Mammal Societies

Mammal Societies

Tim Clutton-Brock

Department of Zoology
University of Cambridge

Library of Congress Cataloging-in-Publication Data

Names: Clutton-Brock, T. H.
Title: Mammal societies / Tim Clutton-Brock.
Description: Chichester, West Sussex : John Wiley & Sons, Inc., 2016. |
    Includes bibliographical references and index.
Identifiers: LCCN 2015040498 | ISBN 9781119095323 (cloth)
Subjects: LCSH: Mammals.
Classification: LCC QL703 .C57 2016 | DDC 599–dc23 LC record available at http://lccn.loc.gov/2015040498

A catalogue record for this book is available from the British Library.

Wiley also publishes its books in a variety of electronic formats. Some content that appears in print may not be available in electronic books.

Cover image: Seal image courtesy of Mike Fedak © Mike Fedak; Meerkat Charge image courtesy of Rob Sutcliffe © Rob Sutcliffe

Set in Meridien 8.5/12pt by Thomson Digital, India
Printed and bound in Singapore by Markono Print Media Pte Ltd

1   2016

Contents

Preface

The core of Darwin's theory of natural selection is the realisation (based on Malthus's demographic projections) that the rate of population increase will inevitably lead to competition between individuals for resources and reproductive opportunities, and that competition will favour individuals that are well adapted to the environments they live in, with the result that their heritable characteristics will increase in future generations. Empirical research on population ecology since Malthus has provided extensive evidence to support Darwin's argument, while research in evolutionary biology and population genetics has confirmed that favourable mutations are likely to spread.

Natural selection adapts animals to the ecological niches that they occupy. Early explorations of animal adaptation mostly examined relationships between anatomical traits and the challenges imposed on different species by their physical environments. More recently, research has documented the impact of the social environment on the selection pressures operating on both sexes and on the evolution of behavioural, physiological and anatomical adaptations. In addition, an increasing range of studies have explored the consequences of contrasts in social organisation and the adaptations they generate for ecological processes within and between species, as well as for other areas of biology, including population genetics, epidemiology and conservation biology.

While there have been excellent reviews of the social behaviour of particular Orders of mammals, there have been few attempts to integrate research across mammalian groups and more extensive reviews of social organisation are available for ants and birds than for mammals. My aim in writing this book was to produce an integrated account of the evolution of mammalian societies within the framework provided by theoretical and empirical research on social evolution. As one of the main reasons for studying animal societies is to provide a general perspective on studies of the evolution of human societies, I particularly wanted to integrate research on primates with studies of other mammalian groups in order to explore the extent to which studies of non-human mammals (including non-primates as well as primates) provide insight into the evolution of hominin societies and human behaviour. Finally, as well as exploring the evolution of variation in the structure and organisation of mammalian societies, I wanted to examine what is known of the consequences of variation in social behaviour and breeding systems for ecological processes.

In contrast to birds, most mammals are polygamous and the structure of mating systems has important effects on the selection pressures operating on females and males and the distribution of sex differences in anatomy, physiology and behaviour. In many mammalian species, different factors determine the distribution of the two sexes: while the need to maintain access to adequate resources and to avoid predation commonly structures the distribution and behaviour of females, selection to maximise access to females often has a more important influence on the distribution and behaviour of males. As a result, although the behaviour of the two sexes coevolves, it is often useful to consider the behaviour of females and males separately. I have consequently organised the book to focus first on the behaviour and reproductive strategies of females and then on those of males. The first chapter provides a brief review of the body of theory relevant to the evolution of social behaviour that has built up over the last 40 years. Chapters 2–9 deal with different aspects of female behaviour in non-human mammals, including sociality, the kinship structure of female groups, mate choice, maternal care, social development, communication and the distribution of competition and cooperation. Subsequently, Chapters 10–16 cover similar topics in males. Chapter 17 then examines the evolution of cooperative breeding systems and Chapter 18 explores the evolution of sex differences in behaviour, physiology and anatomy. Finally, Chapters 19 and 20 provide an introduction to related research on the evolution of breeding systems and social behaviour in hominins and humans.

Acknowledgements

I owe an enormous debt to the large number of people that contributed in different ways to this book. In particular, I am deeply indebted to my wife, Dafila Scott, who has provided continuous support and encouragement; to Katy McAuliffe, who produced syntheses of material for six chapters and comments on others as well as vital encouragement at a time when progress had stalled; and to Penny Roth, who has typed, corrected and retyped all chapters multiple times.

For reading and for generous commenting on drafts that allowed me to improve them, I am grateful to Matt Bell (Chapter 2), Monique Borgerhoff Mulder (Chapters 19 and 20), Andrew Bourke (Chapter 1), Mike Cant (Chapter 17), Alecia Carter (Chapter 6), Ben Dantzer (Chapters 5 and 6), Alan Dixson (Chapter 18), Claudia Feh (Chapter 15), Robert Foley (Chapter 19), Sarah Hrdy (Chapters 9, 19 and 20), Elise Huchard (Chapters 8–11 and 18), Dieter Lukas (Chapters 9, 10 and 14), Katy McAuliffe (Chapters 6 and 20), Karen McComb (Chapter 7), Marta Manser (Chapter 7), Martin Muller (Chapter 16), Ryne Palombit (Chapter 15), Oliver Schülke (Chapter 14), Joan Silk (Chapters 9 and 19), Chuck Snowdon (Chapter 16) and Stuart West (Chapter 1). Any mistakes that remain are, of course, my own. I am also grateful to a large number of people for helping me to locate suitable photographs or for allowing me to use their images.

Dieter Lukas regularly provided relevant papers and useful discussion. In the later stages of production, Bonnie Metherell, Rebecca Stanley, Alex Thompson, Tom Houslay, Gabrielle Davidson, Samantha Leivers and Andrew Szopa-Comley helped me to find photographs and references, and obtain permissions to use images and figures. A large number of colleagues helped me to locate relevant photographs, including Karen Strier, John Hoogland, Carel van Schaik, Tim Caro, Sarah Hrdy, Nigel Bennett, Elise Huchard, Robert Seyfarth, Claudia Fichtel, Dan Rubenstein and Kelvin Matthews. At Wiley-Blackwell, I am grateful to Ward Cooper who originally commissioned the book; to Jolyon Phillips for his meticulous copy-editing; to Kelvin Matthews for his role in organising permissions and in guiding the development of the manuscript; to Kathy Syplywczak for organising the text; and to Emma Strickland for arranging the cover design.

My academic debts extend widely in time and space and I am well aware of how much I owe to my colleagues, collaborators and students. Over the last 35 years while I have been a member of the Zoology Department at Cambridge my work has been supported by the Heads of Department (Gabriel Horn, Pat Bateson, Malcolm Burrows and Michael Akam) and several Departmental Secretaries (including John Andrews, Milly Bodfish and Julian Jacobs). Colleagues on the staff of the Department (including Nick Davies, William Foster, Bill Amos, Andrew Balmford, Bryan Grenfell, Andrea Manica, Claire Spottiswoode and Rebecca Kilner) all contributed in one way or another to its inception and development. So, too, did colleagues at the University of Zurich, where I spent half a year working on the manuscript, including Marta Manser, Barbara Koenig, Gustl Anzenberger and Gerald Kerth.

My debts extend beyond the UK and beyond Europe. For the last 20 years, I have been involved in research on the evolution and ecology of cooperative breeding systems in the southern Kalahari. The late John Skinner and Nigel Bennett have both provided generous help, while Marta Manser has co-directed the meerkat project with me for the last 10 years. Research on meerkats in the Kalahari has involved a substantial number of students, post-docs and collaborators including Christine Drea, Dave Gaynor, Andrew Bateman, Ben Dantzer, Elise Huchard, Alecia Carter, Kirsty MacLeod, Peter Santema, Sinead English, Johanna Nielsen, Raff Mares, Matt Bell, Stu Sharp, Sarah Leclaire, Dom Cram, Constance Dubuc, Ashleigh Griffin, Jack Thorley, Joah Madden, Alex Thornton, Sarah Hodge, Andrew Young, Nobu Kutsukake, Tom Flower, Arpat Ozgul, Julian Drewe, Goran Spong, Neil Jordan, Lynda Sharpe, Andy Russell, Anne Carlson, Pete Brotherton, Andrew MacColl, Michael Scantlebury, Justin O'Riain, Ruth Kansky and Franck Courchamp, as well as a large number of volunteers.

Josephine Pemberton and Loeske Kruuk have also been continuously involved.

Over the same period we began work on other cooperative breeders. Rosie Woodroffe, Mike Cant, Jason Gilchrist, Matt Bell and Sarah Hodge joined me to establish and run a long-term comparative study of banded mongooses in Uganda which now continues under Mike Cant's leadership. Markus Zottl, Philippe Vullioud and Jack Thorley joined me to work on Damaraland mole-rats, in conjunction with Nigel Bennett at Pretoria. Amanda Ridley came to work on Arabian babblers in Israel and then, with Nikki Raihani, came to work on pied babblers in the Kalahari. All of these people have left an indelible mark on my research, my thinking – and this book.

So, too, have the large number of students, post-docs and collaborators that worked with me previously on red deer and Soay sheep or on lek-breeding ungulates, including fallow deer, Uganda kob and Kafue lechwe. My involvement with research on red deer began soon after I had finished my PhD in 1972 when, with support and guidance from Roger Short, Gerald Lincoln and John Fletcher, Fiona Guinness and I started a long-term project that continues today under the leadership of Josephine Pemberton and Loeske Kruuk. Over the 38 years that I led the red deer project, many PhD students and post-docs contributed to the development of the work including Fiona Guinness, Steve Albon, Marion Hall, Rosemary Cockerill, Michael Reiss, Robert Gibson, Michael Appleby, Iain Gordon, Chris Thouless, Callan Duck, Glen Iason, Paul Marrow, Tim Coulson, Owen Price, Martin Major, Karen Rose, Karen McComb, Larissa Conradt, Kelly Moyes, Kathreen Ruckstuhl, Josephine Pemberton, Loeske Kruuk and Dan Nussey. The research on red deer on Rum led on to related projects on Soay sheep on St Kilda in collaboration with Peter Jewell, Steve Albon, Josephine Pemberton, Loeske Kruuk, Mick Crawley, Ken Wilson, Frances Gulland, Bryan Grenfell and Ian Stevenson, as well as to research on fallow deer and other lek-breeding ungulates with Andrew Balmford, Mariko Hiraiwa-Hasegawa, Karen McComb, David Green, James Deutsch, Rory Nefdt, Richard Stillman, Marco Apollonio and Bill Sutherland. Some of the results of these projects are described in relevant chapters but whether I have included them or not, all played some role in the development of my perception of mammal societies and their ecological consequences. So, too, did a range of other collaborators and students: Geoff Parker, Michael Taborsky, Petr Komers, Virpi Lummaa, Claudia Feh, Marco Festa-Bianchet, Andrew Illius, Peter Langley, Amanda Vincent, Sigal Balshine, Márcio Ayres, Kavita Isvaran, Anthony Rylands and Peter Kappeler.

It is also a pleasure to thank people who helped me at the beginning of my career and gave me opportunities without which I would never have been able to make progress. As an undergraduate reading anthropology at Cambridge, I was lucky to be supervised by David Pilbeam who kindled my interest in human evolution and primate societies. I remain enormously grateful to Robert Hinde who agreed to take me on as a PhD student despite my lack of formal scientific qualifications, helped me to make the transition from anthropology to zoology, and taught me to tackle academic questions head-on. Like many others in the field, my interest in the causes and consequences of variation in animal societies was stimulated by the work of John Crook at Bristol and David Lack at Oxford. Subsequently, when I moved to the University of Sussex, John Maynard Smith and Paul Harvey introduced me to evolutionary biology and population genetics.

Finally, I am grateful to the bodies that have funded my long-term research projects, including the Natural Environment Research Council and the Biotechnology and Biological Sciences Research Council of the UK, the Newton Trust and the European Research Council. This book was completed as one component of an ERC Advanced Grant (No. 294494) which supported our long-term research on cooperative breeding systems.

CHAPTER 1

Social evolution

Nothing in biology makes sense except in the light of evolution.

Theodosius Dobzhansky

1.1 Origins

Life is full of dangers, competition for resources and reproductive opportunities is universal and all life forms need to be well adapted to the physical and social environments they occupy in order to grow, survive and breed. While an appreciation of the adaptedness of animal behaviour extends back into antiquity, the modern understanding of adaptation as a consequence of natural selection originates with the work of Darwin (1859, 1871) and Wallace (1870, 1878). In the *Origin of Species* Darwin reviews the diversity of animal adaptations for survival while in *The Descent of Man* he focuses to a greater extent on the evolution of reproductive adaptations as well as on human evolution.

The fundamental importance of Darwin's theory in explaining variation in the morphology, physiology and behaviour of animals was quickly appreciated by his contemporaries. 'If you ask whether we shall call this the century of iron, or of steam, or of electricity', wrote Ludwig Boltzman in 1886, 'then I can answer at once with complete conviction: it will be called the century of the mechanistic understanding of Nature – the century of Darwin' (Boltzman 1905).

But it wasn't. After Darwin's death, scientific attention focused on developmental questions rather than functional ones and his holistic view of biological adaptation was eclipsed by the growth of other biological sub-disciplines. As a result, the true century of Darwin was delayed for nearly 100 years, and is not yet over.

When functional questions were considered in the years following Darwin's death, they mostly related to anatomical adaptations to the physical environment. Before the 1930s, systematic studies of the behaviour and ecology of animals in natural populations were scarce and most were the work of naturalists, sociologists or philosophers who lacked Darwin's theoretical structure, his compelling interest in principles and his readiness to confront apparent exceptions. In many cases, they were satisfied with accurate descriptions of the biology of particular species coupled with ad hoc explanations of the function of particular traits. One important exception was the work of entomologists, like Fabre, who could not ignore the social behaviour of insects and who began to describe the form and structure of colonies and speculate about the mechanisms that maintained them (Fabre 1879; Cézilly 2008).

Only after 1930 did a substantial number of professional biologists start detailed studies of the behaviour and ecology of animals in their natural habitats and, when they did, their principal objective was seldom to explain their evolution or to account for their diversity. They fell into four main groups. First, there were systematists and taxonomists whose principal interest was in phylogeny and development but who found themselves confronted with the obvious diversity of animal societies. Second, there were the founding fathers of animal behaviour, including Julian Huxley, Konrad Lorenz, Niko Tinbergen, Karl von Frisch, T.C. Schneirla and Bill Thorpe. Though their research sometimes encompassed functional aspects of behaviour (especially foraging behaviour), with the exception of Niko Tinbergen, their primary focus was on questions concerning the control and development of behaviour. Third, there were animal ecologists, including Luc Tinbergen, David Lack and A.F. Skutch, whose interests included the regulation of animal populations and the evolution of

Mammal Societies, First Edition. Tim Clutton-Brock.
© 2016 Tim Clutton-Brock. Published 2016 by John Wiley & Sons, Ltd.

life-history parameters and who faced the need to explore the role of territoriality and competition between breeding pairs. Finally, there were the population geneticists, including Ronald Fisher and J.B.S. Haldane, whose principal focus was on the operation of natural selection and the evolution of genetic systems but whose interests inevitably included dispersal and the genetic structure of local populations as well as the evolution of demographic measures. Unlike many of the others, they were well aware of the evolutionary problems raised by social behaviour, though these were tangential to their main interests.

The development of field research after 1930 rapidly revealed the diversity of breeding systems and social behaviour and raised questions about the adaptive significance of these differences. Many of the earliest studies involved insects or birds, since they were relatively easy to observe and their nests are often accessible (Lack 1935; Skutch 1935; Tinbergen 1935). Most birds are monogamous and biparental so that the diversity of social organisation was not a topic of immediate interest. The first professional studies of social behaviour in mammals also date from the 1930s (Figure 1.1). Zuckerman (1929, 1932) explored the social and sexual behaviour of captive baboons and related these to physiological processes, while Fraser Darling's studies of red deer and grey seals (Darling 1937a,b, 1943) and C.R. Carpenter's research on howler monkeys, macaques and gibbons (Carpenter 1934, 1935, 1942) described the size and

Figure 1.1 Early studies of the behaviour of mammals. In the 1930s, (a) Frank Fraser Darling investigated the social and reproductive behaviour of red deer and grey seals, (b) Solly Zuckerman explored the sexual behaviour of captive baboons, and (c) Clarence Ray Carpenter established field studies of several primates, including howler monkeys, spider monkeys, rhesus macaques and gibbons. *Sources*: (a) © http://littletoller.co.uk/authors/frank-fraser-darling/; (b) Reproduced with permission of Zuckerman Archive, University of East Anglia; (c) © Smithsonian Institution Archives. Image SIA Acc. 90-105 [SIA2008-0362].

structure of groups and the reproductive behaviour of individuals and were more concerned with contrasts in ecology.

After 1945, studies of animal ecology and animal behaviour proliferated. In America, which still possessed extensive state forests and national parks, a larger proportion of ecological research was directed towards wildlife management, while in Europe the primary objectives of ecological research were more fundamental in nature. Ecological research focused on foraging behaviour, on the mechanisms regulating population density and on the proximate and ultimate factors influencing life-history parameters, including clutch size, laying data and survival (Lack 1954, 1966). While a substantial proportion of behavioural research was directed at investigating the causation and development of behaviour (Lorenz 1950; Tinbergen 1951; Hinde 1966), a substantial number of studies (mostly of birds) examined feeding behaviour, foraging strategies, territoriality and the benefits of sociality (Tinbergen 1952; Gibb 1954; Hinde 1956), laying the foundations for later work on optimal foraging behaviour (Krebs 1978). However, few studies had yet

monitored the breeding success of individuals throughout their lifespans and little was known of the extent or causes of variation in fitness in natural populations in either sex.

As field studies of birds developed and proliferated, it became obvious that there were striking contrasts in their social behaviour which were consistently related to ecological differences (Orians 1961; Lack 1968). In particular, J.H. Crook's research on weaverbirds showed that there were systematic relationships between variation in social behaviour and contrasts in ecology: species living in open savannah or semi-arid habitats formed the

Figure 1.3 (a–d) In one of the first systematic comparative studies of vertebrate social behaviour, Crook (1964) showed that there were consistent relationships between the size and structure of the colonies of African weaver birds and the type of habitat they lived in: forest-dwelling species mostly breed in pairs or in small colonies while the largest colonies are found in species living in arid savannahs. *Source*: (a–d) From Lack (1968). Reproduced with permission from Taylor & Francis.

Figure 1.2 John Crook, founding father of socio-ecology. *Source*: © Simon Child.

largest breeding colonies while forest-dwelling species mostly lived in pairs or small groups (Figures 1.2 and 1.3). Crook argued that relationships between interspecific differences in social behaviour and contrasts in ecology were a consequence of adaptive responses to variation in the distribution of food resources, nesting sites and predation (Crook 1962, 1964, 1965).

Studies of social behaviour in birds stimulated similar research on mammals. Most European mammals are solitary and nocturnal, so they were less promising targets for field studies than birds but, by the early 1960s, relatively cheap air travel was opening up possibilities for research on diurnal mammals in tropical Africa and Asia. Many of them lived in stable social groups of varying size and structure and the primary aim of many studies of mammals that were established during this period was, for the first time, to describe their social behaviour and the structure of their societies. Since

one motive was to explore the biological origins of human society, many of the earliest field studies of mammals focused on primates, including macaques (Imanishi 1957; Itani 1959; Southwick *et al.* 1965), baboons (DeVore 1965), patas monkeys (Hall 1965) and the African apes (Figure 1.4).

Over the following decade, similar studies began to investigate social behaviour in other groups of mammals, including carnivores (Kruuk 1972; Schaller 1972), rodents (Armitage 1962), ungulates (Walther 1964; Leuthold 1966; Geist 1971), marsupials (Kaufmann 1975; Russell 1984) and cetaceans (Norris 1966; Whitehead 1983; Connor and Smolker 1985) (Figure 1.5). In addition, there was a large increase in field studies of other diurnal primates, including lemurs, New World monkeys and colobines while nocturnal species, which were far harder to observe, did not attract the same level of attention. To make it possible to collect regular

(a)

(c)

(b)

Figure 1.4 Pioneers of long-term primate field studies: (a) Jane Goodall with alpha male Figan in Gombe National Park, Tanzania; (b) George Schaller in the Virungas; and (c) Dian Fossey with Digit in Rwanda. *Sources*: (a) © the Jane Goodall Institute/by Derek Bryceson; (b) © Terrence Spencer/The LIFE Images Collection/Getty Images; (c) © K.J. Stewart and A.H. Harcourt.

Figure 1.5 A selection of mammals that are the subject of continuing long-term, individual-based field studies: (a) rhesus macaques, Puerto Rico (© Alexander Georgiev); (b) chimpanzees, Tanzania (© Ian Gilby); (c) yellow-bellied marmots, USA (© Kenneth Armitage); (d) spotted hyenas, Tanzania and Kenya (© Tim Clutton-Brock); (e) mountain gorillas, Rwanda (© K.J. Stewart and A.H. Harcourt); (f) African lions, Tanzania (© Craig Packer); (g) orangutans, Indonesia (© Anna Marzec, Tuanen Orang Research Project); (h) yellow baboons, Kenya (© Jeanne Altmann); (i) bighorn sheep, Canada (© Fanie Pelletier); (j) red deer, Scotland (© Clutton-Brock); (k) African elephants, Tanzania (© Vicki Fishlock); (l) black-tailed prairie dogs, USA (© Elaine Miller Bond); (m) cheetah, Tanzania (© Dom Cram); (n) muriquis, Brazil (©Thiago Cavalcante Ferreira); (o) sifakas, Madagascar (© Claudia Fichtel); (p) Soay sheep, Scotland (© Arpat Ozgul); (q) white-faced capuchins, Costa Rica (© Katherine MacKinnon); (r) Kalahari meerkats, South Africa (© Tim Clutton-Brock); (s) banded mongooses, Uganda (© Jennifer Sanderson); (t) red-fronted lemurs, Madagascar (© Claudia Fichtel); (u) striped mice, South Africa (© Carsten Schradin). Dates against each species show the approximate time when current long-term studies tracking the life histories of individuals began, though not all studies have maintained continuous records since they started.   *(continued over)*

Figure 1.5 (*Continued*).

observations, and to recognise individuals, it was often necessary to habituate study animals to observation by humans and, once this was done, they often became increasingly trusting, making it possible to observe them from close quarters (Figure 1.6). Techniques for quantifying behaviour in captive and field populations also improved rapidly, making it possible to compare the structure of relationships between individuals and to explore the mechanisms that controlled their development (Hinde 1970, 1973, 1983).

Until the mid 1960s, research on ecology, evolutionary biology and animal behaviour developed independently and there were limited connections between these three areas: for example, neither of Niko Tinbergen's two synthetic books, *The Study of Instinct* (Tinbergen 1951) and *Social Behaviour in Animals* (Tinbergen 1953), cite either Darwin or Fisher. But, by 1960, both theoretical and empirical research began to turn to topics which overlapped behaviour, ecology and evolutionary biology, including the evolution of life histories and social behaviour (Cole 1954; Williams 1957, 1966; Wilson 1971). In the early 1960s, two developments acted as catalysts for the rapid changes in the study of animal societies and reproductive strategies that occurred over the next decade and which are still continuing today. The first was the publication of Wynne-Edwards' monumental book *Animal Dispersion in Relation to Social Behaviour* (Wynne-Edwards 1962). Wynne-Edwards claimed that many animals cooperated to limit their numbers in advance

Figure 1.6 A group of male chimpanzees grooming each other in the Gombe National Park in 1969 while an observer collects data on a check sheet. *Source*: © Tim Clutton-Brock.

of resource shortage in order to improve the probability that groups or populations would survive. Group displays had evolved, he suggested, to allow their members to assess population density and to adjust their reproductive output so as to avoid over-exploitation of their food supplies. Other aspects of social behaviour, including territoriality and dominance hierarchies were, he argued, also involved in the regulation of animal numbers *and had evolved for this purpose*.

Wynne-Edwards' assertion that social behaviour had evolved through selection operating between groups or populations was clearly stated and was contrary to Darwin's persistent emphasis on individual competition as the keystone of evolution as well as to the view that animal populations were limited by the availability of resources (Lack 1954, 1966). Both population geneticists and ecologists rose to the challenge. Ecologists contested the view that social mechanisms regulated population density in advance of resource shortage and showed that density-dependent changes in fecundity and survival were associated with changes in resource availability, predation and disease (Lack 1966, 1968). Formal evolutionary models of Wynne-Edwards' concept of group selection showed that it would only be likely to work where all group members were genetically identical or where there was complete suppression of competition between group members (Maynard Smith 1964) and its general application was explored and refuted (Hamilton 1963; Maynard Smith 1964; Lack 1966; Williams 1966). The controversy drew attention to the fact that many functional explanations of social behaviour relied on benefits to groups or populations and led to a critical re-evaluation of these ideas, initiated by G.C. Williams' influential critique of evolutionary explanations of adaptation (Williams 1966).

The second development was the construction of a coherent body of theory capable of explaining the evolution of social behaviour, reproductive strategies and life histories and the interrelationships between them. The two most important components were the development of the concept of kin selection and inclusive fitness theory (Hamilton 1964), which provided a framework for explanations of the evolution of cooperative and eusocial breeding systems, and the introduction of game theory models to explore the competitive strategies of individuals (Maynard Smith 1974; Parker 1974). Other developments included theoretical models of the evolution of group living (Hamilton 1971) and of breeding systems (Bradbury and Vehrencamp 1977; Emlen and

Oring 1977), of reproductive competition (Trivers 1972), life-history parameters (Parker 1974; Stearns 1977), sperm competition (Parker 1970), mate choice (O'Donald 1962), parental care (Trivers 1974; Maynard Smith 1977), cooperation between unrelated individuals (Trivers 1971), communication (Zahavi 1975) and punishment (Clutton-Brock and Parker 1995).

The framework of theory, based on the theoretical papers of Hamilton, Trivers, Maynard Smith and Parker (Figure 1.7), provided the first satisfactory explanations of variation in animal social behaviour, breeding systems and life histories and emphasised the extent to which the characteristics of an individual's social environment affected its fitness and the selection pressures operating on it. One insight that emerged from this was an understanding that the evolutionary interests of individuals belonging to the same group could diverge as well as converge. While early studies of animal behaviour had seen relationships between males and females, between parents and young and between members of the same social group as harmonious interactions generating social structures that maximised benefits to all, the new framework emphasised the extent to which the interests of individuals differed, leading to conflicts between them, to the evolution of manipulative or exploitative strategies and to social structures that were the outcome of conflicts of interest and which did not necessarily maximise the fitness of all group members (Davies 1992; Arnqvist and Rowe 2005; Bourke 2011). Although they recognised that shared interests could predominate in some cases, they showed that even the most cooperative relationships contained the seeds of conflict.

Another important development was an understanding of the contrasting selection pressures operating on females and males and the role of social behaviour in causing these differences. A seminal paper by Emlen and Oring (1977) showed how the distribution of females was usually related to the distribution of resources and the risk of predation, while the distribution of males commonly depended on the distribution of females. While it came to be appreciated that there were exceptions to this generalisation and that the strategies adopted by males can influence the distribution and reproductive behaviour of females and vice versa, their argument emphasised the need to consider the reproductive strategies of the two sexes separately.

Reviewing the new field that was emerging from the integration of studies of behaviour, ecology and

Figure 1.7 Some of the architects of social evolution theory: (a) Robert Trivers and Bill Hamilton wrestling with a problem at Harvard; (b) John Maynard Smith in his garden; (c) Geoff Parker in 1980; (d) E.O. Wilson. *Sources*: (a) © Sarah Hrdy; (b) © Corbin O'Grady Studio/Science Photo Library; (c) © Geoff Parker; (d) © Jim Harrison (PLoS) https://commons.wikimedia.org/wiki/File%3APlos_wilson.jpg. Used under CC BY 2.5 http://creativecommons.org/licenses/by/2.5/.

population genetics, in 1974 E.O. Wilson had named it sociobiology and predicted that, by the year 2000, it would have become closely allied with population biology and genetics, while traditional ethology and comparative psychology would have been integrated with neurophysiology. Others disagreed: 'I see no signs or probability of this happening and if it did, it could, I believe, be a considerable disaster for biology' wrote W. H. Thorpe. In practice, the first part of Wilson's prediction came about within a few years of the publication of his book and there was a rapid expansion of research, though for studies of non-human animals, sociobiology was gradually abandoned in favour of behavioural ecology (Klopfer 1973; Krebs and Davies 1978).

Over the 40 years since 1975, a combination of theoretical and empirical research has extended and refined our understanding of animal breeding systems and social behaviour (Danchin *et al.* 2008; Székely *et al.* 2010; Davies *et al.* 2012). There has been a substantial improvement in quantitative methods (Martin and Bateson

Figure 1.8 Kalahari meerkats can be habituated to close observation by humans, so it is possible to train them to climb onto electronic balances with small rewards of food or water. *Source*: © Tim Clutton-Brock.

1993) and a progressive refinement of experiments involving both wild and captive animals (Krebs and Davies 1981; Davies *et al.* 2012). Long-term studies that have tracked the life histories of large numbers of individual animals over decades and documented their behaviour and reproductive success have generated quantitative measures of individual differences in fecundity, rearing success and longevity and the factors that affect them, providing access to questions about the costs and benefits of variation in behaviour and reproductive strategies that were previously unavailable (MacColl 2011; Cockburn 2014). In some species, it is possible to habituate large numbers of animals to humans, making it feasible to monitor changes in weight and growth and to collect regular samples of blood, urine and faeces for hormonal and genetic analysis (Figure 1.8). The development of DNA fingerprinting and associated techniques has made it possible to measure the breeding success of males, establish pedigrees and explore the heritability of traits (Jeffreys *et al.* 1985; Charmantier *et al.* 2014). In addition, quantitative comparative studies developed from their initial use as a descriptive tool (Lack 1968) to provide quantitative tests of the generality of specific predictions concerning relationships between ecological, behavioural and anatomical traits that controlled for the effects of phylogeny (Clutton-Brock and Harvey 1977b; Harvey and Pagel 1991). More recently, the advent of gene-based phylogenetic super-trees has made it possible to document sequences of evolutionary events and to identify the ancestral states from which particular traits evolved (Pagel 1994).

Theoretical models of evolutionary processes have continued to explore the operation of selection at different levels. Following extensive critiques of Wynne-Edwards' book, it was initially widely accepted that group selection was only likely to be an important evolutionary process under restrictive conditions (Maynard Smith 1976). However, a subsequent reformulation of the process suggested that selection could operate at multiple levels and that selection operating between groups might, after all, play an important role in the evolution of social behaviour in non-human animals (Wilson 1977; Wilson and Wilson 2007; Nowak *et al.* 2010; Nowak and Allen 2015; Akcay and Van Cleve 2016). Others disagree and have argued that the evolutionary processes described by these models do not differ substantively from Hamilton's concept of kin selection operating through variation in inclusive fitness, and that the two approaches represent alternatives ways of accounting fitness (Gardner *et al.* 2011; Marshall 2011, 2015; Frank 2013).

Recent arguments about differences between models of group and kin selection and the relative importance of these two processes have focused on whether or not high levels of relatedness between group members are necessary for the evolution of eusociality and obligate sterility in insects (Liao *et al.* 2015; Nowak and Allen 2015; Queller *et al.* 2015). While there is no final resolution to this discussion, comparative studies suggest that the initial evolution of eusocial breeding systems has been confined to groups where relatedness between group members is unusually high, though levels of average relatedness may subsequently decline (Hughes *et al.* 2008; Boomsma 2009). Further support for the suggestion that high levels of kinship are necessary for the initial evolution of extensive altruistic cooperation comes from comparative studies of birds and mammals which show that the evolution of cooperative breeding systems has also been associated with unusually high levels of kinship between group members (see Chapters 9 and 17), though humans are an important exception (see Chapter 20). Moreover, unlike models of group selection, the theoretical framework provided by inclusive fitness theory provides a basis for a wide range of predictions about other evolutionary consequences of variation in kinship, and many of them have now been confirmed by empirical studies (Abbot *et al.* 2011).

One reason why arguments about the role of group selection and kin selection in the evolution of cooperative behaviour are important is that they can affect the way in which colony structure and individual behaviour are interpreted. Some proponents of group selection argue that social groups are 'super-organisms' whose size and structure are adapted to maximising survival or breeding success at the group level (Wilson and Sober 1989). Explanations of this kind are most prevalent in studies of social insects, where conflicts of interest between individuals are limited by the suppression of reproduction in other females by the queen or queens (Wilson 1971; Ghiselin 1974) and colonies can show a level of 'functional organisation' resembling the integrated organisation of different parts of the bodies of individual organisms (Wilson and Sober 1989). While this approach may sometimes help to generate useful hypotheses about variation in colony size and structure (Seeley 2001; Hölldobler and Wilson 2009), conflicts of interest between colony members are never eliminated entirely and functional analogies between the most specialised insect societies and individual organisms have important limitations (West-Eberhard 1975; Starr 1979; Gardner and Grafen 2009). In non-human vertebrates, where all group members are potential breeders, conflicts of interest are widespread and intense and treating groups as adapted units offers few insights and is usually misleading (Kitchen and Packer 1999; Clutton-Brock 2009a).

Two related semantic issues concerning the process of evolution need mention. While some evolutionary biologists (including many population geneticists) use 'natural selection' (or 'selection') to refer to relationships between fitness (or components of fitness) and heritable traits, others (including some population geneticists and many sociobiologists and behavioural ecologists) use

natural selection to refer to cases where there are consistent relationships between phenotypic variation and fitness (or its components), distinguishing between selection on phenotypic traits and responses to selection, which vary with their heritability. The acceptance of correlations between phenotypic variation and fitness as a measure of selection is sometimes criticised by geneticists on the grounds that selection pressures operating on phenotypic variation do not necessarily reflect those operating on genetic variation, while behavioural ecologists often respond with the argument that correlations between phenotypic variation and fitness are likely to reflect the selection pressures that operated before heritable traits reached equilibrium.

Contrasts in the usage of 'selection' are often associated with differences in the use of 'adaptation'. Biologists working on the process of evolution commonly use 'adaptation' to refer to changes in gene frequency that increase fitness, while those interested in explaining phenotypic diversity often use it to refer to variation in phenotypic traits that increases fitness, whether or not it has been shown to have a heritable basis, and refer to fitness-enhancing strategies acquired by individuals in the course of their lives through individual or social learning as adaptive. It is particularly important to recognise the presence of differences in usage in discussions of the adaptive significance of social strategies in higher vertebrates and humans, where adaptive tactics that improve the fit of individuals to their social environment (and so increase their fitness) commonly develop as a consequence of individual or social learning, and many differences in behaviour may not be heritable. Like many other behavioural ecologists, I distinguish between selection and the evolutionary response to selection and use 'adaptation' to refer to phenotypic traits in non-human animals that help to fit individuals to their ecological or social environments and so increase their fitness, whether they have been shown to be heritable or not.

In the rest of this chapter, I provide a brief introduction to the development of the main areas of evolutionary theory relevant to understanding contrasts in sociality, reproductive competition, mate choice, parental care, communication and cooperation. Sections 1.2 and 1.3 examine the evolution of female sociality and its consequences for the evolution of mating systems and the form and intensity of reproductive competition in both sexes. Sections 1.4 and 1.5 review our understanding of mate choice and parental care in females and males. Section

1.6 examines the evolution of cooperation and of cooperative breeding systems. Finally, section 1.7 warns about the use of intentional language and the dangers of loaded labels.

1.2 Sociality and mating systems

Early field studies of social behaviour in insects, birds and mammals quickly focused attention on the reasons why many animals live in groups and showed that social behaviour could reduce the risk of predation: for example, research on colonies of black-headed gulls showed that synchronised mobbing deterred predators and that larger numbers of individuals were more effective than smaller ones (Kruuk 1964). Empirical studies led to the development of the first formal models of group-living. In a characteristically original paper, W.D. Hamilton showed that, where predators attack groups and are only likely to take a single animal per attack, individuals gain benefits by aggregating because this increases their per-capita chances of survival (Hamilton 1971). Other studies explored the effects of sociality on the probability that individuals would be detected by predators (Vine 1973; Treisman 1975) and extended the range of ways in which aggregation might reduce the per-capita risk of predation, including effects on the probability that individuals will detect dangers, confuse attackers or defend themselves (Krause and Ruxton 2002).

The potential benefits of sociality in finding and catching food were also recognised. Ward and Zahavi (1973) suggested that the aggregation of birds into flocks might allow individuals to exchange information and might facilitate the location of widely distributed food sources. In addition, comparisons of the hunting success of predators showed that they were commonly more successful when hunting in pairs or small groups than when hunting alone (Wyman 1967; Kruuk 1972; Schaller 1972). Other potential benefits included the sharing of information about the relative probability of different foods and the enhancement of exploitation efficiency.

As the range of animals studied increased, additional benefits of sociality were explored (Krause and Ruxton 2002). These included the retention of heat (especially in animals that hibernate), reductions in the risk of desiccation and improvements in efficiency of movement. In addition, it became apparent that group-living could provide a range of important social benefits, including

the ability of larger groups to displace competitors (Wrangham 1980), to limit immigration or to reduce the risk that take-overs by either sex would lead to infanticide (Packer *et al.* 1990). In cooperative breeders, where a single female monopolises reproduction, studies showed that group-living also increased the reproductive success of the breeding female and ensured the continuity of breeding groups consisting of relatives (Wilson 1971, 1974).

At the same time, the potential costs of sociality came to be recognised. Studies of birds and mammals showed that increases in group size within and across species were commonly associated with increases in territory size, home-range area and day-range length and associated energetic costs of movement (McNab 1963; Schoener 1968; Clutton-Brock and Harvey 1977a,b). In addition, a wide range of field studies showed that increasing group size was often correlated with increased rates of feeding interference or aggression and with reductions in foraging efficiency (Goss-Custard 1970; Jarman 1979; Selman and Goss-Custard 1988) and, in some cases, with increases in parasite load (Hoogland and Sherman 1976; Hoogland 1979; Brown and Brown 1986) or increased risks of detection by predators (Vine 1973; Lindström 1989). It also became apparent that, in plural breeders (species where groups contained multiple breeding females), increases in group size often raised the incidence of reproductive interference between group members and reduced fecundity and juvenile survival (Hoogland 1981; van Schaik 1983) and that relatively large groups sometimes fissioned into smaller ones (Chepko-Sade and Sade 1979).

As field studies multiplied and contrasts between species became clearer, synthetic papers examined the relationship between species differences in sociality and variation in ecological parameters. Crook and Gartlan (1966) compared the social organisation of primates living in contrasting habitats while Jarman (1974) explored the ecological correlates of variation in group size between different species of African antelope. A similar 'socio-ecological' approach was used to explore the causes of intraspecific variation in social behaviour (Richard 1974, 1978; Lott 1991). In addition, related studies began to explore interspecific associations between social behaviour and morphological and physiological adaptations, as well as life-history parameters and relative brain size (Lack 1968; Western 1979;

Clutton-Brock and Harvey 1980; Harvey and Clutton-Brock 1985).

Theoretical studies of the evolution of sociality investigated the effects of increasing group size on the costs and benefits of sociality to individuals. In particular, an important paper by Sibley argued that where the fitness of solitary individuals is lower than that of individuals living in groups, observed group size will commonly exceed the value that maximises the average fitness of group members since solitaries will keep joining groups until the average fitness of their members is equal to that of solitaries (Sibly 1983). Subsequent models examined the extent to which observed group sizes were likely to deviate from the size that optimised average fitness, and showed that variation in the relative fitness of solitaries, in the size of units that transfer between groups and in the relatedness of group members can all affect the probability that group size will deviate from optimal values (Giraldeau and Gillas 1985; Higashi and Yamamura 1993; Kramer 1995; Giraldeau and Caraco 2000; Krause and Ruxton 2002). In addition, other theoretical studies began to explore the effects of group dynamics on the distribution of group sizes (Cohen 1971, 1975).

Comparative and empirical studies also described variation in the kinship structure of groups. Contrasts in kinship between group members are partly caused by variation in fecundity and survival and partly by contrasts in dispersal. Studies of a number of mammals showed that females avoid breeding with close relatives (Packer 1979) and an influential review by Greenwood (1980) demonstrated that, in species which form stable groups, one sex usually disperses to breed elsewhere. Greenwood showed that, in mammals, males were typically the dispersing sex while, in birds, females often dispersed further than males and suggested that this contrast was related to variation in the role of males in defending breeding territories, though recent studies have shown that sex differences in dispersal are more variable and have suggested other explanations for contrasts between birds and mammals (see Chapters 3 and 12).

Most early studies of the evolution of animal sociality considered the average costs or benefits to group members and either disregarded contrasts in the effects of variation in group size on females and males or focused implicitly on females. Important reviews of mating systems in birds and mammals in the late 1970s emphasised the need to consider the separate interests of females and males (Bradbury and Vehrencamp 1977; Emlen and Oring 1977). Empirical tests confirmed that female distributions were closely related to resources while the distribution of males was usually governed primarily by that of females (Ims 1988; Davies 1989).

The recognition that it was necessary to consider the separate interests of females and males had far-reaching consequences. First, it suggested that polygyny was associated with ecological conditions favouring the aggregation of females in stable groups defensible by males, while social monogamy was associated with conditions favouring solitary, widely distributed females (see Chapter 10). Second, it made an important contribution to explanations of the evolution of sex differences in the intensity of reproductive competition and the distribution of associated sex differences in weaponry and body size (see Chapter 18). One extension to this framework was the recognition that multi-male multi-female groups were likely to be found where group size was so large or the reproductive cycles of females were so highly synchronised that more than one female was often receptive at the same time, so that reproductive competition between males was reduced (Altmann 1962; Emlen and Oring 1977; Altmann *et al.* 1996) (see Chapter 11). Third, it led to comparisons of life histories and variation in reproductive success in the two sexes and to the recognition that intense reproductive competition between males is often associated with costs to male survival at several stages of the lifespan (Trivers 1974; Clutton-Brock 1988) (see Chapter 18). And, fourth, it showed that the interests of females and males were frequently in conflict, especially in systems where females are likely to maximise their fitness by mating with multiple males (see Chapter 4) while males are likely to maximise theirs by limiting female opportunities to mate with other partners (Davies 1985, 1989) (see Chapter 15).

Subsequent research on animal breeding systems has refined and extended these generalisations and demonstrated that there are important exceptions to these trends and that the reproductive tactics of each sex can have important consequences for selection on members of the other sex (see Chapters 10 and 15). However, the recognition that the distribution of resources plays a fundamental role in determining the distribution of females and that this, in turn, affects the distribution of males, their opportunities to monopolise multiple partners and the intensity of reproductive competition between them is still of central importance in explaining the diversity of animal societies.

1.3 Reproductive competition

Between males

In the *Origin of Species*, Darwin was principally concerned with explaining the evolution of traits that increased the survival of individuals, but he appreciated that many characteristics of animals, like the elaborate plumage of many male birds, were unlikely to increase an individual's chances of acquiring food or escaping predators (Figure 1.9). *The Descent of Man* provides an explanation of the evolution of these 'secondary' sexual characters and argues that they are adaptations that increase the chance that individuals will acquire breeding opportunities or mates. Darwin identified two ways by which individuals can compete for access to the opposite sex: by direct competition with other members of the same sex for access to mates and the resources necessary for reproduction (such as breeding territories); and by competition to attract breeding partners and induce them to mate. He realised that direct intrasexual competition for breeding opportunities was commonly more intense among males than among females and argued that this was why males commonly showed greater development of traits associated with fighting or other forms of direct competition.

Darwin's recognition of the greater intensity of competition between males (and the sex differences in size and weaponry associated with it) posed a fundamental question. Why do males compete more intensely for females than females do for males? Part of the answer was supplied by analysis of the distribution of breeding success by males and females in fruit flies. In 1948, Bateman showed that variance in breeding success in *Drosophila* was greater in males than females and that breeding success increased more rapidly in relation to the number of mating partners in males than females (Bateman 1948). Sex differences in relationships between fitness and the number of mating partners (Bateman gradients) have now been demonstrated in a number of polygamous species (Clutton-Brock 1988, 2010; Jones *et al.* 2000, 2002) and selection for traits that influence competitive ability is often stronger in males than females (Andersson 1994; Lorch *et al.* 2008). However, the situation is more complex than Bateman appreciated and sex differences in Bateman gradients may often be smaller than was initially supposed (Sutherland 1985; Tang-Martinez and Ryder 2005; Roughgarden and Akçay 2010). In some animals (including the species of *Drosophila* that Bateman worked with) female fitness also increases with partner number (Tang-Martinez and Ryder 2005). In addition, stochastic factors commonly contribute to individual differences in breeding success in both sexes and some models predict that their influence is likely to be greater in males than in females (Sutherland 1985; Gowaty and Hubbell 2005).

But why does partner number have a stronger influence on mating success in males and why is competition for mates usually more intense among males than

(a)   (b)

Figure 1.9 Secondary sexual characters of males include complex weaponry, for example (a) the antlers of red deer, and elaborate ornaments, for example (b) the trains of peacocks. In general, male weaponry is more highly developed in mammals while male ornamentation is more highly developed in birds, reflecting the contrasting importance of intrasexual and intersexual selection in the two groups. *Sources*: (a) © Tim Clutton-Brock; (b) © Roslyn Dakin.

females? In a seminal paper that built on Bateman's work, Trivers argued that it is the relative expenditure by males and females on gametes and parental care ('parental investment') that determines the relative intensity of competition for breeding partners in the two sexes (Trivers 1972). Sex differences in parental investment affect the time necessary to complete a successful breeding attempt or their 'time out' of competition for breeding partners and this limits the potential rate at which males and females can complete breeding attempts, their *potential reproductive rate* or PRR (Clutton-Brock and Parker 1992; Parker and Simmons 1996). Sex differences in 'time out' and PRR in turn affect the relative numbers of each sex that are ready to breed at any point in time (the *operational sex ratio*, or OSR) which, in many systems, is the principal factor determining the relative intensity of intrasexual competition in the two sexes (Trivers 1972; Emlen and Oring 1977; Clutton-Brock and Parker 1992). For example, among species where males care for the young, they typically compete more intensely than females for mating opportunities in species where they can care for multiple clutches of eggs simultaneously and their PRR exceeds that of females, whereas females compete more intensely than males for mates in species where males can only care for a single clutch at a time and their PRR exceeds that of males (Clutton-Brock and Vincent 1991; Ahnesjö *et al.* 2001).

While the OSR establishes the competitive arena in which both sexes compete for breeding partners, estimating the OSR and predicting the relative intensity of selection for traits that increase the competitive ability in the two sexes is not straightforward. It is frequently difficult to decide which individuals should be included in estimates of the OSR and stochastic variation in male success may increase as the OSR rises, weakening the intensity of selection for traits related to competitive ability in males (Sutherland 1985; Klug *et al.* 2010, 2012; Rios Moura and Peixoto 2013). Moreover, in multiparous species, annual breeding success often trades off against the effective breeding lifespan of males (see Chapter 13) and much of the observed variation in male success within years is often the result of age differences between individuals (Clutton-Brock 1983, 1988). As a result, standardised variance in lifetime breeding success among males does not necessarily increase with the degree of polygyny and is not always much greater in males than in females (Lukas and Clutton-Brock 2014). To predict how much members

of each sex should invest in traits that affect their competitive success (their *scope for competitive investment*, or SCI), it is necessary to consider both the OSR and Bateman gradients, as well as the social and ecological factors affecting the costs and benefits of investment in breeding competition. An integrative model constructed by Kokko and her collaborators incorporates these different factors and shows how variation in the OSR can affect Bateman gradients and why contrasts in the OSR do not always predict sex differences in competitive behaviour (Kokko *et al.* 2012).

One general conclusion emerging from research on sexual selection is that species differences in the development of male secondary sexual characters associated with competitive success and the extent of sex differences in these traits may be more closely related to variation in the frequency of fighting and the competitive tactics of the two sexes than to sex differences in reproductive variance or to differences in the OSR. While variation in the frequency of fights and the competitive tactics of the two sexes may be loosely related to differences in the OSR as well as to variance in male breeding success, these relationships may not be close.

Studies of male competition also raise important questions about the evolution of fighting tactics. Why are all-out fights often uncommon? How long and hard should individuals fight? And how might individuals minimise the costs of fighting? Empirical studies of competition show that fighting often has substantial costs to survival in males and is likely to reduce the duration of effective breeding (Geist 1971; Clutton-Brock *et al.* 1979, 1982). Maynard Smith and Parker introduced game theory models to explore the evolution of fighting tactics and showed that high levels of aggression would not necessarily be the most successful tactic (Maynard Smith 1974; Parker 1974): as more aggressive individuals ('hawks') come to predominate, more pacific strategies ('doves') may be favoured, so that both tactics persist. Subsequently, they examined different ways in which individuals might minimise the costs of fighting. Where fights consist of prolonged contests which end when one party is exhausted ('wars of attrition'), they showed that individuals should give up as soon as it becomes clear that they are unlikely to win, so that fights between disparate opponents should be relatively short while those between well-matched ones should be relatively long (Parker 1974; Maynard Smith and Parker 1976). Subsequent models argued that an even better course would be to discourage opponents from pursuing challenges by signalling their superior strength or commitment to winning (Maynard Smith 1982, 1991). They raised questions about the 'honesty' and reliability of signals (Johnstone 1997) which led to a substantial field of theory dealing with the evolution of competitive signalling systems (Maynard Smith and Harper 2003; Searcy and Nowicki 2005) (see Chapter 7).

As empirical studies of reproductive behaviour in natural populations proliferated, the diversity and complexity of male reproductive strategies became apparent. In some societies, dominant males allow one or more subordinate males to remain in the group and subordinates may assist dominants in repelling neighbouring groups or potential rivals (see Chapter 11). Where female groups are large and include several adult breeding males, individual males often form alliances to compete with each other, and may attempt to disrupt the formation of alliances by rivals (see Chapter 14). In many species, males use force, harassing tactics, or intimidation to coerce females to mate with them, generating arms races between the sexes and, in some species, males frequently attempt to kill dependent infants fathered by their competitors (see Chapter 15). In some species, particular males form 'friendships' with particular females, providing them and their infants with some protection from attacks by rival males and benefiting from this by increased reproductive access to their female 'friends'.

Moreover, competition between males does not end at copulation. Where females commonly mate with more than one male per breeding attempt, sperm from more than one male compete within the reproductive tracts of females and adaptations in males that increase the probability that their sperm will inseminate females or reduce the probability of successful fertilisation by subsequent mating partners are common (Parker 1970, 1984, 1998; Simmons 2001). Early studies of mating competition often assumed that the sperm supplies of individual males were virtually unlimited, but more recent work has shown that this is often not the case and that males may limit their allocation of sperm to different mating partners so as to maximise their breeding success (Parker *et al.* 1997; Parker 1998, 2000; Wedell *et al.* 2002). The risk of sperm competition and associated male tactics often affects the intensity and duration of mate guarding bonds between the sexes and the form and duration of competition for mating opportunities (Birkhead and Møller 1992; Tregenza and Wedell 1998; Simmons 2001; Bjork and Pitnick 2006).

Between females

Partly because interactions between females less frequently involve escalated fights and partly because females rarely possess such elaborate weaponry or ornaments, research on reproductive competition initially focused principally on males. However, in many animal societies, females also compete intensely between themselves for rank, access to breeding territories or other resources necessary for conception or rearing offspring and are commonly aggressive towards each other's offspring, sometimes with lethal consequences (Clutton-Brock 2007; Rosvall 2011; Stockley and Bro-Jørgensen 2011; Clutton-Brock and Huchard 2013a). Like males, females may also compete for access to mates and, in some cases, reproductive competition is more intense between females than between males. For example, in polyandrous birds where males care for eggs or offspring and single females can monopolise access to multiple males, OSRs can be biased towards females and females can be more competitive and more ornamented than males (Figure 1.10) (Emlen and Oring 1977; Oring *et al.* 1991a,b). Where males produce unusually large sperm, this too can lead to competition between females for mates (Bjork and Pitnick 2006). In other cases, males may bring nuptial gifts to females that increase their fecundity and females gain direct benefits by mating with multiple partners and so compete for mating partners (Simmons and Gwynne 1993; Simmons 1995; Kvarnemo and Simmons 1999). Similar reversals in the usual pattern of sex differences in reproductive competition also occur in some singular cooperative breeders where variance in female breeding success is larger in females than in males (see Chapter 17).

While sex differences in parental care sometimes mean that the OSR is consistently biased towards one sex, in some species the direction of biases in the OSR changes throughout the breeding season and the relative intensity of reproductive competition in males and females also varies (Forsgren *et al.* 2004; Gowaty and Hubbell 2005). For example, in two-spotted gobies, the relative intensity of competition for mating partners in the two sexes varies throughout the breeding cycle as the relative number of receptive females and males changes (Amundsen and Forsgren 2001; Forsgren *et al.* 2004). It is also important to appreciate that intrasexual competition between females is not confined to cases where OSRs are

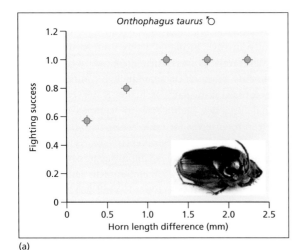

(a)

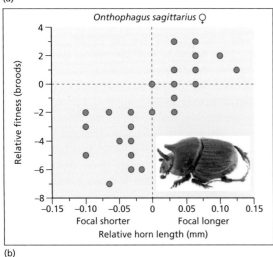

(b)

Figure 1.10 Horn length and fighting success in male and female *Onthophagus* beetles. (a) Relationship between male fighting success and horn length in *Onthophagus taurus*. Graph adapted from Moczek and Emlen (2000). (b) Relationship between relative horn length and relative fitness of competing females in *O. sagittarius*. Positive values on the *x*- and *y*-axes represent cases where, in pairs of competing individuals matched for body size, the focal animal had a larger horn and produced more broods relative to her competitor. *Source*: Graph from Watson and Simmons (2010), reproduced with permission from the Royal Society. *Photo source*: (a) © Tim Murray; (b) © Udo Smidt.

male biased and can occur wherever females can increase their fitness by competing for access to particular males, multiple mating partners or resources necessary for breeding.

While there are fundamental similarities in reproductive competition in males and females, there are also general differences. Where females compete for resources while males compete for access to breeding partners, the fitness benefits of winning particular contests are often likely to be lower in females than in males. In addition, the relative intensity of intrasexual competition (and the development of traits that increase their competitive success) may be more strongly influenced by variation in resource distribution and less by variation in the form of mating systems than in males. The costs of competition are often likely to be higher to females if agonistic interactions endanger the survival of fetuses or dependent infants. As a result, females are often likely to invest less in attempts to win competitive contests than males, though research on singular cooperative breeders shows that this is not always the case (see Chapter 17).

One consequence of the common tendency for females to invest less in winning contests than males is that the survival costs of traits associated with competitive success are seldom as large in females as in males. For example, while the evolution of increased body size in males is often associated with higher juvenile mortality and reduced longevity in adult males compared to females (Clutton-Brock *et al.* 1985; Clutton-Brock and Coulson 2002), there is little evidence that sex differences in survival are reversed in species where reproductive competition is more intense or secondary sexual characters are more highly developed in females (Clutton-Brock 2007). An additional reason for this difference may be that the costs of expenditure on reproductive competition or ornamentation by females depress fecundity or parental investment, constraining the development of secondary sexual characters below the level at which they have measurable costs to female survival (LeBas 2006). For example, elevated levels of testosterone may have adverse effects on the fecundity of females or on the development of their offspring which constrain the evolution of further increases in female competitiveness (Drea *et al.* 2002; Knickmeyer and Baron-Cohen 2006).

As in males, the characteristics that affect the ability of individual females to acquire breeding opportunities, high status or mates and to rear young successfully include the individual's age, weight and hormonal status (see Chapter 8). Maternal status and support in social interactions can also be important, as can the number and identity of allies (Hrdy 1981; Chapais 1992; Walters and Seyfarth 1997). For example, in some social primates, where females from the same matriline support each other in competitive interactions with members of other matrilines, the social rank of individuals as well as their reproductive success depends on the rank of their matriline (see Chapter 8). Where similar characteristics determine competitive ability in the two sexes, selection often favours the development of similar secondary sexual characters in males and females. For example, in *Onthophagus* beetles, selection operating through the effective fighting success on reproduction has led to the evolution of horns in both sexes (Watson and Simmons 2010) (see Figure 1.10). Similarly, intrasexual reproductive competition has led to similar behavioural strategies adapted to the acquisition of rank in both sexes in some primates (see Chapters 8 and 13).

Evidence that reproductive competition can lead to the evolution of secondary sexual characters in females as well as in males has sparked a debate over whether or not this should be regarded as a form of sexual selection. Research on sexual selection and secondary sexual characters has focused principally on males and sexual selection has come to be defined as operating exclusively through competition for mates, mating opportunities or access to gametes produced by the opposite sex (Andersson 1994; Kokko and Jennions 2008). This generates a semantic difficulty in describing the evolution of secondary sexual characters in females, since they typically compete for breeding opportunities rather than mating opportunities. If sexual selection is confined to selection operating through variation in mating opportunities, many examples of selection operating through reproductive competition between females (and the adaptations it generates) would have to be excluded and fundamentally similar evolutionary processes operating in males and females would have to be ascribed to different evolutionary processes (Clutton-Brock and Huchard 2013b). For example, if a narrow definition of sexual selection is adopted, the evolution of horns in male and female *Onthophagus* beetles (Figure 1.10) would have to be attributed to different evolutionary processes. Similarly, selection operating through competition for dominance status in male primates would be regarded as an example of sexual selection (because it enhances mating success) while selection operating through competition for dominance status in females would have to be treated as a consequence of some

other form of selection (because it improves opportunities to breed but may not influence the chance of mating).

Several ways out of this dilemma have been suggested. One possible solution is to recognise some additional category of selection operating through intrasexual competition for resources other than mates, such as 'social selection' (Crook 1972; West-Eberhard 1979, 1983) and this has been advocated in some recent reviews (Roughgarden *et al.* 2006; Lyon and Montgomerie 2012; Tobias *et al.* 2012). However, in social species, distinguishing between 'social' selection and 'sexual' selection is difficult since virtually all selection pressures are affected by social interactions and relationships (Clutton-Brock and Huchard 2013b). Another approach is to abandon the attempt to distinguish between sexual and natural selection and to categorise selection pressures on the basis of whether or not they differ between the sexes (Carranza 2009, 2010), although few selection pressures are unaffected by sex. A third is to broaden the current definition of sexual selection to include all selection processes operating through intrasexual competition for breeding opportunities in either sex (Clutton-Brock 2007; Cornwallis and Uller 2010), although this can lead to problems in distinguishing between sexual selection and natural selection (Shuker 2010). A final approach is to abandon any formal attempt to distinguish between natural and sexual selection and to concentrate instead on comparisons of the relative intensity of selection operating through different components of fitness and on different phenotypic characters (Clutton-Brock 1983, 2004). Whichever solution to these semantic problems is adopted, this discussion underlines the qualitative similarity in the evolutionary mechanisms operating in the two sexes.

1.4 Mate choice

In *The Descent of Man*, Darwin (1871) described examples of female mating preferences and argued that sexual selection operating through consistent preferences for males with particular characteristics or ornaments could lead to the evolution of male secondary sexual characters. However, he did not explain the origins of female choice in any detail and his idea attracted criticism (Wallace 1889; Huxley 1938a,b). 'Shall we assume', wrote T.H. Morgan in 1903, that still another process of selection is going on, that those females whose taste has soared a little higher than that of the average (a variation of this sort having appeared) select males to correspond, and thus the two continue heaping up the ornaments on one side and the appreciation of these ornaments on the other? No doubt an interesting fiction could be built up along these lines, but would anyone believe it, and, if he did, could he prove it?

Today, Morgan's scepticism appears wonderfully dated, for the process he describes was shown to be feasible by Fisher (1930) and empirical studies have confirmed the presence and benefits of consistent female preferences for mating with partners with a variety of anatomical, physiological, behavioural and genetic traits in a wide variety of animals (Andersson 1994). One of the earliest studies that provided clear evidence of female mating preferences based on male ornaments was Andersson's study of long-tailed widowbirds where he showed that he could increase (or decrease) the attractiveness and mating success of males by altering the length of their tail feathers (Andersson 1982), but there are now several other examples where experimental studies have demonstrated female mating preferences based on the characteristics of male ornaments (Andersson 1994; Wilkinson and Reillo 1994; Ryan 1997).

Female mating preferences may provide direct fitness benefits if their choice of mating partner affects their own survival, fecundity or rearing success. Direct benefits of mate choice are widespread and include reduced risks of predation, harassment or disease transmission while mating, improved access to resources defended by males and increased paternal investment (Danchin and Cézilly 2008; Pizzari and Bonduriansky 2010) (see Chapter 4). They can lead to the evolution of signals or ornaments in males that advertise their fertility or their ability to invest (Andersson 1994) (see Chapters 13 and 18).

Alternatively (or additionally), female mating preferences can generate indirect fitness benefits that affect the fitness of their offspring. Several different mechanisms for the evolution of mating preferences through indirect benefits have been suggested. Fisher (1930) argued that, if heritable choice variation in male ornaments associated with fitness arises by chance, females preferring ornamented males will produce sons with superior mating success and alleles favouring female discrimination and those controlling the development of male ornaments will spread together (Fisher 1930; O'Donald 1962, 1967).

Subsequent models confirmed that female preferences and male traits can coevolve in a runaway fashion leading to the evolution of exaggerated male characteristics and strong female mating preferences even if they have substantial costs to survival (Lande 1980, 1987; Kirkpatrick 1982; Lande and Arnold 1985). Empirical studies have confirmed the presence of these costs: for example, in guppies, there is a negative genetic correlation between male attractiveness and the survival of their offspring (Brooks 2000). A second possibility is that conspicuous, costly male traits indicate the overall fitness of their carriers and that females mating with males bearing these traits produce offspring of both sexes that have relatively high fitness. Models of this kind have come to be known as 'good genes' models and are currently regarded as providing the most plausible explanation of the benefits of female mate choice and the evolution of male ornaments through female choice (Andersson 1994; Ryan 1997; Danchin and Cézilly 2008).

The expression of secondary sexual traits is usually strongly condition-dependent and so may allow males to signal their genetic quality to females (see Chapter 4). One suggestion is that females may be selected to favour males with conspicuous ornaments because the high costs of ornaments guarantee the genetic quality of their mates (Zahavi 1975, 1977; Johnstone and Grafen 1993). Although this idea was initially regarded with scepticism (Kirkpatrick 1986), later modelling that combined heritable differences in viability with mating advantages showed that this process might contribute to the evolution of male ornaments, especially if their size depends on the individual's phenotypic condition (Grafen 1990; Andersson 1994).

Where females select partners on the basis of genetic differences, it is necessary to explain why heritable variation in male quality persists (Kirkpatrick and Ryan 1991). One of the first suggestions was that this is a result of continuing coevolution between parasites and their hosts (Hamilton and Zuk 1982), though attempts to test this possibility have produced mixed results (Danchin and Cézilly 2008). Alternatively, variation in male quality may be maintained by the accumulation of deleterious mutations, interactions between selection at different loci, fluctuating selection pressures and variation in the degree of inbreeding (Danchin and Cézilly 2008).

Recent research on female mating preferences has also explored the possibility that females preferentially mate with males whose genotype complements their own (Mays and Hill 2004). Laboratory studies of a variety of organisms suggest that heterozygosity at specific loci can improve individual fitness and that the degree of male ornamentation is sometimes correlated with levels of heterozygosity (von Schantz *et al.* 1997). The best examples of effects of this kind come from studies of the effects of heterozygosity at the major histocompatibility complex (MHC) locus, which is involved in immune function (Jordan and Bruford 1998). Genes at this locus are extremely variable, both within and across species (Zinkernagel and Doherty 1974; Klein 1986; Potts and Wakeland 1990), and individuals which are heterozygous at particular MHC loci are often more resistant to infections and diseases (Gabriel *et al.* 1994; Comings and MacMurray 2000; Penn *et al.* 2002). They can also show high levels of ornamentation, display or social status (Yamaguchi *et al.* 1981; Roberts and Gosling 2003) and several studies have produced empirical evidence of disassociative mating for MHC genotype (see Chapter 4).

As well as selecting particular partners, females may gain both direct and indirect fitness benefits by mating with more than one male (Andersson 1994; Zeh and Zeh 1996). Direct benefits include improved fecundity arising from the avoidance of monopolisation by infertile males or by access to multiple 'nuptial' gifts, as well as improved survival of offspring as a consequence of improved protection and reduced rates of infanticide (see Chapter 4). Indirect benefits include improved viability of offspring caused by avoidance of the negative consequences of genetic incompatibility or selfish genetic elements (Tregenza and Wedell 2002; Price *et al.* 2008).

In some animals, females can also control the probability that their eggs will be fertilised by different males through post-copulatory mate choice. In insects and birds, which commonly store sperm, females can control the paternity of their offspring by storing sperm at different sites and subsequently varying its release (Birkhead and Møller 1992; Ward 1993; Andersson 1994; Simmons 2001). In some mammals, females also store sperm (Birkhead and Møller 1993), but a more common way by which they can manipulate paternity is by controlling the number and identity of males they mate with (see Chapter 4). Females

commonly mate with more than one male. Potential benefits include direct fitness benefits, such as confusing paternity and reducing the risk of infanticide, and indirect benefits associated with increases in genetic variability within litters (Simmons 2001).

While theoretical research tends to contrast different mechanisms maintaining female mating preferences, in reality, several evolutionary mechanisms may often be involved in maintaining the same ornament (Kokko 2003; Kokko *et al.* 2003). For example, where females select males on the basis of direct benefits, their choice may also have indirect benefits, and where females select males on the basis of indirect benefits, their decisions may also benefit their own fitness directly. Indirect benefits may commonly include benefits to survival as well as to attractiveness, and selection pressures outside the context of mate choice may often affect both male traits and female preferences (see Chapter 4).

Where OSRs are biased towards males, mate choice is typically more highly developed in females than males (Trivers 1972; Emlen and Oring 1977). However, this does not preclude the evolution of mating preferences in males and an increasing number of empirical studies have found evidence of mate choice in both sexes (Edward and Chapman 2011). Like female competition for males, the relative choosiness of males varies with the availability of partners and can differ between local populations (Simmons and Gwynne 1993) as well as between stages of the breeding cycle (Amundsen and Forsgren 2001).

Like female preferences, male preferences vary qualitatively and quantitatively. In some species, males prefer familiar partners, while in others they prefer novel partners (Orrell and Jenssen 2002). In some cases, males prefer older, larger or more dominant partners, while in others they prefer younger partners (Werner and Lotem 2003; Wong and Jennions 2003; Herdman *et al.* 2004; Kvarnemo *et al.* 2007). In several species, males also show consistent preferences for partners that have not mated recently (see Chapter 15). Where females are ornamented or brightly coloured, males often show a preference for brighter or more highly ornamented females (Andersson 1994). In addition, males, like females, sometimes copy each other's choice of partners, reinforcing the effects of individual choice (Dugatkin 1992; Widemo 2006).

Figure 1.11 Sexually selected ornaments in females. Scanning electron micrograph of pinnate leg scales on the leg of a female empidid dance fly (*Rhamphomyia longipes*). In most dance fly species, males bring nuptial gifts to females, who compete for suitors; the size of scales on the legs of females is related in some species to their fecundity, and males mate preferentially with females with large leg scales. *Source*: From LeBas *et al.* (2003). Reproduced with permission from the Royal Society.

Individual differences in fecundity between females caused by intrasexual competition for resources are likely to strengthen selection on males to identify and prefer superior partners and, on females, to signal changes and individual differences in fecundity (Berger 1989; Reinbold *et al.* 2002; Clutton-Brock 2007). Male preferences often appear to favour female characteristics associated with fecundity, suggesting that they may generate direct fitness benefits (see Chapter 15). For example, in empididid dance flies, where females are ornamented with large pinnate leg scales whose size reflects their fecundity, males preferentially mate with highly ornamented partners (Cumming 1994; LeBas *et al.* 2003) (Figure 1.11), while sexual swellings in female primates (Figure 1.12) appear to advertise temporal changes in their fertility (see Chapter 7).

1.5 Parental care

Males and females

The extent of parental care in animal societies ranges from species where parents abandon eggs shortly after they are laid to species where they associate with and protect their offspring throughout much of their lives (Clutton-Brock 1991). The first attempts to account for these differences focused not on the evolution of care but on the evolution of egg size and clutch size (Perrins 1965; Lack 1968; Smith and Fretwell 1974). Theoretical treatments argued that the survival of individual embryos should increase rapidly with egg size before approaching an asymptote and used the marginal value theorem to identify the egg size that would provide parents with the greatest fitness return per unit of egg weight or investment (Smith and Fretwell 1974; see also Chapter 5). In most cases, this is likely to be lower than the egg size at which offspring fitness is maximised, favouring parents that produce intermediate-sized eggs. In addition, the evolution of egg size is also likely to be modified by trade-offs between egg size and clutch size or fecundity (Charnov and Krebs 1974). The same model can be used to predict the amount or duration of parental care, leading to the general prediction that, where neonates and juveniles face adverse physical or social environments, the benefits of extended parental care are likely to increase. Empirical studies of a wide range of animals confirm that this is often the case (Clutton-Brock 1991).

Figure 1.12 Sexually selected ornaments in female primates: (a) female baboon and (b) female chimpanzee at mid-cycle. *Source*: (a) © Phyllis Lee; (b) © Michael Wilson.

Figure 1.13 The South American titi monkey is monogamous and produces single young. In contrast to other monotocous, monogamous species, males carry dependent infants for a larger proportion of daytime than females. *Source*: © Kathy West.

Among vertebrates that care for their offspring, patterns of care vary from groups where care is usually restricted to females (including many mammals), through groups where biparental care is normal (as in most birds and a few mammals, including titi monkeys; see Figure 1.13), to groups where exclusive male care is common (as in some fish and amphibia). One obvious question is why patterns of parental care vary so widely? The first theoretical explanations of the evolution of parental care focused on the benefits that each sex would gain by deserting mating partners after copulation, and predicted that uniparental care should be most likely to evolve in whichever sex benefits most from caring for its offspring when the costs of care and the response of the other sex to desertion are taken into account (Maynard Smith 1977; Houston

et al. 2013). Trivers (1972) originally argued that because anisogamy generates male-biased OSRs and more intense mating competition among males than among females, it predisposes females to care and males to invest in competition for mates, and this argument was subsequently accepted as a general explanation of the prevalence of female care (Emlen and Oring 1977; Maynard Smith 1977). However, the prevalence of uniparental male care in some groups suggests that other factors must be involved. Recent models of the initial evolution of parental care have pointed out that where males routinely desert partners after mating, the OSR will become strongly male-biased and the probability that deserting fathers will find additional mates is likely to fall, increasing the relative benefits to males of guarding mates and contributing to parental care (Queller 1997; Houston and McNamara 2002; Kokko and Jennions 2008; Houston *et al.* 2013). Under these conditions, anisogamy may still favour the evolution of care by males rather than females, though contrasts in ecology, life histories and the structure of breeding systems are also likely to have important effects and have the potential to reverse this bias in particular cases (Kokko and Jennions 2008).

Comparisons of patterns of care in different animal groups support the suggestion that ecological factors play an important role in the evolution of sex differences in parental care. For example, in many demersal-breeding fish, intrasexual competition between males has led to the defence of preferred breeding sites. As the presence of eggs attracts further mating partners and males can care for multiple clutches simultaneously, the marginal costs of care may often be lower and the marginal benefits may be higher for males than for females, leading to the evolution of male care (Clutton-Brock and Vincent 1991). Since parents do not usually need to maintain the temperature of eggs and rarely provision young directly, biparental care is relatively rare in fish. In contrast, in birds, the need to brood and feed young with scarce high-energy foods generates strong selection for biparental care and female fecundity is often constrained by the number of eggs that can be brooded. In mammals, the commitment of females to gestation and lactation constrains female opportunities for increasing fecundity even further and (in most species) reduces the marginal benefits of contributing to parental care in males, while the relatively high density of females and their tendency to form groups favours the evolution of polygyny (Clutton-Brock 2009a). While interspecific contrasts in ecology appear to be responsible for major contrasts in patterns of care, the relative costs and benefits of care to the two sexes also vary and, in some cases, can generate changes in the sex responsible for care within or between populations (Alonzo and Klug 2012).

A related question is how partners in biparental species would be expected to divide the workload associated with raising young. Where both parents are involved in care, reductions in the level of contributions by one partner are likely to cause its mate to increase their contributions, though rising costs of care may prevent them from compensating fully (Houston and Davies 1985; Parker et al. 2002; Johnstone and Hinde 2006). Situations of this kind can generate conflicts of interest between partners, and may lead to a process of 'negotiation' over their contributions to care, which may occur both within particular breeding seasons and over evolutionary time (see Chapter 16). As long as individuals do not overcompensate for reductions in their partner's contributions, a stable equilibrium is reached when the reaction curves of the partners intersect and neither partner can improve its fitness by altering its contributions (Parker 1985; Westneat and Sargent 1996; McNamara et al. 2003). In many birds and mammals, males have greater opportunities than females to increase their fitness by extra-pair mating, so that contributions to care

have higher fitness costs to males than females which may explain why females contribute more to parental care than males in many biparental species (Clutton-Brock 1991).

Parent–offspring conflict

Early theories explaining the evolution of parental care usually assumed that parents were free to allocate the resources at their disposal among offspring so as to maximise their own fitness. However, in sexually reproducing animals, offspring will often be selected to extract more care from parents than the amount that would maximise the parent's fitness, generating conflicts of interest between parents and offspring over the level of parental investment (Trivers 1974). For example, where parents are related to their offspring by 0.5, they would be likely to 'disagree' with their offspring over the continuation of parental care from the point when the benefit–cost ratio equals 1.0 to the point when it equals 0.5 (see Chapter 5). Depending on the rate at which the benefit–cost ratio declines with increasing offspring age, this may either be a relatively short period or a relatively long one. In a subsequent model, Trivers also showed that parents and offspring are likely to disagree over the amount of care at particular stages of development. Trivers' argument was subsequently challenged by Alexander (1974) on the grounds that the evolution of strategies in offspring that reduced their parents' fitness would be selected against when the offspring themselves became parents and were similarly exploited by their own offspring. However, subsequent models of parent–offspring conflict showed that, in sexual organisms, 'conflictor' genes are able to spread, though this can be inhibited if their cost to parental fitness is high or parents are able to control the distribution of care (Parker and MacNair 1979; Parker 1985).

Conflicts of interest between parents and offspring are likely to affect all aspects of parental investment, but testing predictions is often complex because of the difficulty of identifying separate optima for parents and offspring. Some of the most convincing evidence that parent–offspring conflict affects patterns of investment comes from studies of the sex ratio in eusocial insects (Trivers and Hare 1976; Charnov 1982; Boomsma and Grafen 1991; West 2009) or from cases where conflicts between offspring reduce brood size below the parent's optimum (Godfray

1994; Mock and Parker 1997). In many cases, conflicts of interest have been settled long ago and the current situation is a consequence of past conflict but there are some interactions where overt behavioural conflicts are played out in real time over the course of each breeding cycle (see Chapter 5). Some of the best examples are of interactions between offspring and parents in nidiculous birds and insects, where young signal or beg for food and parents commonly adjust both their frequency of feeding and the distribution of food in relation to begging frequency (Kilner and Johnstone 1997; Smiseth and Moore 2002; Hinde *et al.* 2009). Though it has been suggested that begging could represent a way of blackmailing parents to supply food in order to reduce wasteful solicitation or the risk of predation (Zahavi 1977; Parker and MacNair 1979), most of the available evidence suggests that it is often an honest signal of need with tangible costs to chicks which maintain its reliability as a signal and prevents exploitation of parents by manipulative offspring (Kilner 2002).

Offspring sex ratios

Especially in polygynous or promiscuous species where males compete intensely for breeding opportunities, parental investment is often likely to affect the fitness of sons more than daughters. In an early paper, Trivers and Willard (1973) argued that this should often cause parents to invest more heavily in individual sons than daughters and that phenotypically superior mothers that can afford the expenditure of energy necessary to rear a son might be expected to specialise in producing male offspring while inferior females would be expected to produce females. Other aspects of parental phenotype that affect the fitness of offspring of one sex more than that of the other (such as their social status or their attractiveness to the opposite sex) would also be expected to affect the sex ratio of progeny, as should temporal fluctuations in the resources necessary for parental care (Clutton-Brock 1991; Leimar 1996) (see Chapter 18). While a number of studies have demonstrated trends in the expected direction, these are often inconsistent and there is still disagreement over the extent to which empirical studies support theoretical predictions (Charnov 1982; Lessells 2002; West 2009).

Trivers and Willard's predictions concern variation in the relative numbers of sons and daughters that different mothers should produce rather than the overall (or average) ratio of males to females. In many insects, average sex ratios vary widely, though in birds and mammals most species produce approximately equal numbers of male and female offspring (Williams 1979; Clutton-Brock 1991). This puzzled some early advocates of group selection, who argued that selection would be expected to favour female-biased sex ratios that maximised the productivity of groups or populations. However, the problem had already been discussed by Darwin (1871, p. 316) and solved independently by Düsing (1883, 1884a,b) and Fisher (1930). Fisher pointed out that because all individuals require a mother and father, if parents (on average) produced a preponderance of one sex, this would increase the average fitness of individuals of the other sex, generating frequency-dependent selection to increase the number of offspring of the rarer sex. Eventually, he argued, this would lead to offspring sex ratios close to parity, unless there are sex differences in survival of juveniles before the end of parental care, in which case sex ratios at the beginning of parental care should be biased towards the less viable sex and biases should be reversed by the end of the period of parental care so as to maintain equal (total) investment in the two sexes (Fisher 1930; Shaw and Mohler 1953; Leigh 1970).

More recently, a range of other factors that can affect the average sex ratio have been identified (West 2009). These include sex differences in parental investment of the kind predicted by Trivers and Willard (1973) and sex differences in the intensity of competition or cooperation between relatives that influence the relative costs of producing sons and daughters to the parents' fitness (Malcolm and Marten 1982; Gowaty and Lennartz 1985; Griffin *et al.* 2005). A range of other mechanisms, including trade-offs between the sex ratio and potential litter size and conflicts of interest between parents or between parents and offspring, may also affect the average sex ratio (Frank 1987; Seger and Stubblefield 2002). In addition, sex determining mechanisms may generate deviations from Fisher's principle and may themselves be affected by selection operating on the sex ratio (Kraak and Pen 2002; Pen and Weissing 2002). Unfortunately, the diversity of evolutionary theories concerning sex ratios means that it is often difficult to attribute the presence or absence of observed deviations from Fisher's principle to particular mechanisms with confidence and there is still disagreement over their relative importance (Cockburn *et al.* 2002).

1.6 Cooperation

Definitions

The evolution of cooperation presents evolutionary biologists with the intriguing problem of explaining why individuals should assist other animals at some cost to their own fitness (Gardner and Foster 2008). To explain cooperation, it is important to be clear about how it is defined. Evolutionary biologists view interactions between individuals as having four potential outcomes for the direct fitness of the participants (Figure 1.14): they can be mutually beneficial or 'mutualistic' if both partners assist each other and both gain 'direct' fitness benefits that increase their survival or breeding success (+/+); altruistic, if one partner assists another at some net cost to its own direct fitness (−/+); selfish, if one partner increases its own direct fitness at some cost to that of its partner (+/−); or spiteful, if one individual's actions reduce both its own fitness and that of its partner (−/−).

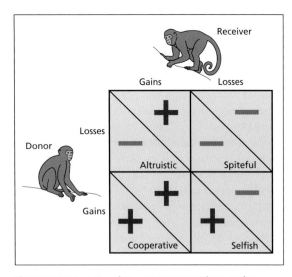

Figure 1.14 Interactions between group members can be allocated to four categories: selfish (+/−); mutually beneficial or mutualistic (+/+); altruistic (−/+); and spiteful (−/−)
Source: From Hauser *et al.* (2009). Reproduced with permission from the Royal Society.

Cooperation is usually taken to include behaviour that generates direct fitness benefits to cooperators through its benefits to others as well as altruistic behaviour that is maintained by indirect benefits (West *et al.* 2007). The inclusion of a requirement that the effects of cooperative behaviour on recipients affect selection operating on the cooperator is necessary because there are many cases where the actions of individuals generate coincidental, unselected benefits to others, which are usually referred to as 'by-product mutualisms' (Brown 1987). For example, individuals that react to approaching predators may alert other group members but their behaviour may be adapted only to maximising their own chances of survival. Although interactions of this kind are sometimes regarded as examples of cooperation, this is misleading since selection is not operating through behaviour that provides benefits to others.

If by-product mutualisms are excluded, explanations of cooperative behaviour in non-human animals fall into six main categories, described in the following sections.

Prestige

Some models of cooperation suggest that cooperative activities represent costly displays or 'handicaps' that increase the 'prestige' of individuals and attract mates or intimidate rivals (Zahavi and Zahavi 1997). While this is feasible, there is little empirical evidence that cooperative behaviour in non-human animals operates in this way (Wright *et al.* 2001; Nomano *et al.* 2013): individuals seldom compete to perform cooperative activities; dominants seldom contribute consistently more than subordinates; and this explanation does not coincide with our current understanding of the causal basis of cooperative behaviour (see Chapter 17).

Induced assistance

In some cases, the selfish behaviour of one (or more) individual induces others into providing assistance at some cost to their own net fitness. For example, stronger or more powerful individuals may coerce other individuals to assist them, using either direct force or the threat of punishment, or needy individuals may harass other group members until they provide assistance (Clutton-Brock and Parker 1995; Cant and Johnstone 2006).

Pseudo-reciprocity

In other cases, individuals may derive direct fitness benefits by assisting or stimulating others, whose responses are purely selfish (Connor 1995). Some inter-specific interactions provide examples of this kind of cooperative behaviour. For example, some ants provide food for fungus colonies whose growth provides food and so contributes to the ants' breeding success, but is not (necessarily) adapted to provide benefits to ants (Höll-dobler and Wilson 1990). In other cases, individuals may modify their behaviour to take advantage of the gener-alised responses of conspecifics: for example, by regularly associating with dominant individuals and grooming them repeatedly, subordinates may habituate them to their presence, gaining shelter from competition and interference by rank neighbours as a consequence (Bar-rett *et al*. 2002; Watts 2002; Cheney and Seyfarth 2008). Cases of this kind are referred to as 'pseudo-reciprocity' (Connor 1995; Leimar and Connor 2003) since the behaviour of one partner is adapted to providing benefits while its partner's responses are purely selfish.

Shared benefits

Both the *Origin of Species* and *The Descent of Man* show that Darwin was aware that social animals commonly coop-erate with each other. In *The Descent of Man*, he describes how 'wolves and some other beasts of prey hunt in packs, and aid one another in attacking their victims', how pelicans 'fish in concert' and 'social animals mutually defend each other'. Subsequent studies show that there are a number of birds and mammals where several breeding females pool their young and care for them together (Figure 1.15) and gain benefits by doing so (see Chapter 17).

Mutually beneficial interactions of this kind are common in social animals and are modelled in 'collective goods' or 'public goods' games (Hawkes 1992; Mester-ton-Gibboons and Dugatkin 1992; Sandler 1992; Dugatkin 1998; Nunn and Lewis 2001; Johnstone and Rodrigues 2016). The simplest models consider two unrelated players and only differ from Prisoner's Dilemma games in their pay-off structure while, in N-person games, all individuals have access to collective goods. Utilisation of collective goods can either have no effect on their availability or can reduce it, leading to competition between contributors; situations of the first kind are described as *non-rival* while those of the second are described as *rival* (Nunn and Lewis 2001).

Models seek to identify the conditions where cooper-ative behaviour is maintained and is not eroded by exploitative 'free-riding' strategies or other collective action problems (CAPs) (Ostrom 1990, 2003; Nunn and Lewis 2001). The exploitative strategies that they consider vary and include cases where free-riders con-tribute nothing to the creation of collective goods as well as cases where they contribute less than the benefits they receive (Marwell and Ames 1980; Sandler 1992). While cooperative behaviour involving non-kin that generates shared mutualistic benefits can be viewed as being main-tained through selection operating on individuals, it is also possible to argue that it reduces the fitness of indi-viduals relative to other group members but increases the fitness of groups of cooperators and so is maintained by group selection (Wilson 1975, 1977).

Cooperators may be able to reduce the risk of exploi-tation in 'public goods' games in a number of different ways. For example, they may associate selectively with each other or they may assist other group members only if they have previously received assistance from any other member of their group, a scenario known as *generalised reciprocity or contingent cooperation* (Pfeiffer *et al*. 2005; Rankin and Taborsky 2009; Taborsky *et al*. 2009 a,b). Alternatively, group members might monitor the frequency of cooperation between third parties (Nowak and Sigmund 1998a,b) or might punish defec-tors by reducing their own level of investment (Houston and Davies 1985) or by imposing some penalty on them (Clutton-Brock and Parker 1995; Andreoni *et al*. 2003).

In non-human animals, cooperation maintained by shared benefits is most likely to be evolutionarily stable if group size is small (Sandler 1992; Nunn and Lewis 2001) and synergistic effects of the presence of other individuals are strong (Kokko *et al*. 2001). For example, the relative importance of shared benefits may increase if there is some degree of labour division or role com-plementarity which helps to increase the efficiency of individuals working together (Olson 1965; Nunn and Lewis 2001). In addition, cooperation can be main-tained in 'public goods' games if group members police each other's activities and punish individuals that fail to cooperate (Boyd and Richerson 1992; Frank 1995; Ruxton and van der Meer 1997; Fehr and Gächter 2002), though individuals might also be expected to avoid contributing to punishing defectors and to free-ride on the efforts of others (Heckathorn 1989; Henrich and Boyd 2001) and empirical studies suggest that 'policing' behaviour in non-human animals is largely confined to groups consisting of close relatives or breed-ing partners.

(a)

(b)

Figure 1.15 Two communal breeders: in both (a) acorn woodpeckers and (b) African lions, several breeding females jointly provide care for their young. *Sources*: (a) © Bruce Lyon; (b) © Dave Hamman.

Reciprocity

In 'public goods' games, cooperating group members derive immediate benefits from assisting each other. However, in many animal societies, individuals assist each other at different times, sometimes giving assistance and sometimes receiving it. Where there are time delays between giving and receiving assistance, individuals might be expected to attempt to minimise their investment in assisting their partners while maximising the amount of assistance they receive, leading to selection for 'free riding'. This will, in turn, favour the evolution of strategies that minimise the risk that cooperation will be exploited by their partners.

One way for cooperators to avoid being exploited is to adjust the amount of assistance they give to particular partners the amount they receive from them. In 1971, Trivers published a classic paper arguing that cooperation between non-kin could be maintained where individuals that give assistance to others at some cost to their own fitness receive assistance from them in turn, so that although giving assistance has temporary costs, these are exceeded by subsequent benefits and both (or all) partners gain when net fitness benefits are calculated across several interactions. Trivers referred to this process as *reciprocal altruism* and used it to explain the evolution of interspecific cooperation as well as of intraspecific cooperation between unrelated individuals. Today, it is usually referred to as 'direct' or 'cost-counting' reciprocity.

As Trivers pointed out, reciprocal altruism is analogous to theoretical games devised by economists called the Prisoner's Dilemma (von Neumann and Morgenstein 1953). In these games, two individuals that cooperate both gain higher pay-offs than if they refuse to assist each other (defect), but the highest pay-off is gained by individuals that defect when their partners cooperate, while the lowest pay-off is to individuals that cooperate when their partners defect (Box 1.1 and Figure 1.16). In 'one-shot' encounters, these conditions always favour defection but, where partners interact repeatedly (as in iterated Prisoner's Dilemma, or IPD, games), strategies that involve copying each other's responses (such as tit-for-tat or TFT) can be stable.

Following the publication of Trivers' paper, the evolution of cooperation in Prisoner's Dilemma games was widely investigated (Axelrod 1984; Dugatkin 1997). Several models showed that the most successful versions of TFT strategies are slightly 'generous' ones, where individuals copy the previous behaviour of their

		Player 2: Cooperate	Player 2: Cheat (defect)
Player 1	Cooperate	R 1-year prison term	S 5-year prison term
	Cheat (defect)	T 0-year prison term	P 3-year prison term

Figure 1.16 Pay-offs for Player 1 in the original Prisoner's Dilemma game. If neither prisoner snitches on his mate ('cooperate'), they both get 1-year prison terms; if both snitch, they both get 3-year terms ('defect'). However, if one prisoner defects while the other cooperates, the defector goes free while the cooperater gets 5 years. The game is defined by the inequalities T > R > P > S and 2 R > T + S. *Source*: From Dugatkin (1997). Reproduced with permission of Oxford University Press, USA.

partners, cooperating when they do and forgiving occasional lapses but eventually responding to defection by ceasing to provide assistance (Nowak 2006). In general, direct reciprocity is most likely to be stable where the costs of assistance are low; where there is a high certainty of repayment; where the potential benefits of the exchange are high; and where the interval between exchanges is short (Trivers 1971, 2006; Wilkinson 1984).

A number of additional tactics that individuals involved in Prisoner's Dilemma type situations might use have been suggested. One way of reducing the chance of assisting non-cooperators is to monitor interactions between third parties, and cooperate selectively with frequent cooperators (Sugden 1986; Boyd and Richerson 1989; Nowak and Sigmund 1998a,b, 2005; Riolo *et al.* 2001; Wedekind and Braithwaite 2002). Cooperation maintained by third party effects (including models of 'reputation' and 'image scoring') is sometimes referred to as *indirect reciprocity* and has been the focus of recent experiments with humans (Milinski *et al.* 2002; McElreath 2003). Although indirect reciprocity will usually strengthen the benefits of cooperating (Nowak and

Box 1.1 Reciprocity and the Prisoner's Dilemma

The Prisoner's Dilemma game envisages a situation where two or more players can choose to cooperate with each other or can refuse to do so (von Neumann and Morgenstein 1953; Axelrod and Hamilton 1981; Dugatkin 1997; Trivers 2006) Refusals to cooperate are usually referred to as 'defecting' or 'cheating'. The example originally used to illustrate this situation is one where two suspects of a crime are interviewed by the police in separate rooms. If both suspects cooperate and keep silent, the police only have enough evidence to send them to prison for 1 year each but if both snitch on their partners or 'defect', the police can send them both down for 3 years (see Figure 1.17). However, if one snitches (defects) but the other doesn't (cooperates), the snitch walks free while the cooperator gets 5 years (Dugatkin 1997).

What is the best strategy trade-off under these conditions? In 'one-shot' encounters where the two players interact once, defection is the best strategy, but in repeated (iterated) versions of the same game, cooperative strategies can generate higher pay-offs than purely selfish ones (Axelrod and Hamilton 1981). A conditionally cooperative strategy, where individuals initially cooperate and subsequently imitate the response of their partner (tit-for-tat or TFT), is sometimes the best solution. TFT is successful because it combines three characteristics (Axelrod 1984): it is never the first to defect; it is swift to retaliate; and it is forgiving of past defection, in the sense that its memory does not extend back beyond the previous move.

Axelrod and Hamilton's initial work led to a spate of papers exploring factors affecting the outcome of different strategies in the Prisoner's Dilemma game (Trivers 2006). While many of them confirm that TFT is the most successful strategy under a range of conditions, some have shown that simple TFT can be beaten by TFT-like strategies that are slightly more forgiving (Boyd and Richerson 1989; Nowak and Sigmund 1992, 1993) or that a mixture of TFT and defecting strategies can be the stable outcome of iterated games (Peck and Feldman 1986; Dugatkin and Wilson 1991; Pollock and Dugatkin 1992). Others show that where TFT-like strategies predominate, a policy of 'win-stay, lose-shift' which repeats successful strategies and abandons less successful ones is favoured (Nowak and Sigmund 1993).

Theoretical studies have also explored a range of factors affecting the outcome of Prisoner's Dilemma games (Dugatkin 1997) and have shown the following.

- Increasing group size may hinder the evolution of cooperation (Boyd and Richerson 1988; Trivers 2006).
- Subdivisions within populations can increase the chance that cooperative strategies will resist invasion by defectors (Pollock 1989).
- Stochastic variation in responses can lead to cycles of cooperation and defection (Nowak and Sigmund 1993).
- Individual variation in behaviour can help to maintain cooperation (McNamara *et al.* 2004).
- Where players alternate between donor and recipient roles, this favours generous TFT over win-stay, lose-shift strategies (Nowak and Sigmund 1994).
- Kinship between players is likely to promote cooperative strategies (Queller 1985).
- Dispersing or 'roving' defectors will hinder the evolution of cooperation (Dugatkin and Wilson 1991).
- Punishment of defectors promotes cooperation (Boyd and Richerson 1992).
- Stochastic 'mistakes' in the responses of players can affect the outcome of games, sometimes hindering the evolution of cooperative strategies (Hirshleifer and Coll 1988; Boyd and Richerson 1989; Stephens *et al.* 1997).
- Monitoring the responses of other group members in reactions with third parties and adjusting responses to them on the basis of their 'reputation' ('observer TFT') can invade populations of individuals playing simple TFT (Pollock and Dugatkin 1992).
- Competition for cooperative partners can help to maintain and increase cooperative behaviour (Roberts 1998).

Related work has explored the effects of more fundamental changes in the assumptions underlying models of the Prisoner's Dilemma (see Trivers 2006). Strategies that can vary between individuals may replace fixed inflexible ones (Johnstone and Godfray 2002). Similarly, strategies that initially invest little in cooperative behaviour but increase their investment after they interact with other cooperators are often favoured (Roberts and Sherratt 1998). The division of investment in cooperative behaviour into small units can limit the potential costs of defectors and favour the evolution of prolonged exchanges of small acts of assistance (Fischer 1988; Trivers 2006). Finally, some recent models envisage situations intermediate between those considered by traditional Prisoner's Dilemma models and models of generalised reciprocity or group augmentation. For example, if purely selfish actions by one individual generate independent, coincidental benefits to another, selection can favour coordinated cooperative investment (Hauert *et al.* 2002; Hauert and Doebeli 2004).

Sigmund 1998a), minor differences in assumptions can lead to different outcomes (Leimar and Hammerstein 2001; Leimar and Connor 2003). For example, the presence of individuals that cannot afford to contribute to cooperative activities broadens the range of conditions under which cooperation is maintained as a result of selection against both defectors and unconditional assistance.

Examples of direct reciprocity are common in humans (see Chapter 20) but are rare in non-human animals and there is a growing perception that the situation considered by IPD models ignores many important features of cooperative interactions and relationships among animals (Noë and Hammerstein 1994; Clutton-Brock 2002, 2009b; Hammerstein 2003; Trivers 2006). These include communication between partners concerning their intentions (Smuts and Watanabe 1990), the capacity of individuals to modify their behaviour in the course of repeated interactions (McNamara *et al.* 1999), variation in partner quality (Noë 1992), the ability of individuals to choose between several alternative partners (Enquist and Leimar 1993) and the effects of different tactics on the individual's subsequent ability to attract or retain partners (Smuts and Watanabe 1990; McNamara *et al.* 1999).

Moreover, it is questionable how often cooperative interactions between animal partners involve an alternation of costs and benefits since fitness benefits should be accounted prospectively and it is frequently unclear that there really are delays between incurring costs and gaining benefits, as models of reciprocal cooperation based on IPD assume. Many cooperative interactions between non-kin occur in the context of well-established supportive relationships that benefit both parties (see Chapter 9) and repeated failure to provide assistance to a regular ally may cause them to engage in punishing tactics or lead to supportive relationships unravelling. Not only may defectors subsequently have difficulty in finding new partners and establishing cooperative relationships with them (Hirshleifer and Rasmusen 1989; Pollock and Dugatkin 1992) but previous partners may form new alliances and re-emerge as competitors. Where providing assistance reduces these dangers, it may be more realistic to account it as providing immediate net benefits rather than net costs that are subsequently offset by greater net benefits. Arguments of this kind suggest that many interaction that have been interpreted as examples of reciprocity may, in practice, represent cases where interacting individuals gain immediate shared benefits from their actions and may be more realistically modelled as restricted 'public goods' games (see Chapter 9).

The direct benefits of cooperating with particular individuals (whether they are kin or non-kin) are likely to be affected both by the partner's ability to contribute to cooperative activities and by their intrinsic skill or power. For example, where individuals provide each other with support in competitive interactions with other group members, alliances with dominant partners are likely to yield larger benefits than alliances with subordinates (see Chapters 9 and 14). Since the availability of desirable partners will usually be limited, individuals might be expected to compete for the best allies and that they appear to do so in several social mammals (Noë and Hammerstein 1994, 1995; Roberts 1998; Noë *et al.* 2001). Some of the most detailed evidence of competition comes from studies of social primates where group members use grooming to establish alliances and tolerate larger asymmetries in grooming when interacting with higher ranking animals (Harcourt 1988, 1989, 1992; Cords 2000).

Cases where competition for social partners generates asymmetries in the provision of assistance are often referred to as 'market effects' to emphasise their similarity to exchanges of services between humans, where the amount that individuals will pay for a resource depends on its availability relative to the number of individuals that need it (Noë and Hammerstein 1994, 1995). Although the presence of market effects is sometimes presented as a prediction of models of cooperation based on reciprocity, they are likely to be common wherever the productivity of different potential partners varies or opportunities for cooperation are limited by the supply of partners. For example, individuals might be expected to compete to form mutualistic relationships with powerful individuals or to coerce individuals whose behaviour is most likely to influence their fitness. As a result, the presence of market effects provides no strong evidence that direct reciprocity is involved or that interactions between individuals resemble the exchanges that occur in human markets.

Altruistic assistance

In some animal societies, a single female in each group monopolises reproduction and her offspring are reared by non-breeding helpers of either or both sexes. Cooperative breeding systems of this kind occur in insects, birds and mammals (Figures 1.15, 1.17 and 1.18). In

Figure 1.17 In some birds, like pied babblers, social groups consist of a single breeding female and her mate, that are assisted in raising young by non-breeding helpers of both sexes. In this photograph, an adult helper brings food to a dependent juvenile. *Source*: © Alex Thompson.

some cooperative vertebrates, like naked mole rats, and many social insects, breeding females and subordinate helpers differ in size and shape as well as in physiology and behaviour (Figure 1.19), while in some social insects (which live in far larger colonies than any cooperative vertebrate), workers are obligately sterile (Figure 1.20). Some authorities refer to species where breeders and helpers differ in morphology as 'eusocial' while others only use the term for species where helpers are obligately sterile. On the first definition, a number of mammals that breed cooperatively would be classed as eusocial while, on the second, no vertebrates would be eusocial (see Chapter 17).

Although most of the examples of cooperative behaviour that Darwin discusses in *The Descent of Man* are examples of shared benefits (see above), he also recognised the existence of cases where individuals provide assistance but do not receive it and appreciated that these represented a serious problem for the theory of natural selection. In the *Origin of Species* (Chapter VIII, p. 228) he describes how the problem of the evolution of sterile females in social insects initially appeared to be 'insuperable, and actually fatal to my whole theory', but then goes on to explain how the problem 'disappears,

when it is remembered that selection may be applied to the family, as well as to the individual, and may thus gain the desired end' (p. 230), presaging developments in our understanding of social evolution that did not occur for another 100 years.

Just over 100 years later, W.D. Hamilton produced the first formal models of animal altruism, laying one of the cornerstones of our current understanding of animal societies. In contrast to many of his contemporaries working on vertebrates, Hamilton was familiar with formal evolutionary theory and population genetics and his thinking owed much to Fisher. In an account of his early work (Hamilton 1988), he describes how his interest in the evolution of insect societies

> began for me while I was an undergraduate reading natural sciences at the University of Cambridge in 1958. I discovered R.A. Fisher's *The Genetical Theory of Natural Selection* in the St John's College Library and immediately realised that this was the key to the understanding of evolution that I had long wanted. I became a Fisher freak and neglected whole courses in my efforts to grasp the book's extremely compressed style and reasoning. I quickly noticed, however, that Fisher's arguments implied a basically different interpretation of adaptation from what I was hearing from most of my

Figure 1.18 In Kalahari meerkats, most groups consist of a single breeding pair of adults and a variable number of non-breeding helpers of both sexes. *Source*: © Tim Clutton-Brock.

lecturers and reading in other books. Was adaptation mainly for the benefit of species (the lecturers' view) or for the benefit of individuals (Fisher's view)? Clearly it was Fisher who had thought out his Darwinism properly; where interpretations differed, therefore, he must be right – but were the others *always* wrong? I started on what seemed the key theme in this puzzle – *altruism*. Did it exist? Could one evolve it in a model?

Figure 1.19 Some of the more specialised cooperative breeders, like naked mole rats live in large colonies; breeding females ('queens') and helpers differ in size and shape, as well as in physiology and behaviour. *Source*: © Lorna Ellen Faulkes.

In 1963, Hamilton published a brief paper arguing that altruism could evolve if it benefited the fitness of relatives and the following year published more formal models of this process (Hamilton 1964). To account for the evolution of cooperation between relatives, he introduced the concept of 'inclusive' fitness, consisting of the 'direct' fitness individuals derive from producing descendants and the 'indirect' fitness that they derive from helping non-descendant relatives minus any benefits received from them. Subsequently Maynard Smith (1964) named the process by which indirect benefits accrue as *kin selection* to distinguish it from group selection. The mathematics incorporated in Hamilton's original papers are complex, but the kernel of his theory can be presented as a simple equation (usually referred to as Hamilton's Rule) which states that a gene will increase in frequency whenever $B/C > 1/r$ where B represents the benefit of the trait that the gene codes for derived by the recipient of assistance, C the costs of assistance to the 'donor' and r Wright's coefficient of relatedness. Mechanisms capable of generating increased levels of r and important indirect fitness benefits include limited dispersal (Hamilton 1964; West *et al.* 2002) and kin discrimination based either on environmental cues (such as prior association) or on genetic mechanisms (such as the recognition of shared

Figure 1.20 Much larger breeding groups occur in social insects, where sterile workers rear the offspring of one or more breeding females, as in yellow meadow ants. The large size of colonies has led to the evolution of specialised workers and larger, specialised queens that are able to produce large numbers of eggs. *Source*: © Tom Houslay.

heritable odours or other characteristics) (Mateo 2002; West *et al.* 2007).

Hamilton's Rule makes intuitive sense. Under most conditions, relatives should be most likely to assist each other where coefficients of relatedness are high, the costs of providing assistance are low and the potential benefits of assistance to recipients are large (Wilson 1974; Dugatkin 1997). However, there are difficulties in incorporating inclusive fitness in formal models of population genetics, especially where natural selection is strong (Cavalli-Sforza and Feldman 1978; Uyenoyama 1984; Mueller and Feldman 1988; Dugatkin 1997). Measuring the indirect benefits and costs of cooperation is also more difficult than it might initially appear since actions that are likely to increase the fitness of non-descendant kin may also contribute to the helper's own fitness or to the fitness of its descendants (Brown 1987; Creel 1990). For example, estimating the costs of cooperation to resident workers or helpers in cooperative societies makes it necessary to decide whether individuals delay dispersal in order to assist relatives (in which case costs may be large) or whether they remain in their natal group to

maximise their survival and their chance of breeding and costs should be measured as the marginal costs of helping per se (in which case costs may be small) (see Chapter 17). In many cooperative species, helpers are the parents of some of the offspring being raised, so that helping may also increase the helper's direct fitness. Finally, assistance may affect the fitness of donors and recipients throughout their entire lifespans, so that estimates of *B* based only on short-term effects (such as the survival of nestlings) are likely to underestimate their real magnitude. Despite these problems, the components of Hamilton's equation have now been measured successfully in several natural systems and its predictions have been tested and verified (Bourke 2014; Hatchwell *et al.* 2014).

Theoretical research since 1964 has clarified the measurement of inclusive fitness and its assumptions (Queller 1996; Frank 2013; West and Gardner 2013) and explored its links to other branches of evolutionary theory (Taylor 1996; Gardner *et al.* 2007), though some theoreticians continue to be critical of its assumptions and argue that inclusive fitness will only be maximised by selection under limited circumstances (Nowak *et al.* 2010; Allen *et al.* 2013). However, the large number of cases where inclusive fitness theory has generated testable predictions about the evolution of social or reproductive strategies that have subsequently been verified by empirical research provides substantial evidence of the insights that it has provided (Abbot *et al.* 2011).

The central importance of kinship in understanding the evolution of animal societies has been recognised in a wide range of empirical studies and emphasised by many reviews (Trivers and Hare 1976; Grafen 1984; Bourke and Franks 1995; Silk 2009; West and Gardner 2013). Across animal species, costly forms of cooperative behaviour are largely confined to species where group members are, on average, closely related, and even in species where members of one sex live in groups consisting of relatives while members of the other sex associate with unrelated individuals, the sex that usually associates with relatives typically shows a greater development of cooperative behaviour (Boomsma 2009; Clutton-Brock 2009b). In addition, where non-breeding individuals have opportunities to choose between joining and assisting close relatives, distant relatives or unrelated individuals, they typically show a strong preference for joining close relatives (Emlen and Wrege 1988; Hatchwell *et al.* 2001; Hatchwell 2007), and where groups include a mixture of related and unrelated animals, individuals are often more cooperative to close kin than to non-kin or distant relatives (Boncoraglio and Saino 2008; Hatchwell 2009; Silk 2009).

Cheats and defectors

Cooperative behaviour invites exploitation and selection is often likely to favour the evolution of strategies that are adapted to exploit cooperative behaviour in other animals (Brembs 1996; West *et al.* 2007). The existence of cheating strategies that exploit the behaviour of cooperators has been demonstrated in bacterial populations (West *et al.* 2006). For example, in the pathogenic bacterium *Pseudomonas aeruginosa*, some individuals produce iron-scavenging agents (siderophores) that benefit local populations at a cost to their own fitness while others do not (Griffin *et al.* 2004).

As the frequency of cheats increases in populations, the density of individuals that can be exploited falls and selection is likely to strengthen the defences of cooperators so that the benefits of cheating strategies are likely to be negatively frequency dependent. If this eventually leads to a situation where the fitness of cheats is lower than that of cooperators, a stable equilibrium may develop; however, if the relative fitness of cheats continues to exceed that of cooperators as their frequency rises, cooperation will become extinct (Ross-Gillespie *et al.* 2007; Ghoul *et al.* 2014). Consequently, where cheating strategies and cooperators both persist in the same populations, we can expect the fitness of cheats to show negative frequency dependence and several studies support this prediction (Ghoul *et al.* 2014).

Large individual differences in contributions to cooperative behaviours are also common in many cooperative vertebrates. However, in most cases, these do not appear to be associated with contrasts in development or reproductive strategies and they are commonly related to contrasts in age, size, sex, reproductive status or nutrition. Many of them probably reflect variation in the costs and benefits of providing assistance (see Chapter 17). For example, in many cooperative breeders, hungry helpers that are in relatively poor condition give a smaller proportion of the food they find to dependent young and the fitness costs they incur for the amount of provisioning they do are not necessarily lower than those of better-fed individuals that contribute more and could be similar or even higher.

Unfortunately, variation in cooperative behaviour is often interpreted as evidence of cheating and individuals that contribute relatively little are often referred to as 'cheats' or 'defectors' when their behaviour may be a result of individual differences in the costs or benefits of contributing to cooperative activities, so that the incidence of cheating in natural populations is likely to be overestimated. So how common is cheating in cooperative societies? Few studies of non-human mammals have yet produced unequivocal evidence of cheating strategies and several which explored behaviour that initially looked like examples of cheating have concluded that the behaviour is a consequence of tactical decisions based on last-minute assessments of the needs of juveniles rather than of cheating strategies (see Chapter 17). One possible explanation is that, where cooperation is highly developed, cooperative behaviour usually provides net benefits that are sufficiently large that cheating is not favoured, while another is that cooperative species have evolved mechanisms that safeguard individuals against potential cheats.

1.7 Loaded labels

Like 'cheating' strategies, many descriptions of social strategies derived from theoretical models carry implications about the intentions of the individuals involved and the functions of their behaviour: examples include teaching, exchanges, punishment, policing, reconciliation and consolation. The use of these labels provides a convenient shorthand description of strategies and can indicate the way in which evolutionary processes are likely to operate (West *et al.* 2007; Ghoul *et al.* 2014) but also has disadvantages. The implication that strategies are intentional has led to arguments between biologists and psychologists (see Chapter 6), though evolutionary biologists are usually aware of this problem and avoid it by adopting functional definitions of strategies that avoid implications about intentions. A more serious problem is that strategy labels (like many of those described) carry implications about the functions of behaviour. As a result, their application to empirical examples often assumes answers to precisely the questions that need to be asked. 'Cheating' is an obvious case where the uncritical application of a strategy label to observed behaviour is often likely to be misleading. Similarly, it can be misleading to refer to examples of mutual assistance as exchanges unless there is real evidence of reciprocity and an alternation of costs and benefits (see Chapter 9). There are many other examples and it is not possible to avoid using strategy labels altogether, but it is important to question the reality of their implications and to find less misleading labels for observed behaviour where possible.

SUMMARY

1. The structure of animal societies and breeding systems exerts profound effects on almost all evolutionary and ecological processes so that an understanding of their diversity and distribution is of central importance in research on virtually all aspects of organismal biology.
2. Although the existence of animal societies has been recognised since classical times, systematic descriptions of animal societies and social behaviour were uncommon before 1960. However, during the 1960s and the 1970s, long-term field studies that were able to recognise individuals and track their life histories began to provide detailed insight into the diversity of social behaviour and social organisation and its causes and consequences.
3. While early explanations of contrasts in social behaviour among animals frequently suggested that they served to increase the survival of groups or populations, theoretical research between 1960 and 1980 demonstrated the extent to which the interests of individuals differ and laid the basis for our current understanding of social evolution based on selection operating at the level of individuals or genes.
4. Research on breeding competition between individuals shows that individual differences in breeding success are often large in both sexes and that individuals compete intensely for breeding opportunities, though individuals may benefit by avoiding fights and settling competitive interactions in other ways where the costs of escalated fights are high. Empirical studies have shown how the factors affecting breeding success commonly differ between the sexes, with female breeding success often depending primarily on access to resources and male success depending principally on access to females.
5. A combination of theoretical and empirical research (which continues to be extended) now provides a framework for explaining the evolution of mate choice and parental care. Early studies showed how the sex that invests most heavily in individual offspring is often more selective in its choice of mating partners than the sex that invests less heavily, while the latter competes more intensely for access to mates than the former. More recent research has refined this generalisation and shown how both sexes frequently compete for mates and are selective of mating partners and how sex differences in competitiveness and choosiness can vary between and within populations and individuals.
6. A combination of theoretical and empirical research has also provided a basis for understanding the evolution of animal cooperation. While several evolutionary mechanisms have the capacity to favour the evolution of cooperative behaviour, most examples of animal cooperation are either cases where both (or all) cooperating individuals gain net fitness benefits from assisting each other or cases where individuals provide assistance to descendent or

non-descendent relatives and so gain net benefits to indirect components of their fitness or (in many cases) a combination of both processes. Most examples of animal cooperation involving the provision of assistance that has substantial costs involve interactions between relatives, while cooperation between non-relatives seldom involves costly actions.

7. Strategy labels generated by theoretical studies are commonly used to describe observed differences in behaviour. While this can stimulate thinking and lead to novel lines of research, it frequently assumes answers to the precise functional questions that need to be asked. It is also important to remember that different theoretical models commonly generate similar predictions, so that a match between the general predictions of a model and empirical results does not necessarily indicate that the model is realistic or that evolution has operated in the way that it suggests.

References

Abbot, P., *et al.* (2011) Inclusive fitness theory and eusociality. *Nature* **471**:E1–E4.

Ahnesjö, I., *et al.* (2001) Using potential reproductive rates to predict mating competition among individuals qualified to mate. *Behavioral Ecology* **12**:397–401.

Alexander, R.D. (1974) The evolution of social behavior. *Annual Review of Ecology and Systematics* **5**:325–383.

Allen, J., *et al.* (2013) Network-based diffusion analysis reveals cultural transmission of lobtail feeding in humpback whales. *Science* **340**:485–488.

Alonzo, S.H. and Klug, H. (2012) Paternity, maternity, and parental care. In: *The Evolution of Parental Care* (eds N.J. Royle, P. T. Smiseth and M. Kölliker). Oxford: Oxford University Press, 189–205.

Altmann, J., *et al.* (1996) Behaviour predicts genetic structure in a wild primate group. *Proceedings of the National Academy of Sciences of the United States of America* **93**:5797–5801.

Altmann, S.A. (1962) A field study of sociobiology of rhesus monkeys, *Macaca mulatta*. *Annals of the New York Academy of Sciences* **102**:338–435.

Amundsen, T. and Forsgren, E. (2001) Male mate choice selects for female coloration in a fish. *Proceedings of the National Academy of Sciences of the United States of America* **98**:13155–13160.

Andersson, M. (1982) Female choice selects for extreme tail length in a widowbird. *Nature* **299**:818–820.

Andersson, M. (1994) *Sexual Selection*. Princeton, NJ: Princeton University Press.

Andreoni, J., *et al.* (2003) The carrot or the stick: rewards, punishments, and cooperation. *American Economic Review* **93**:893–902.

Armitage, K.B. (1962) Social behaviour of a colony of the yellow-bellied marmot (*Marmota flaviventris*). *Animal Behaviour* **10**:319–331.

Arnqvist, G. and Rowe, L. (2005) *Sexual Conflict*. Princeton, NJ: Princeton University Press.

Axelrod, R. (1984) *The Evolution of Cooperation*. New York: Basic Books.

Axelrod, R. and Hamilton, W.D. (1981) The evolution of cooperation. *Science* **211**:1390–1396.

Barrett, L., *et al.* (2002) A dynamic interaction between aggression and grooming reciprocity among female chacma baboons. *Animal Behaviour* **63**:1047–1053.

Bateman, A.J. (1948) Intra-sexual selection in *Drosophila*. *Heredity* **2**:349–368.

Berger, J. (1989) Female reproductive potential and its apparent evaluation by male mammals. *Journal of Mammalogy* **70**:347–358.

Birkhead, T.R. and Møller, A.P. (1992) *Sperm Competition in Birds: Evolutionary Causes and Consequences*. London: Academic Press.

Birkhead, T.R. and Møller, A.P. (1993) Sexual selection and the temporal separation of reproductive events: sperm storage data from reptiles, birds and mammals. *Biological Journal of the Linnean Society* **50**:295–311.

Bjork, A. and Pitnick, S. (2006) Intensity of sexual selection along the anisogamy–isogamy continuum. *Nature* **441**:742–745.

Boltzman, L. (1905) *Populare Schriften*. Leipzig: J.A. Barth.

Boncoraglio, G. and Saino, N. (2008) Barn swallow chicks beg more loudly when brood mates are unrelated. *Journal of Evolutionary Biology* **21**:256–262.

Boomsma, J.J. (2007) Kin selection versus sexual selection: why the ends do not meet. *Current Biology* **17**:R673–R683.

Boomsma, J.J. (2009) Lifetime monogamy and the evolution of eusociality. *Philosophical Transactions of the Royal Society B: Biological Sciences* **364**:3191–3207.

Boomsma, J.J. and Grafen, A. (1991) Colony-level sex ratio selection in the eusocial Hymenoptera. *Journal of Evolutionary Biology* **4**:383–407.

Bourke, A. (2011) *Principles of Social Evolution*. Oxford: Oxford University Press.

Bourke, A. (2014) Hamilton's rule and the causes of social evolution. *Philosophical Transactions of the Royal Society B: Biological Sciences* **369**:20130362.

Bourke, A.F.G. and Franks, N.R. (1995) *Social Evolution in Ants*. Princeton, NJ: Princeton University Press.

Boyd, R. and Richerson, P.J. (1988) The evolution of reciprocity in sizable groups. *Journal of Theoretical Biology* **132**:337–356.

Boyd, R. and Richerson, P.J. (1989) The evolution of indirect reciprocity. *Social Networks* **11**:213–236.

Boyd, R. and Richerson, P.J. (1992) Punishment allows the evolution of cooperation (or anything else) in sizable groups. *Ethology and Sociobiology* **13**:171–195.

Bradbury, J.W. and Vehrencamp. S.L. (1977) Social organization and foraging in emballonurid bats. III. Mating systems. *Behavioral Ecology and Sociobiology* **2**:1–17.

Brembs, B. (1996) Chaos, cheating and cooperation: potential solutions to the Prisoner's Dilemma. *Oikos* **76**:14–24.

Brooks, R. (2000) Negative genetic correlation between male sexual attractiveness and survival. *Nature* **406**:67–70.

Brown, C.R. and Brown, M.B. (1986) Ectoparasitism as a cost of coloniality in cliff swallows (*Hirundo pyrrhonota*). *Ecology* **67**:1206–1218.

Brown, J.L. (1987) *Helping and Communal Breeding in Birds*. Princeton, NJ: Princeton University Press.

Cant, M.A. and Johnstone, R.A. (2006) Self-serving punishment and the evolution of cooperation. *Evolutionary Biology* **19**:1383–1385.

Carpenter, C.R. (1934) A field study of the behavior and social relations of howling monkeys (*Alouatta palliata*). *Comparative Psychology Monographs* **10**:1–168.

Carpenter, C.R. (1935) Behavior of red spider monkeys in Panama. *Journal of Mammalogy* **16**:171–180.

Carpenter, C.R. (1942) Sexual behaviour of free-ranging rhesus monkeys *Macaca mulatta*. I. Specimens, procedures and behavioral characteristics of estrus. *Journal of Comparative Psychology* **33**:113–142.

Carranza, J. (2009) Defining sexual selection as sex-dependent selection. *Animal Behaviour* **77**:749–751.

Carranza, J. (2010) Sexual selection and the evolution of evolutionary theories. *Animal Behaviour* **79**:e5–e6.

Cavalli-Sforza, L.L. and Feldman, M.W. (1978) Darwinian selection and altruism. *Theoretical Population Biology* **14**:268–280.

Cézilly, F. (2008) A history of behavioural ecology. In: *Behavioural Ecology* (eds E. Danchin and L. A. Giraldeau) Oxford: Oxford University Press, 3–27.

Chapais, B. (1992) The role of alliances in social inheritance of rank among female primates. In: *Coalitions and Alliances in Humans and Other Animals* (eds A. H. Harcourt and F. B. M. de Waal) Oxford: Oxford University Press, 29–60.

Charmantier, A., *et al.* (eds) (2014) *Quantitative genetics in the wild*. Oxford University Press, Oxford.

Charnov, E.L. (1982) *The Theory of Sex Allocation*. Princeton, NJ: Princeton University Press.

Charnov, E.L. and Krebs, J.R. (1974) On clutch size and fitness. *Ibis* **116**:217–219.

Cheney, D.L. and Seyfarth, R.M. (2008) *Baboon Metaphysics: The Evolution of a Social Mind*. Chicago: University of Chicago Press.

Chepko-Sade, B.D. and Sade, D.S. (1979) Patterns of group splitting within matrilineal kinship groups. *Behavioral Ecology and Sociobiology* **5**:67–86.

Clutton-Brock, T.H. (1983) Selection in relation to sex. In: *Evolution from Molecules to Men* (ed. B.J. Bendall). Cambridge: Cambridge University Press, 457–481.

Clutton-Brock, T.H. (1988) *Reproductive success. In: Reproductive Success* (ed. T.H. Clutton-Brock). Chicago: University of Chicago Press, 472–486.

Clutton-Brock, T.H. (1991) *The Evolution of Parental Care*. Princeton, NJ: Princeton University Press.

Clutton-Brock, T.H. (2002) Breeding together: kin selection and mutualism in cooperative vertebrates. *Science* **296**:69–72.

Clutton-Brock, T.H. (2004) What is sexual selection? In: *Sexual Selection in Primates: New and Comparative Perspectives* (eds P.M. Kappeler and C. P. van Schaik) Cambridge: Cambridge University Press, 24–36.

Clutton-Brock, T.H. (2006) Cooperative breeding in mammals. In: *Cooperation in Primates and Humans* (eds P. M. Kappeler and C. P. van Schaik) Berlin: Springer Verlag, 173–190.

Clutton-Brock, T.H. (2007) Sexual selection in males and females. *Science* **318**:1882–1885.

Clutton-Brock, T.H. (2009a) Structure and function in mammalian societies. *Philosophical Transactions of the Royal Society B: Biological Sciences* **364**:3229–3242.

Clutton-Brock, T.H. (2009b) Cooperation between non-kin in animal societies. *Nature* **462**:51–57.

Clutton-Brock, T.H. (2010) We do not need a Sexual Selection 2.0 – nor a theory of Genial Selection. *Animal Behaviour* **79**: e7–e10.

Clutton-Brock, T.H. and Coulson, T.N. (2002) Ungulate population dynamics: the devil is in the detail. *Philosophical Transactions of the Royal Society B: Biological Sciences* **357**:1285–1298.

Clutton-Brock, T.H. and Harvey, P.H. (1977a) Species differences in feeding and ranging behaviour in primates. In: *Primate Ecology: Studies of Feeding and Ranging Behaviour in Lemurs, Monkeys and Apes* (ed. T.H. Clutton-Brock). London: Academic Press, 557–584.

Clutton-Brock, T.H. and Harvey, P.H. (1977b) Primate ecology and social organisation. *Journal of Zoology* **183**:1–39.

Clutton-Brock, T.H. and Harvey, P.H. (1980) Primates, brains and ecology. *Journal of Zoology* **190**:309–323.

Clutton-Brock, T.H. and Huchard, E. (2013a) Social competition and its consequences in female mammals. *Journal of Zoology* **289**:151–171.

Clutton-Brock, T.H. and Huchard, E. (2013b) Social competition and selection in males and females. *Philosophical Transactions of the Royal Society B: Biological Sciences* **368**:20130074.

Clutton-Brock, T.H. and Parker, G.A. (1992) Potential reproductive rates and the operation of sexual selection. *Quarterly Review of Biology* **67**:437–456.

Clutton-Brock, T.H. and Parker, G.A. (1995) Punishment in animal societies. *Nature* **373**:209–216.

Clutton-Brock, T.H. and Vincent, A.C.J. (1991) Sexual selection and the potential reproductive rates of males and females. *Nature* **351**:58–60.

Clutton-Brock, T.H., *et al.* (1979) The logical stag: adaptive aspects of fighting in red deer (*Cervus elaphus* L.). *Animal Behaviour* **27**:211–225.

Clutton-Brock, T.H., *et al.* (1982) *Red Deer: The Behaviour and Ecology of Two Sexes*. Chicago: University of Chicago Press.

Clutton-Brock, T.H., *et al.* (1985) Parental investment and sex differences in juvenile mortality in birds and mammals. *Nature* **313**:131–133.

Cockburn, A. (2014) Behavioural ecology as big science: 25 years of asking the same questions. *Behavioral Ecology* **25**:1283–1286.

Cockburn, A., *et al.* (2002) Sex ratios in birds and mammals: can the hypotheses be disentangled? In: *Sex Ratios, Concepts and Research Methods* (ed. I.C.W. Hardy). Cambridge: Cambridge University Press, 266–286.

Cohen, J.E. (1971) *Casual Groups of Monkeys and Men: Stochastic Models of Elemental Social Systems*. Cambridge, MA: Harvard University Press.

Cohen, J.E. (1975) The size and demographic composition of social groups of wild orang-utans. *Animal Behaviour* **23**:543–550.

Cole, L.C. (1954) The population consequences of life history phenomena. *Quarterly Review of Biology* **29**:103–137.

Comings, D.E. and MacMurray, J.P. (2000) Molecular heterosis: a review. *Molecular Genetics and Metabolism* **71**:19–31.

Connor, R.C. (1995). Altruism among non-relatives: alternatives to the 'prisoner's dilemma'. *Trends in Ecology and Evolution* **10**:84–86.

Connor, R.C. and Smolker, R.S. (1985) Habituated dolphins (*Tursiops* sp.) in western Australia. *Journal of Mammalogy* **66**:398–400.

Cords, M. (2000) Grooming partners of immature blue monkeys (*Cercopithecus mitis*) in the Kakamega Forest, Kenya. *International Journal of Primatology* **21**:239–254.

Cornwallis, C.K. and Uller, T. (2010) Towards an evolutionary ecology of sexual traits. *Trends in Ecology and Evolution* **25**:145–152.

Creel, S. (1990) How to measure inclusive fitness. *Proceedings of the Royal Society of London. Series B: Biological Sciences* **241**:229–231.

Crook, J.H. (1962) The adaptive significance of pair formation types in weaver birds. *Symposium of the Zoological Society, London* **8**:57–70.

Crook, J.H. (1964) The evolution of social organisation and visual communication in the weaver birds (Ploceinae). *Behaviour Supplement* **10**:1–178.

Crook, J.H. (1965) The adaptive significance of avian social organisations. *Symposium of the Zoological Society, London* **14**:181–218.

Crook, J.H. (1972) Sexual selection, dimorphism and social organization in the primates. In: *Sexual Selection and the Descent of Man* (ed. B. Campbell). Chicago, IL: Aldine, 231–281.

Crook, J.H. and Gartlan, J.S. (1966) Evolution of primate societies. *Nature* **210**:1200–1203.

Cumming, J.M. (1994) Sexual selection and the evolution of dance fly mating systems (Diptera: Empididae; Empidinae). *Canadian Entomologist* **126**:907–920.

Danchin, E. and Cézilly, F. (2008) Sexual selection: another evolutionary process. In: *Behavioural Ecology* (eds E. Danchin, L.-A. Giraldeau and F. Cézilly). Oxford: Oxford University Press, 363–426.

Danchin, E., *et al.* (eds) (2008) *Behavioural Ecology*. Oxford: Oxford University Press.

Darling, F.F. (1937a) *A Herd of Red Deer*. Oxford: Oxford University Press.

Darling, F.F. (1937b) Habits of wild goats in Scotland. *Journal of Animal Ecology* **6**:21–22.

Darling, F.F. (1943) *Island Farm*. G. Bell.

Darwin, C. (1859) *On The Origin of Species By Means of Natural Selection*. London: John Murray.

Darwin, C. (1871) *The Descent of Man, and Selection in Relation to Sex*. London: John Murray.

Davies, N.B. (1985) Cooperation and conflict among dunnocks, *Prunella modularis*, in a variable mating system. *Animal Behaviour* **33**:628–648.

Davies, N.B. (1989) Sexual conflict and the polygamy threshold. *Animal Behaviour* **38**:226–234.

Davies, N.B. (1992) *Dunnock Behaviour and Social Evolution*. Oxford: Oxford University Press.

Davies, N.B., *et al.* (2012) *An Introduction to Behavioural Ecology*. Oxford: Wiley-Blackwell.

DeVore, I. (1965) *Primate Behavior: Field Studies of Monkeys and Apes*. New York: Holt, Rinehart and Winston.

Drea, C.M., *et al.* (2002) Exposure to naturally circulating androgens during fetal life is prerequisite for male mating but incurs direct reproductive costs in female spotted hyenas. *Proceedings of the Royal Society of London. Series B: Biological Sciences* **269**:1981–1987.

Dugatkin, L.A. (1992) Sexual selection and imitation: females copy the mate choice of others. *American Naturalist* **139**:1384–1389.

Dugatkin, L.A. (1997) *Cooperation Among Animals: An Evolutionary Perspective*. Oxford: Oxford University Press.

Dugatkin, L.A. (1998) A model of coalition formation in animals. *Proceedings of the Royal Society of London. Series B: Biological Sciences* **265**:2121–2125.

Dugatkin, L.A. and Wilson, D.S. (1991) Rover: a strategy for exploiting cooperators in a patchy environment. *American Naturalist* **138**:687–701.

Düsing, C. (1883) Die Factoren welche die Sexualität entscheiden. *Jenaische Zeitschrift für Naturwissenschaft* **16**:428–464.

Düsing, C. (1884a) Die regulierung des geschlechtsverhältnisses bei der vermehrung der menschen, thiere und pflanzen. *Jenaische Zeitschrift für Naturwissenschaft* **17**:593–940.

Düsing, C. (1884b) *Die regulierung des geschlechtsverhältnisses bei der vermehrung der menschen, thiere und pflanzen*. Jena, Fischer.

Edward, D.A. and Chapman, T. (2011) The evolution and significance of male mate choice. *Trends in Ecology and Evolution* **26**:647–654.

Emlen, S.T. and Oring, L.W. (1977) Ecology, sexual selection, and the evolution of mating systems. *Science* **197**:215–223.

Emlen, S.T. and Wrege, P.H. (1988) The role of kinship in helping decisions among white-fronted bee-eaters. *Behavioral Ecology and Sociobiology* **23**:305–315.

Enquist, M. and Leimar, O. (1993) Evolution of cooperation in mobile organisms. *Animal Behaviour* **45**:747–757.

Fabre, J.-H. (1879) *Souvenirs Entomologiques. Etudes sur l'Instinct et les Moeurs des Insectes*. Paris: Editions Delagrave.

Fehr, E. and Gächter, S. (2002) Altruistic punishment in humans. *Nature* **415**:137–140.

Fischer, E.A. (1988) Simultaneous hermaphroditism, Tit for Tat, and the evolutionary stability of social systems. *Ethology and Sociobiology* **9**:119–136.

Fisher, R.A. (1930) *The Genetical Theory of Natural Selection*. Oxford: Clarendon Press.

Forsgren, E., *et al.* (2004) Unusually dynamic sex roles in a fish. *Nature* **429**:551–554.

Frank, S.A. (1987) Individual and population sex allocation patterns. *Theoretical Population Biology* **31**:47–74.

Frank, S.A. (1995) Mutual policing and repression of competition in the evolution of cooperative groups. *Nature* **377**:520–522.

Frank, S.A. (2013) Natural selection. VII. History and interpretation of kin selection theory. *Journal of Evolutionary Biology* **26**:1151–1184.

Gabriel, S.E., *et al.* (1994) Cystic fibrosis heterozygote resistance to cholera toxin in the Cystic fibrosis mouse model. *Science* **266**:107–109.

Gardner, A. and Foster, K.R. (2008) The evolution and ecology of cooperation: history and concepts. In: *Ecology of Social Evolution* (eds J. Korb and J. Heinze) Berlin: Springer-Verlag.

Gardner, A. and Grafen, A. (2009) Capturing the superorganism: a formal theory of group adaptation. *Journal of Evolutionary Biology* **22**:659–671.

Gardner, A., *et al.* (2007) The relation between multilocus population genetics and social evolution theory. *American Naturalist* **169**:207–226.

Gardner, A., *et al.* (2011) The genetical theory of kin selection. *Journal of Evolutionary Biology* **24**:1020–1043.

Geist, V. (1971) *Mountain Sheep: A Study in Behavior and Evolution*. Chicago: University of Chicago Press.

Ghiselin, M.T. (1974) *The Economy of Nature and the Evolution of Sex*. Berkeley: University of California Press.

Ghoul, M., *et al.* (2014) Toward an evolutionary definition of cheating. *Evolution* **68**:318–331.

Gibb, J. (1954) Feeding ecology of tits, with notes on treecreeper and goldcrest. *Ibis* **96**:513–543.

Giraldeau, L.-A. and Caraco, T. (2000) *Social Foraging Theory*. Princeton, NJ: Princeton University Press.

Giraldeau, L.-A. and Gillas, D. (1985) Optimal group size can be stable: a reply to Sibly. *Animal Behaviour* **33**:666–667.

Godfray, C. (1994) *Parasitoids: Behavioural and Evolutionary Ecology*. Princeton, NJ: Princeton University Press.

Goss-Custard, J.D. (1970) Feeding dispersion in some overwintering wading birds. In: *Social Behaviour in Birds and Mammals* (ed. J.H. Crook). London: Academic Press, 3–35.

Gowaty, P.A. and Hubbell, S.P. (2005) Chance, time allocation and the evolution of adaptively flexible sex role behavior. *Integrative and Comparative Biology* **45**:931–944.

Gowaty, P.A. and Lennartz, M.R. (1985) Sex ratios of nestling and fledgling red-cockaded woodpeckers (*Picoides borealis*) favor males. *American Naturalist* **126**:347–353.

Grafen, A. (1984) Natural selection, kin selection and group selection. In: *Behavioural Ecology: An Evolutionary Approach*, 2nd edn (eds J. R. Krebs and N. B. Davies). Oxford: Blackwell Scientific Publications, 62–84.

Grafen, A. (1990) Biological signals as handicaps. *Journal of Theoretical Biology* **144**:517–546.

Greenwood, P.J. (1980) Mating systems, philopatry and dispersal in birds and mammals. *Animal Behaviour* **28**:1140–1162.

Griffin, A.S., *et al.* (2004) Cooperation and competition in pathogenic bacteria. *Nature* **430**:1024–1027.

Griffin, A.S., *et al.* (2005) Cooperative breeders adjust offspring sex ratios to produce helpful helpers. *American Naturalist* **166**:628–632.

Hall, K.R.L. (1965) Behaviour and ecology of the wild patas monkey, *Erythrocebus patas*, in Uganda. *Proceedings of the Zoological Society of London* **148**:15–87.

Hamilton, W.D. (1963) The evolution of altruistic behavior. *American Naturalist* **97**:354–356.

Hamilton, W.D. (1964) The genetical evolution of social behaviour. I. *Journal of Theoretical Biology* **7**:1–16.

Hamilton, W.D. (1971) Geometry for the selfish herd. *Journal of Theoretical Biology* **31**:295–311.

Hamilton, W.D. (1988) The genetic evolution of social behaviour. *Citation Classic in Current Contents* **40**:16.

Hamilton, W.D. and Zuk, M. (1982) Heritable true fitness and bright birds: a role for parasites? *Science* **218**:384–387.

Hammerstein, P. (2003) Why is reciprocity so rare in social animals? A protestant appeal. In: *Genetic and Cultural Evolution of Cooperation* (ed. P. Hammerstein). Cambridge, MA: MIT Press, 83–93.

Harcourt, A.H. (1988) Alliances in contests and social intelligence. In: *Machiavellian Intelligence* (eds R. Byrne and A. Whiten). Oxford: Oxford University Press, 132–152.

Harcourt, A.H. (1989) Sociality and competition in primates and non-primates. In: *Comparative Socioecology* (eds V. Standen and R. A. Foley). Oxford: Blackwell Scientific Publications, 223–242.

Harcourt, A.H. (1992) Coalitions and alliances: are primates more complex than non-primates? In: *Coalitions and Alliances in Humans and Other Animals* (eds A.H. Harcourt and F. B. M. de Waal). Oxford: Oxford University Press, 445–471.

Harvey, P.H. and Clutton-Brock, T.H. (1985) Life-history variation in primates. *Evolution* **39**:559–581.

Harvey, P.H. and Pagel, M.D. (1991) *The Comparative Method in Evolutionary Biology*. Oxford: Oxford University Press.

Hatchwell, B. (2007) Historical perspectives and a focus on fitness. *Behavioural Processes* **76**:73–74.

Hatchwell, B.J. (2009) The evolution of cooperative breeding in birds: kinship, dispersal and life history. *Philosophical Transactions of the Royal Society B: Biological Sciences* **364**:3217–3227.

Hatchwell, B.J., *et al.* (2014) Helping in cooperatively breeding long-tailed tits: a test of Hamilton's rule. *Philosophical Transactions of the Royal Society B: Biological Sciences* **369**:20130565.

Hatchwell, B.J., *et al.* (2001) Kin discrimination in cooperatively breeding long-tailed tits. *Proceedings of the Royal Society of London. Series B: Biological Sciences* **268**:885–890.

Hauert, C. and Doebeli, M. (2004) Spatial structure often inhibits the evolution of cooperation in the snowdrift game. *Nature* **428**:643–646.

Hauert, C., *et al.* (2002) Volunteering as Red Queen mechanism for cooperation in public goods games. *Science* **296**:1129–1132.

Hauser, M., *et al.* (2009) Evolving the ingredients for reciprocity and spite. *Philosophical Transactions of the Royal Society B: Biological Sciences* **364**:3255–3266.

Hawkes, K. (1992) Sharing and collective action. In: *Evolutionary Ecology and Human Behavior* (eds E. A. Smith and B. Winterhalder). Hawthorne, NY: Aldine de Gruyter, 269–300.

Heckathorn, D.D. (1989) Collective action and the second-order free-rider problem. *Rationality and Society* **1**:78–100.

Henrich, J. and Boyd, R. (2001) Why people punish defectors: weak conformist transmission can stabilize costly enforcement of norms in cooperative dilemmas. *Journal of Theoretical Biology* **208**:79–89.

Herdman, E.J.E., *et al.* (2004) Male mate choice in the guppy (*Poecilia reticulata*): do males prefer larger females as mates? *Ethology* **110**:97–111.

Higashi, M. and Yamamura, N. (1993) What determines animal group size? Insider–outsider conflict and its resolution. *American Naturalist* **142**:553–563.

Hinde, C.A., *et al.* (2009) Prenatal environmental effects match offspring begging to parental provisioning. *Proceedings of the Royal Society of London. Series B: Biological Sciences* **276**:2787–2794.

Hinde, R.A. (1956) The biological significance of territories in birds. *Ibis* **98**:340–369.

Hinde, R.A. (1966) *Animal Behavior: A Synthesis of Ethology and Comparative Psychology*. London: McGraw-Hill.

Hinde, R.A. (1970) *Animal Behaviour: A Synthesis of Ethology and Comparative Physiology*. New York: McGraw Hill.

Hinde, R.A. (1973) On the design of check-sheets. *Primates* **14**:393–406.

Hinde, R.A. (1983) *Primate Social Relationships: An Integrated Approach*. Oxford: Blackwell Science.

Hirshleifer, D. and Rasmusen, E. (1989) Cooperation in the repeated prisoner's dilemma with ostracism. *Journal of Economic Behavior and Organization* **12**:87–106.

Hirshleifer, J. and Coll, J.C.M. (1988) What strategies can support the evolutionary emergence of cooperation. *Journal of Conflict Resolution* **32**:367–398.

Hölldobler, B. and Wilson, E.O. (1990) *The Ants*. Cambridge, MA: Harvard University Press.

Hölldobler, B. and Wilson, E.O. (2009) *The Superorganism: The Beauty, Elegance, and Strangeness of Insect Societies*. New York: W. W. Norton and Co.

Hoogland, J.L. (1979) Aggression, ectoparasitism, and other possible costs of prairie dog (Sciuridae: *Cynomys* spp.) coloniality. *Behaviour* **69**:1–35.

Hoogland, J.L. (1981) The evolution of coloniality in white-tailed and black-tailed prairie dogs (Sciuridae: *Cynomys leucurus* and *C. ludovicianus*). *Ecology* **62**:252–272.

Hoogland, J.L. and Sherman, P.W. (1976) Advantages and disadvantages of bank swallow (*Riparia riparia*) coloniality. *Ecological Monographs* **46**:33–58.

Houston, A.I. and Davies, N.B. (1985) The evolution of cooperation and life-history in the dunnock *Prunella modularis*. In: *Behavioural Ecology: Ecological Consequences of Adaptive Behaviour* (eds R. M. Sibley and R. H. Smith). Oxford: Blackwell Scientific Publications, 471–487.

Houston, A.I. and McNamara, J.M. (2002) A self-consistent approach to paternity and parental effort. *Philosophical Transactions of the Royal Society B: Biological Sciences* **357**:351–362.

Houston, A.I., *et al.* (2013) The parental investment models of Maynard Smith: a retrospective and prospective view. *Animal Behaviour* **86**:667–674.

Hrdy, S.B. (1981) *The Woman That Never Evolved*. Cambridge, MA: Harvard University Press.

Hughes, W.O.H., *et al.* (2008) Ancestral monogamy shows kin selection is key to the evolution of eusociality. *Science* **320**:1213–1216.

Huxley, J.S. (1938a) Darwin's theory of sexual selection and the data subsumed by it, in the light of recent research. *American Naturalist* **72**:416–433.

Huxley, J.S. (1938b) The present standing of the theory of sexual selection. In: *Evolution (Essays Presented to E.S. Goodrich)* (ed. G.R. de Beer). New York: Oxford University Press, 11–42.

Imanishi, K. (1957) Learned behavior of Japanese monkeys. *Japanese Journal of Ethnology* **21**:185–189.

Ims, R.A. (1988) Spatial clumping of sexually receptive females induces space sharing among male voles. *Nature* **235**:541–543.

Itani, J. (1959) Paternal care in the wild Japanese monkey, *Macaca fuscata*. *Primates* **2**:61–93.

Jarman, M.V. (1979) *Impala Social Behaviour: Territory, Hierarchy, Mating and the Use of Space*. Berlin: Verlag Paul Parey.

Jarman, P.J. (1974) The social organisation of antelope in relation to their ecology. *Behaviour* **48**:215–267.

Jeffreys, A.J., *et al.* (1985) Hypervariable 'minisatellite' regions in human DNA. *Nature* **314**:67–73.

Johnstone, R.A. (1997) The evolution of animal signals. In: *Behavioural Ecology: An Evolutionary Approach*, 4th edn (eds

J. R. Krebs and N. B. Davies). Oxford: Blackwell Science, 155–178.

Johnstone, R.A. and Godfray, H.C. (2002) Models of begging as a signal of need. In: *The Evolution of Nestling Begging: Competition, Cooperation and Communication* (eds J. Wright and M. L. Leonard). Dordrecht: Kluwer, 1–20.

Johnstone, R.A. and Grafen, A. (1993) Dishonesty and the handicap principle. *Animal Behaviour* **46**:759–764.

Johnstone, R.A. and Hinde, C.A. (2006) Negotiation over offspring care: how should parents respond to each other's efforts. *Behavioral Ecology* **17**:818–827.

Johnstone, R.A. and Rodrigues, A.M.M. (2016) Cooperation and the common good. *Philosophical Transactions of the Royal Society B: Biological Sciences*, DOI: 10.1098/rstb.2015.0086. [Accessed 18 January 2016]

Jones, A.G., et al. (2000) The Bateman gradient and the cause of sexual selection in a sex-role-reversed pipefish. *Proceedings of the Royal Society of London. Series B: Biological Sciences* **267**:677–680.

Jones, A.G., et al. (2002) Validation of Bateman's principles: a genetic study of sexual selection and mating patterns in the rough-skinned newt. *Proceedings of the Royal Society of London. Series B: Biological Sciences* **269**:2533–2539.

Jordan, W.C. and Bruford, M.W. (1998) New perspectives on mate choice and the MHC. *Heredity* **81**:127–133.

Kaufmann, J.H. (1975) Field observations of the social behaviour of the eastern grey kangaroo, *Macropus giganteus*. *Animal Behaviour* **23**:214–221.

Kilner, R.M. (2002) The evolution of complex begging displays. In: *The Evolution of Nestling Begging: Competition, Cooperation and Communication* (eds J. Wright and M. L. Leonard). Dordrecht: Kluwer, 87–106.

Kilner, R.M. and Johnstone, R.A. (1997) Begging the question: are offspring solicitation behaviours signals of need? *Trends in Ecology and Evolution* **12**:11–15.

Kirkpatrick, M. (1982) Sexual selection and the evolution of female choice. *Evolution* **36**:1–12.

Kirkpatrick, M. (1986) The handicap mechanism of sexual selection does not work. *American Naturalist* **127**: 222–240.

Kirkpatrick, M. and Ryan, M.J. (1991) The evolution of mating preferences and the paradox of the lek. *Nature* **350**:33–38.

Kitchen, D.M. and Packer, C.R. (1999) Complexity in vertebrate societies. In: *Levels of Selection in Evolution* (ed. L. Keller). Princeton, NJ: Princeton University Press, 176–196.

Klein, J. (1986) How many class-II immune-response genes: a commentary on a reappraisal. *Immunogenetics* **23**:309–310.

Klopfer, P.H. (1973) *Behavioral Aspects of Ecology*. London: Prentice-Hall.

Klug, H., et al. (2010) The mismeasurement of sexual selection. *Journal of Evolutionary Biology* **23**:447–462.

Klug, H., et al. (2012) Theoretical foundation of parental care. In: *The Evolution of Parental Care* (eds N.I.J. Royle, P. T. Smiseth and M. Kölliker). Oxford: Oxford University Press, 21–39.

Knickmeyer, R.C. and Baron-Cohen, S. (2006) Topical review: Fetal testosterone and sex differences in typical social development and in autism. *Journal of Child Neurology* **21**:825–845.

Kokko, H. (2003) Review Paper: Are reproductive skew models evolutionarily stable? *Proceedings of the Royal Society of London. Series B: Biological Sciences* **270**:265–270.

Kokko, H. and Jennions, M.D. (2008) Parental investment, sexual selection and sex ratios. *Journal of Evolutionary Biology* **21**:919–948.

Kokko, H., et al. (2001) The evolution of cooperative breeding through group augmentation. *Proceedings of the Royal Society of London. Series B: Biological Sciences* **268**:187–196.

Kokko, H., et al. (2003) The evolution of mate choice and mating biases. *Proceedings of the Royal Society of London. Series B: Biological Sciences* **270**:653–664.

Kokko, H., et al. (2012) Unifying cornerstones of sexual selection: operational sex ratio, Bateman gradient and the scope for competitive investment. *Ecology Letters* **15**:1340–1351.

Kraak, S.B.M. and Pen, I. (2002) Sex determining mechanisms in vertebrates. In: *Sex Ratios: Concepts and Research Methods* (ed. I.C.W. Hardy). Cambridge: Cambridge University Press, 158–177.

Kramer, D.L. (1995) Are colonies supraoptimal groups? *Animal Behaviour* **33**:1031–1032.

Krause, J. and Ruxton, G.D. (2002) *Living in Groups*. Oxford: Oxford University Press.

Krebs, J.R. (1978) Optimal foraging: decision rules for predators. In: *Behavioural Ecology* (eds J. R. Krebs and N. B. Davies). Oxford: Blackwell Scientific Publications, 23–63.

Krebs, J.R. and Davies, N.B. (1978) *Behavioural Ecology: An Evolutionary Approach*. Oxford: Blackwell Scientific Publications.

Krebs, J.R. and Davies, N.B. (1981) *An Introduction to Behavioural Ecology*. Oxford: Blackwell Scientific Publications.

Kruuk, H. (1964) Predators and anti-predator behaviour of the black-headed gull (*Larus ridibundus* L). *Behaviour Supplement* **11**:1–129.

Kruuk, H. (1972) *The Spotted Hyena: A Study of Predation and Social Behaviour*. Chicago: University of Chicago Press.

Kruuk, L.E.B. (2004) Estimating genetic parameters in natural populations using the 'animal model'. *Philosophical Transactions of the Royal Society of London. Series B: Biological Sciences* **359**:873–890.

Kvarnemo, C., et al. (2007) Sexually selected females in the monogamous Western Australian seahorse. *Proceedings of the Royal Society of London. Series B: Biological Sciences* **274**:521–525.

Kvarnemo, C. and Simmons, L.W. (1999) Variance in female quality, operational sex ratio and male mate choice in a bush cricket. *Behavioral Ecology and Sociobiology* **45**:245–252.

Lack, D. (1935) Territory and polygamy in a bishop-bird, *Euplectes hordeacea hordeacea* (Linn.). *Ibis* **77**:817–836.

Lack, D. (1954) *The Natural Regulation of Animal Numbers*. Oxford: Clarendon Press.

Lack, D. (1966) *Population Studies of Birds*. Oxford: Oxford University Press.

Lack, D. (1968) *Ecological Adaptation for Breeding in Birds*. London: Methuen.

Lande, R. (1980) Sexual dimorphism, sexual selection, and adaptation in polygenic characters. *Evolution* **34**:292–305.

Lande, R. (1987) Genetic correlations between the sexes in the evolution of sexual dimorphism and mating preferences. In: *Sexual Selection: Testing the Alternatives* (eds J. W. Bradbury and M. B. Andersson). Chichester: John Wiley & Sons Ltd, 83–94.

Lande, R. and Arnold, S.J. (1985) Evolution of mating preference and sexual dimorphism. *Journal of Theoretical Biology* **117**:651–664.

LeBas, N.R. (2006) Female finery is not for males. *Trends in Ecology and Evolution* **21**:170–173.

LeBas, N.R., *et al.* (2003) Nonlinear and correlational sexual selection on 'honest' female ornamentation. *Proceedings of the Royal Society of London. Series B: Biological Sciences* **270**:2159–2165.

Leigh, E.G. (1970) Sex ratio and differential mortality between the sexes. *American Naturalist* **104**:205–210.

Leimar, O. (1996) Life-history analysis of the Trivers and Willard sex-ratio problem. *Behavioral Ecology* **7**:316–325.

Leimar, O. and Connor, R.C. (2003) By-product benefits, reciprocity and pseudoreciprocity in mutualism. In: *Genetic and Cultural Evolution of Cooperation* (ed. P. Hammerstein). Cambridge, MA: MIT Press, 203–222.

Leimar, O. and Hammerstein, P. (2001) Evolution of cooperation through indirect reciprocity. *Proceedings of the Royal Society of London. Series B: Biological Sciences* **268**:745–753.

Lessells, C.M. (2002) Parental investment in relation to offspring sex. In: *The Evolution of Begging* (eds J. Wright and M. C. Leonard). Dordrecht: Kluwer, 65–85.

Leuthold, W. (1966) Variations in territorial behavior of Uganda kob Adenota kob thomasi (Neumann 1896). *Behaviour* **27**:215–258.

Liao, X., *et al.* (2015) Relatedness, conflict, and the evolution of eusociality. *PLOS Biology* **13** (3): e1002098.

Lindström, Å. (1989) Finch flock size and the risk of hawk predation at a migratory stopover site. *The Auk* **106**:225–232.

Lorch, P.D., *et al.* (2008) Quantifying the potential for sexual dimorphism using upper limits on Bateman gradients. *Behaviour* **145**:1–24.

Lorenz, K. (1950) The comparative method in studying innate behavioural patterns. In: *Physiological Mechanisms in Animal Behaviour*. Symposia of the Society for Experimental Biology No. IV. Oxford: Academic Press, 221–254.

Lott, D.F. (1991) *Intraspecific Variation in the Social Systems of Wild Vertebrates*. Cambridge: Cambridge University Press.

Lukas, D. and Clutton-Brock, T.H. (2014) Costs of mating competition limit male lifetime breeding success in polygynous mammals. *Proceedings of the Royal Society of London. Series B: Biological Sciences* **281**:20140418.

Lyon, B.E. and Montgomerie, R. (2012) Sexual selection is a form of social selection. *Philosophical Transactions of the Royal Society B: Biological Sciences* **367**:2266–2273.

MacColl, A.D. (2011) The ecological causes of evolution. *Trends in Ecology and Evolution* **26**:514–522.

McElreath, R. (2003) Reputation and the evolution of conflict. *Journal of Theoretical Biology* **220**:345–357.

McNab, B.K. (1963) Bioenergetics and the determination of the home range size. *American Naturalist* **97**:133–140.

McNamara, J.M., *et al.* (1999) Incorporating rules for responding to evolutionary games. *Nature* **401**:368–371.

McNamara, J.M., *et al.* (2003) Should young ever be better off with one parent than two? *Behavioral Ecology* **14**:301–310.

McNamara, J.M., *et al.* (2004) Variation in behaviour promotes cooperation in the Prisoner's Dilemma game. *Nature* **428**:745–748.

Malcolm, J.R. and Marten, K. (1982) Natural selection and the communal rearing of pups in African wild dogs, *Lycaon pictus*. *Behavioral Ecology and Sociobiology* **10**:1–13.

Marshall, J.A.R. (2011) Group selection and kin selection: formally equivalent approaches. *Trends in Ecology and Evolution* **26**:325–332.

Marshall, J.A.R. (2015) *Social Evolution and Inclusive Fitness Theory: An Introduction*. Princeton, NJ: Princeton University Press.

Martin, P. and Bateson, P.P.G. (1993) *Measuring Behaviour: An Introductory Guide*. Cambridge: Cambridge University Press.

Marwell, G. and Ames, R. (1980) Experiments on the provision of public-goods II: Provision points, stakes, experience and the free rider problem. *American Journal of Sociology* **85**:926–937.

Mateo, J.M. (2002) Kin-recognition abilities and nepotism as a function of sociality. *Proceedings of the Royal Society of London. Series B: Biological Sciences* **269**:721–727.

Maynard Smith, J. (1964) Group selection and kin selection. *Nature* **201**:1145–1147.

Maynard Smith, J. (1974) The theory of games and the evolution of animal conflicts. *Journal of Theoretical Biology* **47**:209–221.

Maynard Smith, J. (1976) Group selection. *Quarterly Review of Biology* **51**:277–283.

Maynard Smith, J. (1977) Parental investment: a prospective analysis. *Animal Behaviour* **25**:1–9.

Maynard Smith, J. (1982) *Evolution and the Theory of Games*. Cambridge: Cambridge University Press.

Maynard Smith, J. (1991) Honest signalling: the Philip Sidney game. *Animal Behaviour* **42**:1034–1035.

Maynard Smith, J. and Harper, D. (2003) *Animal Signals*. Oxford: Oxford University Press.

Maynard Smith, J. and Parker, G.A. (1976) The logic of asymmetric contests. *Animal Behaviour* **24**:159–175.

Mays, H.L. and Hill, G.E. (2004) Choosing mates: good genes versus genes that are a good fit. *Trends in Ecology and Evolution* **19**:554–559.

Mesterton-Gibbons, M. and Dugatkin, L.A. (1992) Cooperation among unrelated individuals: evolutionary factors. *Quarterley Review of Biology* **67**:267–281.

Milinski, M., *et al.* (2002) Reputation helps solve the 'tragedy of the commons'. *Nature* **415**:424–426.

Mock, D.W. and Parker, G.A. (1997) *The Evolution of Sibling Rivalry*. Oxford: Oxford University Press.

Moczek, A.P. and Emlen, D.J. (2000) Male horn dimorphism in the scarab beetle, *Onthophagus taurus*: do alternative reproductive tactics favour alternative phenotypes? *Animal Behaviour* **59**:459–466.

Morgan, T.H. (1903) Recent theories in regard to the determination of sex. *Popular Science Monthly* **64**:97–116.

Mueller, T.H. and Feldman, M.W. (1988) The evolution of altruism by kin selection: new phenomena with strong selection. *Etthology and Sociobiology* **9**:223–239.

Noë, R. (1992) Alliance formation among male baboons: shopping for profitable partners. In: *Coalitions and Alliances in Humans and Other Animals* (eds A. H. Harcourt and F. B. M. de Waal). Oxford: Oxford University Press, 285–321.

Noë, R. and Hammerstein, P. (1994) Biological markets: supply and demand determine the effect of partner choice in cooperation, mutualism and mating. *Behavioral Ecology and Sociobiology* **35**:1–11.

Noë, R. and Hammerstein, P. (1995) Biological markets. *Trends in Ecology and Evolution* **10**:336–339.

Noë, R., *et al.* (2001) *Economics in Nature: Social Dilemmas, Mate Choice and Biological Markets*. Cambridge: Cambridge University Press.

Nomano, F.Y., *et al.* (2013) Feeding nestlings does not function as a signal of social prestige in cooperatively breeding chestnut-crowned babblers. *Animal Behaviour* **86**:277–289.

Norris, K.S. (1966) *Whales, Dolphins, and Porpoises*. Berkeley: University of California Press.

Nowak, M.A. (2006) Five rules for the evolution of cooperation. *Science* **314**:1560–1565.

Nowak, M.A. and Allen, B. (2015) Inclusive fitness theorizing invokes phenomena that are not relevant for the evolution of eusociality. *PLOS Biology* **13** (4): e1002134.

Nowak, M.A. and Sigmund, K. (1992) Tit for tat in heterogeneous populations. *Nature* **355**:250–253.

Nowak, M.A. and Sigmund, K. (1993) A strategy of win-stay, lose-shift that outperforms tit-for-tat in the Prisoner's Dilemma game. *Nature* **364**:56–58.

Nowak, M.A. and Sigmund, K. (1994) The alternating prisoners dilemma. *Journal of Theoretical Biology* **168**:219–226.

Nowak, M.A. and Sigmund, K. (1998a) Evolution of indirect reciprocity by image scoring. *Nature* **393**:573–577.

Nowak, M.A. and Sigmund, K. (1998b) The dynamics of indirect reciprocity. *Journal of Theoretical Biology* **194**:561–574.

Nowak, M.A. and Sigmund, K. (2005) Evolution of indirect reciprocity. *Nature* **437**:1291–1298.

Nowak, M.A., *et al.* (2010) The evolution of eusociality. *Nature* **466**:1057–1062.

Nunn, C.L. and Lewis, R.J. (2001) Cooperation and collective action in animal behaviour. In: *Economics in Nature: Social Dilemmas, Mate Choice and Biological Markets* (eds R. Noë, J.

A. R. A. M. van Hooff and P. Hammerstein). Cambridge: Cambridge University Press, 42–66.

O'Donald, P. (1962) The theory of sexual selection. *Heredity* **17**:541–552.

O'Donald, P. (1967) A general model of sexual and natural selection. *Heredity* **22**:499–518.

Olson, M. (1965) *The Logic of Collective Action: Public Goods and the Theory of Groups*. Cambridge, MA: Harvard University Press.

Orians, G.H. (1961) The ecology of blackbird (*Agelaius*) social systems. *Ecological Monographs* **31**:285–312.

Oring, L.W., *et al.* (1991a) Lifetime reproductive success in the spotted sandpiper (*Actitis macularia*): sex-differences and variance-components. *Behavioral Ecology and Sociobiology* **28**:425–432.

Oring, L.W., *et al.* (1991b) Factors regulating annual mating success and reproductive success in spotted sandpipers (*Actitis macularia*). *Behavioral Ecology and Sociobiology* **28**:433–442.

Orrell, K.S. and Jenssen, T.A. (2002) Male mate choice by the lizard *Anolis carolinensis*: a preference for novel females. *Animal Behaviour* **63**:1091–1102.

Ostrom, E. (1990) *Governing the Commons: the Evolution of Institutions for Collective Action*. Cambridge: Cambridge University Press.

Ostrom, E. (2003) Toward a behavioral theory linking trust, reciprocity, and reputation. In: *Trust and Reciprocity: Interdisciplinary Lessons from Experimental Research* (eds E. Ostrom and J. Walker). New York: Russell Sage Foundation, 9–79.

Packer, C. (1979) Inter-troop transfer and inbreeding avoidance in *Papio anubis*. *Animal Behaviour* **27**:1–36.

Packer, C., *et al.* (1990) Why lions form groups: food is not enough. *American Naturalist* **136**:1–19.

Pagel, M. (1994) Detecting correlated evolution on phylogenies: a general method for the comparative analysis of discrete characters. *Proceedings of the Royal Society of London. Series B: Biological Sciences* **255**:37–45.

Parker, G.A. (1970) Sperm competition and its evolutionary consequences in the insects. *Biological Reviews* **45**:525–567.

Parker, G.A. (1974) Assessment strategy and the evolution of fighting behaviour. *Journal of Theoretical Biology* **47**:223–243.

Parker, G.A. (1984) Sperm competition and the evolution of animal mating strategies. In: *Sperm Competition and the Evolution of Animal Mating Systems* (ed. R.L. Smith). New York: Academic Press, 1–60.

Parker, G.A. (1985) Models of parent–offspring conflict. V. Effects of the behaviour of two parents. *Animal Behaviour* **33**:519–533.

Parker, G.A. (1998) Sperm competition and the evolution of ejaculates: towards a theory base. In: *Sperm Competition and Sexual Selection* (eds T. R. Birkhead and A. P. Møller). London: Academic Press, 3–54.

Parker, G.A. (2000) Sperm competition games between related males. *Proceedings of the Royal Society of London. Series B: Biological Sciences* **267**:1027–1032.

Parker, G.A. and MacNair, M.R. (1979) Models of parent–offspring conflict. IV. Suppression: evolutionary retaliation of the parent. *Animal Behaviour* **27**:1210–1235.

Parker, G.A. and Simmons, L.W. (1996) Parental investment and the control of sexual selection: predicting the direction of sexual competition. *Proceedings of the Royal Society of London. Series B: Biological Sciences* **263**:315–321.

Parker, G.A., *et al.* (1997) Sperm competition games: a prospective analysis of risk assessment. *Proceedings of the Royal Society of London. Series B: Biological Sciences* **264**:1793–1802.

Parker, G.A., *et al.* (2002) Intrafamilial conflict and parental investment: a synthesis. *Philosophical Transactions of the Royal Society B: Biological Sciences* **357**:295–307.

Peck, J.R. and Feldman, M.W. (1986) The evolution of helping behavior in large, randomly mixed populations. *American Naturalist* **127**:209–221.

Pen, I. and Weissing, F.J. (2002) Optimal sex allocation: steps towards a mechanistic theory. In: *Sex Ratios: Concepts and Research Methods* (ed. I.C.W. Hardy). Cambridge: Cambridge University Press, 26–47.

Penn, D.J., *et al.* (2002) MHC heterozygosity confers a selective advantage against multiple-strain infections. *Proceedings of the National Academy of Sciences of the United States of America* **99**:11260–11264.

Perrins, C.M. (1965) Population fluctuations and clutch-size in the Great Tit, Parus major L. *Journal of Animal Ecology* **34**:601–647.

Pfeiffer, T., *et al.* (2005) Evolution of cooperation by generalized reciprocity. *Proceedings of the Royal Society of London. Series B: Biological Sciences* **272**:1115–1120.

Pizzari, T. and Bonduriansky, R. (2010) Sexual behaviour: conflict, cooperation and coevolution. In: *Social Behaviour: Genes, Ecology and Evolution* (eds T. Szekely, A. J. Moore and J. Komdeur). Cambridge: Cambridge University Press, 230–266.

Pollock, G.B. (1989) Evolutionary stability of reciprocity in a viscous lattice. *Social Networks* **11**:175–212.

Pollock, G. and Dugatkin, L.A. (1992) Reciprocity and the emergence of reputation. *Journal of Theoretical Biology* **159**:25–37.

Potts, W.K. and Wakeland, E.K. (1990) Evolution of diversity at the major histocompatibility complex. *Trends in Ecology and Evolution* **5**:181–187.

Price, T.A.R., *et al.* (2008) Selfish genetic elements promote polyandry in a fly. *Science* **322**:1241–1243.

Queller, D.C. (1985) Kinship, reciprocity and synergism in the evolution of social behavior. *Nature* **318**:366–367.

Queller, D.C. (1996) The measurement and meaning of inclusive fitness. *Animal Behaviour* **51**:229–232.

Queller, D.C. (1997) Why do females care more than males? *Proceedings of the Royal Society of London. Series B: Biological Sciences* **264**:1555–1557.

Queller, D.C., *et al.* (2015) Some agreement on kin selection and eusociality? *PLOS Biology* **13**:e1002133.

Rankin, D.J. and Taborsky, M. (2009) Assortment and the evolution of generalized reciprocity. *Evolution* **63**:1913–1922.

Reinbold, K., *et al.* (2002) Cryptic male choice: sperm allocation, strategies when female quality varies. *Journal of Evolutionary Biology* **115**:201–209.

Richard, A. (1974) Intraspecific variation in the social organization and ecology of *Propithecus verreauxi*. *Folia Primatologica* **22**:178–207.

Richard, A.F. (1978) *Behavioral Variation: Case Study of a Malagasy Lemur*. Cranbury, NJ: Bucknell University Press.

Riolo, R.L., *et al.* (2001) Evolution of cooperation without reciprocity. *Nature* **414**:441–443.

Rios Moura, R. and Peixoto, P.E.C. (2013) The effect of the operational sex ratio on the opportunity for sexual selection: a meta-analysis. *Animal Behaviour* **86**:675–683.

Roberts, G. (1998) Competitive altruism: from reciprocity to the handicap principle. *Proceedings of the Royal Society of London. Series B: Biological Sciences* **265**:427–431.

Roberts, G. and Sherratt, T.N. (1998) Development of cooperative relationships through increasing investment. *Nature* **394**:175–179.

Roberts, S.C. and Gosling, L.M. (2003) Genetic similarity and quality interact in mate choice decisions by female mice. *Nature Genetics* **35**:103–106.

Ross-Gillespie, A., *et al.* (2007) Frequency dependence and cooperation: theory and a test with bacteria. *American Naturalist* **170**:331–342.

Rosvall, K.A. (2011) Intrasexual competition in females: evidence for sexual selection? *Behavioral Ecology* **22**:1131–1140.

Roughgarden, J. and Akçay, E. (2010) Do we need a Sexual Selection 2.0? *Animal Behaviour* **79**:e1–e4.

Roughgarden, J., *et al.* (2006) Reproductive social behavior: cooperative games to replace sexual selection. *Science* **311**:965–969.

Russell, E.M. (1984) Social behaviour and social organization of marsupials. *Mammal Review* **14**:101–154.

Ruxton, G.D. and van der Meer, J. (1997) Policing: it pays the strong to protect the weak. *Trends in Ecology and Evolution* **12**:250–251.

Ryan, M.J. (1997) Sexual selection and mate choice. In: *Behavioural Ecology: An Evolutionary Approach*, 4th edn (eds J. R. Krebs and N. B. Davies). Oxford: Blackwell Science, 179–202.

Sandler, T. (1992) *Collective Action*. Ann Arbor: University of Michigan Press.

Schaller, G.B. (1972) *The Serengeti Lion: A Study of Predator–Prey Relations*. Chicago: Chicago University Press.

Schoener, T.W. (1968) Sizes of feeding territories among birds. *Ecology* **49**:123–141.

Searcy, W.A. and Nowicki, S. (2005) *The Evolution of Animal Communication*. Princeton, NJ: Princeton University Press.

Seeley, T.D. (2001) Decision making in superorganisms: how collective wisdom arises from the poorly informed masses. In:

Bounded Rationality: The Adaptive Toolbox (eds G. Gigerenzer and R. Selten). Cambridge, MA: MIT Press, 249–261.

Seger, J. and Stubblefield, W.J. (2002) Models of sex ratio evolution. In: *Sex Ratios: Concepts and Research Methods* (ed. I.C.W. Hardy). Cambridge: Cambridge University Press, 1–25.

Selman, J. and Goss-Custard, J.D. (1988) Interference between foraging redshank *Tringa totanus*. *Animal Behaviour* **36**: 1542–1544.

Shaw, R.F. and Mohler, J.D. (1953) The selective significance of the sex ratio. *American Naturalist* **87**:337–342.

Shuker, D.M. (2010) Sexual selection: endless forms or tangled bank? *Animal Behaviour* **79**:e11–e17.

Sibly, R.M. (1983) Optimal group size is unstable. *Animal Behaviour* **31**:947–948.

Silk, J.B. (2009) Nepotistic cooperation in non-human primate groups. *Philosophical Transactions of the Royal Society B: Biological Sciences* **364**:3243–3254.

Simmons, L.W. (1995) Relative parental expenditure, potential reproductive rates, and the control of sexual selection in katydids. *American Naturalist* **145**:797–808.

Simmons, L.W. (2001) *Sperm Competition and Its Evolutionary Consequences in Insects*. Princeton, NJ: Princeton University Press.

Simmons, L.W. and Gwynne, D.T. (1993) Reproductive investment in bushcrickets: the allocation of male and female nutrients to offspring. *Proceedings of the Royal Society of London. Series B: Biological Sciences* **252**:1–5.

Skutch, A.F. (1935) Helpers at the nest. *The Auk* **52**:257–273.

Smiseth, P.T. and Moore, A.J. (2002) Does resource availability affect offspring begging and parental provisioning in a partially begging species? *Animal Behaviour* **63**:577–585.

Smith, C.C. and Fretwell, S.D. (1974) The optimal balance between size and number of offspring. *American Naturalist* **108**:499–506.

Smuts, B.B. and Watanabe, J.M. (1990) Social relationships and ritualized greetings in adult male baboons (*Papio cynocephalus anubis*). *International Journal of Primatology* **11**:147–172.

Southwick, C.H., *et al.* (1965) Rhesus monkeys in north India. In: *Primate Behavior: Field Studies of Monkeys and Apes* (ed. I. DeVore). New York: Holt, Rinehart and Winston, 111–159.

Starr, C.K. (1979) Origin and evolution of insect sociality: a review of modern theory. In: *Social Insects* (ed. H.R. Hermann). New York: Academic Press, 35–79.

Stearns, S.C. (1977) The evolution of life history traits: a critique of the theory and a review of the data. *Annual Review of Ecology and Systematics* **8**:145–171.

Stephens, D.W., *et al.* (1997) On the spurious occurrence of tit-for-tat in pairs of predator-approaching fish. *Animal Behaviour* **53**:113–131.

Stevens, J.R. and Hauser, M.D. (2004) Why be nice? Psychological constraints on the evolution of cooperation. *Trends in Cognitive Sciences* **8**:60–65.

Stockley, P. and Bro-Jørgensen, J. (2011) Female competition and its evolutionary consequences in mammals. *Biological Reviews* **86**:341–366.

Sugden, A. (1986) Trends in ecology and evolution. *Trends in Ecology and Evolution* **1**:2.

Sutherland, W.J. (1985) Chance can produce a sex difference in variance in mating success and explain Bateman's data. *Animal Behaviour* **33**:1349–1352.

Székely, T., *et al.* (eds) (2010) *Social Behaviour: Genes, Ecology and Evolution*. Cambridge: Cambridge University Press.

Taborsky, M., et al. (2016a) The evolution of cooperation based on direct fitness benefits. *Philosophical Transactions of the Royal Society* B. DOI:10.1098/rstb.2015.0472. [Accessed 18 January 2016]

Taborsky, M., et al. (2016b) Correlated pay-offs are key to cooperation. *Philosophical Transactions of the Royal Society B: Biological Sciences* B. DOI:10.1098/rstb.2015.0084. [Accessed 18 January 2016]

Tang-Martinez, Z. and Ryder, T.B. (2005) The problem with paradigms: Bateman's worldview as a case study. *Integrative and Comparative Biology* **45**:821–830.

Taylor, P.D. (1996) Inclusive fitness arguments in genetic models of behaviour. *Journal of Mathematical Biology* **34**:654–674.

Tinbergen, N. (1935) Field observations of East Greenland birds. 1. The behavior of the Red-necked Phalarope (*Phalaropus lobatus* L.) in spring. *Ardea* **24**:1–42.

Tinbergen, N. (1951) *The Study of Instinct*. Oxford: Clarendon Press.

Tinbergen, N. (1952) On the significance of territory in the herring gull. *Ibis* **94**:158–159.

Tinbergen, N. (1953) *Social Behaviour in Animals*. London: Methuen.

Tobias, J.A., *et al.* (2012) The evolution of female ornaments and weaponry: social selection, sexual selection and ecological competition. *Philosophical Transactions of the Royal Society B: Biological Sciences* **367**:2274–2293.

Tregenza, T. and Wedell, N. (1998) Benefits of multiple mates in the cricket *Gryllus bimaculatus*. *Evolution* **52**:1726–1730.

Tregenza, T. and Wedell, N. (2002) Polyandrous females avoid costs of inbreeding. *Nature* **415**:71–73.

Treisman, M. (1975) Predation and the evolution of gregariousness. I. Models for concealment and evasion. *Animal Behaviour* **23**:779–800.

Trivers, R.L. (1971) The evolution of reciprocal altruism. *Quarterly Review of Biology* **46**:35–57.

Trivers, R.L. (1972) Parental investment and sexual selection. In: *Sexual Selection and the Descent of Man, 1871–1971* (ed. B. Campbell). Chicago, IL: Aldine, 136–179.

Trivers, R.L. (1974) Parent–offspring conflict. *American Zoologist* **14**:249–264.

Trivers, R.L. (2006) Reciprocal altruism: 30 years later. In: *Cooperation in Primates and Humans* (eds P. M. Kappeler and C. P. van Schaik). Berlin: Springer-Verlag, 67–83.

Trivers, R.L. and Hare, H. (1976) Haploidploidy and the evolution of the social insect. *Science* **191**:249–263.

Trivers, R.L. and Willard, D.E. (1973) Natural selection of parental ability to vary the sex ratio of offspring. *Science* **179**:90–92.

Uyenoyama, M.K. (1984) Inbreeding and the evolution of altruism under kin selection: effects on relatedness and group structure. *Evolution* **38**:778–795.

van Schaik, C.P. (1983) Why are diurnal primates living in groups? *Behaviour* **87**:120–144.

Vine, I. (1973) Detection of prey flocks by predators. *Journal of Theoretical Biology* **40**:207–210.

von Neumann, J. and Morgenstein, O. (1953) *Theory of Games and Economic Behaviour*. Princeton, NJ: Princeton University Press.

von Schantz, T., *et al.* (1997) Mate choice, male condition-dependent ornamentation and MHC in the pheasant. *Hereditas* **127**:133–140.

Wallace, A.R. (1870) *Contributions to the Theory of Natural Selection*. London: Macmillan.

Wallace, A.R. (1878) *Tropical Nature, and Other Essays*. London: Macmillan.

Wallace, A.R. (1889) *Darwinism: An Exposition of the Theory of Natural Selection with Some of Its Applications*. London: Macmillan.

Walters, J.R. and Seyfarth, R.M. (1997) Conflict and cooperation. In: *Primate Societies* (eds B.B. Smuts, D.L. Cheney, R.M. Seyfarth, R. W. Wrangham and T. T. Struhsaker). Chicago: University of Chicago Press, 306–317.

Walther, F.R. (1964) Einige Verhaltensbeobachtungen an Thomsongazellen (*Gazella thomsoni* Günther, 1884) im Ngorongoro-Krater. *Zeitschrift fur Tierpsychologie* **21**:871–890.

Ward, P. and Zahavi, A. (1973) The importance of certain assemblages of birds as 'information centres' for food finding. *Ibis* **115**:517–534.

Ward, P.I. (1993) Females influence sperm storage and use in the yellow dung fly *Scathophaga stercoraria* (L.). *Behavioral Ecology and Sociobiology* **32**:313–319.

Watson, N.L. and Simmons, L.W. (2010) Reproductive competition promotes the evolution of female weaponry. *Proceedings of the Royal Society of London. Series B: Biological Sciences* **277**:2035–2040.

Watts, D.P. (2002) Reciprocity and interchange in the social relationships of wild male chimpanzees. *Behaviour* **139**:343–370.

Wedekind, C. and Braithwaite, V.A. (2002) The long-term benefits of human generosity in indirect reciprocity. *Current Biology* **12**:1012–1015.

Wedell, N., *et al.* (2002) Sperm competition, male prudence and sperm-limited females. *Trends in Ecology and Evolution* **17**:313–320.

Werner, N.Y. and Lotem, A. (2003) Choosy males in a haplochromine cichlid: first experimental evidence for male mate choice in a lekking species. *Animal Behaviour* **66**:293–298.

West, S.A. (2009) *Sex Allocation*. Princeton, NJ: Princeton University Press.

West, S.A. and Gardner, A. (2013) Adaptation and inclusive fitness. *Current Biology* **23**:R577–R584.

West, S.A., *et al.* (2002) Cooperation and competition between relatives. *Science* **296**:72–75.

West, S.A., *et al.* (2006) Social evolution theory for microorganisms. *Nature Reviews Microbiology* **4**:597–607.

West, S.A., *et al.* (2007) Social semantics: altruism, cooperation, mutualism, strong reciprocity and group selection. *Journal of Evolutionary Biology* **20**:415–432.

West-Eberhard, M.J. (1975) The evolution of social behaviour by kin selection. *Quarterly Review of Biology* **50**:1–33.

West-Eberhard, M.J. (1979) Sexual selection, social competition and evolution. *Proceedings of the American Philosophical Society* **123**:222–234.

West-Eberhard, M.J. (1983) Sexual selection, social competition and speciation. *Quarterly Review of Biology* **58**:155–183.

Western, D. (1979) Size, life history and ecology in mammals. *African Journal of Ecology* **17**:185–204.

Westneat, D.F. and Sargent, R.C. (1996) Sex and parenting: the effects of sexual conflict and parentage on parental strategies. *Trends in Ecology and Evolution* **11**:87–91.

Whitehead, H. (1983) Structure and stability of humpback whale groups off Newfoundland. *Canadian Journal of Zoology* **61**:1391–1397.

Widemo, M.S. (2006) Male but not female pipefish copy mate choice. *Behavioral Ecology* **17**:255–259.

Wilkinson, G.S. (1984) Reciprocal food sharing in the vampire bat. *Nature* **308**:181–184.

Wilkinson, G.S. and Reillo, P.R. (1994) Female choice response to artificial selection on an exaggerated male trait in a stalk-eyed fly. *Proceedings of the Royal Society of London. Series B: Biological Sciences* **255**:1–6.

Williams, G.C. (1957) Pleiotropy, natural selection, and the evolution of senescence. *Evolution* **11**:398–411.

Williams, G.C. (1966) *Adaptation and Natural Selection: A Critique of Some Current Evolutionary Thought*. Princeton, NJ: Princeton University Press.

Williams, G.C. (1979) The question of adaptive sex ratio in outcrossed vertebrates. *Proceedings of the Royal Society of London. Series B: Biological Sciences* **205**:567–580.

Wilson, D.S. (1975) A theory of group selection. *Proceedings of the National Academy of Sciences of the United States of America* **72**:143–146.

Wilson, D.S. (1977) Structured demes and the evolution of group-advantageous traits. *American Naturalist* **111**:157–185.

Wilson, D.S. and Sober, E. (1989) Reviving the superorganism. *Journal of Theoretical Biology* **136**:337–356.

Wilson, D.S. and Wilson, E.O. (2007) Rethinking the theoretical foundation of sociobiology. *Quarterly Review of Biology* **82**:327–348.

Wilson, E.O. (1971) *The Insect Societies*. Cambridge, MA: Belknap Press.

Wilson, E.O. (1974) *Sociobiology, the New Synthesis*. Cambridge, MA: Harvard University Press.

Wong, B.B.M. and Jennions, M.D. (2003) Costs influence male mate choice in a freshwater fish. *Proceedings of the Royal Society of London. Series B: Biological Sciences* **270**:36–38.

Wrangham, R.W. (1980) An ecological model of female-bonded primate groups. *Behaviour* **75**:262–300.

Wright, J., *et al.* (2001) Cooperative sentinel behaviour in the Arabian babbler. *Animal Behaviour* **62**:973–979.

Wyman, J. (1967) The jackals of the Serengeti. *Animals* **10**:79–83.

Wynne-Edwards, V.C. (1962) *Animal Dispersion in Relation to Social Behaviour*. Edinburgh: Oliver and Boyd.

Yamaguchi, M., *et al.* (1981) Distinctive urinary odors governed by the major histocompatibility locus of the mouse. *Proceedings of the National Academy of Sciences of the United States of America* **78**:5817–5820.

Zahavi, A. (1975) Mate selection: a selection for a handicap. *Journal of Theoretical Biology* **53**:205–214.

Zahavi, A. (1977) The cost of honesty: further remarks on the handicap principle. *Journal of Theoretical Biology* **67**:603–605.

Zahavi, A. and Zahavi, A. (1997) *The Handicap Principle: A Missing Piece of Darwin's Puzzle*. New York: Oxford University Press.

Zeh, J.A. and Zeh, D.W. (1996) The evolution of polyandry. I. Intragenomic conflict and genetic incompatibility. *Proceedings of the Royal Society of London. Series B: Biological Sciences* **263**:1711–1717.

Zinkernagel, R.M. and Doherty, P.C. (1974) Immunological surveillance against altered self components by sensitized T-lymphocytes in lymphocytic choriomeningitis. *Nature* **251**:547–548.

Zuckerman, S. (1929) The social life of the primate. *Realist* **1**:72.

Zuckerman, S. (1932) *The Social Life of Monkeys and Apes*. London: Routledge.

CHAPTER 2
Female sociality

2.1 Introduction

An alien observer landing on the earth would quickly notice that life forms were not evenly spread over its surface. Not only are they concentrated in areas where resources are abundant but, within patches of suitable habitat, individuals are commonly clustered in groups. A monospecific stand of ironwood, a gannet colony and a hunting pack of wild dogs all invite the question: why do individuals aggregate in groups? A closer inspection of groups would show that they vary in their structure and stability as well as in their size. Some consist entirely of individuals of one sex, while others contain males as well as females. In some cases most individuals are of similar ages, while in others they include juveniles, adolescents and adults. In some species all adult females breed regularly, while in others one adult female in each group monopolises reproduction. Some groups are stable in membership, others are continually changing as parties of individuals join and leave them; some wander widely throughout their habitat, others defend an exclusive territory. An acute observer would also notice that some group members more frequently associate with each other and interact in a more friendly way than others.

In many social mammals, males guard prospective mating partners and distribute themselves in relation to the distribution of females (see Chapter 1), so that understanding the reasons for contrasts in female sociality is of fundamental importance in understanding the evolution of mammalian breeding systems. In addition, the size and stability of female groups has important consequences both for the evolution of female behaviour (see Chapters 8 and 9) and for the demography and dynamics of populations.

This chapter reviews the distribution and evolution of female groups. Section 2.2 describes different types of female groups and sections 2.3 and 2.4 review the benefits and costs of sociality to females, while section 2.5 examines relationships between group size and fitness. Section 2.6 then describes contrasts in female sociality in different taxa. Subsequently, section 2.7 examines the way in which group members coordinate their behaviour and section 2.8 explores some of the consequences of variation in female sociality.

2.2 Contrasts in female sociality

Although group size and stability often vary widely within species in some mammals, females generally forage alone while in others they usually forage in groups.

Solitary foragers

In some mammals, breeding females are intolerant of each other and females occupy separate territories which they share with their dependent young but not with breeding adults of either sex. North American tree squirrels of the genus *Tamiasciurus* provide one of the best-documented examples of a society of this kind (Figure 2.1). They are central-place foragers who store food in caches for use in winter and the occupation of solitary territories is probably enforced by the need to minimise range area and journey length (Smith 1968).

In a larger proportion of species where females generally forage on their own, they occupy ranges that overlap partially or wholly with those of neighbouring females. Systems of this kind are common in rodents (Michener 1983; Lacey and Sherman 2007), carnivores (Gittleman

Mammal Societies, First Edition. Tim Clutton-Brock.
© 2016 Tim Clutton-Brock. Published 2016 by John Wiley & Sons, Ltd.

Figure 2.1 In North American red squirrels, both sexes defend separate territories against members of the opposite sex as well as against members of their own sex. *Source*: © Ryan W. Taylor.

1989; Sandell 1989) and nocturnal primates (Bearder 1987; Müller and Thalmann 2000). Although females generally forage alone, they frequently maintain complex social networks and the diversity of these systems may be comparable to that of diurnal mammals (Müller *et al.* 2007), although they are considerably less well known as a result of the difficulties of observing small nocturnal mammals. In some of these species (including

Figure 2.2 Male and female fork-tailed lemurs live in stable pairs but forage separately. *Source*: © Roland Hilgarten.

Figure 2.3 Dik dik live in stable pairs, sometimes foraging alone and sometimes together. *Source*: © Peter Brotherton.

several of the terrestrial sciurids), several females (who are usually close relatives) occupy a common range which they defend against neighbouring groups (Michener 1983; Hare and Murie 2007).

Relationships between females and males vary. In a few species, females defend their territories against males as well as against females. In others, females share their range or territory with a single mature male, either foraging separately, as in fork-marked lemurs (Schülke and Kappeler 2003) (Figure 2.2), sometimes together and sometimes apart, as in dik dik (Figure 2.3), or together most of the time, as in owl monkeys (Figure 2.4).

In many species, males occupy ranges or territories larger than those of females and these overlap the ranges of several females (Figure 2.5), and breeding systems of this kind may be the ancestral condition for many mammalian groups (Lukas and Clutton-Brock 2013). Where female ranges overlap partially, male territories usually overlap the ranges of several females and mating is promiscuous, whereas in species where the ranges of several females coincide, a single male may defend access to multiple breeding partners and breeding systems resemble the harem groups of social foragers.

The breeding systems of rodents illustrate the diversity of social organisation among solitary foragers (Michener 1981; Wolff and Sherman 2007). In some species,

Figure 2.4 Owl monkeys live in monogamous pairs and forage together, with their offspring. *Source*: © Margaret Corley.

(a)   (b)

Figure 2.8 In multi-level societies, individuals live in stable breeding groups which aggregate in larger herds or 'super-groups': (a) hamadryas baboon; (b) gelada baboon. *Sources*: (a) © Larissa Swedell; (b) © Noah Snyder-Mackler.

Species living in groups of this kind are commonly referred to as living in 'fission–fusion groups'. In some of these species, group members are hostile to members of neighbouring groups.

Finally, in some mammals, females forage together in social groups but do not have long-term bonds to each other and group membership is unstable. Groups of this kind are relatively common in ungulates and

also occur in some cetaceans as well as in some bats (Figure 2.12).

Networks

In most animal societies, there are also substantial differences in sociality between group members. Some individuals tend to be loners and spend much of their time on the periphery of groups, while others spend more of their

Figure 2.9 In plains zebras, most breeding units consist of a single mature stallion and several associated mares. As in hamadryas baboons, multiple breeding units aggregate in large herds of unstable membership. *Source*: © Tim Clutton-Brock.

Figure 2.10 In spotted hyenas, individuals live in stable groups ('clans') but forage in smaller, unstable parties. *Source*: © Kay Holekamp.

time in the centre; some interact with others relatively frequently while others do so comparatively rarely; some have close social bonds with particular group members while others associate and interact with other group members more equally. Network analysis provides a way of describing the relationships between group members and is now widely used to identify the pattern and strength of social connections between group members (Croft *et al.* 2008; Krause *et al.* 2009; Farine and Whitehead 2015). While it is most commonly applied to proximity data, it can also be used to document the relative frequency with which individuals interact in different ways, for example the relative frequency with which they groom each other or direct aggression at each other (Lusseau and Newman 2004; Wey *et al.* 2008). In addition, it can be used to explore the extent to which social connections affect breeding success and survival (see section 2.6) as well as other behavioural parameters (such as dominance status, frequency of leadership and personality) and developmental processes (such as reliance on social learning).

Since its introduction, network analysis has been widely used to explore social connections in a wide range of social mammals, including cetaceans, carnivores, equids, primates and rodents (Lusseau 2003; Ozgul *et al.* 2006; Sundaresan *et al.* 2007; Kasper and Voelkl 2009; Madden *et al.* 2011) as well as in domestic animals (Webb 2005) and humans (Rand *et al.* 2011) and has shed new light on variation in social structure within and between species (Sueur *et al.* 2008, 2011; Sundaresan *et al.* 2007). It does, however, need to be used with caution for many of its metrics are not straightforward to interpret and social connections between individuals may change with time, making it difficult to collect sufficient data to achieve stable estimates before changes occur (Henzi *et al.* 2009; James *et al.* 2009).

2.3 Benefits of grouping

Reducing risks of predator attack

The simplest explanation of female sociality is that, in saturated environments, the chances of successful dispersal are very low, so that adolescent offspring maximise their chances of survival by remaining in their natal group until a vacant territory is available (Emlen 1991; Lacey and Sherman 2007). Dominant individuals may increase their inclusive fitness by allowing close relatives to remain in the

Figure 2.11 Chimpanzees live in stable groups ('communities') and forage in unstable parties within the range of their community. *Source*: © Jane Goodall Institute / By Derek Bryceson.

group rather than imposing the costs of dispersal by evicting them. Under these circumstances, negative correlations between group size and breeding success would be expected. However, studies of a wide range of species show that sociality can have diverse benefits and that their relative importance varies with their ecology.

In many species, the instantaneous risk of predation declines with increasing group size, though several different mechanisms can be involved. In some species, sociality helps to reduce the per-capita risk of predation because predators only take one individual at a time (Hamilton 1971). This mechanism is sometimes called the 'dilution' effect because it suggests that, by aggregating, individuals are diluting the risk that they will be the predator's target. The dilution effect depends on the assumption that the rate of attacks by predators does not increase in proportion to the size of the group, which

often appears to be the case (Mech 1970; Kruuk 1972; Schaller 1972; Malcolm and van Lawick 1975). For example, great white sharks prey on seals (Figure 2.13) and experiments in which Styrofoam seals were trolled behind boats in 'groups' of different sizes show that the per-capita risk of attack declines as group size increases (De Vos and O'Riain 2010) (Figure 2.14). Dilution effects may help to explain the formation of large herds in migratory ungulates where breeding is highly synchronised and resident carnivores can only kill a very small proportion of individuals (Fryxell and Sinclair 1988; Fryxell 1995).

A third possibility is that living in groups reduces the ability of predators to catch prey (Caro 2005), for example group size increases the chance that at least one member of a group will detect an approaching predator and warn the rest and this rises with increasing group size. This effect has been demonstrated experimentally in insects and birds (Krause and Ruxton 2002) and probably explains the formation of polyspecific feeding or resting groups in many diurnal species. For example, in savannah ungulates, territory-holding males of different species form temporary groups during the middle of the day when the chance of encounters with females is low and respond to the alarm calls of other species as well as to those of conspecifics (Figure 2.15). In some species, group members coordinate their vigilance, increasing the amount of time that they spend looking for predators when the behaviour of other group members suggests that predators are around. For example, in eastern grey kangaroos, the vigilance of individuals increases in response to increases in the vigilance of other group members (Pays *et al.* 2009). Solitary individuals or those in small groups still need to feed and, as the numbers of individuals alert at any one time are usually lower than in large groups, members of small groups are often more likely to be taken by predators.

The presence of several potential prey can also serve to confuse predators, so that they have difficulty in 'locking on' to one individual and may become exhausted before they can make a kill (Morse 1977; Schradin 2000). There is anecdotal evidence of effects of this kind in a number of mammals (Caro 2005) and experimental evidence from some other animals (Neill and Cullen 1974). Finally, in larger species, individuals living in groups can actively defend themselves against predators where individuals would be unable to do so

Figure 2.12 In blue wildebeest, females aggregate in large, unstable herds. *Source*: © David Rose.

(Caro 2005). For example, bison and musk oxen form defensive barriers to predators, with juveniles in the centre and adults on the outside, facing the predator (Mech 1970) (Figure 2.16).

In many social species, individuals living on their own or in small groups appear to be aware of the increased risk of predation. In colonial tuco-tucos, individuals living in their natal groups have higher glucocorticoid

Figure 2.13 Off the coast of South Africa, great white sharks commonly prey on seals which benefit by aggregating. *Source*: © David Jenkins/Getty Images.

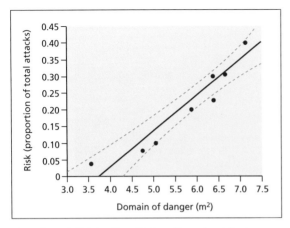

Figure 2.14 'Selfish herd' or dilution effects of sociality in Styrofoam seals presented to great white sharks. Different numbers of Styrofoam seals were attached to poles and trolled in areas where great white sharks commonly feed. Seals are protected from attack by the presence of others outside them so that their 'domain of danger' declines as group size increases. *Source*: From De Vos and O'Riain (2010). Reproduced with permission from the Royal Society.

Figure 2.15 Territorial males of different antelope species whose territories overlap associate with each other in the heat of the day, when no females are around and the chance of mating is low. *Source*: © Morris Gosling.

levels than those that have dispersed and live alone (Woodruff *et al.* 2013), whereas in meerkats, individuals dispersing alone or with one partner have substantially higher levels of circulating cortisol than those dispersing in larger parties (see Chapter 3). Members of small groups often attempt to compensate by raising their level of vigilance (Blumstein and Daniel 2002; Beauchamp

Figure 2.16 In musk oxen, groups that are threatened by predators form a defensive ring with juveniles in the centre. *Source*: © Wayne Lynch/Getty Images.

2008): for example, in white-nosed coatis, vigilance levels when drinking decline with increasing group size and the length of uninterrupted bouts of drinking increases (Burger and Gochfeld 1992).

While most studies of predator benefits have focused on attacks by larger predators, sociality may also have benefits in relation to attacks by parasites. One of the most convincing examples of the dilution effect in mammals comes from a study of attacks on feral horses by blood-sucking tabanid flies in the Camargue. On still summer days, when the horses are most likely to be attacked by flies, they respond by clustering in large groups. Experimental manipulation of group size shows that the rate of fly attacks per horse declines with increasing group size (Duncan and Vigne 1979) (Figure 2.17). Group-living may also facilitate social grooming and so may help to reduce parasite impact or may allow individuals to cleanse wounds and these benefits may offset costs arising from increased rates of disease transmission (Hughes *et al.* 2002; Barrett and Henzi 2006).

Finding and catching food

Living in groups can also help individuals to find food. Where food is cryptic but is distributed in patches that several animals can feed on, individuals are often attracted to sites where others are feeding and sociality can increase intake rates by helping individuals to locate the most productive feeding sites (Krebs *et al.* 1972; Wilkinson 1992). For example, studies of African elephants introduced to a novel environment showed that the proportion of time individuals spent foraging socially were initially correlated with individual differences in body condition and that social foraging declined as individuals got to know their habitat better (Pinter-Wollman *et al.* 2009). In addition, studies of other elephant populations show that the age and experience of mothers and other group members has important consequences for their ability to locate resources during periods of food shortage and affects the survival of calves (Foley *et al.* 2008; Mutinda *et al.* 2011).

Social learning experiments also provide evidence that individuals acquire important information by watching each other (see Chapter 6) and group members may benefit from the signals of other group members (Elgar 1986). For example, chimpanzees which find trees in fruit often give loud pant hoots that attract other members of the same community to the tree (see Chapter 7).

Where food is abundant, callers may obtain immediate benefits from calling because the presence of other individuals reduces their risk of predation or enhances their access to mates. In species that forage in cohesive groups that have to make collective decisions about their movements, like yellow baboons, increased group size can also increase the efficiency of information use and the accuracy of decisions since the proportion of informed individuals necessary for groups to make informed decisions falls as group size rises (Couzin *et al.* 2005).

In some social carnivores, sociality also helps individuals to catch prey and individuals are more likely to be successful in catching food if they hunt in groups than if they hunt alone. Lions, spotted hyenas, jackals and wild dogs can all show higher success rates when hunting some kinds of prey in groups than when hunting alone (Wyman 1967; Kruuk 1972; Schaller 1972; Major 1978; Stander 1992a; Creel and Creel 2002). Increases in group size can also broaden the size range of prey that can be caught and some carnivores adjust their selection of prey to the number of individuals that are hunting together: for example, single lions rarely attempt to kill buffalo while groups often do so (Schaller 1972) and wild dogs in larger packs often hunt adult zebra while small packs specialise in hunting impala or gazelles (Malcolm and van Lawick 1975; Creel and Creel 2002).

While hunting in large groups increases feeding success in some social carnivores, the effects of group size on hunting success do not always compensate for the effects of increased numbers of individuals sharing kills (see section 2.5). One explanation of why individuals continue to hunt in groups under these circumstances is that social hunting may reduce variance in hunting success and the risk of starvation or provides benefits to individuals that are incapacitated and cannot hunt successfully. Another possibility is that sociality helps to reduce the costs of hunting: for example, in wild dogs the average length of successful chases is lower in large packs than small ones and the number of hunts per day increases with pack size, with the result that rates of net energy gain from hunting are larger for members of big packs than for members of small ones (Creel and Creel 2002) (Figure 2.18).

Feeding in groups may also serve to reduce interference by competitors. In many social mammals, larger groups typically displace smaller ones from feeding sites: for example, in wedge-capped capuchin monkeys, large

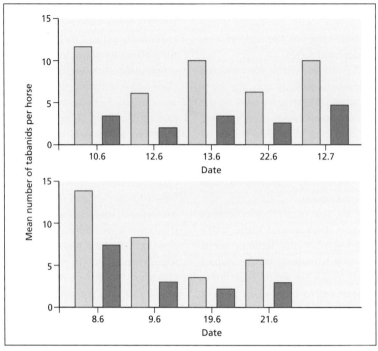

Figure 2.17 Numbers of tabanid flies settling on Camargue horses in small groups (light bars) and large groups (dark bars). The upper figure shows results based on counts of flies on horses in natural groups while the lower one shows them on individuals maintained experimentally in large and small groups. *Source*: From Duncan and Vigne (1979). Reproduced with permission of Elsevier. *Photo source*: © Alison Duncan.

groups consistently displace small groups during inter-group encounters at feeding sites and female breeding success is consistently higher in large groups than in small ones (Robinson 1988, p. 191) (Figure 2.19).

Similarly, large wolf packs displace smaller ones (Cassidy *et al.* 2015). Where larger groups repeatedly displace smaller ones, they may force them to leave their usual range or territory.

Figure 2.18 In cooperative hunters, increased group size often plays an important role in helping group members to catch and defend prey. For example, in African wild dogs, hunting success is strongly influenced by pack size (see Chapter 9). *Source*: © Dave Hamman.

Group size also provides benefits in competition with members of other species. For example, large packs of wolves can defend their kills against scavenging ravens (Vucetich *et al.* 2004), large packs of wild dogs can

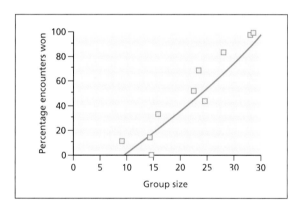

Figure 2.19 The proportion of inter-group contests won against other groups in capuchin monkeys plotted against relative group size. *Source:* From Robinson (1988). Reproduced with permission of Springer Science and Business Media. *Photo source:* © Susan Perry.

defend kills against scavenging hyenas (Creel and Creel 2002) and large groups of hyenas are able to protect their prey against scavenging lions (Smith *et al.* 2008). Similarly, in lions, sociality allows groups to defend group territories and to maintain access to reliable food resources that can be inherited by their offspring (Mosser *et al.* 2015).

Reproductive benefits

In some cases, sociality may reduce the risk of attacks on females by competing males. For example, in some ungulates where females live for most of the year in large, unstable, mixed-sex herds, females that approach oestrus become the target of competition between males and can be killed or injured as a consequence of male competition (see Chapter 10). As oestrus approaches, females leave grazing herds and join males that defend mating territories containing multiple females, where the risk of male harassment is reduced (Clutton-Brock *et al.* 1993).

In other species, female sociality helps to protect dependent juveniles. For example, in southern sea lions, where males defend land-based harems within large aggregations of females, dependent pups are often accidentally killed by territorial males at times when their mother has left temporarily to forage: as harem size increases, the per-capita frequency of interactions between males and females declines and pups are more likely to survive (Campagna *et al.* 1992). In species where immigrant males commonly kill dependent juveniles, female sociality can also contribute to the survival of juveniles because mothers assist each other in defending their young against infanticidal males (see Chapter 15).

In some societies, group-living allows females to share the care of their young with relatives and so increases the survival of juveniles. For example, in black and white colobus monkeys, females with dependent young commonly pass them to other females before descending to the forest floor to forage where they would be at risk (Oates 1977). Similarly, female sperm whales guard the offspring of other females on the surface while their mothers dive deeply (Whitehead and Weilgart 2000). Sociality can also play an important role in protecting juveniles in communal and cooperative breeders (see Chapter 17).

Fitness of emigrants

In social mammals where individuals disperse together, group size may have important effects on fitness through its influence on the size and success of dispersing parties:

individuals in larger parties often show higher rates of survival and are more likely to establish a new range or territory, with the result that the duration of the dangerous period when they are wandering in unfamiliar areas is reduced. Since the number of individuals dispersing at the same time commonly increases with group size, juveniles born to members of larger groups may often be more likely to colonise neighbouring areas than juveniles born to members of small groups. For example, in Kalahari meerkats, dominant females evict older subordinate females between the ages of 2 and 4 years (Clutton-Brock *et al.* 2006, 2010). As group size increases, the number of subordinates dispersing at the same time rises and larger parties (coalitions) of dispersers are derived from larger groups (Figure 2.20a) (Young 2003). Subordinates that are evicted together search for vacant ground but are frequently displaced by resident groups and are subject to high risks of predation. Comparisons show that females in larger parties spend less time vigilant and more time foraging and, as a result, gain more weight during feeding periods (Figure 2.20b–d). They also have lower ectoparasite loads and show reduced levels of circulating cortisol (Figure 2.20e). Individuals that are evicted on their own seldom survive and rarely found breeding groups (Figure 2.20f). Males also disperse from their natal group in parties at around the same age and search for breeding opportunities in other groups or for parties of dispersing females with which they can start new groups and their chances of surviving and establishing new breeding groups also increase with the size of the party they leave in (see Chapter 12).

Group persistence

Most groups are subject to stochastic fluctuations in group size caused by variations in fecundity, mortality and emigration. Where groups defend resources (such as feeding territories or burrows) that are transferred to descendant relatives, there are likely to be indirect fitness benefits in minimising the risk of group extinction. Larger groups are less likely to become extinct as a result of stochastic changes in group size than smaller ones: for example, in prairie dogs, groups that become extinct are smaller than those that persist between years (Figure 2.21). Where group members are close relatives, effects of this kind may make important contributions to the inclusive fitness of individuals.

Individual differences

Individual differences in sociality between group members are often correlated with variation in fitness components. For example, in bighorn sheep, individual females show consistent differences in centrality within social networks that are correlated with lamb survival and female reproductive success and (assuming these correlations reflect causal relationships) the effects of centrality on female fitness may be as large as those of body mass (VanderWaal *et al.* 2009). Similarly, in some social primates, social bonds between females play an important role in establishing and maintaining differences in social status and are correlated with variation in the survival of their offspring (see Chapters 8 and 9). Variation in social relationships can also be correlated with growth, health, survival and longevity.

2.4 Costs of grouping

Predators and parasites

Though there are important benefits to living in groups, there are often substantial costs, too. Several different mechanisms can be involved and it is often difficult to distinguish between their effects. In some mammals, large groups are more likely to be detected by predators: for example, the leks of Uganda kob consistently attract lions and hyenas because they provide a predictable source of prey (Balmford and Turyaho 1992). Large group size also limits opportunities for crypsis as a way of avoiding detection by predators.

Social groups can also facilitate the spread and persistence of parasites or disease (Rand *et al.* 1995; Altizer *et al.* 2003), for example, if one member of an African wild dog group contracts rabies, the disease is rapidly passed to other group members because of the dogs' practice of licking each other's muzzles (Gascoyne *et al.* 1993). Similarly, in Japanese macaques, the centrality of females within networks is related to their parasite load (Macintosh *et al.* 2012). Large groups can also act as reservoirs for parasites and pathogens: in prairie dogs, burrows used by larger groups show increased numbers of ectoparasites compared with burrows used by smaller groups (Hoogland 1995) and studies of bats and social rodents show that levels of ectoparasites increase with group size (Wilkinson 1985). However, positive correlations between group size and the prevalence of parasites and pathogens are not universal. For example, in

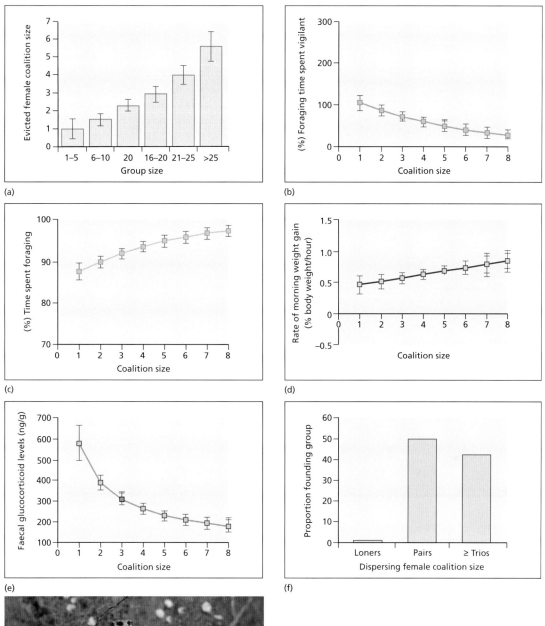

Figure 2.20 Benefits of sociality to dispersing female meerkats: (a) size of parties of females (coalitions) dispersing from groups of different sizes; (b) mean proportion of time individuals in parties of different size spent vigilant; (c) mean proportion of time individuals in parties of different size spent foraging; (d) percentage of body weight gained per hour in parties of different size; (e) faecal glucocorticoid levels in evicted females and dispersing males in relation to party size; (f) proportion of dispersing parties of different sizes that founded new breeding groups. *Source*: From Young (2003). Reproduced with permission of A. Young. *Photo source*: © Tim Clutton-Brock.

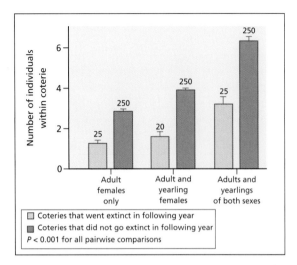

Figure 2.21 Group size and the probability of group extinction in prairie dogs. The figure compares numbers of animals of different categories in coteries (local breeding groups) that went extinct in the following year with numbers in coteries that did not go extinct in the next year. *Source*: From Hoogland (1995). Reproduced with permission of the University of Chicago Press. *Photo source*: © Elaine Miller Bond.

European badgers, the prevalence of tuberculosis is higher in small groups than in large ones (Woodroffe *et al.* 2009).

Resource competition

Sociality can also increase the energetic costs of foraging. Both within and across species, day range length and foraging time commonly increase with group size, while time spent resting declines (Waser 1977; Chapman and Chapman 2000; Majolo *et al.* 2008; Pollard and Blumstein 2008; Groenendijk *et al.* 2015) though, in some cases, groups smaller than average also show increases in day range length and home range size, perhaps as a result of inter-group competition (Markham *et al.* 2015). Across species both day range length and home range area also increase as the metabolic demands of groups rise (Figure 2.22). Increases in home range or territory size as a result of increasing group size are likely to reduce familiarity with foraging areas, with potential effects on foraging efficiency and the risk of predation.

By concentrating the distribution of individuals, sociality generates competition for food and other resources, reducing food intake and the growth and survival of young, especially in subordinate animals (Clutton-Brock *et al.* 1982b; Whitten 1983; Janson 1988; Wrangham *et al.* 1993). Where food is distributed in patches that cannot accommodate all group members, some individuals may either have to forage separately (with attendant risks of predation) or may be forced to forage in suboptimal locations (Chapman and Chapman 2000). In addition, increased rates of aggression between foraging animals and associated physiological responses may reduce foraging efficiency and raise the energetic requirements of individuals (Dunbar 2010). In conjunction, these effects often increase the amount of time that individuals have to devote to foraging and may contribute to limiting group size (Dunbar *et al.* 2009).

Because of the effects of group size on foraging competition, the distribution of resources would be expected to have an important influence on the costs of sociality (Jarman 1974; Kruuk and Macdonald 1985; Janson and Goldsmith 1995). Feeding competition is most likely to generate direct competition between group members and to constrain group size where food is mobile and easily disturbed or where it consists of discrete, widely spaced items of high nutritional value: for example, experiments with captive macaques show that increases in the average size of food items provided in a social context are associated with increased frequencies of overt competition and punishment (Chancellor and Isbell 2008). Increased competition between group members and associated reductions in group size are also likely where the size of food patches limits the number of individuals that can feed simultaneously or where the addition of extra-group members raises the size of group territories that need to be defended against

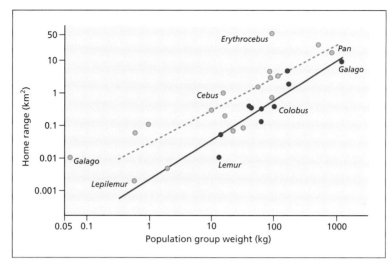

Figure 2.22 As total group weight increases across primate species, the size of group ranges rises. Solid circles and a continuous regression line indicate folivores; open circles and a dotted line indicate insectivores and frugivores. *Source*: From Clutton-Brock and Harvey (1977). Reproduced with permission of John Wiley & Sons, Ltd.

neighbours. Conversely, the effects of increasing group size are likely to be smallest where food is immobile and evenly distributed within large patches or where variation in group size has little effect on territory size because all individuals require access to multiple patches in the course of the year.

Reproductive competition

Increasing group size also generates competition between females for opportunities to reproduce and for access to males, territories, breeding burrows and resources for their offspring (see Chapter 8). Increasing levels of reproductive competition in large groups may also affect decisions about dispersal and play a role in limiting group size. For example, studies of baboons suggest that group size may be limited by attempts by females to maximise their fecundity (Bettridge *et al.* 2010).

In some species, increasing group size has positive effects on some components of fitness and negative ones on others. For example, the size of African wild dog packs is positively related to the breeding success of dominants, but negatively to adult survival (Creel and Creel 2015).

Male infanticide

Finally, increasing group size can raise the risk of male infanticide. In some social mammals where immigrant males kill dependent young (including lions and ursine howler monkeys), males compete for larger groups and

rates of male turnover (and hence infanticide) increase with group size, with the result that female reproductive success declines (Packer 1988; Crockett and Janson 2000) (see Chapter 15).

2.5 Sociality and fitness

Since variation in group size can generate both important benefits and substantial costs, it is unsurprising that contrasts in group size within species are often associated with variation in survival and breeding success. Relationships between group size and individual fitness can take a variety of forms, depending on the relative magnitude of costs and benefits (Figure 2.23). As would be expected, members of social species living alone or in groups substantially smaller than the average for their population commonly show reduced survival or breeding success as a consequence of reductions in their ability to detect predators, combat neighbours or rear young (Clutton-Brock 2009; Ebensperger *et al.* 2012). Some of the clearest examples of relationships of this kind come from studies of cooperative breeders where females are often unable to raise young without other group members to assist them (see Chapter 17).

In many species, fecundity and survival also decline in groups that are substantially larger than the average as a result of increased competition for resources or breeding

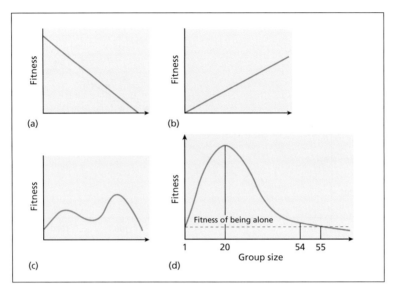

Figure 2.23 Hypothetical relationships between group size and individual fitness: (a–c) differing relationships between group size and individual fitness; (d) fitness of group-living individuals relative to a solitary individual in groups of varying sizes where individual fitness initially increases with group size and then declines. Under these conditions, it will pay solitary individuals to join groups smaller than 55 and group size will commonly exceed the level that maximises individual fitness, as Sibly (1983) demonstrated. *Sources*: Figures (a–c) from Krause and Ruxton (2002). Figure (d) adapted from Sibly (1983) and reproduced with permission of Elsevier.

opportunities and rising levels of interference between group members (Clutton-Brock *et al.* 1982b; van Schaik 1983; Ebensperger *et al.* 2011). For example, in colonial tuco-tucos, females living in their natal groups rear fewer young than dispersers (Lacey 2004). While it might appear surprising that members of larger groups do not move to smaller groups, the risks associated with dispersal are often high and both survival and breeding success are typically low in newly established groups (see Chapter 3), so it may often be in the interests of individuals to remain in established groups that exceed the size that would maximise their fitness (Sibly 1983; Krause and Ruxton 2002; Nunes 2007). As a result, groups may grow in size until the fitness returns gained by additional recruits to established groups are no higher than those of solitaries (Sibly 1983) (see Figure 2.23d). How far group sizes are likely to diverge from the optimum is consequently likely to depend on the relative fitness of solitaries and the shape of fitness functions (Giraldeau and Gillas 1985; Zemel and Lubin 1995; Krause and Ruxton 2002) and may be affected by whether or not individuals can move independently between groups (Kramer 1995), by indirect benefits

associated with joining or tolerating kin (Higashi and Yamamura 1993; Giraldeau and Caraco 2000) and by the relative effects of variation in group size on the fitness of dominants and subordinates (Hamilton 2000).

Attempts to determine whether animals live in groups whose size maximises foraging success or fitness have commonly focused on fission–fusion societies where individuals can leave or join foraging groups freely. Several studies have shown that the average size of foraging parties increases when food is abundant or patch size is relatively large (Wrangham 1987; Symington 1988; Boesch 1996; Campbell 2002; Bradbury *et al.* 2015) but it is usually difficult to determine whether group size approximates to optimal values and different studies have often reached contrasting conclusions (Krause and Ruxton 2002). For example, studies of African lions in Tanzania, where pride members commonly hunt in groups, showed that social hunting increased the chance of catching prey but that the per-capita intake rate of meat was usually maximised by lions hunting alone or in pairs (Caraco and Wolf 1975; Packer *et al.* 1990) while studies of lions in a more arid area of Namibia found that singletons were at a disadvantage in times of food scarcity and small groups were optimal when food was abundant (Stander 1992b). Potential reasons why individuals might hunt in parties larger than those that maximise rates of food intake include the possibility that increasing party size reduces the energetic costs of hunting or reduces the risk of failure (Creel and Creel 1995) as well as social benefits in competitive interactions with neighbouring prides or immigrants (Packer *et al.* 1990) and indirect benefits arising from improvements in the fitness of relatives (Rodman 1981), but it is not yet possible to quantify their effects (Krause and Ruxton 2002).

2.6 Comparative sociality

The ecological factors controlling female sociality differ between groups of mammals but four principal factors are often associated with variation in group size: the relative importance of thermoregulation; the extent to which foraging success is affected by the presence of other individuals; the susceptibility of individuals to predation; and the extent to which breeding females rely on assistance from other group members to rear their young (Alexander 1974; Emlen and Oring 1977; Clutton-Brock and Janson 2012). Contrasting patterns of sociality in different groups of mammals are commonly related to variation in the relative importance of these

four effects. The following sections briefly review the distribution of group size and its correlates in eight different groups of mammals.

Rodents

Sociality in female rodents ranges from species where adult females live in separate ranges, through monogamous pairs and matrilineal groups to species, like capybaras, where multiple breeding groups aggregate in large unstable herds (Wolff 1989; Wolff and Sherman 2007; Hayes *et al.* 2011) (Figure 2.24). All the four main benefits of sociality are probably involved in the evolution of rodent sociality. Thermoregulatory benefits play an important role in species living at high altitudes or extreme latitudes (Armitage 2007). For example, marmots that occupy montane habitats live in colonies and commonly hibernate together in a communal burrow (Armitage *et al.* 2003; Armitage 2007) (Figure 2.25). Their young do not reach adult weight before hibernation and gain important thermoregulatory benefits from hibernating with other group members, which reduces their weight loss and increases their overwinter survival, especially when older animals are siblings (Arnold 1990). In contrast, in woodchucks, which live at lower altitudes, juveniles reach adult weight during their first year and individuals are solitary (Armitage 2007).

Susceptibility to predation generates selection for sociality in diurnal rodents living in open country, like prairie

(a)

(b)

Figure 2.24 (a) Capybaras graze on the flood plains of South American rivers in large herds consisting of multiple breeding unimale units. (b) Colonial tuco-tucos from Argentina. These subterranean ctenomyid rodents live in montane meadows in groups of up to six philopatric females and, in some cases, an immigrant male (Lacey and Ebensperger 2007). Group members share a common burrow and feed on surface vegetation nearby. Like the African mole rats, they are almost entirely subterranean but, unlike the social mole rats, multiple females breed each year, each producing a single litter of precocious young which are kept in a shared breeding chamber. *Source*: (a) © Panoramic Images/Getty Images; (b) © E. Lacey.

Figure 2.25 Alpine marmots live in family groups that hibernate together in winter. *Source*: © DeAgostini/Getty Images.

dogs, ground squirrels and capybaras (Michener 1983; Macdonald *et al.* 2007). In some ground-dwelling sciurids, several related females share overlapping ranges but occupy separate burrows while, in others, they share burrows (Hare and Murie 2007). For example, in black-tailed prairie dogs, several breeding females share several burrows and form 'coteries' and multiple coteries form colonies or 'wards' (Hoogland 1995). Several of these species have highly developed systems of alarm communication that contribute to the survival of individuals (see Chapter 7): for example, in prairie dogs, members of larger wards detect predators more quickly than members of smaller wards and spend less time scanning for danger (Hoogland 1995).

Reproductive cooperation can also be important. Allolactation and cooperative nest defence are widespread in rodents (Lewis and Pusey 1997) and play an important role in the evolution of sociality in some species. For example, female house mice that nest communally with relatives have greater reproductive success than those

nesting alone (König 1994). Cooperative exploitation of resources may also favour increases in group size: among the African mole rats, the largest groups are found in species (like naked mole rats) that use semi-deserts and rely on sparse, heavily clumped supplies of roots and tubers, while species living in more mesic habitats, where food is more abundant and evenly distributed (like the Cape mole rat), either live in pairs or in small groups (see Chapter 17) (Figure 2.26). The suggested explanation (sometimes referred to as the *aridity food-distribution hypothesis*) is that where food sources are widely dispersed and colony members can only extend their tunnels after the soil is softened by rain, sociality is favoured because it permits the rapid expansion of tunnel systems after rain. An association between aridity and sociality is also supported by comparisons of group size between populations of the same species living in habitats of varying aridity (Spinks *et al.* 2000a,b; Nevo 2007).

In many rodents, the ecological mechanisms that favour female sociality also generate costs to individuals that attempt to disperse. Dispersing females are often susceptible to hypothermia, starvation and predation and are seldom able to displace breeders or to join established breeding groups (Wolff 1994; Solomon 2003). As a result, they often show increased levels of glucocorticoids, their probability of survival is often low and selection may favour tolerance of adolescent offspring by breeding females and philopatry in adolescents (Errington, 1963; Nunes 2007; Woodruff *et al.* 2013).

Intraspecific variation in female sociality is common in rodents and is often associated with variation in food availability and population density. The African striped mouse provides one of the most striking examples (Schradin 2005; Schradin and Pillay 2005). In populations living in mesic grasslands, annual survival and population density are low and adult females live alone in exclusive home ranges, which are overlapped by the ranges of males. In contrast, in populations living in arid areas, where succulents are common, survival and population density are relatively high and several breeding females, their adult offspring and one breeding male live in groups of up to 30 individuals that share a common range.

Primates

Most nocturnal prosimians (including the nocturnal lemurs, lorises, bushbabies and tarsiers) forage alone, though in some species related females occupy

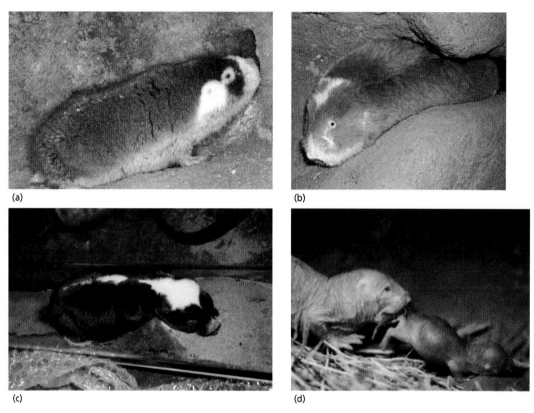

(a)

(b)

(c)

(d)

Figure 2.26 Solitary and social mole rats: solitary species are commonly associated with relatively mesic habitats while the most social species occur in semi-arid or arid habitats. (a) Cape mole rat; (b) silvery mole rat; (c) Damaraland mole rat; (d) naked mole rat. *Sources*: (a) © Nigel Bennett; (b) © Chris Faulkes; (c) © Markus Zöttl; (d) © Lorna Ellen Faulkes.

overlapping ranges and sleep in groups (Bearder 1987; Kappeler 1997; Kappeler and van Schaik 2002) (see Figure 2.5) while the only nocturnal monkeys, the South American night monkeys, live in socially monogamous pairs with dependent young and forage together (Jack 2007) (see Figure 2.4). Nocturnal primates commonly avoid predators by hiding and solitary foraging probably represents an adaptation to minimising predation rates while the formation of daytime resting groups may have thermoregulatory or social benefits (Bearder 1987).

In some diurnal primates, adult females are intolerant of each other and groups consist of socially monogamous pairs and their dependent offspring (van Schaik and Kappeler 2003). Pair-living has evolved independently in all the major primate radiations and is often associated with specialised diets and intolerance between breeding females (Chivers and Raemaekers 1986; Lukas and Clutton-Brock 2013). Although it is also associated with paternal care, phylogenetic reconstructions suggest that this is a consequence rather than a cause of social monogamy (see Chapter 16).

In many of the diurnal primates, multiple adult females live and breed in stable groups that typically consist of matrilineal relatives born in the group (Figure 2.27). Groups are cohesive and individuals commonly forage and sleep in close proximity to each other. The number of breeding females that associate with each other ranges from two to five in many of the arboreal monkeys to more than twenty in some groups of squirrel monkeys, colobine monkeys, baboons and macaques (Clutton-Brock and Harvey 1977; Strier 2011). In some societies, groups of breeding females aggregate in unstable herds or super-groups that can include more than 100 individuals, like the desert-living hamadryas baboons, the montane geladas

Figure 2.27 In contrast to most nocturnal species, ring-tailed lemurs (and many other diurnal lemurs) live in cohesive groups. *Source*: © Cyril Ruoso/Minden Pictures, Minden Pictures/Getty Images.

and the forest-living drills and mandrills (Stammbach 1987; Setchell *et al.* 2002; Patzelt *et al.* 2014). In many of these species, individuals forage primarily on the ground and are (or have been) exposed to predation by large carnivores.

Among the great apes, breeding female orangutans usually forage alone or with their dependent offspring in separate home ranges (Figure 2.28a) (van Schaik and van Hooff 1996; Singleton and van Schaik 2001). Their diet consists primarily of ripe fruit and social

(a)   (b)

Figure 2.28 Contrasts in sociality in the great apes: (a) female orangutans have separate ranges and forage alone while mature males are also solitary and have larger, overlapping ranges; (b) female mountain gorillas live in cohesive groups that include a dominant male, several breeding females and, in some cases, a subordinate male too. *Sources*: (a) © Ralonso/Getty Images; (b) © Alexander Harcourt/Kelly Stewart/Anthrophoto.

foraging increases travel time while their large body size and arboreal niche protect them from predation (Mitani 1989). In contrast, in mountain gorillas, which are herbivorous and primarily terrestrial, several breeding females live together in stable groups with one or more adult males (Watts 1996); feeding competition is rare and variation in group size is not related to obvious differences in foraging effort (Watts 1996) (Figure 2.28b). Leopards occur throughout most of their range and have been known to attack and kill adult gorillas (Schaller 1963), so juveniles are presumably vulnerable and sociality may serve to increase their survival.

Both bonobos and chimpanzees live in groups of ten to more than eighty adults that include multiple breeding females as well as multiple mature males and forage in unstable fission–fusion parties, though groups are larger and more cohesive in bonobos (Boesch 1994; White 1996; Stumpf 2007). They typically travel on the ground and, like gorillas, are exposed to the risk of predation. Individual female chimpanzees occupy distinct ranges and sometimes resist other females that attempt to settle in the same area (Williams *et al.* 2002; Pusey and Schroepfer-Walker 2013) while spatial structure of this kind appears to be absent in bonobos (White 1996). In both species, feeding competition occurs when the size of food patches (usually fruiting trees) is small and the size of feeding parties varies with patch size and food availability (Wrangham *et al.* 1992; White 1996). Patch size and food availability may not vary as much throughout the year in bonobos, which could account for the larger size of parties and the stronger social bonds between females (White 1996).

Comparative analyses of variation in group size in diurnal primates suggest that group size rises as the risk of predation increases and declines as foraging competition increases (Wrangham *et al.* 1993; Janson and Goldsmith 1995; Sterck *et al.* 1997). Within frugivorous species, day range length typically increases with group size, suggesting that individuals in larger groups pay higher foraging costs (Waser 1977; Wrangham *et al.* 1993). The relative strength of these effects (measured by estimating the increase in day range length caused by an additional group member, scaled to the notional day range length of a solitary individual) is correlated with differences in group size (Janson and Goldsmith 1995).

Contrary to prediction, folivorous species tend to live in relatively small groups, and relationships between group size and day range length are often weak or absent. It is sometimes suggested that this is because competition for resources is weak in folivores, but evidence that group size is related to habitat quality and that breeding success often declines in larger groups suggest this may not be the case (Snaith and Chapman 2007). Several possible explanations have been suggested, including the possibility that the low energetic value of foliage leads to tighter constraints on energy expenditure in folivores than frugivores and the possibility that the risk of male infanticide associated with increasing group size is greater (Snaith and Chapman 2007).

Carnivores

Thermoregulation, resource distribution, predation and reproductive benefits all play a significant role in the evolution of sociality in carnivores. In most of the smaller, nocturnal carnivores, females occupy individual territories or ranges and rarely forage in groups even where several individuals share a common territory (Kleiman and Eisenberg 1973; Macdonald 1983; Gittleman 1989). Many of these species live on mobile prey that are easily disturbed, and females forage alone partly to minimise direct competition for resources and partly because they rely on crypsis to avoid predation themselves (Rood 1986; Sandell 1989). Group-living and social foraging are more common among diurnal species, especially those living in relatively open habitats where the risk of predation is high (Stankowich *et al.* 2014). For example, among the mongooses, most nocturnal species are solitary or live in pairs while diurnal species living in open habitats, like dwarf mongooses and meerkats, live and forage in cohesive groups (Figure 2.29). In open habitats, the risk of predation can be extremely high and, in several species, group members take turns to act as sentinels, alerting other group members to approaching predators with a complex system of graded alarm calls (see Chapter 17).

In many medium-sized carnivores that feed on prey substantially smaller than themselves, including European badgers, honey badgers, foxes, jackals, otters and civets, adult females seldom forage together though several females may share a common range and may occupy the same burrow. Contrasts in group size within and between species appear to be related to the structure of the habitat: where resources are aggregated in large clumps and their overall abundance is high, females are commonly social, while in species where females depend

(a)

(b)

(c)

(d)

Figure 2.29 (a) Female slender mongooses usually forage and live alone while (b) female yellow mongooses live in small groups but forage alone or with dependent young. In contrast, (c) dwarf mongooses and (d) meerkats live and forage in cohesive groups. *Sources*: (a, b, d) © Tim Clutton-Brock; (c) © Amy Morris-Drake.

on evenly spaced or sparse resources, they seldom associate with each other, though they sometimes forage in mixed-sex pairs (Macdonald 1983; Kruuk and Macdonald 1985). One suggested explanation, the *resource dispersion hypothesis* (Macdonald 1983; Johnson *et al.* 2002), argues that group-living has multiple benefits and that its evolution is constrained by contrasts in competition caused by variation in the distribution of resources.

In the larger terrestrial carnivores, prey size and hunting technique both play an important role in determining the size of social groups (Gittleman 1989). The larger canids, like wolves, African wild dogs and dholes, which regularly pursue and kill prey larger than themselves, usually live and hunt in cohesive packs (see Figure 2.18). Both hunting success and the energetic costs of hunting are often affected by group size (see Chapter 9) and increased numbers can raise the chance that packs are

able to defend their prey against scavengers or competitors (Fanshawe and Fitzgibbon 1993). In contrast, many of the smaller canids that live on small prey or carrion, like most foxes and jackals, live in pairs or small groups and often forage alone (Gittleman 1989; Moehlman 1989). Similar differences in sociality occur among hyenas: species which commonly pursue and kill prey larger than themselves, like spotted hyenas, live in large clans and hunt in groups (Smith *et al.* 2008), while species that depend on carrion or small prey, like striped hyenas and aardwolves, live in pairs or small groups and normally forage alone (Kruuk 1972; Mills 1990).

Unlike the canids and hyenas, the cats, which typically hunt by stealth, mostly hunt alone even if they regularly kill animals larger than themselves, and social groups seldom include more than one adult female (Packer 1986). For example, female mountain lions, leopards, jaguars and tigers all regularly kill prey larger than themselves but hunt on their own and seldom share territories with other adult females (Gittleman 1989). African lions (which live in fission–fusion prides that include several breeding females and several adult males) usually hunt in groups and are an exception. While the traditional view is that hunting in groups increases hunting success or allows lions to tackle larger prey (Schaller 1972; Stander 1992a), other studies suggest that hunting in groups does not immediately maximise per-capita energetic returns and that sociality may have other benefits, including the defence of group territories and reductions in the risk of infanticide by males (Packer *et al.* 1990; Mosser and Packer 2009; Mosser *et al.* 2015) (see section 2.6).

Like terrestrial carnivores, most seals and sea lions forage independently, though they may feed in the same areas and often rest in unstable groups on rocks or ice (Renouf 1990; Berta *et al.* 2005). During the breeding season, females of many species that breed on land or on land-fast ice, aggregate in colonies that range in size from less than ten females to several thousand, and males defend harems of females (Stirling 1983; Cassini 1999). Aggregation may offer protection against terrestrial and aquatic predators as well as against male harassment (Campagna *et al.* 1992; Cassini 1999, 2000) but contrasts in female sociality may also be associated with the relative availability of suitable breeding sites. In contrast, in species breeding on pack-ice, like harp seals, hooded seals and crab-eaters, females are more widely dispersed alone or in small parties and males often defend a single female at a time.

Ungulates

Most ungulates are susceptible to predation by carnivores and their strategies for avoiding predation, combined with the distribution of their food supplies, play an important part in determining the size of the groups they live in (Jarman 1974; Caro 2005). Ruminants that live in forests or closed-canopy woodlands or isolated patches of cover, like the South American brocket deer, the Asian muntjacs, the African duikers and the okapi, typically live alone or in pairs. They escape predation by rapid flight, followed by immobility and this probably constrains the evolution of sociality (Caro 2005). Most are selective browsers and the distribution of food may increase the probability of feeding interference and also constrain the evolution of sociality (Jarman and Jarman 1973; Jarman and Kruuk 1996). One

exception to the generalisation that forest-dwelling ungulates forage alone or with dependent young are the forest-dwelling pigs and peccaries, which typically forage
in mixed-sex groups or large herds (Eisenberg 1981; Macdonald 2004). Unlike forest deer and antelope, they are omnivorous, their food is distributed in relatively large clumps and they rely on sociality and group defence to reduce the risk of predation.

In contrast to species living in closed environments, grazers living in open grasslands and savannahs (including perissodactyls as well as artiodactyls) frequently aggregate in large unstable herds. Where visibility is high, aggregation probably offers individuals the best chance of avoiding predation and solitary individuals often show higher levels of vigilance (see section 2.4). Here, too, contrasts in the distribution of resources and in the risk of feeding interference may also play a role in constraining sociality (Jarman and Jarman 1973; Jarman and Kruuk 1996): in selective browsers that live in open country, like giraffe or gerenuk, females often forage alone or with dependent young, while species that rely on a mixture of browsing and grazing, like impala or Grant's gazelle, forage in small groups which sometimes aggregate in larger herds. The largest groups are found in relatively unselective grazers like wildebeest and buffalo (Figure 2.30), where the value of individual food items is relatively low and resources are distributed in large clumps that can accommodate many individuals: in African buffalo, large numbers of breeding females aggregate in stable mixed-sex herds that can include several hundred individuals and rarely move far from their regular feeding sites (Sinclair 1977; Prins 1996). Equids that rely on widely distributed, seasonally fluctuating resources, like plains zebra and African and Asiatic wild asses, also collect in large unstable herds that can include over 1000 individuals (Rubenstein and Wrangham 1986; Fryxell and Sinclair 1988; Fitzgibbon 1990). Since diet quality and average body size are inversely related, larger species tend to live in larger groups (Figure 2.31). In some ungulates, there do not appear to be any long-term social bonds between females and males while, in others, herds comprise multiple harem groups consisting of unrelated females, as in plains zebra (Rubenstein and Wrangham 1986): benefits related to predation risk probably play an important role in maintaining sociality in these species. In some species where individuals aggregate in

Figure 2.30 Among ungulates, the largest groups are found in relatively unselective grazers, like buffalo and wildebeest. *Source*: © Tim Clutton-Brock.

unstable groups, like reticulated giraffes, it is possible to recognise nested social connections at several levels, based on individual differences in ranges (VanderWaal *et al.* 2014).

Megaherbivores

Predation and food distribution both play a dominant role in the evolution of sociality in the megaherbivores (Owen-Smith 1988). Because of their size, megaherbivores are

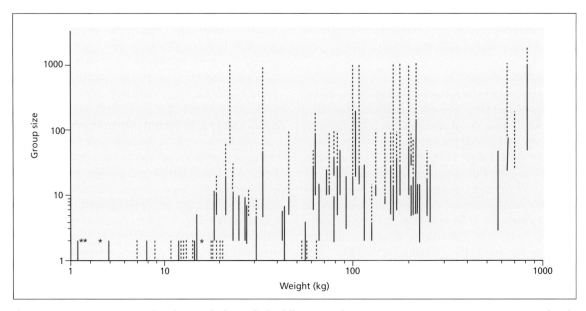

Figure 2.31 Average group size plotted against body weight for different ungulate species. *Source*: From Jarman (1974). Reproduced with permission of Brill.

Figure 2.32 Because of their large size, megaherbivores are less susceptible to predation and many live and forage alone, like white rhinoceros. *Source*: © Norman Owen-Smith.

less susceptible to predation than ungulates, though lions will attack their juveniles and, in the past, larger carnivores may have regularly killed adults. Freed from the need of protection, black, white and Indian rhinoceroses can forage on their own and live in relatively restricted ranges (Figure 2.32) (Owen-Smith 1988). In contrast, elephants range more widely and females associate with matrilineal relatives in groups which sometimes aggregate in larger herds (Figure 2.33), while males disperse to join bachelor groups before establishing individual ranges (Dublin 1983;

Figure 2.33 Female African elephants live in matrilineal groups which sometimes aggregate in large herds, while adult males live alone or in bachelor groups. *Source*: © Vicki Fishlock.

Archie and Chiyo 2011). Predation on young animals by male lions has been documented in several populations and may help to account for the evolution of sociality (McComb *et al.* 2014).

Cetaceans

While many of the cetaceans are social (Connor 2000; Connor *et al.* 2000), the difficulty of identifying individuals and recording their movements mean that less is known of their social organisation than is the case for terrestrial mammals. Among the larger, baleen whales (Mysticetes), humpback whales have been studied most extensively. In the northern hemisphere, humpbacks spend the summer feeding in small unstable groups on traditional feeding grounds in the North Atlantic or the North Pacific (Clapham 2000; Connor 2000; Connor *et al.* 2000), while in winter they migrate to lower latitudes and females from different subpopulations aggregate at sheltered breeding grounds where they are followed by males, who compete for mating opportunities. The other baleen whales are also most commonly seen alone or in small unstable parties and none aggregate in large stable groups like land mammals, though their capacity for long-distance communication means that it is difficult to be sure that groups do not have some multi-level structure. Variation in group size in the baleen whales appears to be associated with diet and foraging behaviour: the larger species, which are mostly filter feeders, aggregate in smaller groups than smaller species which feed on fish (Tershy 1992), possibly because the latter cooperate to corral and concentrate fish schools. Since the energetic costs of locomotion are unusually low for cetaceans, constraints on group size may be weaker than in many terrestrial mammals (Connor *et al.* 2000).

The toothed whales (Odontocetes) feed extensively on fish and squid and many of them live in larger groups than the baleen whales (Connor 2000). Like the baleen whales, sperm whales are inhabitants of the open oceans and females are most commonly found in waters over 1000 m deep, though in some places, males use much shallower waters (Whitehead and Weilgart 2000). Females aggregate in small, apparently matrilineal groups which collect in unstable schools of twenty to forty or more individuals in areas where upwellings generate high primary productivity (the traditional 'grounds' of whalers) to feed on fish, squid and crustaceans. Males disperse from their family units when they

are about 6 years old, often moving to higher latitudes before adopting an annual pattern of migration to breeding 'grounds' where they consort with females and compete for mating opportunities (Best 1979; Whitehead and Weilgart 2000).

Larger and more complex social groups are found in some of the smaller toothed whales. In some populations of bottlenose dolphins, unstable herds of females form the core of communities of 100 or more individuals whose members interact more frequently with each other than with members of similar units in neighbouring areas (Wells *et al.* 1987). In other populations, there is less evidence of spatial structure and females belonging to the same population occupy a mosaic of overlapping home ranges, aggregating temporarily with neighbours in loose associations (Connor *et al.* 2000). In killer whales, some populations are relatively sedentary while others are nomadic (Baird 2000). In sedentary populations, individuals feed principally on fish and live in stable matrineal groups of up to ten individuals; several matrilineal groups form a sub-pod whose members spend more than 95% of their time travelling with other individuals of the same sub-pod; and one to three sub-pods make up a pod whose members share the same range and travel together for at least 50% of the time (Baird 2000). Individuals remain in their natal pod throughout their lives and mating occurs when pods meet, often in areas where food is abundant. In contrast, in nomadic populations, individuals feed extensively on marine mammals (Figure 2.34) and live in smaller groups of two to four, possibly because relatively small groups are more efficient for hunting marine mammals (Baird and Dill 1996). In both populations, mothers appear to have stronger and longer-lasting bonds with their sons than their daughters.

Cetaceans do not defend territories so it is unlikely that benefits to inter-group competition play an important role in the evolution of sociality. Juveniles of all species and adults of some are subject to predation by killer whales and the larger sharks and sociality may often represent an adaptation to minimising the risk of predation. Communal care of young may also favour sociality. For example, in sperm whales, dependent calves are typically left on the surface when their mothers leave to feed and are sometimes suckled by other group members (Baird 2000) and females cooperate to defend other females and their offspring from attacks by predators.

Figure 2.34 Killer whales live in stable 'pods' which may aggregate in larger groups and individuals of both sexes commonly remain in their natal pod throughout their lives. *Source*: © Rus Hoelzel.

There are records of females attacking whaling boats that had fastened to another female and of attempts to break harpoon lines (Caldwell and Caldwell 1966) and whalers frequently made use of the tendency for females to protect calves cooperatively by harpooning (but not killing) calves and then harpooning females that came to protect them.

Sociality may also help the smaller cetaceans to find and catch prey (Connor 2000). When parties of dolphins are searching for fish, individuals will leave larger groups to search for fish schools, calling to other members of their group when they find them. Groups that are pursuing fish schools often form crescents which help to prevent fish escaping and group members will cooperate to herd prey against barriers: in Mauritania, local fishermen that locate schools of mullet attract bottlenose dolphins and humpback whales that drive the fish towards their nets to the benefit of both parties (Busnel 1973). In open water, dolphins often attempt to compress and subdivide fish schools, sometimes using bursts of bubbles to corral them. While some members prevent fish from escaping, others swim through the school at speed, often capturing several fish at a time. Cooperative hunting is also well developed in killer whales (see Chapter 9).

Insectivores and bats

In most terrestrial insectivores, females forage alone or with dependant young and social groups that include more than one breeding female are rare. In some species, males and females defend separate territories while, in others, males occupy ranges overlapping those of females or the sexes form socially monogamous pairs. For example, among the shrews and related species, around 40% of recorded species may live in mixed-sex pairs (Valomy *et al.* 2015).

The largest mammalian aggregations of all are found in bats, where colonies of some cave-dwelling Microchiroptera can include up to 20 million individuals (Bradbury and Vehrencamp 1977; Kunz and Fenton 2003; Kerth 2008) (Figure 2.35). Many of these species feed on ephemeral and unpredictable swarms of insects at high altitudes where the size of foraging patches is very large and the risk of feeding interference or scramble competition is low. Tree-roosting frugivorous Megachiroptera also feed on widely dispersed, variable food supplies and are often migratory and colonies can number over 100,000 individuals (Kunz and Lumsden 2003). In contrast, insectivorous bats that feed close to ground level have more localised feeding sites and commonly roost in hollow trees in smaller groups that change

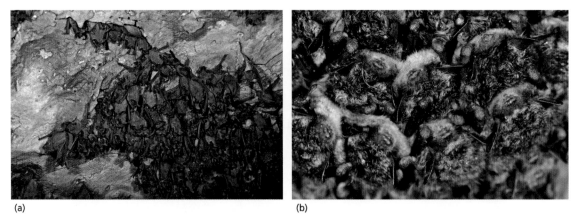

(a)   (b)

Figure 2.35 (a) Mexican free-tailed bats hanging inside bracken cave; (b) hibernating Brandt's bats (*Myotis brandtii*). *Sources*: (a) © S. J. Krasemann/Getty Images; (b) © Yves Adams/Getty Images.

in membership and forage alone (Kerth 2008; Kerth *et al.* 2011). In some species, females forage in the centre of their colony's territory, while males forage alone on the periphery; in others, colony members have individual feeding sites close to their sleeping sites and daughters inherit their mother's site (Kerth *et al.* 2001a,b). The smallest colonies are found in carnivorous bats that roost alone or in pairs as well as in species that use ephemeral roosts, like the sucker-footed bats, that sleep inside furled leaves.

The sex and age structure of roosting colonies varies widely. In some temperate Microchiroptera that roost in small groups, colonies consist of breeding females belonging to one or more matrilines and males may be solitary, form groups of their own or defend female groups (Dwyer 1966; Ransome 1990; Kerth 2008, 2010). In several tropical species, a single male roosts with a harem of females and males compete for harems (Ortega and Arita 2002; Altringham and Senior 2005).

In some species that live in larger colonies, there appears to be no spatial structure within the colony while, in others, females form distinct breeding groups (Kerth 2008). For example, in grey-headed flying foxes, males and females form sexually segregated 'camps' before parturition, with males and females occupying different trees or different levels in the canopy. As young are weaned and females are ready to mate again, males form and defend small territories within camps and attract one or more females while, after mating, both sexes again form large single-sex groups.

The stability of female roosting groups also varies. Many tropical bats form groups throughout the year but, in some temperate species, sociality varies between seasons (Kunz and Lumsden 2003). Individuals often have several different roosting sites within their range and roosting groups vary in membership from night to night (Kerth 2008). In some species, colonies have a fission–fusion structure: for example, Bechstein's bat colonies consist of several stable subunits, each including several matrilineal groups whose members forage independently and sleep in unstable subgroups (Popa-Lisseanu *et al.* 2008; Kerth 2010).

As in other groups of mammals, sociality probably has multiple benefits in bats and their relative importance varies between species (Kerth 2008). Most species cannot build roosts themselves and individuals may be forced to aggregate because the availability of suitable roosting sites is limited. However, limitations in the availability of roosting sites do not provide a satisfactory general explanation of variation in sociality since many tree-roosting species (who have no shortage of potential roosting sites) and species that can construct their own roosting sites (like the tent-making bats) aggregate in groups (Kunz and Lumsden 2003). In some cases, social roosting may generate thermoregulatory benefits: individuals that cluster together in roosts can reduce their heat loss (Speakman and Thomas 2003) and large colonies that live in relatively confined spaces can raise the local ambient temperature (Altringham and Senior 2005). However, thermoregulation is also unlikely to account for all cases of sociality in bats: many tree-roosting tropical bats form roosting groups and so, too, do a number of bats that roost in termite mounds, which provide a warm and stable environment (Kerth 2008). In some bats, predation may play an important role in maintaining sociality, especially in the larger frugivorous species that roost in trees and in cave-roosting species where colony members disperse from the roost in synchrony, saturating any waiting predators (Altringham and Senior 2005). Finally, colonial living may facilitate information transfer about available foraging sites and allow individuals to follow other group members to productive sites (Wilkinson 1992, 1995) or to avoid sites that have already been utilized (Altringham and Senior 2005).

Marsupials and monotremes

Less is known of the social organisation of monotremes and marsupials (including the American didelphids and the Australian dasyurids and macropods) than of the social organisation of eutherian mammals, but there are clearly close parallels between the two radiations (Jarman 1991; Jarman and Kruuk 1996; Winter 1996). Like eutherian carnivores, most of the smaller omnivores or carnivorous marsupials, like quolls, are active at dusk or during the night and many are solitary foragers, as are all three monotremes (Grant 1983) (Figure 2.36). In many of the didelphids and dasyurids, females live in separate but overlapping ranges while males have larger ranges overlapping those of several females, though their social organisation appears to be less structured than that of eutherians: ranges are not

defended, individuals rarely contact each other, allogrooming does not occur, mating is promiscuous and there are no long-term bonds between adults of the same or opposite sexes (Charles-Dominique 1983; Russell 1984). In some folivorous marsupials, like the koalas, females also forage alone and their home ranges overlap little, while males occupy ranges overlapping those of one or more females. In others (including the sap-feeding gliders and possums and the herbivorous wombats) females live in family units consisting either of a single female and her dependent young or in colonies consisting of several breeding females living in separate burrows but sharing a common range that may be defended against neighbours, as in wombats (Johnson and Crossman 1991; Taylor 1993).

The larger herbivorous macropods that live in open country, like grey and red kangaroos, typically forage in groups and sometimes aggregate in large unstable mobs (Figure 2.37). Like ungulates, foraging kangaroos rarely prevent other individuals from joining them, groups are open and unstable and multiple females share overlapping ranges. In more sedentary populations, females frequently remain where they were born for much of their life, occupying overlapping undefended ranges while males disperse more widely (Jarman and Southwell 1986; Jarman 2000). Oestrous females roam widely, often pursued by multiple males who compete for mating access (Jarman 2000). Regular competition between males with overlapping ranges leads to well-defined, size-related dominance relationships and the highest ranking male in a locality typically obtains a high

(a)

(b)

Figure 2.36 Among the smaller crepuscular and nocturnal marsupials, carnivores, like quolls (a), are usually solitary, while some herbivores, like wombats (b), live in colonies. *Sources*: (a) © Craig Dingle/Getty Images; (b) © Visuals Unlimited, Inc. /Dave Watts /Getty Images.

(a)   (b)

Figure 2.37 Many of the larger diurnal macropods are social. Some, like rock wallabies (a), live in pairs which share a single cave or burrow, while in others, like grey kangaroos (b), related females have overlapping ranges and commonly associate with each other, and males live alone or in bachelor groups. *Sources*: (a) © Jouan Rius/Nature Picture Library; (b) © Tim Clutton-Brock.

proportion of available mating, though average tenure of alpha status is little more than a year (Jarman 2000). Today, the larger macropods have few predators apart from humans and dingos (which only invaded Australia around 3500 years ago) but they evolved with a suite of marsupial predators and female sociality may represent an anti-predator strategy.

2.7 The distribution of female sociality

This brief review of female sociality shows that there are consistent relationships between environmental parameters and group size across mammals as well as differences between taxonomic groups in relationships between ecology and sociality. For example, many terrestrial mammals living in grassland or savannah environments that are active by day (including rodents, primates, small carnivores, ungulates and macropods) are exposed to high predation risks and rely on sociality as a defence, while many nocturnal species (including insectivores, carnivores, primates, ungulates and marsupials) rely on crypsis to avoid predation and forage alone or in small groups. Variation in the risk of feeding interference and competition also has similar effects in different groups of mammals. Feeding interference is particularly likely in carnivores that feed on relatively mobile prey and social foraging is uncommon in both eutherian carnivores and in carnivorous marsupials. Where it does occur, it is often associated either with

cooperative hunting techniques that do not rely on stealth (as in the social canids and smaller cetaceans) or with dependency on locally abundant, relatively immobile prey (as in several of the diurnal insectivorous mongooses) (Rood 1986). Frugivores that rely on relatively scarce food items of high nutritional quality, like ripe fruit, also tend to forage in smaller groups than species that live on more abundant resources (Wrangham 1980; Kappeler and van Schaik 2002). Among herbivores (including both eutherians and marsupials), browsing species tend to feed in relatively small groups compared to selective grazers while the largest, densest aggregations are found in relatively unselective grazers (Jarman 1974, 1991; Jarman and Kruuk 1996).

There are also a substantial number of species that diverge from general trends and where the reasons for differences in sociality remain uncertain, despite detailed research. For example, among carnivores, lions are a striking exception to the generalisation that predators which hunt by stealth forage alone (Packer 1986) and there is still disagreement about why they are social (see earlier discussion). Similarly, it is not clear why, among diurnal primates, folivorous species tend to live in smaller groups than frugivorous ones, despite the lower biomass of their resources (Janson and Goldsmith 1995; Snaith and Chapman 2007). In both these cases, it has been suggested that increases in the risk of male infanticide with rising group size may constrain numbers, illustrating the potential importance of social mechanisms and reproductive strategies.

Although there are similar correlations between ecological variables and female sociality in different groups of mammals, consistent differences in social behaviour and social organisation between phyla are also common (Henzi and Barrett 2005; Chapman and Rothman 2009). In some cases, these appear to be related to contrasts in life-history parameters, but many differences are difficult to explain and it is hard to avoid the conclusion that phylogenetic history exerts an important influence on behaviour. However, even if phylogenetic inertia constrains the evolution of adaptations in contemporary species, the origins of these differences probably lie in evolution of alternative adaptations in ancestral populations and the development of gene-based phylogenies now provides opportunities to explore the origin of these differences.

2.8 Group coordination

In social animals, group members often need to synchronise or coordinate their activities and movements if they are to remain together (Conradt and Roper 2003, 2005; Kerth 2010). For example, individuals have to decide when and where to move, what to feed on and where and when to rest or nest. The need for coordination raises important questions about the way in which social decisions are achieved, the extent to which different individuals contribute to them and the relative benefits of involving different numbers of individuals in decision processes (Conradt and Roper 2003, 2009; Kerth 2010; Couzin *et al.* 2005).

Where group decisions have to be made rapidly, they are often based on simple 'self-organising' processes, as in cases where group members avoid predators by copying the responses of their closest neighbours (Laland 2004; Couzin *et al.* 2005). However, copying can lead to incorrect decisions or to ones that benefit the first individuals that change their behaviour but not those that decide later and, where decisions do not need to be made so quickly, there can be advantages in drawing on the information available to several group members before a consensus is established.

Studies of several social mammals suggest that individuals often observe the preferences of other group members before making their decisions. For example, in African buffalo, regular herd movements from resting areas to one of several grazing grounds occur in the early evening (Prins 1996). Once resting herds become active and start to move, the activity and direction of movement of herd members is closely coordinated. Before herds move off, individuals commonly stand, gaze intently in one direction and then may lie down again and continue to rest. There is a close agreement between the direction of gaze in the majority of individuals and the direction of the feeding ground that the herd eventually moves to.

A similar 'voting system' has been described in hamadryas baboons (Stolba 1979). After descending from their sleeping cliffs in the early morning, dominant males descend to ridges and persistently gaze at the horizon. Initially, individuals may gaze in several directions but, as time passes, more individuals gaze in the same direction. Eventually, multiple harem groups move off in a common direction to search for feeding sites. Subgroups may subsequently fragment and forage independently, but they often reconvene at noon at the same waterhole before foraging again and, later, returning to the sleeping cliffs.

Group members commonly rely partly on the decisions of others and partly on information gained independently when deciding when and where to forage or rest. For example, in Bechstein's bats, where females are philopatric and live in stable groups of ten to forty-five and individuals roost in temporary subgroups, colony members rely partly on the behaviour of other group members and partly on information that they have acquired themselves when deciding when to switch roosts (Kerth *et al.* 2006; Kerth 2010). In a series of field experiments, Kerth provided different colony members with conflicting information about the relative suitability of different roosting sites and demonstrated that this increased the frequency of fissioning (Kerth *et al.* 2006).

Individual differences in sex, age, condition and nutritional requirements often generate contrasts in the needs of different group members and conflicts of interest between them (Conradt and Rober 2009). In some cases, the interests of group members diverge to such an extent that, for some individuals, the costs of following consensus decisions exceed the benefits of group cohesion, leading to temporary or permanent separation of some individuals from the group. Variation of this kind may be responsible for the temporary segregation of lactating females in some ungulates as well as for the segregation of males and females (see Chapter 18) and the evolution of fission–fusion societies.

Where the interests of individuals do not diverge to such an extent, it may be in their interests to reach a shared or 'combined' decision. For example, in baboons, individuals follow the lead of multiple initiators (Strandburg-Peshkin *et al.* 2015). When some initiators lead in one direction and some in another, following animals choose one of the two directions if the angle between them is large and an intermediate direction if it is small. The optimal numbers of individuals that should be allowed to contribute to such decisions depend on the frequency-dependent costs and benefits of different choices. In many cases, group decisions are initially combined but, after group members reach a quorum, the commitment of additional group members to the same decision increases quickly until a consensus decision is made. One of the best examples of this process is the choice of nesting sites in honeybees (Seeley and Buhrman 1999) but similar processes probably occur in many mammals (Conradt and Roper 2003).

In theory, the extent to which decisions are shared among group members can range from cases where they depend on the preferences of a single individual to cases where all members contribute equally to the eventual decision. However, in practice, both fully dictatorial systems and fully egalitarian ones are rare and group decisions are usually affected by the preferences of several individuals. The relative number of individuals that contribute to decisions often ranges from more than 50% to less than 20% of group members (Kerth 2010). The optimal number of individuals involved is likely to be affected both by the speed at which decisions need to be made and by the distribution of information across group members (Franks *et al.* 2003; Couzin *et al.* 2005).

In some societies, all adult group members contribute equally to group decisions but, in others, some individuals contribute more than others (Kerth 2010; Smith *et al.* 2016). In some cases, groups are led by the neediest individuals (Conradt *et al.* 2009). In others older or more experienced individuals may exert a disproportionate influence on group decisions to the benefit of all group members. For example, in African elephants, groups are led by the eldest female, who is most likely to have the greatest knowledge of the distribution of resources and the relative dangers posed by competitors or predators (McComb *et al.* 2001, 2011). The relative rank of individuals can also affect their ability to influence group decisions. For example, in hamadryas baboons, where males are dominant to females, adult males control group decisions about the timing and direction of travel (Kummer 1968). Conversely, in some lemurs where females are dominant to males, females usually lead groups (Erhart and Overdorff 1999).

Situations where a single individual initiates or controls group movements are unlikely to maximise average benefits for all group members. For example, in chacma baboons, dominant males lead groups to concentrated food patches which they can monopolise, to the disadvantage of subordinates, who are often unable to forage effectively there (King *et al.* 2008). Despite this, subordinates often follow dominant males to concentrated food patches, though their tendency to do so depends on the strength of their social relationships with the dominant male.

2.9 Consequences of female sociality

Group cohesion

Intraspecific variation in group size affects the cohesiveness of social groups. As the number of individuals rises, the dispersion of group members increases and subgroups can become separated from each other. For example, members of large meerkat groups are dispersed over a larger area than those of small groups and group members can lose communication with each other, generating temporary splits. Sometimes, subgroups locate each other and groups re-form; in others, they can remain separate for days or weeks and may remain as two separate groups. These processes have been most extensively investigated in primates where similar mechanisms are involved in group fission (see Chapter 4).

Where group size is large, the need for group members to communicate with each other over considerable distances has important consequences for the way in which individuals communicate and the structure of signals (see Chapter 7). Across primate species, the repertoire of calls and visual signals and the range of facial expressions increases with group size, which may reflect the need for greater signal diversity in large groups (McComb and Semple 2005; Dunbar 2012).

Increase group size may also affect the structure of social networks and social bonds. Here, too, research has focused extensively on social primates, where allogrooming plays an important role in the maintenance of social bonds and supportive coalitions (see Chapter 9).

Some comparative studies of primates have shown that, across species, the proportion of daytime spent allogrooming and the number of regular grooming partners increase with group size and have suggested that grooming plays an essential role in maintaining group cohesion and that the need to maintain grooming relationships with all other group members may even constrain group size (Dunbar 1991; Lehmann and Dunbar 2009; Dunbar and Lehmann 2013). However, other analyses suggest that associations between group size and grooming time in primates occur because both increase in terrestrial species (Grueter *et al.* 2013). In addition, the maintenance of large stable groups in many mammals that rarely groom each other argues against this suggestion.

Reproductive strategies in males

Female sociality has profound consequences for males and much of the second half of this book is devoted to exploring these effects. Female group size influences the number of mating partners that single males can defend and also affects the reproductive strategies of males (Chapter 10), the numbers of males that associate with female groups (Chapter 11), the form and intensity of mating competition and the duration of effective breeding by males (Chapter 13). Reproductive competition between males in turn affects social relationships between them (Chapter 15) as well as relationships between males and females (Chapter 16) and commonly influences the evolution of sex differences in anatomy, physiology and behaviour (Chapter 18).

Genetic structure of groups and populations

The size and stability of female groups has important consequences for the genetic structure of populations. Where groups include a single breeding female, a high proportion of natal animals are usually half or full sibs and average relatedness among members of whichever sex is philopatric is likely to be high, while variance in relatedness is relatively low. In contrast, as the number of breeding females increases, the proportion of natal animals that are full sibs declines and average relatedness falls, while variance in relatedness increases (Lukas and Vigilant 2005) (Figure 2.38). The number of breeding females can also affect relatedness between group members through its influence on the number of breeding males and the duration of their tenure. As female group size rises, the average tenure of breeding males falls and the number of associated breeding males increases, reducing patrilineal kinship and paternal contributions to relatedness between group members (see Chapter 13). In conjunction, these effects can generate large differences in average relatedness and variance in kinship between juveniles born into the same group as well as

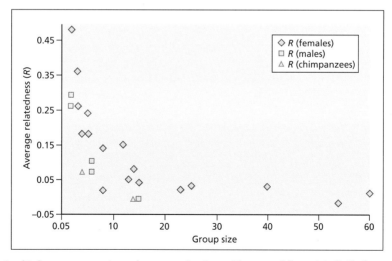

Figure 2.38 The relationship between group size and average relatedness (*R*) among philopatric individuals in primates, separated for species where females are philopatric (diamonds) and where males are philopatric (squares); triangles show the values for chimpanzees. In both categories, relatedness values fall as group size increases. *Source*: From Lukas *et al.* (2005). Reproduced with permission of John Wiley & Sons.

between philopatric adults. For example, in meerkats (where a single dominant female is responsible for most breeding attempts in each group), average relatedness between philopatric females is around 0.4 while among savannah baboons (where multiple females breed) average relatedness between females may usually be little higher than 0.10.

The number of breeding females is not, of course, the only factor contributing to variation in relatedness between natal animals. Other factors that will tend to reduce relatedness between natals include the production of single offspring at long intervals, the length of female breeding lifespans and prolonged retention of natal animals in the group. Contrasts in kinship between female group members may have important consequences for the evolution of female behaviour: high levels of kinship between group members appear to be a necessary precondition for the evolution of altruistic cooperation and cooperative breeding (see Chapter 17) while high variance in kinship is often associated with the formation of kin-based coalitions that are used in competition between subgroups that consist of maternal relatives (see Chapter 8).

The size and stability of social groups also influences the effective population size (N_e) and the genetic structure of populations (Chepko-Sade and Halpin 1987; Dobson et al. 2004). Where social structure inhibits dispersal and increases variance in male success, it may reduce N_e, increasing the loss of genetic diversity though where one sex habitually disperses after adolescence, the division of populations into genetically distinct groups may have opposite effects (see Chapter 4).

Cognitive development and brain size

The increasing number of social partners in large groups may also affect the complexity of social relationships which, in turn, may reinforce selection for cognitive development and brain size, an idea commonly referred to as the *social brain hypothesis* (Humphrey 1976; Dunbar and Shultz 2007a). Interspecific comparisons have demonstrated relationships between differences in cognitive abilities and relative brain size or neocortex size (Reader and Laland 2002; Whiten and van Schaik 2007): for example, across a sample of mammalian carnivores, the ability to extract food from simple puzzle boxes was associated with variation in relative brain size (Holekamp et al. 2015). In addition, selection experiments with fish have demonstrated correlations between brain size and simple cognitive abilities associated with

survival (Kotrschal et al. 2013, 2015). In some taxonomic groups, species differences in relative brain volume or neocortex size are correlated with social variables, though relationships differ between groups and these contrasts are often difficult to explain (Dunbar and Schultz 2007b; MacLean et al. 2009). For example, among the higher primates, relative brain and neocortex size are correlated with average group size (Dunbar and Shultz 2007a,b) and changes in relative brain size and sociality are associated with each other (Pérez-Barbería et al. 2007). Among primates, relative brain or neocortex size has also been shown to be positively associated with the size of social networks (Kudo and Dunbar 2001) and the incidence of coalitionary behaviour (Dunbar and Schultz 2007b) and negatively with indices of male–male breeding competition (Schillaci 2008). In contrast, among non-primate mammals, relative brain size is seldom closely correlated with group size and appears to be associated with pair-bonded social monogamy (Pérez-Barbería et al. 2007; Schultz and Dunbar 2007) as well as with the incidence of care of infants and juveniles by individuals other than their mother (Isler and van Schaik 2012). While it has been suggested that these contrasts may be a consequence of greater complexity in social relationships in pair-bonded species and their extension to larger social groups in primates but not in most other social mammals (Schultz and Dunbar 2007), it is not clear either that social relationships are more complex in pair-bonded mammals than in other mammalian societies (see Chapters 8 and 9) or that there are consistent qualitative differences in social relationships between primates and other social mammals (see Chapter 19).

While the existence of some relationship between social organisation, cognitive abilities and brain size in mammals seems likely, there are fundamental difficulties in measuring both social complexity and generalised cognitive ability (Freeberg et al. 2012). Although group size increases the number of different animals that each group member may encounter, a high proportion of each individual's social interactions are with a relatively small number of other group members (see Chapter 8 and 9) so that the total number of other group members may not provide a reliable measure of social complexity. The number and stability of differentiated social relationships might provide a more realistic index of complexity (Dunbar and Schultz 2007b) but it is not obvious how this should be assessed and satisfactory comparative data are unavailable. The measurement of generalised cognitive abilities also presents problems since performance on different cognitive tasks varies between species and scores on different tasks are often weakly interrelated (Shettleworth 2010; Call and Santos 2012). While species differences in contrasting cognitive abilities may represent adaptations to different social or ecological niches and can provide important insights into the evolution of behaviour, their presence complicates attempts to investigate the distribution and causes of general cognitive abilities.

A further problem is that correlations between brain size and social variables differ between taxonomic groups and are usually difficult to interpret (Healy and Rowe 2007). Within some taxonomic groups, there is no evidence of associations between brain size and contrasts in social organisation: for example, among lemurs, variation in brain size is not related either to group size or to pair-bonding and is instead correlated with differences in diet and activity timing (MacLean *et al.* 2009) and similar relationships between brain size and ecological differences have been demonstrated across broader taxonomic groups by earlier studies (Clutton-Brock and Harvey 1980; Harvey *et al.* 1980; Mace *et al.* 1981). Other analyses have shown that variation in brain size is correlated with contrasts in life-history parameters (Harvey and Bennett 1983; Barton and Capellini 2011) as well as with specific behavioural propensities, including the capacity for self-control (MacLean *et al.* 2014). Interspecific differences in ecological parameters, life-history variables and social behaviour are extensively intercorrelated and these relationships may cause or obscure relationships between brain size and social or ecological parameters (Harvey and Pagel 1991). Moreover, variation in group size is often closely correlated with other parameters likely to affect selection for cognitive abilities and brain size, like home range size (see Figure 2.22), male breeding tenure and reproductive skew (see Chapters 13 and 14) and average relatedness between group members (see Figure 2.38). As a result, it is usually difficult to attribute differences in brain size to particular selection pressures with confidence.

Despite these problems, the evolution of cognitive abilities and brain size and their impact both on social relationships and on technological development raises questions of fundamental importance to our understanding of social evolution in general and the evolution of human societies in particular (Dunbar and Shultz 2007a,

b). Quantitative comparisons offer insights into the evolution of cognitive abilities that are inaccessible to other approaches, new analytical and experimental techniques offer improved ways of distinguishing between hypotheses (Dunbar and Schultz 2007b; MacLean *et al.* 2012, 2014) and the investigation of the distribution of cognitive abilities and brain size and their relation to contrasts in social organisation and ecology remains an important topic for research.

Population dynamics and demography

Female sociality can also have important consequences for population demography and the regulation of population density although these remain largely unexplored. Increased competition for resources in large groups can affect density-dependent processes at the group level, generating reductions in breeding success or survival in younger, weaker or more subordinate individuals and increasing individual differences in breeding success (Dittus 1977, Dittus 1979; van Schaik *et al.* 1983; Clutton-Brock *et al.* 1984). In general, the tendency for sociality to focus intra-group competition on the weakest individuals is likely to intensify density-dependent changes in survival and breeding success at the population level and to enhance population stability in social species (Clutton-Brock 2001), though there may also be cases where sociality has the opposite effect (Dobson and Poole 1998; Courchamp *et al.* 2008).

Sociality may also affect population dynamics through its influence on interactions between groups. In many social vertebrates, larger groups can displace smaller ones and, in some cases, can drive them from their territories (Wrangham 1980). Where the benefits of large group size exceed the effects of increased competition between group members, breeding success and survival may increase with group size, leading to inverse density dependence or *Allee effects* at the group level (Dobson and Poole 1998; Courchamp *et al.* 2008). These effects are particularly likely to occur in cooperative breeders where females cannot rear young without assistance and recruitment rises in relation to the number of available helpers and may have consequences for the stability of group size and the probability of group extinctions (see Chapter 17).

The effects of sociality on patterns of dispersal can also have important consequences for population dynamics. Species that live in stable groups often show low rates of successful dispersal because emigrants are prevented from joining established groups and dispersers are unlikely to establish new breeding units successfully if local environments are saturated. Under these conditions, dispersing females often show high mortality and are reluctant to leave their natal group, only dispersing when evicted by dominant breeders or when they can disperse in small groups (see Chapter 4) and this may explain why, in some social mammals, populations are slow to re-colonise territories where groups have become extinct, leading to spatial variation in population density (Courchamp *et al.* 1999).

Finally, group size often affects susceptibility of animals to disturbance or hunting by humans. Mammals that live in large groups, like woolly monkeys or bearded pigs, are more conspicuous than those that live in small groups or forage alone and they are more likely to be located and larger numbers are more likely to be killed by hunters (Pfeffer and Caldecott 1986; Bennett and Robinson 2000). Moreover, the increased range size of large groups raises the size of reserves necessary to protect them and increases the chance that individuals will move into neighbouring unprotected areas. As a result, interspecific differences in group size and home range area are often positively correlated with local extinction rates (Woodroffe and Ginsberg 1998).

SUMMARY

1. This chapter reviews contrasts in female sociality. In some mammals (including many nocturnal species), breeding females are intolerant of each other's presence and occupy separate ranges or territories that overlap little, while in others they forage alone or with dependent offspring but share part or all of their territories with other females. In other species, several breeding females and their dependent young forage together and, in some, stable or semi-stable groups of breeding females and associated males form large unstable herds. In some species, groups are unstable and change in membership throughout the day (as in many ungulates), while in others they are cohesive and stable (as in some primates and carnivores). In a few species, like chimpanzees, multiple breeding females live in stable communities but forage in smaller parties of unstable membership.

2. Female sociality often increases the capacity of females to detect or defend themselves against predators, to locate or catch food or to rear young. However, it commonly increases competition for resources and breeding opportunities and increases the risk of interference in breeding attempts by other females.

3. In many social species, females living alone or in groups that are substantially smaller than the population average show reduced breeding success or survival. Under these conditions, the upper limits of group size may often exceed the optimal level because solitary females may continue to attempt to join groups until the fitness of their members approaches that of a solitary female.

4. Four main factors are associated with species differences in female sociality in mammals: the relative importance of thermoregulation; the susceptibility of individuals to predation and the nature of their anti-predator tactics; the extent to which foraging success or breeding success is reduced by the presence of competitors; and the extent to which the females rely on assistance from other group members to raise their young.

5. Within groups, individuals coordinate their activities by copying each other and their behaviour is often closely synchronised. Changes in activity are often initiated by younger group members, though older and more experienced individuals can exert a stronger influence on the direction of group travel.

6. The evolution of female sociality has important consequences for the reproductive strategies of males, the extent of polygyny and the evolution of associated adaptations in both sexes. By increasing the extent of polygyny and the intensity of breeding competition between males, increases in female group size are also likely to reduce the breeding tenure and longevity of males.

7. In some groups of mammals, differences in group size are correlated with variation in absolute or relative brain size and it has been suggested that these associations may have a causal basis. However, variation in group size is correlated with many ecological, social and demographic parameters and several other explanations are also feasible.

8. Female sociality can have important effects on population demography and dynamics and can generate both negative and positive correlations between group size (or local population density) and survival and breeding success.

References

Abbeglen, J.-J. (1984) *On Socialization in Hamadryas Baboons: A Field Study*. Cranbury, NJ: Associated University Press.

Alexander, R.D. (1974) The evolution of social behavior. *Annual Review of Ecology and Systematics* **5**:325–383.

Altizer, S., *et al.* (2003) Social organization and parasite risk in mammals: integrating theory and empirical studies. *Annual Review of Ecology, Evolution and Systematics* **34**:517–547.

Altringham, J.D. and Senior, P. (2005) Social systems and ecology of bats. In: *Sexual Segregation in Vertebrates* (eds K. E. Ruckstuhl and P. Neuhaus). Cambridge: Cambridge University Press, 280–302.

Archie, E.A. and Chiyo, P.I. (2011) Elephant behaviour and conservation: social relationships, the effects of poaching, and genetic tools for management. *Molecular Ecology* **21**:765–778.

Armitage, K.B. (2007) Evolution of sociality in marmots: it begins with hibernation. In: *Rodent Societies: An Ecological and Evolutionary Perspective* (eds J. O. Wolff and P.W. Sherman) Chicago: University of Chicago Press, 356–367.

Armitage, K.B., *et al.* (2003) Energetics of hibernating yellow-bellied marmots (*Marmota flaviventris*). *Comparative Biochemistry and Physiology Part A: Molecular and Integrative Physiology* **134**:101–114.

Arnold, W. (1990) The evolution of marmot sociality: II. Costs and benefits of joint hibernation. *Behavioral Ecology and Sociobiology* **27**:239–246.

Aureli, F. (2008) Fission–fusion dynamics. *Current Anthropology* **49**:627–641.

Baird, R.W. (2000) The killer whale: foraging specializations and group hunting. In: *Cetacean Societies: Field Studies of Dolphins and Whales* (eds J. Mann, R. Connor, P. Tyack and H. Whitehead). Chicago: University of Chicago Press, 127–153.

Baird, R.W. and Dill, L.M. (1996) Ecological and social determinants of group size in transient killer whales. *Behavioral Ecology* **7**:408–416.

Balmford, A.P. and Turyaho, M. (1992) Predation risk and lek-breeding in Uganda kob. *Animal Behaviour* **44**:117–127.

Barrett, L. and Henzi, S.P. (2006) Monkeys, markets and minds: biological markets and primate sociality. In: *Cooperation in Primates and Humans* (eds P. M. Kappeler and C. P. van Schaik). Berlin: Springer, 209–232.

Barton, R.A. and Capellini, I. (2011) Maternal investment, life histories, and the costs of brain growth in mammals. *Proceedings of the National Academy of Sciences of the United States of America* **108**:6169–6174.

Bearder, S.K. (1987) Lorises, bushbabies, and tarsiers: diverse societies in solitary foragers. In: *Primate Societies* (eds B.B. Smuts, D.L. Cheney, R.M. Seyfarth, R.W. Wrangham and T. T. Struhsaker). Chicago: University of Chicago Press, 12–24.

Beauchamp, G. (2008) What is the magnitude of the group-size effect on vigilance? *Behavioral Ecology* **19**:1361–1368.

Bennett, E.L. and Robinson, J.G. (2000) *Hunting of Wildlife in Tropical Forests: Implications for Biodiversity and Forest Peoples*. Washington, DC: The World Bank.

Berta, A., *et al.* (2005) *Marine Mammals: Evolutionary Biology*. New York: Academic Press.

Best, P.B. (1979) Social organisation of sperm whales, *Physeter macrocephalus*. In: *Behaviour of Marine Animals: Current Perspectives in Research*, Vol. **3** (eds H. E. Winn and B. C. Olla). New York: Plenum Press, 227–289.

Bettridge, C., *et al.* (2010) Trade-offs between time, predation risk and life history, and their implications for biogeography: a systems modelling approach with a primate case study. *Ecological Modelling* **221**:777–790.

Blumstein, D.T. and Daniel, J.C. (2002) Isolation from mammalian predators differentially affects two congeners. *Behavioral Ecology* **13**:657–662.

Boesch, C. (1994) Chimpanzees–red colobus monkeys: a predator–prey system. *Animal Behaviour* **47**:1135–1148.

Boesch, C. (1996) Social grouping in Taï chimpanzees. In: *Great Ape Societies* (eds W.C. McGrew, L. F. Marchant and T. Nishida). Cambrdge: Cambridge University Press, 101–113.

Bradbury, J.W. and Vehrencamp, S.L. (1977) Social organization and foraging in emballonurid bats. III. Mating systems. *Behavioral Ecology and Sociobiology* **2**:1–17.

Bradbury, J.W., *et al.* (2015) The ideal free antelope: foraging dispersions. *Behavioral Ecology* **26**:1303–1313.

Burger, J. and Gochfeld, M. (1992) Effect of group size on vigilance while drinking in the coati, *Nasua narica* in Costa Rica. *Animal Behaviour* **44**:1053–1057.

Busnel, R.G. (1973) Symbiotic relationship between man and dolphins. *Transactions of the New York Academy of Sciences* **35**:112–131.

Caldwell, M.C. and Caldwell, D.K. (1966) Epimeletic (caregiving) behavior in Cetacea. In: *Whales, Dolphins and Porpoises* (ed. K.S. Norris). Berkeley and Los Angeles: University of California Press, 755–789.

Call, J. and Santos, L.R. (2012) Understanding other minds. In: *The Evolution of Primate Societies* (eds J. C. Matani and J. Call). Chicago: University of Chicago Press, 664–681.

Campagna, C., *et al.* (1992) Group breeding in sea lions: pups survive better in colonies. *Animal Behaviour* **43**:541–548.

Campbell, C.J. (2002) The influence of a large home range on the social structure in free-ranging spider monkeys (*Ateles geoffroyi*) on Barro Colorado Island, Panama. *American Journal of Physical Anthropology* **534**:51–52.

Caraco, T. and Wolf, L.L. (1975) Ecological determinants of group sizes of foraging lions. *American Naturalist* **109**:343–352.

Caro, T.M. (2005) *Antipredator Defenses in Birds and Mammals*. Chicago: University of Chicago Press.

Cassidy, K.A., *et al.* (2015) Group composition effects on aggressive interpack interactions of gray wolves in Yellowstone National Park. *Behavioral Ecology* **26**:1352–1360.

Cassini, M.H. (1999) The evolution of reproductive systems in pinnipeds. *Behavioral Ecology* **10**:612–616.

Cassini, M.H. (2000) A model on female breeding dispersion and the reproductive systems of pinnipeds. *Behavioural Processes* **51**:93–99.

Chancellor, R.L. and Isbell, L.A. (2008) Punishment and competition over food in captive rhesus macaques, *Macaca mulatta*. *Animal Behaviour* **75**:1939–1947.

Chapman, C.A. and Chapman, L.J. (2000) Determinants of group size in primates: the importance of travel costs. In: *On the Move: How and Why Animals Travel in Groups* (eds S. Boinski and P. A. Garber). Chicago: University of Chicago Press, 24–41.

Chapman, C.A. and Rothman, J.M. (2009) Within-species differences in primate social structure: evolution of plasticity and phylogenetic constraints. *Primates* **50**:12–22.

Charles-Dominique, P. (1983) Ecology and social adaptations in didelphid marsupials: comparison with eutherians of similar ecology. In: *Advances in the Study of Mammalian Behavior* (eds J. F. Eisenberg and D. G. Kleiman). American Society of Mammalogists Special Publication No. 7. Stillwater, OK: American Society of Mammalogists, 395–422.

Chepko-Sade, B.D. and Halpin, Z.T. (1987) *Mammalian Dispersal Patterns: The Effects of Social Structure on Population Genetics*. Chicago: University of Chicago Press.

Chivers, D.J. and Raemaekers, J.J. (1986) Natural and synthetic diets of Malayan gibbons. In: *Primate Ecology and Conservation*, Vol. **2** (eds J. Else and P. Lee). Cambridge: Cambridge University Press, 39–56.

Clapham, P.J. (2000) The humpback whale: seasonal feeding and breeding in a baleen whale. In: *Cetacean Societies* (eds J. Mann, R. Connor, P. Tyack and H. Whitehead). Chicago: University of Chicago Press, 173–198.

Clutton-Brock, T.H. (2001) Sociality and population dynamics. In: *Ecology: Achievement and Challenge* (eds M.C. Press, N.J. Huntly and S. Levin). Oxford: Blackwell Science, 47–66.

Clutton-Brock, T.H. (2009) Structure and function in mammalian societies. *Philosophical Transactions of the Royal Society B: Biological Sciences* **364**:3229–3242.

Clutton-Brock, T.H. and Albon, S.D. (1989) *Red Deer in the Highlands*. Oxford: Blackwell Scientific Publications.

Clutton-Brock, T.H. and Harvey, P.H. (1977) Primate ecology and social organisation. *Journal of Zoology* **183**:1–39.

Clutton-Brock, T.H. and Harvey, P.H. (1980) Primates, brains and ecology. *Journal of Zoology* **190**:309–323.

Clutton-Brock, T.H. and Janson, C.J. (2012) Primate socioecology at the crossroads: past, present, and future. *Evolutionary Anthropology: Issues, News, and Reviews* **21**:136–150.

Clutton-Brock, T.H., *et al.* (1982a) *Red Deer: The Behaviour and Ecology of Two Sexes*. Chicago: University of Chicago Press.

Clutton-Brock, T.H., *et al.* (1982b) Competition between female relatives in a matrilocal mammal. *Nature* **300**:178–180.

Clutton-Brock, T.H., *et al.* (1984) Individuals and populations: the effects of social behaviour on population dynamics in deer. *Proceedings of the Royal Society of Edinburgh. Section B. Biological Sciences* **82**:275–290.

Clutton-Brock, T.H., *et al.* (1987) Sexual segregation and density-related changes in habitat use in male and female red deer (*Cervus elaphus*). *Journal of Zoology* **211**:275–289.

Clutton-Brock, T.H., *et al.* (1996) Multiple factors affect the distribution of females in lek-breeding ungulates: a rejoinder to Carbone and Taborsky. *Behavioral Ecology* **7**:373–378.

Clutton-Brock, T.H., *et al.* (1999) Predation, group size and mortality in a cooperative mongoose, *Suricata suricatta*. *Journal of Animal Ecology* **68**:672–683.

Clutton-Brock, T.H., *et al.* (2006) Intrasexual competition and sexual selection in cooperative mammals. *Nature* **444**:1065–1068.

Clutton-Brock, T.M., *et al.* (1993) The evolution of ungulate leks. *Animal Behavior* **46**:1121–1138.

Clutton-Brock, T.M., *et al.* (2010) Adaptive suppresssion of subordinate reproduction in cooperative mammals. *The American Naturalist* **176**:664–673.

Connor, R.C. (2000) Group living in whales and dolphins. In: *Cetacean Societies: Field Studies of Dolphins and Whales* (eds J. Mann, R.C. Connor, P. L. Tyack and H. Whitehead). Chicago: University of Chicago Press, 199–218.

Connor, R.C., *et al.* (2000) The bottlenose dolphin: social relationships in a fission–fusion society. In: *Cetacean Societies: Field Studies of Dolphins and Whales* (eds J. Mann, R. Connor, P. Tyack and H. Whitehead). Chicago: University of Chicago Press, 91–125.

Conradt, L. and Roper, T.J. (2003) Group decision-making in animals. *Nature* **421**:155–158.

Conradt, L. and Roper, T.J. (2005) Consensus decision making in animals. *Trends in Ecology and Evolution* **20**:449–456.

Conradt, L., *et al.* (2009) Leading according to need in animal groups. *American Naturalist* **173**:304–312.

Conradt, L. and Roper, T.J. (2009) Conflicts of interest and the evolution of decision sharing. *Philosophical Transactions of the Royal Society of London B* **364**:807–819.

Courchamp, F., *et al.* (1999) Inverse density dependence and the Allee effect. *Trends in Ecology and Evolution* **14**:405–410.

Courchamp, F., *et al.* (2008) *Allee Effects in Ecology and Conservation*. Oxford: Oxford University Press.

Couzin, I.D., *et al.* (2005) Effective leadership and decision-making in animal groups on the move. *Nature* **433**:513–516.

Creel, S. and Creel, N.M. (1995) Communal hunting and pack size in African wild dogs, *Lycaon pictus*. *Animal Behaviour* **50**:1325–1339.

Creel, S. and Creel, N.M. (2002) *The African Wild Dog: Behavior, Ecology, and Conservation*. Princeton, NJ: Princeton University Press.

Creel, S. and Creel, N.M. (2015) Opposing effects of group size on reproduction and survival in African wild dogs. *Behavioral Ecology* **26**:1414–1422.

Crockett, C.M. and Janson, C.H. (2000) Infanticide in red howlers: female group size, male membership and a possible link to folivory. In: *Infanticide by Males and Its Implications* (eds C. P. van Schaik and C. H. Janson). Cambridge: Cambridge University Press, 75–98.

Croft, D.B., *et al.* (2008) *Exploring Animal Social Networks*. Princeton, NJ: Princeton University Press.

De Vos, A. and O'Riain, J.M. (2010) Sharks shape the geometry of a selfish seal herd: experimental evidence from seal decoys. *Biology Letters* **6**:48–50.

Dittus, W.P.J. (1977) The social regulation of population density and age-sex distribution in the toque monkey. *Behaviour* **63**:281–322.

Dittus, W.P.J. (1979) The evolution of behavior regulating density and age-specific sex ratios in a primate population. *Behaviour* **69**:265–302.

Dobson, A. and Poole, J. (1998) Conspecific aggregation and conservation biology. In: *Behavioural Ecology and Conservation Biology* (ed. T. Caro). Oxford: Oxford University Press, 193–208.

Dobson, F.S., *et al.* (2004) The influence of social breeding groups on effective population size in black-tailed prairie dogs. *Journal of Mammalogy* **85**:58–66.

Douglas-Hamilton, I. (1972) *On the ecology and behaviour of the African elephant*. PhD thesis, University of Oxford, Oxford.

Dublin, H.T. (1983) Cooperation and reproductive competition among female African elephants. In: *Social Behaviour of Female Vertebrates* (ed. S.K. Warner). New York: Academic Press, 291–313.

Dunbar, R.I.M. (1991) Functional significance of social grooming in primates. *Folia Primatologica* **57**:121–131.

Dunbar, R.I.M. (2010) Brain and behaviour in primate evolution. In: *Mind the Gap: Tracing the Origins of Human Universals* (eds P. M. Kappeler and J. B. Silk). Berlin: Springer-Verlag, 315–330.

Dunbar, R.I.M. (2012) Bridging the bonding gap: the transition from primates to humans. *Philosophical Transactions of the Royal Society B: Biological Sciences* **367**:1837–1846.

Dunbar, R.I.M. and Lehmann, J. (2013) Grooming and social cohesion in primates: a comment on Grueter *et al*. *Evolution and Human Behavior* **34**:453–455.

Dunbar, R.I.M. and Shultz, S. (2007a) Evolution in the social brain. *Science* **317**:1344–1347.

Dunbar, R.I.M. and Schultz, S. (2007b) Understanding primate brain evolution. *Philosophical Transactions of the Royal Society B: Biological Sciences* **362**:649–658.

Dunbar, R.I.M., *et al.* (2009) Time as an ecological constraint. *Biological Reviews* **84**:413–429.

Duncan, P. and Vigne, N. (1979) The effect of group size in horses on the rate of attacks by blood-sucking flies. *Animal Behaviour* **27**:623–625.

Dwyer, P.D. (1966) The population pattern of *Miniopterus schrebersii* (Chiroptera) in north-eastern New South Wales. *Australian Journal of Zoology* **14**:1073–1137.

Ebensperger, L.A., *et al.* (2011) Sociality, glucocorticoids and direct fitness in the communally rearing rodent, *Octodon degus*. *Hormones and Behavior* **60**:346–352.

Ebensperger, L.A., *et al.* (2012) Direct fitness of group living mammals varies with breeding strategy, climate and fitness estimates. *Journal of Animal Ecology* **81**:1013–1023.

Eisenberg, J.F. (1981) *The Mammalian Radiations: An Analysis of Trends in Evolution, Adaptation, and Behavior*. Chicago: University of Chicago Press.

Elgar, M.A. (1986) House sparrows establish foraging flocks by giving chirrup calls if the resources are divisible. *Animal Behaviour* **34**:169–174.

Emlen, S.T. (1991) Evolution of cooperative breeding in birds and mammals. In: *Behavioural Ecology: An Evolutionary Approach*, 3rd edn (eds J. R. Krebs and N. B. Davies). Oxford: Blackwell Scientific Publications, 301–337.

Emlen, S.T. and Oring, L.W. (1977) Ecology, sexual selection, and the evolution of mating systems. *Science* **197**: 215–223.

Erhart, E.M. and Overdorff, D.J. (1999) Female coordination of group travel in wild *Propithecus* and *Eulemur*. *International Journal of Primatology* **20**:927–940.

Errington, P.L. (1963) *Muskrat Populations*. Ames, IA: Iowa State University Press.

Fanshawe, J.H. and Fitzgibbon, C.D. (1993) Factors influencing the hunting success of an African wild dog pack. *Animal Behaviour* **45**:479–490.

Farine, D.R. and Whitehead, H. (2015) Constructing, conducting and interpreting animal social network analysis. *Journal of Animal Ecology* **84**:1144–1163.

Fitzgibbon, C.D. (1990) Mixed-species grouping in Thomson's and Grant's gazelles: the antipredator benefits. *Animal Behaviour* **39**:1116–1126.

Foley, C.A.H., *et al.* (2008) Severe drought and calf survival in elephants. *Biology Letters* **4**:541–544.

Franks, N.R., *et al.* (2003) Speed versus accuracy in collective decision making. *Proceedings of the Royal Society of London. Series B: Biological Sciences* **270**:2457–2463.

Freeberg, T.M., *et al.* (2012) Social complexity as a proximate and ultimate factor in communicative complexity DOI: 10.1098/rstb.2011.0213. [Accessed 07 January 2016].

Fryxell, J.M. (1995) Aggregation and migration by grazing ungulates in relation to resources and predators. In: *Serengeti II: Dynamics, Management and Conservation of an Ecosystem* (eds A. R. E. Sinclair and P. Arcese). Chicago: University of Chicago Press, 257–273.

Fryxell, J.M. and Sinclair, A.R.E. (1988) Seasonal migration by white-eared kob in relation to resources. *African Journal of Ecology* **26**:17–31.

Gascoyne, S.C., *et al.* (1993) Rabies in African wild dogs (*Lycaon pictus*) in the Serengeti region. *Journal of Wildlife Diseases* **29**:396–402.

Giraldeau, L.-A. and Caraco, T. (2000) *Social Foraging Theory*. Princeton, NJ: Princeton University Press.

Giraldeau, L.-A. and Gillas, D. (1985) Optimal group size can be stable: a reply to Sibly. *Animal Behaviour* **33**: 666–667.

Gittleman, J.L. (1989) Carnivore group living: comparative trends. In: *Carnivore Behaviour, Ecology and Evolution* (ed. J.L. Gittleman). Ithaca, NY: Cornell University Press, 183–207.

Grant, T.R. (1983) The behavioral ecology of the monotremes. In: *Advances in the Study of Mammalian Behavior* (eds J. F. Eisenberg and D. G. Kleiman). American Society of Mammalogists Special Publication No 7. Stillwater, OK: American Society of Mammalogists, 360–394.

Groenendijk et al. (2015) Effects of territory size on the reproductive success and social system of the Giant Otter in southeastern Peru. *Journal of Zoology* **296**:153–160.

Grueter, C.C., *et al.* (2013) Grooming and group cohesion in primates: implications for the evolution of language. *Evolution and Human Behavior* **34**:61–68.

Hamilton, I.M. (2000) Recruiters and joiners: using optimal skew theory to predict group size and the division of resources within groups of social foragers. *American Naturalist* **155**:684–695.

Hamilton, W.D. (1971) Geometry for the selfish herd. *Journal of Theoretical Biology* **31**:295–311.

Hare, J.F. and Murie, J.O. (2007) Ecology, kinship and ground squirrel sociality: insights from comparative analysis. In: *Rodent Societies: An Ecological and Evolutionary Perspective* (eds J. O. Wolff and P. W. Sherman). Chicago: University of Chicago Press, 345–355.

Harvey, P.H. and Bennett, P.M. (1983) Brain size, energetics, ecology and life history patterns. *Nature* **306**:314–315.

Harvey, P.H. and Pagel, M.D. (1991) *The Comparative Method in Evolutionary Biology*. Oxford: Oxford University Press.

Harvey, P.H., *et al.* (1980) Brain size and ecology in small mammals and primates. *Proceedings of the National Academy of Sciences of the United States of America* **77**:4387–4389.

Hayes, L.D., *et al.* (2011) Towards an integrative model of sociality in caviomorph rodents. *Journal of Mammalogy* **92**:65–77.

Healy, S.D. and Rowe, C. (2007) A critique of comparative studies of brain size. *Proceedings of the Royal Society of London. Series B: Biological Sciences* **274**:453–464.

Henzi, S.P. and Barrett, L. (2005) The historical socioecology of savanna baboons (*Papio hamadryas*). *Journal of Zoology* **265**:215–226.

Henzi, S.P., *et al.* (2009) Cyclicity in the structure of female baboon social networks. *Behavioral Ecology and Sociobiology* **63**:1015–1021.

Higashi, M. and Yamamura, N. (1993) What determines animal group size? Insider–outsider conflict and its resolution. *American Naturalist* **142**:553–563.

Holekamp, K.E., *et al.* (2015) Brains, brawn and sociality: a hyaena's tale. *Animal Behaviour* **103**:237–248.

Hoogland, J.L. (1995) *The Black-tailed Prairie Dog: Social Life of a Burrowing Mammal*. Chicago: University of Chicago Press.

Hughes, W.O., *et al.* (2002) Trade-offs in group living: transmission and disease resistance in leaf-cutting ants. *Proceedings of the Royal Society of London. Series B: Biological Sciences* **269**:1811–1819.

Humphrey, N.K. (1976) The social function of intellect. In: *Growing Points in Ethology* (eds P. P. G. Bateson and R. A. Hinde). Cambridge: Cambridge University Press, 303–317.

Isler, K. and van Schaik, C.P. (2012) Allomaternal care, life history and brain size evolution in mammals. *Journal of Human Evolution* **63**:52–63.

Jack, K. (2007) The cebines. In: *Primates in Perspective* (eds C.J. Campbell, A. Fuentes, K. MacKinnon, M. Panger and S. Bearder). Oxford: Oxford University Press, 107–123.

James, R., *et al.* (2009) Potential banana skins in animal social network analysis. *Behavioral Ecology and Sociobiology* **63**:989–997.

Janson, C.H. (1988) Food competition in brown capuchin monkeys (*Cebus apella*): quantitative effects of group size and tree productivity. *Behaviour* **105**:53–76.

Janson, C.H. and Goldsmith, M.C. (1995) Predicting group size in primates: foraging costs and predation risks. *Behavioral Ecology* **6**:326–336.

Jarman, P.J. (1974) The social organisation of antelope in relation to their ecology. *Behaviour* **48**:215–267.

Jarman, P.J. (1991) Social behaviour and organisation in the Macropodoidea. In: *Advances in the Study of Behavior*, Vol. **20** (eds P.J.B. Slater, J.S. Rosenblatt, C. Beer and M. Milinski). San Diego, CA: Academic Press, 1–50.

Jarman, P.J. (2000) Males in macropod society. In: *Primate Males: Causes and Consequences of Variation in Group Composition* (ed. P.M. Kappeler). Cambridge: Cambridge University Press, 21–33.

Jarman, P.J. and Jarman, M.V. (1973) Social behaviour, population structure and reproductive potential in impala. *African Journal of Ecology* **11**:329–338.

Jarman, P.J. and Kruuk, H. (1996) Phylogeny and spatial organisation in mammals. In: *Comparison of Marsupial and Placental Behaviour* (eds D. B. Croft and U. Ganslosser). Fürth, Germany: Filander Verlag, 80–101.

Jarman, P.J. and Southwell, C. (1986) Grouping, associations and reproductive strategies in eastern grey kangaroos. In: *Ecological Aspects of Social Evolution: Birds and Mammals* (eds D. I. Rubenstein and R. W. Wrangham). Princeton, NJ: Princeton University Press, 399–428.

Johnson, C.N. and Crossman, D.G. (1991) Dispersal and social organization of the northern hairy-nosed wombat *Lasiorhinus krefftii*. *Journal of Zoology* **225**:605–613.

Johnson, D.D., *et al.* (2002) Does the resource dispersion hypothesis explain group living? *Trends in Ecology and Evolution* **17**:563–570.

Kappeler, P.M. (1997) Determinants of primate social organisation: comparative evidence and new insights from Malagasy lemurs. *Biological Reviews* **72**:111–151.

Kappeler, P.M. and van Schaik, C.P. (2002) Evolution of primate social systems. *International Journal of Primatology* **23**:707–740.

Kasper, C. and Voelkl, B. (2009) A social network analysis of primate groups. *Primates* **50**:343–356.

Kerth, G. (2008) Causes and consequences of sociality in bats. *BioScience* **58**:737–746.

Kerth, G. (2010) Group decision-making in animal societies. In: *Animal Behaviour: Evolution and Mechanisms* (ed. P.M. Kappeler). Berlin: Springer, 241–265.

Kerth, G., *et al.* (2001a) Roosting together, foraging apart: information transfer about food is unlikely to explain sociality in female Bechstein's bats (*Myotis bechsteinii*). *Behavioral Ecology and Sociobiology* **50**:283–291.

Kerth, G., *et al.* (2001b) Day roost selection in female Bechstein's bats (*Myotis bechsteinii*): a field experiment to determine the influence of roost temperature. *Oecologia* **126**:1–9.

Kerth, G., *et al.* (2006) Group decision making in fission–fusion societies: evidence from two-field experiments in Bechstein's bats. *Proceedings of the Royal Society of London. Series B: Biological Sciences* **273**:2785–2790.

Kerth, G., *et al.* (2011) Bats are able to maintain long-term social relationships despite the high fission–fusion dynamics of their groups. *Proceedings of the Royal Society of London. Series B: Biological Sciences* **278**:2761–2767.

King, A.J., *et al.* (2008) Dominance and affiliation mediate despotism in a social primate. *Current Biology* **18**:1833–1838.

Kleiman, D.G. and Eisenberg, J.F. (1973) Comparisons of canid and felid social systems from an evolutionary perspective. *Animal Behaviour* **21**:637–659.

Kotrschal, A., *et al.* (2013) The benefit of evolving a larger brain: big-brained guppies perform better in a cognitive task. *Animal Behaviour* **86**:e4–e6.

Kotrschal, A., *et al.* (2015) Brain size affects female but not male survival under predation threat. *Ecology Letters* **18**:646–652.

König, B. (1994) Fitness effects of communal rearing in house mice: the role of relatedness versus familiarity. *Animal Behaviour* **48**:1449–1457.

Kramer, D.L. (1995) Are colonies supraoptimal groups? *Animal Behaviour* **33**:1031–1032.

Krause, J. and Ruxton, G.D. (2002) *Living in Groups*. Oxford: Oxford University Press.

Krause, J., *et al.* (2009) Animal social networks: an introduction. *Behavioral Ecology and Sociobiology* **63**:967–973.

Krebs, J.R., *et al.* (1972) Flocking and feeding in the great tit *Parus major*: an experimental study. *Ibis* **114**:507–530.

Kruuk, H. (1972) *The Spotted Hyena: A Study of Predation and Social Behaviour*. Chicago: University of Chicago Press.

Kruuk, H. and Macdonald, D.W. (1985) Group territories of carnivores: empires and enclaves. In: *Behavioural Ecology: Ecological Consequences of Adaptive Behaviour* (eds R. M. Sibly and R. H. Smith). Oxford: Blackwell Scientific Publications, 521–536.

Kudo, H. and Dunbar, R.I.M. (2001) Neocortex size and social network size in primates. *Animal Behaviour* **62**:711–722.

Kummer, H. (1968) *Social Organization of Hamadryas Baboons*. Chicago: University of Chicago Press.

Kunz, T.H. and Fenton, M.B. (2003) *Bat Ecology*. Chicago: University of Chicago Press.

Kunz, T.H. and Lumsden, L.F. (2003) Ecology of cavity and foliage roosting bats. In: *Bat Ecology* (eds T. H. Kunz and M. B. Fenton). Chicago: Univerisity of Chicago Press, 3–89.

Lacey, E.A. (2004) Sociality reduces individual direct fitness in a communally breeding rodent, the colonial tuco-tuco (*Ctenomys sociabilis*). *Behavioral Ecology and Sociobiology* **56**:449–457.

Lacey, E.A. and Sherman, P.W. (2007) The ecology of sociality in rodents. In: *Rodent Societies: An Ecological and Evolutionary Perspective* (eds J. O. Wolff and P. W. Sherman). Chicago: University of Chicago Press, 243–254.

Laland, K.N. (2004) Social learning strategies. *Learning Behaviour* **32**:4–14.

Lehmann, J. and Dunbar, R.I.M. (2009) Network cohesion, group size and neocortex size in female-bonded Old World primates. *Proceedings of the Royal Society of London. Series B: Biological Sciences* **276**:4417–4422.

Lewis, S.E. and Pusey, A.E. (1997) Factors influencing the occurrence of communal care in plural breeding mammals. In: *Cooperative Breeding in Mammals* (eds N. G. Solomon and J. A. French). Cambridge: Cambridge University Press, 335–363.

Lukas, D. and Clutton-Brock, T.H. (2013) The evolution of social monogamy in mammals. *Science* **341**:526–530.

Lukas, D. and Vigilant, L. (2005) Reply: Facts, faeces and setting standards for the study of MHC genes using noninvasive samples. *Molecular Ecology* **14**:1601–1602.

Lukas, D., *et al.* (2005) To what extent does living in a group mean living with kin? *Molecular Ecology* **14**:2181–2196.

Lusseau, D. (2003) The emergent properties of a dolphin social network. *Proceedings of the Royal Society of London. Series B: Biological Sciences* **270**:S186–S188.

Lusseau, D. and Newman, M.E.J. (2004) Identifying the role that animals play in their social networks. *Proceedings of the Royal Society of London. Series B: Biological Sciences* **271**:S477–S481.

MacIntosh, A.J.J., *et al.* (2012) Monkeys in the middle: parasite transmission through the social network of a wild primate. Plos one DOI: 10.1371/journal.pone.0051144. [Accessed 07 January 2016].

Madden, J.R., *et al.* (2011) The social network structure of a wild meerkat population: 3. Position of individuals within networks. *Behavioral Ecology Sociobiology* **65**:1857–1871.

Markham, A.C., *et al.* (2015) Optimal group size in a higher social mammal. *PNAS* **12**:14882–14887.

McComb, K. and Semple, S. (2005) Coevolution of vocal communication and sociality in primates. *Biology Letters* **1**:381–385.

McComb, K., *et al.* (2001) Matriarchs as repositories of social knowledge in African elephants. *Science* **292**:491–494.

McComb, K., *et al.* (2011) Leadership in elephants: the adaptive value of age. *Proceedings of the Royal Society of London. Series B: Biological Sciences* **282**:3270–3276.

McComb, K., *et al.* (2014) Elephants can determine ethnicity, gender, and age from acoustic cues in human voices. *Proceedings of the National Academy of Sciences of the United States of America* **111**:5433–5438.

Macdonald, D.W. (1983) The ecology of carnivore social behaviour. *Nature* **301**:379–384.

Macdonald, D.W. (2004) *The New Encyclopedia of Mammals*. Oxford: Oxford University Press.

Macdonald, D.W., *et al.* (2007) Social organisation and resource use in capybaras and maras. In: *Rodent Societies: An Ecological and Evolutionary Perspective* (eds J. O. Wolff and P. W. Sherman). Chicago: University of Chicago Press, 393–402.

Mace, G.M., *et al.* (1981) Brain size and ecology in small mammals. *Journal of Zoology* **193**:333–354.

MacLean, E.L., *et al.* (2009) Sociality, ecology, and relative brain size in lemurs. *Journal of Human Evolution* **56**:471–478.

MacLean, E.L., *et al.* (2012) How does cognition evolve? Phylogenetic comparative psychology. *Animal Cognition* **15**:223–238.

MacLean, E.L., *et al.* (2014) The evolution of self control. *Proceedingd of the National Academy of Sciences of the United States of America* **20**:E2140–E2148.

Madden, J.R. and Clutton-Brock, T.H. (2011) Experimental peripheral administration of oxytocin elevates a suite of cooperative behaviours in a wild social mammal. *Proceedings of the Royal Society of London. Series B: Biological Sciences* **278**:1189–1194.

Majolo, B., *et al.* (2008) Costs and benefits of group-living in primates: group size effects on behaviour and demography. *Animal Behaviour* **76**:1235–1247.

Major, P. (1978) Predator–prey interactions in two schooling fishes *Caranx ignobilis* and *Stolephorus purpureus*. *Animal Behaviour* **26**:760–777.

Malcolm, J.R. and van Lawick, H. (1975) Notes on wild dogs (*Lycaon pictus*) hunting zebras. *Mammalia* **39**:231–240.

Mech, L.D. (1970) *The Wolf: The Ecology and Behavior of an Endangered Species*. New York: Natural History Press.

Michener, G.R. (1981) Ontogeny of spatial relationships and social behavior in juvenile Richardson's ground squirrels. *Canadian Journal of Zoology* **59**:1666–1676.

Michener, G.R. (1983) Kin identification, matriarchies, and the evolution of sociality in ground-dwelling sciurids. In: *Advances in the Study of Mammalian Behavior* (eds J. F. Eisenberg and D. G. Kleiman). American Society of Mammalogists Special Publication No 7. Stillwater, OK: American Society of Mammalogists, 528–572.

Mills, M.G.L. (1990) *Kalahari Hyaenas: Comparative Behavioural Ecology of Two Species*. London: Unwin Hyman.

Mitani, J.C. (1989) Orangutan activity budgets: monthly variations and the effects of body size, parturition and sociality. *American Journal of Primatology* **18**:87–100.

Moehlman, P.D. (1989) Intraspecific variation in canid social systems. In: *Carnivore Behavior, Ecology and Evolution* (ed. J.L. Gittleman). Ithaca, NY: Cornell University Press, 143–163.

Morse, D.H. (1977) Feeding behavior and predator avoidance in heterospecific groups. *BioScience* **27**:332–339.

Moss, C.J. and Lee, P.C. (2011) Female social dynamics: fidelity and flexibility. In: *The Amboseli Elephants: A Long-term Perspective on a Long-lived Mammal* (eds C.J. Moss, H. Croze and P. C. Lee). Chicago: University of Chicago Press, 205–223.

Mosser, A. and Packer, C. (2009) Group territoriality and the benefits of sociality in the African lion, *Panthera leo*. *Animal Behaviour* **78**:359–370.

Mosser, A.A., *et al.* (2015) Landscape heterogeneity and behavioral traits drive the evolution of lion group territoriality. *Behavioral Ecology* doi: 10.1093/beheco/arv046.

Müller, A.E. and Thalmann, U. (2000) Origin and evolution of primate social organisation: a reconstruction. *Biological Reviews of the Cambridge Philosophical Society* **75**:405–435.

Müller, M.N., *et al.* (2007) Male coercion and the costs of promiscuous mating for female chimpanzees. *Proceedings of the Royal Society of London. Series B: Biological Sciences* **274**:1009–1014.

Mutinda, H., *et al.* (2011) Decision making and leadership in using the ecosystem. In: *The Amboseli Elephants: A Long-term Perspective on a Long-lived Mammal* (eds C.J. Moss, H. Croze and P. C. Lee). Chicago: University of Chicago Press, 246–259.

Neill, S. and Cullen, J.M. (1974) Experiments on whether schooling by their prey affects the hunting behaviour of cephalopods and fish predators. *Journal of Zoology* **172**:549–569.

Nevo, E. (2007) Mosaic evolution of subterranean mammals: tinkering, regression, progression, and global convergenc. In: *Subterranean Rodents: News from Underground* (eds S. Begall, H. Burda and C. E. Schleich). Berlin: Springer, 375–388.

Nunes, S. (2007) Dispersal and philopatry. In: *Rodent Societies: An Ecological and Evolutionary Perspective* (eds J. O. Wolff and P. W. Sherman). Chicago: University of Chicago Press, 150–163.

Oates, J.F. (1977) Social life of a black-and-white colobus monkey, *Colobus guereza*. *Zeitschrift fur Tierpsychologie* **45**:1–60.

Ortega, J. and Arita, H.T. (2002) Subordinates in harem groups of Jamaican fruit-eating bats (*Artibeus jamaicensis*): satellites or sneaks? *Ethology* **108**:1077–1091.

Owen-Smith, R.N. (1988) *Megaherbivores: The Influence of Very Large Body Size on Ecology*. Cambridge: Cambridge University Press.

Ozgul, A., *et al.* (2006) Effects of patch quality and network structure on patch occupancy dynamics of a yellow-belled marmot population. *Journal of Animal Ecology* **75**:191–202.

Packer, C. (1986) The ecology of sociality in felids. In: *Ecological Aspects of Social Evolution in Birds and Mammals* (eds D. I. Rubenstein and R. W. Wrangham). Princeton, NJ: Princeton University Press, 429–451.

Packer, C. (1988) Constraints on the evolution of reciprocity: lessons from cooperative hunting. *Ethology and Sociobiology* **9**:137–147.

Packer, C., *et al.* (1990) Why lions form groups: food is not enough. *American Naturalist* **136**:1–19.

Patzelt, A., *et al.* (2014) Male tolerance and male–male bonds in a multilevel primate society. *Proceedings of the National Academy of Sciences of the United States of America* **111**:14740–14745.

Pays, O., *et al.* (2009) The effect of social facilitation on vigilance in the eastern gray kangaroo, *Macropus giganteus*. *Behavioral Ecology* **20**:469–477.

Pérez-Barbería, F.J., *et al.* (2007) Evidence for coevolution of sociality and relative brain size in three orders of mammals. *Evolution* **61**:2811–2821.

Pfeffer, P. and Caldecott, J.O. (1986) The bearded pig (*Sus barbatus*) in East Kalimantan and Sarawak. *Journal of the Malaysian Branch of the Royal Asiatic Society* **59**:81–100.

Pinter-Wollman, N., *et al.* (2009) The relationship between social behaviour and habitat familiarity in African elephants (*Loxodonta africana*). *Proceedings of the Royal Society of London. Series B: Biological Sciences* **276**:1009–1014.

Pollard, K.A. and Blumstein, D.T. (2008) Time allocation and the evolution of group size. *Animal Behaviour* **76**:1683–1699.

Popa-Lisseanu, A.G., *et al.* (2008) Highly structured fission–fusion societies in an aerial-hawking, carnivorous bat. *Animal Behaviour* **75**:471–482.

Prins, H.H.T. (1996) *Ecology and Behaviour of the African Buffalo: Social Inequality and Decision Making*. London: Chapman and Hall.

Pusey, A.E. and Schroepfer-Walker, K. (2013) Female competition in chimpanzees. *Philosophical Transactions of the Royal Society B: Biological Sciences* **368**:20130077.

Rand, D.A., *et al.* (1995) Invasion, stability and evolution to criticality in spatially extended, artificial host-pathogen ecologies. *Proceedings of the Royal Society of London. Series B: Biological Sciences* **259**:55–63.

Rand, D.G., *et al.* (2011) Dynamic social networks promote cooperation in experiments with humans. *Proceedings of the National Academy of Sciences of the United States of America* **108**:19193–19198.

Ransome, R.D. (1990) *The Natural History of Hibernating Bats*. London: Christopher Helm.

Reader, S.M. and Laland, K.N. (2002) Social intelligence, innovation, and enhanced brain size in primates. *Proceedings of the National Academy of Sciences of the United States of America* **99**:4436–4441.

Renouf, D. (1990) *The Behaviour of Pinnipeds*. London: Chapman and Hall.

Robinson, J.G. (1988) Group size in wedge-capped capuchin monkeys *Cebus olivaceus* and the reproductive success of males and females. *Behavioral Ecology and Sociobiology* **23**:187–197.

Rodman, P.S. (1981) Inclusive fitness and group size with a reconsideration of group sizes in lions and wolves. *American Naturalist* **118**:275–283.

Rood, J.P. (1986) Ecology and social evolution in the mongooses. In: *Ecological Aspects of Social Evolution* (eds D. I. Rubenstein and R. W. Wrangham). Princeton, NJ: Princeton University Press, 131–152.

Rubenstein, D.I. and Wrangham, R.W. (eds) (1986) *Ecological Aspects of Social Evolution: Birds and Mammals*. Princeton, NJ: Princeton University Press.

Russell, E.M. (1984) Social behaviour and social organization of marsupials. *Mammal Review* **14**:101–154.

Sandell, M. (1989) The mating tactics and spacing patterns of solitary carnivores. In: *Carnivore Behavior, Ecology and Evolution* (ed. J.L. Gittleman). London: Chapman and Hall, 164–182.

Schaller, G.B. (1963) *The Mountain Gorilla: Ecology and Behavior*. Chicago: University of Chicago Press.

Schaller, G.B. (1972) *The Serengeti Lion: A Study of Predator–Prey Relations*. Chicago: Chicago University Press.

Schillaci, M.A. (2008) Primate mating systems and the evolution of neocortex size. *Journal of Mammalogy* **89**:58–63.

Schradin, C. (2000) Confusion effect in a reptilian and a primate predator. *Ethology* **106**:691–700.

Schradin, C. (2005) When to live alone and when to live in groups: ecological determinants of sociality in the African striped mouse (*Rhabdomys pumilio*, Sparrman 1974). *Belgian Journal of Zoology* **135**:77–82.

Schradin, C. and Pillay, N. (2005) The influence of the father on offspring development in the striped mouse. *Behavioral Ecology* **16**:450–455.

Schülke, O. and Kappeler, P.M. (2003) So near and yet so far: territorial pairs but low cohesion between pair partners in a nocturnal lemur, *Phaner furcifer*. *Animal Behaviour* **65**:331–343.

Seeley, T.D. and Buhrman, S.C. (1999) Group decision making in swarms of honey bees. *Behavioral Ecology and Sociobiology* **45**:19–31.

Setchell, J.M., *et al.* (2002) Reproductive parameters and maternal investment in mandrills (*Mandrillus sphinx*). *International Journal of Primatology* **23**:51–68.

Shettleworth, S.J. (2010) *Cognition, Evolution, and Behavior*. New York: Oxford University Press.

Shultz, S. and Dunbar, R.I.M. (2007) The evolution of the social brain: anthropoid primates contrast with other vertebrates. *Proceedings of the Royal Society of London. Series B: Biological Sciences* **274**:2429–2436.

Sibly, R.M. (1983) Optimal group size is unstable. *Animal Behaviour* **31**:947–948.

Sinclair, A.R.E. (1977) *The African Buffalo: A Study of Resource Limitation of Populations*. Chicago: University of Chicago Press.

Singleton, I. and van Schaik, C.P. (2001) Orangutan home range size and its determinants in a Sumatran swamp forest. *International Journal of Primatology* **22**:877–886.

Smith, J.E., *et al.* (2008) Social and ecological determinants of fission–fusion dynamics in the spotted hyaena. *Animal Behaviour* **76**:619–636.

Smith, J.E., *et al.* (2016) Leadership in mammalian societies: emergence, distribution, power and payoff. *Trends in Ecology and Evolution* **31**:54–66.

Smith, M.C. (1968) Red squirrel responses to spruce cone failure in interior Alaska. *Journal of Wildlife Management* **32**:305–317.

Snaith, T.V. and Chapman, A.P. (2007) Primate group size and interpreting socioecological models: do folivores really play by different rules? *Evolutionary Anthropology* **16**:94–106.

Solomon, N.G. (2003) A reexamination of factors influencing philopatry in rodents. *Journal of Mammalogy* **84**:1182–1197.

Speakman, J.R. and Thomas, D.W. (2003) Physiological ecology and energetics of bats. In: *Bat Ecology* (eds T. H. Kunz and M. B. Fenton). Chicago: University of Chicago Press, 430–490.

Spinks, A.C., *et al.* (2000a) Circulating LH levels and the response to exogenous GnRH in the common mole-rat: implications for reproductive regulation in this social, seasonal breeding species. *Hormones and Behavior* **37**:221–228.

Spinks, A.C., *et al.* (2000b) Comparative patterns of philopatry and dispersal in two common mole-rat populations: implications for the evolution of mole-rat sociality. *Journal of Animal Ecology* **69**:224–234.

Stammbach, E. (1987) Desert, forest and montane baboons: multi-level societies. In: *Primate Societies* (eds B.B. Smuts, D.L. Cheney, R.M. Seyfarth, R. W. Wrangham and T. T. Struhsaker). Chicago: University of Chicago Press, 112–120.

Stander, P.E. (1992a) Cooperative hunting in lions: the role of the individual. *Behavioral Ecology and Sociobiology* **29**:445–454.

Stander, P.E. (1992b) Foraging dynamics of lions in a semi-arid environment. *Canadian Journal of Zoology* **70**:8–21.

Stankowich, T., *et al.* (2014) Ecological drivers of antipredator defenses in carnivores. *Evolution* **68**:1415–1425.

Sterck, E.H.M., *et al.* (1997) The evolution of female social relationships in nonhuman primates. *Behavioral Ecology and Sociobiology* **41**:291–309.

Stirling, I. (1983) The evolution of mating systems in pinnipeds In: *Advances in the Study of Mammalian Behavior* (eds J. F. Eisenberg and D. G. Kleiman). American Society of Mammologists Special Publication No 7. Stillwater, OK: American Society of Mammalogists, 489–527.

Stolba, A. (1979) *Entscheidungsfindung in verbänden von* Papio hamadryas. PhD thesis, University of Zurich, Zurich.

Strier, K.B. (2011) *Primate Behavioral Ecology*. Upper Saddle River, NJ: Prentice-Hall.

Strandburg-Peshkin, A., *et al.* (2015) Shared decision-making drives collective movement in wild baboons. *Science* **348**:1358–1361.

Stumpf, R. (2007) Chimpanzees and bonobos: diversity within and between species. In: *Primates in Perspective* (eds C.J. Campbell, A. Fuentes, K.C. MacKinnon, N. Panger and S. K. Bearder). Oxford: Oxford University Press, 321–344.

Sueur, C. and Petit, O. (2008) Organization of group members at departure is driven by social structure in Macaca. *International Journal of Primatology* **29**:1085–1098.

Sueur, C., *et al.* (2011) A comparative network analysis of social style in macaques. *Animal Behaviour* **82**:845–852.

Sundaresan, S.R., *et al.* (2007) Network metrics reveal differences in social organization between two fission–fusion species, Grevy's zebra and onager. *Oecologia* **151**:140–149.

Swedell, L. (2002) Affiliation among females in wild hamadryas baboons (*Papio hamadryas hamadryas*). *International Journal of Primatology* **23**:1205–1226.

Symington, M.M. (1988) Food competition and foraging party size in the black spider monkey (*Ateles paniscus chamek*). *Behaviour* **105**:117–134.

Taylor, R.J. (1993) Observations on the behaviour and the ecology of the common wombat *Vombatus ursinus* in northeast Tasmania. *Australian Mammalogy* **16**:1–7.

Tershy, B.R. (1992) Body size, diet, habitat use, and social behavior of *Balaenoptera* whales in the Gulf of California. *Journal of Mammalogy* **73**:477–486.

Thouless, C.R. and Guinness, F.E. (1986) Conflict between red deer hinds: the winner always wins. *Animal Behaviour* **34**:1166–1171.

VanderWaal, K.L., *et al.* (2009) Optimal group size, dispersal decisions and postdispersal relationships in female African lions. *Animal Behaviour* **77**:949–954.

VanderWaal, K.L., *et al.* (2014) Multilevel social organization and space use in reticulated giraffe (*Giraffa camelopardalis*). *Behavioral Ecology* **25**:17–26.

van Schaik, C.P. (1983) Why are diurnal primates living in groups? *Behaviour* **87**:120–144.

van Schaik, C.P. and Kappeler, P.M. (2003) The evolution of social monogamy in primates. In: *Monogamy: Mating Strategies and Partnerships in Birds, Humans and Other Mammals* (eds U. Reichard and C. Boesch). Cambridge: Cambridge University Press, 59–80.

van Schaik, C.P. and van Hooff, R.A.M. (1996) Toward an understanding of the orangutan's social system. In: *Great Ape Societies* (eds W.C. McGrew, L. F. Marchant and T. Nishida). Cambridge: Cambridge University Press, 3–15.

van Schaik, C.P., *et al.* (1983) The effect of group size on time budgets and social behaviour in wild long-tailed macaques (*Macaca fascicularis*). *Behavioral Ecology and Sociobiology* **13**:173–181.

Volumy, M., *et al.* (2015) Social organization in Eulipotyphla: evidence for a social shrew. *Biology Letters* DOI: 10.1098/rsbl.2015.0825. [Accessed 07 January 2016].

Vucetich, J.A., *et al.* (2004) Raven scavenging favours group foraging in wolves. *Animal Behaviour* **67**:1117–1126.

Waser, P.M. (1977) Feeding, ranging and group size in the mangabey *Cercocebus albigena*. In: *Primate Ecology* (ed. T.H. Clutton-Brock). London: Academic Press, 183–222.

Watts, D.P. (1996) Comparative socio-ecology of gorillas. In: *Great Ape Societies* (eds W.C. McGrew, L. F. Marchant and T. Nishida). Cambridge: Cambridge University Press, 16–28.

Webb, C.R. (2005) Farm animal networks: unraveling the contact structure of the British sheep population. *Preventive Veterinary Medicine* **68**:3–17.

Wells, R.S., *et al.* (1987) The social structure of free-ranging bottlenose dolphins. In: *Current Mammalogy* (ed. H.H. Genoways). New York: Plenum Press, 247–305.

Wey, T., *et al.* (2008) Social network analysis of animal behaviour: a promising tool for the study of sociality. *Animal Behaviour* **75**:333–344.

White, F.J. (1996) Comparative socio-ecology of *Pan paniscus*. In: *Great Ape Societies* (eds W.C. McGrew, L. F. Marchant and T. Nishida). Cambridge: Cambridge University Press, 29–41.

Whitehead, H. and Weilgart, L. (2000) The sperm whale: social females and roving males. In: *Cetacean Societies: Field Studies of Dolphins and Whales* (eds J. Mann, R. Connor, P. Tyack and H. Whitehead). Chicago: University of Chicago Press, 154–172.

Whiten, A. and van Schaik, C.P. (2007) The evolution of animal 'cultures' and social intelligence. *Philosophical Transactions of the Royal Society B: Biological Sciences* **362**:603–620.

Whitten, P.L. (1983) Diet and dominance among female vervet monkeys (*Cercopithecus aethiops*). *American Journal of Primatology* **5**:139–159.

Wilkinson, G.S. (1985) The social organization of the common vampire bat. I. Pattern and cause of association. *Behavioral Ecology and Sociobiology* **17**:111–121.

Wilkinson, G.S. (1992) Information transfer at evening bat colonies. *Animal Behaviour* **44**:501–508.

Wilkinson, G.S. (1995) Information transfer in bats. *Symposia of the Zoological Society of London* **67**:345–360.

Williams, J.M., *et al.* (2002) Female competition and male territorial behaviour influence female chimpanzees' ranging patterns. *Animal Behaviour* **63**:347–360.

Winter, J.W. (1996) Australian possums and Madagascan lemurs: behavioural comparisons of ecological equivalents. In: *Comparison of Marsupial and Placental Behaviour* (eds D. B. Croft and U. Ganslosser). Fürth, Germany: Filander Verlag, 263–292.

Wolff, J.O. (1989) Social behavior. In: *Advances in the Study of Peromyscus (Rodentia)* (eds G. L. Kirkland and J. N. Layne). Lubbock, TX: Texas Tech University Press, 271–291.

Wolff, J.O. (1994) More on juvenile dispersal in mammals. *Oikos* **71**:349–352.

Wolff, J.O. and Sherman, P.W. (eds) (2007) *Rodent Societies: An Ecological and Evolutionary Perspective*. Chicago: University of Chicago Press.

Woodroffe, R. and Ginsberg, J.R. (1998) Edge effects and the extinction of populations inside protected areas. *Science* **280**:2126–2128.

Woodroffe, R., *et al.* (2009) Social group size affects *Mycobacterium bovis* infection in European badgers *Meles meles*. *Journal of Animal Ecology* **78**:818–827.

Woodruff, J.A., *et al.* (2013) Effects of social environment on baseline glucocorticoid levels in a communally breeding rodent, the colonial tuco-tuco (*Ctenomys sociabilis*). *Hormones and Behavior* **64**:566–572.

Wrangham, R.W. (1980) An ecological model of female-bonded primate groups. *Behaviour* **75**:262–300.

Wrangham, R.W. (1987) Evolution of social structure. In: *Primate Societies* (eds B.B. Smuts, D.L. Cheney, R.M. Seyfarth, R. W. Wrangham and T. T. Struhsaker). Chicago: University of Chicago Press, 282–296.

Wrangham, R.W., *et al.* (1992) Female social relationships and social organization of Kibale Forest chimpanzees. In: *Topics in Primatology*, Vol. **1** (eds T. Nishida, W.C. McGrew, P. Marler, M. Pickford and F. B. M. de Waal). Tokyo: University of Tokyo Press, 81–98.

Wrangham, R.W., *et al.* (1993) Constraints on group size in primates and carnivores: population density and day range as assays of exploitation competition. *Behavioral Ecology and Sociobiology* **32**:199–209.

Wyman, J. (1967) The jackals of the Serengeti. *Animals* **10**:79–83.

Young, A.J. (2003) *Subordinate tactics in cooperative meerkats: breeding, helping and dispersal*. PhD thesis, University of Cambridge, Cambridge.

Zemel, A. and Lubin, Y. (1995) Intergroup competition and stable group sizes. *Animal Behaviour* **50**:485–488.

CHAPTER 3
Female dispersal and philopatry

3.1 Introduction

In the dry mountains of the Granite Range of Nevada, two groups of mustangs graze on dry brush. In one group, a young mare is restless and is repeatedly herded back into the group by the resident stallion. When the group moves off along a winding trail, she lags behind. While the stallion's attention is temporarily distracted, the stallion from the neighbouring group gallops out and places himself between the erring mare and the rest of the group, then turns and herds her back towards his own group. The first stallion notices the manoeuvre and wheels to attack, but is too late and he soon returns to his other mares. In the space of a few minutes, the young mare has made a decision that will affect the rest of her life, leaving the protection of her mother and father for life in a group of unrelated strangers that she has previously encountered only as rivals for patches of good grazing. She treads cautiously as she approaches the resident females and the stallion guards her closely to prevent her returning to her neighbouring group.

Among social mammals, there are marked contrasts in kinship structure between species. While these are partly a consequence of differences in the number of breeding females and the extent of reproductive skew in both sexes (as the last chapter describes), the kinship structure of female groups is also strongly affected by contrasts in female philopatry, dispersal and immigration. In some social mammals, most adult females belonging to the same group are natals (individuals that have been born in the same group) and are related to each other to varying extents and have grown up in close proximity to each other while, in others, most adult females are unrelated immigrants that were born in different groups. Though it is sometimes suggested that phylogenetic history may explain much of this variation, the presence of marked contrasts in female philopatry and dispersal between

closely related species and the flexibility of mammalian dispersal as well as direct evidence of recent evolutionary change in dispersal frequency in other species (Ronce 2007) argue against this conclusion.

The reasons for contrasting patterns in the frequency of female dispersal and the kinship structure of female groups have attracted interest for more than 30 years, but there is still little agreement over the importance of different evolutionary mechanisms (Greenwood 1980; Moore and Ali 1984; Pusey 1987; Perrin and Mazalov 1999; Clobert et al. 2008; Clutton-Brock and Lukas 2011). One contributory factor to this is that philopatry and dispersal are of interest to scientists working in different areas of biology who have contrasting agendas and consequently define dispersal in different ways (Bowler and Benton 2005). Population geneticists and demographers interested in the structure and dynamics of populations have commonly used dispersal to refer to permanent movements out of the natal *population* or deme and philopatry as continued presence and reproduction within it (see Mayr 1963; Shields 1982; Reed 1993; Laporte and Charlesworth 2002). In contrast, behavioural biologists interested in the kinship structure of female groups and the causes of immigration and emigration have commonly defined dispersal as movements out of the natal *group or range* (Greenwood 1980; Waser and Jones 1983; Waser 1996). Since many individuals that leave their natal groups settle in neighbouring groups, the contrasting definitions used by population geneticists and behavioural biologists have confused comparisons and complicated conclusions.

As a result of contrary definitions of dispersal, some studies classify females as philopatric only if they remain and breed in the group or territory where they were born while others classify them on the basis of whether they move further from their natal territory than males, even if all juveniles of both sexes leave their natal group or

Mammal Societies, First Edition. Tim Clutton-Brock.
© 2016 Tim Clutton-Brock. Published 2016 by John Wiley & Sons, Ltd.

territory to breed and yet others classify them on the basis of whether or not they leave their natal deme (Greenwood 1980; Clarke *et al.* 1997). As a result, where classifications are based on sex differences in dispersal distance, species whose females are classified as philopatric commonly include ones where most or all females leave their natal territory to breed but do not move as far as males. Their inclusion may obscure important patterns since the factors affecting whether individuals leave or remain in their natal range or group often differ from those affecting the distance they move if they do leave. In many social species, social factors operating within their natal group (including the intensity of aggression directed at them by resident breeding females, their need for alliances with relatives and the availability of unrelated breeding partners) play an important role in determining whether or not females leave their natal group and which individuals remain and which stay (Clutton-Brock and Lukas 2011). In contrast, the distances moved by individuals once they have left their natal group are likely to be strongly affected by ecological parameters, including the quality and availability of vacant habitat and breeding partners.

A further problem is that dispersal distances are difficult to measure accurately. In many species, emigrating individuals are notoriously difficult to follow, with the result that individuals that move relatively far from their natal territory are likely to be lost and recorded distributions of dispersal distance are likely to be systematically biased towards shorter distances (van Noordwijk 1995; Koenig *et al.* 1996). Where records of dispersal distance are based on a small proportion of the initial sample (as, for example, in most studies of birds based on ringing records) biases may be strong and their magnitude is usually difficult to assess (Koenig *et al.* 1996). It may often be unsafe to assume that biases affect both sexes equally and comparisons of mean dispersal distance may conceal important differences in variance (Sharp *et al.* 2008). For example, in species where one sex commonly remains and breeds in their natal group (and is consequently classified as philopatric), individuals that do disperse are often unable to join established breeding groups and so move further than individuals of the sex that disperses more frequently (Pope 2000). As a result, comparisons of mean dispersal distance can mask important differences in distribution.

The recent development of extensive genetic sampling to assign parentage represents an important development which has provided new insights into dispersal in a number of rodents (Telfer *et al.* 2003; Aars *et al.* 2006; Waser *et al.* 2006) as well as other animals (Christie *et al.* 2011; Planes and Lemer 2011) but its application to larger, more mobile animals where it is difficult to catch or sample large numbers of individuals is still limited by logistic problems. Moreover, the application of genetic or demographic techniques (Prugnolle and De Meeûs 2002; Lawson Handley and Perrin 2007; Abadi *et al.* 2010) does not solve the problem that, unless the behaviour of recognisable individuals can be monitored, it is usually difficult to gain a reliable understanding of the reasons why individuals leave their natal groups or of the factors affecting how far they move. Without studies of the movements of individuals, it is seldom possible to test the assumptions of theoretical models of dispersal so that there is a danger that models do not reflect the biological processes important in natural populations. For example, some recent models of dispersal have been based on the assumption that local resource competition (Clark 1978) favours the evolution of dispersal from groups that include a high proportion of close kin (Hamilton and May 1977; Perrin and Mazalov 2000), whereas the available evidence suggests that the presence of relatives often increases the probability that individuals will remain in their natal group (Hoogland 2013).

As this book is primarily concerned with the evolution of animal societies, I use 'philopatry' to refer to the continued presence of individuals in their natal range or group, 'natal dispersal' to refer to permanent movement out of this area and 'secondary dispersal' to refer to permanent or semi-permanent movements of breeding adults who have already left their natal groups for other groups. I avoid classifying individuals of one sex as philopatric or dispersive on the basis of the relative distance moved since this confounds dispersal from the natal group with distance moved after leaving. Inevitably, attempts to apply a single definition of dispersal generate some ambiguities (Creel and Creel 2002): for example, in some ungulates, both sexes occupy distinct breeding ranges that are separate from those they use throughout the rest of the year (Clutton-Brock *et al.* 1982a; Nelson and Mech 1987); in other species, individuals of one sex leave their natal group but settle in part of their natal territory with immigrants of the opposite sex; and, in some, groups subdivide and occupy part of their original range or parents abandon part of their range to their offspring. There is no single definition that is entirely satisfactory in all circumstances and, in such cases, I treat individuals that have left their natal group and associate with unfamiliar, unrelated animals of the opposite sex as having dispersed.

Though rates of dispersal evidently differ between species, our understanding is limited by the difficulties of measuring dispersal in adequate numbers of marked individuals in natural populations (Shields 1987; Waser *et al.* 1994). Even where it is possible to monitor the life histories of large numbers of individuals, it is usually difficult to be certain whether those that suddenly disappear have died or dispersed (Waser *et al.* 1994; Creel and Creel 2002). In addition, it is seldom possible to monitor the fitness of all dispersers and estimates of the survival or breeding success of individuals that settle within the vicinity of their natal group will not necessarily reflect the average fitness of dispersers. As a result, there are few mammals for which we have reliable estimates of the costs or benefits of dispersal or a satisfactory understanding of the factors affecting dispersal rates.

This chapter initially describes variation in female dispersal and philopatry (section 3.2) before discussing associated costs and benefits (sections 3.3 and 3.4), the evolution of species differences (section 3.5), and the consequences of contrasts for group structure and population demography (section 3.6). Explanations of the evolution of sex differences in philopatry and dispersal distance are reviewed in Chapter 12.

3.2 Variation in female philopatry and dispersal

Different species of mammals show marked contrasts in the frequency of reproductive philopatry and natal dispersal in females as well as in the incidence of secondary dispersal (Pusey 1987; Nunes 2007). In some social mammals, female dispersal is uncommon and most females remain in their natal groups throughout their lives and associate with their mothers, sisters and other matrilineal relatives during their lifespan. For example, in social mole rats and tuco-tucos, many females remain in their natal groups throughout their lives and groups commonly include related females from multiple generations (Faulkes and Bennett 2007; Lacey and Ebensperger 2007). In several of the large herbivores, many females remain in their natal groups and associate with maternal relatives throughout their lives: for example, in both red deer (Figure 3.1) and African elephants (Figure 3.2) females

Figure 3.1 In red deer, most daughters adopt ranges overlapping those of their mother and associate with her and their sisters and their offspring throughout their lives. *Source*: © Tim Clutton-Brock.

Figure 3.2 Female African elephants associate with their mother and sisters throughout their lives, and, as in red deer, matrilineal groups often aggregate in larger herds. *Source*: © Karen McComb.

associate with matrilineal kin throughout their lives, though several kin groups may form larger herds. In many social primates, most females also remain in their natal group throughout their lives (Figure 3.3), though the proportion of group members that are close relatives is not necessarily high and average coefficients of

relatedness between group members can be low (see Chapter 2).

Where females remain in their natal groups and associate with maternal relatives, they are often intolerant of unrelated and unfamiliar females from other groups and female immigration into breeding groups is either rare or

Figure 3.3 In some social mammals, like chacma baboons, females usually remain in their natal group throughout their lives unless their group fissions. Female members of the same group often resemble each other closely. *Source*: © Alecia Carter.

does not occur. Where females that leave their natal group are seldom able to join other established breeding groups and the formation of new breeding groups is uncommon, larger groups may fission (Altmann *et al.* 1985; Hoogland 1995; Thierry 2007; Kerth 2008a). Group fission often appears to be a consequence of increased group size, for group cohesion often declines as group size increases (Henzi *et al.* 1997; Waterman 2002; Lefebvre *et al.* 2003; Archie *et al.* 2006).

Group fission differs from other patterns of dispersal in several respects: it does not necessarily involve the occupation of unfamiliar areas nor is it necessarily associated with marked reductions in survival, though these can arise if one of the new groups is forced to move into peripheral habitat. New groups typically include animals of both sexes and variable ages that are familiar with each other as well as many relatives for, when groups split, maternal kinship commonly affects the membership of different subgroups so that fissioning can raise coefficients of relatedness between group members (Okamoto 2004; Van Horn *et al.* 2007). In species where paternal kin tend to associate, paternal kinship, too, may influence the membership of subgroups. For example, studies of baboons, macaques and gorillas show that, when troops fission, paternal relatives also tend to join the same subgroup (Smith 2000; Widdig *et al.* 2006; Nsubuga *et al.* 2008). Variation in kinship between neighbouring groups may affect the distance moved. For example, in high-density populations of African lions, females are less likely to disperse from prides surrounded by large numbers of unrelated females and, when prides fission, daughters settle relatively close to their mother's pride (VanderWaal *et al.* 2009).

In most species where females remain and breed in their natal group, males disperse after reaching adolescence and attempt to join and breed in other groups (see Chapter 12). However, in a few species, including killer whales and pilot whales, both females and males commonly remain in their natal group and females typically mate with members of other social groups (Amos *et al.* 1993; Baird 2000). In some species, members of both sexes may remain and breed in their natal group: for example, in naked mole rats and banded mongooses, both sexes may remain in their natal group and natal animals may breed with each other (Honeycutt *et al.* 1991a,b; O'Riain and Jarvis 1997; Cant *et al.* 2013) though females still show a preference for unrelated males (Cisek 2000; Sanderson *et al.* 2015).

At the other extreme, there are a number of social mammals where most females leave their natal group or territory as adolescents or young adults and, if they survive, breed in other territories or groups, so that resident breeding females are usually immigrants born in different groups or ranges from the one they breed in, and few are close kin. Examples include most social equids (Figure 3.4), spider monkeys and muriquis (Figure 3.5), red colobus monkeys, hamadryas baboons, chimpanzees, mountain gorillas (Figure 3.6) and some tropical bats (Wrangham 1980; Berger 1987; Pusey 1987; Strier and Ziegler 2000; Korstjens and Schippers 2003; Hammond *et al.* 2006; Di Fiore and Campbell 2007; Nagy *et al.* 2007; Stumpf 2007; Kerth *et al.* 2011). In many of these species, social bonds between breeding females are weak, groups often disband if the resident male dies or is displaced and secondary dispersal of females can be common.

Between these extremes lie a variety of species where some females breed in their natal groups and others disperse to breed. In some species, many females leave as adolescents or young adults but a proportion remain and breed in their natal group: for example, in meerkats, dominant females evict all subordinates above a threshold age, so that groups include one mature breeding female who has usually been born in the group and several younger subordinates (Clutton-Brock *et al.* 2006) (Figure 3.7). In these systems, too, dispersing females are often prevented from joining established breeding groups by residents so that, unless they can displace resident females, dispersers either establish new groups or die: examples include African lions, meerkats, golden lion tamarins, ursine howler monkeys and black-tailed prairie dogs (Pusey and Packer 1987a; Crockett and Pope 1993; Hoogland 1995; Sillero-Zubiri *et al.* 1996; Creel and Creel 2002; Clutton-Brock *et al.* 2006; VanderWaal *et al.* 2009). In other cases, dispersing females are sometimes able to join established groups, so that breeding groups may include a mixture of related and unrelated females, as in some Asian colobine monkeys (Kirkpatrick 2007) and some ungulates where related females forage independently but share overlapping home ranges (Clutton-Brock *et al.* 1982a). Finally, in many of the more wide ranging semi-migratory or migratory species, including some bats and several of the larger herbivores, bonds between mature females are weak or absent and adolescent females disperse into local populations after separating from their mothers (Estes 1974; Clutton-Brock

Figure 3.4 Breeding groups of Przewalski's horses in the Gobi desert. Both in horses and in plains zebra, females leave their natal breeding groups after reaching sexual maturity and join other groups, with the result that average levels of kinship between females from the same group are low. *Source*: © Claudia Feh.

et al. 2002; Kerth 2008b). However, if populations are geographically divided or opportunities for dispersal are constrained, related females may still encounter each other more frequently than unrelated females (Pratt and Anderson 1985; Bashaw *et al.* 2007; Bradley *et al.* 2007; Kerth 2008a,b).

The proximate causes of female dispersal also vary. In some species, adolescent females or young adults are

Figure 3.5 A female muriqui with her infant. Muriquis live in multi-male groups: females mate with multiple partners, co-resident males are tolerant of each other and there is little sex difference in body size. *Source*: © Thiago Cavalcante Ferreira.

Figure 3.6 Most female mountain gorillas disperse from their natal group soon after reaching adolescence. *Source*: © Alexander Harcourt/Kelly Stewart/Anthrophoto.

evicted from their natal group by adult females who are usually close relatives: for example, in meerkats (Clutton-Brock *et al.* 1998). In others, including many rodents and some primates, dispersing females appear to leave their group voluntarily (Bekoff 1977; Moore and Ali 1984; Wolff 1993; Nunes 2007) or are kidnapped by males from neighbouring groups (Berger 1986; Swedell

et al. 2011). Within species, the relative frequency of dispersal commonly increases with group size (Pope 2000; Clutton-Brock *et al.* 2008; VanderWaal *et al.* 2009) and with local population density (Matthysen 2005). The frequency of dispersal is commonly reduced by experimental reductions in density (Brody and Armitage 1985; Aars and Ims 2000) and is increased

(a)   (b)

Figure 3.7 Female meerkats are evicted from their natal group by the dominant when they are 2–4 years old: (a) evicted female; (b) evicted female showing a large scar at the base of the tail, caused by repeated biting by the dominant female. *Sources*: (a) © Arpat Ozgul; (b) © Gabriele Cozzi.

by the creation of vacant territories nearby: for example, the removal of neighbouring territory holders increases rates of dispersal in female North American red squirrels (Boutin *et al.* 1993).

In many mammals, social dispersal appears to represent a conditional strategy and both the probability that individuals will leave their natal group and the timing of leaving are affected by their relative size, condition and development and, in some cases, also by their genotype (Bowler and Benton 2005; Nunes 2007; Ronce 2007; Clobert *et al.* 2008). For example, female meadow voles that are malnourished during early development but are well nourished as adults show a stronger tendency to disperse (Bondrup-Nielsen 1993). In some cases, smaller and more subordinate individuals are more likely to disperse (Hanski *et al.* 1991; Bowler and Benton 2005) while, in others, it is larger or heavier individuals that are more likely to leave (Nunes *et al.* 1998; Clutton-Brock *et al.* 2002).

Rates of female dispersal and dispersal distances are also commonly associated with variation in hormone levels: for example, in ground squirrels, experimental increases in androgen at the time of birth raise the tendency for females to disperse as adolescents or young adults (Holekamp 1984; Nunes *et al.* 1999). The availability of vacant territories may also be important. For example, studies of several group-living birds and mammals show that the frequency of dispersal increases with the availability of vacant territories or breeding opportunities outside the group (Dobson 1981; Brody and Armitage 1985; Komdeur *et al.* 1995; Ekman *et al.* 2004). Conversely, the probability that individuals will remain in their natal group may be affected by chance of breeding there. For example, as mothers age and their probability of survival falls, the relative benefits of philopatry to their daughters are likely to increase and, in some animals, the offspring of older mothers are less likely to disperse (Ronce *et al.* 1998).

The size of dispersing groups also differs between species. Where females occupy separate ranges or territories, individuals usually disperse alone and independent dispersal is also common in species where females are readily accepted into established breeding groups. In contrast, where dispersing females cannot join established breeding groups, several females often disperse together: for example, in meerkats, several females are commonly evicted at the same time and often disperse together (Clutton-Brock *et al.* 1998). Similarly, in banded mongooses, established breeding females evict multiple younger females in sporadic eviction events and younger females disperse together and either die or join established breeding groups or found new breeding units (Gilchrist 2008; Cant *et al.* 2013). Where females disperse together, new breeding units initially consist of a mixture of descendant and non-descendant relatives, but dominant females may subsequently evict non-descendant relatives, so that established groups eventually consist of a dominant female and her descendant kin (Young *et al.* 2006; Clutton-Brock *et al.* 2008).

The age at which females disperse differs widely between species and these differences can have an important influence on the average size of groups. In plural breeders where female dispersal usually involves groups fissioning, splinter groups commonly involve old as well as young females and dispersal may occur in females of any age (Okamoto 2004). In contrast, where breeding females occupy separate ranges or territories, subordinates usually leave (or are evicted) soon after reaching sexual maturity (Nunes 2007). In other species, females may remain in their natal group as non-breeders for several years before they leave or are evicted: for example, in meerkats, females reach sexual maturity at a year but are seldom evicted from their natal group until they are at least 2 years old, though few are tolerated after they are 4 years old (Clutton-Brock *et al.* 2008).

The distances moved by dispersing females range from less than 100 m in some rodents (Gaines and McClenaghan 1980; Jones 1987; Stenseth and Lidicker 1992), through distances of a few hundred metres in the case of the more sedentary primates (Pope 2000) to 20 km or more in some carnivores (Creel and Creel 2002) and much larger distances in some cetaceans (Whitehead and Weilgart 2000). Intraspecific variation in the distance moved may be affected by the kinship structure of local groups. For example, in high-density populations of African lions, females are less likely to disperse from prides surrounded by large numbers of unrelated females and, if prides fission, daughters settle relatively close to their mothers (VanderWaal *et al.* 2009). Interspecific differences in the distances moved reflect variation in relative mobility (Sutherland *et al.* 2000) as well as differences in the availability of vacant habitat.

3.3 Benefits of philopatry

Survival

Dispersal from the natal group often has high costs unless dispersing females can join established breeding groups rapidly (Berger 1987; Van Vuren and Armitage 1994; Nunes 2007; Ronce 2007; Doligez and Pärt 2008; Strier 2008). In many species, dispersing individuals lack detailed knowledge of the distribution of resources accrued by older residents and their feeding efficiency may be impaired: for example, studies of African elephants feeding in a novel environment show that foraging efficiency is initially low but increases as they get to know the area (Pinter-Wollman *et al.* 2009). Dispersing meerkats experience increased levels of glucocorticoids and their pituitary responsiveness falls (Figure 3.8): they show lower daily weight gain and they suffer higher parasite loads (Young *et al.* 2006; Young and Monfort 2009). As well as affecting survival, the energetic costs of dispersal may delay breeding and reduce reproductive potential (Ronce 2007; Fisher *et al.* 2009).

Individuals that disperse to unfamiliar areas may also be more vulnerable to predators: for example, experimental studies show that dispersing white-footed mice are more susceptible than residents to predation by owls (Metzgar 1967). In addition, dispersers are often more likely to be attacked by members of resident groups, sometimes with fatal consequences (Fritts and Mech 1981; Packer and Pusey 1982; Boydston *et al.* 2001; Creel and Creel 2002). High rates of mortality in dispersers may be particularly common in carnivores, where attacks by residents are often dangerous (Waser 1996): for example, grey wolves making extra-territorial forays die at five times the rate of residents (Messier 1985) while mortality rates in dispersing African wild dogs are also substantially higher than those of residents (Creel and Creel 2002) (Figure 3.9). However, dispersal is also associated with substantial increases in mortality in many herbivores (Errington 1963; Van Vuren and Armitage 1994): for example, in ursine howler monkeys, over 40% of dispersing females are suspected or known to die (Crockett and Pope 1993).

Reproductive success

As well as suffering the direct costs of moving between groups, dispersing females lose the potential benefits of associating with kin and this may have an important

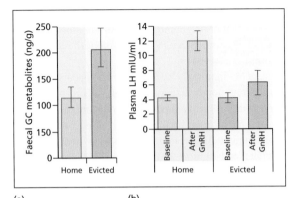

Figure 3.8 Effects of eviction from their natal group on adrenal and reproductive physiology in female meerkats: (a) evicted females show significantly elevated levels of glucocorticoid (GC) metabolites in their faecal samples (predicted means after controlling for significant diel variation); (b) evicted females experience downregulation of their pituitary responsiveness to an exogenous GnRH challenge. Bars represent means ± 1 SE. *Source*: From Young *et al.* (2006) © 2006 National Academy of Sciences, USA. Reproduced with permission. *Photo source*: © Arpat Ozgul.

influence on the probability that they will disperse (Lambin *et al.* 2001; Perrin and Goudet 2001; Silk 2007). Some of the clearest evidence comes from studies of rodents where females occupy individual home ranges that overlap those of neighbouring females and their breeding success or survival is affected by their proximity to kin. For example, in some voles, females show a preference for settling close to relatives and individuals with ranges close to kin breed earlier (Pusenius *et al.* 1998), rear more offspring (Lambin and Yoccoz 1998) and show higher rates of survival in the next breeding season (Lambin and Krebs 1993) than individuals with

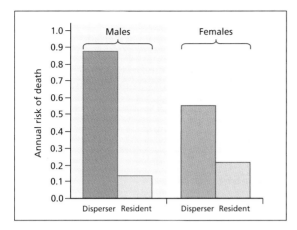

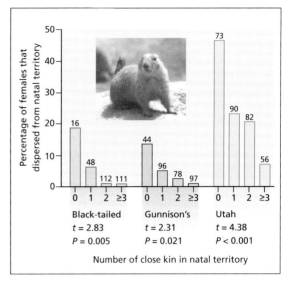

Figure 3.10 Percentage of females that dispersed from groups that contained different numbers of close kin (mother, littermate sister or littermate brother) in the natal territory for three species of prairie dogs. *Source*: From Hoogland (2013). Reproduced with permission from AAAS.

Figure 3.9 Mortality rates in dispersing African wild dogs in the Selous game reserve are substantially higher than in residents: estimated annual risk of death in male and female dispersers compared with residents of the same age. Dispersing dogs may be killed by other predators or by resident packs. *Source*: From Creel and Creel (2002). Reproduced with permission of Scott Creel. *Photo source*: © Krys Jordan.

ranges close to non-kin. Similarly, in alpine marmots, infants are more likely to survive their first winter in hibernation groups consisting largely of close relatives than in groups where most individuals are not closely related (Arnold 1990a,b) while the breeding success of dominant females is depressed by the number of unrelated subordinate females in the group but not by the number of daughters present (Hackländer *et al*. 2003). And, in prairie dogs, where related females cooperate and compete with each other, females without close kin in the group lack support in interactions with

other group members and the proportion that disperse is inversely related to the number of close relatives in the group (Figure 3.10).

There appear to be similar effects of associating with kin in other social mammals, though the evidence is less complete. For example, in grey seals, females that breed in areas where individuals are closely related produce larger and faster-growing pups than females in areas where relatedness is relatively low (Pomeroy *et al*. 2000, 2001). In ursine howler monkeys, females that recruit into their natal group and associate with relatives give birth at earlier ages and have higher breeding success than individuals that disperse and breed in groups consisting largely or exclusively of unrelated females (Crockett and Pope 1993). And, in white-faced capuchins, the lengths of inter-birth intervals are negatively related to the number of matrilineal kin in their group (Fedigan *et al*. 2008).

Where comparisons of breeding success or survival between philopatric individuals and dispersers are based solely on observational data, there is a danger that consistent differences in survival rates in successive generations by local variation in habitat quality produce correlations between breeding success and association with kin that have no causal basis. However, several rodent studies have either manipulated the proportion of relatives in local populations or housed animals with kin versus non-kin and have shown that the presence of kin affects breeding success or survival. For example, experiments with voles show that associating with kin increases fecundity and, in some cases, the survival of young, too (Kawata 1990; Sera and Gaines 1994), though it does not always do so (Boonstra and Hogg 1988; Sera and Gaines 1994; Dalton 2000). Similar effects of the presence of kin have been demonstrated in house mice (König 1994; Dobson et al. 2000; Krackow et al. 2003; Rusu and Krackow 2004): in this case, females living with a familiar sister had a higher chance of breeding successfully than females living with an unfamiliar relative, a familiar unrelated individual or an unfamiliar unrelated one (Figure 3.11). In prairie voles (where communal breeding is common) the presence of additional breeding sisters is associated with reduced weight loss in mothers if food is scarce as well as with improved growth in their pups (Hayes 2000). However, communal breeding is not always associated with improved breeding success: for example, in populations of white-footed mice, some females form communal breeding groups with relatives where population density is relatively high while others breed alone (Wolff 1994) but, in contrast to prairie voles and house mice, there is no obvious effect of communal breeding on reproductive success (Wolff 1994).

There are several different reasons why associating with kin may lead to improved fecundity or breeding success. In some rodents, relatives are more tolerant of each other's presence than unrelated females, rates of aggression between relatives are lower than between unrelated females and home range overlap between relatives is often greater (Sera and Gaines 1994; Mappes et al. 1995). For example, the reproductive success of female voles in enclosures seeded with unrelated females is negatively correlated with their proximity to neighbours but this effect disappears when kin are involved (Mappes et al. 1995; Pusenius et al. 1998). Relatives may also be more tolerant of each other's offspring and the risk of female infanticide may decline where neighbours are close relatives (Lambin and Yoccoz 1998; Dalton 2000), though this does not appear to be the case in all species (Hoogland 1995). In addition, related females may assist each other in defending resources or repelling intruders and younger individuals may benefit from the presence of older

relatives (Silk 2007): for example, in wood rats, females whose mothers are present in their group produce more offspring than those whose mothers are absent (Moses and Millar 1994) and similar effects occur in some primates (Fairbanks and McGuire 1987; Pavelka et al. 2002; Cheney et al. 2004).

3.4 Benefits of dispersing

Competition avoidance

The substantial benefits of philopatry and the high costs that are often associated with dispersing suggest that dispersing females must gain large benefits by leaving their natal range or group. In some cases, dispersal may generate ecological benefits by allowing females to escape from groups or ranges where there is intense competition for resources or breeding opportunities with other individuals and they are unlikely to breed successfully (Perrin and Mazalov 2000; Matthysen 2005; Clobert et al. 2008). In many mammals, subordinate females suffer from competition for resources or breeding opportunities with other group members, leading to reductions in their fecundity and the survival of their offspring, so that dispersing may increase their fitness (Clutton-Brock et al. 1982b, 2006; Hoogland 1995).

Dispersal may also allow females to avoid reproductive competition. Where females defend individual territories, young adult females often become the target of aggression from the resident breeding female (who is often their mother) and are eventually forced to leave (Tilson 1981; Komers and Brotherton 1997; Nunes 2007). Similarly, in some singular cooperative breeders (including meerkats, African wild dogs and tamarins) older subordinate females become the target of aggression from dominant females and eventually leave the group or are evicted (Dietz and Baker 1993; Clutton-Brock et al. 1998; Creel and Creel 2002; Goldizen 2003).

There is considerable evidence that the frequency of dispersal varies in relation to changes in the intensity of competition. In several species, the probability that subordinates will be evicted rises with group size (Clutton-Brock et al. 2008; Cant et al. 2010) and increases at times when dominants or subordinates are attempting to breed (Ebensperger 1998a,b; Clutton-Brock et al. 1998; Creel and Creel 2002). Where younger females appear to leave voluntarily to seek breeding opportunities in other

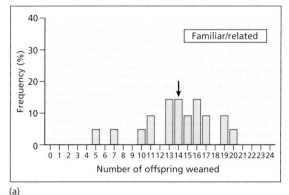

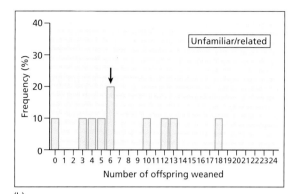

(a)   (b)

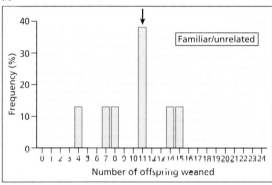

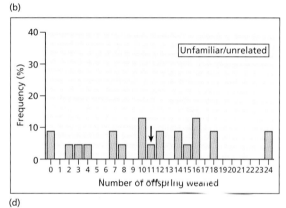

(c)   (d)

Figure 3.11 Effects of the presence of familiar kin on lifetime reproductive success in captive house mice. The four parts of this figure show the distribution of lifetime reproductive success in artificially constructed groups consisting of (a) two familiar related females, (b) two unfamiliar related females, (c) two familiar unrelated females and (d) two unfamiliar unrelated females. Source: (a–d) From König (1994). Reproduced with permission of Blackwell. Photo © Nicolas Perony.

groups, dispersal rates increase if groups exceed the group size that is optimal for their members (Komers and Brotherton 1997; Brockelman *et al.* 1998; Goldizen 2003; McGuire *et al.* 2007). For example, in African lions, female dispersal increases in groups that exceed the size that is optimal for the habitat they occupy (VanderWaal *et al.* 2009). In addition, a number of studies have manipulated the local distribution of resources or competitors and shown that this affects patterns of dispersal: for example, in California ground squirrels, supplementation of local food resources increases rates of emigration from un-supplemented to supplemented areas (Dobson 1979) while in pocket gophers (Sullivan *et al.* 2001), yellow-bellied marmots (Brody and Armitage 1985) and California ground squirrels (Dobson 1981) the experimental removal of individuals or groups has been shown to increase immigration from neighbouring areas. In several social rodents, the probability that females will disperse increases with population density (Lidicker and Patton 1987; Nunes 2007). Increased rates of dispersal are often stimulated by short-term changes in resource abundance (Nunes 2007) and dispersing females frequently move to areas of higher food abundance: for example, in Eurasian red squirrels, dispersing females move to areas where food is relatively abundant (Lurz *et al.* 1997). In some cases, they may also disperse to reduce the chance that their dependent young will be killed by predators (see Chapter 5) or may anticipate the risk of take-over and male infanticide by leaving groups which are not defended effectively by the resident male (Steenbeck 2000; Stokes *et al.* 2003).

While competition for resources or breeding opportunities often leads to increased frequency of dispersal, this is not always so in species where females aggregate with kin. For example, in yellow baboons and red deer, females normally remain in their natal group or range, even if there is strong competition for resources (Alberts and Altmann 1995; Conradt *et al.* 1999) while in some rodents the saturation of local environments is associated with reductions in dispersal and increased retention of daughters (and, in some cases, sons) in the natal range (Gundersen and Andreassen 1998; Lambin *et al.* 2001). In brown bears, too, high population density is associated with reductions in the dispersal distance of females (Støen *et al.* 2006). In these cases the costs of high levels of competition may not be sufficient to offset the costs of dispersing.

Although most functional explanations of female dispersal suggest that females gain direct fitness benefits by dispersing, dispersing females might also gain indirect fitness benefits by reducing the costs of their presence to their kin (Hamilton and May 1977; Gandon 1999; Perrin and Mazalov 2000). However, there is little evidence that the presence of maternal kin stimulates voluntary dispersal after the effects of local density or group size on dispersal frequency have been allowed for (Wolff 1992; Lambin *et al.* 2001). For example, in root voles, adolescent females exposed either to their mother or to an unrelated breeding female were no more likely to show reproductive suppression or to attempt to disperse in the presence of their mother than in the presence of an unrelated female (Le Galliard *et al.* 2007). In some species, the presence of relatives reduces rather than increases the likelihood that females will disperse: for example, in yellow-bellied marmots, the presence of a female's mother reduces the probability that she will disperse (Armitage *et al.* 2011), while, in prairie dogs, the absence of close kin of any kind increases the probability that young females will leave (Hoogland 2013). This is probably because mothers and other relatives provide support that enhances the ability of younger females to compete with other colony members and their absence increases the net benefits of dispersal.

Inbreeding avoidance
Inbreeding costs
While female dispersal is often associated with competition for resources or breeding opportunities, this does not appear to be so in all cases. For example, in wild horses and mountain gorillas, almost all females emigrate from their natal group, even if population density is relatively low and resources are abundant (Berger 1986; Harcourt and Stewart 2007). Examples of habitual female dispersal of this kind suggest that some mechanism other than the avoidance of competition for resources or breeding opportunities must be capable of generating substantial benefits to dispersal in females.

One possible explanation is that dispersal allows females to gain access to unrelated males and to avoid inbreeding with close relatives (Bengtsson 1978; Greenwood 1980; Pusey 1987; Koenig and Haydock 2004). There is extensive evidence that, in outbred populations, breeding with close relatives reduces the fitness of progeny (Morton *et al.* 1956; Charlesworth and Charlesworth 1987; Keller and Waller 2002). The approximate magnitude of inbreeding costs can be predicted from estimates of *genetic load*: the predicted number of 'lethal gene equivalents' (n) carried in recessive form by the average individual (May 1979; Keller 1998), calculated by summing the average probability of occurrence multiplied by the decrease in fitness they produce in a homozygous state for all deleterious recessive genes. Estimates of genetic load for human populations suggest that n is around 2.2 (Schull and Neel 1965; Bodmer and Cavalli-Sforza 1976), indicating that as many as 42% of offspring born to parent–offspring or sister–brother ($r = 0.5$) matings might be expected to die as a consequence of inbreeding depression, while 24% of those born after matings where $r = 0.25$ might be expected to die and 13% where $r = 0.125$ (May 1979).

Empirical estimates of the effects of close inbreeding are now available for a range of mammals either in captivity or in the wild: ungulates (Ralls *et al.* 1979; Slate *et al.* 2000; Marshall *et al.* 2002; Walling *et al.* 2011); rodents (Hill 1974; Haigh 1983; Pugh and Tamarin 1988; Keane 1990; Reeve *et al.* 1990; Krackow and Matuschak 1991; Hoogland 1992); insectivores (Stockley *et al.* 1993); carnivores (Packer and Pusey 1983); primates (Bulger and Hamilton 1988; Dietz and Baker 1993; Alberts and Altmann 1995); and a number of other species (Pusey and Wolf 1996; Frankham *et al.* 2002; Keller and Waller 2002). Most studies show a substantial reduction in the fitness of offspring resulting from matings between close relatives, indicating that individuals are likely to maximise their fitness by avoiding breeding with close relatives. For example, an analysis of data for zoo-bred primates found that inbreeding between known relatives of any level of kinship was associated with a 75% reduction in the number of offspring produced and a 30% increase in juvenile mortality levels (Harcourt and Stewart 2007). Several studies of natural populations have produced similar results (Keller and Waller 2002): for example, research on wild red deer shows that matings between close relatives have a substantial effect on the growth and survival of calves (Walling *et al.* 2011) (Figure 3.12).

The causes of reductions in fitness vary but can include reduced growth and survival, later age at first breeding,

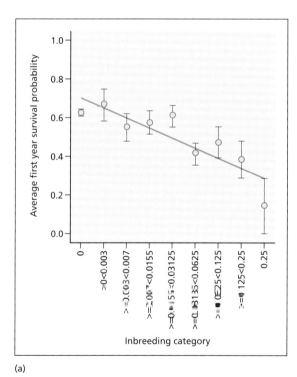

(a)

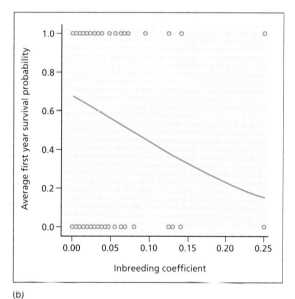

(b)

Figure 3.12 Costs of inbreeding in wild red deer showing the association between the probability of first-year survival in calves and their inbreeding coefficients, with bars representing standard errors. The left-hand plot shows averages of raw data while the right-hand plot shows values of first-year survival predicted by a model including birthweight (solid line) and actual data (open circles). *Source*: From Walling *et al.* (2011). Used under CC BY-2.0. https://creativecommons.org/licenses/by/2.0/uk/. *Photo source*: © Martyn Baker.

lower fecundity, reduced survival of offspring and shorter lifespans. In some cases, increasing susceptibility to disease or parasites may also contribute to these effects. For example, relatively inbred California sea lions show higher frequencies of carcinomas and of helminth infections than less inbred individuals and are less responsive to veterinary treatment (Acevedo-White-house *et al.* 2003). Similarly, Soay sheep showing relatively low levels of heterozygosity show higher loads of intestinal parasites known to affect growth and lower survival than more heterozygous individuals (Coltman *et al.* 1999) with the result that survivors showed significantly higher levels of heterozygosity than those that died (Figure 3.13). Experiments in which some sheep were treated with anthelmintic drugs before the onset of winter showed that, in treated sheep, there was no consistent difference in heterozygosity between individuals that died and those that survived, supporting the suggestion that heterozygosity affects survival through its effects on parasite load.

Although inbreeding is often associated with reduced offspring fitness, the magnitude of effects is variable and some populations show little or no evidence of inbreeding depression. In addition, a substantial number of studies have found no consistent relationship between individual variation in heterozygosity and fitness (Chapman *et al.* 2009). One reason for this may be that past inbreeding has 'purged' populations of deleterious recessive genes (Barrett and Charlesworth 1991; Fowler and Whitlock 1999). For example, inbreeding depression may be slight or absent in black-tailed prairie dogs because a recent population bottleneck led to the purging of deleterious recessives (Hoogland 1992). Similarly, past bottlenecks may explain why inbreeding depression is weak or absent in some populations of dwarf mongooses (Keane *et al.* 1996; Brzeski *et al.* 2014).

Given the reasons for expecting close inbreeding to have negative consequences, evidence of the apparent absence of disadvantages of close inbreeding should be treated with caution. The effects of inbreeding may only be detectable in individuals (or populations) living in challenging environments and comparisons show that moderate to high levels of inbreeding depression are more commonly found in wild populations than in captive animals (Crnokrak and Roff 1999). For example, experiments in which inbred and outbred white-footed mice were reared experimentally and then either maintained in captivity on ad lib food or released into natural populations showed that while the survival of inbred and outbred mice was similar in captivity, in wild populations the survival of inbred mice fell to 56% of that of outbred individuals (Jimenez *et al.* 1994) (Figure 3.14). Similarly, in wild red deer, inbreeding has stronger effects in years when April temperature and food availability are low than in years where they are high and reduces the birthweight of calves born to young and old mothers to a greater extent than those born to mothers in their prime (Coulson *et al.* 1998).

A further reason why the costs of inbreeding may often appear to be lower than they really are is that inbreeding may only affect the survival or breeding success of offspring when they reach adulthood, so that estimates of inbreeding costs based on the number of offspring born or the number that survive to independence are likely to underestimate the real costs of inbreeding (Pusey and Wolf 1996). For example, studies of great tits show that differences in the number of grand-offspring between inbred and outbred individuals are substantially larger than differences in the number of fledglings raised (Szulkin *et al.* 2007). In addition, few studies have yet been able to explore the effects of inbreeding on rates of ageing or longevity in offspring resulting from inbred matings, and it would not be surprising if these prove to be substantial.

Conversely, it is important to remember that, even where inbreeding has substantial costs, it may not always be best for individuals to outbreed since the risks of dispersal are often high. Where the costs of inbreeding avoidance are low relative to the costs of inbreeding, females would be expected to avoid mating with close relatives but where the costs of inbreeding avoidance are unusually high, inbreeding may be favoured (Bengtsson 1978; May 1979; Waser *et al.* 1986; Szulkin *et al.* 2013). For example, the relatively high incidence of inbreeding in naked mole rats may be caused by the difficulties and dangers of dispersing for a subterranean rodent living in an arid environment (Bennett and Faulkes 2000). Outbreeding can also have indirect costs, including the disruption of co-adapted gene complexes or the loss of local genetic adaptations (Lynch 1991; Waser 1993) and, in some cases, individuals may maximise their fitness by avoiding mating either with close relatives or with individuals very dissimilar to themselves (Bateson 1978; Frankham *et al.* 2011; Cohas *et al.* 2008). However, there is limited evidence that effects of this kind are important in natural populations unless interspecific hybridisation

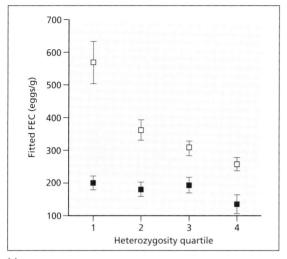

(a)

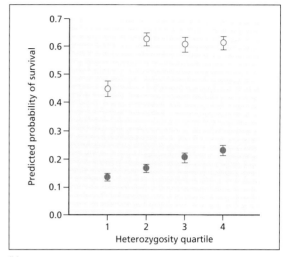

(b)

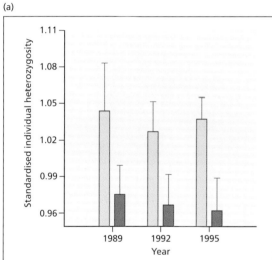

(c)

(d)

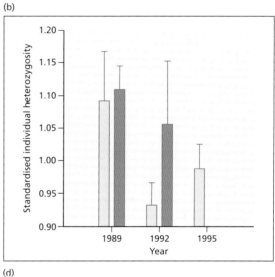

Figure 3.13 Inbreeding, parasite load and survival in Soay sheep: (a) faecal egg count (FEC) of adult sheep in relation to their relative heterozygosity for years of high population density (open squares) and years of low population density (filled squares); (b) overwinter survival probabilities predicted by a generalised linear model of juvenile (filled circles) and adult (open circles) Soay sheep in relation to their relative heterozygosity in three years of high winter mortality; (c) mean standardised heterozygosity of untreated Soay sheep that either died (filled columns) or survived (open columns) in years of high overwinter mortality. Bars represent one standard error of the mean while numbers above each column indicate sample size; (d) mean standardised heterozygosity for Soay sheep treated with anthelmintics that either died (filled columns) or survived (open columns) in years of high overwinter mortality. Bars represent one standard error of the mean while numbers above each column indicate sample size. *Source*: From Coltman *et al.* (1999). Reproduced with permission of John Wiley & Sons. *Photo source*: © Arpat Ozgul.

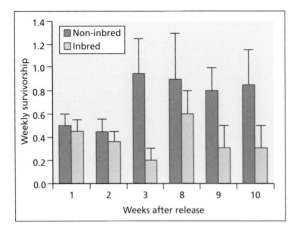

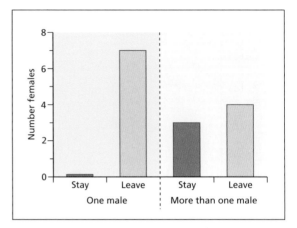

Figure 3.14 Survivorship of inbred and non-inbred deer mice when introduced to a wild population. *Source*: From Jimenez *et al.* (1994). Reproduced with permission from AAAS.

is occurring or dispersal is very limited (Templeton 1986; Waser 1993; Pusey and Wolf 1996).

Mechanisms of inbreeding avoidance

Several different kinds of evidence support the suggestion that inbreeding avoidance plays an important role in the evolution of female dispersal. Where females mature in groups that include their fathers or brothers, they commonly leave the group: for example, in dwarf mongooses and African wild dogs, adolescent females are more likely to disperse from their natal group if their father is still present in the group than if he has emigrated or is dead (Rood 1987; McNutt 1996). In mountain gorillas, females that mature in groups where the only breeding male is their father usually disperse before breeding, while those maturing in groups where there is a second breeding male commonly breed at least once in their natal group (Harcourt and Stewart 2007) (Figure 3.15). In some rodents, experimental studies confirm that the presence of closely related males can stimulate female dispersal (Bollinger *et al.* 1993) although the presence of unfamiliar males in neighbouring territories can also stimulate dispersal in females (McGuire and Getz 1991).

In species where females commonly reach sexual maturity when their father or brothers are still actively breeding in their group, they frequently avoid mating with them (see Chapter 40). In addition, the presence of closely related males in the same group can delay sexual development or conception in females. For example, female

Figure 3.15 Inbreeding avoidance in mountain gorillas. When female gorillas reach breeding age, they leave groups that contain a single mature male that is often their father. In contrast, a substantial proportion of females remain and breed in their natal group in groups that include at least one additional male that was not the dominant male at the time they were immature. *Source*: From Harcourt and Stewart (2007). Reproduced with permission of the University of Chicago Press. *Photo source*: © Alexander Harcourt/Kelly Stewart/Anthrophoto.

white-footed mice whose fathers are still resident in their territory are less likely to become sexually mature before dispersing than those whose mothers are present while those whose mothers are present are no less likely to mature than those lacking both parents (Figure 3.16). Experimental removal of fathers causes daughters to remain longer in their natal territory while the presence of mothers in the natal territory has similar effects on sons.

The relative importance of the different benefits of dispersal is still widely debated (Waser and Jones 1983; Bowler and Benton 2005; Ronce 2007; Clobert *et al.* 2008; Szulkin *et al.* 2013). While there is general

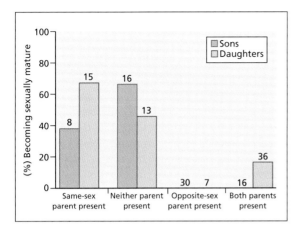

Figure 3.16 The presence of opposite-sex parents suppresses sexual development and stimulates dispersal in juvenile white-footed mice. The figure shows the percentage of sons and daughters that became sexually mature in the presence or absence of their mothers and fathers in their home range. *Source*: From Wolff (1993). Reproduced with permission of Macmillan Publishing Ltd.

agreement that females commonly disperse to avoid competition or persecution, there is still disagreement about the importance of inbreeding avoidance as an ultimate cause of female dispersal. Although many argue that it plays an important role (Pusey 1987; Clutton-Brock 1989; Roze and Rousset 2005; Nunes 2007; Szulkin and Sheldon 2008), some are more sceptical (Moore and Ali 1984; Guillaume and Perrin 2006; Clobert *et al.* 2008), and some theoretical models suggest that inbreeding avoidance is unlikely on its own to account for the evolution of dispersal, though it may modify other mechanisms favouring dispersal (Perrin and Mazalov 1999; Perrin and Goudet 2001).

Disagreements over the importance of inbreeding avoidance in stimulating dispersal often reflect variation in definitions of dispersal and philopatry (see section 3.1). Many reviews of variation in female philopatry by behavioural ecologists focus on the avoidance of mating with first-order relatives in species that live in stable social groups and conclude that the risk of close inbreeding is a common cause of female dispersal (Pusey 1987; Clutton-Brock 1989; Clutton-Brock and Lukas 2011). In contrast, analyses of dispersal distance by evolutionary ecologists are often sceptical of the role of inbreeding avoidance and emphasise the importance of ecological parameters in affecting how far individuals move (Clobert *et al.* 2008). Both may well be correct, and

inbreeding avoidance may often play an important role in stimulating the dispersal of females from their natal groups in social species, while ecological and demographic factors may determine how far they move after leaving (Wolff 1992; Szulkin and Sheldon 2008).

Despite the costs of inbreeding, several studies have found little evidence that dispersers avoid joining relatives or mating with kin (Keller and Arcese 1998; Daniels and Walters 2000; Szulkin and Sheldon 2008). For example, studies of reindeer and red deer have found no evidence that oestrous females avoid related males (Holand *et al.* 2007; Stopher *et al.* 2011). There may be several reasons why dispersing individuals seldom avoid breeding with relatives. Where individuals rely on familiar olfactory cues, continued recognition of relatives that have left their natal group and now belong to groups that include unrelated individuals may be difficult to maintain. Moreover, where the risks associated with dispersal are high, strong selection pressures are likely to favour rapid settlement and breeding. In relatively mobile species, the chances that dispersers will encounter close relatives may consequently be low and the distance moved by dispersing females may depend primarily on the availability of resources, vacant habitat and breeding opportunities, though this may not be the case in relatively sedentary species and research on some small mammals has found that dispersing females that have left their home range or group continue to avoid mating with familiar kin (Dewsbury 1988; Ferkin 1990; Pusey and Wolf 1996; Le Galliard *et al.* 2007). One possible explanation is that these differences are a consequence of interspecific contrasts in mobility: in more mobile species, the chance that dispersing individuals will encounter close relatives as potential breeding partners may be low, while in relatively sedentary species where individuals live in small discontinuous populations, the chance that dispersing females will do so may be substantially higher, generating selection for continued discrimination.

3.5 Species differences in female philopatry

Arguments concerning the role of inbreeding in stimulating female dispersal have focused attention on the evolution of species differences in philopatry and dispersal. Evidence that female philopatry is common and dispersal often has substantial costs suggests that remaining on the natal range or territory is usually the optimal strategy for females. Explaining the evolution of habitual female dispersal in singular breeders presents little difficulty since subordinate females either leave voluntarily to find opportunities to breed or are evicted by the dominant female. In contrast, the evolution of female dispersal in plural breeders where females habitually disperse at adolescence is less easy to explain. Although it is sometimes suggested that phylogenetic history may explain much of this variation (Perrin and Mazalov 1999), the presence of marked contrasts in dispersal between closely related species, the flexibility of mammalian dispersal and evidence of recent evolutionary changes in dispersal in other species argue against this conclusion (Clutton-Brock and Lukas 2011).

While several ecological mechanisms can lead to differences in dispersal, two main explanations for the evolution of habitual female dispersal in plural breeders have been suggested. One, originally suggested by Wrangham (1980) for primates, and later developed by other primatologists (van Schaik 1989; Sterck et al. 1997; Koenig 2002; Isbell 2004), argues that habitual female dispersal occurs in species where the value of individual food items is low and contest competition within groups is either rare or has little effect on fitness. Under these conditions (it is suggested), linear hierarchies and supportive coalitions may be unlikely to develop, the benefits of associating with kin may be low and females may commonly disperse to avoid competing for resources with relatives. Although this explanation of contrasts in dispersal was initially widely accepted, recent reviews have stressed its shortcomings (Janson and van Schaik 2000; Koenig and Borries 2006; Thierry 2008; Clutton-Brock and Janson 2012). In particular, it is not clear that competition for resources is less frequent or less intense among primates where females habitually disperse than among species where females are philopatric (Snaith and Chapman 2007; Thierry 2008) and there does not appear to be any close association between female philopatry and diet quality or food distribution, either across primate species or across other mammals (Clutton-Brock and Janson 2012). While linear female dominance hierarchies and supportive coalitions may be more frequent in species where females are philopatric, this may reflect the fact that females typically remain in the same group for most or all of their lives.

The second suggestion is that habitual female dispersal occurs in species where the breeding tenure of individual

males or male kin groups commonly exceeds the age at which most females are ready to breed, stimulating females to leave their natal groups to locate unrelated partners (Berger 1986; Pusey 1987; Clutton-Brock 1989; Harcourt and Stewart 2007). Across primate genera, females commonly breed in their natal groups and rates of female emigration are usually relatively low where most males are immigrants, while habitual dispersal by females is common where males frequently remain and breed in their natal groups (Pusey and Packer 1987b) (Figure 3.17). Quantitative analyses confirm that habitual female dispersal is frequently associated with average male breeding tenures that exceed the average age at which females reach sexual maturity (Lukas and Clutton-Brock 2011) (Figure 3.18). The same explanation may also help to account for the contrast in dispersal patterns between mammals and birds: in most group-living birds, both sexes are relatively long-lived (Arnold and Owens 1998) and females are often mature by their second year of life, when their father is often likely to be still active in their group and female-biased dispersal is common (Clutton-Brock 2009; Clutton-Brock and McAuliffe 2009).

While the available evidence suggests that habitual female dispersal is associated with the risk of inbreeding, the sample of plural breeding mammals where females habitually disperse to breed is small and there are several species where females commonly remain in their natal

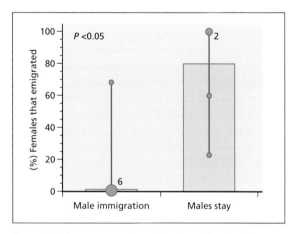

Figure 3.17 Average proportion of females emigrating for primate genera in which most males immigrate from their natal groups versus those where most males remain in their natal group. *Source*: From Harcourt and Stewart (2007). Reproduced with permission of the University of Chicago Press.

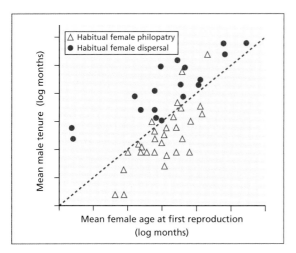

Figure 3.18 Average male tenure plotted against average age of females at first reproduction for 29 mammals where breeding females are usually natals compared with 17 species where females habitually disperse to breed. In 86% of species where females are philopatric, average male tenure is shorter than the average age of females at first breeding, whereas in all species in this sample where females habitually emigrate, the average tenure of individual breeding males or male kin groups exceeds female age at first breeding. *Source:* From Lukas & Clutton-Brock (2011). Reproduced with permission of John Wiley & Sons.

group despite the presence of their father or brothers and either breed with more distantly related males or with males from other groups (including capuchin monkeys, killer whales and banded mongooses) (Baird 2000; Perry *et al.* 2008; Nichols *et al.* 2010, 2015; Sanderson *et al.* 2015). It is not yet clear why this strategy is not more widely adopted but a possible explanation is that females habitually disperse in species where resident males are able to control their access to extra-group partners, and are more likely to remain where groups are large and include some natal males that are not close relatives or where dominant males are unable to restrict female access to intruding males from other groups.

This explanation of habitual female dispersal emphasises the need to understand the causes of variation in male breeding tenure. Several different mechanisms are probably involved (see Chapters 10 and 13). Male tenure is typically longer in socially monogamous vertebrates than in polygynous ones, and declines as the degree of polygyny increases, probably because increased breeding competition between males increases rates of turnover in breeding males in polygynous species (Clutton-Brock and Isvaran 2007). Variation in extrinsic rates of male

mortality may also be important. For example, mortality rates are relatively low in male hamadryas baboons and gorillas as well as in the social equids (Sigg *et al.* 1982; Berger 1987; Harcourt and Stewart 2007). Life-history adaptations, too, may be involved: for example, in African lions, the synchronous production of relatively large numbers of cubs generates dispersing male subgroups capable of displacing resident groups of males, leading to relatively high rates of turnover in resident male groups (Packer *et al.* 1988), with the result that the risk of father–daughter mating is reduced and females can remain and breed in their natal group (Clutton-Brock 1989). A similar outcome occurs in some langurs where emigrating males form bachelor groups that intermittently invade resident groups and evict the resident breeding male (Sugiyama 1965a, Sugiyama 1965a,b; Hrdy 1977; Borries and Koenig 2000). In contrast, in chimpanzees, rates of reproduction are low, males rarely leave their natal community and male kin groups appear to have indeterminate tenure in their community, so that natal females need to leave to gain regular access to unrelated males (Goodall 1986; Wrangham 1986) and similar patterns of social organisation occur in some bats (Nagy *et al.* 2007).

3.6 Social and ecological consequences of female philopatry

Contrasts in female dispersal patterns have important consequences for the structure of social groups (Clutton-Brock 2001; Creel and Creel 2002; Hill 2004; Strier 2008). Where most breeding females are philopatric, female group members are usually maternal relatives, while in species where females habitually disperse to breed, female group members may either be unrelated (if dispersal occurs independently) or may be maternal relatives (if females disperse with other females from their natal group).

Contrasts in the kinship structure of groups in turn affect social relationships between group members. Where most female group members are maternal relatives, strong social bonds and cooperative relationships between group-related members are common (see Chapters 8 and 9). In contrast, where females habitually disperse at adolescence and female group members are seldom close relatives, social bonds between female group members are weaker and cooperative interactions are less frequent (Dunbar 1988). Often-cited examples of the effects of variation in kinship structure on female social relationships include gelada baboons (where females live in unimale harem groups that typically include one male and several matrilineally related females) and hamadryas baboons (where females also live in unimale harem units but males establish relationships with females from other units and coerce females into leaving their natal unit and female unit members are seldom close relatives) (Dunbar 1988; Colmenares 2004).

Contrasts in the kinship structure of female groups may also affect inter-group relations and the frequency of female immigration. In many species where breeding females are usually natals and female groups consist of familiar, related individuals, female immigration into established breeding groups is rare or does not occur (Clutton-Brock and Lukas 2011). In some mammals where breeding females are mostly natal relatives, including lions and spotted hyenas, resident females commonly attack intruders and may kill them for example, in meerkats, groups commonly intrude into the territories of their neighbours, inspect their sleeping burrows and attempt to kill any pups or babysitters that they encounter (see Chapter 17). In contrast, in several species where females habitually disperse to breed in other groups and most resident females are unrelated (as in social equids and gorillas) recent female immigrants may receive aggression from resident females but this is seldom intense or protracted and they are rarely prevented by other females from joining the group (Berger 1986; Harcourt and Stewart 2007). Systematic comparisons of intra- and inter-group relations between females of species where breeding females are usually related natals and those where resident females are mostly unrelated have not yet been attempted but could be revealing.

Contrasts in the strength of social bonds between females can have important consequences for the cohesiveness of female groups and for their persistence across generations as well as for the evolution of male reproductive strategies. For example, where females habitually disperse and harem groups consist primarily of unrelated or distantly related females, the death of resident males often leads to the fragmentation of female groups; in contrast, in species where harem groups consist of matrilineal relatives, the membership of female

groups is often unaffected by the death of resident males (Kummer 1968; Dunbar 1984, 1988; Berger 1987; Stammbach 1987). Differences in the cohesiveness of female groups in turn affect the evolution of male strategies and mating systems (Auld and Rubio de Casas 2013). Opportunities for polygyny increase where female philopatry increases the number of females sharing overlapping ranges or associating in cohesive groups, raising the relative intensity of reproductive competition between males (see Chapter 10), with downstream consequences for their breeding tenure and survival (see Chapter 13). Where, if philopatric females avoid mating with related males or with males reared in the same group (see Chapter 4), female philopatry can cause males to disperse to find willing mating partners (see Chapter 12).

Intense male competition for access to stable female groups in turn affects aspects of female strategies, including the extent to which females are able to choose mating partners and the benefits of mate choice. Female choice appears to be more highly developed and to be associated with more elaborate male ornamentation in animals where females habitually disperse from their natal range and can choose freely between alternative mating partners than in species where males compete to monopolise access to pre-existing female groups and female opportunities to select mating partners are more limited (see Chapter 4). The relative benefits of mate choice may also be reduced where females remain in their natal groups since intense intra-sexual competition between males for access to breeding groups is likely to favour individuals with superior phenotypes, so that female mate choice is redundant.

Variation in dispersal and in the kinship structure of female groups also affects the genetic structure of populations. Where females are philopatric and female groups persist across generations, the spatial structuring of populations is likely to be strong (Hoezel *et al.* 2004). Effective population size may be reduced, though strong structure can also minimise the loss of genetic diversity (see Chapter 2). In contrast, where females habitually disperse to breed, effective population size is likely to be larger and the effects of social divisions on genetic structure may be reduced.

Finally, contrasts in female dispersal are likely to affect the dynamics of groups and populations (Clutton-Brock 2001; Clobert *et al.* 2008). In many social mammals, increases in local density or group size eventually reduce breeding success and survival, especially in younger or weaker individuals. As a result, group-living may contribute to the sensitivity of density-dependent reductions in recruitment and may increase population stability (see Chapter 2). However, where the benefits of sociality are so large that fitness declines in small groups, recruitment and population density may be positively correlated (see Chapter 17). Variation in dispersal may also interact with the effects of increasing density to produce unexpected effects. For example, in some rodents, increased local densities of females can depress the breeding success of unrelated females but have little effect on the recruitment of kin, so that increasing density leads to the formation of kin clusters and to the temporary enhancement of recruitment, introducing time delays in the effects of rising population density and contributing to oscillations in population size (Krebs *et al.* 2007).

SUMMARY

1. In many social mammals, females remain and breed in their natal groups throughout their lives but, in a few, most females disperse after adolescence and few individuals remain and breed in their natal group.
2. Dispersing females commonly show high mortality, so dispersal presumably has substantial benefits.
3. In some cases, females disperse to avoid competition for resources or breeding opportunities, while in others they disperse to locate unrelated mating partners.
4. Habitual dispersal of females after reaching adolescence is usually associated with male breeding tenures that exceed female age at first breeding.
5. Contrasts in the frequency of female dispersal have important consequences for the kinship structure of social groups, for social relationships between group members and for the genetic structure of populations.

References

Aars, J. and Ims, R.A. (2000) Population dynamics and genetic consequences of spatial density-dependent dispersal in patchy populations. *American Naturalist* **155**:252–265.

Aars, J., *et al.* (2006) Widespread gene flow and high genetic variability in populations of water voles *Arvicola terrestris* in patchy habitats. *Molecular Ecology* **15**:1455–1466.

Abadi, F., *et al.* (2010) Estimation of immigration rate using integrated population models. *Journal of Applied Ecology* **47**:393–400.

Acevedo-Whitehouse, K., *et al.* (2003) Disease susceptibility in California sea lions. *Nature* **422**:35.

Alberts, S.C. and Altmann, J. (1995) Balancing costs and opportunities: dispersal in male baboons. *American Naturalist* **145**:279–306.

Altmann, J., *et al.* (1985) Demography of Amboseli baboons, 1963–1983. *American Journal of Primatology* **8**:113–125.

Amos, B., *et al.* (1993) Social structure of pilot whales revealed by analytical DNA profiling. *Science* **260**:670–672.

Archie, E.A., *et al.* (2006) The ties that bind: genetic relatedness predicts the fission and fusion of social groups in wild African elephants. *Proceedings of the Royal Society of London. Series B: Biological Sciences* **273**:513–522.

Armitage, K.B., *et al.* (2011) Proximate causes of dispersal in yellow-bellied marmots, Marmota flaviventris. *Ecology* **92**:218–227.

Arnold, K.E. and Owens, I.P.F. (1998) Cooperative breeding in birds: a comparative test of the life history hypothesis. *Proceedings of the Royal Society of London. Series B: Biological Sciences* **265**:739–745.

Arnold, W. (1990a) The evolution of marmot sociality: I. Why disperse late? *Behavioral Ecology and Sociobiology* **27**:229–237.

Arnold, W. (1990b) The evolution of marmot sociality: II. Costs and benefits of joint hibernation. *Behavioral Ecology and Sociobiology* **27**:239–246.

Auld, J.R. and Rubio de Casas, R. (2013) The correlated evolution of dispersal and mating-system traits. *Evolutionary Biology* **40**:185–193.

Baird, R.W. (2000) The killer whale: foraging specializations and group hunting. In: *Cetacean Societies: Field Studies of Dolphins and Whales* (eds J. Mann, R. Connor, P. Tyack and H. Whitehead) Chicago: University of Chicago Press, 127–153.

Barrett, S.C.H. and Charlesworth, D. (1991) Effects of a change in the level of inbreeding on the genetic load. *Nature* **352**:522–524.

Bashaw, M.J., *et al.* (2007) The structure of social relationships among captive female giraffe (*Giraffa camelopardalis*). *Journal of Comparative Psychology* **121**:46–53.

Bateson, P.P.G. (1978) Sexual imprinting and optimal outbreeding. *Nature* **273**:659–660.

Bekoff, M. (1977) Mammalian dispersal and the ontogeny of individual behavioral phenotypes. *American Naturalist* **111**:715–732.

Bengtsson, B.O. (1978) Avoiding inbreeding: at what cost? *Journal of Theoretical Biology* **73**:439–444.

Bennett, N.C. and Faulkes, C.G. (2000) *African Mole-rats: Ecology and Eusociality.* Cambridge: Cambridge University Press.

Berger, J. (1986) *Wild Horses of the Great Basin*. Chicago: University of Chicago Press.

Berger, J. (1987) Reproductive fates of dispersers in a harem-dwelling ungulate: the wild horse. In: *Mammalian Dispersal Patterns* (eds B.D. Chepko-Sade and Z.T. Halpin) Chicago: University of Chicago Press, 41–54.

Bodmer, W.F. and Cavalli-Sforza, L.L. (1976) *Genetics, Evolution and Man*. San Francisco: W.H. Freeman.

Bollinger, E.K., *et al.* (1993) Inbreeding avoidance increases dispersal movements of the meadow vole. *Ecology* **74**:1153–1156.

Bondrup-Nielsen, S. (1993) Early malnutrition increases emigration of adult female meadow voles, *Microtus pennsylvanicus*. *Oikos* **67**:317–320.

Boonstra, R. and Hogg, I. (1988) Friends and strangers: a test of the Charnov–Finerty hypothesis. *Oecologia* **77**:95–100.

Borries, C. and Koenig, A. (2000) Infanticide in hanuman langurs: social organization, male migration, and weaning age. In: *Infanticide by Males and its Implications* (eds C.P. van Schaik and C.H. Janson) Cambridge: Cambridge University Press, 99–122.

Boutin, S., *et al.* (1993) Post-breeding dispersal by female red squirrels (*Tamiasciurus hudsonicus*): the effect of local vacancies. *Behavioral Ecology* **4**:151–155.

Bowler, D.E. and Benton, T.G. (2005) Causes and consequences of animal dispersal strategies: relating individual behaviour to spatial dynamics. *Biological Reviews* **80**:205–225.

Boydston, E.E., *et al.* (2001) Sex differences in territorial behaviour exhibited by the spotted hyena (Hyaenidae: *Crocuta crocuta*). *Ethology* **107**:369–385.

Bradley, B.J., *et al.* (2007) Potential for female kin associations in wild western gorillas despite female dispersal. *Proceedings of the Royal Society of London. Series B: Biological Sciences* **274**:2179–2185.

Brzeski, K.E. (2014) Inbreeding and inbreeding depression in endangered red wolves (*Canis rufus*). *Journal of Molecular Ecology* **23**:4241–4255.

Brockelman, W.Y., *et al.* (1998) Dispersal, pair formation and social structure in gibbons (*Hylobates lar*). *Behavioral Ecology and Sociobiology* **42**:329–339.

Brody, A.K. and Armitage, K.B. (1985) The effects of adult removal on dispersal of yearling yellow-bellied marmots. *Canadian Journal of Zoology* **63**:2560–2564.

Bulger, J. and Hamilton, W.J. (1988) Inbreeding and reproductive success in a natural chacma baboon, *Papio cynocephalus ursinus*, population. *Animal Behaviour* **36**:574–578.

Cant, M.A., *et al.* (2010) Reproductive control via eviction (but not the threat of eviction) in banded mongooses. *Proceedings of the Royal Society of London. Series B: Biological Sciences* **277**:2219–2226.

Cant, M.A., *et al.* (2013) Demography and social evolution of banded mongooses. In: *Advances in the Study of Behavior*, (eds H.J. Brockmann, T.J. Roper, M. Naguib *et al.*) Vol. **45** San Diego, CA: Academic Press, 407–445.

Chapman, J.R., *et al.* (2009) A quantitative review of heterozygosity–fitness correlations in animal populations. *Molecular Ecology* **18**:2746–2765.

Charlesworth, D. and Charlesworth, B. (1987) Inbreeding depression and its evolutionary consequences. *Annual Review of Ecology and Systematics* **18**:237–268.

Cheney, D.L., *et al.* (2004) Factors affecting reproduction and mortality among baboons in the Okavango Delta, *Botswana*. *International Journal of Primatology* **25**:401–428.

Christie, M.R., *et al.* (2011) Who are the missing parents? Grandparentage analysis identifies multiple sources of gene flow into a wild population. *Molecular Ecology* **20**:1263–1276.

Cisek, D. (2000) New colony formation in the highly inbred eusocial naked mole-rat: outbreeding is preferred. *Behavioral Ecology* **11**:1–6.

Clark, A.B. (1978) Sex ratio and local resource competition in a prosimian primate. *Science* **201**:163–165.

Clarke, A.L., *et al.* (1997) Sex biases in avian dispersal: a reappraisal. *Oikos* **79**:429–438.

Clobert, J., *et al.* (2008) Evolution of dispersal. In: *Behavioural Ecology* (eds E. Danchin, L.A. Giraldeau and F. Cezilly) Oxford: Oxford University Press, 323–359.

Clutton-Brock, T.H. (1989) Female transfer and inbreeding avoidance in social mammals. *Nature* **337**:70–72.

Clutton-Brock, T.H. (2001) Sociality and population dynamics. In: *Ecology: Achievement and Challenge* (eds M.C. Press, N.J. Huntly and S. Levin) Oxford: Blackwell Science, 47–66.

Clutton-Brock, T.H. (2009) Structure and function in mammalian societies. *Philosophical Transactions of the Royal Society B: Biological Sciences* **364**:3229–3242.

Clutton-Brock, T.H. and Isvaran, K. (2007) Sex differences in ageing in natural populations of vertebrates. *Proceedings of the Royal Society of London. Series B: Biological Sciences* **274**:3097–3104.

Clutton-Brock, T.H. and Janson, C.J. (2012) Primate socioecology at the crossroads: past, present, and future. *Evolutionary Anthropology: Issues, News, and Reviews* **21**:136–150.

Clutton-Brock, T.H. and Lukas, D. (2011) The evolution of social philopatry and dispersal in female mammals. *Molecular Ecology* **21**:472–492.

Clutton-Brock, T.H., *et al.* (1982a) *Red Deer: The Behaviour and Ecology of Two Sexes*. Chicago: University of Chicago Press.

Clutton-Brock, T.H., *et al.* (1982b) Competition between female relatives in a matrilocal mammal. *Nature* **300**:178–180.

Clutton-Brock, T.H., *et al.* (1998) Infanticide and expulsion of females in a cooperative mammal. *Proceedings of the Royal Society of London. Series B: Biological Sciences* **265**:2291–2295.

Clutton-Brock, T.H., *et al.* (2002) Sex differences in emigration and mortality affect optimal management of deer populations. *Nature* **415**:633–637.

Clutton-Brock, T.H., *et al.* (2006) Intrasexual competition and sexual selection in cooperative mammals. *Nature* **444**: 1065–1068.

Clutton-Brock, T.H., *et al.* (2008) Group size and the suppression of subordinate reproduction in Kalahari meerkats. *Animal Behaviour* **76**:689–700.

Cohas, A. *et al.* (2008) The genetic similarity between pair members influences the frequency of extrapair paternity in alpine marmots. Animal Behaviour **76**:87–95.

Colmenares, F. (2004) Kinship structure and its impact on behaviour in multilevel societies. In: *Kinship and Behaviour in Primates* (eds B. Chapais and C.M. Berman) Oxford: Oxford University Press, 242–270.

Coltman, D.W., *et al.* (1999) Parasite-mediated selection against inbred Soay sheep in a free-living island population. *Evolution* **53**:1259–1267.

Conradt, L., *et al.* (1999) Lifetime reproductive success in relation to habitat use in female red deer. *Oecologia* **120**:218–224.

Coulson, T.N., *et al.* (1998) Genotype by environment interactions in winter survival in red deer. *Journal of Animal Ecology* **67**:434–445.

Creel, S. and Creel, N.M. (2002) *The African Wild Dog: Behavior, Ecology, and Conservation.* Princeton, NJ: Princeton University Press.

Crnokrak, P. and Roff, D.A. (1999) Inbreeding depression in the wild. *Heredity* **83**:260–270.

Crockett, C.M. and Pope, T.R. (1993) Consequences of sex differences in dispersal for juvenile red howler monkeys. In: *Juvenile Primates: Life History, Development, and Behavior* (eds M.E. Pereira and L.A. Fairbanks) New York: Oxford University Press, 104–118.

Dalton, C.L. (2000) Effects of female kin groups on reproduction and demography in the gray-tailed vole, *Microtus canicaudus. Oikos* **90**:153–159.

Daniels, S.J. and Walters, J.R. (2000) Inbreeding depression and its effects on natal dispersal in red-cockaded woodpeckers. *Condor* **102**:482–491.

Dewsbury, D.A. (1988) Kin discrimination and reproductive behavior in muroid rodents. *Behavior Genetics* **18**:525–536.

Dietz, J.M. and Baker, A.J. (1993) Polygyny and female reproductive success in golden lion tamarins, *Leontopithecus rosalia. Animal Behaviour* **46**:1067–1078.

Di Fiore, A. and Campbell, C.J. (2007) The Atelines: variation in ecology, behaviour and social organisation. In: *Primates in Perspective* (eds C.J. Campbell, A. Fuentes, K.C. MacKinnon, M. Panger and S.K. Bearder) New York: Oxford University Press, 155–185.

Dobson, F.S. (1979) An experimental study of dispersal in the California ground squirrel. *Ecology* **60**:1103–1109.

Dobson, F.S. (1981) An experimental examination of an artificial dispersal sink. *Journal of Mammalogy* **62**:74–81.

Dobson, F.S., *et al.* (2000) An experimental test of kin association in the house mouse. *Canadian Journal of Zoology* **78**:1806–1812.

Doligez, B. and Pärt, T. (2008) Estimating fitness consequences of dispersal: a road to 'know-where'? Non-random dispersal and the underestimation of dispersers' fitness. *Journal of Animal Ecology* **77**:1199–1211.

Dunbar, R.I.M. (1984) *Reproductive Decisions: An Economic Analysis of the Social Strategies of Gelada Baboons.* Princeton, NJ: Princeton University Press.

Dunbar, R.I.M. (1988) *Primate Social Systems.* London: Chapman & Hall.

Ebensperger, L.A. (1998a) Strategies and counterstrategies to infanticide in mammals. *Biological Reviews* **73**:321–346.

Ebensperger, L.A. (1998b) Do female rodents use promiscuity to prevent male infanticide? *Ethology Ecology and Evolution* **10**:129–141.

Ekman, J., *et al.* (2004) Delayed dispersal. In: *Ecology and Evolution of Cooperative Breeding in Birds* (eds W. Koenig and J. Dickinson) Cambridge: Cambridge University Press, 35–47.

Errington, P.L. (1963) *Muskrat Populations.* Ames, IA: Iowa State University Press.

Estes, R.D. (1974) Social organization of the African Bovidae. *IUCN New Series* **24**:166–205.

Fairbanks, L.A. and McGuire, M.T. (1987) Mother–infant relationships in vervet monkeys: response to new adult males. *International Journal of Primatology* **8**:351–366.

Faulkes, C.G. and Bennett, N.C. (2007) African mole-rats: social and ecological diversity. In: *Rodent Societies: An Ecological and Evolutionary Perspective* (eds J.O. Wolff and P.W. Sherman) Chicago: University of Chicago Press, 427–437.

Fedigan, L.M., *et al.* (2008) Predictors of reproductive success in female white-faced capuchins (*Cebus capucinus*). *American Journal of Physical Anthropology* **137**:82–90.

Ferkin, M.H. (1990) Kin recognition and social behavior in microtine rodents. In: *Social Systems and Population Cycles in Voles* (eds R.H. Tamarin, R.S. Ostfeld, S.R. Pugh and G. Bujalska) Basel: Birkhaeuser, 11–24.

Fisher, D.O., *et al.* (2009) Experimental translocation of juvenile water voles in a Scottish lowland metapopulation. *Population Ecology* **51**:289–295.

Fowler, K. and Whitlock, M.C. (1999) The variance in inbreeding depression and the recovery of fitness in bottlenecked populations. *Proceedings of the Royal Society of London. Series B: Biological Sciences* **266**:2061–2066.

Frankham, R., *et al.* (2002) *Introduction to Conservation Genetics.* Cambridge: Cambridge University Press.

Frankham, R., *et al.* (2011) Predicting the probability of outbreeding depression. *Conservation Biology* **25**:465–475.

Fritts, S.H. and Mech, L.D. (1981) Dynamics, movements, and feeding ecology of a newly protected wolf population in northwestern Minnesota. *Wildlife Monographs* **80**: 3–79.

Gaines, M.S. and McClenaghan, L.R. (1980) Dispersal in small mammals. *Annual Review of Ecology and Systematics* **11**:163–196.

Gandon, S. (1999) Kin competition, the cost of inbreeding and the evolution of dispersal. *Journal of Theoretical Biology* **200**:345–364.

Gilchrist, J.S. (2008) Aggressive monopolization of mobile carers by young of a cooperative breeder. *Proceedings of the Royal Society of London. Series B: Biological Sciences* **275**:2491–2498.

Goldizen, A.W. (2003) Social monogamy and its variations in callitrichids: do these relate to the costs of infant care? In: *Monogamy: Mating Strategies and Partnerships in Birds, Humans and Other Mammals* (eds U.H. Reichard and C. Boesch) Cambridge: Cambridge University Press, 232–247.

Goodall, J. (1986) Social rejection, exclusion, and shunning among the Gombe chimpanzees. *Ethology and Sociobiology* **7**:227–236.

Greenwood, P.J. (1980) Mating systems, philopatry and dispersal in birds and mammals. *Animal Behaviour* **28**:1140–1162.

Guillaume, F. and Perrin, N. (2006) Joint evolution of dispersal and inbreeding load. *Genetics* **173**:497–509.

Gundersen, G. and Andreassen, H.P. (1998) Causes and consequences of natal dispersal in root voles, *Microtus oeconomus*. *Animal Behaviour* **56**:1355–1366.

Hackländer, K., *et al.* (2003) Reproductive suppression in female Alpine marmots, *Marmota marmota*. *Animal Behaviour* **65**:1133–1140.

Haigh, G.R. (1983) Effects of inbreeding and social-factors on the reproduction of young female *Peromyscus maniculatus bairdii*. *Journal of Mammalogy* **64**:48–54.

Hamilton, W.D. and May, R.M. (1977) Dispersal in stable habitats. *Nature* **269**:578–581.

Hammond, R.L., *et al.* (2006) Genetic evidence for female-biased dispersal and gene flow in a polygynous primate. *Proceedings of the Royal Society of London. Series B: Biological Sciences* **273**:479–484.

Hanski, I.A., *et al.* (1991) Natal dispersal and social dominance in the common shrew *Sorex araneus*. *Oikos* **62**:48–58.

Harcourt, A.H. and Stewart, K.J. (2007) *Gorilla Society*. Chicago: University of Chicago Press.

Hayes, L.D. (2000) To nest communally or not to nest communally: a review of rodent communal nesting and nursing. *Animal Behaviour* **59**:677–688.

Henzi, S.P., *et al.* (1997) Fission and troop size in a mountain baboon population. *Animal Behaviour* **53**:525–535.

Hill, D.A. (2004) The effects of demographic variation on kinship structure and behaviour in cercopithecines. In: *Kinship and Behaviour in Primates* (eds B. Chapais and C.M. Berman) Oxford: Oxford University Press, 132–150.

Hill, J.L. (1974) *Peromyscus*: effect of early pairing on reproduction. *Science* **186**:1042–1044.

Hoezel, G.A., *et al.* (2004) Dispersal and the population genetics of primate species. In: *Kinship and Behaviour in Primates* (eds B. Chapais and C.M. Berman) Oxford: Oxford University Press, 109–131.

Holand, Ø., *et al.* (2007) No evidence of inbreeding avoidance in a polygynous ungulate: the reindeer (*Rangifer tarandus*). *Biology Letters* **3**:36–39.

Holekamp, K.E. (1984) Natal dispersal in Belding's ground squirrels (*Spermophilus beldingi*). *Behavioral Ecology and Sociobiology* **16**:21–30.

Honeycutt, R.L., *et al.* (1991a) Systematics and evolution of the family Bathyergidae. In: *The Biology of the Naked Mole-rat* (eds P.W. Sherman, J.U.M. Jarvis and R.D. Alexander) Princeton, NJ: Princeton University Press, 66–96.

Honeycutt, R.L., *et al.* (1991b) Genetic variation within and among populations of the naked mole-rat: evidence from nuclear and mitochondrial genomes. In: *The Biology of the Naked Mole-rat* (eds P.W. Sherman, J.U.M. Jarvis and R.D. Alexander) Princeton, NJ: Princeton University Press, 195–208.

Hoogland, J.L. (1992) Levels of inbreeding among prairie dogs. *American Naturalist* **139**:591–602.

Hoogland, J.L. (1995) *The Black-tailed Prairie Dog: Social Life of a Burrowing Mammal*. Chicago: University of Chicago Press.

Hoogland, J.L. (2013) Prairie dogs disperse when all close kin have disappeared. *Science* **339**:1205–1207.

Hrdy, S.B. (1977) *The Langurs of Abu: Female and Male Strategies of Reproduction*. Cambridge, MA: Harvard University Press.

Isbell, L.A. (2004) Is there no place like home? Ecological bases of female dispersal and philopatry and their consequences for the formation of kin groups In: *Kinship and Behaviour in Primates* (eds B. Chapais and C.M. Berman) Oxford: Oxford University Press, 71–108.

Janson, C.H. and van Schaik, C.P. (2000) The behavioral ecology of infanticide by males. In: *Infanticide by Males and its Implications* (eds C.P. van Schaik and C.H. Janson) Cambridge: Cambridge University Press, 469–494.

Jimenez, J.A., *et al.* (1994) An experimental study of inbreeding depression in a natural habitat. *Science* **266**:271–273.

Jones, W.T. (1987) Dispersal patterns in kangaroo rats. In: *Mammalian Dispersal Patterns: The Effects of Social Structure on Population Genetics* (eds B.D. Chepko-Sade and Z.T. Halpin) Chicago: University of Chicago Press, 119–127.

Kawata, M. (1990) Fluctuating populations and kin interaction in mammals. *Trends in Ecology and Evolution* **5**:17–20.

Keane, B. (1990) The effect of relatedness on reproductive success and mate choice in the white-footed mouse, *Peromyscus leucopus*. *Animal Behaviour* **39**:264–273.

Keane, B., *et al.* (1996) No evidence of inbreeding avoidance or inbreeding depression in a social carnivore. *Behavioral Ecology* **7**:480–489.

Keller, L.F. (1998) Inbreeding and its fitness effects in an insular population of song sparrows (*Melospiza melodia*). *Evolution* **52**:240–250.

Keller, L.F. and Arcese, P. (1998) No evidence for inbreeding avoidance in a natural population of song sparrows (*Melospiza melodia*). *American Naturalist* **152**:380–392.

Keller, L.F. and Waller, D.M. (2002) Inbreeding effects in wild populations. *Trends in Ecology and Evolution* **17**:230–241.

Kerth, G. (2008a) Animal sociality: bat colonies are founded by relatives. *Current Biology* **18**:R740–R742.

Kerth, G. (2008b) Causes and consequences of sociality in bats. *BioScience* **58**:737–746.

Kerth, G., *et al.* (2011) Bats are able to maintain long-term social relationships despite the high fission–fusion dynamics of their groups. *Proceedings of the Royal Society of London. Series B: Biological Sciences* **278**:2761–2767.

Kirkpatrick, R.C. (2007) Asian colobines: diversity among leaf-eating monkeys. In: *Primates in Perspective* (eds C.J. Campbell, A. Fuentes, K.C. MacKinnon, N. Panger and S.K. Bearder) New York: Oxford University Press, 186–200.

Koenig, A. (2002) Competition for resources and its behavioral consequences among female primates. *International Journal of Primatology* **23**:759–783.

Koenig, A. and Borries, C. (2006) The predictive power of socio-ecological models: a reconsideration of resource characteristics, agonism and dominance hierarchies. In: *Feeding Ecology in Apes and Other Primates: Ecological, Physiological and Behavioural Aspects* (eds G. Hohmann, M.M. Robbins and C. Boesch) Cambridge: Cambridge University Press, 263–284.

Koenig, W. and Haydock, J. (2004) Incest and incest avoidance. In: *Ecology and Evolution of Cooperative Breeding in Birds* (eds W. Koenig and J.L. Dickinson) Cambridge: Cambridge University Press, 142–156.

Koenig, W.D., *et al.* (1996) Detectability, philopatry, and the distribution of dispersal distances in vertebrates. *Trends in Ecology and Evolution* **11**:514–517.

Komdeur, J., *et al.* (1995) Transfer experiments of Seychelles warblers to new islands: changes in dispersal and helping behaviour. *Animal Behaviour* **49**:695–708.

Komers, P.E. and Brotherton, P.N.M. (1997) Female space use is the best predictor of monogamy in mammals. *Proceedings of the Royal Society of London. Series B: Biological Sciences* **264**:1261–1270.

König, B. (1994) Fitness effects of communal rearing in house mice: the role of relatedness versus familiarity. *Animal Behaviour* **48**:1449–1457.

Korstjens, A.H. and Schippers, E.P. (2003) Dispersal patterns among olive colobus in Taï National Park. *International Journal of Primatology* **24**:515–539.

Krackow, S. and Matuschak, B. (1991) Mate choice for non-siblings in wild house mice: evidence from a choice test and a reproductive test. *Ethology* **88**:99–108.

Krackow, S., *et al.* (2003) Sexual growth dimorphism affects birth sex ratio in house mice. *Proceedings of the Royal Society of London. Series B: Biological Sciences* **270**:943–947.

Krebs, C.J., *et al.* (2007) Social behavior and self-regulation in murid rodents. In: *Rodent Societies: An Ecological and Evolutionary Perspective* (eds J.O. Wolff and P.W. Sherman) Chicago: University of Chicago Press, 173–181.

Kummer, H. (1968) *Social Organization of Hamadryas Baboons.* Chicago: University of Chicago Press.

Lacey, E.A. and Ebensperger, L.A. (2007) Social structure in Octodontid and Ctenomyid rodents. In: *Rodent Societies: An Ecological and Evolutionary Perspective* (eds J.O. Wolff and P.W. Sherman) Chicago: University of Chicago Press, 403–415.

Lambin, X. and Krebs, C.J. (1993) Influence of female related-ness on the demography of female Townsend's vole populations in the spring. *Journal of Animal Ecology* **62**:536–550.

Lambin, X. and Yoccoz, N.G. (1998) The impact of population kin-structure on nestling survival in Townsend's voles *Microtus townsendii*. *Journal of Animal Ecology* **67**:1–16.

Lambin, X., *et al.* (2001) Dispersal, intraspecific competition, kin competition and kin facilitation: a review of the empirical evidence. In: *Dispersal* (eds J. Clobert, E. Danchin, A.A. Shondt and J.D. Nichols) Oxford: Oxford University Press, 110–122.

Laporte, V. and Charlesworth, B. (2002) Effective population size and population subdivision in demographically structured populations. *Genetics* **162**:501–509.

Lawson Handley, L.J. and Perrin, N. (2007) Advances in our understanding of mammalian sex-biased dispersal. *Molecular Ecology* **16**:1559–1578.

Lefebvre, D., *et al.* (2003) Modelling the influence of demographic parameters on group structure in social species with dispersal asymmetry and group fission. *Behavioral Ecology and Sociobiology* **53**:402–410.

Le Galliard, J.F., *et al.* (2007) Mother–offspring interactions do not affect natal dispersal in a small rodent. *Behavioral Ecology* **18**:665–673.

Lidicker, W.Z. and Patton, J.L. (1987) Patterns of dispersal and genetic structure in populations of small rodents. In: *Mammalian Dispersal Patterns: Effects of Social Structure on Population Genetics* (eds B.D. Chepko-Sade and Z.T. Halpin) Chicago: University of Chicago Press, 144–161.

Lukas, D. and Clutton-Brock, T.H. (2011) Group structure, kinship, inbreeding risk and habitual female dispersal in plural-breeding mammals. *Journal of Evolutionary Biology* **22**:2624–2630.

Lurz, P.W.W., *et al.* (1997) Effects of temporal and spatial variation in habitat quality on red squirrel dispersal behaviour. *Animal Behaviour* **54**:427–435.

Lynch, M. (1991) The genetic interpretation of inbreeding depression and outbreeding depression. *Evolution* **45**:622–629.

McGuire, B. and Getz, L.L. (1991) Response of young female prairie voles (*Microtus ochrogaster*) to nonresident males: implications for population regulation. *Canadian Journal of Zoology* **69**:1348–1355.

McGuire, B., *et al.* (2007) Sex differences, effects of male presence and coordination of nest visits in prairie voles (*Microtus ochrogaster*) during the immediate postnatal period. *American Midland Naturalist* **157**:187–201.

McNutt, J.W. (1996) Sex biased dispersal in African wild dogs, *Lycaon pictus*. *Animal Behaviour* **52**:1067–1077.

Mappes, T., *et al.* (1995) Reproductive costs and litter size in the bank vole. *Proceedings of the Royal Society of London. Series B: Biological Sciences* **261**:19–24.

Marshall, T.C., *et al.* (2002) Estimating the prevalence of inbreeding from incomplete pedigrees. *Proceedings of the Royal Society of London. Series B: Biological Sciences* **269**:1533–1539.

Matthysen, E. (2005) Density-dependent dispersal in birds and mammals. *Ecography* **28**:403–416.

May, R.M. (1979) When to be incestuous. *Nature* **279**:192–194.

Mayr, E. (1963) *Animal Species and Evolution*. Cambridge, MA: Harvard University Press.

Messier, F. (1985) Solitary living and extraterritorial movements of wolves in relation to social status and prey abundance. *Canadian Journal of Zoology* **63**:239–245.

Metzgar, L.H. (1967) An experimental comparison of screech owl predation on resident and transient white-footed mice (*Peromyscus leucopus*). *Journal of Mammalogy* **48**:387–391.

Moore, J. and Ali, R. (1984) Are dispersal and inbreeding avoidance related? *Animal Behaviour* **32**:94–112.

Morton, N.E., *et al.* (1956) An estimate of the mutational damage in man on data from consanguineous matings. *Proceedings of the National Academy of Sciences of the United States of America* **42**:855–863.

Moses, R.A. and Millar, J.S. (1994) Philopatry and mother–daughter associations in bushy-tailed woodrats: space use and reproductive success. *Behavioral Ecology and Sociobiology* **35**:131–140.

Nagy, M., *et al.* (2007) Female-biased dispersal and patrilocal kin groups in a mammal with resource-defence polygyny. *Proceedings of the Royal Society of London. Series B: Biological Sciences* **274**:3019–3025.

Nelson, M.E. and Mech, L.D. (1987) Demes within a Northeastern Minnesota deer population. In: *Mammalian Dispersal Patterns: Effects of Social Structure on Population Genetics* (eds B.D. Chepko-Sade and Z.T. Halpin) Chicago: University of Chicago Press, 27–40.

Nichols, H.J., *et al.* (2010) Top males gain high reproductive success by guarding more successful females in a cooperatively breeding mongoose. *Animal Behaviour* **80**:649–657.

Nichols, H.J., *et al.* (2015) Adjustment of costly extra-group paternity according to inbreeding risk in a cooperative mammal. *Behavioral Ecology* **26**: doi: 10.1093/beheco/arv095

Nsubuga, A.M., *et al.* (2008) Patterns of paternity and group fission in wild multimale mountain gorilla groups. *American Journal of Physical Anthropology* **135**:263–274.

Nunes, S. (2007) Dispersal and philopatry. In: *Rodent Societies: An Ecological and Evolutionary Perspective* (eds J.O. Wolff and P.W. Sherman) Chicago: University of Chicago Press, 150–163.

Nunes, S., *et al.* (1998) Body fat and time of year interact to mediate dispersal behaviour in ground squirrels. *Animal Behaviour* **55**:605–614.

Nunes, S., *et al.* (1999) Energetic and endocrine mediation of natal dispersal behavior in Belding's ground squirrels. *Hormones and Behavior* **35**:113–124.

Okamoto, K. (2004) Patterns of group fission. In: *Macaque Societies* (eds B. Thierry, N. Singh and W. Kaumanns) Cambridge: Cambridge University Press, 112–116.

O'Riain, M.J. and Jarvis, J.U.M. (1997) Colony member recognition and xenophobia in the naked mole-rat. *Animal Behaviour* **53**:487–498.

Packer, C. and Pusey, A.E. (1982) Cooperation and competition within coalitions of male lions: kin selection or game theory? *Nature* **296**:740–742.

Packer, C. and Pusey, A.E. (1983) Dispersal, kinship, and inbreeding in African lions. In: *The Natural History of Inbreeding and Outbreeding: Theoretical and Empirical Perspectives* (ed. N.W. Thornhill). Chicago: University of Chicago Press, 375–391.

Packer, C., *et al.* (1988) Reproductive success in lions. In: *Reproductive Success: Studies of Individual Variation in Contrasting Breeding Systems* (ed. T.H. Clutton-Brock). Chicago: University of Chicago Press, 363–383.

Pavelka, M.S.M., *et al.* (2002) Availability and adaptive value of reproductive and postreproductive Japanese macaque mothers and grandmothers. *Animal Behaviour* **64**:407–414.

Perrin, N. and Goudet, J. (2001) Inbreeding, kinship and the evolution of natal dispersal. In: *Dispersal* (eds J. Clobert, E. Danchin, A.A. Dhondt and J.D. Nichols) Oxford: Oxford University Press, 123–142.

Perrin, N. and Mazalov, V. (1999) Dispersal and inbreeding avoidance. *American Naturalist* **154**:282–292.

Perrin, N. and Mazalov, V. (2000) Local competition, inbreeding, and the evolution of sex-biased dispersal. *American Naturalist* **155**:116–127.

Perry, S., *et al.* (2008) Kin-biased social behaviour in wild adult female white-faced capuchins, *Cebus capucinus. Animal Behaviour* **76**:187–199.

Pinter-Wollman, N., *et al.* (2009) The relationship between social behaviour and habitat familiarity in African elephants (*Loxodonta africana*). *Proceedings of the Royal Society of London. Series B: Biological Sciences* **276**:1009–1014.

Planes, S. and Lemer, S. (2011) Individual-based analysis opens new insights into understanding population structure and animal behaviour. *Molecular Ecology* **20**:187–189.

Pomeroy, P.P., *et al.* (2000) Philopatry, site fidelity and local kin associations within grey seal breeding colonies. *Ethology* **106**:899–919.

Pomeroy, P.P., *et al.* (2001) Reproductive performance links to fine-scale spatial patterns of female grey seal relatedness. *Proceedings of the Royal Society of London. Series B: Biological Sciences* **268**:711–717.

Pope, T.R. (2000) The evolution of male philopatry in neotropical monkeys. In: *Primate Males: Causes and Consequences of Variation in Group Composition* (ed. P.M. Kappeler). Cambridge: Cambridge University Press, 219–235.

Pratt, D.M. and Anderson, V.H. (1985) Giraffe social behaviour. *Journal of Natural History* **19**:771–781.

Prugnolle, F. and De Meeûs, T. (2002) Inferring sex-biased dispersal from population genetic tools: a review. *Heredity* **88**:161–165.

Pugh, S.R. and Tamarin, R.H. (1988) Inbreeding in a population of meadow voles, *Microtus pennsylvanicus. Canadian Journal of Zoology* **66**:1831–1834.

Pusenius, J., *et al.* (1998) Matrilineal kin clusters and their effect on reproductive success in the field vole *Microtus agrestis. Behavioral Ecology* **9**:85–92.

Pusey, A.E. (1987) Sex-biased dispersal and inbreeding avoidance in birds and mammals. *Trends in Ecology and Evolution* **2**:295–299.

Pusey, A.E. and Packer, C. (1987a) The evolution of sex-biased dispersal in lions. *Behaviour* **101**:275–310.

Pusey, A.E. and Packer, C.R. (1987b) Dispersal and philopatry. In: *Primate Societies* (eds B.B. Smuts, D.L. Cheney, R.M. Seyfarth, R.W. Wrangham and T.T. Struhsaker) Chicago: University of Chicago Press, 250–266.

Pusey, A.E. and Wolf, M. (1996) Inbreeding avoidance in animals. *Trends in Ecology and Evolution* **11**:201–206.

Ralls, K., *et al.* (1979) Inbreeding and juvenile mortality in small populations of ungulates. *Science* **206**:1101–1103.

Reed, J.M. (1993) A parametric method for comparing dispersal distances. *Condor* **95**:716–718.

Reeve, H.K., *et al.* (1990) DNA fingerprinting reveals high levels of inbreeding in colonies of the eusocial naked mole-rat. *Proceedings of the National Academy of Sciences of the United States of America* **87**:2496–2500.

Ronce, O. (2007) How does it feel to be like a rolling stone? Ten questions about dispersal evolution. *Annual Review of Ecology, Evolution and Systematics* **38**:231–253.

Ronce, O., *et al.* (1998) Natal dispersal and senescence. *Proceedings of the National Academy of Sciences of the United States of America* **95**:600–605.

Rood, J.P. (1987) Dispersal and intergroup transfer in the dwarf mongoose. In: *Mammalian Dispersal Patterns: The Effects of Social Structure on Population Genetics* (eds B.D. Chepko-Sade and Z.T. Halpin) Chicago: University of Chicago Press, 85–103.

Roze, D. and Rousset, F. (2005) Inbreeding depression and the evolution of dispersal rates: a multilocus model. *American Naturalist* **166**:708–721.

Rusu, A.S. and Krackow, S. (2004) Kin-preferential cooperation, dominance-dependent reproductive skew and competition for males in communally nesting female house-mice. *Behavioral Ecology and Sociobiology* **56**:298–305.

Sanderson, J.L., *et al.* (2015) Banded mongooses avoid inbreeding when mating with members of the same natal group. *Molecular Ecology* **24**:3738–3751.

Schull, W.J. and Neel, J.V. (1965) *The Effects of Inbreeding on Japanese Children*. New York: Harper and Row.

Sera, W.E. and Gaines, M.S. (1994) The effect of relatedness on spacing behavior and fitness of female prairie voles. *Ecology* **75**:1560–1566.

Sharp, S.P., *et al.* (2008) Natal dispersal and recruitment in a cooperatively breeding bird. *Oikos* **117**:1371–1379.

Shields, W.M. (1982) *Philopatry, Inbreeding and the Evolution of Sex*. Albany, NY: State University of New York Press.

Shields, W.M. (1987) Dispersal and mating systems: investigating their causal connections. In: *Mammalian Dispersal Patterns: The Effects of Social Structure on Population Genetics* (eds B.D. Chepko-Sade and Z.T. Halpin) Chicago: University of Chicago Press, 3–24.

Sigg, H., *et al.* (1982) Life history of hamadryas baboons: physical development, infant mortality, reproductive parameters and family relationships. *Primates* **23**:473–487.

Silk, J.B. (2007) The adaptive value of sociality in mammalian groups. *Philosophical Transactions of the Royal Society B: Biological Sciences* **362**:539–559.

Sillero-Zubiri, C., *et al.* (1996) Male philopatry, extra pack copulations and inbreeding avoidance in Ethiopian wolves (*Canis simensis*). *Behavioral Ecology and Sociobiology* **38**:331–340.

Slate, J., *et al.* (2000) Inbreeding depression influences lifetime breeding success in a wild population of red deer (*Cervus elaphus*). *Proceedings of the Royal Society of London. Series B: Biological Sciences* **267**:1657–1662.

Smith, K. (2000) Paternal kin matter: the distribution of social behavior among wild, adult female baboons. PhD thesis, University of Chicago.

Snaith, T.V. and Chapman, A.P. (2007) Primate group size and interpreting socioecological models: do folivores really play by different rules? *Evolutionary Anthropology* **16**:94–106.

Stammbach, E. (1987) Desert, forest and montane baboons: multi-level societies. In: *Primate Societies* (eds B.B. Smuts, D.L. Cheney, R.M. Seyfarth, R.W. Wrangham and T.T. Struhsaker) Chicago: University of Chicago Press, 112–120.

Steenbeck, R. (2000) Infanticide by males and female choice in Thomas's langurs. In: *Infanticide by Males and its Implications* (eds C.P. van Schaik and C.H. Janson). Cambridge: Cambridge University Press, 153–177.

Stenseth, N.C. and Lidicker, W.Z. (eds) (1992) *Animal Dispersal: Small Mammals as a Model*. New York: Chapman & Hall.

Sterck, E.H.M., *et al.* (1997) The evolution of female social relationships in nonhuman primates. *Behavioral Ecology and Sociobiology* **41**:291–309.

Stockley, P., *et al.* (1993) Female multiple mating behaviour in the common shrew as a strategy to reduce inbreeding. *Proceedings of the Royal Society of London. Series B: Biological Sciences* **254**:173–179.

Støen, O.-G., *et al.* (2006) Inversely density-dependent natal dispersal in brown bears *Ursus arctos*. *Oecologia* **148**:356–364.

Stokes, E.J., *et al.* (2003) Female dispersal and reproductive success in wild western lowland gorillas (*Gorilla gorilla gorilla*). *Behavioral Ecology and Sociobiology* **54**:329–339.

Stopher, K.V., *et al.* (2011) The red deer rut revisited: female excursions but no evidence females move to mate with preferred males. *Behavioral Ecology* **22**:808–818.

Strier, K.B. (2008) The effects of kin on primate life histories. *Annual Review of Anthropology* **37**:21–36.

Strier, K.B. and Ziegler, T.E. (2000) Lack of pubertal influences on female dispersal in muriqui monkeys, *Brachyteles arachnoides*. *Animal Behaviour* **59**:849–860.

Stumpf, R. (2007) Chimpanzees and bonobos: diversity within and between species. In: *Primates in Perspective* (eds C.J. Campbell, A. Fuentes, K.C. MacKinnon, N. Panger and S.K. Bearder) Oxford: Oxford University Press, 321–344.

Sugiyama, Y. (1965a) Behavioral development and social structure in two troops of hanuman langurs (*Presbytis entellus*). *Primates* **6**:213–247.

Sugiyama, Y. (1965b) On the social change of hanuman langurs (*Presbytis entellus*) in their natural conditions. *Primates* **6**:381–417.

Sullivan, T.P., *et al.* (2001) Reinvasion dynamics of northern pocket gopher (*Thomomys talpoides*) populations in removal areas. *Crop Protection* **20**:189–198.

Sutherland, G.D., *et al.* (2000) Scaling of natal dispersal distances in terrestrial birds and mammals. *Conservation Ecology* **4**:16.

Swedell, L., *et al.* (2011) Female dispersal in hamadryas baboons: transfer among social units in a multilevel society. *American Journal of Physical Anthropology* **145**:360–370.

Szulkin, M. and Sheldon, B.C. (2008) Dispersal as a means of inbreeding avoidance in a wild bird population. *Proceedings of the Royal Society of London. Series B: Biological Sciences* **275**:703–711.

Szulkin, M., *et al.* (2007) Inbreeding depression along a life-history continuum in great tits. *Journal of Evolutionary Biology* **20**:1531–1543.

Szulkin, M., *et al.* (2013) Inbreeding avoidance, tolerance, or preference in animals? *Trends in Ecology and Evolution* **28**:205–211.

Telfer, S., *et al.* (2003) Parentage assignment detects frequent and large scale dispersal in water voles. *Molecular Ecology* **12**:1939–1949.

Templeton, A.R. (1986) Coadaptation and outbreeding depression. In: *Conservation Biology: The Science of Scarcity and Diversity* (ed. M. Soulé) Sunderland, MA: Sinauer Associates, 105–116.

Thierry, B. (2007) The macaques: a double-layered social organisation. In: *Primates in Perspective* (eds C.J. Campbell, A. Fuentes, K.C. MacKinnon, M. Panger and S.K. Bearder) New York: Oxford University Press, 224–239.

Thierry, B. (2008) Primate socioecology, the lost dream of ecological determinism. *Evolutionary Anthropology* **17**:93–96.

Tilson, R.L. (1981) Family formation strategies of Kloss's gibbons. *Folia Primatologica* **35**:259–287.

VanderWaal, K.L., *et al.* (2009) Optimal group size, dispersal decisions and postdispersal relationships in female African lions. *Animal Behaviour* **77**:949–954.

Van Horn, R.C., *et al.* (2007) Divided destinies: group choice by female savannah baboons during social group fission. *Behavioral Ecology and Sociobiology* **61**:1823–1837.

van Noordwijk, A.J. (1995) On bias due to observer distribution in the analysis of data on natal dispersal in birds. *Journal of Applied Statistics* **22**:683–694.

van Schaik, C.P. (1989) The ecology of social relationships amongst female primates. In: *Comparative Socioecology: The Behavioural Ecology of Humans and Other Mammals* (eds V. Standen and R.A. Foley) Oxford: Blackwell Scientific Publications, 195–218.

Van Vuren, D. and Armitage, K.B. (1994) Survival of dispersing and philopatric yellow-bellied marmots: what is the cost of dispersal? *Oikos* **69**:179–181.

Walling, C.A., *et al.* (2011) Inbreeding depression in red deer calves. *Evolutionary Biology* **11**:318.

Waser, N.M. (1993) Population structure, optimal outbreeding, and assortative mating in angiosperms. In: *Natural History of Inbreeding and Outbreeding: Theoretical and Empirical Perspectives* (ed. N.W. Thornhill). Chicago: University of Chicago Press, 173–199.

Waser, P.M. (1996) Patterns and consequences of dispersal in gregarious carnivores. In: *Carnivore Behavior, Ecology and Evolution* (ed. J.L. Gittleman) Ithaca, NY: Cornell University Press, 267–295.

Waser, P.M. and Jones, W.T. (1983) Natal philopatry among solitary mammals. *Quarterly Review of Biology* **58**:355–390.

Waser, P.M., *et al.* (1986) When should animals tolerate inbreeding? *American Naturalist* **128**:529–537.

Waser, P.M., *et al.* (1994) Death and disappearance: estimating mortality risks associated with philopatry and dispersal. *Behavioral Ecology* **5**:135–141.

Waser, P.M., *et al.* (2006) Parentage analysis affects cryptic precapture dispersal in a philopatric rodent. *Molecular Ecology* **15**:1929–1937.

Waterman, J.M. (2002) Delayed maturity, group fission and the limits of group size in female Cape ground squirrels (Sciuridae: *Xerus inauris*). *Journal of Zoology* **256**:113–120.

Whitehead, H. and Weilgart, L. (2000) The sperm whale: social females and roving males. In: *Cetacean Societies: Field Studies of Dolphins and Whales* (eds J. Mann, R. Connor, P. Tyack and H. Whitehead) Chicago: University of Chicago Press, 154–172.

Widdig, A., *et al.* (2006) Consequences of group fission for the patterns of relatedness among rhesus macaques. *Molecular Ecology* **15**:3825–3832.

Wolff, J.O. (1992) Parents suppress reproduction and stimulate dispersal in opposite sex juvenile white-footed mice. *Nature* **359**:409–410.

Wolff, J.O. (1993) What is the role of adults in mammalian juvenile dispersal? *Oikos* **68**:173–176.

Wolff, J.O. (1994) Reproductive success of solitarily and communally nesting white-footed mice and deer mice. *Behavioral Ecology* **5**:206–209.

Wrangham, R.W. (1980) An ecological model of female-bonded primate groups. *Behaviour* **75**:262–300.

Wrangham, R.W. (1986) Ecology and social evolution in two species of chimpanzees. In: *Ecology and Social Evolution: Birds and Mammals* (eds D.I. Rubenstein and R.W. Wrangham) Princeton, NJ: Princeton University Press, 352–378.

Young, A.J. and Monfort, S.L. (2009) Stress and the costs of extra-territorial movement in a social carnivore. *Biology Letters* **5**:439–441.

Young, A.J., *et al.* (2006) Stress and the suppression of subordinate reproduction in cooperatively breeding meerkats. *Proceedings of the National Academy of Sciences of the United States of America* **103**:12005–12010.

CHAPTER 4
Female mating decisions

4.1 Introduction

In a corner of Sussex downland, twenty fallow deer bucks defend territories in a cluster. The individual territories are small, while the entire cluster of territories (the 'lek') is little more than 100 m across (Figure 4.1). Several territories contain females, who stand uneasily, staring around them, herded by the resident male. Two females suddenly leave the territory they are on and run to a neighbouring territory that already contains half a dozen other females and a large male. They are pursued by their original male but the new territory holder runs over and chases him off. Close by, a mixed herd of over a hundred females and non-territorial males grazes peacefully. One female, pestered by several young males, runs out of the herd and gallops to the lek pursued by several males. She stops briefly at one of the peripheral territories and then runs to join the largest group of females.

Lek breeding systems like those of fallow deer are not common in mammals though they raise important questions about the mating decisions of females (see Chapter 10). Why do receptive females move to leks before mating? Why do they concentrate on particular territories? Do they share a preference for particular territories, or for the males that defend them? In either case, what cues are they using? Do they copy each other's preferences for territories or mating partners? And what benefits do they gain by mating with particular partners? Or with different numbers of partners?

Though these questions are brought into sharp focus by lek breeding systems, similar questions need to be asked about female mating decisions in other mammalian societies. In some mammals, females may distribute themselves in relation to food and safety, forming stable groups which males compete to monopolise, so that opportunities for female mating preferences are limited

(see Chapter 10). However, opportunities for females to influence the identity of their mating partners exist in most systems and the ability of males to constrain female mating preferences has probably been overestimated (Smale 1988; Andersson 1994; Drea 2005).

Female mating decisions have been the focus of intensive interest for more than 30 years and a large number of studies have explored female mating preferences in mammals. Before describing the evidence of female choice, it is worth clarifying exactly what I mean when I say that females choose or show a preference for particular mating partners. I use female choice to mean cases where females show an active preference for mating with particular categories of males, whether or not matings lead to conception. Unlike some other definitions, this one carries the implication that not all forms of female behaviour that affect the distribution of mating success in males should necessarily be regarded as mating preferences. For example, in Grevy's zebra, females in the later stages of pregnancy restrict their usually large ranges and move to areas where water is available and they can drink on a daily basis during the period of lactation (Rubenstein 1986) (Figure 4.2). As in other equids, females have a post-partum oestrus and males establish territories in the vicinity of water sources and mate with receptive females visiting sources in their territories. Although female movements affect the distribution of male mating success, differences in mating success between males may simply be a consequence of female habitat preferences. Similarly, in some populations of fallow deer, males hold mating territories under oak trees and their mating success is predicted by variation in the abundance of acorns, which attract females (see Chapter 10). I refer to cases of this kind as 'coincidental' mate choice to emphasise that female preferences have not necessarily evolved through a process of sexual selection.

Mammal Societies, First Edition. Tim Clutton-Brock.
© 2016 Tim Clutton-Brock. Published 2016 by John Wiley & Sons, Ltd.

Figure 4.1 A fallow deer lek in Petworth Park, Sussex, at the height of the annual rut. *Source*: © Tim Clutton-Brock.

In many cases where males defend resource-based territories, their mating success is correlated both with characteristics of their territory and with aspects of their phenotype. For example, in puku (a riverine antelope related to kob), males hold resource-based territories in areas of grassland favoured by grazing females (Rosser 1987, 1992) (Figure 4.3). Females (including receptive and non-receptive individuals) spend more time grazing

Figure 4.2 In Grevy's zebra, females have a post-partum oestrus and males defend mating territories close to water sources that are visited by females in the weeks immediately after parturition. *Source*: © Tui De Roy/Minden Pictures/Getty Images.

(a)

(b)

Figure 4.3 (a) A male puku with females on his territory in swale vegetation on an old river meander in Luangwa Valley, Zambia; (b) a mature male puku guarding his territory. *Source*: (a,b) © Tim Clutton-Brock.

on territories where the risk of predation is low and food is abundant and males defending preferred territories get more matings than those with less preferred ones, though male mating success is independently correlated with indices of male phenotype (neck patch intensity) (Rosser 1992) (Figure 4.4). Similar relationships occur in populations of topi where males defend resource-based territories (Balmford *et al.* 1992) and studies suggest that male success depends partly on territory quality and partly on female preferences for mating with particular males. However, it is not possible to be certain that correlations between male mating success and male phenotype are a consequence of female mating preferences, for they could also be a by-product of associations between male phenotype and the relative ability of males to defend females effectively (see Chapter 10).

The problems of distinguishing between the effects of female mating preferences and those of male–male competition are not confined to species where males hold resource-based territories. In a number of mammals, females are more mobile around the time of mating and this tendency is sometimes interpreted as evidence of attempts by females to broaden their choice of mating partners (Richard *et al.* 2008; Stopher *et al.* 2011). However, as oestrus approaches, females are often increasingly harassed by males (see Chapter 15), and an alternative possibility is that increased mobility in females close to oestrus is a by-product of increased harassment by males. Even though females that move further are more likely to breed with males of superior quality, this does not necessarily eliminate the effects of competition between males.

For example, studies of the movements of female fur seals between harems guarded by static males show that females that move further to mate are more likely to breed with relatively heterozygous males (Hoffman *et al.* 2007). However, heterozygosity is a significant predictor of male phenotypic quality in fur seals (Hoffman *et al.* 2004) and one possible interpretation of these results is that males of superior quality or condition are more likely than others to be able to retain oestrous females within their territory (Stillman *et al.* 1996).

Because of the difficulties of identifying female mating preferences and their effects on male success, studies of female choice have often focused on lek breeding populations where adult males defend territories that contain no significant resources where they are visited by females that are ready to mate. For example, in many deer and antelope, males defend mating territories on leks when they are visited by receptive females (Figure 4.5) and mating rates vary widely between individual males. The most successful males are commonly large, in good condition and in their prime and often have well-developed secondary sexual characters, like horns or antlers (Clutton-Brock *et al.* 1993; Fiske *et al.* 1998), so that mating success and the development of secondary sexual characters are often correlated. It is often assumed that these correlations provide evidence of consistent female preferences for mating with well-developed males in these species, but detailed studies show that the situation is more complex (Clutton-Brock *et al.* 1996). Despite the absence of significant resources on leks, females often show preferences for particular territories and much of

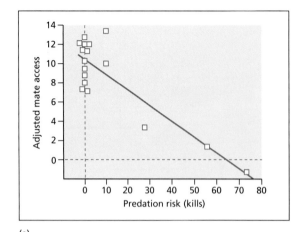

(a)

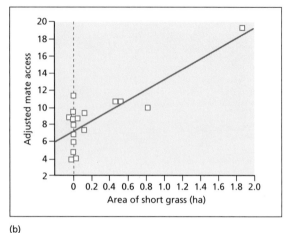

(b)

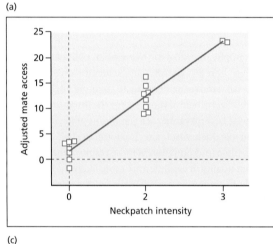

(c)

Figure 4.4 In species where males defend resource-based mating territories, their access to females often depends on the characteristics of their territory as well as their own phenotype. Plots show independent relationships between the number of female puku visiting the territories of different males in Luangwa Valley (Zambia) in relation to (a) the percentage of observed kills that occurred on the territory, (b) the area of short swale grass (a preferred sward), and (c) the intensity of the male's neck patch (a signal of condition and quality). *Source*: From Rosser (1992). Reproduced with permission of Oxford University Press.

the variation in mating success between males often appears to be a consequence of the characteristics of the territories they defend rather than of the male's phenotype (see Chapter 10). As males fight for access to territories preferred by females (see Figure 4.5), larger, stronger males are more likely to gain access to the most popular territories, generating correlations between mating success and male characteristics that may have no causal basis. As a result, it is usually difficult to assess the extent to which female mating preferences contribute to variation in male success.

The problems associated with attempts to separate the effects of male–male competition and female mating preferences emphasise the need for experiments which are able to control for the effects of competition between males, with the result that much of the most convincing evidence of female mating preferences comes from

studies of captive animals. However, experiments with captive animals face other problems since it is usually difficult to assess the extent to which preferences demonstrated in captive groups play an important role in determining the distribution of male breeding success in natural populations (Clutton-Brock and McAuliffe 2009). Moreover, because of the difficulties of manipulating mate choice, many studies examine effects of male characteristics on female attraction or attention, rather than on copulation or paternity and although the responses of females to different males may sometimes reflect mating preferences, this is not always the case. For example, the extent to which females gaze at or approach different males may be affected by the chance that males will respond aggressively as well as by attraction and even measures of mating frequency do not necessarily solve this problem, for some mammals

(a)

(h)

Figure 4.5 (a) Male topi on an East African lek; (b) males fight for access to territories and females. *Source*: (a,b) © Jakob Bro-Jørgensen.

use non-ejaculatory copulations as part of courtship (Dewsbury 1988; Penn and Potts 1998a). The reproductive status of females used in experiments and the stage of the reproductive cycle at which they are conducted can also have an important influence on the strength and direction of mate choice (Widemo and Sæther 1999): in many systems, female movements or mating decisions are affected by the behaviour of other individuals and females seldom make fully independent decisions (see section 4.5).

This chapter reviews evidence of female mating preferences both in natural systems and in captive mammals. The next two sections review studies of the direct benefits of mate choice to female fitness (section 4.2) and of genetic benefits to the fitness of their offspring (section 4.3). Sections 4.4 and 4.5 examine empirical evidence of female preferences for different male traits, and of mate choice copying by females. Sections 4.6 and 4.7 examine the reasons for contrasts in the number of mating partners and for variation in mate choice while section 4.8 describes some of the consequences of variation in female mating preferences.

4.2 Direct benefits of mate choice to females

The potential benefits of mate choice vary widely and have been extensively examined (Andersson 1994). They fall into two main groups: non-genetic or 'direct' benefits, where mate choice contributes to the female's own survival or fecundity; and genetic benefits, where mate choice affects the fitness of a female's progeny through its effects on their genotype (see Andersson 1994).

At least four different ways in which mate choice can generate direct benefits to the fitness of females have been suggested. In some cases, mate choice may contribute to the safety of females. In polygynous societies where males compete intensely for females, they commonly adopt coercive tactics that can injure or kill females (see Chapter 15) and receptive females can gain substantial benefits by joining or mating with males capable of providing effective protection against rivals (Rubenstein 1986; Smuts and Smuts 1993). In some species, dominant males punish females that associate or mate with subordinates and females may gain direct benefits by associating or mating with dominant males.

Mate choice can also contribute to the fecundity of females by reducing the frequency of infertile matings. Though it is often assumed that sperm are always readily available, the sperm supplies of successful males can become depleted (Smale 1988). For example, in Soay sheep, larger males become sperm depleted in the later stages of the rut and the breeding success of smaller males increases (Hogg and Forbes 1997; Preston et al. 2001). Selection is likely to favour female preferences for males with adequate supplies of sperm and this may explain why females often show little interest in males that do not display actively (McComb 1991; Byers et al. 1994).

Mate choice can also generate social benefits. By mating with dominant males, females may gain improved access to food resources because dominant males tolerate their presence and allow them to feed in close proximity. For example, in capuchin monkeys, females solicit copulations with the most dominant male in the group and, by doing so, gain improved access to food resources when food abundance is low (Janson 1984). Similarly, in chimpanzees, females that mate with males are

sometimes allowed to share kills (Stanford 1999). Studies of a number of other primates suggest that affiliative or sexual interactions between females and males commonly influence the extent to which females are tolerated at feeding sites when food is scarce (Barrett and Henzi 2006) (see Chapter 15).

Finally, females may increase their fitness by selecting mates that are able or willing to help to invest in rearing or protecting their progeny if this increases their breeding success or survival or that of their progeny. For example, where males contribute to parental care, females might be expected to favour individuals that are capable and ready to invest in their young and competing males would be expected to advertise their capacity for investment (Heywood 1989; Hoelzer 1989). Similarly, where immigrant males are commonly infanticidal, females may be able to reduce the risk of infanticide either by copulating with immigrant males or by forming close 'friendships' with males that are able to protect their offspring against infanticide (Hrdy 1977) (see Chapter 15).

While exercising mate choice may generate benefits to females, it can also have substantial costs. In some cases, the energetic costs of sampling mating territories may be high: for example, studies of pronghorn antelope (Figure 4.6) suggest that the energetic costs to females of sampling different leks may be substantial (Byers et al. 2005). Other potential costs are likely to include those associated with male harassment or punishment or with reductions in investment by regular partners (see Chapter 15).

4.3 Genetic benefits of mate choice to females

Mate choice can also generate genetic benefits that increase the fitness of the progeny of females that mate with males of superior quality. There are two main explanations of the evolution of female mating preferences (Andersson 1994).

1 *Good genes:* in many animals, females may be able to increase the fitness of their offspring by mating with males with superior genotypes, generating selection pressure on males for traits ('indicators') that advertise their genetic quality (Fisher 1915; Williams 1966; Zahavi 1975). The evolution of male traits through female selection or 'good genes' operates most effectively where breeding success and indicator traits both

Figure 4.6 In pronghorn antelope, males defend females on leks and females commonly visit several leks before mating. *Source*: © Eric Dragesco/Ardea.

show high levels of heritability (Andersson 1994) and studies of a range of animals provide evidence of the heritability of male secondary sexual characters (Alatalo *et al.* 1997). As well as mating selectively with males carrying 'good genes', females should avoid mating with males that are likely to carry genes that detract from the fitness of their offspring. In particular, mating with close relatives lowers average levels of heterozygosity among offspring and, in many animals that usually outbreed, increases the chance that off-spring will be homozygous for deleterious recessive alleles likely to reduce fitness (see Chapter 5). These effects are likely to occur throughout the range of parental relatedness and are not restricted to matings between close relatives (see Slate *et al.* 2000).

2 *Sexy sons*: an alternative model of the evolution of female mating preferences operating through genetic benefits suggests that females favour males that are attractive to other females since they are likely to father sons who are also attractive to females. This idea is usually referred to as the 'sexy sons' hypothesis and was also first outlined by Fisher (1930), who envisaged an initial situation where some aspect of male phenotype (such as tail length or face colour) was

heritable and was associated with a small advantage in male viability, so that females that mated with males carrying the trait were more likely to produce sons with increased viability (see Andersson 1994). Assuming that indicator traits and female preferences both have some heritable basis, preferences for the trait would then spread through the population and males carrying the preferred trait would show increased mating success, generating progressively stronger selection for female preferences for partners that carry it. If females select males on the relative development of the trait (rather than those above an absolute threshold), this could lead to 'runaway' processes, where both the relative development of the indicator and the strength of female preferences coevolve until the increasing costs of indicator traits to viability offset the benefits of further increases in their development (O'Donald 1962, 1967; Andersson 1994; Bakker and Pomiankowski 1995).

An important distinction between 'sexy sons' and 'good genes' models of mate choice is that, in 'sexy sons' models, female mating preferences for attractive partners should increase the fitness of sons but not that of daughters, while 'good genes' selection should benefit the fitness of daughters **as** well as sons. Several studies of the effects of female choice on offspring fitness show that the development of sexually selected characters are correlated with other indicators of fitness, suggesting that they act as indicators of the genetic quality of males (Reynolds and Gross 1992; Nicoletto 1995; Wilkinson 1998). However, studies that have manipulated the extent to which females can choose their partners show that mate choice can increase the fitness of offspring of both sexes. For example, experiments with house mice that allow females to choose between two males and then pair them with either preferred or with non-preferred males show that offspring of both sexes fathered by preferred males have larger ranges, build better nests and are more likely to survive to 60 days (Drickamer *et al.* 2000).

Both 'good genes' and 'sexy sons' models of sexual selection face a number of problems. First, it is necessary to explain why selection has not eroded heritable variation in male quality or male attractiveness. Empirical studies of natural populations show that many traits are partly heritable, though those closely associated with fitness tend to have lower heritabilities than those that have little effect on fitness (Fisher 1930; Gustafsson 1986; Mousseau and Roff 1987). Several possible solutions have been suggested why this should be the case, including temporal and spatial variation in selection pressures (Chaine and Lyon 2008), mutation pressure and pleiotropic effects of particular genes (see Andersson 1994). In addition, Hamilton and Zuk (1982) proposed that male indicators of genetic quality might reflect resistance to parasites or disease, and that arms races between parasites and their hosts may maintain heritable variation in male traits.

A second problem is that female preferences for males with highly developed ornaments or indicators might be expected to favour the evolution of cheats that signal genetic qualities that they do not possess. One possible solution is that females select males on the basis of indicators that are sufficiently costly that inferior males cannot imitate them without depressing their fitness (Zahavi 1975, 1977). While Zahavi's 'handicap' principle was initially criticised, it is now generally accepted that high costs can provide a guarantee of signal honesty (Grafen 1990; Maynard Smith and Harper 2003), though distinguishing between handicaps and costly indices presents problems (see Chapter 7). Another possibility is that females select males on the basis of multi-component signals that are difficult to fake (Rowe 1999) or of traits that are inevitably linked to fitness by physical or genetic constraints.

A third difficulty is that mating preferences may involve costs to females which can have an important influence on the evolution of female choice (Parker 1983; Pomiankowski 1988). Few empirical studies have been able to measure the fitness costs of female choice and it is often unclear whether these are large or small (Andersson 1994). In many systems, the costs of choice may be masked by direct benefits. For example, where males with highly developed ornaments can provide more effective protection of females against other males, these may obscure the costs of mate searching (Smuts and Smuts 1993).

Though different models of the evolution of female mate choice and male ornaments are often presented as alternatives, multiple processes are likely to be involved in the evolution of female mating preferences and are frequently difficult to separate (Balmford and Read 1991; Kokko 2003; Kokko *et al.* 2003). For example, in many cases where females appear to select males to gain direct fitness benefits, their choice may also have genetic benefits to their offspring, while in cases where females appear to select males on the basis of indirect benefits, their decisions may also generate direct fitness benefits. Moreover, selection pressures outside the context of mate choice may often affect the evolution of male traits and female preferences, which may evolve independently of each other (Partridge and Hurst 1998; Ryan 1998).

4.4 Female mating preferences

Females can select mating partners on the basis of a wide range of criteria and this section describes evidence of female preferences for ten different male traits.

Maturity
Where younger or less powerful males are likely to lose mating opportunities to older or stronger individuals unless they copulate quickly, selection often favours coercive strategies, especially in younger or more subordinate males (see Chapter 15). The costs of male coercion to females can be high, especially in systems where multiple males compete simultaneously for the same female (Smuts and Smuts 1993) and, in many of these

species, females avoid mating with immature males that are unable to provide effective defence against rivals (Fox 2002). Elephant seals provide some of the best examples: they breed in large dense colonies; females are commonly courted by multiple males; and, unless one male can control access to them, both females and their pups can be injured (Reiter *et al.* 1981; Le Boeuf and Mesnick 1991). Mature males are usually able to monopolise access to females, providing effective protection against harassment by other males, while younger males are unable to do so. Females show a preference for joining larger harems where they (and their pups) are likely to be safer from male harassment (Pistorius *et al.* 2001; McMahon and Bradshaw 2004), and females that are being courted by younger males give calls that attract the attention of dominant males, who often chase away younger suitors (Figure 4.7) (see Chapter 7).

Costs of male harassment and coercion to females are also high in other systems. For example, in ungulates that live in unstable herds, oestrous females risk being killed by competing males unless they join males capable of protecting them against rivals and often show preferences for joining larger groups (Clutton-Brock *et al.* 1993). In some social primates, females avoid mating with immature partners and, by doing so, may reduce the risk of dangerous harassment or

punishment by older males (Manson 1994a,b; Fox 2002). Like female elephant seals, females often incite competition between males and may gain direct benefits from this (Kuester and Paul 1992). An additional reason why females avoid mating with younger partners may sometimes be that immature males are less likely to be fertile (Cox and Le Boeuf 1977; Small 1988; Byers *et al.* 1994).

Dominance

In a number of mammals, females show a consistent preference for mating with dominant males; however, in natural populations, it is often difficult to exclude the possibility that an increase in female matings with dominant males is a consequence of competition between males or of coercive tactics. In some rodents, females offered a choice of dominant or subordinate partners preferentially mate with more dominant males (Huck and Banks 1982; Shapiro and Dewsbury 1986; Drickamer 1992; Solomon and Keane 2007). Similarly, in natural populations of social primates and carnivores, females often appear to favour dominant partners (Seyfarth 1978; Robinson 1982; Silk and Boyd 1983; Janson 1984, West and Packer 2002) though this is not always the case and, especially in larger groups, dominant males may obtain a smaller share of matings

Figure 4.7 A female elephant seal vocalising while being mounted. *Source*: © Nature Picture Library.

than priority of access models would predict (Fedigan 1983; Manson 1992; Soltis and McElreath 2001; Fox 2002; Alberts *et al.* 2003; Manson 2006). In some lek-breeding ungulates, too, females appear to show a preference for dominant males. For example, studies of topi, where dominant males defending central territories on leks obtain a high proportion of matings, show that females mate more rapidly after arrival on a central territory than on a peripheral one (Bro-Jørgensen 2002; Bro-Jørgensen and Durant 2003), though another interpretation of these results is that females in full oestrus collect in the centre of leks as a consequence of successive movements between territories, leading to reductions in latency to mating on central territories (see Chapter 10).

Selective mating with dominant males can provide direct as well as genetic benefits to the fitness of progeny that can complicate attempts to identify the genetic benefits of mate choice. For example, in some cercopithecine primates that live in multi-male groups, fathers are more likely to protect or support their own offspring than those of other males (see Chapter 15). As high-ranking males can provide more effective care for infants than low-ranking males (Alexander 1970; Stein 1984; Agrell *et al.* 1998), offspring survival may differ between males that vary in social rank and female preferences for dominant males may affect the survival of their progeny as a result of direct fitness benefits.

Fertility

Although males of some polygynous species are capable of fertilising substantial numbers of females, the ejaculates of males that mate repeatedly over a short period often show reduced sperm counts and fertility (Austin and Dewsbury 1986; Huck *et al.* 1986; Preston *et al.* 2001), and their capacity to impregnate females may be temporarily limited (Dewsbury 1982a–c). Several studies of rodents show that females mate preferentially with unmated males (Kramer and Mastromatteo 1973; Huck and Banks 1982; Pierce and Dewsbury 1991). This tendency may be more pronounced in monogamous species, where males commonly care for their offspring and the costs of pairing with an infertile partner may be particularly high, than in polygynous species (Pierce and Dewsbury 1991; Salo and Dewsbury 1995; though see Solomon and Keane 2007). In addition, in some polygynous mammals, oestrous females compete for access to males (Bro-Jørgensen 2002; Bebié and McElligott 2006),

and one interpretation of these observations is that the sperm supplies of 'popular' partners are limited (see Preston *et al.* 2001).

Investment

Where males contribute to rearing or protecting their offspring, females might be expected to favour partners that are prepared to invest heavily. In several baboons and macaques that live in multi-male groups, females develop close affiliative relationships or 'friendships' with particular males (see Chapter 15). 'Friends' are often males that previously held high status in the group, though they seldom include the current alpha male, and females are often more likely to mate with male 'friends' than with other males (Seyfarth 1978; Rasmussen 1983; Smuts 1985) though this is not always the case (Hemelrijk *et al.* 1999; Manson 2006). Males may benefit by forming friendships with females because this increases their mating success or their access to infants used in triadic interactions or because this allows them to contribute to the safety or development of their own offspring. Friendships may benefit females because their 'friends' support them in conflicts with other females or because they increase the chance that males will support their offspring at subsequent stages of development. In addition, friendships with particular males may help to protect a female's offspring against attacks by infanticidal males (see Chapter 15).

Where males invest in their offspring, they may signal their capacity or readiness to provide care or protection of their offspring to prospective mates. In some biparental birds, males feed their mates during courtship and this may allow females to assess the extent to which they are likely to invest in subsequent offspring (Nisbet 1973; Tasker and Mills 1981). Similarly, in some tamarins where multiple males often help to carry young produced by a single female, infant carrying by males may represent a form of courtship and females may be more likely to mate with males that contribute to carrying previous litters (Price 1990; Ferrari 1992), though direct evidence of this is lacking (Baker *et al.* 1993; Tardif and Bales 1997). Studies of captive vervet monkeys show that subordinate males are more likely to behave affiliatively to infants if they are aware that their mother can monitor the interaction, while dominant males show no significant tendency to alter their responses to infants in relation to the presence of their mothers (Figure 4.8a). In addition,

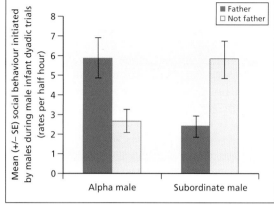

(a)

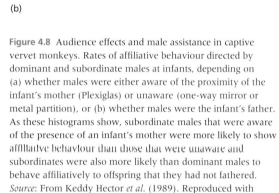

(b)

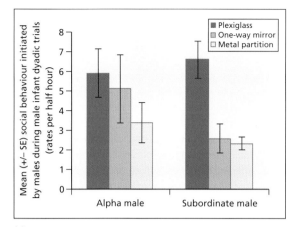

Figure 4.8 Audience effects and male assistance in captive vervet monkeys. Rates of affiliative behaviour directed by dominant and subordinate males at infants, depending on (a) whether males were either aware of the proximity of the infant's mother (Plexiglas) or unaware (one-way mirror or metal partition), or (b) whether males were the infant's father. As these histograms show, subordinate males that were aware of the presence of an infant's mother were more likely to show affiliative behaviour than those that were unaware and subordinates were also more likely than dominant males to behave affiliatively to offspring that they had not fathered. *Source*: From Keddy Hector *et al.* (1989). Reproduced with permission of Elsevier. *Photo source*: © Dorothy Cheney.

subordinate males that are not the father of the infant are more likely to direct affiliative interactions towards infants than dominant males (Figure 4.8b).

Weaponry

In many polygynous mammals, strong sexual selection favouring competitive ability in males has led to the evolution of sexually dimorphic weaponry (see Chapter 18). Sexually dimorphic traits frequently have relatively low growth priority (Huxley 1932), with the result that the relative size of male weaponry is often correlated with individual differences in growth, body size, competitive ability and mating success (Clutton-Brock 1982; Alvarez 1990; Prichard *et al.* 1999; Coltman *et al.* 2002; Mysterud *et al.* 2005). As a result, it is frequently suggested that females show a preference for mating with males with highly developed weapons and that female choice has played an important role in the evolution of male weaponry: for example, studies of

ungulates have commonly shown that individual differences in the breeding success of males are correlated with variation in the absolute or relative size of their horns or antlers and have interpreted this as evidence that female mating preferences have played an important role in the evolution of horn or antler size (Espmark 1964; Geist 1971; Hitchkott *et al.* 2001; Vanpé *et al.* 2007). While this could be the case, associations between the relative development of male weapons and male mating success provide no firm indication that male weaponry is used by females in selecting potential mating partners since they could be a consequence of correlations between weapon size and success in intrasexual competition. In order to show that females prefer males with large weapons it is necessary to manipulate weapon size and to show that correlations between the size of male weapons and male mating success are a consequence of female mating preferences. As yet, all attempts to do this have either been unsuccessful (Lincoln 1994) or have failed to demonstrate that females select males on the basis of the relative size of weapons. For example, experiments where female fallow deer were experimentally induced into oestrus and were offered a choice of males with antlers and males that had had their antlers removed found no evidence of a preference for antlered males (see Chapter 10). Despite this, correlations between horn or antler size and male mating success in ungulates continues to be cited as evidence that females prefer mating with males with large antlers and that female mating preferences have played an important role in the evolution of antlers and horns.

Another suggestion is that low levels of fluctuating asymmetry in the size or structure of male weapons are an indicator of developmental stability and genetic quality (Møller and Pomiankowski 1993; Møller and Swaddle 1997) and some studies of ungulates have suggested that there may be an association between the mating success of males and the symmetry of their horns (Møller *et al.* 1996). However, there is little evidence that this is the case: for example, detailed studies of red deer provide no evidence that antler symmetry is related to mating success and its heritability is low (Kruuk *et al.* 2003).

Coloration

In a substantial number of mammals, males with relatively high testosterone levels are both more frequently aggressive and darker or more distinctively coloured than males with lower testosterone levels, and are commonly preferred as mating partners by females (Vandenbergh 1965; Gerald 2001, 2003; Setchell and Dixson 2001). For example, in African lions, males with dark manes (Figure 4.9a) have higher testosterone levels than light-maned individuals and are more likely to lead in encounters with other prides (West and Packer 2002).

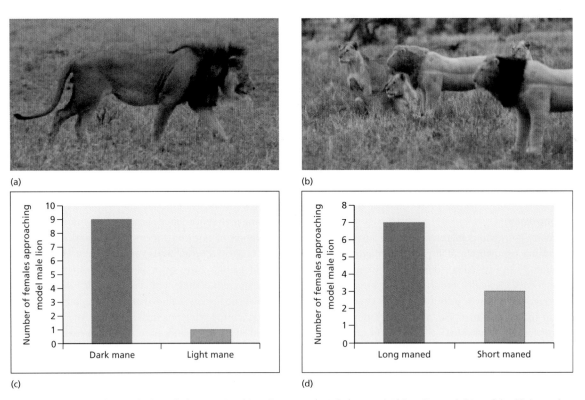

(a)   (b)

(c)   (d)

Figure 4.9 Female preference for long dark manes in African lionesses: (a) a dark-maned African lion and (b) models of light- and dark-maned lions with approaching lionesses. Results showed that lionesses were more likely to approach dark-maned models (c) and models with relatively long manes (d) than light- or short-maned males. *Source*: From West and Packer (2002). Reproduced with permission from AAAS.

Receptive lionesses commonly mate with the darkest-maned male in their group and experiments in which lionesses were presented with life-sized models of light- and dark-maned lions (Figure 4.9b) showed that lionesses were more likely to approach lions with dark manes (Figure 4.9c) as well as males with relatively long manes (Figure 4.9d) (West and Packer 2002). Dark manes retain heat more than light ones and mane colour could be a handicap, providing females with a reliable indication of male quality (Zahavi 1975), though an alternative explanation is that females gain direct benefits from choosing males with dark manes since they have longer breeding tenures with the result that the risk of male infanticide is reduced (Packer *et al.* 1988).

Unlike many other groups of placental mammals, most primates possess trichromatic colour vision (Jacobs 1993)

and, in some species, including vervet monkeys (Gartlan and Brain 1968; Gerald 2001), gelada baboons (Dunbar 1984) and mandrills (Setchell and Dixson 2001), males of high status have more brightly coloured faces, perineal regions, or testes than subordinates (Gerald 2001) (Figure 4.10). Females are often attracted to the brightest males: for example, in mandrill groups, which include large numbers of males, one is typically brighter than all others, and females interact, groom and mate more frequently with him than with other mature males (Setchell 2005) (Figure 4.11). Although brightly coloured males tend to have high status within their groups, the effects of male coloration are stronger than those of rank and remain when the influence of male rank is allowed for (Setchell and Dixson 2001). Similarly, in rhesus macaques, the colour of male faces darkens in

(a)   (b)

Figure 4.10 A bright male mandrill (a) and a paler male (b). *Source*: © Joanna M. Setchell, Centre International de Recherches Medicales de Franceville.

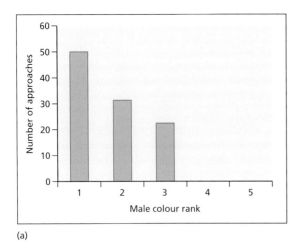

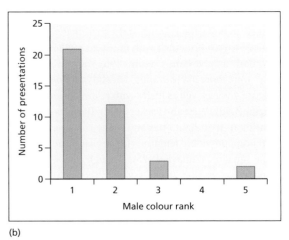

(a)
(b)

Figure 4.11 Male coloration and female preferences in mandrills: (a) female responsibility for proximity, calculated as the number of approaches made by peri-ovulatory females divided by the total number of approaches between males and peri-ovulatory females for males of different colour rank; (b) number of presentations received from peri-ovulatory females by males of different colour rank. *Source*: From Setchell (2005). Reproduced with permission of Springer Science and Business Media.

the breeding season and varies between individuals (Figure 4.12) and females direct a higher proportion of socio-sexual behaviour towards darker males (Dubuc *et al.* 2014) (Figure 4.13). In this species there is no consistent relationship between the skin colour of males and their dominance status (Higham *et al.* 2013) so it is unlikely that male coloration has evolved as an intra-sexual signal.

One possible explanation of female preferences for brightly coloured males is that they increase the chance that a female's daughters will inherit genes providing effective resistance to parasites (Hamilton and Zuk 1982;

Folstad and Karter 1992). This argument is based on the fact that bright skin colouring is commonly associated with relatively high testosterone levels, which suppress immune function (Vandenbergh 1965), so that male brightness may signal an individual's resistance to infection. While this may be the case, there are several other possible reasons why females pay more attention to brightly coloured males. For example, if bright males are likely to have low parasite levels, mating with them may reduce the female's own chances of becoming infected (Loehle 1997). Alternatively, females may pay greater attention to brighter males because they are more

Figure 4.12 Male rhesus monkeys vary in facial colour during the breeding season, ranging from pale pink to dark red. *Source*: © Constance Dubuc.

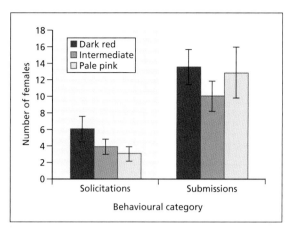

Figure 4.13 Female rhesus macaques respond to variation in the facial colour of males. Variation among three male categories in the average number of females ± SEM that exhibited proceptive and submissive behaviours toward them (data presented without transformation). *Source*: From Dubuc *et al.* (2014). Reproduced with permission of Springer Science and Business Media.

visible ('sensory exploitation') or because they are more likely to be aggressive and so pose greater risks to females who ignore their advances.

Vocal displays

In many mammals, male displays play an important role in the mating decisions of females and can have direct effects on hormonal levels in females and influence the timing of ovulation. For example, in domestic sheep, male courtship affects gonadotropin secretion in females and the timing of the pre-ovulatory surge of luteinising hormone, which leads to ovulation (Pearce and Oldham 1984). Similarly, in red deer, exposure to the roaring displays of males advances the timing of oestrus in females (McComb 1987). Male displays may be used by female mammals in the selection of mating partners: in red deer and fallow deer, the repeated roars of harem-holding stags attract the attention of receptive females, who are more likely to join the harems of males with high roaring rates (McComb 1991; McElligott and Hayden 2001) and similar benefits of high calling rates to males have been demonstrated in some lemurs (Craul *et al.* 2004).

Specific acoustic properties of male calls may also be important in attracting females (Davidson and Wilkinson 2004). For example, the formant frequencies of red deer roars decrease with increasing body size, and females preferentially approach speakers playing roars with lower formant values (Charlton *et al.* 2007). In white-lined bats, males that defend mating territories in tree buttresses produce complex vocalisations that attract females (Catchpole 1980; Hiebert *et al.* 1989), and males with more complex 'songs' have more females on their territories (Davidson and Wilkinson 2004), though it has not yet been shown that the complexity of male songs has a direct effect on female mating preferences or male success. Although these studies support the suggestion that male displays are important to female mate choice, the same displays commonly discourage potential challengers (see Chapter 13) so that it is often difficult to assess the extent to which female preferences are responsible for variation in male success.

Olfactory displays

Olfactory cues also play an important role in female mating preferences and may be involved in female preferences for many phenotypic and genetic characteristics in males (Gosling and Roberts 2001a,b). Females commonly investigate male scent marks, which frequently reflect male condition and other male characteristics (Moore and Marchinton 1974; Sawyer *et al.* 1989; Gosling and Roberts 2001a). For example, in several rodents, females are less attracted to odours from males infected with parasites than to those of uninfected males, and females given a choice between infected and uninfected males are less likely to mate with infected males (Clayton 1991; Kavaliers and Colwell 1995a,b; Penn and Potts 1998a; Klein *et al.* 1999; Willis and Poulin 2000; Ehman and Scott 2001). Experiments with house mice also show that females given a choice between males that had been infected with intestinal nematodes and uninfected males are less likely to mate with infected males and produce fewer male offspring if they did so (Ehman and Scott 2002) (Figure 4.14), possibly because parasites depress testosterone levels and influence male odours (Hillgarth and Wingfield 1997; Barnard *et al.* 1998). Since resistance to parasites can be heritable (Enriquez *et al.* 1988), and parasite load can affect male status (Hausfater and Watson 1976; Freeland 1981; Gosling and Roberts 2001a), female preferences for uninfected males may have genetic benefits for their offspring.

Female preferences for mating with dominant partners may often be caused by olfactory cues, for dominant

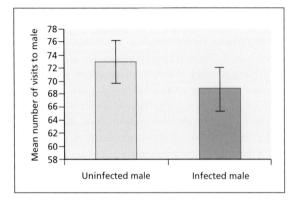

(a)

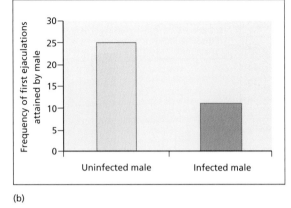

(b)

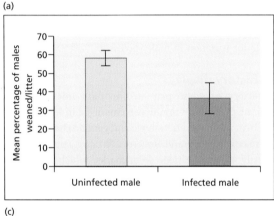

(c)

Figure 4.14 Female preferences in captive house mice for uninfected males versus males infected with intestinal nematodes: (a) mean number of visits to males; (b) frequency of first ejaculations per mate; (c) mean number of male offspring weaned per litter. *Source*: From Ehman and Scott (2002). Reproduced with permission of Cambridge University Press.

individuals commonly scent mark more frequently than subordinates, over-marking the scent marks of rivals (Gosling and Roberts 2001a,b). In some species, females appear to use the frequency of scent marking by males, or the frequency with which they over-mark the scent of rivals, as a basis for assessing potential mates (Desjardins *et al.* 1973; Johnston *et al.* 1997; Rich and Hurst 1998, 1999; Gosling and Roberts 2001a,b).

The odours produced by dominant males also differ qualitatively from those produced by subordinates and often reflect their higher androgen levels (Novotny *et al.* 1990), and the compounds present in their scent marks may play an important role in attracting females (Yamaguchi *et al.* 1981; Hayashi 1990; Mossman and Drickamer 1996; Rich and Hurst 1998; Humphries *et al.* 1999; Thomas 2002). The size of male scent glands, the frequency of scent marking, and the chemical structure of scent marks are all heritable (Horne and Ylönen 1998; Roberts and Gosling 2003) and, in some rodents, females

show a consistent preference not only for dominant males but for their sons as well (Drickamer 1992).

Relatedness

Breeding with close relatives commonly depresses the number and fitness of offspring (see Chapter 3), and females use several different tactics to avoid mating with close relatives (Solomon and Keane 2007) (Table 4.1) In some species, females delay sexual maturity if their father is resident in their natal group. For example, in prairie dogs, the presence of fathers is associated with delays in the onset of sexual maturity in their daughters (Hoogland 1982) though this effect disappears when groups contain a second, unrelated male as well as their father (Hoogland 1995) (Figure 4.15). A similar effect occurs in yellow-bellied marmots (Figure 4.16) while in alpine marmots, females are more likely to breed with males other than their mate if levels of genetic similarity between them are unusually high (Cohas *et al.* 2008).

Table 4.1 Alternative strategies of outbreeding.

1. Females disperse from natal group or area, e.g. horses, chimpanzees
2. Females living in their natal group delay onset of sexual maturity if their fathers are still present in the group, e.g. black-tailed prairie dogs, deer mice
3. Females living in their natal groups mate selectively with males from other groups, e.g. killer whales, Ethiopian wolves
4. Females living in their natal groups selectively mate with males that have immigrated into the group, e.g. Damaraland mole-rats, olive baboons
5. Females avoid mating with related partners both within their natal group and when mating outside the group, e.g. house mice and some other rodents where both sexes are sedentary

Where individuals of both sexes remain in their natal groups after reaching breeding age, females commonly avoid mating with first-order relatives. For example, in killer whales, both females and males commonly remain in their natal group throughout their lives and both sexes mate with partners from other groups (Amos *et al.* 1993; Baird 2000). A similar phenomenon occurs in some carnivores: for example, subordinate female meerkats living in groups that include their father and several mature brothers typically mate with roving males from neighbouring groups (Griffin *et al.* 2003; Young 2003; Young *et al.* 2007) and dominant females paired with relatively closely related males more frequently breed outside the group (Leclaire *et al.* 2013).

Where breeding groups include a mixture of related males and immigrants (as in savannah baboons and banded mongooses), females often show little sexual interest in natal males. For example, in olive baboons, breeding females readily approach, groom and mate with males that have recently immigrated into their group but show little sexual interest in natal males and rarely mate with them (Packer 1979). In banded mongooses, females usually avoid breeding with closely related males, too, though they do so occasionally (Nichols *et al.* 2014, 2015; Sanderson *et al.* 2015).

In many cases where females avoid mating with related males, they appear to be responding to their familiarity rather than to kinship *per se*. For example, in spotted hyenas, subordinate females show little sexual interest in natal males or in immigrants that were already present in the group when they were born but will mate with unrelated rovers or more recent immigrants from other groups (Smale *et al.* 1997; Young 2003; Höner *et al.* 2007).

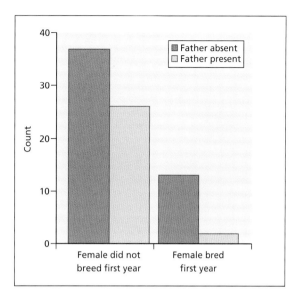

Figure 4.15 Female tactics for avoiding breeding with close relatives in prairie dogs. The figure shows numbers of female prairie dogs that bred or did not breed in their first year when their father was present or absent (data from Hoogland 1982). *Source:* From Blouin and Blouin (1988). Reproduced with permission of Elsevier. *Photo source:* © Elaine Miller Bond.

Cross-fostering experiments in rodents show that unrelated individuals reared together subsequently avoid breeding with each other while relatives reared apart breed readily, indicating that inbreeding avoidance is based on avoiding mating with familiar individuals (Dewsbury 1988; Pusey and Wolf 1996). Humans may also use similar cues in selecting males, for individuals commonly show an aversion to breeding with childhood associates, whether or not they are related (see Chapter 20).

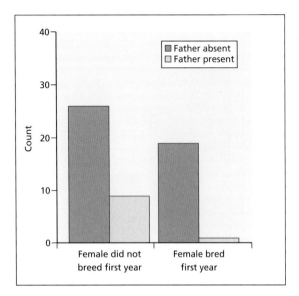

Figure 4.16 Female tactics for avoiding breeding with close relatives in yellow-bellied marmots. The figure shows numbers of female yellow-bellied marmots that bred or did not breed in their first year when their father was present or absent (data from Armitage 1984). *Source*: From Blouin and Blouin (1988). Reproduced with permission of Elsevier. *Photo source*: © Ben Husley.

Finally, in a number of mammals where the breeding tenure of individual males or male kin groups commonly exceeds the age at which females reach sexual maturity, females normally leave their natal group after reaching breeding age and either establish new breeding units or join other established groups (see Chapter 3). After

leaving their natal group, there is limited evidence of selection against related males and females will usually breed with related males if they encounter them. The absence of discrimination against related males may reflect the low probability that dispersing females will encounter close male relatives in wide-ranging species for, in some relatively sedentary species, females avoid breeding with close relatives encountered outside their natal group or territory, whether or not they are familiar with them (Dewsbury 1988; Ferkin 1990). For example, in house mice and in some voles, females are apparently able to discriminate between odours from siblings and unrelated individuals and avoid mating with siblings (Gilder and Slater 1978; Gavish *et al.* 1984; Winn and Vestal 1986; Bolhuis *et al.* 1988; Krackow and Matuschak 1991). In such cases, individuals may use cues derived either from their own phenotype or from the phenotype of parents or littermates to form a template which is subsequently used in comparisons with other individuals (Beauchamp *et al.* 1988).

While outbreeding generally increases the fecundity of females and the fitness of their offspring, it can also have costs if it disrupts co-adapted gene complexes and some studies suggest that individuals avoid breeding both with close relatives and with entirely unrelated partners (Bateson 1983). For example, in alpine marmots, the frequency with which females breed with extra-group partners increases when levels of genetic similarity between the breeding pair are unusually high or unusually low (Cohas *et al.* 2008). Similarly, studies of captive white-footed mice have shown that females mate preferentially with cousins, though studies of several other rodents have found no effects of this kind (Solomon and Keane 2007) and it is not yet clear how common they are.

Heterozygosity, genetic dissimilarity and compatibility

Where heterozygous males have higher fitness, females might increase the fitness of their progeny by selectively mating with relatively heterozygous partners. Studies of alpine marmots have found positive correlations between multilocus heterozygosity and juvenile survival (Cohas *et al.* 2008) and in several other in rodents, females show a preference for more heterozygous partners over more homozygous ones (Potts *et al.* 1991; von Schantz *et al.* 1997; Penn and Potts 1998a,b; Foerster *et al.* 2003). In fur seals, females that have moved further to mate are more likely to breed with relatively heterozygous males

Box 4.1 Disassociative mating and the MHC locus

Genes at the MHC locus are involved in immune function and are extremely variable at two polymorphic loci, both within and across species (Zinkernagel and Doherty 1974; Klein 1986; Potts and Wakeland 1990). Several studies show that individuals which are heterozygous at particular MHC loci are more resistant to particular infections or diseases, including malaria and cholera (Comings and MacMurray 2000; Penn 2002), and that they also show high levels of ornamentation, display or social status (von Schantz *et al.* 1996; Roberts and Gosling 2003). In some populations of mammals, MHC homozygous genotypes are rarer and heterozygous ones are commoner than would be expected by chance, suggesting that females mate selectively with males that have different MHC genotypes to their own (Potts *et al.* 1991; Jordan and Bruford 1998; Amos *et al.* 2001). For example, house mice living in semi-natural enclosures produce offspring that are more heterozygous at the MHC locus than would be expected by chance (Carroll and Potts 2007), and much of this excess appears to be a consequence of extra-territorial matings by females, suggesting that females favour matings with genetically dissimilar partners.

Experiments with fish, mice and humans show that females can detect MHC-related odours and confirm that they are often attracted to individuals with genotypes dissimilar from their own (Potts *et al.* 1991; Penn 2002; Milinski 2003), in some cases favouring partners carrying specific MHC alleles that complement their own genotype (Egid and Brown 1989). MHC-based odour preferences may also be involved in kin recognition and in the avoidance of close inbreeding (Yamazaki *et al.* 1988; Penn and Potts 1999).

Evidence for disassociative mating for MHC genotype in mice and other vertebrates could suggest that females have a reference genotype that is used to compare the genotype of potential mates with the individual's own genotype (self-inspection or the 'armpit effect') or with that of close kin (negative familial imprinting). However, cross-fostering experiments show that female mice avoid mating with individuals carrying the MHC genes of the family in which they are reared rather than with mice carrying their own MHC genes (Penn and Potts 1998b) (see Figure 4.17), possibly because this provides a more effective way of identifying close relatives than self-inspection (Penn and Potts 1998b).

(Hoffman *et al.* 2007). Some studies have found positive correlations between multilocus heterozygosity in juveniles and survival (Cohas *et al.* 2008). However, correlations between heterozygosity and fitness are seldom strong (Chapman *et al.* 2009), the offspring of relatively heterozygous fathers do not necessarily show high levels of heterozygosity (Cohas *et al.* 2007a; Leclaire *et al.* 2013) and preferences for heterozygous partners may be maintained by direct fitness benefits rather than genetic ones.

Females may also increase the heterozygosity and fitness of their progeny by mating with partners whose genotypes are dissimilar to their own. Some of the best evidence of disassociative mating is provided by studies of the effects of major histocompatibility complex (MHC) genotype on mate choice (Box 4.1 and Figure 4.17).

Evidence that individuals favour genetically dissimilar mates raises the question of how these preferences interact with selection for good genes (Colegrave *et al.* 2002; Mays and Hill 2004). One possibility is that there is a hierarchy of cues used in mating preferences: for example, individuals may make an initial choice on the basis of 'good genes' and then use compatibility criteria to select a mate from among acceptable males (Roberts and Gosling 2003). Experiments with mice provide some evidence for hierarchies of this kind: females prefer males who scent mark frequently (a trait that is associated with both androgen levels

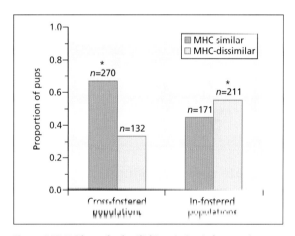

Figure 4.17 Evidence for familial imprinting in house mice using wild-derived mouse pups. Neonate pups were fostered into families with dissimilar MHC genotypes (cross-fostering) or into families with the same genotype as that of their parents (in-fostering) and then tested to see whether cross- and in-fostered mice differed in the frequency with which they mated and bred with individuals of similar versus dissimilar MHC genotype to their own. Cross-fostered females avoided mating with males carrying MHC genes characteristic of their foster family and were more likely to mate with individuals whose MHC genotype resembled their own than in-fostered animals, suggesting that they respond to olfactory cues from their foster family. *Source*: From Penn and Potts (1998b). Reproduced with permission from the Royal Society.

and dominance and may be an indicator of genetic quality), but where variation in scent marking rate is small, they show a preference for males with MHC genotypes unlike their own (Roberts and Gosling 2003). The relative importance of selection for good genes versus compatibility in mates may also vary between individuals. For instance, in species where pair-bonds are enforced by female aggression, dominant females may choose (and monopolise) highly ornamented males, while subordinate females may be forced to choose among less highly ornamented males on the basis of compatibility (Mays and Hill 2004).

MHC-based preferences may be most highly developed in species where females use olfactory cues to choose between males, and MHC-related mate choice may be more highly developed in fish and mammals than in birds because of their advanced olfactory capabilities (Mays and Hill 2004). MHC-based mate choice could have initially evolved as a mechanism allowing individuals to identify and avoid mating with close kin; if so, MHC-related preferences might be expected to be weak or absent in species where one sex habitually disperses at adolescence and the chance that dispersers will encounter close relatives is low (Penn and Potts 1998a,b, 1999). This could explain why there is no evidence of disassociative mating for the MHC genotype in Soay sheep, where male competition appears to determine the identity of mating partners, though there are several other possible explanations (Paterson and Pemberton 1997; Pemberton et al. 2004).

4.5 Mate choice copying

One tactic that females might be expected to use to select mates is to copy the mating choices of other females and studies of several social animals suggest that breeding females commonly copy each other's choice of mates and may gain indirect benefits by doing so (Dugatkin 1992). However, producing firm evidence that females copy each other's choice of mating partners is difficult since female mammals also copy each other's movements for reasons unconnected to mating preferences, so that evidence that females follow each other does not necessarily indicate that they are copying each other's choice of mating partners. For example, in some ungulates where males defend mating territories, females commonly follow each other between territories, aggregating in territories of particular males (Clutton-Brock et al. 1993). Although this could occur as a result of oestrous females copying each other's mating preferences, females (and their offspring) are often subjected to lower rates of male harassment when they are in larger groups (Carranza and Valencia 1999; McMahon and Bradshaw 2004) and receptive females may aggregate for this reason (McComb and Clutton-Brock 1994; Nefdt and Thirgood 1997). Attempts to avoid this problem using controlled experiments have so far produced mixed results. While female rats offered a choice between a male that had copulated recently and one that had not done so consistently preferred males that had recently copulated (Galef et al. 2008), experiments with fallow deer found that oestrous females show no preference for males that they have seen mating over those that had not mated (Clutton-Brock and McComb 1993; McComb and Clutton-Brock 1994).

In a number of mammals where females live in matrilineal groups, related females commonly breed with the same male (Pemberton et al. 1992). For example, both in red deer and in greater horseshoe bats, related females often join the same mating partners (Rossiter et al. 2005; Stopher et al. 2012). It is sometimes suggested that these trends are a consequence of female preferences for mating with the same male as their relatives in order to increase coefficients of relatedness within social groups, and to reduce competition and increase the potential for kin-selected cooperation. However, a simpler explanation is that related females tend to associate with each other and gain direct fitness benefits by doing so.

4.6 Partner number and post-copulatory mate choice

The number of males that females mate with in each breeding attempt varies widely, both within species and between them. In some socially monogamous species, females mate principally or exclusively with a single male, while in others extra-pair mating is common (Clutton-Brock and Isvaran 2006; Isvaran and Clutton-Brock 2007; Cohas and Allainé 2009). Similar differences occur among polygynous species: in some species females typically mate with a single male, while in others they commonly mate with multiple partners, often in rapid succession (Jolly 1966; Birdsall and Nash 1973; Tutin 1979; Cords et al. 1986; Packer et al. 1991; Pereira and Weiss 1991; East et al. 2003) and, in some species, many litters show multiple paternity (Hohoff *et al.* 2003). Intraspecific variation in partner number is also common: for example, in one population of thirteen-lined ground squirrels, the frequency of multiple paternity within litters ranged from 0 to 50% between years (Schwagmeyer and Brown 1983; Schwagmeyer and Parker 1987).

In some mammals, mating with multiple partners appears to be a consequence of competition between males, and females make no obvious attempt to instigate it. For example, in Soay sheep, oestrous females commonly mate with large numbers of males, but they show little evidence of active mate choice and often attempt to avoid pursuing males (Clutton-Brock *et al.* 2004; Pemberton *et al.* 2004). Similarly, in thirteen-lined ground squirrels, most females copulate with all males that locate and attempt to mate with them (Schwagmeyer 1984, 1986; Schwagmeyer and Woontner 1985).

Variation in the capacity of males to control reproductive access in females often has an important influence on interspecific differences in partner number (Clutton-Brock and Isvaran 2006). For example, in alpine marmots, the number of extra-pair paternities increases with the number of subordinate males present in groups (Cohas *et al.* 2007b, 2009). Across species, mating with multiple partners is also more frequent where groups include more than one adult male or where resident males cannot prevent visiting males gaining access to females and comparative studies show that rates of extra-group paternity increase with group size as well as with the duration of breeding seasons (see Chapter 13). Variation in the costs of male coercive tactics to females may also affect the frequency with which females mate with multiple partners. For example, in some ungulates where males possess dangerous weapons and the coercive strategies of males can have lethal consequences to females, female oestrus often terminates soon after mating and most females copulate only once or twice per season, usually with a single partner (Clutton-Brock *et al.* 1988).

In some mammals, females appear to go out of their way to instigate matings with more than one partner. For example, in ring-tailed lemurs and in blue monkeys, oestrous females often solicit mating from several partners from outside their group in the course of a single day (Jolly 1966; Cords *et al.* 1986). Similarly, in brown capuchin monkeys, as well as in several baboons and macaques that live in multi-male groups, females make active attempts to mate with several different males (Taub 1980; Janson 1984). It is often difficult to be sure that females are not coerced into mating with multiple partners, but some studies are able to exclude this possibility. For example, in yellow-toothed cavies, oestrous females race between males, making it difficult for single males to monopolise them, and experiments in which receptive females were allowed to choose between different males in an apparatus that prevented harassment or monopolisation of females showed that 90% of mating females actively solicited copulations with more than one male, and that females preferred heavier males and those that courted more frequently (Schwarz-Weig and Sachser 1996; Hohoff *et al.* 2003).

The costs of mating with multiple partners may include the energy costs of locating suitable partners, the risk of harassment, punishment or reduced investment by dominant males (see Chapter 15), increased risks of acquiring sexually transmissible diseases, reduced control of paternity and increased competition between offspring born in the same litters, though comparative studies of mammals have not detected any consistent relationship between partner number and female mortality (Lemaître and Gaillard 2013). Several different groups of benefits that could offset any costs of mating with multiple partners have been suggested (Table 4.2), though their relative importance is still uncertain. Mating with

Table 4.2 Suggested benefits to females of mating with multiple partners.

| **Direct benefits** |
| Reduced risk of infertility |
| Increased protection of infants against conspecifics or predators |
| Ejaculate utilisation by females for nutritional purposes or increased numbers of nuptial gifts |
| Increased tolerance of mother by males |
| Reduced risk of infanticide by males |
| Increases in total investment by males in dependent young |
| |
| **Indirect benefits** |
| Improved flexibility of mate choice ('trading up') |
| Reduced risk of fertilisation by close relatives |
| Improved sperm quality |
| Improved opportunities to bias paternity (cryptic mate choice) |
| Genetic diversity within litters, leading to reduced competition for resources between littermates; reduced disease transmission; reduced chances of total breeding failure |

Source: data from Dewsbury (1982a,b), Jennions and Petrie (2000) and Hosken and Blanckenhorn (1999)

multiple partners may increase the chance that females will conceive (Schwagmeyer 1984; Ridley 1988, 1990): for example, studies of artificial insemination in domestic mammals show that the use of sperm from more than one male can increase the probability of conception (Hess *et al*. 1954; Beatty 1960). However, not all studies show that multiple mating increases fecundity and, in some rodents, females that mate with multiple partners show reduced conception rates (Schwagmeyer 1986). For example, female deermice and Djungarian hamsters that mate with multiple partners are less likely to become pregnant than those that mate with a single partner (Dewsbury and Baumgardner 1981; Dewsbury 1982a, b,c; Wynne-Edwards and Lisk 1984), perhaps because the risk of male infanticide increases when multiple males have access to a female's territory and because it is beneficial to females to avoid conception (Dewsbury 1982a,b,c). In some cases, mating with multiple males may be a response to previous breeding failure (Ens *et al*. 1993). For example, in rock wallabies, which form monogamous pairs, females are more likely to breed with males other than their mate if their previous off-spring failed to survive to emergence from the pouch (Spencer *et al*. 1998).

Second, mating with multiple partners may increase the survival of fetuses to term as a result of sperm competition. For example, in (captive) yellow-toothed cavies, females that are allowed to mate with multiple partners have fewer stillbirths and wean significantly more of their pups (Figure 4.18). Since the same males were used to fertilise single versus multiple mating females in this experiment, and young were suckled by multiple females (Künkele and Hoeck 1995), this effect was unlikely to be a consequence of differences in male quality or of differences in parental care, and the most likely explanation is that sperm competition weeded out qualitatively inferior gametes (Sivinski 1984; Keil and Sachser 1998).

Third, mating with multiple males may increase the survival of juveniles by increasing the amount of care or protection they receive from males and reducing the risks of infanticide (see Chapter 15) or females may gain tolerance or support from males by mating with them, which may contribute to their fecundity and survival. In Japanese macaques, males are eight times more likely to attack infants if they have not previously mated with their mothers, and infants born to females that mate with multiple males attract less aggression (Soltis *et al*. 2000).

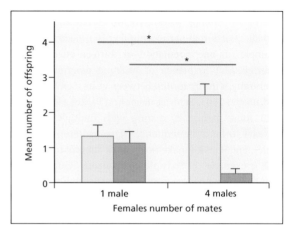

Figure 4.18 Effects of mate number on breeding success in yellow-toothed cavies. The figure shows the number of pups that are stillborn (dark bars) and the number weaned (hatched bars) by females that were restricted to mating with a single male compared with females allowed to mate with four males. *Source*: From Keil and Sachser (1998). Reproduced with permission of John Wiley & Sons. *Photo source*: © Matthias Asher.

Similarly, among some rodents in which male infanticide is common, controlled experiments show that resident males are less likely to attack neonates if they have previously had sexual experience with the mother (vom Saal 1984; Ebensperger 1998a,b).

A final possibility is that mating with multiple males contributes to the genetic or phenotypic quality of surviving offspring, either by increasing the chance that a female will breed with a higher-quality male (trading-up) or by increasing competition between sperm in the female reproductive tract (Schwagmeyer 1984; Eberhard 1996). In alpine marmots, offspring that are a product of extra-pair matings show increased survival during their first two years of life and, in females, increased breeding success in adulthood compared with those that are the result of matings between partners (Cohas *et al.* 2007b).

4.7 Variation in mate choice and partner number

Within species

Although most studies of female mate choice have investigated whether there are consistent female preferences for particular characteristics in males, the benefits and costs of particular choices are likely to vary in space and time and this might be expected to lead to variation in sampling tactics, mating preferences and the degree of choosiness (Zeh and Zeh 1996; Widemo and Sæther 1999; Fawcett and Johnstone 2003). There is increasing evidence that this is the case: studies of birds show that the strength and, in some cases, the direction of female mating preferences can vary widely (Chaine and Lyon 2008), and several studies indicate that the strength of female mating preferences also varies in mammals. For example, in mice, preferences for MHC-dissimilar males, which can be detected in oestrous females, disappear when females are not in oestrus (Egid and Brown 1989) and several studies of humans also suggest that MHC-related odour preferences as well as female preferences for masculinity in males vary with the stage of the reproductive cycle or the hormonal status of subjects (Jordan and Bruford 1998; Penton-Voak and Perrett 2001).

The probability that females will mate with multiple partners can also change throughout the reproductive cycle. In several monkeys, females initiate copulation more frequently around the time of ovulation than they do during other times of the cycle and sexual behaviour declines during the luteal phase (Manson 2006). In female chimpanzees, individuals commonly copulate with multiple partners during the early stages of the oestrus cycle but are more likely to copulate repeatedly with high-ranking males during the later stages, when conception is most likely (Matsumoto-Oda 1999). However, changes in the probability that females will conceive are also likely to affect the behaviour of males, and it is often difficult to be sure that changes in the number of partners are a consequence of female decisions rather than changes in male behaviour.

Between species

Evidence of intraspecific variation in the strength and direction of mate choice suggests that interspecific differences in mate choice are likely to be widespread and large. In general, it is reasonable to suppose that female mating preferences are likely to be more highly developed in species where females have the opportunity to choose freely between multiple mating partners and males invest heavily in parental care, so that females gain direct as well as indirect benefits from mate selection, than in species where females typically remain and breed in their natal groups and males compete to monopolise access to female groups and do not contribute to parental care. As a result, it is unfortunate that a high proportion of studies of mate choice in mammals have focused on polygynous species where males defend access to female groups and are not involved in parental care rather than on monogamous species where males contribute substantially to parental care (see Chapter 16). Not only may the benefits of mate choice to females be smaller in polygynous species than in monogamous societies where males are involved in parental care but male competition is typically intense and may often constrain female choice. In addition, it is often difficult to distinguish its effects from those of female choice unless controlled experiments are feasible.

Relatively few male mammals show the kind of sexually dimorphic ornamentation that is common in birds (see Chapter 18). One possible explanation is that the common tendency for males to establish breeding territories where they are subsequently joined by females favours the evolution of male ornamentation in birds, whereas the common tendency in mammals for males to join and defend pre-existing female groups favours the evolution of fighting ability and weaponry in males. Alternatively, it may be that the evolution of viviparity is associated with a reduction in the relative importance of pre-copulatory female mate choice and an increase in post-zygotic mechanisms of sexual selection (Zeh and Zeh 2000; Pollux *et al.* 2014). However, we still understand relatively little about olfactory communication (see Chapter 6) and visual indicators of male quality may be less highly developed in mammals than birds because olfaction plays a more important role in mammalian mate choice. It is an intriguing possibility that, in some mammals, males may possess the olfactory equivalent of the peacock's tail.

There is also a historical contrast in the focus of research between studies of birds and mammals: while studies of birds have commonly investigated female preferences for male coloration or plumage structure and have tended to assume that they are maintained by indirect fitness benefits (Andersson 1994), studies of mammals (and primates in particular) have more commonly focused on female preferences for dominant males and have tended to assume that they are maintained by direct fitness benefits (Janson 1984; Small 1989). Although it is conceivable that the direct fitness consequences of mating decisions are larger in mammals than birds because male coercive tactics are more highly developed in mammals, relatively few studies of birds have yet attempted to investigate the direct benefits of mating with highly ornamented males while relatively few studies of mammals have investigated the indirect benefits that females may gain by mating with dominant males.

4.8 Consequences of female mating preferences

The evolutionary consequences of female mating preferences are pervasive. In particular, the avoidance of close relatives as mating partners by females is probably the principal cause of male dispersal in many social mammals (see Chapter 12) and of female dispersal in some (see Chapter 3). Female preferences for male characteristics are probably responsible for the evolution of many visual, auditory and olfactory displays in males and the phenotypic traits associated with them (Andersson 1994). In addition, female preferences for familiar partners and for individuals that are likely to contribute to defending young have probably played an important role in the evolution of social relationships between individual males and females (see Chapter 15) and paternal care (see Chapter 16).

The evolution of exaggerated male ornaments has also had a wide range of evolutionary and ecological consequences. Since male ornaments maintained by female choice are likely to evolve to a point where their benefits in terms of male attractiveness to

females are balanced by costs to growth or survival (Fisher 1930; Iwasa et al. 1991; Pomiankowski et al. 1991) they may often have substantial costs to male growth or survival. There is anecdotal evidence that exaggerated ornaments increase the risk of predation on males (Magnhagen 1991) and experimental manipulation of tail length in sexually dimorphic birds shows that exaggerated tails affect the foraging success and survival of males (Møller and de Lope 1994; Møller et al. 1995; Saino et al. 1997). In addition, comparative studies show that when resource availability is low, sex differences in growth and body size are commonly associated with sex differences in survival (Clutton-Brock et al. 1985; Promislow 1992; Promislow et al. 1994) as well as with differential costs of rearing sons and daughters to maternal survival (see Chapter 18).

The costs of secondary sexual characters to males and females suggest that species where males perform elaborate displays or possess exaggerated ornaments may suffer reductions in mean population fitness compared to species where sexual selection is weaker and the secondary sexual characteristics of males are less developed (Møller 2000; Møller and Danchin 2008) and the risk of local or global extinction may be elevated (Tanaka 1996). Although this issue has not yet been systematically investigated in mammals, studies of birds provide some evidence of this effect. A continent-wide survey of bird populations in North America shows that sexually dichromatic species have a 23% higher rate of local extinction than monochromatic species (Doherty et al. 2003). In addition, studies of bird species introduced to oceanic islands show that sexually dichromatic ones are less likely to become established than monochromatic species (McClain et al. 1995; Sorci et al. 1998). However, polygyny and sexual dimorphism are often associated with coercive tactics in males that may also lower the fitness of females (see Chapter 15) and it is difficult to distinguish between the effects of associated behaviour and those of secondary sexual characters.

Strong female preferences for particular partners and large differences in mating success between males may have other evolutionary consequences. In particular, where they increase reproductive skew in males. For example, in red deer, around one-fifth of females in each year breed with males that they have bred with previously, leading to higher than expected levels of inbreeding (Stopher et al. 2012). As a result they are likely to reduce the number of males contributing to each generation, leading to lower genetic diversity within populations (Shields 1987) and increasing the rate at which populations diverge and the frequency of speciation (Fisher 1930; Lande 1981; Schluter and Price 1993).

SUMMARY

1. By mating selectively with particular males, females can gain immediate benefits to their direct fitness as well as genetic benefits that affect the fitness of their offspring.

2. While female preferences can increase the mating success of particular males, it is often difficult to distinguish between the contributions of female mating preferences to variation in male mating success and those of variation in male competitive success. Moreover, the characteristics of males that are commonly suggested to affect female mating preferences (including their size and the relative development of secondary sexual characters) are also likely to affect male competitive success.

3. As a result, neither evidence of variation in male mating success nor correlations between male success and male phenotype necessarily indicate that female mating preferences are involved and it is usually difficult to produce unequivocal evidence for female mating preferences without experiments that control for the effects of male competition. Since experiments of this kind are seldom feasible with wild mammals, much of the evidence for female mate choice in non-human mammals relies either on experiments with captive animals under restricted conditions or on correlational evidence that is open to other interpretations.

4. In natural populations, female mammals often appear to join and mate with mature dominant males. In many cases, this may help to reduce dangerous harassment by other males, though it may also provide indirect fitness benefits to females as a result of improvements in the survival or breeding success of their offspring.

5. It is often suggested that females mate preferentially with males with well-developed weapons (including horns, antlers and teeth), but there is little convincing evidence that this is the case or that the evolution of male weapons is strongly affected by benefits other than their effectiveness in fights with other males. In contrast, there is stronger evidence that female preferences for particular males is affected by their visual, vocal and olfactory displays.

6. Females can gain substantial benefits to the fitness of their offspring by avoiding breeding with close relatives and there is substantial evidence that they avoid mating with close relatives or with individuals that they were familiar with as juveniles or adolescents. There is also some evidence that female mating preferences may be affected by their genetic similarity and compatibility with different males.

7. The mating preferences of other females may provide a reliable indication of the relative benefits of mating with different partners and females often appear to copy each other's choice of mating partners. However, it is often difficult to determine whether females are copying each other's mating preferences or each other's movements, which are not necessarily related to mating preferences.

8. In some mammals, females usually mate with a single partner in each breeding attempt, while in others they typically mate with multiple partners. Where females mate with multiple partners, males often appear to be unable to monopolise access to females effectively, though in some species females actively solicit matings from multiple partners. Potential benefits of mating with multiple partners include increases in fertility and in the number of males contributing to parental care and reductions in the risk of male infanticide.

9. Female mating preferences exert an important influence on patterns of dispersal in both sexes as well as on the evolution of competitive behaviour in males and the development of secondary sexual characters.

References

Agrell, J., *et al.* (1998) Counter-strategies to infanticide in mammals: costs and consequences. *Oikos* **83**:507–517.

Alatalo, R.V., *et al.* (1997) Heritabilities and paradigm shifts. *Nature* **385**:402–403.

Alberts, S.C., *et al.* (2003) Queuing and queue-jumping: long-term patterns of reproductive skew in male savannah baboons, *Papio cynocephalus*. *Animal Behaviour* **65**:821–840.

Alexander, B.K. (1970) Parental behaviour of adult male Japanese monkeys. *Behaviour* **36**:270–285.

Alvarez, F. (1990) Horns and fighting in male Spanish ibex, *Capra pyrenaica*. *Journal of Mammalogy* **71**:608–616.

Amos, B., *et al.* (1993) Social structure of pilot whales revealed by analytical DNA profiling. *Science* **260**:670–672.

Amos, W., *et al.* (2001) Do female grey seals select genetically diverse mates? *Animal Behaviour* **62**:157–164.

Andersson, M. (1994) *Sexual Selection*. Princeton, NJ: Princeton University Press.

Armitage, K.B., *et al.* (2011) Proximate causes of dispersal in yellow-bellied marmots, *Marmota flaviventris*. *Ecology* **92**:218–227.

Austin, D. and Dewsbury, D.A. (1986) Reproductive capacity of male laboratory rats. *Physiology and Behaviour* **37**:627–632.

Baird, R.W. (2000) The killer whale: foraging specializations and group hunting. In: *Cetacean Societies: Field Studies of Dolphins and Whales* (eds J. Mann, R. Connor, P. Tyack and H. Whitehead) Chicago: University of Chicago Press, 127–153.

Baker, A.J., *et al.* (1993) Behavioral evidence for monopolization of paternity in multimale groups of golden lion tamarins. *Animal Behaviour* **46**:1091–1103.

Bakker, T.C.M. and Pomiankowski, A. (1995) The genetic basis of female mate preferences. *Journal of Evolutionary Biology* **8**:129–171.

Balmford, A.P. and Read, A.F. (1991) Testing alternative models of sexual selection through female choice. *Trends in Ecology and Evolution* **6**:274–276.

Balmford, A.P., *et al.* (1992) Correlates of female choice in resource-defending antelope. *Behavioral Ecology and Sociobiology* **31**:107–114.

Barnard, C.J., *et al.* (1998) The role of parasite-induced immunodepression, rank and social environment in the modulation of behaviour and hormone concentration in male laboratory mice (*Mus musculus*). *Proceedings of the Royal Society of London. Series B: Biological Sciences* **265**:693–701.

Barrett, L. and Henzi, S.P. (2006) Monkeys, markets and minds: biological markets and primate sociality. In: *Cooperation in Primates and Humans* (eds P.M. Kappeler and C.P. van Schaik) Berlin: Springer, 209–232.

Bateson, P. (1983) Optimal outbreeding. In: *Mate Choice* (ed. P. Bateson). Cambridge: Cambridge University Press, 257–309.

Beatty, R.A. (1960) Fertility of mixed semen from different rabbits. *Journal of Reproduction and Fertility* **1**:52–60.

Beauchamp, G.K., *et al.* (1988) Preweaning experience in the control of mating preferences by genes in the major histocompatibility complex of the mouse. *Behavior Genetics* **18**:537–547.

Bebié, N. and McElligott, A.G. (2006) Female aggression in red deer: does it indicate competition for mates? *Mammalian Biology* **71**:347–355.

Birdsall, D.A. and Nash, D. (1973) Occurrence of successful multiple insemination of females in natural populations of deer mice (*Peromyscus maniculatus*). *Evolution* **27**:106–110.

Blouin, S.F. and Blouin, M. (1988) Inbreeding avoidance behaviors. *Trends in Ecology and Evolution* **3**:230–233.

Bolhuis, J.J., *et al.* (1988) Preferences for odors of conspecific non-siblings in the common vole, *Microtus-arvalis*. *Animal Behaviour* **36**:1551–1553.

Bro-Jørgensen, J. (2002) Overt female mate competition and preference for central males in a lekking antelope. *Proceedings of the National Academy of Sciences of the United States of America* **99**:9290–9293.

Bro-Jørgensen, J. and Durant, S.M. (2003) Mating strategies of topi bulls: getting in the centre of attention. *Animal Behaviour* **65**:585–594.

Byers, J.A., *et al.* (1994) Pronghorn females choose vigorous mates. *Animal Behaviour* **47**:33–43.

Byers, J.A., *et al.* (2005) A large cost of female mate sampling in pronghorn. *American Naturalist* **166**:661–668.

Carranza, J. and Valencia, J. (1999) Red deer females collect on male clumps at mating areas. *Behavioral Ecology* **10**:525–532.

Carroll, L.S. and Potts, W.K. (2007) Sexual selection: using social ecology to determine fitness difference. In: *Rodent Societies: An Ecological and Evolutionary Perspective* (eds J.O. Wolff and P.W. Sherman) Chicago: University of Chicago Press, 57–67.

Catchpole, C.K. (1980) Sexual selection and the evolution of complex songs among European warblers of the genus *Acrocephalus*. *Behaviour* **74**:149–166.

Chaine, A.S. and Lyon, B.E. (2008) Adaptive plasticity in female mate choice dampens sexual selection on male ornaments in the lark bunting. *Science* **319**:459–462.

Chapman, J.R., *et al.* (2009) A quantitative review of heterozygosity–fitness correlations in animal populations. *Molecular Ecology* **18**:2746–2765.

Charlton, B., *et al.* (2007) Female red deer prefer the roars of larger males. *Biology Letters* **3**:382–385.

Clayton, D.H. (1991) The influence of parasites on host sexual selection. *Parasitology Today* **7**:329–334.

Clutton-Brock, T.H. (1982) The functions of antlers. *Behaviour* **79**:108–125.

Clutton-Brock, T.H. and Isvaran, K. (2006) Paternity loss in contrasting mammalian societies. *Biology Letters* **2**:513–516.

Clutton-Brock, T.H. and McAuliffe, K. (2009) Female mate choice in mammals. *Quarterly Review of Biology* **84**:3–27.

Clutton-Brock, T.H. and McComb, K. (1993) Experimental tests of copying and mate choice in fallow deer. *Behavioral Ecology* **4**:191–193.

Clutton-Brock, T.H., *et al.* (1982) *Red Deer: The Behaviour and Ecology of Two Sexes*. Chicago: University of Chicago Press.

Clutton-Brock, T.H., *et al.* (1985) Parental investment and sex differences in juvenile mortality in birds and mammals. *Nature* **313**:131–133.

Clutton-Brock, T.H., *et al.* (1988) Passing the buck: resource defence, lek breeding and mate choice in fallow deer. *Behavioral Ecology and Sociobiology* **23**:281–296.

Clutton-Brock, T.H., *et al.* (1993) The evolution of ungulate leks. *Animal Behaviour* **46**:1121–1138.

Clutton-Brock, T.H., *et al.* (1996) Multiple factors affect the distribution of females in lek-breeding ungulates: a rejoinder to Carbone and Taborsky. *Behavioral Ecology* **7**:373–378.

Clutton-Brock, T.H., *et al.* (2004) The sheep of St Kilda. In: *Soay Sheep: Dynamics and Selection in an Island Population* (eds T.H. Clutton-Brock and J.M. Pemberton) Cambridge: Cambridge University Press, 17–51.

Cohas, A., *et al.* (2006) Extra-pair paternity in the monogamous alpine marmot (*Marmota marmota*): The roles of social setting and female mate choice. *Behavioral Ecology and Sociobiology* **55**:597–605.

Cohas, A., *et al.* (2007a) Extra-pair paternity in alpine marmot (Marmota marmota): Genetic quality and genetic diversity effects. *Behavioral Ecology and Sociobiology* **61**:1081–1092.

Cohas, A., *et al.* (2007a) Are extra-pair young better than within-pair young? A comparison of survival and dominance in alpine marmot. *Journal of Animal Ecology* **76**:771–781.

Cohas, A., *et al.* (2008) The genetic similarity between pair members influences the frequency of extrapair paternity in alpine marmots. *Animal Behaviour* **76**:87–95.

Cohas, A. and Allainé, D. (2009) Social structure influences extra-pair paternity in socially monogamous mammals. *Biology Letters* DOI:10.1098/rsbl.2008.0760. [Accessed 08 January 2016]

Colegrave, N., *et al.* (2002) Mate choice or polyandry: reconciling genetic compatibility and good genes sexual selection. *Evolutionary Ecology Research* **4**:911–917.

Coltman, D.W., *et al.* (2002) Age-dependent sexual selection in bighorn rams. *Proceedings of the Royal Society of London. Series B: Biological Sciences* **269**:165–172.

Comings, D.E. and MacMurray, J.P. (2000) Molecular heterosis: a review. *Molecular Genetics and Metabolism* **71**:19–31.

Cords, M., *et al.* (1986) Promiscuous mating among blue monkeys in the Kakamega forest, Kenya. *Ethology* **72**:214–226.

Cox, C.R. and Le Boeuf, B.J. (1977) Female incitation of male competition: a mechanism in sexual selection. *American Naturalist* **111**:317–335.

Craul, M., *et al.* (2004) First experimental evidence for female mate choice in a nocturnal primate. *Primates* **45**:271–274.

Davidson, S.M. and Wilkinson, G.S. (2001) Function of male song in the greater white-lined bat, *Saccopteryx bilineata*. *Animal Behaviour* **67**:883–891.

Desjardins, C., *et al.* (1973) Social rank in house mice: differentiation revealed by ultraviolet visualization of urinary marking patterns. *Science* **182**:939–941.

Dewsbury, D.A. (1982a) Dominance rank, copulatory behavior and differential reproduction. *Quarterly Review of Biology* **57**:135–159.

Dewsbury, D.A. (1982b) Ejaculate cost and male choice. *American Naturalist* **119**:601–610.

Dewsbury, D.A. (1982c) Pregnancy blockage following multiple-male copulation or exposure at the time of mating in deer mice *Peromyscus maniculatus*. *Behavioral Ecology and Sociobiology* **11**:37–42.

Dewsbury, D.A. (1988) Kin discrimination and reproductive behavior in muroid rodents. *Behavior Genetics* **18**:525–536.

Dewsbury, D.A. and Baumgardner, D.J. (1981) Studies of sperm competition in two species of muroid rodents. *Behavioral Ecology and Sociobiology* **9**:121–133.

Ditchkoff, S.S., *et al.* (2001) Major-histocompatibility-complex-associated variation in secondary sexual traits of white-tailed deer (*Odocoileus virginianus*): evidence for good-genes advertisement. *Evolution* **55**:616–625.

Doherty, P.F.J., *et al.* (2003) Sexual selection affects local extinction and turnover in bird communities. *Proceedings of the National Academy of Sciences of the United States of America* **100**:5858–5862.

Drea, C.M. (2005) Bateman revisited: The reproductive tactics of female primates. *Integrative and Comparative Biology* **45**:915–923.

Drickamer, L.C. (1992) Oestrous female house mice discriminate dominant from subordinate males and sons of dominant from sons of subordinate male by odour cues. *Animal Behaviour* **43**:868–870.

Drickamer, L.C., *et al.* (2000) Free female mate choice in house mice affects reproductive success and offspring viability and performance. *Animal Behaviour* **59**:371–378.

Dubuc, C., *et al.* (2014) Is male rhesus macaque red color ornamentation attractive to females? *Behavioral Ecology and Sociobiology* **68**:1215–1224.

Dugatkin, L.A. (1992) Sexual selection and imitation: females copy the mate choice of others. *American Naturalist* **139**:1384–1389.

Dunbar, R.I.M. (1984) *Reproductive Decisions: An Economic Analysis of the Social Strategies of Gelada Baboons*. Princeton, NJ: Princeton University Press.

East, M.L., *et al.* (2003) Sexual conflicts in spotted hyenas: male and female mating tactics and their reproductive outcome with respect to age, social status and tenure. *Proceedings of the Royal Society of London. Series B: Biological Sciences* **270**:1247–1254.

Ebensperger, L.A. (1998a) Strategies and counterstrategies to infanticide in mammals. *Biological Reviews* **73**:321–346.

Ebensperger, L.A. (1998b) Do female rodents use promiscuity to prevent male infanticide? *Ethology, Ecology and Evolution* **10**:129–141.

Eberhard, W.G. (1996) *Female Control: Sexual Selection by Cryptic Female Choice*. Princeton, NJ: Princeton University Press.

Egid, K. and Brown, J.L. (1989) The major histocompatibility complex and female mating preferences in mice. *Animal Behaviour* **38**:548–550.

Ehman, K.D. and Scott, M.E. (2001) Urinary odour preferences of MHC congenic female mice, *Mus domesticus*: implications for kin recognition and detection of parasitized males. *Animal Behaviour* **62**:781–789.

Ehman, K.D. and Scott, M.E. (2002) Female mice mate preferentially with non-parasitized males. *Parasitology* **125**:461–466.

Enriquez, F.J., *et al.* (1988) *Nematospiroides dubius*: genetic control of immunity to infections of mice. *Experimental Parasitology* **67**:12–19.

Ens, B.J., *et al.* (1993) Divorce in the long-lived and monogamous oystercatcher, *Haematopus ostralegus*: incompatibility or choosing the better option? *Animal Behaviour* **45**:1199–1217.

Espmark, Y. (1964) Rutting behaviour in reindeer *Rangifer tarandus* L. *Animal Behaviour* **12**:159–163.

Fawcett, T.W. and Johnstone, R.A. (2003) Mate choice in the face of costly competition. *Behavioral Ecology* **14**:771–779.

Fedigan, L.M. (1983) Dominance and reproductive success in primates. *Yearbook of Physical Anthropology* **26**:91–129.

Ferkin, M.H. (1990) Kin recognition and social behavior in microtine rodents. In: *Social Systems and Population Cycles in Voles* (eds R.H. Tamarin, R.S. Ostfeld, S.R. Pugh and G. Bujalska) Basel: Birkhaeuser, 11–24.

Ferrari, S.F. (1992) The care of infants in a wild marmoset (*Callithrix flaviceps*) group. *American Journal of Primatology* **26**:109–118.

Fisher, R.A. (1915) The evolution of sexual preference. *The Eugenics Review* **7**:184–192.

Fisher, R.A. (1930) *The Genetical Theory of Natural Selection.* Oxford: Clarendon Press.

Fiske, P., *et al.* (1998) Mating success in lekking males: a meta-analysis. *Behavioral Ecology* **9**:328–338.

Foerster, K., *et al.* (2003) Females increase offspring heterozygosity and fitness through extra-pair matings. *Nature* **425**:714–717.

Folstad, I. and Karter, A.J. (1992) Parasites, bright males and the immunocompetence handicap. *American Naturalist* **139**:603–622.

Fox, E.A. (2002) Female tactics to reduce sexual harassment in the Sumatran orangutan (*Pongo pygmaeus abelii*) *Behavioral Ecology and Sociobiology* **52**:93–101.

Freeland, W.J. (1981). Parasitism and behavioral dominance among male mice. *Science* **213**:461–462.

Galef, B.G., *et al.* (2008) Evidence of mate choice copying in Norway rats, *Rattus norvegicus. Animal Behaviour* **75**:1117–1123.

Gartlan, J.S. and Brain, C.K. (1968) Ecology and social variability in *Cercopithecus aethiops* and *C. mitis*. In: *Primates: Studies in Adaptation and Variability* (ed. P.C. Phyliss). New York: Holt, Rinehart and Winston, 253–292.

Gavish, L., *et al.* (1984) Sibling recognition in the prairie vole, *Microtus ochrogaster. Animal Behaviour* **32**:362–366.

Geist, V. (1971) *Mountain Sheep: A Study in Behavior and Evolution.* Chicago: University of Chicago Press.

Gerald, M.S. (2001) Primate colour predicts social status and aggressive outcome. *Animal Behaviour* **61**:559–566.

Gerald, M.S. (2003) How color may guide the primate world: possible relationships between sexual selection and sexual dichromatism. In: *Sexual Selection and Reproductive Competition in Primates: New Perspectives and Directions* (ed. C.B. Jones). Norman, OK: American Society of Primatologists, 141–171.

Gilder, P.M. and Slater, P.J.B. (1978) Interest of mice in conspecific male odours is influenced by degree of kinship. *Nature* **274**:364–365.

Gosling, L.M. and Roberts, S.C. (2001a) Scent-marking by male mammals: cheat-proof signals to competitors and mates. In: *Advances in the Study of Behavior*, Vol. **30** (eds P.J.B. Slater, J.S. Rosenblatt, C.T. Snowdon and T.J. Roper) San Diego, CA: Academic Press, 169–217.

Gosling, L.M. and Roberts, S.C. (2001b) Testing ideas about the function of scent marks in territories from spatial patterns. *Animal Behaviour* **62**:F7–F10.

Grafen, A. (1990) Biological signals as handicaps. *Journal of Theoretical Biology* **144**:517–546.

Griffin, A.S., *et al.* (2003) A genetic analysis of breeding success in the cooperative meerkat (*Suricata suricatta*). *Behavioral Ecology* **14**:472–480.

Gustafsson, L. (1986) Lifetime reproductive success and heritability: empirical support for Fisher's fundamental theorem. *American Naturalist* **128**:761–764.

Hamilton, W.D. and Zuk, M. (1982) Heritable true fitness and bright birds: a role for parasites? *Science* **218**:384–387.

Hausfater, G. and Watson, D.F. (1976) Social and reproductive correlates of parasite ova emissions by baboons. *Natures* **262**:688–689.

Hayashi, S. (1990) Social condition influences sexual attractiveness of dominant male mice. *Zoological Science* **7**:889–894.

Hemelrijk, C.K., *et al.* (1999) 'Friendship' for fitness in chimpanzees? *Animal Behaviour* **58**:1223–1229.

Hess, E.A., *et al.* (1954) Some of the influences of mixed ejaculates upon bovine fertility. *Journal of Dairy Science* **37**:649–650.

Heywood, J.S. (1989) Sexual selection by the handicap mechanism. *Evolution* **43**:1387–1397.

Hiebert, S.M., *et al.* (1989) Repertoire size, territory acquisition and reproductive success in the song sparrow. *Animal Behaviour* **37**:266–273.

Higham, J.P., *et al.* (2013) The endocrinology of male rhesus macaque social and reproductive status: a test of the challenge and social stress hypotheses. *Behavioral Ecology and Sociobiology* **67**:19–30.

Hillgarth, N. and Wingfield, J.C. (1997) Parasite-mediated sexual selection: endocrine aspects. In: *Host–Parasite Evolution: General Principles and Avian Models* (eds D.H. Clayton and J. Moore) Oxford: Oxford University Press, 78–104.

Hoelzer, G.A. (1989) The good parent process of sexual selection. *Animal Behaviour* **38**:1067–1078.

Hoffman, J.I., *et al.* (2004) Exploring the relationship between parental relatedness and male reproductive success in an Antarctic fur seal *Arctocephalus gazella. Evolution* **58**:2087–2099.

Hoffman, J.I., *et al.* (2007) Female fur seals show active choice for males that are heterozygous and unrelated. *Nature* **445**:912–914.

Hogg, J.T. and Forbes, S.H. (1997) Mating in bighorn sheep: frequent male reproduction via a high-risk unconventional tactic. *Behavioral Ecology and Sociobiology* **41**:33–48.

Hohoff, C., *et al.* (2003) Female choice in a promiscuous wild guinea pig, the yellow-toothed cavy (*Galea musteloides*). *Behavioral Ecology and Sociobiology* **53**:341–349.

Höner, O.P., *et al.* (2007) Female mate-choice drives the evolution of male-biased dispersal in a social mammal. *Nature* **448**:798–801.

Hoogland, J.L. (1982) Prairie dogs avoid extreme inbreeding. *Science* **215**:1639–1641.

Hoogland, J.L. (1995) *The Black-tailed Prairie Dog: Social Life of a Burrowing Mammal.* Chicago: University of Chicago Press.

Horne, T.J. and Ylönen, H. (1998) Heritabilities of dominance-related traits in male bank voles (*Clethrionomys glareolus*). *Evolution* **52**:894–899.

Hosken, D.J. and Blanckenhorn, W.U. (1999) Female multiple mating, inbreeding avoidance and fitness: it is not only the magnitude of costs and benefits that counts. *Behavioral Ecology* **10**:462–464.

Hrdy, S.B. (1977) *The Langurs of Abu: Female and Male Strategies of Reproduction.* Cambridge, MA: Harvard University Press.

Huck, U.W. and Banks, E.M. (1982) Male dominance status, female choice and mating success in the brown lemming, *Lemmus trimucronatus. Animal Behaviour* **30**:665–675.

Huck, U.W., *et al.* (1986) Determinants of mating success in the golden hamster (*Mesocricetus auratus*): social dominance and mating tactics under seminatural conditions. *Animal Behaviour* **34**:971–989.

Humphries, R.E., *et al.* (1999) Unravelling the chemical basis of competitive scent marking in house mice. *Animal Behaviour* **58**:1177–1190.

Huxley, J.S. (1932) *Problems of Relative Growth.* London: Methuen.

Isvaran, K. and Clutton-Brock, T.H. (2007) Ecological correlates of extra-group paternity in mammals. *Proceedings of the Royal Society of London. Series B: Biological Sciences* **274**:219–224.

Iwasa, Y., *et al.* (1991) The evolution of costly mate preferences. II. The handicap principle. *Evolution* **45**:1431–1442.

Jacobs, G.H. (1993) The distribution and nature of color vision among the mammals. *Biological Reviews* **68**:413–471.

Janson, C.H. (1984) Female choice and mating system of the brown capuchin monkey *Cebus apella* (Primates: Cebidae). *Zeitschrift für Tierpsychologie* **65**:177–200.

Jennions, M.D. and Petrie, M. (2000) Why do females mate multiply? A review of the genetic benefits. *Biological Reviews of the Cambridge Philosophical Society* **75**:21–64.

Johnston, R.E., *et al.* (1997) Scent counter-marking by male meadow voles: females prefer the top-scent male. *Ethology* **103**:443–453.

Jolly, A. (1966) *Lemur Behavior.* Chicago: University of Chicago Press.

Jordan, W.C. and Bruford, M.W. (1998) New perspectives on mate choice and the MHC. *Heredity* **81**:127–133.

Kavaliers, M. and Colwell, D.D. (1995a) Discrimination by female mice between the odours of parasitized and non-parasitized males. *Proceedings of the Royal Society of London. Series B: Biological Sciences* **261**:31–35.

Kavaliers, M. and Colwell, D.D. (1995b) Odours of parasitized males induce aversive responses in female mice. *Animal Behaviour* **50**:1161–1169.

Keddy Hector, A.C., *et al.* (1989) Intraspecific variation in the tuning of female cricket frogs. *American Zoologist* **29**:A16.

Keil, A. and Sachser, N. (1998) Reproductive benefits from female promiscuous mating in a small mammal. *Ethology* **104**:897–903.

Klein, J. (1986) How many class-II immune-response genes: a commentary on a reappraisal. *Immunogenetics* **23**:309–310.

Klein, S.L., *et al.* (1999) *Trichinella spiralis* infection in voles alters female odor preference but not partner preference. *Behavioral Ecology and Sociobiology* **45**:323–329.

Kokko, H. (2003) Review Paper: Are reproductive skew models evolutionarily stable? *Proceedings of the Royal Society of London. Series B: Biological Sciences* **270**:265–270.

Kokko, H., *et al.* (2003) The evolution of mate choice and mating biases. *Proceedings of the Royal Society of London. Series B: Biological Sciences* **270**:653–664.

Krackow, S. and Matuschak, B. (1991) Mate choice for non-siblings in wild house mice: evidence from a choice test and a reproductive test. *Ethology* **88**:99–108.

Kramer, L. and Mastromatteo L.A. (1973) Role of olfactory stimuli during copulation in male and female rats. *Journal of Comparative and Physiological Psychology* **85**:528–535.

Kruuk, L.E.B., *et al.* (2003) Fluctuating asymmetry in a secondary sexual trait: no associations with individual fitness, environmental stress or inbreeding, and no heritability. *Journal of Evolutionary Biology* **16**:101–113.

Kuester, J. and Paul, A. (1992) Influence of male competition and female mate choice on male mating success in Barbary macaques (*Macaca sylvanus*). *Behaviour* **120**:192–217.

Künkele, J. and Hoeck, H.N. (1995) Communal suckling in the cavy *Galea musteloides. Behavioral Ecology and Sociobiology* **37**:385–391.

Lande, R. (1981) Models of speciation by sexual selection on polygenic traits. *Proceedings of the National Academy of Sciences of the United States of America* **78**:3721–3725.

Le Boeuf, B.J. and Mesnick, S. (1991) Sexual behavior of male northern elephant seals. I. Lethal injuries to adult females. *Behaviour* **116**:143–162.

Leclaire, S., *et al.* (2013) Mating strategies in dominant meerkats: evidence for extra-pair paternity in relation to genetic relatedness between pair mates. *Journal of Evolutionary Biology* **26**:1499–1507.

Lemaître, J.F. and Gaillard, J.M. (2013) Polyandry has no detectable mortality cost in female mammals. *PLOS ONE* **18**:e66670.

Lincoln, G.A. (1994) Teeth, horns and antlers: the weapons of sex. In: *The Differences Between the Sexes* (eds R.V. Short and E. Balaban) Cambridge: Cambridge University Press, 131–158.

Loehle, C. (1997) The pathogen transmission avoidance theory of sexual selection. *Ecological Modelling* **103**:231–250.

McClain, D.K., *et al.* (1995) Sexual selection and the risk of extinction in introduced birds on oceanic islands. *Oikos* **74**:27–34.

McComb, K. (1987) Roaring by red deer stags advances the date of oestrus in hinds. *Nature* **330**:648–649.

McComb, K. (1991) Female choice for high roaring rates in red deer, *Cervus elaphus*. *Animal Behaviour* **41**:79–88.

McComb, K. and Clutton-Brock, T.H. (1994) Is mate choice copying or aggregation responsible for skewed distributions of females on leks? *Proceedings of the Royal Society of London. Series B: Biological Sciences* **255**:13–19.

McElligott, A.G. and Hayden, T.J. (2001) Postcopulatory vocalizations of fallow bucks: who is listening? *Behavioral Ecology* **12**:41–46.

McMahon, C.R. and Bradshaw, C.J.A. (2004) Harem choice and breeding experience of female southern elephant seals influence offspring survival. *Behavioral Ecology and Sociobiology* **55**:349–362.

Magnhagen, C. (1991) Predation risk as a cost of reproduction. *Trends in Ecology and Evolution* **6**:183–186.

Manson, J.H. (1992) Measuring female mate choice in Cayo Santiago rhesus macaques. *Animal Behaviour* **44**:405–416.

Manson, J.H. (1994a) Male aggression: a cost of female mate choice in Cayo Santiago rhesus macaques. *Animal Behaviour* **48**:473–475.

Manson, J.H. (1994b) Mating patterns, mate choice, and birth season heterosexual relationships in free-ranging rhesus macaques. *Primates* **35**:417–433.

Manson, J.H. (2006) Mate choice. In: *Primates in Perspective* (eds C.J. Campbell, A. Fuentes, K.C. MacKinnon, M. Panger and S.K. Bearder) Oxford: Oxford University Press, 447–463.

Matsumoto-Oda, A. (1999) Female choice in the opportunistic mating of wild chimpanzees (*Pan troglodytes schweinfurthii*) at Mahale. *Behavioral Ecology and Sociobiology* **46**:258–266.

Maynard Smith, J. and Harper, D. (2003) *Animal Signals*. Oxford: Oxford University Press.

Mays, H.L. and Hill, G.E. (2004) Choosing mates: good genes versus genes that are a good fit. *Trends in Ecology and Evolution* **19**:554–559.

Milinski, M. (2003) The function of mate choice in sticklebacks: optimizing MHC genetics. *Journal of Fish Biology* **63**:1–16.

Møller, A.P. (2000) Sexual selection and conservation. In: *Behaviour and Conservation* (eds L.M. Gosling and W.J. Sutherland) Cambridge: Cambridge University Press, 161–171.

Møller, A.P. and Danchin, E. (2008) Behavioural ecology and conservation. In: *Behavioural Ecology* (eds E. Danchin, L.A. Giraldeau and F. Cezilly) Oxford: Oxford University Press, 647–656.

Møller, A.P. and de Lope, F. (1994) Differential costs of a secondary sexual character: an experimental test of the handicap principle. *Evolution* **48**:1676–1683.

Møller, A.P. and Pomiankowski, A. (1993) Fluctuating asymmetry and sexual selection. *Genetica* **89**:267–279.

Møller, A.P. and Swaddle, J.P. (1997) *Asymmetry, Developmental Stability and Evolution*. Oxford: Oxford University Press.

Møller, A.P., *et al.* (1995) Foraging costs of a tail ornament: experimental evidence from two populations of barn swallows *Hirundo rustica* with different degrees of sexual size dimorphism. *Behavioral Ecology and Sociobiology* **37**:289–295.

Møller, A.P., *et al.* (1996) Horn asymmetry and fitness in gemsbok, *Oryx g. gazella*. *Behavioral Ecology* **7**:247–253.

Moore, W.G. and Marchinton, R.L. (1974) Marking behaviour and its social function in white-tailed deer. In: *The Behaviour of Ungulates and its Relation to Management* (eds V. Geist and F. Walther) Gland, Switzerland: International Union for the Conservation of Nature Publications, 447–456.

Mossman, C.A. and Drickamer, L.C. (1996) Odor preferences of female house mice (*Mus domesticus*) in seminatural enclosures. *Journal of Comparative Psychology* **110**:131–138.

Mousseau, T.A. and Roff, D.A. (1987) Natural selection and the heritability of fitness components. *Heredity* **59**:181–197.

Mysterud, A., *et al.* (2005) Climate-dependent allocation of resources to secondary sexual traits in red deer. *Oikos* **111**:245–252.

Nefdt, R.J.C. and Thirgood, S.J. (1997) Lekking, resource defence and harassment in two subspecies of lechwe antelope. *Behavioral Ecology* **8**:1–9.

Nichols, H.J., *et al.* (2014) Evidence for frequent incest in a cooperatively breeding mammal. *Biology Letters* **10**:20140898.

Nichols, H.J., *et al.* (2015) Adjustment of costly extra-group paternity according to inbreeding risk in a cooperative mammal. *Behavioral Ecology* **26**:doi: 10.1093/beheco/arv095

Nicoletto, P.F. (1995) Offspring quality and female choice in the guppy, *Poecilia reticulata*. *Animal Behaviour* **49**:377–387.

Nisbet, I.C.T. (1973) Terns in Massachusetts: present numbers and historical changes. *Bird Banding* **44**:27–55.

Novotny, M., *et al.* (1990) Chemistry of male dominance in the house mouse, *Mus domesticus*. *Experientia* **46**:109–113.

O'Donald, P. (1962) The theory of sexual selection. *Heredity* **17**:541–552.

O'Donald, P. (1967) A general model of sexual and natural selection. *Heredity* **22**:499–518.

Packer, C. (1979) Inter-troop transfer and inbreeding avoidance in *Papio Anubis*. *Animal Behaviour* **27**:1–36.

Packer, C., *et al.* (1988) Reproductive success in lions. In: *Reproductive Success: Studies of Individual Variation in Contrasting Breeding Systems* (ed. T.H. Clutton-Brock). Chicago: University of Chicago Press, 363–383.

Packer, C., *et al.* (1991) A molecular genetic analysis of kinship and cooperation in African lions. *Nature* **351**:562–565.

Parker, G.A. (1983) Mate quality and mating decisions. In: *Mate Choice* (ed. P. Bateson). Cambridge: Cambridge University Press, 141–164.

Partridge, L. and Hurst, L.D. (1998) Sex and conflict. *Science* **281**:2003–2008.

Paterson, S. and Pemberton, J.M. (1997) No evidence for major histocompatibility complex-dependent mating patterns in a free-living ruminant population. *Proceedings of the Royal Society of London. Series B: Biological Sciences* **264**:1813–1819.

Pearce, D.T. and Oldham, C.M. (1984) The ram effect, its mechanism and application to the management of sheep. In: *Reproduction in Sheep* (eds D.R. Lindsay and D.T. Pearce) New York: Cambridge University Press, 26–34.

Pemberton, J.M., *et al.* (1992) Behavioural estimates of male mating success tested by DNA fingerprinting in a polygynous mammal. *Behavioral Ecology* **3**:66–75.

Pemberton, J.M., *et al.* (2004) Mating patterns and male breeding success. In: *Soay Sheep: Dynamics and Selection in an Island Population* (eds T.H. Clutton-Brock and J.M. Pemberton) Cambridge: Cambridge University Press, 166–189.

Penn, D.J. (2002) The scent of genetic compatibility: sexual selection and the major histocompatibility complex. *Ethology* **108**:1–21.

Penn, D.J. and Potts, W.K. (1998a) Chemical signals and parasite-mediated sexual selection. *Trends in Ecology and Evolution* **13**:391–396.

Penn, D.J. and Potts, W.K. (1998b) MHC-disassortative mating preferences reversed by cross-fostering. *Proceedings of the Royal Society of London. Series B: Biological Sciences* **265**:1299–1306.

Penn, D.J. and Potts, W.K. (1999) The evolution of mating preferences and major histocompatibility complex genes. *American Naturalist* **153**:145–164.

Penton-Voak, I.S. and Perrett, D.I. (2001) Male facial attractiveness: perceived personality and shifting female preferences for male traits across the menstrual cycle. In: *Advances in the Study of Behavior*, Vol. 30 (eds P.J.B. Slater, J.S. Rosenblatt, C.T. Snowdon and T.J. Roper) San Diego, CA: Academic Press, 219–259.

Pereira, M.E. and Weiss, M.L. (1991) Female mate choice, male migration and the threat of infanticide in ringtailed lemurs. *Behavioral Ecology and Sociobiology* **28**:141–152.

Pierce, J.D. and Dewsbury, D.A. (1991) Female preference for unmated versus mated males in two species of voles (*Microtus ochrogaster* and *Microtus montanus*). *Journal of Comparative Psychology* **105**:165–171.

Pistorius, P.A., *et al.* (2001) Pup mortality in southern elephant seals at Marion Island. *Polar Biology* **24**:828–831.

Pollux, B.J.A., *et al.* (2014) The evolution of the placenta drives a shift in sexual selection in livebearing fish. *Nature* **513**:233–236.

Pomiankowski, A. (1988) The evolution of female mate preferences for male genetic quality. *Oxford Surveys in Evolutionary Biology* **5**:136–184.

Pomiankowski, A., *et al.* (1991) The evolution of costly mate preferences. II. Fisher and biased mutation. *Evolution* **45**:1422–1430.

Potts, W.K. and Wakeland, E.K. (1990) Evolution of diversity at the major histocompatibility complex. *Trends in Ecology and Evolution* **5**:181–187.

Potts, W.K., *et al.* (1991) Mating patterns in seminatural populations of mice influenced by MHC genotype. *Nature* **352**:619–621.

Preston, B.T., *et al.* (2001) Dominant rams lose out by sperm depletion. *Nature* **409**:681–682.

Price, E.C. (1990) Infant carrying as a courtship strategy of breeding male cotton-top tamarins. *Animal Behaviour* **40**:784–786.

Prichard, A.K., *et al.* (1999) Factors affecting velvet antler weights in free-ranging reindeer in Alaska. *Rangifer* **19**:71–76.

Promislow, D.E.L. (1992) Costs of sexual selection in natural populations of mammals. *Proceedings of the Royal Society of London. Series B: Biological Sciences* **247**:203–210.

Promislow, D.E.L., *et al.* (1994) Sexual selection and survival in North American waterfowl. *Evolution* **48**:2045–2050.

Pusey, A.E. and Wolf, M. (1996) Inbreeding avoidance in animals. *Trends in Ecology and Evolution* **11**:201–206.

Rasmussen, K.L.R. (1983) Influence of affiliative preferences upon the behaviour of male and female baboons during sexual consortships. In: *Primate Social Relationships: An Integrated Approach* (ed. R.A. Hinde). Oxford: Blackwell Science, 116–120.

Reiter, J.R., *et al.* (1981) Female competition and reproductive success in northern elephant seals. *Animal Behaviour* **29**:670–687.

Reynolds, J.D. and Gross, M.R. (1992) Female mate preference enhances offspring growth and reproduction in a fish, *Poecilia reticulata*. *Proceedings of the Royal Society of London. Series B: Biological Sciences* **250**:57–62.

Rich, T.J. and Hurst, J.L. (1998) Scent marks as reliable signals of the competitive ability of mates. *Animal Behaviour* **56**:727–735.

Rich, T.J. and Hurst, J.L. (1999) The competing countermarks hypothesis: reliable assessment of competitive ability by potential mates. *Animal Behaviour* **58**:1027–1037.

Richard, E., *et al.* (2008) Ranging behaviour and excursions of female roe deer during the rut. *Behavioural Processes* **79**:28–35.

Ridley, M. (1988) Mating frequency and fecundity in insects. *Biological Reviews* **63**:509–549.

Ridley, M. (1990) The control and frequency of mating in insects. *Functional Ecology* **4**:75–84.

Roberts, S.C. and Gosling, L.M. (2003) Genetic similarity and quality interact in mate choice decisions by female mice. *Nature Genetics* **35**:103–106.

Robinson, J.G. (1982) Intrasexual competition and mate choice in primates. *American Journal of Primatology* **3**:131–144.

Rosser, A.M. (1987) *Resource defence in an African antelope, the puku* (Kobus vardoni). PhD thesis, University of Cambridge, Cambridge

Rosser, A.M. (1992) Resource distribution, density, and determinants of mate access in puku. *Behavioral Ecology* **3**:13–24.

Rossiter, S.J., *et al.* (2005) Mate fidelity and intra-lineage polygyny in greater horseshoe bats. *Nature* **437**:408–411.

Rowe, C. (1999) Receiver psychology and the evolution of multicomponent signals. *Animal Behaviour* **58**:921–931.

Rubenstein, D.I. (1986) Ecology and sociality in horses and zebras. In: *Ecological Aspects of Social Evolution* (eds D. I. Rubenstein and R. W. Wrangham) Princeton, NJ: Princeton University Press, 282–302.

Ryan, M.J. (1998) Sexual selection, receiver biases, and the evolution of sex differences. *Science* **281**:1999–2003.

Saino, N., *et al.* (1997) Immunocompetence, ornamentation, and viability of male barn swallows (*Hirundo rustica*). *Proceedings of the National Academy of Sciences of the United States of America* **94**:549–552.

Salo, A.L. and Dewsbury, D.A. (1995) Three experiments on mate choice in meadow voles (*Microtus pennsylvanicus*). *Journal of Comparative Psychology* **109**:42–46.

Sanderson, J.L., *et al.* (2015) Banded mongooses avoid inbreeding when mating with members of the same natal group. *Molecular Ecology* **24**:3738–3751.

Sawyer, T.G., *et al.* (1989) Response of female white-tailed deer to scrapes and antler rubs. *Journal of Mammalogy* **70**:431–433.

Schluler, D. and Price, T. (1993) Honesty, perception and population divergence in sexually selected traits. *Proceedings of the Royal Society of London. Series B: Biological Sciences* **253**:117–122.

Schwagmeyer, P.L. (1984) Multiple mating and intersexual selection in thirteen-lined ground squirrels. In: *The Biology of Ground-dwelling Squirrels* (eds J.O. Murie and G.R. Michener) Lincoln, NE: University Press of Nebraska, 275–293.

Schwagmeyer, P.L. (1986) Effects of multiple mating on reproduction in female thirteen-lined ground squirrels. *Animal Behaviour* **34**:297–298.

Schwagmeyer, P.L. and Brown, C.H. (1983) Factors affecting male–male competition in thirteen-lined ground squirrels. *Behavioral Ecology and Sociobiology* **13**:1–6.

Schwagmeyer, P.L. and Parker, G.A. (1987) Queuing for males in thirteen-lined ground squirrels. *Animal Behaviour* **35**:1015–1025.

Schwagmeyer, P.L. and Woontner, S.J. (1985) Mating competition in an asocial ground squirrel, Spermophilus tridecemlineatus. *Behavioral Ecology and Sociobiology* **17**:291–296.

Schwarz-Weig, E. and Sachser, N. (1996) Social behaviour, mating system and testes size in Cuis (*Galea musteloides*). *Zeitschrift für Säugetierkunde* **61**:25–38.

Setchell, J.M. (2005) Do female mandrills prefer brightly colored males? *International Journal of Primatology* **26**:715–735.

Setchell, J.M. and Dixson, A.F. (2001) Changes in the secondary adornments of male mandrills (*Mandrillus sphinx*) are associated with gain and loss of alpha status. *Hormones and Behavior* **39**:177–184.

Seyfarth, R.M. (1978) Social relationships among adult male and female baboons. II. Behaviour throughout the female reproductive cycle. *Behaviour* **64**:227–247.

Shapiro, L.E. and Dewsbury, D.A. (1986) Male dominance, female choice and male copulatory behavior in two species of voles (*Microtus ochrogaster* and *Microtus montanus*). *Behavioral Ecology and Sociobiology* **18**:267–274.

Shields, W.M. (1987) Dispersal and mating systems: investigating their causal connections. In: *Mammalian Dispersal Patterns: The Effects of Social Structure on Population Genetics* (eds B.D. Chepko-Sade and Z.T. Halpin) Chicago: University of Chicago Press, 3–24.

Silk, J.B. and Boyd, R. (1983) Female cooperation, competition and mate choice in matrilineal macaque groups. In: *Social Behavior of Female Vertebrates* (ed. S.K. Wasser). New York: Academic Press, 316–348.

Sivinski, J. (1984) Sperm in competition. In: *Sperm Competition and the Evolution of Animal Mating Systems* (ed. R.L. Smith). London: Academic Press, 86–115.

Slate, J., *et al.* (2000) Inbreeding depression influences lifetime breeding success in a wild population of red deer (*Cervus elaphus*). *Proceedings of the Royal Society of London. Series B: Biological Sciences* **267**:1657–1662.

Smale, L. (1988) Influence of male gonadal hormones and familiarity on pregnancy interruption in prairie voles. *Biology of Reproduction* **39**:28–31.

Smale, L., *et al.* (1997) Sexually dimorphic dispersal in mammals: patterns, causes and consequences. In: *Advances in the Study of Behavior*, Vol. **26** (eds C. Snowdon, P.J.B. Slater, M. Milinski and J.S. Rosenblatt) San Diego, CA: Academic Press, 181–250.

Small, M.F. (1988) Female primate sexual behavior and conception: are there really sperm to spare. *Current Anthropology* **29**:81–100.

Small, M.F. (1989) Female choice in non-human primates. *American Journal of Physical Anthropology* **32**:103–127.

Smuts, B.B. (1985) *Sex and Friendship in Baboons*. New York: Aldine.

Smuts, B.B. and Smuts, R.W. (1993) Male aggression and sexual coercion of females in nonhuman primates and other mammals: evidence and theoretical implications. In: *Advances in the Study of Animal Behaviour*, Vol. **22** (eds P.J.B. Slater, J.S. Rosenblatt, C.T. Snowdon and M. Milinski) San Diego, CA: Academic Press, 1–63.

Solomon, N.G. and Keane, B. (2007) Reproductive strategies in female rodents. In: *Reproductive Strategies in Female Rodents* (eds J.O. Wolff and P.W. Sherman) Chicago: University of Chicago Press, 42–56.

Soltis, J. and McElreath, R. (2001) Can females gain extra paternal investment by mating with multiple males? A game theoretic approach. *American Naturalist* **158**:519–529.

Soltis, J., *et al.* (2000) Infanticide by resident males and female counter-strategies in wild Japanese macaques (*Macaca fuscata*). *Behavioral Ecology and Sociobiology* **48**:195–202.

Sorci, G., *et al.* (1998) Plumage dichromatism in birds predicts introduction success in New Zealand. *Journal of Animal Ecology* **67**:263–269.

Spencer, P.B.S., *et al.* (1998) Enhancement of reproductive success through mate choice in a social rock-wallaby, *Petrogale assimilis* (Macropodidae) as revealed by microsatellite markers. *Behavioral Ecology and Sociobiology* **43**:1–9.

Stanford, C.B. (1999) *The Hunting Apes*. Princeton, NJ: Princeton University Press.

Stein, D.M. (1984) Ontogeny of infant–adult male relationships during the first year of life for yellow baboons (Papio cynocephalus) In: *Primate Paternalism* (ed. D. M. Taub). New York: Van Nostrand Reinhold, 213–243.

Stillman, R.A., *et al.* (1996) Black hole models of ungulate lek size and distribution. *Animal Behaviour* **52**:891–902.

Stopher, K.V., *et al.* (2011) The red deer rut revisited: female excursions but no evidence females move to mate with preferred males. *Behavioral Ecology* **22**:808–818.

Stopher, K.V., *et al.* (2012) Re-mating across years and intralineage polygyny are associated with greater than expected levels of inbreeding in wild red deer. *Journal of Evolutionary Biology* **25**:2457–2469.

Tanaka, Y. (1996) Social selection and the evolution of animal signals. *Evolution* **50**:512–523.

Tardif, S.D. and Bales, K. (1997) Is infant-carrying a courtship strategy in callitrichid primates? *Animal Behaviour* **53**:1001–1007.

Tasker, C.R. and Mills, J.A. (1981) A functional analysis of courtship feeding in the red-billed gull. *Behaviour* **77**:222–241.

Taub, D.M. (1980) Female choice and mating strategies among wild Barbary macaques (Macaca sylvanus) In: *The Macaques: Studies in Ecology, Behavior and Evolution* (ed. D.G. Lindburg). New York: Van Nostrand Reinhold, 287–344.

Thomas, S.A. (2002) Scent marking and mate choice in the prairie vole, *Microtus ochrogaster*. *Animal Behaviour* **63**:1121–1127.

Tutin, C.E.G. (1979) Mating patterns and reproductive strategies in a community of wild chimpanzees (*Pan troglodytes schweinfurthii*). *Behavioral Ecology and Sociobiology* **6**:29–39.

Vandenbergh, J.G. (1965) Hormonal basis of the sex skin in male rhesus monkeys. *General and Comparative Endocrinology* **5**:31–34.

Vanpé, C., *et al.* (2007) Antler size provides an honest signal of male phenotypic quality in roe deer. *American Naturalist* **169**:402–193.

vom Saal, F.S. (1984) Proximate and ultimate causes of infanticide and parental behavior in male mice. In: *Infanticide: Comparative and Evolutionary Perspectives* (eds G. Hausfater and S.B. Hrdy). New York: Aldine, 401–424.

von Schantz, T., *et al.* (1996) MHC genotype and male ornamentation: genetic evidence for the Hamilton–Zuk model. *Proceedings of the Royal Society of London. Series B: Biological Sciences* **263**:265–271.

von Schantz, T., *et al.* (1997) Mate choice, male condition-dependent ornamentation and MHC in the pheasant. *Hereditas* **127**:133–140.

West, P.M. and Packer, C. (2002) Sexual selection, temperature, and the lion's mane. *Science* **297**:1339–1343.

Widemo, F. and Sæther, S.A. (1999) Beauty is in the eye of the beholder: causes and consequences of variation in mating preferences. *Trends in Ecology and Evolution* **14**:26–31.

Wilkinson, G.S. (1998) Male eye span in stalk-eyed flies indicates genetic quality by meiotic drive suppression. *Nature* **391**:276–279.

Williams, G.C. (1966) *Adaptation and Natural Selection: A Critique of Some Current Evolutionary Thought*. Princeton, NJ: Princeton University Press.

Willis, C. and Poulin, R. (2000) Preference of female rats for the odours of non-parasitised males: the smell of good genes? *Folia Parasitologica* **47**:6–10.

Winn, B.E. and Vestal. B.M. (1986) Kin recognition and choice of males by wild female house mice (*Mus musculus*). *Journal of Comparative Psychology* **100**:72–75.

Wynne-Edwards, K.E. and Lisk, R.D. (1984) Djungarian hamsters fail to conceive in the presence of multiple males. *Animal Behaviour* **32**:626–628.

Yamaguchi, M., *et al.* (1981) Distinctive urinary odors governed by the major histocompatibility locus of the mouse. *Proceedings of the National Academy of Sciences of the United States of America* **78**:5817–5820.

Yamazaki, K. (1988) Familial imprinting determines H-2 selective mating preferences. *Science* **240**:1331–1332.

Young, A.J. (2003) Subordinate tactics in cooperative meerkats. breeding, helping and dispersal. PhD thesis, University of Cambridge, Cambridge.

Young, A.J., *et al.* (2007) Subordinate male meerkats prospect for extra-group paternity: alternative reproductive tactics in a cooperative mammal. *Proceedings of the Royal Society of London. Series B: Biological Sciences* **274**:1603–1609.

Zahavi, A. (1975) Mate selection: a selection for a handicap. *Journal of Theoretical Biology* **53**:205–214.

Zahavi, A. (1977) The cost of honesty: further remarks on the handicap principle. *Journal of Theoretical Biology* **67**:603–605.

Zeh, D.W. and Zeh, J.A. (2000) Reproductive mode and speciation: the viviparity driven conflict hypothesis. *BioEssays* **22**:938–946.

Zeh, J.A. and Zeh, D.W. (1996) The evolution of polyandry. I. Intragenomic conflict and genetic incompatibility. *Proceedings of the Royal Society of London. Series B: Biological Sciences* **263**:1711–1717.

Zinkernagel, R.M. and Doherty, P.C. (1974) Immunological surveillance against altered self components by sensitized T-lymphocytes in lymphocytic choriomeningitis. *Nature* **251**:547–548.

CHAPTER 5
Maternal care

5.1 Introduction

In a grove of fever trees, a group of vervet monkeys are foraging for fallen pods. Males displace females from the most productive feeding sites and females displace adolescents and juveniles. Mothers help to shelter their offspring from aggression from other group members and the sons and daughters. On the edge of the group, an orphan female forages warily, one eye on her closest neighbour. Lacking any support from a mature female, she is often threatened and frequently displaced. Her chances of survival are low and her breeding success if she lives is unlikely to be high.

Extensive maternal care is a conspicuous feature of all mammals and the extent and duration of care has far-reaching benefits to offspring (Figure 5.1). But all components of maternal care have substantial costs to mothers, reducing their capacity to invest in other offspring, lowering their subsequent fecundity and, in some cases, depressing their condition and survival. The presence of these effects raises a wide range of important questions about the evolution and distribution of maternal care and the effects of interactions with other individuals on the mother's investment strategies. What are the relative costs and benefits of gestation and lactation? For how long should mothers gestate their offspring and for how long should they allow them to nurse? Are optimal birth and weaning dates similar for parents and their offspring or are there conflicts of interest about the length of these crucial stages of development? How should mothers distribute their resources among their offspring and how should they respond to conflict between them?

This chapter examines the evolutionary causes and consequences of maternal care and the ways in which maternal investment is distributed. Section 5.2 briefly reviews five areas of theory that have structured our understanding of variation in maternal investment: optimal strategies of investment per offspring; trade-offs between maternal investment at different stages of development; investment in different offspring; the effects of conflicts of interest between siblings; and conflicts between parents and offspring. The next two sections examine empirical studies of variation in maternal investment before birth (section 5.3) and its effects on offspring development (section 5.4) while the following two sections describe variation in maternal investment during lactation (section 5.5) and after weaning (section 5.6). Mothers might be expected to adjust the extent to which they invest in offspring in relation both to the variation in the costs of parental care and its relative benefits to offspring and section 5.7 reviews evidence of adaptive strategies of this kind. In some cases, mothers may be free to adjust their levels of investment to the level that maximises their own fitness but conflicts of interest between sibs as well as between parents and their offspring can lead to evolutionary conflicts and, in some cases, may prevent parents from optimising their level of investment: section 5.8 examines the extent and effects of competition between siblings while section 5.9 reviews evidence of parent–offspring conflict and its consequences. Finally, section 5.10 briefly reviews some of the evolutionary consequences of maternal care.

5.2 The evolution of maternal care

Investment and maternal survival

In all animals where mothers invest substantially in their offspring and breed more than once, there are trade-offs between maternal investment in particular breeding attempts and the mother's subsequent survival (see Chapter 1). Increased energetic expenditure on offspring

Mammal Societies, First Edition. Tim Clutton-Brock.
© 2016 Tim Clutton-Brock. Published 2016 by John Wiley & Sons, Ltd.

Figure 5.1 In many social mammals, offspring are dependent on their mothers for comfort and social support until they reach maturity. A female rhesus macaque comforts her adolescent offspring, closely guarded by an adult male. *Source*: © Alexander Georgiev.

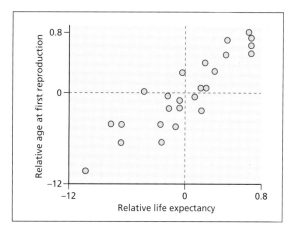

(a)

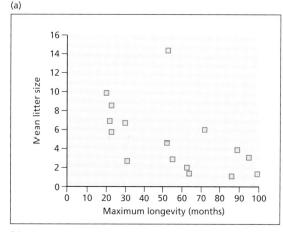

(b)

Figure 5.2 (a) Relative age at first reproduction plotted against relative life expectancy at birth for natural populations of mammals. Relative values refer to deviations from logarithmic regression lines for each variable plotted on adult female body size. Numbers refer to different mammalian genera. (b) Mean litter size plotted against maximum longevity in mammals. *Source*: From Harvey and Zammuto (1985). Reproduced with permission of Macmillan Publishers Ltd. (b) From Eisenberg (1981). Reproduced with permission of the University of Chicago Press.

usually reduces the mother's subsequent survival or her ability to invest in other offspring (or both). Energetic costs are likely to generate fitness costs, though these can be affected by a wide range of ecological and social factors and relationships between energetic costs and fitness costs may be non-linear (Clutton-Brock 1991; Alonso-Alvarez and Velando 2012). In general, high risks of extrinsic mortality (such as high rates of starvation caused by unpredictable environments or high rates of predation) are likely to favour rapid and frequent reproduction at the expense of maternal survival while low rates of extrinsic mortality will favour smaller clutches or litters, and lower breeding frequency and may lead to prolonged breeding lifespans in females (Pianka 1974; Stearns 1992; Roff 2002). As expected, species differences in female age at first breeding increase the average life expectancy (Figure 5.2a) while litter size is negatively correlated with longevity (Eisenberg 1981; Harvey and Zammuto 1985) (Figure 5.2b).

Safe harbours and offspring development

In many animals, the relative susceptibility of offspring to environmental factors that can affect their growth, survival or eventual breeding success varies between different stages of development. For example, in many oviparous animals, eggs and juveniles are relatively safe in nests which are guarded by adults; however,

once juveniles start to disperse on their own, they are at immediate risk of predation or starvation and mortality rates increase (Clutton-Brock 1991). Other things being equal, parents might be expected to maximise the amount of time that dependent offspring spend in the safest stages of development, an idea known as the *safe harbour* hypothesis (Shine 1978), and several empirical studies of fish support this hypothesis (Shine 1978;

Sargent *et al.* 1987; Rüber *et al.* 2004). Where parental resources are limited, selection can also favour deferring differential investment in offspring until the later stages of development when there is a higher chance that offspring will survive to maximise the benefits of care to the parent's fitness (Clutton-Brock 1991). However, while both these arguments are reasonable, it is important to remember that they depend on the initial premise that other things are equal, which often is not the case. For example, if the effects of investment on parental survival vary between the stages of offspring development, parents may attempt to minimise the time that offspring spend in stages that are associated with the highest risks to their own survival.

Allocation of resources to different offspring

There is also an (almost) inevitable trade-off between the extent to which mothers invest in individual offspring and offspring number or breeding frequency. The first quantitative model of this trade-off by Smith and Fretwell (1974) used the marginal value theorem to predict the optimal level of investment in eggs (Box 5.1 and Figure 5.3) but the same argument applies to other forms of maternal investment. Harsh environments, high rates of predation early in life or intense competition with conspecifics are all likely to increase the effects of maternal investment on offspring survival and to favour increases in the extent and duration of care (McGinley *et al.* 1987; Roff 2002).

Box 5.1 Trade-offs between offspring size and number

The first quantitative model of the trade-offs between the size and the number of offspring was produced by Smith and Fretwell (1974) and has formed the basis of nearly all subsequent work (Clutton-Brock and Godfray 1991). Smith and Fretwell asked how a parent should distribute a fixed amount of resources (M) amongst an indefinite number of young. They had in mind a mother whose only investment in her offspring lay in provisioning eggs so that the only trade-off involved was between the size and number of eggs, though their basic argument also applies equally well to investment in dependent young. Their approach assumes that the way in which parental resources are distributed within particular breeding attempts does not affect the risk of the whole clutch being lost or the parent's subsequent survival or breeding success.

Suppose the mother produces eggs of size s which survive to maturity with probability $k \cdot f(s)$. The probability of survival has two components: one, $f(s)$, influenced by parental investment and a second component, k, which summarises all the other mortality risks that are unaffected by egg size. The number of eggs produced by the mother is simply M/s, the total resources available divided by the amount invested in each individual. The fitness of the mother is thus $(M/s) \cdot k \cdot f(s)$, the number of offspring multiplied by their own individual fitness. Optimal egg size is found by maximising fitness with respect to investment per egg and is $f(s)/f'(s)$ where the prime refers to the fitness of an offspring. The solution to this problem is of a marginal value form (a ratio of a function and its derivative) and can also be derived graphically (Figure 5.3a). The optimal size of offspring is defined by the point at which a straight line from the origin is tangent to the offspring fitness curve ($f'(s)$) for, at this point, the parent maximises its fitness returns per unit of investment. Where the fitness of an offspring rises with increasing size but on a decelerating curve (Figure 5.3b), a mother should produce the maximum number of the smallest viable individuals, whereas if the offspring fitness curve always accelerates (Figure 5.3c), all resources should be invested in a single large offspring.

Since 1974, a substantial number of models have extended Smith and Fretwell's treatment (Clutton-Brock 1991; Alonso-Alvarez and Velando 2012). Relationships between parental expenditure and offspring fitness need not be as simple as those shown in Figure 5.3a and multiple optima may occur, favouring multimodal distributions of egg or offspring size. Such effects can exaggerate differences in offspring size among parents because some parents are unable to 'reach' higher optima and effects of this kind may generate individual differences in offspring sex ratios in some dimorphic species (see Chapter 18). Where optimal propagule size varies between environments, conditional strategies for varying propagule size in relation to parental environment might be expected (Lloyd 1984; Haig and Westoby 1989). In multiparous organisms, optimal propagule size may change if offspring differ in quality (Temme 1986), if parents vary in their ability to care for propagules (Sargent *et al.* 1987), or if resource levels vary but propagule number is predetermined (Lloyd 1987; McGinley and Charnov 1988). In some models, reduced juvenile survival can also select for increased propagule size (Kolding and Fenchel 1981; Sibly and Calow 1986), while in others propagule size may be affected by selection pressures on litter size (Parker and Begon 1986). In addition, while most models of optimal propagule size assume that the way in which parental resources are distributed within clutches or litters does not affect the average fitness of the entire clutch or the parent's subsequent survival or breeding success, in some situations these assumptions are invalid. For example, if competitive interactions between nestlings attract predators but are minimised by differences in size between siblings, selection may favour parents that produce young of different sizes within the same litter (Slagsvold *et al.* 1984; Magrath 1988). Similarly, in species where offspring assist their parents' subsequent breeding attempts or compete with subsequent (or previous) sibs, the form of these interactions may affect the optimal distribution of parental expenditure among progeny.

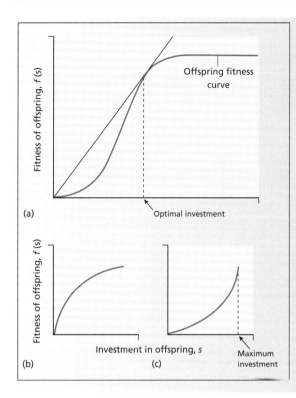

Figure 5.3 Estimating the optimal size of propagules. (a) Where offspring fitness initially increases slowly with increasing parental investment, then accelerates and finally increases more slowly again, parents obtain maximal returns on investment at the point where a line from the origin is tangential to the offspring fitness curve. Other relationships between offspring fitness and parental investment generate different results: (b) where increases in offspring fitness relative to parental investment decline with increasing investment, selection will favour low levels of parental investment; (c) where increases in offspring fitness relative to parental investment continue to rise with increasing investment, selection will favour maximal levels of investment in each offspring. *Source*: From Smith and Fretwell (1974).

Comparative studies of relative egg size among invertebrates and fish support this prediction (Clutton-Brock and Godfray 1991). Similarly, among mammals, species that bear their young above ground, where they are immediately exposed to predation, typically produce smaller litters of offspring that are relatively large and precocious, while species that keep their offspring in burrows or tree cavities produce large litters of relatively small neonates. For example, rabbits (which breed in burrows) produce larger litters of relatively altricial young while hares (which breed above ground) produce small numbers of relatively precocial young (Swihart 1984; Boyd 1985; Hackländer *et al.* 2002).

Conflicts between sibs

Where offspring demand more resources from their mothers than the latter are able (or willing) to supply, sibs may compete with each other to maximise their share of parental resources (Mock and Parker 1997). The extent of conflict is likely to be affected by ecological and social parameters, including family size, methods of food distribution used by adults, and asymmetries in age or size between sibs (Mock and Parker 1997; Roulin and Dreiss 2012). Differences in kinship between littermates would also be expected to affect the intensity of competition between siblings, for the converse of Hamilton's rule predicts that the spread of selfish strategies should be restricted to situations where the benefits of selfish behaviour exceed their costs to victims, multiplied by the coefficient of relatedness between interacting parties (see Chapter 1). As a result, genetic relatedness might be expected to reduce competition between sibs (Johnstone and Roulin 2003; Johnstone 2004), although high relatedness among littermates may also increase the similarity of their requirements and so may increase competition (Le Cam *et al.* 2009; Aguirre and Marshall 2012; Roulin and Dreiss 2012).

Models of non-lethal competition between siblings have explored a range of different situations, including scramble competition between sibs and the formation of consistent dominance relations with or without negotiation between competitors or parental intervention. Under scramble competition, the intensity of sibling competition is affected by whether the costs of competition are restricted to individuals that compete or whether they are shared between all littermates; where there are dominance hierarchies, outcomes are affected by assumptions about the extent of competition between different individuals (Godfray and Parker 1991; Mock and Parker 1997). Models of lethal competition have explored the conditions under which selection operating on parents and on their offspring favours brood reduction through siblicide, filial infanticide and suicide: they suggest that siblicide can evolve more easily than either filial infanticide or suicide (O'Connor 1978).

Under some conditions, cooperative strategies between littermates may also be favoured, especially if relatedness is high. For example, individuals may avoid the costs of competing with each other over their share of parental resources by negotiating their share before parents arrive to feed them (Roulin 2002; Roulin and Dreiss 2012). Similarly, where the tendency of parents (or other carers) to provide food is controlled by signals of need generated by dependants, offspring might be expected to cooperate to maintain parental contributions by adjusting their signalling behaviour to maximise its effects on parents (Johnstone 2004).

Parent–offspring conflict

Early accounts of maternal investment considered offspring as passive entities and assumed that mothers controlled the allocation of resources. However, in diploid animals, parents and their progeny should commonly 'disagree' about the level of parental investment (see Chapter 1), and offspring would often be expected to do their best to bias maternal investment away from their parents' optimum level towards their own (Box 5.2 and Figure 5.4): conflict may arise either over the amount of resources allocated to different broods (*inter-brood conflict*) or over allocation to different members of the same brood (*intra-brood conflict*). Although Trivers' argument was initially contested (Alexander 1974), subsequent models confirmed that conflicts of interest between parents and offspring are likely to be widespread and explored the factors affecting the intensity of conflict and the extent to which parents would be expected to retaliate to manipulation by offspring (MacNair and Parker 1979; Godfray 1995a,b).

Where there are conflicts of interest between parents and their offspring over the extent of parental care, selection is likely to favour offspring that give signals that exaggerate their need for resources and parents that attempt to make a realistic assessment of their needs (Trivers 1974; Kilner and Hinde 2012). Where offspring have private information about their needs which parents cannot access, they should exploit this asymmetry in information (Grodzinski and Lotem 2007). Under these circumstances, the genetic mechanisms controlling offspring signals of need and those controlling the supply of resources by parents are likely to coevolve, leading to genetic correlations between these two sets of traits (Kölliker *et al.* 2005). Whether these are positive or negative is likely to be affected by the role of parents and offspring in controlling the upper limits of provisioning (Kilner and Hinde 2012): where parents control provisioning, positive genetic correlations between parent and offspring traits are expected; where offspring exert greater control than parents, these should be replaced by negative genetic correlations (Kölliker *et al.* 2005, 2012). In general, the available data for vertebrates appears to support both these predictions (Kilner and Hinde 2012).

A second important issue is whether conflicts between parents and their offspring are likely to be resolved so that they have stable outcomes that favour either parents or offspring, or whether struggles are likely to persist with the result that females differ in how closely they achieve their optimal level of investment (Smiseth *et al.* 2008; Kilner and Hinde 2012). Both kinds of outcome are possible (Parker and MacNair 1979; Godfray 1995b) and which is most likely frequently depends on the form of interactions between parental strategies and offspring strategies, the costs of begging to offspring and the extent to which parents and offspring control provisioning (Godfray 1995b; Parker *et al.* 2002; Kilner and Hinde 2012).

Unfortunately, the predictions of theoretical models of parent–offspring conflict are difficult to test since it is usually hard to identify levels of investment that are optimal for parents or to compare them with levels that are optimal for offspring in order to determine the extent to which the behaviour of offspring causes parents to modify their optimal level of investment. However, using long-term datasets that combine genetic data with accurate measures of fitness, some studies have now provided direct evidence of conflicts of interest between parents and offspring (see section 5.9) and recent studies of insects and birds have investigated how evolution proceeds in the presence of conflict and what the eventual outcome is likely to be (Kilner and Hinde 2012).

Conflicts between carers

Where males contribute to parental care, their input may allow females to reduce their level of investment

Box 5.2 Parent–offspring conflict

Trivers' initial model of parent–offspring conflict envisaged a situation where mothers produced a single young each year, where subsequent offspring were full sibs, and where fathers did not invest in their offspring after conception. His graphical model (Figure 5.4a) envisages a situation where the amount of parental expenditure per day is fixed and the benefits of parental care to the current offspring decline as it grows older, while the costs to the parent either remain constant or increase, so that the ratio of benefit to cost (*B/C*) shows a progressive decline. Basing his argument on Hamilton's concept of inclusive fitness, Trivers argued that because the coefficient of relatedness between the offspring and its future sibs will be 0.5 while the offspring's relatedness to itself is 1.0, the offspring should favour the mother's continuing parental care until the costs of care to the mother's fitness exceed twice the benefit to itself. Mothers and offspring should consequently 'disagree' over the continuation of parental care from the time when the *B/C* ratio equals 1.0 to the time when it equals 0.5 (Figure 5.4a). Depending on the rate at which *B/C* declines with increasing offspring age, this may either be a relatively short period (curve I) or a relatively long one (curve II). Where offspring can raid parental resources directly or can deceive parents into exceeding their own optimal level of investment (for example, by begging frequently for food), Trivers predicted that they should do so until the *B/C* ratio falls to 0.5 if future offspring are full sibs or 0.25 if they are half-sibs. Periods of evolutionary conflict between parents and their offspring might be expected to give rise to overt behavioural conflict, such as the apparent 'disagreements' over weaning between female mammals and their young (Trivers 1974). In addition, he showed that evolutionary conflicts between parents and offspring are likely to arise over levels of investment at any stage of the period of parental care: since the cost of parental expenditure to the adult will be double the cost that is suffered by its offspring, the latter will be selected to favour a level of parental expenditure that maximises the differences between *B* and *C*/2, while parents will be selected to favour a level that maximises the difference between *B* and *C* (Figure 5.4b). As both the parent and the current offspring will be related to the parent's future progeny by 0.5 (assuming that they have the same father), subsequent treatments of parent–offspring conflict have argued that the *costs* of expenditure to the parent and its current offspring will be identical and that it is the *benefits* to parents and offspring that will differ (Lazarus and Inglis 1986).

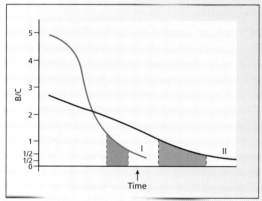

(a)

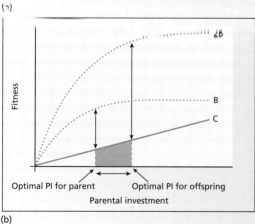

(b)

Figure 5.4 (a) The benefit–cost ratio (*B/C*) of a parental care act (such as nursing) toward an offspring as a function of time. Benefit is measured in units of the reproductive success of the current offspring, and cost in comparable units of the reproductive success of the mother's future offspring. Two species are plotted: in species I the *B/C* ratio decays quickly; in species II, it does so more slowly. Shaded areas indicate times during which parent and offspring are in conflict over whether the parental care should continue. Future sibs are assumed to be full sibs. If future sibs were half-sibs, the shaded areas would have to be extended until *B/C* = ¼. (b) Parent–offspring conflict at a particular point in time. Line C shows the fitness costs experienced by parents and offspring as a consequence of parental investment; line B denotes the fitness benefits gained by parents from the provision of parental investment; line 2B shows the fitness benefits experienced by offspring as their parents supply investment (assuming a monogamous mating system in a diploid species). The optimal levels of investment for each party can be found at the point at which they experience greatest benefit for least cost. The horizontal arrow indicates the disparity between optima, which is the source of parent–offspring conflict. *Sources*: (a) From Trivers (1974). Reproduced with permission of Oxford University Press. (b) From Lazarus and Inglis (1986), redrawn by Hinde and Kilner (2012). Reproduced with permission of Elsevier.

(Maynard Smith 1977) (see Chapter 16). Where partners compensate for reductions in the extent of investment by their mates, selection on both parents is likely to favour individuals that induce their partners to increase their level of investment so that they can reduce their own input and conflicts of interest between carers are likely to be common (Houston and Davies 1985; Barta *et al.* 2002; Alonzo and Klug 2012). To predict the relative investment of fathers and mothers in parental care in biparental species, it is necessary to consider both the relative costs of care to males and females and the likely responses of their partners to changes in their level of expenditure (see Chapter 16).

5.3 Prenatal investment

Gestation

All living mammals are viviparous, with the exception of the two families of monotremes: the platypuses (Ornithorhynchidae) and the spiny anteaters (Tachyglossidae). Viviparity is an effective way of increasing the survival of offspring and may also accelerate their growth, and comparative studies of its distribution among invertebrates, reptiles and fish show that it is associated with harsh conditions, high levels of predation or parasitism or intense competition, especially between juveniles and adults (Clutton-Brock 1991).

The evolution of endothermy probably allowed the first mammals to adopt a nocturnal way of life and facilitated the evolution of fossorial habits, maternal brooding and lactation (Hopson 1973; Hayssen 1993). Parental provisioning may subsequently have encouraged a reduction in egg size, allowing the development of egg retention and viviparity (Pond 1977, 1983; Lillegraven *et al.* 1987; Dunbrack and Ramsay 1989). Marsupials and eutherians subsequently developed alternative strategies of prenatal development: contemporary marsupials possess a breeding system where prenatal expenditure and gestation length are minimised, and offspring growth can be manipulated by delayed implantation or embryonic diapause (Low 1978), while most eutherians have longer and less flexible gestation periods and produce larger neonates that are weaned relatively quickly.

Gestation periods differ between species, from less than 15 days in some marsupials and smaller rodents to nearly 2 years in elephants (Hayssen 1993). Interspecific differences in gestation length are partly a consequence of variation in maternal size: across mammals, gestation length (in days) increases as (maternal weight in grams)$^{0.26}$ (Blueweiss *et al.* 1978), though shallower slopes and lower correlation coefficients are found within orders (Clutton-Brock 1991). Like the relative duration of other life-history stages, the relative duration of gestation is inversely associated with mortality rates, probably because a combination of high juvenile and adult mortality favours rapid development and reproduction (Gaillard *et al.* 1989; Harvey *et al.* 1989a,b; Roff 2002). Since differences in habitat and diet often affect mortality rates, they are commonly associated with variation in rates of development, including relative gestation length. For example, rodents that are food specialists have longer gestation periods, are weaned later, reach sexual maturity later and have longer lifespans than folivores (Mace 1979), while forest-dwelling primates develop more slowly for their body size than savannah-dwelling species and show a lower maximum rate of reproduction (Ross 1988; Harvey *et al.* 1989a,b).

Some differences in relative gestation length are a consequence of trade-offs between the relative development of offspring before versus after birth. As the safe harbour hypothesis predicts, eutherian mammals that give birth in comparatively safe sites underground or in tree holes produce comparatively light, altricial neonates after relatively short gestation periods, whereas those that bear their young in open environments, where they are exposed to attacks from predators, produce relatively precocial nenonates (Gaillard *et al.* 1989; Derrickson 1992) and similar associations between the weight of neonates and the risk of infant mortality occur in seals (Kovacs and Lavigne 1986).

In the early stages of gestation energy costs to mothers are typically low, but during the last two quarters they increase rapidly before eventually reaching a ceiling (Johnson *et al.* 2001). For example, in pregnant rodents, increases in mean daily calorie intake over nonreproductive rates range between 18 and 25% and rising maternal weight can reduce the mother's foraging success and increase her susceptibility to predation (Gittleman and Thompson 1988; Clutton-Brock 1991).

While the energetic costs of the later stages of gestation are considerable, they are substantially lower than those of lactation and the efficiency of energy conversion into the growth of offspring tissue is also substantially higher during gestation than lactation (Speakman 2008). For example, energetic studies of guinea pigs show that a mother's average increase in energy intake above the non-reproductive level is only 16% during gestation compared with 92% during lactation, while the efficiency of energy conversion is nearly twice as high during gestation as during lactation (Künkele 2000). The higher costs of lactation are commonly associated with increased costs to maternal fitness relative to the costs of gestation. For example, compared to females that give birth but lose their calves shortly afterwards, female red deer that rear a calf successfully are significantly more likely to die in the following winter, are less likely to breed again the following year and produce lighter calves with lower survival chances if they do so (Clutton-Brock *et al.* 1989).

In eutherian mammals, mothers are able to vary the duration of gestation to a rather limited extent. For example, in red deer, relatively high spring temperatures and improved levels of nutrition are associated with a reduction of 0.77 days in gestation length for every degree increase in mean temperature (Asher *et al.* 2005; Clements *et al.* 2011). Social factors may also be important, especially in species where births are highly synchronised and young born out of synchrony are less likely to survive. For example, in banded mongooses (Figure 5.5), where breeding females from the same group typically give birth on the same day, pups born later are usually killed and late-conceiving mothers give birth after slightly shorter gestation periods than those that conceive early (Cant 2000) (Figure 5.6).

Competition between neonates and conflicts of interest between parents and offspring may also affect the evolution of gestation length and precociality. Where competition between neonates is intense and larger individuals are more likely to be successful, there is likely to be strong selection for increases in prenatal growth rates, which may raise the costs of gestation to mothers (Stockley and Parker 2002). Across species, the average number of offspring per maternal nipple (after controlling for body mass) provides a possible index of the intensity of competition between neonates and comparative analyses show that this is positively correlated with prenatal growth rates and negatively with postnatal growth. Stockley and Parker argue that mothers may respond to increased costs of prenatal growth by reducing gestation length and that selection on offspring to increase prenatal growth and on parents to restrict costs by shortening gestation may lead to coevolutionary increases in prenatal growth. This could explain the existence of the altricial–precocial dichotomy in mammals and the tendency for altricial species to have

Figure 5.5 In banded mongooses, which live in large groups that include several breeding adults of each sex, several females commonly give birth on the same day. *Source*: © Tim Clutton-Brock.

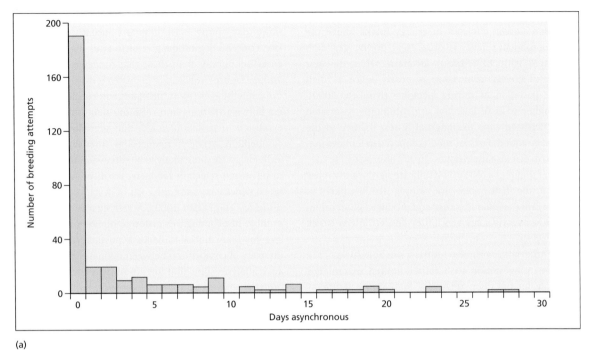

(a)

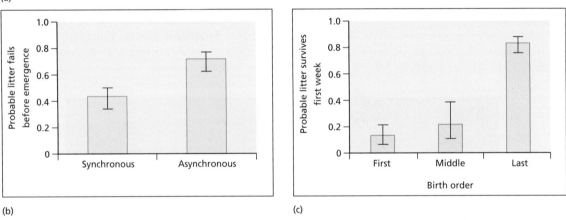

(b)   (c)

Figure 5.6 Extreme birth synchrony and infanticide in banded mongooses. In this species, most adult females breed in each breeding attempt and typically give birth on the same day. (a) Frequency histogram of synchronous and asynchronous breeding attempts. Several females give birth on the same day in 63% of breeding attempts. (b) Synchronous communal litters are less likely to fail in the den. (c) In litters that are asynchronous, litters born to females that give birth first are more likely to fail in the first week compared to those that give birth last: the dependency of early pup survival on the pregnancy status of co-breeders is probably a consequence of infanticide. Extreme birth synchrony in this species appears to be an adaptation to avoid infanticide and minimise competitive disparities between young. *Source*: Adapted from Hodge *et al.* (2011). Reproduced with permission from the Royal Society.

relatively high prenatal growth rates. However, the extent to which the duration of gestation is controlled by the mother is unclear, postnatal growth rates are also likely to have substantial costs to mothers, and a

considerable number of ecological factors may affect trade-offs between prenatal and postnatal growth. As a result, the relative importance of parent–offspring conflict in the evolution of precociality is uncertain.

Species differences

Across mammalian species, neonatal weight increases as (maternal weight)$^{0.66-0.72}$ (Blueweiss et al. 1978) though shallower slopes and lower correlation coefficients are found within Orders (Harvey and Clutton-Brock 1985; Gittleman 1986). As a result of the negative allometry of birthweight, females of larger species give birth to off-spring that are a smaller proportion of their own weight than those produced by smaller species and the proportion of a mother's daily energy budget expended on reproduction declines with increasing body size.

Contrasts in the relative costs of gestation to mothers may exert an important influence on prenatal growth. In several groups of mammals, interspecific differences in neonatal size are associated with variation in diet quality: for example, folivorous rodents and primates tend to produce smaller neonates relative to maternal weight than specialised fruit or seed eaters (Mace 1979; Harvey and Clutton-Brock 1985). Neonatal size relative to maternal size tends to decline in species where life expectancy of adults is relatively low, possibly because heavy birthweights require longer gestation periods, which are selected against where the chance that the mother will die before her offspring reach independence is high (Harvey et al. 1989a,b; Clutton-Brock 1991).

Variation in the effects of birthweight on neonatal survival is clearly important, too. Across some taxonomic groups, the size of neonates increases and litter size declines as environmental conditions deteriorate or the risk of predation to newborn young increases, leading to variation in the relative development of neonates along an altriciality–precociality axis (Martin and MacLarnon 1985, Clutton-Brock 1991). However, multiple factors (including contrasts in phylogeny) affect relationships between the length of gestation and lactation periods and the extent of prenatal and postnatal growth, and there is no simple relationship between them (Stearns 1983; Derrickson 1992). For example, the length of lactation period is extended in precocial primates but shortened in precocial rodents. These correlations between gestation length, birthweight and the risk of mortality are sometimes cited as in line with the predictions of the safe harbour hypothesis (Shine 1978; Sargent et al. 1987), though long gestation periods and relatively early weaning may also have energetic benefits to mothers if energy transfer during gestation is more efficient than during lactation.

Intraspecific variation

During the early stages of gestation, fetal growth rates are slow but they typically accelerate at around the midpoint of the gestation period. In monotocous species, the weight of young at birth is usually correlated with the mother's weight, size and condition as well as with ecological factors that affect maternal characteristics (Hastings and Testa 1998; Pomeroy et al. 1999; Wilson and Festa-Bianchet 2009). For example, in red deer, birthweights are positively correlated with temperature and grass growth in the last 2 months of the gestation period (Albon et al. 1987; Moyes et al. 2011) (Figure 5.7), while in white-tailed deer they are negatively correlated with winter snowfall (Mech et al. 1991). In polytocous species, relationships between maternal condition and offspring birthweight are complicated by trade-offs between birthweight and litter size and there is often no simple association between maternal condition and neonatal weight (Russell et al. 2002; Stevenson et al. 2004; Wilson and Festa-Bianchet 2009). For example, in mammals that commonly produce twins, like roe deer and Soay sheep, low maternal weight is associated with reductions in litter size rather than birthweight, weakening correlations between the weight of mothers and the birthweight of offspring (Clutton-Brock et al. 2004; Wilson and Festa-Bianchet 2009).

Variation in maternal hormones can also have an important influence on the prenatal development of offspring (Mateo 2009; Dantzer et al. 2013). For example, socially dominant female spotted hyenas show higher levels of faecal androgen metabolites during pregnancy than subordinate females and their cubs are more frequently aggressive than those born to mothers with lower androgen levels (Dloniak et al. 2006). Increased aggressiveness in infants may contribute to their eventual rank and breeding success, though social support of juveniles by their mothers appears to play an important role in establishing status (East et al. 2009).

In some mammals, elevated levels of glucocorticoids in pregnant females increase neonatal growth and size: for example in North American red squirrels, mothers that have relatively high levels of glucocorticoids produce offspring that show enhanced growth (Dantzer et al. 2013). Glucocorticoid levels in pregnant females can also influence the development of sex differences among offspring: for example, the male offspring of rats reared in crowded stressful conditions show increased behavioural feminisation (Dahlöf et al. 1977), while in guinea pigs

Figure 5.7 On the island of Rum in the Inner Hebrides, researchers catch a high proportion of deer calves each year soon after birth by netting them when they are left on the hill by their mothers. *Source*: © Tim Clutton-Brock.

elevated cortisol levels in pregnant females are associated with behavioural masculinisation in daughters and infantilisation in sons (Sachser and Kaiser 1996; Kaiser and Sachser 2001; Kaiser *et al.* 2003). Proximity to male and female fetuses in the uterus can also affect the hormonal environment that fetuses experience and their subsequent development (see Chapter 18).

Selection for prenatal growth

In many mammals there is strong selection for prenatal growth. Studies of ungulates where researchers have been able to catch large numbers of infants soon after birth and monitor the subsequent survival and breeding success of different individuals from the same birth cohorts have demonstrated the importance of prenatal growth. For example, in Soay sheep, the probability that a lamb will survive its first winter is related to its birth date, birthweight and subsequent growth (Figure 5.8) (Clutton-Brock *et al.* 1992). Positive associations between neonatal weight and juvenile survival have also been found in bighorn and Dall's sheep (Bunnell 1980; Festa-Bianchet 1988), mountain goats (Côté and Festa-Bianchet 2001), roe deer (Gaillard *et al.* 1997), red deer (Guinness *et al.* 1978a,b), white-tailed deer (Mech *et al.* 1991), moose (Keech *et al.* 2000) and bison

(Green and Rothstein 1991). These effects may operate both through the susceptibility of juveniles to starvation and through susceptibility to disease or predation (Mech *et al.* 1991).

In several species, heavy-born individuals also show increased reproductive performance as adults. For example, heavy-born female red deer calves produce relatively heavy offspring throughout their reproductive lives and lose a relatively low proportion of their progeny before they mature (Guinness *et al.* 1978b; Albon *et al.* 1987). In some cases, relationships between birthweight and juvenile survival among members of the same cohort vary with population density and are only apparent in particular categories of individuals or in years when population density is high (Festa-Bianchet *et al.* 1996, 1997; Côté and Festa-Bianchet 2001). In others, inter-annual variation in climate generates contrasts in growth, survival and reproductive performance between cohorts (Albon *et al.* 1987; Mech *et al.* 1991).

5.4 Maternal effects

As a result of the impact of maternal characteristics on the level and timing of investment, the age and identity

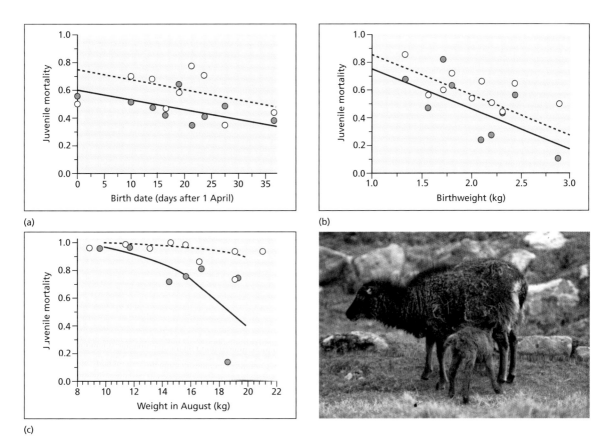

(a)

(b)

(c)

Figure 5.8 Survival of Soay sheep lambs on St Kilda depends on their birth date and birthweight. (a) Winter mortality of cohorts of lambs born in different years, controlling for the effects of population size (filled circles are for females, open circles are for males); (b) winter mortality of lambs plotted on birthweight, allowing for effects of population size (filled circles are female, open circles are male); (c) juvenile winter mortality plotted on weight at 4 months, allowing for the effect of population size (filled circles are female, open circles are male). In all plots, lines are fitted equations through all data while displayed points show mean mortality for records in different categories. *Source*: From Clutton-Brock *et al.* (1992). Reproduced with permission of John Wiley & Sons. *Photo Source*: Tim Clutton-Brock.

of mothers often has an important influence on the development and fitness of their offspring (Bernardo 1996; Uller 2012). For example, in European rabbits, offspring born to females in their prime are more likely to survive and and to breed successfully than those of young and old mothers (Rodel *et al.* 2009) and similar effects have been found in ungulates (Clutton-Brock and Albon 1989; Clutton-Brock *et al.* 2004).

Maternal effects can be generated by several different mechanisms. In some cases, they are a consequence of the direct impact of ecological or social factors. For example, in rodents, the experimental manipulation of maternal diets or the injection of flavourants into the amnion shows how maternal physiology can affect the feeding preferences of neonates and juveniles, generating similarities in feeding behaviour between mothers and their offspring (Galef and Henderson 1972; Schaal *et al.* 1995) (see Chapter 6). Where environmental factors affect the prenatal growth of offspring they can have a substantial impact on both their survival as juveniles and their reproductive success as adults and can lead to transgenerational effects. For example, long-term records of red deer show that cold weather in late winter and early spring which retards plant growth is associated with depressed fetal growth and relatively low birthweights of calves (Albon *et al.* 1983). Over the rest of their lifespan, females born in cold springs produced relatively light calves and the average birthweight of calves born to

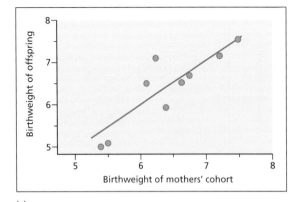

(a)

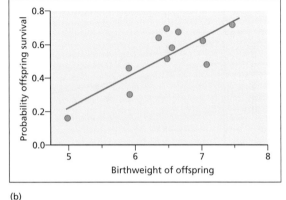

(b)

Figure 5.9 Correlations between birthweight, offspring birthweight, and offspring survival in red deer calves from different cohorts. (a) Shows the mean birthweight of offspring born to different cohorts of mothers across their lifespans plotted against the mean birthweight of the mothers' cohort; (b) shows the mean survival of calves born to members of different cohorts plotted against the mean birthweight of the cohort. *Source*: From Albon *et al*. (1987). Reproduced with permission of John Wiley & Sons. *Photo source*: Tim Clutton-Brock.

different cohorts of females differed by up to 40% between cohorts, while calf survival over the first 2 years of life ranged from under 10% to over 60% (Figure 5.9). Female nutrition has also been shown to be associated with the growth and survival of their grand-offspring in white-tailed deer (Mech *et al*. 1991). Environmental effects operating through mothers can also affect the reproductive performance and parental behaviour of their offspring as adults, leading to trans-generational effects (Champagne and Curley 2011). In banded mongooses, which live in groups that include multiple breeding females, heavy offspring are more likely to win competitive interactions with their littermates and to obtain improved access to resources (Hodge *et al*. 2009), while in meerkats the extent to which individuals contribute to allo-parental care is associated with their own experience of care as juveniles (English *et al*. 2010).

Variation in the physical or social environment of pregnant females may also affect the development of young through its effects on maternal hormone levels, including sex hormones and glucocorticoids (Mateo 2009; Dantzer *et al*. 2013). For example, if pregnant females experience adverse climatic conditions, their responses can affect the hypothalamic–pituitary–adrenal functioning of their developing offspring as well as their responses to acute stress and novel objects and can influence some forms of learning (Maccari *et al*. 1995; Champagne and Curley 2011; Siegeler *et al*. 2011). Social conditions that raise glucocorticoid levels in pregnant females can have long-lasting consequences for their offspring. For example, the daughters of female cavies exposed to unstable social environments during pregnancy show higher testosterone levels and are permanently masculinised while their sons are infantilised and show retarded development of the hypothalamic–pituitary–adrenocortical axis, reduced aggression and lower attentiveness but enhanced body weight as adolescents (Sachser and Kaiser 1996; Kaiser and Sachser 1998, 2001; Kaiser *et al*. 2003; Siegeler *et al*. 2011).

Maternal androgen levels during gestation and lactation may also have an important influence on the development of offspring. In some marmosets, maternal androgen levels during the first trimester of pregnancy are negatively correlated with prenatal growth and birthweight as well as with postnatal catch-up growth (Smith *et al.* 2010). In other species, there is evidence that maternal androgen levels can affect the behaviour of infants and juveniles. For example, in spotted hyenas dominant females have relatively high androgen levels and their offspring are rather more frequently aggressive than those of subordinate mothers (Dloniak *et al.* 2006) (see Chapter 6).

In some cases, maternal effects can be caused by interactions between genes and environmental factors that affect gene expression (Champagne 2008; Champagne and Curley 2009, 2012; Shea *et al.* 2011). For example, experiments with mice and rats show that the development of mother–infant relationships affects the responsiveness of offspring to stress as well as their maternal behaviour as adults as a consequence of changes in DNA methylation and gene expression that affect the promoter regions of steroid receptor genes and these effects can be transferred across generations (Curley *et al.* 2008; Champagne and Curley 2009). Contrasts in the environment experienced during early development can also lead to changes in gene expression and affect the behaviour of individuals as adults (Curley *et al.* 2008; Leshem and Schulkin 2012).

Maternal effects are not confined to processes operating before birth and may operate through the quantity or quality of milk that mothers produce or through the extent and regularity that mothers feed their offspring before they achieve nutritional independence (see section 5.5). In other cases, they may operate through the social conditions that developing individuals encounter, which are often strongly influenced by their mother's social rank as well as by the size and age structure of their group.

Since genetic differences between mothers and between offspring can modify the form and extent to which mothers affect the development of their offspring, maternal effects are likely to be subject to selection (Chevrud and Wolf 2009; Uller 2012). One important question is how commonly maternal effects are adaptive, helping to adjust the development of offspring to the ecological conditions they are likely to encounter as adults and generating *predictive adaptive responses* or PARs (Mateo 2009; Gluckman *et al.* 2005a,b). The available evidence is mixed. For example, it has been suggested that the production of small neonates and juveniles by mothers living in environments where resources are scarce might help to adapt their offspring to similar conditions, an idea that is sometimes referred to as the *thrifty phenotype* hypothesis (Barker 1998; Gluckman 2005a,b; Bateson and Gluckman 2011). However, an alternative possibility is that differences in juvenile development are the result of developmental, ecological or social constraints that limit the capacity of mothers to invest in their offspring and generate non-adaptive contrasts in development. One way of distinguishing between these explanations is to determine whether individuals that develop under adverse conditions are more successful than those reared under optimal conditions when they experience adverse conditions in adulthood (as PAR models would predict) or less successful (as 'constraint' models would suggest). Studies of roe deer (Douhard *et al.* 2014), baboons (Lea *et al.* 2015) and humans (Hayward and Lumaa 2013), as well as a range of other organisms (Uller 2012) all suggest that individuals reared under adverse conditions are less resilient to adverse conditions in adulthood than those that develop under more favourable conditions, as 'constraint' models would predict and, as yet, there is little convincing evidence to support the thrifty phenotype hypothesis.

While evidence favouring the thrifty phenotype hypothesis is scarce, other studies indicate that PARs can occur in mammals. For example, research on natural populations of North American red squirrels shows that maternal effects operating through variation in glucocorticoid levels in pregnant females can help to adapt the development of offspring to the social conditions they are likely to face (Dantzer *et al.* 2013). In this study, pregnant female squirrels were exposed either to regular playback of territorial conspecifics, mimicking high population density, or to playback of white noise. Mothers exposed to frequent territorial calls showed heightened cortisol levels and produced offspring that showed increased rates of growth. Experiments in which breeding females were provisioned with cortisol confirmed that their offspring grew significantly faster (Dantzer *et al.* 2013) (Figure 5.10). Other studies of red squirrels show that increased growth rates enhance the competitive success of individuals and increase the chance that they will be

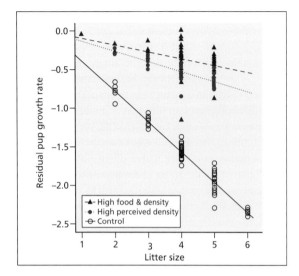

Figure 5.10 Adaptive plasticity in growth in North American red squirrels. Plots show the relative growth rates of pups that were born to mothers who experienced: (1) high population density and high food availability during gestation; (2) high perceived density (because they were exposed to levels of conspecific calls indicating high density but with lower levels of food availability); and (3) controls. The experiment shows that pups born to mothers exposed to high perceived levels of population density showed levels of growth similar to those that experienced high food availability and high population density. *Source*: From Dantzer *et al.* (2013). Reproduced with permission from AAAS. *Photo source*: © Ryan W. Taylor.

able to acquire territories when competition is intense (McAdam and Boutin 2003), indicating that these changes are likely to increase the fitness of offspring.

The extent of maternal effects and their potential role in evolutionary processes has only recently been recognised and credible evidence supporting adaptive

explanations is usually difficult to obtain. Although it is usually possible to construct adaptive explanations for maternal effects, the data to test them are often unavailable. For example, the tendency for social instability to infantilise the development of males in cavies could be beneficial in younger males queuing for reproductive opportunities in high-density populations where competition is common and social relationships are frequently unstable, but the effects of infantilisation on male fitness under natural conditions are unknown (Siegeler *et al.* 2011). Theoretical studies suggest that the strength of maternal effects could be affected by conflicts of interest between mothers and their offspring (Uller and Pen 2011) but whether there is any consistent association between parent–offspring conflict and the distribution of PARs is also unknown. As a result, it is not yet clear how commonly maternal effects are adaptive and how often they are unselected by-products of ecological or social constraints.

5.5 Lactation and infant care

Duration

All female mammals suckle their young with milk produced by specialised mammary glands that are originally thought to be derived from sweat or skin glands (Hayssen 1993; Balshine 2012). With the exception of monotremes, where milk oozes from their skin and is licked from the mother's milk-soaked fur by their offspring (Brawand *et al.* 2008), young receive milk through specialised nipples that may be derived from hair follicles. Lactation may have had substantial advantages for the first insectivorous mammals because it allowed mothers to continue to feed their offspring when food supplies were unpredictable (Dall and Boyd 2004).

Steroid and thyroid hormones and hormones of the prolactin and growth hormone family control the development of mammary glands and the differentiation of mammary cells and stimulate associated changes in the mother's physiology (Tucker 1994; Akers 2002, 2006). A surge in prolactin occurs at parturition and can stimulate lactation, though changes in growth hormone, insulin, renal corticoids and progesterone can also be involved (Forsyth 1986). The hormonal mechanisms controlling the maintenance of lactation vary. In ruminants, growth hormone is particularly important (Peel *et al.* 1983), while in other mammalian groups (including primates) changes in prolactin secretion and neural mechanisms operating through the frequency or duration of nursing play an important role, both in maintaining lactation and in delaying subsequent conceptions (McNeilly *et al*. 1982, 1983).

Lactation duration varies widely between species (Hayssen 1993). In some eutherians, young can feed themselves from the time they are born and lactation lasts for less than a week while, in others, young derive a proportion of their resources from their mothers for several years. For example, in African elephants, mothers often continue to suckle their offspring for over 2 years and calves whose mothers die when they are less than 2 years old rarely survive (Moss and Lee 2011). These differences are associated both with contrasts in the abundance and timing of resources and with differences in reproductive behaviour.

The pinnipeds provide some of the best examples of relationships between environmental factors and inter-specific differences in lactation and development (Kovacs and Lavigne 1992; Schulz and Bowen 2005). They fall into three main groups: the relatively small-bodied otar-iids (fur seals and sea lions), the larger phocids (true seals) and the walrus (Figure 5.11). The otariids breed on land, have lengthy lactation periods lasting 4–12 months and females intersperse lactation with foraging trips to sea, when they abandon their pups for up to 10 days (Schulz and Bowen 2005). The fat content of the milk of otariid species is related to the number of days females spend at sea between suckling periods while the relative length of lactation periods is negatively correlated with latitude, possibly because breeding conditions at low latitudes are more prone to unpredictable environmental oscillations that delay maturation, such as El Nino effects (Werhy and Trillman 1996)

Female phocids breed on land or ice and can feed their pups from resources that have been stored in the form of fat and, unlike otariids, remain close to their pups until they are weaned (Kovacs and Lavigne 1986; Schulz and Bowen 2005) (see Figure 5.11). Partly as a result of continuous maternal attendance, the duration of lacta-tion is shorter in phocids than otariids, though it varies with breeding habit and phocid species whelping on the pack ice (which commonly breaks up in storms) have shorter lactation periods than those breeding on land or fast ice. For example, the hooded seal weans its pups in an astonishing 4 days, during which pups increase their weight from 22 to 45 kg (Bowen *et al*. 1985; Boness and

Bowen 1996). Predation, too, may be involved: the northern pack ice breeders, like the hooded seal, are subject to land-based predation by polar bears and have shorter lactation periods than the southern pack ice breeders that are not exposed to terrestrial predators (Oftedal *et al*. 1987). Among northern seals that breed on fast ice, the birthweight of pups is lower in species that shelter their young in caves or fissures where they are relatively safe from predators than in species that keep their young in the open (Kovacs and Lavigne 1986). Unlike the other seals, walrus have an extended lactation period and pups accompany their mothers and suckle at sea.

Interspecific differences in lactation period are associ-ated with variation in milk content, milk yield and energy output. All seals produce milk that is unusually high in energy content, but species with very short lactation periods have the highest fat contents. For exam-ple, the hooded seal produces milk that has an average fat content of over 60%, the highest known for any mam-mal. Species with very short lactation periods and high milk energy content also have unusually large amounts of mammary tissue, very high daily milk yields (both in absolute terms and relative to the mother's energy bud-get), and relatively high weight gain per unit energy expended on pups. Despite this, they show reduced total expenditure on lactation. For example, the combined weight of mammary glands in female hooded seals exceeds 5 kg, daily milk production is around 8.5 kg/day (or 4.8% of maternal body weight) and pups gain 5.7 kg/day during lactation, but the mother's total energy expenditure on lactation is less than that of species with extended lactation periods (Oftedal *et al*. 1987; Skibiel *et al*. 2013).

Energetic costs

In most mammals, lactation typically has energetic costs that exceed those of gestation by a substantial margin and often affect maternal condition (Oftedal 1985; Speakman 2008). Across species, maternal energy intake increases with mothers' body mass and is two and a half to five times higher in lactating females than in those that are pregnant or are not breeding (Figure 5.12a). Energy intake is higher in species with relatively large litters, after the effects of maternal weight are controlled (Figure 5.12b). A similar rela-tionship between maternal energy needs and litter size occurs within species, generating negative correlations

(a)

(b)

Figure 5.11 Land and ice-breeding pinnipeds. (a) Grey seals breed on land and have extended lactation periods while (b) harp seals breed on pack ice or ice-flows and have shorter lactation periods. *Sources*: (a) © Westend61/Getty Images; (b) © Visuals Unlimited, Inc./Gerard Lacz/Getty Images.

between litter size and the growth of infants. For example, in mice, the food intake of mothers increases with the size of their litters, while the weight of their pups declines (Figure 5.13).

The daily milk yield and the gross energy output of mothers at peak lactation rises with (maternal weight)$^{0.83}$ (Oftedal 1981, 1984a,b), while daily milk production as a percentage of maternal weight declines

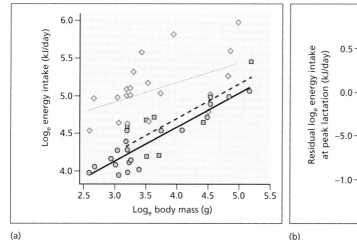

(a)   (b)

Figure 5.12 Energetic costs of breeding in different species of small rodents. (a) Plot of energy intake of females, when not breeding (circles), pregnant (squares) and lactating (diamonds). There is no significant elevation in intake during pregnancy, but intake during lactation is significantly elevated compared with non-breeders and pregnant individuals. (b) Residual energy intake during lactation is significantly associated with variation in litter size. *Source*: From Speakman (2008). Reproduced with permission from the Royal Society.

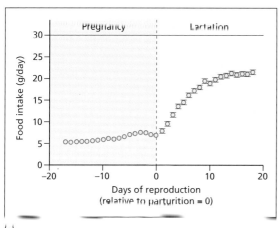

(a)

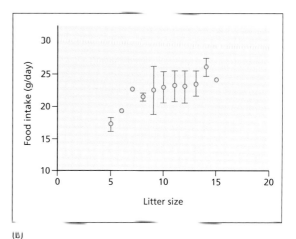

(B)

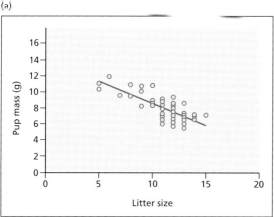

(c)

Figure 5.13 Energy costs of reproduction in mice.
(a) Relationship between mean food intake and day of reproduction: intake increases over days 1–10, but then reaches a plateau at days 10–18. (b) Asymptotic food intake (average over days 10–18) plotted against litter size: bigger litters require greater energy demands but, for litters of more than ten, the total intake is capped at 23 g/day. (c) Pup body mass at weaning in relation to litter size. In all cases $n = 71$ unmanipulated litters. *Source*: From Speakman and Król (2005). Reproduced with permission of Springer Science and Business Media.

with increasing maternal size, from 28% in pygmy shrews to 1.25% in elephants (Hanwell and Peaker 1977). However, as a result of the increased duration of lactation in large species, total expenditure on lactation per reproductive event or per lifespan increases with (maternal body weight)$^{0.8-1.14}$ (Gordon 1989; Reiss 1989) and, after the effects of body weight are controlled, daily milk yields differ between species by a factor of four (Oftedal 1984a,b). For example, pigs produce nearly twice as much milk as predicted, while reindeer, baboons and humans produce only half as much (Oftedal 1981). Females of litter-bearing species show consistently higher gross energy and protein yield than monotocous species and there is a close correlation between the mother's relative energy output in milk and the ratio of litter metabolic mass to mother's mass when the effects of variation in relative metabolic rate have been allowed for (Oftedal 1981, 1984a,b). Similar increases in gross energy production with litter size occur within species: for example, sheep with twin lambs or triplets show substantially higher milk yields than those with single-tons, and the proportion of their total energy turnover allocated to reproduction increases (Milne 1987). Among monotocous species, daily milk production varies in relation to the size and growth rate of offspring (Clutton-Brock 1991; Speakman 2008).

Milk content also varies widely between species. Comparisons of milk composition between species are complicated by the presence of pronounced changes in milk quality over the lactation period, but there are substantial differences between species that are associated with contrasts in phylogeny, ecology and behaviour (Oftedal 1984a,b; Cockburn 1989; Skibiel *et al*. 2013). Carnivores produce milk that includes a higher percentage of fat, dry matter, protein and energy than herbivores and omnivores (Figure 5.14). For example, mammals that suckle their young infrequently, like many phocid seals, tend to have more concentrated milk than those that suckle at shorter intervals, including most precocial species (Ben Shaul 1962). In addition, species in which lactating females feed little during lactation produce milk that is low in sugar but high in fat, minimising the need for gluconeogenesis while transferring energy to their young (Oftedal 2000). In other species, the costs to mothers of carrying milk affects its content: for example, in bats, flight places a premium on the reduction of weight, and milk is high in fat and dry matter (Jenness and Sloan 1970).

Species living on food supplies that are rich or relatively abundant during the breeding season tend to have high daily milk yields for their body weight, while species living on less abundant resources extend the duration of the lactation period and reduce daily expenditure. For example, studies of eastern wood rats which inhabit relatively poor ground show that they breed comparatively slowly, extend their lactation period and reduce their daily energy expenditure (McClure 1987). Among carnivores, species with herbivorous or folivorous diets, like the black bear, the giant panda and the red panda, also show relatively low daily energy outputs during lactation compared with more carnivorous species (Gittleman 1986). In contrast, in species where several adults cooperate to rear young, like coyotes, red foxes, dholes and meerkats, infants and dependent juveniles show relatively high total growth rates in relation to maternal body weight (Case 1978; Gittleman and Oftedal 1987).

In response to the energetic costs of lactation, females commonly increase foraging time at the expense of other activities and the average energy intake of lactating females is substantially higher than that of pregnant or non-breeding females and varies with food availability and quality (Felton *et al*. 2009). In some species, pregnant and lactating females consistently select foods of higher nutritional quality compared with the food selected by non-breeding females or males (Staines and Crisp 1978; Murray *et al*. 2009). Lactating females also show a range of physiological and anatomical adjustments, including growth of the alimentary tract and associated organs, and changes in immune function (Speakman 2008). In some species, females also appear to be able to adjust their fat reserves during the early stages of lactation in anticipation of subsequent costs. For example, in North American red squirrels, females whose litters were experimentally increased in the early stages of lactation accumulated larger reserves than controls and lost them during the later stages (Humphries and Boutin 1996) (Figure 5.15).

Intraspecific variation

Both the duration of lactation and the milk yields of females vary widely between populations and between individuals. Lactating mothers adjust milk production in relation to variation in their condition and access to resources, with important consequences for the development and survival of their offspring. For example, in harbour seals, the (absolute) weight loss of mothers

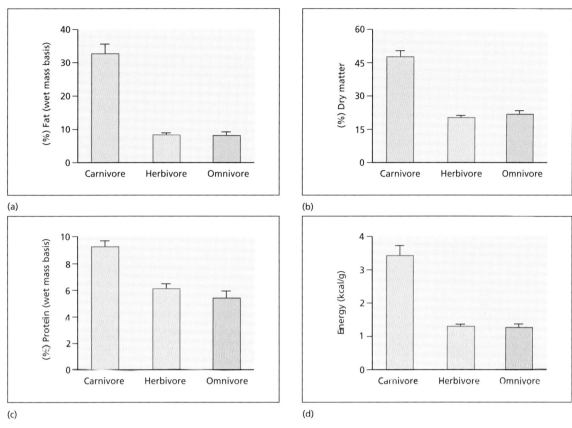

Figure 5.14 Comparison of milk composition between mammals with different diets: (a) milk fat concentration; (b) milk dry matter concentration; (c) milk protein concentration; (d) milk energy density. Error bars are standard errors. Sample sizes for each milk constituent: fat, $n = 130$; protein, $n = 130$; dry matter, $n = 124$; energy density, $n = 116$. *Source*: From Skibiel *et al*. (2013). Reproduced with permission of John Wiley & Sons.

during lactation (a measure of energy transferred) increases with the mother's initial body mass, and pups born to heavy mothers gain more weight and show higher rates of survival than those born to light ones (Bowen *et al.* 2001). Milk quality also varies with the availability and quality of resources and can have an important influence on the development of offspring. For example, reduction in the protein content of food fed to laboratory mice reduces the protein content of milk, affecting both prenatal and postnatal growth rates (Derrickson and Lowas 2007). Variation in offspring demand also affects milk quality and offspring growth. For example, female cotton rats that rear relatively large litters produce relatively dilute, energy-poor milk and their offspring have lower growth rates than those in

small litters (Rogowitz 1996). In some cases, mothers adjust the duration of lactation to minimise costs. For example, in southern elephant seals, mothers that give birth to young relatively late in the season show higher rates of energy transfer to their pups but wean them at younger ages than mothers that give birth early in the season (Arnbom *et al.* 1997).

Milk production, milk quality and offspring growth rates also show pronounced differences between individuals. For example, in grey seals, there are consistent differences between females in daily milk output, milk composition and duration of lactation which are associated with differences in the weight of pups at weaning (Lang *et al.* 2009). Individual differences in milk yield probably make an important contribution to variation in

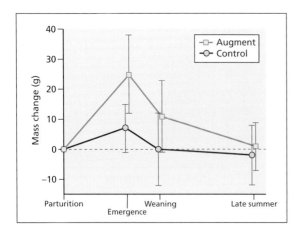

Figure 5.15 Strategic changes in body mass in North American red squirrels. During the early stages of lactation, females whose litters were experimentally increased ('augment') laid down larger reserves which they lost during the later stages of lactation. Emergence refers to days 35–45 following parturition, when most juveniles emerged; weaning refers to days 60–70 when most juveniles were fully weaned; and late summer refers to a 5-day period in August. Bars indicate SD. *Source*: From Humphries and Boutin (1996). Reproduced with permission of John Wiley & Sons. *Photo source*: © Ryan W. Taylor.

growth rate between offspring born to different mothers, which is pronounced in many wild mammals and has been shown to be partly heritable in red squirrels (see section 5.4) and feral sheep (Wilson *et al.* 2009; Childs *et al.* 2011).

The mechanisms constraining milk production are still uncertain. While increased food availability is often associated with increases in litter size, energy intake and milk production, this is not the case in all circumstances and increased food availability can instead be associated with reductions in ranging and time spent foraging (Akbar and Gorman 1993, 1996), suggesting that some intrinsic mechanism may limit food intake.

There have been several suggestions as to the mechanisms that might be involved, including central control of the digestive system, limits of heat dissipation and constraints on the neuroendocrine processes that control feeding behaviour and their relative importance is still unclear (Speakman and Król 2005). It is also uncertain whether intrinsic mechanisms impose unavoidable constraints over evolutionary time and an alternative possibility is that lactation is constrained by costs to other components of fitness, including the mother's survival and subsequent breeding success.

Fitness costs

Despite associated increases in food intake and the adjustment of milk production to resource availability, lactation often leads to loss of maternal body weight and reductions in body condition, thermogenesis, immune function and physical activity (Speakman 2008). In extreme cases, changes can include hypertrophy of the liver, kidney and digestive organs.

Changes in maternal condition as a consequence of the costs of lactation are often associated with reduction in the subsequent fecundity of females. For example, in red deer, mothers that raise calves successfully are more likely to fail to breed the following year than those that either do not produce calves or whose calves die shortly after birth (Clutton-Brock *et al.* 1983, 1989). Similarly, wild chimpanzee females that have successfully raised offspring may not resume oestrous cycling for at least 2 years (Emery Thompson *et al.* 2012). Under harsh environmental conditions, lactation is also associated with substantial increases in female mortality. In Soay sheep, mothers that have raised twins are substantially less likely to survive in years of high mortality than those that have raised single offspring or have failed to breed, especially if they are breeding for the first time or are relatively light, though these effects disappear in years when food is abundant and mortality is low (Figure 5.16). These results emphasize the need for estimates of fitness costs to be based on more than one season.

5.6 Post-weaning investment

Post-weaning care

In short-lived mammals, maternal care often terminates at or soon after weaning, while in longer-lived species

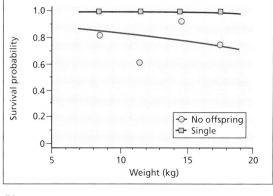

(a)

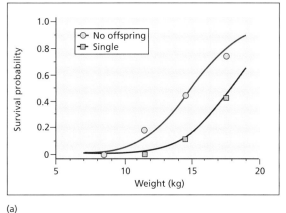

(b)

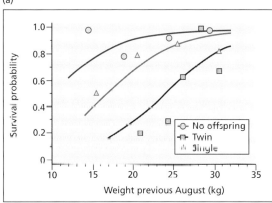

(c)

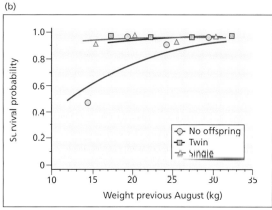

(d)

Figure 5.16 Variation in the survival costs of breeding in female Soay sheep on St Kilda in relation to their age, weight and litter size in years of high and low mortality. (a) Overwinter survival of juveniles (animals <1 year old) in relation to their weight in summer and whether or not they reared a lamb in years when population density and mortality were high. (b) As for (a) but in years of low population density and mortality. (c) Similar figures for adult females (>12 months old), distinguishing between mothers that failed to breed, those that reared a single lamb and those that reared twin lambs in years of high mortality. (d) As for (c) but in years of low density and mortality. *Source*: From Clutton-Brock *et al.* (1996). Reproduced with permission of John Wiley & Sons. *Photo source*: © Arpat Ozgul.

mothers often continue to provide some form of care to weaned offspring throughout their lives. In some carnivores, mothers help their offspring to find or catch food: for example, cheetah cubs are reliant on kills made by their mothers throughout their first year of life and their food intake declines once they separate from their mothers (Caro 1994). Provisioning adolescent offspring has substantial energetic costs to mothers, since sharing access to kills reduces their own food intake and the presence of pups can interfere with their foraging efficiency (Caro 1994). Similarly, in several of the mongooses, mothers and other group members feed infants and juveniles with prey (mostly insects and their larvae, arachnids and small vertebrates) that they have caught until they are able to forage effectively on their own and mothers also help their young to learn how to catch or handle difficult or dangerous prey (see Chapter 6).

Inheritance of resources

In some mammals, mothers continue to invest in their progeny in other ways. In a wide range of species, mothers tolerate the presence of weaned offspring within their territory for periods ranging from a few weeks to several years or, in some species, for life (see Chapter 3). Continuing maternal tolerance often has substantial benefits to offspring since the risks associated with dispersal are often high (see Chapter 3) but it can have costs to their mothers, whose access to resources, condition and breeding success commonly declines as the number of resident offspring increases (see Chapter 2). As might be expected, mothers often discriminate between adolescent group members and are more likely to evict individuals that are larger, older or less closely related (see Chapters 3 and 17).

One consequence of maternal tolerance of adult offspring is that offspring often inherit their mother's territory or breeding position on her death. For example, in meerkats, a subordinate female's best chances of surviving and breeding is to inherit her mother's position following her death (Clutton-Brock *et al.* 2006). Where males remain in their mother's group, they can also inherit important resources from their mothers. For example, in chimpanzees, males inherit their mother's ranges and ranging patterns (Murray *et al.* 2008).

While the inheritance of maternal resources can be a by-product of maternal tolerance of adult offspring, in some species mothers appear to adjust their territorial behaviour to provide their offspring with continuing access to their natal area or to resources that they have established in it. For example, in North American red squirrels, adults defend exclusive territories containing a food cache that is necessary for survival and juveniles must establish their own independent territory and food cache in early autumn if they are to survive the winter (Larsen and Boutin 1994; Berteaux and Boutin 2000; Boon *et al.* 2008). Before conceiving, adult females establish additional food caches in their territories and subsequently relinquish or 'bequeath' these to their adolescent offspring (Boutin *et al.* 2000). In

around 30% of cases, mothers allow their offspring to occupy the breeding territory where they were born and to use the food cache site that it contains, relocating to vacant territories nearby themselves (Price 1992; Boutin and Larsen 1993; Boutin *et al.* 1993; Price and Boutin 1993). Evidence of bequeathal of resources by parents to their offspring is available in a number of other rodents that rely on renewable resources, including Columbian ground squirrels, kangaroo rats and dusky-footed woodrats (Harris and Murie 1984; Jones 1986; Cunningham 2005). Bequeathal would be expected to evolve where recruits are at a disadvantage relative to established adults and dispersal costs to juveniles are high.

Social support

In some group-living species, mothers play an important role in sheltering their offspring from competition with other group members and establishing their relative dominance rank (see Chapter 8). As a result, the presence of mothers often has an important influence on the growth, survival and breeding success of their offspring. For example in bushy-tailed wood rats, which live in small groups on isolated rocky outcrops, juvenile females are more likely to survive their first winter if their mother is present in the group, though this effect disappears if the density of adult females is experimentally reduced (Moses and Millar 1994). The effects of the loss of mothers commonly decline with the increasing age of their offspring at the time of their mother's death. For example, in long-tailed macaques, few offspring that lose their mother when they are less than a year old survive while around 70% of those that lose them when they are 2–4 years old do so (van Noordwijk and van Schaik 1999) (Figure 5.17).

In some species, the absence of mothers affects one sex of offspring more than the other. In red deer, where females continue to associate with maternal relatives throughout their lives while males disperse, calves orphaned before or after weaning show reduced chances of surviving which, in females (but not males), persist throughout their lives (Andres *et al.* 2013). In other species, the loss of their mothers affects sons more than daughters. For example, in killer whales, where adult offspring of both sexes may continue to associate with their mothers, a mother's death has a stronger effect on the fitness of sons than on the survival of daughters (Foster *et al.* 2012) (Figure 5.18).

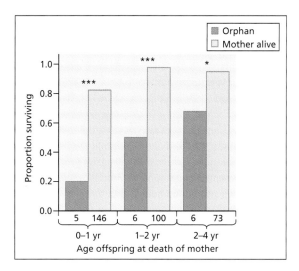

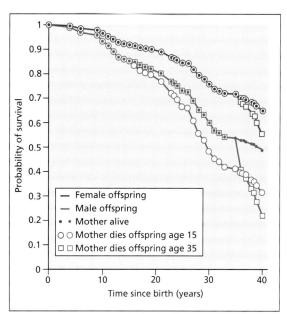

Figure 5.17 Age-related survival of infant and juvenile long-tailed macaques whose mothers were alive or dead. The figure shows the proportion of infants surviving to the next age category after the deaths of their mothers compared with the survival of immatures of the same age whose mothers were living. *Source*: From van Noordwijk and van Schaik (1999). Reproduced with permission of Springer Science and Business Media. *Photo source*: © Anna Marzec, Tuanen Orang Research Project.

Figure 5.18 Effects of their mother's presence on the survival of male and female offspring in killer whales. The figure shows survival curves (derived from a Cox proportional hazards model) for male and female offspring that experienced their mother's death at different ages. (*Photo*) Adult sons (i and ii) travelling with their post-reproductive mother (iii). *Source* and *Photo*: From Foster *et al.* (2012). Reproduced with permission from AAAS.

Post-reproductive investment

In some long-lived species, females commonly survive for several years after ceasing to breed: for example, in killer whales, females commonly live for several years after ceasing to breed and play an important role in leading the group, especially when food is scarce (Ward *et al.* 2009; Brent *et al.* 2015). The existence of post-reproductive lifespans in females has been interpreted in different ways. First, it may merely be a by-product of greater stochastic variation in survival than in fecundity (Packer *et al.* 1998). Alternatively, if offspring born to elderly females are unlikely to achieve independence before their mother dies, selection to prolong female fecundity may be weaker than selection to prolong female survival, leading to an earlier onset of senescence in fecundity (Ricklefs *et al.* 2003). Third, selection may favour the evolution of post-reproductive lifespans if elderly females achieve greater fitness returns by ceasing to breed themselves and investing in their grand-offspring rather than in additional offspring. In this case, we might expect to find positive correlations between the presence of post-reproductive females and the breeding success of their offspring (known as *grand-mother effects*) and a high proportion of females should survive beyond the edge of reproduction (Alvarez 2000; Hawkes 2003). In addition, where generations of females overlap, reproductive competition between generations may favour the premature termination of reproduction by older individuals as a result of asymmetries in relat-edness (Cant and Johnstone 2008; Johnstone and Cant 2010) (see Chapter 20).

The evolution of grandmother effects has attracted much attention as a consequence of interest in the evolution of the post-reproductive lifespans in women (see Chapter 20). In several human populations, the survival of children is improved by the presence of their grandmothers (see Chapter 20) and similar effects occur in some other mammals. For example, in African ele-phants, the presence of grandmothers is associated with increased survival of calves (see section 6.2). However, there is little evidence of grandmother effects in several other mammals.

5.7 Investment strategies

Maternal investment and fitness costs

Mothers would be expected to adjust the extent to which they invest in their offspring in relation to both the costs of care to their own fitness and the relative benefits to offspring. Where resources are limited or the mother's condition is depressed, mothers reduce investment in their offspring, birthweights fall, and they invest less in protection of their young and, in extreme cases, pregnant females may abort or resorb fetuses before birth and mothers may desert or kill their offspring (Smith 1987; Clutton-Brock 1991). However, it is frequently difficult to determine whether reductions in maternal investment represent adaptive strategies that adjust maternal expen-diture in relation to likely costs to the fitness of mothers or are simply a direct consequence of energetic restric-tions and both mechanisms may be common. One line of evidence which suggests that mothers are making stra-tegic decisions about investment is that, in some rodents, mothers terminate investment in unusually small litters (Lee and Cockburn 1985). Similarly, after the death of one of their cubs, female brown bears sometimes aban-don the surviving cubs and calculations show that moth-ers are likely to increase their reproductive success by

abandoning lone cubs if this allows them to re-conceive a larger litter immediately (Tait 1980) (Figure 5.19).

Age-related patterns of maternal investment have also been interpreted as evidence of adaptive adjustment of maternal investment to variation in the costs of care (Clutton-Brock 1991; Klug *et al.* 2012). Younger or primiparous females are often less likely to rear young successfully than older females, and their survival can be more strongly affected by breeding. As would be expected, fecundity is often influenced to a greater extent by environmental conditions in young females than in those in their prime (Schneider and Wade 1989; Schino and Troisi 2005). In addition, where annual survival falls with increasing age, old females whose likelihood of future reproduction ('reproductive value') is low would be expected to transfer a higher proportion of their remaining resources to their offspring and some evidence suggests they may do so. For example, in red deer, where maternal condition falls with increasing age, hinds over 12 years old allow their calves to suck for longer and their offspring show improved condition and overwinter sur-vival compared with those of middle-age hinds, though the costs of breeding to the mother's survival increase with her age (Clutton-Brock *et al.* 1987). The improvement in reproductive performance occurs towards the end of the mother's lifespan, so it seems unlikely that it is caused by improved reproductive or feeding skills. A similar increase in offspring birthweight with maternal age has been found in moose (Ericsson *et al.* 2001). In contrast, studies of humans and captive chimpanzees have shown that, after parity has been controlled, there is a negative effect of age on birthweight (Fessler *et al.* 2005).

Maternal investment and offspring needs

Throughout lactation and post-weaning care, mothers respond to the needs of their offspring. Milk yield is stimulated by nursing, causing mothers to adjust their production in relation to variation in the demands of offspring (Loudon and Racey 1987; Speakman 2008): as offspring grow, their demands increase and maternal milk production rises to meet them, until the needs of offspring eventually outstrip their mother's potential to produce milk and they gradually transfer to solid food (Loudon and Racey 1987). Dependent infants and juve-niles frequently use vocal signals to solicit feeding and parents often adjust their level of investment to these calls (see Chapter 7). Offspring vary in their begging behaviour while parents differ in their responses to

Figure 5.19 Female brown bear and three cubs. Field studies show that females may abandon lone cubs if this allows them to re-conceive a larger litter. *Source*: © Mint Images/Art Wolfe/Getty Images.

offspring begging and these differences can have a genetic basis, indicating that they are likely to coevolve (Kölliker and Richner 2001; Kilner and Hinde 2012).

Maternal investment and offspring quality

Parents may also adjust their behaviour to differences in the age and quality of their offspring that affect their chances of surviving and breeding successfully. The probability of surviving to breeding age is generally higher in older juveniles than in younger offspring, both because older juveniles are closer to maturation and because the instantaneous rate of juvenile mortality usually declines with increasing age. Older offspring are consequently of greater value to parents, who might be expected to take greater risks to ensure their survival, so that parental investment might be expected to increase with offspring age (Andersson *et al.* 1980; Winkler 1987). However, an alternative argument is that parental investment is likely to have a greater effect on the fitness of younger, smaller or weaker offspring and, if so, parents would be expected to bias their investment towards younger or smaller offspring (Clutton-Brock 1991).

In practice, it is often difficult to establish whether contrasts in investment or growth between offspring are a consequence of parental discrimination or whether they are by-products of competition between offspring. For example, female marmosets tend to carry larger offspring more frequently than their smaller sibs but larger juveniles are able to gain priority of access to their mothers, who do not appear to discriminate between their offspring (Tardif *et al.* 2002). Where there is evidence of parental discrimination, it usually appears to favour younger, smaller or weaker offspring (Clutton-Brock 1991). For example, female fur seals attempt to protect their pups against older offspring that are not yet weaned (Trillmich and Wolf 2008). Similarly, experiments with mice show that mothers of inbred litters are more attentive to their offspring immediately after birth than mothers of outbred litters and, as a result, survival of inbred pups does not differ from that of outbred ones (Margulis 1997). In goats and rhesus macaques, mothers with twins provide more care for the weaker member of the pair (Spencer-Booth 1969; Klopfer and Klopfer 1977) and in Japanese macaques, mothers with malformed young support their infants more than mothers with healthy offspring (Turner *et al.* 2005).

It is also possible that parental strategies are more flexible than is commonly supposed. Long-term field studies that can measure changes in juvenile survival show that the effects of variation in growth differ between years. For example, in Soay sheep, the effects of body weight on juvenile survival are much stronger in years when population density is high than in years when it is low (Clutton-Brock *et al.* 1996, 2004). In North American red squirrels, too, fast-growing offspring are favoured by selection in some years but not in others (McAdam and Boutin 2003) and mothers appear to be able to adjust their investment in an adaptive fashion (Dantzer *et al.* 2013).

Non-offspring nursing

Since the costs of lactation are high, mothers might be expected to restrict nursing to their own offspring. However, non-offspring nursing occurs in a substantial number of mammals, including both communal and cooperative breeders (see Chapter 17). In species that do not breed cooperatively, non-offspring nursing is most frequent in litter-bearing (polytocous) species that live in small kin-based groups (MacLeod and Lukas 2014), like house mice. Communal nursing can reduce costs to mothers by spreading peak energy loads and may also generate benefits to the fitness of relatives (Packer *et al.* 1992; König 1994).

In monotocous species, non-offspring nursing more frequently appears to be a consequence of milk theft by unrelated pups or misdirected parental care (Reiter *et al.* 1978; McCracken 1984; Boness 1990) and is most frequently found in species living in large groups where individuals live in close proximity to each other and confusion is common. In several of these species, lactating females threaten or attack unrelated juveniles that attempt to nurse (Reiter *et al.* 1978; Packer *et al.* 1992; Lewis and Pusey 1997).

5.8 Relationships between siblings

Sibling competition

As maternal milk supplies are rarely super-abundant in naturally regulated populations, competition between littermates for access to lactating females is common (Clutton-Brock 1991; Hudson and Trillmich 2008). In some cases, this takes the form of scramble competition. For example, in rabbits, where nursing is confined to a short period each day, littermates extract milk as fast as possible from their mother until supplies are exhausted and larger juveniles are able to extract more milk, direct

competition for access to nipples is uncommon, and there is no consistent tendency for larger juveniles to monopolise the most productive nipples (Hudson and Distel 1982). In other species, nursing young compete for access to particular nipples (Bautista *et al.* 2005): for example, newborn domestic pigs compete aggressively for access to the most productive nipples using sharp teeth that develop rapidly and larger or more dominant individuals obtain priority to preferred nipples (McBride 1963; Fraser and Thompson 1991). Similarly, meerkat pups compete directly for access to helpers that are most likely to feed them (see Chapter 17). In some birds, aggressive competition between brood mates increases where multiple paternity reduces average relatedness within broods (Boncoraglio *et al.* 2009) and siblicide and cannibalism are more often directed towards half-siblings than full siblings (Pfennig 1997), but it is not yet clear whether similar effects occur in mammals.

Competition between littermates can be so intense that smaller siblings are excluded and, in some species, larger siblings may also attempt to kill smaller competitors (Mock and Parker 1997). For example, in spotted hyenas, cubs are born with fully erupted front teeth (Frank *et al.* 1991), and compete aggressively for dominance from the first day of life, sometimes killing subordinate littermates before they are weaned (Hofer and East 1997, 2008; Wahaj and Holekamp 2006) (Figure 5.20). The intensity of competition between littermates is greater in same-sex than mixed-sex litters and is related to factors affecting their mother's nutrition and her rank (Golla *et al.* 1999), increasing at times when food is in short supply and pups are fed less frequently and grow more slowly (Hofer and East 2008) (Figure 5.21). One unusual feature of competition between hyena cubs is that mortality is most frequent in all-female litters (Hofer and East 1997), which may reflect the unusual intensity of competition between females in this species (see Chapter 8). Among birds, the most intense aggression between siblings is found in species where brood size is small, rearing periods are prolonged and parents do not pass food directly to their chicks (Gonzalez-Voyer *et al.* 2007) but, as yet, there has been no systematic comparative study of sibling aggression in mammals.

Successive offspring can also compete for access to lactating females. In many mammals, variation in offspring growth rates affects the duration of lactation and the period for which juveniles associate with their mothers (Trillmich 1990; Dahle and Swenson 2003). This can

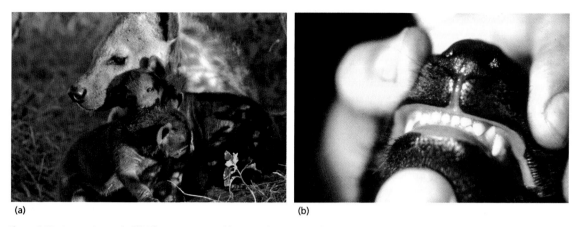

(a) (b)

Figure 5.20 Aggression and siblicide among spotted hyena cubs. (a) Twin hyena cubs at the natal den. (b) Dentition of spotted hyena on day of birth: canines are already 6–7 mm long, incisors 2–4 mm long. *Sources*: (a) © Kay Holekamp; (b) © Steve Glickman.

lead to increases in the periods when yearlings and infants born to the same mother overlap, generating competition for maternal resources that is often won by larger older offspring. For example, in fur seals and sea lions, poorly developed yearlings often compete to suckle with pups born in the following breeding season (Figure 5.22), depressing the growth of pups and

sometimes causing them to starve (Trillmich and Wolf 2008) (Figure 5.23). Changes in environmental conditions or food availability can exacerbate these effects. For example, reductions in food availability associated with fluctuations in climatic factors cause female Galapagos fur seals to travel further to find food and this increases intervals between foraging trips and

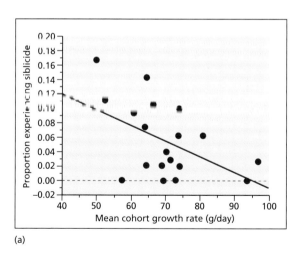

(a)

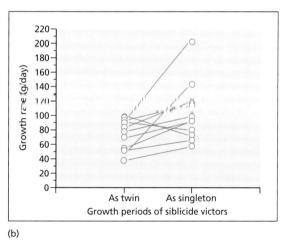

(b)

Figure 5.21 (a) The proportion of spotted hyena cubs per cohort experiencing siblicide in twin litters in relation to the mean growth rate per cohort. For illustrative purposes, the line of $y = 0$ (no siblicide) is marked with a dashed line. Each data point is labelled with the year of the cohort. (b) Comparison of growth rates of dominant cubs in siblicidal litters during the twin period and the subsequent singleton period after their littermate died of siblicide. *Source*: (a,b) From Hofer and East (2008). Reproduced with permission of Springer Science and Business Media.

(a)   (b)

Figure 5.22 (a) Galapagos sea lion pup and yearling compete for the mother's milk. (b) A mother defends her pup from competition from her yearling. *Source*: (a,b) © Fritz Trillmich.

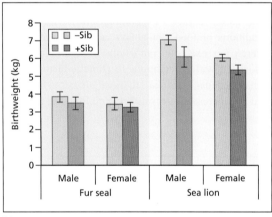

(a)

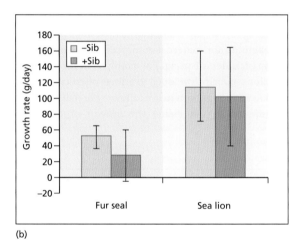

(b)

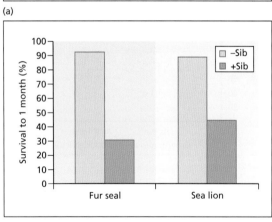

(c)

Figure 5.23 Birthweight, growth rate and survival in fur seals and sea lion pups when their mother is supporting an older sibling (+ sib) or is not doing so (− sib). *Source*: From Trillmich and Wolf (2008). Reproduced with permission of Springer Science and Business Media.

reduces the growth rates of their offspring (Trillmich 1990). As a result, there is increased overlap between successive offspring and the growth and survival of pups is reduced.

Sibling cooperation

Under some circumstances, siblings might be expected to cooperate rather than to compete with each other (Johnstone 2004; Roulin and Dreiss 2012). For example, where siblings disperse together, individuals that selfishly exclude other brood members from access to resources may be likely to reduce their own fitness (see Chapter 3). In addition, where littermates are close relatives, excessive expenditure on competition may reduce the inclusive fitness of individuals, favouring the evolution of negotiation between brood members over their share of resources (Roulin 2002; Johnstone and Roulin 2003). However, active cooperation between siblings would only be expected where the indirect genetic benefits that it generates exceed the costs involved (West *et al.* 2002) and this may rarely be the case.

Some studies of birds suggest that kin selection can moderate competition between siblings. For example, in blue-footed boobies (where siblicide between brood mates is common), dominant chicks allow subordinate chicks to feed during short-term periods of food shortage (Anderson and Ricklefs 1995). Other studies have suggested that fledglings may negotiate their share of resources within their brood: for example, in barn owls, hungry chicks give regular calls during periods when parents are hunting and their siblings appear to refrain from competing with them when parents next bring food (Dreiss *et al.* 2010). However, an alternative interpretation is that the cost–benefit ratios of competing for limited resources are substantially higher for hungry chicks than for well-fed ones and that by signalling their intention to compete while parents are absent, hungry chicks discourage well-fed ones from competing and well-fed ones avoid competition to maximise their own direct fitness.

Dependent juveniles may also cooperate to manipulate parents or other carers. Where the tendency of adults to invest in juveniles is related to the combined rate of begging of all pups, individuals might be expected to cooperate to signal their need for additional investment, sharing the costs of high begging rates (Wilson and Clark 2002; Johnstone 2004). If so, individual rates of begging might be expected to decline in large litters and some empirical studies of mammals have demonstrated precisely this pattern. In banded mongooses, where pups born to several breeding females are provisioned by 'dedicated' helpers, pups in large litters beg less than those in small ones (Bell 2007). Similarly, in meerkats, experimental reductions in the number of begging pups lead to increases in per-capita begging rates (Madden *et al.* 2009). However, interpretation is complicated, for pups avoid begging when other pups are vocalising nearby, which may allow them to maximise their chances of being fed per vocalisation given. Where per-capita begging rates decline in larger litters (as in banded mongooses), this may be because there are fewer opportunities for pups in large litters to call at a time when no other pups are vocalising, with the result that average begging rates are reduced. If so, reductions in begging rates in large litters may be a consequence of scramble competition to attract the attention of carers rather than of cooperation and studies of meerkats suggest that this is likely to be the case (Madden *et al.* 2009).

Sibling competition does not terminate at weaning and frequently persists as long as siblings remain in the same group. Direct competition between siblings is particularly intense in singular breeders where a single female in each group monopolises the production of offspring and subordinates queue for breeding opportunities in their natal group (see Chapter 17).

5.9 Parent–offspring conflict

While conflicts of interest between parents and their offspring are probably widespread (see section 5.2), there are relatively few cases where it is possible to measure and compare evolutionary optima for parents and their offspring, so there are few firm empirical examples of parent–offspring conflict. However, there are several cases where enough is known about the interests of parents and their offspring to be sure that their interests diverge (Shea *et al.* 2011; Kilner and Hinde 2012). For example, long-term data for Soay sheep show that mothers benefit by producing twins rather than singletons while the fitness of offspring is maximised if lambs develop on their own, so that selection on mothers favours larger litters than selection acting on offspring (Wilson *et al.* 2005). Another case where there is strong evidence of evolutionary conflict involves studies of the effects of imprinted genes whose expression depends on which parent they are transmitted in (Box 5.3).

There are also numerous examples of overt conflicts between parents and offspring in situations where there is circumstantial evidence of conflicts of interest. In many mammals, suckling delays the resumption of ovulatory cycles and prolongs inter-birth intervals (Short 1976; McNeilly *et al.* 2006) and offspring may attempt to manipulate this effect to prolong maternal investment. For example, juvenile rhesus macaques increase the frequency of their attempts to suck at the onset of the next mating season, when their mothers are likely to conceive again (Gomendio 1989). Attempts by mothers to prevent juveniles from suckling may generate weaning tantrums in which offspring kick, scream and even attempt to attack their mothers and these tantrums appear to make mothers nervous and tense and they often give in to their offspring's demands (Goodall 1986). In polytocous species, mothers commonly attempt to ensure that larger or more vociferous offspring do not monopolise resources (see previous section) and this can lead to overt conflicts with offspring. Similar conflicts also occur in monotocous species where yearlings and offspring are dependent on their mothers at the same time (see section 5.8).

Conflicts between parents and offspring are not confined to lactation periods. Conflicts over the period for which offspring remain in the parent's range or territory are also common and are frequently associated with rising levels of aggression directed by mothers at their offspring until the latter disperse (see Chapter 3). The young of some rodents actively resist being removed from the breeding burrow by their mother, beating her off with their forepaws and refusing to be picked up (Daly and Daly 1975). However, behavioural disputes between parents and offspring do not necessarily indicate that there are evolutionary conflicts between them (Mock and Forbes 1992; Kilner and Hinde 2012). In some cases, mothers may control the distribution of parental resources and maximise their own fitness by adjusting investment to the responses of their offspring. For example, detailed studies of weaning in grey seals suggest that mothers control the timing of weaning but adjust it to information on the pup's hunger state derived from begging behaviour (Smiseth and Lorentsen 2001). In other cases, acts of aggression between young may promote maternal fitness by facilitating brood reduction (Trillmich and Wolf 2008; Kilner and Hinde 2012). Developing juveniles may also use parental responses to modify their behaviour in an adaptive fashion (Kölliker *et al.* 2012). For example, yearling rhesus macaques respond to changes in the frequency with which mothers reject their attempts to nurse by reducing time spent in proximity to their mother and increasing independence, and interactions between mothers and offspring may coincide more

Box 5.3 Parent–offspring conflict and genetic imprinting

In many mammals, there are likely to be conflicts between mothers and their offspring over the extent of prenatal growth: individual offspring are likely to maximise their fitness by exceeding the rate of growth that is optimal for their mother (see Box 5.2). Fathers (or genes inherited from fathers) would also be expected to favour higher levels of maternal investment than those that would be optimal for mothers, since they do not pay the costs involved in additional care. One case where there is strong evidence of conflict involves imprinted genes whose expression depends on which parent they were transmitted by. This unusual pattern of inheritance may be a consequence of conflict between maternally and paternally inherited genes over the provision of maternal investment arising from the fact that fathers do not suffer the costs of increased maternal investment (Haig and Westoby 1989; Haig 2004). Experiments on *Igf2* and *Igf2r* genes in mice provide some support for this idea. *Igf2* is a paternally imprinted gene which encodes an insulin-related polypeptide that controls the extraction of resources from pregnant mothers by their offspring. Experimental inactivation of the paternal allele of *Igf2* causes females to produce offspring 40% smaller at birth than their usual size while inactivation of the female allele does not have any similar effect (Haig 1997). Conversely, inactivation of the maternal allele of the maternally imprinted gene *Igf2r*, which reduces the influence of *Igf2* on resource transfer, leads to increases in the birthweight of offspring of 20–30% while, in this case, inactivation of the paternal allele has little effect (Haig 1997). Whether many maternally and paternally imprinted genes have similar effects is not yet clear (Kilner and Hinde 2012).

closely with models of dynamic assessment processes than with models of parent–offspring conflict (Devinney *et al.* 2001).

It is also important to appreciate that not all differences in maternal behaviour are necessarily adaptive. For example, detailed studies of maternal behaviour in rhesus macaques show that some females treat their offspring roughly and are frequently aggressive to them throughout their period of dependence (Maestripieri 1998). Abusive mothers do not expend less effort on parental behaviour and are often unusually controlling in their relationships with their offspring. Macaques are monotocous, so maternal abuse is unlikely to represent an adaptive attempt to prevent particular offspring from monopolising maternal investment and it seems more likely that they are an example of pathological behaviour (Horsfall 1984; Leonard *et al.* 1988).

5.10 Consequences of maternal care

The commitment of (almost) all female mammals to gestation and lactation has evolutionary consequences that affect most aspects of ecology, life histories and behaviour and are so pervasive that it is beyond the scope of this section to explore them fully. The retention of young within the mother's body and their subsequent provisioning from maternal resources during lactation allows mammals to occupy environmental niches that are inaccessible to oviparous vertebrates. However, the high energetic costs of maternal care accentuate the effects of environmental conditions and individual differences in phenotype on the development of offspring, leading to pronounced contrasts in development between cohorts and strong maternal effects.

The commitment of female mammals to parental care also has far-reaching consequences for males. The long periods that females are absent from the reproductive arena bias operational sex ratios towards males, favouring the evolution of pre-copulatory mate guarding by males, intensifying reproductive competition between them and strengthening selection for the evolution of competitive ability in males and traits associated with it (see Chapter 13). Intense reproductive competition between males has further consequences: it is likely to raise the potential costs of paternal care and to discourage its evolution (see Chapter 16) and it commonly shortens the breeding tenure and effective breeding lifespans of males and can lead to the evolution of infanticidal behaviour in males (see Chapter 15), with associated effects on both the evolution of male care and female mating preferences (see Chapter 4).

SUMMARY

1. All female mammals invest heavily in their offspring from conception to weaning. In many longer-lived species, mature daughters continue to live with their mothers and maternal care extends until the mother's death.

2. Heavy investment in individual offspring trades off against maternal survival as well as against the extent to which mothers can invest in other offspring. Species living in ecologically or socially challenging environments commonly produce relatively large offspring.

3. The high costs of maternal care are frequently associated with adaptive maternal strategies that maximise female fitness. These can involve either extensions of maternal care or its premature termination where offspring are unlikely to survive.

4. The characteristics of mothers often exert strong effects on the development of their offspring. In some cases, mothers may adjust the development of their offspring to the environment they are likely to experience after they are weaned but it is frequently difficult to distinguish between the effects of adaptive maternal strategies and the unselected consequences of ecological variation.

5. Both conflicts of interest between siblings and conflicts between offspring and their parents are likely to influence the distribution of maternal care. Mothers often intervene to protect weaker offspring while offspring and their parents are frequently involved in agonistic interactions over the duration of care.

References

Aguirre, J.D. and Marshall, D.J. (2012) Does genetic diversity reduce sibling competition? *Evolution* **66**:94–102.

Akbar, Z. and Gorman, M.L. (1993) The effect of supplementary feeding upon the sizes of the home ranges of woodmice *Apodemus sylvaticus* living on a system of maritime sand-dunes. *Journal of Zoology* **231**:233–237.

Akbar, Z. and Gorman, M.L. (1996) The effect of supplementary food upon the activity patterns of wood mice, *Apodemus sylvaticus*, living on a system of maritime sand-dunes. *Journal of Zoology* **238**:759–768.

Akers, R.M. (2002) *Lactation and the Mammary Gland*. Ames, IA: Iowa State Press.

Akers, R.M. (2006) Major advances associated with hormone and growth factor regulation of mammary growth and lactation in dairy cows. *Journal of Dairy Science* **89**:1222–1234.

Albon, S.D., *et al.* (1983) The influence of climatic variation on the birth weights of red deer (*Cervus elaphus*). *Journal of Zoology* **200**:295–298.

Albon, S.D., *et al.* (1987) Early development and population dynamics in red deer. II. Density-independent effects and cohort variation. *Journal of Animal Ecology* **56**:69–81.

Alexander, R.D. (1974) The evolution of social behavior. *Annual Review of Ecology and Systematics* **5**:325–383.

Alonso-Alvarez, C. and Velando, A. (2012) Benefits and costs of parental care. In: *The Evolution of Parental Care* (eds N.J. Royle, P.T. Smiseth and M. Kölliker). Oxford: Oxford University Press, 40–61.

Alonzo, S.H. and Klug, H. (2012) Paternity, maternity, and parental care. In: *The Evolution of Parental Care* (eds N.J. Royle, P.T. Smiseth and M. Kölliker). Oxford: Oxford University Press, 189–205.

Alvarez, H.P. (2000) Grandmother hypothesis and primate life histories. *American Journal of Physical Anthropology* **113**:435–450.

Anderson, D.J. and Ricklefs, R.E. (1995) Evidence of kin-selected tolerance by nestlings in a siblicidal bird. *Behavioral Ecology and Sociobiology* **37**:163–168.

Andersson, M., *et al.* (1980) Parental defence of offspring: a model and an example. *Animal Behaviour* **28**:536–542.

Andres, D., *et al.* (2013) Sex differences in the consequences of maternal loss in a long-lived mammal, the red deer (*Cervus elaphus*). *Behavioral Ecology and Sociobiology* **67**:1249–1258.

Arnbom, T., *et al.* (1997) Factors affecting maternal expenditure in southern elephant seals during lactation. *Ecology* **78**:471–483.

Asher, G.W., *et al.* (2005) Influence of level of nutrition during late pregnancy on reproductive productivity of red deer: 2. Adult hinds gestating wapiti × red deer crossbred calves. *Animal Reproduction Science* **86**:285–296.

Balshine, S. (2012) Patterns of parental care in vertebrates. In: *The Evolution of Parental Care* (eds N.J. Royle, P. T. Smiseth and M. Kölliker). Oxford: Oxford University Press, 62–80.

Barker, D.J. (1998) *Mothers, Babies and Health in Later Life*. Edinburgh: Churchill Livingstone.

Barta, Z., *et al.* (2002) Sexual conflict about parental care: the role of reserves. *American Naturalist* **159**:687–705.

Bateson, P. and Gluckman, P. (2011) *Plasticity, Robustness, Development and Evolution*. Cambridge: Cambridge University Press.

Bautista, A., *et al.* (2005) Scramble competition in newborn domestic rabbits for an unusually restricted milk supply. *Animal Behaviour* **70**:1011–1021.

Bell, M.B. V. (2007) Cooperative begging in banded mongoose pups. *Current Biology* **17**:717–721.

Ben Shaul, D.M. (1962) The composition of the milk of wild animals. *International Zoo Yearbook* **4**:333–342.

Bernardo, J. (1996) Maternal effects in animal ecology. *American Zoologist* **36**:83–105.

Berteaux, D. and Boutin, S. (2000) Breeding dispersal in female North American red squirrels. *Ecology* **81**:1311–1326.

Blueweiss, L., *et al.* (1978) Relationships between body size and some life-history parameters. *Oecologia* **37**:257–272.

Boncoraglio, G., *et al.* (2009) Fine-tuned modulation of competitive behaviour according to kinship in barn swallow nestlings. *Proceedings of the Royal Society of London. Series B: Biological Sciences* **276**:2117–2123.

Boness, D.J. (1990) Fostering behavior in Hawaiian monk seals: is there a reproductive cost? *Behavioral Ecology and Sociobiology* **27**:113–122.

Boness, D.J. and Bowen, W.D. (1996) The evolution of maternal care in pinnipeds. *BioScience* **46**:645–654.

Boon, A.K., *et al.* (2008) Personality, habitat use, and their consequences for survival in North American red squirrels *Tamiasciurus hudsonicus*. *Oikos* **117**:1321–1328.

Boutin, S. and Larsen, K.W. (1993) Does food availability affect growth and survival of males and females differently in a promiscuous small mammal, *Tamiasciurus hudsonicus*? *Journal of Animal Ecology* **62**:364–370.

Boutin, S., *et al.* (1993) Post-breeding dispersal by female red squirrels (*Tamiasciurus hudsonicus*): the effect of local vacancies. *Behavioral Ecology* **4**:151–155.

Boutin, S., *et al.* (2000) Anticipatory parental care: acquiring resources for offspring prior to conception. *Proceedings of the Royal Society of London. Series B: Biological Sciences* **267**:2081–2085.

Bowen, W.D., *et al.* (1985) Birth to weaning in 4 days: remarkable growth in the hooded seal, *Cystophora cristata*. *Canadian Journal of Zoology* **63**:2841–2846.

Bowen, W.D., *et al.* (2001) Foraging effort, food intake and lactation performance depend on maternal mass in a small phocid seal. *Functional Ecology* **15**:325–334.

Boyd, I.L. (1985) Investment in growth by pregnant wild rabbits in relation to litter size and sex of the offspring. *Journal of Animal Ecology* **54**:137–147.

Brawand, D., *et al.* (2008) Loss of egg yolk genes in mammals and the origin of lactation and placentation. *PLOS Biology* **6**:e63.

Brent, L.J.N., *et al.* (2015) Ecological knowledge, leadership, and the evolution of menopause in killer whales. *Current Biology* **25**:746–750.

Bunnell, F.L. (1980) Factors controlling lambing period of Dall's sheep. *Canadian Journal of Zoology* **58**:1027–1031.

Cant, M.A. (2000) Social control of reproduction in banded mongooses. *Animal Behaviour* **59**:147–158.

Cant, M.A. and Johnstone, R.A. (2008) Reproductive conflict and the separation of reproductive generations in humans. *Proceedings of the National Academy of Sciences of the United States of America* **105**:5332–5335.

Caro, T.M. (1994) *Cheetahs of the Serengeti Plains: Group Living in an Asocial Species*. Chicago: University of Chicago Press.

Case, T.J. (1978) On the evolution and adaptive significance of postnatal growth rates in the terrestrial vertebrates. *Quarterly Review of Biology* **53**:243–282.

Champagne, F.A. (2008) Epigenetic mechanisms and the transgenerational effects of maternal care. *Frontiers in Neuroendocrinology* **29**:386–397.

Champagne, F.A. and Curley, J.P. (2009) Epigenetic mechanisms mediating the long-term effects of maternal care on development. *Neuroscience and Biobehavioral Reviews* **33**:593–600.

Champagne, F.A. and Curley, J.P. (2011) Epigenetic influence of the social environment. In: *Brain, Behavior and Epigenetics* (eds A. Petronis and J. Mill). Berlin. Springer, 185–208.

Champagne, F.A. and Curley, J.P. (2012) Genetics and epigenetics of parental care. In: *The Evolution of Parental Care* (eds N.J. Royle, P. T. Smiseth and M. Kölliker). Oxford: Oxford University Press, 304–324.

Chevrud, J.M. and Wolf, J.B. (2009) Genetics and evolutionary consequences of maternal effects. In: *Maternal Effects in Mammals* (eds D. Maestripieri and J.M. Mateo). Chicago: University of Chicago Press, 11–37.

Childs, D.Z., *et al.* (2011) Predicting traits and measuring selection in complex life histories: reproductive allocation decisions in Soay sheep. *Ecology Letters* **14**:985–992.

Clements, M.N., *et al.* (2011) Gestation length variation in a wild ungulate. *Functional Ecology* **25**:691–703.

Clutton-Brock, T.H. (1991) *The Evolution of Parental Care* Princeton, NJ: Princeton University Press.

Clutton-Brock, T.H. and Albon, S.D. (1989) *Red Deer in the Highlands*. Oxford: Blackwell Scientific Publications.

Clutton-Brock, T.H. and Godfray, H.C.J. (1991) Parental investment. In: *Behavioural Ecology*, 3rd edn (eds J. R. Krebs and N. B. Davies). Oxford: Blackwell Scientific Publications, 234–262.

Clutton-Brock, T.H., *et al.* (1983) The costs of reproduction to red deer hinds. *Journal of Animal Ecology* **52**:367–383.

Clutton-Brock, T.H., *et al.* (1987) Interactions between population density and maternal characteristics affecting fecundity and survival in red deer. *Journal of Animal Ecology* **56**:857–871.

Clutton-Brock, T.H., *et al.* (1989) Fitness costs of gestation and lactation in wild mammals. *Nature* **337**:260–262.

Clutton-Brock, T.H., *et al.* (1992) Early development and population fluctuations in Soay sheep. *Journal of Animal Ecology* **61**:381–396.

Clutton-Brock, T.H., *et al.* (1996) Population fluctuations and life-history tactics in female Soay sheep. *Journal of Animal Ecology* **65**:675–689.

Clutton-Brock, T.H., *et al.* (2004) The sheep of St Kilda. In: *Soay Sheep: Dynamics and Selection in an Island Population* (eds T. H. Clutton-Brock and J. M. Pemberton). Cambridge: Cambridge University Press, 17–51.

Clutton-Brock, T.H., *et al.* (2006) Intrasexual competition and sexual selection in cooperative mammals. *Nature* **444**:1065–1068.

Cockburn, A. (1989) Adaptive patterns in marsupial reproduction. *Trends in Ecology and Evolution* **4**:126–130.

Côté, S.D. and Festa-Bianchet, M. (2001) Birth date, mass and survival in mountain goat kids: effects of maternal characteristics and forage quality. *Oecologia* **127**:230–238.

Cunningham, S. (2005) *Dispersal and inheritance in the dusky-footed woodrat* Neotoma fuscipes. PhD thesis, University of California at Berkeley.

Curley, J.P., *et al.* (2008) Transgenerational effects of impaired maternal care on behaviour of offspring and grandoffspring. *Animal Behaviour* **75**:1551–1561.

Dahle, B. and J.E. Swenson (2003) Factors influencing length of maternal care in brown bears (*Ursus arctos*) and its effect on offspring. *Behavioral Ecology and Sociobiology* **54**:352–358.

Dahlöf, L.-G., *et al.* (1977) Influence of maternal stress on offspring sexual behaviour. *Animal Behaviour* **25**:958–963.

Dall, S.R.X. and Boyd, I.L. (2004) Evolution of mammals: lactation helps mothers to cope with unreliable food supplies. *Proceedings of the Royal Society of London. Series B: Biological Sciences* **271**:2049–2057.

Daly, M. and Daly, S. (1975) Socio-ecology of Saharan gerbils, especially *Meriones libycus*. *Mammalia* **39**:289–312.

Dantzer, B., *et al.* (2013) Density triggers maternal hormones that increase adaptive offspring growth in a wild mammal. *Science* **340**:1215–1217.

Derrickson, E.M. (1992) Comparative reproductive strategies of altricial and precocial eutherian mammals. *Functional Ecology* **6**:57–65.

Derrickson, E.M. and Lowas, S.R. (2007) The effects of dietary protein levels on milk protein levels and postnatal growth in laboratory mice (*Mus musculus*). *Journal of Mammalogy* **88**:1475–1481.

Devinney, B.J., *et al.* (2001) Changes in yearling rhesus monkeys' relationships with their mothers after sibling birth. *American Journal of Primatology* **54**:193–210.

Dloniak, S.M., *et al.* (2006) Rank-related maternal effects of androgens on behaviour in wild spotted hyenas. *Nature* **440**:1190–1193.

Douhard, M., *et al.* (2014) Fitness consequences of environmental conditions at different life stages in a long-lived vertebrate. *Proceedings of the Royal Society of London. Series B: Biological Sciences* **281**:20140276.

Dreiss, A., *et al.* (2010) How siblings adjust sib–sib communication and begging signals to each other. *Animal Behaviour* **80**:1049–1055.

Dunbrack, R.L. and Ramsay, M.A. (1989) The evolution of viviparity in amniote vertebrates: egg retention versus egg sise reduction. *American Naturalist* **133**:138–148.

East, M.L., *et al.* (2009) Maternal effects on offspring social status in spotted hyenas. *Behavioral Ecology* **20**:478–483.

Eisenberg, J.F. (1981) *The Mammalian Radiations: An Analysis of Trends in Evolution, Adaptation, and Behavior.* Chicago: University of Chicago Press.

Emery Thompson, M., *et al.* (2012) The energetics of lactation and the return to fecundity in wild chimpanzees. *Behavioral Ecology* **23**:1234–1241.

English, S., *et al.* (2010) Consistent individual differences in cooperative behaviour in meerkats (*Suricata suricatta*). *Journal of Evolutionary Biology* **23**:1597–1604.

Ericsson, G., *et al.* (2001) Age-related reproductive effort and senescence in free-ranging moose, *Alces alces. Ecology* **82**:1613–1620.

Felton, A.M., *et al.* (2009) Protein content of diets dictates the daily energy intake of a free-ranging primate. *Behavioral Ecology* **20**:685–690.

Fessler, D.M.T., *et al.* (2005) Examining the terminal investment hypothesis in humans and chimpanzees: associations among maternal age, parity and birth weight. *American Journal of Physical Anthropology* **127**:95–104.

Festa-Bianchet, M. (1988) Birthdate and survival in bighorn lambs (*Ovis canadensis*). *Journal of Zoology* **214**:653–661.

Festa-Bianchet, M., *et al.* (1996) The development of sexual dimorphism: seasonal and lifetime mass changes in bighorn sheep. *Canadian Journal of Zoology* **74**:330–342.

Festa-Bianchet, M., *et al.* (1997) Body mass and survival of bighorn sheep. *Canadian Journal of Zoology* **75**: 1372–1379.

Forsyth, I.A. (1986) Variation among species in the endocrine control of mammary growth and function: the roles of prolactin, growth hormone, and placental lactogen. *Journal of Dairy Science* **69**:886–903.

Foster, E.A., *et al.* (2012) Adaptive prolonged postreproductive life span in killer whales. *Science* **337**:1313.

Frank, L.G., *et al.* (1991) Fatal sibling aggression, precocial development, and androgens in neonatal spotted hyenas. *Science* **252**:702–704.

Fraser, D. and Thompson, B.K. (1991) Armed sibling rivalry among suckling piglets. *Behavioral Ecology and Sociobiology* **29**:9–15.

Gaillard, J.-M., *et al.* (1989) An analysis of demographic tactics in birds and mammals. *Oikos* **56**:59–76.

Gaillard, J.-M., *et al.* (1997) Early survival in roe deer: causes and consequences of cohort variation in two contrasted populations. *Oecologia* **112**:502–513.

Galef, B.G. and Henderson, P.W. (1972) Mother's milk: a determinant of the feeding preferences of weaning rat pups. *Journal of Comparative and Physiological Psychology* **78**:213–219.

Gentry, R.L. and Kooyman, G.L. (eds) (1986) *Fur Seals: Maternal Strategies on Land and at Sea.* Princeton, NJ: Princeton University Press.

Gittleman, J.L. (1986) Carnivore life history patterns: allometric, phylogenetic, and ecological associations. *American Naturalist* **127**:744–771.

Gittleman, J.L. and Oftedal, O.T. (1987) Comparative growth and lactation energetics in carnivores. *Symposia of the Zoological Society of London* **57**:41–77.

Gittleman, J.L. and Thompson, S.D. (1988) Energy allocation in mammalian reproduction. *American Zoologist* **28**:863–876.

Gluckman, P.D., *et al.* (2005a) Predictive adaptive responses and human evolution. *Trends in Ecology and Evolution* **20**:527–533.

Gluckman, P.D., *et al.* (2005b) Environment influences during development and their later consequences for health and disease: implications for the interpretation of empirical studies. *Proceedings of the Royal Society of London. Series B: Biological Sciences* **272**:671–677.

Godfray, H.C.J. (1995a) Evolutionary theory of parent–offspring conflict. *Nature* **376**:133–138.

Godfray, H.C.J. (1995b) Signaling of need between parents and young: parent–offspring conflict and sibling rivalry. *American Naturalist* **146**:1–24.

Godfray, H.C.J. and Parker, G.A. (1991) Clutch sise, fecundity and parent–offspring conflict. *Philosophical Transactions of the Royal Society B: Biological Sciences* **332**:67–79.

Golla, W., *et al.* (1999) Within-litter sibling aggression in spotted hyaenas: effect of maternal nursing, sex and age. *Animal Behaviour* **58**:715–726.

Gomendio, M. (1989) Suckling behaviour and fertility in rhesus macaques (*Macaca multatta*). *Journal of Zoology* **217**:449–467.

Gonzalez-Voyer, A., *et al.* (2007) Why do some siblings attack each other? Comparative analysis of aggression in avian broods. *Evolution* **61**:1946–1955.

Goodall, J. (1986) *The Chimpanzees of Gombe: Patterns of Behavior.* Cambridge, MA: Belknap Press.

Gordon, I.J. (1989) The interspecific allometry of reproduction: do larger species invest relatively less in their offspring? *Functional Ecology* **3**:285–288.

Green, W.C.H. and Rothstein, A. (1991) Sex bias or equal opportunity? Patterns of maternal investment in bison. *Behavioral Ecology and Sociobiology* **29**:373–384.

Grodzinski, U. and Lotem, A. (2007) The adaptive value of parental responsiveness to nestling begging. *Proceedings of the Royal Society of London. Series B: Biological Sciences* **274**:2449–2456.

Guinness, F.E., *et al.* (1978a) Factors affecting calf mortality in red deer (*Cervus elaphus*). *Journal of Animal Ecology* **47**:817–832.

Guinness, F.E., *et al.* (1978b) Factors affecting reproduction in red deer (*Cervus elaphus*) hinds on Rhum. *Journal of Reproduction and Fertility* **54**:325–334.

Hackländer, K., *et al.* (2002) Postnatal development and thermoregulation in the precocial European hare (*Lepus europaeus*). *Journal of Comparative Physiology B* **172**:183–190.

Haig, D. (1997) Parental antagonism, relatedness asymmetries and genomic imprinting. *Proceedings of the Royal Society of London. Series B: Biological Sciences* **264**:1657–1662.

Haig, D. (2004) Genomic imprinting and kinship: how good is the evidence? *Annual Review of Genetics* **38**:553–585.

Haig, D. and Westoby, M. (1989) Parent-specific gene expression and the triploid endosperm. *American Naturalist* **134**:147–155.

Hanwell, A. and Peaker, M. (1977) Physiological effects of lactation on the mother. In: *Comparative Aspects of Lactation* (ed. M. Peaker). London: Zoological Society of London, 297–312.

Harris, M.P. and Murie, J.O. (1984) Inheritance of nest sites in female Columbian ground squirrels. *Behavioral Ecology and Sociobiology* **15**:97–102.

Harvey, P.H. and Clutton-Brock, T.H. (1985) Life-history variation in primates. *Evolution* **39**:559–581.

Harvey, P.H. and Zammuto, R.M. (1985) Patterns of mortality and age at first reproduction in natural populations of mammals. *Nature* **315**:319–320.

Harvey, P.H., *et al.* (1989a) Causes and correlates of life history differences among mammals. In: *Comparative Socioecology. The Behavioural Ecology of Humans and Other Mammals* (eds V. Standen and R. Foley). Oxford: Blackwell Scientific Publications, 305–318.

Harvey, P.H., *et al.* (1989b) Life history variation in placental mammals: unifying the data with theory. *Oxford surveys in evolutionary biology* **6**:13–31.

Hastings, K.K. and Testa, J.W. (1998) Maternal and birth colony effects on survival of Weddell seal offspring from McMurdo Sound, Antarctica. *Journal of Animal Ecology* **67**:722–740.

Hawkes, K. (2003) Grandmothers and the evolution of human longevity. *American Journal of Human Biology* **15**:380–400.

Hayssen, V. (1993) Empirical and theoretical constraints on the evolution of lactation. *Journal of Dairy Science* **76**:3213–3233.

Hayward, A.D. and Lummaa, V. (2013) Testing the evolutionary basis of the predictive adaptive response hypothesis in a preindustrial human population. *Evolution, Medicine, and Public Health* **2013**:106–117.

Hodge, S.J., *et al.* (2009) Maternal weight, offspring competitive ability, and the evolution of communal breeding. *Behavioral Ecology* **20**:729–735.

Hodge, S.J., *et al.* (2011) Reproductive competition and the evolution of extreme birth synchrony in a cooperative mammal. *Biology Letters* **7**:54–56.

Hofer, H. and East, M.L. (1997) Skewed offspring sex ratios and sex composition of twin litters in Serengeti spotted hyaenas (*Crocuta crocuta*) are a consequence of siblicide. *Applied Animal Behaviour Science* **51**:307–316.

Hofer, H. and East, M.L. (2008) Siblicide in Serengeti spotted hyenas: a long-term study of maternal input and cub survival. *Behavioral Ecology and Sociobiology* **62**:341–351.

Hopson, J.A. (1973) Endothermy, small size and the origin of mammalian reproduction. *American Naturalist* **107**:446–452.

Horsfall, J.A. (1984) Brood reduction and brood division in coots. *Animal Behaviour* **32**:216–225.

Houston, A.I. and Davies, N.B. (1985) The evolution of cooperation and life-history in the dunnock *Prunella modularis*. In: *Behavioural Ecology: Ecological Consequences of Adaptive Behaviour* (eds R. M. Sibley and R. H. Smith). Oxford: Blackwell Scientific Publications, 471–487.

Hudson, R. and Distel, H. (1982) The pattern of behaviour of rabbit pups in the nest. *Behaviour* **79**:255–271.

Hudson, R. and Trillmich, F. (2008) Sibling competition and cooperation in mammals: challenges, developments and prospects. *Behavioral Ecology and Sociobiology* **62**:299–307.

Humphries, M.M. and Boutin, S. (1996) Reproductive demands and weight gains: a paradox in female red squirrels (*Tamiasciurus hudsonicus*). *Journal of Animal Ecology* **65**:332–338.

Jenness, R. and Sloan, R.E. (1970) The composition of milks of various species: a review. *Dairy Science Abstracts* **32**:599–612.

Johnson, M.S., *et al.* (2001) Limits to sustained energy intake. III. Effects of concurrent pregnancy and lactation in *Mus musculus*. *Journal of Experimental Biology* **204**:1947–1956.

Johnstone, R.A. (2004) Begging and sibling competition: how should offspring respond to their rivals? *American Naturalist* **163**:388–406.

Johnstone, R.A. and Cant, M.A. (2010) The evolution of menopause in cetaceans and humans: the role of demography. *Proceedings of the Royal Society of London. Series B: Biological Sciences* **277**:3765–3771.

Johnstone, R.A. and Roulin, A. (2003) Sibling negotiation. *Behavioral Ecology* **14**:780–786.

Jones, W.T. (1986) Survivorship in philopatric and dispersing kangaroo rats *Dipodomys spectabilis*. *Ecology* **67**:202–207.

Kaiser, S. and Sachser, N. (1998) The social environment during pregnancy and lactation affects the female offsprings' endocrine status and behaviour in guinea pigs. *Physiology and Behavior* **63**:361–366.

Kaiser, S. and Sachser, N. (2001) Social stress during pregnancy and lactation affects in guinea pigs the male offsprings' endocrine status and infantilizes their behaviour. *Psychoneuroendocrinology* **26**:503–519.

Kaiser, S., *et al.* (2003) Early social stress in female guinea pigs induces a masculinization of adult behavior and corresponding changes in brain and neuroendocrine function. *Behavioural Brain Research* **144**:199–210.

Keech, M.A., *et al.* (2000) Life-history consequences of maternal condition in Alaskan moose. *Journal of Wildlife Management* **64**:450–462.

Kilner, R.M. and Hinde, C.A. (2012) Parent–offspring conflict. In: *The Evolution of Parental Care* (eds N.I. Royle, P. T. Smiseth and M. Kölliker). Oxford: Oxford University Press, 119–132.

Klopfer, P.H. and Klopfer, M. (1977) Compensatory responses of goat mothers to their impaired young. *Animal Behaviour* **25**:286–291.

Klug, H., *et al.* (2012) Theoretical foundation of parental care. In: *The Evolution of Parental Care* (eds N.J. Royle, P. T. Smiseth and M. Kölliker). Oxford: Oxford University Press, 21–39.

Kolding, S. and Fenchel, T.M. (1981) Patterns of reproduction in different populations of five species of the amphipod genus *Gammarus. Oikos* **37**:167–172.

Kölliker, M. and Richner, H. (2001) Parent–offspring conflict and the genetics of offspring solicitation and parental response. *Animal Behaviour* **63**:395–407.

Kölliker, M., *et al.* (2005) The coadaptation of parental supply and offspring demand. *American Naturalist* **166**:506–516.

Kölliker, M., *et al.* (2012) Parent–offspring co-adaptation. In: *The Evolution of Parental Care* (eds N.J. Royle, P. T. Smiseth and M. Kölliker). Oxford: Oxford University Press, 285–303.

König, B. (1994) Components of lifetime reproductive success in communally and solitarily nursing house mice: a laboratory study. *Behavioral Ecology and Sociobiology* **34**:275–283.

Kovacs, K.M. and Lavigne, D.M. (1986) Maternal investment and neonatal growth in phocid seals. *Journal of Animal Ecology* **55**:1035–1051.

Kovacs, K.M. and Lavigne, D.M. (1992) Maternal investment in otariid seals and walruses. *Canadian Journal of Zoology* **70**:1953–1964.

Künkele, J. (2000) Energetics of gestation relative to lactation in a precocial rodent, the guinea pig (*Cavia porcellus*). *Journal of Zoology* **250**:533–539.

Lang, S.L.C., *et al.* (2009) Repeatability in lactation performance and the consequences for maternal reproductive success in gray seals. *Ecology* **90**:2513–2523.

Larsen, K.W. and Boutin, S. (1994) Movements, survival, and settlement of red squirrel (*Tamiasciurus hudsonicus*) offspring. *Ecology* **75**:214–223.

Lazarus, J. and Inglis, I.R. (1986) Shared and unshared parental investment, parent–offspring conflict and brood size. *Animal Behaviour* **34**:1791–1804.

Lea, A.J., *et al.* (2015) Developmental constraints in a wild primate. *American Naturalist* **185**:809–821.

Le Cam, S., *et al.* (2009) Fast versus slow larval growth in an invasive marine mollusc: does paternity matter? *Journal of Heredity* **100**:455–464.

Lee, A.K. and Cockburn, A. (1985) Spring declines in small mammal populations. *Acta Zoologica Fennica* **173**:75–76.

Leonard, M.L., *et al.* (1988) Parent–offspring aggression in moorhens. *Behavioral Ecology and Sociobiology* **23**:265–270.

Leshem, M. and Schulkin, J. (2012) Transgenerational effects of infantile adversity and enrichment in male and female rats. *Developmental Psychobiology* **54**:169–186.

Lewis, S.E. and Pusey, A.E. (1997) Factors influencing the occurrence of communal care in plural breeding mammals. In: *Cooperative Breeding in Mammals* (eds N. G. Solomon and J. A. French). Cambridge: Cambridge University Press, 335–363.

Lillegraven, J.A., *et al.* (1987) The origin of eutherian mammals. *Biological Journal of the Linnean Society* **32**:281–336.

Lloyd, D.G. (1984) Variation strategies of plants in heterogeneous environments. *Biological Journal of the Linnean Society* **21**:357–385.

Lloyd, D.G. (1987) Selection of offspring size at independence and other size-versus-number strategies. *American Naturalist* **129**:800–817.

Loudon, A.S.I. and Racey, P.A. (1987) *Reproductive Energetics in Mammals.* London: Symposia of the Zoological Society of London.

Low, B.S. (1978) Environmental uncertainty and the parental strategies of marsupials and placentals. *American Naturalist* **112**:197–213.

McAdam, A.G. and Boutin, S. (2003) Effects of food abundance on genetic and maternal variation in the growth rate of juvenile red squirrels. *Journal of Evolutionary Biology* **16**:1249–1256.

McBride, G. (1963) The teat order and communication in young pigs. *Animal Behaviour* **11**:53–56.

Maccari, S., *et al.* (1995) Adoption reverses the long-term impairment in glucocorticoid feedback induced by prenatal stress. *Journal of Neuroscience* **15**:110–116.

McClure, P.A. (1987) The energetics of reproduction and life histories of cricetinerodents (*Neotoma floridana* and *Sigmodon hispidus*). *Symposia of the Zoological Society of London* **57**:241–258.

McCracken, G.F. (1984) Communal nursing in Mexican free-tailed bat maternity colonies. *Science* **223**:1090–1091.

Mace, G.M. (1979) *The evolutionary ecology of small mammals.* PhD thesis, University of Sussex.

McGinley, M.A. and Charnov, E.L. (1988) Multiple resources and the optimal balance between size and number of offspring. *Evolutionary Ecology* **2**:77–84.

McGinley, M.A., *et al.* (1987) Parental investment in offspring in variable environments: theoretical and empirical considerations. *American Naturalist* **130**:370–398.

MacLeod, K.J. and Lukas, D. (2014) Revisiting non-offspring nursing: allonursing evolves when the costs are low. *Biology Letters* **10**:20140378.

MacNair, M.R. and Parker, G.A. (1979) Models of parent–offspring conflict. III. Intra-brood conflict. *Animal Behaviour* **27**:1202–1209.

McNeilly, A.S., *et al.* (1982) Evidence for direct inhibition of ovarian function by prolactin. *Journal of Reproduction and Fertility* **65**:559–569.

McNeilly, A.S., *et al.* (1983) Release of oxytocin and prolactin in response to suckling. *British Medical Journal* **286**:257–259.

McNeilly, A.S., *et al.* (2006) Physiological mechanisms underlying lactational amenorrhea. *Annals of the New York Academy of Science* **709**:145–155.

Madden, J.R., *et al.* (2009) Calling in the gap: competition or cooperation in littermates' begging behaviour. *Proceedings of* the Royal Society of London. Series B: Biological Sciences **276**:1255–1262.

Maestripieri, D. (1998) Social and demographic influences on mothering style in pigtail macaques. Ethology **104**:379–385.

Magrath, R.D. (1988) Hatching asynchrony in altricial birds: nest failure and adult survival. American Naturalist **131**:893–900.

Margulis, S.W. (1997) Inbreeding-based bias in parental responsiveness to litters of oldfield mice. Behavioral Ecology and Sociobiology **41**:177–184.

Martin, R.D. and MacLarnon, A.M. (1985) Gestation period, neonatal size and maternal investment in placental mammals. Nature **313**:220–223.

Mateo, J.M. (2009) Maternal influences on development, social relationships, and survival behaviors. In: Maternal Effects in Mammals (eds D. Maestripieri and J. M. Mateo). Chicago: University of Chicago Press, 133–158.

Maynard Smith, J. (1977) Parental investment: a prospective analysis. Animal Behaviour **25**:1–9.

Mech, L.D., et al. (1991) Effects of maternal and grandmaternal nutrition on deer mass and vulnerability to wolf predation. Journal of Mammalogy **72**:146–151.

Milne, J.A. (1987) The effect of litter and maternal size on reproductive performance of grazing ruminants. Symposia of the Zoological Society of London **57**:189–201.

Mock, D.W. and Forbes, L.S. (1992) Parent–offspring conflict: a case of arrested development. Trends in Ecology and Evolution **7**:409–413.

Mock, D.W. and Parker, G.A. (1997) The Evolution of Sibling Rivalry. Oxford: Oxford University Press.

Moses, R.A. and Millar, J.S. (1994) Philopatry and mother–daughter associations in bushy-tailed woodrats: space use and reproductive success. Behavioral Ecology and Sociobiology **35**:131–140.

Moss, C.J. and Lee, P.C. (2011) Female reproductive strategies: individual life histories. In: The Amboseli Elephants (eds C.J. S Moss, H. Croze and P. C. Lee). Chicago: University of Chicago Press, 187–204.

Moyes, K., et al. (2011) Advancing breeding phenology in response to environmental change in a wild red deer population. Global Change Biology **17**:2455–2469.

Murray, C.M., et al. (2008) Adult male chimpanzees inherit maternal ranging patterns. Current Biology **18**:20–24.

Murray, C.M., et al. (2009) Reproductive energetics in free-living female chimpanzees (Pan troglodytes schweinfurthii). Behavioral Ecology **20**:1211–1216.

O'Connor, R.J. (1978) Brood reduction in birds: selection for fratricide, infanticide and suicide. Animal Behaviour **26**:79–96.

Oftedal, O.T. (1981) Milk, protein and energy intakes of suckling mammalian young: a comparative study. PhD thesis, Cornell University, Ithaca, NY.

Oftedal, O.T. (1984a) Milk composition, milk yield and energy output at peak lactation: a comparative review. Symposia of the Zoological Society of London **51**:33–85.

Oftedal, O.T. (1984b) Body size and reproductive strategy as correlates of milk energy yield in lactating mammals. Acta Zoologica Fennica **171**:183–186.

Oftedal, O.T. (1985) Pregnancy and lactation. In: Bioenergetics of Wild Herbivores (eds R. J. Hudson and R. G. White). Boca Raton, FL: CRC Press, 215–238.

Oftedal, O.T. (2000) Use of maternal reserves as a lactation strategy in large mammals. Proceedings of the Nutrition Society **59**:99–106.

Oftedal, O.T., et al. (1987) The behavior, physiology, and anatomy of lactation in the pinnipedia. In: Current Mammalogy (ed. H.H. Genoways). New York: Plenum Press, 175–245.

Packer, C., et al. (1992) A comparative analysis of non-offspring nursing. Animal Behaviour **43**:265–281.

Packer, C., et al. (1998) Reproductive cessation in female mammals. Nature **392**:807–811.

Parker, G.A. and Begon, M. (1986) Optimal egg size and clutch size: effects of environment and maternal phenotype. American Naturalist **128**:573–592.

Parker, G.A. and MacNair, M.R. (1979) Models of parent–offspring conflict. IV. Suppression: evolutionary retaliation of the parent. Animal Behaviour **27**:1210–1235.

Parker, G.A., et al. (2002) Intrafamilial conflict and parental investment: a synthesis. Philosophical Transactions of the Royal Society B: Biological Sciences **357**:295–307.

Peel, C.J., et al. (1983) Effect of exogenous growth hormone in early and late lactation on lactational performance of dairy cows. Journal of Dairy Science **66**:776–782.

Pfennig, D.W. (1997) Kinship and cannibalism. BioScience **47**:667–675.

Pianka, E.R. (1974) Evolutionary Ecology. New York: Harper and Row.

Pomeroy, P.P., et al. (1999) Consequences of maternal size for reproductive expenditure and pupping success of grey seals at North Rona, Scotland. Journal of Animal Ecology **68**:235–253.

Pond, C.M. (1977) The significance of lactation in the evolution of mammals. Evolution **31**:177–199.

Pond, C.M. (1983) Parental feeding as a determinant of ecological relationships in Mesozoic terrestrial ecosystems. Acta Palaeontologica Polonica **20**:215–224.

Price, K. (1992) Territorial bequeathal by red squirrel mothers: a dynamic model. Bulletin of Mathematical Biology **54**:335–354.

Price, K. and Boutin, S. (1993) Territorial bequeathal by red squirrel mothers. Behavioral Ecology **4**:144–150.

Reiss, M.J. (1989) The Allometry of Growth and Reproduction. Cambridge: Cambridge University Press.

Reiter, J., et al. (1978) Northern elephant seal development: the transition from weaning to nutritional independence. Behavioral Ecology and Sociobiology **3**:337–367.

Ricklefs, R.E., et al. (2003) Age-related patterns of fertility in captive populations of birds and mammals. Experimental Gerontology **38**:741–745.

Rodel, H.G., *et al.* (2009) Family legacies: short- and long-term fitness consequences of early-life conditions in female European rabbits. *Journal of Animal Ecology* **78**:789–797.

Roff, D.A. (2002) *Life History Evolution*. Sunderland, MA: Sinauer.

Rogowitz, G.L. (1996) Trade-offs in energy allocation during lactation. *Integrative and Comparative Biology* **36**:197–204.

Ross, C. (1988) The intrinsic rate of natural increase and reproductive effort in primates. *Journal of Zoology* **214**:199–219.

Roulin, A. (2002) The sibling negotiation hypothesis. In: *The Evolution of Begging: Competition, Cooperation and Communication* (eds J. Wright and M. L. Leonard). Dordrecht: Kluwer Academic, 107–126.

Roulin, A. and Dreiss, A.N. (2012) Sibling competition and cooperation over parental care. In: *The Evolution of Parental Care* (eds N.I. Royle, P. T. Smiseth and M. Kölliker). Oxford: Oxford University Press, 133–149.

Rüber, L., *et al.* (2004) Evolution of mouthbrooding and life-history correlates in the fighting fish genus *Betta*. *Evolution* **58**:799–813.

Russell, A.F., *et al.* (2002) Factors affecting pup growth and survival in co-operatively breeding meerkats *Suricata suricatta*. *Journal of Animal Ecology* **71**:700–709.

Sachser, N. and Kaiser, S. (1996) Prenatal social stress masculinizes the females' behaviour in guinea pigs. *Physiology and Behavior* **60**:589–594.

Sargent, R.C., *et al.* (1987) Parental care and the evolution of egg size in fishes. *American Naturalist* **129**:32–46.

Schaal, B., *et al.* (1995) Olfactory preferences in newborn lambs: possible influence of prenatal experience. *Behaviour* **132**:351–365.

Schino, G. and Troisi, A. (2005) Neonatal abandonment in Japanese macaques. *American Journal of Physical Anthropology* **126**:447–452.

Schneider, J.E. and Wade, G.N. (1989) Effects of maternal diet, body weight and body composition on infanticide in Syrian hamsters. *Physiology and Behavior* **46**:815–821.

Schulz, T.M. and Bowen, W.D. (2005) The evolution of lactation strategies in pinnipeds: a phylogenetic analysis. *Ecological Monographs* **75**:159–177.

Shea, N., *et al.* (2011) Three epigenetic information channels and their different roles in evolution. *Journal of Evolutionary Biology* **24**:1178–1187.

Shine, R. (1978) Propagule size and parental care: the 'safe harbor' hypothesis. *Journal of Theoretical Biology* **75**:417–424.

Short, R.V. (1976) Lactation: the central control of reproduction. *CIBA Foundation Symposium* **45**:73–86.

Sibly, R. and Calow, P. (1986) Why breeding earlier is always worthwhile. *Journal of Theoretical Biology* **123**:311–319.

Siegeler, K., *et al.* (2011) The social environment during pregnancy and lactation shapes the behavioral and hormonal profile of male offspring in wild cavies. *Developmental Psychobiology* **53**:575–584.

Skibiel, A.L., *et al.* (2013) The evolution of the nutrient composition of mammalian milks. *Journal of Animal Ecology* **82**:1254–1264.

Slagsvold, T., *et al.* (1984) On the adaptive value of intraclutch egg-size variation in birds. *The Auk* **101**:685–697.

Smiseth, P.T. and Lorentsen, S.-H. (2001) Begging and parent–offspring conflict in grey seals. *Animal Behaviour* **62**:273–279.

Smiseth, P.T., *et al.* (2008) Parent–offspring conflict and co-adaptation: behavioural ecology meets quantitative genetics. *Proceedings of the Royal Society of London. Series B: Biological Sciences* **275**:1823–1830.

Smith, A.S., *et al.* (2010) Maternal androgen levels during pregnancy are associated with early-life growth in Geoffroy's marmosets, *Callithrix geoffroyi*. *General and Comparative Endocrinology* **166**:307–313.

Smith, C.C. and Fretwell, S.D. (1974) The optimal balance between size and number of offspring. *American Naturalist* **108**:499–506.

Smith, W.P. (1987) Maternal defense in Columbian white-tailed deer: when is it worth it? *American Naturalist* **130**:310–316.

Speakman, J.R. (2008) The physiological costs of reproduction in small mammals. *Philosophical Transactions of the Royal Society B: Biological Sciences* **363**:375–398.

Speakman, J.R. and Król, E. (2005) Limits to sustained energy intake IX: a review of hypotheses. *Journal of Comparative Physiology B* **175**:375–394.

Spencer-Booth, Y. (1969) The effects of rearing rhesus monkey infants in isolation with their mothers on their subsequent behaviour in a group situation. *Mammalia* **33**:80–86.

Staines, B.W. and Crisp, J.M. (1978) Observations on food quality in Scottish red deer (*Cervus elaphus*) as determined by chemical analysis of rumen contents. *Journal of Zoology* **185**:253–259.

Stearns, S.C. (1983) The influence of size and phylogeny on patterns of covariation among life-history traits in the mammals. *Oikos* **41**:173–187.

Stearns, S.C. (1992) *The Evolution of Life Histories*. Oxford: Oxford University Press.

Stevenson, I.R., *et al.* (2004) Adaptive reproductive strategies. In: *Soay Sheep: Dynamics and Selection in an Island Population* (eds T. H. Clutton-Brock and J. M. Pemberton). Cambridge: Cambridge University Press, 243–275.

Stockley, P. and Parker, G.A. (2002) Life history consequences of mammal sibling rivalry. *Proceedings of the National Academy of Sciences of the United States of America* **99**:12932–12937.

Swihart, R.K. (1984) Body size, breeding season length, and life history tactics of lagomorphs. *Oikos* **43**:282–290.

Tait, D.E.N. (1980) Abandonment as a reproductive tactic: the example of grizzly bears. *American Naturalist* **115**:800–808.

Tardif, S.D., *et al.* (2002) Can marmoset mothers count to three? *Effect of litter size on mother–infant interactions*. *Ethology* **108**:825–836.

Temme, D.H. (1986) Seed size variability: a consequence of variable genetic quality among offspring? *Evolution* **40**:414–417.

Trillmich, F. (1990) The behavioral ecology of maternal effort in fur seals and sea lions. *Behaviour* **114**:3–20.

Trillmich, F. and Wolf, J.B.W. (2008) Parent–offspring and sibling conflict in Galapagos fur seals and sea lions. *Behavioral Ecology and Sociobiology* **62**:363–375.

Trivers, R.L. (1974) Parent–offspring conflict. *American Zoologist* **14**:249–264.

Tucker, H.A. (1994) Lactation and its hormonal control. In: *The Physiology of Reproduction* (eds E. Knobil and J. D. Neill). New York: Raven Press.

Turner, S.E., *et al.* (2005) Maternal behavior and infant congenital limb malformation in a free-ranging group of *Macaca fuscata* on Awaji Island, Japan. *International Journal of Primatology* **26**:1435–1457.

Uller, T. (2012) Parental effects in development and evolution. In: *The Evolution of Parental Care* (eds N.J. Royle, P. T. Smiseth and M. Kölliker). Oxford: University of Oxford Press, 247–266.

Uller, T. and Pen, I. (2011) A theoretical model of the evolution of maternal effects under parent–offspring conflict. *Evolution* **65**:2075–2084.

van Noordwijk, M.A. and van Schaik, C.P. (1999) The effects of dominance rank and group size on female lifetime reproductive success in wild long-tailed macaques, *Macaca fascicularis*. *Primates* **40**:105–130.

Wahaj, S.A. and Holekamp, K.E. (2006) Functions of sibling aggression in the spotted hyaena, *Crocuta crocuta*. *Animal Behaviour* **71**:1401–1409.

Ward, E.J., *et al.* (2009) The role of menopause and reproductive senescence in a long-lived social mammal. *Frontiers in Zoology* **6**:1–10.

West, S.A., *et al.* (2002) Cooperation and competition between relatives. *Science* **296**:72–75.

Wilson, A.J. and Festa-Bianchet, M. (2009) Maternal effects in wild ungulates. In: *Maternal Effects in Mammals* (eds D. Maestripieri and J. Mateo). Chicago: University of Chicago Press, 83–103.

Wilson, A.J., *et al.* (2005) Selection on mothers and offspring: whose phenotype is it and does it matter. *Evolution* **59**:451–463.

Wilson, A.J., *et al.* (2009) Trading offspring size for number in a variable environment: selection on reproductive investment in female Soay sheep. *Journal of Animal Ecology* **78**:354–364.

Wilson, D.S. and Clark, A.B. (2002) Begging and cooperation. In: *The Evolution of Begging: Competition, Cooperation and Communication* (eds J. Wright and M. L. Leonard). Dordrecht: Kluwer Academic, 43–64.

Winkler, D.W. (1987) A general model for parental care. *American Naturalist* **130**:526–543.

CHAPTER 6
Social development

6.1 Introduction

In the hard sunlight of a Kalahari morning, adult meerkats forage in the dry litter of a fallen tree. Scattered among them, four 8-week-old pups beg for food or dig desultorily themselves. A young adult unearths a large scorpion, which backs away, its claws raised and its sting curled above its head. In an instant, the meerkat seizes it across the thorax, bites off the venomous tip of its tail and carries it, still struggling angrily, to the nearest pup. Without damaging it further, it drops it in front of the pup, but remains nearby guarding the pup and the injured scorpion from interference by other group members. Unused to mobile food of this size, the pup pats it uncertainly and tries, unsuccessfully, to bite its head. Quickly, the adult steps in, biting the scorpion several times and leaving it struggling feebly and unable to defend itself further. Faced with an immobile prey, the pup begins to eat the tail and gradually works its way up to the head and claws until none is left.

In animals where parents do not care for their young, developing juveniles have limited opportunities to learn technical or social skills from conspecifics and, where learning occurs, it is typically based on trial and error, increasing the risk of mistakes and their associated costs. In contrast, in social animals, individuals can learn complex skills and acquire information both from their parents and from other group members. Interactions with peers, older juveniles and adults allow them to learn how to respond in an adaptive fashion without experiencing the costs of mistakes. Bouts of play with other juveniles and adolescents may help them both to hone their motor skills and to practice a variety of social responses (Figure 6.1).

In many longer-lived species (especially those living in unpredictable environments) individuals are intermittently exposed to extreme environmental events that threaten their survival (or that of their offspring) and may draw on the cumulative knowledge of older group members. Information shared between group members can also be transferred across generations, and can lead to behavioural differences between groups which are inherited by subsequent generations, producing traditional differences in behaviour which may not be directly related to current variation in ecology or social organisation.

Our understanding of the extent and pattern of species differences in social development is more fragmentary than our knowledge of many other aspects of social behaviour and a large proportion of research in this area has been carried out on captive animals maintained on ad lib food under restrictive conditions. In addition, field research has concentrated on species likely to show advanced cognitive abilities, including the cercopithecine monkeys and the great apes. As a result, while our understanding of the cognitive capacities of non-human mammals is increasing steadily, we still lack the data necessary for a quantitative assessment of interspecific differences or of their relationship to variation in ecological and social factors.

This chapter explores the effects of group-living on social development. Section 6.2 describes evidence of social learning while section 6.3 reviews the process of social development, and section 6.4 examines the consequences of play between juveniles and adolescents. Section 6.5 reviews recent research on self-awareness and social knowledge in wild mammals and its consequences. Section 6.6 explores the origins, extent and consequences of individual differences in behaviour

Mammal Societies, First Edition. Tim Clutton-Brock.
© 2016 Tim Clutton-Brock. Published 2016 by John Wiley & Sons, Ltd.

Figure 6.1 Sub-adult meerkats playing. *Source*: © Russell Venn.

and personality while section 6.7 describes the extent and causes of differences between groups.

6.2 Social learning

Where young are neither guarded nor protected by their parents, the initial development of survival skills depends primarily on interactions between genetically programmed propensities and individual experience. In contrast, in animals with prolonged parental care, associations between juveniles and their mothers or other adults allow juveniles to learn from adults to identify opportunities and dangers and to exploit or avoid them (Galef and Giraldeau 2001; Griffin 2004; Moore 2004; Hoppitt and Laland 2013). Naive individuals can acquire information by vertical transmission from parents, by horizontal transmission from their sibs or peers or by oblique transmission from unrelated adults (Galef and Laland 2005; Whiten 2012). Juveniles and sub-adults are often likely to use older or more dominant individuals as models: for example, in wild chimpanzees, juveniles pay more attention to the feeding techniques of individuals older than themselves than to those of younger individuals (Biro *et al*. 2003) (Figure 6.2) while vervet monkeys

pay more attention to the behaviour of the dominant female in their group than to that of other animals and are most likely to learn from her (van de Waal *et al.* 2010).

In many societies, younger animals are most likely to learn behavioural traits or skills from individuals that they associate with on a regular basis (Coussi-Korbel and

Figure 6.2 A juvenile chimpanzee observes the foraging techniques of its mother. *Source*: © Joshua Leonard.

Fragaszy 1995) and the distribution of socially learned traits reflects the structure of social networks (Cantor and Whitehead 2013). For example, in bottlenose dolphins, the tendency for individuals to learn to beg for food from recreational fishermen is associated with the amount of time they spend with other individuals that beg (Donaldson *et al.* 2012). Similarly, experiments with big brown bats in which individuals were taught a solution to a novel foraging task show that other members of their group commonly adopt the same approach (Wright *et al.* 2011).

Social learning can involve different processes, including social facilitation (where an individual's motivational state is affected by the presence of a conspecific) and local stimulus enhancement (where learning results from an individual's attention being drawn to a particular location or object) as well as emulation (where individuals copy the goals of others) and imitation (where they copy their actions as well as their goals) (Tomasello 1996, 1999; Zentall 2001). While social learning is often attributed to imitation, relatively few studies (especially those of free-ranging animals) can definitely exclude the possibility that changes in behaviour are a consequence of simpler

mechanisms, including social facilitation or local enhancement (Heyes and Ray 2000; Heyes 2001).

Social learning often plays an important role in the development of feeding behaviour (Rapaport and Brown 2008; Hoppitt and Laland 2013). For example, experiments with ground squirrels show that the presence of their mothers increases both the rate at which juveniles develop preferences for particular foods and their rate of food intake (Peacock and Jenkins 1988). Similarly, in black rats living in pine forests in Israel, young animals learn how to strip seeds from pine cones from older animals and, if they fail to learn this technique when young, they seldom learn on their own (Zohar and Terkel 1996).

Age-structured groups provide adolescents with access to older individuals that possess experience of the distribution of resources and the risks associated with interactions with competitors or predators. For example, in African elephants (Figure 6.3), matriarchs possess a detailed knowledge of the distribution of resources in outlying parts of their range, which is of crucial importance to survival in drought years (Moss and Lee 2011; Mutinda *et al.* 2011). As a result of their greater experience, older females are also more likely to respond

Figure 6.3 In African elephants, all members of matrilineal female groups help to protect calves born in the group. *Source*: © Phyllis Lee.

appropriately in risky situations: for example, older matriarchs are more likely to distinguish and respond to the roars of male lions, which are more likely than lionesses to attack young elephants (McComb *et al.* 2011). As in several other social mammals (see Chapter 5) the absence of mothers in elephants is associated with reduced survival of their calves (Figure 6.4a), while the presence of grandmothers and the number of allo-mothers are associated with increased calf survival (Figure 6.4b,c). Similar effects may well be common in many other wide-ranging long-lived species, including cetaceans (McAuliffe and Whitehead 2005; Johnstone and Cant 2010).

Social learning also plays an important role in other contexts (Hoppitt and Laland 2013). Juveniles and sub-adults commonly learn to identify dangerous predators from the reactions of other group members (Griffin 2004). For example, experiments with tammar wallabies in which individuals that had been exposed to a 'trained' tutor that was either fearful of a model fox or indifferent to it showed that individuals in the first group responded more strongly to the presentation of the model than

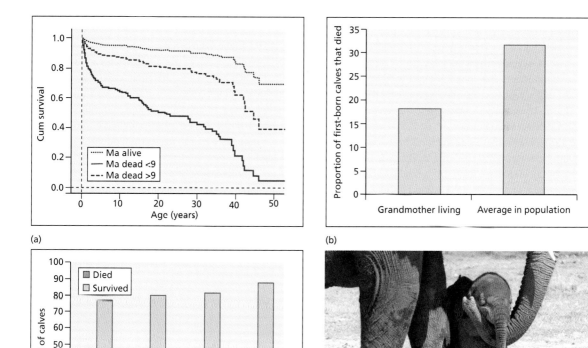

(a)

(b)

(c)

Figure 6.4 Effects of maternal experience on the survival of African elephant calves. (a) The survival of daughters in African elephants whose mothers were still alive (dotted line), whose mothers died when they were at least 9 year old (dashed line), and whose mothers died when they were less than 9 years old (solid line). Effects of the absence of mothers persist into the later years of their daughters' lifespans and influence both the survival of their daughters and the survival of their grand-daughters. (b) Proportion of first-born calves whose grandmothers were still living that died compared with the average for first-born calves in the population. (c) The percentage of calves that die or survive the first 2 years of life in relation to the number of allo-mothers. *Source*: From Moss and Lee (2011). Reproduced with permission of the University of Chicago Press. *Photo source*: © Karen McComb.

those in the second (Griffin and Evans 2003). Similarly, experiments with captive rhesus monkeys show that individuals can acquire a fear of snakes through viewing the responses of other animals to videotapes, though they are apparently unable to learn similar associations when alarm calls are paired with other objects (Cook and Mineka 1990). The responses of juveniles to alarm calls may also be acquired by social learning (Seyfarth and Cheney 1980; Hollén and Manser 2006; Hollén *et al.* 2008).

Social learning is also important in the acquisition of social skills. It is often involved in the development of calling behaviour (Janik and Slater 1997, 2000): for example, studies of greater spear-nosed bats show that the development of group-specific differences in call structure is dependent on social learning and that, as group composition changes, females modify the structure of their calls to resemble those of other group members (Boughman 1998). Similar processes may occur in cetaceans (Yurk *et al.* 2002). For example, male humpback whales using the same breeding grounds sing similar songs which develop in parallel during the course of the breeding season, presumably because individuals adjust their songs in relation to those given by other males at the same breeding site (Payne 1999; Garland *et al.* 2011; Cantor and Whitehead 2013). In many social mammals, juveniles and adolescents also learn their relative status and rank within their groups by observing the reactions and responses of close kin and may also acquire a range of other social skills by observing interactions between other group members (Galef and Laland 2005; Hoppitt and Laland 2013).

While reliance on social learning can have advantages because it enables individuals to avoid the costs of trial and error learning, it can be disadvantageous if conditions change with the result that individuals learn suboptimal or irrelevant responses (Laland and Williams 1998; Giraldeau *et al.* 2002). The benefits of social learning are consequently likely to vary with the relative stability of environments (Galef 1995; Whitehead 2007): in very stable environments, genetic control of development may be the cheapest and most effective way to ensure adaptation; in locally unstable environments, individual learning may be most effective; in environments that vary on large spatial or temporal scales, social learning may be best. Some empirical evidence supports these predictions. For example, experiments with rats show that the presence of

'demonstrators' has less effect on the food choice of their cage mates in animals maintained under unpredictable feeding regimes than in animals fed at predictable times for predictable periods (Galef and Whiskin 2004).

Since the individual's experience of environmental variability depends on its lifespan, the extent to which individuals rely on social learning is also likely to vary with longevity. Short-lived organisms may rarely experience environmental changes during their lives and may consequently be able to rely on genetic determinism, while in long-lived species, like many large mammals, individuals may commonly experience environmental changes during their lives and social learning may consequently be relatively advantageous (Boyd and Richerson 1985).

Sociality also provides opportunities for individuals to train or teach juveniles and adolescents (Ewer 1969; Hoppit *et al.* 2008; Thornton and Raihani 2008). The usual definition of teaching (Caro and Hauser 1992) requires that an experienced individual modifies its behaviour at some cost to itself when one or more naive individuals are nearby and that these changes increase the rate or precision with which the naive animal learns an appropriate skill or response. Teaching may either involve *opportunity teaching* (where changes in the teacher's behaviour put the pupil in a situation where they are likely to develop a new skill) or *coaching*, where teachers adjust their behaviour in response to the pupil's actions, using various forms of encouragement or punishment to affect the development of skills (Caro and Hauser 1992; Hoppit *et al.* 2008). For example, in golden lion tamarins, adults use calls that they give to encourage younger juveniles to take proffered food to direct sub-adults to foraging sites that contain hidden prey (Rapaport 2011).

The most convincing evidence of teaching in non-human mammals comes from studies of carnivores where group members teach juveniles complex techniques needed to catch and dismember prey. For example, in cheetah, mothers of cubs less than 2 months old typically suffocate their prey (principally hares and Thomson's gazelles), while mothers of 3–4 month old cubs commonly bring back prey to their cubs and release it, repeatedly calling to their cubs, and giving them an opportunity to chase it before they intervene and kill it (Figure 6.5). As pups develop, mothers catch, pursue and knock over their prey and then leave their cubs to catch it and disembowel it. Subsequently, they sometimes run

Figure 6.5 Female cheetah provide their cubs with opportunities to learn how to handle prey: a female cheetah with her dependent cubs and a juvenile gazelle. *Source*: © Richard Du Toit/Minden Pictures/Getty Images.

relatively slowly after prey and allow their cubs to overtake them and be the first to knock down prey themselves (Caro 1994). Meerkat mothers (as well as other group members) help dependent juveniles to learn how to handle and kill scorpions by initially presenting them with incapacitated prey whose stings have been removed and, as pups age, gradually reducing the damage they inflict on the scorpion before giving it to the pup (Thornton and McAuliffe 2006). In this case, it was possible to show experimentally that mothers not only adjusted the way in which they handled prey to pups of different ages at some cost to their own rate of foraging but that this increased the rate at which pups learned to handle scorpions (Box 6.1 and Figure 6.6).

Box 6.1 Teaching in meerkats

One of the few studies that provides evidence that meets the criteria for teaching set by Caro and Hauser (1992) involves wild meerkats where parents and helpers commonly feed scorpions to dependent pups (Figure 6.6a,b). While meerkats have some immunity to scorpion stings, severe stings can kill pups or generate prolonged immobility and inappetence. Before feeding scorpions to young pups, helpers and parents usually kill or disable them by removing their stings but, as pups grow older, adults become increasingly likely to feed them intact scorpions (Figure 6.6a–c). The proportion of scorpions that are fed intact can be increased by playing parents and helpers the calls of pups older than those they are feeding (Figure 6.6d) while the proportion fed dead can be increased by playing them the calls of younger pups (Figure 6.6e).

Experiments in which pups were experimentally provisioned with live or dead scorpions confirm that provisioning with live prey increases the pup's skill in handling and reduces handling time relative to that of pups experimentally fed with dead prey, showing that the behaviour of helpers improves the acquisition of handling skills by pups. Feeding intact scorpions to pups requires helpers to monitor the pup while they are handling the scorpion and can consequently increase costs to helpers.

The extent to which helpers are involved in teaching varies with their age: younger helpers who are still growing contribute less than older ones but neither previous experience of helping nor individual differences in nutritional status are associated with individual involvement in teaching, suggesting that it may have little cost (Thornton and Raihani 2008). As well as teaching juveniles feeding skills, parents and helpers commonly use mild (or not so mild) threats to discourage adolescents from dangerous or disruptive activities. For example, in some primates, mothers discourage dependent juveniles from straying outside the group by mild threats or punishments (MacKinnon 2011). Teaching is not confined to mammals and has been described in insects (Franks and Richardson 2006) and birds (Raihani and Ridley 2008).

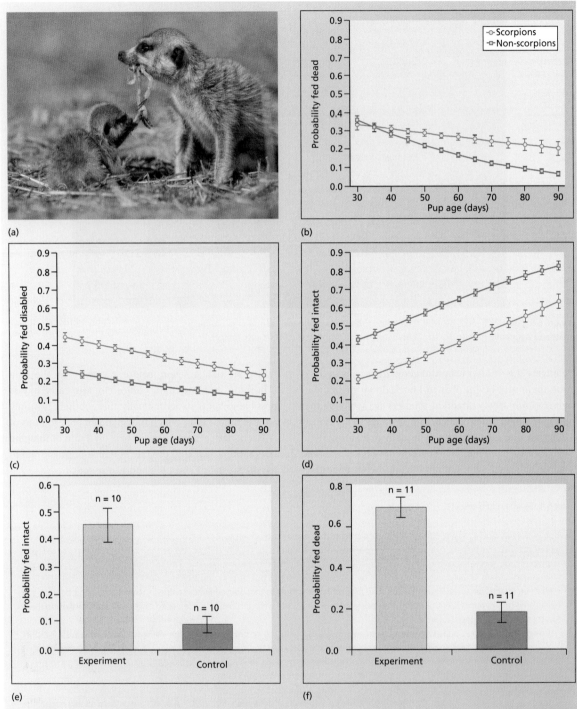

Figure 6.6 Teaching in wild meerkats. (a) In meerkats, parents and helpers provision dependent pups with invertebrates and small vertebrates, including scorpions, whose stings can severely affect pups. As the age of pups increases, the probability that they will be given (b) dead or (c) disabled prey declines and the probability that they will be given intact prey (d) increases. At all ages, scorpions are less likely to be fed intact than other prey. If parents and helpers are played the calls of pups older than those they are feeding, they (e) increase the proportion of intact scorpions fed to pups, whereas if they are played the calls of pups younger than those they are feeding, they (f) increase the proportion fed dead. *Source*: Graphs (b–f) from Thornton and McAuliffe (2006). Reproduced with permission from AAAS. *Photo source*: © Robin Hoskyns.

There is a continuing debate as to whether behaviour of the kind described in this section should be referred to as 'teaching'. Some psychologists define teaching as the intentional transfer of information and, as evidence of intentionality is usually impossible to provide, are sceptical that teaching occurs in wild animals. In contrast, behavioural ecologists typically adopt functional definitions and define teaching as involving costly modifications of adult behaviour that are adjusted to the age or experience of developing juveniles and increase the rate at which they acquire important skills, and although firm evidence that meets these criteria is uncommon, teaching clearly occurs in some species.

How widely teaching is distributed is not yet clear. For example, rat dams do not teach their young to select appropriate foods (Galef et al. 2005) and there is limited evidence of active teaching of foraging skills in higher primates (Hoppit et al. 2008). For example, in chimpanzees, where daughters apparently learn how to fish with grass stems or twigs by observing the behaviour of their mothers, there is no indication that mothers adjust their behaviour to help them to learn this technique (Lonsdorf et al. 2004). Teaching may be less common in omnivores and frugivores than in carnivores because they seldom have to handle mobile prey and similar processes may operate through local enhancement or the provision of opportunities to learn, so that changes in the teacher's behaviour are hard to detect.

6.3 Social development

In most social mammals, social development is not a steady process and there are sensitive periods when social responses develop rapidly (Schnehla and Rosenblatt 1963; Bateson 1979). Variation in opportunities during these periods often has long-lasting effects. For example, social and ecological factors affecting pregnant and lactating females can have an important influence on the development of their offspring, with downstream consequences for their behaviour and reproductive performance as adults (see Chapter 5). The timing and duration of sensitive periods often appear to be adapted to contrasts in the social environment that neonates encounter (Bateson 1979). For example, the neonates of altricial species (which are unlikely to be separated from their mothers) learn to recognise them relatively slowly, whereas the young of precocial species (including

many ungulates) rapidly learn to distinguish between their mothers and other group members and their responses are difficult to reverse (Klopfer and Gamble 1966; Vince 1993).

Early experience of social interactions can also affect adult behaviour and physiology. For example, in laboratory mice, males that suckle more on anterior teats and gain weight more rapidly are more likely to attain high rank than those that suckle on posterior ones (Barnard et al. 1998). In rats, licking and grooming by mothers affects the expression of oxytocin receptor genes in female offspring, with downstream effects on the maternal skills of their offspring in adulthood (Meaney 2001). Contrasts in experience frequently affect the extent of changes in behaviour and physiology at adolescence and the timing of exposure to social interactions can be critical (Sachser et al. 2013). For example, in some rodents, exposure to aggression from adults during early adolescence increases their tendency to be aggressive as adults, while exposure during late adolescence has the opposite effect (Delville et al. 1998; Terranova et al. 1999).

The social environment in which individuals are raised can also have important effects on development (Taborsky and Oliveira 2012). Variation in group size and local population density commonly affects glucocorticoid levels in mothers and their offspring, with downstream effects on their behaviour as adults (Creel et al. 2013). The structure of groups in which individuals develop can also have important consequences for their behaviour as adults. For example, in guinea pigs, males reared in mixed-sex colonies establish stable dominance relations with little overt aggression, and are able to integrate into novel groups of unfamiliar animals without intense aggression, while males maintained with a single female from early adolescence are more overtly aggressive and do not easily integrate into new social groups (Wolf a and Sachser 2009; Sachser et al. 2011, 2013). This is apparently because adolescent males that are involved in frequent interactions during early adolescence show high testosterone levels, which reduce the reactivity of the hypothalamic–pituitary–adrenal axis, leading to low levels of aggressive behaviour, whereas infrequent social interactions during early adolescence are associated with low testosterone levels, high hypothalamic–pituitary–adrenal reactivity and high levels of aggressive behaviour.

Although social and ecological influences on development have often been interpreted as non-adaptive, they may also represent adaptive adjustments that match individuals to the conditions they are likely to encounter in later life (Sachser *et al.* 2013). For example, rearing male guinea pigs in groups may mimic the social environment that individuals would encounter in high-density populations and their ability to establish relationships may increase their fitness. However, distinguishing between the non-adaptive consequences of adverse environments and adaptive plasticity is usually difficult unless it is possible to investigate the effects of adjustments on subsequent fitness under conditions that approximate to natural, which is rarely feasible (Bateson and Gluckman 2011; Dantzer *et al.* 2013).

6.4 Play

In many social animals, juveniles and adolescents often learn important skills through playful interactions with their peers (Fagen 1974) (Figures 6.1, 6.7). Some form of play involving juveniles and adolescents is ubiquitous in animal societies (Bekoff 1972), though its frequency and form vary widely between species (Pellis and Iwaniuk 1999). Within species, the frequency and form of play usually varies with the age of individuals and play is normally more frequent between littermates than between other individuals (Chalmers 1984; Bekoff and Allen 1998) and sex differences are also common (see Chapter 18).

Play is energetically costly and energy expenditure on play can affect growth: for example, in wild Assamese macaques, individual differences in time spent playing

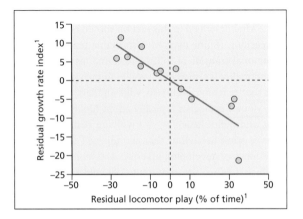

Figure 6.8 Trade-off between time spent playing and an index of growth in infant Assamese macaques. Residuals are deviations from the average percentage of time spent playing, controlled for average food availability and lactation category. *Source*: From Berghanel *et al.* (2015). Reproduced with permission from AAAS. Used under CC BY-NC http://creativecommons.org/licenses/by-nc/4.0/. *Photo source*: © Julia Ostner.

are negatively correlated with growth (Figure 6.8). As might be expected, the frequency and duration of bouts of play commonly decline when food is scarce (Lee 1984; Barrett *et al.* 1992) and several studies have shown that time spent playing can be increased by experimental provisioning: for example, in wild meerkats, individuals whose food is experimentally increased double their rate of play (Sharpe *et al.* 2002).

The high energetic costs of play are often taken to suggest that it must serve an important function and two main types of benefits have been suggested (Bekoff 1977; Bekoff and Allen 1998). First, play may help juveniles to establish close social bonds that increase the chance that

Figure 6.7 A young adult chacma baboon playing with a juvenile. *Source*: © Alecia Carter.

they will disperse together, or may help to improve the individual's fighting skills or their ability to establish dominance relationships (Bekoff 1977; Bekoff and Allen 1998; Blumstein et al. 2013). An alternative possibility is that its principal benefit lies in its contributions to the development of motor skills, coordination or the integration of the central nervous system (Pellegrini et al. 2007). For example, in Assamese macaques, individual differences in play are positively correlated with differences in the acquisition of motor skills (Berghanel et al. 2015). Across species, high rates of play tend to occur at stages of development when cerebellar synaptogenesis and muscle differentiation are most malleable and neural plasticity in brain regions associated with play peak at around the same time (Byers and Walker 1995; Fairbanks 2000).

Unfortunately, it is difficult to test alternative explanations of play and the selection pressures maintaining play are still unresolved (Burghardt 2005). Some studies support the suggestion that it has important consequences for subsequent social behaviour. For example, long-term studies of yellow-bellied marmots show that the directional outcome of playful interactions is correlated with later dominance rank (Blumstein et al. 2013). However, others do not: for example, studies of meerkats have found no evidence that individual differences in involvement in play are associated with reductions in rates of aggression, increased fighting success or enhanced social cohesion and the frequency with which individuals play together is not correlated with the probability that they will disperse together (Sharpe and Cherry 2003; Sharpe 2005).

6.5 Social knowledge

Studies of social development raise important questions about the extent to which social animals understand each other and can predict the likely responses of others to their actions. In many social mammals, individuals can recognise other individuals belonging to their group from olfactory, vocal or visual cues. For example, female African elephants can distinguish between the calls and faeces of individuals belonging to their family or their bond group (a collection of related females) and those of other females (McComb et al. 2000; Bates et al. 2008). Information of this kind is not restricted to the receiver's own species: playback experiments with elephants show that they can also distinguish differences in ethnicity, age and sex between groups of humans (McComb et al. 2014).

Social animals are often aware of the absence of other group members as well as of their presence. For example, horses react more strongly to playback of the whinny of a group member who is temporarily absent from the group than to those of an individual who is present but out of sight (Proops et al. 2009) (Figure 6.9), while elephants respond more strongly to dung from family members that are not present at the time or are walking behind them than to dung from family members that are present in their group or are walking ahead of them (Bates et al. 2008).

In some species, individuals also appear to be aware of their own identity. One common way of testing for self-awareness is to mark individuals in some obvious way and then to present them with a mirror and assess whether they can connect the mark in their mirror image with themselves. In chimpanzees (Tomasello and Call 1997), dolphins (Reiss and Marino 2001) and elephants (Plotnik et al. 2006) some individuals connect mirror images with themselves (Figures 6.10 and 6.11) but most other mammals that have been tested (including several Old World monkeys) fail this test, suggesting that their level of social awareness is not as highly developed. However, many species that fail the mirror test still behave as if they have knowledge of their place in a network of social relations. Seyfarth and Cheney (2000), who have pioneered research in this area, conclude that monkeys have a well-developed sense of their 'social selves' which plays an important role in guiding their responses but which differs from human self-awareness.

In some social mammals, individuals of both sexes also recognise relationships between third parties and use this knowledge to adjust their own behaviour. For example, female vervet monkeys, baboons and chimpanzees appear to be able to recognise dominance relations and kinship connections and social bonds between third parties (Cheney and Seyfarth 1990, 2008; Seyfarth and Cheney 2012; Wittig et al. 2014), while female spotted hyenas routinely support dominant individuals over subordinate opponents in fights between third parties, even if both individuals are subordinate to them (Engh et al. 2005).

The capacity to recognise relationships between third parties allows individuals to respond to interactions involving relatives or allies (Cheney and Seyfarth

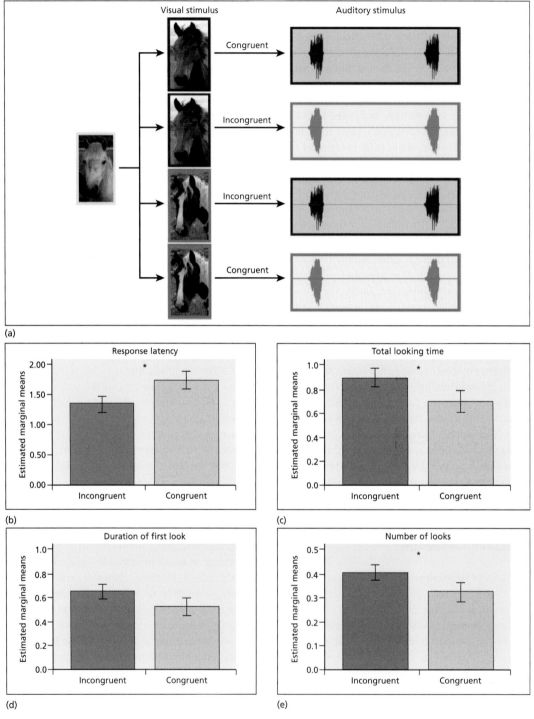

Figure 6.9 Horses played the vocalisations (whinnies) of known individuals present in the same paddock but temporarily out of sight (congruent) react less strongly than those played the whinnies of known individuals that are not present in the same paddock (incongruent). (a) Experimental structure; (b–e) responses of subjects to congruent versus incongruent whinnies. *Source*: From Proops *et al.* (2009). Reproduced with permission of the National Academy of Sciences, USA.

Figure 6.10 Chimpanzees given access to mirrors appear to recognise that the mirror presents an image of themselves. *Source*: © Xavier Hubert-Brierre.

Figure 6.11 An Asian elephant closely inspects her reflection in a mirror. *Source*: © Joshua Plotnik.

1990, 2008; Seyfarth and Cheney 2012). For example, male chacma baboons that have friendly relations with particular females appear to recognise the risk that potentially infanticidal males pose to the offspring of their 'friends' and take steps to protect them (see Chapter 15). Females are able to recognise the screams and threat grunts of their own close relatives as well as those of the kin of other individuals and to identify circumstances that are likely to involve risk to their own kin (Cheney and Seyfarth 1999). In a remarkable series of playback experiments, pairs of unrelated female baboons (one dominant and one subordinate) were played sequences of calls that mimicked fights between their relatives while, as controls, they were played sequences that either mimicked fights between individuals unrelated to either of them or sequences that involved the dominant's relatives in contact with an unrelated partner. When the playback mimicked a dispute between relatives of both females or a dispute between a relative of the dominant female and another animal, both females looked toward the speaker (Figure 6.12a) and at each other (Figure 6.12b) for longer than when the playback was of a dispute between two females unrelated to either of the subject animals. After playbacks of fights in which relatives of both subjects were involved, dominant females were more likely to supplant subordinate partners than after playbacks that did not involve their kin (Figure 6.12c, DS) and were less likely to act in a friendly way towards them (Figure 6.12c, DA) while subordinates were more likely to approach dominants and initiate a friendly interaction (Figure 6.12c, SA).

Female baboons are also able to integrate knowledge of kin structure and relative dominance within groups and respond more strongly to playbacks of interactions that violate existing dominance relations than to those that do not. For example, when chacma baboons were played a threat grunt from one animal followed by a scream from another that mimicked either established dominance relations or reversals of the usual pattern, between members of the same matriline or between members of a different matriline (an even less common event), subjects responded more strongly to playback of reversals between members of the same matriline than to playback of interactions between members of the same matriline that were in line with the established hierarchy (Cheney and Seyfarth 2008). In addition, they responded even more strongly to playback of reversals involving members of different matrilines. Recent interactions also affect the responses of individuals to each other (Bergman et al. 2003). For example, female baboons that have been threatened by another individual react more strongly if played a recording of their aggressor's threat grunt shortly afterwards than if they are played a recording of a grunt from another individual (Seyfarth and Cheney 2012). Similarly, females respond less strongly to threat grunts of other females with whom they have recently exchanged grooming than to threat grunts of females that have recently threatened them (Engh et al. 2006).

In higher primates (and in chimpanzees, in particular), individuals sometimes appear to use their knowledge of other individuals to manipulate or deceive them (Byrne and Whiten 1992). For example, experiments with macaques using one-way mirrors show that subordinate females are more likely to supplant and threaten the juvenile offspring of dominant females at times when the juvenile is out of sight of its dominant relatives (Cheney and Seyfarth 1990). Similarly, rhesus macaques given the opportunity to 'steal' grapes from a noisy source or a source from which they can be extracted silently are more likely to choose the silent source if the experimenter is looking away (and so might be alerted by a noise) than if the experimenter is looking at them (Santos et al. 2006). Subordinate chimpanzees are more likely to retrieve food placed outside rather than within the view of dominants but show no discrimination if the watching dominant is replaced with another individual (Hare et al. 2000). Similar tactics are also used by females to avoid conflicts with males (see Chapter 7).

In some mammals, individuals may also be aware of the extent of each other's knowledge. Following individual chimpanzees in Budongo Forest, experimenters placed a stuffed model of a viper in their path and recorded their behaviour when other chimpanzees approached the area (Crockford et al. 2012). Chimps that encountered the viper were more likely to give warning calls if approaching individuals were unaware of the snake, while their own proximity to the snake was unimportant. Whether experiments of this kind indicate that individuals have an understanding of the mental states of others (a *theory of mind*) or whether they are reacting to relatively simple cues that are correlated with the knowledge that others' possess is uncertain and the extent to which non-human mammals understand the knowledge or reactions of others is still unclear (Heyes

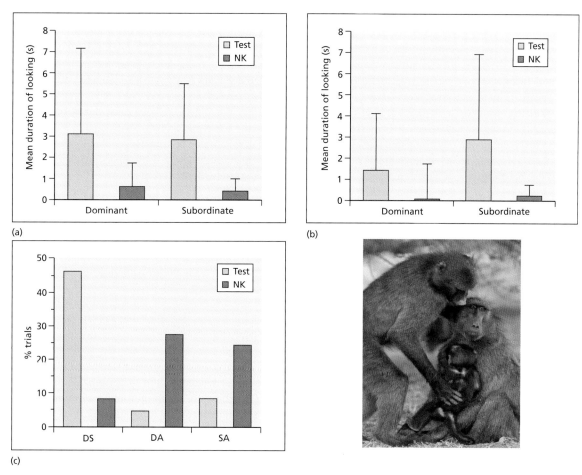

Figure 6.12 Social knowledge in female chacma baboons. (a) The mean duration that dominant and subordinate subjects looked towards the speaker following playback of sequences that included the calls of both of their close relatives (Test condition) or neither of the subjects' relatives (NK condition). Histograms show means ± SD for dominant and subordinate subjects in each condition. (b) The mean duration that dominant and subordinate subjects looked towards each other following playback of sequences that included the calls of both their close relatives (Test condition) or neither of the subjects' relatives (NK condition). (c) The proportion of subjects' first interactions with their partners that took different forms following each trial condition. DS, dominant subject supplanted her subordinate partner or, DA, approached and interacted in a friendly manner with her subordinate partner, SA, subordinate subject approached her dominant partner and interacted in a friendly manner. *Source.* From Cheney and Seyfarth (1999). Reproduced with permission of Elsevier. *Photo source.* © Alecia Carter.

1998; Penn and Povinelli 2007; Cheney and Seyfarth 2008). Experiments with several other primates, including capuchin monkeys, have found no evidence that individuals are sensitive to the extent of each other's knowledge (Hare *et al.* 2003) and one interpretation of these contrasts is that they reveal the unusual cognitive development of chimpanzees.

So why do cognitive abilities vary so widely between species? Evidence of advanced cognitive abilities in

several large-brained mammals (including primates, cetaceans and elephants) suggests that cognitive development is associated with relative brain size (Dunbar 1992; Dunbar and Shultz 2007), though different cognitive abilities are not always closely correlated either with each other or with brain size and the strength of correlations differs between taxa (Shultz and Dunbar 2007; MacLean *et al.* 2013). Brain size, in turn, has been shown to be correlated to a variety of ecological and behavioural parameters, including differences in diet, manipulative skills, tool use and innovativeness (Clutton-Brock and Harvey 1980; Mace *et al.* 1980; Reader and Laland 2002; Hartwig *et al.* 2011) as well as with variation in alloparental care (Isler and van Schaik 2012), group size and measures of social complexity (see Chapter 2). Unfortunately, many of these traits are inter-correlated, making it difficult to identify which of them is responsible for the evolution of differences in brain size and, not all mammals possessing relatively large brains or showing well-developed cognitive abilities are social (Chalmeau *et al.* 1997; Fox *et al.* 1999). It seems unlikely that the evolution of cognitive abilities or brain size depends on any single set of ecological or social variables and more probable that a range of different environmental scenarios requires individuals to integrate different kinds of information and to make complex opportunistic decisions requiring improved cognitive abilities and increases in brain development.

A final question is why social awareness and cognitive abilities are so highly developed in primates. Relative to other groups of mammals, primates have relatively large brains (Martin 1990), rely extensively on extractive foraging and appear to be adept at recognising and classifying items according to their relationships to each other (Tomasello and Call 1997; Shettleworth 2010). One possible answer is that a combination of large group size, stability of group membership and low average in relatedness between group members may increase the complexity of social relationships among group members and, in conjunction with extractive foraging, this may have led to selection favouring an understanding of cause and effect, improved cognitive abilities and enhanced brain size (see Chapter 19).

6.6 Individual differences and personality

Watch any group of individually recognisable mammals closely for any period of time and you start to be struck by the presence of consistent differences in behaviour between them. Some are shy, others bold; some are frequently aggressive while others avoid agonistic interactions; some are frequently involved in affiliative or cooperative interactions while others less frequently interact with other group members. Part of this variation is related to differences in age, sex and social rank, but,

after these effects are allowed for, there are still large individual differences in behaviour which are often consistent over time (Sih *et al.* 2004; Komdeur 2006; English *et al.* 2010). For example, studies of spotted hyenas show that individuals vary in assertiveness, excitability, human-directed behaviour, sociability and curiosity and these differences are not a consequence of contrasts in sex or age (Gosling 1998). Similar differences have been demonstrated in primates (Capitanio 1999, 2011a; Gosling and John 1999), dogs (Jones and Gosling 2005) and ungulates (Bergvall *et al.* 2011).

Systematic comparisons of the extent and structure of personality differences across species are not yet available. However, there are suggestions that contrasts in personality may be particularly highly developed in the great apes, and that personality differences among chimpanzees show affinities with those found in humans (Weiss *et al.* 2012).

Individual differences in particular forms of behaviour are often inter-correlated, so that individuals show contrasting syndromes of behaviour or 'personalities' (Carere and Eens 2005). In wild female baboons, Seyfarth and his co-workers (Seyfarth *et al.* 2012) identified three distinctive dimensions of personality: 'nice' individuals (who were relatively friendly to other females and responded benignly to grunts from lower-ranking individuals); 'aloof' animals, who were less friendly and responded most frequently to grunts from higher-ranking individuals; and 'loners', who were relatively unfriendly and were often on their own. These differences were associated with indices of sociality: 'nice' females had high sociality scores whereas 'aloof' females had lower sociality scores but more stable partner preferences and 'loners' had lower sociality scores, unstable partner preferences and higher glucocorticoid levels. Studies of captive male rhesus macaques also show that individuals vary widely in their sociability (Capitanio 1999) and that relatively unsociable animals can be further divided into 'loners' that appear to be satisfied with their lot and 'dissatisfied' individuals which attempt to establish closer social relationships with other group members but are often rejected (Capitanio *et al.* 2014; Hawkley and Capitanio 2015).

Contrasts in personality are associated with differences in other behavioural responses and processes. For example, in rodents, relatively aggressive individuals show more active responses to defeats, controllable shocks and uncontrollable tasks than less aggressive ones, which often show a tendency to withdraw (Benus *et al.* 1991). They may also affect the rate of social learning. For example, in chacma baboons, there are consistent differences in the rate of social learning between personality types: anxious animals are more likely to show higher rates of improvement in task solving after watching a demonstrator than less anxious individuals (Carter *et al.* 2014).

Personality differences are frequently associated with contrasts in social status, growth and development, cortisol levels and responses to social stress (Capitanio *et al.* 2004; Réale *et al.* 2007; Biro and Stamps 2008; Smith and Blumstein 2008) and can also be correlated with variation in feeding behaviour, health and susceptibility to infectious diseases (Bergvall *et al.* 2011; Capitanio 2011b; Capitanio and Cole 2015; Kappeler *et al.* 2015). For example, studies of captive rhesus macaques show that individuals which have low levels of sociability and appear to be dissatisfied with their status are particularly susceptible to disease and show higher levels of lymphoid tissue innervation and nerve growth factor, indicating that differences have a molecular basis (Sloan *et al.* 2008; Capitanio *et al.* 2014).

In natural populations, differences in personality are commonly associated with variation in survival, fecundity, rearing success and longevity (Réale *et al.* 2007; Biro and Stamps 2008; Smith and Blumstein 2008). For example, recent research has demonstrated correlations between components of fitness or fitness-related traits and individual differences in aggressiveness (Boon *et al.* 2008), boldness (Fairbanks *et al.* 2004a; Monestier *et al.* 2015), predator responses (Runyan and Blumstein 2004), courtship (Scuett *et al.* 2010), sociality (Silk 2003) and parental behaviour (Altmann 2001). In some cases, there is evidence of contrasting effects on different components of fitness. For example, in red squirrels, more active females have offspring that grow faster and are more likely to bequeath them part of their territory but show lower rates of overwinter survival than less active females (Boon *et al.* 2007, 2008). The benefits of differences in personality may also vary with environmental conditions. For example, a study of chipmunks, found that bolder individuals had higher reproductive success when food availability is low but lower success when food was plentiful (Lecoeur *et al.* 2015). Changes in personality with age may also reflect changes in reproductive opportunities: for example, in mouse lemurs, older males which have a high probability of breeding successfully are bolder than young males who

have a lower chance of breeding in the near future but high probability of future fitness (Dammhahn 2012).

Recent research has started to explore the causes of individual differences in behaviour and personality. Effects of the prenatal environment are often important (see Chapter 5) and early interactions with parents, peers and other adults can also have long-lasting effects on subsequent development and adult behaviour (see section 6.3). For example, rapid early growth may lead to increased risk-taking by faster-growing individuals which may improve their breeding success at some cost to their survival (Stamps 2007; Smith and Blumstein 2010). Individual differences in personality can also have a genetic basis (van Oers *et al.* 2005; Wilson and Réale 2006) and individual differences in aggressiveness, infanticidal behaviour, maternal behaviour, sociability and cooperation among mammals have all been shown to be heritable (Svare *et al.* 1984; Benus *et al.* 1991; Perrigo *et al.* 1993; Peripato and Cheverud 2002; Fairbanks *et al.* 2004b). Correlations between traits can also be heritable: for example, selection experiments on mice show that males selected for high nest-building behaviour were more aggressive and had shorter attack latencies than those selected for low nest-building behaviour (Sluyter *et al.* 1995).

Contrasts in personality may represent adaptive responses to variation in environmental or social conditions and trade-offs between the effects of different personality traits may maintain variation in behaviour in natural populations (Benus *et al.* 1991; Carere and Eens 2005; Dingemanse *et al.* 2007). For example, in great tits, less exploratory females survive better in years when food availability is low but show reduced survival when it is high, and fluctuating environmental conditions may maintain heritable variation in exploratory behaviour in populations (Dingemanse *et al.* 2001). Differences in personality may also adjust the behaviour of individuals to their capacity to exploit resources or compete for breeding opportunities and 'state-dependent' models show that processes of this kind are capable of maintaining consistent variation within populations (Dall *et al.* 2004). Another possibility is that contrasts in personality allow individuals to adopt contrasting social roles and exploit different social niches (a process usually referred to as *social niche specialisation*) and that variation is maintained by frequency-dependent selection (Bergmüller and Taborsky 2010; Wolf and McNamara 2012; Montiglio *et al.* 2013). However, as yet, there is little concrete evidence for effects of this kind in social mammals, possibly because stochastic factors usually exert such a strong effect on the eventual status of individuals that selection may not favour 'prospective' adjustments of development (Carter *et al.* 2014).

While the possibility that contrasts in personality reflect adaptive variation in behaviour has attracted much attention, personality differences can also be non-adaptive by-products of the effects of environmental challenges on perturbations or development. For example, both starvation and acute social stress have effects on the physical and behavioural development of individuals that are associated with reductions rather than increases in their fitness and it would not be surprising if they generated contrasts in behaviour, too. Both adaptive and non-adaptive contrasts in personality may well be common and, as yet, it is usually difficult to distinguish clearly between them or to assess their relative frequency in natural populations (Dingemanse and Réale 2005; Trillmich and Hudson 2011). Neither evidence that personality traits are heritable nor demonstrations that they are associated with components of fitness necessarily indicate that they represent adaptive responses, for adverse environments commonly generate variation in fitness which is partly mediated by genetic differences. As in the case of maternal effects, direct evidence that differences in personality are adaptive either requires evidence that personality traits which are more frequent in one environment than another cause individuals to have higher fitness in the first environment than the second, or that changes in selection pressures lead to predictable changes in the relative frequency of different personality traits (see Chapter 5). Producing evidence of either kind in mammals under conditions that approximate to natural faces major logistical obstacles and is seldom feasible, while non-experimental studies suffer from the difficulty that they are seldom able to determine whether correlations between personality traits and fitness have a causal basis.

6.7 Traditions

Where social learning plays an important role in the development of behaviour, contrasts in the experience of individuals can establish group-specific differences in behaviour that are learned by subsequent generations, leading to persistent differences in behaviour between groups or subpopulations which are not necessarily correlated with variation in current environmental factors or with contrasts in genotype (Galef 2009; Whiten 2012; Hoppitt and Laland 2013). Zoologists often refer to variation of this kind as 'cultural' to emphasise its similarity to human culture but some anthropologists and sociologists prefer to define culture in ways that restrict its application to humans and this has generated arguments about the existence of culture in non-human animals (McGrew 1998; Castro and Toro 2003; Laland and Hoppitt 2003; Laland and Galef 2009). To avoid confusion, I refer to consistent differences in behaviour between groups or populations that are a consequence of some form of social learning and are not a direct consequence of environmental or genetic differences as 'traditions' rather than 'cultural behaviour' although, like most other zoologists, I regard the two as synonymous.

Many of the best-documented examples of traditions in natural populations of non-human animals involve variation in feeding or foraging behaviour. One of the first detailed descriptions of traditional behaviour in mammals was of the spread and maintenance of washing food (sweet potatoes) to remove sand in provisioned groups of Japanese macaques (Kawai 1965) (Figure 6.13). Persistent differences in foraging behaviour or food handling techniques between groups or local populations that persist across generations have now been documented in a wide range of mammals, including chimpanzees (Boesch and Tomasello 1998; Lycett *et al.* 2007), orangutans (van Schaik and Pradhan 2003), vervet monkeys (van de Waal *et al.* 2013), capuchin monkeys (Phillips 1998; Moura and Lee 2004), macaques (Sinha 2005), killer whales (Yurk *et al.* 2002), sperm and humpback whales and dolphins (Whitehead and Rendell 2004). In some species, there are also traditional differences in the timing of foraging. For example, in wild meerkats, there are consistent inter-group differences in the time at which individuals start to forage that appear to be vertically transmitted and are not associated with contrasts in food selection or foraging success (Thornton *et al.* 2010). Some long-term studies of particular populations have been able to document the origin of new traditions and shown that their diffusion through local populations follows the pattern of social networks. For example, studies of humpback whales, which often use curtains of bubbles to concentrate fish schools, have shown how the spread of a new

(a)

(b)

Figure 6.13 (a) Japanese macaques on Koshima Island foraging on the beach. (b) Japanese macaque eating a sweet potato after washing it in the ocean, Kojima, Japan. *Sources*: (a) © Nobu Kutsukake. (b) © Cyril Ruoso/Minden Pictures/Getty Images.

technique – lobtail feeding (Figure 6.14) – coincided with social networks (Allen *et al.* 2013).

Contrasting traditions are also particularly well developed in the use of tools to catch or process food. Some of the most extensive evidence of variation in tool use comes from studies of chimpanzees (Whiten *et al.* 1999; Boesch 2003; Whiten 2005). Surveys of different communities of chimpanzees in the Congo Basin show that communities differ both in the form of their tools and in the materials used to make them

Figure 6.15 Chimpanzees in Bossou (Guinea) frequently use hammer-stones to crack hard fruit. *Source:* © Kathelijne Koops.

Figure 6.14 Some populations of humpback whales in the Gulf of Maine have developed a technique of feeding known as lobtail feeding, which consists of striking the water surface one to four times (a), followed by the use of bubbles to concentrate fish schools (b) and subsequently swimming through the school to catch their prey (c) (see Allen *et al.* 2013). *Source:* © Jennifer Allen.

(Sanz *et al.* 2004) and patterns of tool use also vary on a wider geographical scale (Whiten *et al.* 1999; Lycett *et al.* 2007). For example, chimpanzees in several West African sites regularly use hammer-stones to crack hard fruit, while those in Kibale Forest and Gombe in East Africa seldom if ever do so, though they commonly modify twigs to fish for ants (Whiten *et al.* 1999) (Figure 6.15). Variation in traditions of tool use is also common in other animals. In wild capuchin monkeys living in dry forest habitats, individuals use sticks (which they commonly modify) to probe for insects, and leaves as cups to retrieve water from tree cavities (Phillips 1998; Moura and Lee 2004). Like chimpanzees, they also use hammer-stones to crack hard nuts (Figure 6.16) and appear to have a well-developed understanding of the properties of different tools and their capacity to crack nuts of varying hardness (Fragaszy *et al.* 2010). Traditions of tool use have also developed in some cetaceans. For example, in one population of bottlenose dolphins, members of some matrilines use sponges as foraging tools and genetic analysis shows that this technique is transmitted from mothers to daughters (Krützen *et al.* 2005) (Figure 6.17).

While the frequency and extent of tool use and the types of tool used differ widely between populations (Humle and Matsuzawa 2002), it is often difficult to be certain that natural variation in environmental conditions plays no part in the development of these

Figure 6.16 Capuchin monkeys in dry forests use stones to crack hard nuts. *Source*: © Noemi Spagnoletti.

differences. Individual learning also often appears to be involved and studies of the development of tool use show that individuals can spontaneously manufacture and use tools without contact with adults of their own species (Kenward *et al.* 2005) and hone their skills by sophisticated processes of trial and error (Fragaszy *et al.* 2013).

Behavioural traditions occur in many other contexts (Byrne *et al.* 2004; McGrew 2004). Some of the best evidence of traditional variation in mammalian behaviour comes from comparisons of communication. For example, in chimpanzees, the use of particular postures and clasps in grooming interactions differs between populations (McGrew 1998, 2004) (Figure 6.18) while members of some groups of capuchin monkeys have developed elaborate games that differ between groups and may serve to test the strength of social bonds (see Chapter 9). Studies of vocal communication also emphasise the importance of traditional differences. For example, in killer whales (where members of both sexes typically remain in their natal clan throughout their lives), different clans have distinct vocal dialects (Yurk *et al.* 2002). In sperm whales, too, different clans have distinct vocal repertoires which are probably transmitted from parents to their offspring (Whitehead 1998; Rendell

Figure 6.17 In the Shark Bay dolphin population in Western Australia, members of some matrilines use sponges as foraging tools. *Source*: © Simon Allen, The Dolphin Innovation Project.

Figure 6.18 Behavioural variation in wild chimpanzees: while social grooming is common in all populations of chimpanzees, the hand-clasp shown in this picture is used only in some populations. *Source*: © Ian Gilby.

and Whitehead 2003; Whitehead and Rendell 2004) while in humpback whales, males belonging to the same population sing songs that conform to the version current in their population and new song types spread in a unidirectional manner, like cultural ripples (Garland *et al.* 2011). Traditional differences can also arise in agonistic behaviour and the effects of social rank (Sapolsky and Share 2004; Sinha 2005). For example, in a troop of olive baboons, the death of the most aggressive males from tuberculosis left a cohort of unusually unaggressive survivors and subsequent recruits to the same group adopted a similar pacific approach to social interactions (Sapolsky and Share 2004).

A number of studies have explored the behavioural mechanisms underlying the development of traditions (van Leeuwen and Haun 2013, 2015). Horizontal transmission is often involved and the distribution of traditions is often closely correlated with variation in the

frequency and proximity of association between individuals. For example, the spread of socially learned begging behaviour within a population of bottlenose dolphins was closely related to the structure of social networks and individuals preferentially associated with individuals that used the same socially learned foraging techniques (Cantor and Whitehead 2013). Similarly, the spread of the use of moss sponges to collect water by wild chimpanzees coincided with changes in the structure of social networks (Hobaiter *et al.* 2014). Experiments with wild meerkats that involved training one animal in each group to remove food from an apparatus when it was out of the sight of other group members and then retesting in their presence show that rapid horizontal transmission can generate local traditions for using particular techniques (Thornton and Malapert 2009; Thornton and Clutton-Brock 2011). In wild vervet monkeys, individuals were offered artificial foods coloured either pink or blue and were trained to avoid pink foods in one group and to avoid blue foods in the other, with the result that individuals which changed groups quickly switched their preferences to the new local norm (van de Waal *et al.* 2013) (Figure 6.19).

Vertical transmission is also common (Whiten 1989). For example, experiments with captive groups of rats, where all group members were taught an arbitrary food preference and experienced individuals were gradually replaced with naive individuals, show that individuals adopt the practices of their predecessors (Galef and Allen 1995). In many groups of animals, the same traditions persist in the same groups across generations and can exert an important influence on habitat use and ranging behaviour. For example, in chimpanzees, specific techniques of tool use are often transferred from mother to daughter (Lonsdorf *et al.* 2004), while in red deer on Rum one female learned how to access the island graveyard to graze and so too did her offspring and grand-offspring (Clutton-Brock *et al.* 1982). Similarly, in wild chimpanzees, adult males adopt foraging ranges that coincide with those of their mother at the time when they were dependant juveniles (Murray *et al.* 2006) (see Figure 6.2).

Traditions may help to increase the probability that inexperienced individuals will develop responses that are well adapted to local conditions (Galef 1995, 1996), though firm evidence of fitness benefits associated with traditional behaviour is scarce. In addition,

Figure 6.19 Experimental set-up illustrating preferential foraging in vervet monkeys. Maize corn dyed either pink or blue was provided intermittently in two adjacent containers. The photograph shows an infant sitting on the colour earlier made distasteful to its mother, as it eats the colour currently preferred by its mother and the rest of the group. *Source*: From van de Waal *et al.* (2013). Reproduced with permission from AAAS.

rapid changes in environmental or social conditions may also lead to situations where traditional responses are non-adaptive (Hoppitt and Laland 2013). Inappropriate traditions should consequently be quickly abandoned and experiments with food preferences in captive rats suggest that this is often the case (Galef and Whiskin 1998).

Traditional behaviour may be most likely to develop when traditional innovation is common: for example, across primate species, the frequency of behavioural innovation and social learning appear to be correlated with each other (Figure 6.20). Some studies suggest that traditions are unusually common among cetaceans and primates but evidence is sparse and there is a danger that attempts to investigate the frequency of behavioural traditions have focused selectively on long-lived species while systematic investigations of the relative extent of traditional behaviour in non-primate mammals are not yet available (Thornton and Clutton-Brock 2011).

Since traditions commonly affect ranging behaviour, they can have important consequences for the frequency with which individuals associate and the structure of populations (Cantor and Whitehead 2013). In bottlenose dolphin populations, for example, individuals preferentially associate with animals that have learned the same foraging skills (Mann *et al.* 2012). The establishment of traditions may increase the isolation of local populations, contributing to contrasts in gene frequency as a result of selection and drift (West-Eberhard 2003, Richerson and Boyd 2008).

In some cases, traditions may enable populations to persist in changing environments, while in others they may allow them to occupy new ones (Crispo 2007; Badyaev 2009). They may also affect (and be affected by) selection pressures operating on specific traits and may have an important influence on particular propensities – a process known as gene-culture co-evolution (Feldman *et al.* 1985; Richerson and Boyd 2005) that has probably played a central role in human evolution (see Chapter 19 and 20). In addition, where traditions represent adaptive responses to particular environments, underlying genetic propensities to perform the specific behaviours involved may become

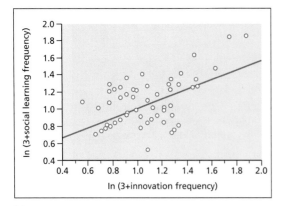

(a)

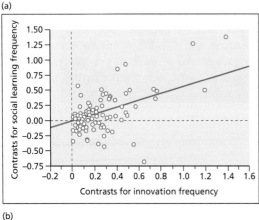

(b)

Figure 6.20 Frequencies of innovation and social learning, corrected for research effort, co-vary across non-human primates: (a) raw data, with points representing one species; (b) independent contrasts. *Source*: From Reader and Laland (2002). © 2002 National Academy of Sciences, USA.

canalised through selection operating on the developmental system leading to the evolution of heritable

responses, a process known as *genetic assimilation* (Waddington 1953).

Traditional behaviour can also have important consequences for the dynamics and management of populations. In many social mammals, individuals probably learn territory boundaries, escape routes, foraging paths and techniques and migration routes from other members of their group (Byrne 2000; Garber 2000; Garber *et al.* 2009). In migratory species, these traditions may affect the timing of seasonal movements, the routes used and the location of wintering or breeding grounds (Whitehead and Weilgart 2000). Natural ecological processes or changes in management regimes that alter the access of individuals to traditional ranges or to routes between them may consequently have substantial effects on reproduction and survival, and may even reduce the viability of populations (Caro 1998, 2007; Tuyttens and Macdonald 2000; Gosling 2003).

The division of local population units by traditional boundaries is also likely to have important demographic consequences by restricting colonisation of vacant habitat and population growth (Durant 2000; Woodroffe 2003) so that populations are slow to re-occupy ranges left vacant by groups that have disbanded or died out as a result of stochastic processes. Divisions of this kind also have important implications for management. For example, in spatially subdivided populations of deer, maximum sustainable yields, based on overall productivity, may overestimate potential yields in subpopulations whose fecundity is lower than the average. As a result, the imposition of culls based on average population productivity is likely to reduce less productive subpopulations and may eventually extinguish them (Clutton-Brock and Albon 1989).

SUMMARY

1. One of the consequences of sociality is that individuals can learn responses and skills from observing or interacting with other group members. While social learning enables individuals to avoid the costs of trial and error learning, it can have disadvantages if social or ecological conditions change and individuals learn inappropriate responses.

2. Sociality also provides opportunities for parents or helpers to teach juveniles particular responses or skills. Recent studies of carnivores provide evidence that adults teach juveniles to catch and dismember prey but firm examples of other forms of teaching are scarce.

3. Variation in the social and ecological environment of neonates and their mothers often has effects on the development of their behaviour that can persist throughout the lives of individuals, affecting their survival and eventual breeding success.

4. In many social animals (and especially in primates), individuals are aware of interactions and relationships between third parties. However, evidence that individuals are aware of the extent of knowledge that others possess (a 'theory of mind') is limited and there appear to be both qualitative and quantitative differences in social awareness between humans and other animals.

5. Individuals differ in their responses to social and ecological challenges and show contrasting personalities. The personality of individuals often has some genetic basis and differences are frequently associated with variation in components of fitness.

6. Social learning is often responsible for the development of traditions or cultural differences between groups. Cultural differences in foraging or feeding skills are often well developed and include pronounced differences in the use of tools. Traditional behaviour can accelerate learning and reduce costs associated with trial and error and can lead to interactions between selection and traditional behaviour that have probably played a particularly important role in the evolution of human social behaviour.

References

Allen, J., et al. (2013) Network-based diffusion analysis reveals cultural transmission of lobtail feeding in humpback whales. *Science* **340**:485–488.

Altmann, J. (2001) *Baboon Mothers and Infants*. Chicago: University of Chicago Press.

Badyaev, A.V. (2009) Evolutionary significance of phenotypic accommodation in novel environments: an empirical test of the Baldwin effect. *Philosophical Transactions of the Royal Society B. Biological Sciences* **364**:1125–1141.

Barnard, C.J., et al. (1998) Maternal effects on the development of social rank and immunity trade-offs in male laboratory mice (*Mus musculus*). *Proceedings of the Royal Society of London. Series B: Biological Sciences* **265**:2087–2093.

Barrett, L., et al. (1992) Environmental influences on play behaviour in immature gelada baboons. *Animal Behaviour* **44**:111–115.

Bates, L.A., et al. (2008) African elephants have expectations about the locations of out-of-sight family members. *Biology Letters* **4**:34–36.

Bateson, P.P.G. (1979) How do sensitive periods arise and what are they for? *Animal Behaviour* **27**:470–486.

Bateson, P.P.G. and Gluckman, P. (2011) *Plasticity, Robustness, Development and Evolution*. Cambridge: Cambridge University Press.

Bekoff, M. (1972) The development of social interaction, play, metacommunication in mammals: an ethological perspective. *Quarterly Review of Biology* **47**:412–434.

Bekoff, M. (1977) Mammalian dispersal and the ontogeny of individual behavioral phenotypes. *American Naturalist* **111**:715–732.

Bekoff, M. and Allen, C. (1998) Intentional communication and social play: how and why animals negotiate and agree to play. In: *Animal Play: Evolutionary, Comparative, and Ecological Perspectives* (eds M. Bekoff and J.A. Byers). Cambridge: Cambridge University Press, 97–114.

Benus, R.F., et al. (1991) Behavioural differences between artificially selected aggressive and non-aggressive mice: response to apomorphine. *Behavioural Brain Research* **43**:203–208.

Berghanel, A., et al. (2015) Locomotor play drives motor skill acquisition at the expense of growth: a life history trade-off. *Science Advances* **1**:e1500451.

Bergman, T.J., et al. (2003) Hierarchical classification by rank and kinship in baboons. *Science* **302**:1234–1236.

Bergmüller, R. and Taborsky, M. (2010) Animal personality due to social niche specialisation. *Trends in Ecology and Evolution* **25**:504–511.

Bergvall, U.A., et al. (2011) Personality and foraging decisions in fallow deer, *Dama dama*. *Animal Behaviour* **81**:101–112.

Biro, D., et al. (2003) Cultural innovation and transmission of tool use in wild chimpanzees: evidence from field experiments. *Animal Cognition* **6**:213–223.

Biro, P.A. and Stamps, J.A. (2008) Are animal personality traits linked to life-history productivity? *Trends in Ecology and Evolution* **23**:361–368.

Blumstein, D.T., et al. (2013) Early play may predict later dominance relationships in yellow-bellied marmots (*Marmota flaviventris*). *Proceedings of the Royal Society of London. Series B: Biological Sciences* **280**:20130485.

Boesch, C. (2003) Is culture a golden barrier between human and chimpanzee? *Evolutionary Anthropology: Issues, News, and Reviews* **12**:82–91.

Boesch, C. and Tomasello, M. (1998) Chimpanzee and human cultures. *Current Anthropology* **39**:591–614.

Boon, A.K., et al. (2007) The interaction between personality, offspring fitness and food abundance in North American red squirrels. *Ecology Letters* **10**:1094–1104.

Boon, A.K., et al. (2008) Personality, habitat use, and their consequences for survival in North American red squirrels *Tamiasciurus hudsonicus*. *Oikos* **117**:1321–1328.

Boughman, J.W. (1998) Vocal learning by greater spear-nosed bats. *Proceedings of the Royal Society of London. Series B: Biological Sciences* **265**:227–233.

Boyd, R. and Richerson, P.J. (1985) *Culture and the Evolutionary Process*. Chicago: University of Chicago Press.

Burghardt, G.M. (2005) *The Genesis of Animal Play: Testing the Limits*. Cambridge, MA: MIT Press.

Byers, J.A. and Walker, C. (1995) Refining the motor training hypothesis for the evolution of play. *American Naturalist* **146**:25–40.

Byrne, R.W. (2000) How monkeys find their way: leadership, coordination, and cognitive maps of African baboons. In: *On the Move: How and Why Animals Travel in Groups* (eds S. Boinski and P.A. Garber). Chicago: University of Chicago Press, 491–518.

Byrne, R.W. and Whiten, A. (1992) Cognitive evolution in primates: evidence from tactical deception. *Man* **27**:609–627.

Byrne, R.W., *et al.* (2004) Understanding culture across species. *Trends in Cognitive Sciences* **8**:341–346.

Cantor, M. and Whitehead, H. (2013) The interplay between social networks and culture: theoretically and among whales and dolphins. *Philosophical Transactions of the Royal Society B: Biological Sciences* **368**:20120340.

Capitanio, J.P. (1999) Personality dimensions in adult male rhesus macaques: prediction of behaviors across time and situation. *American Journal of Primatology* **47**:299–320.

Capitanio, J.P. (2011a) Individual differences in emotionality: social temperament and health. *American Journal of Primatology* **73**:507–515.

Capitanio, J.P. (2011b) Health and social relationships in non-human primates: toward a comparative health psychology. In: *Oxford Handbook of Health Psychology* (ed. H.S. Friedman). New York: Oxford University Press, 860–883.

Capitanio, J.P., *et al.* (2004) Personality characteristics and basal cortisol concentrations in adult male rhesus macaques (*Macaca mulatta*). *Psychoneuroendocrinology* **29**:1300–1308.

Capitanio, J.P., *et al.* (2014) A behavioral taxonomy of loneliness in humans and rhesus monkeys (*Macaca mulatta*). *PLOS ONE* **9**(10): e110307.

Capitanio, J.P. and Cole, S.W. (2015) Social instability and immunity in rhesus monkeys: the role of the sympathetic nervous system. *Philosophical Transactions of the Royal Society B: Biological Sciences* **370** (1669), doi: 10.1098/rstb.2014.0104. [Accessed 10 January 2016].

Carere, C. and Eens, M. (2005) Unravelling animal personalities: how and why individuals consistently differ. *Behaviour* **142**:1149–1157.

Caro, T.M. (1994) *Cheetahs of the Serengeti Plains: Group Living in an Asocial Species*. Chicago: University of Chicago Press.

Caro, T.M. (1998) *Behavioral Ecology and Conservation Biology*. Oxford: Oxford University Press.

Caro, T.M. (2007) Behavior and conservation: a bridge too far? *Trends in Ecology and Evolution* **22**:394–400.

Caro, T.M. and Hauser, M.D. (1992) Is there teaching in non-human animals? *Quarterly Review of Biology* **67**:151–174.

Carter, A.J., *et al.* (2014) Personality predicts the propensity for social learning in a wild primate. *PeerJ* **2**:e283.

Castro, L. and Toro, M.A. (2003) The evolution of culture: from primate social learning to human culture. *Proceedings of the National Academy of Sciences of the United States of America* **101**:10235–10240.

Chalmeau, R., *et al.* (1997) Cooperative problem solving by orangutans (*Pongo pygmaeus*). *International Journal of Primatology* **18**:23–32.

Chalmers, N.R. (1984) Social play in monkeys: theories and data. In: *Play in Animals and Humans* (ed. P.K. Smith). Oxford: Basil Blackwell, 119–141.

Cheney, D.L. and Seyfarth, R.M. (1990) *How Monkeys See the World: Inside the Mind of Another*. Chicago: University of Chicago Press.

Cheney, D.L. and Seyfarth, R.M. (1999) Recognition of other individuals' social relationships by female baboons. *Animal Behaviour* **58**:67–75.

Cheney, D.L. and Seyfarth, R.M. (2008) *Baboon Metaphysics: The Evolution of a Social Mind*. Chicago: University of Chicago Press.

Clutton-Brock, T.H. and Albon, S.D. (1989) *Red Deer in the Highlands*. Oxford: Blackwell Scientific Publications.

Clutton-Brock, T.H. and Harvey, P.H. (1980) Primates, brains and ecology. *Journal of Zoology* **190**:309–323.

Clutton-Brock, T.H., *et al.* (1982) *Red Deer: The Behaviour and Ecology of Two Sexes*. Chicago: University of Chicago Press.

Cook, M. and Mineka, S. (1990) Selective associations in the observational conditioning of fear in rhesus monkeys. *Journal of Experimental Psychology: Animal Behavior Processes* **16**:372–389.

Coussi-Korbel, S. and Fragaszy, D.M. (1995) On the relation between social dynamics and social learning. *Animal Behaviour* **50**:1441–1453.

Creel, S., *et al.* (2013) The ecology of stress: effects of the social environment. *Functional Ecology* **27**:66–80.

Crispo, E. (2007) The Baldwin effect and genetic assimilation: revisiting two mechanisms of evolutionary change mediated by phenotypic plasticity. *Evolution* **61**:2469–2479.

Crockford, C., *et al.* (2012) Wild chimpanzees inform ignorant group members of danger. *Current Biology* **22**:142–146.

Dall, S.R.X., *et al.* (2004) The behavioural ecology of personality: consistent individual differences from an adaptive perspective. *Ecology Letters* **7**:734–739.

Dammhahn, M. (2012) Are personality differences in a small iteroparous mammal maintained by a life-history trade-off? *Proceedings of the Royal Society of London. Series B: Biological Sciences* **279**:2645–2651.

Dantzer, B., *et al.* (2013) Density triggers maternal hormones that increase adaptive offspring growth in a wild mammal. *Science* **340**:1215–1217.

Delville, Y., *et al.* (1998) Behavioral and neurobiological consequences of social subjugation during puberty in golden hamsters. *Journal of Neuroscience* **18**:2667–2672.

Dingemanse, N.J. and Réale, D. (2005) Natural selection and animal personality. *Behaviour* **142**:1159–1184.

Dingemanse, N.J., *et al.* (2004) Fitness consequences of avian personalities in a fluctuating environment. *Proceedings of the Royal Society of London. Series B: Biological Sciences* **271**:847–852.

Dingemanse, N.J., *et al.* (2007) Behavioural syndromes differ predictably between 12 populations of three-spined stickleback. *Journal of Animal Ecology* **76**:1128–1138.

Donaldson, R., *et al.* (2012) The social side of human–wildlife interaction: wildlife can learn harmful behaviours from each other. *Animal Conservation* **15**:427–435.

Dunbar, R.I.M. (1992) Neocortex size as a constraint on group size in primates. *Journal of Human Evolution* **22**:469–493.

Dunbar, R.I.M. and Shultz, S. (2007) Evolution in the social brain. *Science* **317**:1344–1347.

Durant, S. (2000) Dispersal patterns, social organization and population viability. In: *Behaviour and Conservation* (eds L.M. Gosling and W.J. Sutherland). Cambridge: Cambridge University Press, 172–197.

Engh, A.L., *et al.* (2005) Patterns of alliance formation and postconflict aggression indicate spotted hyaenas recognize third-party relationships. *Animal Behaviour* **69**:209–217.

Engh, A.L., *et al.* (2006) Who, me? Can baboons infer the target of vocalizations? *Animal Behaviour* **71**:381–387.

English, S., *et al.* (2010) Consistent individual differences in cooperative behaviour in meerkats (*Suricata suricatta*). *Journal of Evolutionary Biology* **23**:1597–1604.

Ewer, R.F. (1969) The instinct to teach. *Nature* **222**:698.

Fagen, R. (1974) Selective and evolutionary aspects of animal play. *American Naturalist* **108**:850–858.

Fairbanks, L.A. (2000) The developmental timing of primate play: a neural selection model. In: *Biology, Brains and Behavior: The Evolution of Human Development* (eds S.T. Parker, J. Langer and M.L. McKinney). Santa Fe, NM: School of American Research Press, 131–158.

Fairbanks, L.A., *et al.* (2004a) Adolescent impulsivity predicts adult dominance attainment in male vervet monkeys. *American Journal of Primatology* **64**:1–17.

Fairbanks, L.A., *et al.* (2004b) Genetic contributions to social impulsivity and aggressiveness in vervet monkeys. *Biological Psychiatry* **55**:642–647.

Fox, E.A., *et al.* (1999) Intelligent tool use in wild Sumatran orangutans. In: *The Mentality of Gorillas and Orangutans* (eds S.T. Parker, R.W. Mitchell and H.L. Miles). Cambridge: Cambridge University Press, 99–116.

Fragaszy, D.M., *et al.* (2010) How wild bearded capuchin monkeys select stones and nuts to minimize the number of strikes per nut cracked. *Animal Behaviour* **80**:205–214.

Fragaszy, D.M., *et al.* (2013) The fourth dimension of tool use: temporally enduring artefacts aid primates learning to use tools. *Philosophical Transactions of the Royal Society B: Biological Sciences* **368**:20120410.

Franks, N.R. and Richardson, T. (2006) Teaching in tandem-running ants. *Nature* **439**:153.

Galef, B.G. (1995) Why behaviour patterns that animals learn socially are locally adaptive. *Animal Behaviour* **49**:1325–1334.

Galef, B.G. (1996) The adaptive value of social learning: a reply to Laland. *Animal Behaviour* **52**:641–644.

Galef, B.G. (2009) Culture in animals? In: *The Question of Animal Culture* (eds K.N. Laland and B.G. Galef). Cambridge, MA: Harvard University Press, 222–246.

Galef, B.G. and Allen, C. (1995) A new model system for studying behavioural traditions in animals. *Animal Behaviour* **50**:705–717.

Galef, B.G. and Giraldeau, L. (2001) Social influences on foraging in vertebrates: causal mechanisms and adaptive functions. *Animal Behaviour* **61**:3–15.

Galef, B.G. and Laland, K.N. (2005) Social learning in animals: empirical studies and theoretical models. *BioScience* **55**:489–499.

Galef, B.G. and Whiskin, E.E. (1998) Limits on social influence on food choices of Norway rats. *Animal Behaviour* **56**:1015–1020.

Galef, B.G. and Whiskin, E.E. (2004) Effects of environmental stability and demonstrator age on social learning of food preferences by young Norway rats. *Animal Behaviour* **68**:897–902.

Galef, B.G., *et al.* (2005) A new way to study teaching in animals: despite demonstrable benefits, rat dams do not teach their young what to eat. *Animal Behaviour* **70**:91–96.

Garber, P.A. (2000) Evidence for the use of spatial, temporal, and social information by primate foragers. In: *On the Move: How and Why Animals Travel in Groups* (eds S. Boinski and P.A. Garber). Chicago: University of Chicago Press, 261–298.

Garber, P.A., *et al.* (2009) Primate cognition: integrating social and ecological information in decision-making. In: *South American Primates: Comparative Perspectives in the Study of Behavior, Ecology, and Conservation* (eds P.A. Garber, A. Estrada, J.C. Bicca-Marques, E.W. Heymann and K.B. Strier). Developments in Primatology: Progress and Prospects. New York: Springer, 365–385.

Garland, E.C., *et al.* (2011) Dynamic horizontal cultural transmission of humpback whale song at the ocean basin scale. *Current Biology* **21**:687–691.

Giraldeau, L.-A., *et al.* (2002) Potential disadvantages of using socially acquired information. *Philosophical Transactions of the Royal Society B: Biological Sciences* **357**:1559–1566.

Gosling, L.M. (2003) Adaptive behavior and population viability. In: *Animal Behavior and Wildlife Conservation* (eds M. Festa-Bianchet and M. Apollonio). Washington, DC: Island Press, 13–30.

Gosling, S.D. (1998) Personality dimensions in spotted hyenas (*Crocuta crocuta*). *Journal of Comparative Psychology* **112**:107–118.

Gosling, S.D. and John, O.P. (1999) Personality dimensions in nonhuman animals. *Current Directions in Psychological Science* **8**:69–75.

Griffin, A.S. (2004) Social learning about predators: a review and prospectus. *Animal Learning and Behavior* **32**:131–140.

Griffin, A.S. and Evans, C.S. (2003) Social learning of anti-predator behaviour in a marsupial. *Animal Behaviour* **66**:485–492.

Hare, B., *et al.* (2000) Chimpanzees know what conspecifics do and do not see. *Animal Behaviour* **59**:771–785.

Hare, B., *et al.* (2003) Do capuchin monkeys, *Cebus apella,* know what conspecifics do and do not see? *Animal Behaviour* **65**:131–142.

Hartwig, W., *et al.* (2011) Relative brain size, gut size, and evolution in New World monkeys. *Anatomical Record* **294**:2207–2221.

Hawkley, L.C. and Capitanio, J.P. (2015) Perceived ritual isolation, evolutionary fitness and health outcomes: a lifespan approach. *Philosophical Transactions of the Royal Society B: Biological Sciences* **307**, doi: 10.1098/rstb.2014.0114. [Accessed 10 January 2016].

Heyes, C.M. (1998) Theory of mind in nonhuman primates. *Behavioral and Brain Sciences* **21**:101–114.

Heyes, C.M. (2001) Causes and consequences of imitation. *Trends in Cognitive Sciences* **5**:253–261.

Heyes, C.M. and Ray, E.D. (2000) What is the significance of imitation in animals? In: *Advances in the Study of Behavior*, Vol. 29 (eds P.J.B. Slater, J.S. Rosenblatt, C.T. Snowdon and T.J. Roper). San Diego, CA: Academic Press, 215–245.

Hobaiter, C., *et al.* (2014) Social network analysis shows direct evidence for social transmission of tool use in wild chimpanzees. *PLOS Biology* **12** (9): e1001960.

Hollén, L.I. and Manser, M.B. (2006) Ontogeny of alarm call responses in meerkats, *Suricata suricatta*: the roles of age, sex and nearby conspecifics. *Animal Behaviour* **72**:1345–1353.

Hollén, L.I., *et al.* (2008) Ontogenetic changes in alarm-call production and usage in meerkats (*Suricata suricatta*): adaptations or constraints? *Behavioral Ecology and Sociobiology* **62**:821–829.

Hoppitt, W. and Laland, K.N. (2013) *Social Learning: An Introduction to Mechanisms, Methods, and Models*. Princeton, NJ: Princeton University Press.

Hoppit, W.J.E., *et al.* (2008) Lessons from animal teaching. *Trends in Ecology and Evolution* **23**:486–493.

Humle, T. and Matsuzawa, T. (2002) Ant-dipping among the chimpanzees of Bossou, Guinea, and some comparisons with other sites. *American Journal of Primatology* **58**:133–148.

Isler, K. and van Schaik, C.P. (2012) Allomaternal care, life history and brain size evolution in mammals. *Journal of Human Evolution* **63**:52–63.

Janik, V.M. and Slater, P.J.B. (1997) Vocal learning in mammals. In: *Advances in the Study of Behavior*, Vol. **26** (eds P.J.B. Slater, C.T. Snowdon, J.S. Rosenblatt and M. Milinski). San Diego, CA: Academic Press, 59–99.

Janik, V.M. and Slater, P.J.B. (2000) The different roles of social learning in vocal communication. *Animal Behaviour* **60**:1–11.

Johnstone, R.A. and Cant, M.A. (2010) The evolution of menopause in cetaceans and humans: the role of demography. *Proceedings of the Royal Society of London. Series B: Biological Sciences* **277**:3765–3771.

Jones, A.C. and Gosling, S.D. (2005) Temperament and personality in dogs (*Canis familiaris*): a review and evaluation of past research. *Applied Animal Behaviour Science* **95**:1–53.

Kaiser, S. and Sachser, N. (2009) Effects of prenatal social stress on offspring development. *Pathology or adaptation? Current Directions in Psychological Science* **18**:118–121.

Kappeler, P.M., *et al.* (2015) Sociality and health: impacts on disease susceptability and transmission. *Philosophical Transactions of the Royal Society B: Biological Sciences* **370** (1669), DOI: 10.1098/rstb.2014.0116. [Accessed 10 January 2016].

Kawai, M. (1965) Newly-acquired pre-cultural behavior of the natural troop of Japanese monkeys on Koshima islet. *Primates* **6**:1–30.

Kenward, B., *et al.* (2005) Tool manufacture by naive juvenile crows. *Nature* **433**:121.

Klopfer, P.H. and Gamble, J. (1966) Maternal imprinting in goats: the role of chemical senses. *Zeitschrift fur Tierpsychologie* **23**:588–592.

Komdeur, J. (2006) Variation in individual investment strategies among social animals. *Ethology* **112**:729–747.

Krützen, M., *et al.* (2005) Cultural transmission of tool use in bottlenose dolphins. *Proceedings of the National Academy of Sciences of the United States of America* **102**:8939–8943.

Laland, K.N. and Galef, B.G. (2009) *The Question of Animal Culture*. Cambridge, MA: Harvard University Press.

Laland, K.N. and Hoppitt, W. (2003) Do animals have culture? *Evolutionary Anthropology: Issues, News, and Reviews* **12**:150–159.

Laland, K.N. and Williams, K. (1998) Social transmission of maladaptive information in the guppy. *Behavioral Ecology* **9**:493–499.

Le Coeur, C., *et al.* (2015) Temporarily fluctuating selection on a personality trait in a wild rodent population. *Behavourial Ecology* **26**:1285–1291.

Lee, P.C. (1984) Ecological constraints on the social development of vervet monkeys. *Behaviour* **91**:245–262.

Lonsdorf, E.V., *et al.* (2004) Sex differences in learning in chimpanzees. *Nature* **428**:715–716.

Lycett, S.J., *et al.* (2007) Phylogenetic analyses of behavior support existence of culture among wild chimpanzees. *Proceedings of the National Academy of Sciences of the United States of America* **104**:17588–17592.

McAuliffe, K. and Whitehead, H. (2005) Eusociality, menopause and information in matrilineal whales. *Trends in Ecology and Evolution* **20**:650.

McComb, K., *et al.* (2000) Unusually extensive networks of vocal recognition in African elephants. *Animal Behaviour* **59**:1103–1109.

McComb, K., *et al.* (2011) Vocal communication and social knowledge in African elephants. In: *The Amboseli Elephants: A Long-term Perspective on a Long-lived Mammal* (eds C.J. Moss, H. Croze and P.C. Lee). Chicago: University of Chicago Press, 162–173.

McComb, K., *et al.* (2014) Elephants can determine ethnicity, gender, and age from acoustic cues in human voices. *Proceedings of the National Academy of Sciences of the United States of America* **111**:5433–5438.

Mace, G.M., *et al.* (1980) Is brain size an ecological variable? *Trends in Neurosciences* **3**:193–196.

McGrew, W.C. (1998) Culture in nonhuman primates? *Annual Review of Anthropology* **27**:301–328.

McGrew, W.C. (2004) *The Cultured Chimpanzee*. Cambridge: Cambridge University Press.

MacKinnon, K.C. (2011) Social beginnings: the tapestry of infant and adult interactions. In: *Primates in Perspective* (eds C.J. Campbell, A. Fuentes, K.C. MacKinnon, S.K. Bearder and R.M. Stumpf). Oxford: Oxford University Press, 440–455.

MacLean, E.L., *et al.* (2013) Group size predicts social but not nonsocial cognition in lemurs. *PLOS ONE* **8**:e66359.

Mann, J., *et al.* (2012) Social networks reveal cultural behaviour in tool-using dolphins. *Nature Communications* **3**:980.

Martin, R.D. (1990) *Primate Origins and Evolution: A Phylogenetic Reconstruction*. Princeton, Princeton University Press.

Meaney, M.J. (2001) Maternal care, gene expression, and the transmission of individual differences in stress reactivity across generations. *Annual Review of Neuroscience* **24**:1161–1192.

Monestier, C., *et al.* (2015) Is a proactive mum a good mum? A mother's coping style influences early fawn survival in roe deer. *Behavioural Ecology* **26**:1395–1403.

Montiglio, P.-O., *et al.* (2013) Social niche specialization under constraints: personality, social interactions and environmental heterogeneity. *Philosophical Transactions of the Royal Society B: Biological Sciences* **368**:20120343.

Moore, A.J. (2004) Behavioural genetics: all in the family. *Nature* **429**:517–518.

Moss, C.J. and Lee, P.C. (2011) Female social dynamics: fidelity and flexibility. In: *The Amboseli Elephants: A Long-term Perspective on a Long-lived Mammal* (eds C.J. Moss, H. Croze and P.C. Lee). Chicago: University of Chicago Press, 205–223.

Moura, A.C. de A. and Lee, P.C. (2004) Capuchin stone tool use in Caatinga dry forest. *Science* **306**:1909.

Murray, C.M., *et al.* (2006) Foraging strategies as a function of season and rank among wild female chimpanzees (*Pan troglodytes*). *Behavioral Ecology* **17**:1020–1028.

Mutinda, H., *et al.* (2011) Decision making and leadership in using the ecosystem. In: *The Amboseli Elephants: A Long-term Perspective on a Long-lived Mammal* (eds C.J. Moss, H. Croze and P.C. Lee). Chicago: University of Chicago Press, 246–259.

Payne, K. (1999) The progressively changing songs of humpback whales: a window on the creative process in a wild animal. In: *The Origins of Music* (eds N.T. Wallin, B. Merker and S. Brown). Cambridge, MA: MIT Press, 135–150.

Peacock, M.M. and Jenkins, S.H. (1988) Development of food preferences: social learning by Belding's ground squirrels *Spermophilus beldingi*. *Behavioral Ecology and Sociobiology* **22**:393–399.

Pellegrini, A.D., *et al.* (2007) Play in evolution and development. *Developmental Review* **27**:261–276.

Pellis, S.M. and Iwaniuk, A.N. (1999) The roles of phylogeny and sociality in the evolution of social play in muroid rodents. *Animal Behaviour* **58**:361–373.

Penn, D.C. and Povinelli, D.J. (2007) On the lack of evidence that non-human animals possess anything remotely resembling a 'theory of mind'. *Philosophical Transactions of the Royal Society B: Biological Sciences* **362**:731–744.

Peripato, A.C. and Cheverud, J.M. (2002) Genetic influences on maternal care. *American Naturalist* **160**:S173–S185.

Perrigo, G., *et al.* (1993) Genetic mediation of infanticide and parental behavior in male and female domestic and wild stock house mice. *Behavior Genetics* **23**:525–531.

Phillips, K.A. (1998) Tool use in wild capuchin monkeys (*Cebus albifrons trinitatis*). *American Journal of Primatology* **46**:259–261.

Plotnik, J.M., *et al.* (2006) Self-recognition in an Asian elephant. *Proceedings of the National Academy of Sciences of the United States of America* **103**:17053–17057.

Proops, L., *et al.* (2009) Cross-modal individual recognition in domestic horses (*Equus caballus*). *Proceedings of the National Academy of Sciences of the United States of America* **106**:947–951.

Raihani, N.J. and Ridley, A.R. (2008) Experimental evidence for teaching in wild pied babblers. *Animal Behaviour* **75**:3–11.

Rapaport, L.G. (2011) Progressive parenting behavior in wild golden lion tamarins. *Behavioral Ecology* **22**:745–754.

Rapaport, L.G. and Brown, G.R. (2008) Social influences on foraging behavior in young nonhuman primates: learning what, where, and how to eat. *Evolutionary Anthropology* **17**:189–201.

Reader, S.M. and Laland, K.N. (2002) Social intelligence, innovation, and enhanced brain size in primates. *Proceedings of the National Academy of Sciences of the United States of America* **99**:4436–4441.

Réale, D., *et al.* (2007) Integrating animal temperament within ecology and evolution. *Biological Reviews* **82**:291–318.

Reiss, D. and Marino, L. (2001) Mirror self-recognition in the bottlenose dolphin: a case of cognitive convergence. *Proceedings of the National Academy of Sciences of the United States of America* **98**:5937–5942.

Rendell, L.E. and Whitehead, H. (2003) Vocal clans in sperm whales (*Physeter macrocephalus*). *Proceedings of the Royal Society of London. Series B: Biological Sciences* **270**:225–231.

Richerson, P.J. and Boyd, R. (2008) *Not by Genes Alone: How Culture Transformed Human Evolution*. Chicago: University of Chicago Press.

Runyan, A.M. and Blumstein, D.T. (2004) Do individual differences influence flight initiation distance? *Journal of Wildlife Management* **68**:1124–1129.

Sachser, N., *et al.* (2011) Adaptive modulation of behavioural profiles by social stress during early phases of life and adolescence. *Neuroscience and Biobehavioral Reviews* **35**:1518–1533.

Sachser, N., *et al.* (2013) Behavioural profiles are shaped by social experience: when, how and why. *Philosophical Transactions of the Royal Society B: Biological Sciences* **368**:20120344.

Santos, L.R., *et al.* (2006) Rhesus monkeys, *Macaca mulatta*, know what others can and cannot hear. *Animal Behaviour* **71**:1175–1181.

Sanz, C., *et al.* (2004) New insights into chimpanzees, tools, and termites from the Congo Basin. *American Naturalist* **164**:567–581.

Sapolsky, R.M. and Share, L.J. (2004) A pacific culture among wild baboons: its emergence and transmission. *PLOS Biology* **2**: e106.

Schneirla, T.C. and Rosenblatt, J.S. (1963) Critical periods in the development of behavior. *Science* **139**:1110–1115.

Scuett, W., *et al.* (2010) Sexual selection and animal personality. *Biological Reviews* **85**:217–246.

Seyfarth, R.M. and Cheney, D.L. (1980) The ontogeny of vervet monkey alarm calling behavior: a preliminary report. *Zeitschrift fur Tierpsychologie* **54**:37–56.

Seyfarth, R.M. and Cheney, D.L. (2000) Social awareness in monkeys. *American Zoologist* **40**:902–909.

Seyfarth, R.M. and Cheney, D.L. (2012) Knowledge of social relations. In: *The Evolution of Primate Societies* (eds J.C. Mitani, J. Call, P. Kappeler, R.A. Palombit and J.B. Silk). Chicago: University of Chicago Press, 628–642.

Seyfarth, R.M., *et al.* (2012) Variation in personality and fitness in wild female baboons. *Proceedings of the National Academy of Sciences of the United States of America* **109**:16980–16985.

Sharpe, L.L. (2005) Frequency of social play does not affect dispersal partnerships in wild meerkats. *Animal Behaviour* **70**:559–569.

Sharpe, L.L. and Cherry, M.I. (2003) Social play does not reduce aggression in wild meerkats. *Animal Behaviour* **66**:989–997.

Sharpe, L.L., *et al.* (2002) Experimental provisioning increases play in free-ranging meerkats. *Animal Behaviour* **64**:113–121.

Shettleworth, S.J. (2010) *Cognition, Evolution, and Behavior*. New York: Oxford University Press.

Shultz, S. and Dunbar, R.I.M. (2007) The evolution of the social brain: anthropoid primates contrast with other vertebrates. *Proceedings of the Royal Society of London. Series B: Biological Sciences* **274**:2429–2436.

Sih, A., *et al.* (2004) Behavioral syndromes: an ecological and evolutionary overview. *Trends in Ecology and Evolution* **19**:372–378.

Silk, J.B. (2003) Cooperation without counting: the puzzle of friendship. In: *Genetic and Cultural Evolution of Cooperation* (ed. P. Hammerstein). Berlin: Dahlem University Press, 37–54.

Sinha, A. (2005) Not in their genes: phenotypic flexibility, behavioural traditions and cultural evolution in wild bonnet macaques. *Journal of Biosciences* **30**:51–64.

Sloan, E.K., *et al.* (2008) Social temperament and lymph node innervation. *Brain, Behavior and Immunity* **22**:717–726.

Sluyter, F., *et al.* (1995) A comparison between house mouse lines selected for attack latency or nest-building: evidence for a genetic basis of alternative behavioral strategies. *Behavior Genetics* **25**:247–252.

Smith, B.R. and Blumstein, D.T. (2008) Fitness consequences of personality: a meta-analysis. *Behavioral Ecology* **19**:448–455.

Smith, B.R. and Blumstein, D.T. (2010) Behavioral types as predictors of survival in Trinidadian guppies (*Poecilia reticulata*). *Behavioral Ecology* **21**:919–926.

Stamps, J.A. (2007) Growth–mortality tradeoffs and 'personality traits' in animals. *Ecology Letters* **10**:355–363.

Svare, B., *et al.* (1984) Infanticide: accounting for genetic variation in mice. *Physiology and Behavior* **33**:137–152.

Taborsky, B. and Oliveira, R.F. (2012) Social competence: an evolutionary approach. *Trends in Ecology and Evolution* **27**:679–688.

Terranova, M.L., *et al.* (1999) Behavioral and hormonal effects of partner familiarity in periadolescent rat pairs upon novelty exposure. *Psychoneuroendocrinology* **24**:639–656.

Thornton, A. and Clutton-Brock, T.H. (2011) Social learning and the development of individual and group behaviour in mammal societies. *Philosophical Transactions of the Royal Society B: Biological Sciences* **366**:978–987.

Thornton, A. and McAuliffe, K. (2006) Teaching in wild meerkats. *Science* **313**:227–229.

Thornton, A. and Malapert, A. (2009) Experimental evidence for social transmission of food acquisition techniques in wild meerkats. *Animal Behaviour* **78**:255–264.

Thornton, A. and Raihani, N.J. (2008) The evolution of teaching. *Animal Behaviour* **75**:1823–1836.

Thornton, A., *et al.* (2010) Multi-generational persistence of traditions in neighbouring meerkat groups. *Proceedings of the Royal Society of London. Series B: Biological Sciences* **277**:3623–3629.

Tomasello, M. (1996) Do apes ape? In: *Social Learning in Animals: The Roots of Culture* (eds C.M. Heyes and B.G. Galef Jr). San Diego, CA: Academic Press, 319–346.

Tomasello, M. (1999) *The Cultural Evolution of Human Cognition*. Cambridge, MA: Harvard University Press.

Tomasello, M. and Call, J. (1997) *Primate Cognition*. New York: Oxford University Press.

Trillmich, F. and Hudson, R. (2011) The emergence of personality in animals: the need for a developmental approach. *Developmental Psychobiology* **53**:505–509.

Tuyttens, F.A.M. and Macdonald, D.W. (2000) Consequences of social perturbation for wildlife management and conservation. In: *Behavior and Conservation* (eds L.M. Gosling and W.J. Sutherland). Cambridge: Cambridge University Press, 315–329.

van de Waal, E., *et al.* (2010) Selective attention to philopatric models causes directed social learning in wild vervet monkeys. *Proceedings of the Royal Society of London. Series B: Biological Sciences* **277**:2105–2111.

van de Waal, E., *et al.* (2013) Potent social learning and conformity shape a wild primate's foraging decisions. *Science* **340**:483–485.

van Leeuwen J.C. and Huan, D.B.M. (2013) Conformity in nonhuman primates: fad or fact. *Evolution and Human Behaviour* **34**:1–7.

van Leeuwen J.C. and Huan, D.B.M. (2014) Conformity without majority? The case for demarcating social from majority influences. *Animal Behaviour* **96**:187.

van Oers, K., *et al.* (2005) Contribution of genetics to the study of animal personalities: a review of case studies. *Behaviour* **142**:9–10.

van Schaik, C.P. and Pradhan, G.R. (2003) A model for tool-use traditions in primates: implications for the coevolution of culture and cognition. *Journal of Human Evolution* **44**:645–664.

Vince, M.A. (1993) Newborn lambs and their dams: the interactions that lead to suckling. In: *Advances in the Study of Behavior*, Vol. 22 (eds P.J.B. Slater, M. Milinski, C.T. Snowdon and J.S. Rosenblatt). San Diego, CA: Academic Press, 239–268.

Waddington, C.H. (1953) Genetic assimilation of an acquired character. *Evolution* **7**:118–126.

Weiss, A., *et al.* (2012) All too human? Chimpanzee and orangutan personalities are not anthropomorphic projections. *Animal Behaviour* **83**:1355–1365.

West-Eberhard, M.J. (2003) *Developmental Plasticity and Evolution*. Oxford: Oxford University Press.

Whitehead, H. (1998) Cultural selection and genetic diversity in matrilineal whales. *Science* **282**:1708–1711.

Whitehead, H. (2007) Learning, climate and the evolution of cultural capacity. *Journal of Theoretical Biology* **245**:341–350.

Whitehead, H. and Rendell, L. (2004) Movements, habitat use and feeding success of cultural clans of South Pacific sperm whales. *Journal of Animal Ecology* **73**:190–196.

Whitehead, H. and Weilgart, L. (2000) The sperm whale: social females and roving males. In: *Cetacean Societies: Field Studies of Dolphins and Whales* (eds J. Mann, R. Connor, P. Tyack and H. Whitehead). Chicago: University of Chicago Press, 154–172.

Whiten, A. (1989) Transmission mechanisms in primate cultural evolution. *Trends in Ecology and Evolution* **4**:61–62.

Whiten, A. (2005) The second inheritance system of chimpanzees and humans. *Nature* **437**:52–55.

Whiten, A. (2012) Social learning, traditions, and culture. In: *The Evolution of Primate Societies* (eds J.C. Mitani, J. Call, P.M. Kappeler, R.A. Palombit and J.B. Silk). Chicago: University of Chicago Press, 682–700.

Whiten, A., *et al.* (1999) Cultures in chimpanzees. *Nature* **399**:682–685.

Wilson, A.J. and Réale, D. (2006) Ontogeny of additive and maternal genetic effects: lessons from domestic mammals. *American Naturalist* **167**:E23–E38.

Wittig, R.M., *et al.* (2014) Triadic social interactions operate across time: a field experiment with wild chimpanzees. *Proceedings of the Royal Society of London. Series B: Biological Sciences* **281**:20133155.

Wolf, M. and McNamara, J.M. (2012) On the evolution of personalities via frequency-dependent selection. *American Naturalist* **179**:679–692.

Woodroffe, R. (2003) Dispersal and conservation: a behavioural perspective on metapopulation persistence. In: *Animal Behavior and Wildlife Conservation* (eds M. Festa-Bianchet and M. Apollonio). Washington, DC: Island Press, 33–48.

Wright, G.S., *et al.* (2011) Social learning of a novel foraging task by big brown bats, *Eptesicus fuscus*. *Animal Behaviour* **82**:1075–1083.

Yurk, H., *et al.* (2002) Cultural transmission within maternal lineages: vocal clans in resident killer whales in southern Alaska. *Animal Behaviour* **63**:1103–1119.

Zentall, T.R. (2001) Imitation in animals: evidence, function, and mechanisms. *Cybernetics and Systems* **32**:53–96.

Zohar, O. and Terkel, J. (1996) Social and environmental factors modulate the learning of pine-cone stripping techniques by black rats, *Rattus rattus*. *Animal Behaviour* **51**:611–618.

CHAPTER 7

Communication

7.1 Introduction

What's in a grunt?

Olive baboons are foraging in long grass on the slopes above Lake Tanganyika. As they walk, females intermittently grunt, sometimes softly, sometimes more loudly. Occasionally they seem to answer each other while, sometimes, the timing of grunts given appears to be independent of grunts heard. If you listen carefully, you can hear that grunts differ in pitch and structure between individuals, and if you count them you soon see that the rate of grunting varies too.

Why do baboons grunt? Is it to keep contact with other group members? Or to make sure they don't approach too closely? Do mothers with dependent young grunt more than those without offspring? Can individuals recognise each other by their grunts? Are regular calls of this kind confined to species where group members benefit from each other's presence? And what's in a grunt? Do grunts simply convey 'indexical' information about the individual's internal state or do they convey 'referential' information about the external world? Do grunts also incorporate information about the grunter's health, dominance or hormonal status? And are grunts honest signals or do individuals use them to mislead rivals?

Many of these questions are not specific to grunting and can be asked about other vocalisations, including calls used to warn other group members of danger, to deter rivals, reassure juveniles or attract males. Many of the same questions can also be asked about other forms of female communication, including visual, tactile and olfactory signals, though one important advantage of studying vocal signals is that it is often possible to record specific calls from specific individuals and use them in playback experiments to investigate their function (Figure 7.1).

In contrast to birds, nocturnal habits and terrestrial niches represent the ancestral condition for mammals and their communication systems reflect this. Olfactory signals are widely used to mark regularly used pathways and delineate territories. They have the advantage over visual and vocal signalling that they operate when individuals are absent and persist for considerable periods. Although birds also use olfactory signals (Leclaire *et al.* 2011), most mammals rely to a greater extent on olfactory signals and both their capacity to generate olfactory signals and their ability to discriminate between them is more highly developed than in most other vertebrates (Roberts 2007). Conversely, mammals rely to a lesser extent on visual signals than birds, and visual displays and permanent ornaments are not as highly developed (Clutton-Brock and Huchard 2013).

Mammals also use vocal signals to defend territories, attract mates and threaten rivals, warn and reassure other group members, initiate friendly interactions and maintain contact. Vocal signals are easier to measure and quantify accurately than olfactory or visual signals and can be recorded, modified and played back to individuals, so that it is possible to design realistic experiments to explore the information they convey and investigate their function. As a result, much of our understanding of mammalian communication is based on studies of vocal signals and our knowledge of the content and function of olfactory and visual signals is more limited. However, it is important to recognise that this bias does not necessarily reflect the relative importance of different types of signal. Moreover, vocal, olfactory and visual signals are often used in conjunction with each other and different signalling modalities can modify, extend or amplify the information conveyed in other components of the signal.

Paradoxically, studies of communication have been impeded by variation in the meanings attached to different terms. In this chapter I use the terminology adopted by Maynard Smith and Harper (2003) (Table 7.1).

Mammal Societies, First Edition. Tim Clutton-Brock.
© 2016 Tim Clutton-Brock. Published 2016 by John Wiley & Sons, Ltd.

Figure 7.1 In meerkats, it is possible to record calls from close quarters for playback experiments. *Source*: © Tim Clutton-Brock.

The evolution of signalling systems

Animals commonly respond to cues, which provide a guide to the signaller's future action (Hasson 1994), as well as to signals, which are traits that have evolved as a

Table 7.1 Definition of terms used in this chapter.

Signal: an act or structure that alters the behaviour of another organism, which evolved because of that effect, and which is effective because the receiver's response has also evolved
Cue: a feature of the world, animate or inanimate, that can be used by an animal as a guide to future action
Ritualisation: evolutionary process whereby a cue may be converted into a signal
Handicap: a signal whose reliability is ensured because its cost is greater than required by efficacy requirements; the signal may be costly to produce, or have costly consequences (e.g. vulnerability cost)
Cost: loss of fitness resulting from making a signal. Includes:
Efficacy cost: cost needed to ensure that the information can be reliably perceived
Strategic cost: cost needed by the handicap principle to ensure honesty
Index: a signal whose intensity is causally related to the quality being signalled, and which cannot be faked
Minimal-cost signal: a signal whose reliability does not depend on its cost (i.e. not a handicap) and which can be made by most members of a population (i.e. not an index)

Source: Maynard Smith and Harper (2003).

result of their effects on other individuals and have coevolved with the reactions of receivers (Maynard Smith and Harper 2003). Early studies of animal communication assumed that signals evolved to maximise the efficiency of information transfer, which may be the case where callers and recipients both benefit from the transfer of information. However, individuals also communicate with rivals and here selection is likely to favour signals that cause recipients to behave in a way that increases the signaller's fitness, while counteracting selection pressures will favour recipients that react in a way that is likely to maximise their own fitness (Dawkins and Krebs 1978; Krebs and Dawkins 1984; Searcy and Nowicki 2005). As a result, rather than evolving as cooperative interactions in which both participants adjust their behaviour to achieve a common goal, many forms of communication may resemble 'arms races' where senders attempt to manipulate receivers to their own advantage and receivers do their best to anticipate the strategies of signallers and adjust their responses to maximise their own fitness (Dawkins and Krebs 1978; Krebs and Dawkins 1984; Rendall *et al.* 2009). Nevertheless, both parties must still benefit from the interaction, for if senders maximised their fitness by avoiding signalling or receivers did so by not responding, this would favour the abandonment of the signalling system (Maynard Smith and Harper 2003). As a result, signals must convey some accurate information, for otherwise it would be in the interests of receivers to ignore them (Seyfarth *et al.* 2010).

Signal reliability

The recognition that signalling systems involving competitors may be the result of evolutionary arms races raises important questions about the mechanisms that maintain the reliability of signals to receivers. There are four main ways in which reliability can be maintained. First, signal reliability can be maintained if receivers that detect dishonest signals punish frauds so severely that dishonest signalling reduces the signaller's fitness (Rohwer and Rohwer 1978; Tibbetts and Dale 2004). Both the magnitude of punishment and the probability that fraudulent signals will be detected are likely to influence the effectiveness of this process.

Second, signals may be inevitably correlated with the size, condition or intention of the sender so that they convey reliable information and cannot be misleading. Many of the repetitive displays used by males to attract mates or repel competitors inevitably contain information that accurately reflects the signallers competitive ability or resource-holding potential (RHP) even if selection has favoured individuals that exaggerate their potential (Clutton-Brock and Albon 1979). For example, territorial tigers mark the trunks of trees with their claws, reaching as high as they can, while dwarf mongooses will stand on their front legs to leave anal secretions on the highest vegetation that they can reach (Figure 7.2). The height of marks provides an unfakeable signal of the body size of the territory owner (Thapar 1986) and individuals respond more to higher marks than to lower ones (Sharpe 2015). Similarly, the rate of roaring that can be sustained by male red deer in confrontations with rivals during the autumn rut may be inevitably related to their fighting prowess because roaring rate signals how rapidly they would run out of breath in a prolonged pushing encounter. Cases where individuals give signals that are inevitably correlated with the signaller's size, quality or competitive prowess for biophysical reasons are sometimes referred to as examples of 'honest advertisement' (Clutton-Brock and Albon 1979) and signals of this kind are usually referred to as *indices* (Arak and Enquist 1995; Maynard Smith and Harper 2003). Although indices sometimes involve single modalities, they often involve more than one modality (either simultaneously or in succession) which may help to maintain their 'honesty' (Enquist and Leimar 1983, 1990). Many indices were probably adopted as signals because they coincided with the sensory modalities of receivers and attracted their attention as well as being indicators of the signaller's power or RHP. Selection may then have exaggerated them until the costs of further elaboration exceeded any benefit, with the result that they now have energy and fitness costs that contribute to their reliability (Higham 2013).

Third, signals may be reliable indicators of an individual's quality or competitive ability because they have such large costs to the signaller's fitness that only superior individuals can afford to give them or because high-quality individuals are able to produce more elaborate signals for a given fitness cost. Cost-related theories of signal reliability were first proposed by Zahavi (1975, 1993) who suggested that selection is likely to have favoured the evolution of particular actions or traits as signals because they had initially high fitness costs and were consequently unfakeable by inferior individuals. Zahavi drew analogies between the evolution of costly signals and the handicaps imposed on sportsmen and argued that conspicuous wastage of energy on signalling by superior individuals maintained signal honesty since inferior individuals would be unable to afford similar extravagance (Figure 7.3). Zahavi's theory was initially rejected (Kirkpatrick 1986) before being shown to be feasible by Grafen (1990a,b), who demonstrated that signal honesty can be maintained if, for low-quality individuals, the fitness costs of giving a signal exceed the benefits while, for high-quality individuals, benefits exceed costs. However, while it is possible for signal reliability to be maintained by the increasing costs of signalling to inferior individuals, as Zahavi suggested, more recent models show that it is also possible for signal reliability to be maintained by quality-dependent trade-offs between the characteristics of signals and their costs to signallers (Lachmann *et al.* 2001; Getty 1998a,b, 2006; Higham 2013): for example, high-quality individuals may be more efficient at converting advertising into fitness than low-quality ones and so do not need to incur absolutely higher fitness costs for signals to be reliable (Getty 2006). In contrast to handicap models of signal reliability, these models do not suggest that sexually selected signals should necessarily be wasteful or excessively extravagant (Biernaskie *et al.* 2014).

(a)

(b)

Figure 7.2 The height of marks provides an unfakeable signal of the size of the marker: (a) a tiger marks on a tree with its claws; (b) a dwarf mongoose leaves an anal mark on vegetation. *Sources*: (a) © Steve Winter/National Geographic/Getty Images; (b) © Lynda Sharpe.

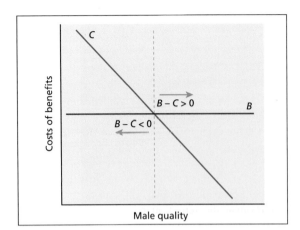

Figure 7.3 A summary of the assumptions of the handicap principle. The benefit (*B*) of producing a costly signal or ornament is assumed to be roughly equal for all males. The cost (*C*) of producing the signal or ornament is assumed to be lower for high-quality males, because they are in better condition and have additional resources to invest in ornament production. In this case, the benefit of producing the ornament only outweighs the cost (*B* – *C* >0) for high-quality males, so only high-quality males are selected to produce the ornament, making the ornament a reliable signal of male quality. More generally, the handicap principle requires that the cost–benefit ratio is lower for individuals giving stronger signals. *Source*: From Davies *et al.* (2012). Reproduced with permission of John Wiley & Sons.

Signal function

While many signals convey information about the signaller's competitive ability, body size, condition or motivation (Taylor and Reby 2010), signals can also reflect specific aspects of their situation or environment (Seyfarth and Cheney 2003; Rendall *et al.* 2009; Manser 2010). For example, the waggle dance of bees provides receivers with information about the type of food that is available, how far away it is and which direction it lies in (Seeley 1985). Similarly, some mammalian alarm calls reflect the type of predator involved, how close it is and how great a risk it poses (Fischer *et al.* 2001; Frederiksen and Slobodchikoff 2007; Townsend and Manser 2013). Signals that convey specific information about the social or physical environment are referred to as *functionally referential* to distinguish them from *indexical* signals that provide information about the signaller's own characteristics, such as their size, condition or motivational state (Seyfarth and Cheney 1990; Bradbury and Vehrencamp 1998). However, while it is convenient to contrast them, both types of information are commonly transmitted in

the same signal (Manser *et al.* 2002; Seyfarth and Cheney 2003).

When signals are used to transfer information about the external environment to relatives and allies, their evolution may correspond more closely to 'public goods' games than to arms races (see Chapter 17). Selection processes may favour the development of signals that can be given by all individuals (referred to as *minimal-cost signals*; see Table 7.1) and are unlikely to favour the evolution of signals that have large costs to the direct fitness of signallers unless they generate large indirect benefits.

Signal complexity

Another important question concerns the complexity of signalling systems. Both total repertoire size and the number of calls serving similar functions vary widely between species and the consistency of particular calls also varies (Bradbury and Vehrencamp 1998; Searcy and Nowicki 2005). Why? It is sometimes suggested that species differences in repertoires of calls provide a measure of the complexity of communication systems. But is this really the case? For example, some species give graded calls that reflect the caller's circumstances and have relatively few discrete calls and it is not clear that their systems are any less complex than those of species with larger repertoires (Manser *et al.* 2014). So if repertoire size is not a good indicator of signalling complexity, what is?

This chapter reviews our understanding of signalling systems in females, focusing principally on vocal signals. Section 7.2 examines nine different categories of signals: contact, feeding and moving calls between group members; calls between mothers and dependent young; alarm calls and sentinel calls; sexual signals; threats and reassurance; and territorial signals while the evolution of male displays is explored in Chapters 13 and 15. Section 7.3 returns to some of the general questions raised at the start of this chapter.

7.2 Types of signal

Contact calls

In many social mammals, mature individuals of both sexes give regular calls while they forage which help group members to remain in contact with each other (McComb and Reby 2005). For example, foraging meerkats give gentle calls every few seconds (Townsend *et al.* 2010), foraging baboons give regular grunts

(Cheney *et al.* 1996; Rendall *et al.* 1999) and foraging dolphins give regular clicks and whistles (Janik and Slater 1998, 2000; Quick and Janik 2012) (Figure 7.4).

(a)

(b)

(c)

Figure 7.4 (a, b, c) Foraging dolphins surround and concentrate fish schools, making individual fish easier to catch. In some populations, they stir up mud around fish schools to concentrate them while, in others, they cooperate to drive fish on shore and then beach to catch them. *Source*: © Richard Connor.

In most cases, contact calls are used over relatively short distances ('close' calls), but in some they are used to maintain contact over considerable distances. For example, African elephants give very low frequency rumbles that can be transmitted over more than a kilometre and help to maintain contact and cohesion between widely separated individuals (Poole 1999; McComb *et al.* 2003; Soltis *et al.* 2005a,b; Leighty *et al.* 2008), while in some whales vocalisations between group members may allow them to remain in contact over many kilometres (Tyack 2000). The amplitude and structure of contact calls varies with the usual distance between foraging animals.

Individual differences in contact calls allow group members to identify each other. For example, bottlenose dolphins have 'signature' whistles which they use when separated from their group (Janik and Slater 1998). Playback experiments with elephant rumbles show that the responses of receivers vary according to the caller's kin group, individual identity and familiarity and that the same individuals exchange calls repeatedly (McComb *et al.* 2003; Soltis *et al.* 2005b). In rhesus macaques, too, individuals discriminate between contact calls given by different individuals and are more likely to respond to calls given by close kin (Rendall *et al.* 1996). Individual meerkats discriminate between the calls of the dominant female and those of other adult females and display submission to calls from the dominant female if they are at risk of being evicted (Reber *et al.* 2013).

Contact calls can also convey information about the individual's activity, reproductive state and motivation. For example, elephants can distinguish the rumbles of males in musth from those of other males (Poole 1999; Soltis *et al.* 2005b) and baboons give distinctive grunts when they are about to move across an open area, where group cohesion and coordination is likely to be important to all individuals (Rendall *et al.* 1999). As in some other animals, different components of contact calls can carry different information; for example, contact calls of banded mongooses contain two acoustically different segments, one of which is stable and differs between individuals and the other varies with the individual's activity (Jansen *et al.* 2012).

Contact calling helps to maintain group cohesion and synchronise the activity of group members and is most frequent in species where individuals forage in stable groups but are often out of sight of each other. Within species, individuals tend to give and answer contact calls (and signature calls in particular) in situations where they are likely to be separated from other group members. For example, bottlenose dolphins give signature calls when they are likely to become separated from other group members and rarely give these calls when they are close together (Janik and Slater 1998).

While contact calls may increase the safety or foraging efficiency of group members, they can also have costs, especially in species that live on mobile prey. As would be expected, regular contact calling is usually absent in social carnivores that hunt by stealth but is common in pack-living cursorial hunters. Similarly, resident killer whales, which live on shoaling fish with relatively poor hearing ability, give frequent pulsed calls that help to maintain contact between group members while transient killer whales, which feed on mammals with much greater auditory acuity, seldom give pulsed calls until after a successful hunt (Deecke *et al.* 2005).

How do individuals benefit by giving contact calls? One possibility is that giving contact calls elicits responses from neighbours and allows individuals foraging in cover to distinguish between the movements of other group members and those of predators, or reduces the chance that the signaller will become isolated from their group (Manser 1998; Townsend *et al.* 2011). For example, female baboons are more likely to answer barks given by close relatives at times when they are separated from

other females or at the end of group progressions (Cheney *et al.* 1996). Alternatively, or additionally, giving contact calls may benefit the foraging efficiency of individuals if it reduces the chance that nearby animals will intrude on the caller's foraging space.

As well as giving regular contact calls, individuals that have become separated from their group and are attempting to re-establish contact may give specific 'lost' calls that attract the attention of other group members and may cause them to vocalise in response. For example, in meerkats, individuals that get separated from the group initially emit soft 'lost' calls, which turn into loud barks if they continue to be unable to see or hear other group members (Manser 1998). Lost calls often cause other group members to stand up bipedally or to move in the direction of the lost animal.

Moving calls

Calls are also used to coordinate group movements: one or more individuals gives a specific 'moving' call that initiates and synchronises transitions between one activity and another or movements between feeding sites. For example, in meerkats, any group member may initiate a movement from one feeding patch to another by giving a moving call (Manser 1998; Bousquet *et al.* 2011) (Figure 7.5). If no

Figure 7.5 A meerkat group on the move. Leading individuals (which are often subordinates) inspect areas into which they are moving before other group members commit themselves. *Source*: © Tim Clutton-Brock.

other group members give the same call, the group typically remains in the same patch, but if several others give moving calls, the probability that the group will move to another feeding patch increases sharply.

Decisions about activity transitions or movements seldom appear to be controlled by particular individuals (Turbé 2006; Perony and Townsend 2013). For example, when dominant and subordinate meerkats were trained to expect food in opposite directions, trained individuals initiated movements in the direction where they expected food but there was no stronger tendency for groups to be more likely to follow the direction of movements initiated by dominants than those initiated by subordinates (Bousquet and Manser 2011). Similar 'voting' systems play an important role in coordinating group movement in other species (see Chapter 2) but not all involve specialised moving calls (Fischer and Zinner 2011; King and Sueur 2011). For example, in chacma baboons, group members frequently vocalise before groups move but the frequency of calling does not appear to influence the probability that an initiation of movement will occur (Sueur 2011).

Vocalisations are not the only signal that is used by group members to remain in contact with each other: where individuals forage in subgroups, olfactory marks may also play an important role in maintaining the cohesion of group members (Gosling and Roberts 2001). In addition, many species have physical traits that help to maintain visual contact, including distinctive marks on their tails which help individuals to remain in visual contact when moving in cover. For example, meerkats have prominent black tail tips and individuals hold their tails erect when the group is moving in cover, with the result that the visibility of individuals is increased

Feeding calls

In some mammals, individuals that find good foraging sites give loud calls that attract other group members to the same site (Clay et al. 2012). For example, spider monkeys and chimpanzees that find food sources give loud calls that attract other individuals (Chapman and Lefebvre 1990; Clark and Wrangham 1994; Slocombe et al. 2010). As well as calling, group members sometimes give other auditory signals: for example, chimpanzees drum on the buttress roots of forest trees (Sabater-Pí 1979; Goodall 1986).

Feeding calls often convey information about the quality of resources that have been located (Clay et al.

2012). For example, spider monkeys are more likely to give feeding calls in large fruiting trees than in small ones as well as in trees where fruit is relatively abundant (Chapman and Lefebvre 1990). Similarly, chimpanzees more frequently give feeding calls if food patches are large and food cannot be monopolised (Slocombe et al. 2010) and the structure of their calls also varies with the size of food patches (Kalan et al. 2015). Experiments with captive chimpanzees show that they use different calls when they find high-preference foods from those used when they encounter common low-preference foods (Slocombe and Zuberbühler 2005, 2006). Rhesus macaques that find high-quality rare foods also give different calls from those that find low-quality common foods (Hauser 1998).

Food calls are frequently interpreted as a form of cooperative behaviour and this is a likely interpretation in some cases. For example, the food calls given by male chimpanzees help to inform other group members of the location of potential feeding sites (Slocombe et al. 2010) and individuals are more likely to call it there is an individual with which they have a close social bond in the vicinity (Schel et al. 2013). However, giving feeding calls may also generate more direct benefits. In some cases, they may attract potential mating partners: for example, dominant male chimpanzees are more likely to give feeding calls if there are oestrous females nearby (Kalan and Boesch 2015). Attracting additional group members may also reduce the chance that finders will be displaced by rival groups with overlapping ranges or by competing species (Chapman and Lefebvre 1990; Holekamp et al. 2000; Wilson and Wrangham 2003, 2007). Calling may also announce the individual's possession of a food source (Cheney and Seyfarth 2008). For example, white-faced capuchin monkeys commonly call when they encounter high-quality foods and observations show that individuals that call are subsequently less likely to be approached by other group members who were in sight when they discovered the food than those that kept silent (Gros-Louis 2004). Finally, feeding calls may slow down the forward movement of the group, giving finders the chance to exploit feeding sites that they locate without being left behind.

Mother–offspring signals

Communication between offspring and their mothers often involves a combination of visual, vocal and olfactory signals and their relative importance varies with social and ecological conditions. In birds and mammals where breeding females aggregate in large groups and mothers need to identify each other from some distance, they commonly rely on vocal signals (Aubin and Jouventin 1998; Jouventin *et al*. 1999). For example, Mexican free-tailed bats are reared in crèches that can include millions of individuals at densities of up to 5000/m^2 (McCracken 1984). Playback experiments show that mothers recognise and respond to the calls of their offspring while pups do not appear to respond selectively to the calls of their mothers (Balcombe 1990) (Figure 7.6).

Where mothers do not need to recognise their offspring over long distances, they often rely on a combination of vocal and olfactory cues and their relative importance can change throughout the period of development. In some mammals, infants learn to identify their mother's odour from amniotic cues before birth (Wells and Hepper 2006), while in others they learn to recognise their mother's calls during the first days after they are born. For example, in fur seals, which breed in large colonies, pups learn to recognise their mother's calls when they are 2–5 days old and mothers do not leave the colony to forage until pups have reached this age (Insley 2000; Charrier *et al*. 2001).

In many mammals, dependent young give begging calls whose rate signals their hunger and need for food. For example, experiments with piglets show that hungry juveniles give begging calls at a higher rate than well-fed individuals and that mothers are more likely to respond to their calls (Weary and Fraser 1995; Weary *et al*. 1996). Hungry or slow-growing piglets also continue to massage their mother's udders after milk ejection, which may help to increase their mother's milk production (Jensen *et al*. 1998; Dostalkova *et al*. 2002; Torrey and Widowski 2007). In meerkats, frequent begging calls are given by pups and juveniles that are foraging with adults and food deprivation increases the call rates of pups (Manser and Avey 2000; Carlson *et al*. 2006; Manser *et al*. 2008). High begging rates stimulate helpers and breeders to distribute more food to pups and to feed individuals that call longer and more intensely (see Chapter 17).

If begging calls are honest signals, high rates of begging might be expected to have appreciable energetic costs. Attempts to estimate the costs of begging in birds have produced mixed results (Kilner 2001; Horn and Leonard 2002; Smiseth and Moore 2004). While some studies show that the energetic costs of begging typically comprise a very small part of the total energy budget of nestlings (Chappell and Bachman 2000), experiments with captive canaries suggest that frequent begging can depress the growth of chicks (Kilner 2001), indicating that the costs of begging may be higher in natural populations.

Alarm signals

In many mammals that live in stable social groups, individuals that detect potential danger give alarm signals involving vocal, visual or olfactory cues that alert other group members to the presence of danger (Klump and Shalter 1984; Zuberbühler 2009). For example, group-living ungulates that are alarmed by predators give a variety of visual signals that alert conspecifics, including tail flagging, bounding, stotting and foot-stamping, often associated with snorting or whistling (Caro *et al*. 2004). In some species, alarm displays utilise physical structures like the white tails of some deer and the erectile rump patches of springbok (Kiley-Worthington 1976; Caro *et al*. 2004). Alarmed individuals may also give olfactory signals. For example, springbok and pronghorn antelope both produce olfactory signals from specialised glands when alarmed (Moy 1970; Bigalke 1972). Similarly, stressful conditions or events cause cattle to release alarm substances in their urine that induce changes in the behaviour of other group members (Boissy 1995). Rodents, too, often give alarm signals that involve multiple modalities (Klump and Shalter 1984; Blumstein 2007).

Interspecific differences in the diversity and structure of alarm signals are related both to the need to avoid detection and to the distance over which individuals need to communicate (Marler 1955). The frequency with which alarm calls are given and the repertoire of calls are typically higher in diurnal species than in nocturnal ones, which commonly rely on crypsis to avoid predation (Blumstein 2007), though some social nocturnal species, like the plains viscacha, can have large repertoires of alarm calls (Branch *et al*. 1993). In many small mammals, alarm calls are usually of high frequency and can be ultrasonic, which may serve to reduce the chance that predators will be able to locate their source (Caro 2005). For example, ground squirrels give ultrasonic alarm calls that limit their range and which are unlikely to be detectable by their principal predators (Wilson and Hare 2004). In contrast, the alarm calls of

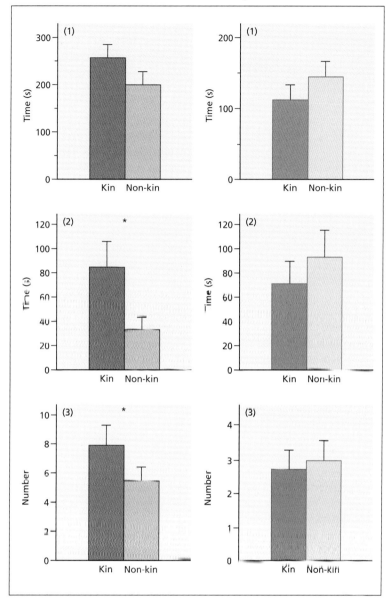

Figure 7.6 Mothers respond to the vocalisations of their pups in Mexican free-tailed bats. Left column shows responses of mothers to playbacks of isolation calls given by their own pups versus unrelated pups: (1) time spent on each side of arena; (2) time spent in contact with models; (3) number of separate contacts with models. Right column shows responses of pups to playbacks of echolocation calls given by their mothers or by unrelated females: (1) time spent on each side of arena; (2) time spent in contact with models; (3) number of separate contacts with models. *Source*: From Balcombe (1990). Reproduced with permission of Elsevier.

larger mammals, including primates and ungulates, are often louder and relatively low in frequency (Caro 2005), reflecting increased selection for communication over longer distances and reduced need to avoid detection.

Contrasts in call structure may also be related to inter-specific differences in habitats: for example, forest-dwelling rodents have relatively low-frequency calls which are less subject to scattering by leaves and

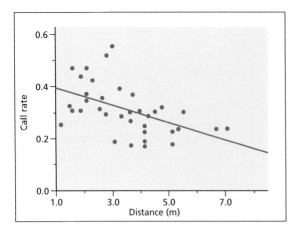

Figure 7.7 Rates of calling by Richardson's ground squirrels at different distances from a model predator. *Source*: From Warkentin *et al.* (2001). Reproduced with permission of NRC Research Press.

branches and so attenuate less than high-frequency ones (Emmons 1978; Blumstein 2007).

Within species, the rate at which alarm signals are given, the duration of visual signals, the amplitude of calls and the number of individuals calling commonly increase as the perceived danger rises (Caro 2005). For example, experiments with ground squirrels show that the rate of calling increases as the distance between the caller and the predator declines (Warkentin *et al.* 2001) (Figure 7.7). The amplitude and structure of calls can also change with the increasing urgency of threats (Macedonia and Evans 1993): for example, the amplitude of calls given by California ground squirrels and meerkats to potential predators increases with the risk that they pose to callers (Leger *et al.* 1984; Manser 2001; Furrer and Manser 2009).

In some mammals, structurally different calls are used for different kinds of predators (Caro 2005; Townsend and Manser 2013). For example, a classic study by Seyfarth and Cheney showed that vervet monkeys give structurally different alarm calls for eagles, snakes and leopards: playback of different calls caused individuals on the ground to look up for eagle calls, stand bipedally and inspect the ground and nearby grass for snake calls and run into the trees for leopard calls (Seyfarth *et al.* 1980a,b; Seyfarth and Cheney 1990) (Figure 7.8). Responses given to the same call by individuals already in the trees also differed from those on the ground (Figure 7.9). Subsequent research has shown that several other species, including red squirrels, meerkats and several social primates give different calls to

different categories of predators (Zuberbühler *et al.* 1997; Greene and Meagher 1998; Manser 2001; Fichtel and Kappeler 2002; Fichtel and Hammerschmidt 2003). In Gunnison's prairie dogs, individuals even give different calls to humans wearing different coloured T-shirts and differently shaped clothes (Slobodchikoff *et al.* 1991, 2009a,b).

Although referential alarm calls are widespread, they are not universal and have not been detected in a number of other social species, including yellow-bellied marmots and several cercopithecine monkeys (Hohmann 1989; Blumstein and Armitage 1997a; Fischer 1998; Fischer *et al.* 2001). Their apparent absence in these species suggests that referential calls may evolve in species where individuals have to rely on different tactics to avoid or deter different predators (Macedonia and Evans 1993; Fichtel and Kappeler 2002; Furrer and Manser 2009), indicating that there may not be a close association between the occurrence of referential alarm calls and relative cognitive abilities and that, contrary to some suggestions, there is little reason to suppose that they represent a precursor to the evolution of language (Wheeler and Fischer 2012).

In most species where alarm calls are given, receivers react rapidly to them, adjusting their responses to the type and urgency of calls or to the number of individuals calling (Caro 2005; Blumstein 2007). In some cases, the magnitude and duration of their responses is positively related to cortisol levels (Voellmy *et al.* 2014). Since responding to alarm calls is likely to reduce the immediate foraging success of receivers and may have other costs, including effects on their glucocorticoid levels (Caro 2005; Blumstein 2007), receivers might be expected to adjust their responses to the reliability of callers and several recent studies suggest that they do so. For example, juveniles commonly give inappropriate alarm calls and adults often disregard alarm calls given by younger animals or scan their surroundings for potential danger rather than immediately fleeing to safety (Seyfarth and Cheney 1990; Ramakrishnan and Coss 2000; Hollén and Radford 2009). Contrasts in the reactions of adults to the calls of younger versus older animals may be a result of a generalised tendency to monitor the reliability of calls and to adjust responses to their likely accuracy. For example, experiments with ground squirrels in which the alarm calls of particular individuals were made more or less reliable by pairing them with or without a model predator showed that the

(a)   (b)

Figure 7.8 (a,b) Contrasting responses of vervet monkeys to playback of different alarm calls. When they are played calls that signal danger from aerial predators, they look up, while when they are played alarm calls given to snakes, they look down. *Source*: © Robert Seyfarth.

animals quickly learned to discriminate between reliable and unreliable signals (Hare and Atkins 2001).

Alarm calling may benefit callers in at least two different ways (Caro 2005; Blumstein 2007). First, it may generate direct benefits to the fitness of callers if the responses of receivers improve the caller's own survival chances. In some cases, alarm calls may inform predators that they have been detected and encourage them to pursue other prey, reducing the chance that callers will be attacked (Woodland *et al.* 1980; Bildstein 1983; Caro *et al.* 2004; Caro 2005) and several empirical studies suggest that alarm calls can work in this way. For example, the alarm snorts of ungulates and the alarm calls of arboreal monkeys that have detected predators both appear to reduce the risk that callers will be selected

by predators (Zuberbühler *et al.* 1999; Caro 2005). Similarly, individual differences in the frequency and duration of stotting in Thompson's gazelles are related to their condition and to the chance that they will be caught by cursorial predators and appear to encourage wild dogs to select other prey (Fitzgibbon and Fanshawe 1988; Caro 1994, 2005). Alarm calls may also encourage predators that hunt by stealth to leave the area, reducing the need for their prey to monitor their movements (Tilson and Norton 1981; Zuberbühler *et al.* 1999). A final possibility is that alarm calling benefits callers because it induces responses in other group members that increase the chance that they will be selected by predators (Charnov and Krebs 1975), but there is little evidence of effects of this kind. For example, a detailed study of chipmunks

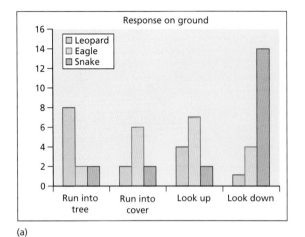

(a)

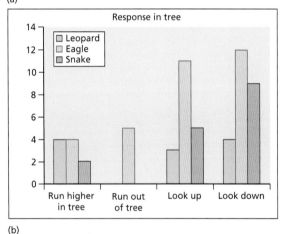

(b)

Figure 7.9 Responses of vervet monkeys to playbacks of alarm calls given to aerial predators, terrestrial predators and snakes: (a) responses to playbacks of calls given to leopards, eagles and snakes by animals on the ground; (b) similar responses given by animals already in the trees. *Source*: From Seyfarth *et al.* (1980b). Reproduced with permission of Elsevier.

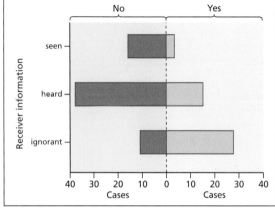

Figure 7.10 Alarm calling by chimpanzees in relation to a presented snake. Individuals that had already seen a snake were more likely to call if others were not aware of the danger ('ignorant') than if they showed some concern but could have seen the snake ('heard') or if they were obviously aware of the snake ('seen'). Dark bars indicate no alarm calls produced; light bars indicate at least one alarm call produced. 'Receiver information' indicates receiver ignorance or knowledge from the perspective of the subject, divided into the following three categories. 'Seen' indicates knowledgeable receivers: the subject had seen all receivers see the snake model. 'Heard' indicates partially knowledgeable receivers: the subject had heard an alarm call when all receivers were within 50 m of the snake model but could not have seen all receivers see the snake model. 'Ignorant' indicates that the subject could not have seen all receivers see the snake and had not heard an alert hoo when all current receivers were within earshot (50 m) of the alert hoo. *Source*: From Crockford *et al.* (2012). Reproduced with permission of Elsevier. *Photo source*: © Kathelijne Koops.

found no evidence that calling increased the survival of callers as a result of the responses of receivers (Smith 1978).

Second, alarm calling may alert other group members to the presence of danger and reduce the chance that they will be taken by predators and there is substantial evidence not only that they do so but that they are adapted to have this effect (see Figure 7.10). Alarm calls typically cause other group members to scan for danger or run to take shelter in burrows or trees and solitary individuals are frequently more likely to be killed by predators than those in groups (see Chapter 2). In addition, variation in the rate at which alarm calls are given, their amplitude and structure are often associated with the urgency or magnitude faced by receivers rather than by callers (Zuberbühler 2009). For example, in blue monkeys, the frequency with which males give alarm calls is related to the proximity of other group members to a potential predator rather than to the caller's own proximity (Papworth *et al.* 2008), while in Thomas' langurs breeding males continue to give alarm calls to predators until all other group members have responded with an alarm call and are presumably aware of the danger (Wich and de Vries 2006). In chimpanzees, alarm calls are frequently directed at 'friends' or allies, callers monitor the reactions of individuals and calling ceases once individuals have located the risk (Figure 7.10): for example, group members are more likely to give alarm calls if they perceive snakes and other individuals are still ignorant of their danger. Finally, both interspecific and intraspecific comparisons indicate that alarm calling increases with group size and that solitary animals either seldom give alarm calls or do not do so at all (Blumstein and Armitage 1997b; Blumstein 2007). For example, in yellow mongooses (a facultatively social species where individuals either forage alone or with one or two others) as well as in meerkats, individuals only give alarm calls while foraging in groups (Le Roux *et al.* 2008).

Where alarm calls are maintained by selection operating through their effects on other group members, several different mechanisms may be involved. First, alarm calling may represent an extended form of parental care and calling may be maintained by its effects on the fitness of offspring. In several species, alarm calls are given most frequently when vulnerable juveniles are with the group (Schwagmeyer 1980). For example, in capuchin monkeys, the likelihood that females will give alarm calls increases when dependent offspring are present (Wheeler 2008), while in yellow-bellied marmots alarm calling rates differ between adults and are associated with the presence of dependent young rather than with variation in relatedness between group members (Blumstein and Armitage 1997b; Blumstein and Daniel 2004).

Second, indirect fitness benefits may be involved. For example, in Belding's ground squirrels, females living with their daughters or sisters give alarm calls more frequently than those living with more distant relatives (Sherman 1980, 1981). Similarly, in prairie dogs (Figure 7.11), females without resident offspring but with non-descendent relatives in the group called as

Figure 7.11 In black-tailed prairie dogs, the probability that adults of either sex will give alarm calls is related to the presence of kin in the group. *Photo source*: © Elaine Miller Bond.

frequently as those with offspring, suggesting that benefits to non descendent kin are likely to be involved (Hoogland 1995) (Figure 9.2). Comparisons of sex differences in alarm calling also support the suggestion that indirect fitness benefits play a role. In species where females live in kin-based groups while males spend much of the year in open groups consisting primarily of unrelated individuals (as in many ungulates and some social rodents), males either give alarm calls less frequently than females or do not do so at all (Hirth and McCullough 1977; Sherman 1980, 1981; Clutton-Brock *et al.* 1982; Hoogland 1995).

A third possibility is that where the fitness of individuals increases with group size, alarm calling is maintained by mutualistic benefits or by some form of group selection (Kokko *et al.* 2001). While alarm calling appears to be most highly developed where relatives are involved, alarm calls are also given in groups consisting of unrelated animals and, in kin-based groups, by unrelated immigrants as well as natal animals, suggesting that mutualistic benefits operating through either group augmentation or some form of group selection could be important (see Chapter 1). As more data become available, it would be useful to investigate whether the frequency of alarm calls is consistently higher in species where individual fitness increases with group size compared with those where it declines (see Chapter 2).

A final suggestion is that giving alarm calls is an evolutionary handicap to callers enabling them to gain social prestige and to impress competitors or potential breeding partners (Zahavi and Zahavi 1997). If so, alarm calling would be expected to involve substantial risks and alarm calls should be given principally by phenotypically superior animals. However, there is little evidence that directly supports this suggestion or which indicates that alarm calling serves to increase mating success (Clutton-Brock *et al.* 1999; Wright *et al.* 2001). It is also uncertain that giving alarm calls involves substantial costs to the callers. Evidence that the structure of calls in smaller species is adapted to minimise the chance that predators will detect the caller suggests that there must be costs associated with giving alarm calls and, in some mammals, individuals only give alarm calls from positions of safety: for example, klipspringers that have been chased by jackals give loud alarm calls after reaching safety (Tilson and Norton 1981). However, direct evidence that alarm calling increases the risk that callers will be caught by predators is scarce. One of the few exceptions involves Belding's ground squirrels, where individuals that give alarm calls are more likely to be captured by terrestrial predators than individuals that do not call (Sherman 1985) (Figure 7.12). Whether the absence of such other studies suggests that giving alarm calls seldom has substantial costs or whether it merely reflects the difficulty of measuring variation in the rate of attacks by predators is uncertain.

Sentinel calls

In some birds and mammals that breed cooperatively, including babblers, dwarf mongooses and meerkats, group members take turns to watch for danger from elevated positions while others feed. Vigilance commonly increases in potentially dangerous situations and is positively correlated with glucocorticoid levels (Voellmy *et al.* 2014). In some of these species, sentinels also give repeated calls that inform others that they are vigilant (Rasa 1986; Ridley *et al.* 2013) (see Chapter 17) (Figure 7.13). These calls help to increase the foraging

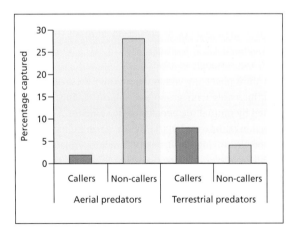

Figure 7.12 Costs of giving alarm calls in Belding's ground squirrel. While individuals that called to aerial predators were less likely to be captured by predators than those that did not call, individuals that called in response to terrestrial predators were more likely to be captured. *Source*: Data from Caro (2005) and Sherman (1985).

Figure 7.13 In meerkats, one group member commonly acts as a sentinel while the rest of the group are foraging. If they detect predators, sentinels give alarm calls which contain information about the type of danger and the urgency of the situation. *Source*: © Tim Clutton-Brock.

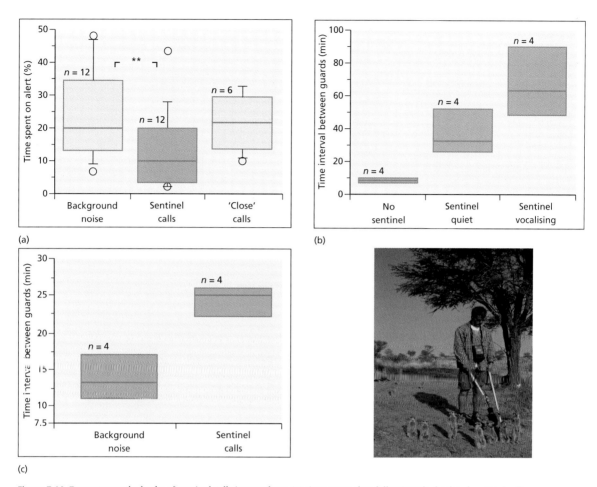

(a)

(b)

(c)

Figure 7.14 Response to playbacks of sentinel calls in meerkats: (a) time spent alert following playbacks of sentinel calls versus contact calls; (b) interval between times when foraging individuals scanned their surroundings (guards), depending on whether there was no sentinel up, a sentinel that was not vocalising, or a sentinel emitting calls; (c) interval between guards when playbacks of background noise or sentinel calls were broadcast. The box plots show 25th, 50th and 75th percentiles. *Source*: From Manser (1999). Reproduced with permission from the Royal Society. *Photo source*: © Tim Clutton-Brock.

efficiency of other group members: for example, playback of sentinel calls to foraging meerkats shows that sentinel calls reduce the amount of time that foraging animals spend alert (Figure 7.14) (Manser 1999) and studies of babblers have produced similar results (Hollén *et al.* 2008). In dwarf mongooses, the probability that sentinels will vocalise when on 'sentinel duty' varies in relation to the habitat their group is foraging in and sentinels vocalise more frequently when groups are foraging in dense cover where visibility is restricted (Figure 7.15a) and when group members are relatively widely separated (Figure 7.15b) (Kern and Radford

2013). Sentinels are also particularly likely to vocalise shortly after an alarm call has been given (Figure 7.15c), perhaps signalling to other group members that the danger is past and they can resume foraging. Individuals vary their rate of calling, and reduce their rate when the risk of predation is high.

Sexual signals

Although females seldom engage in such elaborate displays to attract breeding partners as males do (see Chapter 13), female signals of reproductive status or motivation that attract potential mating partners are

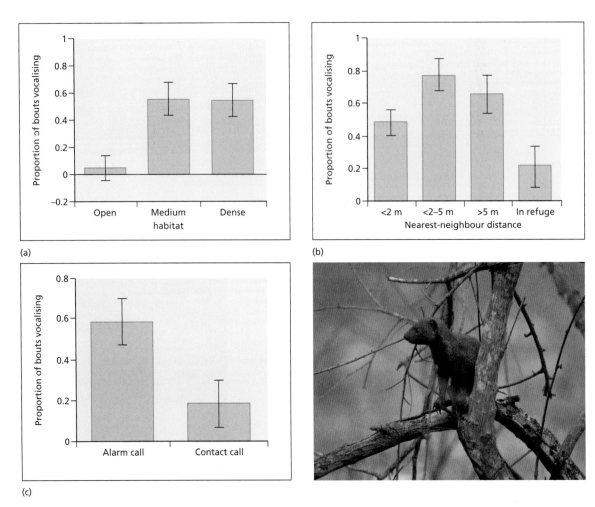

Figure 7.15 Dwarf mongooses adjust sentinel calls to the risk faced by other group members. Histograms show the proportion of bouts when sentinels gave calls when groups were in open areas versus cover (a) when other group members were nearby or in a refuge versus times when they were further away (b), after an alarm call had been recently played to them compared to periods after they had been played a control call (c). *Source*: From Kern and Radford (2013). Reproduced with permission of Elsevier. *Photo source*: © Amy Morris-Drake.

common among mammals (Clutton-Brock and Huchard 2013). Olfactory signals are widespread and commonly exert an important influence on the behaviour of adults of both sexes (Johnston 1986; Kappeler 1998; Slade *et al.* 2003). For example, female ring-tailed lemurs treated with contraceptives show significant changes in genital odorants and become less attractive to males (Crawford *et al.* 2011). Olfactory signals also provide a basis for recognition of individuals (Hurst 1990; Swaisgood *et al.* 1999; Hurst *et al.* 2001; Nevison *et al.* 2003) as well as

information on pair-wise relatedness and variation in heterozygosity (Charpentier *et al.* 2010; Leclaire *et al.* 2013).

Visual signals of sexual status are also common (Tobias *et al.* 2012; Clutton-Brock and Huchard 2013). In some primates, the skin colour of females changes throughout the breeding cycle. For example, in rhesus macaques, the facial colouring of females changes throughout the breeding cycle and females with brighter faces are more likely to attract males (Dubuc *et al.* 2009). In other species, females have developed specialised ornaments that signal changes in receptivity or fertility. For example, in gelada baboons, females have prominent chest patches whose size and colour changes throughout the breeding cycle (Figure 7.16) while some primates that live in multi-male groups have developed large perineal swellings that vary in size and colour throughout the breeding cycle (Box 7.1 and Figures 7.17 and 7.18). Detailed

(a) (b) (c) (d)

Figure 7.16 Cyclical changes in the chest patches of female gelada baboons: (a) an immature female prior to her first sexual swelling of the chest; (b) a lactating female with a relatively pale chest patch; (c) a nulliparous female with her first chest swelling (note there are no 'vesicles' surrounding the patch); (d) a nulliparous female after several chest swellings (note the 'vesicles' surrounding the patch). *Source*: © Jacintha Beehner.

Box 7.1 Sexual swellings in primates

In some primates, breeding females have distinctive swellings of the perineal region that change in size and coloration throughout the reproductive cycle (see Figure 7.17). Perineal swellings are most commonly found in species that usually live in groups that include multiple breeding females and multiple breeding males, where females often mate with multiple partners (Clutton-Brock and Harvey 1976; Zinner *et al.* 2004). In these species, females may increase their fitness by mating with multiple males, either because this increases their fecundity (see Chapter 4) or because it increases the number of males that tolerate or support their offspring (see Chapter 15). The size and brightness of a female's swelling usually increases as she approaches ovulation, though temporal changes do not signal proximity to ovulation precisely and there is considerable interspecific variation in the relationship between changes in swelling size and ovulation (Deschner *et al.* 2003, 2004; Gesquiere *et al.* 2007; Higham *et al.* 2008).

(a)

(b)

(c)

Figure 7.17 Cyclical changes in sexual swellings in yellow baboons: photographs show changes in the size of the swelling on the same female through the reproductive cycle. *Source*: © Courtney Fitzpatrick.

Some studies of baboons and macaques have suggested that individual differences in the size of female swellings (see Figure 7.18) are correlated with individual differences in fecundity and have suggested that swellings may have originated as a signal of receptivity and subsequently evolved to signal differences in individual quality (Domb and Pagel 2001), an idea referred to by primatologists as the *reliable indicator hypothesis*. However, other studies have found no consistent relationships between individual differences in the size of female swellings and fecundity and little evidence that males consistently favour females with relatively large swellings (Zinner *et al.* 2002; Engelhardt *et al.* 2005). For example, a detailed study of mating behaviour in yellow baboons found that females with relatively large swellings do not have higher reproductive success and are not consistently preferred as partners by dominant males (Fitzpatrick *et al.* 2015). Instead, female fertility increased in the course of successive cycles and dominant males preferred partners who had also raised offspring recently. Since the relative size of swellings increases with cycle number, one possibility is that perineal swellings have evolved to signal temporal changes in female fertility rather than individual differences. This need not, of course, suggest that female primates have not evolved a range of other signals that reflect their receptivity or relative fertility, including olfactory signals and changes in behaviour.

(a)   (b)

Figure 7.18 (a,b) Individual differences in the size of sexual swellings in yellow baboons. Two individuals, both photographed at maximal swelling size for that cycle, representing the smallest swelling and the largest swelling in the group. *Source*: © Courtney Fitzpatrick.

studies show that the size and colour of the swelling is associated with changes in fertility, though relationships are not always close and vary widely between species (see Box 7.1).

Copulation calls

In some mammals, including elephant seals, lions, elephants, several cercopithecine monkeys and chimpanzees, copulating females give loud calls that frequently attract other males and may increase both the number of males they mate with and the relative status of their partners (Cox and Le Boeuf 1977; Poole 1989; Pradhan *et al.* 2006). Like perineal swellings, these calls are found in species where females live in social groups where they have access to more than one potential mating partner: for example, among primates they occur in many species that live in multi-male groups (Maestripieri and Roney 2005). In some cases, copulation calls signal the female's breeding status: for example, in yellow baboons, the copulation calls of females indicate the caller's proximity to ovulation (Semple *et al.* 2002), though other studies have found no close relationship between the incidence of copulation calls and changes in female fertility (O'Connell and Cowlishaw 1994; Pfefferle *et al.* 2008; Townsend *et al.* 2008).

Copulation calls may have several different functions and, once again, it is difficult to distinguish between them. They may alert and attract dominant males or consorts, and so may reduce the risk of dangerous harassment by multiple males or increase the genetic quality of mating partners (Pradhan *et al.* 2006). For example, in elephant seals, females vocalise when mounted by younger males and their calls frequently attract the attention of the dominant males (Cox and Le Boeuf 1977). In some primates, the copulation calls of particular females are individually recognisable and calls may stimulate consorting males to guard the female more closely (Semple 2001). Copulation calls may also serve to attract additional mates and so may enhance sperm competition or reduce paternity certainty and increase the number of males that will protect the female's offspring (O'Connell and Cowlishaw 1994) (see Chapter 15). For example, playback experiments with Barbary macaques show that they increase the probability that other males will approach females and shorten latency between matings (Semple 1998). One possible explanation of why copulation calls are not always closely associated with female fertility is that it may pay females to avoid signalling ovulation so precisely that males are able to assess their probability of paternity accurately and avoid protecting young that are not their own (Pradhan *et al.* 2006). Another is that although there may be no close association between female fertility and whether or not copulation calls are given, details of the structure of calls provide an accurate signal of the female's proximity to ovulation (Semple 1998).

Threats and reassurance

While escalated fights between females are typically less frequent than fights between males, they occur intermittently in most mammalian societies and are relatively frequent in some (see Chapter 8). However, even where reproductive competition is more intense between females than males, females seldom possess more elaborate weaponry than males (Young and Bennett 2013). A possible explanation is that females use threats to deter or displace competitors more commonly than males since escalated contests are more likely to affect female fitness because of their effects on abortion or on the survival of offspring, with the result that selection for aggression and weapon development is not as strong in females as in males (Cant and Johnstone 2009; Cant et al. 2010; Cant and Young 2013).

In some species, members of both sexes combine threatening gestures with facial expressions that signal their intention to attack if recipients do not withdraw or modify their behaviour. Among primates, visual signals of this kind are often associated with vocal signals (Ghazanfar and Logothetis 2003) and muscle action (Parr et al. 2007) and are particularly well developed in chimpanzees (Figure 7.19) (Burrows et al. 2006).

Facial expressions are commonly monitored by nearby group members (Goossens et al. 2008) and can affect the attention they pay to interactions between third parties. Females also give threat calls in support of relatives or unrelated allies: for example, in wild baboons, females give threat calls when observing disputes involving their relatives and playback experiments suggest this helps to provide support for relatives and deters their opponents from escalating aggression (Wittig et al. 2007).

As well as giving threats, females give reassuring calls that signal their peaceable intentions (see Chapter 8). For example, in baboons, dominant females that approach subordinates give grunts that indicate that they do not intend to attack or displace them (Cheney et al. 1995; Cheney and Seyfarth 2008). Similarly, in rhesus macaques, giving reassuring grunts increases the chance that females will be allowed to handle the infants of other females (Silk 2001), while in baboons reassuring grunts also play an important role in reconciling individuals that have recently been involved in aggressive interactions with each other.

Territorial signals

In socially monogamous mammals, both females and males are commonly involved in territorial displays. For example, in dik dik and klipspringers, females as well as males leave scent marks on prominent sites in

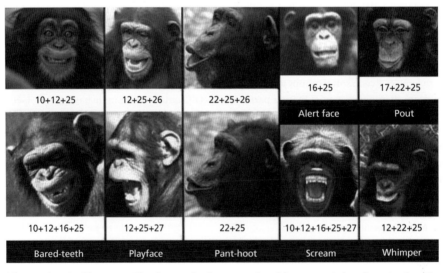

Figure 7.19 Facial expressions in chimpanzees. The photographs show examples of the prototypical configuration for chimpanzee facial expressions identified using discriminant function analysis. *Source*: From Parr *et al.* (2007). Reproduced with permission of L.A. Parr.

Figure 7.20 Wolves howling in chorus. Choruses of this kind are common in larger mammals where groups defend sizeable territories. They often help group members to locate each other as well as signalling the number of individuals present in groups. *Source*: © Juniors Bildarchiv GmbH/Alamy Stock Photo.

their territories (Komers and Brotherton 1997; Roberts and Dunbar 2000). Females mark more frequently when they are sexually receptive than at other times, though they continue to mark throughout the year. In several socially monogamous primates, females are also involved in vocal displays. For example, in resident mated pairs of gibbons, females duet with their male partners and playback of female songs elicits calling by both sexes and approaches to the playback site that are led by females (Mitani 1984; Cowlishaw 1992).

Females also contribute to territorial defence in some polygynous species. For example, females as well as males contribute to vocal choruses in wolves, lions and howler monkeys (Figures 7.20, 7.21 and 7.22) (Harrington and Mech 1979; Sekulic 1982; Raemaekers and Raemaekers 1984; East and Hofer 1991). In African lions, females advertise their presence and recruit other group members by synchronised roaring choruses which serve to deter intruders (McComb *et al.* 1994) and playback experiments have shown that individuals are able to assess the number of different individuals involved in choruses (Wilson *et al.* 2001; Benson-Amram *et al.* 2011).

Resident lionesses without cubs usually approach playbacks of groups of intruders that they outnumber by at least two members but seldom approach larger groups (Figure 7.23). Similarly, parties of spotted hyenas are usually reluctant to approach playbacks of larger parties than their own (Figure 7.24).

In some mammals, females also play a proactive role in competition with neighbouring groups, probing the ranges of their neighbours and attacking or killing dependent juveniles or solitary adults if they come across them (Pusey and Packer 1993). Attacks are not confined to dependent young: for example, groups of female hyenas that encounter solitary adults from neighbouring clans commonly attack and sometimes kill them (Kruuk 1972; Boydston *et al.* 2001) and female meerkats and chimpanzees can also be involved in lethal attacks on parties of adults from neighbouring groups (Bygott 1979; Clutton-Brock and Manser 2015). Since larger groups usually displace smaller ones in these species (see Chapter 2), tactics that prevent neighbouring groups from breeding or which reduce their size are likely to increase the fitness of attackers.

Figure 7.21 In lions, both sexes commonly roar in chorus, signalling their presence and the number of individuals in the group. *Source*: © Gallo Images/Dave Hamman. Digital Vision/Getty Images.

Figure 7.22 Choruses are also common in primates where groups defend territories, as in ursine (red) howler monkeys. *Source*: © Carolyn M. Crockett.

(a)

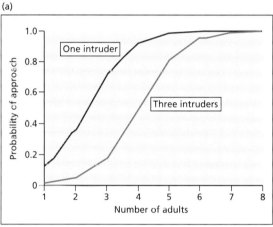

(b)

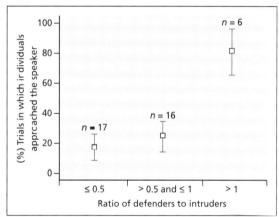

Figure 7.24 Assessment of intruding groups in spotted hyenas. Proportion of playback experiments in which spotted hyenas approached speakers playing calls of fewer individuals than that of the resident group, of similar numbers to the resident group and of larger numbers of intruders. *Source*: From Benson-Amram *et al.* (2011). Reproduced with permission of Elsevier. *Photo source*: © Marion Fast

Figure 7.23 Assessment of intruding groups in African lionesses. (a) Lionesses listening to playback of roars simulating an intruding group. (b) Probability that residents approached the source of playbacks of one and three intruders in relation to the number of adults present. *Source*: From McComb *et al* (1994). Reproduced with permission of Elsevier. *Photo source*: © Dave Hamman.

In territorial species, individuals may benefit from the ability to recognise the level of threat posed by intruders and to discriminate between neighbours and intruders as well as between kin and non-kin. Several studies show that individuals discriminate the signals of neighbours from those of unknown intruders, sometimes reacting more strongly to unfamiliar individuals (Zenuto 2010) and sometimes more to familiar ones (Müller and Manser 2007). In some mammals, individuals also discriminate between kin and non-kin by phenotype matching, using templates based either on their own phenotype or on that of relatives (Holmes and Sherman 1982; Mateo and Johnston 2000). Both olfactory and vocal cues can be used for this purpose so long as they show high levels of heritable variation (Tibbetts and Dale 2007) and several studies have now shown that signals used to distinguish kin from non-kin have relatively high heritability (Scherrer and Wilkinson 1993). For example, North American red squirrels use individually variable 'rattle' calls to claim ownership of their territory, responding more strongly to calls given by non-kin than to those given by relatives, and analyses confirm that variation in the structure of these calls has a large heritable component (Wilson *et al.* 2015).

7.3 Signalling in theory and practice

Repertoire size and signal variability

As research on mammalian signalling systems has evolved, the subtlety and complexity of signals has gradually come to be appreciated. Signals often convey information about the signaller's age, condition and emotional state as well as about the specific circumstances it has encountered (Cheney and Seyfarth 1990; Manser 2009, 2010). In some cases, they can involve a single modality, but they commonly involve several modalities that interact to provide precise information (Proops *et al.* 2009) and the sequence in which different signals are given can affect their meaning and information content (Kershenbaum *et al.* 2014). Calls often differ between individuals and animals that exchange signals frequently (including mothers and offspring, mated pairs, members of stable groups and territorial neighbours) are often able to recognise each other's signals, allowing them to respond appropriately to the calls of different individuals (Rendall 2003; Townsend *et al.* 2014). In at least some species, individuals also recognise and monitor signals between third parties, providing more extensive information about potential competitors and allies than could be gained from direct experience alone (Cheney and Seyfarth 1999, 2008).

The size of signal repertoires varies widely between species (Bradbury and Vehrencamp 1998). One explanation of these differences, sometimes referred to as the *social complexity hypothesis*, argues that the complexity of signalling systems increases in species where individuals interact with many different individuals in different contexts and comparative studies provide some support for this idea (Blumstein 2007; Freeberg *et al.* 2012; Krams *et al.* 2012). For example, across primate species, repertoire size increases with average group size (McComb and Semple 2005) (Figure 7.25a), while in mongooses and mole rats repertoire size is larger in social than in solitary species (though, among social species, the number of distinct call types does not increase with group size) (Knotkova *et al.* 2009; Manser *et al.* 2014).

The diversity of visual and olfactory signals may also be related to group size. For example, across samples of

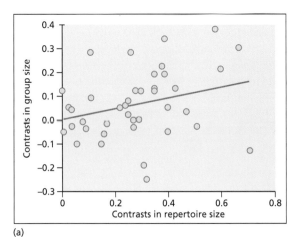

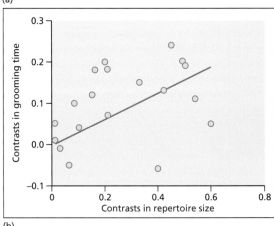

Figure 7.25 Repertoire size and group size in different primate species. Relationships between (a) contrasts in repertoire size and contrasts in group size and (b) contrasts in repertoire size and contrasts in time spent grooming. Repertoire size was square root transformed, group size was log transformed and percentage grooming time was log + 1 transformed. *Source:* From McComb and Semple (2005). Reproduced with permission from the Royal Society.

primate species, the diversity of facial expressions increases with average group size, (Dobson 2009) and in rodents the complexity of chemical signals also rises with group size (delBarco-Trillo *et al.* 2011). In contrast, there is little evidence that the structure of mating systems affects the complexity of signals (Shultz and Dunbar 2007): for example, among primates, repertoires of vocal and visual signals do not appear to be larger in polygamous species than in monogamous ones (Freeberg *et al.* 2012).

Group size is unlikely to be the only social parameter that affects repertoire size. Across different species of ground squirrels, the number of distinct alarm calls rises with indices of the complexity of social roles rather than with group size (Blumstein and Armitage 1997b; Pollard and Blumstein 2012), whereas vocal repertoire size in primates is correlated both with group size and with the proportion of time spent allo-grooming (McComb and Semple 2005) (see Figure 7.25b). Ecological factors may also be important: although few systematic attempts have yet been made to investigate relationships between ecology and repertoire size in mammals, studies of other animal groups (including birds and frogs) suggest that contrasts in habitat, body size and phylogeny may have a stronger influence on the complexity of communication systems than variation in social organisation (Ord and Garcia-Porta 2012). While this does not necessarily imply that the same is true for mammals, it emphasises the need for studies of signal complexity to determine whether correlations between signal complexity and social organisation persist when the effects of ecological differences are allowed for.

Contrasts in social organisation may affect the need for individuals to recognise each other's signals and this may have consequences for the variability of signals. For example, across different species of sciurids, the individuality of calls increases with group size (Pollard and Blumstein 2012) (Figure 7.26). Similarly in bats, are positively associated

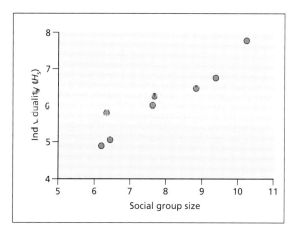

Figure 7.26 Individuality of calls increases with group size across sciurid rodents. The plot shows the relationship between an estimate of alarm call individuality and average group size. Points show raw data but a phylogenetically controlled analysis of independent contrasts produced similar results. *Source*: From Pollard and Blumstein (2012). Reproduced with permission from the Royal Society.

with differences in the individuality of isolation calls given by juveniles interspecific differences in colony size (Wilkinson 2003). In addition, individuals may meet the need for more specific or more complex communication by varying the complexity of sequences of calls (Kershenbaum *et al.* 2014).

Variation in social complexity may also affect the extent to which signals vary within individuals, for studies of an increasing number of signalling systems suggest that individuals commonly adjust their signals in relation to their target audiences. For example, male rock hyraxes give complex sequences of calls and modify the complexity and duration of their 'songs' in relation to the number of other individuals that are listening (Dermatsev *et al.* 2014). Similarly, captive orangutans appear to adjust their gestural signals in relation to the extent to which they are being understood by human observers, narrowing the range of signals when observers show partial understanding and elaborating their range of gestures when they are not understood (Cartmill and Byrne 2007).

Early studies of animal signals commonly suggested that the sophistication, content and complexity of signals increase in relation to species differences in cognitive abilities (see Chapter 6). Satisfactory comparisons of the complexity of signalling systems are problematic but, as studies of communication have progressed, it has become increasingly doubtful whether this is the case. For example, it is not clear that the vocal signals of chimpanzees are more sophisticated or more complex than those of monkeys or some social birds. Instead of reflecting variation in cognitive abilities, variation in the content and complexity of animal signals may reflect the relative needs of individuals to convey different forms of information to different receivers or audiences.

Honesty and deception

In contrast to humans, where social pressures and third party punishment encourages individuals to tell the truth (see Chapter 19), animals have no commitment to give honest signals, and we should expect individuals both to conceal information and to give misleading signals when it is in their interests to do so (Dawkins and Krebs 1978; Krebs and Dawkins 1984). Studies of animal signalling systems have produced substantial evidence that individuals often avoid providing information that is likely to jeopardise their interests and that 'factual concealment' is common (Whiten and Byrne 1988; de Waal 1992). Individuals also adjust their actions to the presence or absence of other individuals: for example, male vervet monkeys vary their rates of affiliative and aggressive behaviour towards infants depending on the presence of their mothers (Keddy Hector *et al.* 1989) and female chimpanzees avoid giving copulation calls when dominant females are nearby (Townsend *et al.* 2008). Females sometimes use similar tactics to avoid male control. For example, in gelada baboons, females involved in extra-pair copulations are often punished by dominant males and are more likely to mate when the resident male is some distance away and are less likely to vocalise when they do so (Le Roux *et al.* 2013) (Figure 7.27).

Individuals might also be expected to attempt to deceive other group members where this is to their advantage (Dawkins and Krebs 1978; Krebs and Dawkins 1984) and several cases of deceptive communication have been described. For example, vervet monkeys sometimes give deceptive alarm calls to intruding neighbours (Cheney and Seyfarth 1990). Some primates appear to react to non-existent external events in order to redirect the attention of competitors (de Waal 1992). For example, juvenile baboons which are close to subordinate adults that are in possession of valued food items sometimes give screams similar to those usually uttered by infants that have been attacked and, by doing so, provoke their mothers to attack the subordinate, allowing them to steal the food item (Byrne and Whiten 1985, 1997). In putty-nosed monkeys, males give alarm calls that are usually used to respond to leopards to prompt their group to change activity or to move on (Arnold and Zuberbühler 2006a,b) and, in topi, males give alarm snorts that help to retain receptive females in their territories when no predators are around (Bro-Jørgensen and Pangle, 2010).

While deceptive behaviour has been most extensively investigated in primates, it is not confined to primates and its distribution does not appear to be closely associated with cognitive abilities. For example, grey squirrels appear to mislead competitors or potential scroungers over the location of their caching sites by covering a proportion of sites without burying food there (Steele *et al.* 2008). Some of the most complex examples of deception involve birds that mimic alarm calls of other species. For example, drongos often parasitise foraging meerkats, giving false alarm calls and snatching food that adults have given to dependent pups when they run to shelter (Figure 7.28). Some individuals have learned to mimic meerkat alarm calls and intersperse them with their own alarm calls, reducing the rate at which

meerkats habituate to misleading calls (Flower 2011; Flower and Gribble 2012).

While there are a number of well-known cases of dishonest signalling, examples are not common and evidence of honest signalling is more abundant. One possible explanation is that individuals that give misleading signals are punished by animals that detect them (Rohwer 1977, 1982). However, although misleading threats are sometimes punished, there is limited evidence that individuals that give inaccurate signals become a target of aggression. A more likely explanation is that in species where the same animals interact repeatedly, selection has favoured an ability to disregard unreliable signals, with the result that individuals rapidly learn to identify inaccurate signallers and cease reacting to them. This may explain why many of the clearest examples of deception come from studies of communication between individuals from different groups or species (Caro 2014), where the same individuals do not interact frequently, so that cheats may be hard to identify. A final possibility is that we may overestimate the extent of conflict between signallers and recipients. While some differences between the interests of senders and recipients is usual, most of the call types reviewed in this chapter are not overtly competitive and several are likely to increase the inclusive fitness of the caller as well as that of recipients, so that selection for dishonesty is weak.

Signal reliability

The extent to which interest in signal evolution has focused on the competitive signals of males has tended to obscure the fact that many signals are likely to benefit both signallers and recipients. Under these circumstances, signals do not necessarily need to be costly to be credible (Silk *et al.* 2000; Maynard Smith and Harper 2003), their structure may not be closely correlated with the individual's quality or competitive ability and they may not have high fitness costs (Maynard Smith and Harper 2003).

Where there are conflicts of interest between signallers and recipients, some mechanism presumably maintains signal reliability and discourages dishonest signalling. There is little evidence that dishonest or inaccurate signallers are punished but, where individuals live in close proximity and interact repeatedly, it may be difficult for individuals to deceive each other effectively and signal reliability may be constrained by

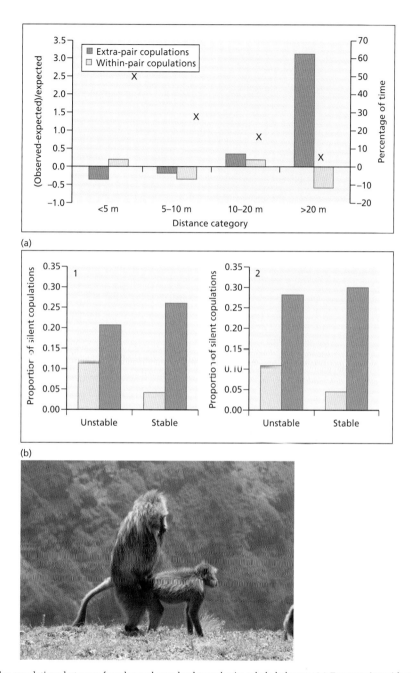

(a)

(b)

Figure 7.27 Stealthy copulations between females and non-leader males in gelada baboons. (a) Frequencies with which copulations occurred at different distances between the leader and follower male: extra-pair copulations occur significantly more often when leader males were >20 m away from the copulating pair ('X' indicates the percentage of time that leader and follower males spend at different distances from each other). (b) Proportion of copulations in which (1) the female and (2) the male remains silent when engaged in extra-pair and within-pair copulations during stable and unstable periods. During both periods, both sexes were significantly more likely to remain silent when engaged in an extra-pair copulation. *Source*: From Le Roux *et al.* (2013). Reproduced with permission of Macmillan Publishers Ltd. *Photo source*: © Clay Wilton.

Figure 7.28 A drongo watches a foraging meerkat pup closely, waiting for an opportunity to steal any prey it finds or is given. *Source*: © Tim Clutton-Brock.

the ineffectiveness of deception (see previous section). In some cases, signal reliability may be maintained by unfakeable relationships between signal characteristics and individual quality. For example, the structure of roars in red deer stags provides an index of their body size (see Chapter 13). Similarly, territorial choruses in lions and spotted hyenas probably provide reliable indices of their group's competitive ability because they reveal the number of individuals present (McComb *et al.* 1994; Benson-Amram *et al.* 2011).

Some female signals could be handicaps. For example, anecdotal evidence suggests that the perineal swelling of female primates may have substantial costs, since swellings may be lacerated and can become infected and also elicit harassment by competing males (see Chapter 15). Perhaps the high costs of perineal swellings favoured their evolution as signals of quality and

their reliability is currently maintained because only superior females can afford the costs of maintaining large swellings to advertise their reproductive status, as handicap models suggest. However, producing firm evidence that signals evolved as handicaps and are maintained by their costs to inferior individuals is problematic. To show that a signal is a handicap, it is not sufficient to show that the signal has energetic costs, for many repetitive signals have high short-term energetic costs which do not necessarily involve substantial fitness costs (Clutton-Brock and Albon 1979). Nor is it sufficient to show that a signal has fitness costs for, where signallers are competing with each other, we can expect selection to exaggerate indices of quality or competitive ability to a point where the costs of their further development would offset any benefits so that, at equilibrium, competitive signals are often likely to have substantial costs that limit their further development. It is also insufficient to show that the fitness costs of the same signal are greater to weaker individuals than to stronger ones since it is (almost) inevitable that traits that have appreciable costs will depress the fitness of inferior individuals more than that of superior ones, so that the presence of disparities in cost need not indicate that they have played an important role in signal evolution (Maynard Smith and Harper 2003). Finally, signal reliability may commonly be maintained by quality-dependent trade-offs between signal characteristics and viability and differences in the efficiency of individuals in converting advertising into fitness (Getty 2006). Though this explanation of the maintenance of signal reliability appears feasible and the predictions of these models are not restricted to cases where the fitness consequences of signals are additive, realistic comparative measures of the viability costs of signals to individuals of varying quality are not yet available.

As an empirical test of the handicap principle, it is consequently worth asking whether there are convincing examples of competitive signals in non-human animals that are unnecessarily extravagant and wasteful, as handicap models of signalling predict. For example, do we observe powerful individuals in the opening stages of conflicts intentionally putting themselves at a disadvantage? Are signals frequently more costly than they need to be to maximise their efficacy? And where individuals are signalling to potential breeding partners, do they perform actions whose only conceivable function is to increase the strategic costs of signalling? There is little positive evidence of such tactics and many traits used in competitive signalling that initially appeared to be extravagant and wasteful have proved, on closer inspection, to be well adapted to exploiting the sensory modalities of the opposite sex or to be effective weapons (see Chapter 13). Since it is now clear that the handicap

mechanism proposed by Zahavi is not the only way by which signal reliability can be maintained, a broader approach to signal evolution is needed that incorporates quality-dependent viability trade-offs and explores the consequences of individual variation in the efficiency of signal production.

SUMMARY

1. Mammals use a wide range of signalling modalities, often combining visual, vocal and olfactory signals. While signallers and receivers sometimes gain mutual benefits from the transfer of information, other exchanges resemble arms races, where senders attempt to manipulate receivers to their own advantage and the latter attempt to anticipate the strategies of signallers and adjust their responses to maximise their own fitness.

2. Where the interests of signallers and receivers diverge, signals may be selected to give exaggerated or deceptive signals. Signal honesty may be preserved either if signals are indices and their frequency or structure is inevitably associated with the characteristics of signallers or if signals are handicaps and are so costly that only superior animals can afford to give them.

3. Group members commonly stay in contact with each other by giving regular contact calls. In addition, they use moving calls to encourage other group members to join them in moving to new resource patches and feeding calls to attract them to particular food supplies. Offspring often solicit food from their mothers with regular begging calls.

4. In many social mammals, individuals that detect predators give alarm calls that alert other group members to the danger. In some cases, these signal both the urgency of the risk and the category of danger. In some cooperative breeders, group members take turns to act as sentinels and give calls that inform other individuals that they are on guard.

5. Females also use visual and vocal signals to attract breeding partners, threaten rivals and deter intruders from other groups.

6. Species differ in the size of their signal repertoire and the variability of signals. Species differences in repertoire size and in signal variability both appear to correlate with variation in group size and structure.

7. While individuals often withhold information, deceptive signals directed at other group members do not appear to be common and many of the best examples of deception involve members of different groups.

8. In many cases, signals are likely to have evolved as indices and signal honesty is maintained through associations between the structure and frequency of signals and the characteristics of signallers. While theoretical studies show that signal honesty could be maintained through 'strategic' costs and signals could have evolved as handicaps, there is limited empirical evidence that this evolutionary pathway played an important role in signal evolution.

References

Alak, A. and Enquist M. (1995) Conflict, receiver bias and the evolution of signal form. *Philosophical Transactions of the Royal Society B: Biological Sciences* **349**:337–344.

Arnold, K. and Zuberbühler, K. (2006a) Language evolution: semantic combinations in primate calls. *Nature* **441**:303.

Arnold, K. and Zuberbühler, K. (2006b) The alarm-calling system of adult male putty-nosed monkeys *Cercopithecus nictitans martini*. *Animal Behaviour* **72**:643–653.

Aubin, T. and Jouventin, P. (1998) Cocktail-party effect in king penguin colonies. *Proceedings of the Royal Society of London. Series B: Biological Sciences* **265**:1665–1673.

Balcombe, J.P. (1990) Vocal recognition of pups by mother Mexican free-tailed bats, *Tadarida brasiliensis mexicana*. *Animal Behaviour* **39**:960–966.

Benson-Amram, S., et al. (2011) Numerical assessment and individual call discrimination by wild spotted hyaenas, *Crocuta crocuta*. *Animal Behaviour* **82**:743–752.

Biernaskie, J.M., et al. (2014) The evolution of index signals to avoid the cost of dishonesty. *Proceedings of the Royal Society of London. Series B: Biological Sciences* **281**:20140876.

Bigalke, R.C. (1972) Observations on the behaviour and feeding habits of the springbok, *Antidorcas marsupialis*. *Zoologica Africana* **7**:333–359.

Bildstein, K.L. (1983) Why white-tailed deer flag their tails. *American Naturalist* **121**:709–715.

Blumstein, D.T. (2007) The evolution of alarm communication in rodents: structure, function and the puzzle of apparently altruistic calling. In: *Rodent Societies: An Ecological and Evolutionary Perspective* (eds J.O. Wolff and P.W. Sherman). Chicago: University of Chicago Press, 317–327.

Blumstein, D.T. and Armitage, K.B. (1997a) Alarm calling in yellow-bellied marmots: I. The meaning of situationally variable alarm calls. *Animal Behaviour* **53**:143–171.

Blumstein, D.T. and Armitage, K.B. (1997b) Does sociality drive the evolution of communicative complexity? A comparative test with ground-dwelling sciurid alarm calls. *American Naturalist* **150**:179–200.

Blumstein, D.T. and Daniel, J.C. (2004) Yellow-bellied marmots discriminate between the alarm calls of individuals and are more responsive to calls from juveniles. *Animal Behaviour* **68**:1257–1265.

Boissy, A. (1995) Fear and fearfulness in animals. *Quarterly Review of Biology* **70**:165–191.

Bousquet, C.A.H. and Manser, M.B. (2011) Resolution of experimentally induced symmetrical conflicts of interest in meerkats. *Animal Behaviour* **81**:1101–1107.

Bousquet, C.A.H., *et al.* (2011) Moving calls: a vocal mechanism underlying quorum decisions in cohesive groups. *Proceedings of the Royal Society of London. Series B: Biological Sciences* **278**:1482–1488.

Boydston, E.E., *et al.* (2001) Sex differences in territorial behaviour exhibited by the spotted hyena (Hyaenidae: *Crocuta crocuta*). *Ethology* **107**:369–385.

Bradbury, J.W. and Vehrencamp, S.L. (1998) *Principles of Animal Communication*. Sunderland, MA: Sinauer Associates.

Branch, L.C., *et al.* (1993) Recruitment, dispersal, and group fusion in a declining population of the plains vizcacha (*Lagostomus maximus*; Chinchillidae). *Journal of Mammalogy* **74**:9–20.

Bro-Jørgensen, J. and Pangle, W.M. (2010) Male tope antelopes alarm snort deceptively to retain females for mating. *American Naturalist* **176**:E33–E39.

Burrows, A.M., *et al.* (2006) Muscles of facial expression in the chimpanzee (*Pan troglodytes*): descriptive, comparative and phylogenetic contexts. *Journal of Anatomy* **208**:153–167.

Bygott, J.D. (1979) Agonistic behaviour, dominance, and social structure in wild chimpanzees of the Gombe National Park. In: *Great Apes* (eds D.A. Hamburg and E.R. McCown). Menlo Park, CA: Benjamin Cummings, 405–427.

Byrne, R.W. and Whiten, A. (1985) Tactical deception of familiar individuals in baboons (*Papio ursinus*). *Animal Behaviour* **33**:669–673.

Byrne, R.W. and Whiten, A. (1997) Machiavellian intelligence. In: *Machiavellian Intelligence II: Extensions and Evaluations* (eds A. Whiten and R.W. Byrne). Cambridge: Cambridge University Press, 1–23.

Cant, M.A. and Johnstone, R.A. (2009) How threats influence the evolutionary resolution of within-group conflict. *American Naturalist* **173**:759–771.

Cant, M.A. and Young, A.J. (2013) Resolving social conflict among females without overt aggression. *Philosophical Transactions of the Royal Society B: Biological Sciences* **368**:20130076.

Cant, M.A., *et al.* (2010) Reproductive control via eviction (but not the threat of eviction) in banded mongooses. *Proceedings of the Royal Society of London. Series B: Biological Sciences* **277**:20092097.

Carlson, A.A., *et al.* (2006) Cortisol levels are positively associated with pup-feeding rates in male meerkats. *Proceedings of the Royal Society of London. Series B: Biological Sciences* **273**:571–577.

Caro, T.M. (1994) Ungulate antipredator behaviour: preliminary and comparative data from African bovids. *Behaviour* **128**:189–228.

Caro, T.M. (2005) *Antipredator Defenses in Birds and Mammals*. Chicago: University of Chicago Press.

Caro, T.M. (2014) Antipredator deception in terrestrial vertebrates. *Current Zoology* **60**:16–25.

Caro, T.M., *et al.* (2004) Adaptive significance of antipredator behaviour in artiodactyls. *Animal Behaviour* **67**:205–228.

Cartmill, E.A. and Byrne, R.W. (2007) Orangutans modify their gestural signaling according to their audience's comprehension. *Current Biology* **17**:1345–1348.

Chapman, C.A. and Lefebvre, L. (1990) Manipulating foraging group size: spider monkey food calls at fruiting trees. *Animal Behaviour* **39**:891–896.

Chappell, M.A. and Bachman, G.C. (2000) Energetic costs of begging behaviour. In: *The Evolution of Begging: Competition, Cooperation and Communication* (eds J. Wright and M.L. Leonard). Dordrecht: Kluwer Academic, 143–162.

Charnov, E.L. and Krebs, J.R. (1975) The evolution of alarm calls: altruism or manipulation? *American Naturalist* **109**:107–112.

Charpentier, M.J.E., *et al.* (2010) Message 'scent': lemurs detect the genetic relatedness and quality of conspecifics via olfactory cues. *Animal Behaviour* **80**:101–108.

Charrier, I., *et al.* (2001) Mother's voice recognition by seal pups *Nature* **412**:873.

Cheney, D.L. and Seyfarth, R.M. (1990) *How Monkeys See The World: Inside the Mind of Another*. Chicago: University of Chicago Press.

Cheney, D.L. and Seyfarth, R.M. (1999) Recognition of other individuals' social relationships by female baboons. *Animal Behaviour* **58**:67–75.

Cheney, D.L. and Seyfarth, R.M. (2008) *Baboon Metaphysics: The Evolution of a Social Mind*. Chicago: University of Chicago Press.

Cheney, D.L., *et al.* (1995) The role of grunts in reconciling opponents and facilitating interactions among adult female baboons. *Animal Behaviour* **50**:249–257.

Cheney, D.L., *et al.* (1996) The function and mechanisms underlying baboon 'contact' barks. *Animal Behaviour* **52**:507–518.

Clark, A.P. and Wrangham, R.W. (1994) Chimpanzee arrival pant-hoots: do they signify food or status? *International Journal of Primatology* **15**:185–205.

Clay, Z., *et al.* (2012) Food-associated vocalizations in mammals and birds: what do these calls really mean? *Animal Behaviour* **83**:323–330.

Clutton-Brock, T.H. and Albon, S.D. (1979) The roaring of red deer and the evolution of honest advertisement. *Behaviour* **69**:145–170.

Clutton-Brock, T.H. and Harvey, P.H. (1976) Evolutionary rules and primate societies. In: *Growing Points in Ethology* (eds P.P.G.

Bateson and R.A. Hinde). Cambridge: Cambridge University Press, 195–237.

Clutton-Brock, T.H. and Huchard, E. (2013) Social competition and its consequences in female mammals. *Journal of Zoology* **289**:151–171.

Clutton-Brock, T.H. and Manser, M.B. (in press) Kalahari meerkats. In: *Cooperative Breeding in Vertebrates: Studies of Ecology, Evolution and Behaviour* (eds W.D. Koenig and J.L. Dickinson). Cambridge: Cambridge University Press.

Clutton-Brock, T.H., *et al.* (1982) *Red Deer: The Behaviour and Ecology of Two Sexes*. Chicago: University of Chicago Press.

Clutton-Brock, T.H., *et al.* (1999) Selfish sentinels in cooperative mammals. *Science* **284**:1640–1644.

Cowlishaw, G. (1992) Song function in gibbons. *Behaviour* **121**:131–153.

Cox, C.R. and Le Boeuf, B.J. (1977) Female incitation of male competition: a mechanism in sexual selection. *American Naturalist* **111**:317–335.

Crawford, J.C., *et al.* (2011) Smelling wrong: hormonal contraception in lemurs alters critical female odour cues. *Proceedings of the Royal Society of London. Series B: Biological Sciences* **278**:122–130.

Davies, N.B., *et al.* (2012) *An Introduction to Behavioural Ecology*. Chichester, West Sussex: Wiley-Blackwell.

Dawkins, R. and Krebs, J.R. (1978) Animal signals: information or manipulation. In: *Behavioural Ecology: An Evolutionary Approach* (eds J.R. Krebs and N.B. Davies). Oxford: Blackwell Science, 282–309.

Deecke, V.B., *et al.* (2005) The vocal behaviour of mammal-eating killer whales: communicating with costly calls. *Animal Behaviour* **69**:395–405.

delBarco-Trillo, J., *et al.* (2011) Night and day: the comparative study of strepsirrhine primates reveals socioecological and phylogenetic patterns in olfactory signals. *Journal of Evolutionary Biology* **24**:82–98.

Dermatsev, V., *et al.* (2014) Male hyraxes increase song duration and syntax complexity in the presence of an alert individuals. *Behavioral Ecology* doi: 10.1093/beheco/aru155.

Deschner, T., *et al.* (2003) Timing and probability of ovulation in relation to sex skin swelling in wild West African chimpanzees, *Pan troglodytes verus*. *Animal Behaviour* **66**:551–560.

Deschner, T., *et al.* (2004) Female sexual swelling size, timing of ovulation, and male behavior in wild West African chimpanzees. *Hormones and Behavior* **46**:204–215.

de Waal, F.B.M. (1992) Intentional deception in primates. *Evolutionary Anthropology: Issues, News, and Reviews* **1**:86–92.

Dobson, S.D. (2009) Socioecological correlates of facial mobility in nonhuman anthropoids. *American Journal of Physical Anthropology* **139**:413–420.

Domb, L.G. and Pagel, M. (2001) Sexual swellings advertise female quality in wild baboons. *Nature* **410**:204–206.

Dostalkova, I., *et al.* (2002) Begging for milk: evolution of teat massaging in suckling pigs. *Journal of Theoretical Biology* **215**:321–332.

Dubuc, C., *et al.* (2009) Sexual skin color contains information about the timing of the fertile phase in free-ranging *Macaca mulatta*. *International Journal of Primatology* **30**:777–789.

East, M.L. and Hofer, H. (1991) Loud calling in a female-dominated mammalian society: II. Behavioural contexts and functions of whooping of spotted hyaenas, *Crocuta crocuta*. *Animal Behaviour* **42**:651–669.

Emmons, L.H. (1978) Sound communication among African rainforest squirrels. *Zeitschrift fur Tierpsychologie* **47**:1–49.

Engelhardt, A., *et al.* (2005) Female sexual behavior, but not sex skin swelling, reliably indicates the timing of the fertile phase in wild long-tailed macaques (*Macaca fascicularis*). *Hormones and Behavior* **47**:195–204.

Enquist, M. and Leimar, O. (1983) Evolution of fighting behaviour: decision rules and assessment of relative strength. *Journal of Theoretical Biology* **102**:387–410.

Enquist, M. and Leimar, O. (1990) The evolution of fatal fighting. *Animal Behaviour* **39**:1–9.

Fichtel, C. and Hammerschmidt, K. (2003) Responses of squirrel monkeys to their experimentally modified mobbing calls. *Journal of the Acoustical Society of America* **113**:2927–2932.

Fichtel, C. and Kappeler, P.M. (2002) Anti-predator behavior of group-living Malagasy primates: mixed evidence for a referential alarm call system. *Behavioral Ecology and Sociobiology* **51**:262–275.

Fischer, J. (1998) Barbary macaques categorize shrill barks into two call types. *Animal Behaviour* **55**:799–807.

Fischer, J. and Zinner, D. (2011) Communicative and cognitive underpinnings of animal group movement. In: *Coordination in Human and Primate Groups* (eds M. Boos, M. Kolbe, P.M. Kappeler and T. Ellwart). Berlin: Springer, 229–244.

Fischer, J., *et al.* (2001) Acoustic features of female chacma baboon barks. *Ethology* **107**:33–54.

Fitzgibbon, C. and Fanshawe, J.H. (1988) Stotting in Thomson's gazelles: an honest signal of condition. *Behavioral Ecology and Sociobiology* **23**:69–74.

Fitzpatrick, C.L., *et al.* (2015) Exaggerated estrous swellings and male mate choice in primates: testing the reliable indicator hypothesis in the Amboseli baboons. *Animal Behaviour* **104**:175–185.

Flower, T. (2011) Fork-tailed drongos use deceptive mimicked alarm calls to steal food. *Proceedings of the Royal Society of London. Series B: Biological Sciences* **278**:1548–1555.

Flower, T.P. and Gribble, M. (2012) Kleptoparasitism by attacks versus false alarm calls in fork-tailed drongos. *Animal Behaviour* **83**:403–410.

Frederiksen, J.K. and Slobodchikoff, C.N. (2007) Referential specificity in the alarm calls of the black-tailed prairie dog. *Ethology Ecology and Evolution* **19**:87–99.

Freeberg, T.M., *et al.* (2012) Social complexity as a proximate and ultimate factor in communicative complexity. *Philosophical Transactions of the Royal Society B: Biological Sciences* **367**:1785–1801.

Furrer, R. and Manser, M. (2009) The evolution of urgency-based and functionally referential alarm calls in ground-dwelling species. *American Naturalist* **173**:400–410.

Gesquiere, L.R., *et al.* (2007) Mechanisms of sexual selection: sexual swellings and estrogen concentrations as fertility indicators and cues for male consort decisions in wild baboons. *Hormones and Behavior* **51**:114–125.

Getty, T. (1998a) Handicap signalling: when fecundity and viability do not add up. *Animal Behaviour* **56**:127–130.

Getty, T. (1998b) Reliable signalling need not be a handicap. *Animal Behaviour* **56**:253–255.

Ghazanfar, A.A. and Logothetis, N.K. (2003) Neuroperception: facial expressions linked to monkey calls. *Nature* **423**:937–938.

Goodall, J. (1986) *The Chimpanzees of Gombe: Patterns of Behavior.* Cambridge, MA: Belknap Press.

Goossens, B.M.A., *et al.* (2008) Gaze following in monkeys is modulated by observed facial expressions. *Animal Behaviour* **75**:1673–1681.

Gosling, L.M. and Roberts, S.C. (2001) Testing ideas about the function of scent marks in territories from spatial patterns. *Animal Behaviour* **62**:F7–F10.

Grafen, A. (1990a) Biological signals as handicaps. *Journal of Theoretical Biology* **144**:517–546.

Grafen, A. (1990b) Sexual selection unhandicapped by the Fisher process. *Journal of Theoretical Biology* **144**:473–516.

Greene, E. and Meagher, T. (1998) Red squirrels, *Tamiasciurus hudsonicus,* produce predator-class specific alarm calls. *Animal Behaviour* **55**:511–518.

Gros-Louis, J. (2004) The function of food-associated calls in white-faced capuchin monkeys, *Cebus capucinus,* from the perspective of the signaller. *Animal Behaviour* **67**:431–440.

Hare, J.F. and Atkins, B.A. (2001) The squirrel that cried wolf: reliability detection by juvenile Richardson's ground squirrels (*Spermophilus richardsonii*). *Behavioral Ecology and Sociobiology* **51**:108–112.

Harrington, F.H. and Mech, L.D. (1979) Wolf howling and its role in territory maintenance. *Behaviour* **68**:207–249.

Hasson, O. (1994) Cheating signals. *Journal of Theoretical Biology* **167**:223–238.

Hauser, M.D. (1998) Functional referents and acoustic similarity: field playback experiments with rhesus monkeys. *Animal Behaviour* **55**:1647–1658.

Higham, J.P. (2013) How does honest costly signaling work? *Behavioral Ecology* **25**:8–11.

Higham, J.P., *et al.* (2008) Baboon sexual swelling: information content of size and color. *Hormones and Behavior* **53**:452–462.

Hirth, D.H. and McCullough, D.R. (1977) Evolution of alarm signals in ungulates with special reference to white-tailed deer. *American Naturalist* **111**:31–42.

Hohmann, G. (1989) Vocal communication of wild bonnet macaques (*Macaca radiata*). *Primates* **30**:325–345.

Holekamp, K.E., *et al.* (2000) Group travel in social carnivores. In: *On the Move: How and Why Animals Travel in Groups* (eds S. Boinski and P.A. Garber). Chicago: University of Chicago Press, 587–627.

Hollén, L.I. and Radford, A.N. (2009) The development of alarm call behaviour in mammals and birds. *Animal Behaviour* **78**:791–800.

Hollén, L.I., *et al.* (2008) Cooperative sentinel calling? Foragers gain increased biomass intake. *Current Biology* **18**:576–579.

Holmes, W.G. and Sherman, P.W. (1982) The ontogeny of kin recognition in two species of ground squirrels. *American Zoologist* **22**:491–517.

Hoogland, J.L. (1995) *The Black-tailed Prairie Dog: Social Life of a Burrowing Mammal.* Chicago: University of Chicago Press.

Horn, A.G. and Leonard, M.C. (2002) Efficacy and the design of begging signals. In: *The Evolution of Begging* (eds J. Wright and M. Leonard). Dordrecht: Kluwer Academic, 127–141.

Hurst, J.L. (1990) Urine marking in populations of wild house mice *Mus domesticus* Rutty. I. Communication between males. *Animal Behaviour* **40**:209–222.

Hurst, J.L., *et al.* (2001) Individual recognition in mice mediated by major urinary proteins. *Nature* **414**:631–634.

Insley, S.J. (2000) Long-term vocal recognition in the northern fur seal. *Nature* **406**:404–405.

Janik, V.M. and Slater, P.J.B. (1998) Context-specific use suggests that bottlenose dolphin signature whistles are cohesion calls. *Animal Behaviour* **56**:829–838.

Janik, V.M. and Slater, P.J.B. (2000) The different roles of social learning in vocal communication. *Animal Behaviour* **60**:1–11.

Jansen, D.W.A.M., *et al.* (2012) Segmental concatenation of individual signatures and context cues in banded mongoose (*Mungos mungo*) close calls. *BMC Biology* **10**:97.

Jensen, P.E.R., *et al.* (1998) Teat massage after milk ingestion in domestic piglets: an example of honest begging? *Animal Behaviour* **55**:779–786.

Johnston, R.E. (1986) Effects of female odors on the sexual behavior of male hamsters. *Behavioral and Neural Biology* **46**:168–188.

Jouventin, P., *et al.* (1999) Finding a parent in a king penguin colony: the acoustic system of individual recognition. *Animal Behaviour* **57**:1175–1183.

Kalan, A.K. and Boesch, C. (2015) Audience effects in chimpanzee food calls and their potential for recruiting others. *Behavioral Ecology and Sociobiology* **69**:1701–1712.

Kalan, A.K., *et al.* (2015) Wild chimpanzees modify food call structure with respect to tree size for a particular food species. *Animal Behaviour* **101**:1–9.

Kappeler, P.M. (1998) To whom it may concern: the transmission and function of chemical signals in *Lemur catta.* *Behavioral Ecology and Sociobiology* **42**:411–421.

Keddy Hector, A.C., *et al.* (1989) Male parental care, female choice and the effect of an audience in vervet monkeys. *Animal Behaviour* **38**:262–271.

Kern, J.M. and Radford, A.N. (2013) Call of duty? Variation in use of the watchman's song by sentinel dwarf mongooses, *Helogale parvula. Animal Behaviour* **85**:967–975.

Kershenbaum, A., *et al.* (2014) Acoustic sequences in non-human animals: a tutorial review and prospectus. *Biological Reviews* doi: 10.1111/brv.12160

Kiley-Worthington, M. (1976) The tail movements of ungulates, canids and felids with particular reference to their causation and function as displays. *Behaviour* **56**:69–115.

Kilner, R.M. (2001) A growth cost of begging in captive canary chicks. *Proceedings of the National Academy of Sciences of the United States of America* **98**:11394–11398.

King, A.J. and Sueur, C. (2011) Where next? Group coordination and collective decision making by primates. *International Journal of Primatology* **32**:1245–1267.

Kirkpatrick, M. (1986) The handicap mechanism of sexual selection does not work. *American Naturalist* **127**:222–240.

Klump, G.M. and Shalter, M.D. (1984) Acoustic behaviour of birds and mammals in the predator context. I. Factors affecting the structure of alarm signals. II. The functional significance and evolution of alarm signals. *Zeitschrift fur Tierpsychologie* **66**:189–226.

Knotkova, E., *et al.* (2009) Vocalisations of the silvery mole-rat: comparison of vocal repertoires in subterranean rodents with different social systems. *Bioacoustics* **18**:241–257.

Kokko, H., *et al.* (2001) The evolution of cooperative breeding through group augmentation. *Proceedings of the Royal Society of London. Series B: Biological Sciences* **268**:187–196.

Komers, P.E. and Brotherton, P.N.M. (1997) Dung pellets used to identify the distribution and density of dik-dik. *African Journal of Ecology* **35**:124–132.

Krams, I., *et al.* (2012) Linking social complexity and vocal complexity: a parid perspective. *Philosophical Transactions of the Royal Society B: Biological Sciences* **367**:1879–1891.

Krebs, J.R. and Dawkins, R. (1984) Animal signals: mind-reading and manipulation. In: *Behavioural Ecology: An Evolutionary Approach*, 2nd edn (eds J.R. Krebs and N.B. Davies). Oxford: Blackwell Science, 380–402.

Kruuk, H. (1972) *The Spotted Hyena: A Study of Predation and Social Behaviour*. Chicago: University of Chicago Press.

Lachmann, M., *et al.* (2001) Cost and conflict in animal signals and human language. *Proceedings of the National Academy of Sciences of the United States of America* **98**:13189–13194.

Leclaire, S., *et al.* (2011) An individual and a sex odor signature in kittiwakes? Study of the semiochemical composition of preen secretion and preen down feathers. *Naturwissenschaften* **98**:615–624.

Leclaire, S., *et al.* (2013) Odour-based kin discrimination in the cooperatively breeding meerkat. *Biology Letters* **9**:20121054.

Leger, D.W., *et al.* (1984) Vocalizations of Belding's ground squirrels (*Spermophilus beldingi*). *Animal Behaviour* **32**:753–754.

Leighty, K.A., *et al.* (2008) Rumble vocalizations mediate interpartner distance in African elephants, *Loxodonta africana*. *Animal Behaviour* **76**:1601–1608.

Le Roux, A., *et al.* (2008) The audience effect in a facultatively social mammal, the yellow mongoose, *Cynictis penicillata*. *Animal Behaviour* **75**:943–949.

Le Roux, A., *et al.* (2013) Evidence for tactical concealment in a wild primate. *Nature Communications* **4**:1462.

McComb, K. and Reby, D. (2005) Vocal communication networks in large terrestrial mammals. In: *Animal Communication Networks* (ed. P.K. McGregor). Cambridge: Cambridge University Press, 372–389.

McComb, K. and Semple, S. (2005) Coevolution of vocal communication and sociality in primates. *Biology Letters* **1**:381–385.

McComb, K., *et al.* (1994) Roaring and numerical assessment in contests between groups of female lions, *Panthera leo*. *Animal Behaviour* **47**:379–387.

McComb, K., *et al.* (2003) Long-distance communication of acoustic cues to social identity in African elephants. *Animal Behaviour* **65**:317–329.

McCracken, G.F. (1984) Communal nursing in Mexican free-tailed bat maternity colonies. *Science* **223**:1090–1091.

Macedonia, J.M. and Evans, C.S. (1993) Essay on contemporary issues in ethology: variation among mammalian alarm call systems and the problem of meaning in animal signals. *Ethology* **93**:177–197.

Maestripieri, D. and Roney, J.R. (2005) Primate copulation calls and postcopulatory female choice. *Behavioral Ecology* **16**:106–113.

Manser, M.B. (1998) *The evolution of auditory communication in suricates* Suricata suricatta. PhD thesis, University of Cambridge, Cambridge.

Manser, M.B. (1999) Response of foraging group members to sentinel calls in suricates, *Suricata suricatta*. *Proceedings of the Royal Society of London. Series B: Biological Sciences* **266**:1013–1019.

Manser, M.B. (2001) The acoustic structure of suricates' alarm calls varies with predator type and the level of response urgency. *Proceedings of the Royal Society of London. Series B: Biological Sciences* **268**:2315–2324.

Manser, M.B. (2009) What do functionally referential alarm calls refer to? In: *Cognitive Ecology* (eds R. Dukas and J.M. Ratcliffe). Chicago: University of Chicago Press, 229–248.

Manser, M.B. (2010) Generation of functionally referential and motivational vocal signals in mammals. In: *Handbook of Mammalian Vocalization* (ed. S. Brudzynski). San Diego, CA: Academic Press, 477–486.

Manser, M.B. and Avey, G. (2000) The effect of pup vocalisations on food allocation in a cooperative mammal, the meerkat (*Suricata suricatta*). *Behavioral Ecology and Sociobiology* **48**:429–437.

Manser, M.B., *et al.* (2002) Suricate alarm calls signal predator class and urgency. *Trends in Cognitive Sciences* **6**:55–57

Manser, M.B., *et al.* (2008) Signals of need in a cooperatively breeding mammal with mobile offspring. *Animal Behaviour* **76**:1805–1813.

Manser, M.B., *et al.* (2014) Vocal complexity in meerkats and other mongooses. In: *Advances in the Study of Behavior*, Vol. **46** (eds M Naguib et al.). San Diego, CA: Academic Press, 281–310.

Marler, P. (1955) Characteristics of some animal calls. *Nature* **176**:6–8.

Mateo, J.M. and Johnston, R.E. (2000) Retention of social recognition after hibernation in Belding's ground squirrels. *Animal Behaviour* **59**:491–499.

Maynard Smith, J. and Harper, D. (2003) *Animal Signals*. Oxford: Oxford University Press.

Mitani, J.C. (1984) The behavioral regulation of monogamy in gibbons (*Hylobates muelleri*). *Behavioral Ecology and Sociobiology* **15**:225–229.

Moy, R.F. (1970) Histology of the subauricular and rump glands of the pronghorn (*Antilocapra americana* Ord). *American Journal of Anatomy* **129**:65–87.

Müller, C. and Manser, M.B. (2007) 'Nasty neighbours' rather than 'dear enemies' in a social carnivore. *Proceedings of the Royal Society of London. Series B: Biological Sciences* **274**:959–965.

Nevison, C.M., *et al.* (2003) The ownership signature in mouse scent marks is involatile. *Proceedings of the Royal Society of London. Series B: Biological Sciences* **270**:1957–1963.

O'Connell, S.M. and Cowlishaw, G. (1994) Infanticide avoidance, sperm competition and mate choice: the function of copulation calls in female baboons. *Animal Behaviour* **48**:687–694.

Ord, T.J. and Garcia-Porta, J. (2012) Is sociality required for the evolution of communicative complexity? Evidence weighed against alternative hypotheses in diverse taxonomic groups. *Philosophical Transactions of the Royal Society B: Biological Sciences* **367**:1811–1828.

Papworth, S., *et al.* (2008) Male blue monkeys alarm call in response to danger experienced by others. *Biology Letters* **4**:472–475.

Parr, L.A., *et al.* (2007) Classifying chimpanzee facial expressions using muscle action. *Emotion* **7**:172–181.

Perony, N. and Townsend, S.W. (2013) Why did the meerkat cross the road? Flexible adaptation of phylogenetically-old behavioural strategies to modern-day threats. *PLOS Biology* **8**: e52834.

Pfefferle, D., *et al.* (2008) Female Barbary macaque (*Macaca sylvanus*) copulation calls do not reveal the fertile phase but influence mating outcome. *Proceedings of the Royal Society of London. Series B: Biological Sciences* **275**:571–578.

Pollard, K.A. and Blumstein, D.T. (2012) Evolving communicative complexity: insights from rodents and beyond. *Philosophical Transactions of the Royal Society B: Biological Sciences* **367**:1869–1878.

Poole, J.H. (1989) Mate guarding, reproductive success and female choice in African elephants. *Animal Behaviour* **37**:842–849.

Poole, J.H. (1999) Signals and assessment in African elephants: evidence from playback experiments. *Animal Behaviour* **58**:185–193.

Pradhan, G.R., *et al.* (2006) The evolution of female copulation calls in primates: a review and a new model. *Behavioral Ecology and Sociobiology* **59**:333–343.

Proops, L., *et al.* (2009) Cross-modal individual recognition in domestic horses (*Equus caballus*). *Proceedings of the National Academy of Sciences of the United States of America* **106**:947–951.

Pusey, A.E. and Packer, C. (1993) Infanticide in lions: consequences and counterstrategies. In: *Infanticide and Parental Care* (eds S. Parmigiani and F.S. vom Saal). Chur, Switzerland: Harwood Academic Press, 277–299.

Quick, N.J. and Janik, V.M. (2012) Bottlenose dolphins exchange signature whistles when meeting at sea. *Proceedings of the Royal Society of London. Series B: Biological Sciences* **279**:2539–2545.

Raemaekers, J.J. and Raemaekers, P.M. (1984) The Ooaa duet of the gibbon (*Hylobates lar*): a group call which triggers other groups to respond in kind. *Folia Primatologica* **42**:209–215.

Ramakrishnan, U. and Coss, R.G. (2000) Age differences in the responses to adult and juvenile alarm calls by bonnet macaques (*Macaca radiata*). *Ethology* **106**:131–144.

Rasa, O.A.E. (1986) Coordinated vigilance in dwarf mongoose family groups: the watchman's song hypothesis and the cost of guarding. *Ethology* **71**:340–344.

Reber, S.A., *et al.* (2013) Social monitoring via close calls in meerkats. *Proceedings of the Royal Society of London. Series B: Biological Sciences* **280**:20131013.

Rendall, D. (2003) Acoustic correlates of caller identity affect intensity in the vowel-like grunt vocalizations of baboons. *Journal of the Acoustical Society of America* **113**:3390–3402.

Rendall, D., *et al.* (1996) Vocal recognition of individuals and kin in free-ranging rhesus monkeys. *Animal Behaviour* **51**:1007–1015.

Rendall, D., *et al.* (1999) The meaning and function of grunt variants in baboons. *Animal Behaviour* **57**:583–592.

Rendall, D., *et al.* (2009) What do animal signals mean? *Animal Behaviour* **78**:233–240.

Ridley, A.R., *et al.* (2013) Is sentinel behaviour safe? An experimental investigation. *Animal Behaviour* **85**:137–142.

Roberts, S.C. (2007) Scent marking. In: *Rodent Societies: An Ecological and Evolutionary Perspective* (eds J.O. Wolff and P.W. Sherman). Chicago: University of Chicago Press, 255–267.

Roberts, S.C. and Dunbar, R.I.M. (2000) Female territoriality and the function of scent-marking in a monogamous antelope (*Oreotragus oreotragus*). *Behavioral Ecology and Sociobiology* **47**:417–423.

Rohwer, S. (1977) Status signaling in Harris sparrows: some experiments in deception. *Behaviour* **61**:107–129.

Rohwer, S. (1982) The evolution of reliable and unreliable badges of fighting ability. *American Zoologist* **22**:531–546.

Rohwer, S. and Rohwer, F.C. (1978) Status signalling in harris sparrows: experimental deceptions achieved. *Animal Behaviour* **26**:1012–1016.

Sabater-Pí, J. (1979) Feeding behaviour and diet of chimpanzees (*Pan troglodytes troglodytes*) in the Okorobiko Mountains of Rio Muni (West Africa). *Zeitschrift fur Tierpsychologie* **50**:265–281.

Schel, A.M., *et al.* (2013) Chimpanzee food calls are directed at specific individuals. *Animal Behaviour* **86**:955–965.

Scherrer, J.A. and Wilkinson, G.S. (1993) Evening bat isolation calls provide evidence for heritable signatures. *Animal Behaviour* **46**:847–860.

Schwagmeyer, P.L. (1980) Alarm calling behavior of the thirteen-lined ground squirrel, *Spermophilus tridecemlineatus*. *Behavioral Ecology and Sociobiology* **7**:195–200.

Searcy, W.A. and Nowicki, S. (2005) *The Evolution of Animal Communication*. Princeton, NJ: Princeton University Press.

Seeley, T.D. (1985) *Honeybee Ecology: A Study of Adaptation in Social Life*. Princeton, NJ: Princeton University Press.

Sekulic, R. (1982) The function of howling in red howler monkeys (*Alouatta seniculus*). *Behaviour* **81**:38–54.

Semple, S. (1998) The function of Barbary macaque copulation calls. *Proceedings of the Royal Society of London. Series B: Biological Sciences* **265**:287–291.

Semple, S. (2001) Individuality and male discrimination of female copulation calls in the yellow baboon. *Animal Behaviour* **61**:1023–1028.

Semple, S., et al. (2002) Information content of female copulation calls in yellow baboons. *American Journal of Primatology* **56**:43–56.

Seyfarth, R.M. and Cheney, D.L. (1990) The assessment by vervet monkeys of their own and another species' alarm calls. *Animal Behaviour* **40**:754–764.

Seyfarth, R.M. and Cheney, D.L. (2003) Signalers and receivers in animal communication. *Annual Review of Psychology* **54**:145–173.

Seyfarth, R.M., et al. (1980a) Monkey responses to three different alarm calls: evidence of predator classification and semantic communication. *Science* **210**:801–803.

Seyfarth, R.M., et al. (1980b) Vervet monkey alarm calls: semantic communication in a free-ranging primate. *Animal Behaviour* **28**:1070–1094.

Seyfarth, R.M., et al. (2010) The central importance of information in studies of animal communication. *Animal Behaviour* **80**:3–8.

Sharpe, L.L. (2015) Handstand scent marking: height matters to dwarf mongooses. *Animal Behaviour* **105**:173–179.

Sherman, P.W. (1980) The limits of ground squirrel nepotism. In: *Sociobiology: Beyond Nature/Nurture?* (eds G.W. Barlow and J. Silverberg). Boulder, CO: Westview Press, 505–544.

Sherman, P.W. (1981) Kinship, demography and Belding's ground squirrel nepotism. *Behavioral Ecology and Sociobiology* **8**:251–259.

Sherman, P.W. (1985) Alarm calls of Belding's ground squirrels to aerial predators: nepotism or self-preservation? *Behavioral Ecology and Sociobiology* **17**:313–323.

Shultz, S. and Dunbar, R.I.M. (2007) The evolution of the social brain: anthropoid primates contrast with other vertebrates. *Proceedings of the Royal Society of London. Series B: Biological Sciences* **274**:2429–2436.

Silk, J.B. (2001) Grunts, girneys, and good intentions: the origins of strategic commitment in nonhuman primates. In: *Evolution and the Capacity for Commitment* (ed. R. Nesse). New York: Russell Sage Foundation, 138–158.

Silk, J.B., et al. (2000) Cheap talk when interests conflict. *Animal Behaviour* **59**:423–432.

Slade, B.E., et al. (2003) Oestrous state dynamics in chemical communication by captive female Asian elephants. *Animal Behaviour* **65**:813–819.

Slobodchikoff, C.N., et al. (1991) Semantic information distinguishing individual predators in the alarm calls of Gunnison's prairie dogs. *Animal Behaviour* **42**:713–719.

Slobodchikoff, C.N., et al. (2009a) Prairie dog alarm calls encode labels about predator colors. *Animal Cognition* **12**:435–439.

Slobodchikoff, C.N., et al. (2009b) *Prairie Dogs: Communication and Community in an Animal Society.* Cambridge, MA: Harvard University Press.

Slocombe, K.E. and Zuberbühler, K. (2005) Functionally referential communication in a chimpanzee. *Current Biology* **15**:1779–1784.

Slocombe, K.E. and Zuberbühler, K. (2006) Food-associated calls in chimpanzees: responses to food types or food preferences? *Animal Behaviour* **72**:989–999.

Slocombe, K.E., et al. (2010) Production of food-associated calls in wild male chimpanzees is dependent on the composition of the audience. *Behavioral Ecology and Sociobiology* **64**:1959–1966.

Smiseth, P.T. and Moore A.J. (2004) Signalling of hunger when offspring forage by both begging and self-feeding. *Animal Behaviour* **67**:1083–1088.

Smith, S.F. (1978) Alarm calls, their origin and use in *Eutamias sonomae*. *Journal of Mammalogy* **59**:888–893.

Soltis, J., et al. (2005a) African elephant vocal communication I: antiphonal calling behaviour among affiliated females. *Animal Behaviour* **70**:579–587.

Soltis, J., et al. (2005b) African elephant vocal communication II: rumble variation reflects the individual identity and emotional state of callers. *Animal Behaviour* **70**:589–599.

Steele, M.A., et al. (2008) Cache protection strategies of a scatter-hoarding rodent: do tree squirrels engage in behavioural deception? *Animal Behaviour* **75**:705–714.

Sueur, C. (2011) Group decision-making in chacma baboons: leadership, order and communication during movement. *BMC Ecology* **11**:26.

Swaisgood, R.R., et al. (1999) Giant pandas discriminate individual differences in conspecific scent. *Animal Behaviour* **57**:1045–1053.

Taylor, A.M. and Reby, D. (2010) The contribution of source–filter theory to mammal vocal communication research. *Journal of Zoology* **280**:221–236.

Thapar, V. (1986) *Tiger, Portrait of a Predator.* London: Collins.

Tibbetts, E.A. and Dale, J. (2004) A socially enforced signal of quality in a paper wasp. *Nature* **432**:218–222.

Tibbetts, E.A. and Dale, J. (2007) Individual recognition: it is good to be different. *Trends in Ecology and Evolution* **22**:529–537.

Tilson, R.L. and Norton, P.M. (1981) Alarm duetting and pursuit deterrence in an African antelope. *American Naturalist* **118**:455–462.

Tobias, J.A., et al. (2012) The evolution of female ornaments and weaponry: social selection, sexual selection and ecological competition. *Philosophical Transactions of the Royal Society B: Biological Sciences* **367**:2274–2293.

Torrey, S. and Widowski, T.M. (2007) Relationship between growth and non-nutritive massage in suckling pigs. *Applied Animal Behaviour Science* **107**:32–44.

Townsend, S.W. and Manser, M.B. (2013) Functionally referential communication in mammals: the past, present and the future. *Ethology* **119**:1–11.

Townsend, S.W., *et al.* (2008) Female chimpanzees use copulation calls flexibly to prevent social competition. *PLOS ONE* 3:2431.

Townsend, S.W., *et al.* (2010) Meerkat close calls encode group-specific signatures, but receivers fail to discriminate. *Animal Behaviour* 80:133–138.

Townsend, S.W., *et al.* (2011) All clear? Meerkats attend to contextual information in close calls to coordinate vigilance. *Behavioral Ecology and Sociobiology* 65:1927–1934.

Townsend, S.W., *et al.* (2014) Acoustic cues to identity and predator context in meekat barks. *Animal Behaviour* 94:143–149.

Turbé, A. (2006) *Habitat use, ranging behaviour and social control in meerkats.* PhD thesis, University of Cambridge, Cambridge.

Tyack, P.L. (2000) Functional aspects of cetacean communication. In: *Cetacean Societies: Field Studies of Dolphins and Whales* (ed. J. Mann). Chicago: University of Chicago Press, 270–307.

Voellmy, I.K., *et al.* (2014) Mean fecal glucocorticoid metabolites are associated with vigilance, whereas immediate cortisol levels better reflect acute anti-predator responses in meerkats. *Hormones and Behavior* 66:759–765.

Warkentin, K.J., *et al.* (2001) Repetitive calls of juvenile Richardson's ground squirrels (*Spermophilus richardsonii*) communicate response urgency. *Canadian Journal of Zoology* 79:569–573.

Weary, D.M. and Fraser, D. (1995) Calling by domestic piglets: reliable signals of need? *Animal Behaviour* 50:1047–1055.

Weary, D.M., *et al.* (1996) Sows show stronger responses to isolation calls of piglets associated with greater levels of piglet need. *Animal Behaviour* 52:1247–1253.

Wells, D.L. and Hepper, P.G. (2006) Prenatal olfactory learning in the domestic dog. *Animal Behaviour* 72:681–686.

Wheeler, B.C. (2008) Selfish or altruistic? An analysis of alarm call function in wild capuchin monkeys, *Cebus apella nigritus*. *Animal Behaviour* 76:1465–1475.

Wheeler, B.C. and Fischer, J. (2012) Functionally referential signals: a promising paradigm whose time has passed. *Evolutionary Anthropology* 21:195–205.

Whiten, A. and Byrne, R.W. (1988) Tactical deception in primates. *Behavioral and Brain Sciences* 11:233–244.

Wich, S.A. and de Vries, H. (2006) Male monkeys remember which group members have given alarm calls. *Proceedings of the Royal Society of London. Series B: Biological Sciences* 273:735–740.

Wilkinson, G.S. (2003) Social and vocal complexity in bats. In: *Animal Social Complexity: Intelligence, Culture, and Individualized Societies* (eds F.B.M. de Waal and P.L. Tyack). Cambridge, MA: Harvard University Press, 322–341.

Wilson, D.R. and Hare, J.F. (2004) Animal communication: ground squirrel uses ultrasonic alarms. *Nature* 430:523.

Wilson, D. R., *et al.* (2015) Red squirrels use territorial vocalizations for kin discrimination. *Animal Behaviour* 107:79–85.

Wilson, M.L. and Wrangham, R.W. (2003) Intergroup relations in chimpanzees. *Annual Review of Anthropology* 32:363–392.

Wilson, M.L. and Wrangham, R.W. (2007) Chimpanzees (*Pan troglodytes*) modify grouping and vocal behaviour in response to location-specific risk. *Behaviour* 144:1621–1653.

Wilson, M.L., *et al.* (2001) Does participation in intergroup conflict depend on numerical assessment, range location, or rank for wild chimpanzees? *Animal Behaviour* 61:1203–1216.

Wittig, R.M., *et al.* (2007) Kin-mediated reconciliation substitutes for direct reconciliation in female baboons. *Proceedings of the Royal Society of London. Series B: Biological Sciences* 274:1109–1115.

Woodland, D.J., *et al.* (1980) The pursuit deterrent function of alarm signals. *American Naturalist* 115:748–753.

Wright, J., *et al.* (2001) Cooperative sentinel behaviour in the Arabian babbler. *Animal Behaviour* 62:973–979.

Young, A.J. and Bennett, N.C. (2013) Intra-sexual selection in cooperative mammals and birds: why are females not bigger and better armed? *Philosophical Transactions of the Royal Society B: Biological Sciences* 368:20130075.

Zahavi, A. (1975) Mate selection: a selection for a handicap. *Journal of Theoretical Biology* 53:205–214.

Zahavi, A. (1993) The fallacy of conventional signalling. *Philosophical Transactions of the Royal Society B: Biological Sciences* 340:227–230.

Zahavi, A. and Zahavi, A. (1997) *The Handicap Principle: A Missing Piece of Darwin's Puzzle.* New York: Oxford University Press.

Zenuto, R.R. (2010) Dear enemy relationships in the subterranean rodent *Ctenomys talarum*: the role of memory of familiar odours. *Animal Behaviour* 79:1247–1255.

Zinner, D., *et al.* (2002) Significance of primate sexual swellings. *Nature* 420:142–143.

Zinner, D., *et al.* (2004) Sexual selection and exaggerated sexual swellings of female primates. In: *Sexual Selection in Primates* (eds P.M. Kappeler and C.P. van Schaik). Cambridge: Cambridge University Press, 71–89.

Zuberbühler, K. (2009) Survivor signals: the biology and psychology of animal alarm calling. In: *Advances in the Study of Behavior*, Vol. 40 (eds M. Naguib, K. Zuberbühler, N.S. Clayton and V.M. Janik). San Diego, CA: Academic Press, 277–322.

Zuberbühler, K., *et al.* (1997) Diana monkey long-distance calls: messages for conspecifics and predators. *Animal Behaviour* 53:589–604.

Zuberbühler, K., *et al.* (1999) The predator deterrence function of primate alarm calls. *Ethology* 105:477–490.

CHAPTER 8
Competition between females

8.1 Introduction

In a forest clearing in Sri Lanka a middle-aged, low-ranking female macaque rapidly stuffs her cheek pouches with fallen fruit while her daughter forages around her, careful to keep out of her way. A juvenile female from a more dominant matriline – 10 years younger and half her weight – feeds nearby. Suddenly, the juvenile runs across to the matriarch, sits in front of her and reaches out to touch her lips. With great determination, she inserts her fingers between the matriarch's lips and tries to pull her jaw down. The matriarch turns her head away but neither threatens nor attacks the juvenile. After repeated attempts, the juvenile gets her fingers into the matriarch's mouth, pulls down her jaw, reaches into her cheek pouch and pulls out a handful of fruit, which she quickly stuffs into her own mouth before scampering back to her mother who is watching the interaction from the lower branches of a tree fifty metres away. The low-ranking matriarch abandons her feeding site and moves away to feed out of sight of the dominant female and her daughters.

Although competition to reproduce is one of the cornerstones of the theory of natural selection (see Chapter 1), the first detailed studies of breeding competition focused almost exclusively on males. Compared with competition between males, female competition less frequently involves escalated contests and is less often associated with the evolution of exaggerated secondary sexual characters. Moreover, individual differences in breeding success among females are less immediately obvious: whereas measures of breeding success across a single season are often sufficient to reveal large individual differences among males, it is usually necessary to monitor the success of females over several breeding attempts to appreciate the magnitude of

Figure 8.1 Aggressive interactions between females over resources or rank in red-fronted lemurs can lead to the eviction of subordinate females. *Source*: © Claudia Fichtel.

Mammal Societies, First Edition. Tim Clutton-Brock.
© 2016 Tim Clutton-Brock. Published 2016 by John Wiley & Sons, Ltd.

individual differences and to identify their causes (Clutton-Brock 1988, 1991a). As a result, only when long-term studies were able to track the life histories of large samples of individual females was the magnitude of individual differences in reproductive success apparent.

The immediate causes of competition between females vary widely. As a result of the high energetic demands of gestation and lactation (see Chapter 5), the reproductive success of females is often constrained by the availability of resources. As a result, females often compete directly for food, threatening or attacking other individuals that forage close to them and their dominance rank can have an important influence on their access to resources (Clutton-Brock 2009; Stockley and Bro-Jørgensen 2011). In other species, females compete for access to space, territories, breeding burrows, breeding partners or dominance rank (Hoogland 1995; Clutton-Brock 2009). The size and distribution of resources can have an important influence on the frequency and intensity of competitive interaction (see Chapter 2).

Competition between females can be intense in singular as well as in plural breeders and varies with the availability of resources or reproductive opportunities and the number of individuals competing for them

(Figures 8.1 and 8.2). For example, in prairie dogs, the frequency of fights, chases and territorial disputes between members of different subgroups increases in larger groups (Figure 8.3a,b) and the reproductive success of females declines (Hoogland 1995). Similarly, in long-tailed macaques, increases in group size are associated with increases in the length of day-journeys, rising social tension and reductions in female breeding success (van Noordwijk and van Schaik 1999) (Figure 8.4). Effects of increasing group size on the frequency of competitive interactions between females have been found in many other social mammals, including primates, ungulates and carnivores (Clutton-Brock *et al.* 1982, 2008; Silk 2007; Clutton-Brock 2009). Demographic parameters affected by increasing group size commonly include rates of growth and development, age at first breeding, breeding frequency, breeding success and longevity (Hoogland 1995; Pride 2005; Stockley and Bro-Jørgensen 2011; Lacey 2004).

As well as competing for access to resources, females commonly compete directly for breeding opportunities or mates. In some social species, they compete to become sexually mature (see Chapter 17) and, even where males compete intensely for mating partners, females may still compete for males at particular stages of the breeding

Figure 8.2 Aggressive interactions over rank are common in female baboons. *Source*: © Alecia Carter.

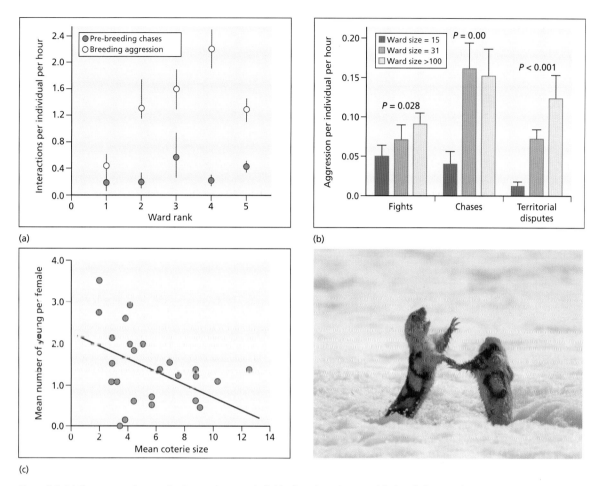

Figure 8.3 (a) Frequency of aggressive interactions per individual per hour between black-tailed prairie dogs in wards of different sizes. (b) Frequencies of fights, chases and territorial disputes in wards of different sizes. (c) Reproductive success of females in family groups (coteries) of different sizes. *Source*: (a,c) Data from Hoogland (1981); (b) From Hoogland (1995). Reproduced with permission of the University of Chicago Press; *Photo source*: © Elaine Miller Bond.

cycle (Stockley and Bro-Jørgensen 2011). For example, in topi, where males defend mating territories during a well-defined mating season, there can be more than one receptive female in a male's harem on the same day and, under these circumstances, females often compete for the attention of males (Bro-Jørgensen 2002, 2011) (Figure 8.5).

Competing for mates may help females to ensure that they are mated by one or more males within the time frame of their reproductive cycles (Ball and Parker 1996; Parker and Ball 2005). In strongly polygynous or promiscuous systems, the sperm supplies of successful males can become depleted (Preston *et al.* 2001) or males may strategically conserve sperm for subsequent mating opportunities. For example, in ungulates, the frequency of overt female competition for mating partners appears to increase in populations where adult sex ratios are strongly biased towards females and there is a high degree of reproductive synchrony (Milner-Gulland *et al.* 2003; Stockley and Bro-Jørgensen 2011).

Finally, females commonly compete to raise offspring, to protect their access to resources and to establish their status within the group or to prevent them being evicted by other females (Clutton-Brock 1991b; Silk 2007; Stockley and Bro-Jørgensen 2011). Competition of this kind is particularly intense in plural breeders that live in stable groups that include several distinct matrilines, like

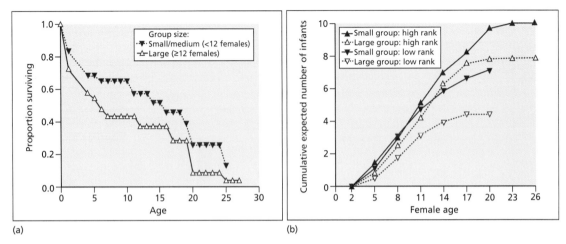

(a)   (b)

Figure 8.4 Costs of sociality to females in long-tailed macaques: (a) age-specific survival rates of females in large groups compared with those in small or medium-sized groups; (b) cumulative number of offspring surviving to at least 1 year of age produced per female in small or medium-sized groups compared to large (≥12 females) groups for females with high or low dominance rank. *Source*: From van Noordwijk and van Schaik (1999). Reproduced with permission of Springer Science and Business Media.

many of the baboons and macaques and spotted hyenas. In some of these species, the size of matrilineal subgroups affects their relative dominance and breeding success and female members of dominant matrilines may direct aggression at female recruits to subordinate matrilines, who represent potential competitors.

This chapter examines direct competition between female group members and describes the competitive strategies used by females and their consequences.

Section 8.2 describes the tactics used by females in competitive interactions; section 8.3 describes the role of dominance and the factors affecting the acquisition of rank in females. Frequent competition generates costs to winners as well as to losers and section 8.4 describes some of strategies used by females to minimise the costs of competition. Finally, section 8.5 explores some of the consequences of female competition.

Figure 8.5 Oestrous female topi competing for access to a male. *Source*: © Jakob Bro-Jørgensen.

8.2 Competitive tactics

Fighting between female mammals is not uncommon, though it is usually less frequent than between males. In some species, there is intense competition between female juveniles. For example, in spotted hyenas, food shortages stimulate intense aggression and occasional siblicide between cubs and aggression between female littermates is more intense than between males, leading to higher rates of mortality in female-biased litters (Hofer and East 1997, 2008). In singular breeders, adult females commonly fight over access to breeding territories (Fernandez-Duque and Huck 2013), while in plural breeders females often fight when important resources are at stake: for example, female prairie dogs commonly fight for access to breeding burrows (Hoogland 1995). Fights are also common where females attempt to evict other females (or their offspring) from breeding groups. For example, in ursine howler monkeys, fights between females often occur when they attempt to evict each other (or their offspring) from breeding groups (Crockett 1984; Crockett and Pope 1988). Similarly, in banded mongooses, fights between females are common when coalitions of older dominant females attempt to evict groups of younger females from breeding groups (Gilchrist 2006; Cant et al. 2010). In singular cooperative breeders, where reproductive skew is typically large, the death of the breeding female is often followed by intense fighting between her daughters and the death or eviction of unsuccessful competitors (see Chapter 17).

As yet, detailed studies of fighting tactics have been almost totally confined to studies of males (see Chapter 13). However, accounts of fights between females suggest that their distribution and duration coincide with the predictions of theoretical models (see Chapter 1). For example, fights between females appear to be most frequent and most intense where the fitness benefits of winning or the costs of losing are large and to be longest when the power (resource-holding potential or RHP) of contestants is approximately similar (see Chapter 17). In some plural breeders, the frequency and intensity of aggression is lower between close relatives than distant relatives or unrelated individuals (Holekamp and Smale 2000; Silk 2007) and related females often support each other in competitive interactions with other group members (see Chapter 9). Despite this, closely related females will readily engage in prolonged and sometimes lethal fights when

competing for breeding opportunities: some of the best examples involve cooperative breeders, but intense competition between females also occurs in plural breeders (Hoogland 1985; Stockley and Bro-Jørgensen 2011).

One contrast between intrasexual conflicts in females and males is that competition between females is less commonly associated with extravagant vocal or visual displays than competition between males. Why this should be the case is not immediately clear, for the net benefits of winning are not necessarily smaller in contests between females. One possible explanation is that the energetic costs of reproduction are higher and more prolonged in females, with the result that extravagant displays are more likely to compromise their fitness.

Threats, punishment and harassment

While conflicts between females sometimes lead to direct fighting, the majority of aggressive interactions between group members involve threats rather than physical attacks. Threats allow individuals to modify the behaviour of potential competitors without incurring the costs and risks associated with escalated fights (Maynard Smith 1974) but are only likely to be effective where individuals that give them have the capacity to damage or impose other costs on recipients (Parker 1974; Cant and Johnstone 2009).

In many societies, dominant individuals also punish subordinates that infringe their interests, inflicting fitness costs that offset the benefits of repeating the same behaviour (Clutton-Brock and Parker 1995). For example, captive chimpanzees will punish individuals that steal their food (Jensen et al. 2007; Jensen 2010; Riedl et al. 2012). Punishing tactics are often used by dominants to reduce the incidence of feeding competition with subordinates, to constrain their access to social partners or to coerce them into cooperative behaviour. Subordinates that repeatedly infringe the interests of the same dominant individual commonly receive progressively larger punishments and may, eventually, be evicted from the group or even killed (Clutton-Brock and Parker 1995).

Conflicts of interest between group members also lead to regular harassment. For example, where a subordinate female is competing with a dominant for a divisible resource, repeated attempts to gain access by the subordinate may eventually raise the costs of continued defence to the dominant to a point where the net benefits of maintaining exclusive access are lower than the costs of defence. Situations of this kind resemble a 'war of attrition' where the ability of two individuals to win a contest depends on the period for which they are ready to persist in competing for it and the winner is the individual that can afford to persist for the longest time (Clutton-Brock and Parker 1995). Persistent harassment can occur in a variety of circumstances. In some societies, dominant females harass subordinates or their offspring, sometimes directing unprovoked attacks at them which may raise glucocorticoid levels, sap their confidence, discourage retaliatory attacks and, in extreme cases, cause them to leave the group (Stockley and Bro-Jørgensen 2011; Clutton-Brock and Huchard 2013). Harassment is also used by subordinates to modify the behaviour of dominants. For example, hungry individuals sometimes harass successful foragers or hunters for a share of the food that they have acquired (see Chapter 9) and adolescents of either sex may harass copulating couples (Clutton-Brock and Harvey 1976).

In most animal societies, dominant individuals more commonly attempt to modify or control the behaviour of other individuals by threats, punishment or harassment than by rewards or inducements (see Chapter 9). There are likely to be good reasons why punishment is a more common tactic than rewarding in animal societies (Clutton-Brock and Parker 1995). Where power asymmetries are well established, the costs of punishing to dominants may be relatively low whereas the costs of rewarding subordinates will often be substantial. As a result, the costs of repeated punishment to punishers may often be negligible whereas the costs of giving repeated rewards are likely to be substantial. Second, a dominant animal will often be able to reduce the fitness of a subordinate to a much greater extent through repeated punishment than it could increment it through inducements or rewards, so that rewards are likely to be less effective than punishments in deterring individuals from actions (such as challenging for the dominant position) that may generate substantial benefits. A third advantage of punishment as a controlling tactic is that it is likely to reduce the power or RHP of the recipient and to lower its capacity to challenge the punisher in the future whereas rewards may strengthen the power of subordinates, increasing the risk of subsequent competition. Finally, punishment may have the effect of deterring other subordinates from similar actions whereas rewarding potential competitors may either encourage them to repeat their behaviour or stimulate others to adopt the same tactics (the *Danegeld effect*; see Box 20.4). One prediction based on this argument is that dominants should engage in conspicuous displays when punishing subordinates to ensure that they attract the attention of as many individuals as possible but should be relatively secretive in their allocation of rewards. All four benefits of punishment as a means of control are exemplified in many human societies and have been recognised by political scientists (see Chapter 20), though the relative benefits of control via rewards versus punishment have received surprisingly little attention from evolutionary theorists.

Reproductive suppression

In many social mammals, female group members (who are often close kin) compete with each other to breed and raise young. Regular aggression directed by dominant females at subordinates or at their offspring is common (especially where groups are large enough to include females belonging to several distinct matrilines) and often depresses the breeding success or survival of subordinates (Abbott 1993; Silk 2007; Beehner and Lu 2013), disrupting their reproductive cycles and causing them to downregulate their reproductive systems (Wasser and Barash 1983; Young 2009). Some form of reproductive interference occurs in a substantial number of mammals. For example, in yellow baboons, dominant females direct frequent aggression at cycling subordinate females in the follicular phase, and these attacks can increase the number of cycles before conception (Wasser and Starling 1988; Huchard and Cowlishaw 2011). In some rodents, carnivores and primates, subordinates are often temporarily infertile (Young 2009), especially in cooperative breeders, where a single female in each group monopolises reproduction and her offspring are reared by other group members (see Chapter 17). However, reproductive suppression of subordinate females is not confined to these systems and occurs in a wide range of species with other breeding systems, including both social and solitary breeders (Beehner and Lu 2013). For example, in subordinate hamsters, interactions with dominant females shortly after mating increase implantation failures while interactions later in pregnancy lead to increased rates of fetal mortality (Huck *et al.* 1988a,b). Recent studies of several species where subordinates show evidence of suppression have demonstrated that suppression increases where resources are limited. For example, in Damaraland mole rats, physiological suppression of subordinate females is most pronounced in the dry season and eases during the annual rains, when ecological constraints are relaxed (Young *et al.* 2010).

Early studies of reproductive suppression suggested that it was caused by chronic elevation of glucocorticoid adrenal hormones as a result of social 'stress' induced by regular aggression from dominants (Wasser and Barash 1983; Abbott *et al.* 2003). However, in several species, cortisol levels either do not vary consistently between subordinates and dominants, or subordinates show lower glucocorticoid levels than dominants (Creel 2001; Starling *et al.* 2010; Creel *et al.* 2013). In addition, cues signalling the presence of dominant females can prevent subordinate females conceiving in the absence of direct interactions with dominant females (Young *et al.* 2006).

An alternative interpretation is that reproductive suppression is a result of reproductive restraint by subordinates rather than of active interference by dominants (Saltzman *et al.* 2009). Several reasons why it might benefit subordinates to defer breeding have been suggested, including reduced foraging skills and associated energetic constraints, the absence of unrelated breeding partners, negative effects of breeding at the same time as dominants on the fitness of their own offspring, and costs to indirect components of their fitness if dominants are close relatives (Young 2009). In particular, the absence of unrelated breeding partners in the group commonly affects the development of subordinates and the replacement of related dominant males with unrelated males can prompt subordinate females to upregulate their reproductive systems and compete for the breeding role (Cooney and Bennett 2000).

Whether reproductive suppression in subordinates is most realistically interpreted as a consequence of reproductive constraint by dominants or of reproductive restraint by subordinates is still debated (Young 2009) and a clear distinction between these two explanations is often difficult to draw. Evidence that subordinates forgo breeding in the absence of dominant females does not necessarily preclude a role for constraint by dominants, since dominants commonly kill offspring born to subordinates or evict subordinates that attempt to breed and the threat of interference by dominant females may be sufficient to induce reproductive suppression (Johnstone and Cant 1999). Moreover, lower levels of glucocorticoids in subordinates than dominants do not necessarily suggest that increased cortisol levels play no role in reproductive suppression. For example, subordinates that have downregulated their reproductive physiology and are no longer the target of regular attacks may show low cortisol levels while increased levels of aggression

from dominants directed at subordinates that attempt to breed may lead to temporary increases in glucocorticoid levels large enough to suppress reproduction. For example, in banded mongooses, elevated glucocorticoid levels in pregnant subordinate females reduce their reproductive success (Sanderson *et al.* 2015) and increased rates of pregnancy loss in subordinate females exposed to stressful conditions occur in other social species, too (Clutton-Brock *et al.* 2008; Henry *et al.* 2013). The most likely conclusion is that several interrelated mechanisms contribute to reproductive suppression in subordinates and it is probable that their relative importance differs between species (Young 2009).

Attempts by females to prevent other females from breeding or to reduce their success in rearing offspring are sometimes regarded as examples of spite since they can occur at times when resources are abundant (Stockley and Bro-Jørgensen 2011). However, although this is theoretically possible (Gardner and West 2004), the fitness costs of attacks on subordinates and their offspring may often be low while simultaneous breeding by subordinates may often have substantial costs to dominants and their dependants (Clutton-Brock *et al.* 2010). Consequently, it is more likely that attempts by dominants to suppress reproduction by subordinates are examples of selfish behaviour rather than of spite (see Chapter 9).

Infanticide

In many social mammals, dominant females also direct aggression at the offspring of subordinate females. In some cases, this affects their access to resources and their condition and can lead to serious wounding or death (Wasser and Starling 1986; Lloyd and Rasa 1989; Muroyama and Thierry 1996). In societies where matrilineal female groups compete with each other within a larger group and the relative rank of matrilines is related to their size (as in several baboons and macaques), additional female recruits to competing matrilines represent a threat to competitors and the frequency of attacks on juveniles or sub-adults by members of other matrilines can be sex-dependent. For example, in captive bonnet macaques, dominant females selectively target female juveniles born into low-ranking matrilines, who show low survival compared either to the sons of subordinate mothers or to the daughters of mothers belonging to high-ranking matrilines (Silk *et al.* 1981; Silk 1988). One study of macaques has even produced evidence suggesting that subordinate females pregnant with female offspring are

Figure 8.6 A dominant female meerkat killing one of her grand-offspring. In meerkats, females rarely kill each other's offspring unless they are pregnant. *Source*: © Andy Young.

more likely to be wounded by other group members than those pregnant with males (Sackett 1981). Although these results have not been confirmed by field studies of primates, research on captive lemurs shows that fetal sex affects olfactory cues produced by pregnant females (Crawford and Drea 2015). Effects of regular aggression from other females on the development or survival of offspring are by no means restricted to primates and have been demonstrated in several other plural breeders (Hoogland 1995; Silk 2007; Clutton-Brock *et al.* 2010).

In some social mammals, females commonly attempt to kill newborn offspring born to their competitors if an opportunity arises (Figure 8.6). While infanticide by females has attracted less attention than infanticide by males (see Chapter 15), female infanticide is probably both more widespread and more frequent (Digby 2000). It is more commonly found in litter-bearing species than in monotocous ones (Lukas and Huchard 2014). Heightened levels of circulating testosterone may play an important role in the control of female infanticidal behaviour: females show a pronounced increase in circulating levels of testosterone during the second half of gestation in several social mammals (Clutton-Brock *et al.* 2006) and testosterone levels are known to be correlated with infanticidal behaviour in other mammals (Ebensperger 1998).

In some cases, infanticidal females usually kill young that are unrelated or distantly related to them. For example, in Belding's ground squirrels, infanticidal individuals are usually distant relatives or unrelated to the young they kill (Sherman 1981), while in bank voles familiarity between females decreases their tendency to kill each other's offspring (Ylönen *et al.* 1997). However, in others, females regularly kill the offspring of close relatives. Some of the best examples come from studies of black-tailed prairie dogs, where breeding females commonly kill litters born to other females belonging to the same social group and the frequency of female infanticide increases with the size of breeding groups (Hoogland 1985, 1995) (Figure 8.7). Mothers whose

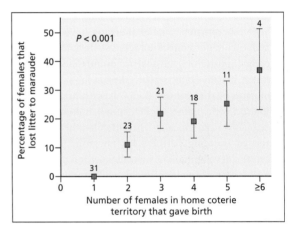

Figure 8.7 In black-tailed prairie dogs, the proportion of females giving birth that lose their litters to infanticide by marauding females (who are often close relatives) increases as the number of breeding females in the coterie rises. *Source*: From Hoogland (1995). Reproduced with permission of the University of Chicago Press. *Photo source*: © Elaine Miller Bond.

pups are killed typically occupy nursery burrows close to the killers and are smaller and lighter than their neighbours and, in many cases, are close relatives of the females that attack their young. In meerkats and marmosets, pregnant dominant females commonly kill the newborn offspring of subordinate females that give birth in the group, which would otherwise be heavier than their own offspring (Clutton-Brock *et al.* 1998; Lazaro-Perea *et al.* 2000; Young *et al.* 2006; Saltzman *et al.* 2009). In meerkats, subordinate females are commonly the daughters of dominants, and pregnant dominant females frequently kill their own grand-offspring

(Clutton Brock *et al.* 1998; Young *et al.* 2006) (see Figure 8.8).

Infanticide can have several different benefits to dominant females. In some cases, it can reduce immediate competition for space or resources (Wolff and Cicirello 1989; Tuomi *et al.* 1997). For example, in meerkats, simultaneous breeding by more than one female reduces the growth of pups (Clutton-Brock *et al.* 2010; Bell *et al.* 2014) and evidence that infanticide is more likely in pregnant than non-pregnant females also suggests that its function is partly to reduce competition for the killer's offspring.

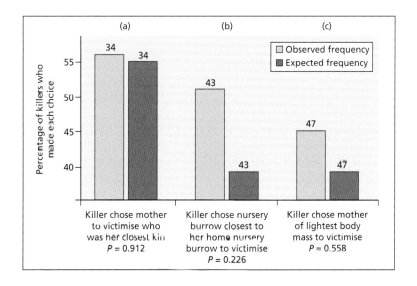

Figure 8.8 Observed and expected frequency of infanticide by marauding female prairie dogs. The figure shows the proportions of marauders that killed the offspring of (a) their closest kin, (b) their closest neighbour and (c) the lightest female in their group. *Source*: From Hoogland (1995). Reproduced with permission of the University of Chicago Press. *Photo source*: © Elaine Miller Bond.

Eviction

In a substantial number of social mammals, competition between resident females leads to evictions or to group splits. In singular breeders, increasing aggression directed by dominant females at older subordinates often builds up until subordinates are chased out of the group by the dominant female. In meerkats, dominant females commonly evict subordinates during the latter half of their (own) gestation period (Clutton-Brock *et al.* 1998) and (virtually) all female subordinates are evicted before they are 4 years old. The probability that subordinates will be evicted increases if they are distant relatives of the dominant female or if they attempt to breed (see Chapter 17).

Eviction of subordinate females by dominants is also common in plural breeders when group size is large. For example, in banded mongooses, coalitions of older dominant females intermittently evict entire cohorts of younger females from their group at times when several older females are in oestrus (Gilchrist 2006; Cant *et al.* 2010). Eviction of younger or lower-ranking females is also common in some primates (Figure 8.9). In red-fronted lemurs, the probability that females will be evicted depends primarily on the total size of their group and is not related to their kinship to other group members (Kappeler and Fichtel 2012), while high-ranking female

ursine howler monkeys frequently evict younger and lower-ranking females from their groups (Pope 2000).

Eviction commonly exposes emigrants to substantial risks (see Chapter 3) and can also raise cortisol levels and induce abortion: for example, in meerkats, pregnant subordinates that are evicted usually abort soon afterwards (Gilchrist 2006; Young *et al.* 2006; Young 2009). In some species, subordinates that become the target of regular aggression from dominants attempt to pacify them. For example, subordinate female meerkats that are at risk of eviction by dominant females engage in frequent submissive gestures and attempts to groom them, which are often rejected (Kutsukake and Clutton-Brock 2006). Experiments in which grooming frequency was experimentally reduced produced increased rates of aggression, suggesting that tactics of this kind may affect the extent to which dominants tolerate subordinates and defer evictions (Madden and Clutton-Brock 2009).

In meerkats, evicted females are often allowed to rejoin their natal groups after the dominant female's pups are several days old and the risk of infanticide has receded (Young *et al.* 2006). Since pregnant subordinates are commonly infanticidal, eviction probably reduces the chance that dominant females will lose their litters (see Chapter 17). In addition, eviction and abortion

Figure 8.9 Eviction of younger or lower-ranking females is common in some social mammals: an evicted female ursine (red) howler monkey with her infant. *Source*: © Carolyn M. Crockett.

may increase the chances that subordinates will suckle pups born subsequently to the dominant female, for females that have recently aborted are more likely to lactate to the offspring of the dominant female than those that have not done so (MacLeod *et al.* 2013).

In plural breeders that live in large groups, increases in group size can generate rising levels of aggression between subgroups of females and weaken social bonds, eventually causing groups to split into two or more separate groups with distinct home ranges. For example, in macaques, increases in group size commonly lead to increased competition between females which eventually lead larger groups to split (Okamoto 2004). Where matrilineal groups split, they often do so along matrilineal lines so that average levels of kinship between group members increase (Van Horn *et al.* 2007). Compared with evictions, the immediate costs of group splitting are usually low, though it may have deferred costs if one of the new groups is small and is forced to occupy an inadequate range or becomes more susceptible to predation (Bettridge *et al.* 2010). However, as yet, few studies have been able to assess how common such effects are.

8.3 Social structure and competition

Dominance systems

Where conflicts over limited resources occur between individuals of contrasting fighting ability, less powerful individuals often benefit by avoiding conflict and allowing their opponents to monopolise resources without fighting (Bernstein 1981; Sapolsky 1993). A high proportion of potential conflicts between group members are resolved because individuals either avoid the proximity of dominants or adjust their behaviour to avoid conflict as soon as they are threatened. Where there are consistent differences in fighting ability or power between individuals, the avoidance of conflict by weaker individuals often generates hierarchies of dominance (or submission) between group members (Rowell 1974; Silk 1993). Some early descriptions of dominance sometimes suggested that hierarchies were adaptations that benefited groups by reducing conflict between their members but a more likely interpretation is that they are consequences of attempts by individuals to avoid the costs of frequent conflicts that they are unlikely to win (Clutton-Brock and Harvey 1976).

While female dominance hierarchies are common in social mammals, the frequency of interactions, the regularity of outcomes and the linearity of hierarchies all differ between species as well as within them (Silk 1993; East and Hofer 2010). In some species, there are well-defined dominance hierarchies in both sexes and subordinate individuals seldom win encounters with competitors of higher rank, as in many of the cercopithecine primates. In others, hierarchies are unstable or there is no regular pattern in the outcome of aggressive interactions between adult females. For example, in red deer, the relative dominance of females is affected by whether or not they are within their usual range (Thouless and Guinness 1986) while in African lions, individuals are seldom displaced from feeding sites and there are no marked differences in the frequency with which individuals give and receive threats (Schaller 1972; Packer *et al.* 2001). Similarly, in Kalahari meerkats, foraging females usually respect each other's access to feeding sites and seldom contest access to food, though the most dominant female in each group occasionally steals food from subordinates (Kutsukake and Clutton-Brock 2006).

The reasons for variation in the consistency of dominance relationships between females are uncertain. In primates, it has been suggested that the presence of strong linear hierarchies in females is associated with reliance on foods that are distributed in patches of high value and with intense direct competition between group members for resources (Wrangham 1980; Sterck *et al.* 1997). Some intraspecific comparisons support this suggestion (Barton and Whiten 1993; Barton *et al.* 1996), although the responses of individuals to changes in the availability or distribution of resources are affected by a number of social factors including group structure and patterns of association within groups and are not consistent (Henzi *et al.* 2013). Differences in dominance relationships between species are also variable and do not appear to be associated with ecological parameters in any simple fashion. In some taxonomic groups, there are contrasts in the prominence of hierarchies which do not appear to be correlated with obvious differences in ecology. For example, the structure and regularity of dominance hierarchies differs among species of macaques and is not obviously associated with variation in ecology and appears to be associated with phylogenetic differences (Thierry 1990; Ménard 2004), while among lemurs similar patterns of social structure are found in species with contrasting feeding ecology (Kappeler 1997). Among other mammals, female hierarchies have been reported in herbivores as well as in carnivores and their prominence and regularity vary between species with similar diets (Wells and von Goldschmidt-Rothschild 1979; Rubenstein and Nuñez 2009). For example, they are weak or absent in lionesses (Packer *et al.* 2001), but are well developed in spotted hyenas (Holekamp *et al.* 1996; East and Hofer 2010).

Attempts to relate variation in dominance relationships to contrasts in ecological or social conditions are constrained by limitations of sample size as well as by the difficulties of comparing hierarchies between populations or species. Most intraspecific comparisons rely on assessments of two or three social groups, when much larger samples of groups are needed to establish whether there are consistent relationships between variation in dominance and social or ecological conditions. Sample size presents less of a problem with interspecific comparisons but there are large differences in the ways in which interactions are sampled and recorded, which complicate quantitative comparisons. In addition, there are pronounced interspecific differences in the frequency of dominance interactions that do not necessarily reflect

contrasts in the structure of hierarchies and further complicate comparisons between species.

Dominance and reproductive success

As longitudinal records of female breeding success have become available, an increasing number of studies have demonstrated positive correlations between dominance and breeding success in females (Stockley and Bro-Jørgensen 2011; Pusey 2012). Studies of several primates show that high-ranking females have priority of access to resources (Holand *et al.* 2004; Murray *et al.* 2006) and that they breed earlier, more frequently and for longer than subordinates (Bulger and Hamilton 1987; Smuts and Nicolson 1989; Wasser *et al.* 2004). That their infants grow faster (Johnson 2003; Altmann and Alberts 2005) and are more likely to survive their first year of life (Silk *et al.* 2003; Pusey 2012). For example, in spotted hyenas, high-ranking females have priority of access at kills, breed at younger ages than subordinates, wean their offspring more rapidly, breed more frequently and produce more surviving offspring (Holekamp *et al.* 1996; Holekamp and Dloniak 2009; East and Hofer 2010) (Figures 8.10 and 8.11). Positive correlations between

Figure 8.10 Dominance interactions between female spotted hyenas are common and a female's social rank has an important influence on her breeding success. *Source*: © Marion East and Heribert Hofer.

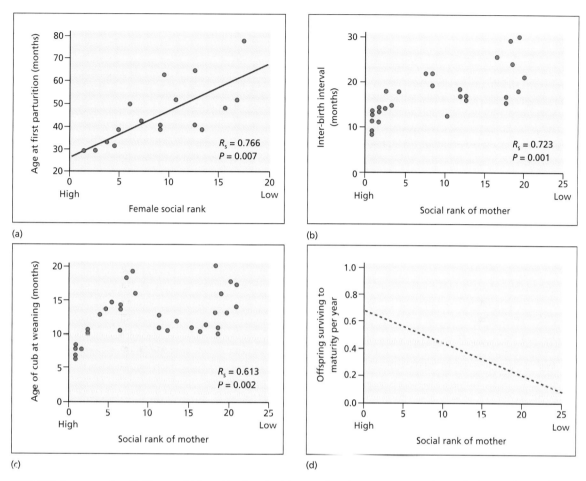

Figure 8.11 In groups of spotted hyenas, high-ranking females (a) start to breed at a younger age, (b) give birth more frequently, (c) wean their offspring more rapidly, and (d) produce more surviving offspring than lower-ranking females. *Source*: From Holekamp *et al.* (1996). Reproduced with permission of BioScientifica Ltd.

female dominance and breeding success have also been found in species living in more open groups, including elephants (Lee 2011) and red deer (Clutton-Brock *et al.* 1988). As well as affecting breeding success, dominance rank can influence the extent to which individuals are exposed to the risk of predation and their relative rates of survival. For example, in long-tailed macaques, high-ranking females are more likely than subordinates to maintain safe, central positions in the group where they are less exposed to predators (van Noordwijk and van Schaik 1987).

Although a substantial number of studies have found positive correlations between dominance and breeding success or survival, these are not universal and

relationships vary with ecological conditions (Stockley and Bro-Jørgensen 2011). For example, studies of provisioned groups of Japanese macaques found no association between female dominance and breeding success (Gouzoules *et al.* 1982) and one study of ring-tailed lemurs found that positive correlations between dominance and breeding success were restricted to large groups (Takahata *et al.* 2008). In some species, correlations between dominance and breeding success are strongest when conditions are poor (Pusey 2012), though extreme food shortage can also weaken relationships between dominance and breeding success (Woodroffe and Macdonald 1995). For example, a study of yellow baboons during a period when the population was declining found no association between dominance and breeding success (Wasser *et al.* 2004). In conjunction, these results suggest that correlations between dominance and reproductive success are closest under conditions of moderate to severe food shortage and may weaken when resources are abundant or when they are extremely scarce. In addition, the effects of dominance on components of fitness may be mediated by the extent of an individual's social connections with other females (and males) from their group (Silk *et al.* 2009; Archie *et al.* 2014) and the importance of these effects probably varies between species.

The acquisition of dominance

Where female dominance and breeding success are closely correlated, strong selection pressures are likely to favour the acquisition of high status by females. In many species, older females establish their dominance over younger individuals and age-related dominance relations occur in a wide range of mammals, including feral ponies (Rutberg and Greenberg 1990), African elephants (Archie *et al.* 2006), mountain goats (Côté 2000), meerkats (Clutton-Brock *et al.* 2006), chimpanzees (Pusey *et al.* 1997) and bottlenose dolphins (Samuels and Gifford 1997). In some species, dominance status is also associated with body mass: for example, experiments with house mice show that body mass predicts subsequent dominance rank (Rusu and Krackow 2004).

In some mammals, the development of weaponry used in intraspecific interactions affects the ability of females to compete. Interspecific comparisons of the distribution of horns and antlers in female ruminants show that their presence is associated with group size (Roberts 1996), while some intraspecific comparisons suggest that they are associated with the frequency of feeding competition as well as with the need for effective defence against predators (Packer 1983). For example in caribou, where both sexes have antlers, females use their antlers in competition for feeding sites at times when the ground is covered with snow: across populations, the proportion of antlered females increases with mean snow depth (Schaefer and Mahoney 2001) and individuals with larger antlers are more successful in obtaining access to limited food (Barrette and Vandal 1986), though this may be a consequence of correlations between female antler size and age or body size. In Soay sheep,

where some females are horned while others are hornless, horned females are more likely to initiate and win aggressive interactions than hornless ones (Robinson and Kruuk 2007), while studies of cattle show that the experimental removal of horns in females leads to reductions in the ability of individuals to dominate competitors in newly established groups (Boussou 1972). As in males, individual differences in horn size in adult females are often correlated both with early growth and with measures of breeding success in adulthood (Rughetti and Festa-Bianchet 2011).

Androgen levels can affect the aggressiveness of females and their probability of acquiring dominant status (Staub and de Beer 1997). In several species where female competition is intense (including meerkats, spotted hyenas and several social lemurs), dominant females show elevated levels of aggression and of circulating testosterone during gestation (Clutton-Brock 2007; Drea 2007) although, within species, relationships between dominance status and androgen levels are often inconsistent and are rarely higher in females than in males (von Engelhardt *et al.* 2000; Goymann *et al.* 2001; Koren *et al.* 2006). There is also evidence that selection for aggressiveness in females increases testosterone levels and reduces fecundity. In breeds of Swiss domestic cattle, where females have been selected to fight in tournaments (Figure 8.12), they show enhanced testosterone levels and a masculine appearance and are usually dominant to females of breeds where females have not been selected to compete (Plusquellec and Boussou 2001). As a result of selection for fighting ability in these breeds, there have been reductions both in milk production and in fecundity (Figure 8.13).

Multiple developmental factors can affect the chance that females will acquire and maintain high status. Both the birthweight of juveniles and their subsequent growth rates often affect their relative rank and these differences can be maintained into adulthood (Clutton-Brock 1991b; Clutton-Brock *et al.* 2006) so that environmental and social factors that influence early growth and development often have an important influence on the probability of acquiring high rank in adulthood (Clutton-Brock 1991b). Where dominant mothers have greater access to resources than subordinates and they produce heavier offspring that grow faster, their offspring often have a higher chance of acquiring high status. For example, in Kalahari meerkats, dominant females are able to displace subordinates from feeding sites and gain more weight

Figure 8.12 In the Italian Western Alps, cow fights are staged in spring and autumn between females that have been selectively bred for their aggressiveness and fighting ability. Females of fighting breeds exhibit a number of masculine traits, including increased levels of testosterone, deeper chests, larger necks and curly hair, and are usually dominant to other breeds. *Source*: © Cristina Sartori.

each day and the daughters of dominant females are heavier at birth, grow faster and are more likely to acquire dominant status as adults than the daughters of subordinates (Clutton-Brock *et al.* 2006). Similarly, in spotted hyenas, the offspring of dominant females have higher circulating levels of insulin-like growth factor (IGF-1), grow faster and are both more likely to survive and to breed successfully than those of subordinate mothers (Holekamp *et al.* 1996; Höner *et al.* 2010).

Variation in hormone levels associated with conflicts over maternal status can also affect the development of offspring. Rank-related differences in the mother's hormonal status during pregnancy can affect fetal development: for example, in spotted hyenas, dominant females have higher androgen levels during the second half of gestation and offspring of mothers with high androgen levels during pregnancy are more aggressive towards other cubs and mount them more frequently than cubs born to mothers with low androgen levels (Dloniak *et al.* 2006; Holekamp *et al.* 2013) (Figure 8.14).

Dependent rank

Social relationships also play an important role in determining and maintaining female status (Silk *et al.* 2009; Lonsdorf and Ross 2012). In several cercopithecine primates, mothers support their daughters in competitive interactions with the offspring of other females and this helps to establish their rank in the group (Chapais 1988, 2004; Silk *et al.* 2006a). Experimental studies confirm that maternal support has important effects in determining female status. As a result, adolescent females whose mothers have died or dispersed from their natal group seldom acquire high rank as adults (Walters 1980; Johnson 1987). For example, in vervet monkeys, individuals that lose their mothers as juveniles or adolescents are usually low ranking as adults (Figure 8.15) and the absence of mothers can increase the chance that their daughters will disperse (see Chapter 3).

Associations between maternal rank and the rank and breeding success of their daughters raise questions about the relative importance of environmental and genetic

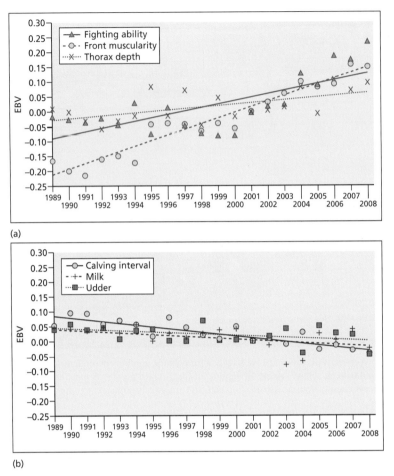

(a)

(b)

Figure 8.13 Human selection for fighting ability in breeds of cattle used for cow fights has led to heritable increases (annual estimated breeding values or EBV) in (a) fighting ability, front muscularity and thorax depth and (b) reductions in milk production, fecundity and udder size.

factors in determining female status. The available evidence suggests that both may be involved, though their relative importance may differ between species. In spotted hyenas, females sometimes adopt cubs born to other members of their clan and long-term data show that their rank as adults depends on the rank of their surrogate mother rather than on the rank of their genetic mother (East *et al.* 2009; East and Hofer 2010) (Figure 8.16). Since social and genetic factors can interact to affect gene expression (Champagne and Curley 2009; Tung *et al.* 2010, 2012), it is also possible that epigenetic mechanisms may play an important role (see Chapter 5). Other evidence suggests that variation in rank can be heritable. For example, selection experiments with captive rodents

have demonstrated genetic variance for dominance (Moore *et al.* 2002) and a quantitative analysis of dominance interactions between wild female red deer, using a multi-generational genetic pedigree, also suggests that dominance is partly heritable (Wilson *et al.* 2011).

Although the relative rank of females often increases with their age, where females live in large stable groups, mothers commonly support their younger daughters against their older sisters and this establishes inverse relationships between the relative age of their offspring and their relative rank, which often persist after the mother's death (Kawai 1958; Holekamp *et al.* 1996; Chapais 2004; East *et al.* 2009) (Figure 8.17). Current data suggest that 'youngest ascendancy' rules of this kind

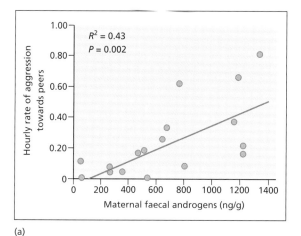

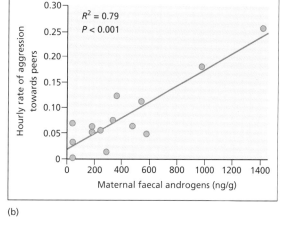

(a)

(b)

Figure 8.14 Maternal effects in spotted hyenas, showing variation in the frequency with which female offspring direct aggression at peers in relation to variation in their exposure to maternal faecal androgen before birth: (a) aggression directed by cubs (2–4 months) at peers at dens; (b) aggression directed by adult offspring (>24 months) towards peers. *Source*: From Holekamp *et al*. (2013). Reproduced with permission from the Royal Society. *Photo source*: © Kay Holekamp.

may be restricted to societies where females live in groups that include several competing matrilines, like baboons and spotted hyenas. In these societies, matriarchs may support younger offspring against their older sisters either because this helps to protect their younger offspring from competition with older sibs, or because, by constraining the status of older daughters, they reduce the risk that coalitions of their older daughters will be able to displace them (Horrocks and Hunte 1983).

In societies where groups include several matrilines, adolescent females may receive support from their sisters and other matrilineal relatives as well as from their mothers. Individuals belonging to relatively high-ranking matrilines benefit from having larger numbers of high-ranking relatives who are more socially active and can help to induce submission in competitors more effectively (Chapais 1992, 2004; Pereira 1992). As a result, females born into high-ranking matrilines commonly show faster growth, higher survival, acquire higher status and have higher fitness than those

belonging to low-ranking matrilines (Silk 2007, 2009). In some species, the relative rank of matrilineal groups is associated with their size, while in others it appears to be determined by the dominance of the group's matriarch (Silk 2007, 2009). The quality of relationships between females and the structure of the supportive networks to which they belong may also be important (Silk *et al*. 2013). 'Dependent rank systems, where the status of individuals is strongly influenced by the rank of their matrilineal group, are common in the baboons and macaques (Chapais 2004; Kapsalis 2004), though they do not occur in all social primates (Nakamichi and Koyama 1997; Sauther *et al*. 1999; Perry *et al*. 2008).

Long-term studies of primates have documented the relative frequency of support given to different categories of relatives and their effects. In general, females are most likely to support close matrilineal kin and support is given to mothers, offspring, grandmothers, grand-offspring and, in some cases, to aunts and nieces, but seldom to more distant relatives whose relatedness is below 0.25

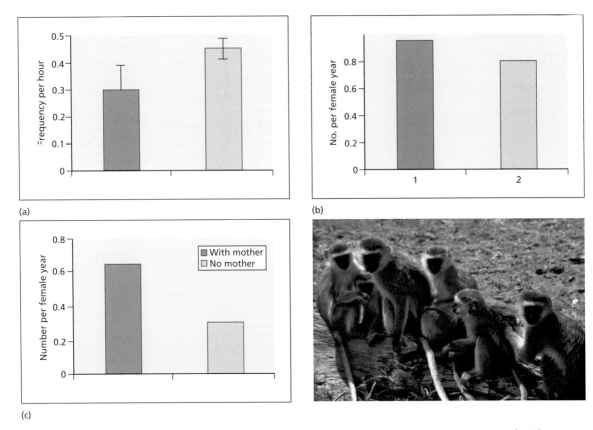

Figure 8.15 Effects of the absence of mothers on (a) the frequency of received aggression, (b) the pregnancy rate and (c) the production of surviving infants in female vervet monkeys. *Source*: From Horrocks and Hunte (1983). Reproduced with permission of Elsevier. *Photo source*: © Dorothy Cheney.

(Kapsalis and Berman 1996; Silk 2009). As yet, it is unclear whether this threshold is a consequence of constraints on the ability to recognise kin or occurs because it becomes more difficult to satisfy the requirements of Hamilton's rule as relatedness declines. Experiments with Japanese macaques show that sisters, grandmothers and great-grandmothers are able to influence rank acquisition by immature females, while aunts, grand-aunts and cousins rarely do so (Chapais *et al.* 2001; Chapais 2004).

Recent studies of baboons and macaques also suggest that patrilineal kinship can affect supportive relationships, though effects are usually weaker than those of matrilineal kinship (Silk 2007, 2009). For example, rhesus macaques avoid intervening against paternal half-sisters (Widdig *et al.* 2006), female baboons form stronger bonds with their paternal half-sisters than with unrelated

individuals when they have few maternal kin in the group (Silk *et al.* 2006a,b) and juvenile mandrills have closer relationships with paternal half-sibs than with unrelated adults (Charpentier *et al.* 2007). However, paternal kinship does not affect the strength of social bonds in all species: for example, white-faced capuchins show no tendency to give preferential treatment to paternal half-sibs over unrelated individuals (Perry *et al.* 2008).

Females also derive support from alliances with members of different matrilines (see Chapter 9). Members of large high-ranking matrilines typically have more extensive alliances with unrelated females than low-ranking females and the effects of these alliances enhance those of matriline size. Relationships between females and males can also influence the acquisition of dominance. For example, in baboons, females frequently develop

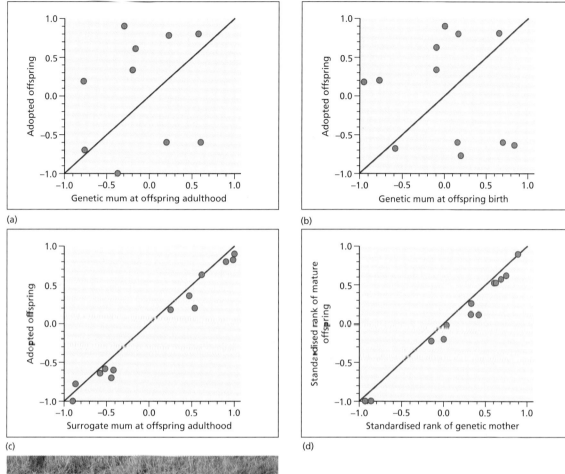

(a)

(b)

(c)

(d)

Figure 8.16 Standardised rank of adopted spotted hyenas at adulthood in relation to (a) the standardised rank of their genetic mother when they reached adulthood, (b) the standardised rank of their genetic mother on the date they were born; (c) the standardised rank of their surrogate mother when the adopted young reached adulthood; (d) the standardised rank of their genetic mothers. *Source*: From East *et al.* (2009). Reproduced with permission of Oxford University Press. *Photo source*: © Sarah Benhaiem.

long-lasting relationships ('friendships') with particular males (Smuts 1985; Silk 2007) and male 'friends' will intervene in competitive interactions involving females with whom they have 'friendships' (see Chapter 15). Relationships between females and their 'friends' of both sexes are recognised by competitors (Cheney and

Seyfarth 1999) and these alliances can influence both the maintenance of rank and the breeding success of females (Silk *et al.* 2009).

As well as influencing the rank and eventual breeding success of juveniles (Maestripieri 2009; Silk 2009), the rank of the matriline that females are born into can

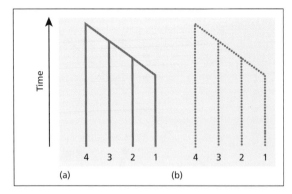

Figure 8.17 Idealised rank systems between females in two groups of sibs. Within each group, the first-born individual (1) has the lowest rank while the last born (4) has the highest rank. All members of one matriline may be dominant to all members of the other.

influence the chance that they will be evicted from the group. For example, when groups of ring-tailed lemurs become large, members of one matriline may attempt to evict individuals from subordinate matrilines (or, in some cases, entire matrilines) and supportive coalitions can play a crucial role in enabling individuals to remain in their natal group or to join other groups (Nakamichi and Koyama 1997; Sauther *et al.* 1999). A similar situation may occur in banded mongooses (see Chapter 17).

Patterns of association and support within social groups commonly vary throughout the year as ecological conditions change. For example, networks of regular associations between female baboons that are evident when food is scarce can disappear when food is plentiful and be replaced by more casual and less predictable patterns of association (Henzi *et al.* 2009). When food becomes scarce again, regular patterns of association re-emerge, though they do not necessarily involve the same individuals, suggesting that there is an important stochastic element to the formation of social networks.

8.4 Conflict proliferation and limitation

Retaliation

In mammal societies where there are large differences in power between dominant females and subordinates, subordinates rarely retaliate to threats or attacks from

dominants for, if they do, this commonly elicits repeated attacks. For example, in meerkats, older subordinates who are attacked by dominant females respond with submissive actions and commonly attempt to groom dominant females that direct aggression at them (Kutsukake and Clutton-Brock 2006). In contrast, where differences in power between resident females are smaller, subordinates are more likely to retaliate to threats or attacks from dominants (Silk 2007). Retaliatory attacks often occur shortly after the initial interaction but may also be delayed for considerable periods. For example, in chimpanzees, attacks sometimes generate retaliation from recipients weeks or even months after the initial attack (de Waal 1982).

Individuals that have been the target of aggression may also threaten or attack relatives of the aggressor rather than the aggressor itself, a tactic described as *redirected aggression*. For example, after female vervet monkeys have been attacked by another group member, they are more likely to attack relatives or allies of their attacker than at other times and their relatives and allies may also attack the original aggressor or its relatives or allies (Figure 8.18). Similar examples of retaliatory aggression involving third parties have been documented in females in several other primates, including Japanese macaques (Figure 8.19). The targets of these retaliatory attacks are often younger and less dominant than the initial victim and aggressors may be unable to monitor or prevent retaliation of this kind.

Reconciliation and reassurance

Where aggressive interactions between pairs of animals generate counter-attacks by their relatives and allies, this may depress their foraging efficiency and reduce the capacity of individuals and their kin to compete effectively with other group members. For example, in some macaques, conflicts are followed by a period of uncertainty when social stress is high, animals commonly scratch and yawn, the heart rates of animals involved in the conflict are elevated, and the probability of the victim of an initial attack being aggressive to another group member is high (Smucny *et al.* 1997; Das *et al.* 1998; Silk 2002). These effects tend to be stronger where the original conflicts involved kin or individuals that previously had strong attractive or supportive relationships (Aureli 1997; Kutsukake and Castles 2001, 2004).

Studies have explored the strategies that individuals use to placate opponents or to limit the social impact of

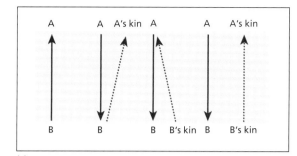

(a)

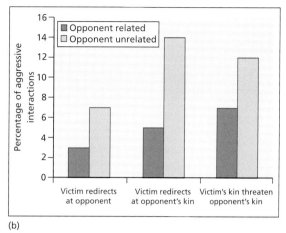

(b)

Figure 8.18 Retaliation after aggression in vervet monkeys:
(a) different types of retaliatory interactions that can occur after
aggressive interactions where A attacks or threatens B;
(b) relative frequency with which individuals that have
recently received aggression and their kin direct threats or
attacks at their opponents or their opponent's kin, depending
on whether opponents were related or unrelated. *Source*: From
Cheney and Seyfarth (1990). Reproduced with permission of
the University of Chicago Press. *Photo source*: © Dorothy
Cheney.

conflicts (de Waal 1989; Aureli and de Waal 2000).
During the minutes immediately after an aggressive
interaction, either of the contestants may initiate an
amicable interaction with its former opponent. For
example, immediately after conflicts between pairs of
long-tailed macaques, the probability that one will initiate an affiliative interaction with the individual that it
has just interacted with is high (Aureli *et al.* 1989)
(Figure 8.20). Similar increases in the probability of
affiliative interactions after contacts have now been
demonstrated in a number of monkeys and apes (de
Waal and van Roosmalen 1979; de Waal and Ren 1988;
de Waal 1993; Aureli and de Waal 2000; Silk 2002) as
well as in some carnivores (Hofer and East 2000; Wahaj
et al. 2001), cetaceans (Samuels and Flaherty 2000) and
ungulates (Cords 1992; Schino 1998). In some cases,
victims initiate these interactions but aggressors may
do so, too (Das *et al.* 1997; Das 2000). Several lines of
evidence suggest that affiliative interactions after conflicts serve to restore relaxed relationships between them
(Das 2000). After affiliative or reassuring interactions
have occurred, the probability of further aggression is
commonly reduced (Kappeler and van Schaik 1992).

For example, if aggression is induced experimentally
between pairs of macaques at experimental feeding sites,
dominants are subsequently more likely to tolerate subordinates if an affiliative interaction has occurred and
subordinates are less likely to avoid dominants than if
there has been no affiliative interaction (Aureli *et al.*
1989; Aureli and de Waal 2000).

Interactions following conflicts often involve vocalisations. For example, in chacma baboons, aggressors commonly approach their victims after an aggressive attack or
threat and grunt at them (Figure 8.21). Victims of attacks
that had been played grunts given by their aggressors were
subsequently more likely to approach their aggressors as
well as to tolerate approaches by them and were also
less likely to be supplanted by their previous opponent
(Cheney and Seyfarth 1997) (Figure 8.22). In contrast,
playbacks of reconciliatory grunts given by victims to
aggressors did not increase the rate at which previous
opponents approached or initiated friendly contact with
them, suggesting that interactions modified the subsequent behaviour of victims rather than aggressors.

Affiliative interactions after conflicts are commonly
described as 'reconciliation', implying that their function

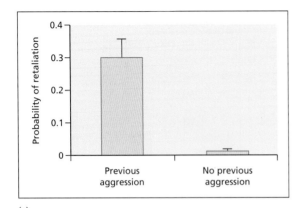

(a)

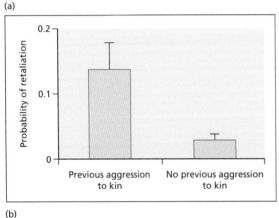

(b)

Figure 8.19 Probability of retaliatory aggression in Japanese macaques by (a) individuals that have been attacked or threatened and (b) the kin of individuals that have been attacked or threatened. *Source*: From Aureli *et al.* (1992). Reproduced with permission of Elsevier.

is to restore relationships that have been disturbed by conflict (Aureli and de Waal 2000; Cords and Aureli 2000), though other explanations have been suggested (Silk *et al.* 1996; Silk 1997; 2000, 2002). Reconciliatory interactions are commoner between individuals that previously had relatively close social relationships (Aureli *et al.* 1989; Aureli 1997; Call *et al.* 1999). For example, in Japanese macaques, individuals that have stronger social relationships with each other are more likely to reconcile with each other than those with weaker relationships (Schino *et al.* 1998). There is also experimental evidence that the strength of previous relationships affects the probability of reconciliation. For example, individual long-tailed macaques that were trained to be tolerant became more likely to initiate affiliative interactions with each other after conflicts (Cords and

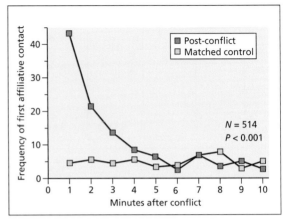

Figure 8.20 The frequency of the first affiliative contact between former opponents during the first 10 minutes after a conflict between them (PC) compared with the frequency of affiliative contacts between matched pairs of control animals not involved in conflicts in captive long-tailed macaques. *Source*: From Aureli *et al.* (1989). Reproduced with permission of John Wiley & Sons.

Thurnheer 1993), indicating that tolerance may increase the probability that individuals will reconcile with each other (Cords and Aureli 2000).

Consolation and intervention

In societies where aggressive interactions may spread and involve the relatives or friends of contestants, other group members sometimes direct affiliative interactions at victims of attacks (Lindburg 1973; de Waal and van Roosmalen 1979; de Waal and Aureli 1996; Call *et al.* 2002). Kin or allies of contestants may also initiate affiliative interactions with each other (York and Rowell 1988; Cheney and Seyfarth 1989; Aureli and van Schaik 1991). For example, in macaques, 'victims' of serious attacks by other group members are more likely to receive affiliative contacts shortly after attacks than at other times (Figure 8.23a). Aggressors are also more likely to receive affiliative contacts from their kin as well as from the kin of their previous opponents (Das *et al.* 1998; Das 2000; Petit and Thierry 2000) but not as frequently as victims (Figure 8.23b).

These interactions are often interpreted as attempts by other group members to 'console' victims and calm aggressors and to limit the social disruption caused by aggression. In some circumstances, individuals also appear to encourage previous contestants to reconcile (Fraser *et al.* 2008). For example, de Waal describes how, after bouts of aggression between male chimpanzees, a

Figure 8.21 A female chacma baboon grunts at another to signal her friendly intentions. *Source*: © Joan Silk.

female would sometimes groom one of the males and then appear to lead him over to his former opponent, eventually leaving the two males to groom each other (de Waal and van Roosmalen 1979).

Several studies have investigated the consequences of third-party interactions after disputes, but their results are mixed. For example, while some studies of macaques and chimpanzees have found that affiliative interactions with third parties reduce signs of stress in individuals that have recently been involved in agonistic encounters, others have found no evidence of effects of this kind (Koski and Sterck 2007; Fraser *et al.* 2008; Das *et al.* 1998; McFarland and Majolo 2012). Moreover, in some cases the probability that victims will be attacked or threatened by other group members increases in the minutes after a dispute, possibly because conflicts provide opportunities for third parties to establish dominance over temporarily weakened rivals.

8.5 Consequences of female competition

Reproductive skew
In societies where some females are able to monopolise access to a disproportionate share of the resources necessary for reproduction, individual differences in breeding success and reproductive skew among females may be large and competition for breeding opportunities is often intense (Clutton-Brock 2009; Stockley and Bro-Jørgensen 2011). Well-defined female dominance hierarchies, like those found in baboons and spotted hyenas, generate relatively high values of reproductive skew, while reproduction is more egalitarian in societies where there are no regular hierarchies, as in African lions (Packer *et al.* 2001) (Figure 8.24). The highest values of reproductive skew are found in singular cooperative breeders where dominant females suppress the fertility of subordinate females and can produce large litters at frequent intervals because their young are protected and fed by other group members (see Chapter 17).

Sexual dimorphism
As in males (see Chapter 13), high levels of reproductive skew in females increase the opportunity for selection and may raise the intensity of selection for traits that increase competitive ability, leading to the evolution of secondary sexual characters in females (see Chapter 1) (Clutton-Brock 2004, 2007; Bro-Jørgensen 2011). For example, in spotted hyenas, the lifetime breeding success of females is correlated with indices of body size (Swanson *et al.* 2011) and females are slightly larger than males. Similarly, in

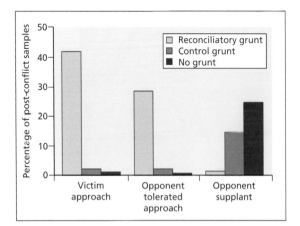

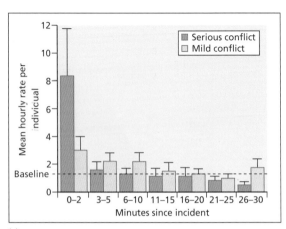

(a)

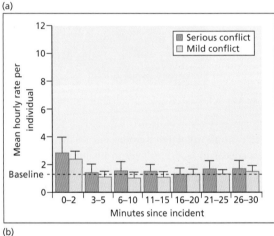

(b)

Figure 8.22 Effects of grunting by dominant aggressors after disputes in chacma baboons on the probability that 'victims' will approach aggressors. In these experiments, the victims of recent attacks were played a 'reconciliatory' grunt given by their aggressors in the minute following attacks and their interactions were then monitored for the next half hour. These treatments were matched with controls in which individuals were either played grunts given by other animals that had not been involved in the conflict or were not played any grunts. The figure compares the proportion of experiments in which victims approached their previous aggressors, tolerated approaches by them or were supplanted by them. *Source*: From Cheney and Seyfarth (1997). Reproduced with permission of Elsevier. *Photo source*: © Joan Silk.

Figure 8.23 'Consolation' in long-tailed macaques: (a) shortly after attacks 'victims' are more likely to receive social contacts from other group members than at other times; (b) aggressors are also more likely to receive contacts but to a lower extent. *Source*: From de Waal and Aureli (1996).

several singular cooperative breeders, dominant females play an important role in preventing subordinates from breeding and are usually the largest individuals in their group and are socially dominant to all other group members, including males (Reeve and Sherman 1991; Faulkes and Abbott 1997; Clutton-Brock *et al.* 2001; Creel and Creel 2002). In meerkats and naked mole rats, females that acquire the breeding position show a period of secondary

growth that is reduced or absent in males and may help them to maintain their status (see Chapter 17).

In some plural breeders where female competition is unusually intense, dominant females can show an unusual development of traits that enhance competitive ability, including larger canine size and heightened testosterone at some stages of the breeding cycle (Plavcan *et al.* 1995; Holekamp and Smale 2000; Drea 2007) though it is seldom the case that testosterone levels are higher in females than in males (Goymann *et al.* 2001). Their genitalia may also show signs of masculinisation (Licht *et al.* 1992, 1998; Drea *et al.* 1998; Glickman *et al.* 1998): spotted hyenas are the best-known example

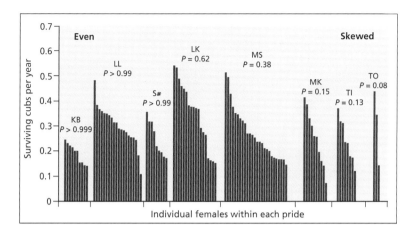

Figure 8.24 Egalitarian reproduction in African lionesses. The figure shows the number of surviving cubs produced per year by adult lionesses in eight prides. The most successful female was the mother of 50% or more of the surviving cubs in less than half of the prides. *Source*: From Packer *et al.* (2001). Reproduced with permission from AAAS. *Photo source*: © Dom Cram.

(Figure 8.25), but genitalia of adult females also show evidence of masculinisation in other species where females compete intensely, including some lemurs and golden moles (Ostner *et al.* 2003; Drea 2007; Drea and Weil 2008). Several explanations of the virilisation of females in these species have been suggested, including the possibility that it is a non-adaptive by-product of elevated testosterone levels or of increased sensitivity to androgens (Racey and Skinner 1979; Frank 1997), or a consequence of selection for precocial and aggressive offspring (Goymann *et al.* 2001). Alternatively, virilisation may be adaptive in these societies if mimicking the characteristics of males helps subordinate females to deflect aggression directed at them by dominant females (Hawkins *et al.* 2002; Hofer and East 2003). For example, the striking pseudo-penis and pseudo-scrotum of female

spotted hyenas closely mimics male genitalia and may allow subordinate females to mimic males and to reduce the amount of aggression they receive from dominant females (Kruuk 1972). Similarly, in fossas, juvenile females develop an enlarged spinescent clitoris supported by an os clitoridis and a pigmented secretion on the fur underparts which appears to mimic the characteristics of adult males and may help to reduce aggression from territorial females (Hawkins *et al.* 2002). This explanation is strengthened by evidence that adolescent males show evidence of transient feminisation in some primates where there is intense competition between males (Kuhn 1972).

As in males, reproductive competition between females for breeding partners has led to the evolution of signals of condition and reproductive status that attract

Figure 8.25 Masculinisation of female genitalia in spotted hyenas. (a) Diagram of the reproductive anatomy of a female spotted hyena. (b) (*left*) Drawing of the external genitalia of an adult female (by R.J. Gordon) and (*right*) photographs of the external genitalia of an adult female and a male. (c) External view of the penis of a male and the clitoris of a female (through which she will give birth). (d) Side view of male and female cubs. *Source*: (a,b) From Cunha *et al.* (2014), adapted from Drea. Reproduced with permission of Elsevier; (c,d) © Steve Glickman.

the opposite sex. For example, female facial coloration in some baboons and macaques is brighter during the fertile phase of their oestrous cycles (Setchell *et al.* 2006; Dubuc *et al.* 2009). In addition, the detailed structure of copulatory calls given by females changes with their stage of oestrus (O'Connell and Cowlishaw 1994; Semple *et al.* 2002) and playback experiments show that, in macaques, males can discriminate between calls given by females at different stages of their cycle and are most attracted to the calls of females in late oestrus (Semple and McComb 2000).

Breeding competition between females can also lead to the development of prominent secondary sexual characters in females. These occur in a substantial number of species and, in many cases, appear to be associated with breeding systems where females gain direct fitness

benefits by mating with multiple partners and consequently compete to attract males (see Chapter 1). One of the most striking examples of female ornaments in mammals are the cyclical perineal swellings found in monkeys and apes that live in multi-male groups where females have access to multiple partners (see Chapter 7). In several of these species, females can gain support and protection for themselves and their offspring from males they consort with and may increase their direct fitness by attracting and mating with multiple partners, and males may increase their direct fitness by mating with females with large swellings. As might be expected, large swellings are more effective in attracting males and may have originated as a signal of receptivity.

In some cases, selection operating through reproductive competition between females has reversed the usual pattern of sex differences in behaviour. For example, in a number of mammals where reproductive competition between females is unusually intense, including spotted hyenas, meerkats and several of the social lemurs, females are more frequently aggressive than males and are usually dominant to them (Kappeler 1993; Clutton-Brock 2009; East and Hofer 2010). One explanation of the evolution of female dominance over males is that it is a consequence of selection on females to obtain priority of access to resources or to protect themselves against male harassment, while another possibility is that it is a by-product of intense selection for competitive success in females (Clutton-Brock and Huchard 2013). Sex role-reversal is probably relatively uncommon in mammals compared to birds because the evolution of lactation and the absence of exclusive male care constrains variation in female breeding success, limiting the potential pay-offs of mating competition in females (see Chapter 1). Its distribution does not appear to be closely related to mating systems and it may occur where ecological or social conditions generate unusually intense reproductive competition between females, as in many singular cooperative breeders (Clutton-Brock 2009).

The evolution of enhanced levels of competitiveness and associated traits in females raises important questions about their costs as well as about the evolutionary mechanisms limiting their development. In males, the development of traits that enhance competitive ability (such as increased adult body size, fast growth rates and aggressiveness) is often associated with increases in the parasite load of males (see Chapter 13) and with reductions in the survival of juveniles, adolescents and adults, especially at times when resources are scarce (see Chapter 18). A recent study of meerkats (where dominant females are more frequently aggressive than other group members) has shown that levels of some parasites are also highest in dominant females (Smyth and Drea 2015) but there is little evidence that sex differences in survival are reversed in species where reproductive competition is more intense or secondary sexual characters are more highly developed in females (Clutton-Brock 2009). One possibility is that the costs of expenditure by females on reproductive competition or ornamentation depress fecundity before they reach a level at which they have measurable costs to female survival, and that costs to fecundity constrain the development of secondary sexual characters (LeBas 2006). For example, elevated levels of testosterone may have adverse effects on the fecundity of females or on the development of their offspring which constrain the evolution of further increases in female competitiveness (Drea *et al.* 2002; Knickmeyer and Baron-Cohen 2006). One line of evidence supporting this suggestion is that Swiss cattle that have been selected for competitive ability show reductions in milk yield and fecundity (see Figure 8.13).

Regulation of group size
Competition between females has important consequences for the size of female groups. In a substantial number of mammals, adult females do not tolerate other adult females in their range or territory and this may reduce the overall density of females. In other species, females tolerate female relatives for several years after they mature but eventually evict them from their group and, in some, dominant females tolerate the presence of mature females of all ages, as in many of the social primates. These differences are associated with contrasts in diet and food distribution: for example, female intolerance of other adults is common in insectivores and in carnivores that depend on mobile prey, while adult females tolerate each other's proximity in many frugivores and herbivores as well as in some insectivores that feed on social insects (see Chapter 2).

Contrasts in female tolerance have important consequences for female group size (see Chapter 2). Where adult females are intolerant of each other, social groups typically consist of a single adult female and her dependent offspring, sometimes accompanied by one or more unrelated males. In contrast, where breeding females tolerate each other's presence, female groups can reach much larger sizes and groups often vary widely in size as a result of stochastic processes (Ozgul *et al.* 2014). However, even in the most social species, increasing group size eventually leads to competition between group members for resources or reproductive opportunities, triggering increased rates of aggression and eviction which limit further increases in group size (see Chapter 3).

Contrasts in female sociality caused by differences in the extent to which breeding females tolerate each other have far-reaching consequences for many other aspects of behaviour. In particular, they have important effects on social relationships between females and the evolution of cooperative behaviour (Chapter 9) as well as on the evolution of mating systems (Chapter 10), mating competition between males (Chapter 13) and the evolution of sex differences (Chapter 18). They are also likely to have important consequences for the regulation of population density and other ecological processes. However, as yet, relatively few studies have been able to quantify relationships between group size and variation in survival, breeding success and emigration and systematic comparisons of these relationships are not yet possible. When empirical data eventually become available, comparisons of these relationships are likely to provide important insights into variation in the regulation of group size and the dynamics of local populations (Clutton-Brock 2001; Clutton-Brock and Sheldon 2010).

SUMMARY

1. Although competition between females has received less attention than competition between males, female mammals commonly compete for resources, breeding opportunities and mating partners, as well as for their dominance status and that of their offspring.

2. Escalated fights between females are typically less frequent than between males, and female competition often involves threats, harassment and punishment, which may explain why secondary sexual characters associated with competition are less highly developed in females than in males. In some mammals, dominant females reduce or suppress the reproductive success or survival of subordinates through frequent aggression, infanticide and eviction.

3. The dominance status of females is commonly affected by their age, size and strength as well as by their social connections and the number and rank of related and unrelated coalition partners and allies that will provide support in competitive interactions with other group members. Although there are exceptions, high-ranking females frequently have priority of access to resources and show higher survival and breeding success than lower-ranking females.

4. In some social mammals, supportive relationships between females can stimulate retaliation against aggressors by the kin of victims or by their unrelated allies and this can lead to the proliferation of conflicts, affecting many other members of the group. In several mammals where females live in large stable groups, victims and aggressors frequently 'reconcile' after agonistic interactions while other group members may 'console' victims and this may limit the repercussions of conflict for other group members.

5. Female competition exerts an important influence on the extent of reproductive skew in females and on the strength of selection in females for traits that affect competitive success. Intense competition and high levels of skew can lead to the evolution in females of traits that improve their competitive ability and resemble those of males in polygynous species. In some cases, it can also be associated with the dominance of females over resident males.

6. Competition between females also has important consequences for female sociality and the size of female groups. Where female competition for resources is intense, females are commonly solitary and social monogamy may evolve, while reductions in resource competition may increase the extent to which breeding females tolerate each other and permit the evolution of sizeable groups, with important consequences for the evolution of mating systems.

7. Reproductive competition between females may also affect the dynamics of groups and the regulation of group size. Where reproductive competition is intense, it may restrict group size in advance of constraints imposed by resource availability.

References

Abbott, D.H. (1993) Social conflict and reproductive suppression in marmosets and tamarin monkeys. In: *Primate Social Conflict* (eds W.A. Mason and S.P. Mendoza) Albany: State University of New York Press, 331–372.

Abbott, D.H., *et al.* (2003) Are subordinates always stressed? A comparative analysis of rank differences in cortisol levels among primates. *Hormones and Behavior* **43**:67–82.

Altmann, J. and Alberts, S.C. (2005) Growth rates in a wild primate population: ecological influences and maternal effects. *Behavioral Ecology and Sociobiology* **57**:490–501.

Archie, E.A., *et al.* (2006) Dominance rank relationships among wild female African elephants, *Loxodonta africana*. *Animal Behaviour* **71**:117–127.

Archie, E.A., *et al.* (2014) Social affiliation matters: both same-sex and opposite-sex relationships predict survival in wild female baboons. *Proceedings of the Royal Society of London. Series B: Biological Sciences* **281**:20141261.

Aureli, F. (1997) Post-conflict anxiety in nonhuman primates: the mediating role of emotion in conflict resolution. *Aggressive Behaviour* **23**:315–328.

Aureli, F. and de Waal, F.B.M. (2000) *Natural Conflict Resolution*. Berkeley: University of California Press.

Aureli, F. and van Schaik, C.P. (1991) Post-conflict behaviour in long-tailed macaques (*Macaca fascicularis*). II. Coping with the uncertainty. *Ethology* **89**:101–114.

Aureli, F., *et al.* (1989) Functional aspects of reconciliation among captive long-tailed macaques (*Macaca fascicularis*). *American Journal of Primatology* **19**:39–51.

Aureli, F., *et al.* (1992) Kin-oriented redirection among Japanese macaques: an expression of a revenge system? *Animal Behaviour* **44**:283–291.

Ball, M.A. and Parker, G.A. (1996) Sperm competition games: external fertilization and adaptive infertility. *Journal of Theoretical Biology* **180**:141–150.

Barrette, C. and Vandal, D. (1986) Social rank, dominance, antler size and access to food in snow-bound wild woodland caribou. *Behaviour* **97**:118–146.

Barton, R.A. and Whiten, A. (1993) Feeding competition among female olive baboons *Papio anubis*. *Animal Behaviour* **46**:777–789.

Barton, R.A., *et al.* (1996) Ecology, feeding competition and social structure in baboons. *Behavioral Ecology and Sociobiology* **38**:321–329.

Beehner, J.C. and Lu, A. (2013) Reproductive suppression in female primates: a review. *Evolutionary Anthropology: Issues, News, and Reviews* **22**:226–238.

Bell, M.B.V., *et al.* (2014) Suppressing subordinate reproduction provides benefits to dominants in cooperative societies of meerkats. *Nature Communications* **5**:4499.

Bernstein, I.S. (1981) Dominance: the baby and the bathwater. *Behavioral and Brain Sciences* **4**:419–457.

Bettridge, C., *et al.* (2010) Trade-offs between time, predation risk and life history, and their implications for biogeography: a systems modelling approach with a primate case study. *Ecological Modelling* **221**:777–790.

Bouissou, M.F. (1972) Influence of body weight and presence of horns in social rank in domestic cattle. *Animal Behaviour* **20**:474–477.

Bro-Jørgensen, J. (2002) Overt female mate competition and preference for central males in a lekking antelope. *Proceedings of the National Academy of Sciences of the United States of America* **99**:9290–9293.

Bro-Jørgensen, J. (2011) Intra- and intersexual conflicts and cooperation in the evolution of mating strategies: lessons learnt from ungulates. *Evolutionary Biology* **38**:28–41

Bulger, J. and Hamilton, W.J. (1987) Rank and density correlates of inclusive fitness measures in a natural chacma baboon (*Papio ursinus*) troop. *International Journal of Primatology* **8**:635–650.

Cant, M.A. and Johnstone, R.A. (2009) How threats influence the evolutionary resolution of within-group conflict. *American Naturalist* **173**:759–771.

Cant, M.A., *et al.* (2010) Reproductive control via eviction (but not the threat of eviction) in banded mongooses. *Proceedings of the Royal Society of London. Series B: Biological Sciences* **277**:20092097.

Call, J., *et al.* (2002) Postconflict third-party affiliation in stumptailed macaques. *Animal Behaviour* **65**:209–216.

Call, J., *et al.* (1999) Reconciliation patterns among stumptailed macaques: a multivariate approach. *Animal Behaviour* **58**:165–172.

Champagne, F.A. and Curley, J.P. (2009) The trans-generational influence of maternal care on offspring gene expression and behavior in rodents. In: *Maternal Effects in Mammals* (eds D. Maestripieri and J.M. Mateo) Chicago: University of Chicago Press, 182–202.

Chapais, B. (1988) Experimental matrilineal inheritance of rank in female Japanese monkeys. *Animal Behaviour* **36**:1025–1037.

Chapais, B. (1992) The role of alliances in social inheritance of rank among female primates. In: *Coalitions and Alliances in Humans and Other Animals* (eds A.H. Harcourt and F.B.M. de Waal) Oxford: Oxford University Press, 29–60.

Chapais, B. (2004) How kinship generates dominance structures: a comparative perspective. In: *Macaque Societies: A Model for the Study of Social Organization* (eds B. Thierry, M. Singh and W. Kaumanns) Cambridge: Cambridge University Press, 186–203.

Chapais, B., *et al.* (2001) Kin selection and the distribution of altruism in relation to degree of kinship in Japanese macaques (*Macaca fuscata*). *Behavioral Ecology and Sociobiology* **49**:493–502.

Charpentier, M.J.C., *et al.* (2007) Kin discrimination in juvenile mandrills, *Mandrillus sphinx*. *Animal Behaviour* **73**:37–45.

Cheney, D.L. and Seyfarth, R.M. (1989) Redirected aggression and reconciliation among vervet monkeys, *Cercopithecus aethiops*. *Behaviour* **110**:258–275.

Cheney, D.L. and Seyfarth, R.M. (1997) Reconciliatory grunts by dominant female baboons influence victims' behaviour. *Animal Behaviour* **54**:409–418.

Cheney, D.L. and Seyfarth, R.M. (1999) Recognition of other individuals' social relationships by female baboons. *Animal Behaviour* **58**:67–75.

Clutton-Brock, T.H. (1988) Reproductive success. In: *Reproductive Success: Studies of Individual Variation in Contrasting Breeding Systems* (ed. T.H. Clutton-Brock). Chicago: University of Chicago Press, 472–486.

Clutton-Brock, T.H. (1991a) Lifetime data and the measurement of selection. *Evolution* **45**:454.

Clutton-Brock, T.H. (1991b) *The Evolution of Parental Care.* Princeton, NJ: Princeton University Press

Clutton-Brock, T.H. (2001) Sociality and population dynamics. In: *Ecology: Achievement and Challenge* (eds M.C. Press, N.J. Huntly and S. Levin) Oxford: Blackwell Science, 47–66.

Clutton-Brock, T.H. (2004) What is sexual selection? In: *Sexual Selection in Primates: New and Comparative Perspectives* (eds P.M. Kappeler and C.P. van Schaik) Cambridge: Cambridge University Press, 24–36.

Clutton-Brock, T.H. (2007) Sexual selection in males and females. *Science* **318**:1882–1885.

Clutton-Brock, T.H. (2009) Sexual selection in females. *Animal Behaviour* **77**:3–11.

Clutton-Brock, T.H. and Harvey, P.H. (1976) Evolutionary rules and primate societies. In: *Growing Points in Ethology* (eds P.P.G.

Bateson and R.A. Hinde) Cambridge: Cambridge University Press, 195–237.

Clutton-Brock, T.H. and Huchard, E. (2013) Social competition and its consequences in female mammals. *Journal of Zoology* **289**:151–171.

Clutton-Brock, T.H. and Parker, G.A. (1995) Punishment in animal societies. *Nature* **373**:209–216.

Clutton-Brock, T.H. and Sheldon, B.C. (2010) Individuals and populations: the role of long-term, individual-based studies in ecology and evolutionary biology. *Trends in Ecology and Evolution* **25**:562–573.

Clutton-Brock, T.H., *et al.* (1982) Competition between female relatives in a matrilocal mammal. *Nature* **300**:178–180.

Clutton-Brock, T.H., *et al.* (1986) Great expectations: dominance, breeding success and offspring sex ratios in red deer. *Animal Behaviour* **34**:460–471.

Clutton-Brock, T.H., *et al.* (1988) Reproductive success in male and female red deer. In: *Reproductive Success: Studies of Individual Variation in Contrasting Breeding Systems* (ed. T.H. Clutton-Brock). Chicago: University of Chicago Press, 325–343.

Clutton-Brock, T.H., *et al.* (1998) Infanticide and expulsion of females in a cooperative mammal. *Proceedings of the Royal Society of London. Series B: Biological Sciences* **265**:2291–2295.

Clutton-Brock, T.H., *et al.* (2001) Cooperation, control and concession in meerkat groups. *Science* **291**:478–481.

Clutton-Brock, T.H., *et al.* (2006) Intrasexual competition and sexual selection in cooperative mammals. *Nature* **444**:1065–1068.

Clutton-Brock, T.H., *et al.* (2008) Group size and the suppression of subordinate reproduction in Kalahari meerkats. *Animal Behaviour* **76**:689–700.

Clutton-Brock, T.H., *et al.* (2010) Adaptive suppression of subordinate reproduction in cooperative mammals. *American Naturalist* **176**:664–673.

Cooney, R. and Bennett, N.C. (2000) Inbreeding avoidance and reproductive skew in a cooperative mammal. *Proceedings of the Royal Society of London. Series B: Biological Sciences* **267**:801–806.

Cords, M. (1992) Post-conflict reunions and reconciliation in long-tailed macaques. *Animal Behaviour* **44**:57–61.

Cords, M. and Aureli, F. (2000) Reconciliation and relationship qualities. In: *Natural Conflict Resolution* (eds F. Aureli and F.B.M. de Waal) Cambridge: Cambridge University Press, 177–198.

Cords, M. and Thurnheer, S. (1993) Reconciling with valuable partners by long-tailed macaques. *Ethology* **93**:315–325.

Côté, S.D. (2000) Dominance hierarchies in female mountain goats: stability, aggressiveness and determinants of rank. *Behaviour* **137**:1541–1566.

Crawford, J.C. and Drea, C.M. (2015) Baby on board: olfactory cues indicate pregnancy and fetal sex in a non-human primate. *Biology Letters* **11**:20140831.

Creel, S. (2001) Social dominance and stress hormones. *Trends in Ecology and Evolution* **16**:491–497.

Creel, S. and Creel, N.M. (2002) *The African Wild Dog: Behavior, Ecology, and Conservation*. Princeton, NJ: Princeton University Press.

Creel, S., *et al.* (2013) The ecology of stress: effects of the social environment. *Functional Ecology* **27**:66–80.

Crockett, C.M. (1984) Emigration by female red howler monkeys and the case for female competition. In: *Female Primates: Studies by Women Primatologists* (ed. M.F. Small). New York: Alan R. Liss, 159–173.

Crockett, C.M. and Pope, T.R. (1988) Inferring patterns of aggression from red howler monkey injuries. *American Journal of Primatology* **15**:289–308.

Cunha, G., *et al.* (2014) Development of the external genitalia: perspectives from the spotted hyena (*Crocuta crocuta*). *Differentiation* **87**:4–22.

Das, M. (2000) Conflict management via third parties: post-conflict affiliation of the aggressor. In: *Natural Conflict Resolution* (eds F. Aureli and F.B.M. de Waal) Berkeley: University of California Press, 263–280.

Das, M., *et al.* (1997) Affiliation between aggressors and third parties following conflicts in long-tailed macaques (*Macaca fascicularis*). *International Journal of Primatology* **18**:159–181.

Das, M., *et al.* (1998) Post-conflict affiliation and stress-related behavior of long-tailed macaque aggressors. *International Journal of Primatology* **19**:53–71.

de Waal, F.B.M. (1982) *Chimpanzee Politics: Power and Sex Among Apes*. New York: Harper & Row.

de Waal, F.B.M. (1989) *Peacemaking Among Primates*. Cambridge, MA: Harvard University Press.

de Waal, F.B.M. (1993) Reconciliation among primates: a review of empirical evidence and unresolved issues. In: *Primate Social Conflict* (eds W.A. Mason and S.P. Mendoza) Albany: State University of New York Press, 111–144.

de Waal, F.B.M. and Aureli, F. (1996) Consolation, reconciliation, and a possible cognitive difference between macaques and chimpanzees. In: *Reaching Into Thought: The Minds of the Great Apes* (eds A.E. Russon, K.A. Bard and S.T. Parker) Cambridge: Cambridge University Press, 80–110.

de Waal, F.B.M. and Ren, R.M. (1988) Comparison of the reconciliation behavior of stumptail and rhesus macaques. *Ethology* **78**:129–142.

de Waal, F.B.M. and van Roosmalen, A. (1979) Reconciliation and consolation among chimpanzees. *Behavioral Ecology and Sociobiology* **5**:55–66.

Digby, L. (2000) Infanticide by female mammals: implication for the evolution of social systems. In: *Infanticide by Males and its Implications* (eds C.P. van Schaik and C.H. Jansen) Cambridge: Cambridge University Press, 423–446.

Dloniak, S.M., *et al.* (2006) Rank-related maternal effects of androgens on behaviour in wild spotted hyenas. *Nature* **440**:1190–1193.

Drea, C.M. (2007) Sex and seasonal differences in aggression and steroid secretion in *Lemur catta*: are socially dominant females hormonally 'masculinized'? *Hormones and Behavior* **51**:555–567.

Drea, C.M. and Weil, A. (2008) External genital morphology of the ring-tailed lemur (*Lemur catta*): females are naturally 'masculinized'. *Journal of Morphology* **260**:451–463.

Drea, C.M., *et al.* (1998) Androgens and masculinisation of genitalia in the spotted hyena *Crocuta crocuta*: effects of prenatal anti-androgens. *Journal of Reproduction and Fertility* **113**:117–127.

Drea, C.M., *et al.* (2002) Exposure to naturally circulating androgens during fetal life is prerequisite for male mating but incurs direct reproductive costs in female spotted hyenas. *Proceedings of the Royal Society of London. Series B: Biological Sciences* **269**:1981–1987.

Dubuc, C., *et al.* (2009) Sexual skin color contains information about the timing of the fertile phase in free-ranging *Macaca mulatta*. *International Journal of Primatology* **30**:777–789.

East, M.L. and Hofer, H. (2010) Social environments, social tactics and their fitness consequences in complex mammalian societies. In: *Social Behaviour* (eds T. Szekely, A.J. Moore and J. Komdeur) Cambridge: Cambridge University Press, 360–390.

East, M.L., *et al.* (2009) Maternal effects on offspring social status in spotted hyenas. *Behavioral Ecology* **20**:478–483.

Ebensperger, L.A. (1998) Strategies and counterstrategies to infanticide in mammals. *Biological Reviews* **73**:321–346.

Faulkes, C.G. and Abbott, D.H. (1997) The physiology of a reproductive dictatorship: regulation of male and female reproduction by a single breeding female in colonies of naked mole-rats. In: *Cooperative Breeding in Mammals* (eds N.G. Solomon and J.A. French) Cambridge: Cambridge University Press, 302–334.

Fernandez-Duque, E. and Huck, M. (2013) Till death (or an intruder) do us part: intrasexual competition in a monogamous primate. *PLOS ONE* **8**:e53724.

Frank, L.G. (1997) Evolution of genital masculinization: why do female hyaenas have such a large 'penis'? *Trends in Ecology and Evolution* **12**:58–62.

Fraser, O.N., *et al.* (2008) Stress reduction through consolation in chimpanzees. *Proceedings of the National Academy of Sciences of the United States of America* **105**:8557–8562.

Gardner, A. and West, S.A. (2004) Spite and the scale of competition. *Journal of Evolutionary Biology* **17**:1195–1203.

Gilchrist, J.S. (2006) Female eviction, abortion and infanticide in banded mongooses (*Mungos mungo*): implications for social control of reproduction and synchronized parturition. *Behavioral Ecology* **11**:664–669.

Glickman, S.E., *et al.* (1998) Androgens and masculinisation of genitalia in the spotted hyaena (*Crocuta crocuta*). 3. Effects of juvenile gonadectomy. *Journal of Reproduction and Fertility* **113**:129–135.

Gouzoules, H., *et al.* (1982) Behavioral dominance and reproductive success in female Japanese monkeys (*Macaca fuscata*). *Animal Behaviour* **30**:1138–1150.

Goymann, W., *et al.* (2001) Androgens and the role of female hyperaggressivness in spotted hyenas (*Crocuta crocuta*). *Hormones and Behavior* **39**:83–92.

Hawkins, C.E., *et al.* (2002) Transient masculinisation in the Fossa *Cryptoprocta ferox* (Carnivora, Viverridae). *Biology of Reproduction* **66**:610–615.

Henzi, S.P., *et al.* (2009) Cyclicity in the structure of female baboon social networks. *Behavioral Ecology and Sociobiology* **63**:1015–1021.

Henry, M.L.D., *et al.* (2013) High rates of pregnancy loss by subordinates leads to high reproductive skew in wild golden lion tamarins (*Leontopithecus rosalia*). *Hormones and Behaviour* **63**:675–683.

Henzi, S.P., *et al.* (2013) Scalar social dynamics in female vervet monkey cohorts. *Philosophical Transactions of the Royal Society B: Biological Sciences* **368**:20120351.

Hofer, H. and East, M.L. (1997) Skewed offspring sex ratios and sex composition of twin litters in Serengeti spotted hyaenas (*Crocuta crocuta*) are a consequence of siblicide. *Applied Animal Behaviour Science* **51**:307–316.

Hofer, H. and East, M.L. (2000) Conflict management in female-dominated spotted hyenas. In: *Natural Conflict Resolution* (eds F. Aureli and F.B.M. de Waal) Berkeley: University of California Press, 232–234.

Hofer, H. and East, M.L. (2003) Behavioral processes and costs of co-existence in female spotted hyenas: a life-history perspective. *Evolutionary Ecology* **17**:315–331.

Hofer, H. and East, M.L. (2008) Siblicide in Serengeti spotted hyenas: a long-term study of maternal input and cub survival. *Behavioral Ecology and Sociobiology* **62**:341–351.

Holand, Ø., *et al.* (2004) Social rank in female reindeer (*Rangifer tarandus*): effects of body mass, antler size and age. *Journal of Zoology* **263**:365–372.

Holekamp, K.E. and Dloniak, S.M. (2009) Maternal effects in fissiped carnivores. In: *Maternal Effects in Mammals* (eds D. Maestripieri and J. Mateo) Chicago: University of Chicago Press, 227–255.

Holekamp, K.E. and Smale, L. (2000) Feisty females and meek males: reproductive strategies in the spotted hyena. In: *Reproduction in Context* (eds K. Wallen and J. Schneider) Cambridge, MA: MIT Press, 257–285.

Holekamp, K.E., *et al.* (1996) Rank and reproduction in the female spotted hyaena. *Journal of Reproduction and Fertility* **108**:229–237.

Holekamp, K.E., *et al.* (2013) Developmental constraints on behavioural flexibility. *Philosophical Transactions of the Royal Society B; Biological Sciences* **386**:20120350.

Höner, O.P., *et al.* (2010) The fitness of dispersing spotted hyaena sons is influenced by maternal social status. *Nature Communications* **1**:60.

Hoogland, J.L. (1981) Nepotism and cooperative breeding in the black-tailed prairie dog (Sciuridae: Cynomys ludovicianus) In: *Natural Selection and Social Behavior* (eds R.D. Alexander and D.W. Tinkle) New York: Chiron Press, 283–310.

Hoogland, J.L. (1985) Infanticide in prairie dogs: lactating females kill offspring of close kin. *Science* **230**:1037–1040.

Hoogland, J.L. (1995) *The Black-tailed Prairie Dog: Social Life of a Burrowing Mammal.* Chicago: University of Chicago Press.

Horrocks, J. and Hunte, W. (1983) Maternal rank and offspring rank in vervet monkeys: an appraisal of the mechanisms of rank acquisition. *Animal Behaviour* **31**:772–782.

Huchard, E. and Cowlishaw, G. (2011) Female–female aggression around mating: an extra cost of sociality in a multimale primate society. *Behavioral Ecology* **22**:1003–1011.

Huck, U.W., *et al.* (1988a) Social dominance and reproductive success in pregnant and lactating golden hamsters (*Mesocricetus auratus*) under seminatural conditions. *Physiology and Behavior* **44**:313–319.

Huck, U.W., *et al.* (1988b) Progesterone levels and socially-induced implantation failure and fetal resorption in golden hamsters (*Mesocricetus auratus*). *Physiology and Behavior* **44**:321–326.

Jensen, K. (2010) Punishment and spite, the dark side of cooperation. *Philosophical Transactions of the Royal Society B: Biological Sciences* **365**:2635–2650.

Jensen, K., *et al.* (2007) Chimpanzees are vengeful but not spiteful. *Proceedings of the National Academy of Sciences of the United States of America* **104**:13046–13050.

Johnson, J.A. (1987) Dominance rank in juvenile olive baboons, *Papio anubis*: the influence of gender, size, maternal rank and orphaning. *Animal Behaviour* **35**:1694–1708.

Johnson, S.E. (2003) Life history and the competitive environment: trajectories of growth, maturation, and reproductive output among chacma baboons. *American Journal of Physical Anthropology* **120**:83–98.

Johnstone, R.A. and Cant, M.A. (1999) Reproductive skew and the threat of eviction: a new perspective. *Proceedings of the Royal Society of London. Series B: Biological Sciences* **266**:275–279.

Kappeler, P.M. (1993) Female dominance in primates and other mammals. In: *Perspectives in Ethology* (eds P.P.G. Bateson, P.H. Klopfer and W.S. Thompson) New York: Plenum, 143–158.

Kappeler, P.M. (1997) Determinants of primate social organisation: comparative evidence and new insights from Malagasy lemurs. *Biological Review* **72**:111–151.

Kappeler, P.M. and Fichtel, C. (2012) Female reproductive competition in *Eulemur rufifrons*: eviction and reproductive restraint in a plurally breeding Malagasy primate. *Molecular Ecology* **21**:685–698.

Kapsalis, E. (2004) Matrilineal kinship and primate behavior. In: *Kinship and Behavior in Primates* (eds B. Chapais and C.M. Berman) Oxford: Oxford University Press, 153–176.

Kapsalis, E. and Berman, C.M. (1996) Models of affiliative relationships among free-ranging rhesus monkeys (*Macaca mulatta*). *Behaviour* **133**:1235–1263.

Kappelar, P. and van Schaik, C.P. (1992) Methodological and evolutionary aspects of reconciliation in primates. *Ethology* **92**: 51–69.

Kawai, M. (1958) On the rank system in a natural group of Japanese monkey (I). *Primates* **1**:111–130.

Knickmeyer, R.C. and Baron-Cohen, S. (2006) Topical review: Fetal testosterone and sex differences in typical social development and in autism. *Journal of Child Neurology* **21**:825–845.

Koren, L., *et al.* (2006) Elevated testosterone levels and social ranks in female rock hyrax. *Hormones and Behavior* **49**:470–477.

Koski, S.E. and Sterck, E.H.M. (2007) Triadic post-conflict affiliation in captive chimpanzees; does consolation console? *Animal Behaviour* **73**:133–142.

Kruuk, H. (1972) *The Spotted Hyena: A Study of Predation and Social Behaviour.* Chicago: University of Chicago Press.

Kuhn, H.J. (1972) On the perineal organ of male *Procolobus badius*. *Journal of Human Evolution* **1**:371–378.

Kutsukake, N. and Castles, D.L. (2001) Reconciliation and variation in post-conflict stress in Japanese macaques (*Macaca fuscata fuscata*): testing the integrated hypothesis. *Animal Cognition* **4**:259–268.

Kutsukake, N. and Castles, D.L. (2004) Reconciliation and post-conflict third-party affiliation among wild chimpanzees in the Mahale Mountains, Tanzania. *Primates* **45**:157–165.

Kutsukake, N. and Clutton-Brock, T.H. (2006) Aggression and submission reflect reproductive conflict between females in cooperatively breeding meerkats *Suricata suricatta*. *Behavioral Ecology and Sociobiology* **59**:541–548.

Lazaro-Perea, C., *et al.* (2000) Behavioral and demographic changes following the loss of the breeding female in cooperatively breeding marmosets. *Behavioral Ecology and Sociobiology* **48**:137–146.

Lacey, E.A. (2004) Sociality reduces individual direct fitness in a communally breeding rodent, the colonial taco-taco (*Ctenomys sociabilis*). *Behaviour Ecology Socialbiology* **56**:449–457.

LeBas, N.R. (2006) Female finery is not for males. *Trends in Ecology and Evolution* **21**:170–173.

Lee, P.C. (2011) Dominance in female elephants. In: *The Amboseli Elephants* (eds C.J. Moss, H. Vroze and P.C. Lee) Chicago: University of Chicago Press, 190–191.

Licht, P., *et al.* (1992) Hormonal correlates of 'masculinization' in female spotted hyaenas (*Crocuta crocuta*). 2. Maternal and fetal steroids. *Journal of Reproduction and Fertility* **95**:463–474.

Licht, P., *et al.* (1998) Androgens and masculinization of genitalia in the spotted hyaena (*Crocuta crocuta*). 1. Urogenital morphology and placental androgen production during fetal life. *Journal of Reproduction and Fertility* **113**:105–116.

Lindburg, D.G. (1973) Grooming behavior as a regulator of social interactions in rhesus monkeys. In: *Behavioral Regulators of Behavior in Primates* (ed. C.R. Carpenter). Cranbury, NJ: Associated University Presses, 124–148.

Lloyd, P.H. and Rasa, O.A.E. (1989) Status, reproductive success and fitness in Cape mountain zebra (*Equus zebra zebra*). *Behavioral Ecology and Sociobiology* **25**:411–420.

Lonsdorf, E.V. and Ross, S.R. (2012) Socialization and development of behavior. In: *The Evolution of Primate Societies* (eds J.C. Mitani, J. Call, P.M. Kappeler, R.A. Palombit and J.B. Silk) Chicago: University of Chicago Press, 245–268.

Lukas, D. and Huchard, E. (2014) The distribution of male infanticide in mammals. *Science* **346**:841–844.

McFarland, R. and Majolo, B. (2012) The occurrence and benefits of postconflict bystander affiliation in wild Barbary macaques, *Macaca sylvanus*. *Animal Behaviour* **84**:583–591.

MacLeod, K.J., *et al.* (2013) Factors predicting the frequency, likelihood and duration of allonursing in the cooperatively breeding meerkat. *Animal Behaviour* **86**:1059–1067.

Madden, J.R. and Clutton-Brock, T.H. (2009) Manipulating grooming by decreasing parasite load causes unpredicted changes in antagonism. *Proceedings of the Royal Society of London. Series B: Biological Sciences* **276**:1263–1268.

Maestripieri, D. (2009) Maternal influences on offspring growth, reproduction and behavior in primates. In: *Maternal Effects in Mammals* (eds D. Maestripieri and J. Mateo) Chicago: University of Chicago Press, 256–291.

Maynard Smith, J. (1974) The theory of games and the evolution of animal conflicts. *Journal of Theoretical Biology* **47**:209–221.

Ménard, N. (2004) Do ecological factors explain variation in social organisation. In: *Macaque Societies: A Model for the Study of Social Organisation* (eds B. Thierry, M. Singh and W. Kaumanns) Cambridge: Cambridge University Press, 237–262.

Milner-Gulland, E.J., *et al.* (2003) Reproductive collapse in saiga antelope harems. *Nature* **422**:135.

Moore, A.J., *et al.* (2002) The evolution of interacting phenotypes: genetics and evolution of social dominance. *American Naturalist* **160**:S186–S197.

Muroyama, Y. and Thierry, B. (1996) Fatal attack on an infant by an adult female Tonkean macaque. *International Journal of Primatology* **17**:219–227.

Murray, C.M., *et al.* (2006) Foraging strategies as a function of season and rank among wild female chimpanzees (*Pan troglodytes*). *Behavioral Ecology* **17**:1020–1028.

Nakamichi, N. and Koyama, N. (1997) Social relationships among ring-tailed lemurs (*Lemur catta*) in two free-ranging troops at Berenty Reserve, Madagascar. *Internal Journal of Primatology* **18**:73–93.

O'Connell, S.M. and Cowlishaw, G. (1994) Infanticide avoidance, sperm competition and mate choice: the function of copulation calls in female baboons. *Animal Behaviour* **48**:687–694.

Okamoto, K. (2004) Patterns of group fission. In: *Macaque Societies: A Model for the Study of Social Organisation* (eds B. Thierry, N. Singh and W. Kaumanns) Cambridge: Cambridge University Press, 112–116.

Ostner, J., *et al.* (2003) Intersexual dominance, masculinized genitals and prenatal steroids: comparative data from lemurid primates. *Naturwissenschaften* **90**:141–144.

Ozgul, A., *et al.* (2014) Linking body mass and group dynamics in an obligate cooperative breeder. *Journal of Animal Ecology* **83**:1357–1366.

Packer, C. (1983) Sexual dimorphism: the horns of African antelopes. *Science* **221**:1191–1193.

Packer, C., *et al.* (2001) Egalitarianism in female African lions. *Science* **293**:690–693.

Parker, G.A. (1974) Assessment strategy and the evolution of fighting behaviour. *Journal of Theoretical Biology* **47**: 223–243.

Parker, G.A. and Ball, M.A. (2005) Sperm competition, mating rate and the evolution of testis and ejaculate sizes: a population model. *Biology Letters* **1**:235–238.

Pereira, M.E. (1992) The development of dominance relations before puberty in Cercopithecine societies. In: *Aggression and Peacefulness in Humans and Other Primates* (eds J. Silverberg and J.P. Gray) New York: Oxford University Press, 117–149.

Perry, S., *et al.* (2008) Kin-biased social behaviour in wild adult female white-faced capuchins, *Cebus capucinus*. *Animal Behaviour* **76**:187–199.

Petit, O. and Thierry, B. (2000) Do impartial interventions in conflicts occur in monkeys and apes? In: *Natural Conflict Resolution* (eds F. Aureli and F.B.M. de Waal) Berkeley: University of California Press, 267–269.

Plavcan, J.M., *et al.* (1995) Competition, coalitions and canine size in primates. *Journal of Human Evolution* **28**:245–276.

Plusquellec, P. and Boussou, M.F. (2001) Behavioural characteristics of two dairy breeds of cows selected (Hérens) or not (Brune des Alpes) for fighting and dominance ability. *Applied Animal Behaviour Science* **72**:1–21.

Pope, T.R. (2000) Reproductive success increases with degree of kinship in cooperative coalitions of female red howler monkeys (*Alouatta seniculus*). *Behavioral Ecology and Sociobiology* **48**:253–267.

Preston, B.T., *et al.* (2001) Dominant rams lose out by sperm depletion. *Nature* **409**:681–682.

Pride, R.E. (2005) Optimal group size and seasonal stress in ring-tailed lemurs *Lemur catta*. *Behavioral Ecology* **16**:550–560.

Pusey, A.E. (2012) Magnitude and sources of variance in female reproductive performance. In: *The Evolution of Primate Societies* (eds J.C. Mitani, J. Call, P.M. Kappeler, R.A. Palombit and J.B. Silk) Chicago: University of Chicago Press, 343–366.

Pusey, A.E., *et al.* (1997) The influence of dominance rank on the reproductive success of female chimpanzees. *Science* **277**:828–831.

Racey, P.A. and Skinner, J.D. (1979) Endocrine aspects of sexual mimicry in spotted hyaenas *Crocuta crocuta*. *Journal of Zoology* **187**:315–326.

Reeve, H.K. and Sherman, P.W. (1991) Intracolonial aggression and nepotism by the breeding female naked mole-rat. In: *The Biology of the Naked Mole-Rat* (eds P.W. Sherman, J.U.M. Jarvis and R.D. Alexander) Princeton, NJ: Princeton University Press, 337–357.

Riedl, K., *et al.* (2012) No third-party punishment in chimpanzees. *Proceedings of the National Academy of Sciences of the United States of America* **109**:14824–14829.

Roberts, S.C. (1996) The evolution of hornedness in female ruminants. *Behaviour* **133**:399–442.

Robinson, M.R. and Kruuk, L.E.B. (2007) Function of weaponry in females: the use of horns in intrasexual competition for resources in female Soay sheep. *Biology Letters* **3**:651–654.

Rowell, T.E. (1974) The concept of social dominance. *Behavioral Biology* **11**:131–154.

Rubenstein, D.I. and Nuñez, C.M. (2009) Sociality and reproductive skew in horses and zebras. In: *Reproductive Skew in*

Vertebrates: Proximate and Ultimate Causes (eds R. Hager and C.B. Jones) Cambridge: Cambridge University Press, 196–226.

Rughetti, M. and Festa-Bianchet, M. (2011) Effects of horn growth on reproduction and hunting mortality in female chamois. *Journal of Animal Ecology* **80**:438–447.

Rusu, A.S. and Krackow, S. (2004) Kin-preferential cooperation, dominance-dependent reproductive skew and competition for males in communally nesting female house-mice. *Behavioral Ecology and Sociobiology* **56**:298–305.

Rutberg, A.T. and Greenberg, S.A. (1990) Dominance aggression frequencies and modes of aggressive competition in feral pony mares. *Animal Behaviour* **40**:322–331.

Sandersen, J.L., *et al.* (2015) Elevated glucocorticoid concentrations during gestation predict reduced reproductive success in subordinate female banded mongooses. *Biology Letters* **11**, DOI: 10.1098/rsbl.2015.0620. [Accessed 10 January 2016].

Sackett, G.P. (1981) Receiving severe aggression correlates with foetal gender in pregnant pigtail monkeys. *Developmental Psychobiology* **14**:267–272.

Saltzman, W., *et al.* (2009) Reproductive skew in female common marmosets: what can proximate mechanisms tell us about ultimate causes? *Proceedings of the Royal Society of London. Series B: Biological Sciences* **276**:389–399.

Samuels, A. and Flaherty, C. (2000) Peaceful conflict resolution in the sea? In: *Natural Conflict Resolution* (eds F. Aureli and F.B.M. de Waal) Berkeley: University of California Press, 229–231.

Samuels, A. and Gifford, T. (1997) A quantitative assessment of dominance relations among bottlenose dolphins. *Marine Mammal Science* **13**:70–99.

Sapolsky, R.M. (1993) The physiology of dominance in stable versus unstable social hierarchies. In: *Primate Social Conflict* (eds W.A. Mason and S.P. Mendoza) Albany: State University of New York Press, 171–204.

Sauther, M.L., *et al.* (1999) The socioecology of the ringtailed lemur: thirty-five years of research. *Evolutionary Anthropology: Issues, News, and Reviews* **8**:120–132.

Schaefer, J.A. and Mahoney, S.P. (2001) Antlers on female caribou: biogeography of the bones of contention. *Ecology* **82**:3556–3560.

Schaller, G.B. (1972) *The Serengeti Lion: A Study of Predator–Prey Relations*. Chicago: Chicago University Press.

Schino, G. (1998) Reconciliation in domestic goats. *Behaviour* **135**:343–356.

Schino, G., *et al.* (1998) Intragroup variation in conciliatory tendencies in captive Japanese macaques. *Behaviour* **135**:897–912.

Semple, S. and McComb, K. (2000) Perception of female reproductive state from vocal cues in a mammal species. *Proceedings of the Royal Society of London. Series B: Biological Sciences* **267**:707–712.

Semple, S., *et al.* (2002) Information content of female copulation calls in yellow baboons. *American Journal of Primatology* **56**:43–56.

Setchell, J.M., *et al.* (2006) Signal content of red facial coloration in female mandrills (*Mandrillus sphinx*). *Proceedings of the Royal Society of London. Series B: Biological Sciences* **273**:2395–2400.

Sherman, P.W. (1981) Reproductive competition and infanticide in Belding's ground squirrels and other animals. In: *Natural Selection and Social Behavior* (eds R.D. Alexander and D.W. Tinkle) New York: Chivon Press, 311–331.

Silk, J.B. (1988) Maternal investment in captive bonnet macaques (*Macaca radiata*). *American Naturalist* **132**:1–19.

Silk, J.B. (1993) The evolution of social conflict among female primates. In: *Primate Social Conflict* (eds W.A. Mason and S.P. Mendoza) Albany: State University of New York Press, 49–83.

Silk, J.B. (1997) The function of peaceful post-conflict contacts among primates. *Primates* **38**:265–279.

Silk, J.B. (2000) The function of peaceful post-conflict interactions: an alternative view. In: *Natural Conflict Resolution* (eds F. Aureli and F.B.M. de Waal) Berkeley: University of California Press, 179–181.

Silk, J.B. (2002) The form and function of reconciliation in primates. *Annual Review of Anthropology* **31**:21–44.

Silk, J.B. (2007) The adaptive value of sociality in mammalian groups. *Philosophical Transactions of the Royal Society B: Biological Sciences* **362**:539–559.

Silk, J.B. (2009) Nepotistic cooperation in non-human primate groups. *Philosophical Transactions of the Royal Society B: Biological Sciences* **364**:3243–3254.

Silk, J.B., *et al.* (1981) Differential reproductive success and facultative adjustment of sex ratios among captive female bonnet macaques (*Macaca radiata*). *Animal Behaviour* **29**:1106–1120.

Silk, J.B., *et al.* (1996) The form and function of post-conflict interactions between female baboons. *Animal Behaviour* **52**:259–268.

Silk, J.B., *et al.* (2003) Social bonds of female baboons enhance infant survival. *Science* **302**:1331–1334.

Silk, J.B., *et al.* (2006a) Social relationships among adult female baboons (*Papio cynocephalus*). I. Variation in the strength of social bonds. *Behavioral Ecology and Sociobiology* **61**:183–195.

Silk, J.B., *et al.* (2006b) Social relationships among adult female baboons (*Papio cynocephalus*). II. Variation in the quality and stability of social bonds. *Behavioral Ecology and Sociobiology* **61**:197–204.

Silk, J.B., *et al.* (2009) The benefits of social capital: close social bonds among female baboons enhance offspring survival. *Proceedings of the Royal Society of London. Series B: Biological Sciences* **276**:3099–3104.

Silk, J., *et al.* (2013) A practical guide to the study of social relationships. *Evolutionary Anthropology* **22**:213–225.

Smucny, D.A., *et al.* (1997) Post-conflict affiliation and stress reduction in captive rhesus macaques. *Advances in Ethology* **32**:157.

Smuts, B.B. (1985) *Sex and Friendship in Baboons*. New York: Aldine.

Smuts, B.B. and Nicolson, N. (1989) Reproduction in wild female olive baboons. *American Journal of Primatology* **19**:229–246.

Smyth, K.N. and Drea, C.M. (2015) Patterns of parasitism in the cooperatively breeding meerkat: a cost of dominance for females. *Behavioral Ecology* doi: 10.1093/beheco/arv132

Starling, A.P., *et al.* (2010) Seasonality, sociality, and reproduction: long-term stressors of ring-tailed lemurs (*Lemur catta*). *Hormones and Behavior* **57**:76–85.

Staub, N.L. and de Beer, M. (1997) The role of androgens in female vertebrates. *General and Comparative Endocrinology* **108**:1–24.

Sterck, E.H.M., *et al.* (1997) The evolution of female social relationships in nonhuman primates. *Behavioral Ecology and Sociobiology* **41**:291–309.

Stockley, P. and Bro-Jørgensen, J. (2011) Female competition and its evolutionary consequences in mammals. *Biological Reviews* **86**:341–366.

Swanson, E.M., *et al.* (2011) Lifetime selection on a hypoallometric size trait in the spotted hyena. *Proceedings of the Royal Society of London. Series B: Biological Sciences* **278**:3277–3285.

Takahata, Y., *et al.* (2008) The relationship between female rank and reproductive parameters of the ringtailed lemur: a preliminary analysis. *Primates* **49**:135–138.

Thierry, B. (1990) Feedback loop between kinship and dominance: the macaque model. *Journal of Theoretical Biology* **145**:511–521.

Thouless, C.R. and Guinness, F.E. (1986) Conflict between red deer hinds: the winner always wins. *Animal Behaviour* **34**:1166–1171.

Tung, J., *et al.* (2010) Evolutionary genetics in wild primates: combining genetic approaches with field studies of natural populations. *Trends in Genetics* **26**:353–362.

Tung, J., *et al.* (2012) Social environment is associated with gene regulatory variation in the rhesus macaque immune system. *Proceedings of the National Academy of Sciences of the United States of America* **109**:6490–6495.

Tuomi, J., *et al.* (1997) On the evolutionary stability of female infanticide. *Behavioral Ecology and Sociobiology* **40**:227–233.

Van Horn, R.C., *et al.* (2007) Divided destinies: group choice by female savannah baboons during social group fission. *Behavioral Ecology and Sociobiology* **61**:1823–1837.

van Noordwijk, M.A. and van Schaik, C.P. (1987) Competition among female long-tailed macaques, *Macaca fascicularis*. *Animal Behaviour* **35**:577–589.

van Noordwijk, M.A. and van Schaik, C.P. (1999) The effects of dominance rank and group size on female lifetime reproductive success in wild long-tailed macaques, *Macaca fascicularis*. *Primates* **40**:105–130.

von Engelhardt, N., *et al.* (2000) Androgen levels and female social dominance in *Lemur catta*. *Proceedings of the Royal Society of London. Series B: Biological Sciences* **267**:1533–1539.

Wahaj, S.A., *et al.* (2001) Reconciliation in the spotted hyena (*Crocuta crocuta*). *Ethology* **107**:1057–1074.

Walters, J. (1980) Interventions and the development of dominance relationships in female baboons. *Folia Primatologica* **34**:61–89.

Wasser, S.K. and Barash, D.P. (1983) Reproductive suppression among female mammals: implications for biomedicine and sexual selection theory. *Quarterley Review of Biology* **58**:513–538.

Wasser, S.K. and Starling, A.K. (1986) Reproductive competition among female yellow baboons. In: *Primate Ontogeny, Cognition and Social Behaviour* (eds J.G. Else and P.C. Lee) Cambridge: Cambridge University Press, 343–354.

Wasser, S.K. and Starling, A.K. (1988) Proximate and ultimate causes of reproductive suppression among female yellow baboons at Mikumi National Park, *Tanzania*. *American Journal of Primatology* **16**:97–121.

Wasser, S.K., *et al.* (2004) Population trend alters the effects of maternal dominance rank on lifetime reproductive success in yellow baboons (*Papio cynocephalus*). *Behavioral Ecology and Sociobiology* **56**:338–345.

Wells, S.M. and von Goldschmidt-Rothschild, B. (1979) Social behaviour and relationships in a herd of Camargue horses. *Zeitschrift fur Tierpsychologie* **49**:363–380.

Widdig, A., *et al.* (2006) Consequences of group fission for the patterns of relatedness among rhesus macaques. *Molecular Ecology* **15**:3825–3832.

Wilson, A.J., *et al.* (2011) Indirect genetics effects and evolutionary constraint: an analysis of social dominance in red deer, *Cervus elaphus*. *Journal of Evolutionary Biology* **24**:772–783.

Wolff, J.O. and Cicirello, D.M. (1989) Field evidence for sexual selection and resource competition infanticide in white-footed mice. *Animal Behaviour* **38**:637–642.

Woodroffe, R. and Macdonald, D.W. (1995) Female/female competition in European badgers, *Meles meles*: effects on breeding success. *Journal of Animal Ecology* **64**:12–20.

Wrangham, R.W. (1980) An ecological model of female-bonded primate groups. *Behaviour* **75**:262–300.

Ylönen, H., *et al.* (1997) Infanticide in the bank vole (*Clethrionomys glareolus*): occurrence and the effect of familiarity on female infanticide. *Annales Zoologici Fennici* **34**:259–266.

York, A.D. and Rowell, T.E. (1988) Reconciliation following aggression in patas monkeys, *Erythrocebus patas*. *Animal Behaviour* **36**:502–509.

Young, A.J. (2009) The causes of physiological suppression in vertebrate societies: a synthesis. In: *Reproductive Skew in Vertebrates: Proximate and Ultimate Causes* (eds R. Hager and C.B. Jones) Cambridge: Cambridge University Press, 397–436.

Young, A.J., *et al.* (2006) Stress and the suppression of subordinate reproduction in cooperatively breeding meerkats. *Proceedings of the National Academy of Sciences of the United States of America* **103**:12005–12010.

Young, A.J., *et al.* (2010) Physiological suppression eases in Damaraland mole-rat societies when ecological constraints on dispersal are relaxed. *Hormones and Behavior* **57**:177–183.

CHAPTER 9

Cooperation between females

9.1 Introduction

In the sloping sun of a Kalahari morning, a group of meerkats leaves its burrow to begin foraging. As the animals fan out to search for food, one awkwardly climbs through the branches of a thorny shrub until it stands, swaying perilously, six feet above the ground. It stares around the horizon, looking faintly worried and starts to give regular calls which announce to other group members that it is on guard, allowing them to feed without interruption. A hunting eagle appears two kilometres away and the sentinel starts to give gentle alarm calls. Other group members look up, then resume feeding. Suddenly a jackal appears two hundred metres away. The sentinel's alarm calls rise in tone, pitch and amplitude, and both the sentinel and the other group members dash into the nearest burrow.

Competition between group members for resources or mates is not surprising, but well-developed cooperative behaviour is more difficult to explain (see Chapter 1). How does a meerkat sentinel benefit from its actions? Is it trying to protect its offspring from danger? Or other relatives in the group? Do unrelated immigrants share guarding duties? And, if so, what prevents some individuals from contributing less than their share? Is it possible that the impression that the sentinel's behaviour is adapted to protecting other group members is misleading? Cooperative behaviour raises many fundamental questions about animal behaviour and has consequently been a major focus of research on animal societies.

Suggested explanations of the evolution of cooperative behaviour include group selection, kin selection and several forms of mutualistic interactions, including shared benefits and direct, indirect and generalised reciprocity (see Chapter 1). Subordinates may sometimes be forced to cooperate by the coercive tactics of dominant individuals. Finally, apparently cooperative behaviour can be a consequence of individuals synchronising or coordinating purely selfish actions which generate coincidental benefits to others.

Cooperation between female mammals takes many different forms, ranging from huddling to minimise heat loss (Arnold 1990) to complex strategies of alliance and counter-alliance (Whiten and Byrne 1988, 1997). This chapter describes different contexts in which group members cooperate with each other, while Chapter 17 examines the evolution of cooperative breeding. Section 9.2 describes nine different forms of cooperative behaviour that are common in social mammals; section 9.3 reviews evidence of 'cheating' strategies and tactics used to control cheating, while section 9.4 reviews the relative importance of different evolutionary mechanisms favouring cooperation and section 9.5 examines some of the consequences of cooperation between females.

9.2 Cooperation in different contexts

Defence against predators

In some social mammals, group members respond collectively to predators, jointly attacking animals much larger than themselves (see Chapter 2). Coordinated defence involving females occurs in several of the smaller social carnivores (Russell 1983; Rasa 1986, 1987a, 1987b; Rood 1986) as well as in cetaceans (Whitehead and Weilgart 2000) and in some herbivores, including musk ox (Mech 1970), eland (Kruuk 1972) and elephants (Dublin 1983; Moss 1988). Where multiple group members engage collectively in predator defence or collaborate to mob predators, it is often difficult to be sure that defence is designed to provide benefits to other individuals. However, in some cases, group members attempt to rescue individuals that have been attacked by predators, providing clear evidence of cooperation.

Mammal Societies, First Edition. Tim Clutton-Brock.
© 2016 Tim Clutton-Brock. Published 2016 by John Wiley & Sons, Ltd.

For example, in one case, a group of banded mongooses, led by the dominant male, attacked a martial eagle that had carried off a group member and successfully rescued it (Rood 1983). Some social mammals support incapacitated individuals. For example, banded mongooses will guard and feed sick or wounded individuals (Rood 1986), while female elephants commonly remain with sick or wounded individuals from the same matrilineal group and defend them against predators (Moss 1988). Behaviour of this kind appears to be restricted to societies where most group members are relatives, so it is likely to be an example of kin-selected assistance.

In some social mammals (including several carnivores and primates), group members collaborate in mobbing predators (Rasa 1987a; Tamura 1989; Fitzgibbon 1994; Kobayashi 1996; Graw 2005) and similar behaviour occurs in some birds (Curio 1978; Frankenberg 1981; Dugatkin and Godin 1992). Costs of mobbing probably include the risk of attack and injury as well as lost time and energy, but why animals benefit from mobbing predators is still unclear. Like sentinel duty, mobbing could be an example of by-product mutualism or of shared benefits since individuals may benefit if predators move away. It is likely to be less risky and more effective to mob at the same time as other individuals, which could help to explain why, in birds, members of several different species often mob predators together (Curio 1978).

Alternatively, studies of mobbing in birds have suggested that it could be an example of reciprocity since individuals that do not contribute to mobbing are likely to share benefits derived from the behaviour of individuals that contribute to mobbing and individual decisions to approach and harass predators often depend on the behaviour of others (Fitzgibbon 1994; Raihani and Bshary 2011). One experimental study of pied flycatchers showed that the probability that pairs will join neighbours in mobbing a predator is affected by whether the same pair had joined them in previous mobbing interactions (Krams *et al.* 2008). In some cases, mobbing may generate benefits to dependent juveniles or other relatives and so could be maintained by kin selection (Graw and Manser 2007). Finally, mobbing could allow individuals to display their qualities as potential breeding partners (Zahavi and Zahavi 1997).

Detailed studies of mobbing provide some insight into factors that affect its distribution but no conclusive answers concerning the evolutionary mechanisms that maintain it. In several species, the duration of mobbing appears to be associated with differences in the threat represented by the predator that is the focus of attention (Kobayashi 1996). For example, meerkats mob Cape cobras (whose bite is deadly) longer and more intensely than less mobile puff adders or non-venomous mole-snakes (Graw 2005) (Figure 9.1). Individual

Figure 9.1 A meerkat group mobbing a Cape cobra. *Source*: © Tim Clutton-Brock.

contributions to mobbing often vary with age: for example, in meerkats, younger adults mob for longer than juveniles and older adults and males mob more than females when age effects are allowed for (Graw 2005). In contrast, in some social birds, females mob more than males (Maklakov 2002). Increased mobbing rates in male meerkats and female babblers (both of which typically disperse to breed in other groups) suggests that it is unlikely that individuals mob in order to advertise their quality as mates and could indicate that mobbing plays a role in forging coalitions among potential dispersers, though there is little evidence that this is the case (Maklakov 2002). Juveniles may mob less than experienced adults because they have not yet learned to identify potential predators: inexperienced vervet monkeys, for example, seldom mob snakes and only learn to do so from older group members (Cheney and Seyfarth 1990).

Why younger adults mob more than older ones is uncertain but suggests that mobbing should not be regarded as an extension of parental care (Graw 2005).

Alarm calls and sentinel behaviour

In many social mammals, group members that detect predators give alarm calls that alert other group members to danger (see Chapter 7). Where giving alarm calls has costs to the caller, individuals that are likely to gain larger direct or indirect benefits from calling would be expected to be more likely to give calls and several studies suggest that individuals are more likely to call if they have close relatives in the group. For example, black-tailed prairie dogs tested with a stuffed badger (Figure 9.2a) were more likely to call if they had relatives in the group (Figure 9.2b) and males (but not females) called more often if they had parents, full siblings or offspring in the

(a)

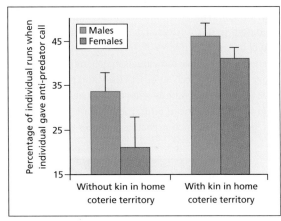

(b)

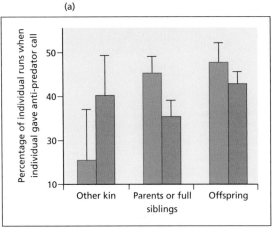

(c)

Figure 9.2 (a) Selective anti-predator alarm calling by prairie dogs tested with a stuffed badger. During simulated predator 'attacks' about 50% of the prairie dogs, most individuals that had either descendent or non-descendent kin within earshot gave anti-predator calls while others watched in silence. (b) Both males and females were more likely to call if they had relatives in the group. (c) Males called more often if they had parents, siblings or offspring in the group while females did not appear to discriminate. *Source*: From Hoogland (1995). Reproduced with permission of the University of Chicago Press. *Photo source*: © Elaine Miller Bond.

group than if only non-distant relatives were present (Figure 9.2c).

Coordinated vigilance, where individuals take turns acting as sentinels, occurs in some mammals as well as in some social birds (Gaston 1977; Wright *et al.* 2001). In dwarf mongooses and meerkats, sentinels stand in elevated positions where they can scan the area around the group (Rasa 1986, 1987a, 1987b; Clutton-Brock *et al.* 1999a; Kern and Radford 2013) and are particularly likely to remain in position in risky conditions. In both species, sentinels give repetitive calls (the 'watchman's song') that inform other group members that there is a

sentinel on guard, allowing them to forage without regularly checking for danger (Figure 9.3). In dwarf mongooses, sentinels are more likely to vocalise in denser habitats and when other group members are widely separated, suggesting that their calls may also help to maintain group cohesion (Kern and Radford 2013). When predators are sighted, meerkat sentinels give graded alarm calls that alert other members of the group to danger (Manser 1999). Bouts of sentinel duty may last for 10 minutes or more and are usually initiated when the previous sentry stands down, so that there is typically only a single individual on guard at a time

(a)

(b)

(c)

Figure 9.3 Sentinel behaviour in meerkats: (a,b) meerkats foraging in Kalahari Gemsbok Park: individuals commonly dig up to 20 cm below ground to find prey and cannot watch for predators while doing so; (c) digging animals frequently stop to glance around them but do so less frequently if a raised guard is vocalising; (d) while the group is foraging, one animal commonly climbs a tree or mound and watches for predators. *Source*: From Clutton-Brock *et al.* (1999a) and Clutton-Brock and Manser (2015). *Photo sources*: © Tim Clutton-Brock.

(Clutton-Brock *et al.* 1999a). In meerkats, most adults contribute to sentinel duty, though breeding females seldom act as sentinels. In larger groups, where there are more individuals to contribute to guarding, there is usually a sentinel on duty for a larger proportion of the day and rates of predation are reduced. Foraging groups with young pups also have a sentinel on duty for more of the day than those without pups (Santema and Clutton-Brock 2013).

It has been suggested that sentinel behaviour could be an example of by-product mutualism rather than of cooperation (Bednekoff 1997; Clutton-Brock *et al.* 1999a; Kitchen and Packer 1999). Sentinels are commonly the first animals to see predators and, once an individual has satisfied its need for food, going on sentinel duty if no other animal is guarding could maximise its chance of detecting danger (Bednekoff 1997). As would be expected, individuals often go on guard after successful foraging bouts and experimental feeding increases both the probability that individuals will go on guard and the duration of their guarding bouts (Figure 9.4a,b). Moreover, sentinels that are experimentally forced to abandon guarding bouts prematurely reduce the time until they next go on guard (Figure 9.4c). However, if sentinel behaviour was purely selfish, it is not obvious why sentinels should give either

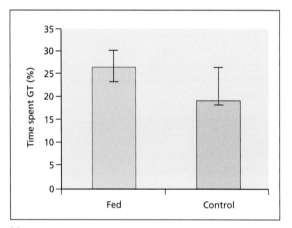

(a)

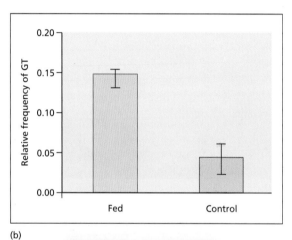

(b)

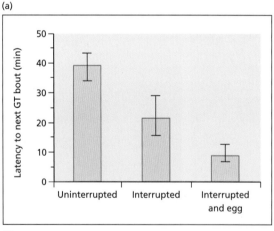

(c)

Figure 9.4 Factors affecting sentinel behaviour in meerkats. (a,b) Individuals often go on raised guard (GT) after successful foraging bouts and experimental feeding increases both the probability that they will go on guard and the duration of their guarding bouts. (c) Forcing sentinels to abandon guarding bouts ('interrupted') prematurely reduces the latency to their next bout while experimental feeding with egg reduces the latency further. *Source*: From Clutton-Brock *et al.* (1999a). Reproduced with permission of AAAS. *Photo source*: © Tim Clutton-Brock.

alarm calls or the watchman's song, or why helpers should spend more time guarding when dependent juveniles are foraging with the group. Both these observations suggest that sentinel behaviour is adapted to minimising the chance that other group members will be caught by predators.

Sentinel behaviour appears to be restricted to species that form stable groups that include a high proportion of close relatives, and sentinels may gain indirect as well as direct benefits by their behaviour. Among meerkats, there is no obvious tendency for group members to adjust their contributions to guarding to their relatedness to other group members and immigrants contribute as much as individuals closely related to the breeding pair (Clutton-Brock et al. 2002; Clutton-Brock and Manser 2015).

Defence against neighbours and immigrants

Social animals where multiple females occupy well-defined ranges often defend them cooperatively against neighbouring groups and, when neighbouring groups encounter each other, their relative size frequently plays an important part in deciding the outcome. For example, in meerkats, relative group size affects both the probability that encounters with other groups will escalate into fights and the chances of winning them if this occurs (Figure 9.5) (Young 2003). Deterring intruders is

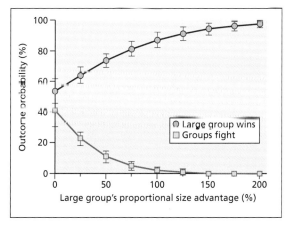

Figure 9.5 Benefits of group size in encounters between rival meerkat groups. The figure shows how a group's proportional size advantage in an interaction with a rival group influences the probability that (i) the group wins the interaction and (ii) that the interaction escalates into a fight. Means and standard errors are predictions from generalized linear models. *Source*: From Young (2003). Reproduced with permission of A. Young.

important both to prevent a gradual loss of parts of the group's territory and, in some cases, to reduce the risk of attacks on juveniles or breeding females (see Chapter 15).

In some animals where females help to defend their group range, they advertise the size of their group by synchronised vocalisations (Harrington and Mech 1979; Sekulic 1982; Raemaekers and Raemaekers 1984; East and Hofer 1991; Zahavi and Zahavi 1997; Radford 2003). For example, in African lions, females advertise their presence and recruit other group members by synchronised roaring choruses which serve to deter or scare away intruders (McComb et al. 1994; Grinnell and McComb 2001) while solitary intruders seldom roar (Grinnell and McComb 2001). Lionesses appear to be able to assess the size of intruding groups relative to the size of their own group and are more reluctant to approach groups larger in size than their own.

In some social carnivores, females also play a proactive role in competition with neighbouring groups, probing the ranges of their neighbours and attacking or killing dependent juveniles if they come across them (Pusey and Packer 1993). For example, meerkat groups go on forays into the territories of neighbours and carefully inspect any burrow systems that they find. If they come across burrows containing pups and a babysitter, they attack them and attempt to chase off the babysitter and kill the pups (Clutton-Brock and Manser 2015). Lethal attacks on members of neighbouring groups are not confined to dependent young. For example, groups of female hyenas that encounter solitary adults from neighbouring clans commonly attack and sometimes kill them (Kruuk 1972; Boydston et al. 2001) and meerkats of both sexes are involved in lethal attacks on dispersing adults from neighbouring groups (Clutton-Brock and Manser 2015).

Coordinated defence can also play an important role in successful dispersal and the establishment of new groups. In many social mammals, females disperse in parties after reaching adolescence, sometimes after they have been ejected from their natal group by resident females (see Chapter 3). Increases in the size of dispersing splinter groups raises their chances of establishing a new territory and breeding successfully. For example, in cheetah, adolescent females that disperse in groups show higher survival than those dispersing alone (Durant et al. 2004), while in meerkats females that disperse on their own are unable to maintain an effective system of predator detection or to feed effectively (Young 2003) and are also unlikely to be joined by dispersing males or to establish new breeding groups. Since only relatively large breeding groups are able to produce sizeable parties of dispersing females, effects of this kind can lead to powerful selection pressures favouring group traits that help to maintain group size (see Chapter 2).

Coordinated defence by females can also play an important role in increasing the survival of offspring where immigrant males kill dependent young (see Chapter 15). For example, after taking over prides, groups of male lions attempt to kill dependent cubs and multiple females are more likely to defend their cubs successfully against intruders than single females (Packer *et al.* 1990). Roaring choruses involving multiple females help to deter male intruders and playback experiments show that intruding males are more

reluctant to approach three females roaring in chorus than a single female (Grinnell and McComb 1996). As territorial groups typically involve a high proportion of relatives, individuals may gain indirect as well as direct benefits from cooperative defence but it is often unclear whether individuals that contribute to group displays are assisting each other or are merely maximising benefits to themselves by synchronising their activities.

Cooperative hunting

Cooperative hunting can take a variety of forms. In some species, group members forage or hunt synchronously but rarely cooperate to catch individual prey: for example, though male cheetahs often hunt prey simultaneously, they seldom cooperate to bring down animals (Caro 1994). Similarly, banded mongooses, dwarf

(a)

(b)

(c)

(d)

Figure 9.6 Cooperative hunting in African wild dogs. (a) A social rally preceding a wild dog hunt. (b) Pack members chase wildebeest herds, selecting one or more likely prey. (c) After prey have turned at bay, some pack members surround them: larger or more dangerous prey are held by multiple pack members until they are demobilised, and dragged to the ground and killed. (d) Pack members eat as much as they can. *Sources*: (a,c,d) © Dave Hamman; (b) © Bruce Davidson/Nature Picture Library.

mongooses and meerkats forage synchronously in stable groups but pursue prey independently (Rood 1986; Clutton-Brock *et al.* 1999b). In other species, individuals hunting together cooperate to catch large prey. For example, Malagasy fossas usually hunt alone but sometimes hunt in pairs or trios, cooperating to catch lemurs and subsequently sharing the meat (Lührs and Dammhahn 2010). Similarly, African lions commonly hunt smaller prey on their own but hunt larger prey, like buffalo, in groups (Schaller 1972; Packer *et al.* 1990).

Cooperative hunting is most highly developed in the larger canids that do not rely on stealth to approach their prey and which often hunt animals larger than themselves (Kruuk 1975; Johnsingh 1982; Creel and Creel 2002) and in the smaller cetaceans (Baird 2000; Connor *et al.* 2000). Studies of African wild dogs provide some of the best examples of cooperative hunting. Hunts are initiated by a greeting ceremony before the group start to move in search of prey, usually spread out over 10–100 m (Figure 9.6a). After prey is located, several dogs may attack simultaneously from different directions in an attempt to force herds to run. Single dogs may incite a charge, then pack-mates quickly separate herd members that lag behind the rest of the group. Several individuals often pursue the same animal in turn, while other group members may intercept it if it circles (Figure 9.6b). Once their prey is run to a standstill, some dogs may attract its attention from the front while others attack it from behind (Figure 9.6c). When wild dogs hunt wildebeest with dependent calves, some pack members distract or attack the mother while others attack its offspring. One or more pack members may hold the head of dangerous prey (such as warthogs) while others start to disembowel it from the rear. Pack members cooperate to defend carcasses against scavenging hyenas or lions (though they are not always successful in doing so) as well as in regurgitating meat for pups at the natal den (Figure 9.6d). Wolves use similar techniques when hunting large prey, like elk or bison (Figure 9.7).

In wild dogs, cooperative hunting both increases efficiency and reduces hunting costs. Hunting success (measured in terms of kills per hunt) is higher in large packs than small ones and prey size as well as the probability that several animals will be killed in the same hunt also increases, so that the amount of meat available per dog per day rises with group size (Figure 9.8a–c). In addition, large packs hunt more frequently (Figure 9.8e) and do not have to chase their prey so far (Figure 9.8f), so that meat gained per kilometre of travel increases (Figure 9.8d). In some species, communal hunting may also reduce variance in hunting success (Stephens and Krebs 1986), though it does not appear to have this effect in wild dogs (Creel and Creel 2002). When all these effects are aggregated, the best estimates suggest that dogs gain substantial energetic benefits by hunting in larger packs (Figure 9.8g) though adult survival is negatively related to pack size (Creel and Creel 2015), perhaps as a result of reproductive competition within packs.

The importance of cooperation in hunting can vary between populations living in different habitats or feeding on different prey. For example, in East African lions, foraging efficiency (calculated in terms of meat per hunt) is usually highest for solitary individuals (Packer *et al.* 1990; Scheel and Packer 1991). In contrast, in the open

(a)

(b)

Figure 9.7 A wolf pack (a) pursues an elk and (b) holds a bison at bay. *Source*: © Douglas W. Smith.

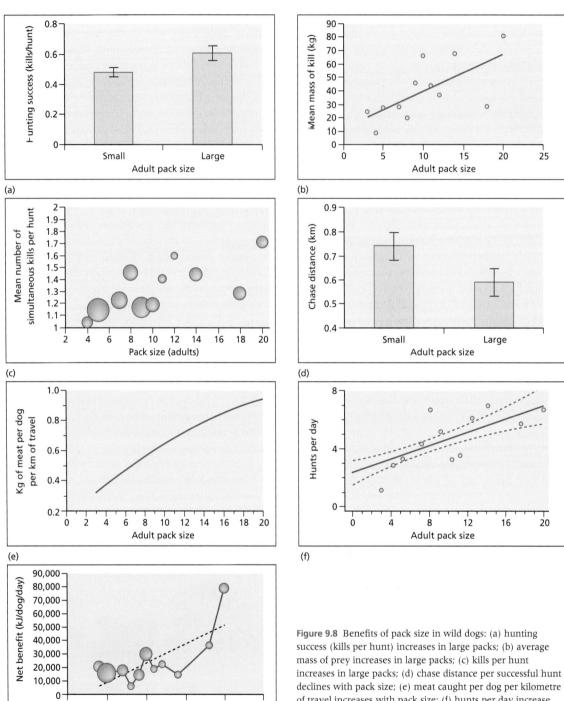

(a)

(b)

(c)

(d)

(e)

(f)

(g)

Figure 9.8 Benefits of pack size in wild dogs: (a) hunting success (kills per hunt) increases in large packs; (b) average mass of prey increases in large packs; (c) kills per hunt increases in large packs; (d) chase distance per successful hunt declines with pack size; (e) meat caught per dog per kilometre of travel increases with pack size; (f) hunts per day increase with pack size; (g) net energetic benefit of hunting per individual increases with pack size. *Source*: From Creel and Creel (2002). Reproduced with permission of Scott Creel.

semi-arid plains of Namibia, hunting success increases with group size and, in the dry season, when prey is scarce and coordinated, cooperative hunting is essential for lions to obtain their daily food requirements (Stander 1992; Stander and Albon 1993). In Namibian lions, hunts often involve coordinated stalks, where some group members ('wings') typically circle and charge prey animals while others ('centres') wait in hiding and coordinated hunts are more consistently successful than hunts where pride members did not coordinate their stalking behaviour. The same individuals usually hunt either as wings or centres and hunts are most successful when individuals are in their preferred positions.

The extent of cooperation in hunting also differs between populations of chimpanzees (Figure 9.9). In the mature rainforests of Taï on the Ivory Coast, male chimpanzees cooperate to catch red colobus monkeys (Boesch and Boesch 1989; Boesch 1994a, 1994b, 1994cBoesch 1994a–c; Boesch *et al.* 2006). Hunting success increases with group size and individuals benefit from hunting in groups, even though some individuals contribute little to catching monkeys (Boesch 1994b). In contrast, at Gombe in Tanzania, chimps often hunt independently as well as in groups, there is limited evidence of cooperation, and the probability that they will capture red colobus monkeys (their main prey) is no higher when they are hunting in groups than when they are hunting alone (Busse 1978; Boesch 1994b). One possible explanation of the differences between Taï and Gombe is that at Gombe the forest canopy is discontinuous and monkeys can be more easily isolated and caught by individual chimps than at Taï where the canopy is higher and more continuous (Boesch *et al.* 2006). However, male chimpanzees also share meat in Ugandan populations where there is limited evidence of active cooperation in hunts (Mitani and Watts 2001).

Shared benefits probably play an important role in the maintenance of cooperative hunting. In many mammals where several individuals cooperate to catch prey, all members of hunting groups often have some access to kills and there is some evidence that variation in the extent to which group members cooperate to catch prey is associated with variation in the extent to which kills are shared. For example, in chimpanzees at Taï, where males hunt cooperatively, meat is shared between members of the hunting party, with hunters receiving more than non-hunters and successful hunters getting the most (Boesch 1994a,c; Boesch and Boesch-Achermann 2000). In contrast, at Gombe, where hunting is less cooperative, the distribution of meat appears to be more strongly affected by rank, with dominant individuals receiving a disproportionate share, either by sharing or by stealing from the successful hunter (Goodall 1986; Stanford 1998; Boesch *et al.* 2006). However, to what extent these contrasts reflect differences in access to meat rather than in willingness to share kills is difficult to resolve. Coercion may also be involved for, in some cases, the sharing of meat appears to represent a response to harassment by individuals begging for food (Feistner and McGrew 1989). For example, persistent begging by chimpanzees after successful hunts often appears to cause hunters to share food to avoid beggars (Teleki

(a)

(b)

Figure 9.9 (a) A male chimpanzee grabs a juvenile colobus while its mother leaps to safety; (b) meat sharing after a group hunt in chimpanzees. *Source*: © Ian Gilby.

1973; Gilby 2006). Similarly, experiments with captive chimpanzees show that food 'owners' are more likely to share food when harassed by beggars and that group members spend more time harassing 'owners' in circumstances where they were more likely to gain access to food by this tactic (Stevens 2004).

In most species where group members hunt cooperatively, there is little evidence that kinship affects either the probability that individuals will hunt together or, if they are successful, that kills will be shared. Chimpanzees tend to hunt with individuals of similar age and rank but show no consistent tendency to hunt with relatives or to favour them in the distribution of meat (Boesch *et al.* 2006). However, this does not necessarily suggest that kin selection has played no part in the evolution of cooperative hunting for many cooperative hunters, including wild dogs and killer whales, live in groups whose members are mostly either relatives or potential breeding partners and the benefits of cooperation with related group members will contribute to inclusive fitness (see Chapter 1).

Food storage and sharing

In some herbivorous mammals, group members collect and store food that is subsequently used by all group members. For example, beavers store food in autumn that is used by all group members in winter (Müller-Schwarz and Sun 2003; Sun 2003). Similarly, in some of the social mole rats, group members store food in chambers connected to the main burrow system (Jarvis and Bennett 1991; Bennett and Faulkes 2000). In most species where food is stored and subsequently used, a majority of group members are relatives or potential breeding partners and storing food probably generates both shared direct benefits and indirect benefits.

In some cases, food sharing includes non-kin as well as kin and it has been suggested that it is maintained by reciprocity. One of the most widely cited examples involves food sharing by vampire bats, which live in stable groups within larger colonies and disperse each night to try to find hosts from which they can suck blood (Wilkinson 1984; Carter and Wilkinson 2013a,b) (Figure 9.10). On any night, a proportion of individuals are unsuccessful and, after returning to the roost, these animals solicit meals of regurgitated blood from other group members (Wilkinson 1985a,b, 1988). The system fulfils most of the requirements for reciprocity: average relatedness between female group members is low (0.02–0.11) and individuals will feed unrelated partners as well as relatives, suggesting that this practice may not be maintained by kin selection; groups are relatively stable, so that the probability that the same individuals will encounter each other repeatedly is high; the benefit of the blood donated is greater than the cost of the gift to a well-fed bat; and vampires are able to recognise each other. Finally, studies of captive bats show that some pairs of individuals both donate and receive more blood from each other than others (Box 9.1 and Figures 9.11 and 9.12) and food sharing is commonly initiated by donors, indicating that it is unlikely to be coerced. However, it is still unclear

(a)　　　　　　　　　　　　　　　　　　(b)

Figure 9.10 Vampire bats live in stable groups whose members share blood meals with each other: (a) a resting group; (b) a pair sharing a blood meal. *Source*: © Gerald Carter.

Box 9.1 Reciprocal altruism in vampire bats

Vampire bats in north-western Costa Rica roost in groups of eight to twelve adult females with an equal number of dependent offspring (Wilkinson 1984, 1990) (see Figure 9.10). As in many other social mammals, females commonly remain in their natal groups throughout their lives, while males disperse at about 1 year. Group composition is stable and groups generally consist of several matrilines although unrelated females are recruited to roosting groups at an average rate of one unrelated female per group every 2 years (Wilkinson 1984). Adults forage at night for animal blood and individuals must consume 50–100% of their body weight to sustain their metabolic requirements and are likely to die of starvation if they fail to feed for three successive days. Foraging is chancey, individuals are often unsuccessful and most nights around one-quarter of all bats return to the roost without having fed. These individuals beg for blood from successful foragers in their group and commonly receive it.

Most observed regurgitations are between a mother and her offspring but a minority involve females feeding each other or feeding each other's young or males feeding young. When kinship effects are allowed for, bats are most likely to feed or be fed by individuals that they associate with regularly and captive bats that are caged together are more likely to share food both with kin and non-kin than with wild ones (Carter and Wilkinson 2013b). Detailed studies of food sharing in captive groups show that individuals are more likely to receive food from individuals that they have given food to or received food from in the past and that the frequency of food sharing is positively correlated with the frequency of mutual allo-grooming, though there is no evidence of a regular alternation of roles and it is not clear whether bats discriminate against non-cooperators (Wilkinson 1984; Carter and Wilkinson 2013b) (see Figure 9.12).

The vampire bat study is frequently cited as evidence of cost-counting reciprocity in non-human animals (Dugatkin 1997) but whether it provides evidence for reciprocity is questionable and depends on the definition of reciprocity that is used (Carter and Wilkinson 2013a). If reciprocity is used in a restrictive sense to include only those cooperative interactions where there is a delay between the costs and benefits of cooperating, and individuals alternate in their roles and discriminate against non-reciprocators, then it is unclear whether reciprocity is involved since it is uncertain whether the immediate costs of sharing blood exceed the immediate benefits or whether bats discriminate against non-cooperators and there is no regular alternation of roles (Carter and Wilkinson 2013b). If reciprocity is taken to include all cases where unrelated individuals gain mutual benefits by assisting each other, then it is almost certainly involved, but a simpler explanation is that it is a consequence of immediate mutualistic benefits. Moreover, as most allo-feeding interactions in natural colonies appear to involve relatives, an alternative explanation is that its occurrence between unrelated individuals is a consequence of limitations in kin discrimination.

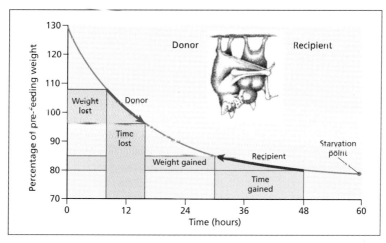

Figure 9.11 Cost–benefit analysis of blood sharing in vampire bats. The author weighed adult females returning to the roost after feeding and then weighed them every hour for the next 24 hours. An individual who had fed might return at 130% of its pre-feeding weight, whereas a bat who failed to feed on two successive nights might return at 80% of its earlier weight; half the weight of a blood meal is lost through urination within the first hour after feeding. Time represents the number of hours for which the bat can survive before it dies of starvation after not feeding for 60 hours. Bats lose weight rapidly in the first 12 hours after feeding and then lose weight more slowly. As a result, an individual that regurgitates a blood meal to hungry roost-mates loses a relatively small amount of time, while the gain to a hungry recipient is much larger. *Source*: From Wilkinson (1990). Reproduced with permission of Patricia J. Wynne.

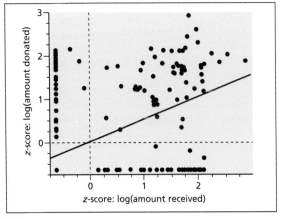

(a)

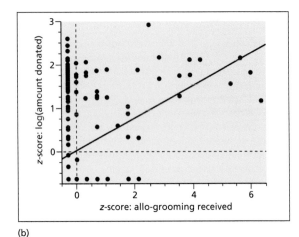

(b)

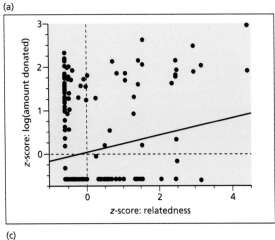

(c)

Figure 9.12 Relationships between the frequency of food donated by vampire bats in captive groups and (a) food received, (b) allo-grooming received and (c) relatedness. The probability that bats would donate food was more accurately predicted by food received and by allo-grooming frequency involving the same individual in the past than it was by relatedness. *Source*: From Carter and Wilkinson (2013b).

whether it is maintained by direct reciprocity, generalised reciprocity or a sharing of immediate benefits of the kind envisaged by public goods games (see Box 9.1).

Experimental studies of other captive mammals also show that the readiness of individuals to share food can be affected by whether or not individuals previously received food from their partners (Stephens *et al.* 2002; Hauser *et al.* 2009), though this is not always the case (Melis *et al.* 2008). For example, experiments with Norway rats show that the frequency with which individuals assist partners in gaining access to food are affected by whether or not they received assistance in previous encounters (Rutte and Taborsky 2007) as well as by the tactics they adopted: rats that played with partners that adopted 'tit for tat' strategies were more likely to provide assistance than those interacting with partners that provided assistance on a random schedule (Viana *et al.* 2010).

Research on captive primates has also produced some evidence of reciprocity. Using captive tamarins, Hauser and his colleagues presented individual tamarins with a situation where they had access to an apparatus that they could manipulate to deliver food to another (unrelated) individual but not for themselves. Animals played against a 'stooge' that either always pulled the tool handle to deliver rewards or against a second stooge that never did so. They were more likely to pull the tool when playing against the first individual, though they cooperated less than 50% of the time and the frequency with which they pulled the tool declined as the experiment progressed (Hauser *et al.* 2003). There is, however, little evidence that similar behaviour occurs in natural populations either of Norway rats or of other primates and these experiments may tell us more about the capacity of individuals to learn and to adopt the most productive strategy in simplified interactions than about the prevalence of cooperation in natural populations.

Stimulated by evidence of prosocial behaviour in humans (see Chapter 20), a number of studies of (mostly) captive primates have investigated the extent to which individuals adjust their behaviour to the needs of others (de Waal 1996; de Waal and Suchak 2010; Amici *et al.* 2014). While primates, like many other mammals, sometimes assist kin and may adjust their behaviour to their needs, they seldom direct costly forms of assistance at non-relatives or adapt their behaviour to their needs (Silk 2005, 2006; Clutton-Brock 2009a). For example, although chimpanzees often engage in cooperative behaviour that is likely to increase their own fitness, and may have some capacity for empathy, they seldom direct costly forms of assistance at unrelated individuals or adjust their behaviour to their needs (Silk 2005b; Amici *et al.* 2014).

Experiments with food sharing in primates have also shown that the responses of partners can be sensitive to inequalities between the rewards they win and those won by their partners. For example, experiments with captive capuchin monkeys in neighbouring cages that were trained to exchange a token for a reward showed that if one individual was given a more attractive reward than the other or was more likely to be rewarded if it failed to respond, the second animal rapidly ceased to participate in the experiment (Brosnan 2006, 2013). Interactions of this kind are sometimes suggested to be the result of 'inequity aversion' (de Waal and Suchak 2010). Similar interactions occur between captive chimpanzees and could indicate that they have a 'sense of fairness', though there is little evidence that individuals are averse to inequalities from which they benefit more than others (Hopper *et al.* 2014; Silk 2005b). Moreover, an alternative interpretation is that they represent examples of loss aversion (Chen and Santos 2006) and may be analogous to jealousy in humans. For example, one possible functional explanation of the tendency for individuals to cease to participate in experiments if they receive smaller rewards than their partner is that it may pay individuals to leave feeding sites if they are aware of other individuals foraging more successfully in nearby areas.

Communal care of young

In some social mammals, females assist with each other's breeding attempts. For example, in the Rodrigues fruit bat, unrelated females assist pregnant mothers during parturition, grooming their ano-vaginal region, helping them to adopt a feet-down birthing position, grooming recently born pups and helping to manoeuvre pups into a suckling position (Kunz *et al.* 1994). Similarly, before female bottlenose dolphins give birth, they commonly associate with other adult females, who subsequently provide assistance during the birth and help to bring newborn young to the surface to breathe (Tavolga and Essapian 1957).

In some species, female group members contribute to carrying each other's infants or to protecting them against predators, other females or potentially infanticidal males (Mitani and Watts 1997). For example, in some colobine monkeys, females other than the mother commonly handle dependent infants and may carry them for short periods (Poirier 1968; Hrdy 1976; Oates 1977). In baboons, too, females show intense interest in young infants and commonly attempt to groom them (Figure 9.13). Allo-mothering is most frequently found in mammals where females live in matrilineal groups, including primates, elephants, chamois, dolphins, killer whales and sperm whales (Tavolga and Essapian 1957; Caldwell *et al.* 1966; Best 1979; Haenel 1986; Lee 1987; Norris and Schilt 1988; Ruckstuhl and Ingold 1998) and there are few records of breeding females assisting each other in species where relatedness between female group members is low (Briga *et al.* 2012). For example, allo-mothering is rare or absent in red colobus monkeys, where females disperse at adolescence and few female group members are close relatives, but is common in black and white colobus monkeys, which live in matrilineal groups (Struhsaker 1975).

Comparative analyses suggest that non-maternal caretaking is often associated with increased rates of infant growth and reduced inter-birth intervals (Mitani and Watts 1997). Detailed studies of particular species suggest that it can increase fitness both in recipients and in donors (Fairbanks 1990). For example, in captive vervet monkeys, inter-birth intervals of individual breeding females decline as the amount of allo-mothering received by their offspring increases (Fairbanks and McGuire 1984). Direct benefits of allo-mothering may include assistance in social interactions, social tolerance or support and experience of parental care: in vervet monkeys, the first infants born to mothers with experience of allo-mothering are more likely to survive than those born to females that have not contributed to allo-mothering. Allo-mothers may also gain indirect benefits, since their

Figure 9.13 Female baboons show intense interest in the young infants of other females, who are often closely protected by their mothers. To gain access to infants, other females often groom their mothers for long periods. *Source*: © Joan Silk.

attentions are frequently directed at the offspring of close relatives (Fairbanks and McGuire 1984; Fairbanks 1990).

While allo-parental care can have important benefits to infants and juveniles, there are also cases where infant handling by females other than mothers can have substantial costs to juveniles. For example, in a number of social primates, including grey langurs, yellow baboons and Barbary macaques, females sometimes kidnap infants from their mothers (Paul *et al.* 1996; Kleindorfer and Wasser 2004). Kidnapped infants are frequently handled in a careless or abusive fashion and infant kidnapping can lead to the infant's death. For example, in Barbary macaques, around one-third of kidnapped infants die as a consequence of being kidnapped (Paul *et al.* 1996). How kidnapping females benefit from handling infants is not certain but one possibility is that infant kidnapping and abusive handling is an expression of reproductive competition between females rather than cooperation (Kleindorfer and Wasser 2004).

Allo-lactation

In some social mammals, breeding females nurse infants other than their own offspring (Koenig 1995; Lewis and Pusey 1997; Hayes 2000). The frequency of allo-lactation varies widely and can differ between populations (Gero *et al.* 2009). In some cases, dependent offspring occasionally steal milk from breeding females other than their mothers: for example, in elephant seals, unrelated offspring may attempt to suck from females other than their mother but are usually driven off when they are detected (McCracken 1984; Boness 1990; Taber and Macdonald 1992). In others, females are tolerant of attempts by infants other than their own to nurse. For example, in African lions, females often breed synchronously, litters are frequently combined and females allow offspring born to other females to suckle, though lactating females discriminate in favour of their own offspring (Packer *et al.* 1992; Lewis and Pusey 1997). In some rodents (including prairie dogs and house mice), breeding females also combine litters and do not discriminate between their own offspring and those of other females (König 1994a, b; Hoogland 1995), while in some populations of sperm whales several females nurse dependent juveniles (Gero *et al.* 2009). Finally, in some cooperative breeders, non-breeding females nurse offspring born to the dominant female or to other subordinate females (see Chapter 17). Comparative analyses suggest that non-offspring nursing is least frequent and most likely to be classified as milk theft in species that produce single young and that its incidence is higher and is more likely to be tolerated in species with large litters (Packer *et al.* 1992). Among polytocous species, non-offspring nursing declines in species where group size is usually large, possibly because relatedness between group members is reduced.

A range of potential benefits of allo-lactation to mothers or their offspring has been suggested. Positive benefits to mothers may include the disposal of excess milk supplies, reduced peak energetic costs as a result of load sharing and indirect benefits from assisting relatives. Positive benefits to offspring include reduction in the risk of starvation or predation through increased growth dilution effects and immunological benefits associated with receiving milk from multiple females (Caraco and Brown 1986; Hoogland *et al.* 1989; Manning *et al.* 1992; Wilkinson 1992; König 1994a, 2006; Dugatkin 1997, Roulin and Heeb 1999; Hayes and Solomon 2004).

Some of the most detailed research on the consequences of allo-lactation has been carried out on captive groups of house mice, where pairs of breeding females commonly share a burrow and pool their litters, suckling their own and other female's infants indiscriminately (Manning *et al.* 1995; König 1994a,b, 1997, 2006). Experimental studies of captive groups show that females which combine litters with sisters show higher fecundity, are more likely to rear their offspring and show higher levels of reproductive success than those that breed on their own, while females that combine litters with unrelated partners show intermediate fitness (König 1994a,c) (Figure 9.14a). Pairs of females do not increase their milk production and König attributes the increased breeding success of pairs over solitary breeders to a reduction in peak energy load caused by asynchrony in the peak needs of the two litters (König 2006). The lower breeding success of pairs made up of unrelated females is associated with a tendency for one female to dominate reproduction, suggesting that related females are more likely to be tolerant of each other's breeding attempts. This effect is apparently caused by familiarity rather than kinship per se, because unrelated but familiar females that breed together are as successful as familiar relatives. As group size increases past two, average breeding success declines, the degree of skew in breeding success increases and reproductive success is less equally shared between participating females (Figure 9.14b). However, some empirical studies have failed to detect any benefit of the number of allo-lactators to the growth

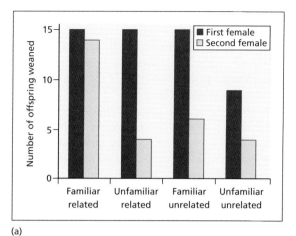

(a)

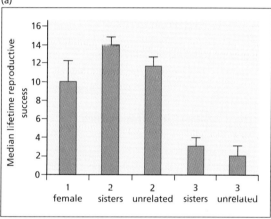

(b)

Figure 9.14 Effects of the presence of kin on the lifetime breeding success of female house mice in artificial groups. (a) The median number of offspring weaned within a standardised lifespan in four types of artificially created social groups. The breeding success of females is highest if they are paired with another female which is both familiar and related. While the first female to start breeding (dark columns) only has a lowered reproductive success when paired with an unfamiliar unrelated female, the second female (pale columns) always produces fewer offspring if not paired with a familiar related female. (b) Effects of different numbers of females. The figure shows the number of offspring weaned during an experimental lifespan of 6 months (median ± SE) in female house mice as a function of group size (number of females per group ranged between one and three) and of genetic relatedness. Sisters were familiar full-sibs that grew up together, while unrelated females were previously unfamiliar. An unrelated adult male was always present. *Source*: Data from König *et al.* (1994a).

and survival of infants. For example, studies of wild meerkats where female helpers commonly lactate to pups born to the dominant female failed to find any consistent relationship between the number of allo-lactators and the growth or survival of the pups (MacLeod and Clutton-Brock 2014).

It seems likely that the evolutionary mechanisms maintaining allo-lactation vary. Where allo-suckling is involuntary, it may occur because the costs of discrimination or prevention exceed the benefits (Pusey and Packer 1994). In contrast, regular suckling by non-breeding females of other females' offspring is probably cooperative and is likely to be maintained either by indirect or mutualistic benefits. As yet there is no evidence that the tendency for females to suckle each other's offspring is contingent on the behaviour of other lactating females and the distribution of regular allo-lactation suggests that it may be maintained by indirect benefits, supplemented by shared direct benefits in species where individual fitness increases with group size (Kokko *et al.* 2001).

Mutual support in competitive interactions

In many mammal societies where females are philopatric and generations overlap, females support their daughters and other close relatives in competitive interactions with other group members (see Chapter 8). Their assistance reduces the chance that offspring will be evicted and helps to establish their rank in the group and mutual support plays an important role in maintaining the rank and breeding success of the female's matriline. It is probably maintained by a combination of direct and indirect fitness benefits.

Though supportive relationships often involve relatives, they can also involve non-kin. For example, in some cercopithecine primates, females commonly develop affiliative relations or alliances both with related females and with unrelated 'friends' (Walters and Seyfarth 1997; Kapsalis 2004; de Waal and Brosnan 2006; Silk 2006). The extent and strength of these relationships can have important consequences for female fitness: for example, in female yellow baboons, individual differences in social connectedness are positively correlated both with individual differences in survival and with the survival of their offspring (Silk 2007; Silk *et al.* 2009; Archie *et al.* 2014). Individuals are often likely to gain more by forming supportive alliances with partners of higher rank than themselves and commonly compete to form supportive

relationships with powerful members of their group (see Chapter 8). In baboons, high-ranking females are more commonly involved in coalitions than low-ranking ones, probably because the costs of assisting allies are lower and the potential benefits are larger (Silk *et al.* 2004), while lower-ranking partners are often more likely to support higher-ranking ones than vice versa (Cheney 1977).

One interpretation of supportive interactions is that they represent examples of reciprocity and involve exchanges of assistance between partners which prevent the evolution of cheating strategies (Trivers 1971; Raihani and Bshary 2011). In some cases, individuals are more likely to support individuals that have assisted or groomed them in the recent past (Hemelrijk and Ek 1991; Cheney *et al.* 2010), as models of reciprocity would predict (Figure 9.15). However, realistic measurements of fitness pay-offs in particular interactions are rarely possible, so it is not clear that particular interactions really involve temporary costs to the fitness of individuals that are subsequently offset by larger benefits when favours are returned. An alternative view is that it is more realistic to measure the costs and benefits of interactions over longer time periods and that, if benefits are accounted in this way, individual supportive actions are likely to generate prospective benefits which exceed any costs involved (Clutton-Brock 2009a). In many social mammals, the fitness of individuals is affected by their relationships with friends and allies, the maintenance of these alliances and friendships generates important benefits (see Chapter 8) and individuals that

Figure 9.15 In female chacma baboons, recent grooming increases the subsequent probability of social support between groomimg parners (Cheney *et al.* 2010). *Photo source:* © Dorothy Cheney.

consistently fail to provide support to allies are likely to be abandoned and their previous partners may form alliances with their rivals and become competitors. As a result, providing support to allies may often represent an individual's best strategy, and it may be unrealistic to assume that providing assistance involves temporary net fitness costs or that defecting is likely to generate benefits in the short term. In situations of this kind, the usual pay-offs of interactions may differ from those usually envisaged by models of the Prisoner's Dilemma and may more closely resemble those in public goods games (see Chapter 1). This need not suggest that individuals would benefit by maintaining cooperative relationships with the same allies indefinitely and they may often be able to enhance their fitness by opportunistic changes of partners, and frequently do so (see Chapter 14).

In species where female group members vary widely in relatedness, like many of the cercopithecine primates, the strength of social bonds and the frequency of affiliative interactions between females and other group members varies widely between individuals and is affected by their age and social status as well as by the number of their immediate kin. These differences are associated with contrasts in the frequency of interactions. For example, in yellow baboons, the frequency of affiliative interactions with other females declines as their age increases and is positively correlated with the size of their families, though it is not consistently related to their social status (Archie *et al.* 2014). In contrast, affiliative interactions with males increase with the female's social status and are consistently related to their age, though individuals with strong social connections to other females have weaker social connections to males.

In some primate species, individuals recognise the existence of bonds between third parties (see Chapter 7). For example, in wild chimpanzees, individuals that have been involved in fights respond more strongly to playbacks of aggressive barks from social partners of their previous opponent than to those of individuals that did not have a close social connection with them (Wittig *et al.* 2014a). Whether individuals also monitor changes in the strength of social bonds between other group members is not yet clear but would appear likely.

Where an individual's success in competitive interactions between group members depends on support from relatives, friends or allies, selection may favour individuals that monitor the strength of social bonds and test their reliability at intervals (Zahavi 1977; Collins 1981, 1993). Bond testing is suggested as a possible explanation of the strange traditional 'games' played between some group members in some monkeys. In some wild groups of capuchins, one individual may place another animal's hand or foot over its face and, with eyes closed, inhale deeply for over a minute (Fedigan 1993; Perry *et al.* 2003). In other cases, pairs of monkeys spend long periods of time sucking each others' toes, ears or tails or inserting their fingers into each others' mouths or ears or even under their eyelids (Perry *et al.* 2003). Some of these 'games' are usually mutual while, in others, partners alternate in roles. 'Games' are usually played repeatedly by the same dyads and are mostly played by a small number of individuals in each group that maintain close social bonds. These games appear to be traditional in particular groups and their occurrence varies between groups. Whether they allow individuals to monitor the reliability of their allies is still unclear but, if so, it is surprising that similar interactions are not more common in other species.

Allo-grooming

Some of the most detailed analyses of cooperative behaviour in mammals have focused on allo-grooming. In many social species, group members spend a substantial amount of time grooming each other, either simultaneously or alternately (Figure 9.16). Allo-grooming serves to remove dirt, flakes of dry skin and parasites (Sparks 1967; Tanaka and Takefushi 1993), grooming frequency and duration vary with parasite abundancy (Mooring 1995; Mooring *et al.* 1996) and groomers often focus their attention on parts of the body that carry heavy parasite loads (Zamma 2002) or on regions that individuals cannot reach themselves (Simpson 1973; Barton 1985). Individuals that live alone, or which lack access to grooming partners for other reasons, commonly show high ectoparasite levels and being groomed reduces parasite load (Akinyi *et al.* 2013) and appears to calm both groomers and groomees, lowering heart rates and raising beta-endorphin levels (Boccia 1987; Schino *et al.* 1988; Keverne *et al.* 1989; Feh and de Mazières 1993). For example, in horses, grooming is concentrated at sites where ganglia lie close beneath the skin and experimental grooming at the same sites reduces the recipient's heart rate while grooming on other parts of the body has little or no effect (Feh and de Mazières 1993).

Figure 9.16 Allo-grooming in (a) Przewalski's horses, (b) grey langurs and (c) meerkats. *Sources*: (a) © Claudia Feh; (b) © Volker Sommer; (c) © Helen Wade.

Unlike many forms of cooperative behaviour, allo-grooming often involves alternating bouts in which individuals both give and receive grooming. It is common both in mammals that live in open social groups as well as in species that live in closed groups. For example, in impala, which live in loosely structured groups whose membership varies from day to day (Jarman 1979), sessions of reciprocal grooming often involve unrelated individuals that are not regular partners and are frequently of different sex and age (Mooring and Hart 1992). One group member initiates a grooming session with another individual by starting to scrape its neck with its incisors (Hart and Hart 1992; Mooring and Hart 1992); the groomer commonly pauses after six to twelve scrapes, and the recipient then grooms the originator a similar number of times. During the course of an allo-grooming session, both partners usually receive approximately the same number of bouts (Figure 9.17).

Among primates, too, allo-grooming sessions are usually divided into a series of alternating sub-bouts involving each partner in turn. This may help to reduce the potential benefits of defection, especially when the benefits of defecting are smaller than the cost of losing subsequent 'parcels' of grooming (Connor 1995). Individuals usually receive approximately the same number of bouts that they give and detailed analysis of grooming interactions shows that partners match the duration as well as the frequency of grooming bouts to those of their partners (Figure 9.18a) (Manson *et al.* 2004), though discrepancies increase when they differ in social rank (Figure 9.18b).

In animals that live in stable groups, some individuals usually groom each other more than others. Network analysis is now commonly used to describe and compare patterns of interaction between individuals within groups and has shown that relationships between

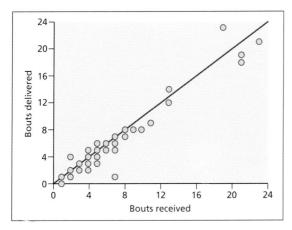

Figure 9.17 Matching of allo-grooming bouts in female impala. *Source*: From Hart and Hart (1992). Reproduced with permission of Elsevier.

individuals are often stable over considerable periods and differ between groups (Croft *et al.* 2008; Madden *et al.* 2009). For example, female baboons have a limited number of regular grooming partners and show considerable fidelity to the same partner (Silk 2012). Regular grooming partners are commonly animals of the same age and sex (Seyfarth and Cheney 1984; Schino 2001) and are often matrilineal relatives (Gouzoules and Gouzoules 1987; Silk 2009). Increased rates of grooming between matrilineal relatives occur partly because members of the same matriline associate with each other but there is still a tendency for females to preferentially groom matrilineal relatives after the frequency of association is allowed for (Chapais 2006; Silk 2009).

In some species, paternal relatedness also appears to affect grooming relationships. Studies of rhesus macaques and baboons show that paternal half-sibs are more likely to interact amicably and to groom each other than unrelated individuals and are less likely to intervene against them in competitive interactions (Widdig *et al.* 2001; Smith *et al.* 2003; Silk 2009). In rhesus macaques, affiliative interactions between paternal half-sibs are less frequent than between maternal half-sibs, while in baboons rates of interaction between paternal and maternal half-sibs are similar in frequency, possibly because the size of matrilines is smaller, so that individuals do not have as many relatives to choose among (Smith *et al.* 2003) (Figure 9.19a). The mechanisms by which individuals discriminate paternal kin are not yet known and could involve visual or olfactory matching of

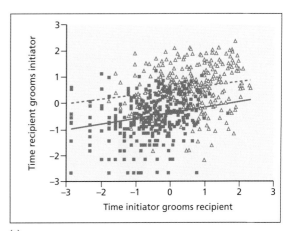

(a)

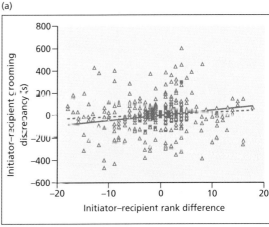

(b)

Figure 9.18 Time-matched grooming in bonnet macaques and white-faced capuchins: (a) duration of grooming by recipients plotted on the duration of grooming by the initiator in bonnet-faced macaques; (b) differences between the grooming times of initiators and recipients as a function of differences in their dominance ranks. *Source*: From Manson *et al.* (2004). Reproduced with permission of Elsevier.

phenotypes (either to the father's phenotype or to their own) though, especially in natural populations, it is usually difficult to exclude the possibility that they are based on familiarity (Potts *et al.* 1991; Parr and de Waal 1999; Widdig 2007). For example, in yellow baboons, paternal sibs are commonly closely matched for age (Figure 9.19b) and this is likely to contribute to increases in association and to rates of interaction (Smith *et al.* 2003).

Grooming is often used by subordinates to placate dominant individuals who are likely to react aggressively

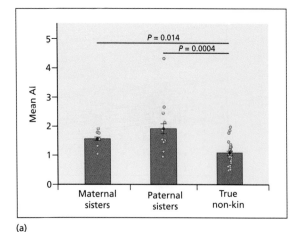

(a)

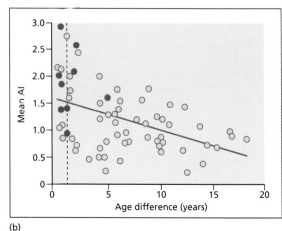

(b)

Figure 9.19 Nepotism in baboons. (a) Relative frequency of affiliative interactions between maternal sisters, paternal sisters and non-kin in wild female baboons. Each open circle represents the mean affiliation index (mean AI, based on five separate measures of affiliation) for one female towards all females within the given kin category; closed circles are means. (b) Effects of age differences on affiliation between non-kin (light blue circles) and paternal sisters (dark blue circles). *Sources*: (a) From Smith *et al.* (2003). Reproduced with permission from the Royal Society. *Photo source*: © Anne Engh.

towards them. For example, in Kalahari meerkats, older subordinate females that become a target of aggression from the dominant female and are likely to be evicted from the group increase the frequency with which they groom her (Kutsukake and Clutton-Brock 2006) and experimental reductions of parasite load (which reduce grooming frequency) lead to increased rates of aggression (Madden and Clutton-Brock 2009).

Grooming may also help to establish and maintain affiliative relationships between individuals which serve to protect them against competition with other group members: in several primates, recent grooming affects the probability that partners will engage in other forms of affiliative behaviour, including the probability that they

will tolerate each other at feeding sites and support each other in competitive interactions with other group members (Weisbard and Goy 1976; Kummer 1978; Henzi and Barrett 1999; Barrett *et al.* 2002; Cheney *et al.* 2010). For example, captive chimpanzees are more likely to share food with individuals that have recently groomed them (de Waal 1997), while in macaques, baboons and vervet monkeys, recent grooming is often associated with increases in the probability that individuals will support each other (Seyfarth and Cheney 1984; Hemelrijk 1994). In some species, these effects are stronger where grooming partners are unrelated than where they are close relatives: for example, in chacma baboons, recent grooming with unrelated individuals increases the chance that females will respond to their recruitment calls while recent grooming with relatives has no similar effect (Seyfarth and Cheney 1984; Cheney *et al.* 2010) (Figure 9.20).

Individuals also compete to groom higher ranking individuals and may intervene to disrupt the development of grooming alliances between potential competitors. For example, in bonnet macaques, females groom higher ranking partners more frequently than lower ranking ones and also compete to groom them more frequently (see Figure 9.21). As dominants are likely to be disadvantaged if lower-ranking animals form alliances against them, dominant females actively try to prevent grooming interactions between subordinates (Cheney and Seyfarth 1990; Silk 1982; de Waal and Luttrell 1986; Harcourt 1992; Kutsukake and Clutton-Brock 2006), and studies of vervet monkeys show that the frequency with which females disrupt grooming pairs

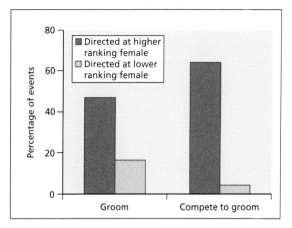

Figure 9.21 Competition for dominant allies in female bonnet macaques. Histograms show the percentage of interactions involving grooming and supplanting from grooming (compete to groom) by ten adult females that involved higher-ranking females (dark bars) or lower-ranking females ('compete to groom', pale bars) than the subject. (Histograms do not sum to 100 because the focal female groomed other group members besides adult females.) *Source*: Adapted from Harcourt (1992) using data from Silk 1982.

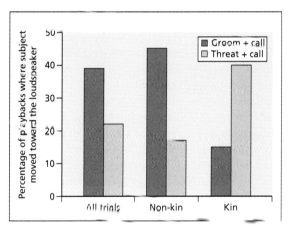

Figure 9.20 Contingent responses to recruitment calls by related and unrelated pairs of wild female baboons in an experiment designed to mimic a situation in which the former grooming partner was threatening another individual and soliciting aid. In the test condition (groom and call) female baboons that had recently exchanged grooming with an unrelated subordinate female were played the threat grunts of a lower-ranking female. Controls (threat and call) were similar experiments conducted on pairs of females where one had previously threatened the other. The figure shows that non-kin were substantially more likely to move towards the playback speaker if they had previously groomed with the caller than if they had previously threatened her while the situation was reversed where members of pairs were related. *Source*: From Cheney *et al.* (2010). Reproduced with permission of the National Academy of Sciences, USA.

depends on the rank of the individual being groomed (Cheney and Seyfarth 1990).

Where individuals compete to form regular grooming relationships, 'market' models that explore the effects of variation in supply and demand often predict how the distribution of grooming is likely to vary in relation to variation in supply and demand. For example, female baboons often compete to gain access to juveniles born to other females by grooming their mothers and, when infants are scarce, females groom their mothers for relatively long periods to gain access to the infant (Henzi and Barrett 2002) (Figure 9.22). Market theory also predicts the distribution of imbalances in grooming time. Where subordinates compete to groom higher-ranking animals, they may induce higher-ranking individuals to allo-groom with them by being less demanding about the extent of reciprocation (Manson *et al.* 2004) with the result that high-ranking individuals obtain a disproportionate amount of grooming relative to the amount that they give (see Figures 9.18 and 9.21).

Ecological conditions can affect the extent of grooming imbalances between dominants and subordinates. In chacma baboons, the extent of discrepancies in grooming

Figure 9.22 When 'black' infants are scarce, female baboons commonly groom their mothers for long periods to obtain access to them. *Source*: © Anne Engh.

between dominants and subordinates increases as the benefits that subordinates are likely to gain from the tolerance or support of dominants rises. Subordinates that groom dominant individuals are more likely to tolerate asymmetries in grooming interactions where tolerance at feeding sites is important because food is short or is distributed in clumps, whereas the contributions of dominants and subordinates are more closely matched when resources are more plentiful (Barrett *et al.* 1999, 2002) (Figure 9.23). However, effects of this kind are not universal and can vary within as well as between species. For example, while vervet monkeys in a Kenyan population responded to food shortage and adverse environmental conditions with increased levels of aggression and a reinforcement of dominance relations, vervets in a South African population did not show the same responses, possibly as a consequence of contrasts in the demographic structure of groups (Henzi *et al.* 2013). These comparisons emphasise the variability of social responses to competition for resources and the wide range of social factors that may influence them.

Policing

In humans, individuals frequently punish individuals whose actions affect the welfare of others or break social conventions (see Chapter 20). Actions of this kind are referred to as *third-party punishment* and play an important role in maintaining human cooperation.

Non-human animals commonly punish individuals whose activities affect their own fitness (see Chapter 8) and will also respond to aggression directed at relatives or friends by threatening or attacking the aggressor(s) (see Chapter 8). In addition, in some primates, dominant males may intervene in aggressive interactions between females that are likely to disrupt the group (see Chapter 15). However, in contrast to humans, non-human animals seldom respond to actions by other group members that are likely to affect the fitness of third parties unless victims are kin or regular allies and regular third-party punishment appears to be a derived trait that is largely confined to human societies (see Chapter 20). For example, captive chimpanzees do not punish individuals that steal food from third parties, even if they are close relatives (Riedl *et al.* 2012).

The causal basis of cooperation

The physiological mechanisms involved in the control of cooperative behaviour include sex hormones (especially testosterone and oestrogen) and stress hormones (especially cortisol) (see Chapter 17). The expression of many forms of behaviour necessary for survival and reproduction (including eating, drinking and copulating) are reinforced by 'natural rewards' (Kelley and Berridge 2002) and there is increasing evidence that social interactions may act as rewards and reinforce the expression of cooperative or affiliative behaviour via the effects on the production of oxytocin and serotonin or their receptors (Ross *et al.* 2009; Crockford *et al.* 2013; Dölen *et al.* 2013). For example, in wild chimpanzees, oxytocin levels increase after bouts of allo-grooming as well as after individuals have shared food (Crockford *et al.* 2013; Wittig *et al.* 2014b). In addition, in cooperative breeders, there is evidence that individual contributions to cooperative activities can be positively as well as negatively related to corticosteroid levels and these effects may be sex-specific (see Chapter 17). As yet, there is no integrated understanding of the causal basis of cooperative behaviour in mammals and it is not yet clear whether different forms of cooperation are under the control of different systems or whether different hormonal systems interact to affect cooperation. Chapter 17 returns to this issue and reviews research on the mechanisms controlling contributions to cooperative activities in cooperative breeders.

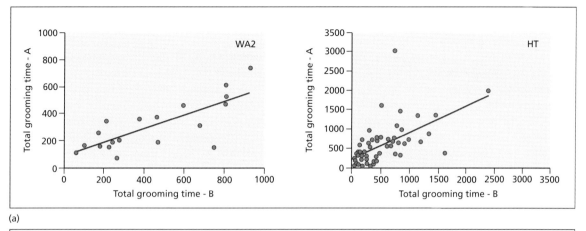

(a)

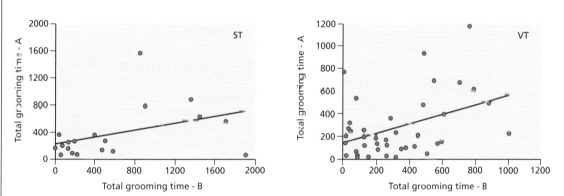

(b)

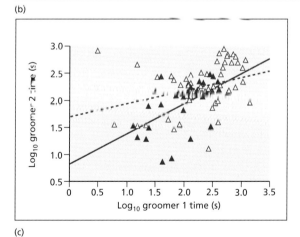

(c)

Figure 9.23 Time matching between grooming partners in allo-grooming sessions in chacma baboons. (a) In two troops of baboons in the Drakensberg Mountains where food was widely distributed (WA2, HT, rates of aggression between females were low and female dominance relationships were not obvious and grooming partners contributed similar amounts of grooming per session whereas (b) in two troops in the Western Cape, where food was more localised (ST, VT), rates of aggression between females were higher, female dominance relationships were well defined and relative contributions to grooming were less equal. (c) Within the population from the Western Cape, inequalities in grooming contributions increased at times when competition for food was most intense and subordinate females gained greater benefits from tolerance by dominants (high competition period indicated by open triangles, dotted line; low competition period indicated by closed triangles, solid line). *Sources*: (a,b) From Barrett *et al.* (1999). Reproduced with permission from the Royal Society. (c) From Barrett *et al.* (2002). Reproduced with permission of Elsevier.

9.3 Cheating in theory and practice

A central issue in understanding the evolution of cooperation is to explain why its benefits are not eroded by
cheating strategies that exploit the benefits provided by
cooperators but avoid the costs of providing reciprocal
assistance (see Chapter 1). In the context of evolution,
cheats (or cheating strategies) are maintained by selection operating through benefits derived from accepting
assistance from cooperators and providing less (or no)
assistance in return. Cheating strategies may involve
deception (and evidence of deception would be a strong
indicator of cheating) but need not do so (Ghoul *et al.*
2014).

So how common is cheating? In many animal societies, some individuals contribute more to cooperative
activities than others (Bergmüller *et al.* 2010). For example, some lionesses consistently hang back in encounters
between prides (Heinsohn and Packer 1995; Legge
1996), female baboons and macaques differ in the probability that they will provide each other with grooming or
support in competitive interactions (Silk 1982; Hauser
et al. 2009) and meerkat helpers vary in the extent to
which they contribute to provisioning young (see
Chapter 17). Differences of this kind are often cited as
evidence of cheating, but an alternative interpretation is
that they are a consequence of individual differences in
the costs or benefits of providing assistance (see
Chapter 1). Where they have been investigated, they
have commonly been found to be closely associated with
individual differences in the costs or benefits of providing
assistance or with phenotypic traits that affect these
(including variation in age, sex, condition and reproductive status). In addition, they often vary through time
within individuals (Russell *et al.* 2003), suggesting that
they may well be a consequence of variation in the costs
or benefits of providing assistance. This emphasises the
point that variation in cooperative behaviour should not
be taken to indicate that cheating is involved unless some
individuals consistently contribute less to cooperative
behaviour *relative to the costs or benefits they derive from
providing assistance* (see Chapter 1). For example, we
would not regard individuals that selectively help close
kin more than unrelated animals as cheating and it is
misleading to refer to individuals that help less because
helping has higher costs to their fitness as cheating either.
Unfortunately, it is usually difficult to be certain that

contrasts in cooperative behaviour are not associated
with variation in the costs or benefits of providing
assistance.

Since it is usually difficult to distinguish between
cheating and the consequences of optimal variation in
investment in cooperative behaviour, empirical studies
have investigated the extent of deception involving cooperative behaviour. Although several models of cooperative behaviour suggest that individuals should attempt to
deceive their partners or other group members over the
extent of their contributions to cooperative activities
(Hirshleifer and Rasmusen 1989; Pollock and Dugatkin
1992; Hammerstein 2003), there is little firm evidence of
deceptive strategies of this kind.

One potential example of deceptive cooperation that
has been investigated in detail is a phenomenon known
as 'false feeding' that occurs in a number of cooperative
birds and mammals where a foraging parent or helper
catches a food item, brings it to dependent young and
then, instead of presenting it to the young, eats it itself.
While one explanation is that 'false feeders' are attempting to deceive their mates or other group members over
the extent to which they are contributing to cooperative
activities (Boland *et al.* 1997), an alternative possibility is
that individuals cannot assess the needs of dependent
young until they are close to them and that false feeding
occurs where individuals decide not to give a food item to
young after they have evaluated the needs of dependants
relative to their own (Canestrari *et al.* 2004). Studies of
false feeding in meerkats (as well as in some birds)
support this second interpretation and show that, in
most cases where false feeding occurs, adults bringing
food initially approached a juvenile that was already
eating or was no longer giving regular begging calls
and do not suggest that any attempt is made to deceive
other group members (Canestrari *et al.* 2004; Clutton-
Brock *et al.* 2005).

Consistent cheats and deceivers could be uncommon
because they are likely to be detected and punished in
dyadic interactions with other group members (Clutton-
Brock and Parker 1995). Punishment is commonly used
to maintain and reinforce dominance relationships in
mammals (see Chapter 8) and is sometimes used to
encourage partners to respond in dyadic interactions.
For example, gorillas sometimes give threatening vocalisations to individuals that have just stopped grooming
them, apparently encouraging them to start again

(Harcourt 1992) while chimpanzees may threaten regular allies that withhold their support (de Waal 1982) or individuals that steal their food (Jensen *et al.* 2007; Riedl *et al.* 2012). However, as group members tolerate large differences in the extent to which individuals contribute to cooperative activities and lazy individuals are rarely punished or threatened, it seems unlikely that punishment plays a major role in maintaining cooperation in mammals (Santema and Clutton-Brock 2012; Riehl and Frederickson 2016).

While unequivocal evidence of cheating strategies is uncommon, there are substantial difficulties in demonstrating that individual differences in cooperative behaviour are not the result of variation in costs or benefits, so that the lack of firm evidence does not necessarily suggest that cheating strategies are rare or that selection favouring cheating is weak. Here, studies of the heritability of variation in contributions to cooperative behaviours may be useful (see Chapter 17). Moreover, even if cheating is uncommon in current populations, this need not indicate that its effects on the evolution of cooperation have been unimportant and the risks of exploitation by cheating strategies may explain why cooperation between unrelated animals is seldom highly developed unless the costs of cooperative behaviour are low or cooperation generates immediate benefits to both parties.

9.4 The evolution of cooperation

Existing studies of cooperation in mammals provide a basis for assessing the role of different evolutionary mechanisms in maintaining cooperative behaviour (see Chapter 1). If cooperation is defined as behaviour adapted to increasing the fitness of other individuals (see Chapter 1), it is clear that most forms of cooperation (and especially those that are energetically costly or risky) are more highly developed where group members are closely related, providing support for the view that kin selection has played a central role in the evolution of cooperation (Abbot *et al.* 2011; Briga *et al.* 2012). In addition, many forms of dyadic cooperation that generate direct benefits to both parties, such as cooperative hunting, allo-grooming or the provision of support in competitive interactions within the group, are also more frequent in groups where most females are related (Clutton-Brock *et al.* 2006).

There is also widespread evidence of nepotism (the selection assistance of kin) within groups. For example,

in many of the cercopithecine primates where females live in large groups that include several matrilines, members of the same matriline selectively support each other in competitive interactions with members of other matrilines (Kurland 1977; Chapais and Berman 2004; Silk 2007, 2009). In some species, affiliative and cooperative interactions are also more frequent between patrilineal relatives than would be expected by chance (Widdig *et al.* 2001, 2006; Smith *et al.* 2003; Silk *et al.* 2006; Widdig 2007). Where most group members are related, cooperators often adjust their behaviour to the needs of recipients (see Chapter 17), while they seldom do so when interacting with unrelated animals and often appear to be indifferent to their welfare or survival (Silk 2005, 2006).

As might be expected, nepotism is more highly developed where the costs of cooperative behaviour are high or benefits are unequal and is weaker in interactions like allo-grooming that generate benefits to both participants (Schino and Aureli 2010). In addition, it appears to be strongest in societies where coefficients of kinship vary widely between group members and is often weak in societies where all group members are close relatives (see Chapter 17): it may be for this reason that many of the best examples of the selective assistance of relatives (Silk 2007, 2009) come from species that live in large groups where relatedness varies widely between group members and average relatedness is low.

While cooperation is most highly developed in societies where some individuals are close kin, it is by no means restricted to relatives and often occurs between unrelated group members as well as between members of different species (Trivers 1971; Clutton-Brock 2009a; Schino and Aureli 2010). In some cases, the selection pressures maintaining the behaviour of individuals may be unaffected by the consequences of their behaviour for others and behaviour that appears to be cooperative may be best explained as an example of by-product mutualism (see Chapter 1). Some examples of social hunting and social defence probably fall in this category and, even when animals hunt cooperatively (as in African wild dogs and some dolphins), close analysis of the hunting behaviour of individuals would be likely to show that 'cooperative' hunts probably consist of a mixture of cooperative and selfish actions.

Other social interactions between unrelated or distantly related individuals, including many examples of affiliative or supportive behaviour, appear to meet the criteria of cooperation (see Chapter 1). Models of reciprocity often interpret cooperative interactions as exchanges and assume that cooperative behaviour involves an initial cost as a result of providing assistance that is subsequently offset by larger benefits when their partners(s) provide them with assistance in turn. In these models, cooperation is preserved from erosion by cheating strategies because cooperators detect cheats and selectively direct cooperative behaviour at individuals (or groups of individuals) that are likely to reciprocate. Several different processes may be involved: individuals may adjust their responses to previous interactions with the same partner ('direct' or 'cost-counting' reciprocity), to observation of their partner's previous behaviour towards third parties ('indirect' reciprocity) or to previous interactions with any member of a group of individuals ('generalised' reciprocity).

Alternatively, cooperation between unrelated individuals can be maintained if cooperation generates immediate shared benefits and individuals that defect or cheat reduce their own fitness by failing to cooperate. Situations of this kind are modelled in public goods games (see Chapter 1). Although models of these processes do not necessarily incorporate conditional responses to the past behaviour of partners, there is no reason to suppose that these may not be involved and selective cooperation would be likely to increase mutualistic benefits (Pfeiffer *et al.* 2005; Clutton-Brock 2009a; Leimar and Hammerstein 2010; Taborsky *et al.* 2016).

The essential difference between explanations of cooperation based on reciprocity and those based on immediate shared benefits is consequently whether there is a delay between the costs suffered by cooperators and the benefits they accrue when their partner(s) reciprocate (as models of reciprocity based on the Prisoner's Dilemma assume) or whether cooperators gain immediate benefits that exceed the costs of defecting. In many cases where unrelated individuals assist each other, they appear likely to gain immediate fitness benefits from their behaviour: examples include many cases of cooperative hunting or foraging as well as the use of coalitions in competitions over resources or reproductive opportunities. Their actions may only generate realised fitness benefits (such as improvements in breeding success or survival) months or even years later, but attempts to assess fitness benefits should be prospective and should incorporate probable effects.

Instead of interpreting cooperative interactions between unrelated animals as exchanges, it may often be useful to view them as attempts by individuals to alter or maintain their social niches (Flack *et al.* 2006; Bergmüller and Taborsky 2007; Montiglio *et al.* 2013). In many group-living animals, the influence of social parameters (such as group size, group composition and the strength and extent of long-lasting social relationships) on individual fitness may often be at least as important as the influence of physical parameters (see Chapter 8). Just as individuals attempt to modify their physical environment by digging burrows or building nests, they commonly attempt to modify their social environment by joining or leaving social groups, evicting potential competitors or interfering with their breeding attempts. The development and maintenance of long-term cooperative relationships both with kin and with non-kin may often represent an extension of this strategy which helps individuals to protect themselves from competition with other group members and allows them to promote their own status and breeding success.

9.5 Consequences of cooperation

The development of cooperation between group members has extensive evolutionary and ecological consequences. Cooperative detection of predators and defence against attacks help to protect individuals and to increase survival. The cooperative construction of burrows and nests, cooperative hibernation and cooperative food storage serve to shelter individuals from the effects of climatic variation and cooperative foraging and hunting expand the range of resources that can be utilised. As a result, cooperation can permit species to colonise and explore ecological niches and environments that would not be accessible to solitary species. For example, many rodents living at high altitudes or latitudes cooperate to dig and maintain communal burrows and hibernate together to preserve heat (Arnold 1990); sentinel behaviour is most frequent in diurnal species living in open environments where predation is common (Clutton-Brock *et al.* 1999c); and cooperative hunting is most highly developed in species that do not hunt by stealth and prey on animals larger than themselves (Caro 1994; Creel and Creel 2002).

Cooperation also plays an important role in protecting individuals from competition between and within groups. In many species, cooperative defence and cooperative displays are involved in the defence of feeding or breeding territories (see Chapter 2). Within groups, cooperation provides protection against competitors as well as opportunities to exploit others. In some species, females compete to conceive or to raise their offspring while competitors attempt to interfere with their breeding attempts or kill their offspring (see Chapter 8). Supportive relationships and alliances between females also play an important role in establishing the social rank of individuals and of matrilines and in preventing younger females from being evicted from the group by older ones.

Cooperative relationships between group members can have far-reaching consequences for the evolution of dispersal. In particular, the cooperative defence of group territories often limits the capacity of individuals that disperse on their own either to join established breeding groups or to found new breeding units and is likely to favour the evolution of social dispersal (see Chapter 3). Where larger parties are more likely to be able to join established breeding groups or found new ones and party size increases with group size, this may in turn favour the evolution of behaviour that contributes to maintaining or increasing group size.

Cooperation between group members can also have important demographic consequences. In species that depend on cooperation to feed or breed, cooperation may initially generate positive correlations between group size and the survival or breeding success of individuals. As group size and population density are often positively correlated, associations between group size and survival or breeding success may lead to negative density-dependence (Allee effects) which can in turn have important consequences for population stability and risk of local extinction (see Chapter 17).

SUMMARY

1. While competition between group members is (almost) inevitable, cooperative behaviour is less easily explained and represents a potential challenge to evolutionary theory. Mechanisms favouring the evolution of cooperation between females include mutualistic benefits shared by cooperators (including immediate shared benefits of the kind modelled by public goods games and direct, indirect and generalised reciprocity) and kin selection. In addition, cooperative behaviour may sometimes be induced by harassment or by coercive strategies.
2. Female mammals cooperate in a wide range of contexts. Female group members commonly assist each other in mobbing predators and in signalling their presence, in defending territories or breeding sites against neighbouring groups, in catching prey and storing food and in raising young. In some cases, successful foragers share food with unsuccessful individuals, who can include unrelated individuals as well as relatives. In larger groups, particular group members may form coalitions or alliances against other group members and maintain them by regular affiliative interactions, including social grooming.
3. The causal mechanisms underlying cooperative behaviour are not yet well understood. Variation in sex hormones, cortisol and peptide hormones (including oxytocin and serotonin) may all be involved.
4. Although theoretical studies suggest that individuals should attempt to take advantage of cooperators by withholding or minimising the assistance they provide, empirical evidence of 'cheating' strategies is scarce. Individuals often differ in their contributions to cooperative behaviour, many of these differences are probably consequences of individual differences in the costs or benefits of cooperation.
5. Most forms of cooperation between females in social mammals typically involve related individuals who are familiar with each other. Where unrelated individuals cooperate, they usually appear to gain shared benefits from assisting each other. Though these are sometimes regarded as reciprocal exchanges involving an alternation of costs and benefits, it may be more realistic to measure the fitness consequences of providing assistance over longer time scales and to regard cooperative interactions between unrelated individuals as components of mutualistic relationships that provide immediate benefits to both partners rather than as sequential exchanges of fitness.
6. In some cases, the evolution of cooperative rearing can allow mammals to occupy habitats that they could not otherwise live in or to exploit resources that they could not otherwise use. Reliance on cooperation may also affect many aspects of population dynamics and demography. In particular, it may reduce the chances of successful dispersal by solitary individuals and encourage the evolution of social dispersal. In extreme cases, it may generate positive correlations between group size or local population density and the fitness of individuals, leading to Allee effects at the group level.

References

Abbot, P., *et al.* (2011) Inclusive fitness theory and eusociality. *Nature* **471**:E1–E4.

Akcay, E., and Van Cleve, J. (2016) There is no fitness but fitness and the lineage is the bearer. *Philosophical Transactions of the Royal Society B.* DOI: 10.1098/rstb.2015.0085. [Accessed 18 January 2016].

Akinyı, M.Y., *et al.* (2013) Role of grooming in reducing tick load in wild baboons (*Papio cynocephalus*). *Animal Behaviour* **85**:559–568.

Amici, F., *et al.* (2014) Lack of prosociality in great apes, capuchin monkeys and spider monkeys: convergent evidence from two different food distribution tasks. *Proceedings of the Royal Society of London. Series B: Biological Sciences* 281: 20141699

Archie, E.A., *et al.* (2014) Social affiliation matters: both same-sex and opposite-sex relationships predict survival in wild female baboons. *Proceedings of the Royal Society of London. Series B: Biological Sciences* 281: 20141261

Arnold, W. (1990) The evolution of marmot sociality: II. Costs and benefits of joint hibernation. *Behavioral Ecology and Sociobiology* **27**:239–246.

Baird, R.W. (2000) The killer whale: foraging specializations and group hunting. In: *Cetacean Societies: Field Studies of Dolphins and Whales* (eds J. Mann, R. Connor, P. Tyack and H. Whitehead) Chicago: University of Chicago Press, 127–153.

Barrett, L., *et al.* (1999) Market forces predict grooming reciprocity in female baboons. *Proceedings of the Royal Society of London. Series B: Biological Sciences* **266**:665–670.

Barrett, L., *et al.* (2002) A dynamic interaction between aggression and grooming reciprocity among female chacma baboons. *Animal Behaviour* **63**:1047–1053.

Barton, R. (1985) Grooming site preferences in primates and their functional implications. *International Journal of Primatology* **6**:519–532.

Bednekoff, P.A. (1997) Mutualism among safe, selfish sentinels: a dynamic game. *American Naturalist* **150**:373–392.

Bennett, N.C. and Faulkes, C.G. (2000) *African Mole-rats: Ecology and Eusociality*. Cambridge: Cambridge University Press.

Bergmüller, R. and Taborsky, M. (2007) Adaptive behavioural syndromes due to strategic niche specialization. *BMC Ecology* **7**:12.

Bergmüller, R., *et al.* (2010) Evolutionary causes and consequences of consistent individual variation in cooperative behaviour. *Philosophical Transactions of the Royal Society B: Biological Sciences* **365**:2751–2764.

Best, P.B. (1979) Social organisation of sperm whales, Physeter macrocephalus. In: *Behaviour of Marine Animals: Current Perspectives in Research, Vol. 3 Cetaceans* (eds H.E. Winn and B.C. Olla) New York: Plenum Press, 227–289.

Boccia, M.L. (1987) The physiology of grooming: a test of the tension reduction hypothesis. *American Journal of Primatology* **12**:330–338.

Boesch, C. (1994a) Hunting strategies of Gombe and Tai chimpanzees. In: *Chimpanzee Cultures* (eds W.C. McGrew, F.B.M. de Waal, R.W. Wrangham and P. Heltne) Cambridge, MA: Harvard University Press, 77–91.

Boesch, C. (1994b) Cooperative hunting in wild chimpanzees. *Animal Behaviour* **48**:653–667.

Boesch, C. (1994c) Chimpanzees–red colobus monkeys: a predator–prey system. *Animal Behaviour* **47**:1135–1148.

Boesch, C. and Boesch, H. (1989) Hunting behavior of wild chimpanzees in the Taï National Park. *American Journal of Physical Anthropology* **78**:547–573.

Boesch, C. and Boesch-Achermann, H. (2000) *The Chimpanzees of the Taï Forest: Behavioural Ecology and Evolution*. Oxford: Oxford University Press.

Boesch, C., *et al.* (2006) Cooperative hunting in chimpanzees: kinship or mutualism? In: *Cooperation in Primates and Humans: Evolution and Mechanisms* (eds P.M. Kappeler and C.P. van Schaik) Berlin: Springer, 139–150.

Boland, C.R.J., *et al.* (1997) Deception by helpers in cooperatively breeding white-winged choughs and its experimental manipulation. *Behavioral Ecology and Sociobiology* **41**:251–256.

Boness, D.J. (1990) Fostering behavior in Hawaiian monk seals: is there a reproductive cost? *Behavioral Ecology and Sociobiology* **27**:113–122.

Boydston, E.E., *et al.* (2001) Sex differences in territorial behaviour exhibited by the spotted hyena (Hyaenidae: *Crocuta crocuta*). *Ethology* **107**:369–385.

Briga, M., *et al.* (2012) Care for kin: within-group relatedness and allomaternal care are positively correlated and conserved throughout the mammalian phylogeny. *Biology Letters* **8**:533–536.

Brosnan, S.F. (2006) Nonhuman species' reactions to inequity and their implications for fairness. *Social Justice Research* **19**:153–185.

Brosnan, S.F. (2013) Conflicts in cooperative social interactions in nonhuman primates. In: *War, Peace, and Human Nature* (ed. D.P. Fry). Oxford: Oxford University Press, 406–420.

Busse, C.D. (1978) Do chimpanzees hunt cooperatively? *American Naturalist* **112**:767–770.

Caldwell, D.K., *et al.* (1966) Behavior of the sperm whale, Physeter catodon L. In: *Whales, Dolphins and Porpoises* (ed. K.S. Norris) Berkeley: University of California Press, 677–717.

Canestrari, D., *et al.* (2004) False-feedings at the nests of carrion crows *Corvus corone corone*. *Behavioral Ecology and Scoiobiology* **55**:477–483.

Caraco, T. and Brown, J.L. (1986) A game between communal breeders: when is food-sharing stable. *Journal of Theoretical Biology* **118**:379–393.

Caro, T.M. (1994) *Cheetahs of the Serengeti Plains: Group Living in an Asocial Species*. Chicago: University of Chicago Press.

Carter, G.G. and Wilkinson, G.S. (2013a) Does food sharing in vampire bats demonstrate reciprocity? *Communicative and Integrative Biology* **6**:e25783.

Carter, G.G. and Wilkinson, G.S. (2013b) Food sharing in vampire bats: reciprocal help predicts donations more than relatedness or harassment. *Proceedings of the Royal Society of London. Series B: Biological Sciences* 280: 20122573

Chapais, B. (2006) Kinship, competence and cooperation in primates. In: *Cooperation in Primates and Humans: Evolution and Mechanisms* (eds P.M. Kappeler and C.P. van Schaik) Berlin: Springer, 47–64.

Chapais, B. and Berman, C.M. (2004) *Kinship and Behavior in Primates*. Oxford: Oxford University Press.

Chen, M.K. and Santos, L.R. (2006) Some thoughts on the adaptive function of inequity aversion: an alternative to Brosnan's social hypothesis. *Social Justice Research* **19**:201–207.

Cheney, D.L. (1977) The acquisition of rank and the development of reciprocal alliances among free-ranging immature baboons. *Behavioral Ecology and Sociobiology* **2**:303–318.

Cheney, D.L. and Seyfarth, R.M. (1990) *How Monkeys See the World: Inside the Mind of Another*. Chicago: University of Chicago Press.

Cheney, D.L., *et al.* (2010) Contingent cooperation between wild female baboons. *Proceedings of the National Academy of Sciences of the United States of America* **107**:9562–9566.

Clutton-Brock, T.H. (2006) Cooperative breeding in mammals. In: *Cooperation in Primates and Humans: Evolution and Mechanisms* (eds P.M. Kappeler and C.P. van Schaik) Berlin: Springer, 173–190.

Clutton-Brock, T.H. (2009a) Cooperation between non-kin in animal societies. *Nature* **462**:51–57.

Clutton-Brock, T.H. (2009b) Structure and function in mammalian societies. *Philosophical Transactions of the Royal Society B: Biological Sciences* **364**:3229–3242.

Clutton-Brock, T.H. and Manser, M.B. (2015) Kalahari meerkats. In: *Cooperative Breeding: Studies of Ecology, Evolution and Behaviour* (eds W.D. Koenig and J.L. Dickinson) Cambridge: Cambridge University Press.

Clutton-Brock, T.H. and Parker, G.A. (1995) Punishment in animal societies. *Nature* **373**:209–216.

Clutton-Brock, T.H., *et al.* (1999a) Selfish sentinels in cooperative mammals. *Science* **284**:1640–1644.

Clutton-Brock, T.H., *et al.* (1999b) Reproduction and survival of suricates (*Suricata suricatta*) in the southern Kalahari. *African Journal of Ecology* **37**:69–80.

Clutton-Brock, T.H., *et al.* (1999c) Predation, group size and mortality in a cooperative mongoose, *Suricata suricatta*. *Journal of Animal Ecology* **68**:672–683.

Clutton-Brock, T.H., *et al.* (2002) Evolution and development of sex differences in cooperative behavior in meerkats. *Science* **297**:253–256.

Clutton-Brock, T.H., *et al.* (2004) Behavioural tactics of breeders in cooperative meerkats. *Animal Behaviour* **68**:1029–1140.

Clutton-Brock, T.H., *et al.* (2005) 'False-feeding' and aggression in meerkat societies. *Animal Behaviour* **69**:1273–1284.

Clutton-Brock, T.H., *et al.* (2006) Intrasexual competition and sexual selection in cooperative mammals. *Nature* **444**:1065–1068.

Collins, R. (1981) On the microfoundations of macrosociology. *American Journal of Sociology* **86**:984–1014.

Collins, R. (1993) Emotional energy as the common denominator of rational action. *Rationality and Society* **5**:203–230.

Connor, R.C. (1995) The benefits of mutualism: a conceptual framework. *Biological Reviews* **70**:427–457.

Connor, R.C., *et al.* (2000) The bottlenose dolphin: social relationships in a fission–fusion society. In: *Cetacean Societies: Field Studies of Dolphins and Whales* (eds J. Mann, R. Connor, P. Tyack and H. Whitehead). Chicago: University of Chicago Press, 91–125.

Creel, S. and Creel, N.M. (2002) *The African Wild Dog: Behavior, Ecology, and Conservation*. Princeton, NJ: Princeton University Press.

Crockford, C., *et al.* (2013) Urinary oxytocin and social bonding in related and unrelated wild chimpanzees. *Proceedings of the Royal Society of London. Series B: Biological Sciences* **280**:20122765.

Croft, D.B., *et al.* (2008) *Exploring Animal Social Networks*. Princeton, NJ: Princeton University Press.

Curio, E. (1978) The adaptive significance of avian mobbing. 1. Teleonomic hypotheses and predictions. *Zeitschrift fur Tierpsychologie* **48**:175–183.

de Waal, F.B.M. (1982) *Chimpanzee Politics: Power and Sex Among Apes*. New York: Harper & Row.

de Waal, F.B.M. (1996) *Good Natured: The Origins of Right and Wrong In Humans and Other Animals*. Cambridge, MA: Harvard University Press.

de Waal, F.B.M. (1997) The chimpanzee's service economy: food for grooming. *Evolution of Human Behaviour* **18**:375–386.

de Waal, F.B.M. and Brosnan, S.F. (2006) Simple and complex reciprocity in primates. In: *Cooperation in Primates and Humans: Evolution and Mechanisms* (eds P.M. Kappeler and C.P. van Schaik) Berlin: Springer, 85–105.

de Waal, F.B.M. and Luttrell, L.M. (1986) The similarity principle underlying social bonding among female rhesus monkeys. *Folia Primatologica* **46**:215–234.

de Waal, F.B.M. and Suchak, M. (2010) Prosocial primates: selfish and unselfish motivations. *Philosophical Transactions of the Royal Society B: Biological Sciences* **365**:2711–2722.

Dölen, G., *et al.* (2013) Social reward requires coordinated activity of the nucleus accumbens oxytocin and serotonin. *Nature* **501**:179–184.

Dublin, H.T. (1983) Cooperation and reproductive competition among female African elephants. In: *Social Behaviour of Female Vertebrates* (ed. S.K. Wasser). New York: Academic Press, 291–313.

Dugatkin, L.A. (1997) *Cooperation Among Animals: An Evolutionary Perspective*. Oxford: Oxford University Press.

Dugatkin, L.A. and Godin, J.G.J. (1992) Prey approaching predators: a cost–benefit perspective. *Annales Zoologici Fennici* **29**:233–252.

Durant, S.M., *et al.* (2004) Factors affecting life and death in Serengeti cheetahs: environment, age and sociality. *Behavioral Ecology* **15**:11–22.

East, M.L. and Hofer, H. (1991) Loud calling in a female-dominated mammalian society: II. Behavioural contexts and functions of whooping of spotted hyaenas, *Crocuta crocuta*. *Animal Behaviour* **42**:651–669.

Fairbanks, L.A. (1990) Reciprocal benefits of allomothering for female vervet monkeys. *Animal Behaviour* **40**:553–562.

Fairbanks, L.A. and McGuire, M.T. (1984) Determinants of fecundity and reproductive success in captive vervet monkeys. *American Journal of Primatology* **7**:27–38.

Fedigan, L.M. (1993) Sex differences and intersexual relations in adult white-faced capuchins (*Cebus capucinus*). *International Journal of Primatology* **14**:853–877.

Feh, C. and de Mazières, J. (1993) Grooming at a preferred site reduces heart rate in horses. *Animal Behaviour* **46**:1191–1194.

Feistner, A.T.C. and McGrew, W.C. (1989) Food-sharing in primates: a critical review. In: *Perspectives in Primate Biology* (eds P. Seth and S. Seth) New Delhi: Today & Tomorrow's Press, 21–36.

Fitzgibbon, C.D. (1994) The costs and benefits of predator inspection behavior in Thomson gazelles. *Behavioral Ecology and Sociobiology* **34**:139–148.

Flack, J.C., *et al.* (2006) Policing stabilizes construction of social niches in primates. *Nature* **439**:426–429.

Frankenberg, E. (1981) The adaptive significance of avian mobbing. IV. Alerting others and perception advertisement in blackbirds facing an owl. *Zeitschrift fur Tierpsychologie* **55**:97–118.

Gaston, A.J. (1977) Social behaviour within groups of jungle babblers (*Turdoides striatus*). *Animal Behaviour* **25**:828–848.

Gero, S., *et al.* (2009) Who cares? Between-group variation in alloparental caregiving in sperm whales. *Behavioral Ecology* **20**:838–843.

Ghoul, M., *et al.* (2014) Toward an evolutionary definition of cheating. *Evolution* **68**:318–331.

Gilby, I.C. (2006) Meat sharing among the Gombe chimpanzees: harassment and reciprocal exchange. *Animal Behaviour* **71**:953–963.

Goodall, J. (1986) *The Chimpanzees of Gombe: Patterns of Behavior*. Cambridge, MA: Belknap Press.

Gouzoules, S. and Gouzoules, H. (1987) Kinship. In: *Primate Societies* (eds B.B. Smuts, D.L. Cheney, R.M. Seyfarth, R.W. Wrangham and T.T. Struhsaker) Chicago: University of Chicago Press, 299–305.

Graw, B. (2005) Cooperative antipredator behaviour in meerkats (*Suricata suricatta*): the function of secondary cue inspection and mobbing. PhD thesis, University of Zurich

Graw, B. and Manser, M. (2007) The function of mobbing in cooperative meerkats. *Animal Behaviour* **74**:507–517.

Grinnell, J. and McComb, K. (1996) Maternal grouping as a defense against infanticide by males: evidence from field playback experiments on African lions. *Behavioral Ecology* **7**:55–59.

Grinnell, J. and McComb, K. (2001) Roaring and social communication in African lions: the limitations imposed by listeners. *Animal Behaviour* **62**:93–98.

Haenel, N.J. (1986) General notes on the behavioral ontogeny of Puget Sound killer whales and the occurrence of allomaternal behavior. In: *Behavioral Biology of Killer Whales* (eds B. Kirkevold and J.S. Lockard) New York: Alan R. Liss, 285–300.

Hammerstein, P. (2003) Why is reciprocity so rare in social animals? A protestant appeal. In: *Genetic and Cultural Evolution of Cooperation* (ed. P. Hammerstein). Cambridge, MA: MIT Press, 83–93.

Harcourt, A.H. (1992) Coalitions and alliances: are primates more complex than non-primates? In: *Coalitions and Alliances in Humans and Other Animals* (eds A.H. Harcourt and F.B.M. de Waal) Oxford: Oxford University Press, 445–471.

Harrington, F.H. and Mech, L.D. (1979) Wolf howling and its role in territory maintenance. *Behaviour* **68**:207–249.

Hart, B.L. and Hart, L. (1992) Reciprocal allogrooming in impala, *Aepyceros melampus*. *Animal Behaviour* **44**:1073–1083.

Hauser, M.D., *et al.* (2003) Give unto others: genetically unrelated cotton-top tamarin monkeys preferentially give food to those who altruistically give food back. *Proceedings of the Royal Society of London. Series B: Biological Sciences* **270**:2363–2370.

Hauser, M.D., *et al.* (2009) Evolving the ingredients for reciprocity and spite. *Philosophical Transactions of the Royal Society B: Biological Sciences* **364**:3255–3266.

Hayes, L.D. (2000) To nest communally or not to nest communally: a review of rodent communal nesting and nursing. *Animal Behaviour* **59**:677–688.

Hayes, L.D. and Soloman, N.G. (2004) Costs and benefits of communal rearing to female prairie voles (*Microtus ochrogaster*). *Behavioral Ecology and Sociobiology* **56**: 585–593.

Heinsohn, R.G. and Packer, C. (1995) Complex cooperative strategies in group-territorial African lions. *Science* **269**:1260–1262.

Hemelrijk, C.K. (1994) Support for being groomed in long-tailed macaques *Macaca fascicularis*. *Animal Behaviour* **48**:479–481.

Hemelrijk, C.K. and Ek, A. (1991) Reciprocity and interchange of grooming and 'support' in captive chimpanzees. *Animal Behaviour* **41**:923–935.

Henzi, S.P. and Barrett, L. (1999) The value of grooming to female primates. *Primates* **40**:47–59.

Henzi, S.P. and Barrett, L. (2002) Infants as a commodity in a baboon market. *Animal Behaviour* **63**:915–921.

Henzi, S.P., *et al.* (2013) Scalar social dynamics in female vervet monkey cohorts. *Philosophical Transactions of the Royal Society B: Biological Sciences* **368**:20120351.

Hirshleifer, D. and Rasmusen, E. (1989) Cooperation in the repeated prisoner's dilemma with ostracism. *Journal of Economic Behavior and Organization* **12**:87–106.

Hoogland, J.L. (1995) *The Black-tailed Prairie Dog: Social Life of a Burrowing Mammal*. Chicago: University of Chicago Press.

Hoogland, J.L., *et al.* (1989) Communal nursing in prairie dogs. *Behavioral Ecology and Sociobiology* **24**:91–95.

Hopper, L.H., *et al.* (2014) Social comparison mediates chimpanzees' responses to loss, not frustration. *Animal Cognition* **17**: 1303–1311.

Hrdy, S.B. (1976) Care and exploitation of nonhuman primate infants by conspecifics other than the mother. In: *Advances in the Study of Behavior*, Vol. **6** (eds J.S. Rosenblatt, R.A. Hinde, E. Shaw and C. Beer) San Diego, CA: Academic Press, 101–158.

Jarman, M.V. (1979) *Impala Social Behaviour: Territory, Hierarchy, Mating and the Use of Space*. Berlin: Verlag Paul Parey.

Jarvis, J.U.M. and Bennett, N.C. (1991) Ecology and behaviour of the Family Bathyergidae. In: *The Biology of the Naked Mole-rat* (eds P.W. Sherman, J.U.M. Jarvis and R.D. Alexander) Princeton, NJ: Princeton University Press, 66–96.

Jensen, K., *et al.* (2007) Chimpanzees are vengeful but not spiteful. *Proceedings of the National Academy of Sciences of the United States of America* **104**:13046–13050.

Johnsingh, A.J.T. (1982) Reproductive and social behaviour of the dhole, *Cuon alpinus* (Canidae). *Journal of Zoology* **198**:443–463.

Johnstone, R.A. and Rodrigues, A.M.M. (2016) Cooperation and the common good. *Philosophical Transactions of the Royal Society B*. DOI: 10.1098/rstb.2015.0086. [Accessed 18 January 2016].

Kapsalis, E. (2004) Matrilineal kinship and primate behavior. In: *Kinship and Behavior in Primates* (eds B. Chapais and C.M. Berman) Oxford: Oxford University Press, 153–176.

Kelley, A.E. and Berridge, K.C. (2002) The neuroscience of natural rewards: relevance to addictive drugs. *Journal of Neuroscience* **22**:3306–3311.

Kern, J.M. and Radford, A.N. (2013) Call of duty? Variation in use of the watchman's song by sentinel dwarf mongooses, *Helogale parvula*. *Animal Behaviour* **85**:967–975.

Keverne, E.B., *et al.* (1989) Beta-endorphin concentrations in cerebrospinal fluid of monkeys are influenced by grooming relationships. *Psychoneuroendocrinology* **14**:155–161.

Kitchen, D.M. and Packer, C.R. (1999) Complexity in vertebrate societies. In: *Levels of Selection in Evolution* (ed. L. Keller). Princeton, NJ: Princeton University Press, 176–196.

Kleindorfer, S. and Wasser, S.K. (2004) Infant handling and mortality in yellow baboons (*Papio cynocephalus*): evidence for female reproductive competition? *Behavioral Ecology and Sociobiology* **56**:328–337.

Kobayashi, T. (1996) The biological function of snake mobbing by Siberian chipmunks: II. Functions beneficial for the mobbers themselves. *Journal of Ethology* **14**:9–13.

Koenig, A. (1995) Group size, composition, and reproductive success in wild common marmosets (*Callithrix jacchus*). *American Journal of Primatology* **35**:311–317.

Kokko, H., *et al.* (2001) The evolution of cooperative breeding through group augmentation. *Proceedings of the Royal Society of London. Series B: Biological Sciences* **268**:187–196.

König, B. (1994a) Fitness effects of communal rearing in house mice: the role of relatedness versus familiarity. *Animal Behaviour* **48**:1449–1457.

König, B. (1994b) Components of lifetime reproductive success in communally and solitarily nursing house mice: a laboratory study. *Behavioral Ecology and Sociobiology* **34**:275–283.

König, B. (1994c) Communal nursing in mammals. *Verhandlungen der Deutschen Zoologischen Gesellschaft* **87**:115–128.

König, B. (1997) Cooperative care of young in mammals. *Naturwissenschaften* **84**:95–104.

König, B. (2006) Non-offspring nursing in mammals: general implications from a study on house mice. In: *Cooperation in Primates and Humans: Mechanisms and Evolution* (eds P.M. Kappeler and C.P. van Schaik) Berlin: Springer, 191–205.

Krams, I., *et al.* (2008) Experimental evidence of reciprocal altruism in the pied flycatcher. *Behavioral Ecology and Sociobiology* **62**:599–605.

Kruuk, H. (1972) *The Spotted Hyena: A Study of Predation and Social Behaviour*. Chicago: University of Chicago Press.

Kruuk, H. (1975) Functional aspects of social hunting in carnivores. In: *Function and Evolution in Behaviour* (eds G. Baerends, C. Beer and A. Manning) Oxford: Clarendon Press, 119–141.

Kummer, H. (1978) On the value of social relationships to nonhuman primates: a heuristic scheme. *Social Science Information* **17**:687–705.

Kunz, T.H., *et al.* (1994) Allomaternal care: helper-assisted birth in the Rodrigues fruit bat, *Pteropus rodricensis* (Chiroptera: Pteropodidae). *Journal of Zoology* **232**:691–700.

Kurland, J.A. (1977) Kin selection in the Japanese monkey. In: *Contributions to Primatology*, Vol. **12** Basel: Karger.

Kutsukake, N. and Clutton-Brock, T.H. (2006) Aggression and submission reflect reproductive conflict between females in cooperatively breeding meerkats *Suricata suricatta*. *Behavioral Ecology and Sociobiology* **59**:541–548.

Lee, P.C. (1987) Allomothering among African elephants. *Animal Behaviour* **35**:278–291.

Legge, S. (1996) Cooperative lions escape the Prisoner's Dilemma. *Trends in Ecology and Evolution* **11**:2–3.

Leimar, O. and Hammerstein, P. (2010) Cooperation for direct fitness benefits. *Philosophical Transactions of the Royal Society B: Biological Sciences* **365**:2619–2626.

Lewis, S.E. and Pusey, A.E. (1997) Factors influencing the occurrence of communal care in plural breeding mammals. In: *Cooperative Breeding in Mammals* (eds N.G. Solomon and J.A. French) Cambridge: Cambridge University Press, 335–363.

Lührs, M.-L. and Dammhahn, M. (2010) An unusual case of cooperative hunting in a solitary carnivore. *Journal of Ethology* **28**:379–383.

McComb, K., *et al.* (1994) Roaring and numerical assessment in contests between groups of female lions, *Panthera leo*. *Animal Behaviour* **47**:379–387.

McCracken, G.F. (1984) Communal nursing in Mexican free-tailed bat maternity colonies. *Science* **223**:1090–1091.

MacLeod, K.J. and Clutton-Brock, T.H. (2014) Low costs of allonursing in meerkats: mitigation by behavioural change? *Behavioral Ecology* doi: 10.1093/beheco/aru205

Madden, J.R. and Clutton-Brock, T.H. (2009) Manipulating grooming by decreasing parasite load causes unpredicted changes in antagonism. *Proceedings of the Royal Society of London. Series B: Biological Sciences* **276**:1263–1268.

Madden, J.R., *et al.* (2009) The social network structure of a wild meerkat population: 2. Intragroup interactions. *Behavioral Ecology and Sociobiology* **64**:81–95.

Maklakov, A.A. (2002) Snake-directed mobbing in a cooperative breeder: anti-predator behaviour or self-advertisement for the formation of dispersal coalitions? *Behavioral Ecology and Sociobiology* **52**:372–378.

Manning, C.J., *et al.* (1992) Communal nesting patterns in mice implicate MHC gene in kin recognition. *Nature* **360**:581–583.

Manning, C.J., *et al.* (1995) Communal nesting and communal nursing in house mice, *Mus musculus domesticus*. *Animal Behaviour* **50**:741–751.

Manser, M.B. (1999) Response of foraging group members to sentinel calls in suricates, *Suricata suricatta*. *Proceedings of the Royal Society of London. Series B: Biological Sciences* **266**:1013–1019.

Manson, J.H., *et al.* (2004) Time-matched grooming in female primates? New analyses from two species. *Animal Behaviour* **67**:493–500.

Mech, L.D. (1970) *The Wolf: The Ecology and Behavior of an Endangered Species*. New York: Natural History Press.

Melis, A.P., *et al.* (2008) Do chimpanzees reciprocate favours? *Animal Behaviour* **76**:951–962.

Mitani, J.C. and Watts, D. (1997) The evolution of non-maternal caretaking among anthropoid primates: do helpers help? *Behavioral Ecology and Sociobiology* **40**:213–220.

Mitani, J.C. and Watts, D.P. (2001) Why do chimpanzees hunt and share meat? *Animal Behaviour* **61**:915–924.

Montiglio, P.-O., *et al.* (2013) Social niche specialization under constraints: personality, social interactions and environmental heterogeneity. *Philosophical Transactions of the Royal Society B: Biological Sciences* **368**:20120343.

Mooring, M.S. (1995) The effect of tick challenge on grooming rate by impala. *Animal Behaviour* **50**:377–392.

Mooring, M.S. and Hart, B.L. (1992) Reciprocal allogrooming in dam-reared and hand-reared impala fawns. *Ethology and Sociobiology* **90**:37–51.

Mooring, M.S., *et al.* (1996) Grooming in impala: role of oral grooming in removal of ticks and effects of ticks in increasing grooming rate. *Physiology and Behavior* **59**:965–971.

Moss, C. (1988) *Elephant Memories*. New York: William Morrow.

Müller-Schwarz, D. and Sun, L. (2003) *The Beaver: Natural History of a Wetlands Engineer*. Ithaca, NY: Cornell University Press.

Norris, K.S. and Schilt, C.R. (1988) Cooperative societies in three-dimensional space: on the origins of aggregations, flocks, and schools, with special reference to dolphins and fish. *Ethology and Sociobiology* **9**:149–179.

Oates, J.F. (1977) Social life of a black-and-white colobus monkey, *Colobus guereza*. *Zeitschrift für Tierpsychologie* **45**:1–60.

Packer, C., *et al.* (1990) Why lions form groups: food is not enough. *American Naturalist* **136**:1–19.

Packer, C., *et al.* (1992) A comparative analysis of non-offspring nursing. *Animal Behaviour* **43**:265–281.

Parr, L.A. and de Waal, F.B.M. (1999) Visual kin recognition in chimpanzees. *Nature* **399**:647–648.

Paul, A., *et al.* (1996) The sociobiology of male–infant interactions in Barbary macaques, *Macaca sylvanus*. *Animal Behaviour* **51**:155–170.

Perry, S., *et al.* (2003) Social conventions in wild white-faced capuchin monkeys: evidence for traditions in a neotropical primate. *Current Anthropology* **44**:241–268.

Pfeiffer, T., *et al.* (2005) Evolution of cooperation by generalized reciprocity. *Proceedings of the Royal Society of London. Series B: Biological Sciences* **272**:1115–1120.

Poirier, F.E. (1968) The Nilgiri langur (*Presbytis johnii*) mother–infant dyad. *Primates* **9**:45–68.

Pollock, G. and Dugatkin, L.A. (1992) Reciprocity and the emergence of reputation. *Journal of Theoretical Biology* **159**:25–37.

Potts, W.K., *et al.* (1991) Mating patterns in seminatural populations of mice influenced by MHC genotype. *Nature* **352**:619–621.

Pusey, A.E. and Packer, C. (1993) Infanticide in lions: consequences and counterstrategies. In: *Infanticide and Parental Care Chur*, (eds S. Parmigiani and F.S. vom Saal) Switzerland: Harwood Academic Press, 277–299.

Pusey, A.E. and Packer, C. (1994) Non-offspring nursing in social carnivores: minimizing the costs. *Behavioral Ecology* **5**:362–374.

Radford, A.N. (2003) Territorial vocal rallying in the green wood hoopoe: influence of rival group size and composition. *Animal Behaviour* **66**:1035–1044.

Raemaekers, J.J. and Raemaekers, P.M. (1984) The Ooaa duet of the gibbon (*Hylobates lar*): a group call which triggers other groups to respond in kind. *Folia Primatologica* **42**:209–215.

Raihani, N.J. and Bshary, R. (2011) Resolving the iterated prisoner's dilemma: theory and reality. *Journal of Evolutionary Biology* **24**:1628–1639.

Rasa, O.A.E. (1986) Coordinated vigilance in dwarf mongoose family groups: the watchman's song hypothesis and the cost of guarding. *Ethology* **71**:340–344.

Rasa, O.A.E. (1987a) The dwarf mongoose: a study of behavior and social structure in relation to ecology in a small social carnivore. In: *Advances in the Study of Behaviour*, Vol. **17** (eds J.S. Rosenblatt, C. Beer, M.-C. Busnel and P.J.B. Slater) San Diego, CA: Academic Press, 121–163.

Rasa, O.A.E. (1987b) Vigilance behaviour in dwarf mongooses: selfish or altruistic? *South African Journal of Science* **83**:587–590.

Riedl, K., *et al.* (2012) No third-party punishment in chimpanzees. *Proceedings of the National Academy of Sciences of the United States of America* **109**:14824–14829.

Riehl, C. and Frederickson, M.E. (2016) Cheating and punishment in cooperative animal societies. *Philosophical Transactions of the Royal Society B*. DOI: 10.1098/rstb.2015.0090. [Accessed 18 January 2016].

Rood, J.P. (1983) Banded mongoose rescues back member from eagle. *Animal Behaviour* **31**:1261–1262.

Rood, J.P. (1986) Ecology and social evolution in the mongooses. In: *Ecological Aspects of Social Evolution* (eds D.I. Rubenstein and R.W. Wrangham) Princeton, NJ: Princeton University Press, 131–152.

Ross, H.E., *et al.* (2009) Characterization of the oxytocin system regulating affiliative behavior in female prairie voles. *Neuroscience* **162**:892–903.

Roulin, A. and Heeb, P. (1999) The immunological function of allosuckling. *Ecology Letters* **2**:319–324.

Ruckstuhl, K.E. and Ingold, P. (1998) Baby-sitting in chamois: a form of cooperation in females? *Mammalia* **62**:125–128.

Russell, A.F., *et al.* (2003) Cost minimization by helpers in cooperative vertebrates. *Proceedings of the National Academy of Sciences of the United States of America* **100**:3333–3338.

Russell, J.K. (1983) Altruism in coati bands: nepotism or reciprocity. In: *Social Behaviour of Female Vertebrates* (ed. S.K. Wasser). New York: Academic Press, 263–290.

Rutte, C. and Taborsky, M. (2007) Generalized reciprocity in rats. *PLOS Biology* **5**:1421–1425.

Santema, P. and Clutton-Brock, T.H. (2012) Dominant female meerkats do not use aggression to elevate work rates of helpers in response to increased brood demand. *Animal Behaviour* **83**: 827–832.

Santema, P. and Clutton-Brock, T.H. (2013) Meerkat helpers increase sentinel behaviour and bipedal vigilance in the presence of pups. *Animal Behaviour* **85**:655–661.

Schaller, G.B. (1972) *The Serengeti Lion: A Study of Predator–Prey Relations.* Chicago: Chicago University Press.

Scheel, D. and Packer, C. (1991) Group hunting behaviour of lions: a search for cooperation. *Animal Behaviour* **41**:697–709.

Schino, G. (2001) Grooming, competition and social rank among female primates: a meta-analysis. *Animal Behaviour* **62**:265–271.

Schino, G. and Aureli, F. (2010) The relative roles of kinship and reciprocity in explaining primate altruism. *Ecology Letters* **13**:45–50.

Schino, G., *et al.* (1988) Allogrooming as a tension reduction mechanism: a behavioral approach. *American Journal of Primatology* **16**:43–50.

Sekulic, R. (1982) The function of howling in red howler monkeys (*Alouatta seniculus*). *Behaviour* **81**:38–54.

Seyfarth, R.M. (1976) Social relationships among female baboons. *Animal Behaviour* **24**:917–938.

Seyfarth, R.M. and Cheney, D.L. (1984) Grooming alliances and reciprocal altruism in vervet monkeys. *Nature* **308**:541–543.

Silk, J.B. (1982) Altruism among female *Macaca radiata*: explanations and analysis of patterns of grooming and coalition-formation. *Behaviour* **79**:162–188.

Silk, J.B., *et al.* (2005) Chimpanzees are indifferent is the welfare of unrelated group members. *Nature* **27**: 1357–1359.

Silk, J.B. (2005) The evolution of cooperation in primate groups. In: *Moral Sentiments and Material Interests: Origins, Evidence, and Consequences* (eds H. Gintis, S. Bowles, R. Boyd and E. Fehr) Cambridge, MA: MIT Press, 43–74.

Silk, J.B. (2006) Practicing Hamilton's Rule: kin selection in primate groups. In: *Cooperation in Primates and Humans: Mechanisms and Evolution* (eds P.M. Kappeler and C.P. van Schaik) Berlin: Springer, 25–46.

Silk, J.B. (2007) Social components of fitness in primate groups. *Science* **317**:1347–1351.

Silk, J.B. (2009) Nepotistic cooperation in non-human primate groups. *Philosophical Transactions of the Royal Society B: Biological Sciences* **364**:3243–3254.

Silk, J.B. (2012) The adaptive value of sociality. In: *The Evolution of Primate Societies* (eds J.C. Mitani, J. Call, P.M. Kappeler, R.A. Palombit and J.B. Silk) Chicago: University of Chicago Press, 554–564.

Silk, J.B., *et al.* (2004) Patterns of coalition formation by adult female baboons in Amboseli, Kenya. *Animal Behaviour* **67**:573–582.

Silk, J.B., *et al.* (2006) Social relationships among adult female baboons (*Papio cynocephalus*). I. Variation in the strength of social bonds. *Behavioral Ecology and Sociobiology* **61**:183–195.

Silk, J.B., *et al.* (2009) The benefits of social capital: close social bonds among female baboons enhance offspring survival. *Proceedings of the Royal Society of London. Series B: Biological Sciences* **276**:3099–3104.

Simpson, M. (1973) Social grooming of male chimpanzees. In: *Comparative Ecology and Behaviour of Primates* (eds J. Crook and R. Michael) London: Academic Press, 411–505.

Smith, K., *et al.* (2003) Wild female baboons bias their social behaviour towards paternal half-sisters. *Proceedings of the Royal Society of London. Series B: Biological Sciences* **270**: 503–510.

Sparks, J. (1967) Allogrooming in primates: a review. In: *Primate Ethology* (ed. D. Morris) London: Weidenfeld and Nicolson: 148–175.

Stander, P.E. (1992) Cooperative hunting in lions: the role of the individual. *Behavioral Ecology and Sociobiology* **29**: 445–454.

Stander, P.E. and Albon, S.D. (1993) Hunting success of lions in a semi arid environment. *Symposium of the Zoological Society of London* **65**:127–143.

Stanford, C.B. (1998) *Chimpanzee and Red Colobus: The Ecology of Predator and Prey.* Cambridge, MA: Harvard University Press.

Stephens, D.W. and Krebs, J.R. (1986) *Foraging Theory.* Princeton, NJ: Princeton University Press.

Stephens, D.W., *et al.* (2002) Discounting and reciprocity in an iterated Prisoner's Dilemma. *Science* **298**:2216–2218.

Stevens, J.R. (2004) The selfish nature of generosity: harassment and food sharing in primates. *Proceedings of the Royal Society of London. Series B: Biological Sciences* **271**:451–456.

Struhsaker, T.T. (1975) *The Red Colobus Monkey.* Chicago: University of Chicago Press.

Sun, L. (2003) Monogamy correlates, socioecological factors and mating systems in beavers. In: *Monogamy: Mating Strategies and Partnerships in Birds, Humans and Other Mammals* (eds U.H. Reichard and C. Boesch) Cambridge: Cambridge University Press, 138–146.

Taber, A.B. and Macdonald, D.W. (1992) Spatial organization and monogamy in the Mara *Dolichotis patagonum*. *Journal of Zoology* **227**:417–438.

Taborsky, M., *et al.* (2016a) The evolution of cooperation based on direct fitness benefits. *Philosophical Transactions of the Royal Society B.* DOI: 10.1098/rstb.2015.0472. [Accessed 18 January 2016].

Taborsky, M., *et al.* (2016b) Correlated pay-offs are key to cooperation. *Philosophical Transactions of the Royal Society B: Biological Sciences B.* DOI: 10.1098/rstb.2015.0084. [Accessed 18 January 2016].

Tamura, N. (1989) Snake-directed mobbing by the Formosan squirrel *Callosciurus erythraeus thaiwanensis. Behavioral Ecology and Sociobiology* **24**:175–180.

Tanaka, I. and Takefushi, H. (1993) Elimination of external parasites (lice) is the primary function of grooming in free-ranging Japanese macaques. *Anthropological Science* **101**:187–193.

Tavolga, C. and Essapian, F.S. (1957) The behavior of the bottle-nosed dolphin (*Tursiops truncatus*): mating, pregnancy, parturition and mother–infant behavior. *Zoologica* **42**:11–31.

Teleki, G. (1973) *The Predatory Behavior of Wild Chimpanzees.* Lewisburg, PA: Bucknell University Press.

Trivers, R.L. (1971) The evolution of reciprocal altruism. *Quarterly Review of Biology* **46**:35–57.

Viana, D.S., *et al.* (2010) Cognitive and motivational requirements for the emergence of cooperation in a rat social game. *PLOS ONE* **5**:e8483.

Walters, J.R. and Seyfarth, R.M. (1997) Conflict and cooperation. In: *Primate Societies* (eds B.B. Smuts, D.L. Cheney, R.M. Seyfarth, R.W. Wrangham and T.T. Struhsaker) Chicago: University of Chicago Press, 306–317.

Weisbard, C. and Goy, R.W. (1976) Effect of parturition and group composition on competitive drinking order in stumptail macaques (*Macaca arctoides*). *Folia Primatologica* **25**:95–121.

Whitehead, H. and Weilgart, L. (2000) The sperm whale: social females and roving males. In: *Cetacean Societies: Field Studies of Dolphins and Whales* (eds J. Mann, R. Connor, P. Tyack and H. Whitehead) Chicago: University of Chicago Press, 154–172.

Whiten, A. and Byrne, R.W. (1988) Tactical deception in primates. *Behavioral and Brain Sciences* **11**:233–244.

Whiten, A. and Byrne, R.W. (eds) (1997) *Machiavellian Intelligence II: Extensions and Evaluations.* Cambridge: Cambridge University Press.

Widdig, A. (2007) Paternal kin discrimination: the evidence and likely mechanisms. *Biological Reviews* **82**:319–334.

Widdig, A., *et al.* (2001) Paternal relatedness and age proximity regulate social relationships among adult female rhesus macaques. *Proceedings of the National Academy of Sciences of the United States of America* **98**:13769–13773.

Widdig, A., *et al.* (2006) Paternal kin bias in the agonistic interventions of adult female rhesus macaques (*Macaca mulatta*). *Behavioral Ecology and Sociobiology* **61**:205–214.

Wilkinson, G.S. (1984) Reciprocal food sharing in the vampire bat. *Nature* **308**:181–184.

Wilkinson, G.S. (1985a) The social organization of the common vampire bat. I. Pattern and cause of association. *Behavioral Ecology and Sociobiology* **17**:111–121.

Wilkinson, G.S. (1985b) The social organization of the common vampire bat: II. Mating system, genetic structure, and relatedness. *Behavioral Ecology and Sociobiology* **17**:123–134.

Wilkinson, G.S. (1988) Reciprocal altruism in bats and other mammals. *Ethology and Sociobiology* **9**:85–100.

Wilkinson, G.S. (1990) Food sharing in vampire bats. *Scientific American* **262**:76–82.

Wilkinson, G.S. (1992) Communal nursing in the evening bat *Nycticeius humeralis. Behavioral Ecology and Sociobiology* **31**:225–235.

Wittig, R.M., *et al.* (2014a) Triadic social interactions operate across time: a field experiment with wild chimpanzees. *Proceedings of the Royal Society of London. Series B: Biological Sciences* **281**:20133155.

Wittig, R.M., *et al.* (2014b) Food sharing is linked to urinary oxytocin levels and bonding in related and unrelated wild chimpanzees. *Proceedings of the Royal Society of London. Series B: Biological Sciences* **281**:20133096.

Wright, J., *et al.* (2001) State-dependent sentinels: an experimental study in the Arabian babbler. *Proceedings of the Royal Society of London. Series B: Biological Sciences* **268**:821–826.

Young, A.J. (2003) Subordinate tactics in cooperative meerkats: breeding, helping and dispersal. PhD thesis, University of Cambridge, Cambridge

Zahavi, A. (1977) Testing of a bond. *Animal Behaviour* **25**:246–247.

Zahavi, A. and Zahavi, A. (1997) *The Handicap Principle: a Missing Piece of Darwin's Puzzle.* New York: Oxford University Press.

Zamma, K. (2002) Grooming site preferences determined by lice infection among Japanese macaques in Arashiyama. *Primates* **43**:41–49.

CHAPTER 10

Mating systems

10.1 Introduction

In an East African rain forest, a group of black and white colobus monkeys are starting to feed. The group consists of four adult females, two of them with juveniles, several adolescents and a mature male, who watches the females closely and is ready to chase away any intruding males. Fifty metres away, a group of sixty red colobus monkeys, including a dozen breeding females and half a dozen adult males, are feeding in the same stand of trees. Both females and males are tolerant of each other, feeding a few feet apart on bunches of newly grown leaves.

The difference in tolerance between males in the two colobus species (Figure 10.1) raises questions about the reasons for such obvious contrasts in reproductive strategies. Why are males intolerant of each other's presence in some species and not in others? When single males defend groups of females, what determines the size of their harems and what consequences does this have for their dynamics? Why, in some species, do males defend mating territories rather than female groups and what affects their size? And why are male mating territories sometimes distributed throughout the areas used by females and sometimes aggregated in tight clusters or 'leks'?

As field studies of mammals accumulated in the second half of the twentieth century, both the extent to which male breeding behaviour varied and the ecological and evolutionary consequences of different mating systems became apparent. In some mammals, most breeding males defend access to a single adult female though either or both sexes sometimes mate with other individuals (Lukas and Clutton-Brock 2013). Some reviews have classified monogamous systems according to the type of evidence that is available, distinguishing between *social monogamy* (breeding adults living in pairs), *sexual monogamy* (pairs of males and females that mate exclusively with each other) and *genetic monogamy* (where genetic evidence confirms

that paired adults do not breed with other conspecifics) (Gowaty 1996; Reichard 2003).

Polygynous and promiscuous mating systems also vary. In some species, males seldom guard females and engage in scramble competition, as in some cavies (Sachser *et al.* 1999; Waterman 2007): lack of mate defence is thought to be associated with high local densities of receptive females, whereas lower densities are predicted to favour pre-copulatory mate guarding (Dobson 1984). In others, males search for receptive females and, when they have found one, guard her (usually for a limited period) before moving on to search for another; successful males may guard and mate with several females in the course of a breeding season but rarely guard more than one female at a time. In some species, males defend territories overlapping the ranges of several females or the entire ranges of female groups (Charles-Dominique 1983; Macdonald 1983; Michener 1983) and show *range-guarding polygyny*: where female density is low, individual males may be unable to guard the range of more than one female; where female ranges overlap each other, males commonly guard access to several females. Systems of this kind (which occur in many rodents, small carnivores, nocturnal primates and ungulates) are also sometimes referred to as *facultative polygyny* (Michener 1981). Where the ranges of female groups are too large to be defended and female groups are stable, males commonly guard access to female groups. Finally, in a minority of species (including many wide-ranging ungulates and at least one carnivore) neither female ranges nor female groups can be economically guarded but females return regularly to particular sites where males guard mating territories smaller than female ranges: I refer to these as *site-guarding polygyny*. In most of these species, males locate their territories on or close to resources that attract females but, in some, they guard small clustered mating territories that contain virtually no

Mammal Societies, First Edition. Tim Clutton-Brock.
© 2016 Tim Clutton-Brock. Published 2016 by John Wiley & Sons, Ltd.

(a)   (b)

Figure 10.1 Arboreal folivores with contrasting breeding systems: (a) black and white colobus live in small groups that typically consist of several philopatric breeding females and a single immigrant male; (b) red colobus live in groups of 20–80, usually consisting of multiple philopatric males and multiple immigrant females. *Sources*: (a) © John Oates; (b) © Tim Clutton-Brock.

resources, forming *leks* similar to those found in a number of polygynous birds (Höglund and Alatalo 1995).

The first attempts to make sense of the diversity of mammalian mating systems treated the structure and composition of breeding groups as species-specific traits and suggested that contrasts in social organisation were advantageous to all individuals (see Chapter 1). Today, we appreciate that male breeding strategies vary widely between populations of the same species as well as between individuals. Evidence that species differences in mating systems are associated with particular ecological conditions (Crook 1965; Lack 1968) does not contradict this view, for these associations are a product of the adaptation of individual behaviour to ecological and social conditions.

Another crucial development in our understanding of the evolution of animal breeding systems is the realisation that determinants of breeding success (and attendant selective pressures) differ consistently between the sexes. The generality of the insight was demonstrated in an important review of animal breeding systems by Emlen and Oring (1977), in which they argued that the distribution of females was controlled by the distribution of resources necessary to produce and raise young and that the distribution of females controlled the distribution and reproductive strategies of males. This proposition has since been tested and confirmed in empirical studies of a range of avian and mammalian breeding systems (Davies *et al.* 2012). For example, Ims (1988) showed that by adding food to ranges of individual female grey-sided voles, he could reduce the size of their

home ranges and cause them to become more aggregated. To show that males were responding to the distribution of females, he placed cages containing females at distances corresponding to the usual distance between home ranges of natural females and moved the cages each day to simulate the natural movements of females. When caged females were widely distributed, males adopted dispersed ranges, overlapping the ranges of particular females, whereas when he placed several female cages close together, males focused their movements on these clusters. When these experiments were repeated with caged males and freely ranging females, females did not respond to changes in the distribution of males. While the distribution of resources is not the only factor affecting female distributions and reproductive strategies and female distributions are not the only factor affecting male distributions and reproductive strategies, there is sufficient generality in Emlen and Oring's proposition that it continues to provide an important component of current attempts to account for the evolution of animal breeding systems.

This chapter reviews our current understanding of the causes and consequences of variation in mammalian mating systems. Section 10.2 examines different forms of monogamy while section 10.3 explores five kinds of polygynous mating system. Section 10.4 reviews evidence of intraspecific differences in mating systems while section 10.5 examines the extent to which contrasts in the behaviour of males and the structure of breeding groups affect variation in reproductive success and paternity among males. Finally, section 10.6 discusses the consequences of variation in breeding systems for varia-tion in male breeding success, selection on males and male life histories, as well as for the genetic structure of groups and local populations.

10.2 Social monogamy

Forms of social monogamy

Monogamous mating systems are widely but unevenly distributed among mammals (Lukas and Clutton-Brock 2013). They are common among carnivores (especially among canids) and around 15% of primate species typi-cally form monogamous pairs, including some lemurs, most or all of the gibbons and, among the New World monkeys, the marmosets, tamarins, titi monkeys and night monkeys (Kleiman 1977; van Schaik and Kappeler 2003). Most insectivores have traditionally been regarded as solitary but a recent survey of shrews and related species suggests that a substantial minority may live in pairs (Valomy *et al.* 2015; Rathbun 1979; Fitz-gibbon 1997). Monogamous societies are more sparsely distributed across other mammalian groups, including ungulates (Dunbar and Dunbar 1980; Kishimoto and Kawamichi 1996; Brotherton *et al.* 1997), rodents (Mich-ener 1981; Ribble 2003; Sommer 2003), bats (Bradbury and Vehrencamp 1977) and marsupials (Brown *et al.* 2007; Runcie 2000). Phylogenetic reconstructions sug-gest that, in most cases, socially monogamous systems appear to have evolved from an ancestral condition where breeding females were solitary (Lukas and Clut-ton-Brock 2013), though it is possible that there may be

some exceptions (Kappeler 2014; Schultz *et al.* 2011; Kappeler and Fichtel 2015).

The relative duration of pair-bonds varies widely. In most monogamous mammals, adolescents pair with a member of the opposite sex soon after dispersing and establish a joint territory (Figure 10.2). In some species, like dik dik or marmosets, pairs commonly persist until one of the pair dies and 'divorce' is rare (Brotherton and Komers 2003; Fietz 2003; Sommer 2003), while in others 'divorce' is more frequent and the same individual com-monly pairs with several different partners in the course of its lifetime: for example, North American beavers commonly pair with two to four different partners in the course of their lives (Müller-Schwarz and Sun 2003; Sun 2003). These distinctions are probably quantitative rather than qualitative and several long-term studies of species previously thought to pair for life (including some gibbons) suggest that individuals may change partners more commonly than was supposed (Brockelman *et al.* 1998; Sommer and Reichard 2000). Eventually, it will be useful to base comparisons on the relative frequency of 'divorce' or on the average number of mates per lifespan but, as yet, available data are too uncommon for system-atic comparisons to be feasible.

In some monogamous societies, territory defence is sex-specific but in others both sexes may repel intruders of either sex. For example, in monogamous ungulates, like dik dik and serow, territory defence is usually sex-specific, whereas in beavers either sex may repel intruding males or females (Brotherton and Komers 2003; Kishimoto 2003; Sun 2003). In some species, pairs spend much of the day together while in others they forage separately (Michener

(a)

(b)

Figure 10.2 Two socially monogamous mammals: (a) dik dik and (b) white-bearded gibbons. *Sources*: (a) © Peter Brotherton; (b) © Thomas Marent/Minden Pictures/Getty Images.

1983; Ostfeld 1990; Fietz 2003; Kishimoto 2003), though the boundaries of their ranges usually coincide (Fietz 2003; Schülke and Kappeler 2003). In a small minority of species, multiple monogamous pairs share a common breeding site and different adults take turns to guard crèches of dependent young: for example, in maras (an antelope-like lagomorph (Figure 10.3) several pairs may share a burrow and different individuals guard young on different days (Macdonald *et al.* 2007).

Although partners occasionally mate outside the pair-bond in most monogamous societies, females commonly show preferences for their partners and a close attachment to their mates in most socially monogamous mammals (Anzenberger 1992; Williams *et al.* 1992) and genetic analyses show that extra-pair paternities are uncommon in most species (Mason 1966; Richardson 1987; Agren *et al.* 1989; Palombit 1994; Richardson *et al.* 2003; Huck *et al.* 2014); however, in a few species including alpine marmots (Goossens *et al.* 1998) and fat-tailed dwarf lemurs (Fietz *et al.* 2000; Fietz 2003), a substantial proportion of young are fathered by extra-group males. In some cases, the incidence of extra-pair paternity may vary with female access to potential partners. For example, in marmosets, females are more likely to have access to extra-group males when population density is high and the incidence of extra-pair paternity increases with population density (Goossens *et al.* 1998).

While extra-pair matings by dominant females are uncommon in many monogamous species, extra-group matings by adolescent or subordinate females who often lack unrelated mating partners in their group are more frequent. For example, in meerkat groups, a high proportion of young born to the dominant female are fathered by the resident dominant male while subordinate females (who are not guarded by the resident male and are often his offspring) usually mate outside the group when they breed (Spong *et al.* 2008). Where differences in the frequency of extra-group mating exist between dominant and subordinate females and subordinate females are not guarded by resident males, the common tendency to combine extra-group paternities for all females is likely to lead to underestimates of the effectiveness of male guarding.

Social monogamy and male care

In a proportion of socially monogamous mammals, males contribute to rearing young they are likely to have fathered, brooding, protecting, carrying or feeding them, and sometimes feeding their mothers too (see Chapter 16). As in biparental birds (Bjorklund and Westman 1986; Bart and Tornes 1989) the presence of males often has an important influence on the growth and survival of the juveniles (see Chapter 16).

Evidence that male care contributes to the reproductive output of females has often been used to suggest that

Figure 10.3 Maras are one of the few mammals that live in monogamous pair within larger colonies. *Source*: © www.bridgetdavey .com/Moment Open/Getty Images.

the evolution of male care has been responsible for the evolution of social monogamy. However, males make no obvious contribution to care in some monogamous mammals and their presence appears to have little effect on the reproductive rate of females or the survival of their offspring (see Chapter 16). For example, in dik dik, males do not contribute to the direct care of offspring, they do not associate more closely with females with dependent young and their presence does not appear to contribute to the ability of females to detect predators (Brotherton and Komers 2003) (Figure 10.4). Phylogenetic reconstructions of the evolution of monogamy and paternal care suggest that social monogamy has evolved more frequently in the absence of paternal care than in its presence, indicating that the evolution of male care in mammals is usually a consequence of monogamous mating rather than a cause (Lukas and Clutton-Brock 2013; Opie *et al.* 2013) (see Chapter 16).

In primates, where male infanticide is common, a similar argument suggests that social monogamy has evolved where the presence of the father is necessary to prevent his offspring being killed by infanticidal males (van Schaik and Dunbar 1990) though, in many monogamous mammals, observations of attempted infanticide are scarce or absent (see Chapter 15). While some phylogenetic reconstructions have suggested that the evolution of social monogamy in primates is associated with the risk of male infanticide (Opie *et al.* 2013, 2014), other analyses have failed to find any consistent association between the evolution of social monogamy and the risk of male infanticide, either across mammals (Lukas and Clutton-Brock 2013) or among primates (Lukas and Clutton-Brock 2014a).

Evidence that the evolution of male care usually succeeds the evolution of social monogamy suggests that other mechanisms explain its evolution. It seems likely that the difficulty of defending multiple mating partners where females occupy separate ranges is important. In most socially monogamous mammals, adult females are intolerant of each other and occupy

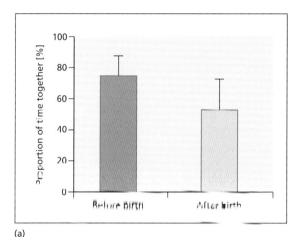

(a)

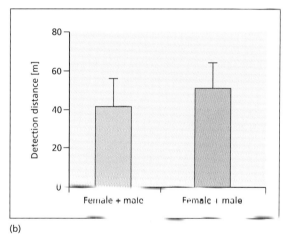

(b)

Figure 10.4 In dik dik, there is little evidence of paternal care: (a) proportion of time spent together by heterosexual pairs before versus after the birth of their offspring; (b) distance at which mothers detected the presence of a potential predator when associated with a male versus times when they were apart. *Sources*: Data from Brotherton and Rhodes (1996) and Brotherton (1994). *Photo source*: © Peter Brotherton.

separate territories, floating males quickly take possession of unguarded territories and the spatial separation of females may prevent males from guarding the range of more than one female at a time. For example, studies of dik dik (Brotherton and Manser 1997) and elephant shrews (Fitzgibbon 1997) show that males that attempt to defend two females and their territories quickly lose one of them to an intruder. An additional characteristic of monogamous mammals is that the relative density of females is usually low: although not all comparative analyses have found a consistent association between monogamous mating systems and population density in mammals (Dobson *et al.* 2010), a recent analysis of a large sample of non-human mammals shows that social monogamy is associated with low female density and occurs where breeding females are intolerant of each other (Figure 10.5) (Lukas and Clutton-Brock 2013).

This explanation of the evolution of social monogamy suggests that polygyny would be the optimal strategy for males if this was not precluded by the separation of females. If so, males would be expected to attempt to prevent females from evicting intruders and often do so (Davies 2000). For example, in dik dik, males do not evict additional females that appear within their territory and may even attempt to prevent their mates from doing so (Brotherton 1994; Brotherton and Rhodes 1996). In other cases, reproductive interference between females may be so intense that the addition of a second breeding partner reduces the combined reproductive output of the two females (and hence the breeding success of the male) and, under these conditions, males as well as females may be intolerant of intruding members of the opposite sex.

Social monogamy may also help to increase paternity certainty or may increase a male's chances of maintaining breeding access to a particular female over several seasons. Genetic studies show that the incidence of extra-pair paternities is relatively low in most mammals where males spend most of the day with their mates (see section 10.5) and, where this is the case, successful males may maximise their lifetime breeding success by prolonging associations with successful females on high-quality territories. Another suggestion is that monogamy may be favoured because it increases the frequency of full sibs within broods, reduces conflict between brood members and increases the total number of surviving young (Peck and Feldman 1988) though it seems unlikely that effects of this

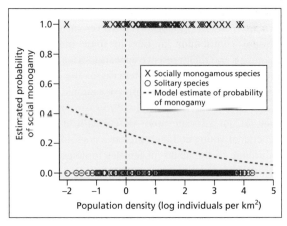

Figure 10.5 Changes in the probability of the evolution of social monogamy with variation in population density (individuals per km²) calculated across mammals. The open circles represent the observed values for solitary species (*n* = 411), the crosses the observed values for socially monogamous species (*n* = 89, 18% of all species). Population density (logarithm of the number of individuals per km²) has a substantial influence on the probability that a species is socially monogamous or solitary. At the highest population densities, there is only a 6% probability that a species will be socially monogamous, whereas the probability rises to 44% at the lowest population densities. Several socially monogamous species showing high population densities are singular cooperative breeders, where regular breeding is restricted to a single female. *Sources*: From Lukas and Clutton-Brock (2013). Reproduced with permission of AAAS.

kind would often be sufficient to offset the benefits of polygyny to males.

Social polyandry

In some mammals that commonly form monogamous pairs, some breeding groups consist of a breeding female and several adult males that contribute to guarding or rearing and are often siblings. Groups of this kind are common in marmosets and tamarins (Goldizen 1987, 2003; Garber 1997; Heymann and Soini 1999; Díaz-Muñoz 2011) as well as in some carnivores, including kinkajous (Kays 2003) and African wild dogs (Creel and Creel 2002). In some of these species, multiple males mate with the breeding female (Rylands 1982; Creel and Creel 2002; Goldizen 2003), but a single male commonly fathers most of the infants born in the group (Baker *et al.* 1993; Huck *et al.* 2005; Anzsenberger and Falk 2012). For example, in African wild dogs, multiple males (who are often brothers) mate with the dominant female but one male is usually responsible for around 90% of paternities (Girman *et al.* 1997) and a similar situation may occur in groups of common marmosets (Nievergelt *et al.* 2000). In these species, dominant males may benefit from the presence of additional males because this either increases their chance of acquiring or maintaining access to the breeding female or raises the number of offspring that can be raised while high coefficients of relatedness between males reduce any costs of sharing paternity. Alternatively, males may be unable to prevent additional males from joining the group. Subordinate males may gain direct benefits from the presence of other males if this provides them with some access to breeding females or increases the chance that they will inherit the group following the death of the dominant male or they may gain indirect benefits if they are related to the dominant male, but firm evidence of this is scarce.

Polyandrous mating systems are often found in cooperative breeders. One possible explanation is that where only a small proportion of adult females breed and there is consequently strong intrasexual competition between males, this favours the formation of competitive male coalitions and reproductive queues (see Chapter 17). Another possibility is that the costs of reproduction to females are unusually high in these species, generating strong selection on females to enlist the support of multiple males (see Chapter 16). Several studies of polyandrous species have produced evidence of substantial costs associated with reproduction: for example, in wild dogs, average litter size is unusually large (seven to eight pups) and the costs of feeding pups are probably high (Creel and Creel 2002). Similarly, female tamarins may lose 10–20% of their body weight during lactation (Garber and Teaford 1986; Nievergelt and Martin 1998) and males, too, show significant weight loss during the period when they are carrying young (Sanchez *et al.* 1999; Achenbach and Snowdon 2002). Comparative studies of primates also suggest that the number of resident males may increase with day range length and home range size (which may reflect the costs of carrying young to females). For example, among marmosets and tamarins, pygmy marmosets (which have the smallest ranges and often park their young when feeding) have the lowest number of adult males per group; marmosets (which have slightly larger ranges) show an increase in the number of males per group; and tamarins (which have the largest ranges) have the highest average numbers of adult males per group

(Goldizen 2003; see also Terborgh and Goldizen 1985; Tardif 1994; Heymann 2000).

10.3 Polygynous systems

Scramble promiscuity

In some social mammals, males show little tendency to guard females or the ranges they occupy, mating is promiscuous and multiple paternity of litters is common. Some of the best examples come from cavies, where promiscuous breeding systems have been described in several species, though in other species males defend resources used by females or adults form monogamous pairs (Adrian and Sachser 2011). For example, in common yellow-toothed cavies, which live in dry habitats where vegetation is sparse and food density is low, females occupy non-overlapping home ranges (Figure 10.6). When they become receptive, they start racing around and attract the attention of males who are unable to monopolize them (Schwarz-Weig and Sachser 1996). Males chase receptive females and attempt to copulate

Figure 10.6 Scramble competition for mating access to females is usual in yellow-toothed cavies and (as expected) testes size is unusually large. *Source*: © Norbert Sachser.

with them. Although copulating males are occasionally displaced by rivals, competing males usually queue peaceably until copulation has finished and then continue to chase the female. Females do not terminate solicitations until all males present have copulated with them and litters of mixed paternity are common. As would be expected (see Chapter 13), the relative size of male testes is among the highest known for mammals (Sachser *et al.* 1999) and males produce relatively large numbers of small sperm (Holt 1977).

Female-guarding polygyny

In a number of mammals, dominant males seek, associate with or guard individual females during part or all of their oestrous period but make little or no attempt to defend either female ranges or groups or particular sites. Systems of this kind are found in a number of large-bodied species where females range widely and their movements are difficult to predict, including elephants, giraffe, red and grey kangaroos, sperm whales and some populations of orangutans (van Schaik and van Hooff 1996; Singleton and Van Schaik 2001) (Figure 10.7, Box 10.1 and Figure 10.8). Many of these species show pronounced sex differences in body size and weight and, in some, males continue to grow throughout their lives (Box 10.2 and Figure 10.9).

(a) (b)

Figure 10.7 Two species where single males follow receptive females for limited periods: (a) African elephants and (b) sperm whales. *Sources*: (a)© Karen McComb; (b) © Hal Whitehead.

Box 10.1 Female-guarding polygyny in African elephants and sperm whales

Male African elephants continue to grow throughout most of their lives, and can reach twice the body weight of females (Poole 1994). Males leave their natal families at around 14 years and begin to associate with other males, though they rarely begin to breed regularly until they are over 30 years old (Lee *et al.* 2011). After reaching breeding age, they start to show an annual period of 'musth' at a predictable time of year (Poole *et al.* 2011). Musth males are aggressive towards other males, especially those in musth and escalated fights are not uncommon (Poole 1994). They show circulating testosterone levels five times higher than their non-musth levels and secrete a viscous liquid from swollen temporal glands as well as a trail of strong-smelling urine (Poole 1994) and making repeated low-frequency calls (Poole 1987, 1989a,b) (Figure 10.8). They feed little and spend their days searching for oestrous females, visiting different female groups in turn. When a dominant male locates a female in oestrus, he may guard her for several days, mating with her repeatedly, and a high proportion of calves are fathered by males in musth (Poole *et al.* 2011). The musth periods of younger and smaller males are relatively short and may be terminated by an attack by an older musth male but, as individuals increase in size and rank, their musth periods extend and coincide progressively with the time of year when most females conceive. Studies of elephant populations that are recovering from episodes of heavy poaching suggest that as the number of mature males in a population increases, the average duration of musth periods declines and the development of sexual behaviour in younger animals is progressively delayed (Poole 1994).

(a) (b)

Figure 10.8 (a) A mature bull elephant in musth. During their rutting period (musth) mature bulls feed little and are unusually aggressive. They secrete liquid from their temporal glands and often maintain partial erections and are intolerant of other adult males in the vicinity. (b) A musth male pursuing a female. *Source*: (a) © Vicki Fishlock; (b) © Phyllis Lee.

Bull sperm whales show mating strategies that may be similar to those of male elephants (see Figure 10.7) (Weilgart *et al.* 1996; Whitehead and Weilgart 2000). Females are commonly found in groups of up to twenty or more, consisting mostly (though not exclusively) of matrilineal relatives (Christal *et al.* 1998) and have ranges spanning distances of around 1000 km (Dufault and Whitehead 1995) in the tropics and subtropics. Males leave their mother's groups at around 6 years and migrate towards the poles, where food is more abundant. In their late twenties, when they are around 13 m long, males return to warmer waters and start to compete for access to receptive females, though they continue growing until their forties (Best 1979; Whitehead and Weilgart 2000). Breeding males are usually solitary and appear to avoid each other (Whitehead 1993), remaining in areas where females concentrate and moving between female groups (Whitehead 1993). When males locate receptive females, they court them and may remain with them for hours or days, before moving on to search for further females. As in elephants, the density of receptive females is low and calculations suggest that roving strategies may be favoured if the time taken to move between groups is less than the duration of a female's oestrus (Whitehead 1990).

Box 10.2 Male competition and continuous growth in kangaroos

In eastern grey kangaroos, multiple females share a common range, aggregating in unstable parties of three to twelve (Jarman and Southwell 1986; Jarman 1991, 2000) and local densities are relatively high (approximately fifteen females and five to six adult males per square kilometre) (see Figure 2.37b). When individual females are in oestrus (usually for a single day), they range more widely than usual and may attract the attention of males by their behaviour or by olfactory cues. Male grey kangaroos grow throughout their lifespans and reach weights of around 90 kg, more than twice the average weight of females (40 kg). Males belonging to the same local population establish a hierarchy by sparring and fighting and rank is closely related to age and weight. In most local populations, there is a recognisable alpha male, who has priority of access to resource and females and is responsible for 60–75% of all matings (Walker 1995, 1996). As males grow larger with increasing age, their rank increases until they are in a position to challenge for the alpha position. If they beat the existing alpha male and assume his status, their mating success increases immediately but they are seldom able to maintain their position for more than a year.

During the breeding season, adult males associate with female groups, moving between groups in search of oestrous females with whom they consort for anything between a few hours and several days. Fights between males are not uncommon and consorting males are likely to be displaced if a larger and more dominant male joins the group. Females appear to be able to determine whether or not they will mate with a particular male, though they usually accept the outcome of competition between males and show little evidence of active mate selection, generally mating one to four times during oestrous periods with up to four different males. Female groups are probably too unstable to be permanently defensible by males and, as females are commonly found in habitats where visibility is limited, the total area that a male can defend effectively may be small. Jarman suggests that the total area that a dominant male can search for receptive females is substantially larger than the area he could defend, so that dominant males maximise their mating opportunities by roving and would restrict their success if they defended territories (Jarman and Southwell 1986; Jarman 2000).

Figure 10.9 Male red kangaroos boxing. *Source*: © Ulrike Kloecker.

Range-guarding polygyny

Where females are aggregated in ranges small enough to be effectively defended, territorial defence may provide the most effective strategy for monopolising access to females. If females are solitary when foraging, males commonly defend ranges or territories that are larger than female ranges and associate with particular females at times when they are receptive. Mating systems of this kind are widespread in rodents (Waterman 2007), monotremes (Grant 1983), small marsupials

(Charles-Dominique 1983; Lee and Cockburn 1985), carnivores (Macdonald 1983; Rood 1986; Gese 2001; Kays 2003), ungulates (Jarman 1974; Dunbar and Dunbar 1979; Barrette 1987; Kishimoto 2003) and nocturnal prosimians (Bearder 1987; Kappeler and van Schaik 2002) and may represent the ancestral condition for all mammalian groups (Lukas and Clutton-Brock 2013).

Species that breed in this way differ widely in the relative size of male territories and female ranges as

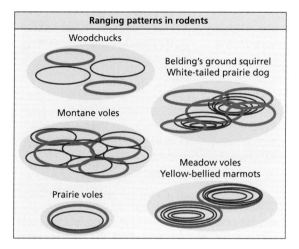

Figure 10.10 Schematic representation of contrasts in range overlap in North American rodents. Heavy line represents male territories or ranges, faint lines female territories or ranges. Range sizes are standardised and do not reflect contrasts between species. *Source:* Adapted from Michener (1983). Reproduced with permission from the American Society of Mammalogists.

well as in the extent to which the boundaries of male territories and female ranges coincide. The mating systems of rodents provide a good example of this diversity (Figure 10.10). In some species, like woodchucks, males and females each defend separate foraging areas and males visit neighbouring females in the breeding season (Michener 1983, Ferron and Ouellet 1989), while in others, like thirteen-lined squirrels and yellow-bellied marmots, males occupy ranges or territories that overlap the ranges of several females but their boundaries do not coincide, so that males usually have access to more than one female (Downhower and Armitage 1971; McLean 1982). Finally, in species where several breeding females share a common foraging area, like black-tailed prairie dogs, males defend territories whose boundaries coincide with those of multiple females and mating systems closely resemble those where males guard access to female groups (Michener 1983; Hoogland 1995).

The number of females that males can access in systems of this kind depends partly on the relative size of male territories and female ranges and partly on the extent to which female ranges overlap (Michener 1983; Kishimoto 2003; Ribble 2003). Where male territories are substantially larger than female ranges, males commonly have breeding access to several females and

can breed polygynously (Cockburn 1988; Clutton-Brock 1989; Ribble 2003)). Conversely, where male territories or ranges are no larger than those of females, males may have regular access to no more than one or two females. Since the number of females that males can access depends on the range size and density of females and these vary widely within species, these systems are often referred to as facultative polygyny.

Interspecific differences in the size of female ranges and the degree of polygyny are related to variation in diet and resource distribution (Ostfeld 1990; Waterman 2007). For example, among microtine rodents, female ranges are commonly clumped and overlapping in species that feed primarily on grasses and successful males often have breeding access to several females. In contrast, in species that depend on more patchily distributed food supplies, such as fruit or forbs, females are often territorial and widely dispersed and males occupy less intensively defended ranges overlapping the territories of multiple females (Ostfeld 1985).

Similar associations between food distribution, female density and mating systems occur within species. For example, in some montane subspecies of deer mice where density is low, males and females are monogamous, whereas in subspecies living at lower altitude, males are polygynous or rove in search of oestrous females (Wolff and Cicirello 1991). Similarly, in oribi, males form monogamous pairs with females in populations where female density is low and females are evenly distributed, but where female density is higher and females are aggregated, males defend access to several females (Brashares and Arcese 2002). Grouping patterns and mating behaviour also vary within populations in relation to contrasts in resource distribution (Kays 2003; Ribble 2003); for example, bushy-tailed wood rats living in small rock outcrops tend to form pairs while those living in larger outcrops form polygynous groups (Escherick 1981). Similarly, in Gunnison's prairie dogs, monogamous pairs are frequent where food distribution is uniform but are replaced by polygynous systems where food is more patchily distributed (Travis *et al.* 1995).

Group-guarding polygyny
Permanent harems
Where groups of females occupy overlapping ranges or forage together, a single mature male commonly associates closely with each group of females and defends

them against rival males. Breeding groups of this kind are usually referred to as 'harem' groups or 'uni-male' groups: some of the best-studied examples of societies of this kind are of leaf-eating monkeys, including the African colobus monkeys and the Asiatic langurs (Figure 10.11) Where female groups occupy well-defined ranges, males commonly defend access to their range as well as to groups themselves or to the burrows, caves or roosting sites regularly used by females (Kerth *et al.* 2011) while in species where females live in stable social groups in ranges that are too large to be effectively defended, males commonly defend access to groups rather than their ranges.

Seasonal harems

Where breeding seasons are long or the timing of breeding is unpredictable, males may defend access to female groups throughout the year, while in species where reproduction is strongly seasonal, defence of female groups is often limited to the breeding season (Figure 10.12). For example, in red deer, males spend most of the year in bachelor groups in areas adjacent to those used most intensively by females (Clutton-Brock *et al.* 1982) but during the weeks before the onset of the breeding season in late September, bachelor groups break up, as males leave their usual ranges and move to areas used principally by females to collect and defend harems (Figure 10.13), mating with

Figure 10.11 Like many other langurs, silvered leaf monkeys live in harem groups that usually include a single mature immigrant male. *Source*: © Perry van Duijnhoven.

Figure 10.12 Greater kudu form temporary harem groups during the breeding season, but during the rest of the year males may either be solitary or live in loose bachelor groups. *Source*: © Norman Owen-Smith.

Figure 10.13 In red deer, males defend seasonal harems that often include females from more than one matrilineal group.
Source: © Martyn Baker.

females (hinds) as they come into oestrus (Gibson and Guinness 1980 (Figure 10.14).

Similar mating systems are also found in some seals where breeding is strongly seasonal, females aggregate on traditional breeding beaches to give birth and males collect and defend harems of females. For example, in elephant seals, females aggregate on particular beaches where they are safe from disturbance throughout the breeding season and males defend large harems that may include more than fifty females (Le Boeuf and Reiter 1988). Across species, the size of harem groups varies in response to variation in the distribution of females. For example, in seals that breed on pack ice, females are widely dispersed and males defend single females or small groups of females, while in most land-breeding species, where females are more densely aggregated, males often defend much larger harems (Stirling 1975; Le Boeuf and Reiter 1988).

Associations of harems

In some species where males defend access to several females, harem groups regularly associate with each other in larger herds. For example, in gelada baboons, multiple harem units consisting of a single breeding male and several breeding females aggregate in large herds (Dunbar, 1984; see Figure 10.15). A similar pattern of social organisation occurs in plains zebra and wild horses. In some species, social connections between harem groups can be identified at several different levels. For example, hamadryas baboons live in one-male units or harems, usually consisting of a breeding male and one to five breeding females (Kummer 1968). Groups of two or three one-male units commonly associate with each other, forming 'clans' whose males are often relatives (Stolba 1979) and several clans plus a number of single males form an independent foraging unit or band. Three or four different bands may use a single set of sleeping cliffs: each morning, when the animals come down from the cliffs, each band moves off in different directions with one-male units belonging to the same clan, travelling close to each other. One-male units form when a sub-adult male leaves his natal unit and forms a close relationship with a juvenile or sub-adult female from one of the other units in his clan, though some males may also attach themselves to an existing unit as a follower (Kummer 1968; Kummer *et al.* 1974). Males subsequently try to attract additional females to their unit, sometimes taking adult females away from mature males

Figure 10.14 A red deer stag repeatedly tries to mount a female, who finally stands, allowing him to mount and ejaculate. *Source*: © Tim Clutton-Brock.

or replacing harem holders altogether, usually during one of the intense battles among adult males belonging to the same band that erupt from time to time. Females often belong to several one-male units at different stages during their lives as a result of take-overs by males, sometimes moving between clans or even between bands (Swedell *et al.* 2011).

Dynamics of harem groups

The dynamics of harem groups vary. Where females are philopatric and breeding groups persist throughout the year, males typically leave their natal group after adolescence and compete for access to female groups (see Chapter 12). In some species, males compete to displace residents on their own; in others, several males (who are often brothers) leave their natal group together and attempt to displace resident males, and, in a few, dispersing males initially join bachelor groups which invade breeding groups and attempt to displace residents and individual males subsequently compete to establish dominance and evict subordinates. For example, in grey langurs, males leave their natal groups as young adults and join bachelor groups whose members invade breeding groups in the mating season and may displace resident males, subsequently fighting among themselves until only one or two are left (Sugiyama *et al.* 1965; Borries 2000).

In many species that live in harem groups, the tenure of males is relatively short and females are likely to breed with several dominant males in the course of their lives (see section 10.6), though the duration of male tenure varies widely between species, ranging from less than a year to more than a decade (Clutton-Brock 1989). In some cases, males intersperse periods when they defend females or territories with periods when they rejoin nonbreeding groups and regain lost condition (Prins 1989; Clutton-Brock *et al.* 1993) while, in others, males that have gained access to a female group and subsequently lost it seldom breed again (Sugiyama 1967; Sommer and Rajpurohit 1989).

In a small number of species, including hamadryas baboons, gorillas and several equids, females usually leave their natal group after adolescence and harem groups are formed by young males that emigrate from their natal group and attract or kidnap one or more females from another group, forming the nucleus of new breeding groups (see Chapter 3). Incipient groups are then joined by emigrant females from other groups, so that few female group members are close relatives. Since natal males are likely to be unrelated to resident females, they may be accepted by females as mating partners and may inherit the breeding position in their natal group if the dominant dies (Harcourt and Stewart 2007) (see Chapter 12).

Subordinate males

In many mammals that breed in harem groups, some groups contain one or more adult natal males which seldom breed within the group, though they may mate

Figure 10.15 Gelada baboons live in single male breeding units that aggregate in large herds. *Source*: © Noah Snyder-Mackler.

with females from neighbouring groups. Subordinate males may either be individuals who have been born in the group and have not yet dispersed or animals that have immigrated from other groups and are tolerated by the dominant male. In most cases, natal males are subordinate to immigrants in social interactions, though this is not invariably the case: for example, in spotted hyenas, immigrant males (which are responsible for almost all breeding) are subordinate to natal males that have not yet dispersed (Holekamp and Smale 2000)

In many species, subordinate males assist dominant males (and, in some cases, dominant females) in repelling intruders or challengers and their presence may help to reduce the frequency of extra-group paternities or prolong the tenure of established males (see Chapter 13). Where females are philopatric, natal males are likely to be related to many resident females and are seldom accepted as mates (see Chapter 4) and eventually disperse and attempt to breed in other groups, rarely inheriting the dominant position in their natal group. Dispersing subordinates may either attack and evict the resident dominant male in an established breeding group or may join an established group as a subordinate and inherit the dominant position after the death of the previous dominant. For example, in

gelada baboons, some dominants acquire their position by immigrating as subordinates and ousting the breeding male while others join breeding groups as subordinates and either inherit the breeding position after the death of the dominant or displace him after some years (Dunbar 1984).

In species where most females habitually disperse to breed (as in gorillas and social equids), subordinate males may also either be immigrants or natals. However, since both natals and immigrants are likely to be unrelated to many resident females, both may breed with resident females and inherit the breeding position. The presence of natal subordinates consequently represents a threat to the breeding success of the dominant male and dominant males often guard receptive females closely (Berger 1986; Harcourt and Stewart 2007).

Multi-male groups

Where females live in relatively large groups that commonly include more than three breeding females, several mature males often associate with groups of females and breed with them. In primates, multi-male groups appear to have evolved directly from systems where females were solitary and uni-male harem groups may represent a derived state (Schultz *et al.* 2011; Dunn *et al.* 2015).

(a)　　　　　　　　　　　　　　　　　　　　(b)

Figure 10.16 Two species that commonly form stable groups that include multiple breeding females and multiple breeding males: (a) sifakas and (b) banded mongoose. *Sources*: (a) © Flávia Koch; (b) © Jennifer Sanderson.

Multi-male groups are more frequent in species where female reproduction is seasonal, so that more than one female group member is often in oestrus at the same time (Nunn 1999) (see Chapter 11). Species that commonly form multi-male groups include African lions (Schaller 1972), banded mongooses (Rood 1986; Cant 1998), ring-tail lemurs (Jolly 1966), sifakas (Richard 1987), howler monkeys (Pope 2000), red colobus monkeys (Struhsaker 1975), spider and woolly monkeys (Di Fiore and Campbell 2007) and African buffalo (Figure 10.16) (Prins 1996). Relationships between breeding males from neighbouring groups are commonly hostile and resident breeding males often cooperate to defend their group against male intruders (Kowalewski and Garber 2015; Zhang *et al.* 2006). Cooperative defence of breeding groups by resident males may help to increase their paternity certainty and may also reduce the risk of infanticide by incoming males. In some species, there is evidence that participation in group defence is increased in males that have sired offspring in the group For example, in black howler monkeys, males that have fathered offspring in the group contribute more to howling 'choruses' than those that have not done so (van Belle *et al.* 2014).

Kinship between resident males varies widely in these systems. In monotocous species where males disperse to breed, like chacma baboons and African buffalo, resident males are usually immigrants that have joined the group independently and are mostly unrelated to each other, whereas in litter-bearing species, like lions and meerkats, several males often disperse and immigrate together and resident males are often close relatives. Resident males are also likely to be related to each other in species where females habitually disperse to breed, like chimpanzees, red colobus monkeys and spider monkeys, though average kinship in these species is not necessarily high (see Chapter 12).

While relationships between males from neighbouring groups are commonly hostile whether males are related or not, they are particularly hostile where defending males are related to each other, as in chimpanzees, spider monkeys, meerkats and lions, and many examples of lethal encounters between males from rival groups involve societies of this kind (Figure 10.17 and Box 10.3). Conversely, relationships between adult males that are resident in the same group appear to be more tolerant and relaxed in species where males are related (see Chapter 14). However, interspecific differences in relatedness between resident males are usually associated with familiarity and an alternative interpretation is that relationships between resident males are most likely to be tolerant and relaxed where they have grown up together, whether they are closely related or not.

Site-guarding polygyny
Resource-based mating territories
Instead of defending the ranges of female groups or female groups themselves, males in some species defend mating territories much smaller than female ranges in areas regularly visited by females and associations between males and particular females are short-lived. For example, in Grevy's zebra, which live in the arid grasslands of northern Kenya, female ranges cover

(a)   (b)

Figure 10.17 Chimpanzees live in fission–fusion communities that typically include multiple immigrant females and a number of natal breeding males, while bonobos live in more cohesive social groups of similar structure. (a) An all-male group of chimpanzees patrol the boundary of their community's range. Patrolling groups listen for intruders and attack single males or parties smaller than their own, sometimes with lethal consequences. (b) A group of bonobos cautiously approaches a food source. Social relationships within and between bonobo communities are less aggressive than those of chimpanzees. *Sources:* chimpanzees © Xavier Hubert-Brierre; bonobos © Takeshi Furuichi.

Box 10.3 Mating systems in chimpanzees and bonobos

Both chimpanzees (Figure 10.17a) and bonobos (pygmy chimpanzees) (Figure 10.17b) live in stable communities that include multiple breeding males as well as multiple breeding females. In both species, individuals forage in unstable subgroups, though groups tend to be larger and more cohesive in bonobos (White 1996, Wrangham *et al.* 1996). One dominant male in each community mates more frequently with females around the time of ovulation than other males, but dominance shows little relationship to mating success among other males (Tutin 1979; Tutin and McGinnis 1981; Hasegawa and Hiraiwa-Hasegawa 1983; Ihobe 1992; Kano 1992; Takahata *et al.* 1996). Within chimpanzee communities, males commonly form competitive coalitions with each other to establish and maintain their social rank (Kano 1996; Takahata *et al.* 1996; White 1996) while in bonobos, males maintain long-term bonds with their mothers, who support their sons in interactions with other males (Ihobe 1992; Kano 1992, 1996). Dominant male chimpanzees are generally tolerant of reproductive activity among subordinates and matings are seldom interrupted though, in some populations, males form consortships with oestrous females and attempt to induce them to travel to less used areas of the community range, presumably because this increases the chance that they can monopolise reproductive access throughout the female's oestrous period (McGinnis 1979; Tutin 1979, 1980). Reproductive competition between males appears to be less frequent and less intense in bonobos than in chimpanzees (Nishida and Hiraiwa-Hasegawa 1987; Kano 1996), possibly as a consequence of reduction in the seasonality of their habitats and food supplies (Wrangham 1986). An alternative suggestion is that breeding competition is reduced because female bonobos are receptive for a larger proportion of the oestrous cycle than female chimpanzees (Savage-Rumbaugh and Wilkerson 1978; Kano 1996) but recent studies do not support this conclusion (Takahata *et al.* 1996).

   While relationships between males chimpanzees belonging to the same community are relatively relaxed, relationships between males from neighbouring communities are often tense or hostile. Resident males form 'patrolling' groups that visit the boundaries of their range or intrude into the ranges of neighbouring communities. If 'patrolling' groups encounter solitary males or parties smaller than their own, they commonly attack them. Attacks are intense and can be prolonged and victims that are unable to escape may be killed (Wilson and Wrangham 2003; Wilson *et al.* 2004). Females from neighbouring communities may also be attacked, especially if they are not sexually receptive (Williams *et al.* 2004) and infants may also be killed (Kutsukake and Matsusaka 2002). These attacks may represent attempts to maintain or increase the size of the group's range: one explanation of the territorial behaviour of male chimpanzees is that they are defending resources used by their community while another is that they are attempting to increase the number of females to which they have access. Analysis of long-term data from Gombe supports the first of these two explanations, for the number of females in communities did not increase with the size of community ranges while measures of foraging behaviour (including time spent in larger foraging parties) and female reproductive rates increase with the size of community ranges (Williams *et al.* 2004). Much less is known of relationships between neighbouring communities in bonobos, but they do not appear to be as hostile as those between communities of chimpanzees (Hohmann, 2001).

hundreds of square kilometres and are too large to be effectively defended my males (Klingel 1974; Rubenstein 1986). Like other equids, female Grevy's zebra have a post-partum oestrus and need access to water during the weeks following the birth of their foal and males defend mating territories close to water sources used by breeding females, mating with those that move through their territories.

Breeding systems of this kind are common in large herbivores that live at comparatively low density in relatively open habitats (Owen-Smith 1975; Leuthold 1978; Gosling 1986a). They also occur in cheetah, where a proportion of males defend territories substantially smaller than female ranges in areas regularly used by gazelles, their main prey (Caro 1994), as well as in some populations of orangutans (Rodman and Mitani 1986) and some bats (Bradbury and Vehrencamp 1977; Kerth 2008). In aseasonal breeders, male mating territories may be defended throughout the year, while in seasonal breeders territory defence is commonly confined to the mating season (Gosling 1986a; Clutton-Brock *et al*. 1993) though males may return to the same territory in successive seasons (Young and Franklin 2004).

The size and distribution of male mating territories vary widely. Where resources used by females are aggregated in areas that regularly attract large numbers of females, males may defend relatively small territories. For example, in black lechwe (a flood-plain antelope belonging to the Reduncinae), large herds of females aggregate on the productive grasslands on the immediate margins of the flood and territories seldom exceed 1–2 ha (Thirgood *et al*. 1992) (Figure 10.18a). In puku (a riverine antelope from southern Africa), males defend contiguous territories of around 5 ha throughout the year on old river meanders which are regularly visited by females in search of annual grasses and forbs (Rosser 1992) (see Chapter 4). In contrast, the size of male mating territories is much larger in antelope living in more arid areas where resources are not so strongly localised, like waterbuck, springbok and gerenuk (Figure 10.18b) (Walther *et al*. 1983).

In several of the gazelles and antelopes where males defend large resource-based territories, they establish one or more stamping grounds within their territories which are marked with urine and faeces (Walther *et al*. 1983; Gosling 1986a; Jackson and Skinner 1998). Males often attempt to herd receptive females onto these stamping grounds and, when no females are in their territories, they return and rest there.

In some cases, territorial males tolerate satellite males within their territories (Owen-Smith 1972, 1975; Wirtz 1981, 1982). Satellites may benefit from joining territorial males either because their breeding frequency (though usually low) is higher than it would be if they were not resident in a territory or because they acquire improved chances of inheriting the territory when the resident male ages or dies, while resident males may benefit from their assistance in expelling intruders (see Chapter 11).

Where males defend resource-based territories, the numbers of females visiting particular territories and the mating success of the males defending them are often related to the resources they contain. For example, in puku, there is a close association between the area of productive forb-rich grassland in the territories of different males, the number of females visiting their territories and observational estimates of their mating success (Rosser 1987) (see Chapter 4). Similarly, in populations of fallow deer in southern England, where males defend territories under oak trees visited by females in search of acorns (Figure 10.19), individual trees differ in the size of their acorn crop and there is a close association between the number of does visiting a male's territory and the acorn production of the tree or trees he is defending (Clutton-Brock *et al*. 1988b). Experiments that manipulated the density of food under oak trees defended by bucks changed the attractiveness of territories and the mating success of the bucks that guarded them (Figure 10.19).

Site-guarding polygyny appears to develop where female ranges are too large to be defensible, female groups are small or unstable and resources are clumped, so that females regularly return to the same sites and habitats are relatively open (Gosling 1986a; Clutton-Brock 1989). For example, in antelope that live in open savannahs, where female ranges are too large to be defended, males commonly defend resource-based territories substantially smaller than female ranges, whereas in smaller species where females do not range as widely (like reedbuck and bushbuck), males typically defend larger territories covering most of the ranges of individual females or female groups.

The relationship between female ranging patterns and the size of male mating territories is well illustrated in the Reduncinae (a group of riverine or flood-plain antelopes that include reedbuck, water buck, kob, puku and lechwe) (Clutton-Brock *et al*. 1993). In reedbuck, females typically live in groups of one to three in small ranges of

(a)

(b)

Figure 10.18 Where female ranges are too large to be defended and females are solitary or live in unstable groups, male ungulates commonly defend mating territories smaller than female ranges in areas where resources regularly attract females. (a) Male black lechwe defend small resource-based territories on the edge of the Bangweulu floodplain in Zambia; (b) while male waterbuck (which, despite their name, live in drier areas than lechwe), defend larger resource-based territories around lakes and rivers. *Source*: © Tim Clutton-Brock.

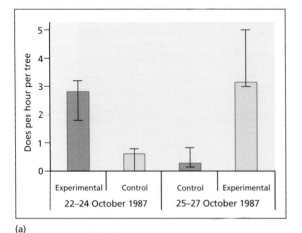

(a)

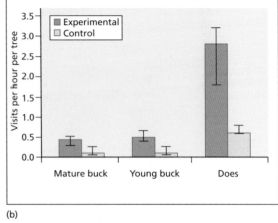

(b)

Figure 10.19 Effects of supplementing food on resource territories under oak trees defended by fallow deer bucks during the rut in southern England. In these experiments, 25 kg of deer pellets were spread at first light under oak trees defended by bucks on each of 3 days, while control territories were visited but no food was distributed. (a) The mean number of does per hour under experimental versus control trees over an initial 3-day period and then over a subsequent 3 days when treatments were reversed on the same territories. (b) The combined number of visits paid to experimental and control territories by mature bucks, young bucks and does. *Source:* From Clutton-Brock *et al.* (1988b). Reproduced with permission of Springer Science and Business Media. *Photo source:* © Ben Pitcher.

less than 100 ha and individual males defend territories covering the entire ranges of female groups for protracted periods. In contrast, female waterbuck have ranges of up to 1000 ha and males defend territories that only cover part of female ranges in areas regularly visited by females. Finally, in Uganda kob, white-eared kob and Kafue lechwe, females have much larger ranges but aggregate in big herds whose movements are unpredictable while males defend very small territories (often less than 0.1 ha in size) on leks for a few weeks at a time (Clutton-Brock *et al.* 1993).

Leks

Lek breeding systems, where males defend very small mating territories that do not contain significant resources in well-defined clusters, occur in some bats (Bradbury 1977), pinnipeds (Fay *et al.* 1984) and ungulates (Clutton-Brock *et al.* 1993), as well as in some birds, fish and insects (Höglund and Alatalo 1995)

(Figure 10.20). In some cases, leks persist throughout the season though individual males frequently replace each other; in others, leks re-form in successive mating seasons, often on the same site. The numbers of males per lek vary widely both within and between species, ranging from three to five males to over 50 or more. Somewhat similar systems involving more mobile males that do not defend well-defined territories occur in some marsupials, procyonids and lemurs. For example, in white-nosed coatis (where females live in matrilineal groups while males are usually solitary) up to a dozen males display around female bands throughout the brief mating season, females commonly mate with multiple partners and reproductive skew among males does not appear to be high (Booth-Binczik *et al.* 2004). In some ways, the mating system of coatis appears to resemble that of ring-tailed lemurs, which are not usually regarded as lek breeders: males are resident in female groups throughout the year and both residents and visiting

Figure 10.20 Part of an Indian blackbuck lek in Rajasthan: note that the only female on the lek is standing on an area that has been heavily used by previous animals. *Source*: © Jayabharathy Ranganathan.

males may mate with females (Jolly 1966; Sauther and Sussman 1993). Both these systems differ from the leks of ungulates since males are not territorial and it is questionable whether either should be regarded as analogous to the leks of birds or ungulates.

Mammalian leks have been most extensively studied in ungulates, where they occur both in seasonal breeders, like fallow deer, and in relatively aseasonal ones, like Uganda kob (Clutton-Brock *et al*. 1993). In both seasonal and aseasonal breeders, leks form on traditional sites that are used in successive years while, between leks, younger or more subordinate males defend larger territories that contain resources used by both sexes. For example, in Uganda kob, mature males defend small mating territories that do not contain obvious resources in clusters or 'territorial grounds' (Figure 10.21). Female kob may visit or rest on leks and receptive females commonly collect there, but females that visit leks spend little time grazing. Between leks, younger or smaller males defend larger mating territories. Most matings occur towards the centre of the lek, while younger or less dominant males

defending larger territories between clusters seldom mate successfully.

Males that defend territories or leks usually feed little and rapidly lose condition, and after 1–2 weeks defending a territory they return to grazing herds to replenish their resources before returning again, though not necessarily to the same territory (Balmford 1990, 1992; Deutsch 1992, Deutsch and Weeks 1992). In all lek-breeding ungulates that have been studied so far, there are typically large differences in mating success between territorial males (Figure 10.22) which are often correlated with individual differences in age and size, as well as with variation in horn length and display rate (Clutton-Brock *et al*. 1993; Höglund and Alatalo 1995; Bro-Jørgensen and Durrant 2003). The period for which males defend their territories is often relatively short and breeding males frequently replace each other on the same territory though in topi, it can last for several years (Bro-Jørgensen 2011; Bro-Jørgensen and Durant 2003).

Among ungulates, lek breeding appears to be associated with the formation of large, unstable, mixed-sex

Figure 10.21 (a) Schematic map of mating territories in Uganda kob. From Leuthold (1966). Reproduced with permission from Brill. Males holding territories in clusters (leks) obtain almost all matings but younger or more subordinate animals hold larger territories between leks but obtain very few matings: (b) female kob in feeding groups; (c) an active lek; (d) a territorial male. *Photo sources*: © Tim Clutton-Brock.

herds that range widely and rely on unpredictable resources (Gosling 1986b; Clutton-Brock *et al*. 1989, 1993). For example, among the reduncine antelopes, males of species occupying relatively mesic areas, like puku, red lechwe and black lechwe, defend resource-based territories in areas regularly visited by females, while males of species that rely on resources that are more widely and less predictably distributed, like Kafue lechwe, Uganda kob and white-eared kob, defend small territories on leks (Clutton-Brock *et al*. 1993; Deutsch 1994). Within species of lek-breeding ungulates, the formation of leks is associated with high local densities of females and is replaced by male defence of isolated

territories in populations where female density is relatively low (Monfort-Braham 1975; Apollonio 1989; Langbein and Thirgood 1989; Balmford 1990; Clutton-Brock *et al*. 1993; Thirgood *et al*. 1999).

Lek breeding in ungulates is of particular interest since, at first sight, it appears to be an exception to the general rule that the distribution of females depends on the distribution of resources, while the distribution of males depends on that of females. The commonest explanation of the evolution of leks is that mating on leks provides females with opportunities to compare mating partners and to select high-quality mates that will maximise their chances of conceiving

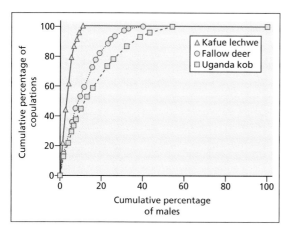

Figure 10.22 Variation in mating success among males in three lek-breeding ungulates: fallow deer, Kafue lechwe and Uganda kob. Plots show the cumulative percentage of copulations on leks involving different percentages of territorial males, ranked on their relative success. *Source*: From Clutton-Brock *et al.* (1993). Reproduced with permission of Elsevier.

and the quality of their offspring (Höglund and Alatalo 1995; Carbone and Taborsky 1996; Bro-Jørgensen 2010). However, this argument provides no explanation of why female mate choice should be particularly

important in species that form leks, nor does it account for the ecological correlates of lek breeding. A further reason for being sceptical is that attempts to demonstrate consistent female mating preferences in lek-breeding mammals have produced mixed results. For example, four studies have investigated whether females on leks are more likely to mate with successful males per unit time spent in their territories (Clutton-Brock *et al.* 1993). Of these, three (of kob, lechwe and fallow deer) found no indication that mating probability per female varies with male success, while a fourth (of topi) found that the latency from the time females entered territories to mating was lower for lek territories than for single territories, as well as for central lek territories versus peripheral ones (Bro-Jørgensen 2002). While this last result could indicate that oestrous female topi prefer to mate with males holding central territories, an alternative possibility is that females that are in or close to oestrus are more likely to aggregate on central territories for other reasons.

Detailed studies of lek-breeding ungulates also show that there are several different processes that contribute to variation in male mating success on leks (Box 10.4). Despite the absence of significant resources on mating territories, females often show consistent preferences for

Box 10.4 Mating on kob and fallow deer leks

The most detailed studies of mating behaviour in mammals involve lek-breeding ungulates, including Uganda kob, fallow deer, Kafue lechwe, Indian blackbuck and topi (Clutton-Brock *et al.* 1993). Uganda kob form large unstable herds, including females and males of different ages. When females approach oestrus, they leave feeding herds and move to leks, where groups of five to fifty males defend small clustered mating territories. There are commonly several leks within a female's range, so that females have to decide which to choose: they tend to prefer those where grass cover is short and avoid those that are covered or surrounded by grass long enough to hide predators. Experiments where leks and the area around them were mowed show that this increased their attractiveness to females and raised the mating success of resident males (Deutsch and Weeks 1992).

Once on the lek, receptive females collect on particular territories and the males that defend attractive territories show high mating success (Leuthold 1966; Balmford *et al.* 1992). Larger or older males typically defend the most popular territories and have higher mating success than smaller or younger territory holders (Balmford *et al.* 1992). Although results of this kind are often interpreted as evidence of mating preferences by females in lek breeders, female kob appear to prefer particular territories rather than the males that occupy them (Balmford 1990; Balmford and Read 1991) and preferred territories are usually those that already contain several females or which have recently been heavily used by females. Experiments involving the transfer of earth between popular and unpopular territories show that this has a strong effect on the number of females that move there as well as on the mating success of the males that defend them (Deutsch 1992) and similar experiments with Kafue lechwe produced identical results (Deutsch and Weeks 1992). In kob, territories that attract females tend to be distant from thickets or other places where predators can hide and are often close to the centre of the lek and proximity to the sites of past predator kills decreases the mating success of males holding territories nearby. Since the most popular territories tend to be held by the largest and most competitive males, preferences for particular territories generate correlations between mating success and male phenotype which may be caused partly or wholly by male competition.

One way of determining whether female preferences for particular partners have an important influence on male success is to experimentally force males to change territories and to determine whether individual differences in mating success persist. Experiments of this kind have been used to assess the role of female mating preferences in fallow deer where (as in kob) mature males defend small mating territories in clusters that attract receptive females. Here, too, there are large differences in mating success between males that are partly a consequence of the characteristics of the territories they occupy (Clutton-Brock *et al.* 1988b). In an attempt to identify the extent to which female mating preferences contributed to variation in male success, samples of territorial males were forced to change territories during the peak period of the rut by pinning sheets of black plastic (which were avoided by females) to the centre of their territories (Clutton-Brock *et al.* 1989). As would be expected if female mating preferences played an important role in determining male success, males that were successful before they were forced to move were also successful on their new territories, while males that had previously been unsuccessful were also unsuccessful on their new territories (Clutton-Brock *et al.* 1989). However, even evidence of this kind does not necessarily indicate that variation in male success is caused by female preferences for particular mating partners since correlations between the success of males on different territories may be generated by individual differences in the capacity of males to retain females on popular territories and therefore do not provide conclusive evidence of female mate choice (Clutton-Brock and McComb 1993).

   An alternative approach, which has been successfully used to demonstrate the role of female mating preferences in birds (Höglund and Alatalo 1995), is to modify the characteristics or attributes of males and assess effects of female movements and mating success. Experiments of this kind with captive female fallow deer that had been induced into full oestrus showed that, as in natural populations, receptive females exhibited a strong tendency to join other females and that the characteristics of males apparently had little effect on their movements (Clutton-Brock and McComb 1993) (Figure 10.25a). Receptive females were as frequently attracted to paddocks containing younger versus older males (Figure 10.25b) and those containing males with and without antlers (Figure 10.25c). In contrast, they showed strong preferences for paddocks containing a male with females (Figure 10.25d) that were absent in anoestrous females (Figure 10.25e). However, they showed no preference for paddocks containing bucks they had not seen mating (Figure 10.25f). Moreover, when they were offered a choice of paddocks containing a male with females and paddocks containing only females, they showed no preference for those containing a male with females (Figure 10.25g), suggesting that the tendency for oestrous females to join larger breeding groups may occur because they are attracted to other females.

moving to particular territories. For example, in Uganda kob, the territories particular males hold play an important part in determining their mating success in different bouts of reproductive activity (Balmford 1991) (Figure 10.23). Predation risk exerts important effects on territory preferences (Figure 10.24) and territories preferred by females tend to be those which are well away from thickets where predators can hide, and the experimental removal of thickets increases female use of particular territories (Deutsch and Weeks 1992). Experiments with natural populations of kob and lechwe also suggest that females are attracted by olfactory cues to areas previously used by other females, while experiments with fallow deer show that females in oestrus have a strong tendency to aggregate with each other and that the distribution of other females may have a stronger effect on female movements than the characteristics of males (Figure 10.25).

   A first question to ask is why females leave feeding groups or herds as they approach oestrus. While it is possible that they do so in order to compare and select potential mating partners, another possibility is that they would otherwise be exposed to high levels of dangerous harassment by multiple males since both the membership of grazing herds and dominance relationships

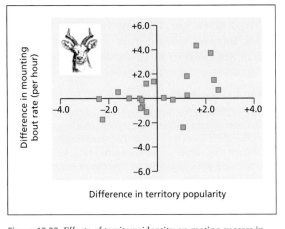

Figure 10.23 Effects of territory identity on mating success in Uganda kob. The figure shows the mating success of the same males on different territories during successive periods of active competition for mates. Differences in mating success (measured as rate of mounting bouts) of the same males on different territories are correlated with differences in the number of females visiting different territories and the mean success of other males on the same territories (territory popularity). When males defend more 'popular' territories, where other males are relatively successful, in successive bouts of territorial defence, their mating success usually increases, while it declines when they move to territories where other males are relatively unsuccessful. *Source*: From Balmford (1991). Reproduced with permission of Elsevier.

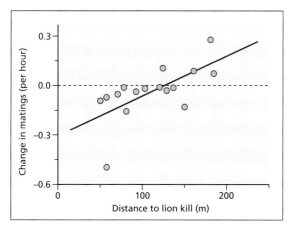

Figure 10.24 The association between changes in the mating success of male kob defending territories on a lek before and after a lion kill and the distance from their territory to the kill, showing data for 16 territories on which at least one mating occurred. *Source*: From Deutsch and Nefdt (1992). Reproduced with permission of Macmillan Publishers Ltd.

between males in these groups are unstable. Observations of females that are approaching oestrus in mixed-sex herds in several lek-breeding ungulates show that they are frequently harassed by multiple males and that mating pairs are commonly attacked by other males and females are sometimes killed (Fryxell 1987; Clutton-Brock *et al.* 1993). While females may be frequently disturbed by resident males while they are in male territories, they are usually safe from dangerous harassment by multiple males and are seldom damaged by territory holders.

So why do receptive female ungulates aggregate on leks rather than on isolated mating territories? Many studies of lek breeding mammals have argued that receptive females move to leks to take advantage of opportunities for choosing mating partners in order to gain indirect fitness benefits from mating with particular males, citing evidence of correlations between male mating success on leks and the phenotypic characteristics of males, such as horn or body size, or between mating success and aspects of male behaviour (Hoglund and Alatalo 1995). However, correlations of this kind do not provide reliable evidence of female mating preferences (see Chapter 4) and, while it is possible

that female ungulates move to leks to increase their choice of mating partners, they may also gain direct fitness benefits by doing so. Females are probably safer from predators with other females on leks than on isolated male territories and there is evidence that they show strong preferences for joining other females and that the risk of predation also has an important influence on their choice of territories. It is also possible that the tendency for receptive females to aggregate on leks is partly a by-product of their frequent moves between neighbouring territories caused by the attentions of territory holders or by intrusions by neighbouring males. A model of random female movements between male territories (Stillman 1993, 1996) shows that if females move repeatedly between neighbouring territories before mating, they will tend to aggregate in the centre of clusters of male territories. In these circumstances, males in their prime will benefit by holding territories in clusters and are likely to compete for central territories, generating correlations between male mating success and phenotypic attributes of males that are related to their competitive ability.

Arguments over the evolution of lek breeding are not yet resolved. One general conclusion that emerges from studies of lek-breeding mammals is that several different processes probably contribute to the movements of receptive females and that it is difficult to separate the effects of male competition, female aggregation and female preferences for particular territories from those of female preferences for mating with particular males. Producing unequivocal evidence of female mating preferences on natural leks is beset with obstacles, so its absence does not necessarily suggest that females do not show strong mating preferences. However, evidence of large differences in mating success between males does not necessarily indicate that mate choice is strong and other mechanisms responsible for the evolution of lek-breeding systems may include avoidance of harassment and predation risk by females and the possibility that the aggregation of receptive females on clustered male territories is a by-product of repeated movements between territories combined with a tendency to move to the nearest male. Further studies are needed that trace the movements of individual females from grazing herds to their eventual mating sites and partners and explore the mechanisms affecting their movements at each stage.

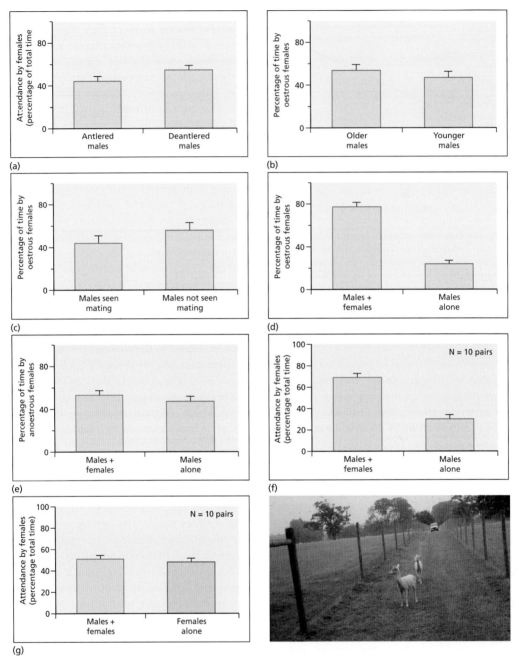

Figure 10.25 Female preferences in captive fallow deer. In these experiments, females that had been induced into oestrus were presented with paddocks containing different combinations of animals and the time they spent next to each paddock was recorded. Bars show the percentage of time spent by oestrous females in front of paddocks containing (a) antlered versus de-antlered male fallow deer; (b) older males (5–7 years) versus younger males (2 years); (c) males they had seen mating versus those they had not seen mating; (d) males with females versus those without; (e) repeats (d) but used anoestrous females who showed no preference for males with females. In a second series of experiments, females were initially offered a choice of males with females and males alone and, once again, favoured males with females (f), but showed no preference for males with females over females on their own (g). *Source*: From Clutton-Brock and McComb (1993). Reproduced with permission of Oxford University Press. *Photo source*: © Tim Clutton-Brock.

10.4 Genetic mating systems

The previous sections of this chapter have been concerned with contrasts in the guarding behaviour of males and the form of mating systems. However, patterns of association between the sexes do not necessarily reflect the distribution of paternity and the real nature of mating systems. The discovery of genetic techniques that make it possible to identify paternity (Jeffreys *et al.* 1985) has led to dramatic changes in our understanding of animal mating systems, especially those of monogamous species. In many mammals, females often attempt to mate with more than one partner and recent studies show that they can gain direct as well as indirect fitness benefits by doing so (see Chapter 4). While guarding males may attempt to prevent females from mating with other partners, they are often unable to do so completely and their success depends on the effectiveness of male guarding strategies as well as on the structure of groups.

The frequency with which females breed with males that are not members of their social group (extra-group paternity or EGP) ranges from less than 5% in some monogamous species such as California deer mice (Ribble 1991, 1992) to over 60% in some primates (Clutton-Brock and Isvaran 2006; Isvaran and Clutton-Brock 2007; Cohas and Allainé 2009). Contrasts in the frequency of EGP between species could be a consequence of variation in the benefits to females of choosing to mate outside their group (Petric and Lipsitch 1994; Slagsvold and Lifjeld 1994), of the availability of alternative partners (Møller and Birkhead 1993) or of the capacity of resident males to monopolise access to receptive females and constrain female choice (Stutchbury and Morton 1995, Clutton-Brock and Isvaran 2006). The number of mammals for which genetic estimates of paternity are available is still small, but the available evidence suggests that variation in the ability of resident males to monopolise access to females is responsible for much of the variation. Rates of EGP increase as the number of breeding females in groups rises and decrease as the length of mating seasons increases (Figure 10.26). In addition, the extent to which the most dominant resident male in each group loses paternity ('extra-dominant paternity') is typically higher in monogamous species than in polygynous ones (Clutton-Brock and Isvaran 2006; Cohas and Allainé 2009) (Figure 10.27a). Rates of extra-group paternity are also higher in species where the sexes are intermittently

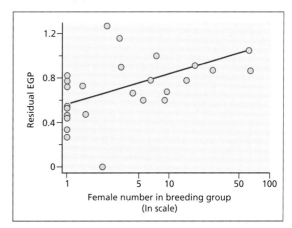

(a)

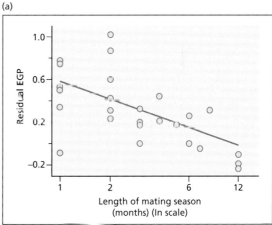

(b)

Figure 10.26 Comparison of the frequency of extra-group paternity in 26 mammals for which genetic estimates of paternity were available. Extra-group paternity (EGP) was (a) positively related to the number of females in breeding groups and (b) negatively related to the length of the breeding season. EGP was significantly related to both variables, in each panel displays partial residual plots of EGP. EGP was arcsine transformed for the analysis and residuals were calculated and best-fit lines drawn based on coefficients from the phylogenetic analysis. *Source*: From Isvaran and Clutton-Brock (2007). Reproduced with permission of the Royal Society.

associated than in those where they are continuously associated (Figure 10.27b), though this difference is no longer significant if comparisons are restricted to group-living species (Cohas and Allaine, 2009).

The presence of EGP raises important questions about the reliability of visual estimates of mating success. Here, differences in the duration of female receptivity and the number of partners may be important. For example, in red

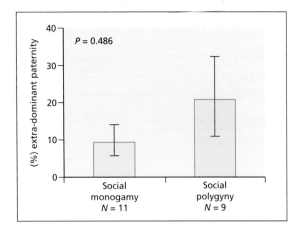

(a)

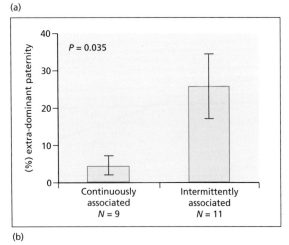

(b)

Figure 10.27 Frequencies of paternity involving males other than the most dominant male in each group, comparing (a) socially monogamous with polygynous species and (b) species where both sexes are continuously associated versus those where the sexes spend considerable periods apart. *Source*: From Clutton-Brock and Isvaran (2006). Reproduced with permission of the Royal Society.

deer and fallow deer, many females mate once per season and, when they mate more than once, it is usually in quick succession and with the same male (Clutton-Brock *et al.* 1982, 1986). As might be expected, there is a close correlation between the number of matings different males were seen to perform and the proportion of calves they fathered and defending stags fertilised a high proportion of the females that were present in their harems at the time of conception (Pemberton *et al.* 1992). In contrast, female Soay sheep mate repeatedly during their period of oestrus, often with different males (Preston *et al.* 2003b; Pemberton

et al. 2004), a substantial proportion of females defended and mated by dominant males produce offspring fathered by other individuals, and observations of mating frequency do not provide a reliable index of male breeding success (Pemberton *et al.* 2004).

10.5 Consequences of polygyny

Male life histories

Polygynous mating systems have far-reaching consequences for the life histories of males. Where mature males can monopolise access to large numbers of available females, sub-adult and young adult males are seldom able to obtain access to receptive females or female groups and start to breed later than females. These differences increase as the size of harems and the degree of reproductive skew among males rises. For example, in ursine howler monkeys, males commonly begin to breed when they are 7 years old while females begin to breed at 5 years old (Crockett and Pope 1993); in red deer, males seldom hold harems before they are 6 or 7 years old while females typically breed in their third or fourth year of life (Clutton-Brock *et al.* 1982) (Figure 10.28); and in elephant seals, males seldom defend harems successfully until they are at least 9 years old, while females begin to breed at 3 or 4 years old (Le Boeuf and Reiter 1988). In some species, the inability of younger males to defend females successfully has led to the development of alternative reproductive tactics (see Chapter 13) but these are rarely as successful as mate guarding.

Competition between males also has important effects on male survival and can lead to substantial differences in longevity between the sexes (see Chapter 18). In territorial species, males spend more time on their own than females, are less vigilant and may be slower to escape from approaching predators because they are reluctant to leave their territories (Estes and Goddard 1979; Fanshawe and Fitzgibbon 1993). Fights between males are frequent in most systems and males are often injured and are sometimes killed. For example, in langurs, males are more likely than females to die from external causes (including fighting and predation) throughout most of their lives (Rajpurohit and Sommer 1991). In addition, loss of condition in the breeding season often means that breeding males enter the next winter (or, in tropical species, the next dry season) with lower fat reserves than females and are more likely to die from starvation.

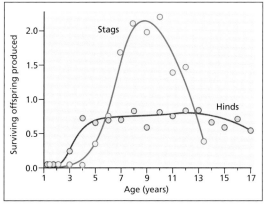

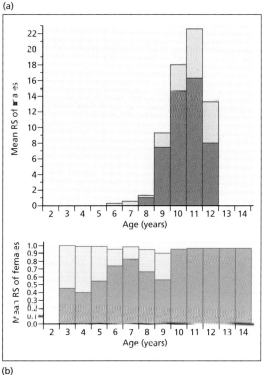

(a)

(b)

Figure 10.28 Age-specific reproductive success in (a) male and female red deer and (b) northern elephant seals. (a) The mean number of surviving offspring produced per year by females (hinds) of different ages, and the estimated number of surviving calves fathered by males (stags) of different ages, based on identification of the male defending the mother of each calf over the period of conception. (b) The mean number of pups born per year to individuals in each age category while estimates of male success are based on observations of copulations by recognisable males. Darker sections show numbers of pups born and weaned. *Source*: (a) From Clutton-Brock *et al.* (1988b). Reproduced with permission of Springer Science and Business Media (b) From Le Boeuf and Reiter (1988).

When large predators are present, they often target injured males or those in poor condition, increasing the costs of mate guarding further. For example, male Thompson's gazelles are more likely to be killed by predators (cheetah, wild dog, spotted hyena, lion and leopard) than are mature females (Fitzgibbon and Lazarus 1995) while in greater kudu, many males that defend breeding territories are killed before the next season, so that a male's chances of breeding successfully in more than two seasons are low (Owen-Smith 1993).

Male adaptations associated with competition also have substantial costs to male survival and higher rates of mortality in males than females also occur in food-limited populations of many polygynous species that are not subject to predation. For example, in island populations of Soay sheep where predators are absent, winter mortality is consistently higher in adult and juvenile males than females in years when overall mortality is high because heavy energy expenditure in the autumn rut leaves males in poor body condition at the beginning of the winter and increases their chances of dying of starvation before the next spring (Grubb 1974). These effects increase with population density (Clutton-Brock *et al.* 2004), while permanent castration of males a few days after birth stops them taking an active part in the rut or reducing feeding time and leads to spectacular increases in adult survival relative both to intact males and to females (Figure 10.29a) (Jewell *et al.* 1986; Jewell 1997; Stevenson *et al.* 2004). In addition, temporary castration of juvenile males during the rut (by treatment with progesterone) prevents them from rutting and leads to consistent increases in their survival through the next winter, especially in lighter animals (Stevenson and Bancroft 1995) (Figure 10.29b).

In many polygynous mammals, males also show faster rates of ageing and reduced longevity compared with females (Vinogradov 1998; Clutton-Brock and Isvaran 2007; Bonduriansky *et al.* 2008). While the cumulative costs of breeding competition may contribute to these differences, similar differences are found in zoo populations and domestic animals where management practices limit opportunities for direct competition between males and a likely explanation is that the compression of effective breeding lifespans of males in polygynous species has reduced the strength of selection for longevity in males, leading to more rapid rates of ageing and reductions in longevity (Clutton-Brock and Isvaran 2007). In line with this, there is some

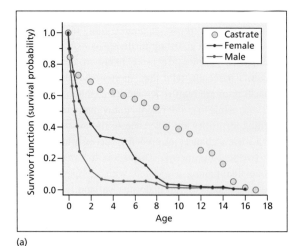

(a)

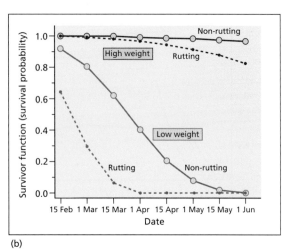

(b)

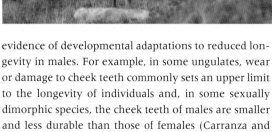

Figure 10.29 Effects of castration on survival of Soay sheep: (a) survival plots for adult castrates, in fact males and females from the same cohorts; (b) survival probabilities of juvenile males excluded from rutting by progestogen treatment (non-rutting) and control males (rutting). Survival plots are shown for heavy (high weight) and light (low weight) juveniles (weights in August preceding the rut are 12 kg and 20 kg respectively). *Source*: (a) Adapted from Jewell (1997) with permission of John Wiley & Sons. (b) From Stevenson *et al.* 2004. Reproduced with permission from Cambridge University Press. *Photo source*: © Arpat Ozgul.

evidence of developmental adaptations to reduced longevity in males. For example, in some ungulates, wear or damage to cheek teeth commonly sets an upper limit to the longevity of individuals and, in some sexually dimorphic species, the cheek teeth of males are smaller and less durable than those of females (Carranza and Pérez-Barbería 2007).

In polygynous mammals, a combination of delays in the onset of effective breeding, high rates of mortality and accelerated ageing rates often restrict the effective breeding lifespans of males to a relatively small number of years. For example, both in red deer and in elephant seals, individual males rarely breed successfully for more than 4 years, though they may survive for several years after they cease to breed effectively. Similarly, in eastern grey kangaroos, where the alpha male in a local population is responsible for 60–75% of all matings, the average tenure of alpha males is around a year (Jarman 2000). In contrast, in monogamous species, the breeding lifespans of males

are usually substantially longer and are comparable to those of females (Clutton-Brock and Isvaran 2007).

Interspecific comparisons show that there is a strong negative correlation between female group size (an indication of the extent of polygyny) and the duration of male tenure, this effect being pronounced in single-male groups and weaker in multi-male groups where males assist each other in defending the group against rivals that may displace them (Lukas and Clutton-Brock 2014) (see Chapter 11). Within species, too, male tenure length appears to decline as the degree of polygyny increases. For example, the tenure of male coalitions in lions is reduced when they are defending large prides of females (Packer *et al.* 1988), while in Thomas' langurs the frequency of male take-overs increases with female group size (Steenbeek *et al.* 2000). In hartebeest, males holding territories in habitats where females are abundant defend smaller territories than those in areas where females are less common but show significantly higher rates of territory changeover (Gosling 1986b). In topi, estimates of male tenure suggest that males holding resource territories occupy them for 4.3 years compared with 2.3 years for males holding peripheral territories on leks and 1.7 years for males holding central lek territories (Bro-Jørgensen and Durant 2003).

Variation in male breeding success

In polygynous species where individual males can monopolise breeding access to large numbers of females, a high proportion of males fail to breed in any season, while a small number of individuals mate with a large number of partners. As average harem size (or the degree of polygyny) grows, both the number of males without regular access to females and variance in mating success among breeding males will increase, especially where the adult sex ratio is relatively equal (Wade and Shuster 2002). While most females breed each year, standardised variance in breeding success calculated within seasons is often substantially lower in females than in males, especially in monotocous species (Clutton-Brock 1988; Lukas and Clutton-Brock 2014b).

While differences in male breeding success within seasons are often large, much of this variation is a consequence of differences in male age and standardised variance in reproduction among males is often smaller where breeding success is calculated over the lifetime than when it is calculated within seasons. Individual differences in lifetime breeding success are also likely to be further reduced by extra-group matings and by negative correlations between male breeding success and reproductive tenure or longevity (Clutton-Brock 1983, 1988; Lukas and Clutton-Brock 2014b). Conversely, where some females are consistently successful (or unsuccessful) at rearing young, standardised variance in female breeding success is likely to be much larger when it is calculated over the lifespan than when it is calculated within seasons and this effect is likely to be particularly large in monotocous species where females have protracted breeding lifespans. Similarly, in monogamous species where males can control access to their partners and have relatively long breeding lifespans, standardised variance in male breeding success is likely to be larger when calculated over the lifespan then when calculated within seasons.

These relationships have important implications. They suggest that sex differences in (standardised) fitness variance in polygynous mammals are often likely to be much smaller when calculated over the lifespan than when calculated within years (Clutton-Brock 1983, 1988). In addition, they indicate that contrasts in (standardised) variance in male success between polygynous and monogamous species and sex differences in (standardised) variance within polygynous species are both likely to be smaller when reproductive success is calculated over the lifespan of individuals than when it is calculated within seasons across animals of unknown age (Clutton-Brock 1983). Both predictions now appear to be supported by empirical data, though studies that are able to measure variation in lifetime reproductive success in both sexes are still scarce. While standardised variance in male breeding success within seasons is usually larger in males than females in polygynous species, sex differences in standardised variance calculated over the lifespan of males are smaller and are inconsistent (Clutton-Brock 1988, 2004; Lukas and Clutton-Brock 2014b) and comparative studies of mammals based on genetic data suggest that standardised variance in male lifetime breeding success may not be consistently larger in polygynous species than in monogamous ones (Lukas and Clutton-Brock 2014).

Selection on male traits

Intense competition for mating opportunities between males in polygynous species is often associated with the evolution of sex differences in body size, weaponry, testes size, physiology and behaviour which have important consequences for survival and longevity and secondary sexual characters are usually more highly developed in species with polygynous breeding systems than in monogamous species (see Chapter 18). Among primates, they are also more developed in species that usually live in uni-male harem groups than in those that typically live in multi-male groups (Dunn et al. 2015), which supports the evidence that uni-male groups represent a derived condition (Schultz et al. 2011). However, relationships between breeding systems and the extent of sexual dimorphism are variable and inconsistent. Contrasts in the effects of breeding systems on variance in male success may contribute to these differences but other factors are also likely to be involved. Part of the observed variation in male success may be a consequence of chance (Sutherland 1985; Gowaty and Hubbell 2005) or of the effects of differences in age (Clutton-Brock 1983). In addition, male fighting techniques vary widely between species (see Chapter 13) with the result that the competitive success of males is affected by different traits in different species, generating contrasts in the selection pressures operating on the same traits in different species (see Chapter 18).

Kinship structure

Reductions in the breeding tenure of males associated with polygyny also have important effects on the kinship structure of groups. Where individual males can monopolise substantial numbers of females within particular seasons, the proportion of paternal half-sibs born into the next generation will be relatively high, increasing coefficients of relatedness among recruits. As a result, effective population size and genetic heterogeneity within cohorts will tend to decline, while heterogeneity between groups may increase (Nunney 1993, Storz *et al.* 2001). Conversely, reductions in the effective breeding lifespans of males in polygynous societies will reduce the proportion of paternal half-sibs and full sibs in successive cohorts and increase variation in relatedness and genetic heterogeneity between cohorts. The relative magnitude of these effects probably varies between

species and will determine the influence of polygyny on effective population size and genetic heterogeneity.

The effects of polygyny on male life histories will also interact with variation in female life histories to affect the kinship structure of social groups. Reductions in male skew will reduce the proportion of juveniles born in the same group in the same season that are full sibs, while reductions in male tenure will reduce the incidence of full sibs born in successive seasons. Where females are philopatric, reductions in male skew or male tenure are also likely to reduce coefficients of relatedness between resident females, with consequences for average coefficients of relatedness between their offspring as well as between all group members. As a result, despite relatively high numbers of paternal half-sibs within cohorts, relatively low average levels of kinship between group members may be typical in groups of polygynous monotocous species, partly as a result of low levels of reproductive skew in females and males and partly because male tenure is usually short (Lukas and Clutton-Brock 2014b).

SUMMARY

1. Contrasts in mammalian mating systems are structured by the distribution of receptive females in time and space and the capacity of males to guard access to them before and after mating.

2. Where female density is high, breeding is closely synchronised and the first male to mate with a female is likely to inseminate her; males seldom guard access to females, searching for receptive partners and leaving them once they have mated. In contrast, where the density of receptive females is very low and their distribution is unpredictable (as in elephants and baleen whales) males search for breeding partners and defend them while they are receptive but seldom defend groups or breeding territories.

3. Where breeding is not closely synchronised, females are intolerant of each other and female density is low, males may associate with and guard a single breeding female at a time. Social monogamy is often associated with extensive paternal care, but this appears to be a consequence rather than a cause of its evolution.

4. Where females live in stable groups, males often guard access to the entire group or its range. Where female group size is large, female groups (or their ranges) are often guarded by several males. In species that have predictable breeding seasons, guarding is often seasonal; where breeding seasons are unpredictable or prolonged, males may guard female groups or their ranges throughout the year, though individuals may rest from active breeding for periods of a few weeks or a few months to regain lost condition.

5. Where female density is relatively high but females forage separately or aggregate in unstable groups (as in many small mammals and nocturnal species), males often defend territories overlapping the ranges of several females. In contrast, where females live in unstable groups and female density is relatively low but females regularly revisit particular feeding sites (as in many ungulates), males often defend resource-based mating territories substantially smaller than female ranges. In some of these species, males defend much smaller territories that do not contain significant resources in clusters (leks) where they are visited by receptive females.

6. The effectiveness of mate guarding varies with the structure of mating groups and the behaviour of males. Rates of extra-pair paternity are low in most monogamous species, while among polygynous species rates of extra-group paternity rise with female group size and breeding synchrony and are higher where males and females forage separately for parts of each day than where they are continuously associated.

7. The structure of mating systems has far-reaching consequences for the biology of both sexes. Polygyny is commonly associated with intense competition between males and high levels of reproductive skew among males within seasons.

However, competition reduces the length of the effective breeding lifespans of males and individual differences in breeding success calculated over the lifespan are similar in the two sexes.

8. Intense competition for access to females in polygynous systems generates stronger selection in males than females for traits that increase competitive success, leading to the evolution of pronounced sex differences in body size, weaponry and ornamentation in many polygynous societies.

9. The structure of mating systems also affects the kinship structure of groups. Especially in litter-bearing species, monogamous breeding pairs generate relatively large numbers of full sibs with the result that average kinship between young born into groups is likely to be relatively high. In contrast, in polygynous systems, average kinship between natal animals is often relatively low even though members of the same age cohorts often share the same father. This is partly because multiple females breed, partly because the breeding tenure of males is short and partly because average relatedness between resident breeding females is unlikely to be high. Monotocy and long inter-birth intervals are also likely to reduce both paternal and maternal relatedness between young born in the same group and so may be associated with relatively low average coefficients of relatedness within groups.

References

Achenbach, G.G. and Snowdon, C.T. (2002) Costs of caregiving: weight loss in captive adult male cotton-top tamarins (*Saguinus oedipus*) following the birth of infants. *International Journal of Primatology* **23**:179–189.

Adrian, O. and Sachser, N. (2011) Diversity of social and mating systems in cavies: a review. *Journal of Mammalogy* **92**:39–53.

Agren, G., *et al.* (1989) Ecology and social behavior of Mongolian gerbils, *Meriones unguiculatus*, at Xilinhot, Inner-Mongolia, China. *Animal Behaviour* **37**:11–27.

Anzenberger, G. (1992) Monogamous social systems and paternity in primates. In: *Paternity in Primates* (eds R.D. Martin, A.F. Dixson and E.J. Wickings). Basel: Karger, 203–224.

Anzenberger, G. and Falk, B. (2012) Monogamy and family life in callitrichid monkeys: deviations, social dynamics and captive management. *International Zoo Yearbook*, **6**:109–122.

Apollonio, M. (1989) Lekking in fallow deer: just a matter of density? *Ethology, Ecology and Evolution* **1**:291–294.

Balmford, A.P. (1990) Lekking in Uganda kob. PhD thesis, University of Cambridge, Cambridge.

Balmford, A.P. (1991) Mate choice on leks. *Trends in Ecology and Evolution* **6**:87–92.

Balmford, A.P. (1992) Social dispersion and lekking in Uganda kob. *Behaviour* **120**:177–191.

Baker, A.J., *et al.* (1993) Behavioural evidence for monopolization of paternity in multi-male groups of golden lion tamarins. *Animal Behaviour* **46**:1091–1103.

Balmford, A.P. and Read, A.F. (1991) Testing alternative models of sexual selection through female choice. *Trends in Ecology and Evolution* **6**:274–276.

Balmford, A.P., *et al.* (1992) Correlates of male mating success and female choice in a lek-breeding antelope. *Behavioral Ecology* **3**:112–123.

Barrette, C. (1987) The comparative ecology of the chevrotains, musk deer, and morphologically conservative deer. In: *Biology and Management of the Cervidae* (ed. C.M. Wemmer). Washington, DC: Smithsonian Institution Press, 200–213.

Bart, J. and Tornes, A. (1989) Importance of monogamous male birds in determining reproductive success. *Behavioral Ecology and Sociobiology* **24**:109–116.

Bearder, S.K. (1987) Lorises, bushbabies, and tarsiers: diverse societies in solitary foragers. *Primate Societies* (eds B.B. Smuts, D.L. Cheney, R.M. Seyfarth, R.W. Wrangham and T.T. Struhsaker), Chicago: University of Chicago Press, 12–24.

Beehler, B.M. and Foster, M.S. (1988) Hotshots, hotspots, and female preference in the organization of lek mating systems. *American Naturalist* **131**:203–219.

Berger, J. (1986) *Wild Horses of the Great Basin*. Chicago: University of Chicago Press.

Best, P.B. (1979) Social organisation of sperm whales, *Physeter macrocephalus*. In: *Behaviour of Marine Animals: Current Perspectives in Research, Vol. 3 Cetaceans* (eds H.E. Winn and B.C. Olla). New York: Plenum Press, 227–289.

Björklund, M. and Westman, B. (1986) Adaptive advantages of monogamy in the great tit (*Parus major*): an experimental test of the polygyny threshold-model. *Animal Behaviour* **34**:1436–1440.

Bonduriansky, R., *et al.* (2008) Sexual selection, sexual conflict and the evolution of ageing and life span. *Functional Ecology* **22**:443–453.

Booth-Binczik, S.D. *et al.* (2004) Lek-like mating in white-nosed coatis (*Nasua narica*): socio-ecological correlates of intraspecific variability in mating systems. *Journal of Zoology* **262**:179–185.

Borries, C. (2000) Male dispersal and mating season influxes in Hanuman langurs living in multi-male groups. In: *Primate Males: Causes and Consequences of Variation in Group Composition* (ed. P.M. Kappeler). Cambridge: Cambridge University Press, 146–158.

Bradbury, J.W. (1977) Lek mating behavior in hammer-headed bat. *Zeitschrift fur Tierpsychologie* **45**:225–255.

Bradbury, J.W. and Vehrencamp, S.L. (1977) Social organization and foraging in emballonurid bats. III. Mating systems. *Behavioral Ecology and Sociobiology* **2**:1–17.

Brashares, J.S. and Arcese, P. (2002) Role of forage, habitat and predation in the behavioural plasticity of a small African antelope. *Journal of Animal Ecology* **71**:626–638.

Brockelman, W.Y., *et al.* (1998) Dispersal, pair formation and social structure in gibbons (*Hylobates lar*). *Behavioral Ecology and Sociobiology* **42**:329–339.

Bro-Jørgensen, J. (2002) Overt female mate competition and preference for central males in a lekking antelope. *Proceedings of the National Academy of Sciences of the United States of America* **99**:9290–9293.

Bro-Jørgensen, J. and Durant, S.M. (2003) Mating strategies of topi bulls: getting in the centre of attention. *Animal Behaviour* **65**:585–594.

Bro-Jørgensen, J. (2010) Intra- and intersexual conflicts and cooperation in the mating strategies: lessons learnt from ungulates. *Evolutionary Biology* **38**:28–41.

Bro-Jørgensen, J. (2011) Queuing in space and time reduces the lek paradox on an antelope lek. *Evolutionary Ecology*, **25**:1385–1395.

Brotherton, P.N.M. (1994) *The evolution of monogamy in the dik-dik*. PhD thesis, University of Cambridge, Cambridge.

Brotherton, P.N.M. and Komers, P.E. (2003) Mate guarding and the evolution of social monogamy in mammals. In: *Monogamy: Mating Strategies and Partnerships in Birds, Humans and other Mammals* (eds U.H. Reichard and C. Boesch). Cambridge: Cambridge University Press, 59–80.

Brotherton, P.N.M. and Manser, M.B. (1997) Female dispersion and the evolution of monogamy in the dik-dik. *Animal Behaviour* **54**:1413–1424.

Brotherton, P.N.M. and Rhodes, A. (1996) Monogamy without bi-parental care in a dwarf antelope. *Proceedings of the Royal Society of London. Series B: Biological Sciences* **263**:23–29.

Brotherton, P.N.M., *et al.* (1997) Genetic and behavioural evidence of monogamy in a mammal, Kirk's dik-dik (*Madoqua kirkii*). *Proceedings of the Royal Society of London. Series B: Biological Sciences* **264**:675–681.

Brown, M., *et al.* (2007) Monogamy in an Australian arboreal marsupial, the yellow-bellied glider (*Petaurus australis*). *Australian Journal of Zoology* **55**:185–195.

Cant, M.A. (1998) *Cooperative breeding in the banded mongoose*, Mungos mungo. PhD thesis, University of Cambridge, Cambridge.

Carbone, C. and Taborsky, M. (1996) Mate choice or harassment avoidance? A question of female control at the lek. *Behavioral Ecology* **7**:370–378.

Caro, T.M. (1994) *Cheetahs of the Serengeti Plains*. Chicago: University of Chicago Press.

Carranza, J. and Pérez-Barbería, F.J. (2007) Sexual selection and senescence: male size-dimorphic ungulates evolved relatively smaller molars than females. *American Naturalist* **170**:370–380.

Charles-Dominique, P. (1983) Ecology and social adaptations in didelphid marsupials: comparison with eutherians of similar ecology. In: *Advances in the Study of Mammalian Behavior* (eds J.F. Eisenberg and D.G. Kleiman). American Society of Mammalogists Special Publication No. 7. Stillwater, OK: American Society of Mammalogists, 395–422.

Christal, J., *et al.* (1998) Sperm whale social units: variation and change. *Canadian Journal of Zoology* **76**:1431–1440.

Clutton-Brock, T.H. (1983) Selection in relation to sex. In: *Evolution from Molecules to Men* (ed. B.J. Bendall). Cambridge: Cambridge University Press, 457–481.

Clutton-Brock, T.H. (1988) Reproductive success. In: *Reproductive Success: Studies of Individual Variation in Contrasting Breeding Systems* (ed. T.H. Clutton-Brock). Chicago: University of Chicago Press, 472–486.

Clutton-Brock, T.H. (1989) Mammalian mating systems. *Proceedings of the Royal Society of London. Series B: Biological Sciences* **236**:339–372.

Clutton-Brock, T.H. (2004) What is sexual selection? In: *Sexual Selection in Primates: New and Comparative Perspectives* (eds P.M. Kappeler and C.P. van Schaik). Cambridge: Cambridge University Press, 24–36.

Clutton-Brock, T.H. and Isvaran, K. (2006) Paternity loss in contrasting mammalian societies. *Biology Letters* **2**:513–516.

Clutton-Brock, T.H. and Isvaran, K. (2007) Sex differences in ageing in natural populations of vertebrates. *Proceedings of the Royal Society of London. Series B: Biological Sciences* **274**:3097–3104.

Clutton-Brock, T.H. and McComb, K. (1993) Experimental tests of copying and mate choice in fallow deer. *Behavioral Ecology* **4**:191–193.

Clutton-Brock, T.H., *et al.* (1982) *Red Deer: The Behaviour and Ecology of Two Sexes*. Chicago: University of Chicago Press.

Clutton-Brock, T.H., *et al.* (1986) Great expectations: dominance, breeding success and offspring sex ratios in red deer. *Animal Behaviour* **34**:460–471.

Clutton-Brock, T.H., *et al.* (1988a) Reproductive success in male and female red deer. In: *Reproductive Success: Studies of Individual Variation in Contrasting Breeding Systems* (ed. T.H. Clutton-Brock). Chicago: University of Chicago Press, 325–343.

Clutton-Brock, T.H., *et al.* (1988b) Passing the buck: resource defence, lek breeding and mate choice in fallow deer. *Behavioral Ecology and Sociobiology* **23**:281–296.

Clutton-Brock, T.H., *et al.* (1989) Mate choice on fallow deer leks. *Nature* **340**:463–465.

Clutton-Brock, T.H., *et al.* (1993) The evolution of ungulate leks. *Animal Behaviour* **46**:1121–1138.

Clutton-Brock, T.H., *et al.* (2004) Population dynamics of Soay sheep. In: *Soay Sheep: Dynamics and Selection in an Island Population* (eds T.H. Clutton-Brock and J.M. Pemberton). Cambridge: Cambridge University Press, 52–88.

Cockburn, A. (1988) *Social Behaviour in Fluctuating Populations*. London: Croom Helm.

Cohas, A. and Allainé, D. (2009) Social structure influences extra-pair paternity in socially monogamous mammals. *Biology Letters* **5**:313–316.

Creel, S. and Creel, N.M. (2002) *The African Wild Dog: Behavior, Ecology, and Conservation*. Princeton, NJ: Princeton University Press.

Crockett, C.M. and Pope, T.R. (1993) Consequences of sex differences in dispersal for juvenile red howler monkeys. In: *Juvenile Primates: Life History, Development, and Behavior* (eds M.E. Pereira and L.A. Fairbanks). New York: Oxford University Press, 104–118.

Crook, J.H. (1965) The adaptive significance of avian social organisations. *Symposium of the Zoological Society, London* **14**:181–218.

Davies, N.B. (2000) Multi-male breeding groups in birds: ecological causes and social conflicts. In: *Primate Males: Causes and consequences of variation on group composition* P.M. Kappeler (ed.). Cambridge University Press, Cambridge, 64–71.

Davies, N.B., *et al.* (2012) *An Introduction to Behavioural Ecology*, 4th edn. Oxford: Wiley-Blackwell.

Deutsch, J.C. (1992) Reproductive strategies in a lek-breeding antelope, the Uganda kob. PhD thesis, University of Cambridge, Cambridge.

Deutsch, J.C. (1994) Lekking by default: female habitat preferences and male strategies in Uganda kob. *Journal of Animal Ecology* **63**:101–115.

Deutsch, J.C. and Nefdt, R.J.C. (1992) Olfactory cues influence female choice in two lek-breeding antelopes. *Nature* **356**:596–598.

Deutsch, J.C. and Weeks, P. (1992) Uganda kob prefer high-visibility leks and territories. *Behavioral Ecology* **3**:223–233.

Díaz-Muñoz, S.L. (2011) Paternity and relatedness in a polyandrous nonhuman primate: testing adaptive hypotheses of male reproductive cooperation. *Animal Behaviour* **82**:563–571.

Di Fiore, A. and Campbell, C.J. (2007) The Atelines: variation in ecology, behaviour and social organisation. In: *Primates in Perspective* (eds C.J. Campbell, A. Fuentes, K.C. MacKinnon, M. Panger and S.K. Bearder). New York: Oxford University Press, 155–185.

Dobson, F.S. (1984) Environmental influences on sciurid mating systems. In: *The Biology of Ground-dwelling Squirrels: Annual Cycles, Behavioral Ecology, and Sociality* (eds J.O. Murie and G.R. Michener). Lincoln: University of Nebraska Press, 229–249.

Dobson, F.S., *et al.* (2010) Spatial dynamics and the evolution of social monogamy in mammals. *Behavioral Ecology* **21**:747–752.

Downhower, J.F. and Armitage, K.B. (1971) Yellow-bellied marmot and the evolution of polygamy. *American Naturalist* **105**:355–370.

Dufault, S. and Whitehead, H. (1995) The geographic stock structure of female and immature sperm whales in the South Pacific. *Report of the International Whaling Commission, Special Issue* **45**:401–405.

Dunbar, R.I.M. (1984) *Reproductive Decisions: An Economic Analysis of the Social Strategies of Gelada Baboons*. Princeton, NJ: Princeton University Press.

Dunbar, R.I.M. and Dunbar, E.P. (1979) Observation on the social organisation of common duiker in Ethiopia. *African Journal of Ecology* **17**:249–252.

Dunbar, R.I.M. and Dunbar, E.P. (1980) Pairbond in klipspringer. *Animal Behaviour* **28**:219–229.

Dunn, J.C., *et al.* (2015) Evolutionary trade-off between vocal tract and testes dimensions in howler monkeys. *Current Biology* **25**:2839–2844.

Emlen, S.T. and Oring, L.W. (1977) Ecology, sexual selection, and the evolution of mating systems. *Science* **197**:215–223.

Escherick, P.C. (1981) *Social Biology of the Bushy-tailed Woodrat, Neotoma cinerea*. University of California Publications in Zoology 110. Berkeley: University of California Press.

Estes, R.D. and Goddard, J. (1979) Prey selection and hunting behaviour of the African wild dog. *Journal of Wildlife Management* **31**:52–70.

Fanshawe, J.H. and Fitzgibbon, C.D. (1993) Factors influencing the hunting success of an African wild dog pack. *Animal Behaviour* **45**:479–490.

Fay, F.H., *et al.* (1984) Time and location of mating and associated behaviour of the Pacific walrus, Odobenus rosmarus divergens Illiger. In: *Soviet–American Cooperative Research on Marine Mammals* (eds F.H. Fay and G.A. Fedoseev). NOAA Technical Report NMFS **12**:89–99.

Ferron, J. and Ouellet, J.P. (1989) Temporal and intersexual variations in the use of space with regard to social organization in the woodchuck (*Marmota monax*). *Canadian Journal of Zoology* **67**:1642–1649.

Fietz, J. (2003) Pair-living and mating strategies in the fat-tailed dwarf lemur (*Cheirogaleus medius*). In: *Monogamy: Mating Strategies and Partnership in Birds, Humans and other Mammals* (eds U.H. Reichhard and C. Boesch). Cambridge: Cambridge University Press, 214–231.

Fietz, J., *et al.* (2000) High rates of extra-pair young in the pair-living fat-tailed dwarf lemur, *Cheirogaleus medius*. *Behavioral Ecology and Sociobiology* **49**:8–17.

Fitzgibbon, C.D. (1997) The adaptive significance of monogamy in the golden-rumped elephant-shrew. *Journal of Zoology* **242**:167–177.

Fitzgibbon, C.D. and Lazarus, J. (1995) Antipredator behavior of Serengeti ungulates: individual differences and population consequences. In: *Serengeti II: Dynamics, Management and Conservation of an Ecosystem* (eds A.R.E. Sinclair and P. Arcese). Chicago: Chicago University Press, 274–296.

Fryxell, J.M. (1987) Lek breeding and territorial aggression in white-eared kob. *Ethology* **75**:211–220.

Garber, P.A. (1997) One for all and breeding for one: cooperation and competition as a tamarin reproductive strategy. *Evolutionary Anthropology: Issues, News, and Reviews* **5**:187–199.

Garber, P.A. and Teaford, M.F. (1986) Body weights in mixed species troops of *Saguinus mystax mystax* and *Saguinus fuscicollis nigrifrons* in Amazonian Peru. *American Journal of Physical Anthropology* **71**:331–336.

Gese, E.M. (2001) Territorial defense by coyotes (*Canis latrans*) in Yellowstone National Park, Wyoming: who, how, where, when, and why. *Canadian Journal of Zoology* **79**:980–987.

Gibson, R.M. and Guinness, F.E. (1980) Behavioral factors affecting male reproductive success in red deer (*Cervus elaphus*). *Animal Behaviour* **28**:1163–1174.

Girman, D.J., *et al.* (1997) A molecular genetic analysis of social structure, dispersal, and interpack relationships of the African wild dog (*Lycaon pictus*). *Behavioral Ecology and Sociobiology* **40**:187–198.

Goldizen, A.W. (1987) Facultative polyandry and the role of infant-carrying in wild saddle-back tamarins (*Saguinus fuscicollis*). *Behavioral Ecology and Sociobiology* **20**:99–109.

Goldizen, A.W. (2003) Social monogamy and its variations in callitrichids: do these relate to the costs of infant care? In:

Monogamy: Mating Strategies and Partnerships in Birds, Humans and Other Mammals (eds U.H. Reichard and C. Boesch). Cambridge: Cambridge University Press, 232–247.

Goossens, B., *et al.* (1998) Extra-pair paternity in the monogamous Alpine marmot revealed by nuclear DNA microsatellite analysis. *Behavioral Ecology and Sociobiology* **43**:281–288.

Gosling, L.M. (1986a) Economic consequences of scent marking in mammalian territoriality. In: *Chemical Signals in Vertebrates 4* (eds D. Duvall, D. Müller-Schwarze and R.M. Silverstein). Berlin: Springer, 385–395.

Gosling, L.M. (1986b) The evolution of mating strategies in male antelopes. In: *Ecological Aspects of Social Evolution* (eds D.I. Rubenstein and R.W. Wrangham). Princeton, NJ: Princeton University Press, 244–281.

Gowaty, P.A. (1996) Battles of the sexes and origins of monogamy. In: *Partnerships in Birds: The Study of Monogamy* (ed. J.M. Black). Oxford: Oxford University Press, 21–52.

Gowaty, P.A. and Hubbell, S.P. (2005) Chance, time allocation and the evolution of adaptively flexible sex role behavior. *Integrative and Comparative Biology* **45**:931–944.

Grant, T.R. (1983) The behavioral ecology of the monotremes. In: *Advances in the Study of Mammalian Behavior* (eds J.F. Eisenberg and D.G. Kleiman). American Society of Mammalogists Special Publication No. 7. Stillwater, OK: American Society of Mammalogists, 360–394.

Grubb, P. (1974) Population dynamics of the Soay sheep. In: *Island Survivors: The Ecology of the Soay Sheep of St Kilda* (eds P.A. Jewell, C. Milner and J.M. Boyd). London: Athlone Press, 242–272.

Harcourt, A.H. and Stewart, K.J. (2007) *Gorilla Society*. Chicago: University of Chicago Press.

Hasegawa, T. and Hiraiwa-Hasegawa, M. (1983) Opportunistic and restrictive mating among wild chimpanzees in the Mahale Mountains, Tanzania. *Journal of Ethology*: 75–85.

Heymann, E.W. (2000) The number of adult males in callitrichine groups and its implication for callitrichine social evolution. In: *Primate Males: Causes and Consequences of Variation in Group Composition* (ed. P.M. Kappeler). Cambridge: Cambridge University Press, 64–71.

Heymann, E.W. and Soini, P. (1999) Offspring number in pygmy marmosets, *Cebuella pygmaea,* in relation to group size and the number of adult males. *Behavioral Ecology and Sociobiology* **46**:400–404.

Höglund, J. and Alatalo, R.V. (1995) *Leks*. Princeton, NJ: Princeton University Press.

Holekamp, K.E. and Smale, L. (2000) Feisty females and meek males: reproductive strategies in the spotted hyena. In: *Reproduction in Context* (eds K. Wallen and J. Schneider). Cambridge, MA: MIT Press, 257–285.

Holt, W.V. (1977) Postnatal development of the testes in the cuis, *Galea musteloides*. *Laboratory Animals* **11**:87–91.

Hoogland, J.L. (1995) *The Black-tailed Prairie Dog: Social Life of a Burrowing Mammal*. Chicago: University of Chicago Press.

Hohmann, G. (2001) Association and social interaction between strangers and residents in bonobos (*Pan paniscus*). *Primates* **42**:91–99.

Huck, M., *et al.* (2005) Paternity and kinship patterns in polyandrous moustached tamarins (*Saguinus mystax*). *American Journal of Physical Anthropology* **127**:449–464.

Huck, M., *et al.* (2014) Correlates of genetic monogamy in socially monogamous mammals: insights from Azara's owl monkeys. *Proceedings of the Royal Society of London. Series B: Biological Sciences* **281**:20140195.

Ihobe, H. (1992) Male male relationships among wild bonobos (*Pan paniscus*) at Wamba, Republic of Zaire. *Primates* **33**:163–179.

Ims, R.A. (1988) Spatial clumping of sexually receptive females induces space sharing among male voles. *Nature* **235**:541–543.

Isvaran, K. and Clutton-Brock, T.H. (2007) Ecological correlates of extra-group paternity in mammals. *Proceedings of the Royal Society of London. Series B: Biological Sciences* **274**:219–224.

Jackson, T.P. and Skinner, J.D. (1998) The role of territoriality in the mating system of the springbok *Anticorcas marsupialis*. *Transactions of the Royal Society of South Africa* **53**:271–282.

Jarman, P.J. (1974) The social organisation of antelope in relation to their ecology. *Behaviour* **48**:215–267.

Jarman, P.J. (1991) Social behaviour and organisation in the Macropodoidea. In: *Advances in the Study of Behavior*, Vol. 20 (eds P.J.B. Slater, J.S. Rosenblatt, C. Beer and M. Milinski). San Diego, CA: Academic Press, 1–50.

Jarman, P.J. (2000) Males in macropod society. In: *Primate Males: Causes and Consequences of Variation in Group Composition* (ed. P.M. Kappeler). Cambridge: Cambridge University Press, 21–33.

Jarman, P.J. and Southwell, C. (1986) Grouping, associations and reproductive strategies in eastern grey kangaroos. In: *Ecological Aspects of Social Evolution: Birds and Mammals* (eds D.I. Rubenstein and R.W. Wrangham). Princeton, NJ: Princeton University Press, 399–428.

Jeffreys, A.J., *et al.* (1985) Hypervariable 'minisatellite' regions in human DNA. *Nature* **314**:67–73.

Jewell, P.A. (1997) Survival and behaviour of castrated Soay sheep (*Ovis aries*) in a feral island population on Hirta, St Kilda, Scotland. *Journal of Zoology* **243**:623–636.

Jewell, P.A., *et al.* (1986) Multiple mating and siring success during natural oestrus in the ewe. *Journal of Reproduction and Fertility* **77**:81–89.

Jolly, A. (1966) *Lemur Behavior*. Chicago: University of Chicago Press.

Kano, T. (1992) *The Last Ape: Pygmy Chimpanzee Behavior and Ecology*. Stanford, CA: Stanford University Press.

Kano, T. (1996) Male rank order and copulation rate in a unit-group of bonobos at Wamba, Zaïre. In: *Great Ape Societies* (eds W.C. McGrew, L.F. Marchant and T. Nishida). Cambridge: Cambridge University Press, 135–145.

Kappeler, P.M. (2014) Lemur behaviour informs the evolution of social monogamy. *Trends in Ecology and Evolution* **29**:591–593.

Kappeler, P.M. and Fichtel, C. (2015) The evolution of Eulemus social organization. *International Journal of Primatology*.

Kappeler, P.M. and van Schaik, C.P. (2002) Evolution of primate social systems. *International Journal of Primatology* **23**:707–740.

Kays, R. (2003) Social polyandry and promiscuous mating in a primate-like carnivore: the kinkajou (*Potos flavus*). In: *Monogamy: Mating Strategies and Partnerships in Birds, Humans and*

Other Mammals (eds U.H. Reichard and C. Boesch). Cambridge: Cambridge University Press, 125–137.

Kerth, G. (2008) Animal sociality: bat colonies are founded by relatives. *Current Biology* **18**:R740–R742.

Kerth, G., *et al.* (2011) Bats are able to maintain long-term social relationships despite the high fission–fusion dynamics of their groups. *Proceedings of the Royal Society of London. Series B: Biological Sciences* **278**:2761–2767.

Kishimoto, R. (2003) Social monogamy and social polygyny in a solitary ungulate, the Japanese serow (*Capricornis crispus*). In: *Monogamy: Mating Strategies and Partnerships in Birds, Humans and Other Mammals* (eds U.H. Reichard and C. Boesch). Cambridge: Cambridge University Press, 147–158.

Kishimoto, R. and Kawamichi, T. (1996) Territoriality and monogamous pairs in a solitary ungulate, the Japanese serow, *Capricornis crispus. Animal Behaviour* **52**:673–682.

Kleiman, D.G. (1977) Monogamy in mammals. *Quarterly Review of Biology* **52**:39–69.

Klingel, H. (1974) Soziale organisation und verhalten des Grevy-Zebras (*Equus grevyi*). *Zeitschrift fur Tierpsychologie* **36**:37–70.

Kowalewski, M.M. and Garber, P.A. (2015) Solving the collective action problem during intergroup encounters: the case of black and gold howler monkeys (*Alouatta caraya*). In: *Howler Monkeys: Behavior, Ecology and Conservation* (eds M.M. Kowalewski, P.A. Garber, L. Cortes-Ortiz, B. Urbani and D. Youlatos). New York: Springer, 165–190.

Kummer, H. (1968) *Social Organization of Hamadryas Baboons.* Chicago: University of Chicago Press.

Kummer, H., *et al.* (1974) Triadic differentiation: inhibitory process protecting pair bonds in baboons. *Behaviour* **49**:62–87.

Kutsukake, N. and Matsusaka, T. (2002) Incident of intense aggression by chimpanzees against an infant from another group in Mahale Mountains National Park, Tanzania. *American Journal of Primatology* **58**:175–180.

Lack, D. (1968) *Ecological Adaptation for Breeding in Birds.* London: Methuen.

Langbein, J. and Thirgood, S.J. (1989) Variation in mating systems of fallow deer (*Dama dama*) in relation to ecology. *Ethology* **83**:195–214.

Le Boeuf, B.J. and Reiter, J. (1988) Lifetime reproductive success in northern elephant seals. In: *Reproductive Success: Studies of Individual Variation in Contrasting Breeding Systems* (ed. T.H. Clutton-Brock). Chicago: University of Chicago Press, 344–362.

Lee, A.K. and Cockburn, A. (1985) *Evolutionary Ecology of Marsupials.* Cambridge: Cambridge University Press.

Lee, P.C., *et al.* (2011) Male social dynamics: independence and beyond. In: *The Amboseli Elephants: A Long-term Perspective on a Long-lived Mammal* (eds C.J. Moss, H. Croze and P.C. Lee). Chicago: University of Chicago Press, 260–271.

Leuthold, W. (1966) Variations in territorial behavior of Uganda kob *Adenota kob thomasi* (Neumann 1896). *Behaviour* **27**:215–258.

Leuthold, W. (1978) On the ecology of gerenuk *Litocranius walleri. Journal of Animal Ecology* **47**:561–580.

Lukas, D. and Clutton-Brock, T.H. (2013) The evolution of social monogamy in mammals. *Science* **341**:526–530.

Lukas, D. and Clutton-Brock, T.H. (2014a) Evolution of social monogamy in primates is not consistently associated with male infanticide. *Proceedings of the National Academy of Sciences of the United States of America* **111**:E1674.

Lukas, D. and Clutton-Brock, T.M. (2014b) Costs of mating competition limit male lifetime breeding success in polygynous mammals. *Proceedings of the Royal Society of London. Series B: Biological Sciences* **281**, DOI: 10.1098. [Accessed 13 January 2016].

Macdonald, D.W. (1983) The ecology of carnivore social behaviour. *Nature* **301**:379–384.

Macdonald, D.W., *et al.* (2007) Social organisation and resource use in capybaras and maras. In: *Rodent Societies: An Ecological and Evolutionary Perspective* (eds J.O. Wolff and P.W. Sherman). Chicago: University of Chicago Press, 393–402.

McGinnis, P.R. (1979) Sexual behaviour in free-living chimpanzees: consort relationships. In: *The Great Apes* (eds D.A. Hamburg and E.R. McCown). Menlo Park, CA: Benjamin/Cummings, 429–439.

McLean, I.G. (1982) The association of female kin in the arctic ground-squirrel *Spermophilus parryii. Behavioral Ecology and Sociobiology* **10**:91–99.

Mason, W.A. (1966) Social organization of the South American monkey, *Callicebus moloch*: a preliminary report. *Tulane Studies in Zoology* **13**:23–30.

Michener, G.R. (1981) Ontogeny of spatial relationships and social behavior in juvenile Richardson's ground squirrels. *Canadian Journal of Zoology* **59**:1666–1676.

Michener, G.R. (1983) Kin identification, matriarchies, and the evolution of sociality in ground-dwelling sciurids. In: *Advances in the Study of Mammalian Behavior* (eds J.F. Eisenberg and D.G. Kleiman). American Society of Mammalogists Special Publication No. 7. Stillwater, OK: American Society of Mammalogists, 528–572.

Møller, A.P. and Birkhead, T.R. (1993) Cuckoldry and sociality: a comparative study of birds. *American Naturalist* **142**:118–140.

Monfort-Braham, N. (1975) Variations dans la structure sociale du topi, *Damaliscus korrigum* Ogilby, au Parc National de l'Akagera, Rwanda. *Zeitschrift fur Tierpsychologie* **39**:332–364.

Müller-Schwarze, D. and Sun, L. (2003) *The Beaver: Natural History of a Wetlands Engineer.* Ithaca, NY: Cornell University Press.

Nievergelt, C.M. and Martin, R.D. (1998) Energy intake during reproduction in captive common marmosets (*Callithrix jacchus*). *Physiology and Behavior* **65**:849–854.

Nievergelt, C.M., *et al.* (2000) Genetic analysis of group composition and breeding system in a wild common marmoset (*Callithrix jacchus*) population. *International Journal of Primatology* **21**:1–20.

Nishida, T. and Hiraiwa-Hasegawa, M. (1987) Chimpanzees and bonobos: cooperative relationships among males. In: *Primate Societies* (eds B.B. Smuts, D.L. Cheney, R.M. Seyfarth, R.W. Wrangham and T.T. Struhsaker). Chicago: University of Chicago Press, 154–165.

Nunn, C.L. (1999) The number of males in primate social groups: a comparative test of the socioecological model. *Behavioral Ecology and Sociobiology* **46**:1–13.

Nunney, L. (1993) The influence of mating system and overlapping generations on effective population size. *Evolution* **47**:1329–1341.

Opie, C., *et al.* (2013) Male infanticide leads to social monogamy in primates. *Proceedings of the National Academy of Sciences of the United States of America* **110**:13328–13332.

Opie, C., *et al.* (2014) Reply to Lukas and Clutton-Brock: Infanticide still drives primate monogamy. *Proceedings of the National Academy of Sciences of the United States of America* **111**:E1675.

Ostfeld, R.S. (1985) Limiting resources and territoriality in microtine rodents. *American Naturalist* **126**:1–15.

Ostfeld, R.S. (1990) The ecology of territoriality in small mammals. *Trends in Ecology and Evolution* **5**:411–415.

Owen-Smith, R.N. (1972) Territoriality: the example of the white rhinoceros. *Zoologica Africana* **7**:273–280.

Owen-Smith, R.N. (1975) The social ethology of the white rhinoceros *Ceratotherium simum* (Burchell 1817). *Zeitschrift fur Tierpsychologie* **38**:337–384.

Owen-Smith, N. (1993) Comparative mortality rates of male and female Kudus: the costs of sexual size dimorphism. *Journal of Animal Ecology* **62**:428–440.

Packer, C., *et al.* (1988) Reproductive success in lions. In: *Reproductive Success: Studies of Individual Variation in Contrasting Breeding Systems* (ed. T.H. Clutton-Brock). Chicago: University of Chicago Press, 363–383.

Palombit, R.A. (1994) Extra-pair copulations in a monogamous ape. *Animal Behaviour* **47**:721–723.

Peck, J.R. and Feldman, M.W. (1988) Kin selection and the evolution of monogamy. *Science* **240**:1672–1674.

Pemberton, J.M., *et al.* (1992) Behavioural estimates of male mating success tested by DNA fingerprinting in a polygynous mammal. *Behavioral Ecology* **3**:66–75.

Pemberton, J.M., *et al.* (2004) Mating patterns and male breeding success. In: *Soay Sheep: Dynamics and Selection in an Island Population* (eds T.H. Clutton-Brock and J.M. Pemberton). Cambridge: Cambridge University Press, 166–189.

Petrie, M. and Lipsitch, M. (1994) Avian polygyny is most likely in populations with high variability in heritable male fitness. *Proceedings of the Royal Society of London. Series B: Biological Sciences* **256**:275–280.

Poole, J.H. (1987) Rutting behavior in African elephants: the phenomenon of musth. *Behaviour* **102**:283–316.

Poole, J.H. (1989a) Announcing intent: the aggressive state of musth in African elephants. *Animal Behaviour* **37**:140–152.

Poole, J.H. (1989b) Mate guarding, reproductive success and female choice in African elephants. *Animal Behaviour* **37**:842–849.

Poole, J.H. (1994) Sex differences in the behaviour of African elephants. In: *The Differences Between the Sexes* (eds R.V. Short and E. Balaban). Cambridge: Cambridge University Press, 331–346.

Poole, J.H., *et al.* (2011) Longevity, competition, and musth: a long-term perspective on male reproductive strategies. In: *The Amboseli Elephants: A Long-term Perspective on a Long-lived Mammal* (eds C.J. Moss, H. Croze and P.C. Lee). Chicago: University of Chicago Press, 272–290.

Pope, T.R. (2000) Reproductive success increases with degree of kinship in cooperative coalitions of female red howler monkeys (*Alouatta seniculus*). *Behavioral Ecology and Sociobiology* **48**:253–267.

Preston, B.T., *et al.* (2003) Soay rams target reproductive activity towards promiscuous females' optimal insemination period. *Proceedings of the Royal Society of London. Series B: Biological Sciences* **270**:2073–2078.

Prins, H.H.T. (1989) Condition changes and choice of social environment in African buffalo bulls. *Behaviour* **108**:297–324.

Prins, H.H.T. (1996) *Ecology and Behaviour of the African Buffalo: Social Inequality and Decision Making*. London: Chapman & Hall.

Rajpurohit, L.S. and Sommer, V. (1991) Sex differences in mortality among langurs (*Presbytis entellus*) of Jodhpur, Rajasthan. *Folia Primatologica* **56**:17–27.

Rathbun, G.B. (1979) The social structure and ecology of elephant shrews. *Advances in Ethology* **20**:1–77.

Reichard, U.H. (2003) Social monogamy in gibbons: the male perspective. In: *Monogamy: Mating Strategies and Partnerships in Birds, Humans and Other Mammals* (eds U.H. Reichard and C. Boesch). Cambridge: Cambridge University Press, 190–213.

Ribble, D.O. (1991) The monogamous mating system of *Peromyscus californicus* as revealed by DNA fingerprinting. *Behavioral Ecology and Sociobiology* **29**:161–166.

Ribble, D.O. (1992) Lifetime reproductive success and its correlates in the monogamous rodent, *Peromyscus californicus*. *Journal of Animal Ecology* **61**:457–468.

Ribble, D.O. (2003) The evolution of social and reproductive monogamy in *Peromyscus*: evidence from *Peromyscus californicus* (the California mouse). In: *Monogamy: Mating Strategies and Partnerships in Birds, Humans and Other Mammals* (eds U.H. Reichard and C. Boesch). Cambridge: Cambridge University Press, 81–92.

Richard, A.F. (1987) Malagasy prosimians: female dominance. In: *Primate Societies* (eds B.B. Smuts, D.L. Cheney, R.M. Seyfarth, R.W. Wrangham and T.T. Struhsaker). Chicago: University of Chicago Press, 15–33.

Richardson, D.S., *et al.* (2003) Avian behaviour: altruism and infidelity among warblers. *Nature* **422**:580.

Richardson, P.R.K. (1987) Aardwolf mating system: overt cuckoldry in an apparently monogamous mammal. *South African Journal of Science* **83**:405–410.

Rodman, P.S. and Mitani, J.C. (1986) Orangutans: sexual dimorphism in a solitary species. In: *Primate Societies* (eds B.B. Smuts, D.L. Cheney, R.M. Seyfarth, R.W. Wrangham and T.T. Struhsaker). Chicago: University of Chicago Press, 146–154.

Rood, J.P. (1986) Ecology and social evolution in the mongooses. In: *Ecological Aspects of Social Evolution* (eds D.I. Rubenstein and R.W. Wrangham). Princeton, NJ: Princeton University Press, 131–152.

Rosser, A.M. (1987) Resource defence in an African antelope, the puku (Kobus vardoni). PhD thesis, University of Cambridge, Cambridge.

Rosser, A.M. (1992) Resource distribution, density, and determinants of mate access in puku. *Behavioral Ecology* **3**:13–24.

Rubenstein, D.I. (1986) Ecology and sociality in horses and zebras. In: *Ecological Aspects of Social Evolution* (eds D.I. Rubenstein and R.W. Wrangham). Princeton, NJ: Princeton University Press, 282–302.

Runcie, M.J. (2000) Biparental care and obligate monogamy in the rock-haunting possum, *Petropseudes dahli*, from tropical Australia. *Animal Behaviour* **59**:1001–1008.

Rylands, A.B. (1982) The behavior and ecology of three species of marmosets and tamarins (Callitrichidae, Primates) in Brazil. PhD thesis, University of Cambridge, Cambridge.

Sachser, N., *et al.* (1999) Behavioural strategies, testis size and reproductive success in two caviomorph rodents with different mating systems. *Behaviour* **136**:1203–1217.

Sanchez, S., *et al.* (1999) Costs of infant-carrying in the cotton-top tamarin (*Saguinus oedipus*). *American Journal of Primatology* **48**:99–111.

Sauther, M.L. and Sussman, R.W. (1993) A new interpretation of the social organisation and mating system of the ringtailed lemur. In: *Lemur Social Systems and Their Ecological Basis* (eds J.U. Ganzhorn and P.M. Kappeler). New York: Springer.

Savage-Rumbaugh, E.S. and Wilkerson, B.J. (1978) Socio-sex ual behaviour in *Pan paniscus* and *Pan troglodytes*: a comparative study. *Journal of Human Evolution* **7**:327–344.

Schaller, G.B. (1972) *The Serengeti Lion*. Chicago: University of Chicago Press.

Schülke, O. and Kappeler, P.M. (2003) So near and yet so far: territorial pairs but low cohesion between pair partners in a nocturnal lemur, *Phaner furcifer*. *Animal Behaviour* **65**:331–343.

Schultz, S., *et al.* (2011) Stepwise evolution of stable sociality in primates. *Nature* **479**:219–222.

Schwarz-Weig, E. and Sachser, N. (1996) Social behaviour, mating system and testes size in Cuis (*Galea musteloides*). *Zeitschrift für Säugetierkunde* **61**:25–38.

Singleton, I. and van Schaik, C.P. (2001) Orangutan home range size and its determinants in a Sumatran swamp forest. *International Journal of Primatology* **22**:877–886.

Slagsvold, T. and Lifjeld, J.T. (1994) Incomplete female knowledge of male quality may explain variation in extra-pair paternity in birds. *Behaviour* **134**:353–371.

Sommer, S. (2003) Social and reproductive monogamy in rodents: the case of the Malagasy giant jumping rat. In: *Monogamy: Mating Strategies and Partnerships in Birds, Humans and Other Mammals* (eds U.H. Reichard and C. Boesch). Cambridge: Cambridge University Press, 109–124.

Sommer, V. and Rajpurohit, L.S. (1989) Male reproductive success in harem troops of Hanuman langurs (*Presbytis entellus*). *International Journal of Primatology* **10**:293–317.

Sommer, V. and Reichard, U. (2000) Rethinking monogamy: the gibbon case. In: *Primate Males: Causes and Consequences of Variation in Group Composition* (ed. P.M. Kappeler). Cambridge: Cambridge University Press, 159–168.

Spong, G.F., *et al.* (2008) Factors affecting reproductive success of dominant male meerkats. *Molecular Ecology* **17**:2287–2299.

Steenbeek, R., *et al.* (2000) Costs and benefits of the one-male, age-graded and all-male phase in wild Thomas's langur

groups. In: *Primate Males: Causes and Consequences of Variation in Group Composition* (ed. P.M. Kappeler). Cambridge: Cambridge University Press, 130–145.

Stevenson, I.R. and Bancroft, D.R. (1995) Fluctuating trade-offs favour precocial maturity in male Soay sheep. *Proceedings of the Royal Society of London. Series B: Biological Sciences* **262**:267–275.

Stevenson, I.R., *et al.* (2004) Adaptive reproductive strategies. In: *Soay Sheep: Dynamics and Selection in an Island Population* (eds T.H. Clutton-Brock and J.M. Pemberton). Cambridge: Cambridge University Press, 243–275.

Stillman, R.A., *et al.* (1993) Black holes, mate retention, and the evolution of ungulate leks. *Behavioral Ecology* **4**:1–6.

Stillman, R.A., *et al.* (1996) Black hole models of ungulate lek size and distribution. *Animal Behaviour* **52**:891–902.

Stirling, I. (1975) Factors affecting the evolution of social behaviour in the pinnipedia. In: *Biology of the Seal: Rapports et Procès-Verbaux des Réunions* Vol. **169** (eds K. Ronald and A. W. Mansfield). Charlottenlund, Denmark: International Council for the Exploration of the Sea, 205–212.

Stolba, A. (1979) *Entscheidungsfindung in verbänden von Papio hamadryas*. PhD thesis, University of Zurich.

Storz, J.F., *et al.* (2001) Genetic consequences of polygyny and social structure in an Indian fruit bat, *Cynopterus sphinx*. I. Inbreeding, outbreeding, and population subdivision. *Evolution* **55**:1215–1223.

Struhsaker, T.T. (1975) *The Red Colobus Monkey*. Chicago: University of Chicago Press.

Stutchbury, B.J.M. and Morton, E.S. (1995) The effect of breeding synchrony on extra-pair mating systems in songbirds. *Behaviour* **132**:675–690.

Sugiyama, Y. (1967) Social organization of hanuman langurs. In: *Social Communication Among Primates* (ed. S. Altmann). Chicago: University of Chicago Press, 221–236.

Sugiyama, Y., *et al.* (1965) Home range, mating season, male group and inter-troop relations in Hanuman langurs (*Presbytis entellus*). *Primates* **6**:73–106.

Sun, L. (2003) Monogamy correlates, socioecological factors and mating systems in beavers. In: *Monogamy: Mating Strategies and Partnerships in Birds, Humans and Other Mammals* (eds U.H. Reichard and C. Boesch). Cambridge: Cambridge University Press, 138–146.

Sutherland, W.J. (1985) Chance can produce a sex difference in variance in mating success and explain Bateman's data. *Animal Behaviour* **33**:1349–1352.

Swedell, L., *et al.* (2011) Female dispersal in hamadryas baboons: transfer among social units in a multilevel society. *American Journal of Physical Anthropology* **145**:360–370.

Takahata, Y., *et al.* (1996) Comparing copulations of chimpanzees and bonobos: do females exhibit proceptivity or receptivity. In: *Great Ape Societies* (eds W.C. McGrew, L.F. Marchant and T. Nishida). Cambridge: Cambridge University Press, 146–158.

Tardif, S.D. (1994) Relative energetic cost of infant care in small-bodied neotropical primates and its relation to infant-care patterns. *American Journal of Primatology* **34**:133–143.

Terborgh, J. and Goldizen, A.W. (1985) On the mating system of the cooperatively breeding saddle-backed tamarin (*Saguinus fuscicollis*). *Behavioral Ecology and Sociobiology* **16**:293–299.

Thirgood, S.J., *et al.* (1992) Mating system and ecology of black lechwe (*Kobus*, Bovidae) in Zambia. *Journal of Zoology* **228**:155–172.

Thirgood, S., *et al.* (1999) Intraspecific variation in ungulate mating strategies: the case of the flexible fallow deer. In: *Advances in the Study of Behavior*, Vol. **28** (eds P.J.B. Slater, J.S. Rosenblatt, T.J. Roper and C.T. Snowdon). San Diego, CA: Academic Press, 333–361.

Travis, S.E., *et al.* (1995) Ecological and demographic effects on intraspecific variation in the social system of prairie dogs. *Ecology* **76**:1794–1803.

Tutin, C.E.G. (1979) Mating patterns and reproductive strategies in a community of wild chimpanzees. *Behavioral Ecology and Sociobiology* **6**:29–39.

Tutin, C.E.G. (1980) Reproductive behavior of wild chimpanzees in the Gombe National Park, Tanzania. *Journal of Reproduction and Fertility Supplement* **28**:43–57.

Tutin, C.E.G. and McGinnis, P.R. (1981) Chimpanzee reproduction in the wild. In: *Reproductive Biology of the Great Apes* (ed. C.E. Graham). New York: Academic Press, 239–264.

van Belle, S., *et al.* (2014) Social and genetic factors mediating male participation in collective group defence in black howler monkeys. *Animal Behaviour* **98**:7–17.

van Schaik, C.P. and Dunbar, R.I.M. (1990) The evolution of monogamy in large primates: a new hypothesis and some crucial tests. *Behaviour* **115**:30–62.

van Schaik, C.P. and Kappeler, P.M. (2003) The evolution of social monogamy in primates. In: *Monogamy: Mating Strategies and Partnerships in Birds, Humans and Other Mammals* (eds U. Reichard and C. Boesch). Cambridge: Cambridge University Press, 59–80.

van Schaik, C.P. and van Hooff, R.A.M. (1996) Toward an understanding of the orangutan's social system. In: *Great Ape Societies* (eds W.C. McGrew, L.F. Marchant and T. Nishida). Cambridge: Cambridge University Press, 3–15.

Vinogradov, A.E. (1998) Male reproductive strategy and decreased longevity. *Acta Biotheoretica* **46**:157–160.

Volomy, M., *et al.* (2015) Social organization in Eulipotyphla: evidence for a social shrew. *Biology Letters*, DOI: 10.1098/rsbl.2015.0825. [Accessed 13 January 2016].

Wade, M.J. and Shuster, S.M. (2002) The evolution of parental care in the context of sexual selection: a critical reassessment of parental investment theory. *American Naturalist* **160**:285–292.

Walker, L.V. (1995) *Mate choice in female eastern grey kangaroos,* Macropus giganteus. PhD thesis, University of New England, Armidale, NSW, Australia.

Walker, L.V. (1996) Female mate choice: are marsupials really so different. In: *Comparison of Marsupial and Placental Behaviour* (eds D.B. Croft and U. Ganslosser). Fürth, Germany: Filander Verlag, 208–225.

Walther, F.R., *et al.* (1983) *Gazelles and Their Relatives: A Study in Territorial Behaviour*. Norwich, NY: Noyes Publications.

Waterman, J.M. (2007) Male mating strategies in rodents. In: *Rodent Societies: An Ecological and Evolutionary Perspective* (eds J.O. Wolff and P.W. Sherman). Chicago: University of Chicago Press, 27–41.

Weilgart, L.S., *et al.* (1996) A colossal convergence. *American Scientist* **84**:278–287.

White, F.J. (1996) Comparative socio-ecology of Pan paniscus. In: *Great Ape Societies* (eds W.C. McGrew, L.F. Marchant and T. Nishida). Cambridge: Cambridge University Press, 29–41.

Whitehead, H. (1990) Rules for roving males. *Journal of Theoretical Biology* **145**:355–368.

Whitehead, H. (1993) The behaviour of mature male sperm whales on the Galapagos breeding grounds. *Canadian Journal of Zoology* **71**:689–699.

Whitehead, H. and Weilgart, L. (2000) The sperm whale: social females and roving males. In: *Cetacean Societies: Field Studies of Dolphins and Whales* (eds J. Mann, R. Connor, P. Tyack and H. Whitehead). Chicago: University of Chicago Press, 154–172.

Williams, J.M., *et al.* (2004) Why do male chimpanzees defend a group range? *Animal Behaviour* **68**:523–532.

Williams, J.R., *et al.* (1992) Development of partner preferences in female prairie voles (*Microtus ochrogaster*): the role of social and sexual experience. *Hormones and Behavior* **26**:339–349.

Wilson, M.L. and Wrangham, R.W. (2003) Intergroup relations in chimpanzees. *Annual Review of Anthropology* **32**:363–392.

Wilson, M.L., *et al.* (2004) New cases of intergroup violence among chimpanzees in Gombe National Park, Tanzania. *International Journal of Primatology* **25**:523–549.

Wirtz, P. (1981) Territorial defense and territory take-over by satellite males in the waterbuck Kobus ellipsiprymnus (Bovidae). *Behavioral Ecology and Sociobiology* **8**:161–162.

Wirtz, P. (1982) Territory holders, satellite males and bachelor males in a high-density population of waterbuck (*Kobus ellipsiprymnus*) and their associations with conspecifics. *Zeitschrift fur Tierpsychologie* **58**:277–300.

Wolff, J.O. and Cicirello, D.M. (1991) Comparative paternal and infanticidal behaviour of sympatric white-footed mice (*Peromyscus leucopus noveboracensis*) and deermice (*P. maniculatus nuibiterrae*). *Behavioral Eoclogy* **2**:38–45.

Wrangham, R.W. (1986) Ecology and social evolution in two species of chimpanzees. In: *Ecology and Social Evolution: Birds and Mammals* (eds D.I. Rubenstein and R.W. Wrangham). Princeton, NJ: Princeton University Press, 352–378.

Wrangham, R.W., *et al.* (1996) Social ecology of Kanyawara chimpanzees: implications for understanding the costs of great ape groups. In: *Great Ape Societies* (eds W.C. McGrew, L.F. Marchant and T. Nishida). Cambridge: Cambridge University Press, 45–57.

Young, J.K. and Franklin, W.L. (2004) Territorial fidelity of territorial male guanacos in the Patagonia of southern Chile. *Journal of Mammalogy* **85**:72–78.

Zhang, P., *et al.* (2006) Social organization of Sichuan snub-nosed monkeys (*Rhinopithecus roxellana*) in the Qinling Mountains, Central China. *Primates* **47**:374–382.

CHAPTER 11

Association between males

11.1 Introduction

On a moonlit night on the Serengeti plains, lions are active. As grazing herds of wildebeest shift uneasily, two young males from a pride pace silently into the territory of a neighbouring group. They disregard the wildebeest around them and pace forward, listening intently. In the distance, a lioness roars softly. The intruding males both respond, initially roaring alternately and then synchronously in a rising chorus. From a kopje a kilometre away there is an immediate response of deep roars, indicating the presence of several resident males. Outnumbered, the intruders listen carefully then, unhurriedly, turn and walk away in the opposite direction, leaving the resident pride to their own devices.

Like females, males can gain important benefits by associating and assisting each other to monopolise access to resources and females, though the presence of several mature males in the same group generates competition for breeding opportunities as well as for access to resources, leading to conflicts between resident males. In some societies, several males who are typically unrelated to each other associate with groups of females but gain no benefits from each other's presence; in others, males who are often relatives cooperate to gain access to female groups and to defend them against intruders or challengers.

This chapter describes associations and relationships between males, focusing on societies where two or more mature males associate with the same females for protracted periods. Section 11.2 describes different kinds of association between males, while sections 11.3 and 11.4 examine evidence of the costs and benefits of association to males. Section 11.5 explores social and genetic relationships between resident males, while section 11.6 examines the factors affecting the number of resident males. Section 11.7 describes contrasts in the distribution

of male breeding success in different groups. Finally, section 11.8 examines the consequences of variation in the size and structure of male groupings for the reproductive strategies of males and females and the genetic structure of groups.

11.2 Contrasts in the formation and structure of male groups

Temporary associations

In many social mammals, unrelated males associate temporarily with each other while attempting to mate with females but derive little or no benefit from each other's presence. For example, male sheep temporarily aggregate around oestrous females and compete for opportunities to mate with them sometimes forming mating queues (Grubb 1974) (Figure 11.1). Similarly, males of several African antelope, like puku, defend mating territories in areas where concentrations of resources attract females (see Chapter 10). Since males compete for space as well as for individual females and clusters of territories may attract predators (Balmford and Turyaho 1992; Fitzgibbon 1994), it is unlikely that males gain from each other's proximity, though it is often difficult to be certain since clusters of males may attract females or help to advance their period of receptivity.

In other cases where males associate temporarily with each other, they evidently benefit from each other's presence. For example, young male red deer collect around the edge of the harems of dominant males during the mating season, often running in and attempting to drive out females when the dominant male's attention is focused elsewhere (Clutton-Brock et al. 1979). Incursions by one or more young males on one side of a harem commonly provoke chases by the dominant male (Figure 11.2),

Mammal Societies, First Edition. Tim Clutton-Brock.
© 2016 Tim Clutton-Brock. Published 2016 by John Wiley & Sons, Ltd.

Figure 11.1 A mating queue in Soay sheep. *Source*: © Brian Preston.

Figure 11.2 Benefits of association to sneakers: young male red deer collect around the harems of stags holding harems and, when dominants chase one away, others have an opportunity to slip into the harem from the opposite side. *Source*: © Tim Clutton-Brock.

leaving the other side of the harem exposed to other males. As a result, when several young males are present, they are more likely to gain access to females. Association can also benefit males by reducing the chance of predation (Hamilton 1971). For example, on the East African plains, territorial males of different antelope species commonly group together in the middle of the day when no females are on their territories, probably reducing their individual chances of being attacked by predators or improving their ability to detect danger (see Chapter 2). In both these examples, males may synchronise their activities and gain fitness benefits from doing so but their actions are not adapted to benefit others and they probably represent examples of by-product mutualism rather than cooperation (see Chapter 1).

Bachelor groups

In a number of mammals, non-breeding males form bachelor groups. All male groups occur in a substantial number of polygynous species, including howler monkeys (Rudran 1979; Crockett and Sekulic 1984), several langurs (Sugiyama 1965; Rudran 1973; Steenbeek 2000; Steenbeek and van Schaik 2001), patas monkeys (Ohsawa *et al.* 1993), gelada baboons (Dunbar 1984),

Japanese macaques (Sugiyama 1976) and gorillas (Robbins 1996; Watts 1996, 2000) and in a number of other social mammals, including equids (Berger 1987), macropods (Macfarlane and Coulson 2005), pinnipeds (Staniland 2005), bats (Altringham and Senior 2005) and cetaceans (Mann *et al.* 2000) (Figures 11.3 and 11.4). They are particularly common in ungulates where they consist either of males that have not yet acquired mating territories, as in kob and lechwe (Leuthold 1966, 1967; Nefdt and Thirgood 1997), or of males that have not yet acquired dominant positions in breeding groups, as in the social equids (Berger 1986; Feh 1999, 2001). In other species, bachelor groups consist of mature males that are 'resting' between periods of mating activity, as in African buffalo (Prins 1996).

In some seasonal breeders, adolescent and mature males form bachelor groups in areas peripheral to concentrations of females during the non-breeding season. For example, in red deer, males form unstable all-male groups during the non-breeding season (Lincoln *et al.* 1970; Gibson and Guinness 1980) but these groups split up at the beginning of the annual mating season when males move to their individual rutting areas (Clutton-Brock *et al.* 1982). Loosely knit associations between

Figure 11.3 Bachelor groups are common among ungulates: in red deer, stags form unstable all-male groups outside the mating season. *Source*: © Laurie Campbell.

Figure 11.4 A bachelor group of gelada baboons. *Source*: © Clay Wilton.

males occur in a number of other deer and antelope, and are commonly (but not universally) associated with sex differences in habitat use and feeding behaviour (see Chapter 18). In African elephants, males leave their natal families at around 14 years and move to areas used largely by bulls, associating with other males in unstable parties (Moss and Poole 1983; Poole 1994). When adult males enter breeding condition, they leave their usual ranges to search for receptive females (Poole 1987, 1989a,b).

Bachelor males may benefit by aggregating for several reasons. In some cases, aggregation probably helps to reduce the risk of predation, which may explain why all-male groups appear to be more common in diurnal terrestrial primates than in arboreal species (Struhsaker 1969). In other cases, males may gain energetic benefits by associating because this allows them to forage more successfully or to defend access to resource patches (see section 11.4). The formation of bachelor groups can also enable males to feed in preferred areas defended by territorial males. For example, in puku and lechwe, all-male groups are able to graze in forb-rich areas defended by territorial males because their numbers swamp the defences of individual territory holders (Nefdt 1992; Rosser 1992). Finally, the formation of bachelor

groups may have reproductive benefits. For example, in grey langurs and gelada baboons, members of all-male groups simultaneously invade and attack resident males. When they are successful in expelling resident males, invading males then fight amongst themselves until, in most cases, only one or two dominant individuals remain (Sugiyama 1965; Hrdy 1977; Dunbar 1984).

Residents and satellites

Longer-term associations of unrelated males occur in species where mature males defend individual territories. In some of these species, males disperse from their natal groups after reaching adulthood and settle within the range of a dominant male but do not associate closely with him, as in waterbuck (Wirtz 1982) and orangutans (van Schaik and van Hooff 1996). In many cases where dominant males tolerate satellite males within their ranges, there is a clear difference in age or development between territorial males and satellite males and satellites maybe evicted before they reach full maturity (van Hooff 2000): for example, in orangutans, dominant males may tolerate younger satellite males within their ranges but evict them if they start to advertise their presence by loud calls (Galdikas 1981; van Schaik and van Hooff 1996; Knott and Kahlenberg 2007) (Figure 11.5). As a result,

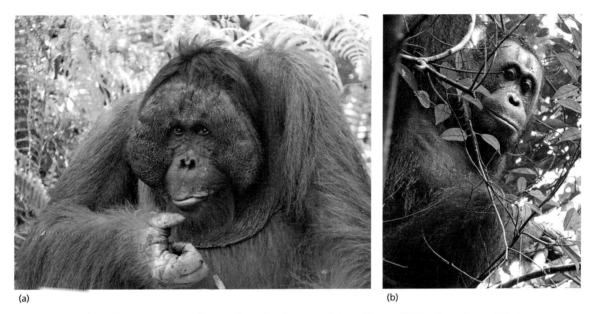

(a)   (b)

Figure 11.5 (a) 'Flanged' male orangutan; (b) an unflanged male. *Source*: © Anna Marzec, Tuanan Orang Research Project.

dominant males and satellites use different mating tactics: while dominants advertise their presence and wait for females to come to them, satellites actively search for females and often attempt to obtain matings by force (see Chapter 15).

Similar associations between a mature male and one or more younger males occur in several species where females are philopatric and live in stable groups that are defended by a single mature male. For example, in Thomas's langurs, groups consist of multiple philopatric females and their juvenile and adolescent offspring of both sexes and a single mature immigrant male (Steenbeek and van Schaik 2001). Subordinate natal males that have not yet dispersed to breed assist resident males (who are often their fathers) to defend their group against intruders. Similar relationships between natal males and resident breeding males occur in howler monkeys (Steenbeek 2000; Pope 2000) and in meerkats (Clutton-Brock *et al*. 1998; Young 2003). Associations between mature dominant males and one or more younger males also occur in societies where females habitually disperse to breed. For example, in both horses and gorillas, groups commonly consist of several immigrant females, one dominant male and one or more subordinate males who may either be natals or unrelated immigrants (Berger 1986; Feh 1999; Harcourt and Stewart 2007).

Multi-male associations

In mammals where breeding females occupy separate ranges, several mature males may share a common range or territory overlapping the ranges of several females (see Chapter 10). For example, in cheetah, several males, who are often brothers, defend a common territory in areas used by females and share mating access to local females (Caro 1994) (Figure 11.6). A similar system occurs in Madagascan fossas, a large (6–11 kg) endemic carnivore that feeds on vertebrates (Lührs and Dammhahn 2010; Lührs and Kappeler 2013). Here, males occupy ranges over 50 km^2 which overlap smaller ranges occupied by individual females. While some males are solitary, littermates sometimes share ranges and hunt together and kill larger prey, such as sifakas, and are physically larger and heavier than solitary males, who do not differ from females in size or weight (Lührs *et al*. 2013). During the month-long mating season, females, who are socially dominant to males, visit traditional mating trees and may share the same mating tree sequentially or simultaneously with other females. They display and attract males and commonly mate with several partners and (Hawkins and Racey 2009; Lührs and Kappeler 2014). Males compete for access to females and larger males are more successful at monopolising females during the peak period of receptivity.

Figure 11.6 In African cheetah, several related males may cooperate to defend large breeding territories against other males. *Source*: © Anup Shah/Nature Picture Library.

Where females live in sizeable groups, several mature males often associate with them and defend access to them. In some species, like Cape ground squirrels (Waterman 1997, 1998), spider monkeys and muriquis (Di Fiore and Campbell 2007), chimpanzees (Boesch 1996; Williams *et al.* 2004) and bottlenose dolphins (Connor 2000), males associated with a group or community of females aggregate in unstable parties. In others, including African buffalo (Prins 1996) and many of the baboons and macaques (Jolly 2007; Thierry 2007), they forage with groups of females in cohesive groups.

Associations of mature males form in different ways as a result of contrasts in female philopatry and litter size (Table 11.1). Where resident females are usually philopatric and produce single young (as in many of the baboons and macaques), males typically disperse on their own and join other breeding groups. Since males leaving the same group may join different breeding groups, many resident males are unrelated to each other although, in some cases, related males from the same natal group may join the same breeding group at different times (Cheney and Seyfarth 1983). In many species

Table 11.1 Contrasts in the formation and structure of multi-male groups.

Origin of males	Relatedness of males to each other	Female dispersal pattern	Relatedness of females to resident males	Examples
Independent immigrants	Mostly unrelated	Females philopatric	Mostly unrelated	Savannah baboons, spotted hyena
Social immigrants	Mostly related	Females philopatric	Mostly unrelated	African lions, meerkats
Natal males	Related (though not necessarily closely)	Females habitually disperse	Mostly unrelated	Chimpanzees, red colobus
Natal males	Related	Females habitually remain in natal group	Mostly related	Killer whales, naked mole rats

where males emigrate and immigrate independently, immigrating males rarely evict previous residents, though they may rapidly rise to dominant positions in their new group (van Noordwijk and van Schaik 2004). Finally, in a small number of monotocous species where females typically remain and breed in their natal groups (including killer whales and, possibly, pilot whales), males often remain in their natal group throughout their lives and both sexes breed outside their group (see Chapter 12), so that resident males are related both to each other and to resident females.

Where females are philopatric but produce sizeable litters, adolescent males (who are often half or full siblings) frequently disperse together and either form new breeding groups with emigrant females or replace resident males in established breeding groups (see Chapter 12). In societies of this kind (which occur in several social rodents as well as in meerkats and lions), resident breeding males are often siblings, and are often tolerant of each other's attempts to breed. Resident males are seldom closely related to resident females, though dispersing males sometimes join groups containing female relatives that emigrated previously from their group (Nielsen *et al.* 2012).

Where females habitually disperse to breed after reaching adolescence, natal males are often unrelated to most breeding females in their group and so are potential mating partners (see Chapter 4). In many of these species (including feral horses, hamadryas baboons, mountain gorillas, banded mongooses and greater sac-winged bats), some males remain and breed in their natal group throughout their lives, while others disperse and join bachelor groups, establish new breeding groups with emigrant females, or immigrate into established breeding groups as primary or secondary males (Berger 1986; Phillips-Conroy *et al.* 1992; Harcourt and Stewart 2007; Nagy *et al.* 2007). Within species, the frequency of male dispersal appears to reflect the distribution of opportunities to breed (see Chapter 12) and contrasts in the chances that dispersing males will be able to locate dispersing females or displace resident males may also have an important influence on species differences in the frequency of male dispersal. For example, the frequency of male dispersal may be lower in monotocous species where females disperse, like chimpanzees and greater sac-winged bats, than in polytocous species, like banded mongooses, because subgroups of dispersing males are seldom likely to be large enough to displace resident males (see Chapter 12).

While there are important interspecific differences in the kinship structure of male groups, exceptions are common. Where females are philopatric and monotocous and males disperse independently, like capuchin monkeys and baboons, some males may remain and breed in their natal group (Perry *et al.* 2008) or male sibs may immigrate independently into the same breeding group (Cheney and Seyfarth 1983). Conversely, in polytocous species where groups of related males emigrate together, small parties may join up with unrelated individuals so that if they eventually immigrate into the same breeding group, some resident males are unrelated to other males in their group, as in lions (Packer and Pusey 1982; Packer *et al.* 1991).

11.3 Costs of association to males

Where males aggregate with each other in single-sex groups, the costs of association are likely to be relatively small and may be offset by benefits derived from improved predator detection or increased access to breeding females (see section 11.4). In contrast, where dominant males tolerate sexually mature subordinate males within breeding groups that are unrelated to the females they are defending, they risk losing mating opportunities to them and the costs of tolerating their presence are potentially large.

Comparative studies of male mating success based on genetic data show that the number of paternities dominant males lose to other males resident in the same group varies widely and is affected both by the capacity of dominant males to guard access to receptive females and by variation in social structure and kinship between subordinate males and resident females (Clutton-Brock and Isvaran 2006; Cohas and Allainé 2009). Where stable groups of breeding females are defended by a single immigrant male and several younger natal males, as in many of the langurs, resident females rarely breed with natal males and dominant males lose few breeding opportunities either to other males in their group or to intruding males from other groups.

A single dominant male can also be responsible for a high proportion of paternities in many species where several immigrant males associate with groups of breeding females. For example, at least two-thirds of young born in groups of capuchin monkeys, howler monkeys, yellow baboons and crab-eating macaques are fathered

Table 11.2 Estimates of the proportion of paternities achieved by the most dominant male in each group.

Meerkats	>90%
Mountain gorilla	85%
Yellow baboon	81%
Capuchin	80%
Red fronted lemurs	71%
Chimpanzee (three studies)	30–67%
Bonobo	30%
Spotted hyena	20%
Muriquis	18%

Source: Clutton-Brock and Isvaran (2006); Cohas and Allainé (2009); Strier *et al.* (2011).

by a single dominant male (Table 11.2). However, the presence of increasing numbers of resident males is commonly associated with an increase in the proportion of paternities lost to other group members by dominant males and the degree of reproductive skew is then frequently reduced. For example, across primates, the proportion of matings achieved by the most successful male in the group falls as the number of resident males increases (Kutsukake and Nunn 2006, 2009; Gogarten and Koenig 2013) (Figure 11.7). Negative correlations between the proportion of offspring fathered by dominant males and the number of males present also occur

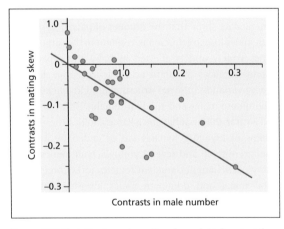

Figure 11.7 Variation in male mating skew plotted against the number of males per group for primate species, showing that the proportion of matings involving the most successful male declines as the number of males per group increases. Estimates of mating skew based on observational data. *Source*: From Kutsukake and Nunn (2006). Reproduced with permission of Springer Science and Business Media.

within a number of other species (Hager 2009; Kutsukake and Nunn 2009; Lardy *et al.* 2012).

Where females habitually disperse to breed, resident females are likely to be unrelated both to immigrant males and to natal males, which are consequently potential mating partners (see Chapter 3). Natal males are typically relatives who have grown up together and relationships between them are frequently more relaxed than those between males in species where resident males are immigrants. Where group size is small and one resident male is older and larger than the others, as in horses and mountain gorillas, the dominant male usually fathers more than two-thirds of young born in the group (Feh 1999, 2001; Bradley *et al.* 2005), but where groups are larger and females are more widely dispersed, dominant males often lose a larger proportion of paternities to other resident males, as in muriquis, chimpanzees and bonobos (Strier *et al.* 2011) (see Table 11.2).

Tolerating other resident males can also have costs to the tenure or longevity of dominant males. The presence of potential competitors in the group may increase the probability of challenges from within the group and the proportion of time spent mate guarding by dominants, with potential costs to their condition and the tenure of the breeding role (see Chapter 13) as well as increased risks of injury. It can also increase their cortisol levels, which may in turn depress their immune function and lead to increases in parasite load. For example, in alpine marmots, the probability of dominant males retaining their position in breeding groups declines as the number of sexually mature subordinates in the group increases (see section 11.7).

11.4 Benefits of association to males

Where the presence of subordinate males is unlikely to have significant costs to dominant males, they may be tolerated because the costs of evicting them exceed the benefits. However, where subordinate males obtain a share of reproduction, the costs of eviction would have to be large to prevent eviction from being the best strategy, suggesting that dominant males may gain important benefits from the presence of subordinates.

Evidence from a considerable number of studies suggests that the presence of subordinate males has benefits to dominant males. For example, in both waterbuck and wild horses, satellite males help dominants to drive away

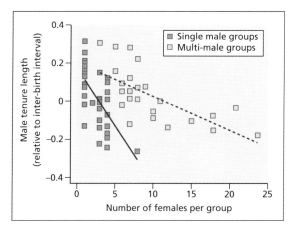

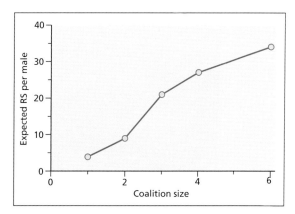

Figure 11.8 Across mammals male breeding tenure declines as the number of females per group increases. Dominant males maintain their positions longer in species in which there are few females per group and, for a given number of females per group, tenure lengths are shorter in species in which groups contain only a single male (dark squares) compared with species in which groups usually contain multiple males (light squares). For comparison, tenure length has been adjusted for the length of the inter-birth interval in these species. *Source:* From Lukas and Clutton-Brock 2014.

intruders, allowing dominants to defend access to females (Wirtz 1982; Stevens 1990; Feh 1999, 2001). Comparative studies suggest that the presence of additional males can have important effects on the breeding tenure of males: comparisons of the average tenure of breeding males across different species of mammals show that the duration of tenure declines more rapidly with increasing numbers of resident females if there is only a single male in the group (Lukas and Clutton-Brock 2014) (Figure 11.8).

Intraspecific comparisons also suggest that the presence of subordinate males can have important benefits to dominants. In capuchin monkeys and gelada baboons, the presence of additional males reduces the risk of takeovers by groups of roving males and extends the tenure of dominants (Fedigan and Jack 2004; Muniz *et al.* 2010; Snyder-Mackler *et al.* 2012). Similarly, in greater sac-winged bats, where groups of natal males defend access to resources used by groups of females, the breeding tenure of males (which accounts for a substantial proportion of variance in male breeding success) increases with the number of males present (Nagy *et al.* 2012). Some of the clearest evidence of the reproductive benefits of association to males comes from studies of lions,

Figure 11.9 The (per capita) lifetime breeding success of male lions increases with the number of males in coalitions. *Source:* From Packer *et al.* (1988). *Photo source:* © Tim Clutton-Brock.

where the probability that parties of males will take over groups of females, the period for which they are likely to defend them and the per-capita breeding success of males all increase with the number of males present (Figure 11.9) (Packer *et al.* 1988). In primates, too, the number of resident males can have an important influence on the outcome of competitive interactions between groups (Wrangham and Peterson 1996; see Figure 11.10).

As well as assisting each other in fights with intruders, resident males (who are often close relatives) contribute to displays, like roaring choruses, that help to discourage intruders: intruding males are able to detect the number of individuals contributing to choruses and commonly avoid escalating contests with larger coalitions (Grinnell and McComb 2001) so that all residents benefit from reductions in the frequency of escalated fights. Similar

Figure 11.10 Males in two groups of rhesus macaques confront each other in a dispute. *Source*: © Kevin Rosenfield.

choruses occur in several other mammals where groups of males defend access to females, including wolves (Harrington and Mech 1979), howler monkeys (Sekulic 1982; Chiarello 1995) and chimpanzees (Wilson *et al.* 2001, 2002).

Male associations can also increase the survival of offspring born to dominant males. In mountain gorillas, subordinate males help to defend groups against intruding males and the presence of co-residents helps to prolong the tenure of breeding males (Sicotte 1995; Harcourt and Stewart 2007). This can have an important influence on the survival of offspring born to dominant males, for when dominants die and no subordinate is resident in the group, incoming males frequently kill dependent juveniles or groups may fragment and females that join other groups may also lose their offspring to infanticide. Where resident subordinate males are present when a dominant dies, they frequently succeed to the dominant position and rarely kill dependent young, who are familiar and may be their own offspring or may be related to them in other ways (Watts 2000; Robbins and Robbins 2005; Harcourt and Stewart 2007). As a result, dominant males gain direct fitness by allowing subordinate males to remain in the group and by not excluding them entirely from reproduction. Where natal males may remain and breed in their natal

group, tolerating them may also allow dominants (who are often their fathers) to ensure that they are succeeded by a close relative rather than an immigrant (Harcourt and Stewart 2007; Henzi *et al.* 2010).

But are benefits to a male's tenure sufficient to offset the costs of tolerating subordinates to the dominant male's breeding success? One of the few studies to attempt to answer this question involves red-fronted lemurs, which live in groups of five to ten with an approximately equal sex ratio (Kappeler and Port 2008; Port *et al.* 2010) (Figure 11.11a). Groups typically include several adult males who may either be natals or immigrants. One male is usually clearly dominant and is responsible for fathering around 70% of juveniles born in the group while 29% are fathered by subordinates. Since dominant males lose nearly one-third of paternities, they might be expected to do their best to evict subordinates but instead they usually tolerate their presence. A likely reason for this is that the presence of subordinates extends the dominant's tenure of the breeding position in the group: long-term data show that the probability that groups will be taken over by intruding males falls as the number of defending males increases until an asymptote is reached where the addition of further resident males has little effect (Figure 11.11b). Calculations show that males defending

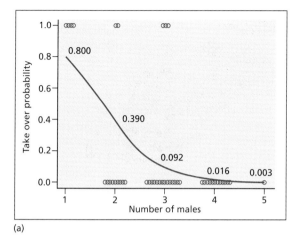

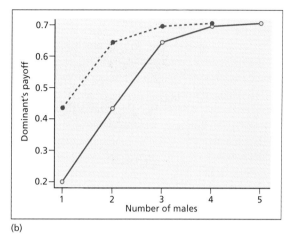

Figure 11.11 Red-fronted lemurs live in multi-male groups of 5–15 animals: (a) estimated risk of a take-over by external males plotted against the number of resident males; (b) the fitness pay-offs to dominant males if they reject (solid line) or accept (dashed line) attempts by subordinates to join them in relation to the number of males present *Source*; From Port *et al.* (2010). Reproduced with permission from the Royal Society. *Photo source*: © Claudia Fichtel.

female groups on their own, as well as those living in groups with a single subordinate, would both gain substantial fitness benefits by allowing additional males to join them. Since the proportion of paternity lost to subordinates does not increase as their numbers rise, dominants should accept additional subordinates, even if the marginal benefits of their presence are small.

Conditional strategies often allow dominant males to minimise the costs and maximise the benefits of tolerating subordinates. In horses, the probability that dominant males will tolerate subordinate males in their group depends on their competitive ability: the most dominant males are usually the sons of dominant mothers and often defend harems on their own, while the sons of subordinate mothers more commonly form alliances with other males (Feh 1999, 2001).

In addition to reproductive benefits, males can gain energetic advantages from associating. For example,

males of several carnivores that take relatively large prey commonly hunt in groups, including lions (Bertram 1978), cheetah (Caro 1994), wild dogs (Creel and Creel 2002) and fossas (Lührs and Dammhahn 2010), as do males in some aquatic species that feed either on large fast moving prey, like killer whales (Baird 2000) and some otters (Blundell *et al.* 2004), or on shoals of fish that can be compressed by cooperative manoeuvres, like several of the smaller cetaceans (Connor 2000). Social hunting by groups of males also occurs in some primates, including chimpanzees (see Chapter 2). Hunting associations can generate substantial benefits: for example, male cheetah that form larger associations hunt larger prey, eat more per unit time and are fatter than solitary individuals (Caro 1994) and similar associations between body weight and the number of males present occur in fossas (Lührs *et al.* 2013).

There is also extensive evidence that subordinate males can gain important fitness benefits from associating with dominants. Where females are philopatric and natal males rarely breed within their natal group, subordinate natal males may gain important survival benefits by remaining in established groups until they reach full adult size (see Chapter 12). In addition, they may increase their fitness by mating opportunistically with roving females or members of other groups and so may achieve some level of reproductive success: for example, in meerkats, subordinate males that have not yet left their natal group often visit other groups and mate with subordinate females while they are there if they get a chance (Young *et al.* 2007). Immigrant subordinates that are unrelated to resident females may acquire a share of paternity within their group, especially in species where females frequently mate with multiple males (Clutton-Brock and Isvaran 2006). Immigrants may also inherit the group following the death or displacement of the dominant male, so that the proportion of their lives spent in breeding groups where they have access to females can be higher than in males that disperse to join bachelor groups.

Finally, where females habitually disperse, males may gain increased mating success by remaining in their natal groups and may also improve their chances of acquiring the breeding role. For example, in mountain gorillas, the likely fitness pay-offs to individuals of joining or remaining in breeding groups as followers are around twice as high as those of males that join bachelor groups (Watts 2000).

11.5 Kinship, familiarity, cooperation and hostility

In many species where males defend access to groups of females, resident males are related and gain indirect as well as direct fitness benefits by associating. Relations between males originating from the same breeding group are often closer and more tolerant than those between males that have immigrated independently into breeding groups as adults and are mostly unrelated. For example, in killer whales, where resident males are natals that breed outside the group, resident males regularly exchange vocal signals and often hunt together (Baird 2000). Similarly, male chimpanzees (who are usually born in the same group but are not necessarily close

relatives) use pant hoots to keep in touch with each other while they are foraging and often maintain close social relationships (Mitani 1996). In contrast, in species where most resident males have immigrated independently and are not close relatives, as in macaques and chacma baboons, relationships between them are more competitive and many of their calls appear to represent competitive displays that reflect their strength and status rather than coordinating signals (Fischer *et al.* 2004; Cheney and Seyfarth 2008).

It is frequently difficult to separate the effects of kinship on affiliative behaviour from those of familiarity. Affiliative behaviour is often more highly developed between males that have grown up in the same group, even if they are not close relatives. For example, adult male chimpanzees commonly engage in mutual grooming sessions with each other, forming large huddles that often include high- and low-ranking males, and frequently support each other in repelling intruders (Simpson 1973; Boesch 1996), even though few are closely related (Lukas and Vigilant 2005). In contrast, where resident males are unrelated immigrants (as in many of the baboons and macaques, as well as in many social ungulates and macropods), mutual grooming and other affiliative interactions between adult males are less frequent, coordinated defence of female groups is less developed though males may attack intruders independently (Cheney and Seyfarth 1977; Harding 1980; Cowlishaw 1995).

Variation in male kinship can also lead to contrasts in social relationships within groups in local populations. For example, relationships between resident males in some multi-male coteries of prairie dogs are consistently hostile, while in others they are variable or consistently amicable (Hoogland 1995). Where males are consistently amicable, the resident males are usually close kin ($r \geq 0.25$) while relationships between males are seldom amicable in coteries where the resident males are not closely related (Figure 11.12). Similar effects of kinship on social relationships between breeding males have been found in other species. In alpine marmots, dominant males suppress the development of subordinates who could not be their offspring ('non-sons') to a greater extent than that of potential sons, and non-sons receive more injuries and show higher corticosteroid levels and lower androgen levels than potential sons (Arnold and Dittami 1997). In howler monkeys, resident males support each other in conflicts with neighbouring groups,

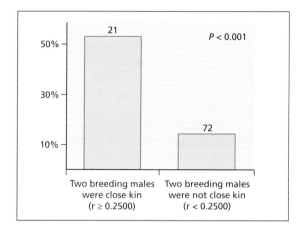

Figure 11.12 Percentage of multi-male prairie dog coteries where interactions between males were primarily amicable in which breeding males were either close kin or not close kin, where interactions between males were primarily amicable. *Source*: From Hoogland (1995). Reproduced with permission of the University of Chicago Press. *Photo source*: © Elaine Miller Bond.

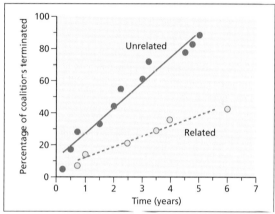

Figure 11.13 Comparative termination rates of coalitions of related versus unrelated male ursine howler monkeys. *Source*: From Pope (2000). Reproduced with permission of Springer Science and Business Media. *Photo source*: © Tim Clutton-Brock.

and coalitions composed of related males persist for longer than those of non-relatives (Pope 2000) (Figure 11.13). In chimpanzees, too, maternally related males are more likely to support each other than the offspring of unrelated females, although not all allies are related (Langergraber *et al.* 2007). Kinship often appears to affect some forms of cooperation more than others: for example, while male chimpanzees selectively support related males in competitive encounters with other males, there is no evidence of kin bias in hunting behaviour and males are no more likely to share meat with maternal kin than with unrelated males (Boesch *et al.* 2006).

Contrasts in kinship also appear to affect relationships between males from neighbouring groups. In species where resident males have immigrated separately and most are unrelated to each other (as in many macaques and baboons), inter-group relationships between males from neighbouring groups are seldom as intensely hostile as in species where groups of related males defend access to female groups, as in lions, meerkats and chimpanzees. For example, in chimpanzees, relationships between males from neighbouring communities are usually intensely hostile and parties of resident males not only patrol the boundaries of their group's range, but also maraud into the territories of neighbouring communities, seeking out and attacking solitary males or smaller parties (Crofoot and Wrangham 2010). Individuals attacked by larger parties are often wounded and are sometimes killed: in a sample of thirty-two known and suspected deaths resulting from attacks at Gombe, over 90% of the victims were male (Wrangham *et al.* 2006). Similar raids into neighboring territories occur in spider monkeys (Aureli *et al.* 2006) as well as in meerkats (Clutton-Brock and Manser 2015).

Raids and attacks may serve to reduce the power of neighbouring groups, allowing males to displace their neighbours and monopolise the females and resources resident within their range (Williams *et al.* 2004; Crofoot and Wrangham 2010). In one case at Gombe, the number of males in one community increased as a result of the provision of artificial food supplies while the number of males in a neighbouring community declined, partly as a result of group members being killed in raids (Goodall 1986). The males from the larger community gradually raided further into their neighbour's territory and eventually drove the remnants of the competing male group out altogether or killed them, extending the boundaries of their own territory to encompass the ranges and resources used by most of the neighbouring group's females. Similar changes in the ranging patterns and breeding success of different groups in relation to the number of males present have been found in other populations of chimpanzees as well as in other primates (Nishida *et al.* 1985; Crofoot and Wrangham 2010). Lethal attacks may be particularly likely in chimpanzees because aggregation in fission–fusion groups commonly leads to encounters between solitaries or pairs of animals and larger parties of neighbours, allowing members of larger parties to engage in lethal attacks at little risk to themselves (Crofoot and Wrangham 2010).

11.6 The size of male associations

Across mammalian societies, the numbers of males associating with breeding groups vary from species where groups commonly consist of a single adult male and several breeding females to groups including several adults of both sexes with an approximately equal sex ratio (see Chapter 10). A considerable number of studies have explored the reasons for these differences, especially in primates (Nunn 1999; Kappeler 2000) but attempts to relate variation in the number of males associated with female groups directly to ecological or social differences have met with limited success (Crook and Gartlan 1966; Clutton-Brock *et al.* 1977; Kappeler 2000) and these differences may be more closely

associated with phylogeny and life-history parameters than with contrasts in ecology (Richard and Dewar 1991; Cords 2000; Dixson 2000; Kappeler 2000). Perhaps this is not surprising, for the costs and benefits to subordinate males of joining or remaining in associations of different size are likely to be affected by many different ecological, social and demographic factors.

To account for variation in the number of resident males in breeding groups among species where females are philopatric, it is necessary to distinguish between natal males that are related to resident females and are unlikely to breed in the group, and immigrants that are potential breeding partners for resident females since different mechanisms are likely to control their numbers. Variation in the number of non-breeding natal males in breeding groups is likely to be partly a consequence of demographic factors (including variation in the fecundity of breeding females and the survival of juveniles) and partly of variation in male dispersal which may be affected by age-related benefits of remaining in the natal group versus dispersing later (see Chapter 12). These may in turn be influenced by a wide range of ecological and social factors, including the effects of age at dispersal on survival and on the probability of acquiring a breeding position, the benefits of delaying dispersal until several males are ready to disperse at the same time, and the costs of age-related changes in the frequency or intensity of aggression received from resident dominants (Dobson 1982; van Hooff 2000). In some species, there are regular patterns in the number of natal males that reflect the individual history of the group. For example, in groups of Thomas's langurs (Figure 11.14), males often take over female groups aggressively and eject the previous males or groups of females may leave to join another male (Steenbeek *et al.* 2000). After a new dominant male is established, groups increase in size and often come to include several mature natal males that do not attempt to breed in the group but assist the resident male (who is often their father) in defending the group against neighbouring males. However, as dominant males age and become less effective at protecting females in their groups against potentially infanticidal attacks by extra-group males, breeding females successively leave the group and transfer to a younger male (Steenbeek *et al.* 2000) (Figure 11.15).

Contrasts in the number of immigrant males in breeding groups are likely to be influenced by variation in the effects of male numbers on the fitness of further immigrants, established dominants and females. Males would

Figure 11.14 Three members of a group of Thomas's langurs. *Source*: © Perry van Duijnhoven.

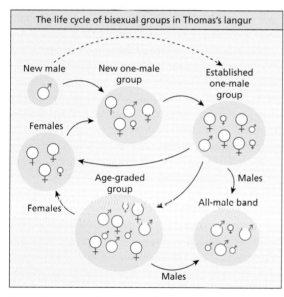

The life cycle of bisexual groups in Thomas's langur

Figure 11.15 Group dynamics in Thomas's langurs. New breeding groups start when females leave an established breeding group to join another male or an immigrant male evicts the resident breeding male in an established group. As the new male's sons mature, the group becomes an 'age-graded' group (Eisenberg *et al.* 1972), though younger males rarely breed with resident females and some may leave to join all-male bands. As the breeding male ages and becomes less able to defend the group against younger immigrant males, females may leave him and either form new groups with males emigrating from other breeding groups or join other established breeding groups. *Source*: From Steenbeek *et al.* (2000).

usually be expected to preferentially join and remain in groups where the ratio of breeding females to resident breeding males is relatively high, and several studies suggest that they do so. For example, both in yellow baboons and in spotted hyenas, males are more likely to join and remain in groups where the ratio of resident females to breeding males is relatively high (Alberts and Altmann 1995; Honer *et al.* 2007). Similarly, in blue monkeys (where males frequently move between groups during the breeding season), the number of males associating with female groups is related to the percentage of days when two or more oestrous females are present (Cords 2000). Where males can move freely between groups, the number of resident males should increase with the number of resident females and their fitness should approximate to an ideal free distribution. As expected, the number of males associated with female groups in different species is positively related to the number of females in the group (Andelman 1986; Altmann *et al.* 1988) (Figure 11.16).

Variation in the ability of resident dominants to monopolise reproduction would also be expected to affect the breeding opportunities of immigrant males and the number of resident males. For a given number of females, the number of resident males would be expected to increase where female reproduction is relatively synchronised so that multiple females are often in oestrus at the same time so that dominant males are

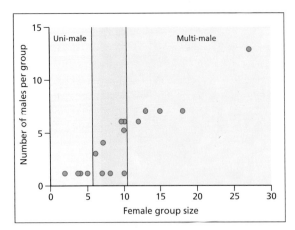

Figure 11.16 Numbers of resident males and females in different cercopithecine primates. Where females live in groups of less than six, single-male harems are usual. Where female group size lies between six and ten, some groups contain more than one adult male although, in many of these species, a single male is responsible for a larger proportion of the matings than other individuals. Where females live in groups of more than ten, several reproductively active males are usually associated with the group. *Source*: From Andelman (1986).

unable to monopolise paternity successfully (Nunn 1999). As predicted, comparative studies of primate groups show that the number of resident adult males in breeding groups increases where reproduction is seasonal and multiple females are often in oestrus at the same time (Ridley 1986; Nunn 1999; Preuschoft and Paul 2000). Other factors that constrain the ability of dominant males to monopolise breeding access to resident females may also affect the relative breeding success of subordinate males and the number of immigrant subordinates males within breeding groups. The extent to which females control the identity of their mates may affect the relative success of subordinate males and so may influence their number. Variation in the average lifespan of males may also be important. For example, reductions in the average tenure of dominant males may raise the pay-offs to subordinate males of queuing within breeding groups and increase the length of queues. So, too, will any positive effects of the duration of residence in breeding groups on the probability that immigrant subordinates will succeed to the dominant position. Contrasts in dominance, resource access, growth and survival between subordinate males in their natal groups and in breeding groups to which

they have immigrated are all likely to affect the chances that immigrants will gain reproductive success and to influence the number of immigrant males queuing in breeding groups and the age at which males disperse (see Chapter 12).

Contrasts in the costs and benefits of additional males to the fitness of resident dominant males may also play an important role in determining the responses of dominants to immigrants and the degree of reproductive skew among males. Where resident males are unable to guard individual females effectively, so that subordinates obtain a substantial share of paternity, residents may be relatively intolerant of the presence of additional subordinate males. Conversely, where dominant males can monopolise breeding access to females effectively and the presence of additional males increases the tenure of breeding males or their ability to exclude intruders or where subordinate males assist in rearing young and their presence increases the output of breeding females (see Chapter 17), dominant males may benefit by tolerating their presence, especially if they are relatives. Several studies suggest that the number of resident males in breeding groups increases in situations where the presence of additional subordinate males is likely to make an important contribution to the fitness of dominants. For example, in ursine howlers, groups including several adult males are uncommon in growing populations, where the density of males is low and competition for breeding access is reduced, while they are common in saturated populations, where male density is high and competition is intense (Pope 2000).

Female mating preferences are also likely to influence the breeding opportunities of immigrants and the number of resident males. For example, in baboons, females commonly show sexual interest in recent immigrants and are less likely to mate with males that have been resident in the group for several years (see Chapter 4), and this may contribute to increases in male immigration rates. Similarly, where males play a major role in caring for young, females are likely to benefit from the presence of more than one male and might be expected to distribute matings so as to increase the chances of paternity among males (see Chapter 15), which may account for the tendency for females to mate with multiple males in some species that show biparental care, including some callitrichids (Goldizen and Terborgh 1989) and African wild dogs (Malcolm 1979; Malcolm and Marten 1982). Where the presence of multiple males provides additional protection and increases the survival of females (Rose 1994; Wrangham and Peterson 1996), females may adjust their behaviour to facilitate male immigration, whereas in species where the size of male associations has adverse effects on female fitness, females may adopt strategies that reduce the chance of immigration. For example, where incoming males kill dependent offspring, females may increase their fitness by discouraging immigrant males or supporting residents in competitive encounters and behaviour of this kind occurs in a number of primates (see Chapter 15).

Interspecific differences in female life histories can also influence the effects of male number on the fitness of additional immigrants, resident dominant males and females, contributing to diversity in adult sex ratios. For example, variation in female reproductive synchrony and in the longevity of breeding males is likely to affect the breeding success of subordinate males and the number of resident males (see Chapter 13). Contrasting patterns of parental care may also influence female mating preferences, with important consequences for the distribution of breeding success within groups and the mating success of subordinate males, while contrasts in litter size may affect the size of dispersing parties of males and the capacity of dominants to defend female groups without assistance and hence their tendency to tolerate additional males.

11.7 Contrasts in reproductive skew

Where several mature resident males are associated with groups of females to which they are not closely related, reproductive competition between males for females is inevitable and is likely to affect the breeding success of males. The magnitude of individual differences in reproductive success between males ('reproductive skew') is affected by the number of years for which individuals can occupy the breeding role, the number and fecundity of females in their group, the proportion of offspring conceived which they father and the survival of their offspring (Clutton-Brock *et al.* 1988; Alberts 2012). Of these four components of breeding success, variation in mating success has attracted more attention than any other and is often associated with variation in the dominance rank of males (see Chapter 14). However, several other factors may also have an important influence on the extent of individual differences in male success.

First, the distribution of receptive females in space and time plays an important role. Where females belonging to the same group are closely associated, dominant males are able to monitor reproductive activity and can often prevent subordinates from mating with other males (see Chapter 10), but where females forage separately (as in some diurnal species, like chimpanzees, as well as in many nocturnal mammals), dominant males may find it difficult to monopolise breeding effectively and subordinates may obtain access to a proportion of females (Clutton-Brock and Isvaran 2006). The distribution of receptive females in time can also constrain the ability of individual males to monopolise breeding access and the degree of skew in males (Altmann 1962): where females breed asynchronously, so that no more than one individual is usually in oestrus at the same time, a dominant male may be able monopolise access to several receptive females in turn and to minimise paternity by other males. In contrast, where females conceive in synchrony and dominant males are unable to guard more than one receptive female at a time, subordinate males are likely to obtain a larger share of paternity (Gogarten and Koenig 2013). One of the first attempts to explore the distribution of mating among males, the *priority of access model*, is based on the assumption that males cannot defend more than one female at a time and argues that the proportion of matings lost by dominant males to other resident males depends on the frequency with which more than one female is in oestrus simultaneously (Altmann 1962) (Box 11.1 and Figure 11.17).

Second, variation in the intrinsic capacity of males to exclude competitors is probably important. For example, where males show continuous growth (as in elephants and many of the large macropods), size differentials between competing males are often relatively large and older heavier males are usually able to monopolise access to females (Jarman and Southwell 1986; Johnson 1989; Poole *et al.* 2011). In contrast, where size differentials are reduced (as in many grazing ungulates and many arboreal monkeys) mating success may be more widely distributed and reproductive skew may be lower, though this has yet to be tested systematically. The nature of male weapons may also play a role: for example the ability of individual males to monopolise access to receptive females may be reduced in species, like lions, where males possess potentially lethal weapons and the costs of escalated contests are high (see Chapter 14). The presence of supportive coalitions between males and

Box 11.1 Priority of access and the distribution of male breeding success

One possible explanation of differences in the degree of reproductive skew among males is that they are a consequence of contrasts in reproductive synchrony among females. The *priority of access model* (Altmann 1962) assumes that where the receptive periods of different females are asynchronous, the most dominant male in the group will be able to guard receptive females in turn and will monopolise mating success but that where two females are in oestrus at the same time, one is likely to be guarded and mated by the second ranking male. If more than two females are in oestrus at the same time, a third male is likely to guard and mate with them and so on. The model predicts that the degree of female synchrony should predict the distribution of mating success across ranked males and that the proportion of matings (or paternities) achieved by subordinates should increase as the synchrony of female cycles rises.

    Both interspecific and intraspecific comparisons of primate data confirm that the breeding success of subordinate males increases with the degree of female synchrony (Pope 1990; Bulger 1993; Say *et al.* 2001; Alberts *et al.* 2006; Kutsukake and Nunn 2006, 2009; Ostner *et al.* 2008) and the priority of access model often predicts the distribution of male mating success with reasonable accuracy. For example, in savannah baboons, there is a reasonably close fit between the proportion of matings achieved by ranked males and proportions predicted by the priority of access model (see Figure 11.18). However, several other mechanisms can affect the distribution of male mating success and may either increase or reduce the degree of reproductive skew. In particular, where females are widely dispersed and dominant males cannot monitor the reproductive state of all individuals, this can reduce the ability of dominant males to monopolise receptive females. Similarly, where females are able to control the identity of their mating partners and do not necessarily prefer to mate with the most dominant male, subordinate males may achieve a larger share of breeding than the priority of access model would predict. For example, in spotted hyenas, where females are widely dispersed and a combination of female dominance over males and their reproductive anatomy allows them to control the identity of their mating partners, reproductive success is more widely distributed among immigrant males than the priority of access model would predict (Engh *et al.* 2002).

related females may also reduce the ability of dominant males to monopolise access to females: for example, male skew is low in both bonobos and muriquis where males receive support in competitive interactions from their female kin (Strier *et al.* 2011; Surbeck *et al.* 2011). Finally,

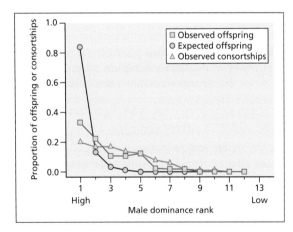

Figure 11.17 Mating success of male yellow baboons of different ranks (based on genetic data) and expected success (based on Altmann's priority of access model). The frequency of observed consortships is also shown. *Source*: From Alberts *et al.* (2006). Reproduced with permission of Elsevier.

the number of resident males can be important. While the presence of more than one male who is unrelated to females typically reduces the proportion of juveniles fathered by the most dominant male (and so reduces skew in breeding success), the presence of further males may be associated with increased skew, since a large proportion seldom breed.

    Third, strong female mating preferences may restrict the ability of individual males to monopolise access to a high proportion of females, especially if different females prefer different males (Cant and Reeve 2002). Female mating preferences are particularly likely to affect the distribution of male breeding success in species where females are dominant to males and so are able to control the identity of their mating partners (Holekamp and Engh 2009). For example, in spotted hyenas, where females are dominant to males and their reproductive anatomy allows them to control mating, male rank has a relatively weak effect on mating success (Engh *et al.* 2002). However, female dominance over males is not associated with low reproductive skew in males in all societies: in some species where females are dominant to males (including red-fronted lemurs and some populations of sifakas) dominant males lose a relatively small proportion of matings to other resident males and father more than two-thirds of young born in their group (Kappeler and Port 2008).

Fourth, the extent to which dominant males rely on assistance from subordinate males to maintain their position can be important. Where dominants rely on subordinate males to assist them in defending groups of females against competing males, they may need to allow them a measure of reproduction in order to dissuade them from dispersing or challenging for the breeding position (Vehrencamp 1983; Johnstone and Cant 2009), an idea derived from transactional models of reproductive skew that were originally constructed to explain the distribution of reproductive skew in females in cooperative breeders (Emlen and Vehrencamp 1983; Vehrencamp 1983; Jamieson and Craig 1987) (see Chapter 17).

Finally, variation in the capacity of males to identify the period when females are most likely to be fertile may be important. For example, where females advertise changes in their reproductive state (see Chapter 8), dominant males may be able to identify and guard most receptive females, whereas cryptic female cycles or weak correlations between female signals and their ovulation status may prevent males from monopolising reproductive access to the same extent. For example, studies of rhesus macaques (where receptive females show changes in facial colour throughout the reproductive cycle but do not have exaggerated sexual swellings) suggest that males may be unable to identify the exact timing of females' fertile phases and this may limit the capacity of dominant males to monopolise reproductive access (Dubuc et al. 2012). Ovulation mode may also be important. For example, in species where ovulation is induced, males may be in a better position to identify the reproductive status of females than where ovulation is spontaneous; as would be expected, interspecific comparisons show that litters with multiple paternity are rarer and dominant males lose a smaller proportion of paternities to other group members in induced ovulators compare to spontaneous ovulators (Soulsbury 2010).

While contrasts in mating success usually make an important contribution to differences in the lifetime breeding success of males, variation in the tenure and lifespan of breeding males is also important (Figure 11.18) (Alberts 2012) and may favour male breeding strategies that minimise the costs of reproductive competition as well as maximising the benefits. Stochastic social or environmental events may often play an important role in determining the breeding lifespans of males (Clutton-Brock 1988; Alberts 2012). However, several studies suggest that males that defend the largest groups of females or the best territories or which have the highest dominance rank attract more competition and spend more energy on reproductive activities and may have shorter breeding lifespans than those with lower annual breeding success (see Chapter 15). In addition, reproductive success early in the lifespan of males may trade off against subsequent growth, breeding success and survival. For example, in red deer, males with high annual rates of reproductive success early in their lifespan breed successfully in fewer seasons than those with lower rates of success (Lemaître et al. 2014).

Individual differences in offspring survival are also likely to contribute to variation in breeding success among males. In polygynous species, like red deer and elephant seals, where males mate with a substantial number of different females throughout the course of their lives, individual differences in offspring survival may contribute a relatively small proportion of individual differences in male lifetime breeding success (Clutton-Brock et al. 1988; Le Boeuf and Reiter 1988; Byers and Dunn 2012). In contrast, where males associate with the same female or group of females for long periods, as in lions, meerkats and baboons, individual differences in offspring survival may contribute a substantial proportion of the observed variation in male reproductive success (Altmann et al. 1988; Packer et al. 1988) and may affect the reproductive strategies of males as well as their choice of mates and their contribution to parental care (see Chapter 16). For example, where male infanticide is common, it may lead to selection for strategies in breeding males that prolong the duration of their tenure of the breeding role if this increases the chance that males will be succeeded by a familiar relative that is unlikely to kill their progeny (Harcourt and Stewart 2007; Henzi et al. 2010). The presence of fathers may also influence the development and breeding success of their offspring (Charpentier et al. 2008) and these effects, too, may also affect the evolution of male competition, behaviour and tolerance.

11.8 Consequences of male association

The size and structure of male associations has important consequences for patterns of male dispersal as well as for

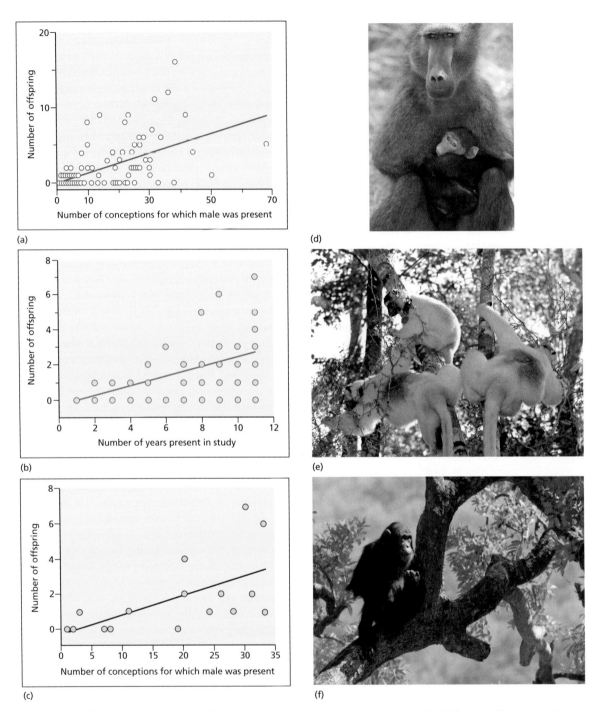

Figure 11.18 Male reproductive success and longevity in three primate species. (a) Baboons. (b) Sifakas. (c) Chimpanzees. *Source*: (a) From Alberts *et al.* (2006). Reproduced with permission of Elsevier; (b) Adapted from Lawler (2007). Reproduced with permission of John Wiley & Sons; (c) Adapted from Wroblewski *et al.* (2009). Reproduced with permission of Elsevier. *Photo sources*: (d) © Alecia Carter; (e) © Claudia Fichtel; (f) © Alecia Carter.

the tenure of successful males. Where females are philopatric, resident males are seldom close relatives and relationships between them are often competitive and intolerant. In many of these species, resident males do not cooperate to defend access to female groups, so that dispersing males are often able to immigrate into established breeding groups. In many of these species, males disperse independently, secondary dispersal is relatively common and male tenure is relatively short, as in baboons (Alberts 2012).

Where groups of closely related males emigrate and immigrate together, as in many polytocous species, dispersing males are often close relatives and their chances of acquiring access to female groups may increase with the size of male groups. Relationships between related males are often tolerant and males commonly cooperate to defend groups of females. In contrast, relations between males from neighbouring groups are intensely hostile, so that males can rarely immigrate into established breeding groups unless they displace the resident males and secondary dispersal is rare. Tolerance between males belonging to the same group and intense hostility between neighbouring groups of males also occurs in species where females disperse to breed and most resident males remain and breed in their natal group, as in muriquis, spider monkeys and chimpanzees. In monotocous species that form groups of this kind (see Chapter 12), the numbers of males ready to disperse at the same time are often smaller than the number of resident males, so that resident males are usually able to prevent take-overs by groups of immigrants, with the result that replacement of resident males is rare. As a result, many males remain and breed in their natal group throughout their lives and the tenure of resident groups of natal males is often very long, as in chimpanzees.

Patterns of male association are also likely to have an important influence on the life histories of males and the structure of populations. The defence of female groups by single males will often favour delays in dispersing by sub-adult males until they reach an age where they can compete successfully with mature males and so may raise the number of non-breeding natal males in breeding groups (see section 11.6). Male tolerance of natal males and factors that affect this (such as the effectiveness of mate guarding or close kinship between resident females and natal males) may also increase delays in male dispersal and increase numbers of natal males in

breeding groups. Since dispersal is commonly associated with increased levels of mortality (see Chapter 12), these processes are likely to increase the representation of young males in the population. Conversely, factors that promote intolerance of sub-adult males by resident dominant males (including increased readiness of females to mate with natal males) would be expected to reduce both the age at which males disperse and the number of resident natal males. These processes, too, are likely to affect adult sex ratios at the population level. For example, early dispersal by males combined with intolerance of male immigrants may contribute to high mortality among dispersing males and may generate relatively low male/female adult ratios at the population level, while the formation of bachelor groups and male philopatry may increase male survival and raise the ratio of mature males to adult females within populations (see Chapter 12).

The presence of multiple males in breeding groups also has important consequences for the reproductive strategies of both sexes. Where several unrelated breeding males associate with breeding groups of females, resident males commonly form coalitions with each other and the ability of dominant males to monopolise mating declines, females are more likely to mate with multiple partners, the incidence of sperm competition increases and selection is likely to favour increases in the size of male ejaculates and testes. For example, across species, relative testis size increases with the frequency of multiple paternities and declines with the proportion of paternities monopolised by alpha males: as would be expected, it is smaller in induced ovulators (where paternity certainly is high) than in spontaneous ovulators (Ramm *et al.* 2005; Soulsbury 2010) (Figure 11.19). Contrasts in the kinship structure of male associations may also influence the frequency of male infanticide. In multi-male groups where resident males are familiar relatives, male infanticide involving resident males is rare or absent, whereas it is common in species where resident males are unrelated immigrants (see Chapter 15). This may in turn affect the evolution of female strategies: increased risks of male infanticide may favour females that mate with multiple partners in each breeding attempt, increasing competition between females to attract mating partners and leading to the evolution of female ornaments that serve to attract mates (see Chapter 8).

The presence of multiple males in breeding groups may also affect selection for cognitive abilities and brain size.

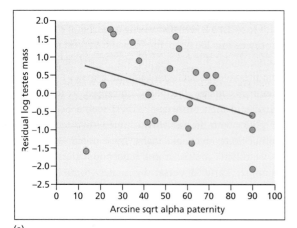

(a)

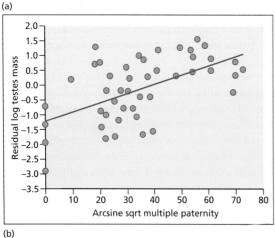

(b)

Figure 11.19 Testes size, alpha male paternity and multiple paternity in mammals. Relative testes mass (a) declines as the proportion of paternity attributed to the alpha male in each group increases and (b) increases as the proportion of litters sharing multiple paternity rises. *Source*: From Soulsbury (2010). Reproduced under the terms of the Open Access licence.

Many of the mammals best known for their advanced cognitive abilities and large brains, like dolphins, spotted hyenas, many of the baboons live in groups or communities that include multiple immigrant males that compete with each other for breeding opportunities. In many of these species, resident males form coalitions with each other to gain or maintain access to females and inter-specific comparisons show that the incidence of coalitionary behaviour is correlated with indices of brain size (see Chapter 14).

Patterns of male association are also likely to have important consequences for the genetic structure of groups (Melnick 1987; Storz 1999; Lukas *et al.* 2005; Di Fiore 2012). Where a single breeding male associates with groups of breeding females and extra-group paternity is low, offspring born in the same cohort are likely to have the same father, increasing coefficients of relatedness within groups (Lukas *et al.* 2005). Social dispersal by groups of related males and successive immigration of closely related males into the same breeding group will tend to increase relatedness within groups, as will social dispersal involving groups of female sibs or fissioning of female groups along matrilineal kinship lines (Olivier *et al.* 1981; Di Fiore 2012). Conversely, independent dispersal of males or females to different breeding groups, reductions in the tenure of breeding males and increases in the number of resident males are likely to reduce relatedness between group members (Lukas *et al.* 2005). Contrasts in female group size may also be important and may reduce kinship between group members, partly because they reduce average maternal relatedness, partly because the number of resident breeding males usually increases with the size of female groups and partly because female group size reduces the breeding tenure of males. As expected, average coefficients of relatedness between philopatric individuals decline rapidly with increasing group size (see Chapter 2).

The effects of contrasting patterns of male association on kinship within groups are also likely to have important consequences for the genetic structure of populations (Sugg *et al.* 1996; Di Fiore 2012). Where females are philopatric, high levels of male skew and high coefficients of relatedness between breeding males are likely to increase genetic differentiation between groups (Chesser 1991a,b). In contrast, frequent immigration by unrelated individuals of either sex, low levels of male skew and relatively short breeding tenure in males are likely to reduce genetic differences and to be associated with relatively weak genetic structure at the population level.

SUMMARY

1. In some social mammals, a single male defends access to multiple females, while in others dominant males tolerate the presence of one or more subordinate males and, in a few, multiple adult males associate with female groups and may defend access to them.

2. Where multiple adult males associate with female groups, they may be related and familiar natals, related and familiar immigrants who have immigrated together, unrelated and unfamiliar males that have immigrated independently or, occasionally, related and (presumably) familiar males that have left their natal group at different times and immigrated independently into the same breeding group.

3. Where dominant males tolerate the presence of other males that are unrelated to resident females within the same breeding group, they risk losing mating opportunities to them: across species, the proportion of offspring fathered by males other than the dominant increases as male number and the synchrony of female breeding rises. In contrast, costs associated with the presence of resident males that are related (and familiar) to resident females are likely to be low since mating between them is unlikely. The proportion of breeding opportunities lost by the most dominant male can also be affected by differentials in competitive ability between males as well as by the capacity of females to exercise mate choice.

4. The presence of additional males can have important benefits to dominants which offset the costs of there presence. Where males defend access to female groups or their ranges, all resident males usually contribute to defence and their presence can increase both the effectiveness of defence and the breeding tenure of dominant males. Subordinate males may benefit by joining or remaining in breeding groups because this provides occasional opportunities to breed before they acquire dominant status or because it increases their chances of inheriting the dominant position in the group.

5. The numbers of males resident in breeding groups depend partly on demographic factors and partly on the adaptive decisions of males concerning when to disperse and which groups to join. As expected, the probability that natal males will leave breeding groups is often related to the availability of breeding opportunities in other groups, while the probability that they will join different breeding groups is related to the ratio of resident males to resident females.

6. Male kinship affects relationships between resident males: resident males that have grown up together and are likely to be related are generally more tolerant of each other's breeding attempts than unrelated males that have immigrated independently into breeding groups. Conversely, relationships between groups of related males living in different groups are frequently more hostile than those between groups of unrelated males that have immigrated independently.

7. The number of resident breeding males and the degree of kinship between them also has important consequences for the incidence of male infanticide and for social relationships between males and females; for the number of males that females mate with per breeding attempt; for the evolution of female ornaments; for sperm competition and related adaptations in males; and for the genetic structure of groups and populations.

References

Alberts, S.C. (2012) Magnitude and sources of variation in male reproductive performances. In: *The Evolution of Primate Societies* (eds J.C. Mitani, J. Call, P.M. Kappeler, R.A. Palombit and J.B. Silk). Chicago: University of Chicago Press, 412–431.

Alberts, S.C. and Altmann, J. (1995) Balancing costs and opportunities: dispersal in male baboons. *American Naturalist* **145**: 279–306.

Alberts, S., *et al.* (2006) Sexual selection in wild baboons: from mating opportunities to paternity success. *Animal Behaviour* **72**: 1177–1196.

Altmann, J., *et al.* (1988) Determinants of reproductive success in savannah baboons, *Papio cynocephalus*. In: *Reproductive Success: Studies of Individual Variation in Contrasting Breeding Systems* (ed. T.H. Clutton-Brock). Chicago: University of Chicago Press, 403–418.

Altmann, S.A. (1962) A field study of sociobiology of rhesus monkeys, *Macaca mulatta*. *Annals of the New York Academy of Sciences* **102**: 338–435.

Altringham, J.D. and Senior, P. (2005) Social systems and ecology of bats. In: *Sexual Segregation in Vertebrates* (eds K.E. Ruckstuhl and P. Neuhaus). Cambridge: Cambridge University Press, 280–302.

Andelman, S. (1986) Ecological and social determinants of cercopithecine mating patterns. In: *Ecological Aspects of Social Evolution: Birds and Mammals* (eds D.I. Rubenstein and R.W. Wrangham). Princeton, NJ: Princeton University Press, 201–216.

Arnold, W. and Dittami, J. (1997) Reproductive suppression in male alpine marmots. *Animal Behaviour* **53**: 53–66.

Aureli, F., *et al.* (2006) Raiding parties of male spider monkeys: insights into human warfare? *American Journal of Physical Anthropology*, **131**:486–497.

Baird, R.W. (2000) The killer whale: foraging specializations and group hunting. In: *Cetacean Societies: Field Studies of Dolphins and Whales* (eds J. Mann, R. Connor, P. Tyack and H. Whitehead). Chicago: University of Chicago Press, 127–153.

Balmford, A.P. and Turyaho, M. (1992) Predation risk and lek-breeding in Uganda kob. *Animal Behaviour* **44**: 117–127.

Berger, J. (1986) *Wild Horses of the Great Basin*. Chicago: University of Chicago Press.

Berger, J. (1987) Reproductive fates of dispersers in a harem-dwelling ungulate: the wild horse. In: *Mammalian Dispersal Patterns* (eds B.D. Chepko-Sade and Z.T. Halpin). Chicago: University of Chicago Press, 41–54.

Bertram, B.C.R. (1978) Living in groups: predators and prey. In: *Behavioural Ecology: An Evolutionary Approach* (eds J.R. Krebs and N.B. Davies). Oxford: Blackwell Science, 64–96.

Blundell, G.M., *et al.* (2004) Kinship and sociality in coastal river otters: are they related? *Behavioral Ecology* **15**: 705–714.

Boesch, C. (1996) Social grouping in Taï chimpanzees. In: *Great Ape Societies* (eds W.C. McGrew, L.F. Marchant and T. Nishida). Cambrdge: Cambridge University Press, 101–113.

Boesch, C., *et al.* (2006) Cooperative hunting in chimpanzees: kinship or mutualism? In: *Cooperation in Primates and Humans* (eds P.M. Kappeler and C.P. van Schaik). Berlin: Springer, 139–150.

Bradley, B.J., *et al.* (2005) Mountain gorilla tug-of-war: silverbacks have limited control over reproduction in multimale groups. *Proceedings of the National Academy of Sciences of the United States of America* **102**: 9418–9423.

Bulger, J.B. (1993) Dominance rank and access to estrous females in male savanna baboons. *Behaviour* **127**: 67–103.

Byers, J. and Dunn, S. (2012) Bateman in nature: predation on offspring reduces the potential for sexual selection. *Science* **338**: 802–804.

Cant, M.A. and Reeve, H.K. (2002) Female control of paternity in cooperative breeders. *American Naturalist* **160**: 602–611.

Caro, T.M. (1994) *Cheetahs of the Serengeti Plains: Group Living in an Asocial Species*. Chicago: University of Chicago Press.

Charpentier, M.J.E., *et al.* (2008) Paternal effects on offspring fitness in a multimale primate society. *Proceedings of the National Academy of Sciences of the United States of America* **105**: 1988–1992.

Cheney, D.L. and Seyfarth, R.M. (1977) Behaviour of adult and immature male baboons during inter-group encounters. *Nature* **269**: 404–406.

Cheney, D.L. and Seyfarth, R.M. (1983) Nonrandom dispersal in free-ranging vervet monkeys: social and genetic consequences. *American Naturalist* **122**: 392–412.

Cheney, D.L. and Seyfarth, R.M. (2008) *Baboon Metaphysics: The evolution of a Social Mind*. Chicago: University of Chicago Press.

Chesser, R.K. (1991a) Gene diversity and female philopatry. *Genetics* **127**: 437–447.

Chesser, R.K. (1991b) Influence of gene flow and breeding tactics on gene diversity within populations. *Genetics* **129**: 573–583.

Chiarello, A.G. (1995) Role of loud calls in brown howlers, *Alouatta fusca*. *American Journal of Primatology* **36**: 213–222.

Clutton-Brock, T.H. (1988) Reproductive success. In: *Reproductive success* (ed. T.H. Clutton-Brock). Chicago: University of Chicago Press, 472–486.

Clutton-Brock, T.H. and Isvaran, K. (2006) Paternity loss in contrasting mammalian societies. *Biology Letters* **2**: 513–516.

Clutton-Brock, T.H., *et al.* (1977) Sexual dimorphism, socionomic sex ratio and body weight in primates. *Nature* **269**: 797–800.

Clutton-Brock, T.H., *et al.* (1979) The logical stag: adaptive aspects of fighting in red deer (*Cervus elaphus* L.). *Animal Behaviour* **27**: 211–225.

Clutton-Brock, T.H., *et al.* (1982) *Red Deer: The Behaviour and Ecology of Two Sexes*. Chicago: University of Chicago Press.

Clutton-Brock, T.H., *et al.* (1988) Reproductive success in male and female red deer. In: *Reproductive Success: Studies of Individual Variation in Contrasting Breeding Systems* (ed. T.H. Clutton-Brock). Chicago: University of Chicago Press, 325–343.

Clutton-Brock, T.H., *et al.* (1993) The evolution of ungulate leks. *Animal Behaviour* **46**: 1121–1138.

Clutton-Brock, T.H., *et al.* (1998) Infanticide and expulsion of females in a cooperative mammal. *Proceedings of the Royal Society of London. Series B: Biological Sciences* **265**: 2291–2295.

Cohas, A. and Allainé, D. (2009) Social structure influences extra-pair paternity in socially monogamous mammals. *Biology Letters* **5**: 313–316.

Connor, R.C. (2000) Group living in whales and dolphins. In: *Cetacean Societies: Field Studies of Dolphins and Whales* (eds J. Mann, R.C. Connor, P.L. Tyack and H. Whitehead). Chicago: University of Chicago Press, 199–218.

Cords, M. (2000) Grooming partners of immature blue monkeys (*Cercopithecus mitis*) in the Kakamega Forest, Kenya. *International Journal of Primatology* **21**: 239–254.

Cowlishaw, G. (1995) Behavioural patterns in baboon group encounters: the role of resource competition and male reproductive strategies. *Behaviour* **132**: 75.

Creel, S. and Creel, N.M. (2002) *The African Wild Dog: Behavior, Ecology, and Conservation*. Princeton, NJ: Princeton University Press.

Crockett, C. and Sekulic, R. (1984) Infanticide in red howler monkeys (Alouatta seniculus). In: *Infanticide: Comparative and Evolutionary Perspectives* (eds G. Hausfater and S.B. Hrdy). New York: Aldine, 173–191.

Crofoot, M.C. and Wrangham, R.W. (2010) Intergroup aggression in primates and humans: the case for a unified theory. In: *Mind the Gap: Tracing the Origins of Human Universals* (eds P.M. Kappeler and J.B. Silk). Berlin: Springer, 171–195.

Crook, J.H. and J.S. Gartlan (1966) Evolution of primate societies. *Nature* **210**: 1200–1203.

Di Fiore, A. (2012) Genetic consequences of primate social organization. In: *The Evolution of Primate Societies* (eds J.C. Mitani, J. Call, P.M. Kappeler, R.A. Palombit and J.B. Silk). Chicago: University of Chicago Press, 269–292.

Di Fiore, A. and Campbell, C.J. (2007) The Atelines: variation in ecology, behaviour and social organisation. In: *Primates in Perspective* (eds C.J. Campbell, A. Fuentes, K.C. MacKinnon,

M. Panger and S.K. Bearder). New York: Oxford University Press, 155–185.

Dixson, A.F. (2000) *Primate Sexuality: Comparative Studies of the Prosimians, Monkeys, Apes, and Human Beings.* Oxford: Oxford University Press.

Dobson, F.S. (1982) Competition for mates and predominant juvenile male dispersal in mammals. *Animal Behaviour* **30**: 1183–1192.

Dubuc, C., *et al.* (2012) Do males time their mate-guarding effort with the fertile phase in order to secure fertilisation in Cayo Santiago rhesus macaques? *Hormones and Behavior* **61**: 696–705.

Dunbar, R.I.M. (1984) *Reproductive Decisions: Economic Analysis of Gelada Baboon Social Strategies.* Princeton, NJ: Princeton University Press.

Eisenberg, J.F., *et al.* (1972) The relation between ecology and social structure in primates. *Science* **176**: 863–874.

Emlen, S.T. and Vehrencamp, S.L. (1983) Cooperative breeding strategies among birds. In: *Perspectives in Ornithology* (ed. A. Brush). Cambridge: Cambridge University Press, 93–120.

Engh, A.L., *et al.* (2002) Reproductive skew among males in a female-dominated mammalian society. *Behavioral Ecology* **13**: 193–200.

Fedigan, L.M. and Jack, K.M. (2004) The demographic and reproductive context of male replacements in *Cebus capucinus. Behaviour* **111**: 755–775.

Feh, C. (1999) Alliances and reproductive success in Camargue stallions. *Animal Behaviour* **57**: 705–713.

Feh, C. (2001) Alliances between stallions are more than just multimale groups: reply to Linklater and Cameron (2000). *Animal Behaviour* **6**: F27–F30.

Fischer, J., *et al.* (2004) Baboon loud calls advertise male quality: acoustic features and their relation to rank, age, and exhaustion. *Behavioral Ecology and Sociobiology* **56**: 140–148.

Fitzgibbon, C.D. (1994) The costs and benefits of predator inspection behavior in Thomson gazelles. *Behavioral Ecology and Sociobiology* **34**: 139–148.

Galdikas, B.M.F. (1981) Orangutan reproduction in the wild. In: *The Reproductive Biology of the Great Apes* (ed. C. Graham). New York: Academic Press, 281–300.

Gibson, R.M. and Guinness, F.E. (1980) Behavioral factors affecting male reproductive success in red deer (*Cervus elaphus*). *Animal Behaviour* **28**: 1163–1174.

Gogarten, J.F. and Koenig, A. (2013) Reproductive seasonality is a poor predictor of receptive synchrony and male reproductive skew among nonhuman primates. *Behavioral Ecology and Sociobiology* **67**: 123–134.

Goldizen, A.W. and Terborgh, J. (1989) Demography and dispersal patterns of a tamarin population: possible causes of delayed breeding. *American Naturalist* **134**: 208–224.

Goodall, J. (1986) *The Chimpanzees of Gombe: Patterns of Behavior.* Cambridge, MA: Belknap Press.

Grinnell, J. and McComb, K. (2001) Roaring and social communication in African lions: the limitations imposed by listeners. *Animal Behaviour* **62**: 93–98.

Grubb, P. (1974) The rut and behaviour of Soay rams. In: *Island Survivors: The Ecology of the Soay Sheep of St Kilda* (eds P.A. Jewell, C. Milner and J.M. Boyd). London: Athlone Press, 195–223.

Hager, R. (2009) Explaining variation in reproductive skew among male langurs: effects of future mating prospects and ecological factors. In: *Reproductive Skew in Vertebrates: Proximate and Ultimate Causes* (eds R. Hager and C.B. Jones). Cambridge: Cambridge University Press, 134–164.

Hamilton, W.D. (1971) Geometry for the selfish herd. *Journal of Theoretical Biology* **31**: 295–311.

Harcourt, A.H. and Stewart, K.J. (2007) *Gorilla Society.* Chicago: University of Chicago Press.

Harding, R.S.O. (1980) Agonism, ranking and the social behavior of adult male baboons. *American Journal of Primatology* **53**: 203–216.

Harrington, F.H. and Mech, L.D. (1979) Wolf howling and its role in territory maintenance. *Behaviour* **68**: 207–249.

Hawkins, C.E. and Racey, P.A. (2009) A novel mating system in a solitary carnivore: the fossa. *Journal of Zoology* **277**: 196–204.

Henzi, S.P., *et al.* (2010) Infanticide and reproductive restraint in a polygynous social mammal. *Proceedings of the National Academy of Sciences of the United States of America* **107**: 2130–2135.

Holekamp, K.E. and Engh, A.L. (2009) Reproductive skew in female-dominated mammalian societies. In: *Reproductive Skew in Vertebrates: Proximate and Ultimate Causes* (eds R. Hager and C.B. Jones). Cambridge: Cambridge University Press, 53–83.

Höner, O.P., *et al.* (2007) Female mate-choice drives the evolution of male-biased dispersal in a social mammal. *Nature* **448**: 798–801.

Hoogland, J.L. (1995) *The Black-tailed Prairie Dog: Social Life of a Burrowing Mammal.* Chicago: University of Chicago Press.

Hrdy, S.B. (1977) *The Langurs of Abu: Female and Male Strategies of Reproduction.* Cambridge, MA: Harvard University Press.

Jamieson, I.G. and Craig, J.L. (1987) Critique of helping behaviour in birds: a departure from functional explanations. In: *Perpectives in Ethology* (eds P.P.G. Bateson and P. Klopfer). New York: Plenum, 79–98.

Jarman, P.J. and Southwell, C. (1986) Grouping, associations and reproductive strategies in eastern grey kangaroos. In: *Ecological Aspects of Social Evolution: Birds and Mammals* (eds D.I. Rubenstein and R.W. Wrangham). Princeton, NJ: Princeton University Press, 399–428.

Johnson, C.N. (1989) Dispersal and philopatry in the Macropodoids. In: *Kangaroos, Wallabies and Rat-Kangaroos* (eds G. Grigg, P. Jarman and I. Hume). Chipping Norton, NSW: Beatty and Sons, 593–601.

Johnstone, R.A. and Cant, M.A. (2009) Models of reproductive skew: outside options and the resolution of reproductive conflict. In: *Reproductive Skew in Vertebrates: Proximate and Ultimate Causes* (eds R. Hager and C.B. Jones). Cambridge: Cambridge University Press, 3–23.

Jolly, C. (2007) Baboons, mandrills and mangabeys: Afropapionis socioecology in a phylogenetic perspective. In:

Primates in Perspective (eds C.J. Campbell, A. Fuentes, K.C. MacKinnon, M. Panger and S.K. Bearder). New York: Oxford University Press, 240–251.

Kappeler, P.M. (2000) *Primate Males: Causes and Consequences of Variation in Group Composition*. Cambridge: Cambridge University Press.

Kappeler, P.M. and Port, M. (2008) Mutual tolerance or reproductive competition? Patterns of reproductive skew among male redfronted lemurs (*Eulemur fulvus rufus*). *Behavioral Ecology and Sociobiology* **62**: 1477–1488.

Knott, C.D. and Kahlenberg, S.M. (2007) Orangutans in perspective: forced copulations and female mating resistance. In: *Primates in Perspective* (eds C.J. Campbell, A. Fuentes, K.C. MacKinnon, M. Panger and S.K. Bearder). New York: Oxford University Press, 290–304.

Kutsukake, N. and Nunn, C.L. (2006) Comparative tests of reproductive skew in male primates: the roles of demographic factors and incomplete control. *Behavioral Ecology and Sociobiology* **60**: 695–706.

Kutsukake, N. and Nunn, C.L. (2009) The causes and consequences of reproductive skew in male primates. In: *Reproductive Skew in Vertebrates: Proximate and Ultimate Causes* (eds R. Hager and C.B. Jones). Cambridge: Cambridge University Press, 165–195.

Langergraber, K.E., *et al.* (2007) The limited impact of kinship on cooperation in wild chimpanzees. *Proceedings of the National Academy of Sciences of the United States of America* **104**: 7786–7790.

Lardy, S., *et al.* (2012) Paternity and dominance loss in male breeders: the cost of helpers in a cooperatively breeding mammal. *PLOS ONE* **7** (1): e29508.

Lawler, R.R. (2007) Fitness and extra-group reproduction in male Verreaux's sifaka: an analysis of reproductive success from 1989–1999. *American Journal of Physical Anthropology* **132**: 267–277.

Le Boeuf, B.J. and Reiter, J. (1988) Lifetime reproductive success in northern elephant seals. In: *Reproductive Success: Studies of Individual Variation in Contrasting Breeding Systems* (ed. T.H. Clutton-Brock). Chicago: University of Chicago Press, 344–362.

Lemaître, J.-F., *et al.* (2014) Early life expenditure in sexual competition is associated with increased reproductive senescence in male red deer. *Proceedings of the Royal Society of London. Series B: Biological Sciences* **281**: 20140792.

Leuthold, W. (1966) Variations in territorial behavior of Uganda kob *Adenota kob thomasi* (Neumann 1896). *Behaviour* **27**: 215–258.

Leuthold, W. (1967) Beobachtungen zum Jugendverhalten von kob-Antilopen. *Zeitschrift für Säugetierkunde* **32**: 59–62.

Lincoln, G.A., *et al.* (1970) The social and sexual behaviour of the red deer stag. *Journal of Reproduction and Fertility Supplement* **11**: 71–103.

Lührs, M.-L. and Dammhahn, M. (2010) An unusual case of cooperative hunting in a solitary carnivore. *Journal of Ethology* **28**: 379–383.

Lührs, M.-L. and Kappeler, P. (2013) Simultaneous GPS tracking reveals male associations in a solitary carnivore. *Behavioral Ecology and Sociobiology* **67**: 1731–1743.

Lührs, M.-L. and Kappeler, P. (2014) Polyandrous mating in treetops: how male competition and female choice interact to determine an unusual carnivore mating system. *Behavioral Ecology and Sociobiology* **68**: 879–889.

Lührs, M.-L., *et al.* (2013) Strength in numbers: males in a carnivore grow bigger when they associate and hunt cooperatively. *Behavioral Ecology* **24**: 21–28.

Lukas, D. and Clutton-Brock, T.H. (2014) Costs of mating competition limit male lifetime breeding success in polygynous mammals. *Proceedings of the Royal Society of London. Series B: Biological Sciences* **281**: 20140418.

Lukas, D. and Vigilant, L. (2005) Reply: Facts, faeces and setting standards for the study of MHC genes using noninvasive samples. *Molecular Ecology* **14**: 1601–1602.

Lukas, D., *et al.* (2005) To what extent does living in a group mean living with kin? *Molecular Ecology* **14**: 2181–2196.

Macfarlane, A.M. and Coulson, G. (2005) Sexual segregation in Australian marsupials. In: *Sexual Segregation in Vertebrates* (eds K.E. Ruckstuhl and P. Neuhaus). Cambridge: Cambridge University Press, 254–279.

Malcolm, J.R. (1979) Social organisation and communal rearing in African wild dogs. PhD thesis, Harvard University, Cambridge, MA.

Malcolm, J.R. and Marten, K. (1982) Natural selection and the communal rearing of pups in African wild dogs, *Lycaon pictus*. *Behavioral Ecology and Sociobiology* **10**: 1–13.

Mann, J., *et al.* (eds) (2000) *Cetacean Societies: Field Studies of Dolphins and Whales*. Chicago: University of Chicago Press.

Melnick, D.J. (1987) The genetic consequences of primate social organization: a review of macaques, baboons and vervet monkeys. *Genetica* **73**: 117–135.

Mitani, J.C. (1996) Comparative studies of African ape vocal behavior. In: *Great Ape Societies* (eds W.C. McGrew, L.F. Marchant and T. Nishida). Cambridge: Cambridge University Press, 241–254.

Moss, C.S. and Poole, J.H. (1983) Relationships and social structure of African elephants. In: *Primate Social Relationships: An Integrated Approach* (ed. R.A. Hinde). Oxford: Blackwell Scientific Publications, 315–325.

Muniz, L., *et al.* (2010) Male dominance and reproductive success in wild white-faced capuchins (*Cebus capucinus*) at Lomas Barbudal, Costa Rica. *American Journal of Primatology* **72**: 1118–1130.

Nagy, M., *et al.* (2007) Female-biased dispersal and patrilocal kin groups in a mammal with resource-defence polygyny. *Proceedings of the Royal Society of London. Series B: Biological Sciences* **274**: 3019–3025.

Nagy, M., *et al.* (2012) Male greater sac-winged bats gain direct fitness benefits when roosting in multimale colonies. *Behavioral Ecology* **23**: 597–606.

Nefdt, R.J.C. (1992) Lek-breeding in Kafue lechwe. PhD thesis, University of Cambridge, Cambridge.

Nefdt, R.J.C. and Thirgood, S.J. (1997) Lekking, resource defence and harassment in two subspecies of lechwe antelope. *Behavioral Ecology* **8**: 1–9.

Nielsen, J.F., *et al.* (2012) Inbreeding and inbreeding depression of early life traits in a cooperative mammal. *Molecular Ecology* **21**: 2788–2804.

Nishida, T., *et al.* (1985) Group extinction and female transfer in wild chimpanzees in the Mahale National Park, Tanzania. *Zeitschrift fur Tierpsychologie* **67**: 284–301.

Nunn, C.L. (1999) The number of males in primate social groups: a comparative test of the socioecological model. *Behavioral Ecology and Sociobiology* **46**: 1–13.

Ohsawa, H., *et al.* (1993) Mating strategy and reproductive success of male patas monkeys (*Erythrocebus patas*). *Primates* **34**: 533–544.

Olivier, T.J., *et al.* (1981) Genetic differentiation among matrilines in social groups of rhesus monkeys. *Behavioral Ecology and Sociobiology* **8**: 279–285.

Ostner, J., *et al.* (2008) Female reproductive synchrony predicts skewed paternity across primates. *Behavioral Ecology* **19**: 1150–1158.

Packer, C. and Pusey, A.E. (1982) Cooperation and competition within coalitions of male lions: kin selection or game theory? *Nature* **296**: 740–742.

Packer, C. *et al.* (1988) Reproductive success of lions. In: *Reproductive Success: Studies of Individual Variation in Contrasting Breeding Systems* (ed. T.H. Clutton-Brock). Chicago: University of Chicago Press, 363–383.

Packer, C., *et al.* (1991) A molecular genetic analysis of kinship and cooperation in African lions. *Nature* **351**: 562–565.

Perry, S., *et al.* (2008) Kin-biased social behaviour in wild adult female white-faced capuchins, *Cebus capucinus*. *Animal Behaviour* **76**: 187–199.

Phillips-Conroy, J.E., *et al.* (1992) Migration of male hamadryas baboons into anubis groups in the Awash National Park, Ethiopia. *International Journal of Primatology* **13**: 455–476.

Poole, J.H. (1987) Rutting behavior in African elephants: the phenomenon of musth. *Behaviour* **102**: 283–316.

Poole, J.H. (1989a) Announcing intent: the aggressive state of musth in African elephants. *Animal Behaviour* **37**: 140–152.

Poole, J.H. (1989b) Mate guarding, reproductive success and female choice in African elephants. *Animal Behaviour* **37**: 842–849.

Poole, J.H. (1994) Sex differences in the behaviour of African elephants. In: *The Differences Between the Sexes* (eds R.V. Short and E. Balaban). Cambridge: Cambridge University Press, 331–346.

Poole, J.H., *et al.* (2011) Longevity, competition, and musth: a long-term perspective on male reproductive strategies. In: *The Amboseli Elephants: A Long-term Perspective on a Long-lived Mammal* (eds C.J. Moss, H. Croze and P.C. Lee). Chicago: University of Chicago Press, 272–290.

Pope, T.R. (1990) The reproductive consequences of cooperation in the red howler monkey: paternity exclusion in multi-male and single-male troops using genetic markers. *Behavioral Ecology and Sociobiology* **27**: 439–446.

Pope, T.R. (2000) Reproductive success increases with degree of kinship in cooperative coalitions of female red howler monkeys (*Alouatta seniculus*). *Behavioral Ecology and Sociobiology* **48**: 253–267.

Port, M., *et al.* (2010) Costs and benefits of multi-male associations in redfronted lemurs (*Eulemur fulvus rufus*). *Biology Letters* **6**: 620–622.

Preuschoft, S. and Paul, A. (2000) Dominance, egalitarianism, and stalemate: an experimental approach to male–male competition in Barbary macaques. In: *Primate Males: Causes and Consequences of Variation in Group Composition* (ed. P.M. Kappeler). Cambridge: Cambridge University Press, 205–216.

Prins, H.H.T. (1996) *Ecology and Behaviour of the African Buffalo: Social Inequality and Decision Making*. London: Chapman & Hall.

Ramm, S.A., *et al.* (2005) Sperm competition and the evolution of male reproductive anatomy in rodents. *Proceedings of the Royal Society of London. Series B: Biological Sciences* **272**: 949–955.

Richard, A.F. and Dewar, R.E. (1991) Lemur ecology. *Annual Review of Ecology and Systematics* **22**: 145–175.

Ridley, M. (1986) The number of males in a primate troop. *Animal Behaviour* **34**: 1848–1858.

Robbins, A.M. and Robbins, M.M. (2005) Fitness consequences of dispersal decisions for male mountain gorillas (*Gorilla beringei beringei*). *Behavioral Ecology and Sociobiology* **58**: 295–309.

Robbins, M.M. (1996) Male–male interactions in heterosexual and all-male wild mountain gorilla groups. *Ethology* **102**: 942–965.

Rose, L.M. (1994) Benefits and costs of resident males to females in white-faced capuchins, *Cebus capucinus*. *American Journal of Primatology* **32**: 235–248.

Rosser, A.M. (1992) Resource distribution, density, and determinants of mate access in puku. *Behavioral Ecology* **3**: 13–24.

Rudran, R. (1973) Adult male replacement in one-male troops of purple-faced langurs (*Presbytis senex senex*) and its effect on population structure. *Folia Primatologica* **19**: 166–192.

Rudran, R. (1979) The demography and social mobility of a red howler (*Alouatta seniculus*) population in Venezuela. In: *Vertebrate Ecology in the Northern Neotropics* (ed. J.F. Eisenberg). Washington, DC: Smithsonian Institution Press, 107–126.

Say, L. *et al.* (2001) Influence of oestrus synchronization on male reproductive success in the domestic cat (*Felis catus* L.) *Proceedings of the Royal Society of London. Series B: Biological Sciences* **268**: 1049–1053.

Sekulic, R. (1982) The function of howling in red howler monkeys (*Alouatta seniculus*). *Behaviour* **81**: 38–54.

Sicotte, P. (1995) Interpositions in conflicts between males in bimale groups of mountain gorillas. *Folia Primatologica* **65**: 14–24.

Simpson, M. (1973) Social grooming of male chimpanzees. In: *Comparative Ecology and Behaviour of Primates* (eds J. Crook and R. Michael). London: Academic Press, 411–505.

Snyder-Mackler, N., *et al.* (2012) Concessions of an alpha male? Cooperative defence and shared reproduction in multi-male primate groups. *Proceedings of the Royal Society of London. Series B: Biological Sciences* **279**: 3788–3795.

Soulsbury, C.D. (2010) Genetic patterns of paternity and testes size in mammals. *PLOS ONE* **5**: e9581.

Staniland, I.J. (2005) Sexual segregation in seals. In: *Sexual Segregation in Vertebrates* (eds K.E. Ruckstuhl and P. Neuhaus). Cambridge: Cambridge University Press, 53–73.

Steenbeek, R. (2000) Infanticide by males and female choice in Thomas's langurs. In: *Infanticide by Males and its Implications* (eds C.P. van Schaik and C.H. Janson). Cambridge: Cambridge University Press, 153–177.

Steenbeek, R. and van Schaik, C.P. (2001) Competition and group size in Thomas's langurs (*Presbytis thomasi*): the folivore paradox revisited. *Behavioral Ecology and Sociobiology* **49**: 100–110.

Steenbeek, R., *et al.* (2000) Costs and benefits of the one-male, age-graded and all-male phase in wild Thomas's langur groups. In: *Primate Males: Causes and Consequences of Variation in Group Composition* (ed. P.M. Kappeler). Cambridge: Cambridge University Press, 130–145.

Stevens, E.F. (1990) Instability of harems of feral horses in relation to season and presence of subordinate stallions. *Behaviour* **112**: 149–161.

Storz, J.F. (1999) Genetic consequences of mammalian social structure. *Journal of Mammalogy* **80**: 553–569.

Strier, K.B., *et al.* (2011) Low paternity skew and the influence of maternal kin in an egalitarian, patrilocal primate. *Proceedings of the National Academy of Sciences of the United States of America* **108**: 18915–18919.

Struhsaker, T.T. (1969) Correlates of ecology and social organization among African cercopithecines. *Folia Primatologica* **11**: 80–118.

Sugg, D.W., *et al.* (1996) Population genetics meets behavioral ecology. *Trends in Ecology and Evolution* **11**: 338–342.

Sugiyama, Y. (1965) Behavioral development and social structure in two troops of hanuman langurs (*Presbytis entellus*). *Primates* **6**: 213–247.

Sugiyama, Y. (1976) Life history of male Japanese macaques. In: *Advances in the Study of Behavior*, Vol. **7** (eds J.S. Rosenblatt, R.A. Hinde, E. Shaw and C. Beer). San Diego, CA: Academic Press, 255–284.

Surbeck, M., *et al.* (2011) Mothers matter! Maternal support, dominance status and mating success in male bonobos (*Pan paniscus*). *Proceedings of the Royal Society of London. Series B: Biological Sciences* **278**: 590–598.

Thierry, B. (2007) The Macaques: a double-layered social organisation. In: *Primates in Perspective* (eds C.J. Campbell, A. Fuentes, K.C. MacKinnon, M. Panger and S.K. Bearder). New York: Oxford University Press, 224–239.

van Hooff, J.A.R.A.M. (2000) Relationships among non-human primate males: a deductive framework. In: *Primate Males: Cause and Consequences of Variation in Group Composition* (ed. P.M. Kappeler). Cambridge: Cambridge University Press, 183–191.

van Noordwijk, A.J. and van Schaik, C.P. (2004) Sexual selection and the careers of primate males: paternity concentration, dominance acquisition tactics and transfer decisions. In: *Sexual Selection in Primates: New and Comparative Perspectives* (eds P.M. Kappeler and C.P. van Schaik). Cambridge: Cambridge University Press, 208–284.

van Schaik, C.P. and van Hooff, R.A.M. (1996) Toward an understanding of the orangutan's social system. In: *Great Ape Societies* (eds W.C. McGrew, L.F. Marchant and T. Nishida). Cambridge: Cambridge University Press, 3–15.

Vehrencamp, S.L. (1983) Optimal degree of skew in cooperative societies. *American Zoologist* **23**: 327–335.

Waterman, J.M. (1997) Why do male Cape ground squirrels live in groups? *Animal Behaviour* **53**: 809–817.

Waterman, J.M. (1998) Mating tactics of male Cape ground squirrels, *Xerus inauris:* consequences of year-round breeding. *Animal Behaviour* **56**: 459–466.

Watts, D.P. (1996) Comparative socio-ecology of gorillas. In: *Great Ape Societies* (eds W.C. McGrew, L.F. Marchant and T. Nishida). Cambridge: Cambridge University Press, 16–28.

Watts, D.P. (2000) Causes and consequences of variation in male mountain gorilla life histories and group membership. In: *Primate Males: Causes and Consequences of Variation in Group Composition* (ed. P.M. Kappeler). Cambridge: Cambridge University Press, 169–179.

Williams, J.M., *et al.* (2004) Why do male chimpanzees defend a group range? *Animal Behaviour* **68**: 523–532.

Wilson, M.L., *et al.* (2001) Does participation in intergroup conflict depend on numerical assessment, range location, or rank for wild chimpanzees? *Animal Behaviour* **61**: 1203–1216.

Wilson, M.L., *et al.* (2002) Chimpanzees and the mathematics of battle. *Proceedings of the Royal Society of London. Series B: Biological Sciences* **269**: 1107–1112.

Wirtz, P. (1982) Territory holders, satellite males and bachelor males in a high-density population of waterbuck (*Kobus ellipsiprymnus*) and their associations with conspecifics. *Zeitschrift für Tierpsychologie* **58**: 277–300.

Wrangham, R.W. and Peterson, D. (1996) *Demonic Males: Apes and the Origins of Human Violence*. New York: Houghton Mifflin Harcourt.

Wrangham, R.W., *et al.* (2006) Comparative rates of violence in chimpanzees and humans. *Primates* **47**: 14–26.

Wroblewski, E.E., *et al.* (2009) Male dominance rank and reproductive success in chimpanzees, *Pan troglodytes schweinfurthii*. *Animal Behaviour* **77**: 873–885.

Young, A.J. (2003) Subordinate tactics in cooperative meerkats: breeding, helping and dispersal. PhD thesis, University of Cambridge, Cambridge.

Young, A.J., *et al.* (2007) Subordinate male meerkats prospect for extra-group paternity: alternative reproductive tactics in a cooperative mammal. *Proceedings of the Royal Society of London. Series B: Biological Sciences* **274**: 1603–1609.

CHAPTER 12

Male dispersal and its consequences

12.1 Introduction

A troop of olive baboons forages among the fishermen's nets beside Lake Tanganyika, searching for scraps of dry fish. A female with a large pink perineal swelling is closely guarded by a dominant male while other males watch closely. Fifty metres behind the group is a younger male who watches the proceedings closely. When females lag behind and the attention of resident males is diverted, he approaches and solicits grooming. He left his own troop a few weeks ago and is attempting to join the beach group by loitering at its edge and interacting with females whenever he can. If he can persist, he will gradually be tolerated by the resident males and will be able to establish closer relationships with the group's females who often show strong interest in strange males. But until he is an accepted group member, he sleeps alone and leads a dangerous existence without the protection that group membership provides.

In olive baboons, as in many other mammals, sex differences in dispersal and philopatry are pronounced: females usually remain and breed in their natal group and seldom disperse to other groups, while males typically leave their natal group as young adults and may change groups several times in the course of their lives (Packer 1979). In contrast, in a few mammals (including greater sac-winged bats, red colobus monkeys and chimpanzees) females habitually disperse from their natal groups after reaching adolescence and males may remain and breed in their natal group throughout their lives (Greenwood 1980; Nagy *et al.* 2007).

Like contrasts in female dispersal, variation in dispersal among males raises important questions about the costs and benefits of leaving their natal group. This chapter examines the causes and consequences of variation in male dispersal and philopatry and the reasons for the marked sex differences in philopatry that are a conspicuous feature of many mammal societies. Section 12.2 describes the proximate causes of dispersal in natal males and variation in dispersal rates, while section 12.3 examines the costs and benefits of dispersal, and sections 12.4 and 12.5 describe the distances moved by dispersing males and the incidence of secondary dispersal from breeding groups. Section 12.6 describes sex differences in philopatry and dispersal in mammals and reviews explanations of their evolution and section 12.7 examines the social and ecological consequences of variation in male dispersal.

As in Chapter 3, I use dispersal to refer to cases where individuals have left their breeding group or range of origin and moved to a separate area or breeding unit; I consequently treat monogamous territorial species, where offspring of both sexes leave their parents' territories, as cases where both sexes disperse from the natal group, even if one sex moves further than the other. In comparisons of dispersal distances, I use estimates of the relative distance moved by individuals after leaving their natal group or range, in order to distinguish questions about the frequency of dispersal from those about the distance moved by dispersers. One reason why it is important to make this distinction is that, in some species, members of one sex may be less likely to disperse from their natal group but may move further if they do so (Clutton-Brock and Lukas 2011).

12.2 Variation in dispersal rates by males

The dispersal of adolescent and post-adolescent males is common among mammals (Greenwood 1980; Dobson 2013), though in a few species (including killer whales, dwarf mongooses, Ethiopian wolves, naked mole rats, gorillas and chimpanzees and some bats), males can

Mammal Societies, First Edition. Tim Clutton-Brock.
© 2016 Tim Clutton-Brock. Published 2016 by John Wiley & Sons, Ltd.

remain and breed in their natal group throughout their lives (Sillero-Zubiri *et al.* 1996; Faulkes and Bennett 2007; Nagy *et al.* 2007). The immediate causes of male dispersal vary. In some species, natal males are evicted from their natal group by intense aggression directed at them by the resident dominant male, by immigrant males or by other group members. For example, in guanacos, territorial males evict adolescent males when they are around 1 year old (Franklin 1983) (Figure 12.1a). Similarly, in red deer, yearling males are chased away from

their mothers' groups by rutting stags (Clutton-Brock *et al.* 1982) (Figure 12.1b). The eviction of adolescent and young adult males also occurs in a number of other mammals, including primates and carnivores (Rajpurohit and Sommer 1993; Robbins 1996).

In other species, males leave their natal group without being forced to do so (Wolff 1993; Smale *et al.* 1997). For example, in spotted hyenas, natal males of 18–24 months leave their natal clans for periods of several weeks or more to visit neighbouring clans, usually returning to

Figure 12.1 In many polygynous ruminants, territorial males chase away juvenile males, breaking their bonds with their mothers: (a) a male guanaco chases away an intruder from its territory; (b) a rutting red deer stag chases away a yearling male but drives a calf (on right of picture) back into its harem. *Sources*: (a) © Hermann Brehm/Nature Picture Library; (b) © Tim Clutton-Brock.

their natal clan several times before finally dispersing permanently (Smale *et al.* 1997; Höner *et al.* 2007, 2010). Since natal males are typically dominant to immigrants, it is unlikely that they have been evicted. In meerkats, males over 12 months old commonly visit neighbouring groups, where they attempt to mate with subordinate females, often returning to sleep with members of their natal group at night (Young 2003; Mares 2012). Agonistic interactions between resident breeding males and natal males are uncommon and there is no evidence that males are forced to leave their natal group. After several forays away from the group, males eventually disperse permanently, typically between the ages of 1.5 and 3 years.

Although it often seems unlikely that males are coerced into leaving, it is usually difficult to be sure that males do not leave their natal groups to escape persecution by resident males. However, in the marsupial mouse, *Antechinus stuartii*, this possibility can be ruled out. In *Antechinus*, there is a concentrated and regular mating season in early spring, when males show high levels of testosterone as well as high corticosteroid levels.

After this, all males die from a variety of stress related conditions, including ulceration of the gastric mucosa, suppression of the immune system and associated increases in parasites and pathogenic microorganisms (Figure 12.2) (Cockburn *et al.* 1985). Young are born around 2 weeks later and spend a further 4–6 weeks obligatorily attached to their mother's teats. Subsequently, the young are kept in a tree hollow and suckled for a further 2–3 months, weaning in December or January. Shortly after weaning, male juveniles disperse from their natal area, moving to join resident females in sites several home-range diameters away, whereas juvenile females mostly remain in their natal area, often continuing to use part of their mother's range (Cockburn *et al.* 1985). Since there are no resident adult males alive at the time when juveniles disperse, the dispersal of adolescent males cannot be a consequence of their actions.

The age of males at dispersal varies widely between species (Figure 12.3). In small mammals, males disperse when they are less than a year old (Gaines and McClenaghan 1980; Cockburn *et al.* 1985; Balloux *et al.* 1998); in larger

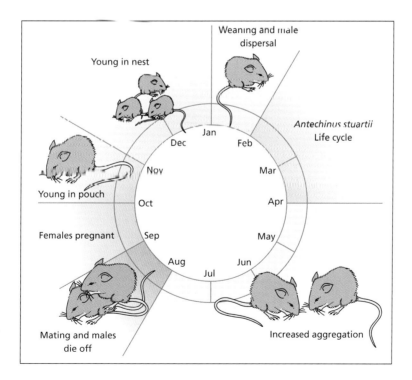

Figure 12.2 The life cycle of *Antechinus stuartii*. *Source*: From Cockburn *et al.* (1985). Reproduced with permission of Elsevier.

Figure 12.3 (a) Number of male and female black-tailed prairie dogs that had dispersed from their natal coterie by different ages. (b) Proportion of different male and female meerkats that dispersed from their natal groups at different ages. (c) Frequency distribution of the age at dispersal in male African wild dogs. (d) Frequency distribution of male dispersal in spotted hyenas. (e) Estimated proportion of natal male yellow baboons dispersing in each age interval based on a survival analysis. *Source*: (a) From Hoogland (1995). Reproduced with permission of the University of Chicago Press; (b) From Clutton-Brock *et al.*, unpublished data; (c) From Creel and Creel (2002). Reproduced with permission of Scott Creel; (d) From Smale *et al.* (1997). Reproduced with permission of Elsevier; (e) From Alberts and Altmann (1995). Reproduced with permission of the University of Chicago Press. *Photo sources*: (a) © Elaine Miller Bond; (b) © Helen Wade; (c) © Neil Jordan; (d) © Tim Clutton-Brock; (e) © Joan Silk.

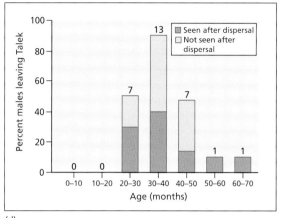

(d)

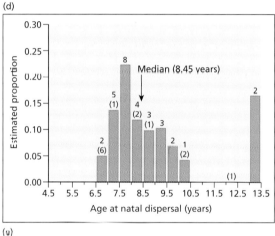

(e)

Figure 12.3 (Continued)

species, including many macropods, ungulates and carnivores, males typically disperse from their mother's range between the ages of 1 and 4 years (Clutton-Brock *et al.* 1982; Berger 1987; Mech 1987; Nelson and Mech 1987; Johnson 1989; Rutberg and Keiper 1993; Waser 1996), while in many of the longer-lived monkeys and apes, dispersal occurs between the ages of 3 and 9 years (Pusey 1987). In the largest and most long-lived mammals, including gorillas and elephants, males are often in their early teens before they leave their mother's range (Harcourt *et al.* 1976; Moss and Poole 1983; Poole 1994).

The relationship between dispersal and maturation differs between species. In some species, males disperse before reaching sexual maturity. For example, male Belding's ground squirrels typically disperse from their natal area as juveniles of between 7 and 12 weeks, almost 2 years before they are fully mature (Holekamp 1984;

Smale *et al.* 1997). Similarly, male prairie dogs disperse as yearlings before reaching sexual maturity at around 21 months (Hoogland 1995). In other species, males do not disperse for several months or even years after reaching sexual maturity. For example, in meerkats, males reach adult body weight at around a year and a half and commonly spend a year or more in their natal group before dispersal. During this time, they frequently go on forays to neighbouring groups in search of receptive females before dispersing permanently (Clutton-Brock and Manser 2015). Similar delays in dispersal occur in many other social mammals, including many primates (Pusey 1987; Pope 2000).

Social factors, too, may be important. Theoretical models predict that the presence of relatives should increase the chance that individuals will disperse in order to reduce competition with their kin (Dobson 2013).

However, in practice, there is often a negative association between the presence of same-sex kin and the probability of dispersal, suggesting that the presence of kin may help to provide some shelter from competition and may consequently delay dispersal. For example, in two species of prairie dogs, females are substantially more likely to disperse in the absence of close relatives than when they are present, apparently because they lack support and cooperation in competitive interactions with other group members if no relatives are present (Hoogland 2013). Similarly, in yellow baboons, males with old mothers or those whose mothers died before they were 6 years old disperse earlier than males with younger mothers (Alberts and Altmann 1995).

Where both sexes typically disperse from their natal area to breed, one sex often leaves before the other, though this is not always the case (Waser and Jones 1983; Smale *et al.* 1997). In some species, males typically disperse before females (Waser and Jones 1983; Smale *et al.* 1997) while this may be reversed in species where females habitually disperse at adolescence, as in equids and mountain gorillas (Harcourt *et al.* 1976; Harcourt 1978; Rutberg and Keiper 1993).

In many species, dispersal appears to represent a discrete phase of development that is associated with changes in behaviour and in hormone levels. For example, while most male prairie dogs do not mature or disperse until their second year of life, individuals that do mature in their first year are likely to disperse before they are a year old (Hoogland 1995). In Belding's ground squirrels, dispersal timing is affected by the body condition of individuals and by endogenous circannual timing mechanisms as well as by perinatal exposure to testosterone (Holekamp and Sherman 1989; Nunes and Holekamp 1996). As male hyenas approach the age of dispersal and begin to range more widely, they exhibit reduced fearfulness of strange objects and increased motor activity (Holekamp 1986; Smale *et al.* 1997).

Individual differences in development often affect the timing of dispersal. For example, in red deer, well-grown adolescents are more likely to disperse from their natal subpopulation than smaller individuals of the same age (Clutton-Brock *et al.* 2002b), though in some other species, smaller or weaker individuals that are less well able to compete for food or mates are more likely to disperse and disperse earlier than larger or more successful individuals (Alberts and Altmann 1995; Solmsen *et al.* 2011).

In some species, there are discrete differences in development and behaviour between individuals that will disperse and those that will remain in their natal group. For example, in naked mole rats, males that are destined to become helpers remain in their natal colony throughout their lives while a minority of males lay down additional fat, contribute little to cooperative activities and attempt to disperse to breed in other colonies (see Chapter 17). Studies of several rodents show that individual variation in the tendency to disperse is heritable (Stenseth 1983), though whether differences in dispersal in naked mole rats also have a genetic basis is not known.

Environmental factors can also have direct effects both on rates of development and on the timing of dispersal. In populations of black bears with access to plentiful food supplies, most males disperse as yearlings, but where food is scarce they disperse as 2 or 3 year olds (Rogers 1987). Similarly, in yellow baboons, the frequency of dispersal differs between populations at high and low density (Alberts and Altmann 1995; though see also Bulger and Hamilton 1988). A number of studies have also shown that males commonly leave their natal group when breeding opportunities occur in neighbouring groups.

Dispersing males often move to groups with a lower ratio of adult males to receptive females than in their previous group, as in baboons (Packer 1979; Manzolillo 1986), olive colobus (Korstjens and Schippers 2003) and macaques (Drickamer and Vessey 1973). For example, in baboons, both natal and non-natal males are more likely to leave groups with a high ratio of males to receptive females and to join groups with a low ratio of males to females (Figure 12.4), though this tendency varies with the individual's competitive ability (Clarke *et al.* 2008). In spotted hyenas, males also selectively immigrate into clans with relatively high numbers of young females and increase their breeding success by doing so (Höner *et al.* 2007, 2010).

In polytocous (litter-bearing) mammals, males often disperse in parties (Waser 1996; Schoof *et al.* 2009). For example, in African lions, males disperse in parties made up of males of the same year-class, some of which are littermates, and compete for access to prides of females (Bygott *et al.* 1979; Packer and Pusey 1982; Packer *et al.* 1988). Parties of males also disperse together in some monotocous ungulates and primates (Crockett and Sekulic 1984; Berger 1987; Pusey and Packer 1987a):

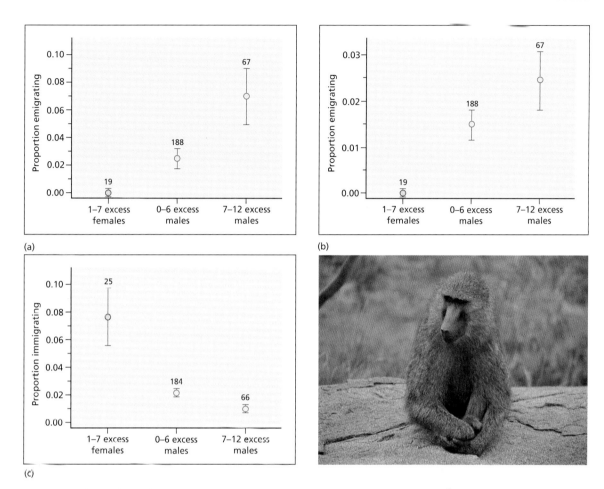

(a)

(b)

(c)

Figure 12.4 Selective dispersal to groups with advantageous male/female ratios by male baboons. Effects of number of excess males on rate of (a) natal emigration, (b) non-natal emigration and (c) immigration. The vertical axis represents the proportion emigrating monthly. The number of excess males or females was computed as the number of adult males minus the number of cycling females. The range was from 7 excess females (equal to −7 excess males) to 12 excess males; this was divided into three categories representing high, medium and low numbers of excess males, indicated on the horizontal axis by the number of excess males or females present. Numbers adjacent to the symbols on the plot indicate the number of group-months in which low, medium or high numbers of excess males were present; bars indicate the standard error of the parameter. *Sources*: From Alberts and Altmann (1995). Reproduced with permission of the University of Chicago Press. *Photo source*: © Joan Silk.

for example, in ursine howler monkeys, males commonly disperse from their natal group in parties of two to four and attempt to displace resident males from other groups (Pope 2000). Single males are seldom able to evict resident groups of males and larger parties of males have an improved chance of gaining access to breeding groups. In lions, too, males dispersing in large parties are generally more successful in winning control of prides and the duration of their tenure in prides is relatively long, so that individuals in larger parties have higher fitness (see Chapter 11).

Dispersing in parties may also reduce the risk of predation and of damaging attacks by members of resident groups (Pusey and Packer 1987a) and may be associated with reduced levels of social stress. For example, in meerkats, males dispersing in larger parties spend less time vigilant, feed more and show lower levels of cortisol than males dispersing in small parties (see Figure 12.5).

Box 12.1 Costs of dispersal in male meerkats

In Kalahari meerkats, natal males disperse voluntarily from their original group between the ages of 20 and 48 months. Initially, males begin to visit other groups, either on their own or in small parties, returning to their natal group at night (Young 2003; Clutton-Brock and Manser 2015). The frequency of 'roving' increases at the beginning of the annual breeding season and continues until males either die or locate and join parties of dispersing females and establish new breeding groups. Roving males are commonly attacked by residents of both sexes and are sometimes killed. Males in large parties spend less time vigilant and forage more successfully than males in small parties (Figure 12.5a) and lose less weight. In addition they show lower levels of glucocorticoids (Figure 12.5b) and have lower ectoparasite levels (Figure 12.5c), and members of larger subgroups are more likely to survive and establish new breeding groups. Similar relationships between the size of dispersing parties and foraging success, condition, glucocorticoid levels and ectoparasite loads occur in females (Young 2003). Since the size of dispersing parties increases with the size of breeding groups, these relationships suggest that subordinates of both sexes gain substantial direct fitness benefits by increasing or maintaining the size of the group they live in.

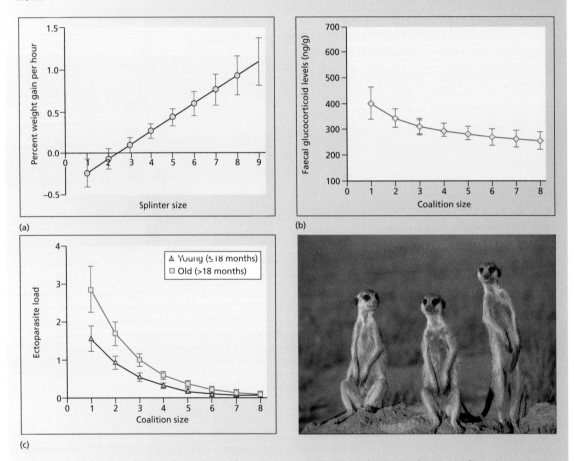

(a)

(b)

(c)

Figure 12.5 Benefits of social dispersal in male meerkats: (a) mean weight gain per hour spent foraging in dispersing parties (splinters) of different sizes; (b) mean levels of faecal glucocorticoids for males in dispersing parties of different sizes; (c) ectoparasite load for young and old individuals in dispersing parties of different sizes. *Sources*: From Young (2003). Reproduced with permission of A. Young. *Photo source*: © Tim Clutton-Brock.

Both the opportunity to disperse in parties and their potential size often depends on the availability of other males of approximately the same age or growth stage in their natal group. As would be expected, social dispersal is commoner in seasonal breeders than in aseasonal breeders (Pusey and Packer 1987a).

Positive relationships between the size of dispersing splinter groups and the fitness of their members are likely to complicate the dispersal decisions of males. Males that join a coalition of older sibs are relatively unlikely to gain the dominant position and so might be expected to delay dispersal until they are able to leave with several younger brothers. However, delaying dispersal is unlikely to be cost-free: not only may it increase the chances that males will die before they begin to breed but, if few males are produced in subsequent litters, individuals either have to delay dispersing beyond the optimal age or may be forced to disperse alone. Factors affecting dispersal decisions have seldom been studied in detail but are likely to include the survival of males in their natal group, their relative probability of acquiring access to females in coalitions of different sizes, the rate at which younger brothers are likely to be produced in their natal group and the possibility of joining up with unrelated males.

12.3 The costs and benefits of dispersal to males

As in females, dispersal is risky for males. Dispersers are at greater risk from predators than residents, as well as from attacks by animals from groups they are attempting to join (Cheney and Seyfarth 1983; Zhao 1996). They feed less effectively and lose body condition: for example, in meerkats, the daily weight gain of males drops after they leave their natal group and they show a progressive loss of weight, an increase in parasite load and elevated levels of corticosteroids, especially if they disperse on their own (Box 12.1 and Figure 12.5). Testosterone levels may also fall during the period of dispersal (Holekamp and Smale 1998; Young 2003).

Many studies of social mammals suggest that a substantial proportion of males that disperse from their natal group die before they find an opportunity to settle and breed (Ralls *et al.* 1980; Waser *et al.* 1994; Isbell and Van Vuren 1996; Smale *et al.* 1997). For example, in yellow baboons, around 13% of males die during dispersal events which typically last around 2–3 months, representing a mortality rate an order of magnitude above the level experienced by males resident in mixed-sex groups (Alberts and Altmann 1995). Males are also unlikely to breed during the 2 months that they spend on their own during the period of dispersal. Where the survival of dispersing individuals cannot be monitored, a conservative estimate of dispersal costs can be calculated by comparing the size of parties that leave breeding groups with those that immigrate into groups (Waser *et al.* 1994). Using this technique, figures for African wild dogs suggest that the daily risk of death is 6.8 times as high in dispersing males as in residents, more than twice the increase suffered by dispersing females (Creel and Creel 2002). Dispersal costs may be higher in males than females in wild dogs because groups are defended by groups of related males that seldom accept male immigrants, whereas in species where female group members are closely related and are hostile to unrelated females (as in meerkats and many social primates) mortality rates associated with dispersal may be higher in females than in males. This may explain why females seldom disperse in many species where females are usually philopatric.

Dispersing males use different strategies to minimise risks and maximise benefits. In some species, they observe or visit other groups before leaving their natal groups (Idani 1991; Isbell and Van Vuren 1996). A number of studies of primates have described how males quietly observe a group before they attempt to immigrate; in meerkats, males often spend part of their time visiting neighbouring groups during the year before they disperse (Clutton-Brock and Manser 2015; Young 2003). Males may also increase their chances of being accepted into breeding groups or of developing cooperative relationships with resident males by attempting to join groups that already contain individuals from their natal group (Schoof *et al.* 2009): for example, both in vervet monkeys and in white-faced capuchins, males commonly join groups that already include previous immigrants from their group and particular pairs of groups tend to exchange males (Cheney and Seyfarth 1983; Wikberg *et al.* 2014). However, patterns of this kind are not universal: for example, male yellow baboons show no tendency to immigrate into groups that contain other members of their previous group (Alberts and Altmann 1995). Moreover, it is not yet clear whether the presence of related previous immigrants eases the acceptance of related males or whether the tendency is a consequence of other factors affecting the movements of dispersing males.

Where the costs of dispersing are high and males leave their natal group of their own volition (see section 12.2), the benefits of dispersal must presumably be large. Three main ways in which males may benefit by dispersing have been suggested (Bengtsson 1978; Greenwood 1980; Moore and Ali 1984; Shields 1987; Moore 1993; Perrin and Mazalov 2000). First, as in females, dispersal may allow males to leave areas where competition for resources is intense, and leaving may generate direct benefits to their own fitness as well as indirect benefits to relatives, who benefit from reductions in local resource competition (Moore and Ali 1984; Perrin and Mazalov 1999, 2000; Dobson 2013).

This is particularly likely to be important in solitary species that defend concentrated food supplies, such as tree squirrels, kangaroo rats and pikas (Smith 1987; Larsen and Boutin 1994). In some ungulates, too, males may move away from their natal area to avoid competition for resources with females (Ruckstuhl and Neuhaus 2001). For example, in red deer, males leave their mother's group in their second or third year of life and move to join male groups in areas peripheral to the major concentrations of females where food resources are more abundant though of lower quality (Staines and Crisp 1978; Clutton-Brock and Albon 1989) and high female density is associated with an increased tendency for males to leave their natal area (Figure 12.6) and a similar pattern occurs in feral horses (Berger 1986). However, it is difficult to be certain that males are not dispersing for other reasons.

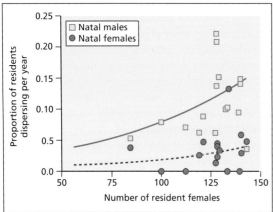

(a)

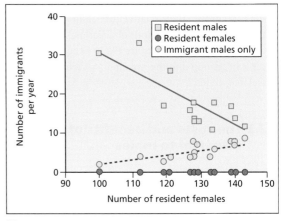

(b)

Figure 12.6 Changes in male emigration and immigration in red deer on Rum associated with increases in female density caused by the cessation of culling: (a) proportion of natal males (blue squares) and females (red circles) of 2–6 years dispersing (permanently) from the study area each year; (b) number of animals of all ages immigrating into the study area per year as permanent residents (blue squares, males; red circles, females) or mating-season immigrants (males only, blue circles) immigrating temporarily during the rut. *Sources*: From Clutton-Brock *et al.* (1997). Reproduced with permission from the Royal Society. *Photo source*: © Tim Clutton-Brock.

Second, dispersal may allow males to escape from breeding competition with resident male competitors or to locate breeding groups where the ratio of males to females is low and mating opportunities are relatively frequent (Greenwood 1980; Dobson 1982; Dobson and Jones 1985). In some species, philopatric males show evidence of reproductive suppression (for example, in striped mice, they have smaller testes, as well as lower levels of testosterone than dispersing males; Schradin *et al.* 2012) and dispersing may allow them to become reproductively active. As in females, dispersal has the potential to increase both the direct fitness of dispersing males and the fitness of male relatives in their group of origin, as a consequence of reductions in local mate competition (Shields 1987).

Third, dispersal often provides males with access to receptive females that are not close relatives (Bengtsson 1978; Greenwood 1980; Shields 1987) and is often associated with the presence of female relatives. Adolescent males frequently disperse from groups or areas where resources and related breeding females are abundant. In several species, neither the provision of food nor female-biased adult sex ratios in breeding groups reduce the proportion of males dispersing at adolescence (Dobson 1979; Dobson and Kjelgaard 1985; Nunes and Holekamp 1996; Smale *et al.* 1997). In many social mammals, resident females are unwilling to mate with males that have been resident since their birth (see Chapter 4), and where males usually disperse at adolescence, they commonly lack access to unrelated females in their natal group (Lukas and Clutton-Brock 2011) and often show lower levels of testosterone and sexual activity than immigrants (Reyer *et al.* 1986; Woodroffe 1993; Holekamp and Smale 1998). For example in spotted hyenas, where natal males are generally dominant to immigrant males (Engh *et al.* 2002), almost all cubs are fathered by immigrants. Natal males also show lower levels of testosterone and lower frequencies of sexual activity than immigrants (Figure 12.7). In contrast, in species where females commonly immigrate and adolescent males have access to unrelated females in their natal range or group (as in banded mongooses, spider monkeys, red colobus, chimpanzees, gorillas and some callitrichids), adolescent males may remain and breed in their natal group throughout their lives (Struhsaker 1975; Harcourt 1978; Goodall 1986; White 1996; Gilchrist 2001; Yamamoto *et al.* 2014). Unfortunately,

systematic comparisons of differences in hormonal status between natal and immigrant males in species where females are commonly philopatric and those where many breeding females are immigrants are not yet available.

While male dispersal from groups where breeding females are mostly natals is often described as a form of inbreeding avoidance, a simpler interpretation is that males disperse from groups where breeding females are relatives because females refuse to mate with familiar or related males (see Chapter 4). Where males invest little in parental care, they would not necessarily be expected to avoid mating with related females unless the costs of inbreeding were high or they were restricted in the number of females they could fertilise (Smith 1979; Waser *et al.* 1986). In practice, it is often difficult to determine whether males disperse to avoid mating with related females or whether they disperse to find willing partners, with the result that firm evidence that males actively avoid breeding with close relatives is scarce. However, some evidence suggests that males may actively avoid breeding with closely related females in some species. In social mammals, males with relatives in the group are more likely to leave than those with few relatives (Kuester and Paul 1999). For example, in dwarf mongooses and African wild dogs, adolescents commonly leave their natal group if the opposite sex breeder is a close relative (Frame *et al.* 1979; Rood 1987; McNutt 1996). Juvenile male white tailed deer that have been orphaned stay longer in their natal range than those whose mothers are alive (Hölzenbein and Marchinton 1992); in white-footed mice, too, sons remain longer in their natal range if their mother is absent than if she is present, though most young males eventually disperse whether or not their mothers are present (Brody and Armitage 1985; Wolff 1992). Experiments with *Antechinus* also show that the removal of mothers increases the chance that their sons will remain in their natal territory rather than dispersing (Cockburn *et al.* 1985).

Although both sexes may avoid breeding with close relatives, this does not necessarily suggest that genetic kinship is the cue that they use to recognise relatives. In many social mammals, familiarity rather than genetic relatedness appears to be used (see Chapter 4), and on the relatively rare occasions when individuals breed with first-order relatives, this is often because they appear to fail to recognise them (Berger and Cunningham 1987).

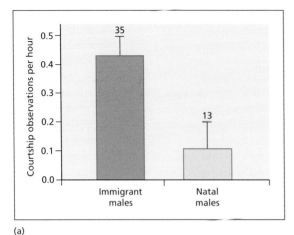

(a)

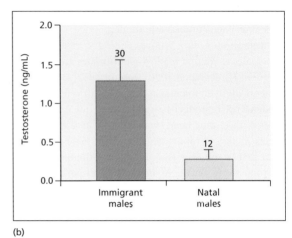

(b)

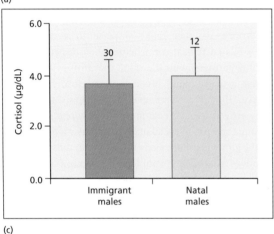

(c)

Figure 12.7 Contrasts in sexual activity and hormone levels between natal and immigrant males in spotted hyenas. (a) Mean hourly rates at which adult natal and immigrant males directed sexual behaviour toward adult resident females of the Talek clan. Sample sizes represent numbers of individuals. Data are expressed as means ± SEM (b,c). Comparisons of testosterone (T) and cortisol levels in immigrant and natal males. *Sources*: From Holekamp and Smale (1998). Reproduced with permission of Elsevier.

12.4 Secondary dispersal by males

In some social mammals, males may disperse more than once during their breeding careers (van Noordwijk and van Schaik 1985; Pusey and Packer 1987a; Cheney *et al.* 1988). The timing and circumstances associated with secondary dispersal vary. In some species, males emigrate to groups where mating opportunities are more abundant than in their own group or the ratio of resident males to females is lower, for example in lions (Pusey and Packer 1987a), baboons (Packer 1979; Manzolillo 1986; Alberts and Altmann 1995; Clarke *et al.* 2008), blue monkeys (Henzi and Lawes 1987), sifakas (Richard *et al.* 2002) and lemurs (Sussman 1992). Coalitions of male lions often abandon smaller female prides when they have an opportunity to take over a larger pride (Pusey and Packer 1987b).

Many of the most detailed studies of secondary dispersal have involved yellow baboons, where the probability that immigrant males will subsequently move to another group appears to be related to changes in breeding success that occur during their period of residency. Male baboons that have recently immigrated are attractive to females and initially show relatively high reproductive success (Altmann *et al.* 1988) but, as time passes, their breeding success declines, apparently because breeding females prefer to mate with fresh immigrants (Altmann *et al.* 1988) and the chance that they will leave the group increases (Figure 12.8). Individuals that are reproductively successful often remain longer than those that are unsuccessful: unsuccessful males stay in groups for an average of 1–2 years while successful males stay for an average of 4.3 years. Rates of secondary dispersal are highest in the first year after immigration, when

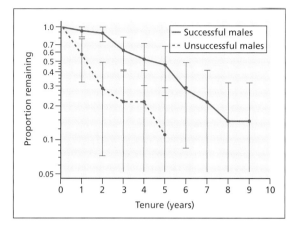

Figure 12.8 Secondary dispersal in yellow baboons is affected by reproductive success in the first group. The graph shows the effects of reproductive success in their group on the chance that immigrant male yellow baboons will remain in the same group. Males were designated as successful based on whether they were above or below the age-specific median for consorting success. Unspecified lower confidence limits indicate lower limits near or equal to zero. *Sources*: From Alberts and Altmann (1995). *Photo source*: © Phyllis Lee.

unsuccessful males typically disperse, and in the sixth year, when a male's first potential daughters reach maturity (Alberts and Altmann 1995). It is not yet clear why females should favour recent immigrants but several factors may be involved, including reduction in the risk that females will breed with their fathers or brothers and increased opportunities for developing supportive relationships with new males (see Chapter 15).

In other cases, secondary dispersal appears to be related to the absence of unrelated females or the presence of close relatives. For example, if dominant female

meerkats die, leaving their male partner without another unrelated female in the group, the dominant male typically emigrates to another group within a few months (Clutton-Brock and Manser 2015). Breeding males may also disperse of there is a risk of breeding with their daughters, even if other, unrelated, breeding partners are present in their groups. For example, in black-tailed prairie dogs, most females reach sexual maturity by the age of 2 years while breeding males may live for

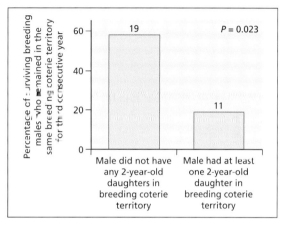

Figure 12.9 Dispersal of older males in black-tailed prairie dogs is related to the likelihood of father–daughter inbreeding. Father–daughter inbreeding is likely when a breeding male has at least one 2-year-old daughter in his breeding coterie during his third year of residency. The number above each bar indicates the number of observed males. The figure shows the percentage of males that remained in the same breeding group for a third year that did or did not have 2 year-old daughters in the group. *Sources*: From Hoogland (1995). Reproduced with permission of the University of Chicago Press. *Photo source*: © Elaine Miller Bond.

4–5 years, with the result that maturing daughters may often be at risk of being mated by their father (Hoogland 1995). Breeding males are less likely to stay in the same colony for more than 2 years if they have one or more mature daughters in the colony than if none of their daughters are present (Figure 12.9).

Sex differences in the frequency of secondary dispersal are also common. In many species, secondary dispersal appears to be less frequent in whichever sex is philopatric, possibly because groups of relatives repel immigrants. The effects of secondary dispersal on the distances individuals move from the group of origin can also differ between the sexes. For example, secondary dispersal by male ground squirrels, red kangaroos and lions commonly increases the distances that males move from their natal group while secondary dispersal by females may involve a movement back towards the natal area or even a return to the group of origin (Holekamp 1984; Oliver 1986; Hanby and Bygott 1987; Dawson 1995; Smale *et al.* 1997).

12.5 Sex differences in philopatry

Studies of dispersal and philopatry in mammals provide extensive evidence both of contrasts between the sexes and of interspecific differences in the extent of these differences (Smale *et al.* 1997). Since dispersal is mediated by hormones and is predictably associated with particular stages of development, these contrasts suggest that sex differences in dispersal may be a consequence of evolved characteristics with an adaptive basis.

A substantial number of different reasons why the costs of dispersing may be lower or the benefits higher in males than in females have been suggested and the functional significance of sex differences in dispersal and philopatry are still debated (Moore and Ali 1984; Lawson Handley and Perrin 2007; Dobson 2013). One possible explanation is that the presence of additional males that are capable of breeding with resident females may represent a greater threat to the breeding success of dominant males than the presence of additional females would to the breeding success of dominant females, with the result that dominant males are less tolerant of adolescent males than dominant females are of adolescent females, generating higher ratios of dispersal in males (Greenwood 1980; Shields 1982; Smale *et al.* 1997). Where local mate competition exerts a stronger

influence on fitness in males than in females, also males may be able to gain greater benefits than females by dispersing to groups with low male/female ratios than females would gain by moving to groups with high male/female ratios (Greenwood 1980; Shields 1987; Perrin and Mazalov 2000).

The energetic costs of dispersal may also be higher in females than in males because of their heavier energetic investment in reproduction and females may gain greater direct benefits than males from remaining in their natal group and associating with close relatives (Sherman 1977; Wrangham 1980; Johnson 1986; Gandon 1999). Experimental studies involving manipulations of food availability or local density confirm that energetic factors often have stronger effects on rates of dispersal in females than in males (Dobson 1979; Dobson and Kjelgaard 1985; Nunes and Holekamp 1996; Smale *et al.* 1997).

While several different factors may contribute to the evolution of sex differences in dispersal and philopatry among mammals, the simplest explanation of the prevalence of female philopatry is that the high costs of dispersal favour philopatry in both sexes (see Chapter 3) and that avoidance of mating with familiar or closely related males by resident females as a result of the costs of inbreeding generates selection on males to disperse to find willing mates (Höner *et al.* 2007; Clutton-Brock and Lukas 2011). Where male tenure is relatively short, so that females are unlikely to mature when their father is still the dominant breeding male in their group, most natal females will have access to unrelated immigrants in their natal group and female philopatry is likely to evolve. However, where dominant males are able to maintain their position for relatively long periods relative to the age of which females begin to breed (see Chapter 3), females may need either to breed outside their group or to disperse to find unrelated partners. Situations of this kind are not common in mammals but occur in several species where females habitually disperse as adolescents or young adults (Harcourt and Stewart 1981; Clutton-Brock 1989a,b; Clutton-Brock and Albon 1989; Sillero-Zubiri *et al.* 1996; Lukas and Clutton-Brock 2011; Yamamoto *et al.* 2014).

Once habitual female dispersal is established, the benefits of philopatry to males may often outweigh the costs, and multiple males may remain and breed in their natal group, forming male kin groups that defend access to female groups or to the resources they rely on, as in greater sac winged bats and chimpanzees (see Chapter 11). The formation of male kin groups may, in turn, prevent single males from immigrating into established breeding groups and reduce the chances of emigrants evicting resident males, reinforcing the costs of dispersal and the relative benefits of philopatry to males and increasing the number of males associated with female groups. Effects of this kind may be stronger in monotocous species, like chimpanzees, than in polytocous ones, like banded mongooses, where dispersing male subgroups may be larger and may consequently have a greater chance of evicting resident males.

In a few mammals, both sexes often breed in their natal group throughout their lives. These appear to fall into two categories. First, there are species where males as well as females typically outbreed with members of other groups when they meet. Breeding systems of this kind include resident groups of killer whales (Heimlich-Boran 1986; Hoelzel and Dover 1991; Baird 2000) and, possibly, pilot whales (Amos *et al.* 1993; Heimlich-Boran 1993). They can also be a temporary phenomenon in species where males habitually disperse. For example, following the death of a dominant male meerkat, the eldest natal male sometimes assumes dominance for several months but does not guard or mate with the dominant female (who is commonly his mother) and the dominant female usually breeds with roving males from other groups until the natal dominant disperses and the group is joined by one or more immigrant males (Clutton-Brock and Manser 2015). A similar situation may explain observations of philopatry in both sexes in Ethiopian wolves (Sillero-Zubiri *et al.* 1996).

Second, there are some relatively sedentary species where both males and females may remain in their natal group throughout their lives and often breed with close relatives. The best-known example of a society of this kind occurs in naked mole rats, where individuals of both sexes may remain in their natal group and breed with close relatives (Alexander 1991; Reeve *et al.* 1990), though the frequency of inbreeding in natural populations is not yet known and females show a preference for unrelated males. High levels of inbreeding may be associated with unusually high costs of dispersal in these species (Braude 2000).

One of the most widely cited contrasts in dispersal is the tendency for dispersal to be female-biased in social birds and male-biased in social mammals (Greenwood 1980; Greenwood and Harvey 1982; Greenwood 1983;

Pusey and Wolf 1996; Clarke *et al.* 1997) and a substantial number of reviews and theoretical papers have sought to explain this difference (Greenwood 1980; Dobson 1982, 2013; Pusey and Wolf 1996; Perrin and Mazalov 2000). Neither Greenwood's original analysis (1980) nor Clarke *et al.*'s more recent one (1997) distinguishes clearly between sex differences in the probability of dispersal from the group or territory of origin and sex differences in the average distance moved by dispersers and both combine data from monogamous species (where members of both sexes typically leave their original breeding unit) with data from group-living species (where members of one sex may breed in their group of origin while members of the other sex typically disperse). However, when comparisons are restricted to animals that breed in groups, they support Greenwood's generalisation that females more commonly breed in their natal group than males in mammals while sex differences in philopatry are reversed in many group-living birds.

Several possible explanations of contrasts in dispersal between birds and mammals have been suggested, though none provide an entirely satisfactory explanation. Greenwood (1980) argued that male birds benefit more than females from remaining close to their original territory (because effective territorial defence is facilitated by local knowledge), whereas male mammals typically breed polygynously and defend groups of females rather than territories so that the benefits of philopatry are reduced (see also Shields 1987). Other suggestions are that female birds may be less tolerant of their daughters because they are capable of adding eggs to clutches that are about to be incubated, while female mammals cannot pursue a similar strategy (Liberg and von Schantz 1985); that the costs of dispersing to females may be higher in mammals than birds as a consequence of the large energetic investment by female mammals in lactation (Johnson 1986); that the relative intensity of local mate competition and local resource competition may favour male dispersal in mammals and female dispersal in birds (Dobson 1982; Perrin and Mazalov 2000); and that as a result of their ability to identify eggs laid by others, females may be able to monopolise breeding success more effectively in birds than in mammals, so that the relative breeding success of subordinates is reduced (Raihani and Clutton-Brock 2010).

A final possibility is that in group-living birds, as in some mammals, prolonged male tenure commonly forces females to disperse to avoid close inbreeding.

The available data on male tenure suggests that this may be the case: in social birds, life expectancy is typically high in both sexes, while females often enter breeding condition in their second year of life and so are likely to mature in groups where the established breeding male is a close relative (Clutton-Brock and Lukas 2011). Like previous explanations, this hypothesis links the contrast between birds and mammals to the incidence of polygyny in mammals, but for different reasons (Clutton-Brock 2009). Instead of polygyny affecting the probability that males will be evicted or the benefits they can derive from moving to new groups, it suggests that polygyny reduces the average tenure of males in breeding groups, so that females rarely need to disperse from their natal group to obtain access to unrelated males. Avoidance by females of mating with familiar males may then lead to the evolution of male dispersal.

While it seems likely that the duration of male tenure and the risk of inbreeding may play a central role in determining which sex typically remains and breeds in its natal group, there are many exceptions to general trends and the probability of dispersing is probably affected by a wide range of ecological and social factors (Gandon and Michelakis 2001). For example, in some species, males are more tolerant of the presence of other adults of the same sex than females, and females are often forced to leave their natal group after adolescence, while males may inherit the breeding role from their fathers (Yamamoto et al. 2014). In addition, the costs and benefits of philopatry may vary both between populations and between groups and these differences may generate variation in the strength or direction of sex differences. For example, in Ethiopian wolves, males typically remain in their natal group and may inherit the dominant breeding role while females usually disperse to other groups; however, in a minority of groups, females inherit the breeding role from their mother and may then remain and breed in their natal group (Sillero-Zubiri et al. 1996). Since many organisms show adaptive plasticity in other aspects of their life histories (Stearns 1992; Lee and Kappeler 2003), it is unsurprising that dispersal, too, is affected by a combination of genetic and environmental factors.

12.6 Dispersal distance

The average distances moved by dispersing males vary from less than 100 m in some rodents (Jones 1987;

Stenseth and Lidicker 1992) or a few hundred metres in the more sedentary primates (Pope 2000), to 20 km or more in some carnivores (Creel and Creel 2002) and much larger distances in cetaceans (Whitehead and Weilgart 2000). As dispersing individuals are usually difficult to monitor systematically, we know less about the distance moved by dispersers than about whether or not they leave their natal group (see Chapter 3).

In most cases, the principal factor governing the distances over which males disperse appears to be the availability of breeding opportunities: where these are available close to the natal territory, males seldom move far; when these are scarce, males may travel long distances from their natal range or territory (Berger 1986; Creel and Creel 2002). For example, dispersing male ursine howler monkeys commonly immigrate into groups close to their natal range if breeding opportunities are available there, but travel further if they are not (Pope 2000) (Figure 12.10). Similarly, young male red deer initially visit different rutting areas until they win access to a harem and then usually return to the same rutting area in successive years (Clutton-Brock et al. 1982). Where dispersing males are prevented from immigrating into established groups, they may have to travel long distances before they settle. For example, in horses, dispersing males commonly move more than 5 km from their natal range before settling (Berger 1986), while in African wild dogs male dispersal distances average more than 25 km (Creel and Creel 2002).

Where dispersal distances are typically short and life-spans are long, males may find themselves surrounded by groups containing relatives (Waser and Jones 1983). As even moderate levels of relatedness between breeders can generate costs to the fitness of progeny (see Chapter 3), males might be expected to adjust their dispersal distance to the local density of relatives so as to avoid the risk of inbreeding, especially in species where paternal care is well developed (Shields 1987). However, although they commonly leave their natal group if no unrelated females are present, there is little concrete evidence that the distance males move is adjusted to the distribution of related females. This could be because few studies of male dispersal have had access to measures of the relatedness of females in neighbouring groups. Alternatively, high annual mortality and restricted overlap between generations may reduce the risk that individuals of either sex will mate with close relatives once they have left the natal group or territory

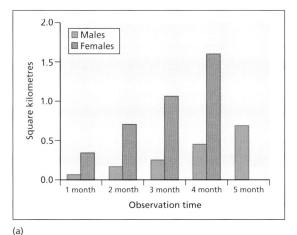

(a)

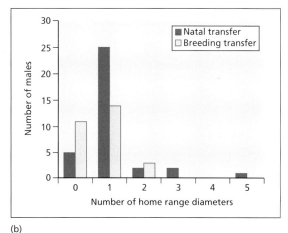

(b)

Figure 12.10 Dispersal in ursine (red) howler monkeys: (a) mean cumulative ranging area increases as a function of observation time for solitary howler males and females; (b) distances travelled by known howler immigrant males from their group of origin. Breeding transfer refers to males transferring from non-natal groups. Zero dispersal distance was assigned to 'non-emigration' events in which males remained in a group long enough to mate with their mother or daughter (i.e. longer than one generation). *Sources*: From Pope (2000). *Photo source*: © Carolyn M. Crockett.

and dispersers may maximise their fitness by taking the first vacant range or breeding opportunity they encounter, irrespective of variation in their kinship to resident females.

In many mammals, average dispersal distances from their natal territory are greater in males than in females (Greenwood 1980; Clarke *et al.* 1997). However, if comparisons are restricted to distances moved by individuals of both sexes that leave their natal group or territory and individuals that remain there are excluded, the pattern of sex differences is often less obvious. Although dispersing males move further than dispersing females in some species (Waser and Jones 1983; McCullough 1985; Gaines and Johnson 1987; Rogers 1987; Smale *et al.* 1997; Radespiel *et al.* 2003), in others there is no sex difference in dispersal distance or dispersing females move further than males (Dice and Howard 1951;

Whitworth and Southwick 1984; Jones 1987; Waser 1996; Balloux *et al.* 1998; Creel and Creel 2002).

Functional explanations of sex differences in dispersal distance often suggest that they are a consequence of the greater benefits that males can gain from moving to groups where the ratio of males to females is relatively low or that they are associated with sex differences in the costs of moving between groups (see section 12.3). While both may be the case, an alternative possibility is that sex differences in dispersal distance are a consequence of sex differences in the availability of breeding opportunities associated with sex differences in breeding tenure or longevity. The relative probability that dispersing males and females will be able to join established breeding groups may also affect dispersal distances. For example, in feral horses (where females habitually disperse after reaching adolescence), dispersing females are readily

The death of dominant breeders may also generate conflicts between parents and offspring of the same sex (Emlen
1997). Where females immigrate and subordinate males
breed in their natal group, the death of their mother may
trigger conflict with their father as soon as an unrelated
female joins the group (Piper and Slater 1993; Zahavi and
Zahavi 1997). Similarly, where subordinate females may
remain and breed in their natal group, the death of the
dominant male and the arrival of immigrant males may
generate conflict between natal females and their mother.

The death of parents may also lead to conflicts between
parents and offspring of the opposite sex over the parent's
continued presence in the group. Where males breed
in their natal group, it may be in the interest of sons to
evict their mother after the death of their father if her
presence prevents unrelated females from joining the
group. Similarly, where females breed in their natal group
if an unrelated male is present (Cooney and Bennett 2000;
O'Riain *et al.* 2000), it may be in the interests of daughters
to evict their father following the death of their mother if
his presence prevents immigrant males from joining the
group. However, in practice, the eviction of parents by
offspring is usually rare in mammals, probably because the
costs of dispersal to parents are high so that they are
reluctant to leave their group.

Population processes that increase relatedness between
group members would also seem likely to favour the
evolution of cooperation. However, they may also
increase competition between relatives (Queller 1992;

West *et al.* 2002; Kümmerli *et al.* 2008) and some models
suggest that these effects may cancel out any tendency to
favour cooperation (Taylor 1992), though others show
that higher rates of dispersal by adults of the sex with
greater variance in reproductive success may restore the
tendency for high levels of relatedness to favour cooperation (Johnstone and Cant 2008; Gardner 2010).

Contrasts in dispersal which lead to variation in the
relative frequency with which sons and daughters compete or cooperate with sibs of the same sex may also
affect strategies of parental investment (Clutton-Brock
1991). In general, parents would be expected to produce
more offspring of whichever sex cooperates most and
competes least with sibs of the same sex. As would be
expected, parents produce more sons than daughters in
some birds and mammals where males assist each other
in gaining access to unrelated mates (see Chapter 18).

Finally, contrasts in dispersal and philopatry can have
important consequences for the regulation of male and
female numbers. Sex differences in dispersal commonly
lead to situations where the numbers of one sex (typically males) are limited by density-dependent changes in
emigration and immigration rates, while those of the
other sex (typically females) are limited by density-
dependent differences in survival and breeding success
in situ (see Chapter 3). In conjunction with sex differences in responses to population density and food shortage, these contrasts can have important implications for
the management of populations (see Chapter 18).

SUMMARY

1. In some social mammals, males are evicted at or just after adolescence, while in others they leave of their own accord.
   'Voluntary' male dispersal is often associated with reluctance by resident females to mate with familiar males.
2. The proportion of males dispersing from their natal groups and their age at dispersal vary between groups and populations
   and, in some species, males may disperse more than once. Both the frequency of male dispersal and the age at which
   males disperse are affected by the rate of male development and the availability of resources, as well as by the proportion
   of resident females that are close relatives.
3. As in females, dispersal has substantial costs to the survival of males, suggesting that the benefits of dispersing must also be
   large. While male dispersal may sometimes allow males to avoid competition for resources or reproductive opportunities,
   in many mammals the principal benefits appear to be increased access to receptive females.
4. In most social mammals, females are more likely than males to remain and breed in their natal group, though in a few
   species females habitually disperse after reaching adolescence. Habitual male dispersal appears to reflect the need of males
   to locate receptive partners while habitual female dispersal appears to occur when males or male kin groups have
   unusually long breeding tenures and females need to leave their natal groups to find unrelated males.
5. Though males more commonly disperse than females, there does not appear to be a consistent sex difference in the
   distance moved, although this is difficult to measure.
6. Contrasts in dispersal affect opportunities for competition and cooperation between group members of both sexes and can
   have an important influence on social relationships between group members as well as on effective population size.

References

Alberts, S.C. and Altmann, J. (1995) Balancing costs and opportunities: dispersal in male baboons. *American Naturalist* **145**:279–306.

Alexander, R.D. (1991) The biology of the naked mole-rat. In: *The Evolution of Eusociality* (eds P.W. Sherman, J.U.M. Jarvis and R.D. Alexander). Princeton, NJ: Princeton University Press.

Altmann, J., *et al.* (1988) Determinants of reproductive success in savannah baboons, *Papio cynocephalus*. In: *Reproductive Success: Studies of Individual Variation in Contrasting Breeding Systems* (ed. T.H. Clutton-Brock). Chicago: University of Chicago Press, 403–418.

Amos, B., *et al.* (1993) Social structure of pilot whales revealed by analytical DNA profiling. *Science* **260**:670–672.

Baird, R.W. (2000) The killer whale: foraging specializations and group hunting. In: *Cetacean Societies: Field Studies of Dolphins and Whales* (eds J. Mann, R. Connor, P. Tyack and H. Whitehead). Chicago: University of Chicago Press, 127–153.

Balloux, F., *et al.* (1998) Breeding system and genetic variance in the monogamous, semi-social shrew, *Crocidura russula*. *Evolution* **52**:1230–1235.

Basset, P., *et al.* (2001) Testing demographic models of effective population size. *Proceedings of the Royal Society of London. Series B: Biological Sciences* **268**:311–317.

Bengtsson, B.O. (1978) Avoiding inbreeding: at what cost? *Journal of Theoretical Biology* **73**:439–444.

Berger, J. (1986) *Wild Horses of the Great Basin*. Chicago: University of Chicago Press.

Berger, J. (1987) Reproductive fates of dispersers in a harem dwelling ungulate: the wild horse. In: *Mammalian Dispersal Patterns: The Effects of Social Structure on Population Genetics* (eds B.D. Chepko-Sade and Z.T. Halpin). Chicago: University of Chicago Press, 41 54.

Berger, J. and Cunningham, C. (1987) Influence of familiarity on frequency of inbreeding in wild horses. *Evolution* **41**:229–231.

Braude, S. (2000) Dispersal and new colony formation in wild naked mole-rats: evidence against inbreeding as the systems of making. *Behaviour Ecology* **11**:7–12.

Brody, A.K. and Armitage, K.B. (1985) The effects of adult removal on dispersal of yearling yellow-bellied marmots. *Canadian Journal of Zoology* **63**:2560–2564.

Bulger, J. and Hamilton, W.J. (1988) Inbreeding and reproductive success in a natural chacma baboon, *Papio-cynocephalus ursinus*, population. *Animal Behaviour* **36**:574–578.

Bush, G.L., *et al.* (1977) Rapid speciation and chromosomal evolution in mammals. *Proceedings of the National Academy of Sciences of the United States of America* **74**:3942–3946.

Bygott, J.D., *et al.* (1979) Male lions in large coalitions gain reproductive advantages. *Nature* **282**:839–841.

Cheney, D.L. and Seyfarth, R.M. (1983) Nonrandom dispersal in free-ranging vervet monkeys: social and genetic consequences. *American Naturalist* **122**:392–412.

Cheney, D.L., *et al.* (1988) Reproductive success in vervet monkeys. In: *Reproductive Success: Studies of Individual Variation in Contrasting Breeding Systems* (ed. T.H. Clutton-Brock). Chicago: University of Chicago Press, 384–402.

Chepko-Sade, B.D. *et al.* (1987) The effects of dispersal and social structure on effective population size. In: *Mammalian Dispersal Patterns* (eds B.D. Chepko-Sade and Z.T. Halpin). Chicago: University of Chicago Press.

Chesser, R.K., *et al.* (1993) Effective sizes for subdivided populations. *Genetics* **135**:1221–1232.

Clarke, A.L., *et al.* (1997) Sex biases in avian dispersal: a reappraisal. *Oikos* **79**:429–438.

Clarke, P.M.R., *et al.* (2008) On the road again: competitive effects and condition-dependent dispersal in male baboons. *Animal Behaviour* **76**:55–63.

Clutton-Brock, T.H. (1989a) Female transfer and inbreeding avoidance in social mammals. *Nature* **337**:70–72.

Clutton-Brock, T.H. (1989b) Review lecture: Mammalian mating systems. *Proceedings of the Royal Society of London. Series B: Biological Sciences* **236**:339–372.

Clutton-Brock, T.H. (1991) *The Evolution of Parental Care*. Princeton, NJ: Princeton University Press.

Clutton-Brock, T.H. (2009) Structure and function in mammalian societies. *Philosophical Transactions of the Royal Society B: Biological Sciences* **364**:3229–3242.

Clutton-Brock, T.H. and Albon, S.D. (1989) *Red Deer in the Highlands*. Oxford: Blackwell Scientific Publications.

Clutton-Brock, T.H. and Lukas, D. (2011) The evolution of social philopatry and dispersal in female mammals. *Molecular Ecology* **21**:472–492.

Clutton-Brock, T.H. and Manser, M.B. (2015) Kalahari meerkats. In: *Cooperative Breeding: Studies of Ecology, Evolution and Behaviour* (eds W.D. Koenig and J.L. Dickinson). Cambridge: Cambridge University Press.

Clutton-Brock, T.H., *et al.* (1982) *Red Deer: The Behaviour and Ecology of Two Sexes*. Chicago: University of Chicago Press.

Clutton-Brock, T.H., *et al.* (2015) Density related changes in sexual selection in red deer. *Proceedings of the Royal Society of London. Series B: Biological Sciences* **264**:1509–1516.

Clutton-Brock, T.H., *et al.* (1998) Infanticide and expulsion of females in a cooperative mammal. *Proceedings of the Royal Society of London. Series B: Biological Sciences* **265**:2291–2295.

Clutton-Brock, T.H., *et al.* (2002a) Evolution and development of sex differences in cooperative behavior in meerkats. *Science* **297**:253–256.

Clutton-Brock, T.H., *et al.* (2002b) Sex differences in emigration and mortality affect optimal management of deer populations. *Nature* **415**:633–637.

Cockburn, A., *et al.* (1985) Inbreeding avoidance and male-biased natal dispersal in *Antechinus* spp. (Marsupialia: Dasyuridae). *Animal Behaviour* **33**:908–915.

Cooney, R. and Bennett, N.C. (2000) Inbreeding avoidance and reproductive skew in a cooperative mammal. *Proceedings* of the Royal Society of London. Series B: Biological Sciences **267**:801–806.

Creel, S. and Creel, N.M. (2002) *The African Wild Dog: Behavior, Ecology, and Conservation*. Princeton, NJ: Princeton University Press.

Crockett, C. and Sekulic, R. (1984) Infanticide in red howler monkeys (Alouatta seniculus). In: *Infanticide: Comparative and Evolutionary Perspectives* (eds G. Hausfater and S.B. Hrdy). New York: Aldine, 173–191.

Dawson, T. (1995) *Kangaroos: Biology of the Largest Marsupials*. Sydney: University of New South Wales Press.

Dice, L.R. and Howard, W.E. (1951) Distances of dispersal by prairie deer mice from birthplaces to breeding sites. Contribution 50, Laboratory of Vertebrate Zoology, University of Michigan.

Di Fiore, A. (2012) Genetic consequences of primate social organization. In: *The Evolution of Primate Societies* (eds J.C. Mitani, J. Call, P.M. Kappeler, R.A. Palombit and J.B. Silk). Chicago: University of Chicago Press, 269–292.

Dobson, F.S. (1979) An experimental study of dispersal in the California ground squirrel. *Ecology* **60**:1103–1109.

Dobson, F.S. (1982) Competition for mates and predominant juvenile male dispersal in mammals. *Animal Behaviour* **30**:1183–1192.

Dobson, F.S. (2013) The enduring question of sex-biased dispersal: Paul J Greenwood's (1980) seminal contribution. *Animal Behaviour* **85**:299–304.

Dobson, F.S. and Jones, W.T. (1985) Multiple causes of dispersal. *American Naturalist* **126**:855–858.

Dobson, F.S. and Kjelgaard, J.D. (1985) The influences of food resources on life history in Columbian ground squirrels. *Canadian Journal of Zoology* **63**:2105–2109.

Dobzhansky, T. and Wright, S. (1943) Genetics of natural populations X. Dispersion rates in *Drosophila pseudoobscura*. *Genetics* **28**:304–340.

Drickamer, L.C. and Vessey, S.H. (1973) Group changing in free-ranging male rhesus monkeys. *Primates* **14**:359–368.

Emlen, S.T. (1997) Predicting family dynamics in social vertebrates. In: *Behavioural Ecology: An Evolutionary Approach*, 4th edn (eds J.R. Krebs and N.B. Davies). Oxford: Blackwell Science, 228–253.

Engh, A.L., *et al.* (2002). Reproductive skew among males in a female-dominated mammalian society. *Behavioral Ecology* **13**:193–200.

Faulkes, C.G. and Bennett, N.C. (2007) African mole-rats: social and ecological diversity. In: *Rodent Societies: An Ecological and Evolutionary Perspective* (eds J.O. Wolff and P.W. Sherman). Chicago: University of Chicago Press, 427–437.

Frame, L.H., *et al.* (1979) Social organization of African wild dogs (Lycaon pictus) on the Serengeti plains, Tanzania 1967–1978. *Zeitschrift fur Tierpsychologie* **50**:225–249.

Franklin, W.L. (1983) Contrasting socioecologies of South America's wild camelids: the vicuña and the guanaco. In: *Advances in the Study of Mammalian Behavior* (eds J.F. Eisenberg

and D.G. Kleiman). American Society of Mammalogists Special Publication No. 7. Stillwater, OK: American Society of Mammalogists, 573–629.

Gaines, M.S. and Johnson, M.L. (1987) Phenotypic and genotypic mechanisms for dispersal in Microtus populations and the role of dispersal in population regulation. In: *Mammalian Dispersal Patterns: The Effects of Social Structure on Population Genetics* (eds B.D. Chepko-Sade and Z.T. Halpin). Chicago: University of Chicago Press, 162–179.

Gaines, M.S. and McClenaghan, L.R. (1980) Dispersal in small mammals. *Annual Review of Ecology and Systematics* **11**: 163–196.

Gandon, S. (1999) Kin competition, the cost of inbreeding and the evolution of dispersal. *Journal of Theoretical Biology* **200**:345–364.

Gandon, S. and Michelakis, Y. (2001) Multiple cause of the evolution of dispersal. In: *Dispersal* (eds J. Clobert, E. Danchin, A.A. Dhondt and J.D. Nichols). Oxford: Oxford University Press, 155–167.

Gardner, A. (2010) Sex-biased dispersal of adults mediates the evolution of altruism among juveniles. *Journal of Theoretical Biology* **262**:339–345.

Gilchrist, J.S. (2001) Reproduction and pup care in the communal breeding banded mongoose. PhD thesis, University of Cambridge, Cambridge.

Goodall, J. (1986) *The Chimpanzees of Gombe: Patterns of Behavior*. Cambridge, MA: Belknap Press.

Greenwood, P.J. (1980) Mating systems, philopatry and dispersal in birds and mammals. *Animal Behaviour* **28**:1140–1162.

Greenwood, P.J. (1983) Mating systems and the evolutionary consequences of dispersal. In: *The Ecology of Animal Movement* (eds I.R. Swingland and P.J. Greenwood,). Oxford: Oxford University Press, 116–131.

Greenwood, P.J. and Harvey, P.H. (1982) The natal and breeding dispersal of birds. *Annual Review of Ecology and Systematics* **13**:1–21.

Hanby, J.P. and Bygott, J.D. (1987) Emigration of subadult lions. *Animal Behaviour* **35**:161–169.

Harcourt, A.H. (1978) Strategy of emigration and transfer by primates, with particular reference to gorillas. *Zeitschrift fur Tierpsychologie* **48**:401–420.

Harcourt, A.H. and Stewart, K.J. (1981) Gorilla male relationships: can differences during immaturity lead to contrasting reproductive tactics in adulthood. *Animal Behaviour* **29**:206–210.

Harcourt, A.H., *et al.* (1976) Male emigration and female transfer in mountain gorilla. *Nature* **263**:226–227.

Heimlich-Boran, J.R. (1986) Cohesive relationships among Puget Sound killer whales. In: *Behavioral Biology of Killer Whales* (eds B.C. Kirkevold and J.S. Lockard). New York: Alan R. Liss, 251–284.

Heimlich-Boran, J.R. (1993) *Social organization of the short-finned pilot whale*, Globicephala macrorhynchus, *with special reference to the comparative social ecology of delphinids*. PhD thesis, University of Cambridge, Cambridge.

Henzi, S.P. and Lawes, M. (1987) Breeding season influxes and the behavior of adult male samango monkeys (*Cercopithecus mitis albogularis*). *Folia Primatologica* **48**:125–136.

Hoelzel, A.R. and Dover, G.A. (1991) Genetic differentiation between sympatric killer whale populations. *Heredity* **66**:191–195.

Holekamp, K.E. (1984) Natal dispersal in Belding's ground squirrels (*Spermophilus beldingi*). *Behavioral Ecology and Sociobiology* **16**:21–30.

Holekamp, K.E. (1986) Proximal causes of natal dispersal in Belding's ground squirrels (*Spermophilus beldingi*). *Ecological Monographs* **56**:365–391.

Holekamp, K.E. and Sherman, P.W. (1989) Why male ground squirrels disperse: a multilevel analysis explains why only males leave home. *American Scientist* **77**:232–239.

Holekamp, K.E. and Smale, L. (1998) Dispersal status influences hormones and behavior in the male spotted hyena. *Hormones and Behavior* **33**:205–216.

Hölzenbein, S. and Marchinton, R.C. (1992) Spatial integration of maturing white-tailed deer in the adult population. *Journal of Mammalogy* **73**:326–324.

Höner, O.P., et al. (2007) Female mate-choice drives the evolution of male-biased dispersal in a social mammal. *Nature* **448**:798–801.

Höner, O.P., et al. (2010) The fitness of dispersing spotted hyaena sons is influenced by maternal social status. *Nature Communications* **1**:60.

Hoogland, J.L. (1995) *The Black-tailed Prairie Dog: Social Life of a Burrowing Mammal*. Chicago: University of Chicago Press.

Hoogland, J.L. (2013) Prairie dogs disperse when all close kin have disappeared. *Science* **339**:1205–1207.

Idani, G. (1991) Social relationships between immigrant and resident bonobo (*Pan paniscus*) females at Wamba. *Folia Primatologica* **57**:83–95.

Isbell, L.A. and Van Vuren, D. (1996) Differential costs of locational and social dispersal and their consequences for female group-living primates. *Behaviour* **133**:1–36.

Johnson, C.N. (1986) Sex-biased philopatry and dispersal in mammals. *Oecologia* **69**:626–627.

Johnson, C.N. (1989) Dispersal and philopatry in the Macropodoids. In: *Kangaroos, Wallabies and Rat-Kangaroos* (eds G. Grigg, P. Jarman and I. Hume). Chipping Norton, NSW: Beatty and Sons, 593–601.

Johnstone, R.A. and Cant, M.A. (2008) Sex differences in dispersal and the evolution of helping and harming. *American Naturalist* **172**:318–330.

Jones, W.T. (1987) Dispersal patterns in kangaroo rats. In: *Mammalian Dispersal Patterns: The Effects of Social Structure on Population Genetics* (eds B.D. Chepko-Sade and Z.T. Halpin). Chicago: University of Chicago Press, 119–127.

Korstjens, A.H. and Schippers, E.P. (2003) Dispersal patterns among olive colobus in Taï National Park. *International Journal of Primatology* **24**:515–539.

Kuester, J. and Paul, A. (1999) Male migration in Barbary macaques (*Macaca sylvanus*) at Affenberg Salem. *International Journal of Primatology* **20**:85–106.

Kümmerli, R., et al. (2008) Limited dispersal, budding dispersal, and cooperation: an experimental study. *Evolution* **63**: 939–949.

Larsen, K.W. and Boutin, S. (1994) Movements, survival, and settlement of red squirrel (*Tamiasciurus hudsonicus*) offspring. *Ecology* **75**:214–223.

Lawson Handley, L.J. and Perrin, N. (2007) Advances in our understanding of mammalian sex-biased dispersal *Molecular Ecology* **16**:1559–1578.

Lee, P.C. and Kappeler, P.M. (2003) Socioecological correlates of phenotypic plasticity of primate life histories. In: *Primate Life Histories and Socioecology* (eds P.M. Kappeler and M.E. Pereira). Chicago: University of Chicago Press, 41–65.

Lewontin, R.C. (1974) *The Genetic Basis of Evolutionary Change*. New York: Columbia University Press.

Liberg, O. and von Schantz, T. (1985) Sex-biased philopatry and dispersal in birds and mammals: the Oedipus hypothesis. *American Naturalist* **126**:129–135.

Lidicker, W.Z. and Patton, J.L. (1987) Patterns of dispersal and genetic structure in populations of small rodents. In: *Mammalian Dispersal Patterns: The Effects of Social Structure on Population Genetics* (eds B.D. Chepko-Sade and Z.T. Halpin). Chicago: University of Chicago Press, 144–161.

Lukas, D. and Clutton-Brock, T.H. (2011) Group structure, kinship, inbreeding risk and habitual female dispersal in plural-breeding mammals. *Journal of Evolutionary Biology* **22**:1337–1343.

Lukas, D. and Clutton-Brock, T.H. (2012) Cooperative breeding and monogamy in mammalian societies. *Proceedings of the Royal Society of London. Series B: Biological Sciences* **279**:2151–2156.

Lukas, D. and Vigilant, L. (2005) Reply: Facts, faeces and setting standards for the study of MHC genes using noninvasive samples. *Molecular Ecology* **14**:1601–1602.

McCullough, D.R. (1985) Variables influencing food habits of white-tailed deer on the George Reserve. *Journal of Mammalogy* **66**:682–692.

McNutt, J.W. (1996) Sex biased dispersal in African wild dogs, *Lycaon pictus*. *Animal Behaviour* **52**:1067–1077.

Manzolillo, D.L. (1986) Factors affecting intertroop transfer by adult male *Papio anubis*. In: *Primate Ontogeny, Cognition, and Social Behaviour* (eds J.G. Else and P.C. Lee). Cambridge: Cambridge University Press, 371–380.

Mares, R. (2012) Extraterritorial prospecting and territory defence in cooperatively breeding meerkats. PhD thesis, University of Cambridge, Cambridge.

Mech, L.D. (1987) Age, season, distance, direction, and social aspects of wolf dispersal from a Minnesota pack. In: *Mammalian Dispersal Patterns: The Effects of Social Structure on Population Genetics* (eds B.D. Chepko-Sade and Z.T. Halpin). Chicago: University of Chicago Press, 55–74.

Moore, J. (1993) Inbreeding and outbreeding in primates: what's wrong with the dispersing sex? In: *The Natural History of Inbreeding and Outbreeding: Theoretical and Empirical Perspectives* (ed. N.W. Thornhill). Chicago: University of Chicago Press, 392–426.

Moore, J. and Ali, R. (1984) Are dispersal and inbreeding avoidance related? *Animal Behaviour* **32**:94–112.

Moss, C.S. and Poole, J.H. (1983) Relationships and social structure of African elephants. In: *Primate Social Relationships: An Integrated Approach* (ed. R.A. Hinde). Oxford: Blackwell Scientific Publications, 315–325.

Nagy, M., *et al.* (2007) Female-biased dispersal and patrilocal kin groups in a mammal with resource-defence polygyny. *Proceedings of the Royal Society of London. Series B: Biological Sciences* **274**:3019–3025.

Nelson, M.E. and Mech, L.D. (1987) Demes within a Northeastern Minnesota deer population. In: *Mammalian Dispersal Patterns: The Effects of Social Structure on Population Genetics* (eds B.D. Chepko-Sade and Z.T. Halpin). Chicago: University of Chicago Press, 27–40.

Nunes, S. and Holekamp, K.E. (1996) Mass and fat influence the timing of natal dispersal in Belding's ground squirrels. *Journal of Mammalogy* **77**:807–817.

Nunney, L. (1999) The effective size of a hierarchically structured population. *Evolution* **53**:1–10.

Oliver, A. (1986) Social organisation and dispersal in the red kangaroo. PhD thesis, Murdoch University, Perth.

O'Riain, M.J., *et al.* (2000) Reproductive suppression and inbreeding avoidance in wild populations of cooperatively breeding meerkats (*Suricata suricatta*). *Behavioral Ecology and Sociobiology* **48**:471–477.

Packer, C. (1979) Inter-troop transfer and inbreeding avoidance in *Papio Anubis*. *Animal Behaviour* **27**:1–36.

Packer, C. and Pusey, A.E. (1982) Cooperation and competition within coalitions of male lions: kin selection or game theory? *Nature* **296**:740–742.

Packer, C., *et al.* (1988) Reproductive success in lions. In: *Reproductive Success: Studies of Individual Variation in Contrasting Breeding Systems* (ed. T.H. Clutton-Brock). Chicago: University of Chicago Press, 363–383.

Perrin, N. and Mazalov, V. (1999) Dispersal and inbreeding avoidance. *American Naturalist* **154**:282–292.

Perrin, N. and Mazalov, V. (2000) Local competition, inbreeding, and the evolution of sex-biased dispersal. *American Naturalist* **155**:116–127.

Piper, W.H. and Slater, G. (1993) Polyandry and incest avoidance in the cooperative stripe-backed wren of Venezuela. *Behaviour* **124**:227–247.

Poole, J.H. (1994) Sex differences in the behaviour of African elephants. In: *The Differences Between the Sexes* (eds R.V. Short and E. Balaban). Cambridge: Cambridge University Press, 331–346.

Pope, T.R. (2000) The evolution of male philopatry in neotropical monkeys. In: *Primate Males: Causes and Consequences*

of Variation in Group Composition (ed. P.M. Kappeler). Cambridge: Cambridge University Press, 219–235.

Pusey, A.E. (1987) Sex-biased dispersal and inbreeding avoidance in birds and mammals. *Trends in Ecology and Evolution* **2**:295–299.

Pusey, A.E. and Packer, C.R. (1987a) Dispersal and philopatry. In: *Primate Societies* (eds B.B. Smuts, D.L. Cheney, R.M. Seyfarth, R.W. Wrangham and T.T. Struhsaker). Chicago: University of Chicago Press, 250–266.

Pusey, A.E. and Packer, C. (1987b) The evolution of sex-biased dispersal in lions. *Behaviour* **101**:275–310.

Pusey, A.E. and Wolf, M. (1996) Inbreeding avoidance in animals. *Trends in Ecology and Evolution* **11**:201–206.

Queller, D.C. (1992) Does population viscosity promote kin selection? *Trends in Ecology and Evolution* **7**:322–324.

Radespiel, L., *et al.* (2003) Patterns and dynamics of sex-biased dispersal in a nocturnal primate, the grey mouse lemur, *Microcebus murinus*. *Animal Behaviour* **65**:709–719.

Raihani, N.J. and Clutton-Brock, T.H. (2010) Higher reproductive skew among birds than mammals in cooperatively breeding species. *Biology Letters* **6**:630–632.

Rajpurohit, L.S. and Sommer, V. (1993) Juvenile male emigration from natal one-male troops in Hanuman langurs. In: *Juvenile Primates: Life History, Development and Behavior, with a New Foreword* (eds M.E. Pereira and L.A. Fairbanks). Oxford: Oxford University Press, 86–103.

Ralls, K., *et al.* (1980) Differential mortality by sex and age in mammals, with specific reference to the sperm whale. *Report of the International Whaling Commission Special Issue* **2**: 223–243.

Rassmann, K., *et al.* (1994) Low genetic variability in a natural alpine marmot population (*Marmota marmota*, Sciuridae) revealed by DNA fingerprinting. *Molecular Ecology* **3**: 347–353.

Reeve, H.K., *et al.* (1990) DNA fingerprinting reveals high levels of inbreeding in colonies of the eusocial naked mole-rat. *Proceedings of the National Academy of Sciences of the United States of America* **87**:2496–2500.

Reyer, H.U., *et al.* (1986) Avian helpers at the nest: are they psychologically castrated? *Ethology* **71**:216–228.

Richard, A.F., *et al.* (2002) Life in the slow lane? Demography and life histories of male and female sifaka (*Propithecus verreauxi verreauxi*). *Journal of Zoology* **256**:421–436.

Robbins, M.M. (1996) Male–male interactions in heterosexual and all-male wild mountain gorilla groups. *Ethology* **102**:942–965.

Rogers, L.L. (1987) Factors influencing dispersal in the black bear. In: *Mammalian Dispersal Patterns: The Effects of Social Structure on Population Genetics* (eds B.D. Chepko-Sade and Z.T. Halpin). Chicago: University of Chicago Press, 75–84.

Rood, J.P. (1987) Dispersal and intergroup transfer in the dwarf mongoose. In: *Mammalian Dispersal Patterns: The Effects of Social Structure on Population Genetics* (eds B.D. Chepko-Sade and Z.T. Halpin). Chicago: University of Chicago Press, 85–103.

Ruckstuhl, K.E. and Neuhaus, P. (2001) Behavioral synchrony in ibex groups: effects of age, sex and habitat. *Behaviour* **138**:1033–1046.

Rutberg, A.T. and Keiper, R.R. (1993) Proximate causes of natal dispersal in feral ponies: some sex differences. *Animal Behaviour* **46**:969–975.

Schoof, V.A.M., *et al.* (2009) What traits promote male parallel dispersal in primates? *Behaviour* **146**:701–726.

Schradin, C., *et al.* (2012) Staying put or leaving home: endocrine, neuroendocrine and behavioral consequences in male African striped mice. *Hormones and Behavior* **63**:136–143.

Sherman, P.W. (1977) Nepotism and the evolution of alarm calls. *Science* **197**:1246–1253.

Sherman, P.W., *et al.* (eds) (1991) *The Biology of the Naked Mole-Rat.* Princeton, NJ: Princeton University Press.

Shields, W.M. (1982) *Philopatry, Inbreeding and the Evolution of Sex.* Albany, NY: State University of New York Press.

Shields, W.M. (1987) Dispersal and mating systems: investigating their causal connections. In: *Mammalian Dispersal Patterns: The Effects of Social Structure on Population Genetics* (eds B.D. Chepko-Sade and Z.T. Halpin). Chicago: University of Chicago Press, 3–24.

Sillero-Zubiri, C., *et al.* (1996) Male philopatry, extra pack copulations and inbreeding avoidance in Ethiopian wolves (*Canis simensis*). *Behavioral Ecology and Sociobiology* **38**: 331–340.

Smale, L., *et al.* (1997) Sexually dimorphic dispersal in mammals: patterns, causes and consequences. In: *Advances in the Study of Behavior*, Vol. **26** (eds C. Snowdon, P.J.B. Slater, M. Milinski and J.S. Rosenblatt). San Diego, CA: Academic Press, 181–250.

Smith, A.T. (1987) Population structure of pikas: dispersal versus philopatry. In: *Mammalian Dispersal Patterns: The Effects of Social Structure on Population Genetics* (eds B.D. Chepko-Sade and Z.T. Halpin). Chicago: Chicago University Press, 128–142.

Smith, R.H. (1979) On selection for inbreeding in polygynous animals. *Heredity* **43**:205–211.

Solmsen, N., *et al.* (2011) Highly asymmetric fine-scale genetic structure between sexes of African striped mice and indication for condition dependent alternative male dispersal factors. *Molecular Ecology* **20**:1624–1634.

Staines, B.W. and Crisp, J.M. (1978) Observations on food quality in Scottish red deer (*Cervus elaphus*) as determined by chemical analysis of rumen contents. *Journal of Zoology* **185**:253–259.

Stearns, S.C. (1992) *The Evolution of Life Histories.* Oxford: Oxford University Press.

Stenseth, N.C. (1983) Causes and consequences of dispersal in small mammals. In: *The Ecology of Animal Movement* (eds I.R. Swingland and P.J. Greenwood,). Oxford: Oxford University Press, 63–101.

Stenseth, N.C. and Lidicker, W.Z. (eds) (1992) *Animal Dispersal: Small Mammals as a Model.* New York: Chapman & Hall.

Struhsaker, T.T. (1975) *The Red Colobus Monkey.* Chicago: University of Chicago Press.

Sussman, R.W. (1992) Male life history and intergroup mobility among ringtailed lemurs (*Lemur catta*). *International Journal of Primatology* **13**:395–413.

Taylor, P.D. (1992) Altruism in viscous populations: an inclusive fitness model. *Evolutionary Ecology* **6**:352–356.

van Noordwijk, A.J. and van Schaik, C.P. (1985) Male migration and rank acquisition in wild long-tailed macaques (*Macaca fascicularis*). *Animal Behaviour* **33**:849–861.

Waser, P.M. (1996) Patterns and consequences of dispersal in gregarious carnivores. In: *Carnivore Behavior, Ecology and Evolution* (ed. J.L. Gittleman). Ithaca, NY: Cornell University Press, 267–295.

Waser, P.M. and Jones, W.T. (1983) Natal philopatry among solitary mammals. *Quarterly Review of Biology* **58**:355–390.

Waser, P.M., *et al.* (1986) When should animals tolerate inbreeding? *American Naturalist* **128**:529–537.

Waser, P.M., *et al.* (1994) Death and disappearance: estimating mortality risks associated with philopatry and dispersal. *Behavioral Ecology* **5**:135–141.

West, S.A., *et al.* (2002) Cooperation and competition between relatives. *Science* **296**:72–75.

White, F.J. (1996) Comparative socio-ecology of Pan paniscus. In: *Great Ape Societies* (eds W.C. McGrew, L.F. Marchant and T. Nishida). Cambridge: Cambridge University Press, 29–41.

Whitehead, H. and Weilgart, L. (2000) The sperm whale: social females and roving males. In: *Cetacean Societies: Field Studies of Dolphins and Whales* (eds J. Mann, R. Connor, P. Tyack and H. Whitehead). Chicago: University of Chicago Press, 154–172.

Whitworth, M.R. and Southwick, C.H. (1984) Sex differences in the ontogeny of social behavior in pikas: possible relationships to dispersal and territoriality. *Behavioral Ecology and Sociobiology* **15**:175–182.

Wikberg, E.C., *et al.* (2014) The effect of male parallel dispersal on the kin composition of groups in white-faced capuchins. *Animal Behaviour* **96**:9–17.

Wilson, M.L. (2013) Chimpanzees, warfare, and the invention of peace. In: *War, Peace, and Human Nature: The Convergence of Evolutionary and Cultural Views* (ed. D.P. Fry). Oxford: Oxford University Press, 361–388.

Wolff, J.O. (1992) Parents suppress reproduction and stimulate dispersal in opposite sex juvenile white-footed mice. *Nature* **359**:409–410.

Wolff, J.O. (1993) What is the role of adults in mammalian juvenile dispersal? *Oikos* **68**:173–176.

Woodroffe, R.W. (1993) Factors affecting reproductive success in the European badger, Meles meles. PhD thesis, University of Oxford, Oxford.

Wrangham, R.W. (1980) An ecological model of female-bonded primate groups. *Behaviour* **75**:262–300.

Wrangham, R.W., *et al.* (2006) Comparative rates of violence in chimpanzees and humans. *Primates* **47**:14–26.

Wright, S. (1932) The role of mutation, inbreeding, crossbreeding and selection in evolution. *Proceedings of the Sixth International Genetics Congress* **1**:356–366.

Wright, S. (1978) *Variability Within and Among Natural Populations*. Chicago: University of Chicago Press.

Yamamoto, M.E., *et al.* (2014) Male and female breeding strategies in a cooperative primate. *Behavioural Processes* **109A**: 27–33.

Young, A.J. (2003) Subordinate tactics in cooperative meerkats: breeding, helping and dispersal. PhD thesis, University of Cambridge, Cambridge.

Zahavi, A. and Zahavi, A. (1997) *The Handicap Principle: A Missing Piece of Darwin's Puzzle*. New York: Oxford University Press.

Zhao, Q.K. (1996) Male–infant–male interactions in Tibetan macaques. *Primates* **37**:135–143.

CHAPTER 13

Reproductive competition among males

13.1 Introduction

Dark from wallowing in a peat bog, a red deer stag stands in the middle of the group of hinds that he is guarding. Across the valley, a second stag appears, heading down to the valley bottom in the direction of the first stag. The resident roars at him, his breath cloudy in the cold air. The approaching stag replies and, in the course of the next quarter hour, the two trade bouts of roars as the new arrival gets steadily closer. When he gets within fifty yards of the hinds, the defending stag runs out to meet him, still roaring. The two stags then move into a stiff parallel walk at right angles to new arrival's direction of approach, ten feet apart and squinting sideways at each other. When this takes them away from the harem they face outwards, roar repeatedly and then turn and walk stiffly back over the same path. As the parallel walk continues, the enthusiasm of the new stag begins to wane and its rate of roaring decreases. Its path eventually diverges and it walks away, still roaring, while the first stag returns to his hinds. Both stags have benefited from the encounter. The intruder, because he avoided a potentially damaging fight that he judged he was unlikely to win; the resident, because he successfully convinced the intruder to withdraw without risking the costs of a fight.

In many species, intense reproductive competition between males is associated with the development of elaborate weapons or ornaments. Darwin was fascinated by the evolution of these 'secondary sexual characters', though he lacked systematic evidence of the benefits and costs of fighting to males or of the extent of female mating preferences. He was particularly interested in the secondary sexual characters of primates: in the discussion on sexual selection in *The Descent of Man* he wrote,

> no case interested and perplexed me so much as the brightly-coloured hinder ends and adjoining parts of certain monkeys.

As these parts are more brightly coloured in one sex than the other, and as they become more brilliant during the season of love, I concluded that the colours has been gained as a sexual attraction. I was well aware that I thus laid myself open to ridicule; though in fact it is not more surprising that a monkey should display his bright-red hinder end than that a peacock should display his magnificent tail.

Darwin (1876)

Since Darwin's era, we know far more about the causes and consequences of reproductive competition between males. This chapter examines competitive strategies in males and describes the tactics that they use to maximise their mating success while minimising the risks associated with competition to their survival and longevity. Section 13.2 reviews evidence of the benefits and costs of fighting while section 13.3 examines the tactics used by males to assess rivals and attract females and section 13.4 explores adaptive variation in fighting behaviour. Subsequently, sections 13.5 and 13.6 examine the benefits and costs of mate guarding and adaptive variation in guarding behaviour while section 13.7 examines alternative male mating tactics and section 13.8 explores research on sperm competition and its effects. Finally, section 13.9 reviews some of the consequences of reproductive competition between males.

13.2 The benefits and costs of fighting

In many mammalian societies, fighting success plays an important role in determining the acquisition of breeding territories and access to females and so has important effects on male fitness. Displacing rivals is often sufficient to ensure access to females and the additional benefits of damaging or killing them are likely to be small, with the result that beaten rivals are rarely pursued for long.

Mammal Societies, First Edition. Tim Clutton-Brock.
© 2016 Tim Clutton-Brock. Published 2016 by John Wiley & Sons, Ltd.

Figure 13.1 Contrasting fighting methods in male mammals. In (a) carnivores and (b) primates, male fights involve biting and males commonly have larger canine teeth than females. *Sources*: (a) © Jonathan & Angela Scott/AWL Images/Getty Images; (b) © Clay Wilton.

Fighting techniques vary widely between species and can involve biting, scratching, kicking, pushing, ramming and stabbing (Figures 13.1, 13.2, 13.3 and 13.4) and the nature of male weaponry reflects these differences (Caro *et al.* 2003; Emlen 2008). As a broad generalisation, males of smaller species often fight by biting or stabbing, while those of larger species commonly engage in pushing contests, where relative body mass often plays an important role in determining the outcome of fights. Where competing males attempt to bite each other, they frequently have larger canine teeth than females (Harvey *et al.* 1978): one of the most extreme examples is found in Chinese water deer, where males compete to bite each other, sometimes killing rivals with bites to the neck or chest (see Chapter 18), and have 6–8 cm long sabre-like canines which are absent in females.

Figure 13.2 In some mammals, fights involve a combination of biting and pushing: (a) Przewalski's stallions fighting; (b) southern elephant seal bulls fighting. *Sources*: (a) © Claudia Feh; (b) © Mike Fedak.

Among ungulates which have evolved horns or antlers, males of smaller monogamous or territorial species typically have relatively simple weapons while more polygynous species have more complex horns or antlers that block dangerous thrusts and lock with those of opponents to provide a secure grip in pushing contests (Caro *et al*. 2003; Emlen *et al*. 2012). These contrasts may reflect differences in the frequency of fighting, which is probably higher in species where single males can defend access to large numbers of females than in those where males defend access to one or two females. Among polygynous ungulates, horn structure has diverged in spectacular fashion and is adapted to protecting the face from opponents, to stabbing through their guard and to

(a)   **(b)**

Figure 13.3 In giraffes, males fight by bludgeoning each other with their heads (a), while male grey kangaroos use their legs to deliver devastating blows (b). *Sources*: (a) © MogensTrolle/Getty Images; (b) © Gunter Ziesler/Photolibrary/Getty Images.

gripping them during pushing contests (Figure 13.5). The diversity of horn structure probably reflects the fact that the benefit of differences in structure depend principally on the traits of other males rather than on ecological factors, with the result that the evolution of horns has followed divergent pathways in different taxonomic groups and there is little or no association between horn structure and ecological parameters.

Although winning fights has important benefits to males, fighting also has substantial costs. Fighting males are usually unable to guard females effectively at the same time and often lose the females they are defending to other males when they are engaged in fights. In addition, fighting is often associated with substantial risks of injury, including the breakage of antlers or horns, damage to eyes and strained or broken legs (Figure 13.6). Few studies have attempted to quantify the risks associated with fighting but long-term studies of red deer suggest that they are often high. On the Isle of Rum, mature males are likely to be involved in at least five fights in the course of a breeding season. A male has approximately a 1 in 20 chance of temporary injury in each fight and a 1 in 100 chance of permanent injury. While the chances that a male will be injured in any one fight are relatively low, they accumulate over the breeding lifespan of individuals and calculations suggest that most male red deer are likely to suffer temporary injury at some point in their breeding lifespan and 20–30% will

suffer permanent injury (Clutton-Brock *et al.* 1979). Studies of the hunting tactics of large predators show that injured males are often killed, so that the fitness costs even of temporary injuries may be very high.

Competition for access to mates has other, less obvious costs. Both temporal changes and individual differences in male competition, aggression and reproductive activity are often associated with increases in testosterone levels (Clutton-Brock *et al.* 1979; Mooradian *et al.* 1987; Moore *et al.* 1998; Pelletier *et al.* 2003; Müller *et al.* 2012). For example, testosterone levels in male chimpanzees rise at times when parous females have maximal swellings and male aggression is common (Müller and Wrangham 2004). Increased testosterone levels commonly raise metabolic rate and may increase growth and competitive success (Gittleman and Thompson 1988; Schradin *et al.* 2009) but they also suppress immune responses and increase parasite load (Shuurs and Verheul 1990; Klein and Nelson 1998) with important effects on male survival. For example in Soay sheep, where individual differences in the load of intestinal parasites affect survival (Gulland *et al.* 1993; Wilson *et al.* 2004), temporary castration of juvenile males reduces overwinter mortality and permanent castration of males as lambs greatly extends their lifespans (Jewell 1997) (see Chapter 10). The effects of testosterone may also be partly responsible for sex differences in parasite load (and associated mortality) in adults, and sex differences in rates of parasitism

Figure 13.4 In some ungulates, including musk oxen (a), males run at each other and use their weight to dislodge smaller rivals, while in others, like red deer (b), males lock horns or antlers and engage in pushing contests, often attempting to stab their rivals. *Sources*: (a) © Paul Oomen/Photographer's Choice/Getty Images; (b) © Tim Clutton-Brock.

increase in sexually dimorphic species (Moore and Wilson 2002) (Figure 13.7).

The development and maintenance of male weapons presumably also imposes energetic costs on males, though these are difficult to measure and it is seldom possible to estimate their consequences for male survival or fitness (Clutton-Brock 1982; Emlen *et al.* 2012). In addition, selection for large body size in males commonly leads to the evolution of sex differences in juvenile growth rates, which are frequently associated with

Figure 13.5 Horn shape and structure vary widely among extant and extinct ruminants. Selection on horn structure may be strongly affected by the traits of rivals and constraints imposed by the effects of environmental factors may be relatively weak, leading to a proliferation of structures. Diversity in horn and antler structure and size in extinct and fossil herbivores: (*a*) Protoceratidae (Tylopoda). 1, *Kyptoceras*∗; 2, *Synthetoceras*∗; 3, *Protoceras*∗; 4, *Paratoceras*∗; 5, *Syndyoceras*∗. (*b*) Giraffidae. 1, *Sivatherium*∗; 2, *Giraffokeryx*∗; 3, *Canthumeryx*∗; 4, *Bramatherium*∗; 5, *Prolibytherium*∗; 6, *Samotherium*∗; 7, *Climacoceras*∗. (*c*) Antilocapridae. 1, *Merriamoceros*∗; 2, *Tetrameryx*∗; 3, *Hexameryx simpsoni*∗; 4, *Paramoceros*∗; 5, *Paracosoryx*∗; 6, *Osbornoceros osborni*∗; 7, *Antilocapra americana*; 8, *Ilingoceros*∗; 9, *Plioceros*∗; 10, *Ramoceros*∗. (*d*) Bovidae. 1, Spanish ibex (*Capra pyrenaica*); 2, dik-dik (*Madoqua kirkii*); 3, Grant's gazelle (*Gazella granti*); 4, kudu (*Tragelaphus strepsiceros*); 5, bighorn sheep (*Ovis canadensis*); 6, waterbuck (*Kobus ellipsiprymnus*); 7, impala (*Aepyceros melampus*); 8, long-horned African buffalo (*Pelorovis antiquus*)∗; 9, Asiatic ibex (*Capra ibex*); 10, chowsingha (*Tetracerus quadricornis*); 11, markhor (*Capra falconeri*). Asterisks denote extinct species. *Source*: From Emlen, 2008). Reproduced with permission from Annual Reviews.

reduced survival in times of resource shortage (see Chapter 18). Finally, heritable traits that increase the fitness of males may have costs to the fitness of females, with the result that daughters fathered by successful males may have reduced fitness (Robinson *et al.* 2006; Foerster *et al.* 2007).

Although in many mammals direct competition and fighting success are important in determining the breeding success of males, the longevity of males and the duration of their access to females also play an important role in determining their reproductive success (see section 13.5). The capacity of males to

Figure 13.6 In many polygynous societies, fighting injuries in males are common and lead to increased rates of mortality (see Chapter 18). A mature red deer stag with a badly injured eye. *Source*: © Tim Clutton-Brock.

attract females may also be important in species where females can control the identity of their mating partners. For example, in spotted hyenas, where the unusual anatomy of the female reproductive tract allows females to control the identity of mating partners, male competitive success and social rank have little effect on breeding success (Engh *et al.* 2002).

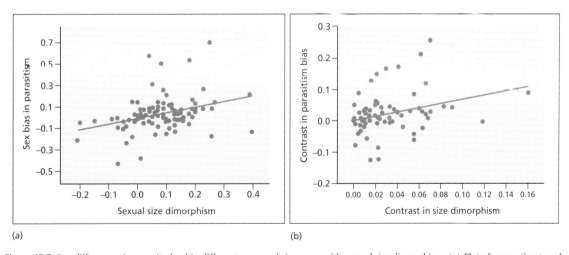

(a)   (b)

Figure 13.7 Sex differences in parasite load in different mammals increase with sexual size dimorphism. (a) Plot of raw estimates of sex bias in parasitism rate (male prevalence – female prevalence) in relation to sexual size dimorphism (logarithm of male mass/ female mass). (b) Plot of independent contrast scores: the lines show the least-squares regression lines; note that the intercept is forced through the origin. *Source*: From Moore and Wilson, 2002. Reproduced with permission from AAAS.

13.3 Assessment and the evolution of male displays

Visual displays

Where the costs of engaging in escalated fights are substantial, individuals would often be expected to avoid fighting (Maynard Smith 1974). Rather than engaging rivals immediately, defending males should threaten opponents in order to convince them to withdraw, while challengers should usually avoid escalated contests that they have little chance of winning (Cant and Johnstone 2009), though there may also be some cases where they should ignore risks or adopt unexpected tactics (Howard 1998; Morrell *et al.* 2005).

Selection on resident males to deter challengers and on challengers to assess residents has led to the evolution of reciprocal displays in which competing males advertise their own size, strength and motivation and assess that of their rivals. Competing males frequently use visual displays, including facial expressions, emphatic postures or gaits, piloerection, penile displays, branch shaking and vegetation thrashing. For example, before male red deer engage in fights, they commonly roar at each other repeatedly and may then perform a parallel walk, in which they eye each other, grind their teeth and roar intermittently (Figure 13.8) (Clutton-Brock and Albon 1979), sometimes breaking off to thrash the vegetation (Poole 1987). In some species, displaying males also pick up vegetation which they appear to use as additional ornaments (Figure 13.9).

In a few polygynous mammals, males have developed elaborate ornaments whose development is correlated with their age, rank or condition. For example, in mandrills, male coloration increases when they gain alpha status and they rapidly become duller if they are beaten in fights (Figure 13.10) and females are attracted to brightly coloured males (Setchell 2005). Similarly, the manes of fully grown male lions signal their status: manes lengthen and darken as individuals mature and

(a) (b) (c) (d)

Figure 13.8 In red deer, males typically engage in competitive displays before fighting: (a) a male approaches a rival stag; (b) stags roaring; (c) parallel walking; (d) fighting. *Sources*: (a, c, d) © Tim Clutton-Brock; (b) © Martyn Baker.

Figure 13.9 Displaying males often pick up vegetation and incorporate it in their displays: (a) fallow deer and (b) river dolphins. *Sources*: (a) © Tim Clutton-Brock; (b) © Projeto Boto.

individual differences in mane size and colouring are correlated with condition and with testosterone levels (West and Packer 2002). Male ornaments are commonly involved in displays directed at potential rivals, as well as at females. For example, experiments with model lions decked out with manes of different length and colour show that males encountering groups of strange males are less likely to approach individuals with long or dark manes than those with short or light ones while the reverse is true for females (West and Packer 2002).

Variation in male colouring may have similar effects in other species. In several lek-breeding antelopes where

(a)

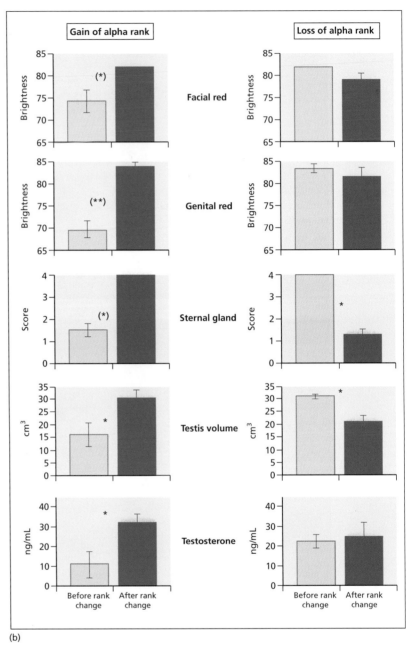

(b)

Figure 13.10 In mandrills, males only acquire full facial coloration when they achieve alpha status and rapidly become duller if they are beaten by rivals: (a) alpha male showing full facial coloration; (b) histograms show changes in coloration, the size of sternal gland and testes volume and testosterone levels in four males that gained alpha status and four males that lost it. *Source*: From Setchell and Dixson (2001). Reproduced with permission of Elsevier. *Photo source*: © Joanna M. Setchell, Centre International de Recherches Medicales de Franceville.

males defend mating territories, dominant males are either darker overall or have more prominent dark markings than subordinates (see Chapter 18) while in mountain sheep, darker rams are dominant to lighter ones and are more likely to guard ewes instead of mating promiscuously (Loehr *et al.* 2008).

Visual displays frequently emphasise male weaponry, such as canine teeth, horns or antlers and, within species, the relative size of male weapons commonly increases with body size (Huxley 1931; Preston *et al.* 2003a; Emlen 2008). As a result, it is often suggested that selection has modified the presence, size or shape of male weapons to enhance their function as signals of fighting ability or genetic quality (Geist 1971; Berglund *et al.* 1996; Bartos and Bahbouh 2006; Emlen 2008; Bergeron *et al.* 2010) or that horns and antlers are costly handicaps that display the quality of males (Zahavi 1975, 1977). However, an alternative possibility is that canine teeth, horns and antlers are functional weapons whose size and shape are principally or exclusively affected by their use in fights rather than their use in displays (Clutton-Brock 1982). Where their relative size is correlated with variation in fighting ability, it is possible that selection for signalling has modified their size or shape, though as fighting ability changes during the course of the breeding season as individuals become exhausted, it would seem more likely that rivals should base their assessment of each other on traits that change over shorter time-spans, like contrasts in display rate, coloration or condition. Empirical studies provide substantial evidence that males pay close attention to cues of this kind: for example, studies of red and fallow deer suggest that males pay close attention to the displays of rivals (Clutton-Brock and Albon 1979; Vannoni and McElligott 2009; Pitcher *et al.* 2014).

As many studies have argued that horns and antlers have evolved to signal the power or quality of their owners, it is worth reflecting on the evidence that would be needed to show that this was the case. Correlations between relative horn or antler size and male status or breeding success are not sufficient since they may arise because the relative size of weapons is associated with individual differences in body size, condition or hormonal status, which in turn may affect fighting ability. Convincing evidence that rivals commonly rely on these traits to assess each other consequently requires a demonstration that experimental increases or changes in weapon size or shape affect assessment. Experiments of this kind face substantial obstacles since it is necessary to maintain the shape and proportion of weapons while modifying their size and although some studies of deer have sometimes attempted to manipulate antler size (Lincoln 1972), they have not yet been successful.

Evidence that would suggest that teeth, horns and antlers have evolved as signals of quality rather than as effective weapons would be a demonstration that they are not used in fights or that their structure precludes their evolution as effective weapons. However, both detailed observations of fighting behaviour and analyses of their mechanical structure suggest that even the most (apparently) unlikely candidates, like the teeth in Chinese water deer, the palmate antlers of fallow deer and the twisted horns of blackbuck, are regularly used in escalated fights and are well adapted to their role as defensive or offensive weapons (Barrette 1977; Kitchener 1985, 1988; Alvarez 1993). In addition, studies of horned beetles (where experiments are more feasible) have found little evidence that females select mating partners on the basis of differences in horn size, structure or symmetry and suggest that horns are functional weapons whose size often affects their effectiveness in conflicts (Brown *et al.* 1985; Hunt and Simmons 1997; Lailvaux *et al.* 2005).

Vocal displays

In many mammals where males defend territories or groups of females, they give loud repetitive calls that signal their presence to potential rivals and can also attract females. In some species, including lions (Grinnell and McComb 2001), bison (Berger and Cunningham 1991; Wyman *et al.* 2008), howler monkeys (Sekulic 1982), hammer-headed bats (Bradbury 1977) and red deer (Reby and McComb 2003), males use low-frequency calls, while in others they use relatively high calls, as in gibbons (Mitani 1984), cetaceans (Payne and McVay 1971; Croll *et al.* 2002) and some rodents (Holy and Guo 2005).

Across species, the fundamental frequency of loud calls is negatively correlated with the length of vocal tracts as well as with body size (Fletcher 2010). Interspecific differences in the form and structure of vocal displays are often associated with variation in social structure. For example, in howler monkeys (where males give loud repetitive roars that are produced by specialised hyoid bones), males of species living in multi-male groups, like mantled howlers, generally roar in chorus and have relatively small hyoid bones and give higher-pitched roars than those of species that live in single-male groups, like ursine howlers, which usually roar alone, have much larger hyoid bones and give deeper roars (Dunn *et al.* 2015) (Figure 13.11).

If loud calls are to be effective in discouraging rivals from persisting with challenges, they need to be reliable indicators of a male's fighting ability (see Chapter 1). Several studies suggest that this is the case. The relative frequency with which males call and the amplitude, fundamental frequency or structure of their calls can reflect their maturity, condition, motivation and body size. For example, the amplitude of the bellows of individual male bison during the breeding season is correlated with their age, size, condition and motivation (Reby *et al.* 1999; Wyman *et al.* 2008). Similarly, in baboons, the 'wahoo' calls of males vary with the male's rank: high-ranking males give 'wa' calls with higher fundamental frequencies and longer 'hoo' syllables than lower-ranking ones and, as males fall in rank, the fundamental frequency of their 'wahoos' declines and the 'hoo' syllable becomes shorter (Fischer *et al.* 2004). In other species, contrasts in the resonant frequencies (formants) of male loud calls are more closely correlated with individual differences in male size or rank than variation in the fundamental frequency of calls (Box 13.1 and Figure 13.12).

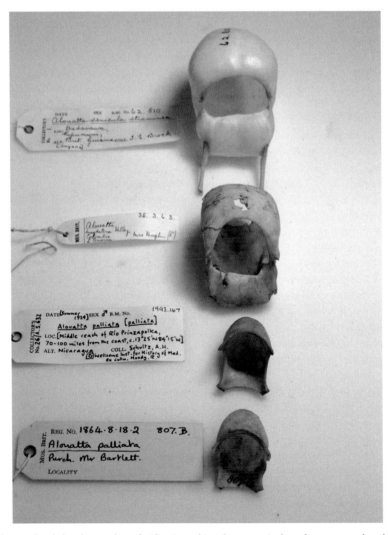

Figure 13.11 Hyoid bones of male howler monkeys that live in multi-male versus single male groups: males of species that usually live in unimale groups (top two specimens) have larger hyoid bones and give lower-pitched roars than males of species that live in multi-male groups (lower two specimens). Top to bottom: *Alouatta macconnellii*, *A. seniculus*, *A. palliata* ×2. *Source*: © Jacob Dunn.

Box 13.1 Call structure and body size in red deer

Studies of several deer species have explored relationships between the structure of loud calls and variation in body size. While interspecific differences in the fundamental frequency of calls are usually associated with body size, intraspecific differences are not necessarily closely related to variation in size or social rank (Reby and McComb 2003; Pfefferle and Fischer 2006; Vannoni and McElligott 2008) and, in some cases, higher-ranking or more successful individuals give higher-pitched loud calls than less successful individuals (Fischer *et al.* 2004; Garcia *et al.* 2013). In contrast, variation in the resonant frequencies of calls (*formants*) are often correlated with individual differences in the length of the vocal tract and so reflect intraspecific variation in body size more closely than fundamental frequency (Reby and McComb 2003; Vannoni and McElligott 2008) (see Figure 13.12). For example, when red deer stags roar, they retract their larynx so that the formant frequencies reflect the entire length of the vocal tract and give an accurate indicator of body size and weight, which are in turn related to measures of reproductive success (Fitch and Reby 2001; Reby and McComb 2003). Presumably, roaring evolved initially in animals that did not retract their larynx and this led to the evolution of cheats that developed the capacity to retract and whose roars overestimated their size, before larynx retraction reached fixation and signal honesty was restored (Davies *et al.* 2012).

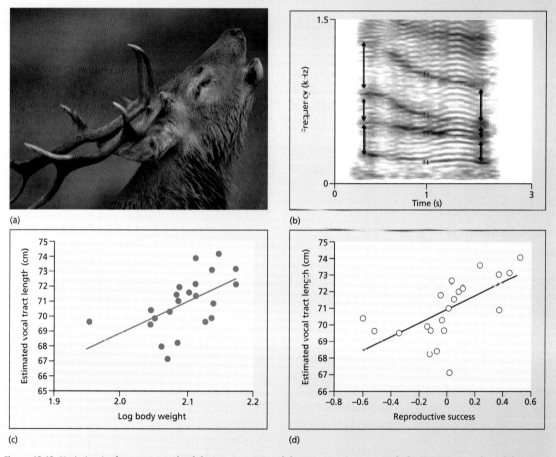

Figure 13.12 Variation in the structure of red deer roars. (a) A red deer stag roaring at a rival. (b) Spectrogram of a red deer roar representing the distribution of the energy (in grey levels). The first four formants are visible as dark bands of energy (labelled F1 to F4), which decrease throughout the vocalisation. The spacing between the formants is shown by the arrowed lines. (c) Across red deer stags, estimated vocal tract length is correlated with body weight as well as with (d) measures of lifetime reproductive success. *Source*: From Reby and McComb (2003). Reproduced with permission of Elsevier. *Photo source*: © Tim Clutton-Brock.

While the harshness of roars is associated with the extent to which they attract the attention of female red deer (Charlton *et al.* 2007; Reby and Charlton 2012), females do not approach harsh roars preferentially and appear to be most attracted to roars with the formant characteristics of young males (Charlton *et al.* 2008), suggesting that intrasexual rather than intersexual selection may have played the main role in the evolution of roar structure. Some other evidence supports this conclusion: for example, playback of roars of unfamiliar red deer stags to harem holders during the rut show that defending stags replied with roars with lower formant spacing than those directed at familiar neighbours and that males extend their vocal tracts more fully when faced with threatening situations (Reby *et al.* 2005).

Several studies provide evidence that rival males use each other's calls as a basis for assessment. For example, in red deer, challenging stags approach harem holders and exchange bouts of roaring in which both males compete to roar more frequently than the other (Clutton-Brock and Albon 1979) (see Figure 13.12). Stags that have been out-roared seldom persist and, if they do, rarely win. Similarly, playback experiments with fallow deer show that younger bucks decrease their rate of calling when they are played the calls of mature males, while mature bucks call more frequently (Komers *et al.* 1997). Similar vocalising contests between rival males also occur in some monogamous territorial species. For example, male gibbons engage in prolonged singing contests with rivals, in which the rate of alternation of calling between residents and their opponents increases as they approach each other (Cowlishaw 1996). During these exchanges, song complexity rises and, as in some birds, is correlated with the male's age, strength and commitment (Mitani 1988).

Loud calls also vary between individuals and allow males to identify themselves. In fallow deer, the structure of groans given by rutting bucks varies widely between individuals (Reby *et al.* 1998) while playback experiments with lions show that females can distinguish the roars of resident males from those of intruders (McComb *et al.* 1993). In some species, males also use exchanges of loud calls to monitor interactions between potential rivals. For example, high-ranking male baboons respond more strongly to vocal exchanges between other high-ranking males than to those between low-ranking males (Kitchen *et al.* 2005).

Olfactory displays

Where males defend breeding territories, they commonly use scent marks to deter intruders or opponents (Ralls 1971; Gosling and Roberts 2001). The source of these marks varies. In some cases, olfactory marks involve urine or faeces (Humphries *et al.* 1999; Hurst *et al.* 2001), while in others they involve secretions from a variety of specialised scent glands which often contain large molecules that operate both as vehicles for volatile constituents and as controlled release systems for long-lasting odours (Gosling and Roberts 2001) (Figures 13.13 and 13.14). Variation in MHC genotype and in major urinary proteins can provide a basis for individual recognition (Hurst *et al.* 2001; Gosling and Roberts 2001) and olfactory signals can also provide information about the age, condition, status, androgen levels and relatedness of resident males, which individuals can compare with their own status. In addition, specific pheromones may be important: for example, studies of elephants in musth show that males produce two 'chiral' forms of the pheromone frontalin and that the ratio of the two forms provides accurate information on the male's age and the relative stage of its musth cycle (Greenwood *et al.* 2005). Scent marks may also identify residents (Gosling 1982) and intruders encountering individuals whose scent matches marks in the locality may be more likely to avoid prolonged contests than those encountering individuals whose scent does not do so (Gosling 1982).

The distribution of scent marks varies both between and within species. In some species, males scent mark at different sites as they move through their territory or range, over-marking the faeces or urine of resident females (Brashares and Arcese 1999; Gosling and Roberts 2001). In others, they maintain a network of established marks that they regularly replenish or use well-defined latrines or middens, which presumably have larger detection ranges than individual marks and are often sited at the perimeter of territories (Müller-Schwarze 1974; Brotherton and Manser 1997; Brashares and Arcese 1999; Roberts and Dunbar 2000). In some social species where groups include multiple breeding males, individuals commonly over-mark each others' scent marks and over-marking rates are correlated with mating success (Jordan *et al.* 2011).

(a)   (b)

Figure 13.13 In dik dik, males mark twigs or grass stems with secretions from their preorbital glands and regularly return to replenish their marks: (a) male dik dik marking; (b) close-up of secretion on twig. *Source*: © Tim Clutton-Brock.

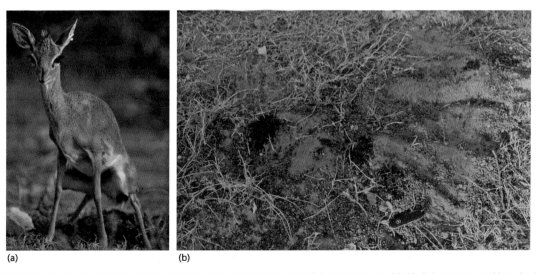

(a)   (b)

Figure 13.14 Dik dik also defecate on middens that signal their occupation of the territory: (a) dik dik defecating on midden; (b) dik dik midden, with a Swiss army knife to show scale. *Source*: © Tim Clutton-Brock.

Scent marks appear to function in the same way as visual and vocal signals, demonstrating the presence of defending males and the likely costs of intruding to potential opponents and encouraging them to withdraw (Gosling and Roberts 2001). For example, male mice avoid areas previously scent-marked by dominant males and are more reluctant to risk prolonged fights with males whose scent indicates that they are territory holders (Jones and Nowell 1989; Hurst 1993; Hurst *et al.* 1994; Gosling *et al.* 1996a,b). Individuals that approach residents on substrates marked by the resident fight at lower intensity than when approaching residents on substrates marked by unknown males (Gosling and McKay 1990). Similarly, when pairs of male rabbits are placed in areas that contain the scent marks of one of the pair, the individual whose scent marks are present is more likely to win any contest (Mykytowycz 1973, 1975). By discouraging intruders, scent marks probably reduce the frequency with which resident males need to fight to maintain their status. For example, territorial male fallow deer that scent mark at high frequencies get involved in fewer agonistic encounters than those that mark less frequently (Stenström 1998).

Scent marks may also help to reinforce the outcome of prior encounters if intruders that have lost contests with residents learn to associate particular individuals with their scent marks and subsequently avoid areas marked by them (Gosling and Roberts 2001). For example, subordinate male mice are more likely to avoid urine odours from dominant individuals that had previously defeated them than odours from unfamiliar males (Carr *et al.* 1970) and several studies show that individuals can remember the scent marks of particular individuals for periods ranging between several weeks and several months or more (Mateo and Johnston 2000).

Scent marks have several advantages over visual or vocal signals. The maintenance of a large network of fresh marks provides an unfakeable signal of the resident's presence and ability to exclude or dominate other males (Gosling and Roberts 2001). In addition, scent marks operate in the resident's absence and can be effective for considerable periods, ranging from 2–3 days for the urine marks of rodents, through periods of 1–2 weeks for the preorbital gland secretions of small antelopes to several months in the case of some anal gland secretions in carnivores (Gosling and Roberts 2001). In addition, marks are effective by night as well as by day and do not betray the immediate location of signallers to predators.

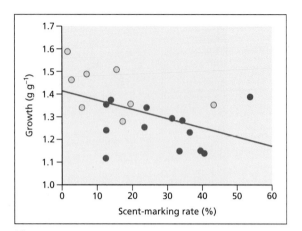

(a)

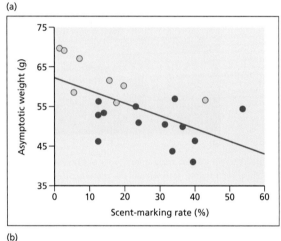

(b)

Figure 13.15 Effects of contrasts in scent-marking rate on growth and asymptotic weight in male mice. Plots show relationships between (a) growth and (b) asymptotic weight and scent-marking frequency. Closed circles denote relatively small males. Proportional growth was calculated by body mass at 25 weeks, divided by initial mass on pairing (9 weeks; ±2 weeks in each case). Asymptotic size is body mass at 25 weeks (the age at which mean population asymptotic size was attained). *Source*: From Gosling *et al.* (2000). Reproduced with permission of Springer Science and Business Media.

Though maintaining a network of fresh scent marks can have substantial benefits, it can also have important costs. Revisiting and replenishing networks of marks requires time and energy: studies of mice show that marking rates vary widely between individuals and litters and that males which mark relatively frequently show reduced growth rates and are smaller as adults (Figure 13.15). Experiments in which male mice were

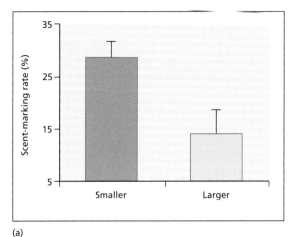

(a)

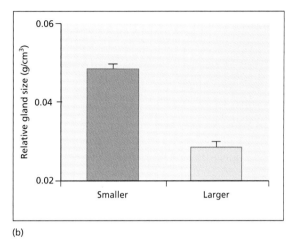

(b)

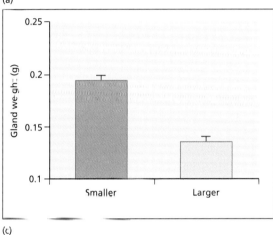

(c)

Figure 13.16 Effects of relative size on investment in scent marking and preputial glands by dominant males in captive mice. Dominant male mice housed with subordinates were either smaller (dark bars) or larger (pale bars) than their subordinate partner. Relatively smaller dominant males have (a) higher signalling frequencies, measured by scent-marking rates and (b, c) both relatively and absolutely larger preputial glands. Preputial gland sizes were obtained from dominant males aged 18 weeks; the weight of empty glands were divided by (femur length)³ to control for body size. Bars represent means + SE. *Source*: From Gosling *et al.* (2000). Reproduced with permission of Springer Science and Business Media.

caged in pairs provide further evidence of a trade-off between marking rates and growth (Gosling and Roberts 2001): dominants that are physically smaller than their cage-mates mark more frequently and have absolutely larger preputial glands than dominants larger than their partners, but have lower growth rates, lower asymptotic weights (Figure 13.16) and are more vulnerable to dominance reversals later in life (Gosling *et al.* 2000).

Scent marks may also attract predators: for example, several diurnal raptors are able to detect the scent marks of voles in ultraviolet light as a result of the proteins they contain and hunt preferentially in areas where scent marks are common (Viitala *et al.* 1995). In some cases, predators may even be able to differentiate between the marks of individuals of different age and sex and adjust their hunting behaviour accordingly (Koivula *et al.* 1999a,b). Scent marks can attract ectoparasites, too.

For example, the scent marks of klipspringer (which are deposited on the ends of low branches) attract ticks, which aggregate around the marks and so gain access to their hosts (Rechav *et al.* 1978; Spickett *et al.* 1981).

The costs of scent marks suggest that males should distribute their marks so as to maximise their impact (Gosling 1981; Guilford and Dawkins 1991, 1993; Roberts and Lowen 1997). Marks are commonly placed at prominent sites close to paths or intersections (Roberts 1997) and are clustered in the area of contested resources (Roper *et al.* 1986; Bel *et al.* 1995; Boero 1995) or where the threat posed by intruders is highest (Gosling and Roberts 2001). For example, in aardwolves, territory holders patrol disputed boundaries and intrusion raises the rate at which existing marks are revisited (Sliwa and Richardson 1998), while in oribi (a small African antelope) territorial males are more likely to mark boundaries shared with neighbouring groups containing several males than those shared with groups containing a single male (Brashares and Arcese 1999).

Where the number of marks that can be regularly replenished is limited, there are likely to be trade-offs between the area that can be marked and the effectiveness of intruder deterrence and territory holders might be expected to distribute their marks so as to optimise the area that they can defend and their effectiveness (Gosling 1981; Gorman 1990; Roberts and Lowen 1997; Gosling and Roberts 2001). Strategies are likely to vary with territory size: where territories are sufficiently small that boundaries can be marked at close intervals so that intruders are unlikely to enter the territory without encountering a mark, boundary marking is likely to be the most effective strategy. In contrast, where territory size is large, and distances between boundary marks are necessarily long, other strategies, such as marking at the core of the territory or marking an 'inner ring' may be more effective (Roberts and Lowen 1997; Gosling and Roberts 2001). Several studies show that the spatial distribution of marks varies with territory size. For example, brown hyenas living in large groups in relatively small territories focus their marks along territory boundaries while in areas where group size is small and territory size is relatively large, a higher proportion of marks are located in the centre of the territory (Mills *et al.* 1980; Mills 1983).

The need to replenish scent marks regularly and the limited volume of faeces (or other secretions) that are available have led to refinements in marking tactics. For example, territorial male oribi deposit less faeces per mark than juvenile males or females and restrict marking to middens on boundaries shared with other males or to deposits of urine or faeces left by females (Brashares and Arcese 1999). Territorial males with relatively large harems, whose territories are likely to attract intruders, mark at higher rates with less faeces per mark than males with small harems, while those with secondary males (who contribute to marking) in their territories reduce their own marking rate and increase the amount of faeces that they deposit on each mark.

Olfactory signalling is not confined to territorial species and is also widely used to signal the condition or status of males in non-territorial species. For example, male African elephants in musth dribble urine continuously through their semi-erect penises, which develop a covering of green algae (the 'green penis' phenomenon) that has a distinctive smell (Poole and Moss 1981; Poole 1987)

(see Chapter 10). Experiments in which urine from musth and non-musth bulls was presented to other males confirm that they can distinguish between the two using their vomeronasal organ system and show that they pay more attention to samples from musth bulls (Hollister-Smith *et al.* 2008).

Honesty and deception in male displays

The use of displays by males to deter rivals raises important questions about their reliability since assessment procedures might be expected to favour cheats whose displays are unrelated to their actual fighting ability (see Chapter 7). Where both signallers do not share common interests, so that either would benefit by deception, the reliability of signals may be maintained by at least four different mechanisms: by unavoidable relationships between male power or strength, so that dishonest signalling is not possible and signals represent reliable indices of male fighting ability (Clutton-Brock and Albon 1979); by quality-dependent trade-offs that allow superior individuals to be more efficient at converting advertising into fitness (Getty 2006); by heightened costs of extravagant signals to inferior males (Zahavi 1975); and by the recognition and punishment of cheats and frauds by other group members (see Chapter 7).

Which of these evolutionary mechanisms is most important? As yet, there is limited evidence of the punishment of frauds, and attempts to deceive may often be ineffective. Some male signals appear likely to be indicators of the individual's strength or quality. For example, the formant structure of red deer stags provides a reliable indicator of their size while the rate at which stags can roar reflects their condition and motivation (see earlier discussion). In sheep, displays involving horns reflect the fighting ability of males because there is a direct connection between horn size and fighting ability (Geist 1971; Preston *et al.* 2003a), while the 'wahoo' calls of male baboons incorporate information on their rank, age and exhaustion (Fischer *et al.* 2004). Many competitive displays involve multiple signalling modalities and contests frequently progress through successive stages that are likely to be increasingly costly and increasingly informative to both contestants (Enquist and Leimar 1983, 1990). For example, in red deer stags, where contests progress from exchanges of roaring, to parallel walks, brief antler clashes and escalated fights (see Figure 13.8), both the costs involved and the accuracy of the information gained probably increase as contests develop.

So how commonly have male displays evolved as handicaps? As in females, there are few if any examples where we can be certain that displays originated as handicaps rather than indices and signal reliability may be maintained by several different mechanisms (see Chapter 7). There is also little evidence that male traits are needlessly wasteful or that males engage in strategic displays that place them at a disadvantage relative to their rivals and many male signals that initially appeared to be extravagant and wasteful have, on closer inspection, proved either to reflect the signallers fighting ability or to be effective in exploiting the sensory modalities of females.

13.4 Adaptive fighting tactics

Where the costs of escalated fights are high, males that commit themselves to fighting would be expected to modify their investment in fighting in relation to variation in the likely benefits and potential costs (Enquist and Leimar 1983, 1987). As would be expected, the frequency and duration of fights varies in relation to the value of resources: for example, fights between red deer stags are more frequent at times when females are receptive, and males are less likely to fight over anoestrous females (Clutton-Brock et al. 1979). In addition, competition for the best mating territories is generally more intense and fights are more frequent and last longer than when males are competing for less popular territories (Le Boeuf 1974; Clutton-Brock et al. 1979).

Individuals also adjust their behaviour to their chances of winning and to the costs of fighting. Males who commit themselves to fights often give up as soon as they are able to judge that their chances of winning are low, while winning males frequently terminate conflicts as soon as it becomes clear that they are the winner (Le Boeuf 1974; Clutton-Brock et al. 1979). Smaller and weaker individuals or those that have been injured commonly avoid fighting altogether and adopt alternative tactics involving sneaking or cryptic approaches to females (Rubenstein 1980), though where they have little to lose they may be more aggressive than larger individuals who have the prospect of greater success in future, a phenomenon sometimes referred to as the 'Napoleon complex' (Morrell et al. 2005).

In many societies, winners seldom continue attacks once rivals flee and lethal fights are usually the result of accidents rather than of persistent attempts to kill rivals. The early ethologists believed that males were inhibited from killing their rivals and explained this as an adaptation that benefited the species (Tinbergen 1953; Lorenz 1966), but a more likely explanation is that the costs of prolonging contests or attempting to damage beaten rivals exceed the benefits, either because doing so involves opportunity losses (such as abandoning a territory or a group of females) or because further aggression increases the risk of retaliation, either from the subject or from its relatives and allies (Enquist and Leimar 1990). For example, when rhesus monkeys have been defeated, they will passively accept incisor bites from the winner but will counter-attack strongly if winners use their canines (Bernstein and Gordon 1974).

Where males can easily damage or kill their opponents, there is little indication that they are inhibited from doing so. For example, in many deer and antelope, males attempt to transfix fallen opponents with their horns or antlers and deaths are not uncommon (Clutton-Brock 1982, Clutton-Brock et al. 1982b). In addition where groups of males compete with each other and there are large disparities in group size, larger groups will pursue and kill solitary individuals or members of smaller groups (Wilson et al. 2001). For example, groups of male chimpanzees patrol the boundaries in their range and attack solitary members of neighbouring communities that they find there and these attacks can lead to serious injuries or even death (Wrangham and Peterson 1996) (see Chapter 11). Similar lethal attacks on outnumbered intruders occur in several social carnivores, including lions, hyenas and meerkats (Kruuk 1972; Schaller 1972; Clutton-Brock and Manser 2015). In many of these species, the size of groups is related to their relative dominance, and the permanent removal of members of neighbouring groups may have protracted benefits (see Chapter 2).

13.5 Benefits and costs of mate guarding

In most polygynous mammals, mate guarding helps to reduce the risk that potential partners will mate with other males and so enhances the breeding success of individual males. For example, in Idaho ground squirrels, territorial males guard females closely for around 3 hours after mating with them, ceasing once females are no longer approached by rival males (Sherman 1989). Where only one male guards a female during her oestrous period, he typically fathers all her pups, while litters born to females that are guarded by several males in turn are usually multiply sired, with the last or longest attending male fathering most of the female's offspring.

Guarding females or breeding territories can increase the breeding success of males for other reasons. Priority of occupation of territories or the establishment of social bonds with females can affect the likely outcome of contests between males (Maynard Smith 1974). In some species, males usually 'respect' each other's ownership of females (Kummer *et al.* 1974) which may explain why, in some seasonal breeders, males often start to guard mating territories or groups of females before the onset of the mating season and why males often return to the same breeding territories in successive seasons (Clutton-Brock *et al.* 1982b; Hoffman *et al.* 2006). Where males and females live in multi-male groups, protracted social bonds between males and particular females may also allow males to protect their progeny (see Chapter 15), while in species where males are involved in parental care, they may increase the effectiveness of care (Scott 1988; van de Pol *et al.* 2006) (see Chapter 16).

While guarding contributes to the mating success of males, it has substantial costs. Males that guard individual females or groups of females are often unable to locate and court other females. In addition, mate guarding commonly prevents effective foraging. For example, male yellow baboons travel less far and have shorter feeding bouts when they are guarding oestrous females (Alberts *et al.* 1996). In species with more predictable breeding seasons, males commonly show regular periods of inappetance during the mating season. For example, in red deer, stags that are defending harems cease feeding almost completely and their condition deteriorates rapidly, eventually limiting their period of breeding activity (Clutton-Brock *et al.* 1982b). In this case, reduction in feeding appears to be caused by hormonal changes at the onset of the breeding season, for a regular period of inappetance combined with reductions in body weight occurs in captive stags maintained separately from females with access to ad lib food (Blaxter *et al.* 1974; Loudon *et al.* 1989).

In other species, guarding constrains the movement of males and prevents them from moving to areas where food is more abundant or of higher quality. For example,

in Kalahari springbok, where the timing of mating seasons is unpredictable, males guard mating territories throughout much of the year and so are forced to remain in areas where food and water are scarce, while females and non-territorial males range more widely in search of food (Jackson *et al.* 1993). Males in the Nossob valley that guard empty territories for weeks on end sometimes become so frustrated that they attempt to copulate with the white posts marking the Botswana–South Africa border which runs down the centre of the riverbed.

The costs of guarding females or territories often increase as the intensity of competition for territories or female groups rises. Across species, the frequency of intrusions and the duration of male tenure on mating territories is typically shorter in lek breeders than in species where males defend resource-based territories (Clutton-Brock *et al.* 1993). Within species, too, male tenure is shorter where competition for females is most intense. For example, male hartebeest and topi that defend mating territories in areas of high female density, where competition is intense, have shorter periods of tenure than those defending territories in areas where females are scarce (Gosling 1986; Bro-Jørgensen and Durant 2003).

In some species, males respond to the high costs of guarding by alternating periods of breeding activity, during which they gradually lose condition, with periods when they abandon their territories and forage alone or with mixed-sex herds. For example, in Uganda kob, males rarely defend territories on leks for more than 2 or 3 weeks at a time and then abandon them and rejoin mixed-sex herds or form bachelor groups before returning to the lek (often on a different territory) for another bout of reproductive activity (Balmford 1992; Clutton-Brock *et al.* 1993). A similar pattern of alternating reproductive activity and non-breeding periods during which males recover condition occurs in African buffalo, where mature males alternate between periods of a few months when they live in mixed-sex herds and compete for access to receptive cows, and periods when they graze alone or in unstable bachelor groups (Prins 1996) (Figure 13.17). While they are guarding cows, their grazing time falls and they gradually lose condition which they regain when feeding outside the herd.

Guarding females or mating territories can also increase the susceptibility of males to predation for several different reasons. Solitary territorial males are less likely to detect predators than those in groups and are

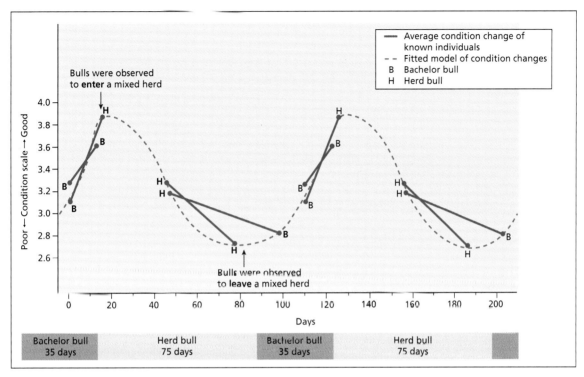

Figure 13.17 Mating behaviour and weight change in African buffalo. When bulls compete for receptive cows and live in mixed-sex groups, they feed little and lose weight, and their competitive ability gradually falls. They subsequently leave breeding herds and live alone or join smaller bachelor groups and their weight and condition recover. *Source*: From Prins (1996). Reproduced with permission from Springer Science and Business Media.

often disproportionately targeted by predators (Owen-Smith 1993; Caro 1994). They may also be easier to stalk since they often persist in defending their territories despite growth of grass long enough to hide predators either on or nearby their territories (Jackson *et al.* 1993). In addition, loss of condition as a result of protracted territory holding or minor injuries may affect the capacity of individuals to escape.

13.6 Adaptive guarding tactics

While access to females plays an important role in determining the success of males, variation in the duration of active breeding and the length of male lifespans is also important in many systems (Clutton-Brock 1988). For example, among territorial male Galapagos sea lions their breeding success is determined principally by the duration of breeding periods (Porschmann *et al.* 2010).

Similarly, in California mice and meerkats, the breeding success of males that attain dominant status is determined principally by the length of their lifespans (Ribble 1992; Clutton-Brock *et al.* 2006).

In many systems, there are likely to be trade-offs between the duration of breeding activity and expenditure of time and effort on guarding females and selection is likely to favour male strategies that maximise fitness returns per unit effort. As would be expected, guarding effort is often adjusted to variation in the likely benefits of monopolising access to females and males guard females more assiduously around the time of conception: their intolerance of intruders increases and the frequency of fighting rises (Clutton-Brock *et al.* 1979; Bercovitch *et al.* 2006; Mysterud *et al.* 2008). For example, in baboons, dominant males guard females more closely during periods when they are likely to conceive and make less effort to prevent them from mating with other males at times when conception is unlikely (Weingrill *et al.* 2003;

Alberts and Fitzpatrick 2012). Similarly, in Soay sheep, dominant rams time their mate guarding to coincide with the period when females are most likely to conceive (Preston *et al*. 2003b; Stevenson *et al*. 2004). In many systems, the presence of receptive females is also associated with increases in the intensity and frequency of competitive displays or with changes in signal structure (Wyman *et al*. 2008).

Interspecific differences in the timing of conception are associated with contrasts in the intensity of guarding behaviour. For example, in horses, seals and mongooses, where females have a post-partum oestrus, males guard females closely during the period when they are likely to conceive but are less restrictive at other times or of individuals that are less likely to conceive (Berger 1986; Gentry 1998; Nichols *et al*. 2010). Similarly, major contrasts in the duration of female receptivity are commonly associated with variation in the duration of male guarding. For example, in fallow deer, where oestrus terminates soon after females have mated, males rarely guard a female with which they have mated for more than half an hour (Clutton-Brock *et al*. 1988). Conversely, where female receptivity is prolonged, last-male sperm precedence is strong or alternative mates are scarce, post-copulatory mate guarding is prolonged. For example, in Idaho ground squirrels, last-male sperm precedence is stronger than in related species and dominant males guard females with which they have mated for relatively long periods (Sherman 1989).

Since the costs of guarding female groups are likely to grow with their size while the efficiency of guarding declines and the per-capita breeding success of females often falls (see Chapter 2), it may not always be to the advantage of males to attempt to monopolise access to as many females as possible. While several studies have shown that males will compete more intensively for large groups of females than for small ones, relatively few have investigated whether males either avoid defending unusually large groups of females or constrain the size of groups that they are defending. One possible exception occurs in vicuña, where males defend resource-based territories of around 20 ha in high-altitude grasslands above 3500 m. Resident males limit the size of the female groups living on their territory, chasing out juvenile females at 10–11 months and preventing immigrant yearlings and adult females from other groups from joining their groups (Koford 1957; Franklin 1983).

Territorial males also adjust their behaviour in relation to the relative risk posed by intruders. For example, territorial males are sometimes more tolerant of neighbours on their boundary than of strange males, who may challenge them for possession of their territory (Getty 1987; Temeles 1994). Conversely, where resident females may attempt to breed with territorial males from neighbouring groups, resident dominant males may be more intolerant of neighbours than of wandering males: recent studies have demonstrated this effect in both striped mice and banded mongooses (Müller and Manser 2007; Schradin *et al*. 2010).

Guarding strategies also vary between individuals. Where young males are unable to compete with adults, they frequently adopt 'alternative' mating tactics that are qualitatively different from those of mature males and do not attempt to defend females or breeding territories but try to sneak copulations when defending males are distracted or temporarily absent (Rubenstein 1980). For example, in land-breeding seals, younger males that are unable to acquire mating territories on the breeding beaches often attempt to mate with females that are joining or leaving the territories of mature males (Lidgard *et al*. 2004). Similarly, in red deer, young males collect on the edge of harems defended by mature stags and quickly run into the harem and attempt to mate when harem holders are fighting rivals or leaving to collect additional females, raising the costs of fighting to harem-holding males (Clutton-Brock *et al*. 1979).

Individual differences in the fighting ability or rank of males are also associated with differences in guarding behaviour. For example, in chimpanzees, males may mate opportunistically, guard oestrous females within the community, or form consortships with oestrous females and accompany them away from the rest of the group (Tutin 1980; Constable *et al*. 2001). Paternity analyses show that dominant males (who are responsible for 45% of paternities) father offspring either as a result of opportunistic mating or by guarding receptive females within the community, while middle- and low-ranking males enhance their probability of mating success by forming consortships with females and leading them away from groups that include other males (Constable *et al*. 2001). While many studies suggest that territorial males that control access to females show higher breeding success than individuals adopting

Figure 13.18 Male Galapagos fur seals defending mating territories. *Source*: © Jana Jeglinski.

alternative mating tactics, this may not always be the case. In fur seals and sea lions, dominant males defend territories (Figure 13.18), and in some South American populations of fur seals, some territorial males may have lower mating success than satellite males that adopt alternative tactics (Franco-Trecu *et al.* 2014).

In many polygynous species, distinct changes in morphology and behaviour occur when males begin to defend mating territories, often in association with increases in testosterone levels. For example, in deer and antelope where mature males defend mating territories, younger males often aggregate with females in unstable herds and do not show either the morphological characteristics of mature males or heightened levels of testosterone until they eventually begin to defend territories themselves (Rosser 1990; Malo *et al.* 2009) (Figure 13.19). Similar changes occur in some primates, including orangutans (Box 13.2 and see Figure 11.5). In some of these species, the age at which individuals attain the characteristics of fully mature males is flexible and differs widely between individuals. Studies of birds show that the costs of delays in switching from sneaking or floating strategies to active territorial defence can be

offset by subsequent benefits to breeding success (Ens *et al.* 1995) and that the form of these trade-offs and the availability and quality of vacant territories affect the age and size at which transitions in breeding strategies should occur (Kokko and Sutherland 1998).

While variation in male strategies is often flexible, some contrasts in breeding strategies are a result of heritable differences in male morphology and behaviour. For example, there are discrete differences in horn shape and size among male Soay sheep which are consistently related to contrasts in breeding behaviour (Box 13.3 and Figure 13.20). While some males have full horns, others have vestigial 'scurred' horns. Both types of males cruise in search of oestrous females but large-horned males that locate receptive females typically defend them for a few hours before moving on, sometimes guarding and copulating with up to a dozen females in a day (Grubb 1974; Stevenson *et al.* 2004), while scurred males and younger horned males do not attempt to defend receptive females but chase them until they stand still and can be mounted. Differences in horn type are associated with variation in longevity and appear to be maintained in the population by heterozygote advantage (Johnston *et al.* 2013).

Figure 13.19 In some mammals, males do not acquire full adult characteristics until they acquire breeding territories: (a) non-territorial male puku; (b) territorial male puku. *Source*: © Tim Clutton-Brock.

Box 13.2 Flexible development in male orangutans

Mature male orangutans are solitary and intolerant of each other (Knott and Kahlenberg 2007). They show a variety of striking secondary sexual characters, including large cheek flanges (see Figure 11.5) and regularly advertise their presence by long calls (Mitani 1985a,b). Younger, unflanged males commonly move between the ranges of flanged males, and their development into flanged males appears to be inhibited by the presence of flanged males. As a result, unflanged males can mature into flanged males at a wide range of ages (van Hooff 2000). Both flanged and unflanged males are reproductively active (Utami *et al.* 2002) but flanged males appear to be preferred by females (Schurmann and van Hooff 1986; Knott *et al.* 2010) and unflanged males more frequently resort to forced matings (see Chapter 15).

Flexible maturation of this kind probably generates benefits to younger males (Mitani 1985a; Utami *et al.* 2002), who may increase their fitness by remaining as satellites to flanged males until a territory holder dies or they have reached an age and size where they have a reasonable chance of displacing a mature male (Pradhan *et al.* 2012). Genetic information on the relative breeding success of flanged and unflanged males is limited and estimates of their relative success vary: while some studies suggest that unflanged males are making the best of a bad job, others indicate that flanged and unflanged males have an equal chance of siring offspring (Müller and Emery Thompson 2012).

Box 13.3 Heritable differences in horn morphology and behaviour in Soay sheep

In Soay sheep on St Kilda, around 85% of mature males have curling horns of varying size while around 15% have small deformed 'scurred' horns. Similar differences occur in females, where around 35% have fully developed horns, 35% have small scurred horns and 30% are completely hornless ('polled') (Clutton-Brock *et al.* 2004) (see Figure 13.20a). Mature, fully horned males defend receptive females and have higher annual mating success than scurred males but do not live as long and so participate in fewer rutting seasons

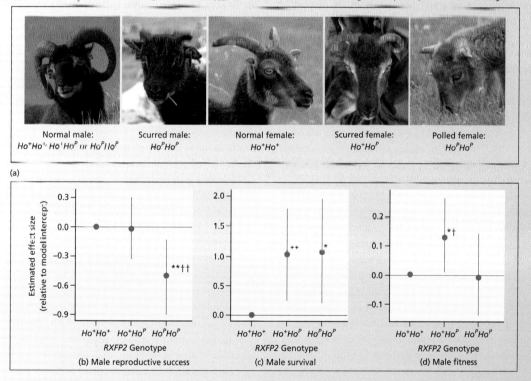

Figure 13.20 Heritable variation in horn development in Soay sheep rams on St Kilda in relation to their genotype at the *Horns* locus (*RXFP2*). (a) *Horns* genotypes and phenotypes in males and females. The estimated effect sizes at each *Horns* genotype for male fitness measures are shown for (b) annual reproductive success in adults, (c) annual survival and (d) overall fitness. Effect sizes were estimated from the posterior mode of a Bayesian GLMM with additional fixed effects of body weight and age. Random effects in the model included individual identity, birth year and measurement year. Effect sizes are given relative to the intercept at Ho^+Ho^+ and the vertical bars indicate the 95% credible interval. * and ** indicate significant difference from the intercept at Ho^+Ho^+ at the 95 and 99% credible intervals respectively. † and †† indicate a significant difference from the intercept at Ho^+Ho^P at the 95 and 99% credible intervals. *Source*: Graphs (b–d) from Johnston *et al.* (2013). Reproduced with permission of Macmillan Publishers Ltd. *Photo sources*: rams and ewes © Arpat Ozgul and Susan Johnston.

(Moorcroft *et al.* 1996; Stevenson *et al.* 2004). Scurred females are heavier than horned females and show higher rates of survival than horned females and higher breeding success, although polled females show considerably lower breeding success than either horned or scurred females (Robinson *et al.* 2008).

Genetic analyses show that variation in horn morphology and size is heritable and that a single gene (*RXFP2*) explains much of the variation in horn type and horn size, both in Soay sheep and in domestic sheep breeds (Johnston *et al.* 2011). Two *RXFP2* alleles at the *Horns* locus, Ho^+ (which leads to larger normal horns) and Ho^P (which leads to smaller horns, scurred horns or no horns), explain horn type in 86% of male Soays and contribute 76% of the additive genetic variation in horn size among normal horned males (Johnston *et al.* 2011). In females, there is no consistent association between genotype at the *Horns* locus and variation in reproductive success or survival, while in males Ho^+Ho^+ and Ho^+Ho^P individuals have higher reproductive success than Ho^PHo^P individuals, and Ho^+Ho^P and Ho^PHo^P males have higher survival than Ho^+Ho^+ males (see Figure 13.20b,c). As a result, only Ho^+Ho^P individuals do not suffer from reductions in either breeding success or survival, suggesting that variation in horn morphology is maintained by heterozygote advantage (Johnston *et al.* 2013). The reasons the *Horns* genotype affects survival in Soay sheep are still unclear. One possibility is that large horns may increase heat loss and so have energetic consequences (Picard *et al.* 1994), while another is that larger normal horned males spend more time defending females and less time grazing during the rut and so are in poorer condition at the onset of winter, generating a trade-off between mating success and survival.

13.7 Alternative tactics

Strong biases in the operational sex ratio in favour of males and intense competition between individuals for access to females commonly excludes younger or smaller males from reproductive competition and encourages the adoption of alternative reproductive tactics (Gross 1996; Taborsky *et al.* 2008). For example, in red deer, younger males hang around the edges of the harems of territorial males and attempt to disrupt them and chase out oestrous females when the male's attention is diverted (Clutton-Brock *et al.* 1979). Genetic analyses suggest that although sneakers have substantially lower breeding success than defending males, they are occasionally successful and their activities reduce the breeding success of territorial males (Pemberton *et al.* 1992).

In other species, there are consistent contrasts in mating tactics between animals of the same age (Rubenstein 1980; Koprowski 2008). In many cases, dominant individuals guard breeding territories or female groups and show higher androgen levels than subordinates, which adopt opportunistic tactics (Moore 1991; Oliveira *et al.* 2008), though in other species, like striped mice, subordinate males that roam in search of females have higher testosterone levels than males that guard breeding territories (Schradin *et al.* 2009). Where males show increased testosterone levels after acquiring dominant status or breeding territories, they may increase in weight as well as in the expression of secondary sexual characters, as in orangutans (see Box 13.2). In some cases, the presence of dominant males is sufficient to delay the development of subordinate males and cause them to adopt opportunistic mating tactics (Gross 1996) (see Chapter 14).

Selection for alternative male strategies may be involved in the evolution of homosexuality. Homosexual behaviour is widespread in non-human mammals and is particularly common in some species, such as bonobos (Vasey 1995; Poiani 2010). In many species, same-sex mounting by males appears early in ontogeny, especially in the context of play, while in females it can be common at particular stages of the oestrous cycle (Dagg 1984). Within species, most forms of homosexual behaviour appear to be most frequent where members of one sex (usually males) lack access to mature individuals of the opposite sex. Numerous explanations of the function of same-sex mounting have been suggested and it may well have multiple functions (Bailey and Zuk 2009; Poiani 2010).

While homosexual behaviour is widespread, preferential or exclusive homosexuality involving individuals that have access to the opposite sex is not. At one time thought to be restricted to humans (Gadpaille 1980), consistent homosexual preferences occur in 5–10% of males in some breeds of sheep (Hulet *et al.* 1964; Price *et al.* 1994). Research on rams showing homosexual preferences suggests that they are responding to male-typical pheromones instead of female-typical pheromones and that their preferences may be associated with modifications of the biochemistry and cellular biology of the olfactory system (Poiani 2010). Experiments involving rearing rams in single-sex groups versus mixed-sex groups indicate that sexual orientation is not strongly affected by experience of the opposite sex early in life (Price *et al.* 1994).

A substantial number of evolutionary explanations of preferential homosexuality have been suggested, including the possibility that homosexuality is a by-product of selection for increased sexual responsiveness in the opposite sex (the sexually antagonistic selection model); that it is a coping strategy in response to interactions with dominant individuals; and that it is a result of selection for cooperative behaviour operating through indirect fitness benefits (Poiani 2010) (see Chapter 20). At the moment there are no strong reasons for preferring any of these explanations to others and it seems possible that more than one mechanism may be involved.

13.8 Sperm competition

One of the principal developments of the theory of sexual selection has been the realisation that where females mate with more than one partner, before conceiving, sperm from different males will compete in the female's tract (see Chapter 1). Under these conditions, the probability that sperm from different males will succeed in fertilising the female is likely to depend on the relative number of their sperm that reach the site of fertilisation, which will usually be related to the number of sperm in their ejaculate (Parker 1982, 1984, 1998), with the result that sperm competition operates in much the same way as a raffle (Stockley 1997; Preston et al. 2003a) and there is strong selection for sperm number. However, some tickets may have a better chance of winning than others and there is growing evidence of substantial individual differences in sperm motility (Gomendio et al. 2007).

The timing of insemination relative to ovulation may also have important effects. In contrast to most birds, where sperm is stored by the female in a bursa and released when eggs are fertilised, the sperm of most mammals swim freely in the female tract and sperm that enters the female tract early in the course of the female's oestrus are often more likely to fertilise ova than those entering later ('first male precedence'), though both mating too long after or too shortly before ovulation can reduce the chances of successful fertilisation (Gomendio et al. 1998; Kraaijeveld-Smit et al. 2002).

Sperm competition is more likely in some breeding systems than others. Where individual females have very short periods of receptivity, as in many of the cervids, sperm competition may be rare, whereas extended oestrous periods are commonly associated with multiple matings and high levels of sperm competition. For example, in Soay sheep, females may copulate with up to seven different males in the course of a 2-day oestrous period and the majority of twins are sired by different fathers (Pemberton et al. 1999). Relatively high levels of sperm competition are also common in species where males do not guard individual females continuously or female groups contain more than one immigrant male who is unrelated to the resident females (see Chapter 10). In contrast, low levels of sperm competition are probably typical of many monogamous mammals where males guard females throughout the day and night and of many polygynous species that live in unimale groups (Clutton-Brock and Isvaran 2006; Isvaran and Clutton-Brock 2007).

Morphological adaptations to sperm competition

Variation in the probability of sperm competition has important consequences for the reproductive strategies of males. In species subject to regular sperm competition, selection would be expected to favour the evolution of large testes since sperm number per ejaculate increases with testis size. Comparative studies of mammals, including primates, rodents and marsupials, have shown that, relative to body size, testes size increases in species where sperm competition is common (Harcourt et al. 1981; Rose et al. 1997; Sachser et al. 1999; Ramm et al. 2005) (Figure 13.21). Contrasts in the relative size of testes between related species have also been shown to be related to variation in the risk of sperm competition between related species. For example, in howler monkeys, species living in single-male groups have relatively smaller testes than those living in multi-male groups (Dunn et al. 2015).

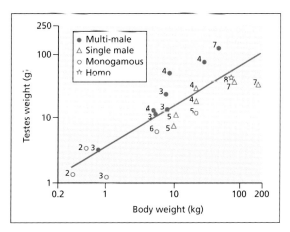

Figure 13.21 Relative to body weight, testes weight is higher in primates that live in multi-male societies (where sperm competition is common) than in species that live in single-male groups or monogamous pairs. The plot shows log combined testes weight (g) versus log body weight (kg) for different primate genera. Filled circles represent multi-male breeding systems, open circles monogamous, and open triangles single male; 8, (☆) *Homo. Source*: From Harcourt et al. (1981). Reproduced with permission of Macmillan Publishers Ltd.

Although comparative studies suggest that sperm competition has favoured the evolution of large ejaculates and large testes in many groups of animals, it is unlikely to be the only factor affecting testes size and aspects of female breeding systems probably also play a role. For example, high levels of breeding synchrony in females may select for the capacity to produce large numbers of sperm, even when sperm competition is rare. As expected, studies of rodents and marsupials suggest that high levels of reproductive synchrony in females are correlated with the evolution of large testes relative to body size, though they differ in whether they attribute this development to breeding synchrony itself or to the high levels of sperm competition that are found in some synchronised breeders (Schradin *et al.* 2009; Fisher *et al.* 2013).

Variation in the likelihood of sperm competition is also related to interspecific differences in other components of male reproductive anatomy. For example, primate species with large testes for their body size also maintain larger sperm reserves in the epididymis (Møller 1988; Müller *et al.* 2012) while some rodents have relatively large prostate glands and seminal vesicles, which produce and store seminal fluids that form vaginal plugs and inhibit the passage of subsequent sperm (Ramm *et al.* 2005) as well as relatively long penises, probably because this allows them to deposit sperm further up the female's tract and so reduce the risk of competition from subsequent matings (Ramm and Stockley 2007). The risk of sperm competition may also have important effects on the chemical structure of ejaculates and comparative studies show that the proteins present in seminal fluid vary widely between species and that those responsible for the construction of copulatory plugs are more abundant in species with large relative testes size, where sperm competition is more likely (Ramm and Stockley 2009) (see Chapter 19).

It has also been suggested that the frequency of sperm competition is associated with contrasts in the size, morphology, motility and competitiveness of sperm, although comparative studies have produced conflicting results and the situation is not yet clear (Gomendio *et al.* 2011). Larger sperm may swim faster and high levels of sperm competition may favour the evolution of large sperm (Gomendio and Roldan 1991). Comparative analyses have mostly focused on sperm midpiece volume, which is associated with increases in the number and volume of the mitochondria that contribute to sperm movement (Anderson and Dixson 2002; Anderson *et al.* 2005), and have shown that interspecific differences in midpiece volume are positively correlated with relative testes size (Anderson

et al. 2005). In addition, attempts to measure the relative swimming speeds of sperm suggest that these are commonly higher in birds that show high levels of extra-pair paternity (Kleven *et al.* 2009) and in mammals that live in multi-male groups (Nascimento 2008).

One intriguing idea is that sperm may cooperate as well as compete. Baker and Bellis (1988) suggested that mammalian ejaculates contained a proportion of deformed 'kamikaze' sperm whose function was to block the passage of sperm from rival males, but neither experimental nor comparative studies support this idea (Harcourt 1991; Moore *et al.* 1999). More recently, experimental studies of woodmice have shown that sperm from the same ejaculate aggregate in 'trains' of cells which have greater motility than individual sperm (Moore *et al.* 2002). Since most of the sperm in these trains undergo a premature acrosome reaction which reduces their capacity to fertilise ova, the formation of trains could be seen as a form of altruism on the part of individual sperm (Moore *et al.* 2002).

Behavioural adaptations to sperm competition

Sperm competition also has important consequences for male mating behaviour. Several studies have confirmed that dominant males time copulations to coincide with the period when females are most likely to conceive. For example, Soay rams time copulations to coincide with the optimal insemination period of females, giving their sperm time to capacitate before ovulation (Preston *et al.* 2003b). The risk of sperm competition may also affect the number of times that males copulate with the same females: in species where several males are likely to have access to females over the period of conception, males often mate repeatedly with the same female, whereas in monogamous species or those where female oestrous periods are short, repeated mating is uncommon (Dewsbury 1982; Stockley and Preston 2004).

Males may also adjust the volume of individual ejaculations in relation to the risk of sperm competition. While it used to be thought that the small size of sperm meant that sperm production was effectively unlimited, it is now clear that the costs of ejaculate production are not trivial (Dewsbury 1982; Partridge and Harvey 1992) and that, in at least some cases, sperm depletion can limit the breeding success of males. For example, in Soay sheep, the number of sperm ejaculated declines as copulation rates increase and (relative to their mounting rate) the siring success of larger males declines during the second half of the annual rut (Preston *et al.* 2001) (Figure 13.22). In this case, total numbers

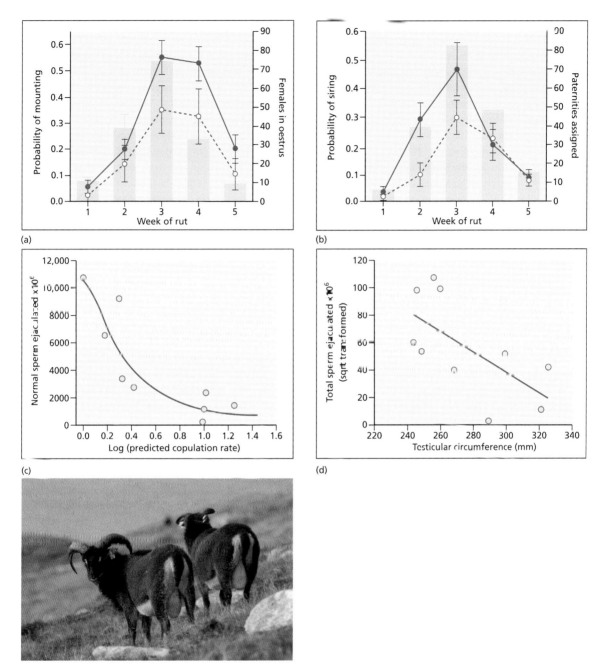

Figure 13.22 Sperm depletion in Soay sheep rams. (a) Temporal variation). in mean probability (± SE) of mounting during a watch in each week of the breeding season for males with above-average (filled circles) or below-average (open circles) hindleg length in each year. Bars indicate mean number of females in oestrus that week (± SE). (b) Temporal variation in mean probability (± SE) of being assigned a paternity in a given week for males with above-average (filled circles) or below-average (open circles) hindleg length in each year. Bars indicate the total number of paternities assigned to lambs conceived during that week. (c) Number of normal sperm ejaculated by rams in relation to their predicted copulation rate: copulation rates were predicted from the male's horn and hindleg length, using parameter estimates from the generalised linear model. (d) Total number of sperm ejaculated (square-root transformed) in relation to testicular circumference. Single ejaculates were collected from natural matings of Soay rams using an intravaginal device. *Source*: From Preston *et al.* (2001). Reproduced with permission of Macmillan Publishers Ltd. *Photo source*: © Arpat Ozgul.

of sperm per ejaculation were negatively correlated with testicular circumference, perhaps as a result of positive correlations between testes size and mating frequency.

Where a male's supply of sperm is likely to become depleted in the course of the breeding season, selection might be expected to favour males that adjust the size of their ejaculations in relation to the perceived intensity of sperm competition and their chances of fertilising offspring relative to other males (Parker 1970, 2006). Recent studies provide evidence of effects of this kind (Stockley and Purvis 1993; Stockley and Preston 2004). For example, male laboratory rats ejaculate 40% more sperm when a rival male is present than when they are caged alone with a receptive female (Pound and Gage 2004). Similarly, male house mice that have experienced high rates of encounters with several other males show higher daily sperm production than those that have experienced low encounter rates with a single other male (Ramm and Stockley 2009), although the size of their ejaculates falls, possibly because of associated changes in the frequency or timing of copulation (Ramm and Stockley 2007, 2009). Where males mate several times with the same female, they might be expected to invest more sperm in the initial copulation in case they are unable to mate again (Reinhold *et al.* 2002) and studies of birds provide evidence of effects of this kind. For example, in domestic fowl, males allocate less sperm to females with whom they have previously mated and more to females with relatively large sexual ornaments (Pizzari *et al.* 2003). While tactics of this kind may appear sophisticated, one possibility is that they are a consequence of variation in the extent to which males are stimulated by competition with other males or by encounters with attractive females.

13.9 Consequences of reproductive competition between males

Intense competition between males is characteristic of many polygynous species and commonly leads to the evolution of secondary sexual characters in males, including increases in body size, testes size, weaponry, vocal apparatus and pelage (see Chapter 18). Sex differences in adult size are commonly associated with sex differences in juveniles and with differences in the costs of rearing sons and daughters to the mother's fitness, which can affect sex ratios at birth (see Chapter 18). In

larger mammals, sexual dimorphism tends to be particularly highly developed in species that live in harem groups where one male monopolises breeding access to several females, while the presence of multiple males is often associated with reductions in the ability of dominant males to monopolise females, reductions in the weaponry and other dimorphic traits and increases in the size of testes. For example, across different species of howler monkeys, the size of hyoid bones used by males to generate roars is negatively correlated with relative testes size and is largest in species living in single-male groups (see Figure 13.11). Across a broad range of primate species, sexual dimorphism in body weight and canine size also tend to be larger in species where a single male defends access to females than where several males are associated with female groups, while relative testes size is reduced (Plavcan 2004; Müller *et al.* 2012) (Figure 13.23) and the relative size of canines in both sexes also tends to be reduced where coalitionary behaviour is common (Müller *et al.* 2012).

Contrasts in reproductive competition also have important consequences for male life histories and reproductive strategies. Intense competition between males for breeding opportunities in polygynous mammals can prevent younger males from breeding, leading to the evolution of alternative mating strategies in younger or smaller males (see Chapter 10). The stress associated with intense intrasexual competition may also affect the rate at which males age: for example, in European badgers, intense intrasexual competition between males is associated with earlier senescence in body mass, which is correlated with breeding success (Beirne *et al.* 2015). Prime males that gain access to females need to guard them closely throughout the day and night and often lose weight rapidly during the course of the breeding season, with the result that few maintain their position for the entire breeding season and individuals displace each other in turn (see Chapter 10). Older males whose condition is deteriorating are often unable to gain access to females or, if they do, can only maintain it for short periods so that the effective breeding lifespans of males is often relatively short and males frequently age faster and die earlier than females (see Chapter 18). In addition, the poor condition of males at the end of the mating season often intensifies effects on survival of fluctuations in resource shortage or harsh weather, leading to higher levels of mortality among males exhausted after the end of breeding seasons. As a result, adult sex ratios in

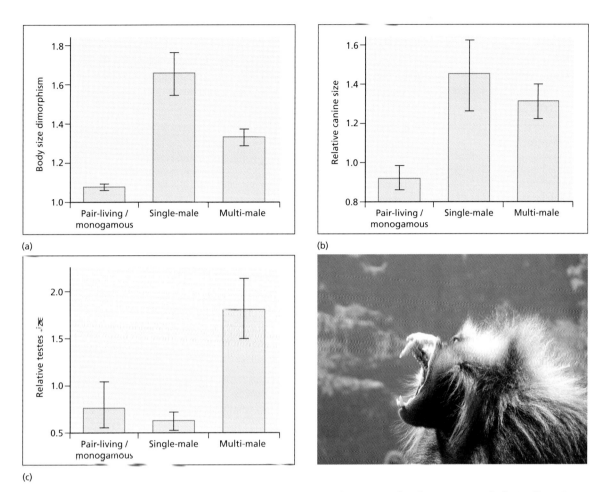

Figure 13.23 Comparative sexual dimorphism between primate genera with contrasting breeding systems: (a) body size dimorphism (adult male divided by adult female weight), (b) relative canine size (a measure of canine size dimorphism; Harvey *et al*. 1978) and (c) relative testes size (a measure of testes size after body size effects have been removed; Harcourt *et al*. 1981) for primate genera with different breeding systems. Sample sizes for P, S and M, respectively: body size dimorphism, *N* = 12, 9, 14; relative canine size, *N* = 4, 7, 9; relative testes size, *N* = 4, 7, 8. Bars indicate one standard error in each direction from the mean. *Source*: From Clutton-Brock and Harvey, 1984. Reproduced with permission of John Wiley & Sons, Ltd. *Photo source*: © Noah Snyder-Mackler.

polygynous societies are frequently biased towards females (see Chapter 18).

Reproductive competition between males also has important consequences for interactions between the sexes as well as for the reproductive tactics of females. Where older or larger males monopolise a high proportion of receptive females, younger or smaller males may adopt coercive tactics to force females they encounter to mate with them before the arrival of older or larger individuals (see Chapter 15). In addition, male exhaustion after intense periods of mating activity is often associated with high rates of turnover in breeding males

leading, in some species, to the evolution of infanticidal behaviour by immigrant males. Both coercive and infanticidal behaviour in males can in turn lead to the evolution of female counter-strategies to minimise their costs and further adaptations in males to overcome them (see Chapter 15).

Finally, high levels of reproductive competition between males and the monopolisation of access to a large proportion of breeding females by a small number of males are likely to affect the genetic structure of populations and may accelerate rates of evolutionary change (Wilson *et al*. 1975; Storz 1999; Solmsen *et al*.

2011). Conversely, low reproductive skew among males and high levels of promiscuity may affect the evolution of the immune system: for example, rates of evolution in immunity genes that interact closely with pathogens have evolved faster in promiscuous primates (Wlasiuk and Nachman 2010). Sperm competition, too, can have important evolutionary consequences and is often associated with the diversification of seminal fluids (Ramm *et al.* 2009), which may accelerate the development of gametic isolation (Martin-Coello *et al.* 2009a,b).

SUMMARY

1. In most mammal societies, males are obliged to fight each other to gain and maintain access to females and male body size and weaponry have an important influence on male fitness. However, escalated fights also have substantial risks and costs to the survival and subsequent breeding success of males.
2. Males seldom persist in attempts to kill rivals once they have chased them away, though when they can easily damage or kill opponents, they are seldom inhibited from doing so.
3. To minimise the costs of fighting, males often assess potential opponents carefully and avoid escalating conflicts with individuals that they are unlikely to beat. There are consequently important benefits to be gained by convincing rivals to avoid persisting in challenges and assessment procedures commonly involve energetically costly displays, including visual, vocal and olfactory signals.
4. Pre-copulatory mate guarding frequently increases the mating success of males, though it can also have substantial costs to their condition, subsequent breeding success and survival. As would be expected, male investment in guarding varies with the size of female groups, the probability that females will conceive and their capacity to defeat rivals.
5. In some polygynous societies, there are discrete differences in guarding tactics between males: larger and older individuals guard access to groups of females, while small or younger individuals attempt to sneak matings when dominant males are absent or engaged elsewhere.
6. In many societies where females frequently mate with multiple partners in each breeding attempt and sperm competition is common, males have developed morphological, physiological and behavioural adaptations that increase their chances of successfully fertilising females. These include larger ejaculates and testes, modifications of sperm structure and increases in the frequency of mating.
7. Competition to guard females often leads to stronger selection in males for traits that increase competitive success, including body size, weaponry and ornamentation, and to increased levels of sexual dimorphism. In addition, it is often associated with reductions in the duration of the effective breeding lifespans of males, with higher mortality throughout much of the lifespan and reduced longevity, as well as with female biases in adult sex ratios.

References

Alberts, S.C. and Fitzpatrick, C.L. (2012) Paternal care and the evolution of exaggerated sexual swellings in primates. *Behavioral Ecology* **23**:699–706.

Alberts, S.C., *et al.* (1996) Mate guarding constrains foraging activity of male baboons. *Animal Behaviour* **51**:1269–1277.

Alvarez, F. (1993) Risks of fighting in relation to age and territory holding in fallow deer. *Canadian Journal of Zoology* **71**:376–383.

Anderson, M.J. and Dixson, A.F. (2002) Sperm competition: motility and the midpiece in primates. *Nature* **416**:496.

Anderson, M.J., *et al.* (2005) Sperm competition and the evolution of sperm midpiece volume in mammals. *Journal of Zoology* **267**:135–142.

Bailey, N.W. and Zuk, M. (2009) Same-sex behavior and evolution. *Trends in Ecology and Evolution* **24**:439–446.

Baker, R.R. and Bellis, M.A. (1988) 'Kamikaze' sperm in mammals. *Animal Behaviour* **36**:936–939.

Balmford, A.P. (1992) Social dispersion and lekking in Uganda kob. *Behaviour* **120**:177–191.

Barrette, C. (1977) Fighting behavior of muntjac and the evolution of antlers. *Evolution* **31**:169–176.

Bartos, L. and Bahbouh, R. (2006) Antler size and fluctuating asymmetry in red deer (*Cervus elaphus*) stags and probability of becoming a harem holder in rut. *Biological Journal of the Linnean Society* **87**:59–68.

Beirne, C., *et al.* (2015) Sex differences in senescence: the role of intra-sexual competition in early adulthood. *Proceedings of the Royal Society of London. Series B: Biological Sciences* **282**:20151086.

Bel, M.-C., *et al.* (1995) Scent deposition by cheek rubbing in the alpine marmot (*Marmota marmota*) in the French Alps. *Canadian Journal of Zoology* **73**:2065–2071.

Bercovitch, F.B., *et al.* (2006) Sociosexual behavior, male mating tactics, and the reproductive cycle of giraffe *Giraffa camelopardalis*. *Hormones and Behavior* **50**:314–321.

Berger, J. (1986) *Wild Horses of the Great Basin*. Chicago: University of Chicago Press.

Berger, J. and Cunningham, C. (1991) Bellows, copulations and sexual selection in bison (*Bison bison*). *Behavioral Ecology* **2**:1–6.

Bergeron, P., *et al.* (2010) Secondary sexual characters signal fighting ability and determine social rank in Alpine ibex (*Capra ibex*). *Behavioral Ecology and Sociobiology* **64**:1299–1307.

Berglund, A., *et al.* (1996) Armaments and ornaments: an evolutionary explanation of traits of dual utility. *Biological Journal of the Linnean Society* **58**:385–399.

Bernstein, I.S. and Gordon, T.P. (1974) The function of aggression in primate societies: uncontrolled aggression may threaten human survival, but aggression may be vital to the establishment and regulation of primate societies and sociality. *American Scientist* **62**:304–311.

Blaxter, K.L., *et al.* (1974) *Farming the Red Deer. The First Report of an Investigation by the Rowett Research Institute and the Hill Farming Research Organisation*, London: HMSO.

Boero, D.L. (1995) Scent deposition behaviour in alpine marmots (*Marmota marmota* L.): its role in territorial defence and social communication. *Ethology* **100**:26–38.

Bradbury, J.W. (1977) Social organization and communication. In: *Biology of Bats*, Vol. **3** (ed. W. Wimsatt). New York; Academic Press, 1–72.

Brashares, J.S. and Arcese, P. (1999) Scent marking in a territorial African antelope: I. The maintenance of borders between male oribi. *Animal Behaviour* **57**:1–10.

Bro-Jørgensen, J. and Durant, S.M. (2003) Mating strategies of topi bulls: getting in the centre of attention. *Animal Behaviour* **65**:585–594.

Brotherton, P.N.M. and Manser, M.B. (1997) Female dispersion and the evolution of monogamy in the dik-dik. *Animal Behaviour* **54**:1413–1424.

Brown, L., *et al.* (1985) Courtship and female choice in the horned beetle *Bolitotherus cornutus* (Panzer) (Coleoptera: Tenebrionidae). *Annals of the Entomological Society of America* **78**:423–427.

Cant, M.A. and Johnstone, R.A. (2009) How threats influence the evolutionary resolution of within-group conflict. *American Naturalist* **173**:759–771.

Caro, T.M. (1994) Ungulate antipredator behaviour: preliminary and comparative data from African bovids. *Behaviour* **128**:189–228.

Caro, T.M., *et al.* (2003) Correlates of horn and antler shape in bovids and cervids. *Behavioral Ecology and Sociobiology* **55**:32–41.

Carr, W.J., *et al.* (1970) Responses of mice to odors associated with stress. *Journal of Comparative Physiological Psychology* **71**:223–228.

Charlton, B.D., *et al.* (2007) Female red deer prefer the roars of larger males. *Biology Letters* **3**:382–385.

Charlton, B.D., *et al.* (2008) Free-ranging red deer hinds show greater attentiveness to roars with formant frequencies typical of young males. *Ethology* **114**:1023–1031.

Clutton-Brock, T.H. (1982) The functions of antlers. *Behaviour* **79**:108–125.

Clutton-Brock, T.H. (1988) Reproductive success. In: *Reproductive Success: Studies of Individual Variation in Contrasting Breeding Systems* (ed. T.H. Clutton-Brock). Chicago: University of Chicago Press, 472–486.

Clutton-Brock, T.H. and Albon, S.D. (1979) The roaring of red deer and the evolution of honest advertisement. *Behaviour* **69**:145–170.

Clutton-Brock, T.H. and Harvey, P.H. (1984) Comparative approaches to investigating adaptation. In: *Behavioural Ecology: An Evolutionary Approach*, 2nd edn (eds J.R. Krebs and N.B. Davies). Oxford: Blackwell Scientific Publications, 7–29.

Clutton-Brock, T.H. and Isvaran, K. (2006) Paternity loss in contrasting mammalian societies. *Biology Letters* **2**:513–516.

Clutton-Brock, T.H., *et al.* (1979) The logical stag: adaptive aspects of fighting in red deer (*Cervus elaphus* L.). *Animal Behaviour* **27**:211–225.

Clutton-Brock, T.H., *et al.* (1982a) Effects of lactation on feeding behaviour and habitat use in wild red deer hinds. *Journal of Zoology* **198**:227–236.

Clutton-Brock, T.H., *et al.* (1982b) *Red Deer: The Behaviour and Ecology of Two Sexes*. Chicago: University of Chicago Press.

Clutton-Brock, T.H., *et al.* (1988) Passing the buck: resource defence, lek breeding and mate choice in fallow deer. *Behavioral Ecology and Sociobiology* **23**:281–296.

Clutton-Brock, T.H., *et al.* (1993) The evolution of ungulate leks. *Animal Behaviour* **46**:1121–1138.

Clutton-Brock, T.H., *et al.* (2004) The sheep of St Kilda. In: *Soay Sheep: Dynamics and Selection in an Island Population* (eds T.H. Clutton-Brock and J.M. Pemberton). Cambridge: Cambridge University Press, 17–51.

Clutton-Brock, T.H., *et al.* (2006) Intrasexual competition and sexual selection in cooperative mammals. *Nature* **444**:1065–1068.

Constable, J.L., *et al.* (2001) Noninvasive paternity assignment in Gombe chimpanzees. *Molecular Ecology* **10**:1279–1300.

Cowlishaw, G. (1996) Sexual selection and information content in Gibbon song bouts. *Ethology* **102**:272–284.

Croll, D.A., *et al.* (2002) Bioacoustics: only male fin whales sing loud songs. *Nature* **417**:809.

Dagg, A.I. (1984) Homosexual behaviour and female–male mounting in mammals: a first survey. *Mammal Review* **14**:155–185.

Darwin, C. (1876) Sexual selection in relation to monkeys. *Nature* **15**:18–19.

Davies, N.B., *et al.* (2012) *An Introduction to Behavioural Ecology*, 4th edn. Oxford: Wiley-Blackwell.

Dewsbury, D.A. (1982) Ejaculate cost and male choice. *American Naturalist* **119**:601–610.

Dunn, J. *et al.* (2015) Evolutionary trade-off between vocal tract and testes dimensions in howler monkeys. *Current Biology* **25**:2839–2844.

Emlen, D.J. (2008) The evolution of animal weapons. *Annual Review of Ecology, Evolution and Systematics* **39**:387–413.

Emlen, D.J., *et al.* (2012) A mechanism of extreme growth and reliable signaling in sexually selected ornaments and weapons. *Science* **337**:860–864.

Engh, A.L., *et al.* (2002) Reproductive skew among males in a female-dominated mammalian society. *Behavioral Ecology* **13**:193–200.

Enquist, M. and Leimar, O. (1983) Evolution of fighting behaviour: decision rules and assessment of relative strength. *Journal of Theoretical Biology* **102**:387–410.

Enquist, M. and Leimar, O. (1987) Evolution of fighting behaviour: the effect of variation in resource value. *Journal of Theoretical Biology* **127**:187–205.

Enquist, M. and Leimar, O. (1990) The evolution of fatal fighting. *Animal Behaviour* **39**:1–9.

Ens, B.J., *et al.* (1995) The despotic distribution and deferred maturity: two sides of the same coin. *American Naturalist* **146**:625–650.

Fischer, J., *et al.* (2004) Baboon loud calls advertise male quality: acoustic features and their relation to rank, age, and exhaustion. *Behavioral Ecology and Sociobiology* **56**:140–148.

Fisher, D.O., *et al.* (2013) Sperm competition drives the evolution of suicidal reproduction in mammals. *Proceedings of the National Academy of Sciences of the United States of America* **110**:17910–17914.

Fitch, W.T. and Reby, D. (2001) The descended larynx is not uniquely human. *Proceedings of the Royal Society of London. Series B: Biological Sciences* **268**:1669–1675.

Fletcher, N.H. (2010) A frequency scaling rule in mammalian vocalization. In: *Handbook of Mammalian Vocalization: An Integrative Neuroscience Approach* (ed. S.M. Brudzynski). San Diego, CA: Academic Press, 51–56.

Foerster, K., *et al.* (2007) Sexually antagonistic genetic variation for fitness in red deer. *Nature* **447**:1107–1110.

Franco-Trecu, V., *et al.* (2014) Sex on the rocks: reproductive tactics and breeding success of South American fur seal males. *Behavioral Ecology* **25**:1513–1523.

Franklin, W.L. (1983) Contrasting socioecologies of South America's wild camelids: the vicuña and the guanaco. In: *Advances in the Study of Mammalian Behavior* (eds J.F. Eisenberg and D.G. Kleiman). American Society of Mammalogists Special Publication No. 7. Stillwater, OK: American Society of Mammalogists, 573–629.

Gadpaille, W.J. (1980) Cross-species and cross-cultural contributions to understanding homosexual activity. *Archives of General Psychiatry* **37**:349–356.

Garcia, M., *et al.* (2013) Do red deer stags (*Cervus elaphus*) use roar fundamental frequency (F0) to assess rivals? *PLOS ONE* **8**: e83946.

Geist, V. (1971) *Mountain Sheep: A Study in Behavior and Evolution.* Chicago: University of Chicago Press.

Gentry, R.L. (1998) *Behaviour and Ecology of the Northern Fur Seal.* Princeton, NJ: Princeton University Press.

Getty, T. (1987) Dear enemies and the prisoner's dilemma: why should territorial neighbors form defensive coalitions? *American Zoologist* **27**:327–336.

Getty, T. (2006) Sexually elected signals are not similar to sports handicaps. *Trends in Ecology and Evolution* **21**:83–88.

Gittleman, J.L. and Thompson, S.D. (1988) Energy allocation in mammalian reproduction. *American Zoologist* **28**:863–876.

Gomendio, M. and Roldan, E.R.S. (1991) Sperm competition influences sperm size in mammals. *Proceedings of the Royal Society of London. Series B: Biological Sciences* **243**:181–185.

Gomendio, M., *et al.* (1998) Sperm competition in mammals. In: *Sperm Competition and Sexual Selection* (eds T.R. Birkhead and A.P. Møller). London: Academic Press, 667–756.

Gomendio, M., *et al.* (2007) Testosterone and male fertility in red deer: response. *Science* **316**:981.

Gomendio, M. *et al.* (2011) Why mammalian lineages respond differently to sexual selection: metabolic rate constrains the evolution of sperm size. *Proceedings of the Royal Society of London. Series B: Biological Sciences* **278**:3135–3141.

Gorman, M.L. (1990) Scent marking strategies in mammals. *Revue Suisse de Zoologie* **97**:3–29.

Gosling, L.M. (1981) Demarkation in a gerenuk territory: an economic approach. *Zeitschrift fur Tierpsychologie* **56**:305–322.

Gosling, L.M. (1982) A reassessment of the function of scent marking in territories. *Zeitschrift fur Tierpsychologie* **60**:89–118.

Gosling, L.M. (1986) The evolution of mating strategies in male antelopes. In: *Ecological Aspects of Social Evolution* (eds D.I. Rubenstein and R.W. Wrangham). Princeton, NJ: Princeton University Press, 244–281.

Gosling, L.M. and McKay, H.V. (1990) Competitor assessment by scent matching: an experimental test. *Behavioral Ecology and Sociobiology* **26**:415–420.

Gosling, L.M. and Roberts, S.C. (2001) Testing ideas about the function of scent marks in territories from spatial patterns. *Animal Behaviour* **62**:F7–F10.

Gosling, L.M., *et al.* (1996a) Avoidance of scent-marked areas depends on the intruder's body size. *Behaviour* **133**:491–502.

Gosling, L.M., *et al.* (1996b) The response of subordinate male mice to scent marks varies in relation to their own competitive ability. *Animal Behaviour* **52**:1185–1191.

Gosling, L.M., *et al.* (2000) Life history costs of olfactory status signalling in mice. *Behavioral Ecology and Sociobiology* **48**:328–332.

Greenwood, D.R., *et al.* (2005) Chemical communication: chirality in elephant pheromones. *Nature* **438**:1097–1098.

Grinnell, J. and McComb, K. (2001) Roaring and social communication in African lions: the limitations imposed by listeners. *Animal Behaviour* **62**:93–98.

Gross, M.R. (1996) Alternative reproductive strategies and tactics: diversity within sexes. *Trends in Ecology and Evolution* **11**:92–98.

Grubb, P. (1974) Mating activity and the social significance of rams in a feral sheep community. In: *The Behaviour of Ungulates and its Relation to Management* (eds V. Geist and F. Walther). IUCN Publication No. 24. Morges, Switzerland: IUCN, 457–476.

Guilford, T. and Dawkins, M.S. (1991) Receiver psychology and the evolution of animal signals. *Animal Behaviour* **42**:1–14.

Guilford, T. and Dawkins, M.S. (1993) Receiver psychology and the design of animal signals. *Trends in Neurosciences* **16**:430–436.

Gulland, F.M.D., *et al.* (1993) Parasite-associated polymorphism in a cyclic ungulate population. *Proceedings of the Royal Society of London. Series B: Biological Sciences* **254**:7–13.

Harcourt, A.H. (1991) Sperm competition and the evolution of nonfertilizing sperm in mammals. *Evolution* **45**:314–328.

Harcourt, A.H., *et al.* (1981) Testis weight, body weight and breeding system in primates. *Nature* **293**:55–57.

Harvey, P.H., *et al.* (1978) Sexual dimorphism in primate teeth. *Journal of Zoology* **186**:475–485.

Hoffman, J.I., *et al.* (2006) Genetic tagging reveals extreme site fidelity in territorial male Antarctic fur seals *Arctocephalus gazella*. *Molecular Ecology* **15**:3841–3847.

Hollister-Smith, J.A., *et al.* (2008) Do male African elephants, *Loxodonta africana*, signal musth via urine dribbling. *Animal Behaviour* **76**:1829–1841.

Holy, T.E. and Guo, Z.S. (2005) Ultrasonic songs of male mice. *PLOS Biology* **3**:e386.

Howard, N. (1998) *n*-person 'soft' games. *Journal of the Operational Research Society* **49**:144–150.

Hulet, C.V., *et al.* (1964) Observations on sexually inhibited rams. *Journal of Animal Science* **23**:1095–1097.

Humphries, R.E., *et al.* (1999) Unravelling the chemical basis of competitive scent marking in house mice. *Animal Behaviour* **58**:1177–1190.

Hunt, J. and Simmons, L.G. (1997) Patterns of fluctuating asymmetry in beetle horns: an experimental examination of the honest signalling hypothesis. *Behavioral Ecology and Sociobiology* **41**:109–114.

Hurst, J.L. (1993) The priming effects of urine substrate marks on interactions between male house mice, *Mus musculus domesticus* Schwarz & Schwarz. *Animal Behaviour* **45**:55–81.

Hurst, J.L., *et al.* (1994) The role of substrate odours in maintaining social tolerance between male house mice, *Mus musculus domesticus*: relatedness, incidental kinship effects and the establishment of social status. *Animal Behaviour* **48**:157–167.

Hurst, J.L., *et al.* (2001) Individual recognition in mice mediated by major urinary proteins. *Nature* **414**:631–634.

Huxley, J.S. (1931) The relative size of antlers in deer. *Proceedings of the Zoological Society of London* **101**:819–864.

Isvaran, K. and Clutton-Brock, T.H. (2007) Ecological correlates of extra-group paternity in mammals. *Proceedings of the Royal Society of London. Series B: Biological Sciences* **274**:219–224.

Jackson, T.P., *et al.* (1993) Some costs of maintaining a perennial territory in the springbok, *Antidorcas marsupialis*. *African Journal of Ecology* **31**:242–254.

Jewell, P.A. (1997) Survival and behaviour of castrated Soay sheep (*Ovis aries*) in a feral island population on Hirta, St Kilda, Scotland. *Journal of Zoology* **243**:623–636.

Johnston, S.E., *et al.* (2011) Genome-wide association mapping identifies the genetic basis of discrete and quantitative variation in sexual weaponry in a wild sheep population. *Molecular Ecology* **20**:2555–2566.

Johnston, S.E., *et al.* (2013) Life history trade-offs at a single locus maintain sexually selected genetic variation. *Nature* **502**:93–95.

Jones, R.B. and Nowell, N.W. (1989) Aversive potency of urine from dominant and subordinate male laboratory mice (*Mus musculus*): resolution of a conflict. *Aggressive Behavior* **15**:291–296.

Jordan, N.R., *et al.* (2011) Scent marking in wild banded mongooses: 2. Intrasexual overmarking and competition between males. *Animal Behaviour* **81**:43–50.

Kitchen, D.M., *et al.* (2005) Male chacma baboons (*Papio hamadryas ursinus*) discriminate loud call contests between rivals of different relative rank. *Animal Cognition* **8**:1–6.

Kitchener, A. (1985) The effect of behaviour and body weight on the mechanical design of horns. *Journal of Zoology* **205**:191–203.

Kitchener, A. (1988) An analysis of the forces of fighting of the blackbuck (*Antilope cervicapra*) and the bighorn sheep (*Ovis canadensis*) and the mechanical design of the horn of bovids. *Journal of Zoology* **214**:1–20.

Klein, S.L. and Nelson, R.J. (1998) Adaptive immune responses are linked to the mating system of arvicoline rodents. *American Naturalist* **151**:59–67.

Kleven, O., *et al.* (2009) Comparative evidence for the evolution of sperm swimming speed by sperm competition and female sperm storage duration in passerine birds. *Evolution* **63**:2466–2473.

Knott, C.D. and Kahlenberg, S.M. (2007) Orangutans in perspective: forced copulations and female mating resistance. In: *Primates in Perspective* (eds C. Campbell, A. Fuentes, K. MacKinnon, M. Panger and S. Bearder). New York: Oxford University Press, 290–304.

Knott, C.D., *et al.* (2010) Female reproductive strategies in orangutans: evidence for female choice and counterstrategies to infanticide in a species with frequent sexual coercion. *Proceedings of the Royal Society of London. Series B: Biological Sciences* **277**:105–113.

Koford, C.B. (1957) The vicuna and the puna. *Ecological Monographs* **27**:153–219.

Koivula, M., *et al.* (1999a) Sex and age-specific differences in ultraviolet reflectance of scent marks of bank voles (*Clethrionomys glareolus*). *Journal of Comparative Psychology A* **185**:561–564.

Koivula, M., *et al.* (1999b) Kestrels prefer scent marks according to species and reproductive status of voles. *Ecoscience* **6**:415–420.

Kokko, H. and Sutherland, W.J. (1998) Optimal floating and queuing strategies: consequences for density dependence and habitat loss. *American Naturalist* **152**:354–366.

Komers, P.E., *et al.* (1997) Age at first reproduction in male fallow deer: age-specific versus dominance-specific behaviors. *Behavioral Ecology* **8**:456–462.

Koprowski, J.L. (2008) Alternative reproductive tactics and strategies of tree squirrels. In: *Rodent Societies: An Ecological* and *Evolutionary Perspective* (eds J.O. Wolff and P.W. Sherman). Chicago: University of Chicago Press, 86–98.

Kraaijeveld-Smit, F.J.L., *et al.* (2002) Factors influencing paternity success in *Antechinus agilis*: last-male sperm precedence, timing of mating and genetic compatibility. *Journal of Evolutionary Biology* **15**:100–107.

Kruuk, H. (1972) *The Spotted Hyena: A Study of Predation and Social Behaviour*. Chicago: University of Chicago Press.

Kummer, H., *et al.* (1974) Triadic differentiation: inhibitory process protecting pair bonds in baboons. *Behaviour* **49**:62–87.

Lailvaux, S.P., *et al.* (2005) Horn size predicts physical performance in the beetle *Euoniticellus intermedius* (Coleoptera: Scarabaeidae). *Functional Ecology* **19**:632–639.

Le Boeuf, B.J. (1974) Male–male competition and reproductive success in elephant seals. *American Zoologist* **14**:163–176.

Lidgard, D.C., *et al.* (2004) The rate of fertilization in male mating tactics of the polygynous grey seal. *Molecular Ecology* **13**:3543–3548.

Lincoln, G.A. (1972) The role of antlers in the behaviour of red deer. *Journal of Experimental Zoology* **182**:233–249.

Loehr, J., *et al.* (2008) Coat darkness is associated with social dominance and mating behaviour in a mountain sheep hybrid lineage. *Animal Behaviour* **76**:1545–1553.

Lorenz, K. (1966) *On Aggression*. New York: Harcourt, Brace & World.

Loudon, A.S.I., *et al.* (1989) A comparison of the seasonal hormone changes and patterns of growth, voluntary food intake and reproduction in juvenile and adult red deer (*Cervus elaphus*) and Père David's deer (*Elaphurus davidianus*) hinds. *Journal of Endocrinology* **122**:733–745.

McComb, K., *et al.* (1993) Female lions can identify potentially infanticidal males from their roars. *Proceedings of the Royal Society of London. Series B: Biological Sciences* **252**:59–64.

Malo, A.F., *et al.* (2009) What does testosterone do for red deer males? *Proceedings of the Royal Society of London. Series B: Biological Sciences* **276**:971–980.

Martin-Coello, J., *et al.* (2009a) Sperm competition promotes asymmetries in reproductive barriers between closely related species. *Evolution* **63**:613–623.

Martin-Coello, J., *et al.* (2009b) Sexual selection drives weak positive selection in protamine genes and high promoter divergence, enhancing sperm competitiveness. *Proceedings of the Royal Society of London. Series B: Biological Sciences* **276**:2427–2436.

Mateo, J.M. and Johnston, R.E. (2000) Retention of social recognition after hibernation in Belding's ground squirrels. *Animal Behaviour* **59**:491–499.

Maynard Smith, J. (1974) The theory of games and the evolution of animal conflicts. *Journal of Theoretical Biology* **47**:209–221.

Mills, M.G.L. (1983) Behavioural mechanisms in territory and group maintenance of the brown hyaena, *Hyaena brunnea*, in the southern Kalahari. *Animal Behaviour* **31**:503–510.

Mills, M.G.L., *et al.* (1980) The scent marking behavior of the brown hyaena *Hyaena brunnes*. *South African Journal of Zoology* **15**:240–248.

Mitani, J.C. (1984) The behavioral regulation of monogamy in gibbons (*Hylobates muelleri*). *Behavioral Ecology and Sociobiology* **15**:225–229.

Mitani, J.C. (1985a) Mating behaviour of male orangutans in the Kutai Game Reserve, Indonesia. *Animal Behaviour* **33**:392–402.

Mitani, J.C. (1985b) Sexual selection and adult male orangutan long calls. *Animal Behaviour* **33**:272–283.

Mitani, J.C. (1988) Male gibbon (*Hylobates agilis*) singing behavior: natural history, song variations and function. *Ethology* **79**:177–194.

Møller, A.P. (1988) Ejaculate quality, testes size and sperm competition in primates. *Journal of Human Evolution* **17**:479–488.

Mooradian, A.D., *et al.* (1987) Biological actions of androgens. *Endocrine Reviews* **8**:1–28.

Moorcroft, P.R., *et al.* (1996) Density-dependent selection in a fluctuating ungulate population. *Proceedings of the Royal Society of London. Series B: Biological Sciences* **263**:31–38.

Moore, H.D.M., *et al.* (1999) No evidence for killer sperm or other selective interactions between human spermatozoa in ejaculates of different males *in vitro*. *Proceedings of the Royal Society of London. Series B: Biological Sciences* **266**:2343–2350.

Moore, H.D.M., *et al.* (2002) Exceptional sperm cooperation in the wood mouse. *Nature* **418**:174–177.

Moore, M.C. (1991) Application of organization-activation theory to alternative male reproductive tactics: a review. *Humans and Behaviour* **25**:154–179.

Moore, M.C., *et al.* (1998) Hormonal control and evolution of alternative male phenotypes: generalizations of models for sexual differentiation. *Integrative and Comparative Biology* **38**:133–151.

Moore, S.L. and Wilson, K. (2002) Parasites as a viability cost of sexual selection in natural populations of mammals. *Science* **297**:2015–2018.

Morrell, L.J., *et al.* (2005) Why are small males aggressive? *Proceedings of the Royal Society of London. Series B: Biological Sciences* **272**:1235–1241.

Müller, C. and Manser, M.B. (2007) 'Nasty neighbours' rather than 'dear enemies' in a social carnivore. *Proceedings of the Royal Society of London. Series B: Biological Sciences* **274**:959–965.

Müller, M.N. and Emery Thompson, M. (2012) Mating, parenting and male reproductive strategies. In: *The Evolution of Primate Societies* (eds J.C. Mitani, J. Call, P.M. Kappeler, R.A. Palombit and J.B. Silk). Chicago: University of Chicago Press, 387–411.

Müller, M.N. and Wrangham, R.W. (2004) Dominance, aggression and testosterone in wild chimpanzees: a test of the 'challenge hypothesis'. *Animal Behaviour* **67**:113–123.

Müller, M.N., *et al.* (2012) Testosterone, development and aging in wild chimpanzees. *American Journal of Physical Anthropology* **147**:220.

Müller-Schwarze, D. (1974) Social functions of various scent glands in certain ungulates and the problems encountered in experimental studies of scent communication. In: *The Behaviour of Ungulates and its Relation to Management* (eds V. Geist and F. Walther). IUCN Publication No. 24. Morges, Switzerland: IUCN, 107–113.

Mykytowycz, R. (1973) Reproduction of mammals in relation to environmental odours. *Journal of Reproduction and Fertility Supplement* **19**:433–446.

Mykytowycz, R. (1975) Activation of territorial behaviour in the rabbit, *Oryctolagus cuniculus,* by stimulation with its own chin gland secretion. In: *International Symposium on Olfaction and Taste*, Vol. **V** (eds D.A. Denton and P. Coghlan). New York: Academic Press, 425–432.

Mysterud, A., *et al.* (2008) The timing of male reproductive effort relative to female ovulation in a capital breeder. *Journal of Animal Ecology* **77**:469–477.

Nascimento, J.M. (2008) *Analysis of sperm motility and physiology using optical tweezers*. PhD thesis, University of California San Diego.

Nichols, H.J., *et al.* (2010) Top males gain high reproductive success by guarding more successful females in a cooperatively breeding mongoose. *Animal Behaviour* **80**:649–657.

Oliveira, R.F., *et al.* (2008) Hormones and alternative reproductive tactics in vertebrates. In: *Alternative Reproductive Tactics: An Integrative Approach* (eds R.F. Oliveira, B. Taborsky and H.J. Brockmann). Cambridge: Cambridge University Press, 132–173.

Owen-Smith, N. (1993) Comparative mortality rates of male and female Kudus: the costs of sexual size dimorphism. *Journal of Animal Ecology* **62**:428–440.

Parker, G.A. (1970) Sperm competition and its evolutionary consequences in the insects. *Biological Reviews* **45**:525–567.

Parker, G.A. (1982) Why are there so many tiny sperm? Sperm competition and the maintenance of two sexes. *Journal of Theoretical Biology* **96**:281–294.

Parker, G.A. (1984) Sperm competition and the evolution of animal mating strategies. In: *Sperm Competition and the Evolution of Animal Mating Systems* (ed. R.L. Smith). New York: Academic Press, 1–60.

Parker, G.A. (1998) Sperm competition and the evolution of ejaculates: towards a theory base. In: *Sperm Competition and Sexual Selection* (eds T.R. Birkhead and A.P. Møller). London: Academic Press, 3–54.

Parker, G.A. (2006) Sexual conflict over mating and fertilisation: an overview. *Philosophical Transactions of the Royal Society B: Biological Sciences* **361**:235–259.

Partridge, L. and Harvey, P.H. (1992) What the sperm count costs. *Nature* **360**:415.

Payne, R.S. and McVay, S. (1971) Songs of humpback whales. *Science* **173**:585–597.

Pelletier, F., *et al.* (2003) Fecal testosterone in bighorn sheep (*Ovis canadensis*): behavioural and endocrine correlates. *Canadian Journal of Zoology* **81**:1678–1684.

Pemberton, J.M., *et al.* (1992) Behavioural estimates of male mating success tested by DNA fingerprinting in a polygynous mammal. *Behavioral Ecology* **3**:66–75.

Pemberton, J.M., *et al.* (1999) Molecular analysis of a promiscuous, fluctuating mating system. *Biological Journal of the Linnean Society* **68**:289–301.

Pfefferle, D. and Fischer, J. (2006) Sounds and size: identification of acoustic variables that reflect body size in hamadryas baboons, *Papio hamadryas*. *Animal Behaviour* **72**:43–51.

Picard, K., *et al.* (1994) Bovid horns: an important site for heat loss during winter? *Journal of Mammalogy* **75**:710–713.

Pitcher, B.J., *et al.* (2014) Fallow bucks attend to vocal cues of motivation and fatigue. *Behavioral Ecology* **25**:392–401.

Pizzari, T., *et al.* (2003) Sophisticated sperm allocation in male fowl. *Nature* **426**:70–74.

Plavcan, J.M. (2004) Sexual selection, measures of sexual selection and sexual dimorphism in primates. In: *Sexual Selection in Primates* (eds P. Kappeler and C. van Schaik). Cambridge: Cambridge University Press, 230–252.

Poiani, A. (2010) *Animal Homosexuality: A Biosocial Perspective.* Cambridge: Cambridge University Press.

Poole, J.H. (1987) Rutting behavior in African elephants: the phenomenon of musth. *Behaviour* **102**:283–316.

Poole, J.H. and Moss, C.J. (1981) Musth in the African elephant, *Loxodonta africana*. *Nature* **292**:830–831.

Porschmann, U., *et al.* (2010) Male reproductive success and its behavioural correlates in a polygynous mammal, the Galapagos sea lion (*Zalophus wollebaeki*). *Molecular Ecology* **19**:2574–2586.

Pound, N. and Gage, M.J.G. (2004) Prudent sperm allocation in Norway rats, *Rattus norvegicus*: a mammalian model of adaptive ejaculate adjustment. *Animal Behaviour* **68**:819–823.

Pradhan, G.R., *et al.* (2012) A model for the evolution of developmental arrest in male orangutans. *American Journal of Physical Anthropology* **149**:18–25.

Preston, B.T., *et al.* (2001) Dominant rams lose out by sperm depletion. *Nature* **409**:681–682.

Preston, B.T., *et al.* (2003a) Overt and covert competition in a promiscuous mammal: the importance of weaponry and testes size to male reproductive success. *Proceedings of the Royal Society of London. Series B: Biological Sciences* **270**:633–640.

Preston, B.T., *et al.* (2003b) Soay rams target reproductive activity towards promiscuous females' optimal insemination period. *Proceedings of the Royal Society of London. Series B: Biological Sciences* **270**:2073–2078.

Price, E.O., *et al.* (1994) Effect of early experience on the sexual performance of yearling rams. *Applied Animal Behaviour Science* **42**:41–48.

Prins, H.H.T. (1996) *Ecology and Behaviour of the African Buffalo: Social Inequality and Decision Making*. London: Chapman & Hall.

Ralls, K. (1971) Mammalian scent marking. *Science* **171**:443–449.

Ramm, S.A. and Stockley, P. (2007) Ejaculate allocation under varying sperm competition risk in the house mouse, *Mus musculus domesticus. Behavioral Ecology* **18**:491–495.

Ramm, S.A. and Stockley, P. (2009) Adaptive plasticity of mammalian sperm production in response to social experience. *Proceedings of the Royal Society of London. Series B: Biological Sciences* **276**:745–751.

Ramm, S.A., *et al.* (2005) Sperm competition and the evolution of male reproductive anatomy in rodents. *Proceedings of the Royal Society of London. Series B: Biological Sciences* **272**:949–955.

Ramm, S.A., *et al.* (2009) Comparative proteomics reveals evidence for evolutionary diversification of rodent seminal fluid and its functional significance in sperm competition. *Molecular Biology and Evolution* **26**:189–198.

Reby, D. and Charlton, B.D. (2012) Attention grabbing in red deer sexual calls. *Animal Cognition* **15**:265–270.

Reby, D. and McComb, K. (2003) Vocal communication and reproduction in deer. In: *Advances in the Study of Behavior*, Vol. **33** (eds P.J.B. Slater, J.S. Rosenblatt, C.T. Snowdown, T.J. Roper and M. Naguib). San Diego, CA: Academic Press, 231–264.

Reby, D., *et al.* (1998) Individuality in the groans of fallow deer (*Dama dama*) bucks. *Journal of Zoology* **245**:79–84.

Reby, D., *et al.* (1999) Spectral acoustic structure of barking in roe deer (*Capreolus capreolus*). Sex-, age- and individual-related variations. *Comptes Rendus de l'Academie des Sciences Serie III. Sciences de la Vie* **322**:271–279.

Reby, D., *et al.* (2005) Red deer stags use formants as assessment cues during intrasexual agonistic interactions. *Proceedings of the Royal Society of London. Series B: Biological Sciences* **272**:941–947.

Rechav, Y., *et al.* (1978) Attraction of the tick *Ixodes neitzi* to twigs marked by the klipspringer antelope. *Nature* **275**:310–311.

Reinhold, K., *et al.* (2002) Cryptic male choice: sperm allocation strategies when female quality varies. *Journal of Evolutionary Biology* **15**:201–209.

Ribble, D.O. (1992) Lifetime reproductive success and its correlates in the monogamous rodent, *Peromyscus californicus. Journal of Animal Ecology* **61**:457–468.

Roberts, S.C. (1997) Selection of scent-marking sites by klipspringers (*Oreotragus oreotragus*). *Journal of Zoology* **243**:555–564.

Roberts, S.C. and Dunbar, R.I.M. (2000) Female territoriality and the function of scent-marking in a monogamous antelope (*Oreotragus oreotragus*). *Behavioral Ecology and Sociobiology* **47**:417–423.

Roberts, S.C. and Lowen, C. (1997) Optimal patterns of scent marks in klipspringer (*Oreotragus oreotragus*) territories. *Journal of Zoology* **243**:565–578.

Robinson, M.R., *et al.* (2006) Live fast, die young: trade-offs between fitness components and sexually antagonistic selection on weaponry in Soay sheep. *Evolution* **60**:2168–2181.

Robinson, M.R., *et al.* (2008) Environmental heterogeneity generates fluctuating selection on a secondary sexual trait. *Current Biology* **18**:751–757.

Roper, T.J., *et al.* (1986) Scent marking with faeces and anal secretion in the European badger (*Meles meles*): seasonal and spatial characteristics of latrine use in relation to territoriality. *Behaviour* **97**:94–117.

Rose, R.W., *et al.* (1997) Testes weight, body weight and mating systems in marsupials and monotremes. *Journal of Zoology* **243**:523–531.

Rosser, A.M. (1990) A glandular neckpatch secretion and vocalizations act as signals of territorial status in male puku (*Kobus vardoni*). *African Journal of Ecology* **28**:314–321.

Rubenstein, D.I. (1980) On the evolution of alternative mating strategies. In: *Limits to Action: The Allocation of Individual Behaviour* (ed. J.E.R. Staddon). New York: Academic Press, 65–100.

Sachser, N., *et al.* (1999) Behavioural strategies, testis size and reproductive success in two caviomorph rodents with different mating systems. *Behaviour* **136**:1203–1217.

Schaller, G.B. (1972) *The Serengeti Lion: A Study of Predator–Prey Relations*. Chicago: Chicago University Press.

Schradin, C., *et al.* (2009) Testosterone levels in dominant sociable males are lower than in solitary roamers: physiological differences between three male reproductive tactics in a sociably flexible mammal. *American Naturalist* **173**:376–388.

Schradin, C., *et al.* (2010) The nasty neighbour in the striped mouse (Rhabdomys pumilio) steals paternity and elicits aggression. *Frontiers in Zoology* **7**:19.

Schurmann, C.L. and van Hooff, J.A. (1986) Reproductive strategies of the orang-utan: new data and a reconsideration of existing sociosexual models. *International Journal of Primatology* **7**:265–287.

Scott, D.K. (1988) Reproductive success in Bewick's swans. In: *Reproductive Success: Studies of Individual Variation in Contrasting Breeding Systems* (ed. T.H. Clutton-Brock). Chicago: University of Chicago Press, 220–236.

Sekulic, R. (1982) The function of howling in red howler monkeys (*Alouatta seniculus*). *Behaviour* **81**:38–54.

Setchell, J.M. (2005) Do female mandrills prefer brightly colored males? *International Journal of Primatology* **26**:715–735.

Setchell, J.M. and Dixson, A.F. (2001) Changes in the secondary adornments of male mandrills (*Mandrillus sphinx*) are associated with gain and loss of alpha status. *Hormones and Behavior* **39**:177–184.

Sherman, P.W. (1989) Mate guarding as paternity insurance in Idaho ground squirrels. *Nature* **338**:418–420.

Shuurs, A.H.W.M. and Verheul, H.A.M. (1990) Effect of gender and sex steroids on the immune response. *Journal of Steroid Biochemistry* **35**:157–172.

Sliwa, A. and Richardson, P.R.K. (1998) Responses of aardwolves, *Proteles cristatus*, Sparrman 1783, to translocated scent marks. *Animal Behaviour* **56**:137–146.

Solmsen, N., *et al.* (2011) Highly asymmetric fine-scale genetic structure between sexes of African striped mice and indication for condition dependent alternative male dispersal factors. *Molecular Ecology* **20**:1624–1634.

Spickett, A.M., *et al.* (1981) *Ixodes (Afrixodes) matopi* n. sp. (Acarina: Ixodidae): a tick found aggregating on pre-orbital gland scent marks of the klipspringer in Zimbabwe. *Onderstepoort Journal of Veterinary Research* **48**:23–30.

Stenström, D. (1998) *Mating behaviour and sexual selection in non-lekking fallow deer (*Dama dama*)*. PhD thesis, Uppsala University.

Stevenson, I.R., *et al.* (2004) Adaptive reproductive strategies. In: *Soay Sheep: Dynamics and Selection in an Island Population* (eds T.H. Clutton-Brock and J.M. Pemberton). Cambridge: Cambridge University Press, 243–275.

Stockley, P. (1997) Sexual conflict resulting from adaptations to sperm competition. *Trends in Ecology and Evolution* **12**:154–159.

Stockley, P. and Preston, B.T. (2004) Sperm competition and diversity in rodent copulatory behaviour. *Journal of Evolutionary Biology* **17**:1048–1057.

Stockley, P. and Purvis, A. (1993) Sperm competition in mammals: a comparative study of male roles and relative investment in sperm production. *Functional Ecology* **7**:560–570.

Storz, J.F. (1999) Genetic consequences of mammalian social structure. *Journal of Mammalogy* **80**:553–569.

Taborsky, M., *et al.* (2008) The evolution of alternative reproductive tactics: concepts and questions. In: *Alternative Reproductive Tactics: An Integrative Approach* (eds R.F. Oliveira, M. Taborsky and H.J. Brockmann). Cambridge: Cambridge University Press, 1–21.

Temeles, E.J. (1994) The role of neighbours in territorial systems: when are they 'dear enemies'? *Animal Behaviour* **47**:339–350.

Tinbergen, N. (1953) *Social Behaviour in Animals*. London: Methuen.

Tutin, C.E.G. (1980) Reproductive behavior of wild chimpanzees in the Gombe National Park, Tanzania. *Journal of Reproduction and Fertility Supplement* **28**:43–57.

Utami, S.S., *et al.* (2002) Male bimaturism and reproductive success in Sumatran orang-utans. *Behavioral Ecology* **13**:643–652.

van de Pol, M., *et al.* (2006) Experimental evidence for a causal effect of pair-bond duration on reproductive performance in oystercatchers (*Haematopus ostralegus*). *Behavioral Ecology* **17**:982–991.

van Hooff, J.A.R.A.M. (2000) Relationships among non-human primate males: a deductive framework. In: *Primate Males: Cause and Consequences of Variation in Group Composition* (ed. P.M. Kappeler). Cambridge: Cambridge University Press, 183–191.

Vannoni, E. and McElligott, A.G. (2008) Low frequency groans indicate larger and more dominant fallow deer (*Dama dama*) males. *PLOS ONE* **3**:e3113.

Vannoni, E. and McElligott, A.G. (2009) Fallow bucks get hoarse: vocal fatigue as a possible signal to conspecifics. *Animal Behaviour* **78**:3–10.

Vasey, P.L. (1995) Homosexual behavior in primates: a review of evidence and theory. *International Journal of Primatology* **16**:173–204.

Viitala, J., *et al.* (1995) Attraction of kestrels to vole scent marks visible in ultraviolet light. *Nature* **373**:425–427.

Weingrill, T., *et al.* (2003) Male consortship behaviour in chacma baboons: the role of demographic factors and female conceptive probabilities. *Behaviour* **140**:405–427.

West, P.M. and Packer, C. (2002) Sexual selection, temperature, and the lion's mane. *Science* **297**:1339–1343.

Wilson, A.C., *et al.* (1975) Social structuring of mammalian populations and rate of chromosomal evolution. *Proceedings of the National Academy of Sciences of the United States of America* **72**:5061–5065.

Wilson, K., *et al.* (2004) Parasites and their impact. In: *Soay Sheep: Dynamics and Selection in an Island Population* (eds T.H. Clutton-Brock and J.M. Pemberton). Cambridge: Cambridge University Press, 113–165.

Wilson, M.L., *et al.* (2001) Does participation in intergroup conflict depend on numerical assessment, range location, or rank for wild chimpanzees? *Animal Behaviour* **61**:1203–1216.

Wlasiuk, G. and Nachman, M.W. (2010) Promiscuity and the rate of molecular evolution at primate immunity genes. *Evolutionary Anthropology* **64**:2204–2220.

Wrangham, R.W. and Peterson, D. (1996) *Demonic Males. Apes and the Origins of Human Violence*. New York: Houghton Mifflin Harcourt

Wyman, M.T., *et al.* (2008) Amplitude of bison bellows reflects male quality, physical condition and motivation. *Animal Behaviour* **76**:1625–1639.

Zahavi, A. (1975) Mate selection: a selection for a handicap. *Journal of Theoretical Biology* **53**:205–214.

Zahavi, A. (1977) The cost of honesty: further remarks on the handicap principle. *Journal of Theoretical Biology* **67**:603–605.

CHAPTER 14

Relationships between males in multi-male groups

14.1 Introduction

In a clearing in an African rain forest, an ageing dominant male chimpanzee sits close to two females feeding in a group of vines. A younger male runs into the clearing, the hair on his head and shoulders bristling and moves across to the two females. One of them immediately presents her rear to the older male, who briefly inspects it while the other continues to feed. The young male hesitates for a moment, glances across at the older male and then attacks the female, tugging her out of the vine, slapping her and sending her sprawling. Screaming with fear, the female presents to the younger male, who briefly pokes her rear end with his finger, sniffs it and then settles down to feed nearby. Throughout the interaction, the dominant male feeds intently, gradually turning away from the younger male and the two females and does not intervene. His fighting ability is waning, the frequency with which he intervenes to prevent younger males from directing aggression at females is falling and so, too, is their tendency to seek his support.

Where more than one male is associated with a group of females, competition for resources and breeding opportunities is inevitable and fights between resident males are likely. Since fighting is costly, strong selection pressures are likely to favour individuals that avoid contests with rivals they are unlikely to beat (see Chapter 13). As a result, where individuals can recognise each other, males commonly avoid fighting with animals that have beaten them and move away when they approach. Where one individual consistently wins or displaces another in agonistic encounters, they are said to be dominant to them and individuals that consistently lose interactions (or are displaced) are said to be subordinate (Schjelderup-Ebbe 1922; Hinde 1976; Drews 1993). Dominance is consequently an attribute of relationships between pairs of individuals, rather than of individuals.

Where one animal can consistently beat or displace another, it can usually beat or displace most of the individuals that are subordinate to it, since the same characteristics tend to determine the outcome of disputes between different pairs of animals. As a result, it is usually possible to identify dominance relationships between males belonging to the same group and dominance hierarchies among males have been described in a wide range of mammals, including macropods, rodents, carnivores, ungulates and primates (Bernstein 1976, 1981; Dewsbury 1982; Jarman 2000).

Where all males interact with each other and relationships are linear (so that Male A interacts with and dominates all individuals that are dominated by Male B who, in turn, dominates all individuals dominated by Male C), simple indices, such as the proportion of males dominated, provide a reliable index of social rank. However, in many groups, some males rarely interact with each other and there are irregularities in the structure of relations between individuals (for example, cases where A dominates B, who dominates C who dominates A but not B) which may be a consequence of the recent history of interactions. The presence of these irregularities complicates the estimation of relative rank (Appleby 1983) and makes it necessary to calculate indices of rank that incorporate a measure of the relative rank of individuals that each male dominates (Whitehead 2008). Several are in use, including David's score (David 1988; Drews 1993) and Elo-rating, a process that incorporates the sequence in which interactions occur (Neumann *et al.* 2011). When hierarchies are linear, different indices generally give similar results, but they diverge when reversals are

Mammal Societies, First Edition. Tim Clutton-Brock.
© 2016 Tim Clutton-Brock. Published 2016 by John Wiley & Sons, Ltd.

common and social structure is more complex (Bayly *et al.* 2006).

Where two or more males are able to deter or displace rivals more effectively than single individuals, selection may favour the development of cooperation between males (Díaz-Muñoz *et al.* 2014). In some mammals, two or more males cooperate to defend access to groups of females (see Chapter 10), while in others resident males form temporary coalitions or longer-lasting alliances which they use in competition for breeding opportunities with rivals. Relationships of this kind commonly lead to complex interactions between resident males as well as to irregularities in dominance relations and it has been suggested that they may favour the evolution of improvements in cognitive abilities.

This chapter examines the causes and consequences of variation in dominance relations between males, primarily in species where breeding males are not closely related to each other. Section 14.2 reviews our understanding of the factors that affect development of dominance relations between males and the causes of individual differences in dominance rank. Sections 14.3 and 14.4 explore the consequences of social rank, while sections 14.5 and 14.6 examine the impact of coalitions and alliances on dominance relationships, social status and reproductive skew among males. Section 14.7 explores the dynamics of supportive relationships and the predictions of market models, while section 14.8 examines the tactics used by males to attract and retain allies and disrupt the alliances of competitors. Finally, section 14.9 reviews some of the consequences of dominance relations for patterns of kinship between group members, for social relations between females and for the evolution of Machiavellian behaviour.

14.2 The development of dominance

In many mammals, male dominance and success in fights are effectively inseparable. For example, in red deer, bighorn sheep, elephants and many social primates where escalated fights are common, the social rank of males depends on their fighting ability and losers often suffer an immediate change in their dominance status relative to the winner (Appleby 1980; Poole 1989; Pelletier and Festa-Bianchet 2006; Alberts 2012). In these species, the rank of males is often associated with traits that affect the fighting ability of males directly. For example, in bighorn sheep, where fights between males involve pushing or ramming contests (Figure 14.1), the dominance rank of males depends on their size and weight (Coltman *et al.* 2002; Pelletier and Festa-Bianchet 2006) (Figure 14.2). Size, weight and condition also have an important influence on male rank and breeding success in many rodents (Hoogland 1995; Asher *et al.* 2008) and pinnipeds (Le Boeuf and Reiter 1988), as well as in some primates (Alberts 2012; Georgiev *et al.* 2015). However, positive correlations between rank and body size or weight are not universal and depend on the way in which males fight. For example, in horses, where individuals fight by biting rather than pushing (see Figure 13.2a), there is no consistent relationship between dominance and body size or weight (Feh 1990, 2001).

The presence of functional weapons, like horns, antlers and canine teeth, can also have an important influence on the fighting success and dominance rank of males. In red deer, some males ('hummels') lack antlers and are unable to fight effectively and so seldom breed

(a)   (b)

Figure 14.1 In bighorn sheep, which fight by charging at each other, horn size and body weight play an important role in determining the outcome of fights. *Source*: © Fanie Pelletier.

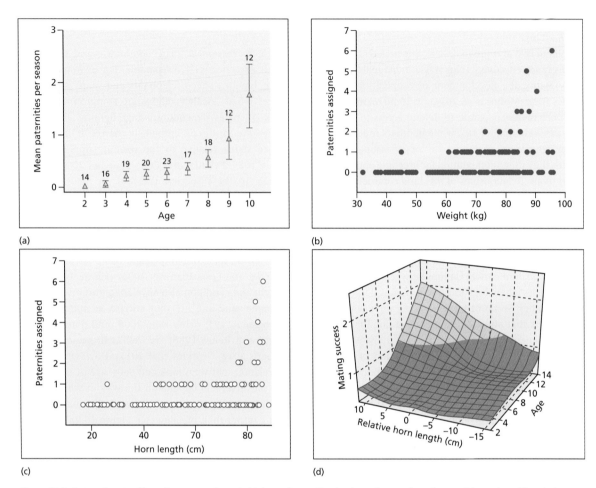

Figure 14.2 Determinants of breeding success in male bighorn sheep. Graphs show the number of paternities assigned in relation to ram age (a), weight (b) and horn length (c); (d) surface plot of fitted mating success predicted by a linear mixed-effects model as a function of age and horn length. *Source*: From Coltman *et al.* (2002). Reproduced with permission from the Royal Society.

successfully (Lincoln *et al.* 1970). Individual differences in antler size are also positively correlated with variation in breeding success in red deer (Kruuk *et al.* 2002), though this does not necessarily indicate that a causal relationship is involved and the relationship may be a result of correlations between antler size and the strength or condition of individuals. The seasonal loss of antlers affects the dominance rank of mature stags and leads to situations where they are temporarily subordinate to younger males who cast their antlers later in the season (Lincoln *et al.* 1970; Lincoln 1994) (Figure 14.3): after losing their antlers, stags revert to boxing with rivals, like hinds (Figure 14.4).

In some social mammals, support from other group members also has an important influence on the

competitive ability and rank of males. For example, in social primates where males disperse and immigrate independently into breeding groups (as in many baboons and macaques), unrelated males often form coalitions with each other which they use in competitive interactions with other males over access to resources (Figure 14.5). The same individuals tend to form coalitions repeatedly and the number and rank of regular allies that a male has can have an important influence on his rank and can help older males retain their status after their fighting ability has begun to wane (de Waal 1985, 1986; Alberts 2012).

Support from maternal relatives can also have an important influence on the relative rank of males in

Figure 14.4 After shedding their antlers, stags resort to boxing with rivals, like hinds. *Source*: © Fiona Guinness.

Figure 14.3 After red deer stags lose their antlers in early spring, they start to regrow their antlers immediately. While their antlers are in velvet, they are sensitive and males that have cast their antlers are unable to fight effectively with those that still retain their antlers. Since older stags cast earlier than younger ones, they often become subordinate to younger males. *Source*: © Tim Clutton-Brock

their natal group. For example, in spotted hyenas, natal males dominate immigrants as well as females ranked below their mother in the clan's hierarchy, though after dispersing and joining another clan, they become subordinate both to resident females and to resident males (Smale *et al.* 1993, 1997). Support from maternal kin can also be important in species where males remain and breed in their group. For example, in bonobos (Figure 14.6), sons commonly remain in the same social group as their mothers and often maintain lifelong social bonds with them (Ihobe 1992). Mothers support their sons in agonistic interactions with other males (Furuichi 1989; Kano 1992) and play an important role in helping them to acquire and maintain high social rank (Kano 1992, 1996). Support from unrelated allies can also play an important role in the acquisition and maintenance of high rank and regular alliances between males occur in many species where several unrelated breeding males live with females in stable groups (see section 14.6).

As among females, there are marked interspecific differences in the frequency and direction of aggression between males, as well as in the consistency of outcomes of agonistic interactions and the extent to which individuals depend on alliances with other group members (Preuschoft *et al.* 1998; van Noordwijk and van Schaik 2004; Thierry 2007). Where resident males are familiar and related, either because they disperse together (as in multi-male groups of lions and meerkats) or because they remain in their natal group (as in muriquis and chimpanzees), relationships between them tend to be relatively tolerant and egalitarian and the outcome of interactions is less consistent than it is in species where resident adult males immigrate independently and are mostly unrelated to each other, as in many of the baboons and macaques.

Figure 14.5 Male rhesus macaques face each other during a competitive encounter. In many of the baboons and macaques, an individual's rank depends on the support it can attract from allies. *Source*: © Kevin Rosenfield.

The determinants of rank also vary between species. In some species, the rank of males depends solely on their capacity to win fights, whereas in others it depends primarily on the extent to which they are supported by allies of either sex. There are also marked differences in dominance style among males that are not captured by descriptions of the structure of hierarchies and the relative rank of individuals (Watts 2010). For example, agonistic interactions between males may be common or rare; aggression may be unidirectional or bidirectional;

Figure 14.6 In bonobos, the rank and breeding success of males depends on maternal support. *Source*: © Pauline Toni.

hierarchies may be relatively linear or may involve many irregularities; and dominance relationships may be stable or constantly changing (Thierry 1985, 2004; de Waal 1989a; Flack and de Waal 2004).

14.3 Dominance and breeding success

Where several unrelated breeding males associate with groups of breeding females, males typically compete to monopolise access to receptive females during the stages of their cycle when they are likely to conceive and the annual breeding success of males is positively correlated with their dominance rank. Positive correlations between social rank and male mating success, based on either observational or genetic measures, have been demonstrated in a wide range of ungulates (Lott 1979; Clutton-Brock *et al.* 1982), rodents (Hoogland 1995) and primates (Ellis 1995; Rodriguez-Llanes *et al.* 2009; Alberts 2012). Both the dominance rank of males and their reproductive success are often correlated with their size and weight (Clutton-Brock *et al.* 1982; Le Boeuf and Reiter 1988; Coltman *et al.* 2002; Asher *et al.* 2008) and where there is no consistent relationship between male size and dominance rank, there is usually no consistent relationship between size and breeding success

Figure 14.7 Two red deer yearling males differing in body size and antler development. In red deer, the antler size of yearling males predicts their breeding success as adults. *Source:* © Tim Clutton-Brock.

either. For example, in horses, meerkats and some lemurs, where males fight by attempting to bite each other, their rank is not consistently related to their size and weight and their breeding success does not depend on their size and weight either (Feh 1990, 2001; Lawler *et al.* 2005; Clutton-Brock *et al.* 2006).

Where the rank and breeding success of males is affected by their size and strength, male rank and breeding success are often also correlated with differences in early development and growth (Lindström 1999; Alberts 2012). For example, in red deer, males that are well grown as yearlings show consistently higher breeding success as adults than smaller individuals (Figure 14.7). In primates, too, variation in early development has an important influence on the age at which individuals attain adult rank and begin to breed as well as on the rank that they attain (Alberts 2012). Variation in early development is commonly correlated with individual differences in birthweight and survival and several studies have shown that heavy-born males are more successful as adults and that light-born ones are less likely to survive and breed successfully (see Chapter 5). For example, in red deer, lifetime breeding success is positively correlated with birthweight in males but not in females (see Figure 14.8).

Where adult rank and breeding success are correlated with early development, the sons of dominant mothers are often more successful than those of subordinates. In red deer, sons born to dominant mothers are more successful in competing for harems during the annual rut than those born to subordinate females and breed more successfully throughout their lives (Figure 14.9) (Clutton-Brock *et al.* 1984, 1986). Similarly, in yellow

baboons and several species of macaques, the sons of dominant mothers grow faster, reach sexual maturity at a younger age and start to breed earlier than those of subordinates (Paul *et al.* 1992; Alberts and Altmann 1995; Altmann and Alberts 2005). Like dominant females (see Chapter 8), dominant males often have priority of access to resources which may reinforce early differences in growth and condition (Post *et al.* 1980; Barrette and Vandal 1986). For example, in red deer, dominant males obtain priority of access to preferred grazing sites and priority access to resources may contribute to their condition and reinforce their status (Appleby 1980).

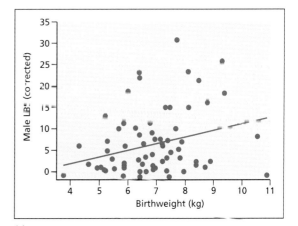

(a)

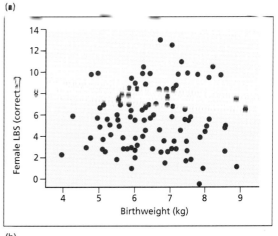

(b)

Figure 14.8 Estimates of lifetime breeding success in male and female red deer plotted on birthweight. Lifetime breeding success (LBS) values have been corrected for possible effects of other variables and the regression lines shown are from generalised linear models. *Source:* From Kruuk *et al.* (1999).

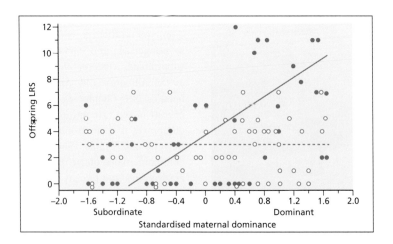

Figure 14.9 Estimates of lifetime reproductive success (LRS) based on observational data for male (solid circles) and female (open circles) red deer in relation to the relative rank of their mother. *Source*: From Clutton-Brock *et al.* (1984). Reproduced with permission of Nature Publishing Group. *Photo source*: © Tim Clutton-Brock.

Correlations between early development and adult reproductive performance may sometimes be a consequence of the effects of maternal rank on the growth of their offspring, though other mechanisms can also be involved. For example, in spotted hyenas, where males disperse to breed, the sons of high-ranking mothers grow faster, are more likely to disperse to clans providing good breeding opportunities, start breeding earlier and have higher reproductive success than sons born to low-ranking mothers (East and Hofer 2001; Engh and Holekamp 2003). However, in this case, it is unlikely that maternal rank influences the fitness of sons through differences in growth, for the size and fighting ability of emigrant males

are not related to their breeding success, which depends principally on the clans they join after dispersing from their natal group as well as on the relationships they develop with individual females (Höner *et al.* 2010).

Social factors that affect the rank of males often also affect their mating success, and where males form coalitions to compete for females, rank and breeding success can be more strongly influenced by social connections than by size or weight. For example, in Assamese macaques, the strength of social bonds between males predicts their future dominance rank, mating success and the number of their offspring (Schülke *et al.* 2010) (Figure 14.10), while in chacma baboons the skill of

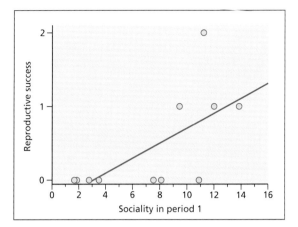

Figure 14.10 Social bonds affect breeding success in male Assamese macaques. On average, males form strong bonds with only three of the ten or eleven co-residents. The strength of the top three bonds ('Sociality in period 1') predicted reproductive success (number of infants sired) measured in the subsequent mating season. The relationship remained significant when the three least-social males that did not sire offspring were excluded. *Source:* From Schülke *et al.* (2010). Reproduced with permission of Elsevier. *Photo source:* © Julia Ostner.

males in managing social relationships and alliances appears to have a stronger influence on their rank than their relative size (Kitchen *et al.* 2003; Cheney and Seyfarth 2008). In chimpanzees, too, males establish their social rank by strategic alliances involving females as well as other males and social connections and alliances play an important role in establishing the rank of males (de Waal 1982; Pusey 1983; Goodall 1986; Langergraber 2012).

Although the breeding success of males is usually positively associated with their dominance rank, the strength of correlations between mating success and rank varies and some studies have found no significant correlation between reproductive success and rank or even (in a small number of cases) a negative correlation (Cowlishaw and Dunbar 1991). In some species where breeding males are familiar relatives, they are tolerant of each other and reproductive skew is relatively low (see Chapter 11). In species where breeding males are unrelated immigrants, the strength of relationships between male mating success and dominance rank can also vary between social groups, as well as within the same group over time (Alberts 2012). For example, within the Amboseli baboon population, where relationships between male mating success and rank have been measured over 32 group-years, correlation coefficients range from +1.0 to –0.7 between group-years (Alberts *et al.* 2003).

Several different mechanisms can affect the ability of males to monopolise females and the strength of relationships between mating success and dominance rank in males. In some species, male mating success depends principally on scramble competition for mating with the maximum number of partners rather than on the number of females that males can guard successfully, and the effects of dominance are reduced: examples include mouse lemurs, Soay sheep and some cavies (Sachser *et al.* 1999; Eberle and Kappeler 2004; Pemberton *et al.* 2004). The distribution of receptive females in time can also affect the ability of males to guard females effectively and influence relationships between the mating success of males and their dominance rank (see Chapter 12). For example, among primates, the mating success of males appears to be more closely related to their dominance rank in aseasonal breeders, like Tonkean macaques, than in more seasonal species, like rhesus macaques, where several females may be in oestrus at the same time and reproductive skew is relatively low (Thierry 2007; Watts 2010).

'Sneaky' tactics adopted by subordinate males can also weaken relationships between social rank and mating success among males, especially in less open environments where females are difficult to guard (Alberts 2012; Gogarten and Koenig 2013) and the strength of female mating preferences can also be important. Although females often show a preference for mating with dominant males (see Chapter 4), this is not universal and female choice can also lead to situations in which there is no relationship between male rank and mating success. For example, in spotted hyenas (where females are dominant to males and are able to control the identity of their mating partners) correlations between dominance rank in immigrant males and breeding success are not close (see Chapter 13) Similarly, in some primates where the dominance rank of older males is maintained by support from allies, female preferences for mating with younger, immigrant males can generate negative correlations between the rank of males and their breeding success. In Japanese macaques, for example, male dominance follows a 'seniority' rule and older immigrant males usually dominate younger ones, but female preferences for mating with younger males often mean that relationships between mating success and dominance rank are inconsistent and can even be negative (Inoue and Takenaka 2008; Alberts 2012).

While most of this section has described systems where aggressive interactions between males are relatively frequent and hierarchies are obvious and well defined, in other species relationships between resident mature males are relaxed and tolerant. In many of these species, breeding males have been reared in the same group and have either remained there (as in spider monkeys, bonobos and killer whales) or have emigrated and joined a new group together (see Chapter 11). For example, in muriquis males and females live in multi-male, multi-female groups that (in some areas) break up into separate foraging groups (Di Fiore and Campbell 2007). Males commonly remain and breed in their natal group and the presence of related females may increase the access of males to potential breeding partners (Strier *et al.* 2011). Females, who typically disperse to breed in other groups (Strier and Ziegler 2000; Strier *et al.* 2002), commonly mate with multiple partners in each breeding cycle (Figure 14.11). Relationships between male muriquis from the same group are tolerant and relaxed and there is little increase in circulating levels of testosterone or cortisol at the onset of the breeding season (Strier 1992; Strier *et al.* 1999). The reasons for the tolerance of males and the lack of sexual dimorphism are uncertain: one suggestion is that this is connected to their arboreal lifestyle and suspensory mode of locomotion, which may prevent males from sequestering females against their will, and increases the capacity of females to choose mating partners (Strier 1992, 1997). In addition,

Figure 14.11 A male muriqui: muriqui groups typically consist of multiple breeding females and multiple males, which are often natals and are usually tolerant of each other. *Source*: © Thiago Cavalcante Ferreira.

suspensory locomotion may increase the costs of increased body weight to males.

14.4 Reproductive skew in multi-male groups

Where males benefit from alliances with other males, they may increase their chances of retaining the support of allies by tolerating their access to breeding females. 'Transactional' models of reproductive skew that predict the extent to which dominants should concede reproductive opportunities to subordinates in order to retain their support were originally developed to provide explanations of variation in degree of reproductive skew among females in cooperative breeders (see Chapter 17) but their predictions are also relevant to the evolution of reproductive skew in males (Keller and Reeve 1994; Clutton-Brock *et al.* 2001). They suggest that, under most conditions, the tendency for dominants to make reproductive concessions to subordinates will increase when the benefits of their assistance to dominants are high and they could easily form a new relationship with another group member.

Several studies of social mammals have produced evidence that is consistent with these predictions, though it is usually difficult to be certain that dominants willingly share reproductive opportunities with subordinates. For example, in gelada baboons, some breeding units contain

Figure 14.12 In gelada baboons, some males form alliances with other adult males (who are often maternal relatives) and jointly defend groups of breeding females. Here, two allied males are involved in a head-to-head dominance display. *Source*: © Clay Wilton.

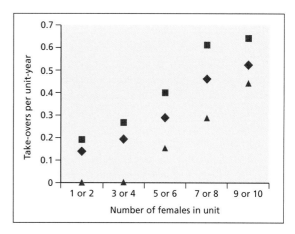

Figure 14.13 In geladas, the presence of 'follower' males within breeding units reduces the risk of the take-over of units by other males and so increases the tenure of breeding males. The figure shows the number of females in units versus take-over rates per unit-year. Diamonds represent all units; squares single-male units; triangles multi-male units. *Source*: From Snyder-Mackler *et al.* (2012). Reproduced with permission from the Royal Society. *Photo source*: © Noah Snyder-Mackler.

subordinate 'follower' males that gain occasional matings and help to defend breeding groups against potential challengers (Snyder-Mackler *et al.* 2012) (Figure 14.12). Units that include 'follower' males are less likely to be taken over by new immigrant males and the tenure of dominants is prolonged by their presence, so it is possible that these benefits affect the readiness of dominants to tolerate subordinate males in their unit (Figure 14.13). One study of chimpanzees has shown that the frequency with which subordinate males supported the alpha male in competitive interactions was correlated with the extent to which the dominant tolerated their attempts to mate with females and frequent supporters enjoyed higher mating success than those which seldom supported the alpha male (Duffy *et al.* 2007). In addition, in chacma baboons, the presence of several breeding males may reduce the risk that the offspring of alpha males will be killed by infanticidal immigrants after their death, which may favour dominant males that allow subordinates to share reproduction and could explain why subordinates often obtain a larger share of breeding attempts than would be expected from the priority of access model (Henzi *et al.* 2010).

A more specific prediction of transactional skew theory is that dominant males should concede a larger share of reproduction to unrelated subordinates to retain their assistance than to relatives (since unrelated males will not gain indirect benefits from their assistance) so that skew should be relatively high where resident males are related to each other (Keller and Reeve 1994). Some intraspecific comparisons appear to support this prediction. For example, in lions, reproductive skew among resident males is generally higher in coalitions of related males than in coalitions of unrelated individuals (Keller and Reeve 1994). However, as Packer has pointed out, coalitions of related males are generally larger than those of non-relatives and skew increases in larger coalitions since breeding is usually shared between coalition partners only when more than one female is in oestrus over the same period, as the priority of access model predicts (Packer *et al.* 1988).

14.5 Dominance, condition and survival

Both the dominance rank and the breeding success of males usually change with age, though not necessarily to the same extent. Where males compete independently, there is often a rapid increase in their breeding success after they attain sexual maturity, followed by a period when they are in their prime. In many polygynous mammals, the start of effective breeding typically occurs several years after individuals reach sexual maturity (see Chapter 10), but the rate at which male rank and reproductive success increase with age can extend over many years and varies widely between species (Poole *et al.* 2011). At some stage after males reach their prime, their social rank and average breeding success begin to fall. In some species, dominance tenure and breeding tenure are synonymous but, in others, previously dominant males retain relatively high levels of breeding success after their fighting ability and social rank have begun to wane. The rate of decline in breeding success with increasing age is often rapid but this, too, varies between species (Clutton-Brock and Isvaran 2007) and appears to be slower in species that live in multi-male groups, where males often form competitive coalitions, than in unimale groups (see Chapter 11).

The period between the age when a male starts to breed effectively and the age at which it ceases to do so is described as its breeding tenure or its effective breeding lifespan and its duration often has a large influence on lifetime breeding success (see Figure 11.18). For example, in spotted hyenas, sifakas, baboons and chimpanzees, individual differences in the breeding tenure of males account for a large proportion of variation in male reproductive success (Alberts 2012) and similar associations have been found in many other species (Clutton-Brock *et al.* 1988, 2006; Hoogland 1995; Pemberton *et al.* 2004). The duration of breeding tenure in males is often related both to the intensity of competition for breeding opportunities and to the costs of mate guarding. High-ranking, successful males often attract more frequent challenges from other group members than low-ranking ones and may be more likely to be wounded (Drews 1996; McCormick *et al.* 2011), increasing the chance that they will be replaced and reducing their tenure. However, effects of this kind may be offset or obscured by positive correlations between the rank of individuals and their health or condition: for example,

one study of rhesus macaques has shown that male rank is positively correlated with the strength of immune responses and negatively with levels of oxidative damage (Georgiev *et al.* 2015)

As well as increasing the risk of injury, competition between males has important physiological consequences. Aggressive competition between group members of either sex commonly leads to heightened levels of glucocorticoid stress hormones that are released into the bloodstream in response to activation of the hypothalamus–pituitary–adrenal axis (Goymann and Wingfield 2004). For example, field studies of baboons show that cortisol levels in males increase at times when dominance relations are uncertain, rank relations are changing or new males have recently entered the group (Alberts *et al.* 1992; Sapolsky 1993; Bergman *et al.* 2005). Glucocorticoids function to mobilise stored resources and increase cardiovascular tone and can reduce costly anabolic processes, including energy storage, growth and immune function, but chronic elevation of glucocorticoid levels can also impair important processes including immune responses and reproduction (Sapolsky *et al.* 2000, 2002).

Studies of relationships between the dominance rank of males and their glucocorticoid levels have produced mixed results. Although some early studies of captive mammals commonly found higher glucocorticoid levels in subordinates than dominants (von Holst 1998), several recent field studies of social species have either found no consistent relationship between male rank and glucocorticoid levels or higher levels in dominants than in subordinate males (Creel and Creel 2002; Creel *et al.* 2013; Goymann and Wingfield 2004; Gesquiere *et al.* 2011). One suggested explanation is that the relationship between male rank and glucocorticoid levels depends on the relative physiological costs of maintaining homeostasis (allostatic load) in dominants and, in particular, on the intensity of overt competition to gain and maintain high social rank (Goymann and Wingfield 2004). Where males have to fight at regular intervals to gain and maintain high rank, glucocorticoid levels are likely to be high and may exceed those of subordinates. In contrast, where dominant status is inherited or males queue peaceably for access to the breeding position and the acquisition and maintenance of high rank does not involve frequent competition, there may be no consistent difference in glucocorticoid levels between dominants and subordinates or subordinates may show higher levels

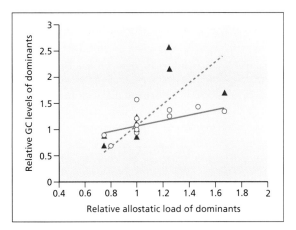

Figure 14.14 Effects of variation in the estimated cost to dominants of both sexes of maintaining their status (relative allostatic load) on the glucocorticoid (GC) levels of dominants relative to those of subordinates, as derived from phylogenetic independent contrasts, for females (triangles, dotted line) and males (circles, solid line). *Source*: From Goymann and Wingfield (2004). Reproduced with permission of Elsevier.

than dominants if they are frequent targets for aggression. Combining various indices of the intensity of conflict necessary to gain and maintain dominance in a sample of birds and mammals, Goymann and Wingfield (2004) show that, in both sexes, the relative glucocorticoid levels of dominants increase with qualitative estimates of the intensity of conflict in both sexes (Figure 14.14). Their results support the suggestion that the effects of rank on relative glucocorticoid levels depend on intensity of social conflict, though other social and ecological processes are also likely to modify the effects of competition on glucocorticoid levels, including the relative nutritional status of dominants and subordinates and variation in kinship between competing males (Rubenstein and Shen 2009; Creel *et al.* 2013).

One reason why understanding the relationship between male dominance rank and glucocorticoid levels is important is because chronically high glucocorticoid levels can depress immune function and lead to increases in parasite load and disease with potential effects on breeding success and survival (Sapolsky *et al.* 2002; Boonstra *et al.* 2007). Several recent studies have shown that dominant males have higher parasite burdens than subordinates. For example, in New Zealand fur seals, territorial males have higher parasite loads than non-territorial males (Negro *et al.* 2010), while in yellow baboons alpha males have higher levels of parasite loads

than subordinates (Hausfater and Watson 1976). Although increased parasite loads do not always depress fitness, they appear to do so in many naturally regulated populations (Gulland 1992; Gulland and Fox 1992; Gulland *et al.* 1993; Wilson *et al.* 2004). As a result, increases in allostatic load are likely to reduce reproductive tenure and longevity in dominant males and to constrain the effects of high skew within seasons on the lifetime breeding success of males.

Evidence of the costs of reproductive competition to males suggests that there may be trade-offs between high dominance rank in males and the period for which individuals maintain their status. Despite increased access to resources, the condition of dominant males is not necessarily higher than that of subordinates and relationships between male rank and breeding tenure or survival also vary. For example, in rabbits, high-ranking males have higher rates of survival than subordinates (von Holst *et al.* 1999), while dominant male prairie dogs (which are typically heavier than subordinates) do not show higher overwinter survival (Hoogland 1995). Evidence that the defence of preferred territories or large groups of females is often associated with reductions in male tenure (see Chapter 13) suggests that high rank may reduce the breeding tenure and longevity of males, though, as yet, few studies have investigated the effect of dominance rank on breeding tenure and longevity and further investigation is needed to identify the usual pattern of relationships.

An associated question is whether males adjust their reproductive behaviour to minimise the costs of maintaining high rank and prolong their tenure. Much of the observed variation in longevity may be a consequence of chance events rather than of individual differences in genotype or phenotype (Clutton-Brock 1988), but some aspects of male behaviour in multi-male groups (including the establishment of alliances with other males and with females) may represent strategies that help to prolong the breeding lifespans of males.

14.6 Coalitions and alliances

The distribution of coalitions

In some social mammals (and especially in primates), males commonly support each other in agonistic interactions with other males (Harcourt and de Waal 1992;

Langergraber 2012). Supportive interactions that are used in competition with other males are usually referred to as coalitions and where the same males assist each other repeatedly, they are said to have formed an alliance. In litter-bearing species, two or more related males (which have often immigrated together) commonly cooperate to defend groups of females (see Chapter 12), while in monotocous species where males immigrate independently into breeding groups, they commonly form coalitions and alliances with unrelated partners against competitors. Coalitionary behaviour between resident males appears to be most highly developed where group membership is stable, the number of resident males is relatively large and few males are closely related to each other (Bissonnette *et al.* 2014). The use of coalitions and alliances between males to gain access to receptive females was described in some of the earliest studies of baboons and macaques. In these species, males typically disperse from their natal group after reaching adolescence, and multiple unrelated males associate with female groups and compete for access to receptive females, More recently, male coalitions and alliances have been described in other social mammals where groups include several adult males and average coefficients of relatedness between males are low, including carnivores (Zabel *et al.* 1992; de Villiers *et al.* 2003) and cetaceans (Connor and Norris 1982; Connor *et al.* 1992a,b).

Although coalitionary behaviour between males appears to be most common in species where resident males are unrelated to each other, male coalitions and alliances also occur in some species where males remain in their natal group, including spider monkeys and chimpanzees (Fedigan and Baxter 1984; Watts 1998). Here, males have the opportunity to ally with close maternal or paternal kin and often do so: for example, studies of wild chimpanzees show that maternal relatives have a tendency to form alliances with each other, though alliances can also involve unrelated males (Langergraber *et al.* 2007; Mitani 2009a,b; Langergraber 2012). Although resident males are likely to be familiar with each other in these species, average relatedness between them is not necessarily high, since resident females are often unrelated, multiple females breed and male skew is often low (Lukas *et al.* 2005). In some mammals where resident males form alliances which they use to sequester and defend receptive females, two or more alliances may associate with each other to

form 'super-alliances': the best-known example concerns bottlenose dolphins from Shark Bay in Australia where alliances of males combine to compete for access to females with rival groupings (Connor *et al.* 1999, 2000).

Early research on primates showed that coalitionary support often affects the rank of males (section 14.2) as well as their reproductive success (section 14.3), leading to 'dependent' rank systems where rank is influenced by alliances with other group members. Several studies have demonstrated that the dependent rank of males is affected by the number, rank and social connections of their allies (Schülke and Ostner 2012). Individuals differ in their capacity to attract and maintain allies, depending partly on their rank, their access to resources and their social connections, which can all help to provide them with 'leverage' in attracting and maintaining allies (Lewis 2002; Flack and de Waal 2004).

Males usually have a small number of regular allies at any one time but alliances change as the relative benefits of forming coalitions with particular individuals declines or coalitions with other partners become more advantageous (Noë 1986; Noë and Sluijter 1995). Shifting alliances are particularly common in the higher primates that live in multi-male groups or communities with unrelated or distantly related individuals, including baboons and chimpanzees (de Waal 1982; Nishida and Hosaka 1996; Cheney and Seyfarth 2008). In many of these species, males compete to form relationships with prospective allies by maintaining close proximity with them, sharing access to resources and supporting them in interactions with other group members (de Waal 1982; Nishida and Hosaka 1996).

In some species, males intervene to disrupt the formation of alliances by competitors or to reduce their support by females (de Waal 1982; Watts 2010). For example, in captive chimpanzees, males that have recently risen in rank and are starting to challenge the alpha male sometimes show persistent aggression towards resident females until they replace the dominant, when their relationships with females immediately become more protective. De Waal (1982) interprets this as a strategy by younger males to weaken supportive relationships between females and the previous dominant and to increase the chance that they will be able to replace him as their principal protector.

The complex social tactics of males have led some primatologists to describe their behaviour as political (de Waal 1982; Flack and de Waal 2004), implying that individuals are aware of relationships between other animals and are capable of manipulating them to their own advantage (Watts 2010). Although this has sometimes attracted criticism, there is increasing evidence that individuals are aware of relationships between third parties in several primates, including capuchin monkeys, vervets, chacma baboons and chimpanzees and that they may attempt to manipulate them to their own advantage (Cheney and Seyfarth 1990, 2008; Perry *et al.* 2004) (see Chapter 6), so that describing them in these terms does not seem inappropriate.

Types of coalitions

Studies of primates have identified different types of coalition. Where resident males are ranked in a hierarchy, coalitions can either be used by two subordinate males to displace a more dominant animal (an 'all-up' coalition) or by two dominant males against a lower-ranking animal (an 'all-down' coalition) or by a dominant male and a subordinate against an individual of intermediate rank (a 'bridging' coalition) (Chapais 1995; van Schaik *et al.* 2006) (Figure 14.15). 'All-up' coalitions are frequently used by middle ranking males to displace more dominant individuals and gain access to receptive females or to establish higher status than males that outranked them on their own (Ehardt and Bernstein 1992; Chapais 1995; van Schaik *et al.* 2006) (Figure 14.16a). For example in some baboon species, middle-ranking males obtain between a quarter and a half of all

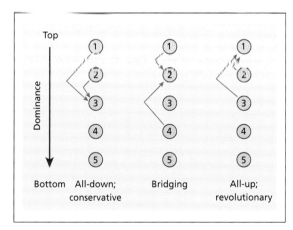

Figure 14.15 Contrasting types of coalitions observed in primates and their consequences. Arrows indicate the direction of aggression. *Source*: From van Schaik *et al.* (2006). Reproduced with permission of Springer Science and Business Media.

consortships by forming 'all-up' coalitions with other males to displace existing consorts (Bercovitch 1986; Noë 1990). Their targets are often high-ranking males that have recently joined the group (Noë 1992) who lose many consorts to coalitions of lower-ranking males (Collins 1981). Middle-ranking males also use 'all-up' coalitions to increase or maintain their dominance rank and coalition members are sometimes dominant to a rival male when both (or all) are present, though they continue to be subordinate when on their own (Figure 14.16b). For example, early studies described how three allied male baboons were able to out-rank all other group members when the three members of the coalition were together, though another male was individually dominant to each of them when they were on their own (Hall and DeVore 1965) and similar situations have been reported in a number of other social mammals (Harcourt and de Waal 1992; Gilby 2012). 'All-up' coalitions can alter the distribution of male breeding success and reduce reproductive skew.

'All-down' coalitions are often used by dominant individuals to suppress subordinate competitors and often do not involve regular allies (Zimen 1976; Noë 1992; Zabel *et al.* 1992). Attacks frequently appear to develop from dyadic interactions where dominant animals attack subordinates and are subsequently joined by some (or all) other group members. Supporting other dominant males in interactions with more subordinate males may help to consolidate dominance relationships between the aggressor and the victim or may reinforce alliances between dominant males (Ehardt and Bernstein 1992). For example, joint attacks on 'scapegoats' may serve to reduce tension between dominant males or may have the effect of reducing the frequency of aggression directed at other group members by younger males (de Waal 1977; Ehardt and Bernstein 1992). 'All-down' coalitions are likely to reinforce pre-existing hierarchies and may increase reproductive skew.

Finally, dominant males may form 'bridging' coalitions with relatively subordinate males against individuals of intermediate status (Chapais 1995; van Schaik *et al.* 2006). Bridging coalitions offer few direct advantages to dominant males, who out-rank the target male on their own and they are often used to assist lower-ranking relatives to increase their rank or to protect them against attacks from higher-ranking individuals that are subordinate to the dominant partner (van Schaik *et al.* 2006) (see Figure 14.16c,d).

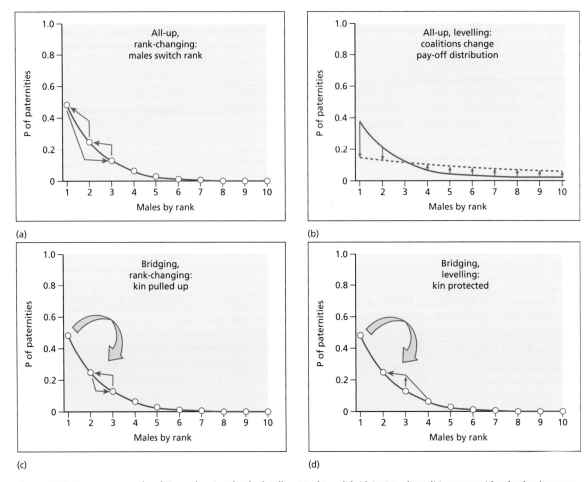

Figure 14.16 Consequences of coalitions, showing that both 'all-up' (a, b) and 'bridging' (c, d) coalitions may either lead to increases in rank or may be levelling (and so reduce differences in rank). *Source*: From van Schaik *et al.* (2006). Reproduced with permission of Springer Science and Business Media.

Comparative analyses suggest that the complexity of coalitionary behaviour is related to group size and to average levels of kinship within groups. Coalitions between males most commonly occur in species where group size is large, multiple males associate with groups and dominance relationships between them are well defined (Olson and Blumstein 2009). A subsequent analysis suggests that species where males form 'revolutionary' coalitions live in larger groups than those that show 'conservative' (all-down) coalitions and that (as a result of the association with group size) average levels of kinship are lowest in species that show revolutionary coalitions and highest in those that do not show coalitionary behaviours.

Evolution of coalitionary behaviour

The evolutionary mechanisms that maintain coalitions and alliances between males are still a matter of debate, fuelled partly by contrasting definitions of evolutionary processes (see Chapter 1). When allies are usually relatives who have remained in their natal group, or dispersed together and immigrated into the same group, or dispersed sequentially to the same breeding group, kin-selected benefits are likely to be involved (see Chapter 14). Relationships between close relatives (especially maternal sibs) are often stronger and more supportive than those between more distant relatives or non-kin (Colmenares 1992; Langergraber 2012) and, in some cases, resident males also support their own offspring more frequently than those of other males (see Chapter 16).

Where males disperse independently and form regular alliances with unrelated individuals (as in many of the baboons and macaques), indirect benefits cannot provide a general explanation of the evolution of alliances, though they may be involved in some cases where related males have immigrated independently into the same group (Langergraber 2012). Alliances between non-kin have frequently been interpreted as examples of reciprocal altruism or reciprocity (Mitani 2006; Hauser *et al.* 2009). For example, in one of the first detailed analyses of coalitionary behaviour between males, Packer (1977) argued that alliances between males were an example of reciprocal altruism (see Chapter 1): male baboons solicit assistance from other individuals by 'head flagging' and individuals that solicit assistance in this way from particular males are likely to be solicited by the same males in future. Packer suggested that as allies took turns to act as beneficiaries (participate and take the female) and altruists (participate without reward), these interactions corresponded to successive moves in an iterated Prisoner's Dilemma game (see Chapter 1). However, subsequent studies based on larger datasets show that after a consorting male has been displaced by a coalition, both members of the coalition commonly attempt to monopolise the female and that the individual that initiated the interaction is no more likely to obtain access to the female than the individual that he solicited (Bercovitch 1988; Noë and Sluijter 1990; Noë 1992).

Several forms of evidence suggest that the tendency for individuals to provide assistance in social interactions depends on the assistance they have received in the past. Although there is rarely any regular alternation of assistance given and received (Clutton-Brock 2009), the repeated failure of regular allies to respond to solicitation commonly leads to the termination of the alliances and the establishment of new ones with other partners (de Waal 1982; Mitani 2006; Gilby 2012). In addition, there is evidence of contingency in supportive and affiliative behaviours over shorter time-spans. For example, in pairs of unrelated female baboons recent grooming with a partner appears to increase the chances that an individual will provide support to the same individual shortly afterwards, especially if the caller has a dependant infant (Cheney *et al.* 2010). In contrast, recent grooming does not increase the responsiveness of related females. Interactions of this kind are commonly referred

to as 'interchanges' and are regarded as examples of reciprocity (Mitani and Watts 1999, 2001; Mitani 2006; Gilby 2012).

Although supportive relationships between individuals and examples of contingency between grooming and social support resemble the interactions considered by models of reciprocity, an alternative interpretation is that grooming and social support represent related components of affiliative relationships between individuals that are maintained by immediate mutual fitness benefits to both participants, even if both individuals do not benefit on every occasion or to the same extent (Bercovitch 1988; Noë and Sluijter 1990; Clutton-Brock 2009; Aureli *et al.* 2012). Mutualistic cooperation would be expected wherever the average, probabilistic pay-offs to both partners in a cooperating pair exceed the pay-off that either can expect from defecting (see Chapter 1). Evidence of contingent responses to partners based on prior experience of their behaviour does not conflict with this interpretation of coalitions, for it would be surprising if individuals were unselective in their choice of coalition partners or allies, since some partners are likely to generate greater benefits than those with others. Consequently, some form of monitoring of the outcome of interactions would be expected in situations where both coalition partners gain immediate benefits from interactions.

The distinction between these two interpretations of the evolution of coalitionary behaviour depends on the way in which fitness benefits are accounted (see Chapter 9). Models of reciprocity usually account the costs and benefits of individual actions immediately after they occur and so envisage situations where actions result in immediate net costs and subsequent larger benefits. They typically assess net benefits within a narrow time frame and assume that the winning strategy is the one that obtains the largest share of resources from a small number of interactions. In contrast, mutualistic explanations often consider the probabilistic effects of an individual's actions on its fitness and recognise that longer-term benefits may exceed immediate costs and should be included in calculations of the immediate fitness consequences of particular actions. Where these benefits are large (for example, because the availability of allies is low and neglected partners are likely to join competing alliances), the provision of assistance may, on average, provide immediate net benefits to both parties and defection may reduce (rather than increase) the fitness of defectors. Explanations of this kind suggest that interactions between allies represent limited public goods games and are maintained by benefits shared by cooperating individuals and that it may be more realistic to interpret the interactions between individuals as attempts to modify their social niches through the establishment and maintenance of supportive relationships than to regard them as exchanges (Clutton-Brock 2009). Unfortunately, which approach to accounting the costs and benefits of alliances is more realistic is not easily resolved by empirical research, since it is seldom feasible to measures changes in fitness over very short time-spans.

The distinction between these two views of the maintenance of coalitions and alliances may look like an example of hair splitting. However, the way in which these behaviours are interpreted affects other predictions and arguments. For example, several of the characteristics of Prisoner's Dilemma models that inhibit the development or maintenance of cooperative behaviour, such as the 'shadow of the future', evaporate if both partners derive immediate net benefits from coalitions (see Chapter 1). More importantly, it affects whether we should interpret supportive interactions between non-human animals as exchanges of social commodities or as attempts by individuals to maintain useful mutualistic relationships with other group members and to modify them to their own advantage.

14.7 Market models and the dynamics of supportive relationships

Both theoretical and empirical studies have also explored the choice of partners and the dynamics of relationships. As a result of differences in their independent dominance rank and their social connections, males differ in their power to affect the outcome of competitive interactions and preferentially seek to form alliances with powerful individuals (Lewis 2002; Watts 2010). 'Market' models of coalition formation developed by economists and psychologists (Kahan and Rapoport 1984) provide an appropriate theoretical framework for exploring their choice of partners (Noë 1992) as well as the distribution of inequalities in the gains derived from successful coalitions. Where two coalition partners are of similar power and both are the optimal allies for a similar number of other males, an approximately equal division of benefits

would be expected (Noë 1992). Conversely, where partners vary in power and one individual is the preferred partner for a larger number of other males and so has the option to join several other alliances, he has greater leverage and may be able to obtain a larger share of the benefits from his partner (Noë et al. 1991; Watts 2010). In some cases, one individual may be the only alliance partner capable of providing a particular benefit for several other males, placing him in a 'veto' position and allowing him to extract substantial concessions from his partners. Market models assume that potential alliance partners can signal their readiness to cooperate and establish an asymmetry in pay-off through some type of 'bargaining' process but, as Noë (1992) points out, this need not be complex: for example, powerful males that acquire a small share of gains after supporting an ally might signal discontent, indicating that they are likely to refuse subsequent invitations to collaborate and subordinates might respond to this by reducing their own share of the gains in subsequent interactions.

Although market-based models provide an intuitively satisfactory framework for understanding the formation of male alliances, many of their assumptions and predictions are difficult to test and the link between theory and empirical observation is still tenuous. It is usually impossible to measure differences in power on an ordinal scale, so that the relative benefits of allying with different partners cannot be predicted with certainty. In addition, evidence that interactions comply with their predictions does not necessarily indicate that they approximate to exchanges or transactions, for individuals are generally more likely to work harder, incur greater costs and accept relatively smaller benefits to obtain access to scarce or valuable resources in situations where no exchanges are involved (see Chapter 1).

14.8 Punishment, retaliation and reconciliation

Where several unrelated males are associated with the same group of breeding females, males often invest considerable time and energy in maintaining their social rank and repeated infringements of the interests of dominant males by intruders or subordinates rapidly attract threats (Figure 14.17), which increase in severity and may culminate in punishing attacks that have substantial costs to recipients (Clutton-Brock and Parker 1995). Punishment

Figure 14.17 A dominant male gelada baboon threatens an intruder by baring its teeth and flipping back its upper lip. If threats are ignored, they may be followed by attacks. *Source*: © Clay Wilton.

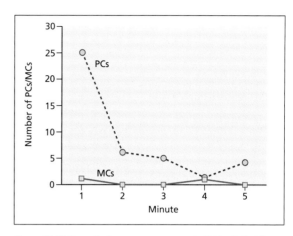

Figure 14.18 In male Japanese macaques, individuals that have been involved in conflicts (PCs) are more likely to interact affiliatively with each other during the next minute than matched controls (MCs). *Source*: From Majolo *et al.* (2005). Reproduced with permission of Springer Science and Business Media.

may take the form of persistent intolerance of the proximity of subordinates, disruption of their attempts to feed or mate, support for their rivals, damaging attacks or, in some cases, ejection from the group. In addition to its immediate costs, punishment may have longer-lasting physiological and psychological consequences for victims: individuals that are the victims of regular attacks often show increased levels of cortisol and other stress-related hormones and reduced levels of testosterone and other steroids (Sapolsky 1992). Associated changes in their signalling capacity and display characteristics may also lead to increases in the probability of challenges by rank neighbours (de Waal 1982; Cheney and Seyfarth 1990) and may affect their supportive relationships with other group members (Preuschoft and van Schaik 2000; Kitchen *et al.* 2003).

As in females, punishing interactions between males can lead to retaliation by victims or their allies against the punisher or his associates. For example, after top-ranking spotted hyenas chase subordinates, the subordinate may subsequently chase a low-ranking animal (Zabel *et al.* 1992). Since group members often join in attacks on victims, redirected aggression may reduce the initial victim's risk of becoming the target of a combined attack. In other cases, it may be more tactical: for example, in some primates, victims redirect aggression at relatives or allies of the original aggressor (Cheney and Seyfarth 1990, 2008).

To avoid protracted disruption, males that have been involved in agonistic interactions may benefit by re-establishing peaceable or affiliative relationships as soon as possible after exchanges of aggression (de Waal 1993;

Aureli *et al.* 2002, 2012). In several social mammals, attacks or fights between males, like those between females, are often followed by 'reconciliatory' interactions, initiated either by the attacker or punisher or by the victim, where one of the two contestants approaches the other and makes an affiliative gesture which is then reciprocated (de Waal 1989b, 1993; Samuels and Flaherty 2000) (Figure 14.18). For example, male chimpanzees that have attacked subordinate males may subsequently approach and hug them, while male macaques may groom individuals they have attacked (de Waal and van Roosmalen 1979; de Waal 1993).

Studies of several primates show that the probability of a further agonistic interaction is reduced after reconciliation has taken place and that individuals with valuable social partners are particularly likely to reconcile (Majolo *et al.* 2005; Fraser *et al.* 2010). For example, long-tailed macaques that had been trained to obtain food in a cooperative task reconciled three times as frequently as individuals that had not been trained to cooperate (Cords and Thurnheer 1993). Reconciliation often has immediate benefits to victims, who show reduced levels of anxiety or stress-related behaviour after reconciliation has occurred and the heart rate of individuals involved in agonistic interactions falls rapidly after reconciliatory exchanges (Aureli and van Schaik 1991; Cords 1992).

'Policing' of aggression between other group members by dominant males who threaten or attack one or both contestants in aggressive exchanges has also been described in a number of monkeys and apes (Das 2000; Petit and Thierry 2000; Watts *et al.* 2000). For example, in some species, dominant males will intervene in support of females or juveniles that are being attacked, threatened or harassed by subordinate males or by females (Ehardt and Bernstein 1992). In some cases where individuals intervene, they do so in support of relatives or allies (Noë 1992; Chapais 1995), but in others there is no close social bond between the intervener and the animal they are protecting (Noë 1992; Cords 1997). Although males sometimes intervene in support of victims, they can also intervene in support of aggressors (Petit and Thierry 1994).

'Policing' of this kind may help to limit the severity or frequency of aggression between females (Ehardt and Bernstein 1992; Watts 1997): for example, in captive macaques, the removal of adult males is commonly followed by an increase in the frequency of intensity of aggression between females (Ehardt and Bernstein 1992). Dominant individuals may also reassure or groom contestants and, in some macaques, 'peaceful' interventions are more frequent and more effective than aggressive ones (Petit and Thierry 1994, 2000). One explanation of these interactions is that they reduce the probability of further aggression being directed at the original victim while an alternative possibility is that they help to prevent disturbance spreading and disrupting the individuals involved in 'policing' (Cords 1997).

In some species, males use their relationships with infants and juveniles to deflect aggression from themselves or to help them to establish alliances with dominant individuals. For example, in Barbary macaques, some males form close associations with particular infants which they sometimes pick up in stressful or dangerous situations and use to deflect aggression from other males (Deag and Crook 1971; Deag 1980) (Figure 14.19). Infant carrying is associated with increases in cortisol levels and may consequently have appreciable costs (Henkel *et al.* 2010). While males are seldom the fathers of the infants they care for or carry (Paul *et al.* 1996), they are often likely to father the next offspring of the females whose offspring they protect, suggesting that this behaviour may also serve to increase mating success (Ménard *et al.* 2001).

Figure 14.19 A male Barbary macaque with an infant. In this species, some males interact frequently with infants and sometimes carry them for protracted periods. *Source*: © Julia Fischer.

14.9 The consequences of male hierarchies

Skew, kinship and genetic structure

Male hierarchies affect the extent to which individual males monopolise paternity in their group, which in turn has important consequences for the extent of reproductive skew, for patterns of kinship within groups and for the genetic structure of populations (Di Fiore 2012; Schülke and Ostner 2012). Where a single despotic male in each group monopolises access to philopatric breeding females, paternal relatedness between natals will be relatively high (Widdig *et al.* 2004; Lukas *et al.* 2005; Schülke and Ostner 2012), partly because an increased number of recruits of both sexes will share the same father and partly because philopatric breeding females are likely to be both paternally and maternally related, so that maternal relatedness between natal animals will also be relatively high (Lukas *et al.* 2005; Widdig 2013). Some of the highest levels of average relatedness in polygynous systems are found in polytocous species where females remain and breed in their natal group while male littermates disperse together and either share breeding access to females in the group they join or acquire the dominant position in turn (Lukas *et al.* 2005). Both low levels of male skew and high rates of male turnover will reduce relatedness between recruits born into the same group in different years.

Through their effects on kinship, contrasts in reproductive skew among males are likely to have an important influence on the genetic structure of populations (Clutton-Brock and Sheldon 2010; Di Fiore 2012). Where reproductive skew is high, male tenure is long and females are usually philopatric, differences in gene frequency between groups are likely to be relatively large (see Chapter 13), whereas low reproductive skew, short male tenure and frequent female emigration would be expected to reduce genetic structure. As yet, comparative estimates of the effects of breeding systems on genetic structure are scarce, but as the number of long-term field studies increases it should be possible to model their effects and to predict variation in the genetic structure of populations (Lukas *et al.* 2005; Di Fiore 2012).

Relationships between females

As a result of their influence on kinship between natals, contrasts in male skew may have important effects on female behaviour. High paternal and maternal relatedness between resident females might be expected to lead to increased levels of tolerance between females and to reductions in the frequency of competitive coalitions and some studies of primates support this prediction. For example, in several species of macaques, paternal kin are more tolerant of each other than they are of non-kin after the effects of maternal relatedness have been controlled (Widdig 2007) and contrasts in the behaviour of females may be associated with interspecific differences in paternal relatedness: for example, in species of macaques where reproductive skew in males is usually high, relationships between females are more tolerant than where male skew is usually low (Schülke and Ostner 2008, 2012; Widdig 2013). High skew and short tenure in males may also lead to increased rates of male infanticide and may favour females that mate with multiple partners (see Chapter 16) and so enhance competition between females to attract males and the development of female ornaments (see Chapter 5).

Machiavellian behaviour and brain size

Many of the best-known examples of opportunistic social strategies involve the establishment, maintenance and termination of coalitions and alliances and the disruption of rival alliances between competing males and other group members of both sexes. As yet, most studies of complex strategies of this kind involve primates, including capuchin monkeys (Perry 1998; Perry and Manson 2008), chacma baboons (Cheney and Seyfarth 2008), macaques (Flack and de Waal 2004) and chimpanzees (Watts 1998, 2002) but similar relationships may also occur in other species, including cetaceans (Connor 2000) and some carnivores (East and Hofer 2010). Behaviour of this kind is often referred to as 'Machiavellian' and is frequent in species living in multi-male groups where breeding males are unrelated immigrants and the frequency of agonistic interactions is relatively high, like many baboons and macaques, as well as in chimpanzees, where males remain in their natal group but are not closely related to each other (Lukas and Vigilant 2005). In contrast, shifting alliances and complex social interactions appear to be less common in polytocous species, like lions and meerkats, where several closely related males defend access to female groups.

It has been suggested that the complex social manoeuvres of unrelated or distantly related individuals living in stable breeding groups have played an important role in the evolution of cognitive abilities and brain structure. The social brain hypothesis (Humphrey 1976) suggests that contrasts in brain development between species are related to the complexity of social relationships and although the assessment of social complexity presents problems (see Chapter 2), comparative studies of primates and other mammals have demonstrated associations between coalitionary behaviour and indices of brain size (Dunbar and Schultz 2007). Several other species recognised for their cognitive abilities, including spotted hyenas and bottlenose dolphins, also form coalitions and alliances and broader comparisons across mammals suggest that species where individuals commonly form coalitions are characterised by relatively large brain sizes as well as by low levels of relatedness between group members. Although it has been suggested that brain size may constrain the ability of species to form coalitions, a more likely explanation of the association between brain size and coalitionary behaviour is that in species where groups are large and stable and most group members are not close relatives, social competition favours the evolution both of cooperative or coalitionary behaviour between males and of increases in brain size (Bissonnette *et al.* 2014; Díaz-Muñoz *et al.* 2014).

SUMMARY

1. Consistent dominance relationships between resident males are common in species where groups contain several breeding males and frequently affect the relative breeding success of individuals.

2. In some species where males compete independently, the rank of males often depends on their fighting ability and is often correlated with their size, weight or condition as well as with the relative development of their secondary sexual characteristics, like horns or antlers. In contrast, where males cooperate with each other in competitive encounters with rivals, the number, rank and condition of their allies may influence their social rank and their individual physical characteristics may be less important.

3. The rank of males is usually positively correlated with their reproductive success, though the strength and stability of correlations varies widely within and between species.

4. High social rank and frequent reproductive activity can have substantial costs to males and may reduce the duration of their effective breeding lifespans, favouring the development of coalitions and alliances.

5. The provision of support between coalition partners is often interpreted as an example of reciprocity. However, partners seldom alternate in providing support and an alternative interpretation is that partners gain immediate benefits by providing support because this maintains useful supportive relationships with their allies.

6. Males commonly compete for powerful allies and the relative rank of coalition partners can affect the division of benefits between them.

7. Like females, males may punish rivals and reconcile with individuals they have threatened or attacked. In some species, they also intervene to limit conflicts between females.

8. Where dominance relationships between males affect the distribution of male breeding success, they can have an important influence on kinship within groups as well as on the genetic structure of populations.

9. The complex social relationships that arise where unrelated individuals form coalitions and alliances may favour the evolution of advanced cognitive abilities and associated developments in brain size and structure.

References

Alberts, S.C. (2012) Magnitude and sources of variation in male reproductive performances. In: *The Evolution of Primate Societies* (eds J.C. Mitani, J. Call, P.M. Kappeler, R.A. Palombit and J.B. Silk). Chicago: University of Chicago Press, 412–431.

Alberts, S.C. and Altmann, J. (1995) Preparation and activation: determinants of age at reproductive maturity in male baboons. *Behavioral Ecology and Sociobiology* **36**:397–406.

Alberts, S.C., *et al.* (1992) Behavioral, endocrine, and immunological correlates of immigration by an aggressive male into a natural primate group. *Hormones and Behavior* **26**:167–178.

Alberts, S.C., *et al.* (2003) Queuing and queue-jumping: long-term patterns of reproductive skew in male savannah baboons, *Papio cynocephalus*. *Animal Behaviour* **65**:821–840.

Altmann, J. and Alberts, S.C. (2005) Growth rates in a wild primate population: ecological influences and maternal effects. *Behavioral Ecology and Sociobiology* **57**:490–501.

Appleby, M.C. (1980) Social rank and food access in red deer stags. *Behaviour* **74**:294–309.

Appleby, M.C. (1983) The probability of linearity in hierarchies. *Animal Behaviour* **31**:600–608.

Asher, M., *et al.* (2008) Large males dominate: ecology, social organization, and mating system of wild cavies, the ancestors of the guinea pig. *Behavioral Ecology and Sociobiology* **62**:1509–1521.

Aureli, F. and van Schaik, C.P. (1991) Post-conflict behaviour in long-tailed macaques (*Macaca fascicularis*). II. Coping with the uncertainty. *Ethology* **89**:101–114.

Aureli, F., *et al.* (2002) Conflict resolution following aggression in gregarious animals: a predictive framework. *Animal Behaviour* **64**:325–343.

Aureli, F., *et al.* (2012) The regulation of social relationships. In: *The Evolution of Primate Societies* (eds J.C. Mitani, J. Call, P.M. Kappeler, R.A. Palombit and J.B. Silk). Chicago: University of Chicago Press, 531–551.

Barrette, C. and Vandal, D. (1986) Social rank, dominance, antler size and access to food in snow-bound wild woodland caribou. *Behaviour* **97**:118–146.

Bayly, K.L., *et al.* (2006) Measuring social structure: a comparison of eight dominance indices. *Behavioural Processes* **73**:1–12.

Bercovitch, F.B. (1986) Male rank and reproductive activity in savanna baboons. *International Journal of Primatology* **7**:533–550.

Bercovitch, F.B. (1988) Coalitions, cooperation and tactics among adult male baboons. *Animal Behaviour* **36**:1198–1209.

Bergman, T.J., *et al.* (2005) Correlates of stress in free-ranging male chacma baboons, *Papio hamadryas ursinus*. *Animal Behaviour* **70**:703–713.

Bernstein, I.S. (1976) Dominance, aggression and reproduction in primate societies. *Journal of Theoretical Biology* **60**:459–472.

Bernstein, I.S. (1981) Dominance: the baby and the bathwater. *Behavioral and Brain Sciences* **4**:419–457.

Bissonnette, A., *et al.* (2014) Socioecology, but not cognition, predicts male coalitions across primates. *Behavioral Ecology* **25**:794–801.

Boonstra, R., *et al.* (2007) The role of the stress axis in life-history adaptations. In: *Rodent Societies: An Ecological and Evolutionary Perspective* (eds J.O. Wolff and P.W. Sherman). Chicago: University of Chicago Press, 139–149.

Chapais, B. (1995) Alliances as a means of competition in primates: evolutionary, developmental, and cognitive aspects. *American Journal of Physical Anthropology* **38**:115–136.

Cheney, D.L. and Seyfarth, R.M. (1990) *How Monkeys See the World: Inside the Mind of Another*. Chicago: University of Chicago Press.

Cheney, D.L. and Seyfarth, R.M. (2008) *Baboon Metaphysics: The Evolution of a Social Mind*. Chicago: University of Chicago Press.

Cheney, D.L., *et al.* (2010) Contingent cooperation between wild female baboons. *Proceedings of the National Academy of Sciences of the United States of America* **107**:9562–9566.

Clutton-Brock, T.H. (1988) Reproductive success. In: *Reproductive Success. Studies of Individual Variation in Contrasting Breeding Systems* (ed. T.H. Clutton-Brock). Chicago: University of Chicago Press, 472–486.

Clutton-Brock, T.H. (2009) Cooperation between non-kin in animal societies. *Nature* **462**:51–57.

Clutton-Brock, T.H. and Isvaran, K. (2007) Sex differences in ageing in natural populations of vertebrates. *Proceedings of the Royal Society of London. Series B: Biological Sciences* **274**:3097–3104.

Clutton-Brock, T.H. and Parker, G.A. (1995) Punishment in animal societies. *Nature* **373**:209–216.

Clutton-Brock, T.H. and Sheldon, B.C. (2010) Individuals and populations: the role of long-term, individual-based studies in ecology and evolutionary biology. *Trends in Ecology and Evolution* **25**:562–573.

Clutton-Brock, T.H., *et al.* (1982) *Red Deer: The Behaviour and Ecology of Two Sexes*. Chicago: University of Chicago Press.

Clutton-Brock, T.H., *et al.* (1984) Maternal dominance, breeding success and birth sex ratios in red deer. *Nature* **308**:358–360.

Clutton-Brock, T.H., *et al.* (1986) Great expectations: dominance, breeding success and offspring sex ratios in red deer. *Animal Behaviour* **34**:460–471.

Clutton-Brock, T.H., *et al.* (1988) Reproductive success in male and female red deer. In: *Reproductive Success: Studies of Individual Variation in Contrasting Breeding Systems* (ed. T.H. Clutton-Brock). Chicago: University of Chicago Press, 325–343.

Clutton-Brock, T.H., *et al.* (1998) Infanticide and expulsion of females in a cooperative mammal. *Proceedings of the Royal Society of London. Series B: Biological Sciences* **265**:2291–2295.

Clutton-Brock, T.H., *et al.* (2001) Cooperation, control and concession in meerkat groups. *Science* **291**:478–481.

Clutton-Brock, T.H., *et al.* (2006) Intrasexual competition and sexual selection in cooperative mammals. *Nature* **444**:1065–1068.

Collins, R. (1981) On the microfoundations of macrosociology. *American Journal of Sociology* **86**:984–1014.

Colmenares, F. (1992) Clans and harems in a colony of hamadryas and hybrid baboons: male kinship, familiarity and the formation of brother-teams. *Behaviour* **121**:61–94.

Coltman, D.W., *et al.* (2002) Age-dependent sexual selection in bighorn rams. *Proceedings of the Royal Society of London. Series B: Biological Sciences* **269**:165–172.

Connor, R.C. (2000) Group living in whales and dolphins. In: *Cetacean Societies: Field Studies of Dolphins and Whales* (eds J. Mann, R.C. Connor, P.L. Tyack and H. Whitehead). Chicago: University of Chicago Press, 199–218.

Connor, R.C. and Norris, K.S. (1982) Are dolphins reciprocal altruists? *American Naturalist* **119**:358–374.

Connor, R.C., *et al.* (1992a) Dolphin alliances and coalitions. In: *Coalitions and Alliances in Humans and Other Primates* (eds A.H. Harcourt and F.B.M. de Waal). Oxford: Oxford University Press, 415–443.

Connor, R.C., *et al.* (1992b) Two levels of alliance formation among male bottlenose dolphins (*Tursiops* sp.). *Proceedings of the National Academy of Sciences of the United States of America* **89**:987–990.

Connor, R.C., *et al.* (1999) Superalliance of bottlenose dolphins. *Nature* **397**:571–572.

Connor, R.C., *et al.* (2000) Male reproductive strategies and social bonds. In: *Cetacean Societies: Field Studies of Dolphins and Whales* (eds J. Mann, R.C. Connor, P.L. Tyack and H. Whitehead). Chicago: University of Chicago Press, 247–269.

Cords, M. (1992) Post-conflict reunions and reconciliation in long-tailed macaques. *Animal Behaviour* **44**:57–61.

Cords, M. (1997) Friendship, alliances, reciprocity, repair. In: *Machiavellian Intelligence II: Extensions and Evaluations* (eds A. Whiten and R.W. Byrne). Cambridge: Cambridge University Press, 24–49.

Cords, M. and Thurnheer, S. (1993) Reconciling with valuable partners by long-tailed macaques. *Ethology* **93**:315–325.

Cowlishaw, G. and Dunbar, R.I.M. (1991) Dominance rank and mating success in male primates. *Animal Behaviour* **41**:1045–1056.

Creel, S. and Creel, N.M. (2002) *The African Wild Dog: Behavior, Ecology, and Conservation*. Princeton, NJ: Princeton University Press.

Creel, S., *et al.* (2013) The ecology of stress: effects of the social environment. *Functional Ecology* **27**:66–80.

Das, M. (2000) Conflict management via third parties: post-conflict affiliation of the aggressor. In: *Natural Conflict Resolution* (eds F. Aureli and F.B.M. de Waal). Berkeley: University of California Press, 263–280.

David, H.A. (1988) *The Method of Paired Comparisons*. London: Charles Griffin.

Deag, J.M. (1980) Interactions between males and unweaned barbary macaques: testing the agonistic buffering hypothesis. *Behaviour* **75**:54–81.

Deag, J.M. and Crook, J.H. (1971) Social behaviour and 'agonistic buffering' in the wild barbary macaque *Macaca sylvana* L. *Folia Primatologica* **15**:183–200.

de Villiers, M.S., *et al.* (2003) Patterns of coalition formation and spatial association in a social carnivore, the African wild dog (*Lycaon pictus*). *Journal of Zoology* **260**:377–389.

de Waal, F.B.M. (1977) The organization of agonistic relations within two captive groups of Java monkeys (*Macaca fascicularis*). *Zeitschrift fur Tierpsychologie* **44**:225–282.

de Waal, F.B.M. (1982) *Chimpanzee Politics: Power and Sex Among Apes*. New York: Harper & Row.

de Waal, F.B.M. (1985) Coalitions in monkeys and apes. In: *Advances in Psychology*, Vol. **24** (ed. H.A.M. Wilke). Amsterdam: Elsevier, 1–27.

de Waal, F.B.M. (1986) The integration of dominance and social bonding in primates. *Quarterly Review of Biology* **61**:459–479.

de Waal, F.B.M. (1989a) Dominance 'style' and primate social organization. In: *Comparative Socioecology: The Behavioural Ecology of Humans and Other Mammals* (eds V. Standen and R.A. Foley). Oxford: Blackwell Scientific Publications, 243–263.

de Waal, F.B.M. (1989b) *Peacemaking Among Primates*. Cambridge, MA: Harvard University Press.

de Waal, F.B.M. (1993) Reconciliation among primates: a review of empirical evidence and unresolved issues. In: *Primate Social Conflict* (eds W.A. Mason and S.P. Mendoza). Albany, NY: State University of New York Press, 111–144.

de Waal, F.B.M. and van Roosmalen, A. (1979) Reconciliation and consolation among chimpanzees. *Behavioral Ecology and Sociobiology* **5**:55–66.

Dewsbury, D.A. (1982) Dominance rank, copulatory behavior and differential reproduction. *Quarterly Review of Biology* **57**:135–159.

Díaz-Muñoz, S.L., *et al.* (2014) Cooperating to compete: altruism, sexual selection and causes of male reproductive cooperation. *Animal Behaviour* **88**:67–78.

Di Fiore, A. (2012) Genetic consequences of primate social organization. In: *The Evolution of Primate Societies* (eds J.C. Mitani, J. Call, P.M. Kappeler, R.A. Palombit and J.B. Silk). Chicago: University of Chicago Press, 269–292.

Di Fiore, A. and Campbell, C.J. (2007) The Atelines: variation in ecology, behaviour and social organisation. In: *Primates in Perspective* (eds C.J. Campbell, A. Fuentes, K.C. MacKinnon, M. Panger and S.K. Bearder). New York: Oxford University Press, 155–185.

Drews, C. (1993) The concept and definition of dominance in animal behavior. *Behaviour* **125**:283–313.

Drews, C. (1996) Contexts and patterns of injuries in free-ranging male baboons (*Papio cynocephalus*). *Behaviour* **133**:443–474.

Duffy, K., *et al.* (2007) Male chimpanzees exchange political support for mating opportunities. *Current Biology* **17**:R586–R587.

Dunbar, R.I.M. and Shultz, S. (2007) Evolution in the social brain. *Science* **317**:1344–1347.

East, M.L. and Hofer, H. (2001) Male spotted hyenas (*Crocuta crocuta*) queue for status in social group dominated by females. *Behavioral Ecology* **12**:558–568.

East, M.L. and Hofer, H. (2010) Social environments, social tactics and their fitness consequences in complex mammalian societies. In: *Social Behaviour* (eds T. Szekely, A.J. Moore and J. Komdeur). Cambridge: Cambridge University Press, 360–390.

Eberle, M. and Kappeler, P.M. (2004) Sex in the dark: determinants and consequences of mixed male mating tactics in *Microcebus murinus*, a small solitary nocturnal primate. *Behavioral Ecology and Sociobiology* **57**:77–90.

Ehardt, C.L. and Bernstein, I.S. (1992) Conflict intervention behaviour by adult male macaques: structural and functional aspects. In: *Coalitions and Alliances in Humans and Other Animals* (eds A.H. Harcourt and F.B.M. de Waal). Oxford: Oxford University Press, 83–111.

Ellis, L. (1995) Dominance and reproductive success among nonhuman animals: a cross-species comparison. *Ethology and Sociobiology* **16**:257–333.

Engh, A. and Holekamp, K.E. (2003) Maternal rank 'inheritance' in the spotted hyena. In: *Social Complexity* (eds F.B.M. de Waal and P. Tyack). Cambridge, MA: Harvard University Press, 149–152.

Fedigan, L.M. and Baxter, M.J. (1984) Sex differences and social organization in free-ranging spider monkeys (*Ateles geoffroyi*). *Primates* **25**:279–294.

Feh, C. (1990) Long-term paternity data in relation to different aspects of rank for Camargue stallions, *Equus caballus*. *Animal Behaviour* **40**:995–996.

Feh, C. (2001) Alliances between stallions are more than just multimale groups: reply to Linklater & Cameron (2000). *Animal Behaviour* **6**:Γ27–F30.

Flack, J.C. and de Waal, F.B.M. (2004) Dominance style, social power, and conflict management: a conceptual framework. In: *Macaque Societies: A Model for the Study of Social Organization* (eds B. Thierry, M. Singh and W. Kaumanns). Cambridge: Cambridge University Press, 157–185.

Fraser, O.N., *et al.* (2010) The function and determinants of reconciliation in *Pan troglodytes*. *International Journal of Primatology* **31**:39–57.

Furuichi, T. (1989) Social interactions and the life history of female *Pan paniscus* in Wamba, Zaire. *International Journal of Primatology* **10**:173–197.

Georgiev, A.V., *et al.* (2015) Male quality, dominance rank, and mating success in free-ranging rhesus macaques. *Behavioral Ecology* doi: 10.1093/beheco/arv008.

Gesquiere, L.R., *et al.* (2011) Life at the top: rank and stress in wild male baboons. *Science* **333**:357–360.

Gilby, I.C. (2012) Cooperation among nonkin: reciprocity, markets and mutualism. In: *The Evolution of Primate Societies* (eds J.C. Mitani, J. Call, P.M. Kappeler, R.A. Palombit and J.B. Silk). Chicago: University of Chicago Press, 514–530.

Gogarten, J.F. and Koenig, A. (2013) Reproductive seasonality is a poor predictor of receptive synchrony and male reproductive skew among nonhuman primates. *Behavioral Ecology and Sociobiology* **67**:123–134.

Goodall, J. (1986) *The Chimpanzees of Gombe: Patterns of Behavior*. Cambridge, MA: Belknap Press.

Goymann, W. and Wingfield, J.C. (2004) Allostatic load, social status and stress hormones: the costs of social status matters. *Animal Behaviour* **67**:591–602.

Gulland, F.M.D. (1992) The role of nematode parasites in Soay sheep (*Ovis aries* L.) mortality during a population crash. *Parasitology* **105**:493–503.

Gulland, F.M.D. and Fox, M. (1992) Epidemiology of nematode infections of Soay sheep (*Ovis aries* L.) on St Kilda. *Parasitology* **105**:481–492.

Gulland, F.M.D., et al. (1993) Parasite-associated polymorphism in a cyclic ungulate population. *Proceedings of the Royal Society of London. Series B: Biological Sciences* **254**:7–13.

Hall, K.R.L. and DeVore, I. (1965) Baboon social behavior. In: *Primate Behavior: Field Studies of Monkeys and Apes* (ed. I. DeVore). New York: Holt, Rhinehart and Winston, 53–110.

Harcourt, A.H. and de Waal, F.B.M. (1992) Cooperation and conflict: from an anthropoidal. In: *Coalitions and Alliances in Humans and Other Animals* (eds A.H. Harcourt and F.B.M. de Waal). Oxford: Oxford University Press, 493–510.

Hauser, M., et al. (2009) Evolving the ingredients for reciprocity and spite. *Philosophical Transactions of the Royal Society B: Biological Sciences* **364**:3255–3266.

Hausfater, G. and Watson, D.F. (1976) Social and reproductive correlates of parasite ova emissions by baboons. *Natures* **262**:688–689.

Henkel, S., et al. (2010) Infants as costly social tools in male Barbary macaque networks. *Animal Behaviour* **79**:1199–1204.

Henzi, S.P., et al. (2010) Infanticide and reproductive restraint in a polygynous social mammal. *Proceedings of the National Academy of Sciences of the United States of America* **107**:2130–2135.

Hinde, R.A. (1976) Interactions, relationships and social structure. *Man* **11**:1–17.

Höner, O.P., et al. (2010) The fitness of dispersing spotted hyaena sons is influenced by maternal social status. *Nature Communications* **1**:60.

Hoogland, J.L. (1995) *The Black-tailed Prairie Dog: Social Life of a Burrowing Mammal*. Chicago: University of Chicago Press.

Humphrey, N.K. (1976) The social function of intellect. In: *Growing Points in Ethology* (eds P.P.G. Bateson and R.A. Hinde). Cambridge: Cambridge University Press, 303–317.

Ihobe, H. (1992) Male male relationships among wild bonobos (*Pan paniscus*) at Wamba, Republic of Zaire. *Primates* **33**:163–179.

Inoue, E. and Takenaka, O. (2008) The effect of male tenure and female mate choice on paternity in free-ranging Japanese macaques. *American Journal of Primatology* **70**:62–68.

Jarman, P.J. (2000) Males in macropod society. In: *Primate Males: Causes and Consequences of Variation in Group Composition* (ed. P.M. Kappeler). Cambridge: Cambridge University Press, 21–33.

Kahan, J.P. and Rapoport, A. (1984) *Theories of Coalition Formation*. Hillsdale, NJ: Lawrence Erlbaum Associates.

Kano, T. (1992) *The Last Ape: Pygmy Chimpanzee Behavior and Ecology*. Stanford, CA: Stanford University Press.

Kano, T. (1996) Male rank order and copulation rate in a unit-group of bonobos at Wamba, Zaïre. In: *Great Ape Societies* (eds W.C. McGrew, L.F. Marchant and T. Nishida). Cambridge: Cambridge University Press, 135–145.

Keller, L. and Reeve, H.K. (1994) Partitioning of reproduction in animal societies. *Trends in Evolution and Ecology* **9**:98–103.

Kitchen, D.M., et al. (2003) Loud calls as indicators of dominance in male baboons (*Papio cynocephalus ursinus*). *Behavioral Ecology and Sociobiology* **53**:374–384.

Kruuk, L.E.B., et al. (1999) Early determinants of lifetime reproductive success differ between the sexes in red deer. *Proceedings of the Royal Society of London. Series B: Biological Sciences* **266**:1655–1661.

Kruuk, L.E.B., et al. (2002) Antler size in red deer: heritability and selection but no evolution. *Evolution* **56**:1683–1695.

Langergraber, K.E. (2012) Cooperation among kin. In: *The Evolution of Primate Societies* (eds J.C. Mitani, J. Call, P.M. Kappeler, R.A. Palombit and J.B. Silk). Chicago: University of Chicago Press, 491–513.

Langergraber, K.E., et al. (2007) The limited impact of kinship on cooperation in wild chimpanzees. *Proceedings of the National Academy of Sciences of the United States of America* **104**:7786–7790.

Lawler, R.R., et al. (2005) Intrasexual selection in Verreaux's sifaka (*Propithecus verreauxi verreauxi*). *Journal of Human Evolution* **48**:259–277.

Le Boeuf, B.J. and Reiter, J. (1988) Lifetime reproductive success in northern elephant seals. In: *Reproductive Success: Studies of Individual Variation in Contrasting Breeding Systems* (ed. T.H. Clutton-Brock). Chicago: University of Chicago Press, 344–362.

Lewis, R.J. (2002) Beyond dominance: the importance of leverage. *Quarterly Review of Biology* **77**:149–164.

Lincoln, G.A. (1994) Teeth, horns and antlers: the weapons of sex. In: *The Differences Between the Sexes* (eds R.V. Short and E. Balaban). Cambridge: Cambridge University Press, 131–158.

Lincoln, G.A., et al. (1970) The social and sexual behaviour of the red deer stag. *Journal of Reproduction and Fertility Supplement* **11**:71–103.

Lindström, J. (1999) Early development and fitness in birds and mammals. *Trends in Ecology and Evolution* **14**:343–348.

Lott, D. (1979) Dominance relations and breeding rate in mature male American bison. *Zeitschrift fur Tierpsychologie* **49**:418–432.

Lukas, D. and Vigilant, L. (2005) Reply: Facts, faeces and setting standards for the study of MHC genes using noninvasive samples. *Molecular Ecology* **14**:1601–1602.

Lukas, D., et al. (2005) To what extent does living in a group mean living with kin? *Molecular Ecology* **14**:2181–2196.

McCormick, H.A., et al. (2011) Male and female aggression: lessons from sex, age, and injury in olive baboons. *Behavioral Ecology* **23**:684–691.

Majolo, B., *et al.* (2005) Postconflict behavior among male Japanese macaques. *International Journal of Primatology* **26**:321–336.

Ménard, N., *et al.* (2001) Is male–infant caretaking related to paternity and/or mating activities in wild Barbary macaques (*Macaca sylvanus*)? *Comptes Rendus de l'Académie des Sciences Series III. Sciences de la Vie* **324**:601–610.

Mitani, J.C. (2006) Reciprocal exchange in chimpanzees and other primates. In: *Cooperation in Primates: Mechanisms and Evolution* (eds P.M. Kappeler and C.P. van Schaik). Berlin: Springer, 107–119.

Mitani, J.C. (2009a) Male chimpanzees form enduring and equitable social bonds. *Animal Behaviour* **77**:633–640.

Mitani, J.C. (2009b) Cooperation and competition in chimpanzees: current understanding and future challenges. *Evolutionary Anthropology: Issues, News, and Reviews* **18**:215–227.

Mitani, J.C. and Watts, D.P. (1999) Demographic influences on the hunting behavior of chimpanzees. *American Journal of Physical Anthropology* **109**:439–454.

Mitani, J.C. and Watts, D.P. (2001) Why do chimpanzees hunt and share meat? *Animal Behaviour* **61**:915–924.

Negro, S.S., *et al.* (2010) Correlation between male social status, testosterone levels, and parasitism in a dimorphic polygynous mammal. *PLOS ONE* **5**:e12507.

Neumann, C., *et al.* (2011) Assessing dominance hierarchies: validation and advantages of progressive evaluation with Elo-rating. *Animal Behaviour* **82**:911–921.

Nishida, T. and Hosaka, K. (1996) Coalition strategies among adult male chimpanzees of the Mahale Mountains, Tanzania. In: *Great Ape Societies* (eds L.F. Marchant and T. Nishida). Cambridge: Cambridge University Press, 114–134.

Noë, R. (1986) Lasting alliances among male savannah baboons. In: *Primate Ontogeny, Cognition and Social Behavior* (eds J.G. Else and P.C. Lee). Cambridge: Cambridge University Press, 381–392.

Noë, R. (1990) A veto game played by baboons: a challenge to the use of the Prisoner's Dilemma as a paradigm for reciprocity and cooperation. *Animal Behaviour* **39**:78–90.

Noë, R. (1992) Alliance formation among male baboons: shopping for profitable partners. In: *Coalitions and Alliances in Humans and Other Animals* (eds A.H. Harcourt and F.B.M. de Waal). Oxford: Oxford University Press, 285–321.

Noë, R. and Sluijter, A.A. (1990) Reproductive tactics of male savanna baboons. *Behaviour* **113**:117–170.

Noë, R. and Sluijter, A.A. (1995) Which adult male savanna baboons form coalitions? *International Journal of Primatology* **16**:77–105.

Noë, R., *et al.* (1991) The market effect: an explanation for pay-off asymmetries among collaborating animals. *Ethology* **87**:97–118.

Olson, M. and Blumstein, D.T. (2009) A trait-bsaed approach to understand the evolution of complex coalitions in male mammals. *Behavioral Ecology* **20**:624–632.

Packer, C. (1977) Reciprocal altruism in olive baboons (*Papio anubis*). *Nature* **265**:441–443.

Packer, C., *et al.* (1988) Reproductive success of lions. In: *Reproductive Success: Studies of Individual Variation in Contrasting Breeding Systems* (ed. T.H. Clutton-Brock). Chicago: University of Chicago Press, 363–383.

Paul, A., *et al.* (1992) Maternal rank affects reproductive success of male Barbary macaques (*Macaca sylvanus*): evidence from DNA fingerprinting. *Behavioral Ecology and Sociobiology* **30**:337–341.

Paul, A., *et al.* (1996) The sociobiology of male–infant interactions in Barbary macaques, *Macaca sylvanus*. *Animal Behaviour* **51**:155–170.

Pelletier, D.L. and Festa-Bianchet, M. (2006) Sexual selection and social rank in bighorn rams. *Animal Behaviour* **71**:649–655.

Pemberton, J.M., *et al.* (2004) Mating patterns and male breeding success. In: *Soay Sheep: Dynamics and Selection in an Island Population* (eds T.H. Clutton-Brock and J.M. Pemberton). Cambridge: Cambridge University Press, 166–189.

Perry, S. (1998) Male–male social relationships in wild white-faced capuchins, *Cebus capucinus*. *Behaviour* **135**:139–172.

Perry, S. and Manson, J.H. (2008) *Manipulative Monkeys: The Capuchins of Lomas Barbudal*. Cambridge, MA: Harvard University Press.

Perry, S., *et al.* (2004) White-faced capuchin monkeys show triadic awareness in their choice of allies. *Animal Behaviour* **67**:165–170.

Petit, O. and Thierry, B. (1994) Aggressive and peaceful interventions in conflicts in Tonkean macaques. *Animal Behaviour* **48**:1427–1436.

Petit, O. and Thierry, B. (2000) Do impartial interventions in conflicts occur in monkeys and apes? In: *Natural Conflict Resolution* (eds F. Aureli and F.B.M. de Waal). Berkeley: University of California Press, 267–269.

Poole, J.H. (1989) Announcing intent: the aggressive state of musth in African elephants. *Animal Behaviour* **37**:140–152.

Poole, J.H., *et al.* (2011) Longevity, competition, and musth: a long-term perspective on male reproductive strategies. In: *The Amboseli Elephants: A Long-term Perspective on a Long-lived Mammal* (eds C.J. Moss, H. Croze and P.C. Lee). Chicago: University of Chicago Press, 272–290.

Post, D.G., *et al.* (1980) Feeding behaviour of yellow baboons (*Papio cynocephalus*): relationship to age, gender and dominance rank. *Folia Primatologica* **34**:170–195.

Preuschoft, S. and van Schaik, C.P. (2000) Dominance and communication. In: *Natural Conflict Resolution* (eds F. Aureli and F.B.M. de Waal). Berkeley: University of California Press, 77–105.

Preuschoft, S., *et al.* (1998) Dominance styles of female and male Barbary macaques (*Macaca sylvanus*). *Behaviour* **135**:731–755.

Pusey, A.E. (1983) Mother–offspring relationships in chimpanzees after weaning. *Animal Behaviour* **31**:363–377.

Rodriguez-Llanes, J.M., *et al.* (2009) Reproductive benefits of high social status in male macaques (*Macaca*). *Animal Behaviour* **78**:643–649.

Rubenstein, D.R. and Shen, S.-F. (2009) Reproductive conflict and the costs of social status in cooperatively breeding vertebrates. *American Naturalist* **173**:650–662.

Sachser, N., *et al.* (1999) Behavioural strategies, testis size and reproductive success in two caviomorph rodents with different mating systems. *Behaviour* **136**:1203–1217.

Samuels, A. and Flaherty, C. (2000) Peaceful conflict resolution in the sea? In: *Natural Conflict Resolution* (eds F. Aureli and F.B.M. de Waal). Berkeley: University of California Press, 229–231.

Sapolsky, R.M. (1992) Neuroendocrinology of the stress-response. In: *Behavioral Endocrinology* (eds J.B. Becker, S.M. Breedlove and D. Crews). Cambridge, MA: MIT Press, 287–324.

Sapolsky, R.M. (1993) The physiology of dominance in stable versus unstable social hierarchies. In: *Primate Social Conflict* (eds W.A. Mason and S.P. Mendoza). Albany, NY: State University of New York Press, 171–204.

Sapolsky, R.M., *et al.* (2000) How do glucocorticoids influence stress responses? Integrating permissive, suppressive, stimulatory, and preparative actions. *Endocrine Reviews* **21**:55–89.

Sapolsky, R.M., *et al.* (2002) The neuroendocrinology of stress and aging: the glucocorticoid cascade hypothesis. *Science of Aging Knowledge Environment* **2002** (38): 21.

Schjelderup-Ebbe, T. (1922) Beiträge zur sozialpsychologie des haushuhns. *Zeitschrift für Psychologie* **88**:225–252.

Schülke, O. and Ostner, J. (2008) Male reproductive skew, paternal relatedness and female social relationships. *American Journal of Primatology* **70**:1–4.

Schülke, O. and Ostner, J. (2012) Ecological and social influences on sociality. In: *The Evolution of Primate Societies* (eds J.C. Mitani, J.C. Call, P.M. Kappeler, R.A. Palombit and J.B. Silk). Chicago: University of Chicago Press, 195–219.

Schülke, O., *et al.* (2010) Social bonds enhance reproductive success in male macaques. *Current Biology* **20**:2207–2210.

Smale, L., *et al.* (1993) Ontogeny of dominance in free-living spotted hyaenas: juvenile rank relations with adult females and immigrant males. *Animal Behaviour* **46**:467–477.

Smale, L., *et al.* (1997) Sexually dimorphic dispersal in mammals: patterns, causes and consequences. In: *Advances in the Study of Behavior*, Vol. **26** (eds C. Snowdon, P.J.B. Slater, M. Milinski and J.S. Rosenblatt). San Diego, CA: Academic Press, 181–250.

Snyder-Mackler, N., *et al.* (2012) Concessions of an alpha male? Cooperative defence and shared reproduction in multi-male primate groups. *Proceedings of the Royal Society of London. Series B: Biological Sciences* **279**:3788–3795.

Strier, K.B. (1992) Causes and consequences of nonaggression in the woolly spider monkey, or muriqui (*Brachyteles arachnoides*) In: *Aggression and Peacefulness in Humans and Other Primates* (eds J. Silverberg and J.P. Gray). Oxford: Oxford University Press, 100–116.

Strier, K.B. (1997) Mate preferences of wild muriqui monkeys (*Brachyteles arachnoides*): reproductive and social correlates. *Folia Primatologica* **68**:120–133.

Strier, K.B. and Ziegler, T.E. (2000) Lack of pubertal influences on female dispersal in muriqui monkeys, *Brachyteles arachnoides*. *Animal Behaviour* **59**:849–860.

Strier, K.B., *et al.* (1999) Seasonal and social correlates of fecal testosterone and cortisol levels in wild male muriquis (*Brachyteles arachnoides*). *Hormones and Behavior* **35**:125–134.

Strier, K.B., *et al.* (2002) Social dynamics of male muriquis (*Brachyteles arachnoides hypoxanthus*). *Behaviour* **139**:315–342.

Strier, K.B., *et al.* (2011) Low paternity skew and the influence of maternal kin in an egalitarian, patrilocal primate. *Proceedings of the National Academy of Sciences of the United States of America* **108**:18915–18919.

Thierry, B. (1985) Patterns of agonistic interactions in three species of macaque (*Macaca mulatta, M. fascicularis, M. tonkeana*). *Aggressive Behavior* **11**:223–233.

Thierry, B. (2004) Social epigenesis. In: *Macaque Societies: A Model for the Study of Social Organization* (eds B. Thierry, M. Singh and V. Kaumanns). Cambridge: Cambridge University Press, 267–289.

Thierry, B. (2007) The macaques: a double-layered social organisation. In: *Primates in Perspective* (eds C.J. Campbell, A. Fuentes, K.C. MacKinnon, M. Panger and S.K. Bearder). New York: Oxford University Press, 224–239.

van Noordwijk, A.J. and van Schaik, C.P. (2004) Sexual selection and the careers of primate males: paternity concentration, dominance-acquisition tactics and transfer decisions. In: *Sexual Selection in Primates: New and Comparative Perspectives* (eds P.M. Kappeler and C.P. van Schaik). Cambridge: Cambridge University Press, 208–284.

van Schaik, C.P., *et al.* (2006) Toward a general model for male–male coalitions in primate groups. In: *Cooperation in Primates and Humans: Mechanisms and Evolution* (eds P.M. Kappeler and C.P. van Schaik). Berlin: Springer, 151–171.

von Holst, D. (1998) The concept of stress and its relevance for animal behavior. In: *Advances in the Study of Behavior*, Vol. **27** (eds A.P. Møller, M. Milinski and P.J.B. Slater). San Diego, CA: Academic Press, 1–131.

von Holst, D., *et al.* (1999) Social rank, stress, fitness, and life expectancy in wild rabbits. *Naturwissenschaften* **86**: 388–393.

Watts, D.P. (1997) Agonistic interventions in wild mountain gorilla groups. *Behaviour* **134**:23–57.

Watts, D.P. (1998) Coalitionary mate guarding by male chimpanzees at Ngogo, Kibale National Park, Uganda. *Behavioral Ecology and Sociobiology* **44**:43–55.

Watts, D.P. (2002) Reciprocity and interchange in the social relationships of wild male chimpanzees. *Behaviour* **139**:343–370.

Watts, D.P. (2010) Dominance, power and politics in non-human and human primates. In: *Mind the Gap: Tracing the Origins of Human Universals* (eds P.M. Kappeler and J.B. Silk). Berlin: Springer, 109–138.

Watts, D.P., *et al.* (2000) Redirection, consolation, and male policing: how targets of aggression interact with bystanders. In: *Natural Conflict Resolution* (eds F. Aureli and F.B.M. de Waal). Berkeley: University of California Press, 281–301.

Whitehead, H. (2008) *Analyzing Animal Societies: Quantitative Methods for Vertebrate Social Analysis*. Chicago: University of Chicago Press.

Widdig, A. (2007) Paternal kin discrimination: the evidence and likely mechanisms. *Biological Reviews* **82**:319–334.

Widdig, A. (2013) The impact of male reproductive skew on kin structure and sociality in multi-male groups. *Evolutionary Anthropology: Issues, News, and Reviews* **22**:239–250.

Widdig, A., *et al.* (2004) A longitudinal analysis of reproductive skew in male rhesus macaques. *Proceedings of the Royal Society of London. Series B: Biological Sciences* **271**:819–826.

Wilson, K., *et al.* (2004) Parasites and their impact. In: *Soay Sheep: Dynamics and Selection in an Island Population* (eds T.H. Clutton-Brock and J.M. Pemberton). Cambridge: Cambridge University Press, 113–165.

Zabel, C.J., *et al.* (1992) Coalition formation in a colony of prepubertal spotted hyenas. In: *Coalitions and Alliances in Humans and Other Animals* (eds A.H. Harcourt and F.B.M. de Waal). Oxford: Oxford University Press, 113–135.

Zimen, E. (1976) On the regulation of pack size in wolves. *Zeitschrift fur Tierpsychologie* **40**:300–341.

CHAPTER 15

Males and females

15.1 Introduction

A herd of over a hundred fallow deer are searching for acorns under oak trees in southern England. The herd consists primarily of does with a considerable number of younger males, for the mature males are defending mating territories elsewhere. One of the does is restless and quickly attracts the attention of a young buck. He sniffs her rump and tries to mount, but the doe slips away. Another young male approaches and sniffs her rear and she runs to the other side of the herd, followed by the two males. She runs again but other males join the chase. She stops for a moment and one of her pursuers mounts, but another buck attacks the pair with his antlers and she dodges away followed by pursuing males. This time, she runs out of the herd and joins an older male who is defending a territory under a separate tree. The male walks out to bring her into the territory and the young males come to a rapid stop, pause for a moment and then move back to the herd. Out of breath, the female stands panting and then lies down close to a clump of reeds. The territorial buck gently sniffs her rear and nuzzles the back of her head but she is not yet ready to mate and does not respond and he lies down nearby.

This sequence of events is the result of a conflict of interest between the sexes. Younger males, who are not yet able to compete successfully for territories with mature animals, are often likely to maximise their fitness by taking every opportunity to mate, while females may maximise their fitness by avoiding mating with them (see Chapter 4). As a result, conflicts between receptive females and males are a common feature of many animal breeding systems. In some cases, one sex has developed a 'winning' or 'controlling' adaptation, with the result that the other sex almost always acquiesces and overt conflicts are rare but, in many cases, both males and females are capable of winning particular contests and both sexes

pursue conflicting strategies: males often adopt coercive tactics that constrain female mating preferences while females attempt to avoid being controlled by males. The responses of females to male coercion can also have important consequences for the mating strategies of males. For example, in ungulates, the tendency for receptive females to move to males that are able to protect them from harassment may explain why males often defend mating territories that contain no resources (see Chapter 10).

While sexual conflicts over mating are particularly intense as a result of the large fitness consequences involved, similar conflicts of interest between males and females occur over other aspects of reproduction (see Chapter 16), as well as over many other activities and resources, including access to food, ranging patterns, communication and group membership (Chapman *et al.* 2003; Arnqvist and Rowe 2005; Lessells 2012; Palombit 2014). One of the starkest examples of a conflict of interest between males and females occurs over the survival of the dependent young of lactating females. When immigrant males replace resident breeding males in stable groups of long-lived mammals with protracted periods of juvenile dependency, they frequently encounter situations where most resident females are lactating and will not be ready to conceive again for several months. Under these circumstances, it can be in the interests of incoming males to kill the dependent offspring of resident females in order to accelerate the return of their mother to receptivity, while it is in the interests of females to prevent them from doing so (Hrdy 1974, 1975). Since either immigrant males or females with dependent young can win these 'contests', and substantial fitness benefits for both sexes are at stake, it is not surprising that they often lead to long and complex social interactions in which both sexes attempt to achieve their goals.

Mammal Societies, First Edition. Tim Clutton-Brock.

Where conflicts between males and females have a genetic basis, they may be a result of sexually divergent optima for alleles that control particular traits (intra-locus conflict) or a result of contrasts in the optimal outcome of interactions for males and females (inter-locus conflict) (Arnqvist and Rowe 2005). Where conflicts occur, females may suffer costs that are unselected by-products of male strategies or males may be selected to inflict costs on females because this increases the chance that they will mate with them (Johnstone and Keller 2000; Lessells 2005). It is also possible that females can increase their fitness by imposing costs on males: for example, it may sometimes be in the interests of females to stimulate aggression between males if this allows them to select the fittest mating partners or reduces the capacity of males to control them (Eberhard 2005), though evidence for effects of this kind is still equivocal (Tregenza *et al.* 2006). Finally, although conflicts of interest between the sexes are common, they are not universal (Roughgarden *et al.* 2006) and there are circumstances where the interests of males and females coincide closely (see Chapters 16 and 17).

This chapter covers three types of interaction between the sexes where conflicts of interest are common. While preferences for mating with particular partners are often more highly developed in females than males, mating preferences are also common in males and are explored in section 15.2. The next two sections examine the evolution of male tactics that increase the chance that females will mate with them. In some cases, the presence or behaviour of males exerts important effects on the physiology or behaviour of females (section 15.3); where females are reluctant to mate with males, it is frequently in the interests of males to coerce them into accepting their advances and males may adopt a range of tactics to induce compliance in females (section 15.4). Coercive behaviour in males is likely to generate counter-selection pressures on females to avoid male control and these have led to a range of adaptations, which are explored in section 15.5. In some species, males can also manipulate the receptivity of females and increase their mating success by killing dependent offspring fathered by their predecessors and male infanticide is sometimes regarded as another form of male coercion: section 15.6 reviews evidence of male infanticide and different interpretations of its evolution, while section 15.7 describes the tactics used by females to avoid their infants being killed. Section 15.8 briefly reviews other forms of post-copulatory sexual conflict

while section 15.9 describes some of the demographic consequences of sexual conflict.

15.2 Male mate choice

Males spend time and energy guarding females and their supply of sperm is finite (see Chapter 13), with the result that they can often increase their mating success and fitness by concentrating their mating effort on accessible females with high potential fecundity. Though male mating preferences would be expected to be particularly strong where parental investment by males exceeds investment by females, consistent mating preferences may increase the fitness of males in a wide range of systems and can evolve in species where males are not involved in parental care or where females invest more than males, as well as in those where males invest more heavily than females (Bonduriansky 2009; Edward and Chapman 2011).

Conditions likely to favour mating preferences in males include heavy investment of time or energy in locating, courting or guarding potential mates, large ejaculate volumes and high variation in female fecundity (Edward and Chapman 2011), which can lead to situations where the direct benefits of mate choice are greater in males than females (Servedio and Lande 2006; Nakahashi 2008; Edward and Chapman 2011). Within populations, the strength of male mating preferences might be expected to be positively correlated with the dominance or power of males and with their prospects of future reproductive success (Galvani and Johnstone 1998; Parker and Pizzari 2010).

Although male mating preferences have received less attention than female preferences, they have now been documented in a wide range of species, including rodents (Schwagmeyer and Parker 1990; Solomon 1993), primates (Keddy-Hector 1992; Müller *et al.* 2006; Parga 2006; Kappeler 2012), ungulates (Berger 1989; Preston *et al.* 2005), carnivores (Szykman *et al.* 2001) and cetaceans (Craig *et al.* 2002; Pack *et al.* 2009). In the majority of cases, males favour females with high potential fecundity, though the characteristics of partners that they prefer vary. For example, in Soay sheep, successful males preferentially guard heavier females during periods of the rut when they are likely to conceive and consortships involving heavier females last longer (Figure 15.1) and, for similar reasons, male humpback whales preferentially guard females without calves (Craig *et al.* 2002).

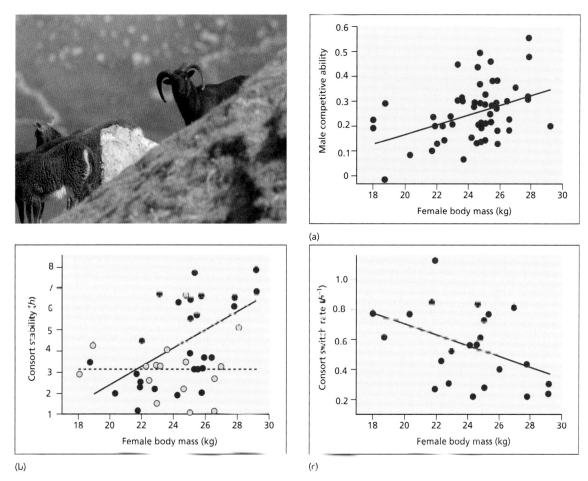

(a)

(b)

(c)

Figure 15.1 Male mate choice in Soay sheep, where rams commonly guard oestrous females during the annual rut: (a) association between female body mass and the competitive ability of the male guarding them, calculated across the rut (the fitted regression line is plotted from the predictions of a model incorporating other factors affecting the competitive ability of males); (b) relationship between consort stability and female body mass during periods of above-average availability of females (filled circles and solid line) and below-average female availability (open circles and dashed line); (c) association between female body mass and the frequency with which the male consorting them changed (consort switch rate). *Source:* From Preston *et al.* (2003). Reproduced with permission from the Royal Society. *Photo source:* © Arpat Ozgul.

In addition, there is a common preference for novel partners, which tend to generate higher levels of arousal in males, more frequent mating and, in some cases, larger ejaculates (see Chapter 13). This is sometimes known as the 'Coolidge effect' after President Coolidge of the USA (Davies *et al.* 2012). When the President and his wife visited a chicken farm, Mrs Coolidge was told by one of the managers that roosters often copulate dozens of times a day, and is said to have replied 'Tell that to the President'. When President Coolidge was told, he asked whether this was with the same hen or with different

hens and was told that many different hens were involved, and replied 'Tell *that* to Mrs Coolidge'.

Males show preferences for different female characteristics in different species. Male mandrills and spotted hyenas favour relatively dominant females (Szykman *et al.* 2001; Setchell and Wickings 2006) while male bison, banded mongooses and chimpanzees prefer older, more fecund females over younger ones (Müller *et al.* 2006; Nichols *et al.* 2010). In some cases, males appear to select partners on the basis of characteristics related to their health. For example, experiments with mice show that males avoid mating with females infected with parasites that reduce their fecundity (Gourbal and Gabrion 2004). Mating preference may also affect the allocation of sperm. For example, males of some bird species allocate more sperm per ejaculation when mating with partners of relatively high fecundity, where sperm competition is likely or where their chances of mating again are relatively low (Pizzari *et al.* 2003; Parker and Pizzari 2010).

One unresolved issue concerns the extent to which males avoid mating with-in. Although male dispersal is commonly described as an example of inbreeding avoidance, in many species inbred matings are rare because females typically reject the advances of familiar or related males, forcing males to disperse from their natal groups to locate willing mating partners (Lehman and Perrin 2003; Höner *et al.* 2007) (see Chapter 12). Whether males also avoid courting or mating with female relatives is unclear, but some evidence suggests that they may do so in some species. For example, bull elephants engage in disproportionately few sexual interactions with either paternal or maternal kin (Archie *et al.* 2007) while male banded mongooses are less likely to guard closely related females than distant relatives or unrelated females (Sanderson *et al.* 2015). In addition, there is some evidence that the presence of female relatives can affect development and dispersal in males. For example, male deer mice delay the development of sexual maturity if their mother is present (Wolff 1992) and male prairie dogs are more likely to disperse if a mature daughter is present in their group (Hoogland 1995) (see Chapter 12).

15.3 Manipulation

In a wide range of mammals, the presence or behaviour of males that are potential breeding partners has important effects on the behaviour or physiology of females. As well as attracting females (see Chapter 4), the presence of males or their displays can exert an important influence on the hormonal status and reproductive physiology of females. For example, in sheep, the presence of rams advances the timing of oestrus and conception in ewes and is widely used by sheep farmers to manipulate reproductive timing (Rosa and Bryant 2002). In red deer, too, male behaviour has an important influence on reproduction in females and playback of male roars to females that are isolated from males advances the timing of conception (McComb 1987) (Figure 15.2).

The presence of males also exerts an important influence on the endocrine status and reproductive activity of females in many rodents (Solomon and Keane 2007). In some social species this varies with the identity of males, and the presence of unrelated males who are potential breeding partners has a stronger effect on the reproductive physiology of females than that of familiar males (see Chapter 4). For example, in Damaraland mole rats, the introduction of unrelated males into breeding groups leads to a rapid increase in reproductive behaviour and competition in females (Cooney and Bennett 2000). Similar effects of the presence of potential breeding partners are probably widespread in other mammals.

15.4 Coercion

Forced copulation

It can often be to the advantage of males to coerce females into mating with them. Attempts to force copulations on females are particularly likely to increase the mating success of males that are unable to defend females or mating territories or that do not possess the resources, phenotype or genetic characteristics to be the optimal mating partner for females (Thornhill 1993). Coercion is particularly likely to increase the fitness of males in situations where females move between male territories, so that individual males cannot be sure that they will be able to monopolise access to particular females for long. Vice versa, when males can be reasonably certain that they will be able to monopolise breeding access to particular females for extended periods (for example, where breeding males and females live in bonded monogamous pairs), there may be little disadvantage to them in waiting to mate and little benefit in attempting to coerce females.

Attempts by males to force copulations on females occur in many animal societies, including many mammals (Ellis 1998; Palombit 2014). For example, young male elephant seals with territories on the periphery of the main breeding assemblies frequently attempt to force females that are crossing their territories to mate with them (Le Boeuf and Mesnick 1991). In orangutans, subordinate satellite males that encounter a receptive female may attempt to force her to copulate before a territorial male appears (Knott 2009; Palombit 2014). Females prefer to mate with prime, flanged males when they are close to ovulation and copulations with

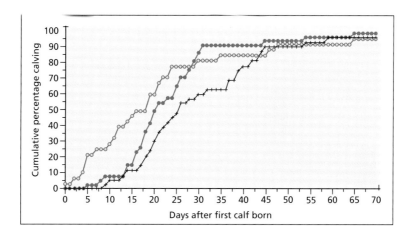

Figure 15.2 Playback of the roars of red deer stags advances the onset of oestrus and conception in female red deer. The figure shows the cumulative percentage of females that have calved in relation to the number of days since the birth of the first calf in the population for three groups exposed to different treatments before being allowed to mate with an intact stag: control [females isolated from males] (cross); playback [females played roars of males] (dark green circles); and exposed to a vasectomised male who could roar and interact with (pale green circles) using values adjusted for variance due to sire stag, sex of calf and sex × sire interaction effects. *Source*: From McComb (1987). Reproduced with permission of Macmillan Publishers Ltd. *Photo source*: © Martyn Baker.

non-territorial males sometimes occur only after fierce female resistance has been violently overcome by the male (Mitani 1985; Utami and van Hooff 2004; Knott 2009) (Figure 15.3). Territorial males, too, will engage in forced copulations, though females are usually more ready to mate with them (Mitani 1985; Palombit 2014). Forced copulation may be particularly common in orangutans because (unlike the other great apes) females are usually solitary and so lack protection from other females (Delgado and van Schaik 2000).

Conflicts of interest between the sexes often resemble 'arms races' where both males and females are likely to increase their fitness by increasing their investment in traits that enhance their chances of winning these encounters (Clutton-Brock and Parker 1995a) (Box 15.1 and Figure 15.4). Since males that successfully coerce females to mate with them can increase the number of offspring they father, while the costs to females of being coerced will, in many cases, lead to a smaller reduction in their fitness, selection pressures favouring coercive tactics in males are often likely to be stronger than those favouring resistance in females, so that 'controlling' traits are often more highly developed in males than females. For example, selection for the ability to punish and intimidate females (and thus to constrain their ability to exercise mate choice) may help to explain the

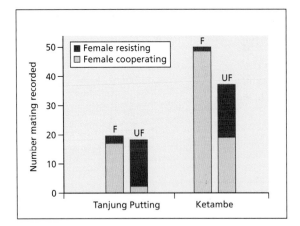

Figure 15.3 Numbers of cooperative and resisted matings by flanged (F) and unflanged (UF) male orangutans at two sites (Tanjung Puting and Ketambe). The photo shows an unflanged male. *Source*: From Utami and van Hoof (2004). Reproduced with permission of Cambridge University Press. *Photo source*: © Anna Marzec, Tuanen Orang Research Project.

evolution of large male body size in cercopithecine primates (Smuts and Smuts 1993).

Where males are substantially larger than females, the energetic costs to females of winning contests with males may be higher than those incurred by males that adopt coercive tactics, so that conflicts are usually won by males (Clutton-Brock and Parker 1995a). However, females would sometimes be expected to win encounters with males and evade their control, otherwise selection would not favour the maintenance of resistance in females. Where the sexes are of similar size, it can be less energetically expensive for a female to prevent mating than for a male to achieve it against female resistance and females

may often win encounters, even if the potential benefits are larger to males. However, lower energetic costs to females do not necessarily imply that females will occupy the winning role, for a variety of other costs may be involved. For example, continued male courtship may prevent the female feeding or engaging in other activities or females may risk injury to themselves or to their offspring by continued resistance. Males may also increase their fitness by raising the costs of resistance to females if this increases the probability that the female will mate with them.

Empirical studies provide extensive evidence of the costs to females of resisting male mating attempts. In northern elephant seals, where males may be up to eight times the weight of females, females run a risk of being crushed by the male and females that leave the (comparative) safety of the harems of dominant bulls are regularly killed by non-territorial males who attempt to control them by force (Le Boeuf and Mesnick 1991; Mesnick and Le Boeuf 1991) (Figure 15.5a). In sea otters, males hold mating partners by the nose or the back of the neck with their teeth or claws, sometimes injuring them and occasionally drowning them (Mestel 1994) (Figure 15.5b). In some ungulates, non-territorial males will attack mating pairs, in an attempt to dislodge males that have already mounted and females are sometimes killed in these struggles (Clutton-Brock *et al.* 1993). Tactics that damage females may increase the fitness of males if they reduce the probability that females will mate again and so lower the risk of sperm competition (Johnstone and Keller 2000). In contrast, females may benefit by behaviour or morphology that raises the costs to males of attempts to mate with them or restricts their access to other partners.

Harassment

In animal societies where males are unable to force females to accept their advances, they may persist in courting them until the costs of continued refusal exceed those of mating with them. In many ungulates, for example, non-territorial males persistently court and attempt to mount females that initially reject their advances (Clutton-Brock *et al.* 1988; Réale *et al.* 1996). Like forced copulation, prolonged sexual harassment can reduce the survival and breeding success of females (Smuts and Smuts 1993; Clutton-Brock and Parker 1995a): experiments with insects that manipulate the frequency of male harassment by varying the adult sex

Box 15.1 Forced copulation and sexual arms races

Where both males and females can invest in a trait that usually allows males to control female mating options or females to evade male control, the eventual outcome or ESS will depend on the relative benefits of winning to each sex and the relative costs of developing the trait. In most mammalian societies, the fitness difference between winning and losing will usually be greater for the male, who gains fertilisations if mating occurs but nothing if it does not (Parker 1984). Under these circumstances there may be strong selection for male traits (such as increased body size) that enhance the male's ability to force copulations on females and these may develop even if they have appreciable costs to male survival. The value of winning for females will generally be smaller than for males since the benefits of mating with one male rather than another will generally be less than the benefits (to males) of an additional mating (Figure 15.4). However, it may be high where the male is genetically incompatible; where females that accept mating attempts by subordinates are likely to suffer substantial direct costs; or where caring males subsequently reduce investment if females are unfaithful and, in these circumstances, there may be stronger selection in females for traits that reduce the risk of forced copulation. For example, it has been suggested that selection to minimise sexual coercion may have been responsible for increased female size in some monomorphic primates (Smuts and Smuts 1993).

The relative costs of developing 'winning' traits to males and females are also likely to affect the outcome of sexual arms races. The important criterion is the marginal cost of increases in armament for each sex (Clutton-Brock and Parker 1995a): in cases where the development of winning traits is more costly for males than females, selection may favour male investment in traits that favour male success at some other stage of the breeding cycle. For example, in flying mammals, the costs to males of increasing body size to a point where they are able to coerce females without difficulty may be prohibitive and selection may focus on aspects of their behaviour. Conversely, in terrestrial species where the energetic constraints limiting body size in males are weaker than in females, selection may favour increases in male size to a point where males can control females. Where both costs and benefits differ between the sexes, the larger of the two net differences is likely to determine whether males or females typically win contests.

In many mammalian societies, sex differences in the benefits of winning reproductive conflicts may be larger than sex differences in the costs of developing winning traits, so that males would be expected to develop traits that enable them to control female choices. For example, selection for the ability to punish and intimidate females (and thus to constrain their ability to exercise mate choice) may help to explain the evolution of large body size in males (Smuts and Smuts 1993). However, the extent of male control is likely to vary in relation to the benefits and costs of winning, and where females benefit substantially more than males by constraining the mate choices of their partners, the outcome of arms races may be reversed.

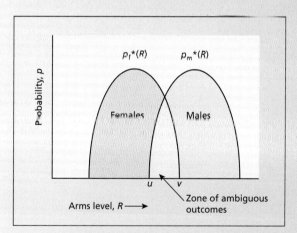

Figure 15.4 Sexual conflict and morphological armaments. The figure shows the ESS probability distributions of arms levels, R, for each sex, $p_m*(R)$ for males and $p_f*(R)$ for females. The ESS prescribes the mean arms level for each sex, and the realised arms level for individuals is randomly distributed about this mean by chance circumstances. In this hypothetical example, males usually win the conflict. They always win if they have an arms level greater than v, and females always lose if their arms level is less than u. Between u and v, the outcome of the contest is ambiguous: females win if their arms level is higher than that of a male they encounter and this maintains selection for female armaments. *Source*: From Clutton-Brock and Parker (1995a), modified from Parker (1983). Reproduced with permission of Elsevier.

(a)

(b)

Figure 15.5 (a) A male elephant seal bites a female during copulation; (b) a male sea otter biting a female. *Source*: (a) © Roberta Olenick/All Canada Photos/Getty Images; (b) © Joe Tomoleoni.

ratio in breeding groups show that strongly male-biased adult sex ratios can lead to substantial reductions in female survival and longevity (Clutton-Brock and Langley 1997). Similar effects probably occur in many mammals. For example, in feral horses, females living in non-territorial harems or multi-male groups where harassment is relatively common are more frequently disturbed while feeding and show lower condition and breeding success than females that associate with territorial males which provide effective protection against other males (Rubenstein 1986; Linklater *et al*. 1999). In some populations of feral sheep, there is high female mortality during the mating season, apparently as a result of male harassment (Réale *et al*. 1996).

Where males are likely to lose mating opportunities unless they copulate rapidly, they may raise the costs of continued refusal to females. For example, in pairs of captive long-tailed macaques, rates of aggression directed by males towards females increase substantially when a second male is present (Zumpe and Michael 1990). Male baboons will pursue cycling females into trees and then drive them out into the terminal branches until they accept matings or fall out of the tree (Figure 15.6). Persistent harassment of receptive females is not confined to primates and occurs in many other species: for example, in bottlenose dolphins coalitions of males will pursue receptive females over considerable distances (Figure 15.7).

Where harassment by males is common and dangerous, females face a quandary. To minimise the costs of harassment, they should mate rapidly with males that harass them, especially if males are less inclined to harass females that have already mated. Conversely, the first males to locate females entering breeding condition are often young or subordinate and females mating with them may suffer direct costs arising from the male's inexperience or from the inability of subordinate males to protect mating partners against attacks by other males while the pair are copulating. Moreover, females that mate with the first males to reach them may fail to take advantage of any genetic benefits arising from the selection of mating partners.

Where females maximise their fitness by refusing to mate with males but it is in the interests of males to induce them to mate with them as rapidly as possible, harassment 'contests' are likely to develop that resemble models of 'asymmetric wars of attrition', where the two individuals compete to establish control by some form of persistence (Maynard Smith 1974; Maynard Smith and Parker 1976). The theoretical structure of contests of this kind has been worked out in relation to contests within each sex (Parker 1979; Hammerstein and Parker 1982) but also applies to contests between the sexes (Clutton-Brock and Parker 1995a). In the simplest case, contests occur between individual males and individual females and each contestant decides at the outset what it is prepared to spend on the contest, which determines for how long the contest will persist and who will win it (Figure 15.8). Since the fitness benefits males gain by coercing females to mate with them are often likely to

Figure 15.6 An adult male chacma baboon has chased a cycling female up a tree and is swatting at her as she hangs from a branch. He subsequently jumped up and down on the branch until she eventually fell from the tree. *Source*: © Dawn Kitchen.

Figure 15.7 Sexual harassment in bottlenose dolphins: three males (closest to camera) in pursuit of a female, who is leading the group. *Source*: © Simon Allen, The Dolphin Innovation Project.

exceed the losses that females suffer by mating with a suboptimal partner, males are likely to be prepared to suffer greater costs and to adopt longer persistence times than females and so are often likely to be the winner (Clutton-Brock and Parker 1995a). However, where females are never likely to win, it may pay them to mate readily with males that harass them in order to minimise the costs of contests. For example, female elephant seals that are leaving the harems of dominant bulls readily accept mating attempts by peripheral males and may reduce the risks of injury by doing so (Mesnick and Le Boeuf 1991). To increase their chances of winning these contests, females may adopt tactics that raise the costs of harassment to males or allow them to avoid male control. For example, in spotted hyenas, the genital anatomy of females prevents mating unless the female cooperates (Frank 1997; East *et al.* 2003), and similar adaptations occur in many insects (Eberhard 1990).

Harassment can take more complex forms, and can have other outcomes. In some species, multiple males often harass females at the same time and the average duration of contests declines as the number of males involved rises (Haigh and Cannings 1989). Interactions of this kind can be dangerous to females, who often do their best to escape. In other cases, contestants may be

able to readjust their persistence time as conflicts progress and they can reassess their chances of winning (Maynard Smith 1976; Enquist and Leimar 1983). Under these circumstances, uncertainty over the likely outcome of contests is likely to extend the length of contests, since both individuals may prefer to suffer the costs of continued assessment in order to obtain a more accurate estimate of the relative power of their opponent.

Although sexual harassment by males frequently appears to coincide with the more general predictions of conflict theory, there have been few quantitative assessments of variation in the behaviour of both sexes in mammals because collecting precise data on male harassment and female responses is usually difficult. As a result, it is not yet clear to what extent empirical results support or conflict with the predictions of different models.

Punishment and intimidation

Male punishment of females that refuse their advances or consort with alternative partners is also common (Clutton-Brock and Parker 1995b) and is sometimes referred to as 'conditioning aggression' (Wrangham and Müller 2009). In the simplest situation, a male threatens or attacks a female who has transgressed his

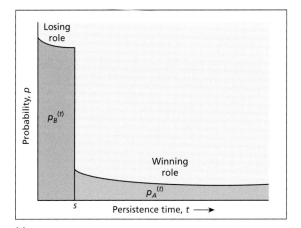

(a)

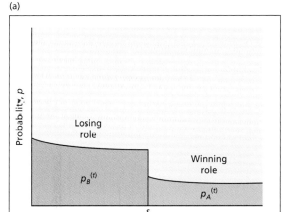

(b)

Figure 15.8 ESS choices of persistence times in asymmetric wars of attrition. At the start of the contest, each player attempts to assess which role it occupied (the losing role B, or the winning role A), and chooses a persistence time from the appropriate distribution, $p_A(t)$ if A, $p_B(t)$ if B. The range of persistence times for role B is continuous between 0 and s, and continuous between s and infinity for role A. One or both contestants may make a mistake in assessing its role: (a) there is a large difference in power between losers and winners and mistakes in role assessment are relatively rare; (b) losers and winners are more equal and mistakes in role assessment are common. Mistakes increase the mean contest costs, and make the outcome more ambiguous for a given degree of asymmetry between contestants. *Source*: Adapted from Hammerstein and Parker (1982). Reproduced with permission of Elsevier.

interests in some way (for example, by refusal to mate with him or by consorting with a rival male) and so increases the chance that she will mate with him. For example, in several primates that live in multi-male

groups, males sometimes attack females that reject their attempts to mate (Figure 15.9), especially at times when they are likely to conceive (Nadler 1982, 1988; Kitchen *et al.* 2009). Male aggression directed at females is also common in chimpanzees (Figure 15.10) and here, too, females receive more aggression from males when they are in oestrus than when they are lactating (Figure 15.11).

A wide range of studies suggest that male aggression directed at receptive or near-receptive females increases the chance that they will mate with aggressors, although it does not always do so (Müller *et al.* 2007, 2009; Swedell and Schreier 2009). For example, male Japanese macaques that direct aggression at females during the mating season are significantly more likely to mate than males that do not (Enomoto 1981). Under these circumstances, punishing tactics are likely to be evolutionarily stable in males as long as the benefits they gain from female compliance exceed the costs of punishing, and the costs of compliance to females are lower than the costs of being punished (Clutton-Brock and Parker 1995b).

Aggression may also have longer-term benefits to males if females can identify particular males and learn to avoid punishment by accepting their attempts to mate (Smuts and Smuts 1993). Data for chimpanzees show that females copulate more frequently with males that direct aggression at them at higher rates than the average (Figure 15.12). Longer-term effects of male aggression may help to explain why, in some species, males often attack females for no apparent reason. For example, 83% of severe male attacks on females in wild chimpanzees observed by Goodall (1986) occurred for no obvious reason and involved cycling females whose sexual swelling had not yet reached full tumescence. In some cases, male aggression towards females may also be a component of intimidatory displays aimed at other males (Kitchen *et al.* 2009). The punishing tactics of males can have substantial fitness costs to females. Serious wounding is not uncommon (Enomoto 1981) and attacks can lead to serious injuries, including broken bones, and can cause abortion, permanent disablement or death (Smuts 1985; Smuts and Smuts 1993): for example, anoestrous female olive baboons are likely to be attacked around five times a week by males and, on average, are seriously wounded by males around once a year; and in one population of wild chimpanzees, over 50% of male and female skeletons show evidence of healed trauma, often involving the face or head (Novak and Hatch 2009).

Figure 15.9 Male aggression towards females and its consequences in chacma baboons. *Source*: © Alecia Carter.

Punishing tactics are often used tactically by males. For example, where males are attempting to guard or sequester females, they commonly punish individuals that stray. In chimpanzees, males will repeatedly attack females in the early stages of consort formation until they become more cooperative and follow more closely

(Goodall 1986), while in hamadryas baboons male aggression appears to play an important role in maintaining the stability of harem groups and is most frequent when a male has taken over a harem from another male or when an established male has recently been joined by a new female immigrant (Swedell and Schreier 2009). If

Figure 15.10 Sexual intimidation in chimpanzees at Gombe. (a) Humphrey, an aggressive high-ranking male, sits and shakes a palm frond as he courts Gilka, a young adult female. (b) Gilka presents her oestrous swelling while staying on the opposite side of a tree. (c) Humphrey, with hair and penis erect, moves around the tree; Gilka tries to keep the tree between them. (d) Humphrey flails the palm frond more vigorously; Gilka screams and clings to the tree but does not fully present for copulation. (e) Humphrey tries again, shaking the branch of a bush; Gilka stays on her side of the tree. (f) Humphrey, still with erection, stands and grabs hold of Gilka, who crouches screaming. (g) Humphrey begins his attack, slapping Gilka on the back. *Source*: From Müller *et al.* (2009). *Photo source*: © David Bygott, Gombe National Park, 1970.

females stray repeatedly, guarding males may increase the severity of punishments: in hamadryas baboons, males that defend harems will initially threaten females that stray with an eyebrow flash, but if they fail to return immediately will bite them on the neck (Kummer 1968; Swedell and Schreier 2009) and frequent punishments are directed at females that stray repeatedly (Figure 15.13). In species living in multi-male groups, dominant males also discourage females from associating or mating with subordinate males by punishing individuals that do so (de Waal 1982; Lindburg 1983; Goodall 1986; Manson and Wrangham 1991; Manson 1994). These attacks can have

some cost to males, since females may retaliate by withholding support for resident or dominant males in interactions with potential usurpers.

The distribution of male aggression directed at females suggests that its function is often to intimidate them. Many of the mammals where males commonly direct aggression at females are species that typically live in multi-male groups where more than one male can gain access to any female, as in baboons and macaques, chimpanzees, spider monkeys, bottlenose dolphins and spotted hyenas (Connor and Vollmer 2009; Kitchen *et al.* 2009; Müller *et al.* 2009). In contrast, similar attacks are

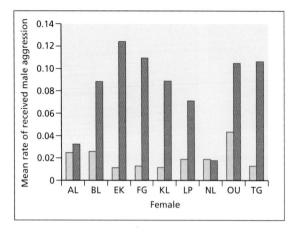

Figure 15.11 Nine different female chimpanzees receive more male aggression when in oestrus (dark bars) than when lactating (pale bars). Letters below histograms refer to the identities of females. *Source*: From Müller *et al.* (2009).

less commonly reported in monogamous species or in those living in harem groups where a single male defends the entire range of a group of females (Smuts and Smuts 1993).

Here, too, the characteristics of both sexes are likely to influence the capacity of males to intimidate females. These include their relative size, the strength of coalitions between females or between females and individual males, and the extent to which males rely on support from females to maintain their tenure of the breeding group (Smuts and Smuts 1993). For example, while male to female aggression is common in chimpanzees, where males are clearly dominant to females, it is relatively uncommon in bonobos, where males rely on support from related females and are not consistently dominant to females (Paoli 2009). The reasons for these differences are uncertain: one suggestion is that contrasts in

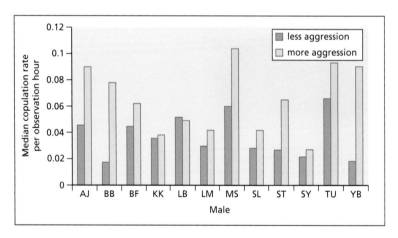

Figure 15.12 In chimpanzees males are more likely to mate with females at which they have previously directed aggression. The figure shows median dyadic rates of sexual coercion and copulation by 13 male chimpanzees with 15 parous females (from a population of chimpanzees in Uganda). Letters refer to different males. For each male, bars indicate median copulation rates with females who received more than (pale blue) or less than (dark blue) the median amount of aggression from that male. *Source*: From Palombit (2014). Reproduced with permission of Elsevier. *Photo source*: © Michael Wilson.

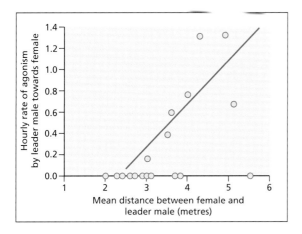

Figure 15.13 Relationships between the average distance between a hamadryas baboon female and her leader male and the overall rate of aggression directed by that male toward the female. *Source*: From Swedell and Schreier (2009). *Photo source*: © Larissa Swedell.

behaviour between chimpanzees and bonobos are a result of variation in the intensity of competition for food (Wrangham 2000), while another is that male intimidation of females is a consequence of the tendency for females to mate with multiple partners in order to confuse paternity and is associated with the risk of male infanticide (van Schaik *et al*. 2004; Clarke *et al*. 2009). However, there are several species where intimidation occurs in the absence of regular male infanticide, so whether there is a general association between male infanticide and male intimidation of females is unclear.

The theoretical basis of several important aspects of punishing tactics remains largely unexplored. For

example, there has been little consideration of the optimal magnitude or frequency of punishments that males should inflict to ensure compliance or of how these would be expected to change if females persistently refuse to comply. Nor do we yet understand the likely effects of variation in the certainty of being punished or the effects of contrasts in the accuracy of information concerning the likely magnitude of punishments. Although inducements or 'bribery' of females is likely to be less effective and more costly to males than punishment, strategies that mix inducements with punishments may often be more effective than either punishing tactics or inducement on their own (see Chapter 8).

15.5 Female counter-strategies to male coercion

While females sometimes respond to male harassment or intimidation by acquiescing to male mating attempts, they often use other tactics to prevent males from constraining their choices or to minimise the costs of male coercion. In some cases, females retreat to positions where they are inaccessible to males. For example, female mountain sheep respond to intruding subordinate rams by backing into inaccessible sites or thick vegetation (Hogg and Forbes 1997). Where males defend mating territories from which young males are excluded, females can often reduce the risk of harassment by joining territorial males. In Grevy's zebras, for example, females are less frequently harassed by males when they associate with territorial males (Sundaresan *et al*. 2007). In some ungulates that live in large unstable herds of mixed membership that lack a consistent male dominance hierarchy, non-territorial males cannot provide effective protection for oestrous females. In some of these species, oestrous females move to clusters of territorial males where they are less likely to be harassed by non-territorial males. Females arriving at these 'leks' often move to male territories where several females are already present (see Chapter 10). Similar behaviour occurs in some primates. In orangutans, females that associate with dominant males suffer less harassment from subordinate males than those that forage on their own (Fox 2002), while in chimpanzees, where males may attack females from neighbouring communities, females may transfer from smaller communities to larger and more powerful ones (Nishida *et al.* 1985; Goodall 1986).

In multi-male societies, females are often at greater risk of attack when consorting with subordinate males and can reduce the risk of harassment by consorting with dominant males. For example, in rhesus macaques, females consorting with dominant males are less likely to be attacked by other males than when they are consorting with low-ranking males (Manson 1992, 1994; Manson and Perry 1993; see also Pope 1990; O'Brien 1991), and in many social mammals females show a preference for associating with high-ranking males (see Chapter 4). In other cases, females commonly form 'friendships' with particular subordinate males that provide protection to them or their offspring, associating preferentially with them throughout the year and mating with them when they enter oestrus (see Chapter 16). Females that are the target of aggression from subordinate males may also solicit support from dominant males, giving calls that attract dominant males, who subsequently chase off the molester(s). The best-known example is in elephant seals, where females call loudly when subordinates attempt to mate with them (Cox and Le Boeuf 1977; Poole 1989), see Figure 15.5.

In some species, females form coalitions to protect each other against males, and female coalitions can even force resident males to emigrate. For example, in chimpanzees, females sometimes gang up on males that have attacked other females or attempted to force them to copulate or threatened their families (Nishida *et al.* 2003) and similar coalitions between females have been observed in bonobos (Smuts and Smuts 1993) as well as in bottlenose dolphins, where females form cooperative coalitions against harassing juvenile males (Connor and Vollmer 2009).

As well as reducing the costs of male harassment, females can adopt tactics that help them to evade male control. These include the avoidance of controlling males and surreptitious matings with males other than the dominant male in their group (Smuts 1985; Clutton-Brock and Huchard 2013). In some species, females have developed physiological or anatomical adaptations that help them to evade male control. For example, it has been suggested that selection on females to evade male control may be responsible for increases in female size and for the evolution of monomorphism in some polygynous primates (Smuts and Smuts 1993). Cryptic ovulation may also represent an adaptation to evading male

control or harassment (Manson 1997; Heistermann *et al.* 2001). Where male harassment is common, the duration of oestrus can also be very short: for example, in fallow deer, females are seldom in oestrus for more than a few hours (Clutton-Brock *et al.* 1988, 1989). Females may also reduce the risk of male coercion by forming long-term bonds with particular males, leading to the formation of long-term 'friendships' between individuals (see Chapter 16).

15.6 Male infanticide

In 1961, a Japanese primatologist, Yukimaru Sugiyama, watched a band of seven male grey langurs (Figure 15.14) expel the resident dominant male from a breeding group (Sugiyama 1965a,b). In the days after the take-over, the new males fought among themselves until only one was left in the group. The new dominant male then bit and killed all the six infants in the group. Soon, afterwards, there was an increase in sexual activity among females and subsequent observations at the same site and at other sites showed that take-overs by males were commonly associated with infant deaths (Sugiyama 1967; Yoshiba 1968; Mohnot 1971). The violence associated with take-overs was surprising, for previous studies of grey langurs had stressed the peaceable and relaxed nature of social relationships between adult males and had contrasted this with the behaviour of male baboons and macaques (Jay 1965). Sugiyama initially suggested that the increase in sexual activity might be caused by the social disruption associated with the take-over (Sugiyama 1965a,b) but later argued that males might kill infants to avoid the delay in female receptivity that occurred when females successfully reared their offspring (Sugiyama 1967).

A common reaction to Sugiyama's results was that infanticide must be a pathological consequence of high local population density, and might resemble the increased frequency of aggression found in laboratory colonies of some rodents kept at unnaturally high densities. Sugiyama's observations stimulated further studies of social dynamics in langurs. In a landmark study, Sarah Hrdy documented further cases where male take-overs were associated with infanticide in a population of grey langurs at Mount Abu in Rajasthan and showed that females whose infants had been killed became sexually

Figure 15.14 A uni-male group of grey langurs. In some populations, multi-male groups are also common. *Source* © Sarah Hrdy

receptive within a few days and copulated with the new dominant male (Hrdy 1974, 1975, 1977a). She argued that the effects of infanticide on female receptivity provided the primary reason why incoming males killed infants, an explanation that has subsequently become known as the *sexual selection hypothesis* for the evolution of male infanticide. Hrdy suggested that the apparent association between infanticide and high local population density occurred because the tenure of dominant males was relatively short in high-density populations where take-overs by extra-group males were relatively common. The gestation length of grey langurs is between 6 and 7 months and is followed by 8–12 months of lactation, so that births are typically spaced 20–30 months apart. Where male tenure is short, the acceleration of female receptivity could, Hrdy suggested, make a substantial contribution to the breeding success of dominant males.

While killing infants may improve the breeding success of individual male langurs, it necessarily reduces the potential growth rate of the group and the population. As a result, Hrdy's explanation of infanticide crystallised the contrast between explanations of social behaviour relying on benefits to all group members and explanations relying only on benefits to particular

individuals (see Chapter 1), generating an immediate controversy with scientists reluctant to accept that the adaptive strategies of individuals might compromise the interests of groups or populations (Sommer 2000). With hindsight, it seems surprising that evidence of infanticide by females (which had been documented in rodents since the 1950s) did not trigger a similar debate: one possible explanation is that female infanticide had been demonstrated mainly in captive rodents, where mothers were known to kill their own offspring under crowded conditions (Fox 1975) and it was not widely appreciated that females in natural populations of many mammals also commonly killed offspring born to their competitors (see Chapter 8).

After the publication of Hrdy's work, theoretical models confirmed the feasibility of the sexual election hypothesis and explored the pay-offs of different strategies to male primates. Early models parameterised for langurs treated infanticide as a fixed strategy and explored its evolution in single-male groups, assuming that infanticidal males would be replaced by other infanticidal males (Chapman and Hausfater 1979; Hausfater *et al.* 1982). They showed that infanticide commonly increases male fitness, though benefits vary with average tenure length and there are potential disadvantages at some tenure lengths. Subsequently, field studies showed that infanticide was also widespread in multi-male groups, and later models investigated its evolution under these conditions, treating it as a conditional strategy (Broom *et al.* 2004) and incorporating the consequences of responses by females and resident males (Boyko and Marshall 2009; Lyon *et al.* 2011). They show that the potential benefits of infanticide to males are affected by the age and likely tenure of males as well as by their probability of fathering the next offspring born to mothers whose infants are killed and the likely responses of females to infanticidal attacks.

Subsequent field studies of other populations of grey langurs (including ones where multi-male groups were common) confirmed the regular occurrence of male infanticide and showed that immigrant males made persistent attempts to kill infants, arguing against suggestions that infant deaths were accidental by-products of social disturbance associated with male immigration

(Figure 15.15). Over the last 40 years, it has become clear that male infanticide occurs in a wide range of social primates as well as in other animals (Freed 1986; Goldstein *et al.* 1986; Veiga 1993; Eggert and Müller 1997) (Figure 15.16). In addition to social primates (including New World monkeys, Old World monkeys and apes), it is common in rodents (especially the Muridae and Sciuridae) and carnivores (especially the Felidae, Mustelidae and Ursidae) (Packer and Pusey 1984; Bartoš and Madlafousek 1994; Blumstein 2000; van Noordwijk and van Schaik 2000). In particular, long-term studies show that male infanticide is common in African lions (Box 15.2 and Figure 15.17) after new males have taken over a pride. Occasional male infanticide has also been reported in several ungulates, including hippopotamus, feral horses, plains zebra and red deer (Berger 1983; Pluháček and Bartoš 2000; Feh and Munkhtuya 2008; Gray 2009), and may occur in some cetaceans (Patterson *et al.* 1998). Infanticide by

Figure 15.15 Male infanticide in grey langurs: (a) a male attacks a female with an infant that he has been pursuing; (b) females cooperate to drive the male away; (c) after a subsequent attack, the female holds her profusely bleeding infant; (d) the infant dies the following day. *Source*: © Volker Sommer.

Figure 15.16 A male lion with a cub that it has recently killed. *Source*: © Tim Caro.

males is not confined to species living in one-male groups or to recent immigrant males: for example, in some macaques or baboons where groups usually include several adult males, infanticide can occur when a new male assumes the dominant position (Kutsukake and Nunn 2006; Palombit 2012).

In some mammals, male infanticide accounts for a substantial proportion of infant mortality: for example,

Box 15.2 Male infanticide in African lions

Some of the best evidence of male infanticide in natural populations of mammals comes from African lions (Bertram 1975; Packer and Pusey 1984). Resident males have relatively short tenure and are usually replaced by rival coalitions in under 2 years. Recent male immigrants have been observed to kill cubs on several occasions and the probability that all cubs will die doubles after a take-over (Figure 15.17). Following infanticide, females whose cubs have been killed breed again after a shorter interval than would have been the case had their cubs survived (Packer and Pusey 1983a,b).

Both juvenile mortality and the frequency of male take-overs are highest in large and small prides. Groups of females with cubs will threaten or attack intruding males and are more successful at driving them off than solitary females (Packer *et al.* 1990). The increased frequency of male take-overs in small prides may occur because these have low reproductive value to males and are frequently defended by males that spend most of their time with a neighbouring larger pride so that male immigration is relatively likely (Packer *et al.* 1988). Increased rates of infanticide in unusually large prides may occur because competition between males for large prides is more intense and male tenure is reduced.

Female lions with cubs appear to avoid situations where infanticide is likely. Playback experiments show that females distinguish between the roars of extra-group males and those of males resident in their pride and avoid interactions with intruding males (McComb *et al.* 1993). Their responses to playbacks of male intruders show a clear contrast with responses to playbacks of female intruders, which are always approached. Females roar in chorus to maintain contact with other pride members and to defend their territory against neighbouring prides, and intruding males are evidently able to distinguish between the roars of solitary females and those of female groups and are more reluctant to approach the latter, especially if they are on their own (McComb *et al.* 1993).

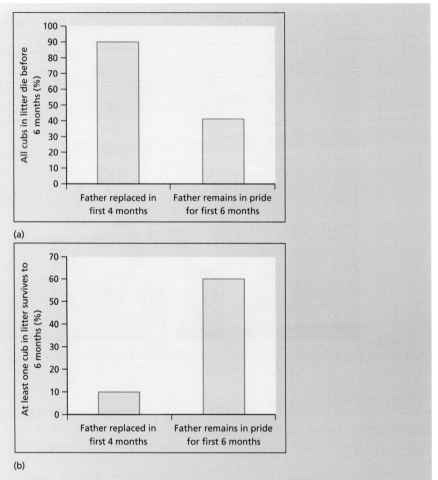

Figure 15.17 Infanticide in African lions: (a) the number of cases where all cubs in litters died before the age of 6 months when the father was replaced during this period or remained in the pride; (b) number of cases where at least one cub in the litter survived to 6 months. *Source*: From Packer and Pusey (1984). Reproduced with permission from Transaction Aldine.

in mountain gorillas, -long-term studies suggest that 37% of infant mortality is a consequence of infanticide (Watts 1989) and it is responsible for a substantial proportion of infant mortality in several other primates (Teichroeb and Sicotte 2008; Palombit 2012) (Figure 15.18).

There is now a large body of evidence that supports the sexual selection hypothesis. Male infanticide is commonly associated with immigration by males into established breeding groups or local populations or, in multi-male groups, with changes in dominance rank among resident males (Wolff and Cicirello 1989, 1991; van Schaik 2000; Palombit 2012). It usually involves either extra-group males or recent immigrants (Packer and Pusey 1984; Perry and Manson 2008): where resident males in multi-male groups kill infants, this is commonly associated with rank changes by males that affect their chances of fathering young (Crockett and Sekulic 1984; Palombit *et al.* 2000; van Schaik 2000). Rates of male infanticide appear to be higher in species where males attain high rank soon after immigrating and to be lower where they enter at the bottom of dominance hierarchies (Broom *et al.* 2004). There is also extensive evidence that male infanticide is intentional and is likely to increase the fitness of males. For example, in some primates, newly dominant immigrant males follow mother–infant pairs for days or even weeks,

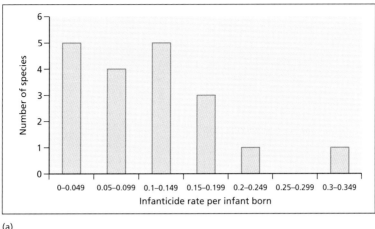

(a)

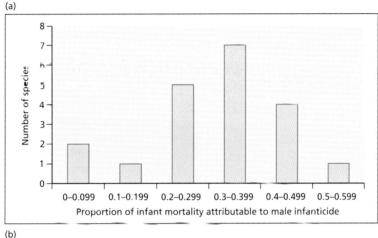

(b)

Figure 15.18 Frequency distribution of male infanticide in different primate species: (a) rate of infanticide per infant born; (b) fraction of infant mortality attributable to male infanticide. *Source*: Data from Janson and van Schaik, 2000.

apparently waiting for a chance to attack the infant (Butynski 1982; Wright 1995).

Both observational and genetic data suggest that victims of infanticide are usually the offspring of previous dominant males or of competing males and are rarely juveniles fathered by the males responsible for infanticide. In addition, there is evidence that sexual experience of specific females commonly reduces the infanticidal tendencies of males (Mallory and Brooks 1978; Labov 1980, 1984; Webster *et al.* 1981). For example, in some strains of mice, previous cohabitation with the mother or exposure to her urine are sufficient to inhibit infanticidal tendencies in males (Huck *et al.* 1982) while, in others, males must have ejaculated with the female and previous exposure to the urine of the mother

is insufficient to inhibit infanticidal tendencies in males (Huck *et al.* 1982). The effects of sexual experience with the mother on the probability that males will be infanticidal vary with the number of days since mating and peak at around 3 weeks, the usual gestation length of mice (Bronson 1979; vom Saal 1984; Perrigo and vom Saal 1994). Studies of primates also suggest that males that have had sexual relationships with females are less likely to kill their offspring. In several species, the infants of pregnant females that have mated with immigrant males are less likely to be killed than those of mothers that have not done so (Hrdy 1974; Struhsaker and Leland 1985; Watts 1989; Borries *et al.* 1999) and where post-conception mating is rare (as in ursine howler monkeys), infants born to females that are pregnant during take-

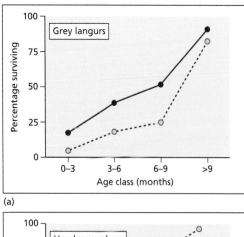

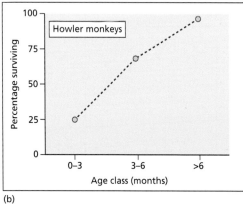

(a)

(b)

Figure 15.19 The probability that immature grey langurs and ursine howlers survive the presence of an infanticidal adult male as a function of their age. In langurs (a), the upper curve refers to cases where attacks or infanticide were witnessed; the lower curve also includes cases where infanticide was considered likely or was presumed. In howlers (b), all 'very probable' infanticides were included. *Source*: Data from Sommer (1994) and Crockett and Sekulic (1984).

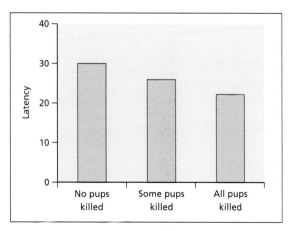

Figure 15.20 Effects of male infanticide in house mice on the latency from the exposure of a female with newborn pups to a newly dominant male to the birth of her next litter, fathered by the introduced male. *Source*: From vom Saal (1984). Reproduced with permission from Transaction Aldine.

overs rarely survive (Packer and Pusey 1983a; Crockett and Sekulic 1984).

Most victims of male infanticide are still at a stage where they are dependent on their mothers, so that their premature death accelerates their mother's return to breeding condition and the death of dependent infants or juveniles advances their mother's next conception. For example, both in grey langurs and in ursine howler monkeys, infants more than 6 months old usually survive the presence of an infanticidal male while over 75% of infants less than 3 months old do not (Figure 15.19). In several primates where male infanticide is common, females whose offspring are killed by incoming males conceive again earlier than those whose offspring survive (van Schaik 2000). Similarly, female laboratory mice whose entire litters are killed by males return to oestrus more quickly than those with surviving pups (Figure 15.20). Infanticidal males are commonly (though not universally) the father of subsequent offspring produced by females whose infants have been killed (Ebensperger 1998a; Bellemain *et al.* 2006) and male infanticide rarely occurs where males are unlikely to have access to the mothers of their victims in subsequent breeding cycles (Palombit 2003).

Research on the hormonal mechanisms underlying male infanticide is also consistent with the sexual selection hypothesis. Most detailed studies of the causal basis of male infanticide have involved rodents, where infanticidal behaviour in males is positively associated with testosterone levels: in some strains of mice, castration reduces the incidence of infanticide by males, and experimentally induced increases in testosterone level raise the probability of infanticide in both sexes (Svare *et al.* 1984). Social factors likely to influence testosterone levels, such as social status, also increase the chance that males will be infanticidal (Huck 1984; vom Saal 1984). For example, subordinate male house mice are less likely to be infanticidal than dominant males but become more likely to kill the offspring of unfamiliar females after attaining the dominant position (vom Saal and Howard 1982; vom Saal 1984). Conversely, when males are defeated in fights, their tendency to be infanticidal declines (vom Saal 1984). Aspects of early development that influence hormonal status during later development can also affect the probability of infanticide. In mice, the position of pups in the uterus (which affects their exposure to androgens) influences their probability of showing infanticidal behaviour as adults (Svare *et al.* 1984; vom Saal 1984). However, these effects are complex and male fetuses that develop with males on either side of them (2M) are less likely to commit infanticide as adults than males with females on either side (0M) (vom Saal 1984), possibly because early exposure to testosterone reduces the strength of subsequent responses to increased testosterone levels (vom Saal 1984). Variation in the frequency of male infanticide may also be partly heritable. Different strains of laboratory mice vary in the frequency with which males kill unrelated pups as well as in the tendency for strange males to induce pregnancy blocks in females and differences in infanticidal tendencies between strains are sometimes related to differences in androgen levels (Svare *et al.* 1984).

The distribution of infanticide across populations of the same species is also consistent with the sexual selection hypothesis. Within several species, male infanticide increases where male replacement rates are relatively high (Janson and van Schaik 2000; Palombit 2012). For example, in blue monkeys, the frequency of male infanticide increases in high-density populations where dominant males are frequently replaced (Butynski 1982, 1990). Similarly, in ursine howler monkeys, Thomas's langurs, some colobus monkeys and lions, large groups of females tend to attract more attempts at take-over than smaller ones and the rate of infanticide is relatively high in large groups (Crockett and Janson 2000; Steenbeek and van Schaik 2001; Teichroeb *et al.* 2012).

Finally, the distribution of infanticide across species is consistent with the sexual selection hypothesis (Janson and van Schaik 2000; Lukas and Huchard 2014). As would be predicted, male infanticide appears to be rare or absent in species where females show a post-partum oestrus and there are few benefits to males in killing dependent young (van Noordwijk and van Schaik 2000). Conversely, it is more frequent in mammals where males are permanently associated with females (and so are likely to have an opportunity to fertilise their subsequent offspring) than in species where bonds between males

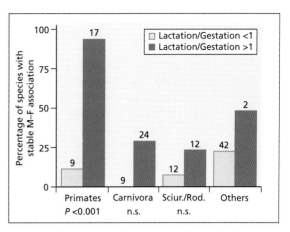

Figure 15.21 Male infanticide has been more commonly reported in mammals where the length of lactation periods exceed the duration of gestation than in those where gestation exceeds lactation. For mammals showing year-round male–female association, the figure shows the proportion of species where the duration of lactation exceeds that of gestation (dark bars) and the proportion where gestation exceeds lactation where male infanticide has been reported (light bars). Numbers show the samples of species involved. *Source:* From Crockett and Janson (2000). Reproduced with permission of Cambridge University Press.

and females are temporary (Lukas and Huchard 2014). The distribution of male infanticide is also positively associated with high reproductive skew among males and short male tenures and negatively with the number of males present in groups (Palombit 2012). It is also associated with aseasonal breeding and is uncommon in species where females breed once a year, and infanticide is unlikely to affect latency to the next breeding attempt, though it occurs in some of these species (Palombit 2012). As the sexual selection hypothesis predicts, it is more frequent in species where the duration of lactation is longer than that of gestation (Janson and van Schaik 2000) (Figure 15.21), though this association could be a consequence of the relationship with aseasonal breeding (Lukas and Huchard 2014).

While much evidence now supports the sexual selection explanation of male infanticide, it does not provide a complete explanation of its distribution and other evolutionary mechanisms are probably involved as well. Not all examples of male infanticide fulfil the conditions required by the sexual selection hypothesis: for example, male infanticide occurs in some seasonal breeders, like prairie dogs, where females breed annually and the death of their offspring is unlikely to have a significant effect on the timing of the next breeding cycle (Coulon *et al.* 1995; Hoogland 1995; Enstam *et al.* 2002). In some cases, infanticidal males are unlikely to be the fathers of the next offspring born to mothers that lost their infants, as in Przewalski's horses (Feh and Munkhtuya 2008). In others, observations suggest that male infanticide is an accidental consequence of male competition: for example, in elephant seals, males appear to kill pups accidentally in the course of fights over access to females (Le Boeuf and Briggs 1977; Doidge *et al.* 1984). Infants or juveniles may also elicit inappropriate responses from males: for example, adolescent male sea lions abduct pups and appear to treat them as substitutes for females, sometimes killing them in the course of attempts to sequester them from other males (Campagna *et al.* 1988).

In some of these cases, male infanticide may increase the fitness of males for reasons other than those suggested by the sexual selection hypothesis. In some cases, it may help males to acquire mates: for example, in Thomas's langurs, male infanticide commonly involves extra-group males that are either solitary or are members of neighbouring groups (Sterck 1997; Steenbeek *et al.* 1999) and, when infants are killed, their mothers often leave their original group to join the infanticidal male (Steenbeek 2000). Similarly, in white-thighed colobus monkeys, intruding extra-group males attack mothers with infants and this may increase their probability of acquiring females (Saj and Sicotte 2005; Teichroeb and Sicotte 2008).

In a few cases, infanticide may help to reduce competition for the male's offspring in the future: for example, in ground squirrels and marmots it may reduce competition for a male's offspring (Sherman 1981; Coulon *et al.* 1995; Hoogland 1995). In chimpanzees, where males defend access to communities of females, killing infants from neighbouring communities reduces competition for ranges or resources and may help to weaken neighbouring groups (Pusey 2001; Williams *et al.* 2002, 2004). There appears to be a male bias among the victims of infanticide in chimpanzees (Clarke 1983; Takahata 1985; Hiraiwa-Hasegawa and Hasegawa 1994; Sommer 1994) and it has been suggested that this is because infanticide helps males to eliminate potential competitors, though another possibility is that these biases are either a consequence of male-biased sex ratios among infants (Nishida *et al.* 2003) or a by-product of sex differences in juvenile behaviour, such as a tendency for male infants to stray further from their mothers (see Chapter 18).

15.7 Female counter-strategies to male infanticide

Responses to intruders

Male infanticide has substantial costs to the fitness of parents of both sexes, and both mothers and fathers might be expected to show adaptations to minimise risks to their offspring. Where immigrant males are commonly infanticidal, females often respond aggressively to potential male immigrants. For example, female territoriality in rodents may help to reduce the risk of infanticide and could have evolved for this reason (Wolff 1985, 1993; Wolff and Peterson 1998). Across species, females tend to respond more aggressively to members of whichever sex are most likely to kill their young. For example, in many rodents where male infanticide is common, females commonly respond aggressively to strange males (Blumstein 2000); in contrast, in species where females are more likely to be infanticidal than males, like golden hamsters, breeding mothers respond more aggressively to female intruders than to males (Wise and Ferrante 1982). Breeding females are often more likely to react aggressively to individuals that are potentially infanticidal than to other members of the same sex. For example, in several primates where immigrant males commonly kill infants, females respond aggressively to potential male immigrants and assist resident males to defend the group against potential immigrants, in some cases helping to drive immigrants out of the group (Mohnot 1971; Fairbanks and McGuire 1987; Wright 1995).

Where females with dependent young are exposed to potentially infanticidal males, they often cooperate to defend young against attacks by males and are sometimes wounded in the process (Borries 1997; Treves 2000). In langurs, several females may cooperate to defend infants and males that attack infants or juveniles are often the target of concerted attacks by other group members, including the infant's mother, her female relatives and, sometimes, one or more males (Hrdy 1977b; Tarara 1987; Agoramoorthy and Rudran 1995; Borries 1997). However, females do not always defend their offspring as vigorously as might be expected and, in some cases, they make little effort to defend them at all. For example, in one reported case, a female blue monkey that had been chased repeatedly by a newly dominant male gave up retreating and pushed her infant towards the male (Böer and Sommer 1992). A possible explanation is that females invest little effort in defence where the costs involved are substantial and the likelihood of success is low (Butynski 1982; Wright 1995).

Females with dependent young often appear to anticipate the risk of infanticide. For example, female langurs with dependent young avoid males that have recently immigrated (Hrdy 1974, 1977a). In chacma baboons, glucocorticoid levels in lactating females rise in response to the arrival of immigrant males, whereas those of cycling and pregnant females do not change (Beehner *et al.* 2005; Engh *et al.* 2006) (Figure 15.22a). Following attempts at infanticide by immigrant males, glucocorticoid levels in lactating females increase further (Figure 15.22b), especially in lactating females that do not have a male 'friend' (Figure 15.22c).

Females are also more likely to be more aggressive towards intruders if they are about to breed or have dependent young. In several mammals, females show an increase in the frequency or intensity of aggression and an increase in testosterone levels during late gestation and early lactation, which may represent an adaptation to the need to defend young from infanticide (Ostermeyer 1983; Maestripieri 1992; Parmigiani and vom Saal 1994; Blumstein 2000). In addition, female aggressiveness during the perinatal period is often associated with the presence of young: for example, experimental studies of female rodents show that the tendency for breeding females to respond aggressively to intruders increases with the size of their litters (Maestripieri and Alleva 1990; Maestripieri and Rossi-Arnaud 1991) and is reduced by

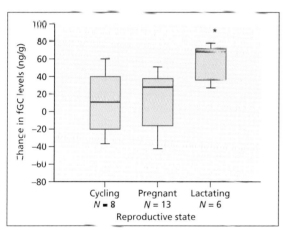

(a)

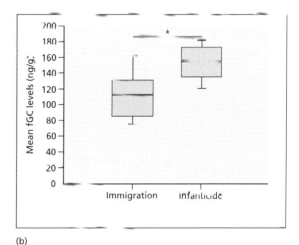

(b)

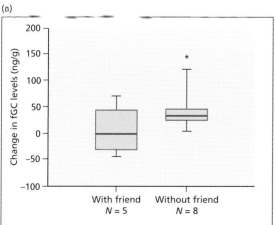

(c)

Figure 15.22 Male infanticide and glucocorticoid (GC) levels in female chacma baboons. (a) Changes in GC levels of cycling, pregnant and lactating females after the immigration of an unfamiliar male. Each box encompasses the 25th through 75th percentiles, with the median represented by an interior line. Extending lines denote 10th and 90th percentiles. An asterisk denotes a significant difference. (b) Mean GC levels of eight lactating females during a period of infanticidal attacks after an unfamiliar dominant male (SO) immigrated into the group. (c) Change in GC levels during a period of infanticidal attacks in lactating females with and without male friends. *Source:* From Engh *et al.* (2006). Reproduced with permission of Elsevier.

removing their litters and restored by returning them (Ebensperger 1998a).

In some species, breeding females commonly leave groups where males are unable to exclude intruders and, by doing so, may reduce the risk of infanticide (see Chapters 3 and 4). For example, in Thomas's langurs, males that have defended a breeding group for several years are often no longer as effective in keeping intruding males out of the group or in protecting infants during attacks as newly dominant individuals and females start to show increased sexual interest in extra-group males, generating increases in the frequency of intrusions (Steenbeek *et al.* 2000). Breeding females may leave the group at this stage and transfer to a new male, sometimes on their own and sometimes together, leaving the previous breeding male and his male offspring as an all-male band. In some cases, females transfer to join extra-group males that have previously killed their offspring: these transfers often occur soon after a female's offspring have reached independence or after a juvenile has been killed by an intruding male, and the last infants born to females before they transfer show relatively high levels of mortality.

Reproductive adaptations

Where mating reduces the risk of infanticide, females might be expected to copulate preferentially with potentially infanticidal males and studies of some mammals where male infanticide is common suggest that they do so (see Chapter 4). In several rodents, females show a preference for mating with dominant males that are particularly likely to kill young (Horne and Ylönen 1996; Agrell *et al.* 1998; Blumstein 2000), while in primates where male infanticide is common, females commonly mate with males that have recently immigrated into their group (Palombit 2012). For example, female grey langurs often mate with recent immigrants, whether or not they are likely to conceive (Hrdy 1979) and similar behaviour has been observed in other primates (Mohnot 1971; Mori and Dunbar 1985; Struhsaker and Leland 1985; van Schaik 2000) as well in other mammals (Jeppsson 1986; Pereira and Weiss 1991; Ebensperger 1998b). However, this is not always the case: for example, some studies of grey langurs found that females were no more likely to mate with recent male immigrants than with previous male incumbents (Sommer 1987) while female meadow voles and house mice show no obvious preference for mating with

infanticidal males over non-infanticidal ones (Ebensperger 1998a).

In many mammals that live in multi-male groups, females normally mate with several different partners in the course of the same reproductive cycle, which may limit the capacity of males to assess paternity. Studies of bank voles, where male infanticide is common, show that polyandry enhances offspring survival (Klemme and Ylönen 2010). In some primates, multiple mating may also serve to attract additional males ('hired guns') that protect females and their infants against other males (Wrangham and Peterson 1996; Pradhan and van Schaik 2008; Henzi *et al.* 2010). For example, in some lemurs, both the probability of group take-overs by external males and the rate of infanticide decline as the number of resident males rises and this may favour female reproductive strategies, like breeding synchrony, that are likely to increase the number of resident males (Ostner and Kappeler 2004). Where females mate with multiple partners and compete for access to males, selection operating through infanticide could have been responsible both for the evolution of large testes in males (Lukas and Huchard 2014) and for the evolution of cyclical perineal swellings in females (Clutton-Brock and Harvey 1976; Nunn 1999).

Where the risk of male infanticide is substantial, females may avoid signalling the precise time of conception. In most mammals, males appear to be unable to recognise their own offspring reliably and their capacity to assess paternity may often depend on their ability to assess the approximate timing of ovulation from oestrogen-related signals in females (Robinson and Goy 1986; Elwood and Kennedy 1994; van Schaik *et al.* 2000). For example, female primates may have developed adaptations that reduce the capacity of males to assess the timing of ovulation while retaining their capacity to attract mates (van Schaik *et al.* 2000). It has been suggested that this could account for the lack of a precise relationship between the cyclical changes in perineal swellings and the timing of ovulation in primates (Nunn 1999) as well as the tendency for females to give copulation calls after matings when they are unlikely to have conceived as well as after matings close to ovulation (Cowlishaw and O'Connell 1996; Pradhan *et al.* 2006).

A similar argument suggests that the risk of male infanticide could be responsible for the evolution of variability in the duration of different stages of the female reproductive cycle. Assuming that long follicular phases permit greater variance in the period between the onset of signals of receptivity and ovulation and help to reduce the capacity of males to identify paternity, females might be expected to have relatively long follicular phases in species where males employ coercive tactics or are likely to be infanticidal (Heistermann *et al.* 1996). As this argument suggests, in Old World monkeys and apes (where sexual dimorphism is pronounced and males commonly use coercive tactics to force females to mate with them) the duration of the follicular phase is longer than in prosimians and New World monkeys, where females usually initiate mating and male coercion is uncommon (van Schaik *et al.* 2000).

Females may also adjust their probability of conception to reduce the risk of male infanticide. Where immigration by new males is associated with a period of instability, when changes in the identity of the dominant male or further immigration by extra-group males are likely and the risk of male infanticide is likely to be high, it may benefit females to delay conception (Hausfater 1984). In several social mammals where male infanticide occurs, females appear to show responses of this kind. For example, lionesses commonly delay conception after group take-overs by immigrant males (Bertram 1975; Packer and Pusey 1983a) and female Thomas's langurs often defer pregnancy for several months after their groups have been taken over by new males, despite mating repeatedly with them (Steenbeck 2000). In some birds where male infanticide occurs, females also stop laying after male replacement and do not resume for some days or weeks (Veiga 1993). In some species, females may also synchronise breeding to increase their ability to mate with multiple partners and so reduce the risk of male infanticide and research on bank voles has confirmed that reproductive synchrony reduces the effect of infanticide on offspring survival (Poikonen *et al.* 2008).

Females whose young are at risk may also terminate investment prematurely by abortion or maternal infanticide if their young are unlikely to survive as a result of male infanticide. A tendency for female mice exposed to strange males to terminate existing pregnancies was first demonstrated in mice by Bruce (1959) and is known as the *Bruce effect*. Subsequent studies of rodents showed that females also respond to androgen-dependent male pheromones associated with urinary proteins (Milligan 1980; Huck 1984) and a similar tendency for

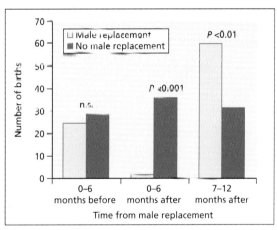

Figure 15.23 Effects of the replacement of breeding males on the birth rate in gelada baboons. The figure shows the number of births to females in the 6 months before the male was replaced, during the 6 months after replacement, and during the 7–12 months after replacement (pale bars) compared with similar-sized groups that did not experience male replacement over the same period (dark bars). Source: From Roberts *et al.* (2012). Reproduced with permission from AAAS.

exposure of pregnant females to strange males to lead to spontaneous abortion has been found in some marmots (Hackländer and Arnold 1999; Pluháček and Bartoš 2000), equids (Berger 1983; Bartoš *et al.* 2011), carnivores (Packer and Pusey 1984) and primates (Pereira 1983; Colmenares and Gomendio 1988; van Noordwijk and van Schaik 2000). For example, following the replacement of the resident male in their breeding group, 80% of pregnant female gelada baboons terminate their pregnancies (Roberts *et al.* 2012) (Figure 15.23).

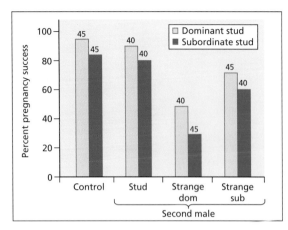

Figure 15.24 Pregnancy block in mice as a function of male social status. Females mated to dominant or subordinate males were subsequently exposed to a strange dominant (Strange dom), a strange subordinate (Strange sub), or the original stud male. Control females remained undisturbed after mating. Numbers of females tested are indicated above bars. Pregnancy success was greater for females exposed to subordinate rather than to dominant strangers and for females mated with dominant as opposed to subordinate studs. *Source*: From Huck (1984). Reproduced with permission from Transaction Aldine.

The Bruce effect was initially interpreted as a pathological response to high density and associated stress, but an alternative explanation is that where the probability of male infanticide is high, spontaneous abortion is advantageous to females because the fitness costs of losing a newly implanted egg or fetus are low while the costs of continuing to invest in a litter that is unlikely to be allowed to survive are high (Hrdy 1979; Huck 1984; Ebensperger 1998a). Some evidence now supports this interpretation. For example, female mice are more likely to fail to breed when exposed to unfamiliar dominant males that they have not encountered before than when exposed to males with whom they have previously mated or to subordinate strangers (Figure 15.24). However, phylogenetic reconstructions of the distribution of the Bruce effect in muroid rodents show that its occurrence is independent of the distribution of male infanticide and suggest that there may not be any close link between male infanticide and the Bruce effect (Blumstein 2000).

Where males are likely to kill dependent offspring, females may also accelerate weaning dependent young. For example, after the experimental introduction of strange males, captive vervet monkeys accelerate weaning (Fairbanks and McGuire 1987). Some aspects of juvenile development may also be adjusted to confuse paternity and reduce the risk of infanticide and there is some evidence that infants or juveniles alter their behaviour after male take-overs so as to deceive males about their age and the extent of their dependence on their mothers (Treves 2000). In addition, it has been suggested that the neonatal coats found in a number of Old World monkeys help to confuse paternity and could also represent an adaptation to reducing the risk of infanticide (Treves 2000).

Social adaptations

Recent models of the evolution of social behaviour in primates have suggested that aspects of social organisation previously explained as adaptations to finding food or avoiding predation may serve to reduce the risk of infanticide (Dunbar 1988; Pradhan and van Schaik 2008). For example, female defence of breeding burrows or territories against conspecifics may reduce the risk of male infanticide and could be adapted to this purpose (Sherman 1981; Ebensperger 1998a; Rödel *et al.* 2008). As might be expected, territorial behaviour is frequently directed at potentially infanticidal individuals of either sex (Ebensperger 1998a) and its intensity often peaks soon after young are born and declines after they have been weaned (Ostermeyer 1983; Maestripieri 1992).

Female sociality may also help to reduce the risk of male infanticide and some recent studies have suggested that protection against infanticide may be one of the principal benefits to females of living in groups (Hrdy 1977a; Crockett and Janson 2000; Nunn and van Schaik 2000; Pradhan and van Schaik 2008). In several social mammals, solitary females are more likely to lose their young to infanticide than those living in groups: for example, in house mice, where several breeding females may nest communally, infanticide is more frequent when females nest alone than when several females use the same nest (Manning *et al.* 1995); in African lions, solitary females suffer higher rates of male take-overs and are more likely to lose their offspring to male infanticide than females living in prides of two to seven (Packer *et al.* 1990). However, while it is conceivable that protection against male infanticide could represent an advantage of sociality in some species, cooperative defence by females is often ineffective in preventing take-overs (Ebensperger 1998a) and females are social in a substantial number of mammals where male infanticide is rare or does not occur (see Chapter 2), suggesting that it must often be maintained for other reasons.

Females may also reduce the risk of infanticide by associating with dominant males (Harcourt and Greenberg 2001; Palombit 2012). For example, in Thomas's langurs, dominant males provide protection against potentially infanticidal intruders and females leave males that are unable to provide effective protection (see Chapter 3). In baboons, females may reduce the risk of infanticide by forming close associations or 'friendships' with particular males, associating with them more frequently than with other males and grooming them frequently (see Chapter 16) (Strum 1984, Smuts 1985; Palombit 1999; Palombit *et al.* 2000). Male friends frequently support females in agonistic interactions with other group members and help to protect dependent offspring (see Chapter 16).

Finally, some studies of primates attribute the evolution of social monogamy to the need for paternal or biparental protection of offspring against conspecifics and support this with evidence of an association between social monogamy and the risk of male infanticide (Opie *et al.* 2013). However, male infanticide appears to be relatively uncommon in monogamous species and other comparative analyses have found no evidence that monogamy is more likely to evolve in lineages where male infanticide is already established, either in primates or in other mammals (Lukas and Clutton-Brock 2013, 2014) (see Chapter 10).

15.8 Post-copulatory sexual conflict

Conflict between the sexes does not end at copulation and extends into the conception and development of offspring, both before and after birth (see Chapter 16). In many animals, females exercise some form of cryptic choice of breeding partners, either through mechanisms that affect the probability that particular matings will lead to conception or through tactics that affect the probability that particular sperm will fertilise their ova (see Chapter 4). Conversely, males may adopt strategies that increase the chance that their own sperm will be successful or which remove the sperm of other males or reduce the probability that females will re-mate (see Chapter 13). In some species, males have developed spiny penises which may cause short-term damage to the female tract and reduce the duration of female receptivity (Stockley 2002). In others, enlarged accessory glands produce substantial amounts of seminal fluids that coagulate and which block the female tract to subsequent sperm, suppress female immunological responses that affect sperm survival, or reduce female receptivity (Palombit 2014). Both in rodents and primates, these accessory glands are larger in species where females commonly mate with multiple partners in the same breeding cycle (Dixson 2002; Dixson and Anderson 2002, Ramm *et al.* 2005), though females can also remove these plugs after mating (Strier 1992; Parga 2003; Setchell and Wickings 2004).

15.9 Demographic consequences of sexual conflict

Conflicts between males and females are often likely to affect the survival of both sexes as well as the fecundity of females. Where the costs that males impose on females are unusually high, these may be sufficient to reduce female numbers or, under extreme conditions, to lead to the extinction of the species (Kokko and Brooks 2003; Le Galliard *et al.* 2005). Some of the best evidence of the demographic consequences of sexual conflict comes from studies of the consequences of male infanticide. Where the risk of group take-overs and infanticide increases with group size, the upper limits of group size may be set by the rising risk of infanticide and associated increases in female dispersal, rather than by competition for food (Steenbeek *et al.* 1999). For example, in ursine howler monkeys, female fecundity does not appear to decline with increasing group size, as it does in many other mammals, possibly because female dispersal induced by the risk of infanticide limits group size in advance of resource constraints (Crockett and Janson 2000).

At the population level, frequent male infanticide is likely to depress recruitment and the potential growth rate of populations and may have important consequences for demography and population dynamics. In ursine howler monkeys, around half of all infant deaths may be caused by male infanticide, and both temporal and spatial variation in the frequency of male take-overs is sufficient to affect average recruitment rates (Crockett and Janson 2000). In bank voles, female breeding success is reduced in infanticidal populations (Poikonen *et al.* 2008) and an experiment which removed resident males (who were subsequently replaced by immigrants) halved the rate of population growth (Andreassen and Gundersen 2006). Theoretical studies suggest that positive feedbacks between male harassment, female mortality and the operational sex ratio can even lead to the extinction of local populations, though this can be prevented by selection favouring counter-strategies in females (Rankin *et al.* 2011).

Several studies also show that male infanticide affects the consequences of management and exploitation. In particular, the removal of mature males by trophy hunters can have important demographic consequences as a result of subsequent increases in the frequency of male infanticide (Swenson 2003). For example, the killing of male bears by hunters increases the risk of infanticide and reduces cub survival and net reproductive output per female by around one-third, with the result that killing one adult male bear has the same effect on the population as killing between 0.5 and 1.0 adult females (Swenson *et al.* 1997) (Figure 15.25). Similarly, in African lions, the hunting of resident males increases the rate of male take-overs and the frequency of male infanticide (Whitman *et al.* 2004) (Figure 15.26) and the effects of infanticide may have a substantial impact on population density. For example, in infanticidal populations of African lions, the removal of males over 3 years old is likely to lead to a substantial reduction in female numbers. However, the magnitude of this effect is reduced if males are culled at older ages, allowing male tenure to increase. Under some conditions, the effects of male infanticide on recruitment may be sufficient to lead to permanent reductions in population density or even to the extinction of populations where male infanticide is common (Kokko and Brooks 2003; Eldakar *et al.* 2009; Palombit 2014).

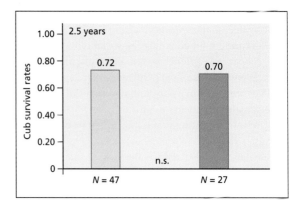

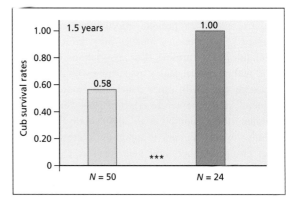

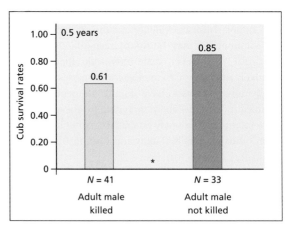

Figure 15.25 Comparison of brown bear cub survival rates in relation to whether adult males were killed (pale bars) or not killed (dark bars) in the area 0.5, 1.5 and 2.5 years before the birth of the cubs. Annual survival is at the top of each bar. *Source*: From Swenson *et al.* (1997). Reproduced with permission of Macmillan Publishers Ltd.

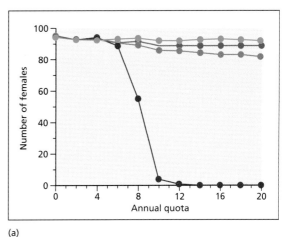

(a)

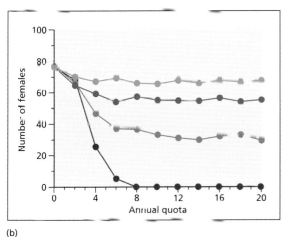

(b)

Figure 15.26 Effects of hunting rates (annual quota) on female numbers in African lions: (a) predicted number of females after 30 years in hypothetical populations where male infanticide did not occur; (b) predicted number of females in infanticidal populations. Figures show the predicted outcome from 100 runs of a simulation model of the population. Colours show the different consequences of shooting males ≥3 years (red), ≥4 years (orange) , ≥5 years (blue) and ≥6 years (green) on population size (number of females). *Source*: From Whitman *et al.* (2004). Reproduced with permission of Macmillan Publishers Ltd. *Photo source*: © Topical Press Agency/Hulton Archive/Getty Images.

SUMMARY

1. In many circumstances, the optimal outcomes of interactions between males and females differ between the sexes, leading to conflicts of interest between them.

2. Like females, males commonly show preferences for particular sexual partners or categories of partner, which are often related to variation in female fecundity.

3. Sexual coercion of females by males is common in mammalian societies and can involve forced copulation, harassment and intimidation of females. Males may increase their chances of mating successfully by increasing the costs of refusal to females and their behaviour can have substantial costs to females.

4. Females often make concerted efforts to avoid coercive males. In some species, the risk of male harassment induces females to aggregate on the territories of males that are able to protect them.

5. Male infanticide is common in mammals and frequently increases the breeding opportunities of infanticidal males. Infants that are killed have rarely been fathered by the male that killed them and their deaths often reduce latency to their mother's next conception. In some cases, male infanticide may also serve to recruit females from other groups or to reduce competition for the male's offspring.

6. Females have evolved tactics which reduce the risk that their infants will be killed or which minimise their investment in offspring that are unlikely to survive. In some cases, this involves the premature abandonment of fetuses or neonates.

7. Where infanticide is an important source of juvenile mortality, management strategies that remove mature males may depress juvenile survival. Infanticide (and other forms of conflict between the sexes) has the capacity to reduce female numbers and, in extreme cases, may even increase the probability of extinction of the population.

References

Agoramoorthy, G. and Rudran, R. (1995) Infanticide by adult and subadult males in free-ranging red Howler monkeys, *Alouatta seniculus*, in Venezuela. *Ethology* **99**:75–88.

Agrell, J., *et al.* (1998) Counter-strategies to infanticide in mammals: costs and consequences. *Oikos* **83**:507–517.

Andreassen, A. and Gundersen, G. (2006) Male turnover reduces population growth: an enclosure experiment on voles. *Ecology* **87**:88–94.

Archie, E.A., *et al.* (2007) Behavioural inbreeding avoidance in wild African elephants. *Molecular Ecology* **16**:4138–4148.

Arnqvist, G. and Rowe, L. (2005) *Sexual Conflict*. Princeton, NJ: Princeton University Press.

Bartoš, L. and Madlafousek, J. (1994) Infanticide in a seasonal breeder: the case of red deer. *Animal Behaviour* **47**:217–219.

Bartoš, L., *et al.* (2011) Promiscuous behaviour disrupts pregnancy block in domestic horse mares. *Behavioral Ecology and Sociobiology* **65**:1567–1572.

Beehner, J.C., *et al.* (2005) The effect of new alpha males on female stress in free-ranging baboons. *Animal Behaviour* **69**:1211–1221.

Bellemain, E., *et al.* (2006) Mating strategies in relation to sexually selected infanticide in a non-social carnivore: the brown bear. *Ethology* **112**:238–246.

Berger, J. (1983) Induced abortion and social factors in wild horses. *Nature* **303**:59–61.

Berger, J. (1989) Female reproductive potential and its apparent evaluation by male mammals. *Journal of Mammalogy* **70**:347–358.

Bertram, B.C.R. (1975) Social factors influencing reproduction of wild lions. *Journal of Zoology* **177**:463–482.

Blumstein, D.T. (2000) The evolution of infanticide in rodents: a comparative analysis. In: *Infanticide by Males and its Implications* (eds C.P. van Schaik and C.H. Janson). Cambridge: Cambridge University Press, 178–197.

Böer, M. and Sommer, V. (1992) Evidence for sexually selected infanticide in captive *Cercopithecus mitis*, *Cercocebus torquatus*, and *Mandrillus leucophaeus*. *Primates* **33**:557–563.

Bonduriansky, R. (2009) Reappraising sexual coevolution and the sex roles. *PLOS Biology* **7**:e1000255.

Borries, C. (1997) Infanticide in seasonally breeding multimale groups of Hanuman langurs (*Presbytis entellus*) in Ramnagar (South Nepal). *Behavioral Ecology and Sociobiology* **41**:139–150.

Borries, C., *et al.* (1999) Males as infant protectors in Hanuman langurs (*Presbytis entellus*) living in multimale groups: defence pattern, paternity and sexual behaviour. *Behavioral Ecology and Sociobiology* **46**:350–356.

Boyko, R.H. and Marshall, A.J. (2009) The willing cuckold: optimal paternity allocation, infanticide and male reproductive strategies in mammals. *Animal Behaviour* **77**:1397–1407.

Bronson, F.H. (1979) The reproductive ecology of the house mouse. *Quarterly Review of Biology* **54**:265–299.

Broom, M., *et al.* (2004) Infanticide and infant defence by males: modelling the conditions in primate multi-male groups. *Journal of Theoretical Biology* **231**:261–270.

Bruce, H.M. (1959) An exteroceptive block to pregnancy in the mouse. *Nature* **184**:105.

Butynski, T.M. (1982) Harem-male replacement and infanticide in the blue monkey (*Cercopithecus mitis stuhlmanni*) in the Kibale Forest, Uganda. *American Journal of Primatology* **3**:1–22.

Butynski, T.M. (1990) Comparative ecology of blue monkeys (*Cercopithecus mitis*) in high- and low-density subpopulations. *Ecological Monographs* **60**:1–26.

Campagna, C., *et al.* (1988) Pup abduction and infanticide in southern sea lions. *Behaviour* **107**:44–60.

Chapman, M. and Hausfater, G. (1979) The reproductive consequences of infanticide in langurs: a mathematical model. *Behavioral Ecology and Sociobiology* **5**:227–240.

Chapman, T., *et al.* (2003) Sexual conflict. *Trends in Ecology and Evolution* **18**:41–47.

Clarke, M.R. (1983) Brief report: Infant-killing and infant disappearance following male takeovers in a group of free-ranging howling monkeys (*Alouatta palliata*) in Costa Rica. *American Journal of Primatology* **5**:241–247.

Clarke, P., *et al.* (2009) Intersexual conflict in primates: infanticide, paternity allocation, and the role of coercion. In: *Sexual Coercion in Primates and Humans: An Evolutionary Perspective on Male Aggression Against Females* (eds M.N. Müller and R.W. Wrangham). Cambridge, MA: Harvard University Press, 42–77.

Clutton-Brock, T.H. and Harvey, P.H. (1976) Evolutionary rules and primate societies. In: *Growing Points in Ethology* (eds P.P.G. Bateson and R.A. Hinde). Cambridge: Cambridge University Press, 195–237.

Clutton-Brock, T.H. and Huchard, E. (2013) Social competition and its consequences in female mammals. *Journal of Zoology* **289**:151–171.

Clutton-Brock, T.H. and Langley, P.A. (1997) Persistent courtship reduces male and female longevity in captive tsetse flies *Glossina morsitans morsitans* Westwood (Diptera: Glossinidae). *Behavioral Ecology* **8**:392–395.

Clutton-Brock, T.H. and Parker, G.A. (1995a) Sexual coercion in animal societies. *Animal Behaviour* **49**:1345–1365.

Clutton-Brock, T.H. and Parker, G.A. (1995b) Punishment in animal societies. *Nature* **373**:209–216.

Clutton-Brock, T.H., *et al.* (1988) Passing the buck: resource defence, lek breeding and mate choice in fallow deer. *Behavioral Ecology and Sociobiology* **23**:281–296.

Clutton-Brock, T.H., *et al.* (1989) Mate choice on fallow deer leks. *Nature* **340**:463–465.

Clutton-Brock, T.H., *et al.* (1993) The evolution of ungulate leks. *Animal Behaviour* **46**:1121–1138.

Clutton-Brock, T.H., *et al.* (2010) Adaptive suppression of subordinate reproduction in cooperative mammals. *American Naturalist* **176**:664–673.

Colmenares, F. and Gomendio, M. (1988) Changes in female reproductive condition following male take-overs in a colony of hamadryas and hybrid baboons. *Folia Primatologica* **50**:157–174.

Connor, R.C. and Vollmer, N. (2009) Sexual coercion in dolphin consortships: a comparison with chimpanzees. In: *Sexual Coercion in Primates and Humans: An Evolutionary Perspective on Male Aggression Against Females* (eds M.N. Müller and R.W. Wrangham). Cambridge, MA: Harvard University Press, 218–243.

Cooney, R. and Bennett, N.C. (2000) Inbreeding avoidance and reproductive skew in a cooperative mammal. *Proceedings of the Royal Society of London. Series B: Biological Sciences* **267**:801–806.

Coulon, J., *et al.* (1995) Infanticide in the Alpine Marmot (*Marmota marmota*). *Ethology Ecology and Evolution* **7**:191–194.

Cowlishaw, G. and O'Connell, S.M. (1996) Male–male competition, paternity certainty and copulation calls in female baboons. *Animal Behaviour* **51**:235–238.

Cox, C.R. and Le Boeuf, B.J. (1977) Female incitation of male competition: a mechanism in sexual selection. *American Naturalist* **111**:317–335.

Craig, I.A.S., *et al.* (2002) Male mate choice and male–male competition coexist in the humpback whale (*Megaptera novaeangliae*). *Canadian Journal of Zoology* **80**:745–755.

Crockett, C.M. and Janson, C.H. (2000) Infanticide in red howlers: female group size, male membership and a possible link to folivory. In: *Infanticide by Males and its Implications* (eds C.P. van Schaik and C.H. Janson). Cambridge: Cambridge University Press, 75–98.

Crockett, C.M. and Sekulic, R. (1984) Infanticide in red howler monkeys (*Alouatta seniculus*) In: *Infanticide: Comparative and Evolutionary Perspectives* (eds G. Hausfater and S.B. Hrdy). New York: Aldine, 173–191.

Davies, N.B., *et al.* (2012) *An Introduction to Behavioural Ecology*, 4th edn. Oxford: Wiley-Blackwell.

Delgado, R.A. and van Schaik, C.P. (2000) The behavioral ecology and conservation of the orangutan (*Pongo pygmaeus*): a tale of two islands. *Evolutionary Anthropology* **9**:201–218.

de Waal, F.B.M. (1982) *Chimpanzee Politics: Power and Sex Among Apes*. New York: Harper & Row.

Dixson, A. (2002) Sexual selection by cryptic female choice and the evolution of primate sexuality. *Evolutionary Anthropology* **11**:195–199.

Dixson, A.L. and Anderson, M.J. (2002) Sexual selection, seminal coagulation and copulatory plug formation in primates. *Folia Primatologica* **73**:63–69.

Doidge, D.W., *et al.* (1984) Density-dependent pup mortality in the Antarctic fur seal *Arctocephalus gazella* at South Georgia. *Journal of Zoology* **202**:449–460.

Dunbar, R.I.M. (1988) *Primate Social Systems*. London: Chapman & Hall.

East, M.L., *et al.* (2003) Sexual conflicts in spotted hyenas: male and female mating tactics and their reproductive outcome with respect to age, social status and tenure. *Proceedings of the Royal Society of London. Series B: Biological Sciences* **270**:1247–1254.

Ebensperger, L.A. (1998a) Strategies and counterstrategies to infanticide in mammals. *Biological Reviews* **73**:321–346.

Ebensperger, L.A. (1998b) Do female rodents use promiscuity to prevent male infanticide? *Ethology Ecology and Evolution* **10**:129–141.

Eberhard, W.G. (1990) Animal genitalia and female choice. *American Scientist* **78**:134–141.

Eberhard, W.G. (2005) Evolutionary conflicts of interest: are females sexual decisions different? *American Naturalist* **165**: S19–S25.

Edward, D.A. and Chapman, T. (2011) The evolution and significance of male mate choice. *Trends in Ecology and Evolution* **26**:647–654.

Eggert, A.-K. and Müller, J.K. (1997) Biparental care and social evolution in burying beetles: lessons from the larder. In: *The Evolution of Social Behavior in Insects and Arachnids* (eds J.C. Choe and B.J. Crespi). Cambridge: Cambridge University Press, 216–236.

Eldakar, O.T., *et al.* (2009) Population structure mediates sexual conflict in water striders. *Science* **326**:816.

Ellis, L. (1998) Neodarwinian theories of violent criminality and antisocial behavior: photographic evidence from nonhuman animals and a review of the literature. *Aggression and Violent Behavior* **3**:61–110.

Elwood, R.W. and Kennedy, H.F. (1994) Selective allocation of parental and infanticidal responses in rodents: a review. In: *Infanticide and Parental Care* (eds S. Parmigiani and F.S. vom Saal). Chur, Switzerland: Harwood Academic Publishers, 397–425.

Engh, A.E., *et al.* (2006) Female hierarchy instability, male immigration and infanticide increase glucocorticoid levels in female chacma baboons. *Animal Behaviour* **71**:1227–1237.

Enomoto, T. (1981) Male aggression and the sexual behavior of Japanese monkeys. *Primates* **22**:15–23.

Enquist, M. and Leimar, O. (1983) Evolution of fighting behaviour: decision rules and assessment of relative strength. *Journal of Theoretical Biology* **102**:387–410.

Enstam, K.L., *et al.* (2002) Male demography, female mating behaviour and infanticide in wild patas monkeys (*Erythrocebus patas*). *International Journal of Primatology* **23**:85–104.

Fairbanks, L.A. and McGuire, M.T. (1987) Mother–infant relationships in vervet monkeys: response to new adult males. *International Journal of Primatology* **8**:351–366.

Feh, C. and Munkhtuya, B. (2008) Male infanticide and paternity analyses in a socially natural herd of Przewalski's horses: sexual selection? *Behavioural Processes* **78**:335–339.

Fox, E.A. (2002) Female tactics to reduce sexual harassment in the Sumatran orangutan (*Pongo pygmaeus abelii*). *Behavioral Ecology and Sociobiology* **52**:93–101.

Fox, L.R. (1975) Cannibalism in natural populations. *Annual Review of Ecology and Systematics* **6**:87–106.

Frank, L.G. (1997) Evolution of genital masculinization: why do female hyaenas have such a large 'penis'? *Trends in Ecology and Evolution* **12**:58–62.

Freed, L.A. (1986) Territory takeover and sexually selected infanticide in tropical house wrens. *Behavioral Ecology and Sociobiology* **19**:197–206.

Galvani, A. and Johnstone, R.A. (1998) Sperm allocation in an uncertain world. *Behavioral Ecology and Sociobiology* **44**:161–168.

Goldstein, H., *et al.* (1986) Infanticide in the Palestine sunbird. *The Condor* **88**:528–529.

Goodall, J. (1986) *The Chimpanzees of Gombe: Patterns of Behavior.* Cambridge, MA: Belknap Press.

Gourbal, B.E.F. and Gabrion, C. (2004) A study of mate choice in mice with experimental *Taenia crassiceps* cysticercosis: can males choose? *Canadian Journal of Zoology* **82**:635–643.

Gray, M.E. (2009) An infanticide attempt by a free-roaming feral stallion (*Equus caballus*). *Biology Letters* **5**:23–25.

Hackländer, K. and Arnold, W. (1999) Male-caused failure of female reproduction and its adaptive value in alpine marmots (*Marmota marmota*). *Behavioral Ecology* **10**:592–597.

Haigh, J. and Cannings, C. (1989) The n-person war of attrition. In: *Evolution and Control in Biological Systems* (eds A.B. Kurzhanski and K. Sigmund). Dordrecht: Kluwer Academic Publishers, 59–74.

Hammerstein, P. and Parker, G.A. (1982) The asymmetric war of attrition. *Journal of Theoretical Biology* **96**:647–682.

Harcourt, A.H. and Greenberg, J. (2001) Do gorilla females join males to avoid infanticide? A quantitative model. *Animal Behaviour* **62**:905–915.

Hausfater, G. (1984) Infanticide in langurs: strategies, counter-strategies, and parameter values. In: *Infanticide: Comparative and Evolutionary Perspectives* (eds G. Hausfater and S.B. Hrdy). New York: Aldine, 257–281.

Hausfater, G., *et al.* (1982) Infanticide as an alternative male reproductive strategy in langurs: a mathematical-model. *Journal of Theoretical Biology* **94**:391–412.

Heistermann, M., *et al.* (1996) Application of urinary and fecal steroid measurements for monitoring ovarian function and pregnancy in the bonobo (*Pan paniscus*) and evaluation of perineal swelling patterns in relation to endocrine events. *Biology of Reproduction* **55**:844–853.

Heistermann, M., *et al.* (2001) Measurement of faecal steroid metabolites in the lion-tailed macaque (*Macaca silenus*): a non-invasive tool for assessing female ovarian function. *Primate Report* **59**:27–42.

Henzi, S.P., *et al.* (2010) Infanticide and reproductive restraint in a polygynous social mammal. *Proceedings of the National Academy of Sciences of the United States of America* **107**:2130–2135.

Hiraiwa-Hasegawa, M. and Hasegawa, T. (1994) Infanticide in non-human primates: sexual selection and local resource competition. In: *Infanticide and Parental Care* (eds S. Parmigiani

and F.S. vom Saal). Chur, Switzerland: Harwood Academic Publishers, 137–154.

Hogg, J.T. and Forbes, S.H. (1997) Mating in bighorn sheep: frequent male reproduction via a high-risk unconventional tactic. *Behavioral Ecology and Sociobiology* **41**:33–48.

Höner, O.P., *et al.* (2007) Female mate-choice drives the evolution of male-biased dispersal in a social mammal. *Nature* **448**:798–801.

Hoogland, J.L. (1995) *The Black-tailed Prairie Dog: Social Life of a Burrowing Mammal.* Chicago: University of Chicago Press.

Horne, T.J. and Ylönen, H. (1996) Female bank voles (*Clethrionomys glareolus*) prefer dominant males: but what if there is no choice? *Behavioral Ecology and Sociobiology* **38**:401–405.

Hrdy, S.B. (1974) Male–male competition and infanticide among the langurs (*Presbytis entellus*). *Folia Primatologica* **22**:19–58.

Hrdy, S.B. (1975) Male and female strategies of reproduction among the langurs of Abu. PhD thesis, Harvard University.

Hrdy, S.B. (1977a) Infanticide as a primate reproductive strategy. *American Scientist* **65**:40–49.

Hrdy, S.B. (1977b) *The Langurs of Abu: Female and Male Strategies of Reproduction.* Cambridge, MA: Harvard University Press.

Hrdy, S.B. (1979) Infanticide among animals: review, classification, and examination of the implications for the reproductive strategies of females. *Ethology and Sociobiology* **1**:13–40.

Huck, U.W. (1984) Infanticide and the evolution of pregnancy block in rodents. In: *Infanticide: Comparative and Evolutionary Perspectives* (eds G. Hausfater and S.B. Hrdy). New York: Aldine, 349–364.

Huck, U.W., *et al.* (1982) Infanticide in male laboratory mice: effects of social status, prior sexual experience, and basis for discrimination between related and unrelated young. *Animal Behaviour* **30**:1158–1165.

Janson, C.H. and van Schaik, C.P. (2000) The behavioral ecology of infanticide by males. In: *Infanticide by Males and its Implications* (eds C.P. van Schaik and C.H. Janson). Cambridge: Cambridge University Press, 469–494.

Jay, P.C. (1965) The common langur of North India. In: *Primate Behavior: Field Studies of Monkeys and Apes* (ed. I. DeVore). New York: Holt, Rinehart and Winston, 197–249.

Jeppsson, B. (1986) Mating by pregnant water voles (*Arvicola terrestris*): a strategy to counter infanticide by males? *Behavioral Ecology and Sociobiology* **19**:293–296.

Johnstone, R.A. and Keller, L. (2000) How males can gain by harming their mates: sexual conflict, seminal toxins and the cost of mating. *American Naturalist* **156**:368–377.

Kappeler, P.M. (2012) Mate choice. In: *The Evolution of Primate Societies* (eds J.C. Mitani, J. Call, P.M. Kappeler, R.A. Palombit and J.B. Silk). Chicago: University of Chicago Press, 343–366.

Keddy-Hector, A.C. (1992) Mate choice in non-human primates. *American Zoologist* **32**:62–70.

Kitchen, D.M., *et al.* (2009) The causes and consequences of male aggression directed at female chacma baboons. In: *Sexual Coercion in Primates and Humans: An Evolutionary Perspective on*

Male Aggression Against Females (eds M.N. Müller and R.W. Wrangham). Cambridge, MA: Harvard University Press, 128–156.

Klemme, I. and Ylönen, H. (2010) Polyandry enhances offspring survival in an infanticidal species. *Biology Letters* 6:24–26.

Knott, C.D. (2009) Orangutans: sexual coercion without sexual violence. In: *Sexual Coercion in Primates and Humans: An Evolutionary Perspective on Male Aggression Against Females* (eds M.N. Müller and R.W. Wrangham). Cambrige, MA: Harvard University Press, 81–111.

Kokko, H. and Brooks, R. (2003) Sexy to die for? Sexual selection and the risk of extinction. *Annales Zoologici Fennici* 40:207–219.

Kummer, H. (1968) *Social Organization of Hamadryas Baboons.* Chicago: University of Chicago Press.

Kutsukake, N. and Nunn, C.L. (2006) Comparative tests of reproductive skew in male primates: the roles of demographic factors and incomplete control. *Behavioral Ecology and Sociobiology* 60.695 706.

Lobov, J.B. (1980) Factors influencing infanticidal behavior in wild male house mice (*Mus musculus*), *Behavioral Ecology and Sociobiology* 6:297–303.

Labov, J.B. (1984) Infanticidal behavior in male and female rodents: sectional introduction and directions for future research. In: *Infanticide: Comparative and Evolutionary Perspectives* (eds G. Hausfater and S.B. Hrdy). New York: Aldine, 323–329.

Le Boeuf, B.J. and Briggs, K.T. (1977) The cost of living in a seal harem. *Mammalia* 41:167–196.

Le Boeuf, B.J. and Mesnick, S. (1991) Sexual behavior of male northern elephant seals: I. Lethal injuries to adult females. *Behaviour* 116:143–162.

Le Galliard, J.-F., et al. (2005) Sex ratio bias, male aggression and population collapse in lizards. *Proceedings of the National Academy of Sciences of the United States of America* 102:18231–18236.

Lehman, L. and Perrin, N. (2003) Inbreeding avoidance through kin recognition: choosy females boost dispersal. *American Naturalist* 162:638–652.

Lessells, C.M. (2005) Why are males bad for females? Models for the evolution of damaging male mating behavior. *American Naturalist* 165:S46–S63.

Lessells, C.M. (2012) Sexual conflict. In: *The Evolution of Parental Care* (eds N.J. Royle, P.T. Smiseth and M. Kölliker). Oxford: Oxford University Press, 150–170.

Lindburg, D.G. (1983) Mating behavior and estrus in the Indian rhesus monkey. In: *Perspectives on Primate Biology* (ed. P.R. Seth). New Delhi: Today & Tomorrow Publishers, 45–61.

Linklater, W.L., et al. (1999) Stallion harassment and the mating system of horses. *Animal Behaviour* 58:295–306.

Lukas, D. and Clutton-Brock, T.H. (2013) The evolution of social monogamy in mammals. *Science* 341:526–530.

Lukas, D. and Clutton-Brock, T.H. (2014) Evolution of social monogamy in primates is not consistently associated with male infanticide. *Proceedings of the National Academy of Sciences of the United States of America* 111:E1674.

Lukas, D. and Huchard, E. (2014) The distribution of male infanticide in mammals. *Science* 346:841–844.

Lyon, J.E., et al. (2011) Mating strategies in primates: a game theoretical approach to infanticide. *Journal of Theoretical Biology* 274:103–108.

McComb, K. (1987) Roaring by red deer stags advances the date of oestrus in hinds. *Nature* 330:648–649.

McComb, K., et al. (1993) Female lions can identify potentially infanticidal males from their roars. *Proceedings of the Royal Society of London. Series B: Biological Sciences* 252:59–64.

Maestripieri, D. (1992) Functional aspects of maternal aggression in mammals. *Canadian Journal of Zoology* 70:1069–1077.

Maestripieri, D. and Alleva, E. (1990) Maternal aggression and litter size in the female house mouse. *Ethology* 84:27–34.

Maestripieri, D. and Rossi-Arnaud, C. (1991) Kinship does not affect litter defence in pairs of communally nesting female house mice. *Aggressive Behavior* 17:223–228.

Mallory, F.F. and Brooks, R.J. (1978) Infanticide and other reproductive strategies in the collared lemming, *Dicrostonyx groenlandicus. Nature* 273:144–146.

Manning, C.J., et al. (1995) Communal nesting and communal nursing in house mice, *Mus musculus domesticus. Animal Behaviour* 50:741 751.

Manson, J.H. (1992) Measuring female mate choice in Cayo Santiago rhesus macaques. *Animal Behaviour* 44:405–416.

Manson, J.H. (1994) Male aggression: a cost of female mate choice in Cayo Santiago rhesus macaques. *Animal Behaviour* 48:473–475.

Manson, J.H. (1997) Primate consortships: a critical review. *Current Anthropology* 38:353–374.

Manson, J.H. and Perry, S.E. (1993) Inbreeding avoidance in rhesus macaques: whose choice? *American Journal of Physical Anthropology* 90:335–344.

Manson, J.H. and Wrangham, R.W. (1991) Intergroup aggression in chimpanzees and human. *Current Anthropology* 32:369 390.

Maynard Smith, J. (1974) The theory of games and the evolution of animal conflicts. *Journal of Theoretical Biology* 47:209–221.

Maynard Smith, J. (1976) Evolution and the theory of games. *American Scientist* 64:41–45.

Maynard Smith, J. and Parker, G.A. (1976) The logic of asymmetric contests. *Animal Behaviour* 24:159–175.

Mesnick, S. and Le Boeuf, B.J. (1991) Sexual behavior of male northern elephant seals: II. Female response to potentially injurious encounters. *Behaviour* 117:262–280.

Mestel, R. (1994) Seamy side of sea otter life. *New Scientist* 141:5.

Milligan, S.R. (1980) Pheromones and rodent reproductive physiology. *Symposia of the Zoological Society of London* 45:251–275.

Mitani, J.C. (1985) Mating behaviour of male orangutans in the Kutai Game Reserve, Indonesia. *Animal Behaviour* 33:392–402.

Mohnot, S.M. (1971) Some aspects of social changes and infant-killing in the hanuman langur, *Presbytis entellus* (Primates: Cercopithecidae), in western India. *Mammalia* 35:175–198.

Mori, U. and Dunbar, R.I.M. (1985) Changes in the reproductive condition of female gelada baboons following the takeover of one-male units. *Zeitschrift fur Tierpsychologie* **67**:215–224.

Müller, M.N., *et al.* (2006) Male chimpanzees prefer mating with old females. *Current Biology* **16**:2234–2238.

Müller, M.N., *et al.* (2007) Male coercion and the costs of promiscuous mating for female chimpanzees. *Proceedings of the Royal Society of London. Series B: Biological Sciences* **274**:1009–1014.

Müller, M.N., *et al.* (2009) Male aggression against females and sexual coercion in chimpanzees. In: *Sexual Coercion in Primates and Humans: An Evolutionary Perspective on Male Aggression Against Females* (eds M.N. Müller and R.W. Wrangham). Cambridge, MA: Harvard University Press, 184–217.

Nadler, R.D. (1982) Reproductive behavior and endocrinology of orang utans. In: *The Orang Utan: Its Biology and Conservation* (ed. L.E.M. de Boer). The Hague: W. Junk, 231–248.

Nadler, R.D. (1988) Sexual aggression in the great apes. *Annals of the New York Academy of Sciences* **528**:154–162.

Nakahashi, W. (2008) Quantitative genetic models of sexual selection by male choice. *Theoretical Population Biology* **74**:167–181.

Nichols, H.J., *et al.* (2010) Top males gain high reproductive success by guarding more successful females in a cooperatively breeding mongoose. *Animal Behaviour* **80**:649–657.

Nishida, T., *et al.* (1985) Group extinction and female transfer in wild chimpanzees in the Mahale National Park, Tanzania. *Zeitschrift fur Tierpsychologie* **67**:284–301.

Nishida, T., *et al.* (2003) Demography, female life history, and reproductive profiles among the chimpanzees of Mahale. *American Journal of Primatology* **59**:99–121.

Novak, S. and Hatch, M. (2009) Intimate wounds: craniofacial trauma in women and female chimpanzees. In: *Sexual Coercion in Primates and Humans: an Evolutionary Perspective on Male Aggression against Females* (eds M.N. Müller and R.W. Wrangham). Cambridge, MA: Harvard University Press, 322–345.

Nunn, C.L. (1999) The evolution of exaggerated sexual swellings in primates and the graded-signal hypothesis. *Animal Behaviour* **58**:229–246.

Nunn, C.L. and van Schaik, C.P. (2000) Social evolution in primates: relative roles of ecology and intersexual conflict. In: *Infanticide by Males and its Implications* (eds C.P. van Schaik and C.H. Janson). Cambridge: Cambridge University Press, 388–412.

O'Brien, T.G. (1991) Female–male social interactions in wedge-capped capuchin monkeys: benefits and costs of group living. *Animal Behaviour* **41**:555–567.

Opie, C., *et al.* (2013) Male infanticide leads to social monogamy in primates. *Proceedings of the National Academy of Sciences of the United States of America* **110**:13328–13332.

Ostermeyer, M. (1983) Maternal aggression. In: *Parental Behaviour in Rodents* (ed. R.W. Elwood). New York: John Wiley & Sons, 151–179.

Ostner, J. and Kappeler, P.M. (2004) Male life history and unusual sex ratio of redfronted lemur, *Eulemur fulvus rufus*, groups. *Animal Behaviour* **67**:249–259.

Pack, A.A., *et al.* (2009) Male humpback whales in the Hawaiian breeding grounds preferentially associate with larger females. *Animal Behaviour* **77**:653–662.

Packer, C. and Pusey, A.E. (1983a) Adaptations of female lions to infanticide by incoming males. *American Naturalist* **121**:716–728.

Packer, C. and Pusey, A.E. (1983b) Dispersal, kinship, and inbreeding in African lions. In: *The Natural History of Inbreeding and Outbreeding: Theoretical and Empirical Perspectives* (ed. N.W. Thornhill). Chicago: University of Chicago Press.

Packer, C. and Pusey, A.E. (1984) Infanticide in carnivores. In: *Infanticide: Comparative and Evolutionary Perspectives* (eds G. Hausfater and S.B. Hrdy). New York: Aldine, 31–42.

Packer, C., *et al.* (1988) Reproductive success of lions. In: *Reproductive Success: Studies of Individual Variation in Contrasting Breeding Systems* (ed. T. H. Clutton-Brock). Chicago: University of Chicago Press, 363–383.

Packer, C., *et al.* (1990) Why lions form groups: food is not enough. *American Naturalist* **136**:1–19.

Palombit, R.A. (1999) Infanticide and the evolution of pair bonds in nonhuman primates. *Evolutionary Anthropology* **7**:117–129.

Palombit, R.A. (2003) Male infanticide in wild savanna baboons: adaptive significance and intraspecific variation. In: *Sexual Selection and Reproductive Competition in Primates: New Perspectives and Directions* (ed. C.B. Jones). Norman, OK: American Society of Primatologists, 367–412.

Palombit, R.A. (2012) Infanticide: male strategies and female counterstrategies. In: *The Evolution of Primate Societies* (eds J.C. Mitani, J. Call, P.M. Kappeler, R.A. Palombit and J.B. Silk). Chicago: University of Chicago Press, 432–468.

Palombit, R.A. (2014) Sexual conflict in nonhuman primates. In: *Advances in the Study of Behavior*, Vol. **46** (eds M. Naguib et al.). San Diego, CA: Academic Press, 191–280.

Palombit, R.A., *et al.* (2000) Male infanticide and defence of infants in chacma baboons. In: *Male Infanticide and its Implications* (eds C.P. van Schaik and C.H. Janson). Cambridge: Cambridge University Press, 123–153.

Paoli, T. (2009) The absence of sexual coercion in bonobos. In: *Sexual Coercion in Primates and Humans: An Evolutionary Perspective on Male Aggression Against Females* (eds M.N. Müller and R.W. Wrangham). Cambridge, MA: Harvard University Press, 410–423.

Parga, J.A. (2003) Copulatory plug displacement evidences sperm competition in *Lemur catta*. *International Journal of Primatology* **24**:889–899.

Parga, J.A. (2006) Male mate choice in *Lemur catta*. *International Journal of Primatology* **27**:107–131.

Parker, G.A. (1979) Sexual selection and sexual conflict. In: *Sexual Selection and Reproductive Competition in Insects* (eds M.S. Blum and N.A. Blum). New York: Academic Press, 123–166.

Parker, G.A. (1983) Mate quality and mating decisions. In: *Mate Choice* (ed. P. Bateson). Cambridge: Cambridge University Press, 141–164.

Parker, G.A. (1984) Sperm competition and the evolution of animal mating strategies. In: *Sperm Competition and the Evolution of Animal Mating Systems* (ed. R.L. Smith). New York: Academic Press, 1–60.

Parker, G.A. and Pizzari, T. (2010) Sperm competition and ejaculate economics. *Biological Reviews* **85**:897–934.

Parmigiani, S. and vom Saal, F.S. (eds), (1994) *Infanticide and Parental Care*. Chur, Switzerland: Harwood Academic Publishers.

Patterson, I.A.P., *et al.* (1998) Evidence for infanticide in bottlenose dolphins: an explanation for violent interactions with harbour porpoises? *Proceedings of the Royal Society of London. Series B: Biological Sciences* **265**:1167–1170.

Pereira, M.E. (1983) Abortion following the immigration of an adult male baboon (*Papio cynocephalus*). *American Journal of Primatology* **4**:93–98.

Pereira, M.E. and Weiss, M.L. (1991) Female mate choice, male migration and the threat of infanticide in ringtailed lemurs. *Behavioral Ecology and Sociobiology* **28**:141–152.

Perrigo, G. and vom Saal, F.S. (1994) Behavioral cycles and the neural timing of infanticide and parental behavior in male house mice. In: *Infanticide and Parental Care* (eds S. Parmigiani and F.S. vom Saal). Chur, Switzerland: Harwood Academic Publishers, 365–396.

Perry, S. and Manson, J.H. (2008) *Manipulative Monkeys: The Capuchins of Lomas Barbudal*. Cambridge, MA: Harvard University Press.

Pizzari, T., *et al.* (2003) Sophisticated sperm allocation in male fowl. *Nature* **426**:70–74.

Pluháček, J. and Bartoš, L. (2000) Male infanticide in captive plains zebra, *Equus burchelli*. *Animal Behaviour* **59**:689–694.

Poikonen, T., *et al.* (2008) Infanticide in the evolution of reproductive synchrony: effects on reproductive success. *Evolution* **62**:612–621.

Poole, J.H. (1989) Mate guarding, reproductive success and female choice in African elephants. *Animal Behaviour* **37**:842–849.

Pope, T.R. (1990) The reproductive consequences of cooperation in the red howler monkey: paternity exclusion in multi-male and single-male troops using genetic markers. *Behavioral Ecology and Sociobiology* **27**:439–446.

Pradhan, G.R. and van Schaik, C.P. (2008) Infanticide-driven intersexual conflict over matings in primates and its effects on social organisation. *Behaviour* **145**:251–275.

Pradhan, G.R., *et al.* (2006) The evolution of female copulation calls in primates: a review and a new model. *Behavioral Ecology and Sociobiology* **59**:333–343.

Preston, B.T., *et al.* (2005) Male mate choice influences female promiscuity in Soay sheep. *Proceedings of the Royal Society of London. Series B: Biological Sciences* **272**:365–373.

Pusey, A.E. (2001) Of genes and apes: chimpanzee social organization and reproduction. In: *Tree of Origin: What Primate Behavior Can Tell Us About Human Social Evolution* (ed. F.B. M. de Waal). Cambridge, MA: Harvard University Press, 613–652.

Ramm, S.A., *et al.* (2005) Sperm competition and the evolution of male reproductive anatomy in rodents. *Proceedings of the Royal Society of London. Series B: Biological Sciences* **272**:949–955.

Rankin, D.J., *et al.* (2011) Sexual conflict and the tragedy of the commons. *American Naturalist* **177**:780–791.

Réale, D., *et al.* (1996) Female-biased mortality induced by male sexual harassment in a feral sheep population. *Canadian Journal of Zoology* **74**:1812–1818.

Roberts, E.K., *et al.* (2012) A Bruce effect in wild geladas. *Science* **335**:1222–1225.

Robinson, J.A. and Goy, R.W. (1986) Steroid hormones and the ovarian cycle. In: *Comparative Primate Biology 3. Reproduction and Development* (eds W.R. Dukelow and J. Erwin). New York: Alan R. Liss, 63–91.

Rödel, H.G., *et al.* (2008) Infanticide and maternal offspring defence in European rabbits under natural breeding conditions. *Ethology* **114**:22–31.

Rosa, H.J.D. and Bryant, M.J. (2002) The 'ram effect' as a way of modifying the reproductive activity in the ewe. *Small Ruminant Research* **45**:1–16.

Roughgarden, J., *et al.* (2006) Reproductive social behavior: cooperative games to replace sexual selection. *Science* **311**:965–969.

Rubenstein, D.I. (1986) Ecology and sociality in horses and zebras. In: *Ecological Aspects of Social Evolution* (eds D.I. Rubenstein and R.W. Wrangham). Princeton, NJ: Princeton University Press, 282–302.

Saj, T.L. and Sicotte, P. (2005) Male takeover in *Colobus vellerosus* at Boabeng-Fiema monkey sanctuary, Central Ghana. *Primates* **46**:211–214.

Sanderson, J.L., *et al.* (2015) Banded mongooses avoid inbreeding when mating with members of the same natal group. *Molecular Ecology* **24**:3738–3751.

Schwagmeyer, P.L. and Parker, G.A. (1990) Male mate choice as predicted by sperm competition in thirteen-lined ground squirrels. *Nature* **348**:62–64.

Servedio, M.R. and Lande, R. (2006) Population genetic models of male and mutual mate choice. *Evolution* **60**:674–685.

Setchell, J.M. and Wickings, E.J. (2004) Sexual swelling in mandrills (*Mandrillus sphinx*): a test of the reliable indicator hypothesis. *Behavioral Ecology* **15**:438–445.

Setchell, J.M. and Wickings, E.J. (2006) Mate choice in male mandrills (*Mandrillus sphinx*). *Ethology* **112**:91–99.

Sherman, P.W. (1981) Reproductive competition and infanticide in Belding's ground squirrels and other animals. In: *Natural Selection and Social Behavior* (eds R.D. Alexander and D.W. Tinkle). New York: Chiron Press, 311–331.

Smuts, B.B. (1985) *Sex and Friendship in Baboons*. New York: Aldine.

Smuts, B.B. and Smuts, R.W. (1993) Male aggression and sexual coercion of females in nonhuman primates and other mammals: evidence and theoretical implications. In: *Advances in the Study of Behavior*, Vol. **22** (eds P.J.B. Slater, J.S. Rosenblatt, C.T. Snowdon and M. Milinski). San Diego, CA: Academic Press, 1–63.

Solomon, N.G. (1993) Body size and social preferences of male and female prairie voles, *Microtus ochrogaster*. *Animal Behaviour* **45**:1031–1033.

Solomon, N.G. and Keane, B. (2007) Reproductive strategies in female rodents. In: *Rodent Societies: An Ecological and Evolutionary Perspective* (eds J.O. Wolff and P.W. Sherman). Chicago: University of Chicago Press, 42–56.

Sommer, V. (1987) Infanticide among free-ranging langurs (*Presbytis entellus*) at Jodhpur (Rajasthan/India): recent observations and a reconsideration of hypotheses. *Primates* **28**:163–197.

Sommer, V. (1994) Infanticide among the langurs of Jodhpur: testing the sexual selection hypothesis with a long-term record. In: *Infanticide and Parental Care* (eds S. Parmigiani and F.S. vom Saal). Chur, Switzerland: Harwood Academic Publishers, 155–198.

Sommer, V. (2000) The holy wars about infanticide: which side are you on and why? In: *Infanticide in Males and its Implications* (eds C.P. van Schaik and C.H. Janson). Cambridge: Cambridge University Press, 9–26.

Steenbeek, R. (2000) Infanticide by males and female choice in Thomas's langurs. In: *Infanticide by Males and its Implications* (eds C.P. van Schaik and C.H. Janson). Cambridge: Cambridge University Press, 153–177.

Steenbeek, R. and van Schaik, C.P. (2001) Competition and group size in Thomas's langurs (*Presbytis thomasi*): the folivore paradox revisited. *Behavioral Ecology and Sociobiology* **49**:100–110.

Steenbeek, R., et al. (1999) Vigilance in wild Thomas's langurs (*Presbytis thomasi*): the importance of infanticide risk. *Behavioral Ecology and Sociobiology* **45**:137–150.

Steenbeek, R., et al. (2000) Costs and benefits of the one-male, age-graded and all-male phase in wild Thomas's langur groups. In: *Primate Males: Causes and Consequences of Variation in Group Composition* (ed. P.M. Kappeler). Cambridge: Cambridge University Press, 130–145.

Sterck, E.H.M. (1997) Determinants of female dispersal in Thomas langurs. *American Journal of Primatology* **42**:179–198.

Stockley, P. (2002) Sperm competition risk and male genital anatomy: comparative evidence for reduced duration of female sexual receptivity in primates with penile spines. *Evolutionary Ecology* **16**:123–137.

Strier, K.B. (1992) *Faces in the Forest: The Endangered Muriqui Monkeys of Brazil*. Cambridge, MA: Harvard University Press.

Struhsaker, T.T. and Leland, L. (1985) Infanticide in a patrilineal society of red colobus monkeys. *Zeitschrift fur Tierpsychologie* **69**:89–132.

Strum, S. (1984) Why males use infants. In: *Primate Paternalism* (ed. D.M. Taub). New York: Van Nostrand Reinhold, 146–185.

Sugiyama, Y. (1965a) Behavioral development and social structure in two troops of hanuman langurs (*Presbytis entellus*). *Primates* **6**:213–247.

Sugiyama, Y. (1965b) On the social change of hanuman langurs (*Presbytis entellus*) in their natural conditions. *Primates* **6**:381–417.

Sugiyama, Y. (1967) Social organization of hanuman langurs. In: *Social Communication Among Primates* (ed. S. Altmann). Chicago: University of Chicago Press, 221–236.

Sundaresan, S.R., et al. (2007) Male harassment influences female movements and associations in Grevy's zebra (*Equus grevyi*). *Behavioral Ecology* **18**:860–865.

Svare, B., et al. (1984) Infanticide: accounting for genetic variation in mice. *Physiology and Behavior* **33**:137–152.

Swedell, L. and Schreier, A. (2009) Male aggression towards females in hamadryas baboons: conditioning, coercion, and control. In: *Sexual Coercion in Primates and Humans: An Evolutionary Perspective on Male Aggression Against Females* (eds M.N. Müller and R.W. Wrangham). Cambridge, MA: Harvard University Press, 244–268.

Swenson, J.E. (2003) Implications of sexually selected infanticide for the hunting of large carnivores. In: *Animal Behavior and Wildlife Conservation* (eds M. Festa-Bianchet and M. Apollonio). Washington, DC: Island Press, 171–189.

Swenson, J.E., et al. (1997) Infanticide caused by hunting male bears. *Nature* **386**:450–451.

Szykman, M., et al. (2001) Association patterns among male and female spotted hyenas (*Crocuta crocuta*) reflect male mate choice. *Behavioral Ecology and Sociobiology* **50**:231–238.

Takahata, Y. (1985) Adult male chimpanzees kill and eat a male newborn infant: newly observed intragroup infanticide in Mahale Mountain National Park, Tanzania. *Folia Primatologica* **44**:161–170.

Tarara, E.B. (1987) Infanticide in a chacma baboon troop. *Primates* **28**:267–270.

Teichroeb, J.A. and Sicotte, P. (2008) Infanticide in ursine colobus monkeys (*Colobus vellerosus*) in Ghana: new cases and a test of the existing hypotheses. *Behaviour* **145**:727–755.

Teichroeb, J.A., et al. (2012) Infanticide risk and male quality influence optimal group composition for *Colobus vellerosus*. *Behavioral Ecology* **23**:1348–1359.

Thornhill, N.W. (ed.), (1993) *The Natural History of Inbreeding and Outbreeding: Theoretical and Empirical Perspectives*. Chicago: University of Chicago Press.

Tregenza, T., et al. (2006) Introduction. Sexual conflict: a new paradigm? *Philosophical Transactions of the Royal Society B: Biological Sciences* **361**:229–234.

Treves, A. (2000) Prevention of infanticide: the perspective of infant primates. In: *Infanticide by Males and its Implications* (eds C.P. van Schaik and C.H. Janson). Cambridge: Cambridge University Press, 223–238.

Utami, S.A. and van Hooff, J.A.R.A.M. (2004) Alternative male reproductive strategies: male bimaturism in orangutans. In: *Sexual Selection in Primates: New and Comparative Perspectives* (eds P.M. Kappeler and C.P. van Schaik). Cambridge: Cambridge University Press, 196–207.

van Noordwijk, M.A. and van Schaik, C.P. (2000) Reproductive patterns in eutherian mammals: adaptations against infanticide? In: *Infanticide by Males and its Implications* (eds

C.P. van Schaik and C.H. Janson). Cambridge: Cambridge University Press, 322–360.

van Schaik, C.P. (2000) Infanticide by male primates: the sexual selection hypothesis revisited In: *Infanticide by Males and its Implications* (eds C.P. van Schaik and C.H. Janson). Cambridge: Cambridge University Press, 27–60.

van Schaik, C.P., *et al.* (2000) Paternity confusion and the ovarian cycles of female primates. In: *Infanticide by Males and its Implications* (eds C.P. van Schaik and C.H. Janson). Cambridge: Cambridge University Press, 361–387.

van Schaik, C.P., *et al.* (2004) A model for within-group coalitionary aggression among males. *Behavioral Ecology and Sociobiology* **57**:101–109.

Veiga, J.P. (1993) Prospective infanticide and ovulation retardation in free-living house sparrows. *Animal Behaviour* **45**:43–46.

vom Saal, F.S. (1984) Proximate and ultimate causes of infanticide and parental behavior in male mice. In: *Infanticide: Comparative and Evolutionary Perspectives* (eds G Hausfater and S.B. Hrdy). New York: Aldine, 401–424.

vom Saal, F.S, and Howard, L.S. (1982) The regulation of infanticide and parental behavior: implications for reproductive success in male mice. *Science* **215**:1270–1272.

Watts, D.P. (1989) Infanticide in mountain gorillas: new cases and a reconsideration of the evidence. *Ethology* **81**:1–18.

Webster, A.B., *et al.* (1981) Infanticide in the meadow vole, *Microtus pennsylvanicus*: significance in relation to social system and population cycling. *Behavioral and Neural Biology* **31**:342–347.

Whitman, K., *et al.* (2004) Sustainable trophy hunting of African lions. *Nature* **428**:175–178.

Williams, J.M., *et al.* (2002) Female competition and male territorial behaviour influence female chimpanzees' ranging patterns. *Animal Behaviour* **63**:347–360.

Williams, J.M., *et al.* (2004) Why do male chimpanzees defend a group range? *Animal Behaviour* **68**:523–532.

Wise, D.A. and Ferrante, F. (1982) Effect of conspecific sex on aggression during pregnancy and lactation in golden hamsters. *Aggressive Behavior* **8**:243–251.

Wolff, J.O. (1985) Maternal aggression as a deterrent to infanticide in *Peromyscus leucopus* and *P. maniculatus*. *Animal Behaviour* **33**:117–123.

Wolff, J.O. (1992) Parents suppress reproduction and stimulate dispersal in opposite sex juvenile white-footed mice. *Nature* **359**:409–410.

Wolff, J.O. (1993) Why are female small mammals territorial? *Oikos* **68**:364–370.

Wolff, J.O. and Cicirello, D.M. (1989) Field evidence for sexual selection and resource competition infanticide in white-footed mice. *Animal Behaviour* **38**:637–642.

Wolff, J.O. and Cicirello, D.M. (1991) Comparative paternal and infanticidal behaviour of sympatric white-footed mice (*Peromyscus leucopus noveboracensis*) and deermice (*P. maniculatus nuibiterrae*). *Behavioral Eoclogy* **2**:38–45.

Wolff, J.O. and Peterson, J.A. (1998) An offspring-defense hypothesis for territoriality in female mammals. *Ethology Ecology and Evolution* **10**:227–239.

Wrangham, R.W. (2000) Why are male chimpanzees more gregarious than mothers? A scramble competition hypothesis. In: *Primate Males: Causes and Consequences of Variation in Group Composition* (ed. P.M. Kappeler). Cambridge: Cambridge University Press, 248–258.

Wrangham, R.W. and Müller, M.N. (2009) Sexual coercion in humans and other primates: the road ahead. In: *Sexual Coercion in Primates and Humans: An Evolutionary Perspective on Male Aggression Against Females* (eds M.N. Müller and R.W. Wrangham). Cambridge, MA: Harvard University Press, 451–468.

Wrangham, R.W. and Peterson, D. (1996) *Demonic Males: Apes and the Origins of Human Violence.* New York: Houghton Mifflin Harcourt.

Wright, P.C. (1995) Demography and life history of free-ranging *Propithecus diadema edwardsi* in Ranomafana national park, Madagascar. *International Journal of Primatology* **16**:835–854.

Yoshiba, K. (1968) Local and intertroop variability in ecology and social behavior of common Indian langurs. In: *Primates: Studies in Adaptation and Variability* (ed. P.C. Jay). New York: Holt, Rinehart and Winston, 217–242.

Zumpe, D. and Michael, R.P. (1990) Effects of the presence of a second male on pair-tests of captive cynomolgus monkeys (*Macaca fascicularis*): role of dominance. *American Journal of Primatology* **22**:145–158.

CHAPTER 16

Paternal care

16.1 Introduction

On the edge of the Okavango Delta, chacma baboons are foraging in long grass, grunting regularly to stay in contact. A female with a black infant feeds nervously, glancing sideways at an adult male. He avoids staring directly at her but edges close to her whenever he can. Suddenly he lunges at her infant, but misses. The female leaps back, hugging her infant closely and screams loudly. From the far side of the group another adult male appears at full gallop, runs directly at the first male and chases him for a hundred metres before relaxing and returning to the female. He sits close to her and grooms her and she relaxes, evidently recognising that he will not attempt to harm her infant.

In contrast to females of other vertebrate groups, all female mammals are committed to care for their offspring after birth (see Chapter 5) and exclusive paternal care does not occur. The lengthy commitment of females to caring for their offspring generates pronounced sex differences in 'time out' from the potential pool of reproductives and strong male biases in the operational sex ratio, which are often associated with strong selection pressures favouring competitive ability (and associated traits) in males (see Chapter 1).

Many mammals have promiscuous mating systems and associations between breeding partners are temporary (see Chapter 10). In these species, the separation of mating, birth and rearing periods restricts opportunities for males to associate with or care for their progeny. Paternal care is more feasible in polygynous systems where breeding partners associate throughout the year but is likely to reduce the capacity of males to invest in mating competition, which is typically intense in these systems (see Chapter 13). In contrast, the fitness costs of parental care to males are likely to be lower in monoga-mous systems where breeding partners associate throughout the year, the operational sex ratio is not as strongly biased towards males and the intensity of competition for mates is reduced.

Despite constraints on the evolution of paternal care, fathers can increase the growth and survival of their offspring in several ways and an increasing number of studies of polygynomous species have produced evidence of some form of paternal care (Elwood 1983; McGuire and Bemis 2007). Where males defend access to breeding females throughout the year, fathers may provide indirect care for their offspring by excluding other males from breeding territories or groups, deter-ring intruders, and helping to build nests or cache food (McGuire and Novak 1984; McGuire and Bemis 2007). In some cases, fathers may also help to protect their offspring against predators or against attacks by poten-tially infanticidal males (Figure 16.1). In contrast, in some monogamous species, males regularly contribute more energetically costly forms of care, including guarding litters of dependent pups at the natal burrow or den or feeding or carrying dependent young (Figure 16.2).

This chapter examines the form and distribution of paternal care in mammals. Section 16.2 compares the protection and care of juveniles (and, in some cases, their mothers) in polygynous and monogamous spe-cies. Section 16.3 then briefly reviews our understand-ing of the mechanisms underlying the expression of male care. Subsequently, section 16.4 explores the benefits and costs of care, while sections 16.5 and 16.6 examine paternal tactics and the effects of conflicts of interest between parents in species showing biparental care. Finally, section 16.7 examines some of the consequences of paternal care for the evolution of mating systems.

Figure 16.1 In chacma baboons, males often protect infants they are likely to have fathered against other resident males
Source: © Dorothy Cheney.

Figure 16.2 In titi monkeys, which live in monogamous pairs, males carry and protect young born to their mate.
Source: © Kathy West.

16.2 The distribution of paternal care

Polygynous mammals

In many polygynous mammals where the same males and females do not associate with each other throughout the year (including many ungulates, pinnipeds and cetaceans) males have little opportunity to invest in their offspring and parental care is restricted to females. However, where males defend territories that overlap female ranges or associate with stable groups of females, they are frequently in a position to influence the survival of their offspring. In species like musk oxen or African buffalo, where adults form defensive formations or attack potential predators, the presence of males may contribute to the survival of other group members (Heard 1992; Prins 1996) and in some ungulates, juvenile mortality is negatively related to the number of mature males in the herd (Mysterud *et al.* 2002). However, these associations may also occur for other reasons, for example because male behaviour advances conception dates and increases female synchrony (see Chapter 10) or because high population density and low food availability have independent effects on juvenile mortality and the adult sex ratio (see Chapter 18).

In polygynous species where males maintain breeding territories throughout the year or live in stable groups

Figure 16.3 A male chacma baboon carries a juvenile. *Source*: © Alecia Carter.

with females, they have greater opportunities to protect young they are likely to have fathered. Recent studies of several polygynous mammals where males defend resource-based territories throughout the year have documented examples of paternal care. For example, in bushbuck, adult males will help to protect calves (Wronkski *et al.* 2006) while in sifakas and snub-nosed monkeys, adult males sometimes carry young (Grieser 1992; Xiang *et al.* 2009) and male brown hyenas sometimes bring food to dependent cubs (Mills 1990). Where males are closely associated with females that they have mated, their presence may also help to increase the survival of dependent infants and juveniles and increase the fecundity of females (Fernandez-Duque *et al.* 2009). In some species, the presence of dominant males can also play an important role in reducing the risk of infanticidal attacks by other males. For example, in harem groups of primates where immigrant males are commonly infanticidal, the continued presence of their fathers is often associated with substantial increases in the survival of juveniles (see Chapter 15). Defending males often appear to be aware of the potential risks of intrusion by other males. For example, in black howler monkeys, playbacks of the calls of strange males are more likely to instigate vocal displays in resident males if vulnerable infants are present in the group (Kitchen 2004).

Males can also play an important role in protecting their offspring against infanticidal attacks in species living in multi-male groups (Figure 16.3). The most detailed evidence of selective protection and support of offspring comes from studies of chacma baboons, where males often form close relationships ('friendships') with particular females, associating and grooming them more frequently than other females and supporting them in agonistic interactions with other group members (Seyfarth and Cheney 1984) (Figure 16.4). Most 'friends' are females with whom the male was known to have copulated in the cycle when an infant was conceived (Palombit 2000) and genetic analyses show that males are often the father of the offspring of their female 'friend' (Huchard *et al.* 2010). Playback experiments also confirm that males respond more strongly to the distress calls of their female friends than to those of other females, especially if these are combined with the calls of a potentially infanticidal male (Palombit *et al.* 1997) (Figure 16.5). In addition 'friendships' between females and males strengthen when females have young infants and weaken or disappear after the death of infants (Figure 16.6). Males appear to anticipate the potential consequences of dangerous events: during periods of instability following the immigration of a dominant male, males involved in friendships with lactating

Figure 16.4 A female chacma baboon grooms her male 'friend'.
Source: © Alecia Carter

females show heightened cortisol levels and reduce time spent in consortships with other females (Cheney *et al.* 2015).

As well as influencing the survival of their offspring, the presence of fathers can affect their growth and maturation. Male baboons appear to be able to differentiate between their own offspring and unrelated juveniles and selectively support their offspring when they are involved in agonistic interactions more frequently than unrelated individuals of the same age (Buchan *et al.* 2003) (Figure 16.7). The presence of fathers in the same group is associated with advanced maturation in their daughters and sons, though this effect is only significant for sons if their father held high rank at the time of their birth and direct evidence of causation is not available (Charpentier *et al.* 2008). Female baboons also appear to be aware both of the dangers associated with male takeovers and of the protection provided by male 'friends', and following the arrival of immigrant males, lactating females with male 'friends' show smaller increases in glucocorticoid levels than those without 'friends' (Figure 16.8) (Beehner *et al.* 2005). An opposite trend occurs in male friends, who have significantly higher glucocorticoid levels during periods of instability than males who do not have friendships with a female (D.L. Cheney, personal communication). In addition, offspring (especially those of subordinate males) play an active role in maintaining associations with their fathers and associate more closely with them when they are feeding in

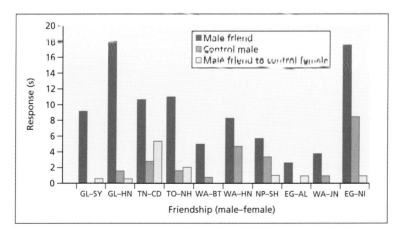

Figure 16.5 Responses of male baboons to playbacks of the screams of females. The figure shows the duration of responses by individual males that are the 'friend' of the female whose screams are played back; those of the same males to playbacks of screams by control females; and the responses of control males to the screams of females that are 'friends' of the first sample of males.
Source: From Palombit *et al.* (1997). Reproduced with permission of Elsevier. *Photo source*: © Alecia Carter.

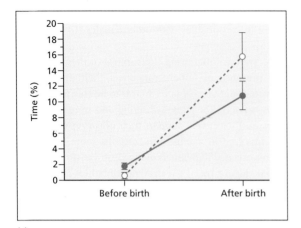

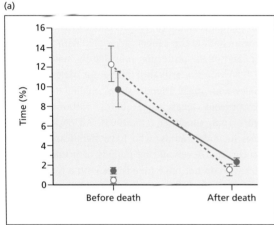

(a)

(b)

Figure 16.6 Changes in proximity (open circles, within 2 m; filled circles, within 2–6m) of male and female 'friends' in baboons: (a) following the birth of an infant to the female friend; (b) following the death of infants belonging to the female friend. *Source*: From Palombit *et al.* (1997). Reproduced with permission of Elsevier.

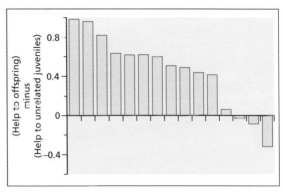

Figure 16.7 Effects of relatedness on care of infants by individual male yellow baboons. Each bar represents the difference, for one male, between the fraction of help given to his genetic offspring and the fraction of help given to unrelated juveniles: bars above the expected value of zero represent males that helped genetic offspring more than unrelated juveniles. *Source*: From Buchan *et al.* (2003). Reproduced with permission of Macmillan Publishers Ltd.

studies suggest that 'friendships' between males and particular females serve primarily to increase the male's reproductive access to that female (van Schaik and Paul 1996; Ménard *et al.* 2001), while others emphasise the

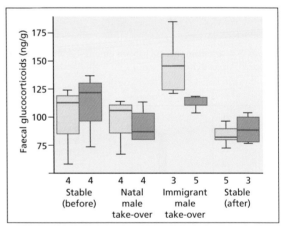

Figure 16.8 Responses of lactating female chacma baboons with male friends (pale bars) and without male friends (dark bars) to take-overs of the alpha position by natal versus immigrant males. The figure compares glucocorticoid levels (ng/g) in lactating females during the 2 months prior to a male take-over of the alpha position; following the take-over of the alpha position by a familiar natal male; following the take-over of the alpha position by an immigrant male; and 3 months after the arrival of immigrant males. Numbers show samples of individuals measured. *Source*: From Beehner *et al.* (2005). Reproduced with permission of Elsevier.

preferred patches or when other adult males are present and their mothers are absent (Huchard *et al.* 2012) (Figure 16.9).

As well as increasing the fitness of offspring, 'friendships' between males and particular females can improve the male's mating success. For example, in vervet monkeys, individual males that tolerate or protect infants when the mother is watching are subsequently more likely to be tolerated by them (Keddy Hector *et al.* 1989) and so may be more likely to mate with the mother subsequently. There is disagreement about the relative importance of these two selection pressures and some

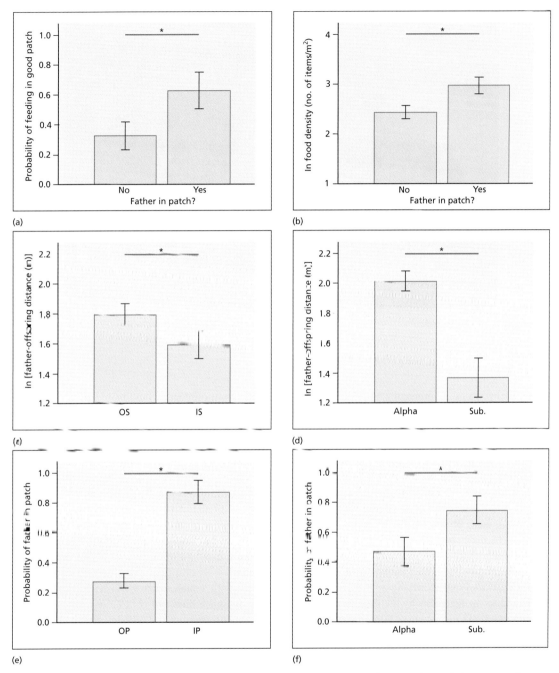

Figure 16.9 Influence of the presence of fathers on the foraging behaviour of juvenile offspring in chacma baboons. (a) Influence on the probability that juvenile offspring would feed in good food patches. (b) Influence on use by juvenile offspring of patches containing a high density of food items. (c) Influence of the presence of a non-father male in sight on $\log_e$ (father–offspring distance); IS, in sight; OS, out of sight. (d) Influence of the father's dominance rank on $\log_e$ (father–offspring distance); Sub, subordinate. (e) Influence of the presence of a non-father male on the probability of father–offspring association in food patches; IP, in patch; OP, out of patch. (f) Influence of the father's dominance rank on the probability of father–offspring association in food patches. Bars represent predicted means and SD computed from the model estimates (and adjusted for other covariates). Asterisk denotes $P < 0.05$. *Source*: From Huchard *et al.* (2012). Reproduced with permission of Oxford University Press.

Figure 16.10 Male Barbary macaques sometimes use their relationships with infants to deflect aggression from other males. *Source*: © Julia Fischer.

role played by males in protecting infants (Palombit 2014).

Males can also gain social benefits from association with particular females and their offspring (Smuts and Gubernick 1992). In Barbary macaques and some other cercopithecine monkeys, males that are involved in agonistic encounters with other males (especially those of higher rank) sometimes pick up and carry infants during the course of the interaction and may pass them to other males (see Chapter 14) (Figure 16.10). Carrying infants during periods of social tension appears to reduce aggression from other males, possibly because escalated attacks on males carrying infants are likely to provoke attacks on aggressors by the infant's kin and is sometimes referred to as 'agonistic buffering' (Deag and Crook 1971). Males frequently use the same infants in successive interactions and an additional reason for them to maintain close relationships with particular infants is that this may improve their mating access to their mothers (Ménard *et al* 2001).

Monogamous mammals

More extensive male care is found in some socially monogamous mammals (Wittenberger and Tilson

1980). Both monogamy and male care are uncommon in marsupials and ungulates (see Chapter 10), but where they occur they are commonly associated. For example, males of the monogamous rock-haunting possum, *Petropseudes dahli*, play an important role in controlling and guarding dependent young (Runcie 2000) and in klipspringer, males share vigilance with their mates and help to detect predators (Dunbar and Dunbar 1980; Dunbar 1985) (Figure 16.11a). However, monogamy is not associated with paternal care in all monogamous species. For example, in dik dik, males play little part in protecting or guarding their offspring: males do not associated more closely with females that have dependent young and their presence does not contribute to the detection of danger (Brotherton and Rhodes 1996; Komers 1996; Komers and Brotherton 1997; Brotherton and Komers 2003) (Figure 16.11b).

Paternal care is also widespread in monogamous rodents and has been extensively described in California mice, mound-building mice, pine voles, prairie voles and some lagomorphs (Dewsbury 1985; Gruder-Adams and Getz 1985; Oliveras and Novak 1986; Woodroffe and Vincent 1994; Getz and Carter 1996; Patris and Baudoin 2000) (Figure 16.12). It also

(a)

(b)

Figure 16.11 In some monogamous antelopes, like klipspringer (a), fathers play an important role in protecting their offspring whereas in others, like dik dik (b), they make little or no contribution to the care of offspring. *Sources*: (a) © Robin Dunbar; dik dik © Peter Brotherton.

occurs in some facultatively polygynous species. For example, in African striped mice, males in high-density populations guard access to breeding groups of three to four females and contribute significantly to parental care and to the development of their offspring (Schradin and Pillay 2003, 2005).

The form of male care varies widely. In some monogamous mammals, males make significant contributions to

(a)   (b)

Figure 16.12 Male care in (a) California mice and (b) African striped mice. *Sources*: (a) © Miles Barton/naturepl.com; (b) © Carsten Schradin.

guarding, incubating and grooming young (McGuire and Bemis 2007). Male incubation appears to be particularly important where litter size (and hence the total mass of pups) is relatively small so that heat retention presents problems, which may account for the evolution of monogamy combined with parental care in some species of deer mice that produce small litters (Ribble 2003). In several monogamous rodents, males and females coordinate the time they spend at the breeding burrow and seldom leave it unattended (McGuire and Novak 1984), and in some voles there are even reports of parents dragging their mate back to the breeding burrow before leaving the nest to feed (Libhaber and Eilam 2002). Males may also assist at the birth of infants: for example, in Djungarian hamsters, males can act as midwives, helping to extract pups from the female's body as they are born (Wynne-Edwards and Reburn 2000) (Figure 16.13). After young are born, male pine voles and muskrats bring food for lactating females (McGuire and Novak 1984; Oliveras and Novak 1986; Marinelli and Messier 1995), while in white-footed mice males accompany weaned young when they start to forage (Schug *et al.* 1992).

Male care is also well developed in monogamous carnivores, including insectivorous species, carrion eaters and hunting species (Figure 16.14). For example in Japanese raccoon dogs, aardwolves and bat-eared foxes, males guard pups at the breeding burrow, giving lactating

females with dependent young an opportunity to forage (Richardson 1987; Yamomoto 1987; Richardson and Coetzee 1988; Wright 2006). In bat-eared foxes, males also bring food to lactating females and dependent young at the breeding burrow (Wright 2006), as do males in many of the more carnivorous foxes and jackals (Moehlman 1983), while in African wild dogs several males (who are often brothers) help to provision lactating females and their pups (Creel and Creel 2002).

In many monogamous or primarily monogamous New World monkeys, including monotocous species, like titis and night monkeys, and twin-bearing ones, like marmosets and tamarins, males usually play an important role in parental care, regularly carrying infants and sometimes transferring food to them (Figure 16.15) (Kleiman 1985; Fuentes 2002; Wolovich *et al.* 2008). For example, in cotton-top tamarins, food is often given to infants both by parents and by older sibs. Infants beg to receive it (Goldizen 2003; Joyce and Snowdon 2007) and those in large groups receive more food than those in small ones (Feistner and Price 1990). In titi monkeys, males are the principal caregivers and infants maintain closer contact with them than with mothers and the temporary removal of their fathers leads to stronger adrenocortical responses in juveniles than the removal of their mothers (Hoffman *et al.* 1995). In some species, males and females play different roles: for example in cotton-top tamarins,

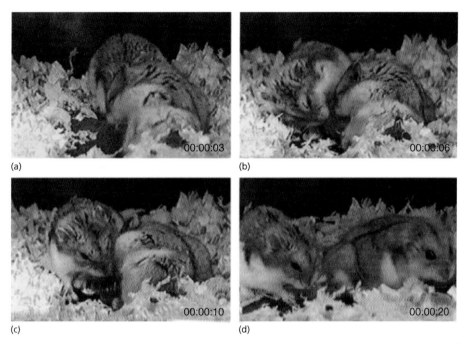

Figure 16.13 Male midwifery in Djungarian hamsters A sequence of four images from a 20-second sequence (times are shown in each frame) during the birth of the second pup in a Djungarian hamster litter. In each image, the male is on the left and the female is on the right. (a) The pup has crowned and is being licked by the female; it is still dark purple in colour and does not have an open airway; the first pup in the litter is visible in the foreground, partially covered by bedding; the male is approaching from the rear. (b) The male is tugging on the head of the crowned pup and is thus mechanically assisting in the birth. (c) The pup is born and the male is clearing the nares of membranes; the pup is just flushed to the bright red colour that indicates haemoglobin oxygenation; the female is engaging in anogenital grooming and pulling on the umbilicus. (d) The female leaves the nest area as she pushes to deliver the placenta; the male is left alone with the neonate and continues to lick and sniff it, removing all membranes, blood and amniotic fluid. *Source*: From Wynne-Edwards and Reburn, 2000. Reproduced with permission of Elsevier. © J. Jones and K.E. Wynne-Edwards.

females carry younger infants more than males while males carry older ones more than females (Tardif *et al.* 1990). In some cases, subordinates (who are often the offspring of the resident pair) also contribute to carrying young, allowing breeders of either or both sexes to reduce their contributions (Wright 1990). In comparison with New World monkeys, paternal care is not as highly developed in the monogamous Old World monkeys and apes. Though males may help to protect their offspring in monogamous as well as polygynous species, they seldom play an important role in carrying or feeding them. In particular, male gibbons appear to play a limited role in parental care, which is only well developed in siamang, where adult males and older juveniles commonly carry infants (Chivers 1974; Lappan 2008).

Differences in the extent of male care between related species are sometimes associated with contrasts in ecology. For example, the difference in male care between klipspringer and dik dik may be related to the nature of the terrain: in klipspringer, females and young are exposed to predators when they leave cliffs to feed in the bed of valleys and males can help to protect them by maintaining vigilance from raised positions, and giving alarm calls if they see predators (Dunbar and Dunbar 1980; Dunbar 1985) while dik dik inhabit scrub and woodland where there is often less relief and they rely on crypsis to avoid predators (Brotherton and Rhodes 1996).

In other cases, contrasts in male care are associated with differences in life-history parameters. In carnivores, the occurrence of extensive guarding is associated with the production of relatively altricial young and the need for lactating females to forage at a distance from the breeding den or burrow (Bekoff *et al.* 1984). In the monotocous titi monkeys and night monkeys, infant

(a)

(b)

Figure 16.14 Monogamous canids with extensive male care: (a) a male bat-eared fox guards a breeding den; (b) a female silver-backed jackal grooming her mate. *Sources*: (a) © Tim Clutton-Brock; (b) © Patricia Moehlman.

dependence is prolonged and females are either pregnant or lactating throughout most of the year (Wright 1990). Finally, in marmosets and tamarins, mothers produce twins more than once a year (Digby *et al.* 2007; Müller and Emery Thompson 2012). In smaller species, where males either play a role in carrying young or are the principal caregivers, the combined weight and size of offspring is a larger fraction of the mother's weight than in the larger species, where females are the principal caregivers (Wright 1990; Tardiff 1994).

The development of extensive paternal care in monogamous mammals raises questions about the reasons for

(a)

(b)

Figure 16.11 In many monogamous primates, males help to care for offspring produced by their mate: (a) male night monkey carrying an infant; (b) male marmoset carrying twins. *Sources*: (a) © Victor Manuel Dávalos, Owl Monkey Project, Formosa-Argentina; (b) © Jeffrey French, Callitrichid Research Center, University of Nebraska at Omaha.

the absence of male lactation. Male lactation has been recorded in some fruit bats (Francis *et al.* 1994; Kunz and Hosken 2009) so its absence in other mammals is puzzling. However, the evolution of functional lactation by males would involve fundamental changes in sexually differentiated ontogeny and the development of a male analogue of the lactogenic effects that occur in late pregnancy in females. Although this does not seem

impossible (Daly 1979), the most likely answer appears to be that the impediments to its evolution have usually been too large for male lactation to develop.

16.3 Control mechanisms

Hormonal mechanisms play an important role in controlling the expression of paternal care. The neuro-endocrine pathways controlling care in males are often the same as those involved in the control of care by females (Kelley 1988; Wynne-Edwards and Reburn 2000) but the relative importance of particular mechanisms often differs between species (Bridges 2008; Müller and Emery Thompson 2012). Prolactin is often associated with the expression of paternal care in males (Wynne-Edwards and Reburn 2000; Ziegler 2000; Smale *et al.* 2005, Wynne-Edwards and Timonin 2007). For example, in California mice, males that assist in caring for young show heightened levels of prolactin, comparable to those found in lactating females (Gubernick and Nelson 1989). Similarly, male marmosets that are carrying twin offspring have higher prolactin concentrations than males that are not doing so (Dixson and George 1982; Mota and Sousa 2000). However, it is not always clear whether variation in prolactin levels is a cause or a consequence of paternal care. For example, in common marmosets, prolactin levels in males increase while they are in contact with offspring, but there are no effects of the birth of infants on male prolactin levels (da Silva Mota *et al.* 2006), and administration of a prolactin blocker to males did not reduce paternal care (Almond *et al.* 2006).

Testosterone can also be involved, though its effects vary. In California mice, castration reduces paternal behaviour, while the replacement of testosterone restores it, possibly because testosterone is converted to oestradiol (Trainor and Marler 2001, 2002; Trainor *et al.* 2003). In other species, there may be a trade-off between male reproductive effort and male parental contributions (Ketterson and Nolan 1999): in some animals, androgen levels are negatively associated with paternal care (Smale *et al.* 2005) and in several species where male care is well developed, testosterone levels typically decline after infants are born and remain low throughout the period when males contribute most to paternal care (Brown *et al.* 1995; Reburn and Wynne-Edwards 1999; Bales *et al.* 2006). For example, in black-tufted-ear marmosets, where males carry older infants more than females, testosterone levels are relatively low in males during the period of maximal paternal care and individuals with the highest testosterone levels contribute least to carrying infants (Nunes *et al.* 2000, 2001) (Figure 16.16).

Once again, the causal mechanisms involved may be complex and variation in testosterone levels may be a consequence as well as a cause of differences in paternal care. For example, research on common marmosets shows that the exposure of fathers to the scent of their own infants during the period that they require care leads to a significant reduction in serum testosterone levels and significant increases in oestradiol (Prudom *et al.* 2008; Ziegler *et al.* 2011). Research on California mice also suggests that increases in testosterone levels may be a precursor of increases in oestradiol (Trainor and Marler 2001, 2002). In addition, not all studies have produced similar results, suggesting that the mechanisms controlling paternal care may be diverse. Differences appear to be present even among closely related species: for example, in white-faced marmosets (unlike black-tufted-ear marmosets), there appears to be no consistent association between paternal care and urinary metabolites of steroid hormones (Cavanaugh and French 2013).

The peptide hormones oxytocin and vasopressin are commonly involved in affiliative behaviour and the formation of pair-bonds and may also play a role in priming or facilitating paternal behaviour (Wynne-Edwards and Reburn 2000; Fernandez-Duque *et al.* 2008). For example, vasopressin levels are positively associated with paternal behaviour in prairie voles (Bales *et al.* 2004) and deer mice (Bester-Meredith *et al.* 1999). In prairie voles, levels of vasopressin receptor that are encoded by the gene *arpr1a* are correlated with space use, fidelity and paternal care (Okhovat *et al.* 2015). The promoter region of genes controlling vasopressin receptors appears to be associated with contrasts in paternal behaviour between monogamous and polygynous voles and transgenic insertions of genes controlling vasopressin receptors from prairie voles into laboratory mice causes them to show patterns of social behaviour and parental care similar to those of prairie voles (Hammock and Young 2005; Donaldson and Young 2008). There is also some evidence that variation in vasopressin receptors affects social behaviour in primates (Rosso *et al.* 2008): for example, the contributions of male marmosets to parental care are associated with increases in the

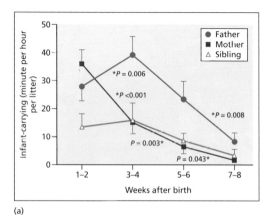

(a)

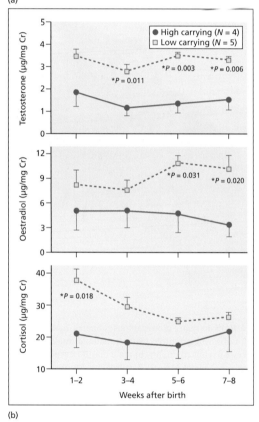

(b)

Figure 16.16 Paternal care in black-tufted-ear marmosets. (a) Rates at which infants were carried by mothers, fathers and siblings in relation to their age; asterisks indicate significant differences between successive time intervals, determined by paired *t*-tests; the sample included nine twin litters. (b) Urinary concentrations of testosterone, oestradiol and cortisol of males following the birth of young; asterisks indicate significant differences between males who carried infants at the highest vs. lowest rates, determined by post-hoc comparisons of means. *Source*: From Nunes *et al.* (2001). Reproduced with permission of Elsevier.

numbers of vasopressin receptors in their brains (Kozorovitskiy *et al.* 2006).

Breeding experience, too, can be important in the development of parental care. In some species, experienced males show substantial changes in their hormonal status several months before their partner gives birth: for example, in cotton-top tamarins, experienced males show significant changes in levels of oestrogens, prolactin and cortisol in the last 2 months of their mate's pregnancy, as well as in their body weight (Ziegler *et al.* 2004). Conditions during early development may also affect the expression of paternal care in adulthood (Nunes *et al.* 2000; Müller *et al.* 2012). For example, in mice, males that develop *in utero* between two male fetuses (2M males) are more likely to show paternal behaviour and less likely to kill infants than males that developed between two female fetuses (0M males) (vom Saal 1983).

The extent of male care may also be influenced by the individual's experience of care during its own development. In African striped mice, the expression of paternal care has a significant non-genetic maternal component and males reared by 'single' mothers contribute more to paternal care of their own litters than males reared by two parents (Rymer and Pillay 2011, 2014). As in the case of maternal effects, it is uncertain whether effects of early experience on offspring development are likely to adjust the development of offspring to the environmental conditions they are likely to face in an adaptive fashion or whether they are by products of variation in paternal investment. Nor is it yet clear whether variation in paternal care generates paternal effects resembling the maternal effects that have been documented in many mammals (see Chapter 5) or whether variation is capable of generating epigenetic effects that cross generations, though recent studies suggest that this is possible (Michel and Tyler 2007; Fernandez-Duque *et al.* 2009).

16.4 Benefits and costs of paternal care

Benefits
Where males regularly contribute to the protection or nourishment of their offspring, male care is often associated with improvements in the growth or survival of infants or juveniles (Woodroffe and Vincent 1994). Where male infanticide is common, the presence of fathers in the group is often associated with substantial

increases in the survival of infants (see Chapter 15), but other forms of male care may also play an important role. For example, in European ground squirrels, the contributions of males to preparing breeding burrows allows females to increase their foraging time and is associated with increases in the weight of pups at emergence (Huber *et al.* 2002). In monogamous species where males contribute to guarding and caring for dependent young, their presence also has an important influence on the survival of infants and juveniles. For example, in bat-eared foxes, paternal attendance at the den is the best predictor of juvenile survival (Wright 2006), while in California mice the experimental removal of fathers reduces offspring survival by 60% (Cantoni and Brown 1997; Gubernick and Teferi 2000) (Figure 16.17), although similar experiments in free-living prairie voles found no similar effects (Getz *et al.* 1993).

In some monogamous rodents, the presence of fathers can also have an important influence on the fecundity of their partners. For example, in California mice, the presence of males is associated with a reduction in inter-birth intervals and an increase in the capacity of females to raise larger litters successfully, and pairs can raise four times as many pups per season as females can rear on their own (Ribble 1992; Cantoni and Brown 1997). Similarly, in marmosets and tamarins, male contributions to carrying and protecting young may allow cooperating pairs to raise more young than single females

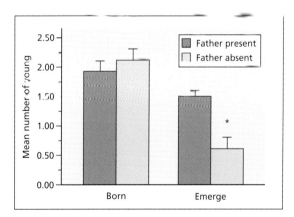

Figure 16.17 Effects of removing fathers on offspring survival in California mice. The figure shows mean number of young reared by female California mice as a consequence of the presence or absence of the male. Dark bars, male present; pale bars, male absent. *Source*: From Gubernick and Teferi, 2000. Reproduced with permission from the Royal Society.

(Goldizen and Terborgh 1989) and female siamang whose mates contribute more than usual to carrying young have shorter inter-birth intervals (Lappan 2008). Studies of singular cooperative breeders also show that helper number is often associated with reduced inter-birth intervals and increased breeding frequency (see Chapter 17). A further benefit of paternal care may be that, if their partners die, males may rear dependent young: for example, in jackals, widowed males sometimes rear their partner's pups successfully (Moehlman 1986).

Costs

Attempts to assess the costs of male care have been largely confined to captive populations of monogamous species where males make regular contributions to carrying or guarding young and have often reached contrasting conclusions. For example, some attempts to estimate the energetic costs of carrying young to males in captive callitrichids have concluded that they are low (Nievergelt and Martin 1998). While other studies have suggested that paternal care may increase energy expenditure in males by as much as 20% above basic

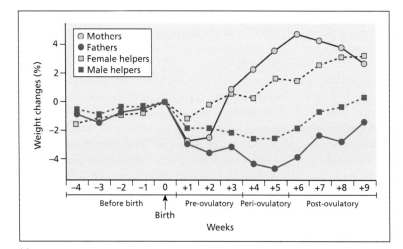

(a)

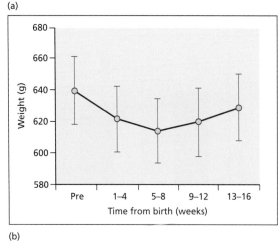

(b)

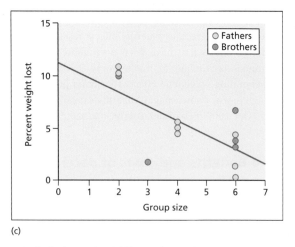

(c)

Figure 16.18 Costs of care to males in cotton-top tamarins. (a) Changes in the body weight of different classes of individuals during the 4 weeks before and the 9 weeks after parturition; the body weight of mothers before birth is not shown. (b) Mean weight (±SE) of males before and after birth of infants. (c) The relationship between male weight loss and group size (excluding infants); pale circles indicate breeding males (father), dark circles indicate sub-adult males (brothers). *Source*: From Sanchez *et al.* (1999). Reproduced with permission of John Wiley & Sons, Ltd. *Photo source*: © Andrew JK Tan/Moment/Getty Images.

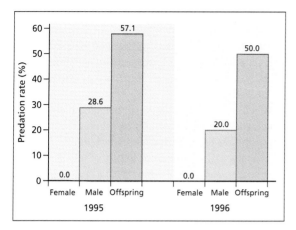

Figure 16.19 Costs of parental care to males in Malagasy giant jumping rats. For each of two years, columns show predation rates on adult females, adult males (which guard breeding burrows) and juveniles. *Source:* From Sommer, 2003. Reproduced with permission of Cambridge University Press.

maintenance costs (Schradin and Anzenberger 2001) and have shown that males lose a substantial proportion of their body weight over the course of breeding attempts (Sanchez *et al.* 1999) and that the extent of weight loss declines as the number of individuals contributing to the care of young increases (Achenbach and Snowdon 2002) (Figure 16.18). Since all these studies involved captive animals with access to plentiful food, larger costs might be expected in natural populations (Goldizen 2003). Paternal care may also expose males to greater risks of predation if carrying young affects the mobility of males (Schradin and Anzenberger 2001): for example, in monogamous Malagasy jumping rats, where males contribute more to guarding offspring than females, they also show higher rates of mortality than females (Sommer 2003) (Figure 16.19).

16.5 Tactical investment

Like maternal care, male care varies in relation to changes in its benefits and costs. For example, across rodents, male care tends to increase in more challenging environments (McGuire and Bemis 2007). Within species, the risks faced by infants or juveniles can also lead to increases in male care: for example, in captive Goeldi's monkeys, the onset of male caregiving was advanced by the experimental

introduction of a potential predator (Schradin and Anzenberger 2003). Increasing costs of male care can reduce male contributions. For example, in hoary marmots, additional mating opportunities (which raise the opportunity costs of care to males) are associated with a reduction in their contributions to parental care (Barash 1975; Marinelli and Messier 1995).

Males would usually be expected to increase their investment if they have a high chance of being the father of a female's offspring and several studies of polygynous species have demonstrated that they are more likely to direct care at their offspring than at other individuals. For example, studies of wild baboons show that males more commonly help to protect their offspring than other infants or juveniles (Buchan *et al.* 2003; Charpentier *et al.* 2008). In some cases, discrimination of this kind may be based on past association or mating experience with particular females: for example, in California mice, caring behaviour in males is maintained if they are exposed to the olfactory cues from their mates, while similar cues from other females do not affect the tendency for males to show paternal behaviour (Gubernick 1990). It is also possible that visual or olfactory cues to paternity may contribute to maintaining male care. For example, studies of rhesus monkeys show that human observers are able to identify the offspring of different fathers so it would not be surprising if male macaques could do so too (Kazem and Widdig 2013).

Few studies of monogamous mammals have been in a position to explore the effects of paternity certainty on male care, but there is some indication that males may reduce care or may abandon their mates if females mate promiscuously or their chances of fathering offspring are low. For example, in aardwolves, males sometimes abandon females that mate promiscuously (Richardson 1987; Richardson and Coetzee 1988). In addition, experiments with meadow voles show that males housed with their mate and her pups spent more time at the nest than those housed with a female rearing young fathered by another individual and that the presence of other males close to breeding females reduces their mate's contribution to parental care (Storey and Snow 1987). However, the extent to which males focus paternal care on their offspring is far from perfect and, in many species, they commonly care for unrelated infants or juveniles as well as for their own offspring. For example, in some monogamous lemurs, extra-pair paternity is common and resident males frequently help to care for young fathered by other males as well as those that they have fathered (Fietz *et al.* 2000; Fietz and Dausmann 2003). Similarly, in common marmosets and tamarins, where most infants are usually fathered by a single male, both fathers and non-fathers contribute to carrying infants (Nievergelt *et al.* 2000; Huck *et al.* 2005).

16.6 Conflicts between parents

In biparental animals where both sexes contribute to the same form of parental care, parents may respond to reductions in parental effort by their partners by reducing their own level of input, though they often do not do so to the same extent (Houston and Davies 1985; Lessells 2012). Under these conditions there are likely to be conflicts of interest between parents, since both would be expected to maximise their fitness by minimising their own level of investment (Barta *et al.* 2002; Chapman *et al.* 2003; Houston *et al.* 2005) and, in some cases, these conflicts may lead to progressive reductions in contributions to care by both parents and may even reduce their combined investment in their offspring (Houston and Davies 1985). Studies of biparental birds involving the experimental removal of one individual show that the responses of their partners to changes in the level of investment by their mates vary widely (Lessells 2012). In some cases, changes in male assistance have little effect on the level of parental care provided by their mates (Slagsvold and Lifjeld 1988; Whittingham 1989), while in others the remaining parent compensates partly or fully for any reductions in input by their mate, leading to negative correlations between the expenditure of partners (Martin 1974; Weatherhead 1979; Smith *et al.* 1982; Breitwisch 1988; Wolf *et al.* 1988).

Relations between the relative contributions of males and females to parental care have not been widely investigated in mammals, though there is some indication that parents compensate for reductions in care by their partners. For example, in siamang, male contributions to care of young vary widely between individuals and females with partners that make relatively large contributions to carrying young reduce their own contributions (Wright 1990; Lappan 2008). Similarly, in several rodents, including muskrats, red-backed voles, rock cavies and gerbils, the presence of males is associated with reductions in maternal

care (McGuire and Bemis 2007). However, responses to changes in care by breeding partners are not present in all species: for example, some studies of house mice, prairie voles and meadow voles have found no significant effects of the presence of fathers on maternal care (McGuire and Bemis 2007).

Where trade-offs in workload between the sexes do occur, they are not necessarily symmetrical. For example, in cotton-top tamarins, the contributions of helpers reduce paternal contributions to care of infants or juveniles to a greater extent than maternal ones (Tardif *et al.* 1990; Zahed *et al.* 2010). In some biparental mammals, the situation is further complicated by qualitative differences in the type of care provided by males and females or by the provision of care to different litters. For example in muskrats, polygynous males only provide care to the young of their first mate and only their first mate reduces their investment in care in response to their contributions (Marinelli and Messier 1995).

Where males and females provide different forms of care of care at different stages of offspring development, there may be selection for complementarity between parents. As yet, few theoretical or empirical studies have investigated the evolution of care where conflicts of interest between caregivers are likely to be reduced by strong selection for complementarity. However, observations of parental behaviour suggest that parents do not necessarily attempt to minimise their level of investment, as models of parental conflict predict. For example, in Goeldi's monkeys, females appear to control the contributions of males to carrying infants and actively prevent them from doing so until infants are 3 weeks old (Schradin and Anzenberger 2003), while in some marmosets group members appear to compete for the opportunity to carry dependent infants (Goldizen 2003).

16.7 Male care and the evolution of mating systems

As section 16.2 describes, paternal care is more frequent and more highly developed in socially monogamous mammals than in polygynous ones. There are two contrasting interpretations of this association. First, paternal care in mammals may be associated with social monogamy because monogamous mating systems reduce opportunities for males to mate with multiple females, lowering reproductive skew among males and reducing the strength of selection in males for traits that contribute to competitive ability. On this argument, social monogamy is a consequence of intolerance between breeding females and the inability of males to defend access to more than one female and the evolution of paternal care is likely either to occur at the same evolutionary stage as the development of monogamy or to succeed it (see Chapter 10).

An alternative possibility is that the need for male care has led to the evolution of social monogamy. One version of this argument suggests that ecological or life-history adaptations in females (such as the production of litters or repeated breeding within seasons) lead to the need for extensive biparental care (Maynard Smith 1977; Klug *et al.* 2012; Trumbo 2012). There is evidence from some other groups of vertebrates that some ecological niches are associated with strong selection for both biparental care and monogamy: for example, among poison-arrow frogs, biparental care and monogamy are associated with the use of small pools where nutrients are inadequate for developing tadpoles (Brown *et al.* 2010). A similar argument (sometimes referred to as the *infant protection hypothesis*) is that where intruding males are commonly infanticidal, resident males need to form protracted bonds with their breeding partners to reduce the risk that their offspring will be killed by other males (van Schaik and Kappeler 1997). The first of these two arguments suggests that the evolution of male care should either precede the evolution of monogamous bonds or should occur at the same time, while the second predicts that the evolution of male infanticide should precede the evolution of monogamy or occur at the same time.

Recent phylogenetic reconstructions have explored the sequential evolution of monogamy and male care in mammals. Both within primates and across a wide array of mammals, the evolution of paternal care most commonly occurs within lineages where social monogamy is already established, though in some cases it occurs at the same point (Lukas and Clutton-Brock 2013; Opie *et al.* 2013). This suggests that the evolution of monogamy facilitates the evolution of paternal care, rather than vice versa.

Investigations of the relationship between the evolution of monogamy and the risk of male infanticide presents problems, for male infanticide is often difficult to detect and so may occur in a wider range of species than existing records suggest. One analysis of the evolution of monogamy in primates (where monogamy has evolved on around six occasions) found a significant association between proxy measures of the incidence of male infanticide and the evolution of monogamy and argued that this supported the infant protection hypothesis (Opie *et al.* 2013). In contrast, an analysis of the association between the evolution of monogamy and the incidence of observed male infanticide across all mammals that identified more than sixty separate evolutions of monogamy found no evidence of an association between monogamy and male infanticide (Lukas and Clutton-Brock 2013). Reanalysis of data for primates collected by the first team using the same categorisation of species but a more conservative analytical approach also found no evidence of a consistent association between monogamy and infanticide (Lukas and Clutton-Brock 2014). Some other evidence also argues against male infanticide as a driver of the evolution of monogamy. For example, the dynamics of intersexual relationships in monogamous species appear to differ from those in species where males protect infants against infanticide (Overdorff 1998). In primates where males provide protection against infanticide, females are commonly responsible for maintaining proximity and for instigating affiliative interactions with males, whereas in several monogamous species, males maintain proximity to females and are the principal instigators of affiliative interactions between the sexes (Palombit 1999, 2000).

While it seems unlikely that selection for male care is responsible for the evolution of monogamy, the evolution of male care may affect many other aspects of mating systems. In several primates that live in groups where mating is usually monopolised by a single male, alpha males that have been defeated and no longer have regular access to breeding females remain in their groups and assist in preventing immigrants from joining the group or in protecting juveniles (see Chapter 15). In addition, where additional males make important contributions to protecting young or prolonging the tenure of the breeding male, dominant males may be more ready to tolerate their presence or to ignore their attempts to mate (see Chapter 10). For example, in some groups of mountain gorillas, dominant breeding males tolerate the presence of a male subordinate, who assists in repelling intruders and challengers and is often one of their sons (Figure 16.20). If resident silverbacks are displaced by immigrant males, the latter frequently

Figure 16.20 In mountain gorillas, some groups include more than one mature male. Subordinate males help to defend groups against immigrant males, who are often infanticidal. In this photograph, the younger male is taking up the more aggressive stance. *Source*: © Sandy Harcourt.

kill dependent young, while the replacement of alpha males by previously resident natal males is rarely associated with male infanticide (Figure 16.21), which may explain why dominant males sometimes tolerate the presence of subordinate natal males and allow them a share of reproduction (Harcourt and Stewart 2007). Effects of this kind may be particularly likely where females routinely emigrate from their natal groups after

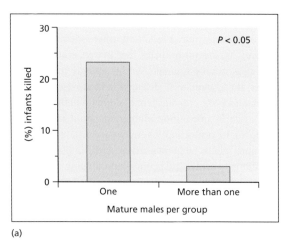

(a)

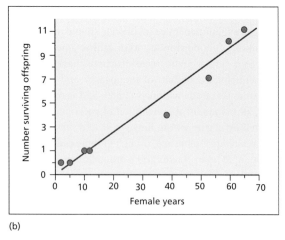

(b)

Figure 16.21 (a) Proportion of mountain gorilla infants killed in infanticidal attacks by males in one-male versus multi-male groups. From Harcourt and Stewart (2007), reproduced with permission from the University of Chicago Press. (b) Number of surviving offspring sired by males as a function of female group size and duration of males' tenure. Female years = mean number of adult females in the group per year of tenure × duration of tenure. *Sources*: Data from (a) Robbins, 1996 and (b) Bradley *et al.* (2005).

reaching sexual maturity and sons can inherit the breeding role in their natal group from their fathers.

Where males help to protect the offspring of mating partners, male care may favour females that mate with multiple partners to induce several males to assist in caring for their young (Davies 1989; Harada and Iwasa 1996). This may increase competition between females for mating partners, which may in turn lead to the evolution of female ornaments that signal female receptivity without providing precise information on the timing of ovulation and attract males. For example, the evolution of perineal swellings in female primates that live in multi-male groups may be a consequence of female competition to attract mating partners and to increase male uncertainty about paternity (Alberts and Fitzpatrick 2012) (see Chapter 7).

SUMMARY

1. Male care is rare or absent in species with promiscuous mating systems where associations between breeding partners are temporary. Males contribute to the protection of juveniles in many polygynous mammals. More extensive male care (including regular carrying and provisioning of infants) is found in many monogamous mammals.
2. Like maternal care, paternal care is associated with increases in circulating levels of prolactin as well as with increased levels of oxytocin and vasopressin, but the mechanisms underlying these associations are not yet well understood. In some rodents, interspecific differences in male care are associated with variation in the promoter region of genes controlling vasopressin.
3. Paternal care commonly contributes to the growth and survival of infants and can increase the fecundity of females. Like maternal care, it has energetic costs and may reduce the condition of males and their opportunities for other reproductive attempts.
4. Contributions to parental care by fathers can allow mothers to reduce their contributions to care. While this suggests that there may be conflicts of interest between parents over the level of investment, the situation is complicated by qualitative differences in the form of care between fathers and mothers and by selection for complementarity of care between parents.
5. Phylogenetic reconstructions of the evolution of paternal care suggest that the evolution of paternal care may often be a consequence of the evolution of monogamy rather than a cause.
6. Less is known of the consequences of paternal than maternal care and our understanding of species differences in its development and form is still very limited. Where male care is highly developed, we can expect to find complex adaptations in the form and extent of paternal care as well as paternal effects on the development of offspring.

References

Achenbach, G.G. and Snowdon, C.T. (2002) Costs of caregiving: weight loss in captive adult male cotton-top tamarins (*Saguinus oedipus*) following the birth of infants. *International Journal of Primatology* **23**:179–189.

Alberts, S.C. and Fitzpatrick, C.L. (2012) Paternal care and the evolution of exaggerated sexual swellings in primates. *Behavioral Ecology* **23**:699–706.

Almond, R.E.A., *et al.* (2006) Suppression of prolactin does not reduce infant care by parentally experienced male common marmosets (*Callithrix jacchus*). *Hormones and Behavior* **49**:673–680.

Bales, K.L., *et al.* (2004) Both oxytocin and vasopressin may influence alloparental behavior in male prairie voles. *Hormones and Behavior* **45**:354–361.

Bales, K.L., *et al.* (2006) Effects of social status, age, and season on androgen and cortisol levels in wild male golden lion tamarins (*Leontopithecus rosalia*). *Hormones and Behavior* **49**:88–95.

Barash, D.P. (1975) Ecology of paternal behavior in the hoary marmot (*Marmota caligata*): an evolutionary interpretation. *Journal of Mammalogy* **56**:613–618.

Barta, Z., *et al.* (2002) Sexual conflict about parental care: the role of reserves. *American Naturalist* **159**:687–705.

Beehner, J.C., *et al.* (2005) The effect of new alpha males on female stress in free-ranging baboons. *Animal Behaviour* **69**:1211–1221.

Bekoff, M., *et al.* (1984) Life history patterns and the comparative social ecology of carnivores. *Annual Review of Ecology and Systematics* **15**:191–232.

Bester-Meredith, J.K., *et al.* (1999) Species differences in paternal behavior and aggression in *Peromyscus* and their associations with vasopressin immunoreactivity and receptors. *Hormones and Behavior* **36**:25–38.

Bradley, B.J., *et al.* (2005) Mountain gorilla tug-of-war: silverbacks have limited control over reproduction in multimale groups. *Proceedings of the National Academy of Sciences of the United States of America* **102**:9418–9423.

Breitwisch, R. (1988) Sex differences in defence of eggs and nestlings by northern mockingbirds, *Mimus polyglottos*. *Animal Behaviour* **36**:62–72.

Bridges, R.S. (2008) *Neurobiology of the Parental Brain*. Burlington, MA: Academic Press.

Brotherton, P.N.M. and Komers, P.E. (2003) Mate guarding and the evolution of social monogamy in mammals. In: *Monogamy: Mating Strategies and Partnerships in Birds, Humans and other Mammals* (eds U.H. Reichard and C. Boesch). Cambridge: Cambridge University Press, 59–80.

Brotherton, P.N.M. and Rhodes, A. (1996) Monogamy without biparental care in a dwarf antelope. *Proceedings of the Royal Society of London. Series B: Biological Sciences* **263**:23–29.

Brown, J.L., *et al.* (2010) A key ecological trait drove the evolution of biparental care and monogamy in an amphibian. *American Naturalist* **175**:436–446.

Brown, R.F., *et al.* (1995) Hormonal responses of male gerbils to stimuli from their mate and pups. *Hormones and Behavior* **29**:474–491.

Buchan, J.C., *et al.* (2003) True paternal care in a multi-male primate society. *Nature* **425**:179–181.

Cantoni, D. and Brown, R.E. (1997) Paternal investment and reproductive success in the California mouse, *Peromyscus californicus*. *Animal Behaviour* **54**:377–386.

Cavanaugh, J. and French, J.A. (2013) Post-partum variation in the expression of paternal care is unrelated to urinary steroid metabolites in marmoset fathers. *Hormones and Behavior* **63**:551–558.

Chapman, T., *et al.* (2003) Sexual conflict. *Trends in Ecology and Evolution* **18**:41–47.

Charpentier, M.J.E., *et al.* (2008) Paternal effects on offspring fitness in a multimale primate society. *Proceedings of the National Academy of Sciences of the United States of America* **105**:1988–1992.

Cheney, D.L., *et al.* (2015) The costs of parental and mating effort for male baboons. *Behavioral Ecology and Sociobiology* **69**:303–312.

Chivers, D.J. (1974) *The Siamang in Malaya: A Field Study of a Primate in a Tropical Rain Forest*. Contributions to Primatology Vol. 4. Basel: Karger.

Creel, S. and Creel, N.M. (2002) *The African Wild Dog: Behavior, Ecology, and Conservation*. Princeton, NJ: Princeton University Press.

Daly, M. (1979) Why don't male mammals lactate? *Journal of Theoretical Biology* **78**:325–345.

da Silva Mota, M.T., *et al.* (2006) Hormonal changes related to paternal and alloparental care in common marmosets *Callithrix jacchus*. *Hormones and Behavior* **49**:293–302.

Davies, N.B. (1989) Sexual conflict and the polygamy threshold. *Animal Behaviour* **38**:226–234.

Deag, J.M. and Crook, J.H. (1971) Social behaviour and 'agonistic buffering' in the wild barbary macaque *Macaca sylvana* L. *Folia Primatologica* **15**:183–200.

Dewsbury, D.A. (1985) Paternal behavior in rodents. *American Zoologist* **25**:841–852.

Digby, L.J., *et al.* (2007) Callitrichines: the role of competition in cooperatively breeding species. In: *Primates in Perspective* (eds C.J. Campbell, A. Fuentes, K.C. Mackinnon, M. Panger and S.K. Bearder). New York: Oxford University Press, 85–106.

Dixson, A.F. and George, L. (1982) Prolactin and parental behaviour in a male New World primate. *Nature* **299**:551–553.

Donaldson, Z.R. and Young, L.J. (2008) Oxytocin, vasopressin, and the neurogenetics of sociality. *Science* **322**:900–904.

Dunbar, R.I.M. (1985) Monogamy on the rocks. *Natural History* **94**:40.

Dunbar, R.I.M. and Dunbar, E.P. (1980) Pairbond in klipspringer. *Animal Behaviour* **28**:219–229.

Elwood, R.W. (1983) Paternal care in rodents. In: *Parental Behaviour of Rodents* (ed. R.W. Elwood). Chichester, West Sussex: John Wiley & Sons Ltd, 235–257.

Feistner, A.T.C. and Price, E.C. (1990) Food-sharing in cottontop tamarins (*Saguinus oedipus*). *Folia Primatologica* **54**:34–45.

Fernandez-Duque, E., *et al.* (2008) Adult male replacement and subsequent infant care by male and siblings in socially monogamous owl monkeys (*Aotus azarai*). *Primates* **49**:81–84.

Fernandez-Duque, E., *et al.* (2009) The biology of paternal care in human and nonhuman primates. *Annual Review of Anthropology* **38**:115–130.

Fietz, J. and Dausmann, K.H. (2003) Costs and potential benefits of parental care in the nocturnal fat-tailed dwarf lemur (*Cheirogaleus medius*). *Folia Primatologica* **74**:246–258.

Fietz, J., *et al.* (2000) High rates of extra-pair young in the pair-living fat-tailed dwarf lemur, *Cheirogaleus medius*. *Behavioral Ecology and Sociobiology* **49**:8–17.

Francis, C.M., *et al.* (1994) Lactation in male fruit bats. *Nature* **367**:691–692.

Fuentes, A. (2002) Patterns and trends in primate pair bonds. *International Journal of Primatology* **23**:953–978.

Getz, L.L. and Carter, C.S. (1996) Prairie-vole partnerships. *American Scientist* **84**:56–62.

Getz, L.L., *et al.* (1993) Social organization of the prairie vole (*Microtus ochrogaster*). *Journal of Mammalogy* **74**:44–58.

Goldizen, A.W. (2003) Social monogamy and its variations in callitrichids: do these relate to the costs of infant care? In: *Monogamy: Mating Strategies and Partnerships in Birds, Humans and Other Mammals* (eds U.H. Reichard and C. Boesch). Cambridge: Cambridge University Press, 232–247.

Goldizen, A.W. and Terborgh, J. (1989) Demography and dispersal patterns of a tamarin population: possible causes of delayed breeding. *American Naturalist* **134**:208–224.

Grieser, B. (1992) Infant development and parental care in two species of sifakas. *Primates* **33**:305–314.

Gruder-Adams, S. and Getz, L.L. (1985) Comparison of the mating system and paternal behavior in *Microtus ochrogaster* and *M. pennsylvanicus*. *Journal of Mammalogy* **66**:165–167.

Gubernick, D.J. (1990) A maternal chemosignal maintains paternal behaviour in the biparental California mouse, *Peromyscus californicus*. *Animal Behaviour* **39**:936–942.

Gubernick, D.J. and Nelson, R.J. (1989) Prolactin and paternal behavior in the biparental California mouse, *Peromyscus californicus*. *Hormones and Behavior* **23**:203–210.

Gubernick, D.J. and Teferi, T. (2000) Adaptive significance of male parental care in a monogamous mammal. *Proceedings of the Royal Society of London. Series B: Biological Sciences* **267**:147–150.

Hammock, E.A.D. and Young, L.J. (2005) Microsatellite instability generates diversity in brain and sociobehavioral traits. *Science* **308**:1630–1634.

Harada, Y. and Iwasa, Y. (1996) Female mate preference to maximise paternal care: a two-step game. *American Naturalist* **147**:996–1027.

Harcourt, A.H. and Stewart, K.J. (2007) *Gorilla Society: Conflict, Compromise, and Cooperation Between the Sexes*. Chicago: University of Chicago Press.

Heard, D.C. (1992) The effect of wolf predation and snow cover on musk ox group size. *American Naturalist* **139**:190–204.

Hoffman, K.A., *et al.* (1995) Responses of infant titi monkeys, *Callicebus moloch*, to removal of one or both parents: evidence for paternal attachment. *Developmental Psychobiology* **28**:399–407.

Houston, A.I. and Davies, N.B. (1985) The evolution of cooperation and life-history in the dunnock *Prunella modularis*. In: *Behavioural Ecology: Ecological Consequences of Adaptive Behaviour* (eds R.M. Sibley and R.H. Smith). Oxford: Blackwell Scientific Publications, 471–487.

Houston, A.I., *et al.* (2005) Conflict between parents over care. *Trends in Ecology and Evolution* **20**:33–38.

Huber, S., *et al.* (2002) Paternal effort and its relation to mating success in the European ground squirrel. *Animal Behaviour* **63**:157–164.

Huchard, E., *et al.* (2010) More than friends? Behavioural and genetic aspects of heterosexual associations in wild chacma baboons. *Behavioral Ecology and Sociobiology* **64**:769–781.

Huchard, E., *et al.* (2012) Paternal effects on access to resources in a promiscuous primate society. *Behavioral Ecology* **24**:229–236.

Huck, M., *et al.* (2005) Paternity and kinship patterns in polyandrous moustached tamarins (*Saguinus mystax*). *American Journal of Physical Anthropology* **127**:449–464.

Joyce, S.M. and Snowdon, C.T. (2007) Developmental changes in food transfers in cotton-top tamarins (*Saguinus oedipus*). *American Journal of Primatology* **69**:955–965.

Kazem, A.J.N. and Widdig, A. (2013) Visual phenotype matching: cues to paternity are present in rhesus macaque faces. *PLOS ONE* **8**:e55846.

Keddy Hector, A.C., *et al.* (1989) Male parental care, female choice and the effect of an audience in vervet monkeys. *Animal Behaviour* **38**:262–271.

Kelley, D.B. (1988) Sexually dimorphic behaviors. *Annual Review of Neuroscience* **11**:225–251.

Ketterson, E.D. and Nolan, V. Jr (1999) Adaptation, exaptation, and constraint: a hormonal perspective. *American Naturalist* **154**:S4–S25.

Kitchen, D.M. (2004) Alpha male black howler monkey responses to loud calls: effect of numeric odds, male companion behaviour and reproductive investment. *Animal Behaviour* **67**:125–139.

Kleiman, D.G. (1985) Paternal care in New World primates. *American Zoologist* **25**:857–859.

Klug, H., *et al.* (2012) Theoretical foundation of parental care. In: *The Evolution of Parental Care* (eds N.J. Royle, P.T. Smiseth and M. Kölliker). Oxford: Oxford University Press, 21–39.

Komers, P.E. (1996) Obligate monogamy without paternal care in Kirk's dik-dik. *Animal Behaviour* **51**:131–140.

Komers, P.E. and Brotherton, P.N.M. (1997) Dung pellets used to identify the distribution and density of dik-dik. *African Journal of Ecology* **35**:124–132.

Kozorovitskiy, Y., *et al.* (2006) Fatherhood affects dendritic spines and vasopressin V1a receptors in the primate prefrontal cortex. *Nature Neuroscience* **9**:1094–1095.

Kunz, T.H. and Hosken, D.J. (2009) Male lactation: why, why not and is it care? *Trends in Ecology and Evolution* **24**:80–85.

Lappan, S. (2008) Male care of infants in a siamang (*Symphalangus syndactylus*) population including socially monogamous and polyandrous groups. *Behavioral Ecology and Sociobiology* **62**:1307–1317.

Lessells, C.M. (2012) Sexual conflict. In: *The Evolution of Parental Care* (eds N.J. Royle, P.T. Smiseth and M. Kölliker). Oxford: Oxford University Press, 150–170.

Libhaber, N. and Eilam, D. (2002) Social vole parents force their mates to baby-sit. *Developmental Psychobiology* **41**:236–240.

Lukas, D. and Clutton-Brock, T.H. (2013) The evolution of social monogamy in mammals. *Science* **341**:526–530.

Lukas, D. and Clutton-Brock, T.H. (2014) Evolution of social monogamy in primates is not consistently associated with male infanticide. *Proceedings of the National Academy of Sciences of the United States of America* **111**:E1674.

McGuire, B. and Bemis, W.E. (2007) Parental care. In: *Rodent Societies: An Ecological and Evolutionary Perspective* (eds J.O. Wolff and P. W. Sherman). Chicago: University of Chicago Press, 231–242.

McGuire, B. and Novak, M. (1984) A comparison of maternal behaviour in the meadow vole (*Microtus pennsylvanicus*), prairie vole (*M. ochrogaster*) and pine vole (*M. pinetorum*). *Animal Behaviour* **32**:1132–1141.

Marinelli, L. and Messier, F. (1995) Parental-care strategies among muskrats in a female-biased population. *Canadian Journal of Zoology* **73**:1503–1510.

Martin, S.G. (1974) Adaptations of polygynous breeding in the Bobolink, *Dolichonyx oryzivorus*. *American Zoologist* **14**:109–119.

Maynard Smith, J. (1977) Parental investment: a prospective analysis. *Animal Behaviour* **25**:1–9.

Ménard, N., *et al.* (2001) Is male–infant caretaking related to paternity and/or mating activities in wild Barbary macaques (*Macaca sylvanus*)? *Comptes Rendus de l'Académie des Sciences Series III. Sciences de la Vie* **324**:601–610.

Michel, G.F. and Tyler, A.N. (2007) Can knowledge of developmental processes illuminate the evolution of parental care? *Developmental Psychobiology* **49**:33–44.

Mills, M.G.L. (1990) *Kalahari Hyaenas: Comparative Behavioural Ecology of Two Species*. London: Unwin Hyman.

Moehlman, P.D. (1983) Socioecology of silverbacked and golden jackals (*Canis mesomelas and Canis aureus*) In: *Advances in the Study of Mammalian Behavior* (eds J.F. Eisenberg and D.G. Kleiman), American Society of Mammalogists Special Publication No. 7. Stillwater, OK: American Society of Mammalogists, 423–453.

Moehlman, P.D. (1986) Ecology of cooperation in canids. In: *Ecological Aspects of Social Evolution in Birds and Mammals* (eds D.I. Rubenstein and R.W. Wrangham). Princeton, NJ: Princeton University Press, 64–86.

Mota, M.T. and Sousa, M.B.C. (2000) Prolactin levels of fathers and helpers related to alloparental care in common marmosets, *Callithrix jacchus*. *Folia Primatologica* **71**:22–26.

Müller, M.N. and Emery Thompson, M. (2012) Mating, parenting and male reproductive strategies. In: *The Evolution of Primate Societies* (eds J.C. Mitani, J. Call, P.M. Kappeler, R.A. Palombit and J.B. Silk). Chicago: University of Chicago Press, 387–411.

Müller, M.N., *et al.* (2012) Testosterone, development and aging in wild chimpanzees. *American Journal of Physical Anthropology* **147**:220.

Mysterud, A., *et al.* (2002) The role of males in the dynamics of ungulate populations. *Journal of Animal Ecology* **71**:907–915.

Nievergelt, C.M. and Martin, R.D. (1998) Energy intake during reproduction in captive common marmosets (*Callithrix jacchus*). *Physiology and Behavior* **65**:849–854.

Nievergelt, C.M., *et al.* (2000) Genetic analysis of group composition and breeding system in a wild common marmoset (*Callithrix jacchus*) population. *International Journal of Primatology* **21**:1–20.

Nunes, S., *et al.* (2000) Variation in steroid hormones associated with infant care behaviour and experience in male marmosets (*Callithrix kuhlii*). *Animal Behaviour* **60**:857–865.

Nunes, S., *et al.* (2001) Interactions among paternal behavior, steroid hormones, and parental experience in male marmosets (*Callithrix kuhlii*). *Hormones and Behavior* **39**:70–82.

Okhovat, M., *et al.* (2015) Sexual fidelity trade-offs promote regulatory variation in the prairie vole brain. *Science* **350**:1371–1374.

Oliveras, D. and Novak, M. (1986) A comparison of paternal behaviour in the meadow vole *Microtus pennsylvanicus*, the pine vole *M. pinetorum* and the prairie vole *M. ochrogaster*. *Animal Behaviour* **34**:519–526.

Opie, C., *et al.* (2013) Male infanticide leads to social monogamy in primates. *Proceedings of the National Academy of Sciences of the United States of America* **110**:13328–13332.

Overdorff, D.J. (1998) Are *Eulemur* species pair-bonded? Social organization and mating strategies in *Eulemur fulvus rufus* from 1988–1995 in southeast Madagascar. *American Journal of Physical Anthropology* **105**:153–166.

Palombit, R.A. (1999) Infanticide and the evolution of pair bonds in nonhuman primates. *Evolutionary Anthropology* **7**:117–129.

Palombit, R.A. (2000) Infanticide and the evolution of male female bonds in animals. In: *Infanticide by Males and its Implications* (eds C.P. van Schaik and C.H. Janson). Cambridge: Cambridge University Press, 239–268.

Palombit, R.A. (2014) Sexual conflict in nonhuman primates. In: *Advances in the Study of Behavior*, Vol. **46** (eds M. Naguib *et al.*). San Diego, CA: Academic Press, 191–280.

Palombit, R.A., *et al.* (1997) The adaptive value of 'friendships' to female baboons: experimental and observational evidence. *Animal Behaviour* **54**:599 614.

Patris, B. and Baudoin, C. (2000) A comparative study of parental care between two rodent species: implications for the mating system of the mound-building mouse *Mus spicilegus*. *Behavioural Processes* **51**:35–43.

Prins, H.H.T. (1996) *Ecology and Behaviour of the African Buffalo: Social Inequality and Decision Making*. London: Chapman & Hall.

Prudom, S.L., *et al.* (2008) Exposure to infant scent lowers serum testosterone in father common marmosets (*Callithrix jacchus*). *Biology Letters* **4**:603–605.

Reburn, C.J. and Wynne-Edwards, K.E. (1999) Hormonal changes in males of a naturally biparental and a uniparental mammal. *Hormones and Behavior* **35**:163–176.

Ribble, D.O. (1992) Lifetime reproductive success and its correlates in the monogamous rodent, *Peromyscus californicus*. *Journal of Animal Ecology* **61**:457–468.

Ribble, D.O. (2003) The evolution of social and reproductive monogamy in *Peromyscus*: evidence from *Peromyscus californicus* (the California mouse). In: *Monogamy: Mating Strategies and Partnerships in Birds, Humans and Other Mammals* (eds U.H. Reichard and C. Boesch). Cambridge: Cambridge University Press, 81–92.

Richardson, P.R.K. (1987) Aardwolf mating system: overt cuckoldry in an apparently monogamous mammal. *South African Journal of Science* **83**:405–410.

Richardson, P.R.K. and Coetzee, M. (1988) Mate desertion in response to female promiscuity in the socially monogamous aardwolf, *Proteles cristatus*. *South African Journal of Zoology* **23**:306–308.

Robbins, M.M. (1996) Male–male interactions in heterosexual and all-male wild mountain gorilla groups. *Ethology* **102**:942–965.

Rosso, L., *et al.* (2008) Mating system and *avpr1a* promoter variation in primates. *Biology Letters* **4**:375–378.

Runcie, M.J. (2000) Biparental care and obligate monogamy in the rock-haunting possum, *Petropseudes dahli*, from tropical Australia. *Animal Behaviour* **59**:1001–1008.

Rymer, T.L. and Pillay, N. (2011) The influence of the early rearing environment on the development of paternal care in African striped mice. *Ethology* **117**:284–293.

Rymer, T.L. and Pillay, N. (2014) Alloparental care in the African striped mouse *Rhabdomys pumilio* is age-dependent and influences the development of paternal care. *Ethology* **120**:11–20.

Sanchez, S., *et al.* (1999) Costs of infant-carrying in the cotton-top tamarin (*Saguinus oedipus*). *American Journal of Primatology* **48**:99–111.

Schradin, C. and Anzenberger, G. (2001) Costs of infant carrying in common marmosets, *Callithrix jacchus*: an experimental analysis. *Animal Behaviour* **62**:289–295.

Schradin, C. and Anzenberger, G. (2003) Mothers, not fathers, determine the delayed onset of male carrying in Goeldi's monkey (*Callimico goeldii*). *Journal of Human Evolution* **45**:389–399.

Schradin, C. and Pillay, N. (2003) Paternal care in the social and diurnal striped mouse (*Rhabdomys pumilio*): laboratory and field evidence. *Journal of Comparative Psychology* **117**:317–324.

Schradin, C. and Pillay, N. (2005) The influence of the father on offspring development in the striped mouse. *Behavioral Ecology* **16**:450–455.

Schug, M.D., *et al.* (1992) Paternal behavior in a natural population of white-footed mice (*Peromyscus leucopus*). *American Midland Naturalist* **127**:373–380.

Seyfarth, R.M. and Cheney, D.L. (1984) Grooming alliances and reciprocal altruism in vervet monkeys. *Nature* **308**:541–543.

Slagsvold, T. and Lifjeld, J.T. (1988) Ultimate adjustment of clutch size to parental feeding capacity in a passerine bird. *Ecology* **69**:1918–1922.

Smale, L., *et al.* (2005) Behavioral neuroendocrinology in non-traditional species of mammals: things the 'knockout' mouse CAN'T tell us. *Hormones and Behavior* **48**:474–483.

Smith, J.N.M., *et al.* (1982) Polygyny, male parental care, and sex ratio in song sparrows: an experimental study. *The Auk* **99**:555–564.

Smuts, B.B. and Gubernick, D.J. (1992) Male–infant relationships in nonhuman primates: paternal investment or mating effort? In: *Father–Child Relations: Cultural and Biosocial Contexts* (ed. B.S. Hewlett) New York: Aldine de Gruyter, 1–30.

Sommer, S. (2003) Social and reproductive monogamy in rodents: the case of the Malagasy giant jumping rat. In: *Monogamy: Mating Strategies and Partnerships in Birds, Humans and Other Mammals* (eds U.H. Reichard and C. Boesch). Cambridge: Cambridge University Press, 109–124.

Storey, A.E. and Snow, D.T. (1987) Male identity and enclosure size affect paternal attendance of meadow voles, *Microtus pennsylvanicus*. *Animal Behaviour* **35**:411–419.

Tardif, S.D. (1994) Relative energetic cost of infant care in small-bodied neotropical primates and its relation to infant-care patterns. *American Journal of Primatology* **34**:133–143.

Tardif, S.D., *et al.* (1990) Infant-care behavior of mothers and fathers in a communal-care primate, the cotton-top tamarin (*Saguinus oedipus*). *American Journal of Primatology* **22**:73–85.

Trainor, B.C. and Marler, C.A. (2001) Testosterone, paternal behavior, and aggression in the monogamous California mouse (*Peromyscus californicus*). *Hormones and Behavior* **40**:32–42.

Trainor, B.C. and Marler, C.A. (2002) Testosterone promotes paternal behaviour in a monogamous mammal via conversion to oestrogen. *Proceedings of the Royal Society of London. Series B: Biological Sciences* **269**:823–829.

Trainor, B.C., *et al.* (2003) Variation in aromatase activity in the medial preoptic area and plasma progesterone is associated with the onset of paternal behavior. *Neuroendocrinology* **78**:36–44.

Trumbo, S.T. (2012) Patterns of parental care in invertebrates. In: *The Evolution of Parental Care* (eds N.J. Royle, P.T. Smiseth and M. Kölliker). Oxford: Oxford University Press, 81–100.

van Schaik, C.P. and Kappeler, P.M. (1997) Infanticide risk and the evolution of male–female association in primates. *Proceedings of the Royal Society of London. Series B: Biological Sciences* **264**:1687–1694.

van Schaik, C.P. and Paul, A. (1996) Male care in primates: does it ever reflect paternity? *Evolutionary Anthropology: Issues, News, and Reviews* **5**:152–156.

vom Saal, F.S. (1983) Variation in infanticide and parental behavior in male mice due to prior intrauterine proximity to female fetuses: elimination by prenatal stress. *Physiology and Behavior* **30**:675–681.

Weatherhead, P.J. (1979) Do savannah sparrows commit the Concorde fallacy? *Behavioral Ecology and Sociobiology* **5**:373–381.

Whittingham, L.A. (1989) An experimental study of paternal behavior in red-winged blackbirds. *Behavioral Ecology and Sociobiology* **25**:73–80.

Wittenberger, J.F. and Tilson, R.L. (1980) The evolution of monogamy: hypotheses and evidence. *Annual Review of Ecology and Systematics* **11**:197–232.

Wolf, L.L., *et al.* (1988) Paternal influence on growth and survival of dark-eyed junco young: do parental males benefit? *Animal Behaviour* **36**:1601–1618.

Wolovich, C.K., *et al.* (2008) Food transfers to young and mates in wild owl monkeys (*Aotus azarai*). *American Journal of Primatology* **70**:211–221.

Woodroffe, R. and Vincent, A. (1994) Mother's little helpers: patterns of male care in mammals. *Trends in Ecology and Evolution* **9**:294–297.

Wright, H.W.Y. (2006) Paternal den attendance is the best predictor of offspring survival in the socially monogamous bat-eared fox. *Animal Behaviour* **71**:503–510.

Wright, P.C. (1990) Patterns of paternal care in primates. *International Journal of Primatology* **11**:89–102.

Wronkski, T., *et al.* (2006) Behavioural repertoire of the bushbuck (*Tragelaphus scriptus*): agonistic interactions, mating behaviour and parent–offspring relations. *Journal of Ethology* **24**:247–260.

Wynne-Edwards, K.E. and Reburn, C.J. (2000) Behavioral endocrinology of mammalian fatherhood. *Trends in Ecology and Evolution* **15**:464–468.

Wynne-Edwards, K.E. and Timonin, M.E. (2007) Paternal care in rodents: weakening support for hormonal regulation of the transition to behavioral fatherhood in rodent animal models of biparental care. *Hormones and Behavior* **52**:114–121.

Xiang, Z.-F., *et al.* (2009) Direct paternal care in black-and-white snub-nosed monkeys. *Journal of Zoology* **278**:157–162.

Yamomoto, I. (1987) Male parental care in the raccoon dog *Nyctereutes procyonoides* during the early rearing period. In: *Animal Societies: Theories and Facts* (eds Y. Ito, J.L. Brown and J. Kikkawa). Tokyo: Japan Scientific Societies, 189–195.

Zahed, S.R., *et al.* (2010) Social dynamics and individual plasticity of infant care behavior in cooperatively breeding cotton-top tamarins. *American Journal of Primatology* **72**:296–306.

Ziegler, T.E. (2000) Hormones associated with non-maternal infant care: a review of mammalian and avian studies. *Folia Primatologica* **71**:6–21.

Ziegler, T.E., *et al.* (2004) Responsiveness of expectant male cotton-top tamarins, *Saguinus oedipus,* to mate's pregnancy. *Hormones and Behavior* **45**:84–92.

Ziegler, T.E., *et al.* (2011) Differential endocrine responses to infant odors in common marmoset (*Callithrix jacchus*) fathers. *Hormones and Behavior* **59**:265–270.

CHAPTER 17

Cooperative breeding

17.1 Introduction

In a dark tunnel, three feet underground, two female mole rats are fighting for domination. The previous queen, their mother, recently died and in the days since her death the resident females have been struggling to establish dominance. Biting, tugging and pushing, the pair move one way and then another until one finally wins, pushes the other back to the end of the tunnel and, eventually, out into the unfamiliar world above ground. The winner returns to the maze of tunnels at the centre of the colony, nuzzling each colony member she encounters. She is already in reproductive condition and before long will produce her first litter of pups, while other members of the colony will maintain and guard the network of foraging tunnels that the group depends on. Over the next months, the new queen will kill or evict all the other females that fought with her for the dominant position until her status is secure.

Some form of cooperation in protecting or rearing young is found in many social mammals. In many monogamous species, both adults contribute to protecting and provisioning their offspring (see Chapter 16). Where several adult females live and breed in stable groups (*plural breeders*), they often cooperate to defend the group's territory against intruders and to defend other individuals against predators, though they typically take care of their own young. Societies of this kind are common among macropods, bats and ungulates as well as in many social primates and carnivores (see Chapter 9). In some species, allo-parental care is more extensive and several breeding females share care of young born in the group and lactating mothers may suckle juveniles born to other females as well as their own offspring. Examples include a number of rodents, including prairie voles, tuco-tucos and degus, several bats (Wilkinson 1987; Solomon and Getz 1997; Hayes and Solomon 2004), a number of social carnivores,

including coatimundis, African lions, and several primates including mouse lemurs (Eberle and Kappelar 2006) and some colobine monkeys (see Figure 17.1) (Bertram 1976; Gompper *et al.* 1997; Xi *et al.* 2008; Pan *et al.* 2014). In these species, females may benefit by sharing care of their young because this allows them for forage, reduces their peak energetic load or increases the effectiveness of defence (see Chapter 9). Species of this kind are commonly referred to as *communal breeders* to distinguish them from *cooperative breeders*, where breeding females are assisted by non-breeding helpers. In many communal breeders, most group members are relatives (though average relatedness between group members is not high). The distribution of allo-parental care is positively correlated with levels of kinship between group members (Briga *et al.* 2012), though cooperation can also occur between non-relatives.

In cooperative breeders, parents are assisted in rearing young by subordinates of either or both sexes which either do not breed in the group or do so irregularly. In most species, a single pair of breeding adults are assisted by several helpers, most of which are their offspring, though groups may also include other natal kin and occasionally immigrants, too. The evolution of societies of this kind appears to have been confined to monogamous ancestors where a high proportion of helpers are the offspring of the breeding pair and most group members are closely related both to each other and the young they help to rear. In contrast to communal breeding systems, which appear to have evolved in polygynous species as well as in monogamous ones, the evolution of cooperative breeding appears to have been confined to monogamous ancestors where a high proportion of helpers are the offspring of the resident breeding pair and are full siblings (see Section 17.8).

In some cooperative breeders only a proportion of breeding groups include non-breeding natal 'helpers' and adult pairs often raise young with assistants. Systems

Figure 17.1 In snub-nosed monkeys (an Asiatic colobine that lives at high altitude), breeding groups consist of multiple single-male units and females from the same unit commonly nurse each other's infants. *Source*: © Zuofu Xiang.

of this kind are usually referred to as *facultative cooperative breeders*: well-studied examples include silver-backed jackals and European foxes (Moehlman 1979, 1989), marmosets and tamarins (Goldizen 1987a; French 1997) and some social rodents (Bennett *et al.* 1994; Solomon and French 1997) (Figure 17.2). In some cases, breeding groups include more than one adult male, as in some of the marmosets and tamarins (French 1997; Digby *et al.* 2007; Díaz-Muñoz 2011) and some groups of meerkats (Clutton-Brock and Manser 2015) and wild dogs (Creel and Creel 2002), though a single male often fathers most offspring born to the dominant female. Multi-male groups may develop where several breeding males have dispersed together from their natal group and have either established a new breeding group or ousted the resident males from an established group (see Chapter 14).

In a smaller number of species, breeding groups are larger and include multiple mature subordinates of one or both sexes who assist a pair of dominant breeders to raise their young. In many of these societies, helpers play a crucial role in raising young and pairs of adults are seldom able to rear young successfully on their own; as a result, they are commonly referred to as *obligate cooperative breeders*. The form of assistance provided by helpers varies. In addition to contributing to defence against predators or

rival groups, helpers may brood young and play an important role in maintaining their temperatures during the winter months, as in some marmots (Arnold 1990a,b) or they may help to maintain foraging tunnels or food stores used by other group members, as in beavers (Sun 2003) and social mole rats (Bennett and Faulkes 2000). In some of the cooperative carnivores (including dwarf mongooses, meerkats and African wild dogs) helpers provision juveniles directly and guard them at the breeding den when their mother forages (Figure 17.3) and female helpers also contribute to suckling young born to the dominant female (Moehlman and Hofer 1997; Creel and Creel 2002; Russell 2004; MacLeod *et al.* 2013). In meerkats, they may also assist dominants by acting as sentinels when the group is foraging (Figure 17.4), renovating sleeping burrows (Figure 17.5), defending the group's range against intruders (Figure 17.6) and mobbing or attacking potential predators (Figure 17.7).

Most mammals that breed cooperatively are *singular breeders*: a single female produces most of the young that are raised successfully, though subordinate females occasionally attempt to breed and are sometimes successful. In many of these species, a single dominant male in each group monopolises reproductive access to the dominant female and fathers almost all her young. In contrast,

(a)

(b)

(c)

(d)

Figure 17.2 Four cooperative breeders: (a) silver-backed jackals; (b) black tufted-ear marmosets; (c) Damaraland mole rats; (d) banded mongooses. *Sources*: (a) © Patricia Moehlman; (b) © Jeffrey French, Callitrichid Research Center, University of Nebraska at Omaha; (c) © Markus Zöttl; (d) © Tim Clutton-Brock.

subordinate females (who are often the daughters of the dominant male and are seldom closely guarded by him) often mate with roving males from other groups. For example, in meerkats one dominant male fathers over 90% of young born to the dominant female in his group, while offspring born to subordinate females are commonly fathered by roving males from other groups (Griffin *et al.* 2003; Spong *et al.* 2008). However, in a few species, multiple individuals of both sexes breed regularly. In some marmosets and tamarins, groups include more than one breeding adult of either or both sexes as well as non-breeding helpers (Nievergelt *et al.* 2000; Heymann, 2000; Sousa *et al.* 2005; Huck *et al.* 2005) while in banded mongooses, multiple females breed, reproductive skew among females is relatively low and offspring are reared by all group members, though subordinate males play a

leading role (Cant *et al.* 2013). Groups of this kind are sometimes referred to as plural *cooperative breeders*.

Patterns of philopatry, dispersal and kinship also vary. As in most other social mammals, breeding females have usually either inherited the breeding role in their natal group or dispersed and founded a new breeding unit and female immigration into established breeding groups is uncommon, though it occurs in some species, including wild dogs (Creel and Creel 2002) and banded mongooses (Cant *et al.* 2001, 2013). Males are typically immigrants, though both sexes can remain and breed in their natal group in some species, including banded mongooses and naked mole rats (Bennett and Faulkes 2000; Cant *et al.* 2013). However, without long-term records of the life histories of individuals, it is easy to misinterpret the nature of these breeding systems. For example, in meerkats, natal

Figure 17.3 In meerkats, helpers of either sex guard dependent pups at the breeding burrow against predators and neighbouring groups for a day at a time while the rest of their group is away foraging. Babysitters are often unable to feed for 24 hours and suffer appreciable weight loss and rarely guard pups on successive days. *Source*: © Tim Clutton-Brock.

Figure 17.4 A meerkat helper on sentry duty. *Source*: © Tim Clutton-Brock.

males can acquire the dominant position temporarily in their natal group following the death of the previous dominant male but they do not guard the breeding female, rarely breed within the group and usually disperse to search for breeding opportunities elsewhere at the start of the next breeding season (Clutton-Brock and Manser 2015).

The breeding systems of obligate cooperative breeders resemble those of primitively eusocial insects where all individuals are capable of breeding. Although these systems are sometimes referred to as eusocial, there is disagreement over the distinction between cooperative breeding and eusociality. E.O. Wilson defined eusocial societies as showing a 'reproductive division of labour with more or less sterile individuals working on behalf of individuals engaged in reproduction' (Wilson 1975) but, depending on the definition of sterility, Wilson's definition can either be used to include or exclude cooperative vertebrates. While some workers favour restricting eusociality to animals where helpers are irreversibly sterile and breeders show a permanent modification for their role (Crespi and Yanega 1995), others argue that many vertebrate species should be classified as eusocial and that different species can be arranged on a continuum of eusociality based either on the degree of reproductive skew or on their reliance on cooperation (Gadagkar 1994; Sherman *et al.* 1995). Both positions have points to recommend them. As Sherman argues, there is a continuum in the development of allo-parental care and in the reliance of breeders on helpers, which extends from communal breeders and facultatively cooperative species through obligate cooperative breeders to eusocial animals where colonies commonly include many thousand individuals and there are irreversible differences in anatomy and behaviour between breeders and workers. However, there are also important differences between the most specialised cooperative vertebrates and the societies of eusocial invertebrates (Crespi and Yanega 1995). In particular, colony size is far smaller in cooperative vertebrates, with the result that the selection pressures operating on breeders and helpers do not diverge to the same extent (Bourke 1999) and, after age effects have been allowed for, physical differences between them are usually small (see Section 17.8).

As the definition of eusociality matters less than the recognition of the similarities and contrasts between social invertebrates and cooperative vertebrates, I avoid using the term 'eusocial', though I regard all cooperative and eusocial societies as falling along a continuum based

Figure 17.5 Meerkats renovating a sleeping burrow. *Source*: © Kerri-Lynn Roelofse-Roodt.

on their reliance on cooperative rearing of young (Clutton-Brock 2009). This seems preferable to a continuum based on reproductive skew, since some social species that do not show extensive allo-parental care show high skew as a result of intense competition between females, while some specialised cooperative breeders with non-breeding helpers show relatively low skew.

Cooperative and eusocial breeding systems are of particular interest to evolutionary biologists because the behaviour of subordinates appears likely to reduce their direct fitness. Four principal explanations for the maintenance of reproductive cooperation have been suggested (Lehmann and Keller 2006; Bergmüller *et al.* 2007). First, that cooperative behaviour is maintained

Figure 17.6 Tails up, a meerkat group charges a neighbouring group that is intruding into its territory. *Source*: © Rob Sutcliffe.

Figure 17.7 A meerkat group mobs a Cape cobra. *Source*: © Tim Clutton-Brock.

because it increases an individual's prestige in its group by demonstrating its capacity to engage in costly activities that benefit others and increases its mating success or social status (Zahavi 1995). Second, that where individuals benefit by remaining in their natal group, cooperative behaviour is maintained by enforcement or coercion as a result of aggression directed at 'lazy' subordinates, a theory that is sometimes referred to as 'pay to stay' (Gaston 1977; Frank 1995; Kokko *et al.* 2002; Bergmüller and Taborsky 2005; Bergmüller *et al.* 2005; Cant and Johnstone 2009). Third, that cooperative behaviour generates direct mutualistic benefits shared by all group members and is maintained either by direct fitness benefits to individuals as a result of generalised reciprocity or group augmentation or by some form of group selection (Kokko *et al.* 2001; Nowak 2006; Wilson and Wilson 2007). And fourth, that cooperative behaviour benefits descendent and non-descendent kin and is maintained by kin selection and indirect fitness benefits (Hamilton 1964; Brown 1987; Emlen 1991). A counter-argument here is that competition with related group members can also offset the benefits of associating with kin (West *et al.* 2002; Lehmann and Rousset 2010).

These four explanations of the evolution of cooperative breeding are non-exclusive, and more than one evolutionary mechanism may often be involved. To assess their relative importance, it is necessary to understand the reasons why subordinates of one or both sexes remain in their natal groups after reaching sexual maturity, why they do not breed and what the consequences of cooperative behaviour are. This chapter deals with each of these three issues in turn. Section 17.2 explores the reasons for delayed dispersal in cooperative breeders, section 17.3 reviews the reasons why subordinate females usually fail to breed successfully and section 17.4 examines the reasons for contrasts in the distribution of breeding success and the extent of reproductive skew. Subsequently, section 17.5 reviews the effects of helpers on other group members and the costs of helping, while sections 17.6 and 17.7 describe the division of labour between helpers and the mechanisms regulating the workloads of individuals. Finally, section 17.8 assesses different explanations of the evolution of cooperative breeding and section 17.9 describes some of its evolutionary and ecological consequences.

17.2 Delayed dispersal

In most cooperative breeders, all female helpers and most males are individuals that have not yet left their natal group. In singular breeders, where one dominant female and one dominant male monopolise reproduction, this raises the question why subordinates should delay dispersing, since delays are likely to shorten their potential breeding lifespans (Solomon and French 1997). As in other social species, subordinates may gain several different benefits by deferring dispersal (see Chapters 3 and 12). In some cases, they may be waiting until reproductive opportunities arise in neighbouring groups: studies of cooperative birds show that the availability of vacant breeding territories or positions in neighbouring groups has an important effect on the probability that individuals of both sexes will leave their 'natal' groups (Koenig and Mumme 1987; Kokko and Ekman 2002). Habitat saturation is probably an important cause of delayed dispersal in cooperative mammals too (Creel and Macdonald 1995; Solomon and Getz 1997). For example, prairie voles living at high population density are less likely to disperse from their natal group than those living at low density (McGuire et al. 1993; Getz et al. 1994) and experimental reduction of density in neighbouring areas increases female rates of dispersal (Lucia et al. 2008).

As in other societies, dispersal entails substantial risks: not only are dispersing animals likely to be attacked by territorial groups but they may be more likely to suffer higher rates of predation because they are unfamiliar with the distribution of refuges. For example, in dwarf mongooses, at least 50% of males and 78% of females that disperse do not survive (Waser et al. 1994; Creel and Waser 1997). Until subordinates have stopped increasing in weight, fighting ability or foraging skills, remaining in their natal group may help to reduce the risks associated with dispersal (Clutton-Brock 2006; Clutton-Brock and Manser 2015). Subordinates may also gain substantial survival benefits by remaining in established breeding groups (Clutton-Brock et al. 1999a,b; Kokko and Ekman 2002) and large groups are also better able to protect their ranges against neighbours more effectively than small groups and so are less likely to be displaced by competitors (Clutton-Brock 2009).

Where the costs of dispersal are high, a female's best chance of surviving, acquiring alpha status and breeding successfully is often to remain in her natal group and to queue for the breeding position (Solomon and French 1997; Kokko and Johnstone 1999; Stephens et al. 2005). Females that inherit the breeding role in established groups are also likely to have higher breeding success than those that establish new breeding groups because established groups are usually larger and increases in group size reduce their contributions to rearing young and raise their breeding success (see section 17.5). Finally, where relatives assist each other, subordinates can gain indirect fitness by remaining in groups that include close relatives and estimates of these benefits show that effects of this kind can be substantial and could play an important part in deciding dispersal decisions (see Section 17.5).

Recent studies have begun to investigate the ecological and social factors affecting the rates of dispersal in cooperative breeders. In some cases, the probability of dispersal rises during or just before the breeding season: for example, in meerkats, both male and female dispersal peaks during the early stages of the breeding season and is associated with increases in rainfall. Subordinate females are often evicted at the start of the breeding season and many males disperse voluntarily at the same time (Mares et al. 2014; Clutton-Brock and Manser 2015). In other cases, dispersal varies in relation to the benefits of remaining in the breeding group or the chances of founding a new group. For example, in marmots, delayed dispersal is more common when the thermodynamic benefits of communal hibernation are high (Armitage 1981, 1999; Blumstein and Armitage 1999). Similarly, several studies of singular breeders suggest that the probability that subordinates will leave their natal group rises as the number of potential competitors for the breeding role increases or their relative chance of acquiring dominant status declines (Clutton-Brock and Lukas 2011). However, not all trends are as expected. For example, in prairie voles, there is apparently no relationship between the probability that individuals will disperse and the number of potential competitors in the group (McGuire et al. 1993).

The direct fitness benefits that individual females gain from remaining in their natal groups and their reluctance to leave them has important implications for the costs of cooperative breeding and the evolution of reproductive skew. Where females maximise their direct fitness by remaining in their natal group, the costs of contributing to cooperative activities should be restricted to the marginal costs of cooperation and should not include the costs of remaining in their natal group and 'forgoing' reproduction. Where individuals are likely to maximise their direct fitness by remaining in their natal group, dominants will not need to make reproductive concessions to subordinates to discourage them from dispersing (Clutton-Brock *et al.* 2001a).

17.3 Reproductive suppression

Costs and benefits

In singular cooperative breeders, subordinate females seldom rear young successfully because dominant females prevent them from conceiving, bearing or rearing young. A likely explanation is that competition between any pups born to subordinates and the dominant female's own pups would reduce the breeding success of dominants and research on meerkats supports this: the growth of pups born to dominants that are rearing pups at the same time as subordinate females is depressed if pups being reared by subordinates are older than those being reared by dominants and these effects increase with the size of the subordinate litter (Figure 17.8) (Hodge *et al.* 2008; Clutton-Brock and Manser 2015). In addition, when all subordinates in some groups were prevented from breeding by contraceptive injections, the foraging success and weight gain of dominants increased and they produced heavier pups, which grew faster (Bell *et al.* 2014) (Figure 17.9).

Subordinate breeding also appears to have costs in banded mongooses, and the post-emergence survival of pups born to dominants is lower when subordinate females breed at the same time (Cant *et al.* 2010, 2014).

While suppressing subordinate reproduction may increase the fitness of dominants, it is likely to have both direct and indirect costs to their fitness. In banded mongooses, mothers that evict subordinates during the course of reproductive events produce lighter pups with lower chances of survival than mothers not involved in evicting other group members (Bell *et al.* 2012) and long-term studies show that pup weight has important effects on the competitive ability and fitness of pups (Hodge *et al.* 2009). Suppressing subordinates is also likely to have indirect fitness costs to dominants, since subordinate females are often close relatives of dominants. As a result, dominants should be most likely to suppress subordinate reproduction at times when the costs of subordinate breeding to dominants are large or the costs of suppression are low. Several studies suggest that this is often the case: for example, in meerkats, suppression of subordinate reproduction increases when dominants are about to breed, and when group size is already large (Clutton-Brock *et al.* 2010). Similar effects occur in banded mongooses, where the extent to which subordinates are suppressed is inversely related to resource availability (Nichols *et al.* 2012).

Fertility suppression

The mechanisms involved in the suppression of subordinate reproduction probably vary between species. In some, subordinate females appear to be unable to conceive. For example, in some singular breeders, subordinate females have lower levels of luteinising hormone (LH), oestrogen or testosterone than dominants, either throughout the breeding season or over the period of oestrus (French 1997; O'Riain *et al.* 2000; Creel and Creel 2002; Carlson *et al.* 2004; Young 2009). Low levels of oestrogen or testosterone combined with reduced levels of LH may indicate that the pituitary is inactive, with the result that the secretion of gonadotrophin-releasing hormone (GnRH), which controls LH production and is produced in the hypothalamus, is low. Whether this is the case can be investigated by 'challenging' individuals with injections of GnRH to see whether this raises levels of LH and leads to increases in levels of circulating oestrogen or testosterone, indicating that the pituitary is active. In meerkats, non-breeding females show

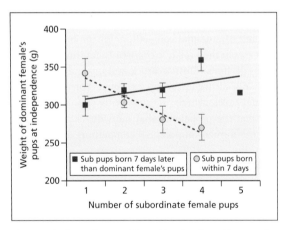

Figure 17.8 Costs of competition with subordinate litters to the growth of meerkat pups born to dominant mothers. The figure shows the weight at independence (3 months) of pups born to dominant females (plotted) on the number of competing subordinate pups, when the subordinate female's pups were born within seven days of the dominant female's pups (open circle, dashed line) versus more than seven days after the dominant female's pups (dark square, solid line). *Source*: From Sarah Hodge.

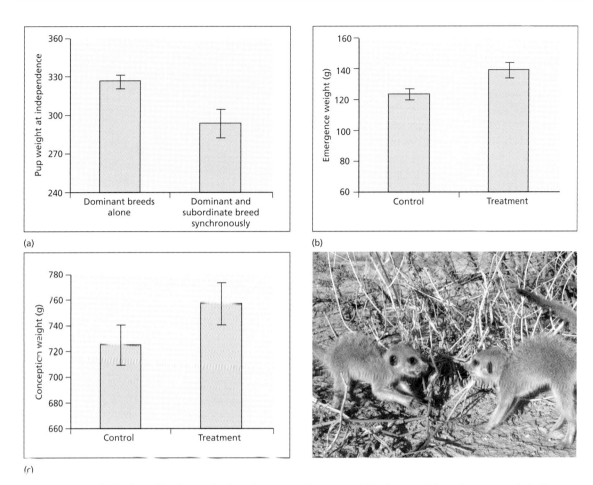

Figure 17.9 Costs of subordinate breeding to dominant female meerkats. (a) Weight of pups at independence (3 months) when dominant females bred alone versus times when at least one other female bred at the same time. (b) Weight of pups at emergence (about 21 days) in groups where all females apart from the dominant female were experimentally prevented from breeding by contraceptive injections (treatment) versus control groups. (c) Weight at conception of dominant females in groups where all other females were prevented from breeding by contraceptive injections (treatment) versus control groups where subordinate females could attempt to breed. *Source:* From Clutton-Brock *et al.* (2010). Reproduced with permission from the University of Chicago Press and T. Clutton-Brock. Plots (b) and (c) from Bell *et al.* (2014). Reproduced under the terms of the Creative Commons Attribution 4.0 licence. CC-BY-4.0. *Photo source:* © Sophie Bell.

lower levels of circulating GnRH than dominants, but when challenged with GnRH they show levels of LH as high as those of dominants, indicating that their pituitaries are active (O'Riain *et al.* 2000). In contrast, in naked mole rats and Damaraland mole rats, differences in LH levels between non-breeding females and breeders persist when individuals are challenged with GnRH and are associated with reductions in the mass of ovaries and with the virtual absence of corpora lutea, indicating a deeper level of suppression (Faulkes and Abbott 1997).

Even where reproductive activity in helpers is relatively deeply suppressed, females can respond quite rapidly to reproductive opportunities. For example, in naked mole rats, repeated hourly doses of GnRH lead to similar LH levels in breeders and non-breeders within 4 hours and non-breeding females that are removed from the colony and paired with males are ready to breed within 8 days (Faulkes *et al.* 1990). In some cases, the extent of suppression varies throughout the year: for example, in Damaraland mole rats, subordinate females show greater sensitivity to GnRH challenges during the rainy season, when breeding usually occurs (Young *et al.* 2010).

Reproductive suppression in subordinates is often attributed to increased levels of glucocorticoids associated with regular aggression directed at them by dominant females (Christian and Davis 1964; Reyer *et al.* 1986; Carter and Roberts 1997; Pottinger and Carrick 2001; Beehner and Lu 2013). In several cooperative breeders, the presence of dominant breeders is sufficient to suppress reproduction in subordinates. For example, in naked mole rats, ovarian function in subordinate females is depressed by the presence of dominant females and is restored if they are removed (Faulkes and Abbott 1997) and a similar effect occurs in cotton-top tamarins (French *et al.* 1984; Widowski *et al.* 1990) (Figure 17.10). If unrelated female pine voles are caged together with an unrelated male, only one female reproduces and the second female commonly dies, apparently as a result of high levels of social stress (Firestone *et al.* 1991). However, reproductive suppression can also occur in the apparent absence of repeated aggression and the relative importance of different processes is still debated (Young 2009; Creel *et al.* 2013). For example, in common marmosets, subordinate females rarely breed, though they receive little aggression from dominants (Saltzman *et al.* 2009). Moreover, several studies of singular breeders have found that dominant females have higher average levels of cortisol than subordinates, which may indicate that glucocorticoid levels are not directly involved in fertility suppression (Creel 2001; Creel and Creel 2002; Creel *et al.* 2013), though an alternative possibility is that subordinates which become the target of aggression from dominants temporarily show very high cortisol levels and that peak (rather than average) cortisol levels are responsible for suppressing reproductive activity in subordinates (Young 2009).

Several other processes may contribute to the suppression of fertility in helpers. Studies of birds show that cooperative breeding is sometimes associated with reliance on complex foraging skills that are slow to develop and are associated with prolonged dependence of juveniles on adults (Du Plessis *et al.* 1995; Du Plessis 2004). Female mammals often have a weight or age threshold below which conception does not occur (Albon *et al.* 1983; Creel and Creel 2002), and in some cooperative societies a proportion of helpers fall below this threshold and so may unable to breed. In addition, some subordinates may lack access to unrelated breeding partners in the same group

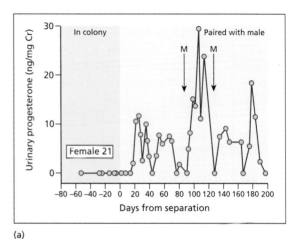

(a)

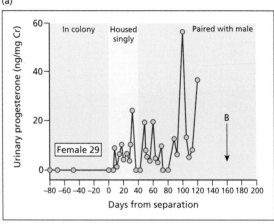

(b)

Figure 17.10 Effects on their progesterone levels of moving two subordinate naked mole rats from a colony which included a queen. *Source*: From Faulkes *et al.* (1990). Reproduced with permission of Bioscientifica.

and may consequently delay reproduction until unrelated males join the group (Greenwood 1980; Pusey 1987; Widowski *et al.* 1990; French 1997; Beehner and Lu 2013). For example, in Damaraland mole rats, experimental introduction of unrelated males to breeding groups causes rapid increases in oestrogen and LH levels, as well as in the frequency of breeding attempts by previously nonbreeding females (Cooney and Bennett 2000).

The presence of dominant females and the absence of unrelated males may often interact to affect the development of reproductive activity in subordinates (Abbott *et al.* 1990; Carter and Roberts 1997; French 1997). For example, both the presence of unrelated males and the absence of the dominant female are necessary to stimulate the onset of organised ovulatory cycles in young female tamarins (Widowski *et al.* 1990; French 1997). In prairie voles, a single drop of male urine on the nose of a female produces rapid changes in the neurotransmitter norepinephrine, as well as luteinising hormone-releasing hormone (LHRH), in the posterior olfactory bulb of the brain (Dluzen *et al.* 1981; Carter and Roberts 1997). Females exposed to male stimuli respond with increased uterine weight (Dluzen *et al.* 1981; Carter and Roberts 1997) but urinary pheromones from other females or the absence of unrelated partners can inhibit reproductive activation.

The mechanisms responsible for suppression of female fertility can also differ between age classes. In particular, the absence of unrelated males or the presence of the dominant female may be more likely to delay the age of sexual maturity in natal females than to suppress reproductive activity in females that have already gained sexual maturity or have attempted to breed (Carter and Roberts 1997; Solomon and French 1997; Solomon and Getz 1997). For example, physiological mechanisms appear to delay ovulation in female wolves that have not yet bred but, once a female has begun to cycle, anoestrus is rare and reproductive failure appears to be caused by failure to copulate (Packard *et al.* 1985). Similar contrasts may occur in meerkats, where sisters of the dominant female (who have commonly gained full sexual maturity and competed with her for alpha status following the death of their mother) more frequently become pregnant and attempt to rear pups than the dominant female's own daughters (Clutton-Brock *et al.* 2001b). Sisters are also more likely to challenge dominant females for alpha status, which probably explains why females that acquire alpha status commonly evict their sisters within a year of acquiring the dominant position (Clutton-Brock and Manser 2015). Similar patterns occur in a number of social insects where several females cooperate to establish colonies and rear workers but subsequently fight until only a single breeding female remains (Bernasconi and Strassmann 1999).

Induced abortion

When subordinates females conceive and attempt to breed, the behaviour of dominant females can prevent them from carrying their fetuses to term. In several cooperative breeders, subordinate females show higher rates of abortion than dominants: for example, in Alpine marmots, aggression directed at fertile subordinate females is associated with increased glucocorticoid levels and reduced levels of progesterone, which may prevent them from maintaining pregnancies. In meerkats, visibly pregnant subordinate females lose a higher proportion of their litters before they give birth (33%) than dominants (18%) (Clutton-Brock *et al.* 2008; Hodge *et al.* 2008) and even larger differences in rates of pregnancy loss between subordinates (64%) and dominants (15%) occur in golden lion tamarins (Henry *et al.* 2013). Estimates of the frequency of induced abortion probably underestimate its actual frequency, for most measures of litter loss relate to subordinate females that have reached the midpoint of gestation and are visibly pregnant and few studies have been able to assess rates of loss during the first half of gestation.

Aggression directed at female subordinates by dominant females (especially pregnant dominants) may often be responsible for the low rates of successful pregnancies in subordinate females. For example, during the second half of gestation, dominant female meerkats commonly evict subordinates from the group, especially if they are pregnant (Young *et al.* 2006) and evicted subordinates show substantial increases in cortisol levels (Young and Monfort 2009) and usually abort their litters (Clutton-Brock *et al.* 2008). In banded mongooses, subordinate females with heightened glucocorticoid levels in late pregnancy lose a relatively high proportion of their young (Sanderson *et al.* 2015). Dominant female callitrichids also direct frequent aggression at subordinates that attempt to breed, which may increase rates of abortion and of the frequency with which newborn infants are abandoned immediately after birth (Henry *et al.* 2013).

Female infanticide

In several cooperative breeders, dominants also kill a high proportion of any litters that subordinates produce. For example, in common marmosets, both dominant and subordinate females will kill each other's litters (Saltzman *et al.* 2009). Similarly, in dwarf mongooses, several females often conceive and give birth in the course of less than a week but litters born to subordinates rarely survive to emergence (Creel and Waser 1997; S.P. Sharp, personal communication). In meerkats, pregnant dominant females commonly kill pups born to subordinates (including their daughters) within 48 hours of their birth if they are pregnant (see Chapter 8); pregnant subordinate females will also kill litters born either to other subordinates or to dominants and the probability that litters born either to dominants or to subordinates will survive to 4 days after birth is reduced by the presence of other pregnant females in the group (Young and Clutton-Brock 2006). Female infanticide is also common in banded mongooses, where several females commonly breed synchronously (Cant *et al.* 2014). In this case, litters are particularly likely to die soon after birth if only a single female breeds, which may indicate that females kill litters which they can be sure do not contain any of their own offspring, whether or not they are pregnant themselves (Cant *et al.* 2014).

Eviction

The final tactic used by dominants to control subordinate reproduction is to evict them from the group (see Chapter 8). Evicting subordinates may not only remove potential competitors but may also discourage subordinates from becoming pregnant and attempting to breed if subordinates are able to defer eviction by avoiding breeding (Cant *et al.* 2010; Cant and Young 2013). In meerkats, pregnant subordinate females are more likely to be evicted by dominants than non-pregnant females (Young *et al.* 2008; Young 2009) though, in banded mongooses, the probability that subordinate females will become pregnant is unrelated to their risk of eviction (Cant *et al.* 2010, 2014).

The upper age at which dominant females tolerate subordinate females varies between species: in some cooperative breeders, subordinate females over a year old become the target of aggression from dominants and are forced to leave the group (Russell 2004; Clutton-Brock and Lukas 2011), while in others dominants will tolerate subordinates for several years but evict them before they reach an age when they attempt to breed on a regular basis. For example, in meerkats, virtually all females are forced to leave the group before they are 30–42 months old (Clutton-Brock and Manser 2015), an age which coincides with a marked increase in the frequency of breeding attempts by subordinates. In addition, dominants are more likely to evict subordinate females that are visibly pregnant than those that are not.

In some species, dominant females will tolerate subordinate females irrespective of their age so long as they do not attempt to breed or to challenge for the breeding role. For example, in naked mole rats, dominant females tolerate the presence of subordinates of all ages, but following a change in the identity of the dominant female, the new dominant subsequently kills or evicts any individuals that have competed with her for the

dominant role (Faulkes and Bennett 2007; P.W. Sherman, personal communication). In some cases, the age of subordinate females at eviction is variable. For example, in banded mongooses, dominant females intermittently evict subordinate females en masse from their group, although they may have previously bred there (Cant *et al.* 2001, 2010, 2013).

Although it is convenient to contrast the different mechanisms responsible for the suppression of reproduction in subordinates, they often interact and their effects can be difficult to distinguish. For example, both the absence of unrelated males and the presence of dominant females may delay the development of subordinate females and prolong the period before they attempt to breed for the first time (French 1997; Clutton-Brock *et al.* 2001a). Discussion of the evolution of facultative sterility in subordinates has often attempted to contrast explanations based on the ability of dominant females to prevent subordinates from breeding ('constraint' models) from theories based on the assumption that subordinates voluntarily terminate breeding ('restraint' models). However, in practice, reproductive restraint in subordinates may often be an evolutionary consequence of constraints imposed by dominants and the two processes may not be easily separable (Young 2009; Cant and Young 2013). For example, the widespread tendency for dominant females to kill offspring born to subordinates may favour the evolution of reproductive restraint by subordinates at times when an established dominant female is present in their group (Cant and Young 2013).

Subordinate counter-strategies

Subordinate females often attempt to evade reproductive suppression. In meerkats, older subordinate females that are at risk of eviction increase the frequency with which they attempt to groom the dominant female and frequently show submissive behaviour towards her (Kutsukake and Clutton-Brock 2006). While they are still in their natal group, subordinate females also attempt to breed at times when the dominant female's control is weak: for example, subordinate female meerkats often conceive and attempt to breed in the months immediately following the acquisition of the dominant position by a new female (Clutton-Brock *et al.* 2001a). In banded mongooses, where multiple females breed in each group, both dominant and subordinate females minimise the risk of infanticide by breeding in close synchrony with each other (Cant *et al.* 2014).

One important question is why, in most cooperative societies, subordinate females rarely challenge the breeding female or displace her. A likely answer is that challengers are unlikely to win contests with breeding females and unsuccessful challenges are likely to lead to the premature eviction of challengers, so that challenging is seldom their best strategy unless the dominant female is sick or wounded. However, an alternative possibility is that subordinates are less likely than dominants to breed effectively or to maintain their status and that as they are commonly offspring of the breeding pair and so are likely to be full sibs of their future offspring, they should exercise restraint and allow the existing dominant to maintain her position. Research on meerkats suggests that subordinates would usually maximise their inclusive fitness by replacing dominants, indicating that their reluctance to challenge dominants is a consequence of their low chance of success (Sharp and Clutton-Brock 2011). This is supported by evidence that, as in other social mammals, subordinate meerkats often challenge and displace dominants if they are sick or wounded.

Reproductive suppression in males

In most cooperative mammals, subordinate males show less evidence of physiological suppression than subordinate females (O'Riain et al. 2000; Young 2009). In meerkats, subordinate males show similar levels of testosterone and LH to those of dominants (O'Riain et al. 2000). In other species, there is evidence of reduced levels of androgens in subordinate males, though this varies with circumstances. For example, in African wild dogs, testosterone levels are not significantly related to dominance during non-mating period but alpha and beta males show higher testosterone levels than other males during mating periods (Creel and Creel 2002). In prairie voles, the testosterone levels of males caged together show an initial increase at around 45 days of age but then decline to low levels if no unrelated females are accessible, increasing again if males are exposed to unrelated females (Carter and Roberts 1997).

Evidence of more extensive suppression in males has been found in Alpine marmots, where dominant males direct more frequent aggression at resident males that are not their own sons than at individuals that they are likely to have fathered and the former show higher levels of corticosteroids and lower levels of androgens (Arnold and Dittami 1997). Among naked mole rats, too,

subordinate males have lower levels of urinary testosterone and reduced levels of plasma LH than breeders and show reduced responses to GnRH challenge (Faulkes and Abbott 1997), their testes are relatively small and their sperm concentrations are low (Faulkes and Abbott 1997). Unusually, their reproductive status appears to be controlled by the presence of the dominant female, for if breeding females are removed from colonies, both non-breeding and breeding males show increased levels of testosterone (Figure 17.11) (Faulkes and Abbott 1997). However, an alternative possibility is that the hormonal

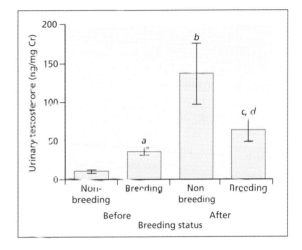

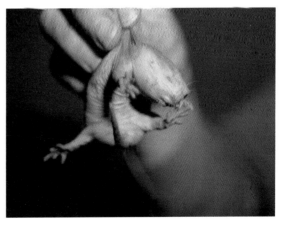

Figure 17.11 Effects of removing the dominant female on the hormonal status of male naked mole rats. The figure shows the mean concentrations of urinary testosterone in breeding and non-breeding males from a colony of 28, before and after removal of the queen. *Source*: From Faulkes and Abbott (1997). Reproduced with permission of Cambridge University Press. *Photo source*: © Tim Clutton-Brock.

changes in males that occur after the removal of dominant females are a response to rapid changes in the
hormonal status of subordinate females rather than to
the absence of dominant females.

17.4 Reproductive skew

Measures

While dominants often prevent subordinates from breeding in cooperative societies, the frequency of breeding by
subordinate females varies widely. In some cooperative
breeders, like naked mole rats, subordinate females
almost never breed while in others, like banded mongooses, most mature females attempt to breed each year
(Cant *et al.* 2013). The distribution of breeding success
within groups is commonly referred to as the degree of
reproductive skew and a number of different indices of
skew have been suggested (Keller and Vargo 1993; Tsuji
and Kasuya 2001). High-skew societies are often referred
to as 'despotic' while low-skew ones are referred to as
'egalitarian'.

Although measures of skew provide a useful measure
of the distribution of reproduction, they have important
limitations. Skew is typically assessed at a particular
point in time and measures of skew often ignore age-
related changes in breeding success. As a result, where
reproductive success varies with age, reproductive skew
measured over the lifetime of individuals may be much
lower than estimates of 'instantaneous' skew would
suggest (see Chapter 13). Another problem is that,
among singular breeders, the magnitude of skew
depends on group size. For example, in societies where
a single dominant female monopolises 100% of breeding, skew will increase as the number of helpers present
rises, though the distribution of breeding remains
unchanged. A third limitation is that measures of
skew can obscure important differences in the way in
which reproduction is distributed, which may help to
indicate the mechanisms involved. Box 17.1 gives some
examples of contrasting ways in which reproduction
might be distributed that could easily be obscured by
focusing on a single index.

Models

Three different groups of theoretical models based on
contrasting assumptions predict the distribution of reproductive skew (Johnstone 2000; Magrath *et al.* 2004;

Reeve and Shen 2013). First, dominant females may
be unable to control subordinate breeding because
they are physically incapable of doing so or the costs
of suppression outweigh the benefits (Reeve *et al.* 1998).
Under these conditions, the division of reproduction
between dominants and subordinates may be a consequence of a 'tug of war', whose outcome is affected by
asymmetries in the power of dominants and subordinates as well as by other variables, including their relatedness to each other (Cant 1998; Magrath *et al.* 2004).

A second group of models (sometimes called 'concession' or 'optimal skew' models) assume that dominants
can control subordinate reproduction by 'inducing' subordinates to remain in the group and assist with rearing
young without challenging them by allowing them to
share in reproduction (Vehrencamp 1979, 1983a,b).
These models predict that the 'incentives' or 'concessions' that dominants must give to subordinates to secure
their cooperation should increase (generating relatively
low levels of reproductive skew) when the subordinate is
more distantly related to the dominant because closely
related subordinates gain greater indirect benefits and so
need less incentive to stay (Vehrencamp 1983b; Reeve
and Ratnieks 1993; Keller and Reeve 1994; Johnstone
2000). Skew might also be expected to decline where
subordinates have more chance of dispersing and breeding successfully (because the level of concession necessary to exceed the pay-off of the subordinate's other
options increases) or where the subordinate has a small
effect on the group's productivity (because this lowers
the subordinate's capacity to increase its indirect fitness
by contributing to the fitness of related dominants and so
increases the share of direct reproduction necessary to
induce it to stay) (Magrath *et al.* 2004).

A third group of models assume that dominants cannot
control the frequency with which subordinates reproduce but can evict them if they claim too large a share of
reproduction, so that subordinates maximise their fitness
by restraining their attempts to breed (Johnstone and
Cant 1999). Under the assumptions of these 'restraint'
models, predictions about the subordinate's share of
reproduction are reversed because the factors that
make staying profitable for subordinates make association profitable for dominants and so increase the level of
subordinate reproduction that dominants will tolerate
(Magrath *et al.* 2004). These models are sometimes
grouped with 'concession' models and referred to as
'transactional' models since, like concession models,

Box 17.1 The distribution of reproduction within groups

It is useful to think of different ways in which reproduction might be distributed within groups. The six matrices shown in Figure 17.12 illustrate different ways in which breeding might be distributed within groups of four potential breeders. The first two matrices illustrate the extremes. In *alpha takes all*, a dominant individual monopolises breeding however many competitors there are in the group. Despite the similarity in the distribution of reproduction, measures of reproductive skew will be higher in groups of four than in groups of two. In contrast, in *egalitarian*, breeding is equally divided among all potential breeders. The other four matrices illustrate intermediate outcomes. In *alpha and beta take all*, two individuals monopolise breeding and other group members get no share, while in *limited dilution*, the share of the dominant animal falls from 100% to 60% as soon as another individual is present, but increasing numbers of subordinates only dilute the subordinate share and do not erode the success of the dominant. In contrast, *continuous dilution* illustrates a situation where the dominant's share is progressively eroded as the number of individuals present increases, while *erosion by sneakers* shows a case where the dominant individual retains a large share of reproduction but additional group members each obtain a small proportion (see Kokko and Johnstone 1999).

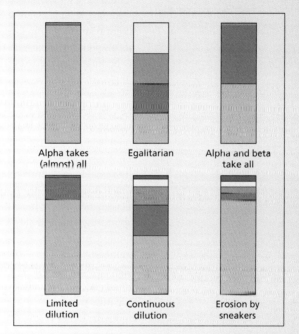

Figure 17.12 Different ways in which male breeding success might be distributed within groups: different colours show the relative breeding success of different individuals. Because of the differences in ways that male success can be distributed, single measures of reproductive skew may often obscure important contrasts.

they depend on the assumption that some form of transaction occurs between dominants and subordinates.

Together, 'concession' and 'restraint' models can be used to identify the boundaries of group stability as well as a zone of conflict in interactions among individuals (Reeve 2000). Figure 17.13 shows the minimum and maximum share of reproduction by subordinates under 'concession' and 'restraint' models as a function of relatedness to the dominant and the degree of ecological constraint on dispersal. The figure shows that, as relatedness (r) increases, the potential breeding success

of a dispersing subordinate (x) falls or the effect of a subordinate on group productivity (k) increases, the concessions that dominants need to make to subordinates to ensure group stability decline, while the maximum reproduction that a subordinate can claim without risking eviction increases. These predictions make intuitive sense (Magrath *et al.* 2004): cooperative groups are likely to be most stable when they consist of close relatives, when constraints on dispersal are strong and when individuals are much more productive when breeding together.

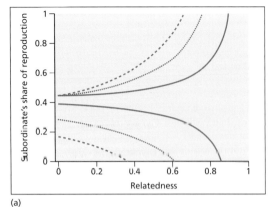

(a)

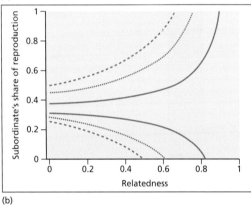

(b)

Figure 17.13 Boundaries of group stability and zones of conflict predicted by concession and restraint models of reproductive skew. (a) Combined effects of concession and restraint models, which delimit the range of sharing and still allow group stability. The figure shows the minimum (concession) and the maximum (restraint) share of reproduction that the subordinate can acquire in a stable group as a function of *r*, the relatedness between the dominant and subordinate. These lower and upper limits are shown for three different values of *x*, the expected reproductive success of a subordinate who disperses to breed independently: 0.7 (solid lines), 0.5 (dotted lines) and 0.3 (dashed lines), all relative to an established lone breeder. In all three cases, *k*, the productivity of the association relative to that of a lone breeder, is equal to 1.8. (b) Combined effects of concession and restraint models, which delimit the range of sharing still allowing group stability. The figure shows the minimum (concession) and the maximum (restraint) share of reproduction that the subordinate can acquire in a stable group as a function of the relatedness between dominant and subordinate. Lower and upper limits are shown for three different values of *k*, the productivity of the association relative to that of a lone breeder: 1.6 (solid lines), 1.8 (dotted lines), and 2.0 (dashed lines). In all three cases, the expected reproductive success of a subordinate who disperses to breed independently relative to an established lone breeder is equal to 0.5. *Source*: (a) Drawn from Johnstone (2000). (b) From Magrath *et al.* (2004).

Tests

Attempts to test the predictions of different skew models are complicated by the fact that similar trends are often predicted by different models. For example, several studies have shown that older subordinate females are both more likely to disperse and more likely to breed in their natal group, as concession models predict (Creel 1997; Creel and Creel 2002), but this could also be because older subordinates are more difficult to control and so are more likely both to breed in situ and to be evicted (Clutton-Brock *et al.* 2001a). Similarly, while several studies of cooperative animals have shown that helpers which are closely related to dominant females are less likely to breed than more distant relatives or unrelated individuals, as transactional models predict (Reeve *et al.* 1998), this could alternatively be because subordinates closely related to dominant females are often the offspring of her mate, so that they lack access to unrelated breeding partners and are more likely to be suppressed.

Evidence currently available suggests that the assumptions of limited control models apply more commonly than those of concession models: dominant females are often unable to control subordinates (as limited control models assume) and the frequency of subordinate reproduction increases where the benefits to dominants of suppressing subordinates are low or the costs are high (Clutton-Brock *et al.* 2001a; Haydock and Koenig 2002; Beekman *et al.* 2003). There is also little evidence that subordinate females that are allowed to breed are subsequently less likely to disperse or more likely to contribute to rearing the dominant's pups in any cooperative vertebrate (Beekman *et al.* 2003): for example, subordinate female meerkats that have bred are no less likely to disperse and no more likely to contribute heavily to rearing pups produced subsequently by the dominant females (Young 2003). Similarly, experimental manipulation of skew in female banded mongooses does not affect rates of eviction or dispersal and experiments provide no support for transactional models (Cant *et al.* 2014). Finally, there is little evidence that subordinate females that have bred are less likely to challenge the dominant and breeding might often be expected to have the opposite effect (Clutton-Brock *et al.* 2001a).

The limited support for transactional skew models in female mammals does not suggest that the strategies of dominants and subordinates do not coevolve. For example, by evicting subordinates that attempt to breed or killing their offspring, dominants may reduce the benefits of breeding attempts to subordinates, lowering the frequency with which they attempt to breed and reducing the costs of controlling them, enhancing their own fitness (Cant and Johnstone 2009; Cant *et al.* 2014). In some circumstances, dominants might even be expected to signal their intentions to evict subordinates or kill their pups if they attempt to breed in order to encourage them to show restraint, though, as yet, there is no evidence of threats of this kind.

To what extent coevolutionary processes exert an important effect on the evolution of reproductive strategies is often hard to assess for it is usually difficult to distinguish between benefits arising directly from the immediate suppression of subordinates and benefits that arise as a result of coevolutionary processes. For example, while it is possible that infanticidal behaviour in female meerkats and mongooses is maintained because it lowers the probability that subordinates will attempt to breed and so reduces the costs of control to dominants, another explanation is that infanticide is maintained by its direct effects in reducing the risk that the offspring of dominant females will have to compete for resources with pups produced by dominants.

Attempts to test the predictions of alternative models of reproductive skew with intraspecific comparisons have tended to divert attention from the distribution of interspecific differences in reproductive skew. Here, recent reviews have demonstrated that, in most mammals that breed cooperatively, either a single female monopolises effective reproduction almost completely or most mature females attempt to breed each year (Lukas and Clutton-Brock 2012). Systems where several mature females breed regularly while others are prevented from breeding are rare, although they occur in some cooperative birds where breeding colonies include several breeding females as well as multiple non-breeders (Cockburn 1998; Vehrencamp and Quinn 2004).

If dominant females gain important benefits by suppressing subordinate reproduction in cooperative societies, why are some cooperative species, like banded mongooses, plural breeders? As yet, it is difficult to answer this question with any certainty since plural cooperative breeders are uncommon among mammals and few have been extensively studied. A possible answer is that singular breeding is associated with strong competition for the resources necessary to support developing juveniles and plural breeding occurs where these are relaxed. For example, among the callitrichid monkeys, reproductive suppression is more highly developed in the marmosets, tamarins and pygmy marmosets of the genera *Callithrix*, *Cebuella* and *Saguinus* where the energetic costs of reproduction are high and the need for alloparental care is greater than in the lion tamarins and Goeldi's monkey (*Leontopithecus* and *Callimico*) where the relative costs of rearing young are lower (Figure 17.14) (French 1997; Diaz-Munoz 2015). Among the mongooses, reproductive

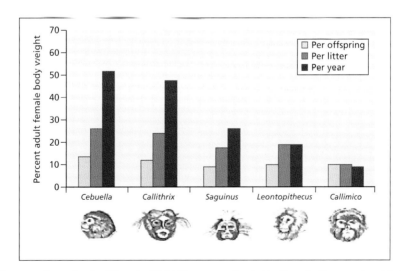

Figure 17.14 Relative costs of gestation in different callitrichid genera, estimated by the ratio (in percentages) between neonate weight and adult female body weight. Estimates are shown per offspring, per litter (offspring × modal litter size) and per annum (litter weight × litters per year). *Source*: From French (1997). Reproduced with permission from Cambridge University Press.

Figure 17.15 In banded mongooses, most pups are fed by dedicated helpers (escorts) that associate with and provision particular pups. Most escorts are subordinate males. *Source*: © Jennifer Sanderson.

suppression and singular breeding occur in species that live in relatively arid environments, like dwarf mongooses and meerkats, while plural breeders, like banded mongooses, occur in more mesic habitats, where food is more abundant and competition for resources may be less intense (Rood 1986; Cant *et al.* 2013) (Figure 17.15). Contrasts in habitat and food availability may also affect the costs of subordinate reproduction to dominants: for example, the pups of banded mongooses are fed by dedicated 'escorts' who are able to provide almost all the food necessary to maintain a single pup; because pups born to dominant mothers obtain priority of access to escorts, dominants might benefit little from attempting to suppress subordinates (Cant *et al.* 2015). Another possibility is that the more abundant food supplies of banded mongooses allow all females to synchronise their breeding cycles, breed in close synchrony and so minimise the risks of female infanticide by confusing the maternity of pups born in the group, while similar levels of synchrony may not be feasible in more arid habitats because large individual differences in condition between females prevent the development of breeding synchrony.

17.5 Benefits and costs of helping

Effects of helpers

While delayed dispersal combined with the suppression of subordinate reproduction by dominants facilitates the evolution of reproductive cooperation, there are many singular breeders that do not rear young cooperatively (see Chapter 2), so additional selection pressures are presumably necessary for cooperative breeding to evolve. To understand the evolution of cooperative breeding, it is obviously important to assess the effects of helpers. Positive correlations between breeding success and helper number have been demonstrated in several cooperative breeders, including silver-backed jackals (Moehlman 1979) (Figure 17.16), meerkats (Clutton-Brock and Manser 2015), dwarf mongooses (Rood 1990) and wild dogs (Creel and Creel 2002), though they are not universal. Improved breeding success in larger groups is frequently associated with increases in food intake, growth and survival of dependent young. For example, in meerkats, increases in the ratio of helpers to pups are associated with increased food intake by pups and increased daily weight gain (Figure 17.17). Increases in the food intake of dependent young are, in turn, associated with relatively high ratios of helpers to juveniles. For example, experiments involving the temporary removal of meerkat helpers or the temporary addition of pups from other litters confirm that increases in helper/pup ratios cause increased daily weight gain in pups (Figure 17.18). Subsequent monitoring of the development of pups showed that natural increases in the daily weight gain of pups are associated with increases in their foraging success and their daily weight gain and survival as juveniles and adolescents, as

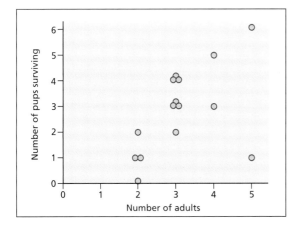

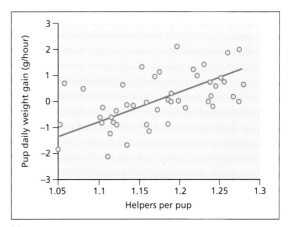

(a)

Figure 17.16 Group size and breeding success in silver-backed jackals in groups containing different numbers of adults. *Source*: From Moehlman (1979). Reprinted with permission of Macmillan Publishers Ltd. *Photo source*: © Patricia Moehlman

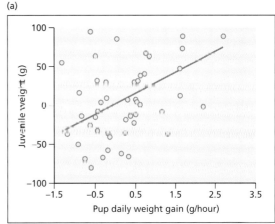

(b)

Figure 17.17 Helper number and pup growth in wild meerkats: (a) correlation between helper/pup ratios in different groups and the daily weight gain (g/hour) of pups aged 35–75 days; (b) correlation between daily weight gain and weight of pups at 3 months. *Source*: From Clutton-Brock *et al.* (2001c). Reproduced with permission of AAAS.

well as with their weight as adults (Clutton-Brock *et al.* 2001c) and individuals that are heavy as pups are also more likely to breed as subordinates and to acquire dominant positions as adults (Russell *et al.* 2007), so that their lifetime reproductive success is higher than that of individuals that are light as pups. Experiments in which samples of pups were artificially provisioned confirmed that early growth affects subsequent survival and breeding success (Russell *et al.* 2007).

Helpers can also increase the reproductive output of dominant females. For example, in meerkats, increases in helper number are associated with reductions in the workload of dominant females and their weight loss during the course of breeding attempts, as well as with

reductions in the intervals between their breeding events and increased frequency of breeding attempts (Russell *et al.* 2003a; Hodge *et al.* 2008). Similar relationships between helper number and the workload, condition and breeding success of dominant females have been documented in several other cooperative mammals (Solomon 1991; Powell and Fried 1992; Sanchez *et al.* 1999; Bales *et al.* 2001) as well as in cooperative birds (Langen 2000).

The presence of multiple helpers may also contribute to the fitness of all group members. For example, in meerkats, the efficiency of vigilance increases as group

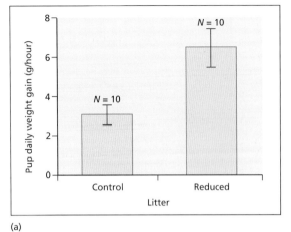

(a)

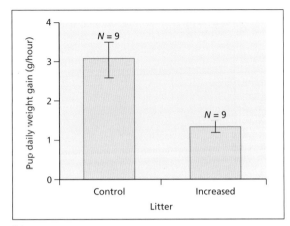

(b)

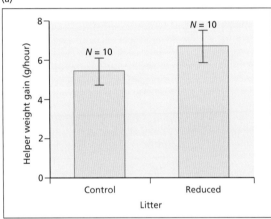

(c)

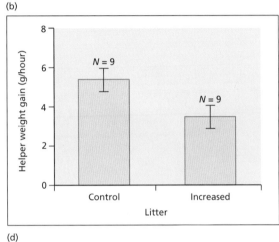

(d)

Figure 17.18 Effects of manipulating helper/pup ratios on daily weight gain in meerkat pups and helpers. In this experiment, helper/pup ratios were either increased by temporarily removing 75% of pups (litter reduced, a) or lowered by temporarily increasing pup numbers by 75% (litter increased, b). Control values were mean measures of daily weight gain in pups and helpers in the same group within 2 days of the experiment. (c, d) Effects of the two treatments on the daily weight gain of helpers. *Source*: From Clutton-Brock *et al.* (2001c). Reproduced with permission of AAAS.
Photo source: © Tim Clutton-Brock.

size rises and mortality falls (Clutton-Brock *et al.* 1999a, b). Large groups are also better able to repel intruding neighbours and to maintain the size of their range (see Chapter 2). In addition, large groups produce larger 'splinter' bands of dispersers, which are able to feed more effectively after leaving their natal group and have a better chance of establishing themselves as new breeding groups than smaller splinters (Young 2003) (see Chapter 3). Finally, large breeding groups are more likely to survive years when rainfall is low and reproduction ceases and to maintain their breeding territories and recover, whereas a high proportion of small groups become extinct during unusually dry years (Clutton-Brock *et al.* 1999c). However, increases in the number of helpers can also generate increased competition for resources and reduce the survival of adults. For example, in African wild dogs, increases in pack size are associated with improvements in pup survival but with a decline in the survival of adults (Creel and Creel 2015).

Average group size across cooperative breeders varies from less than ten to over fifty individuals and these differences appear to be associated with contrasts in the effects of helper number on the fitness of breeders. In some species, the presence of one or two helpers has beneficial effects on the fitness of group members but further increases in helper number either have no further effect or reduce breeding success, suggesting that there is an optimal number of helpers that maximises the breeding success of group members (Goldizen 1987a,b; Komdeur 1994a,b; Allainé and Theuriau 2004). In others, breeding success continues to increase until group size reaches higher levels: for example, in meerkats, the growth and survival of pups continues to increase until group size exceeds twenty and then declines, while the average breeding success of dominants rises throughout the observed range of group size (Clutton-Brock *et al.* 2008). Negative effects of increasing group size on breeding success or survival may eventually constrain group size (Bateman *et al.* 2012; Ozgul *et al.* 2014). Alternatively, group size may be limited by dispersal rather than by reductions in breeding success where the probability of a subordinate female acquiring alpha status is inversely related to the number of older individuals ahead of her in the breeding 'queue' (Kokko and Johnstone 1999), stimulating subordinate females to disperse before groups reach a size at which competition for resources limits breeding success (see section 17.9).

Costs of helping

In many cooperative mammals (as well as in cooperative birds), helping has substantial energetic costs (Heinsohn and Legge 1999; Heinsohn 2004; Russell 2004). For example, in marmosets and tamarins, helpers that carry infants are less mobile, spend less time foraging and suffer reduced calorie intake (Goldizen 1987a,b; Price 1992; Tardif 1997; Schradin and Anzenberger 2001). In meerkats, helpers that babysit pups at the burrow lose

around 1–2% of their total body weight in the course of a day spent babysitting, while those that forage with the group gain around 5% in the course of the day (Clutton-Brock *et al.* 1998b). When dependent pups are foraging with the group, meerkat helpers feeding pups give away around 40% of all the food items they locate and tend to give pups a high proportion of the larger, more valuable items (Brotherton *et al.* 2001). Helpers show reduced growth at times when their group is breeding compared with times when it is not (Russell *et al.* 2003b) and more generous helpers with high levels of contributions show lower rates of growth than less generous ones and temporary reductions in helper/pup ratios are associated with reductions in the daily weight gain of helpers. More generous helpers do not show reduced survival, but generous females are less likely to breed as subordinates, though whether this is a result of the energetic costs of helping or occurs because subordinates that are not primed to breed spend more time helping is impossible to determine (Russell *et al.* 2003b).

Helpers adjust their behaviour to minimise the costs of helping. For example, in meerkats, the extent to which juvenile and sub-adult helpers contribute to cooperative activities increase with their body weight, while the contributions of adult helpers increase with their daily weight gain, though effects differ between the sexes (Figure 17.19). Similar increases in helping behaviour occur if helpers are experimentally provisioned (Figure 7.20). The level of contributions to breeding attempts also varies widely within individuals and animals that have invested heavily in one breeding attempt and are often in poor condition invest relatively little in the next (Russell *et al.* 2003b). Recent studies of other cooperative or communally breeding birds and mammals confirm that helping effort is usually conditional (Wright 1997, Wright *et al.* 2001; Nichols *et al.* 2012). Their results suggest that models of the evolution of cooperation which assume that helpers pay large fixed costs for assisting may be unrealistic and cooperation may be more easily maintained than they indicate (Clutton-Brock 2002).

Allo-lactation

In a number of cooperative carnivores, female helpers contribute to suckling young born to the dominant female (Moehlman and Hofer 1997; Creel and Creel 2002; Russell 2004). Allo-lactation is common in the mongooses (including meerkats, dwarf mongooses and

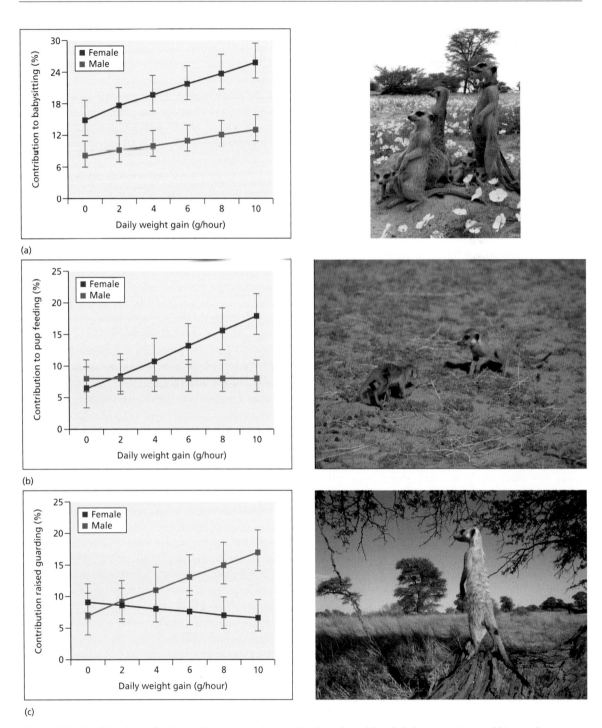

(a)

(b)

(c)

Figure 17.19 Conditional contributions to three cooperative activities by male and female helpers over 1 year old in meerkats. Figures show correlations between individual differences in the hourly weight gain of helpers between dawn and midday and their contributions to (a) babysitting, (b) pup feeding and (c) raised guarding. In females, daily weight gain affected contributions to babysitting and pup feeding but not raised guarding. In males, daily weight gain affected contributions to raised guarding and babysitting but not to pup feeding. *Source*: From Clutton-Brock *et al.* (2002). Reproduced with permission of AAAS. *Photo sources*: (a) © Helen Wade; (b) © Tim Clutton-Brock; (c) © Dave Bell.

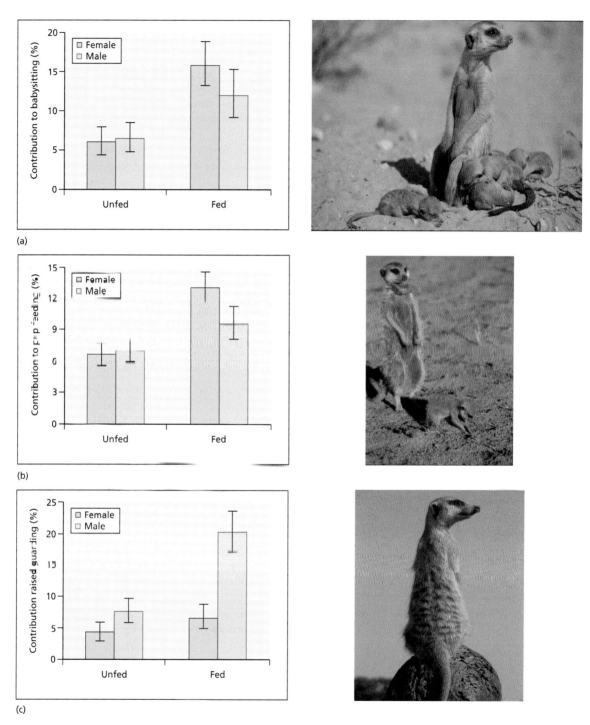

(a)

(b)

(c)

Figure 17.20 Effects of provisioning helpers on their contributions to (a) babysitting, (b) pup feeding and (c) raised guarding. Experimental feeding raised contributions by females to babysitting and pup feeding but not to raised guarding, and contributions by males to babysitting and raised guarding. *Source*: From Clutton-Brock *et al.* (2002). Reproduced with permission of AAAS. *Photo sources*: (a,b) © Tim Clutton-Brock; (c) © Kerry-Lynn Roelofse-Roodt.

handed mongooses) and in the cooperative canids
(where it occurs in wild dogs, wolves, jackals and foxes)
as well as in brown hyenas. In wolves and dwarf mongooses, allo-lactation may follow a pseudopregnancy
that culminates in pseudoparturition and lactation
(Seal *et al.* 1979; Creel *et al.* 1991), while in meerkats
allo-lactators are commonly females that have produced
and lost a litter shortly before the dominant gave birth
(often as a result of infanticide) (MacLeod *et al.* 2013).

Because of the high energetic costs of lactation in
breeding females (see Chapter 5), it is often assumed
that allo-lactation has substantial benefits to pups and
substantial costs to allo-lactators. However, few studies
have yet been able to measure either the benefits or the
costs of allo-lactation. In meerkats, where several subordinate females often contribute to nursing pups, allo-
lactors rarely suckle pups for as long as dominant
females, their milk production and weight loss are considerably lower than that of breeding females (Scantlebury *et al.* 2002; MacLeod *et al.* 2013), and the growth
and survival of pups is not significantly correlated with
the number of females that nurse them. In contrast to
breeding females, allo-lactators do not show increased
overnight weight loss or reduced contributions to other
cooperative activities, though (unlike subordinates of the
same age that do not contribute to allo-lactation) they do
not gain in weight during the course of breeding attempts
(MacLeod and Clutton-Brock 2014).

17.6 Division of labour

Age, size and sex

The distribution of individual contributions to cooperative activities can provide important insights into the
benefits and costs of helping. In all cooperative societies
that have been studied in detail, helpers vary widely in
the extent to which they contribute to rearing young
(Bergmüller *et al.* 2010) and much of this variation
appears to be related to contrasts in costs of helping or
in the benefits that individuals gain from assisting (Barclay and Reeve 2012). Much of the variation in workload
appears to be associated with differences in the age and
body size of helpers (Russell 2004; Clutton-Brock 2006):
the level of contributions initially rises with age, as the
helpers' strength and foraging success increases, and may
subsequently decline in individuals that will shortly
disperse (Owens and Owens 1984; Lacey and Sherman

1991; Tardif *et al.* 1992; Clutton-Brock *et al.* 2002). The
factors responsible for individual differences in the level
of contributions can also change with age: for example,
among meerkats, the workload of helpers less than a year
old increases with body weight rather than foraging
success, while among older helpers foraging success
has a stronger effect than body weight (Clutton-Brock
et al. 2001b).

In species where both sexes help, males and females
often differ in their overall level of contributions (Stacey
and Koenig 1990; Cockburn 1998; Clutton-Brock *et al.*
2002; Cant *et al.* 2013). Members of whichever sex usually
remain and breed in the group (the 'philopatric' sex)
commonly contribute more to rearing young than members of the 'dispersing' sex (Clutton-Brock *et al.* 2002). For
example, in brown hyenas and meerkats (where females
commonly remain and breed in their natal group) female
helpers typically contribute more to rearing young than
males (Owens and Owens 1984; Clutton-Brock *et al.*
2002). Similarly, in meerkats, subordinate females babysit
and feed pups more than subordinate males (Clutton-
Brock *et al.* 2002) and subordinate females are more
responsive than males to increases in begging rate (English
et al. 2008). In contrast, in African wild dogs (where males
often remain and breed in their natal group), males
generally contribute more than females (Malcolm and
Marten 1982) and the number of males in groups exerts
a stronger effect on the breeding success of the dominant
female than group size (McNutt and Silk 2008). In naked
mole rats (where either sex may remain and breed in their
natal group), there are also no obvious sex differences
in contributions to cooperative activities (Lacey and
Sherman 1991; Tardif *et al.* 1992; Schradin and Anzenberger 2001). Similar sex differences occur in cooperative
birds, where female helpers contribute more to rearing
young in species where females may remain and breed in
the group, like Seychelles warblers, while males generally
contribute more than females in species where males
are the philopatric sex (Cockburn 1998; Clutton-Brock
et al. 2002).

In some cooperative breeders, male and female helpers
also differ in their relative contributions to particular
activities. In several cooperative mammals, as well as
some cooperative birds, males tend to contribute disproportionately to sentinel duty while females contribute
disproportionately to guarding and feeding juveniles
(Cockburn 1998; Clutton-Brock *et al.* 2002; Ridley
2003). These differences are accentuated by variation

in food intake. For example, in meerkats (where female helpers contribute more to guarding and feeding pups than males but less to sentinel duty), supplementing the food intake of adult helpers increases contributions to pup feeding and guarding to a greater extent in females than in males (see Figure 17.20), while in banded mongooses (where male helpers contribute more to guarding and feeding pups than females) supplementing the food intake of helpers raises contributions to guarding and feeding pups in males but not females (Cant 2003; Hodge 2007).

Some studies suggest that helpers may discriminate between male and female pups. In evening bats, females are more likely to suckle unrelated female infants than unrelated male infants (Wilkinson 1992). In meerkats, female helpers (but not breeding females or males) show a consistent tendency to feed female pups more than males (Brotherton et al. 2001) and are also less likely to 'false-feed' females than males (Clutton-Brock et al. 2005), while in banded mongooses escorts are more responsive to increases in begging by female pups (Bell 2008).

Individual differences

Although differences in age and sex account for a substantial part of the variation in workload among helpers, large individual differences in workload remain after their effects have been allowed for (Clutton-Brock et al. 2000, 2010; Bergmüller et al. 2010). In some cases, these differences are associated with individual variation in foraging success: for example, in meerkats, individuals that are consistently successful foragers contribute more to cooperative activities than less successful ones and experimental feeding increases the proportion of naturally found food that helpers give to pups (Clutton-Brock et al. 2001b).

The relative contributions of helpers may also be affected by their early development. In meerkats, female pups that receive more food during early development contribute more to pup feeding and babysitting as adolescents than those that receive less food (Clutton-Brock et al. 2002; English et al. 2008). In prairie voles, the extent of allo-parental care varies between populations and these differences, too, may be associated with food availability (Roberts et al. 1998). Several studies suggest that prenatal exposure to sex hormones can affect the extent of parental or allo-parental care. For example, male gerbils positioned between females in the uterus subsequently show relatively high levels of oestrogen,

low testosterone and high levels of cooperative behaviour (Clark and Galef 2000). Similarly, both male and female prairie voles exposed to increased corticosterone levels before birth subsequently tended to spend more time in contact with infants, while postnatal treatment with corticosterone leads to reductions in infant contact (Carter and Roberts 1997). Quantitative genetic analyses based on a multi-generational pedigree show that individual differences to contributions to babysitting and pup feeding in meerkats are partly heritable and are correlated with the extent to which individuals are inbred (Nielsen 2013) and variation in cooperative behaviour has also been shown to be partly heritable in several birds (Charmantier et al. 2007).

Although individual differences in workload are common, there is limited evidence that individuals specialise in particular activities in most cooperative mammals. Early studies of naked mole rats described how smaller helpers were principally involved in the maintenance of tunnels and food collection while larger ones played an increased role in colony defence and speculated that the latter might be a soldier caste (Jarvis 1981). However, subsequent studies suggest that smaller helpers are younger and that contrasting contributions to different activities represent age-related polyethisms (Lacey and Sherman 1991, 1997). Similar age-related changes in cooperative behaviour are present in other species. For example, in meerkats, the youngest helpers contribute relatively little to babysitting (which involves a full day without food) as well as to sentinel duty (which requires experience of the dangers posed by different predators) and relatively more to feeding pups (Clutton-Brock et al. 2001b) while older helpers contribute more to group defence and mobbing predators (Clutton-Brock et al. 2001b; Clutton-Brock and Manser 2015). However, individual differences in contributions to different cooperative activities are positively correlated with each other across individuals and helpers of the same age vary in their overall level of generosity rather than in their relative contributions to different activities (Clutton-Brock et al. 2003).

Since competition for breeding roles is intense in singular cooperative breeders, individuals might be expected to adapt their development to the probability that they will (or will not) acquire a breeding role (Montiglio et al. 2013). There is some evidence of *social niche specialisation* of this kind in naked mole rats, where a small proportion of (mostly male) individuals adopt an alternative growth strategy and attempt to outbreed

(O'Riain *et al.* 1996). O'Riain (personal communication) characterises these potential dispersers as fat, lazy and promiscuous: they lay down more fat than other individuals (Figure 17.21), show elevated levels of LH (indicating that they are sexually primed before leaving the natal burrow) and participate little in cooperative behaviour or colony maintenance but, despite this, do not attract aggression from parents or siblings. They show a tendency to leave their colony and, unlike other individuals, only solicit matings with non-colony members. It is not yet known how dispersers leave their natal colony or whether they immigrate into neighbouring colonies or found new groups (O'Riain *et al.* 1996), nor is it clear whether these differences in growth have a genetic basis nor whether their unusual development is triggered by environmental conditions or by unusually rapid growth

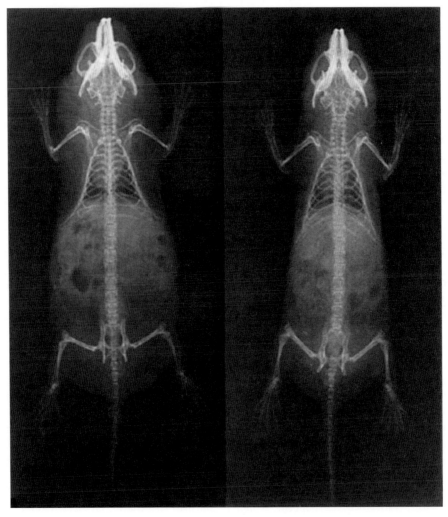

Figure 17.21 X-ray images of disperser (left) and non-disperser (right) siblings in naked mole rats show contrasts in phenotype. Dispersers were heavier than a random sample of non-dispersers from the same age cohort, but there was no significant difference in the skeletal robustness of the two groups and the difference in mass reflects the higher percentage of body fat in dispersers as determined from electromagnetic scanner analyses. Dispersers also had a significantly higher girth-to-body-length ratio, and more fat in the neck region: both features are evident in the soft-tissue outlines of the radiographs. In contrast, colony defenders of a comparable body mass had little subcutaneous fat. *Source*: From O'Riain *et al.* (1996). Reproduced with permission of Macmillan Publishers Ltd.

rates early in life. Specialisation of development also occurs in Damaraland mole rats, where groups include frequent and infrequent helpers who may be of either sex. Frequent helpers are smaller, show little sexual dimorphism and are responsible for most of the cooperative work performed in the colony, while infrequent helpers accumulate larger fat reserves, show greater sexual dimorphism and are more likely to disperse (Bennett and Jarvis 1988; Hazell *et al.* 2000; Scantlebury *et al.* 2006). In contrast, there is no evidence of similar contrasts in developmental strategies in meerkats, possibly because mortality is high and the acquisition of breeding roles is less predictable (Carter *et al.* 2014).

Kinship

Because of the potential importance of indirect fitness benefits in the evolution of cooperative behaviour, studies of cooperative breeders have investigated whether helpers are more likely to assist close relatives than distantly related or unrelated individuals and whether their workload increases when they are helping close relatives. In birds, where helpers are often failed breeders, there is evidence that they are more likely to join and help kin than unrelated individuals (Emlen and Wrege 1988; Russell and Hatchwell 2001), though the relative workload of helpers can be positively related, unrelated or negatively related to their kinship to breeders or juveniles (Komdeur 1994b; Dunn *et al.* 1995; Magrath and Yezerinac 1997; Cockburn 1998). In several social mammals where females live in groups that include non-kin and distant relatives as well as close relatives, females are more likely to support close relatives in competitive interactions (see Chapter 9) and a similar tendency has been found in several communal breeders. For example, female group members preferentially assist close relatives in ground squirrels and prairie dogs (Sherman 1977; Hoogland 1983, 1995). In some communal breeders, females are also more likely to assist close relatives: for example, African lionesses are more likely to suckle offspring that are not their own in groups of close relatives than in groups where females are less closely related (Pusey and Packer 1994).

However, there is evidence of nepotism in some cooperative mammals. In Alpine marmots (where young raised in groups that include closely related helpers have higher overwinter survival than those raised in groups consisting of more distant relatives or non-kin), group members overwintering with close kin lose more weight than those overwintering with less closely related individuals (Arnold 1990a,b). Similarly, some studies of naked mole rats have found that helpers closely related to dependent young require less frequent 'shoving' by the dominant breeder to maintain their levels of activity (Reeve and Sherman 1991), though others have failed to find similar trends (Jacobs and Jarvis 1996). However, in other cooperative breeders there is no consistent relationship between the workload of individual helpers and their proximity of kinship to pups being raised. For example, contributions to caring for young appear to be unrelated to kinship in meerkat (Clutton-Brock *et al.* 2000), although subordinate females are more likely to suckle pups born to the dominant female if they are closely related to her (MacLeod *et al.* 2013).

A possible explanation of variation in the degree of nepotism is that the tendency for individuals to selectively assist kin is confined to species where variance in kinship between group members is high and is absent where all helpers are closely related to breeders (Clutton-Brock 2002). In line with this, an analysis of combined data for cooperative birds and mammals suggests that the extent to which helpers preferentially assist kin declines as average kinship between group members increases (Cornwallis *et al.* 2009).

Social bonds

In several cooperative societies, there is little evidence of persistent bonds between particular helpers and particular juveniles. For example, in meerkats, dependent pups move from helper to helper while they are foraging, quickly abandoning individuals that do not feed them or those that are not foraging actively. Banded mongooses are a striking exception to this generalisation: here, individual pups are escorted and fed by particular helpers for much of their period of dependence (Gilchrist 2004; Gilchrist and Russell 2007) and pups with regular escorts are more likely survive to adulthood (Gilchrist 2004; Hodge 2005). Pups and escorts recognise each other's calls and respond selectively to them (Muller and Manser 2008). Pups defend access to their escorts, beg more intensely and receive more food when they are with their usual escort than when they are with other animals, and escorts are more likely to give away food they find when they are with their usual pup than when they are with other pups (Gilchrist 2004, 2008; Bell 2008). Escorts are usually subordinate males, probably because the costs of feeding pups have little influence on their mating success (which is usually low), whereas most females breed each year and substantial contributions to pup feeding would be likely to affect their fecundity (Hodge 2007). Why persistent social bonds between particular helpers and particular pups occur in banded mongooses but not in meerkats is uncertain but one possibility is that banded mongooses live in more mesic habitats where food is more abundant, so that individual pups can obtain all the food they need from a single helper (Hodge 2003).

Cheating and 'false-feeding'

In some cooperative birds, helpers sometimes visit nests but bring no food to give to the chicks (Clarke 1984; Poiani 1993) or bring food to the nest and then eat it themselves instead of offering it to the young (Doland *et al*. 1997; Canestrari *et al*. 2004). This behaviour is referred to as 'false-feeding' and has been interpreted as a cheating strategy, designed to cause other group members to overestimate the extent to which the individuals contribute to cooperative activities (Boland *et al*. 1997). Similar behaviour occurs in some cooperative mammals. For example, meerkat helpers sometimes catch food items and run across to a begging pup, but then eat the food themselves rather than giving it to the pup (Clutton-Brock *et al*. 2005).

While false-feeding could be a deceptive strategy, there are other possibilities. One is that helpers assess the nutritional needs of juveniles relative to their own immediately before they hand over food, and that false-feeding occurs where they decide that their own needs are more pressing (Canestrari *et al*. 2004). False-feeding is usually relatively uncommon and may occur where conflicting motivational tendencies are finely balanced: in meerkats, false-feeding occurs in no more than 3% of all pup feeding attempts and most frequently involves categories of individuals that are more reluctant to feed pups (including juveniles, males and individuals that have not foraged successfully) than those that are more strongly motivated. In addition, the frequency with which individual pups are false-fed also appears to be inversely related to their need for food and to their attractiveness as targets for helpers.

17.7 Regulation of workload

Like juveniles in many biparental societies, dependent juveniles in cooperative breeders beg for food and helpers adjust their contributions to feeding pups in relation to their needs (Manser and Avey 2000; Horn and Leonard 2002; Kilner 2002). In some callitrichids, adults also give specific calls that encourage juveniles to approach and take food items that are proffered (Rapaport 2011). The effects of pup begging have been investigated by manipulating the hunger of pups, recording their calls and playing them back to adults (Box 17.2 and Figure 17.22). Experiments with meerkats and banded mongooses both show that increases in the begging rate of pups honestly signal their hunger level, and helpers respond to these increases by raising their rate of pup feeding and the proportion of food they find which they give to pups (Bell 2008; English *et al*. 2008; Manser *et al*. 2008). As pups develop and spend more time foraging independently, the rate at which they beg declines and

---

Box 17.2 Pup begging in meerkats and mongooses

In several cooperative mammals where dependent pups accompany foraging adults, they give regular begging calls which stimulate adults to give them food. For example, in meerkats, where pups accompany their group while it is foraging from about 1 month of age, they shuttle between adults, giving begging calls at a rate of around 100 times a minute that elevate cortisol levels in helpers (Carlson *et al*. 2006). Similarly, in banded mongooses, pups follow adults and give regular begging calls which increase the chance that they will be fed (Bell 2008). In both species, the rate at which pups call is associated with their hunger level and can be manipulated by feeding or food-depriving pups (see Figure 17.22) (White 2001; Bell 2008; Manser *et al*. 2008). In meerkats, pups also adjust their calling rates to the probability that they will be fed and call more frequently when close to more generous adults (Kunc *et al*. 2007).

In both species, playback of pups giving begging calls shows that rates of begging affect the frequency with which helpers feed pups. The responses of helpers appear to depend on the cumulative amount of begging they hear and pups reduce their begging rate if litter size is increased (Kunc *et al*. 2007; Bell 2007; Madden *et al*. 2009). While one interpretation of this result is that pups cooperate with each other to manipulate the provisioning rate of helpers (Bell 2007), an alternative explanation is that they prefer to call when other pups are silent because this is most likely to elicit a response from adults and experiments with meerkats suggest that this is the case (Madden *et al*. 2009).

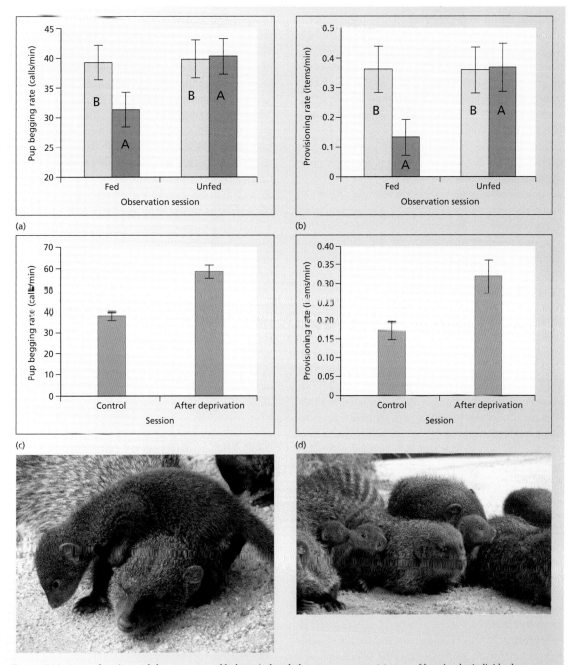

Figure 17.22 Honest begging and the responses of helpers in banded mongoose pups: (a) rates of begging by individual pups before (B) versus after (A) they were experimentally provisioned ('fed') compared with rates of begging by unfed control pups ('unfed'); (b) rates at which pups were given food by escorting helpers before and after they were experimentally provisioned compared with rates for unfed control pups; (c) rates of begging by pups that had been experimentally deprived of food for 10 hours versus controls foraging naturally with groups; (d) rates at which food-deprived and control pups were given food by helpers. *Source*: From Bell (2008). Reproduced with permission from the Royal Society. *Photo sources*: © Jennifer Sanderson.

helpers gradually cease to bring food to them (Manser and Avey 2000). Helpers also respond to qualitative changes in the calls of pups as they increase: in meerkats, playback of the begging calls of young pups increases the frequency with which helpers give food to pups and can induce helpers in groups with older pups that are foraging independently to start feeding them again (Thornton and McAuliffe 2006).

Frequent begging by dependent juveniles induces hormonal changes in helpers that are associated with increases in their contributions to cooperative activities. In some cooperative birds as well as in some cooperative mammals, prolactin levels increase during periods when group members are caring for young and are associated with increased contributions to feeding or guarding young (Gubernick and Nelson 1989; Vleck *et al.* 1991; Schoech *et al.* 1996; Schradin and Anzenberger 2001). Several studies have shown that oxytocin levels are also associated with cooperative or prosocial behaviour (De Dreu *et al.* 2011; Crockford *et al.* 2013): in meerkats, intramuscular injection of oxytocin raises the contributions of individual helpers to guarding, pup feeding and digging, as well as their association with pups (Madden and Clutton-Brock 2011). Changes in cortisol levels may also be involved: for example, in male meerkats, individual differences in the level of contributions to feeding pups are positively correlated with levels of cortisol measured before the onset of pup feeding, and playback of pup begging calls raises cortisol levels as well as pup feeding rates (Carlson *et al.* 2006).

The workload of helpers can also be influenced by the actions of breeders. In some cooperative fish, as well as in some cooperative birds, lazy helpers are sometimes punished by breeders (Mulder and Langmore 1993; Balshine-Earn *et al.* 1998) and coercion may sometimes play a role in maintaining helping behaviour in cooperative mammals, too (Clutton-Brock and Parker 1995; Young 2003). In some cooperative societies, breeders 'stimulate' inactive helpers: for example, during periods of reproductive activity, dominant female naked mole rats shove inactive helpers (especially larger ones and individuals less closely related to her) and the removal of dominant females causes a reduction in the level of helper activity, especially among larger and more distantly related individuals (Reeve and Sherman 1991; Reeve 1992). Similarly, in meerkats, males (but not females) that feed pups relatively infrequently or that

false-feed pups receive more aggression than frequent feeders, mostly from breeders (Clutton-Brock *et al.* 2005). However, in most non-human cooperative societies, aggression directed at lazy helpers by dominants is infrequent and there is little evidence that punishment stimulates collective actions (Riehl and Frederickson 2016). For example, in meerkats, attempts to provoke increased aggression from dominants at helpers by playback of pup begging calls failed to produce any increase in the frequency of threats or attacks (Santema and Clutton-Brock 2012). There is also no indication that recipients of aggression subsequently make larger contributions to cooperative activities (Young *et al.* 2005) or that lazy individuals are more likely to be evicted earlier than generous ones (Clutton-Brock and Manser 2015). Moreover, when dominants direct more aggression at 'lazy' helpers, an alternative interpretation is that they do so because 'lazy' helpers are primed to breed and aggression serves to reduce levels of reproductive hormones. (Wingfield *et al.* 1990; Schoech *et al.* 1996). For example, in naked mole rats, interactions with breeders may inhibit the sexual development of subordinates as well as increasing their contributions to cooperative activities (Faulkes and Abbott 1997; Faulkes and Bennett 2007).

17.8 The evolution of cooperative breeding

So what are the likely evolutionary mechanisms maintaining cooperative breeding in mammals? It is unlikely that cooperative behaviour is commonly maintained by direct benefits to reproductive success operating through prestige: there is little evidence that individuals compete to perform cooperative activities and, when age effects have been allowed for, the most assiduous helpers appear to be less competitive individuals. For example, in naked mole rats, larger more competitive males contribute less to cooperative activities and are more likely to disperse from their natal colony (O'Riain and Braude 2001). Similarly, in meerkats, the largest and oldest helpers frequently contribute less to cooperative activities (Clutton-Brock *et al.* 2002) while individuals that are the product of inbreeding show reduced growth and contribute more than average to cooperative activities (Nielsen 2013).

Nor does it seem likely that helping behaviour is commonly coerced in cooperative mammals, either by dominants or by other subordinates, though aggression from dominants may play an important role in suppressing the sexual development of subordinates and can enhance their contributions to cooperative activities. In most systems, helpers appear to contribute to cooperative activities willingly and rarely attempt to avoid involvement, and although aggression is sometimes directed at lazy helpers, interactions of this kind are uncommon in most cooperative societies (Riehl and Frederickson 2016). Moreover, even where the contributions of helpers are positively correlated with the frequency of aggression received, there is a danger that this is because helpers with relatively high levels of sex hormones both work less and are more likely to breed or to challenge the dominant female and that increases in the amount of aggression they receive are a consequence of the increased risk that their presence poses to dominants rather than of reductions in their contributions to cooperative activities.

This leaves shared mutualistic benefits and kin selection as the two remaining explanations (see section 17.1). Both may often be involved. A range of models have shown that cooperation can be maintained between non-kin by mutualistic benefits though its evolution is facilitated by kinship beteween group members (Kokko *et al.* 2001; Lehmann and Keller 2006; Lehmann *et al.* 2007, Wilson and Wilson 2007; Leggett *et al.* 2011; Nonacs 2011; Riehl and Frederickson 2016) (Figure 17.23) and that it can often generate direct fitness benefits that are shared by all group members (Clutton-Brock 2006; Kingma *et al.* 2011; Johnstone and Rodrigues 2016). A crucial question is consequently whether the initial evolution of cooperative and eusocial breeding systems has been restricted to species where average levels of kinship are high (as explanations based on kin selection would suggest) or whether these systems sometimes evolve where average levels of kinship between group members are low, as some theorists have argued (Nowak 2006). Here, phylogenetic reconstructions of the evolution of cooperative and eusocial societies in insects, birds and mammals appear to provide a clear answer. They show that, in almost all cases, cooperative breeders show unusually high levels of kinship within groups and that the initial evolution of these systems usually occurs in species where monogynous or

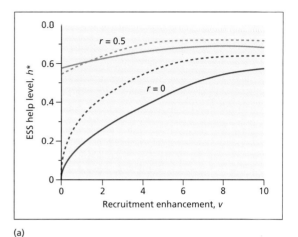

(a)

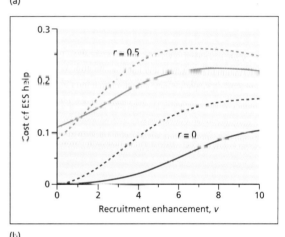

(b)

Figure 17.23 Predicted effects of group augmentation and kinship in a two-player game. The parameter v indicates the impact that helping has on recruitment: $v = 0$ assumes no impact and thus prevents any effect of group augmentation (all offspring disperse). Evolutionarily stable helping is indicated for unrelated ($r = 0$) and related ($r = 0.5$) subordinates as both (a) evolutionarily stable help levels $h*$ and (b) costs expressed as a fractional reduction in survival accepted by subordinates. Solid lines indicate help in the absence of passive benefits and dotted lines assume higher survival in larger groups. The alternative 'no help' equilibrium also exists in all cases where relatedness $r = 0$ (not shown). *Source*: From Kokko *et al.* (2001). Reproduced with permission from the Royal Society.

monogamous mating systems are either already established or where monogamous and cooperative breeding arise at the same stage of evolution (Hughes *et al.* 2008; Boomsma 2009; Cornwallis *et al.* 2010; Lukas and Clutton-Brock 2012). In non-human mammals, the evolution of cooperative breeding systems involving non-breeding helpers appears to have been restricted to species with monogamous mating systems while communal breeding systems (where multiple females breed and share care of their young) appear to have evolved both in monogamous and in polygynous or promiscuous species (Lukas and Clutton-Brock 2012). The apparent absence of kin discrimination in cooperative behaviour does not necessarily suggest that kin selection is weak, for if most groups members are close relatives, selection would not necessarily be expected to favour discrimination (Clutton-Brock 2002).

While cooperative breeding has seldom evolved in groups consisting primarily of unrelated individuals, in many cooperative breeders, groups frequently contain a proportion of individuals that are unrelated to other group members but participate in cooperative activities (Clutton-Brock 2002; Riehl 2013). In some cases, this may be because multiple unrelated males join groups as non-breeding helpers to gain eventual access to breeding females (as in the case of some callitrichids), while in others (like banded mongooses) monogamy appears to have been replaced by polygynous or polygynandrous mating systems after the evolution of cooperative breeding has evolved. Humans (which show a combination of traits found in communal and cooperative breeders but are best regarded as communal breeders) are one of the few exceptions to the general association between the evolution of reproductive cooperation and high levels of kinship among group members, but the mechanisms maintaining human cooperation are sufficiently different from those involved in other animals that it is unsafe to extrapolate from them to non-human animals (see Chapter 20).

Evidence that close kinship between group members is usually necessary for altruistic cooperation to evolve does not indicate that mutualistic benefits are not important and direct fitness benefits shared between group members may play a crucial role in maintaining cooperative breeding in many species (Clutton-Brock 2002; Kingma et al. 2011; Leadbeater et al. 2011; Riehl 2013). In a number of cooperative vertebrates, some group members are not closely related to breeders but contribute as much as close relatives to cooperative activities (Clutton-Brock 2002; Riehl 2013) and cooperation is presumably maintained by direct fitness benefits associated with group-living. Although it is sometimes suggested that mutualistic effects operating through increased group size may be unimportant on the grounds that helping behaviour is not adjusted to variation in group size or its effects (Browning et al. 2012), adjustments of this kind (like adjustments to levels of kinship) would not necessarily be expected if mutualistic benefits are substantial.

The effects of assistance on the fitness of breeders (the b term in Hamilton's rule) and their offspring may also need to be high for cooperative or eusocial systems to evolve (Smaldino et al. 2013). For example, with the exception of humans, the evolution of cooperative breeding in mammals has been restricted to lineages where females produce litters and is usually associated with the production of multiple litters per year (Lukas and Clutton-Brock 2012). A likely explanation of this is that, in monotocous animals, parents are able to rear the young that they produce and helpers are unlikely to increment the success of breeders sufficiently for reproductive cooperation to increase their inclusive fitness. The restriction of cooperative breeding to polytocous species may explain why, among mammals, cooperative breeding is common among carnivores but appears to be absent in cetaceans, despite the reliance of several species on cooperative hunting, as well as why, among non-human primates, cooperative breeding is confined to the callitrichids.

Similar constraints on the evolution of cooperative breeding may help to explain the association between cooperative breeding and environments where rainfall and food availability are low and unpredictable. In birds, cooperative breeding appears to have evolved most frequently in species living in demanding habitats or niches and is commonly associated with arid habitats where rainfall and food availability are low and unpredictable and assistance from helpers is needed to rear offspring (Cockburn and Russell 2011; Jetz and Rubenstein 2011; Rubenstein 2011). Similarly, among mole rats, species living in arid habitats are mostly social and cooperative while those living in more mesic areas are solitary or live in small groups without specialised helpers (Jarvis et al. 1994; Lacey and Ebensperger 2007). Among the mongooses, too, the most extensively cooperative species, like dwarf mongooses and meerkats, are also associated with relatively arid habitats where resources are sparse and the risk of predation is high (Rood 1986; Schneider and Kappeler 2014). However, cooperative breeding is not associated with arid habitats in all species, indicating that different factors may be involved in different groups. For example, most of the marmosets and tamarins live in relatively mesic habitats (Digby et al. 2007).

17.9 Consequences of cooperative breeding

Fecundity and longevity in females

The evolution of cooperative rearing has important consequences for female life histories. Although cooperative breeding in mammals has commonly evolved in lineages where females produce multiple litters per year, the presence of extreme litter sizes and short inter-litter intervals in cooperative breeders living in large groups (including naked mole rats, wild dogs and meerkats) suggests that the evolution of cooperative breeding has led to further increases in female fecundity (Clutton-Brock 2009). Similarly, among higher primates, the marmosets and tamarins (where breeders are commonly assisted by non-breeding helpers and regularly produce twins) show unusually high levels of fecundity (Isler and van Schaik 2012). Evidence that, within species, variation in helper number is negatively correlated with the duration of inter-birth intervals is consistent with the suggestion the cooperative breeding increases fecundity (see section 17.5).

Cooperative rearing can also have important consequence for female longevity. In some eusocial insects where breeding females are provisioned by other group members, they have lifespans that can exceed 20 years, while helpers rarely live for more than a few months (Keller 1998) but it is not yet clear whether similar contrasts in longevity occur in cooperative mammals. Unusually long lifespans in breeding females, which can exceed 20 years, have also been described in groups of some social mole rats (Buffenstein and Jarvis 2002; Sherman and Jarvis 2002; Dammann and Burda 2006; Dammann et al. 2011), and are associated with cellular and physiological adaptations that delay ageing (Buffenstein 2008; Dammann et al. 2012) and reduce susceptibility to cancers (Sherman and Jarvis 2002; Liang et al. 2010; Tian et al. 2013). In some species, helpers appear to have shorter lives than breeders (Dammann and Burda 2011; Schmidt et al. 2013) but it is still uncertain whether extreme longevity in breeders is a characteristic of social mole rats or whether it occurs in all Bathyergidae. In addition, it is not yet clear whether the relatively short lifespans of subordinates are a consequence of earlier and more rapid ageing or whether they are a result of increasing rates of aggression directed at older helpers by breeders of the same sex and costs associated with eviction and dispersal. For example, in meerkats, the life expectancy of helpers is substantially lower than that of breeders, but this is because most helpers are forced to leave their natal groups before they are 4 years old and mortality is high among dispersing animals (Clutton-Brock et al. 2006; Sharp and Clutton-Brock 2009).

Male longevity

In several singular cooperative breeders, males have shorter breeding lifespans than females. For example, in both meerkats and naked mole rats, the average duration of breeding lifespans in males is around half that of breeding females (Lacey and Sherman 1997; Bennett and Faulkes 2000; Clutton-Brock et al. 2006). Several mechanisms may contribute to sex differences in longevity. Dominant males may age faster than dominant females as a result of the energetic costs of guarding the breeding female: for example, in naked mole rats, breeding males (unlike breeding females) lose weight during their period of tenure and can become visibly emaciated even when ad lib food is available (Lacey and Sherman 1997). In meerkats, dominant males may also be more likely to be displaced for, unlike breeding females, their principal competitors are intruding males whose development they cannot control. In addition, following the death of their partners, 'widowers' frequently leave their group to search for access to unrelated females (and so are exposed to the risks associated with dispersal) whereas 'widows' typically remain and breed with immigrant males (Clutton-Brock and Manser 2015).

Variance in breeding success

Differences in the length of breeding lifespans between females and males in singular cooperative breeders affect variance in breeding success in both sexes. The high fecundity and long breeding lifespans of dominant females in singular cooperative breeders, combined with the suppression of reproduction in subordinates, raise variance in female breeding success to levels that exceed those found among males in the most polygynous mammals (Hauber and Lacey 2005; Clutton-Brock 2006). For example, in naked mole rats, successful breeders may produce several hundred offspring during their lifespan (Hauber and Lacey 2005) while dominant female meerkats, despite their shorter lifespans, can rear over eighty surviving pups (Clutton-Brock et al. 2006). Though some females raise large numbers of young, most fail to acquire dominant status and produce no surviving offspring, so that (standardised) variance in female breeding success is unusually high.

In many singular cooperative breeders, males have little opportunity to breed with multiple partners in the same season and their lifetime breeding success depends principally on the fecundity of their partner and their own longevity. Where males have shorter breeding lifespans than females, both maximal lifetime reproductive success and variance in lifetime success are likely to be lower in males than in females: for example, in meerkats, (standardised) variance in breeding success is around half as large in males as in females (Clutton-Brock *et al.* 2006). The shorter breeding lifespans of males also mean that a higher proportion of males than females that reach adulthood breed as dominants at some stage in their lives, and that a higher proportion of females are likely to have multiple breeding partners in the course of their breeding careers.

Reproductive competition

Restrictions on the number of individuals of both sexes that can breed as dominants in singular cooperative breeders is associated with intense competition for dominant status in both sexes. Where breeding females have longer lifespans and show greater variance in lifetime breeding success than males, the frequency and intensity of direct competition would be expected to be greater in females than in males and studies of several cooperative breeders suggest that this may be the case: for example, in several marmosets and tamarins, female–female aggression is both more frequent and more intense than male–male aggression and attacks (which often involve relatives) can lead to severe wounding or death (French 1997; Yanomoto *et al.* 2014) and in one colony of golden lion tamarins, one-third of female–female attacks led to the death of the recipient compared with one-tenth of male–male attacks (Inglett *et al.* 1989). Similarly, in meerkats and naked mole rats, females are more frequently aggressive to each other than are males and sex differences in aggression extend back into the early stages of development (Lacey and Sherman 1991, 1997; Reeve and Sherman 1991; Clutton-Brock *et al.* 2006).

Development

Intense selection to acquire dominant breeding positions in singular cooperative breeders appears to have led to the evolution of developmental strategies that enhance the ability of individuals to acquire or maintain dominant status. Among meerkats, individuals that have recently acquired dominant positions show increased levels of sex hormones and are more frequently aggressive than other group members and reduce their contributions to most cooperative activities (Clutton-Brock *et al.* 1998a, 2004; Russell *et al.* 2004). During the months after acquiring dominant status, they show an accelerated increase in body mass, whose magnitude is inversely related to the difference in mass between them and the heaviest subordinate of the same sex in their group (Russell *et al.* 2004; Huchard, unpublished data). Subsequently, dominants gain weight more slowly than subordinates, possibly as a consequence of the costs of breeding. Since these changes occur in both sexes, it is likely that they represent adaptations to controlling potential competitors rather than adaptations to increasing litter size and fecundity, as has been suggested for similar group patterns in other species (Bennett and Faulkes 2000). After acquiring dominant status, females continue to grow throughout their lives, while males do not, with the result that dominant female meerkats are commonly the largest and heaviest individual in their group and are usually dominant to all other group members.

Pronounced sex differences in growth after acquiring dominant breeding positions also occur in some mole rats (Braude 1991; Jarvis 1991; Faulkes and Abbott 1997). In Damaraland mole rats, females that acquire dominant breeding status develop a more elongate body shape than subordinates (Young and Bennett 2010). While, in naked mole rats, they increase in body weight and size and show increased levels of testosterone (Faulkes and Abbott 1997; Lacey and Sherman 1997). Dominant female naked mole rats become increasingly aggressive to other individuals of the same sex and are usually the most dominant individual in their group. Effects of female competition on the evolution of traits used by females in competitive encounters and reversals in the usual pattern of sex differences in growth, size and behaviour have also been found in some cooperative birds and fish (Aubin-Horth *et al.* 2007; Clutton-Brock and Huchard 2013a,b).

Sex ratio biases

Where offspring of one sex are more likely to assist their parents, the presence of non-breeding helpers can affect patterns of sex allocation (Malcolm and Marten 1982;

Clutton-Brock 1991; Komdeur 2004). Offspring that assist with their parents' subsequent breeding attempts can be regarded as 'repaying' their parents the costs of their production, and where one sex helps more than the other and there are no sex differences in rearing costs, the net costs of producing offspring of the more cooperative sex to their parents are likely to be lower than those of producing the less cooperative sex (Malcolm 1979; Emlen *et al.* 1986). As a result, where sons are more likely to help their parents than daughters, parents might be expected to produce more sons; where daughters help more than sons, the situation may be reversed.

However, predictions about the effects of sex differences in cooperation on sex allocation are complicated by other effects (Pen and Weissing 2000, 2002; Schindler *et al.* 2015). For example, if siblings of one sex compete more intensely with each other for mates or resources, parents would be expected to produce more of the less competitive sex (Hamilton 1967; Clark 1978; Bulmer and Taylor 1980; Gowaty and Lennartz 1985). Or, if siblings of one sex contribute more to each other's fitness (for example because they disperse together and their chances of surviving are related to the size of dispersing subgroups), parents might be expected to produce an excess of that sex (Clutton-Brock 1991). Finally, if individual parents adjust the sex ratio of their offspring to their own need for helpers, the average sex ratio may be unbiased or even biased towards the non-helping sex (Pen and Weissing 2002). Where different models of sex allocation or sex ratio variation generate similar predictions, the relative importance of particular mechanisms is usually difficult to assess (see Chapter 18).

Like many other sex ratio trends (see Chapter 18) those in cooperative breeders are not universal. Some studies of cooperative breeders have produced results that are consistent with predictions that parents should produce an excess of the sex that is more likely to assist them. For example, in some populations of African wild dogs (where sons provide more assistance than daughters), offspring sex ratios shortly after birth are male-biased (Malcolm 1979; Malcolm and Marten 1982) and younger primiparous females, which have relatively high oestrogen levels, produce an excess of sons while multiparous ones produce more daughters (Creel *et al.* 1998; McNutt and Silk 2008). In Alpine marmots, where the presence of subordinate males is associated with reductions in infant overwinter mortality, weaning sex ratios

are also biased towards males (Allainé *et al.* 2000). In captive marmosets and tamarins, species that rely on male helpers to carry young (like golden lion tamarins) produce more male offspring than species that produce single offspring that are usually carried by their parents (like Goeldi's monkey) (Rapaport *et al.* 2013). Studies of Damaraland mole rats (where male and female helpers contribute equally to raising young but males are more likely to disperse) show that the proportion of male pups born increases in relation to the number of female helpers and decreases with the number of male helpers (Lutermann *et al.* 2014). However, other studies of cooperative breeders have found no evidence of similar effects. For example, in meerkats, where subordinate females contribute more to cooperative activities than subordinate males, the sex ratio of pups at emergence is not consistently biased towards females and there is no indication of any consistent relationship between the characteristics of breeding females and the sex ratio of their offspring (MacLeod and Clutton-Brock 2013).

Brain size

The maintenance of large brains involves substantial energetic costs (Isler and van Schaik 2009) and it has been suggested that by reducing the energetic demands on breeding females, allo-parental care and cooperative rearing may permit the evolution of increases in brain size and associated improvements in cognitive development (Burkart and van Schaik 2010; van Schaik and Burkart 2010). Comparative analyses show that, across mammals, there is a positive association between the extent of parental care by individuals other than the mother and indices of brain size and similar correlations are found within several taxonomic groups (Shultz and Dunbar 2007; Isler and van Schaik 2012). However, neither in birds (Iwaniuk and Arnold 2004) nor mammals is there any consistent difference in relative or absolute brain size in adults between cooperative breeders and other species: for example, naked mole rats, which have the most highly developed cooperative system of all mole-rat species, have the (relatively) smallest brains (Kruska and Steffen 2009). While there has been no systematic attempt to compare the cognitive capacities of cooperative breeders with other mammals, there is little indication that cooperative breeders show advanced cognitive abilities (Thornton and Clutton-Brock 2011;

Thornton and Samson 2012). This should not be surprising, for the majority of cooperative breeders live in groups where members of the same sex are typically close relatives, social conflict is rare (except at times when subordinates compete for breeding positions) and complex coalitionary interactions do not occur.

Prosocial behaviour

While cooperative breeders show little evidence of improved cognitive abilities, they show evidence of increased prosocial tendencies compared to other social species. For example, comparisons of callitrichids with other social but non-cooperative New World monkeys suggest that cooperative breeders may pay increased attention to the activities of other group members, greater spatial and temporal coordination of activity, increased tolerance, greater sensitivity to the signals of others and a higher probability of providing assistance (Burkart and van Schaik 2010). Similar tendencies characterise the behaviour of cooperative mongooses, where group members help to defend each other against predators and sometimes bring food to sick animals (Clutton-Brock and Manser 2015), though their behaviour is context-specific and group members rarely hunt cooperatively or, under normal circumstances, share food with other adults. Experiments with siamangs and saki monkeys show that the tendency for individuals to engage in unsolicited prosocial behaviour (in this case, helping another group member access a food reward) is not confined to cooperative breeders, which could suggest that it may occur wherever adults are regularly engaged in alloparental care (Burkart *et al.* 2014) but whether or not this is the case is unknown.

Group size

In some cooperative breeders, increased feeding or breeding competition in large groups may influence the readiness of subordinates to disperse and so regulate group size. For example, where older helpers are consistently more likely to inherit alpha status than younger ones, the rising benefits of dispersal to younger group members as queue length increases in their natal group may eventually exceed those of remaining, even if the costs of dispersal are high. Where subordinate females rarely leave their natal groups of their own volition, the size of groups may be limited by the extent to which

dominant females tolerate subordinates. For example, studies of meerkats suggest that dominant females become increasingly intolerant of the presence of subordinate females once group size exceeds around twenty animals and breeding success begins to fall (Clutton-Brock *et al.* 2008; Clutton-Brock and Manser 2015) and these effects may also play an important role in limiting group size (Bateman *et al.* 2013; Ozgul *et al.* 2014). Similarly, in banded mongooses, the rate of eviction of females increases when the number of resident adult females exceeds eight to ten (Cant *et al.* 2010) and here, too, these effects may contribute to limiting group size.

Interspecific differences in the period for which dominants will tolerate the presence of subordinates appear to play an important role in generating interspecific differences in group size. Where subordinate females are driven out of the group at the end of their first year of life (as in some canids as well as in many cooperative birds), group size is typically small and is likely to be constrained by the annual fecundity of females. Where dominant females tolerate subordinates after the age of sexual maturity and groups include subordinates drawn from several year classes (as in meerkats), group size is larger. Finally, where dominant females tolerate subordinates of all ages, as in naked mole rats, colonies can include several hundred individuals and their size may not be closely regulated (Brett 1991; Faulkes and Bennett 2007).

Population dynamics

The positive effects of helper number on breeding success and survival in singular cooperative breeders may also have important consequences for population dynamics. Where low population density is associated with reductions in average group size, these effects can generate positive correlations between population density and recruitment known as *Allee effects* (Courchamp *et al.* 1999a,b; Angulo *et al.* 2013). If the positive effects of helper number are strong, low population density may be associated with low rates of recruitment per breeding female and populations which have been reduced by density-independent factors (like climatic fluctuations or epidemics) may be slow to recover, with the result that the risk of group or local population extinction may be increased (Courchamp *et al.* 1999a, 2002; Courchamp and Macdonald 2001; Angulo *et al.* 2013) (Figure 17.24).

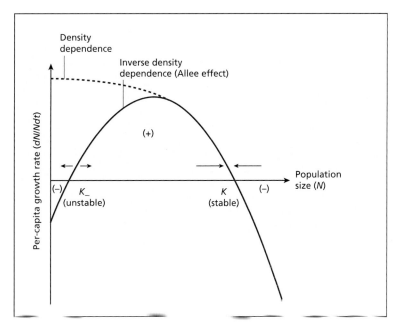

Figure 17.24 Illustration of the Allee effect, from a very simple mathematical model of population dynamics:

$$\frac{dN}{dt} = rN\left(1 - \frac{N}{K}\right)\left(\frac{N}{K_-} - 1\right)$$

The per-capita growth rate (dN/dt) is negative above the carrying capacity (K) and positive below. However, in the presence of an Allee effect, it also decreases below a given population size, and can become negative below a critical population threshold (K_-). When a population displaying this type of population dynamics is driven below the critical threshold, the low per-capita growth rate may lead it to its extinction. *Source*: From Courchamp *et al.* (1999b). Reproduced with permission of Elsevier.

The impact of Allee effects on population survival is likely to vary with ecological conditions. For example, the effects of cooperative breeding on population dynamics have been most extensively investigated in African wild dogs, where both foraging or hunting success and breeding success decline in small groups (Creel 1997, 1998; Courchamp *et al.* 2000; Creel and Creel 2002). Using an individual-based model of wild dog dynamics parameterised with realistic values, Creel showed that the probability of survival of local wild dog populations falls as maximum pack size declines. Allee effects may also reduce the success of re-introduction programmes unless these involve substantial numbers of animals. Wild dogs are adversely affected by the density of lions, which commonly rob them of their prey and may kill their pups (Creel and Creel 2002) and

calculations show that these effects are likely to reduce the chances that local populations will survive and increases the impact of pack size on survival (Courchamp and Macdonald 2001).

While the presence of positive correlations between group size and recruitment may often lead to increased rates of extinction in small groups, this is not necessarily associated with Allee effects at the population level (Gregory *et al.* 2010). For example, in meerkats, where the breeding success of dominant females and the total number of recruits is reduced in small groups, there is no evidence of positive correlations between population density and population growth rates (Bateman *et al.* 2012). These results emphasise that Allee effects may occur at different levels and have different consequences for individuals, groups and populations (Angulo *et al.* 2013).

SUMMARY

1. Cooperative breeders are species where young are raised by breeding adults assisted by non-breeding helpers. In most cooperative breeders, a single male and a single female monopolise reproduction (though they may not do so completely). Non-breeding helpers are usually natal individuals that have not yet dispersed from the group where they were born.

2. Many of these species defend group territories and, when habitats are saturated, delayed dispersal is likely to increase the survival of subordinate animals. However, subordinates are commonly prevented from breeding within the group by the behaviour of dominants, who suppress their fertility, induce abortions, kill offspring and evict older or larger helpers.

3. While subordinates usually have substantially lower breeding success than dominants, they breed occasionally in most cooperative mammals. One possible explanation is that dominants cannot control subordinate breeding entirely, while another is that dominants allow subordinates to breed occasionally to encourage them to stay in the group and assist in rearing infants. Most evidence suggests that subordinate females breed where dominants are unable to prevent them, though males may tolerate reproductive attempts by subordinates if this enhances their tenure.

4. Helping often entails substantial energetic costs to subordinates, though there is limited evidence that these convert into substantial fitness costs. One reason for this is that the workload of helpers varies with their age, size and nutritional status, so that the fitness costs of helping are not necessarily large.

5. Helpers (like breeders) adjust their behaviour to the needs of the infants they are rearing. In several species, infants or juveniles signal their hunger by repeated begging calls and helpers respond to changes in rates of calling.

6. In contrast to communal breeding systems (where several breeding females share care of their offspring), the initial evolution of cooperative breeding systems in mammals appears to have been restricted to monogamous species that produce multiple young per breeding attempt, suggesting that indirect benefits have played a central role in the evolution of cooperative breeding. However, the fitness of group members often increases with group size and direct fitness benefits of helping may often contribute to the maintenance of cooperation, too.

7. Cooperative breeding has far-reaching consequences for life histories, reproductive competition and development as well as for population dynamics and the regulation of group size. Competition for reproductive opportunities is intense in both sexes, especially in females, who often have longer breeding lifespans than males.

References

Abbott, D.H., *et al.* (1990) Pheromonal contraception. In: *Chemical Signals in Vertebrates 5* (eds D.W. Macdonald, D. Müller-Schwarze and S.E. Natynczuk). Oxford: Oxford University Press, 315–328.

Albon, S.D., *et al.* (1983) Fertility and body weight in female red deer: a density dependent relationship. *Journal of Animal Ecology* **52**:969–980.

Allainé, D. and Theuriau, F. (2004) Is there an optimal number of helpers in Alpine marmot family groups? *Behavioral Ecology* **15**:916–924.

Allainé, D., *et al.* (2000) Male-biased sex ratio in litters of alpine marmots supports the helper repayment hypothesis. *Behavioral Ecology* **11**:507–514.

Angulo, E., *et al.* (2013) Do social groups prevent Allee effect related extinctions? *The case of wild dogs. Frontiers in Zoology* **10**:11.

Armitage, K.B. (1981) Sociality as a life-history tactic of ground squirrels. *Oecologia* **48**:36–49.

Armitage, K.B. (1999) Evolution of sociality in marmots. *Journal of Mammalogy* **80**:1–10.

Arnold, W. (1990a) The evolution of marmot sociality: I. Why disperse late? *Behavioral Ecology and Sociobiology* **27**:229–237.

Arnold, W. (1990b) The evolution of marmot sociality: II. Costs and benefits of joint hibernation. *Behavioral Ecology and Sociobiology* **27**:239–246.

Arnold, W. and Dittami, J. (1997) Reproductive suppression in male alpine marmots. *Animal Behaviour* **53**:53–66.

Aubin-Horth, N., *et al.* (2007) Masculinized dominant females in a cooperatively breeding species. *Molecular Ecology* **16**:1349–1358.

Bales, K., *et al.* (2001) Sources of variability in numbers of live births in wild golden lion tamarins (*Leontopithecus rosalia*). *American Journal of Primatology* **54**:211–221.

Balshine-Earn, S., *et al.* (1998) Paying to stay or paying to breed? Field evidence of direct benefits of helping behaviour in a cooperatively breeding fish. *Behavioral Ecology* **9**:432–438.

Barclay, P. and Reeve, H.K. (2012) The varying relationship between helping and individual quality. *Behavioral Ecology* **23**:693–698.

Bateman, A.W., *et al.* (2012) Density dependence in group dynamics of a highly social mongoose, *Suricata suricatta*. *Journal of Animal Ecology* **81**:628–639.

Bateman, A.W., *et al.* (2013) Social structure mediates environmental effects on group size in an obligate cooperative breeder, *Suricata suricatta*. *Ecology* **94**:587–597.

Bednarz, J.C. and Hayden, T.J. (1991) Skewed brood sex ratio and sex-biased hatching sequence in Harris's hawks. *American Naturalist* **137**:116–132.

Beehner, J.C. and Lu, A. (2013) Reproductive suppression in female primates: a review. *Evolutionary Anthropology: Issues, News, and Reviews* **22**:226–238.

Beekman, M., *et al.* (2003) Reproductive conflicts in social animals: who has power? *Trends in Ecology and Evolution* **18**:277–282.

Bell, M.B.V. (2007) Cooperative begging in banded mongoose pups. *Current Biology* **17**:717–721.

Bell, M.B.V. (2008) Strategic adjustment of begging effort by banded mongoose pups. *Proceedings of the Royal Society of London. Series B: Biological Sciences* **275**:1313–1319.

Bell, M.B.V., et al. (2012) The cost of dominance: suppressing subordinate reproduction affects the reproductive success of dominant female banded mongooses. *Proceedings of the Royal Society of London. Series B: Biological Sciences* **279**:20111093.

Bell, M.B.V., et al. (2014) Suppressing subordinate reproduction provides benefits to dominants in cooperative societies of meerkats. *Nature Communications* **5**:4499.

Bennett, N.C. and Faulkes, C.G. (2000) *African Mole-rats: Ecology and Eusociality*. Cambridge: Cambridge University Press.

Bennett, N.C. and Jarvis, J.U.M. (1988) The social structure and reproductive biology of colonies of the mole-rat, *Cryptomys damarensis* (Rodentia, Bathyergidae). *Journal of Mammalogy* **69**:293–302.

Bennett, N.C., et al. (1994) Reproductive suppression in eusocial *Cryptomys damarensis* colonies: socially-induced infertility in females. *Journal of Zoology* **233**:617–630.

Bergmüller, R. and Taborsky, M. (2005) Experimental manipulation of helping in a cooperative breeder: helpers 'pay to stay' by pre-emptive appeasement. *Animal Behaviour* **69**:19–28.

Bergmüller, R., et al. (2005) Helpers in a cooperatively breeding cichlid stay and pay or disperse and breed, depending on ecological constraints. *Proceedings of the Royal Society of London. Series B: Biological Sciences* **272**:325–331.

Bergmüller, R., et al. (2007) Integrating cooperative breeding into theoretical concepts of cooperation. *Behavioural Processes* **76**:61–72.

Bergmüller, R., et al. (2010) Evolutionary causes and consequences of consistent individual variation in cooperative behaviour. *Philosophical Transactions of the Royal Society B: Biological Sciences* **365**:2751–2764.

Bernasconi, G. and Strassmann, J.E. (1999) Cooperation among unrelated individuals: the ant foundress case. *Trends in Ecology and Evolution* **14**:477–481.

Bertram, B.C.R. (1976) Kin selection in lions and in evolution. In: *Growing Points in Ethology* (eds P.P.G Bateson and R.A. Hinde). Cambridge: Cambridge University Press, 281–301.

Blumstein, D.T. and Armitage, K.B. (1999) Cooperative breeding in marmots. *Oikos* **84**:369–382.

Boland, C.R.J., et al. (1997) Deception by helpers in cooperatively breeding white-winged choughs and its experimental manipulation. *Behavioral Ecology and Sociobiology* **41**:251–256.

Boomsma, J.J. (2009) Lifetime monogamy and the evolution of eusociality. *Philosophical Transactions of the Royal Society B: Biological Sciences* **364**:3191–3207.

Bourke, A.F.G. (1999) Colony size, social complexity and reproductive conflict in social insects. *Journal of Evolutionary Biology* **12**:245–257.

Braude, S.H. (1991) Behavior and demographics of the naked mole-rat, Heterocephalus glaber. PhD thesis, University of Michigan, Ann Arbor.

Brett, R.A. (1991) The population structure of naked mole-rat colonies. In: *The Biology of the Naked Mole-rat* (eds P.W. Sherman, J.U.M. Jarvis and R.D. Alexander). Princeton, NJ: Princeton University Press, 97–136.

Briga, M., et al. (2012) Care for kin: within-group relatedness and allomaternal care and positively correlated and conserved throughout the mammalian phylogeny. *Biology Letters* **8**:533–536.

Brotherton, P.N.M., et al. (2001) Offspring food allocation by parents and helpers in a cooperative mammal. *Behavioral Ecology* **12**:590–599.

Brown, J.L. (1987) *Helping and Communal Breeding in Birds*. Princeton, NJ: Princeton University Press.

Browning, L.E., et al. (2012) Kin selection, not group augmentation, predicts helping in an obligate cooperatively breeding bird. *Proceedings of the Royal Society of London. Series B: Biological Sciences* **279**:3861–3869.

Buffenstein, R. (2008) Negligible senescence in the longest living rodent, the naked mole-rat: insights from a successfully aging species. *Journal of Comparative Physiology B* **178**:439–445.

Buffenstein, R. and Jarvis, J.U.M. (2002) The naked mole-rat: a new record for the oldest living rodent. *Science of Aging Knowledge Environment* **2002** (21): 7.

Bulmer, M.G. and Taylor, P.D. (1980) Dispersal and the sex ratio. *Nature* **284**:448–449.

Burkart, J.M. and van Schaik C.P. (2010) Cognitive consequences of cooperative breeding in primates? *Animal Cognition* **13**:1–19.

Canestrari, D., et al. (2004) False-feedings at the nests of carrion crows *Corvus corone corone*. *Behavioral Ecology and Sociobiology* **55**:477–483.

Cant, M.A. (1998) A model for the evolution of reproductive skew without reproductive suppression. *Animal Behaviour* **55**:163–169.

Cant, M.A. (2003) Patterns of helping effort in cooperatively breeding banded mongooses (*Mungos mungo*). *Journal of Zoology* **259**:115–121.

Cant, M.A. (2015) Banded mongooses: demography, life history, and social behavior. In: *Cooperative Breeding: Studies of Ecology, Evolution and Behaviour* (eds W.D. Koenig and J.L. Dickinson). Cambridge: Cambridge University Press.

Cant, M.A. and Johnstone, R.A. (2009) How threats influence the evolutionary resolution of within-group conflict. *American Naturalist* **173**:759–771.

Cant, M.A. and Young, A.J. (2013) Resolving social conflict among females without overt aggression. *Philosophical Transactions of the Royal Society B: Biological Sciences* **368**:20130076.

Cant, M.A., et al. (2001) Eviction and dispersal in cooperatively breeding banded mongooses (*Mungos mungo*). *Journal of Zoology* **254**:155–162.

Cant, M.A., et al. (2010) Reproductive control via eviction (but not the threat of eviction) in banded mongooses. *Proceedings of the Royal Society of London. Series B: Biological Sciences* **277**:20092097.

Cant, M.A., et al. (2013) Demography and social evolution of banded mongooses. In: *Advances in the Study of Behavior*, Vol. **45** (eds H.J. Brockman, et al.). San Diego, CA: Academic Press, 407–445.

Cant, M.A., *et al.* (2014) Policing of reproduction by hidden threats in a cooperative mammal. *Proceedings of the National Academy of Sciences of the United States of America* **111**:326–330.

Carlson, A.A., *et al.* (2004) Hormonal correlates of dominance in meerkats (*Suricata suricatta*). *Hormones and Behavior* **46**:141–150.

Carlson, A.A., *et al.* (2006) Cortisol levels are positively associated with pup-feeding rates in male meerkats. *Proceedings of the Royal Society of London. Series B: Biological Sciences* **273**:571–577.

Carter, A.J., *et al.* (2014) Cooperative personalities and social niche specialisation in female meerkats. *Journal of Evolutionary Biology* **27**:815–823.

Carter, C.S. and Roberts, R.L. (1997) The psychobiological basis of cooperative breeding in rodents. In: *Cooperative Breeding in Mammals* (eds N.G. Solomon and J. A. French). Cambridge: Cambridge University Press, 231–266.

Charmantier, A., *et al.* (2007) First evidence for heritable variation in cooperative breeding behaviour. *Proceedings of the Royal Society of London. Series B: Biological Sciences* **274**:1757–1761.

Christian, J.J. and Davis, D.E. (1964) Endocrines, behavior, and population: social and endocrine factors are integrated in the regulation of growth of mammalian populations. *Science* **146**:1550–1560.

Clark, A.B. (1978) Sex ratio and local resource competition in a prosimian primate. *Science* **201**:163–165.

Clark, M.M. and Galef, B.G. (2000) Why some male Mongolian gerbils may help at the nest: testosterone, asexuality and alloparenting. *Animal Behaviour* **59**:801–806.

Clarke, M.F. (1984) Co-operative breeding by the Australian bell miner *Manorina melanophrys* Latham: a test of kin selection theory. *Behavioral Ecology and Sociobiology* **14**:137–146.

Clarke, M.F., *et al.* (2002) Male-biased sex ratios in broods of the cooperatively breeding bell miner *Manorina melanophrys*. *Journal of Avian Biology* **33**:71–76.

Clutton-Brock, T.H. (1991) *The Evolution of Parental Care*. Princeton, NJ: Princeton University Press.

Clutton-Brock, T.H. (2002) Breeding together: kin selection and mutualism in cooperative vertebrates. *Science* **296**:69–72.

Clutton-Brock, T.H. (2006) Cooperative breeding in mammals. In: *Cooperation in Primates and Humans* (eds P.M. Kappeler and C.P. van Schaik). Berlin: Springer, 173–190.

Clutton-Brock, T.H. (2009) Structure and function in mammalian societies. *Philosophical Transactions of the Royal Society B: Biological Sciences* **364**:3229–3242.

Clutton-Brock, T.H. and Albon, S.D. (1989) *Red Deer in the Highlands*. Oxford: Blackwell Scientific Publications.

Clutton-Brock, T.H. and Huchard, E. (2013a) Social competition and its consequences in female mammals. *Journal of Zoology* **289**:151–171.

Clutton-Brock, T.H. and Huchard, E. (2013b) Social competition and selection in males and females. *Philosophical Transactions of the Royal Society B: Biological Sciences* **368**:20130074.

Clutton-Brock, T.H. and Iason, G.R. (1986) Sex ratio variation in mammals. *Quarterly Review of Biology* **61**:339–374.

Clutton-Brock, T.H. and Lukas, D. (2011) The evolution of social philopatry and dispersal in female mammals. *Molecular Ecology* **21**:472–492.

Clutton-Brock, T.H. and Manser, M.B. (2016) (in press) Kalahari meerkats. In: *Cooperative Breeding: Studies of Ecology, Evolution and Behaviour* (eds W.D. Koenig and J.L. Dickinson). Cambridge: Cambridge University Press.

Clutton-Brock, T.H. and Parker, G.A. (1995) Punishment in animal societies. *Nature* **373**:209–216.

Clutton-Brock, T.H., *et al.* (1998a) Infanticide and expulsion of females in a cooperative mammal. *Proceedings of the Royal Society of London. Series B: Biological Sciences* **265**:2291–2295.

Clutton-Brock, T.H., *et al.* (1998b) Costs of cooperative behaviour in suricates, *Suricata suricatta*. *Proceedings of the Royal Society of London. Series B: Biological Sciences* **265**:185–190.

Clutton-Brock, T.H., *et al.* (1999a) Selfish sentinels in cooperative mammals. *Science* **284**:1640–1644.

Clutton-Brock, T.H., *et al.* (1999b) Predation, group size and mortality in a cooperative mongoose, *Suricata suricatta*. *Journal of Animal Ecology* **68**:672–683.

Clutton-Brock, T.H., *et al.* (1999c) Reproduction and survival of suricates (*Suricata suricatta*) in the southern Kalahari. *African Journal of Ecology* **37**:69–80.

Clutton-Brock, T.H., *et al.* (2000) Individual contributions to babysitting in a cooperative mongoose, *Suricata suricatta*. *Proceedings of the Royal Society of London. Series B: Biological Sciences* **267**:301–305.

Clutton-Brock, T.H., *et al.* (2001a) Cooperation, control and concession in meerkat groups. *Science* **291**:478–481.

Clutton-Brock, T.H., *et al.* (2001b) Contributions to cooperative rearing in meerkats. *Animal Behaviour* **61**:705–710.

Clutton-Brock, T.H., *et al.* (2001c) Effects of helpers on juvenile development and survival in meerkats. *Science* **293**:2446–2449.

Clutton-Brock, T.H., *et al.* (2002) Evolution and development of sex differences in cooperative behavior in meerkats. *Science* **297**:253–256.

Clutton-Brock, T.H., *et al.* (2003) Meerkat helpers do not specialize in particular activities. *Animal Behaviour* **66**:531–540.

Clutton-Brock, T.H., *et al.* (2004) Behavioural tactics of breeders in cooperative meerkats. *Animal Behaviour* **68**:1029–1140.

Clutton-Brock, T.H., *et al.* (2005) 'False-feeding' and aggression in meerkat societies. *Animal Behaviour* **69**:1273–1284.

Clutton-Brock, T.H., *et al.* (2006) Intrasexual competition and sexual selection in cooperative mammals. *Nature* **444**:1065–1068.

Clutton-Brock, T.H., *et al.* (2008) Group size and the suppression of subordinate reproduction in Kalahari meerkats. *Animal Behaviour* **76**:689–700.

Clutton-Brock, T.H., *et al.* (2010) Adaptive suppression of subordinate reproduction in cooperative mammals. *American Naturalist* **176**:664–673.

Cockburn, A. (1998) Evolution of helping behaviour in cooperatively breeding birds. *Annual Review of Ecology and Systematics* **29**:141–177.

Cockburn, A. and Russell, A.F. (2011) Cooperative breeding: a question of climate? *Current Biology* **21**:R195–R197.

Cooney, R. and Bennett, N.C. (2000) Inbreeding avoidance and reproductive skew in a cooperative mammal. *Proceedings of the Royal Society of London. Series B: Biological Sciences* **267**:801–806.

Cornwallis, C.K., et al. (2009) Routes to indirect fitness in cooperatively breeding vertebrates: kin discrimination and limited dispersal. *Journal of Evolutionary Biology* **22**:2445–2457.

Cornwallis, C.K., et al. (2010) Promiscuity and the evolutionary transition to complex societies. *Nature* **466**:969–972.

Courchamp, F. and Macdonald, D.W. (2001) Crucial importance of pack size in the African wild dog *Lycaon pictus*. *Animal Conservation* **4**:169–174.

Courchamp, F., et al. (1999a) Population dynamics of obligate cooperators. *Proceedings of the Royal Society of London. Series B: Biological Sciences* **266**:557–563.

Courchamp, F., et al. (1999b) Inverse density dependence and the Allee effect. *Trends in Ecology and Evolution* **14**:405–410.

Courchamp, F., et al. (2000) Multipack dynamics and the Allee effect in the African wild dog, *Lycaon pictus*. *Animal Conservation* **3**:277–285.

Courchamp, F. et al. (2002) Small pack size imposes a trade-off between hunting and pup-guarding in the painted hunting dog *Lycaon pictus*. *Behavioural Ecology* **13**:20–27.

Creel, S. (1997) Cooperative hunting and group size: assumptions and currencies. *Animal Behaviour* **54**:1319–1324.

Creel, S. (1998) Sizing up the competition. *Natural History* **107**:34–43.

Creel, S. (2001) Social dominance and stress hormones. *Trends in Ecology and Evolution* **16**:491–497.

Creel, S. and Creel, N.M. (2002) *The African Wild Dog: Behavior, Ecology, and Conservation*. Princeton, NJ: Princeton University Press.

Creel, S. and Creel, N.M. (2015) Opposing effects of group size on reproduction and survival in African wild dogs. *Behavioral Ecology* doi: 10.1093/beheco/arv100.

Creel, S. and Macdonald, D. (1995) Sociality, group size, and reproductive suppression among carnivores. In: *Advances in the Study of Behavior, Vol. 24* (eds P.J.B. Slater, J.S. Rosenblatt, C.T. Snowdon and M. Milinski). San Diego, CA: Academic Press, 203–257.

Creel, S. and Waser, P.M. (1997) Variation in reproductive suppression among dwarf mongooses: interplay between mechanisms and evolution. In: *Cooperative Breeding in Mammals* (eds N.G. Solomon and J.A. French). Cambridge: Cambridge University Press, 150–170.

Creel, S., et al. (1991) Spontaneous lactation is an adaptive result of pseudopregnancy. *Nature* **351**:660–662.

Creel, S., et al. (1998) Birth order, estrogens and sex-ratio adaptation in African wild dogs (*Lycaon pictus*). *Animal Reproduction Science* **53**:315–320.

Creel, S., et al. (2013) The ecology of stress: effects of the social environment. *Functional Ecology* **27**:66–80.

Crespi, B.J. and Yanega, D. (1995) The definition of eusociality. *Behavioral Ecology* **6**:109–115.

Crockford, C., et al. (2013) Urinary oxytocin and social bonding in related and unrelated wild chimpanzees. *Proceedings of the Royal Society of London. Series B: Biological Sciences* **280**:20122765.

Dammann, P. and Burda, H. (2006) Sexual activity and reproduction delay ageing in a mammal. *Current Biology* **16**:R117–R118.

Dammann, P., et al. (2011) Extended longevity of reproductives appears to be common in *Fukomys* mole-rats (Rodentia, Bathyergidae). *PLOS ONE* **6**:e18757.

Dammann, P., et al. (2012) Advanced glycation end-products as markers of aging and longevity in the long-lived Ansell's mole-rat (*Fukomys anselli*). *Journals of Gerontology Series A: Biological Sciences and Medical Sciences* **67**:573–583.

De Dreu, C.K.W., et al. (2011) Oxytocin promotes human ethnocentrism. *Proceedings of the National Academy of Sciences of the United States of America* **108**:1262–1266.

Díaz-Muñoz, S.L. (2011) Paternity and relatedness in a polyandrous nonhuman primate: testing adaptive hypotheses of male reproductive cooperation. *Animal Behaviour* **82**:563–571.

Díaz-Muñoz, S.L. (2015) Complex cooperative breeders: Using infant care costs to explain variability in callitrichine social and reproductive behavior. *American Journal of Primatology* DOI: 10.1002/ajp.22431. [Accessed 21 January 2016]

Digby, L.J., et al. (2007) Callitrichines: the role of competition in cooperatively breeding species. In: *Primates in Perspective* (eds C.J. Campbell, A. Fuentes, K.C. MacKinnon, M. Panger and S.K. Bearder). New York: Oxford University Press, 85–106.

Dluzen, D.E., et al. (1981) Male vole urine changes luteinizing hormone releasing hormone and norepinephrine in female olfactory bulb. *Science* **212**:573–575.

Dunn, P.O., et al. (1995) Fairy wren helpers often care for young to which they are unrelated. *Proceedings of the Royal Society of London. Series B: Biological Sciences* **259**:339–343.

Du Plessis, M.A. (2004) Physiological ecology. In: *Ecology and Evolution of Cooperative Breeding in Birds* (eds W.D. Koenig and J.L. Dickinson). Cambridge: Cambridge University Press, 117–127.

Du Plessis, M.A., et al. (1995) Ecological and life-history correlates of cooperative breeding in South African birds. *Oecologia* **102**:180–188.

Eberle, M. and Kappeler, P.M. (2006) Family insurance: kin selection and cooperative breeding in a solitary primate (*Microcebus murinus*). *Behavioral Ecology and Sociobiology* **60**:582–588.

Emlen, S.T. (1991) Evolution of cooperative breeding in birds and mammals. In: *Behavioural Ecology: An Evolutionary Approach* (eds J.R. Krebs and N.B. Davies). Oxford: Blackwell Scientific Publications, 301–337.

Emlen, S.T. and Wrege, P.H. (1988) The role of kinship in helping decisions among white-fronted bee-eaters. *Behavioral Ecology and Sociobiology* **23**:305–315.

Emlen, S.T., *et al.* (1986) Sex-ratio selection in species with helpers-at-the-nest. *American Naturalist* **127**:1–8.

English, S., *et al.* (2008) Sex differences in responsiveness to begging in a cooperative mammal. *Biology Letters* **4**:334–337.

Faulkes, C.G. and Abbott, D.H. (1997) The physiology of a reproductive dictatorship: regulation of male and female reproduction by a single breeding female in colonies of naked mole-rats. In: *Cooperative Breeding in Mammals* (eds N.G. Solomon and J.A. French). Cambridge: Cambridge University Press, 302–334.

Faulkes, C.G. and Bennett, N.C. (2007) African mole-rats: social and ecological diversity. In: *Rodent Societies: An Ecological and Evolutionary Perspective* (eds J.O. Wolff and P.W. Sherman). Chicago: University of Chicago Press, 427–437.

Faulkes, C.G., *et al.* (1990) LH responses of female naked mole-rats, *Heterocephalus glaber*, to single and multiple doses of exogenous GnRH. *Journal of Reproduction and Fertility* **89**:317–323.

Firestone, K.B., *et al.* (1991) Female–female interactions and social stress in prairie voles. *Behavioral and Neural Biology* **55**:31–41.

Frank, S.A. (1995) Mutual policing and repression of competition in the evolution of cooperative groups. *Nature* **377**:520–522.

French, J.A. (1997) Proximate regulation of singular breeding in Callitrichid primates. In: *Cooperative Breeding in Mammals* (eds N.G. Solomon and J.A. French). Cambridge: Cambridge University Press, 34–75.

French, J.A., *et al.* (1984) The effects of social environment on estrogen excretion, scent marking, and sociosexual behavior in tamarins (*Saguinus oedipus*). *American Journal of Primatology* **6**:155–167.

Gadagkar, R. (1994) Why the definition of eusociality is not helpful to understand its evolution and what should we do about it. *Oikos* **70**:485–488.

Gaston, A.J. (1977) Social behaviour within groups of jungle babblers (*Turdoides striatus*). *Animal Behaviour* **25**:828–848.

Getz, L.L., *et al.* (1994) Natal dispersal and philopatry in prairie voles (*Microtus ochrogaster*): settlement, survival, and potential reproductive success. *Ethology Ecology and Evolution* **6**:267–284.

Gilchrist, J.S. (2004) Pup escorting in the communal breeding banded mongoose: behavior, benefits and maintenance. *Behavioral Ecology* **15**:952–960.

Gilchrist, J.S. (2008) Aggressive monopolization of mobile carers by young of a cooperative breeder. *Proceedings of the Royal Society of London. Series B: Biological Sciences* **275**:2491–2498.

Gilchrist, J.S. and Russell, A.F. (2007) Who cares? Individual contributions to pup care by breeders vs non-breeders in the cooperatively breeding banded mongoose *Mungos mungo*. *Behavioral Ecology and Sociobiology* **61**:1053–1060.

Goldizen, A.W. (1987a) Tamarins and marmosets: communal care of offspring. In: *Primate Societies* (eds B.B. Smuts, D.L. Cheney, R.M. Seyfarth, R.W. Wrangham and T.T. Struhsaker). Chicago: University of Chicago Press, 34–43.

Goldizen, A.W. (1987b) Facultative polyandry and the role of infant-carrying in wild saddle-back tamarins (*Saguinus fuscicollis*). *Behavioral Ecology and Sociobiology* **20**:99–109.

Gompper, M.E., *et al.* (1997) Genetic relatedness, coalitions and social behaviour of white-nosed coatis, *Nasua narica*. *Animal Behaviour* **53**:781–797.

Gowaty, P.A. and Lennartz, M.R. (1985) Sex ratios of nestling and fledgling red-cockaded woodpeckers (*Picoides borealis*) favor males. *American Naturalist* **126**:347–353.

Greenwood, P.J. (1980) Mating systems, philopatry and dispersal in birds and mammals. *Animal Behaviour* **28**:1140–1162.

Gregory, S.D. *et al.* (2010) Limited evidence for the demographic Allee effect from numerous species across taxa. *Ecology* **91**:2151–2161.

Griffin, A.S., *et al.* (2003) A genetic analysis of breeding success in the cooperative meerkat (*Suricata suricatta*). *Behavioral Ecology* **14**:472–480.

Griffin, A.S., *et al.* (2005) Cooperative breeders adjust offspring sex ratios to produce helpful helpers. *American Naturalist* **166**:628–632.

Gubernick, D.J. and Nelson, R.J. (1989) Prolactin and paternal behavior in the biparental California mouse, *Peromyscus californicus*. *Hormones and Behavior* **23**:203–210.

Hamilton, W.D. (1964) The genetical evolution of social behaviour. I. *Journal of Theoretical Biology* **7**:1–16.

Hamilton, W.D. (1967) Extraordinary sex ratios. *Science* **156**:477–487.

Hauber, M.E. and Lacey, E.A. (2005) Bateman's principle in cooperatively breeding vertebrates: the effects of non-breeding alloparents on variability in female and male reproductive success. *Integrative and Comparative Biology* **45**:903–914.

Haydock, J. and Koenig, W.D. (2002) Reproductive skew in the polygynandrous acorn woodpecker. *Proceedings of the National Academy of Sciences of the United States of America* **99**:7178–7183.

Hayes, L.D. and Solomon, N.G. (2004) Costs and benefits of communal rearing for female prairie voles *Microtus ochrogaster*. *Behavioral Ecology and Sociobiology* **56**:585–593.

Haymann, E. (2000) The number of adult males in callithrichine groups and its implications for callithrichine social evolution. In: *Primate males* (ed. P.M. Kappeler). Cambridge University Press, Cambridge.

Hazell, R.W.A., *et al.* (2000) Adult dispersal in the co-operatively breeding Damaraland mole-rat (*Cryptomys damarensis*): a case study from the Waterberg region of Namibia. *Journal of Zoology* **252**:19–25.

Heinsohn, R.G. (2004) Parental care, load lightening and costs. In: *Ecology and Evolution of Cooperative Breeding in Birds* (eds W.D. Koenig and J.L. Dickinson). Cambridge: Cambridge University Press, 67–80.

Heinsohn, R.G. and Legge, S. (1999) The cost of helping. *Trends in Ecology and Evolution* **14**:53–57.

Henry, M.D., *et al.* (2013) High rates of pregnancy loss by subordinates leads to high reproductive skew in wild golden lion tamarins (*Leontopithecus rosalia*). *Hormones and Behavior* **63**:675–683.

Hodge, S.J. (2003) The evolution of cooperation in the communal breeding banded mongoose. PhD thesis, University of Cambridge, Cambridge.

Hodge, S.J. (2005) Helpers benefit offspring in both the short and long-term in the cooperatively breeding banded mongoose. *Proceedings of the Royal Society of London. Series B: Biological Sciences* **272**:2479–2484.

Hodge, S.J. (2007) Counting the costs: the evolution of male-biased care in the cooperative breeding banded mongoose. *Animal Behaviour* **74**:911–919.

Hodge, S.J., *et al.* (2008) Determinants of reproductive success in dominant female meerkats. *Journal of Animal Ecology* **77**:92–102.

Hodge, S.J., *et al.* (2009) Food limitation increases aggression in juvenile meerkats. *Behavioral Ecology* **20**:930–935.

Hoogland, J.L. (1983) Nepotism and alarm calling in the black-tailed prairie dog (*Cynomys ludovicianus*). *Animal Behaviour* **31**:472–479.

Hoogland, J.L. (1995) *The Black-tailed Prairie Dog, Social Life of a Burrowing Mammal*. Chicago: University of Chicago Press.

Horn, A.G. and Leonard, M.C. (2002) Efficacy and the design of begging signals. In: *The Evolution of Begging* (eds J. Wright and M. Leonard). Dordrecht: Kluwer Academic, 127–141.

Huck, M. *et al.* (2005) Paternity and kinship patterns in polyandrous moustached tamarins (*Saguinus mystax*). *American Journal Physical Anthropology* **127**: 449–464.

Hughes, W.O.H., *et al.* (2008) Ancestral monogamy shows kin selection is key to the evolution of eusociality. *Science* **320**:1213–1216.

Inglett, B.J., *et al.* (1989) Dynamics of intrafamily aggression and social reintegration in lion tamarins. *Zoo Biology* **8**:67–78.

Isler, K. and van Schaik, C.P. (2009) The expensive brain: a framework for explaining evolutionary changes in brain size. *Journal of Human Evolution* **57**:392–400.

Isler, K. and van Schaik, C.P. (2012) Allomaternal care, life history and brain size evolution in mammals. *Journal of Human Evolution* **63**:52–63.

Iwaniuk, A.N. and Arnold, K.E. (2004) Is cooperative breeding associated with bigger brains? A comparative test in the Corvida (Passeriformes). *Ethology* **110**:203–220.

Jacobs, D.S. and Jarvis, J.U.M. (1996) No evidence for the work-conflict hypothesis in the eusocial naked mole-rat (*Heterocephalus glaber*). *Behavioral Ecology and Sociobiology* **39**:401–409.

Jarvis, J.U.M. (1981) Eusociality in a mammal: cooperative breeding in naked mole-rat colonies. *Science* **212**:571–573.

Jarvis, J.U.M. (1991) Reproduction in naked mole-rats. In: *The Biology of the Naked Mole-rat* (eds P.W. Sherman, J.U.M. Jarvis and R.D. Alexander). Princeton, NJ: Princeton University Press, 426–445.

Jarvis, J.U.M., *et al.* (1994) Mammalian eusociality: a family affair. *Trends in Evolution and Ecology* **9**:47–51.

Jetz, W. and Rubenstein, D.I. (2011) Environmental uncertainty and the global biogeography of cooperative breeding in birds *Current Biology* **21**:72–78.

Johnstone, R.A. (2000) Models of reproductive skew: a review and synthesis. *Ethology* **106**:5–26.

Johnstone, R.A. and Cant, M.A. (1999) Reproductive skew and the threat of eviction: a new perspective. *Proceedings of the Royal Society of London. Series B: Biological Sciences* **266**:275–279.

Johnstone, R.A. and Rodrigues, A.M.M. (2016) Cooperation and the common good. *Philosophical Transactions of the Royal Society of London B* DOI: 10.1098/rstb.2015.0086. [Accessed 21 January 2016]

Keller, L. (1998) Queen lifespan and colony characteristics in ants and termites. *Insectes Sociaux* **45**:235–246.

Keller, L. and Reeve, H.K. (1994) Partitioning of reproduction in animal societies. *Trends in Evolution and Ecology* **9**:98–103.

Keller, L. and Vargo, E.C. (1993) Reproductive structure and reproductive roles in colonies of eusocial insects. In: *Queen Number and Sociality in Insects* (ed. L. Keller). Oxford: Oxford University Press, 16–44.

Kilner, R.M. (2002) The evolution of complex begging displays. In: *The Evolution of Nestling Begging: Competition, Cooperation and Communication* (eds J. Wright and M.L. Leonard). Dordrecht: Kluwer Academic, 87–106.

Kingma, S.A., *et al.* (2011) Multiple benefits drive helping behavior in a cooperatively breeding bird: an integrated analysis. *American Naturalist* **177**:486–495.

Koenig, W.D. and Mumme, R.L. (1987) *Population Ecology of the Cooperatively Breeding Acorn Woodpecker*. Princeton, NJ: Princeton University Press.

Kokko, H. and Ekman, J. (2002) Delayed dispersal as a route to breeding: territorial inheritance, safe havens, and ecological constraints *American Naturalist* **160**:468–484.

Kokko, H. and Johnstone, R.A. (1999) Social queuing in animal societies: a dynamic model of reproductive skew. *Proceedings of the Royal Society of London. Series B: Biological Sciences* **266**:571–578.

Kokko, H., *et al.* (2001) The evolution of cooperative breeding through group augmentation. *Proceedings of the Royal Society of London. Series B: Biological Sciences* **268**:187–196.

Kokko, H., *et al.* (2002) The evolution of parental and alloparental effort in cooperatively breeding groups: when should helpers pay to stay? *Behavioral Ecology* **13**:291–300.

Komdeur, J. (1994a) The effect of kinship on helping in the cooperative breeding Seychelles warbler (*Acrocephalus sechellensis*). *Proceedings of the Royal Society of London. Series B: Biological Sciences* **256**:47–52.

Komdeur, J. (1994b) Experimental evidence for helping and hindering by previous offspring in the cooperative-breeding Seychelles warbler *Acrocephalus sechellensis*. *Behavioral Ecology and Sociobiology* **34**:175–186.

Komdeur, J. (1996) Influence of helping and breeding experience on reproductive performance in the Seychelles warbler: a translocation experiment. *Behavioral Ecology* **7**:326–333.

Komdeur, J. (2004) Sex-ratio manipulation. In: *Ecology and Evolution of Cooperative Breeding in Birds* (eds W.D. Koenig and J.L. Dickinson). Cambridge: Cambridge University Press, 102–116.

Kruska, D.C.T. and Steffen, K. (2009) Encephalization of Bathyergidae and comparison of brain structure volumes between the Zambian mole-rat *Fukomys anselli* and the giant mole-rat *Fukomys mechowii*. *Mammalian Biology* **74**:298–307.

Kunc, H.P., *et al.* (2007) Begging signals in a mobile feeding system: the evolution of different call types. *American Naturalist* **170**:617–624.

Kutsukake, N. and Clutton-Brock, T.H. (2006) Aggression and submission reflect reproductive conflict between females in cooperatively breeding meerkats *Suricata suricatta*. *Behavioral Ecology and Sociobiology* **59**:541–548.

Lacey, E.A. and Ebensperger, L.A. (2007) Social structure in Octodontid and Ctenomyid rodents. In: *Rodent Societies: An Ecological and Evolutionary Perspective* (eds J.O. Wolff and P.W. Sherman). Chicago: University of Chicago Press, 403–415.

Lacey, E.A. and Sherman, P.W. (1991) Social organization of naked mole-rat colonies: evidence for divisions of labor. In: *The Ecology of the Naked Mole-Rat* (eds P.W. Sherman, J.U.M. Jarvis and R.D. Alexander). Princeton, NJ: Princeton University Press, 275–336.

Lacey, E.A. and Sherman, P.W. (1997) Cooperative breeding in naked mole-rats: implications for vertebrate and invertebrate sociality. In: *Cooperative Breeding in Mammals* (eds N.G. Solomon and J.A. French). Cambridge: Cambridge University Press, 267–301.

Langen, T.A. (2000) Prolonged offspring dependence and cooperative breeding in birds. *Behavioral Ecology* **11**:367–377.

Leadbeater, E., *et al.* (2011) Nest inheritance is the missing source of direct fitness in a primitively eusocial insect. *Science* **333**:874–876.

Leggett, H.C., *et al.* (2011) Promiscuity and the evolution of cooperative breeding. *Proceedings of the Royal Society of London. Series B: Biological Sciences* **279**:1404–1411.

Lehmann, L. and Keller, L. (2006) The evolution of cooperation and altruism: a general framework and classification of models. *Journal of Evolutionary Biology* **19**:1365–1378.

Lehmann, L. and Rousset, F. (2010) How life history and demography promote or inhibit the evolution of helping behaviours. *Proceedings of the Royal Society of London. Series B: Biological Sciences* **365**:2599–2617.

Lehmann, L., *et al.* (2007) Group selection and kin selection: two concepts but one process. *Proceedings of the National Academy of Sciences of the United States of America* **104**:6736–6739.

Liang, S., *et al.* (2010) Resistance to experimental tumorigenesis in cells of a long-lived mammal, the naked mole-rat (*Heterocephalus glaber*). *Aging Cell* **9**:626–635.

Lucas, J.R., *et al.* (1997) Dynamic optimization and cooperative breeding: an evaluation of future fitness effects. In: *Cooperative Breeding in Mammals* (eds N.G. Solomon and J.A. French). Cambridge: Cambridge University Press, 171–198.

Lucia, K.E., *et al.* (2008) Philopatry in prairie voles: an evaluation of the habitat saturation hypothesis. *Behavioral Ecology* **19**:774–783.

Lukas, D. and Clutton-Brock, T.H. (2012) Cooperative breeding and monogamy in mammalian societies. *Proceedings of the Royal Society of London. Series B: Biological Sciences* **279**:2151–2156.

Lutermann, H., *et al.* (2014) Sex ratio variation in a eusocial mammal, the Damaraland mole-rat, *Fukomys damarensis*. *Journal of Zoology* **294**:139–145.

McGuire, B., *et al.* (1993) Natal dispersal and philopatry in prairie voles (*Microtus ochrogaster*) in relation to population density, season, and natal social environment. *Behavioral Ecology and Sociobiology* **32**:293–302.

MacLeod, K.J. and Clutton-Brock, T.H. (2013) No evidence for sex ratio variation in the cooperatively breeding meerkat, *Suricata suricatta*. *Animal Behaviour* **85**:645–653.

MacLeod, K.J. and Clutton-Brock, T.H. (2014) Low costs of allonursing in meerkats: mitigation by behavioural change? *Behavioral Ecology* doi: 10.1093/beheco/aru205.

MacLeod, K.J., *et al.* (2013) Factors predicting the frequency, likelihood and duration of allonursing in the cooperatively breeding meerkat. *Animal Behaviour* **86**:1059–1067.

McNutt, J.W. and Silk, J.B. (2008) Pup production, sex ratios, and survivorship in African wild dogs, *Lycaon pictus*. *Behavioral Ecology and Sociobiology* **62**:1061–1067.

Madden, J.R. and Clutton-Brock, T.H. (2011) Experimental peripheral administration of oxytocin elevates a suite of cooperative behaviours in a wild social mammal. *Proceedings of the Royal Society of London. Series B: Biological Sciences* **278**:1189–1194.

Madden, J.R., *et al.* (2009) Calling in the gap: competition or cooperation in littermates' begging behaviour. *Proceedings of the Royal Society of London. Series B: Biological Sciences* **276**:1255–1262.

Magrath, R.D. and Yezerinac, I.S.M. (1997) Facultative helping does not influence reproductive success or survival in cooperatively breeding white-browed scrubwrens. *Journal of Animal Ecology* **66**:658–670.

Magrath, R.D., *et al.* (2004) Reproductive skew. In: *Ecology and Evolution of Cooperative Breeding in Birds* (eds W.D. Koenig and J.L. Dickinson). Cambridge: Cambridge University Press, 157–176.

Malcolm, J.R. (1979) Social organisation and communal rearing in African wild dogs. PhD thesis, Harvard University, Cambridge, MA.

Malcolm, J.R. and Marten, K. (1982) Natural selection and the communal rearing of pups in African wild dogs, *Lycaon pictus*. *Behavioral Ecology and Sociobiology* **10**:1–13.

Manser, M.B. and Avey, G. (2000) The effect of pup vocalisations on food allocation in a cooperative mammal, the meerkat (*Suricata suricatta*). *Behavioral Ecology and Sociobiology* **48**:429–437.

Manser, M.B., *et al.* (2008) Signals of need in a cooperatively breeding mammal with mobile offspring. *Animal Behaviour* **76**:1805–1813.

Mares, R., *et al.* (2014) Timing of predispersal prospecting is influenced by environmental, social and state-dependent factors in meerkats. *Animal Behaviour* **88**:185–193.

Moehlman, P.D. (1979) Jackal helpers and pup survival. *Nature* **277**:382–383.

Moehlman, P.D. (1989) Intraspecific variation in canid social systems. In: *Carnivore Behavior, Ecology and Evolution* (ed. J.L. Gittleman). Ithaca, NY: Cornell University Press, 143–163.

Moehlman, P.D. and Hofer, H. (1997) Cooperative breeding, reproductive suppression and body mass in canids. In: *Cooperative Breeding in Mammals* (eds N.G. Solomon and J.A. French). Cambridge: Cambridge University Press, 76–128.

Montiglio, P.-O., *et al.* (2013) Social niche specialization under constraints: personality, social interactions and environmental heterogeneity. *Philosophical Transactions of the Royal Society B: Biological Sciences* **368**:20120343.

Mulder, R.A. and Langmore, N.E. (1993) Dominant males punish helpers for temporary defection in superb fairy-wrens. *Animal Behaviour* **45**:830–833.

Muller, C.A. and Manser, M. (2008) Mutual recognition of pups and providers in the cooperatively breeding banded mongoose. *Animal Behaviour* **75**:1683–1692.

Nichols, H.J., *et al.* (2012) Resource limitation moderates the adaptive suppression of subordinate breeding in a cooperatively breeding mongoose. *Behavioral Ecology* **23**:635–642.

Nielsen, J.F. (2013) Evolutionary genetics of meerkats (*Suricata suricatta*). PhD thesis, University of Edinburgh.

Nievergelt, C.M. *et al.* (2000) Genetic analysis of group composition and breeding system in a wild common marmoset (*Callithrix jacchus*) population. *International Journal Primatol* **21**:1–20.

Nonacs, P. (2011) Monogamy and high relatedness do not preferentially favor the evolution of cooperation. *BMC Evolutionary Biology* **11**:58.

Nowak, M.A. (2006) Five rules for the evolution of cooperation. *Science* **314**:1560–1565.

O'Riain, M.J. and Braude, S. (2001) Inbreeding versus outbreeding in captive and wild populations of naked mole-rats. In: *Dispersal* (eds J. Clobert, E. Danchin, A.A. Dhondt and J.D. Nichols). New York: Oxford University Press, 143–154.

O'Riain, M.J., *et al.* (1996) A dispersive morph in the naked mole-rat. *Nature* **380**:619–621.

O'Riain, M.J., *et al.* (2000) Reproductive suppression and inbreeding avoidance in wild populations of cooperatively breeding meerkats (*Suricata suricatta*). *Behavioral Ecology and Sociobiology* **48**:471–477.

Owens, D.D. and Owens, M.J. (1984) Helping behaviour in brown hyenas. *Nature* **308**:843–845.

Ozgul, A., *et al.* (2014) Linking body mass and group dynamics in an obligate cooperative breeder. *Journal of Animal Ecology* **83**:1357–1366.

Packard, J.M., *et al.* (1985) Causes of reproductive failure in two family groups of wolves (*Canis lupus*). *Zeitschrift fur Tierpsychologie* **68**:24–40.

Pan, W., *et al.* (2014) Birth intervention and non-maternal infant-handling during parturition in a nonhuman primate. *Primates* **55**:483–488.

Pen, I. and Weissing, F.J. (2000) Sex-ratio optimization with helpers at the nest. *Proceedings of the Royal Society of London. Series B: Biological Sciences* **267**:539–543.

Pen, I. and Weissing, F.J. (2002) Optimal sex allocation: steps towards a mechanistic theory. In: *Sex Ratios: Concepts and Research Methods* (ed. I.C.W. Hardy). Cambridge: Cambridge University Press, 26–47.

Poiani, A. (1993) Social structure and the development of helping behaviour in the bell miner *Manorina melanophrys* Meliphagidae. *Ethology* **93**:62–80.

Pottinger, T.G. and Carrick, T.R. (2001) Stress responsiveness affects dominant–subordinate relationships in rainbow trout. *Hormones and Behavior* **40**:419–427.

Powell, R.A. and Fried, J.J. (1992) Helping by juvenile pine voles (*Microtus pinetorum*), growth and survival of younger siblings, and the evolution of pine vole sociality. *Behavioral Ecology* **3**:325–333.

Price, E.C. (1992) The costs of infant carrying in captive cottontop tamarins. *International Journal of Primatology* **36**:43–55.

Pusey, A.E. (1987) Sex-biased dispersal and inbreeding avoidance in birds and mammals. *Trends in Ecology and Evolution* **2**:295–299.

Pusey, A.E. and Packer, C. (1994) Non-offspring nursing in social carnivores: minimizing the costs. *Behavioral Ecology* **5**:362–374.

Rapaport, L.G. (2011) Progressive parenting behavior in wild golden lion tamarins. *Behavioral Ecology* **22**:745–754.

Rapaport, L.G., *et al.* (2013) Do mothers prefer helpers? Birth sex ratio adjustment in captive callitrichines. *Animal Behaviour* **85**:1295–1302.

Reeve, H.K. (1992) Queen activation of lazy workers in colonies of the eusocial naked mole-rat. *Nature* **358**:147–149.

Reeve, H.K. (2000) A transactional theory of within-group conflict. *American Naturalist* **155**:365–382.

Reeve, H.K. and Ratnieks, F.L.W. (1993) Queen–queen conflicts in polygynous societies: mutual tolerance and reproductive skew. In: *Queen Number and Sociality in Insects* (ed. L. Keller). Oxford: Oxford University Press, 45–85.

Reeve, H.K. and Shen, S.-F. (2013) Unity and disunity in the search for a unified reproductive skew theory. *Animal Behaviour* **85**:1137–1144.

Reeve, H.K. and Sherman, P.W. (1991) Intracolonial aggression and nepotism by the breeding female naked mole-rat. In: *The Biology of the Naked Mole-rat* (eds P.W. Sherman, J.U.M. Jarvis and R.D. Alexander). Princeton, NJ: Princeton University Press, 337–357.

Reeve, H.K., *et al.* (1998) Reproductive sharing in animal societies: reproductive incentives or incomplete control by dominant breeders? *Behavioral Ecology* **9**:267–278.

Reyer, H.U., *et al.* (1986) Avian helpers at the nest: are they psychologically castrated? *Ethology* **71**:216–228.

Ridley, A.R. (2003) The causes and consequences of helping behaviour in the cooperatively breeding Arabian babbler. PhD thesis, University of Cambridge, Cambridge.

Riehl, C. (2013) Evolutionary routes to non-kin cooperative breeding in birds. *Proceedings of the Royal Society of London. Series B: Biological Sciences* **280**:20132245.

Riehl, C. and Frederickson, M.E. (2016) Cheating and punishment in cooperative animal societies. *Philosophical Transactions of the Royal Society of London B* DOI: 10.1098/rstb.2015.0090. [Accessed 21 January 2016]

Roberts, R.L., *et al.* (1998) Cooperative breeding and monogamy in prairie voles: influence of the sire and geographical variation. *Animal Behaviour* **55**:1131–1140.

Rood, J.P. (1986) Ecology and social evolution in the mongooses. In: *Ecological Aspects of Social Evolution* (eds D.I. Rubenstein and R.W. Wrangham). Princeton, NJ: Princeton University Press, 131–152.

Rood, J.P. (1990) Group size, survival, reproduction, and routes to breeding in dwarf mongooses. *Animal Behaviour* **39**:566–572.

Rubenstein, D.I. (2011) Spatiotemporal environmental variation, risk aversion, and the evolution of cooperative breeding as a bet-hedging strategy. *Proceedings of the National Academy of Sciences of the United States of America* **108**:10816–10822.

Russell, A.F. (2004) Mammals: comparisons and contrasts. In: *Ecology and Evolution of Cooperative Breeding in Birds* (eds W. Koenig and J. Dickinson). Cambridge: Cambridge University Press, 210–227.

Russell, A.F. and Hatchwell, B.J. (2001) Experimental evidence for kin-biased helping in a cooperatively breeding vertebrate. *Proceedings of the Royal Society of London. Series B: Biological Sciences* **268**:2169–2174.

Russell, A.F., *et al.* (2003a) Breeding success in cooperative meerkats: effects of helper number and maternal state. *Behavioral Ecology* **14**:486–492.

Russell, A.F., *et al.* (2003b) Cost minimization by helpers in cooperative vertebrates. *Proceedings of the National Academy of Sciences of the United States of America* **100**:3333–3338.

Russell, A.F., *et al.* (2004) Adaptive size modification by dominant female meerkats. *Evolution* **58**:1600–1607.

Russell, A.F., *et al.* (2007) Helpers increase the reproductive potential of offspring in cooperative meerkats. *Proceedings of the Royal Society of London. Series B: Biological Sciences* **274**:513–520.

Saltzman, W., *et al.* (2009) Reproductive skew in female common marmosets: what can proximate mechanisms tell us about ultimate causes? *Proceedings of the Royal Society of London. Series B: Biological Sciences* **276**:389–399.

Sanchez, S., *et al.* (1999) Costs of infant-carrying in the cotton-top tamarin (*Saguinus oedipus*). *American Journal of Primatology* **48**:99–111.

Sanderson, J.L. *et al.* (2015) Elevated glucocorticoid concentrations during gestation predict reduced reproductive success in subordinate female banded mongooses. *Biology Letters* DOI: 10.1098/rsbl.2015.0620. [Accessed 21 January 2016]

Santema, P. and Clutton-Brock, T.H. (2012) Dominant female meerkats do not use aggression to elevate work rates of helpers in response to increased brood demand. *Animal Behaviour* **83**:827–832.

Scantlebury, M., *et al.* (2002) The energetics of lactation in cooperatively breeding meerkats *Suricata suricatta*. *Proceedings of the Royal Society of London. Series B: Biological Sciences* **269**:2147–2153.

Scantlebury, M., *et al.* (2006) Energetics reveals physiologically distinct castes in a eusocial mammal. *Nature* **440**:795–797.

Schneider, T.C. and Kappeler, P.M. (2014) Social systems and life-history characteristics of mongooses. *Biological Reviews of the Cambridge Philosophical Society* **89**:173–198.

Schmidt, C. *et al.* (2013) The long-lived queen: reproduction and longevity in female eusocial Damaraland mole-rats (*Fukomys damarensis*). *African Zoology* **48**:193–196.

Schlinder, S. *et al.* (2015) Sex-specific demography and generalization of the Trivers–Willard theory. *Nature* **526**:249–252.

Schoech, S.J., *et al.* (1996) Prolactin and helping behaviour in the cooperatively breeding Florida scrub-jay, *Apheloma e. coerulesens. Animal Behaviour* **52**:445–456.

Schradin, C. and Anzenberger, G. (2001) Costs of infant carrying in common marmosets, *Callithrix jacchus*: an experimental analysis. *Animal Behaviour* **62**:289–295.

Seal, U.S., *et al.* (1979) Endocrine correlates of reproduction in the wolf. I. Serum progesterone, estradiol and LH during the estrous cycle. *Biology of Reproduction* **21**:1057–1066.

Sharp, S.P. and Clutton-Brock, T.H. (2009) Reproductive senescence in a cooperatively breeding mammal. *Journal of Animal Ecology* **79**:176–183.

Sharp, S.P. and Clutton-Brock, T.H. (2011) Reluctant challengers: why do subordinate female meerkats rarely displace their dominant mothers? *Behavioral Ecology* **22**:1337–1343.

Sherman, P.W. (1977) Nepotism and the evolution of alarm calls. *Science* **197**:1246–1253.

Sherman, P.W. and Jarvis, J.U.M. (2002) Extraordinary life spans of naked mole-rats (*Heterocephalus glaber*). *Journal of Zoology* **258**:307–311.

Sherman, P.W., *et al.* (1995) The eusociality continuum. *Behavioral Ecology* **6**:102–108.

Shultz, S. and Dunbar, R.I.M. (2007) The evolution of the social brain: anthropoid primates contrast with other vertebrates. *Proceedings of the Royal Society of London. Series B: Biological Sciences* **274**:2429–2436.

Smaldino, P.E., *et al.* (2013) Increased costs of cooperation help cooperators in the long run. *American Naturalist* **181**:451–463.

Solomon, N.G. (1991) Current indirect fitness benefits associated with philopatry in juvenile prairie voles. *Behavioral Ecology and Sociobiology* **29**:277–282.

Solomon, N.G. and French, J.A. (eds) (1997) *Cooperative Breeding in Mammals*. Cambridge: Cambridge University Press.

Solomon, N.G. and Getz, L.L. (1997) Examination of alternative hypotheses for cooperative breeding in rodents. In: *Cooperative Breeding in Mammals* (eds N.G. Solomon and J.A. French). Cambridge: Cambridge University Press, 199–230.

Sousa, M.B.C. *et al.* (2005) Behavioral strategies and hormonal profiles of dominant and subordinate common marmoset (*Callithrix jacchus*) females in wild monogamous group. *American Journal of Primatology* **67**:37–50.

Spong, G.F., *et al.* (2008) Factors affecting reproductive success of dominant male meerkats. *Molecular Ecology* **17**:2287–2299.

Stacey, P.B. and Koenig, W.D. (eds) (1990) *Cooperative Breeding in Birds: Long Term Studies of Ecology and Behaviour*. Cambridge: Cambridge University Press.

Stephens, P.A., *et al.* (2005) Dispersal, eviction, and conflict in meerkats (*Suricata suricatta*): an evolutionarily stable strategy model. *American Naturalist* **165**:120–135.

Sun, L. (2003) Monogamy correlates, socioecological factors and mating systems in beavers. In: *Monogamy: Mating Strategies and Partnerships in Birds, Humans and Other Mammals* (eds U.H. Reichard and C. Boesch). Cambridge: Cambridge University Press, 138–146.

Tardif, S.D. (1997) The bioenergetics of parental behaviour and the evolution of alloparental care in marmosets and tamarins In: *Cooperative Breeding in Mammals* (eds N.G. Solomon and J.A. French). Cambridge: Cambridge University Press, 11–33.

Tardif, S.D., *et al.* (1992) Infant-care behaviour of non-reproductive helpers in a communal-care primate, the cotton-top tamarin (*Saguinus oedipus*). *Ethology* **92**:155–167.

Thornton, A. and Clutton-Brock, T.H. (2011) Social learning and the development of individual and group behaviour in mammal societies. *Philosophical Transactions of the Royal Society B: Biological Sciences* **366**:978–987.

Thornton, A. and McAuliffe, K. (2006) Teaching in wild meerkats. *Science* **313**:227–229.

Thornton, A. and Samson, J. (2012) Innovative problem solving in wild meerkats. *Animal Behaviour* **83**:1459–1468.

Tian, X., *et al.* (2013) High-molecular-mass hyaluronan mediates the cancer resistance of the naked mole rat. *Nature* **499**:346–349.

Tsuji, K. and Kasuya, E. (2001) What do the indices of reproductive skew measure? *American Naturalist* **158**:155–165.

van Schaik, C. and Burkart, J.M. (2010) Mind the gap: cooperative breeding and the evolution of our unique features. In: *Mind The Gap: Tracing the Origins of Human Universals* (eds P.M. Kappeler and J.B. Silk). Berlin: Springer, 477–495.

Vehrencamp, S.L. (1979) The roles of individual, kin and group selection in the evolution of sociality. In: *Handbooks of Behavioral Neurobiology*, Vol. **3** (eds P. Marler and J. Vandenbergh). New York: Plenum Press, 351–394.

Vehrencamp, S.L. (1983a) A model for the evolution of despotic versus egalitarian societies. *Animal Behaviour* **31**:667–682.

Vehrencamp, S.L. (1983b) Optimal degree of skew in cooperative societies. *American Zoologist* **23**:327–335.

Vehrencamp, S.L. and Quinn, J.S. (2004) Joint laying systems. In: *Ecology and Evolution of Cooperative Breeding in Birds* (eds W.D. Koenig and J.L. Dickinson). Cambridge: Cambridge University Press, 177–198.

Vleck, C.M., *et al.* (1991) Hormonal correlates of parental and helping behavior in cooperatively breeding Harris' hawks (*Parabuteo unicinctus*). *The Auk* **108**:638–648.

Waser, P.M., *et al.* (1994) Death and disappearance: estimating mortality risks associated with philopatry and dispersal. *Behavioral Ecology* **5**:135–141.

West, S.A., *et al.* (2002) Cooperation and competition between relatives. *Science* **296**:72–75.

White, S.M. (2001) Juvenile development and conflicts of interest in meerkats. PhD thesis, University of Cambridge, Cambridge.

Widowski, T.M., *et al.* (1990) The role of males in the stimulation of reproductive function in female cotton-top tamarins, *Saguinus o. oedipus*. *Animal Behaviour* **40**:731–741.

Wilkinson, G.S. (1987) Altruism and cooperation in bats. In: *Recent Advances in the Study of Bats* (eds M.B. Fenton, P.A. Racey and J.M.V. Rayner). Cambridge: Cambridge University Press, 299–323.

Wilkinson, G.S. (1992) Communal nursing in the evening bat *Nycticeius humeralis. Behavioral Ecology and Sociobiology* **31**:225–235.

Wilson, D.S. and Wilson, E.O. (2007) Rethinking the theoretical foundation of sociobiology. *Quarterly Review of Biology* **82**:327–348.

Wilson, E.O. (1975) *Sociobiology*. Cambridge, MA: Belknap Press.

Wingfield, J.C., *et al.* (1990) The challenge hypothesis: theoretical implications for patterns of testosterone secretion, mating systems, and breeding strategies. *American Naturalist* **136**:829–846.

Wright, J. (1997) Helping-at-the-nest in Arabian babblers: signalling social status or sensible investment in chicks? *Animal Behaviour* **54**:1439–1448.

Wright, J., *et al.* (2001) Cooperative sentinel behaviour in the Arabian babbler. *Animal Behaviour* **62**:973–979.

Xi, W.Z., *et al.* (2008) Benefits to female helpers in wild Rhinopithecus roxellana. *International Journal of Primatology* **29**:593–600.

Yamamoto, M.E., *et al.* (2014) Male and female breeding strategies in a cooperative primate. *Behavioural Processes* **109**:27–33.

Young, A.J. (2003) Subordinate tactics in cooperative meerkats: breeding, helping and dispersal. PhD thesis, University of Cambridge, Cambridge.

Young, A.J. (2009) The causes of physiological suppression in vertebrate societies: a synthesis. In: *Reproductive Skew in Vertebrates: Proximate and Ultimate Causes* (eds R. Hager and C.B. Jones). Cambridge: Cambridge University Press, 397–436.

Young, A.J. and Bennett, N.C. (2010) Morphological divergence of breeders and helpers in Damaraland mole-rats. *Evolution* **64**:3190–3197.

Young, A.J. and Clutton-Brock, T.H. (2006) Infanticide by subordinates influences reproductive sharing in cooperatively breeding meerkats. *Biology Letters* **2**:385–387.

Young, A.J. and Monfort, S.L. (2009) Stress and the costs of extra-territorial movement in a social carnivore. *Biology Letters* **5**:439–441.

Young, A.J., *et al.* (2005) Trade-offs between extraterritorial prospecting and helping in a cooperative mammal. *Animal Behaviour* **70**:829–837.

Young, A.J., *et al.* (2006) Stress and the suppression of subordinate reproduction in cooperatively breeding meerkats. *Proceedings of the National Academy of Sciences of the United States of America* **103**:12005–12010.

Young, A.J., *et al.* (2008) The causes of physiological suppression among female meerkats: a role for subordinate restraint due to the threat of infanticide? *Hormones and Behavior* **53**:131–139.

Young, A.J., *et al.* (2010) Physiological suppression eases in Damaraland mole-rat societies when ecological constraints on dispersal are relaxed. *Hormones and Behavior* **57**:177–183.

Zahavi, A. (1995) Altruism as a handicap: the limitations of kin selection and reciprocity. *Journal of Avian Biology* **26**:1–3.

CHAPTER 18
Sex differences

18.1 Introduction

After completing the *Origin of Species* (1859), which explained animal adaptations in terms of their effects on the survival of individuals, Charles Darwin was well aware that there were still a number of outstanding problems that he needed to deal with. One of the most obvious was that many traits which were more highly developed in one sex than the other, like the antlers of deer or the elaborate plumage of some male birds, looked unlikely to increase survival. Darwin identified these as 'secondary sexual characters' that were not directly involved in reproduction and attributed their evolution to sexual selection, which he saw as depending

> not on a struggle for existence in relation to other organic beings or to external conditions, but on a struggle between the individuals of one sex, generally the males, for the possession of the other sex. The result is not death to the unsuccessful competitor but few or no offspring.

He understood that the development of secondary sexual characters was related to polygamous breeding systems:

> That some relation exists between polygamy and the development of secondary sexual characters, appears nearly certain; and this supports the view that a numerical preponderance of males would be eminently favourable to the action of sexual selection
>
> *The Descent of Man*, Chapter VIII, page 217

He also realised that sexual competition between females can lead to the evolution of secondary sexual characters in females that resemble those of males and, occasionally, to their greater development in females.

The Descent of Man still provides the basis for much of our current understanding of the evolution of sex differences, although it is now clear that sexual selection can operate in ways that Darwin had not appreciated. Where females live in groups and males compete to monopolise access to multiple females, strong selection for competitive ability in males often leads to sexual dimorphism in size and weaponry while selection on males to attract females favours the development of male ornamentation. Where males also deter rivals through displays, selection for the efficacy and reliability of signals may also contribute to the progressive elaboration of secondary sexual characters, further accentuating differences between the sexes (see Chapter 17). In some species, males may coerce females into mating with them, generating additional selection pressures in males for traits that allow them to control potential mating partners as well as counter-selection pressures in females for traits that increase their ability to evade male control (see Chapter 15). Where females mate with more than one partner, there may also be intense competition between sperm in the female tract, leading to the evolution of adaptations in males that increase the probability that their gametes will be successful and, in some cases, to reduced selection for male weaponry and other traits that allow males to monopolise females (Parker 1979).

It is now clear that reproductive competition between females is more intense and more widespread than Darwin realised and that its relative intensity varies widely and can exert an important influence on the characteristics of females and hence on the extent of sex differences (Clutton-Brock 2007; Rosvall 2011; Stockley and Bro-Jørgensen 2011; Clutton-Brock and Huchard 2013a,b). As Chapter 8 describes, increases in the size of social groups and in reproductive skew among females intensify reproductive competition between females, and can increase selection for body size and other traits used in intrasexual competition,

Mammal Societies, First Edition. Tim Clutton-Brock.
© 2016 Tim Clutton-Brock. Published 2016 by John Wiley & Sons, Ltd.

reducing the extent of sex differences in body size. In addition, negative effects of male tactics on females often lead to selection on females to minimise their impact and to the coevolution of male and female tactics.

Contrasts in social organisation affect social competition and the operation of selection in both sexes and so influence the distribution of sex differences. However, the ways in which individuals compete and the intensity of selection on particular traits vary widely (see Chapter 8) and ecological constraints can also modify the evolution of particular tactics and traits. As a result, correlations between particular aspects of social organisation and the distribution or magnitude of sex differences are often weak and differ between taxonomic groups.

This chapter reviews our current understanding of the evolution and distribution of secondary sexual characters and their relation to variation in breeding systems and societies and their consequences. Section 18.2 examines selection on body size in males and females and the distribution of sexual size dimorphism, while sections 18.3 and 18.4 review the evolution of sex differences in weaponry and ornamentation. Section 18.5 then examines sex differences in growth while sections 18.6 and 18.7 explore the consequences of these differences for the behaviour and development of juveniles and adolescents. The next two sections then describe the effects of sexual dimorphism on habitat preferences and feeding ecology in adults (section 18.8), and sex differences in mortality (section 18.9). Finally, sections 18.10 and 18.11 explore the consequences of sex differences in development for sex ratios at birth and among adults, and their implications for the management of populations.

18.2 Body size

As Darwin realised, the relative development of secondary sexual characters in males is associated with polygynous mating systems and frequent conflict between males. In polygynous mammals, like red deer or elephant seals, successful males often have to fight several times per season to retain access to their harems and their breeding success depends on their success in these fights, which is commonly related to their size and weight (see Chapter 13). As a result, selection on

body size operating through variation in reproductive success is often stronger in males than in females and effects of this kind have been demonstrated in a wide range of mammals, from shrews (Bouteiller-Reuter and Perrin 2005) to red deer (Clutton-Brock *et al.* 1982a) and elephants (Poole *et al.* 2011). Conversely, in monomorphic species, male size often has little effect on mating success. For example, in equids, where males fight by biting and there is little sexual dimorphism in body size, there is no consistent association between the size of males and their dominance rank or breeding success (Figure 18.1). Similarly, in sifakas (which are arboreal and sexually monomorphic), there is no evidence of directional selection on size in either sex, though the reproductive success of males is positively correlated with indices of leg shape that are related to locomotor performance (Lawler *et al.* 2005; Lawson Handley and Perrin 2007). Where competition between females is unusually intense, the effects of size on breeding success can be reversed and are sometimes stronger in females than in males (Clutton-Brock *et al.* 2006).

Comparative studies of several mammalian groups, including primates, ungulates and pinnipeds, have shown that sex differences in body size increase with measures of the degree of polygyny and the intensity of reproductive competition between males (Clutton-Brock *et al.* 1977; Alexander *et al.* 1979; Plavcan and van Schaik 1997; Lindenfors *et al.* 2007) (Figure 18.2). However, not all polygynous mammals are dimorphic and interspecific correlations between the degree of polygyny and the extent of sexual size dimorphism are often weak or inconsistent (Müller and Emery Thompson 2012). There are probably many reasons for this. First, male fighting techniques vary widely between species and body size does not lead to increased success in all species (see Chapter 13). Second, contrasts in harem size and variance in annual breeding success among males are likely to overestimate variation in lifetime breeding success since the age of males often has an important influence on their breeding success (see Chapter 13). Third, a wide range of ecological and developmental constraints may affect the evolution of male size, including the extent and impact of increases in the metabolic requirements in males (Key and Ross 1999). For example, in ungulates, browsers show greater levels of sexual dimorphism than grazers (Mysterud 2000; du Toit 2005), possibly because increased body size allows males to access resources that females cannot reach.

(a)

(b)

Figure 18.1 Contrasts in sexual size dimorphism in two polygynous species: males and females in (a) African elephants and (b) plains zebra. *Sources*: (a) © Phyllis Lee; (a) © Manoj Shah/The Image Bank/Getty Images.

A final reason why relationships between mating systems and sexual size dimorphism are often inconsistent is that sex differences in size are affected by selection operating on female size. Competition between females is often strong and the outcome of competitive interactions between females can be affected by their relative body size (see Chapter 8), with effects on the evolution of body size and the extent of sexual dimorphism. For example, in howler monkeys, the social rank of females affects their chances of remaining in their natal group as well as their breeding success (Crockett and Pope 1993) and selection for body size in females could explain why howlers show relatively little sexual dimorphism in adult size. Similarly, a possible explanation of the lack of size dimorphism in

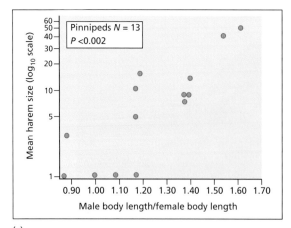

(a)

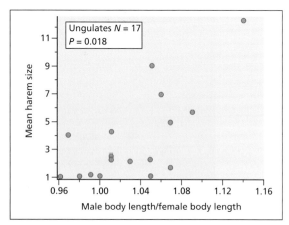

(b)

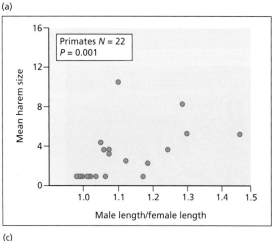

(c)

Figure 18.2 Sexual dimorphism in body size among adults plotted against the mean size of harem groups across species of (a) pinnipeds, (b) ungulates and (c) primates. *Source*: From Alexander *et al.* (1979).

polygynous lemurs (Kappeler 1990) is that strong competition between females for resources during periods of food shortage leads to selection pressures for large size that are as strong in females as males (Schmid and Kappeler 1998; Wright 1999; Richard *et al.* 2002).

18.3 Weaponry

In many mammals with promiscuous or polygynous breeding systems, weapons are more highly developed in males. For example, in Chinese water deer, males have large sabre-like canines that are absent in females (Figure 18.3). In many ruminants, males have horns or antlers which are absent in females, while in species where both sexes carry horns or antlers (as in reindeer

and caribou as well as hartebeest and wildebeest), males typically have longer or thicker horns than females, as well as patches of thickened skin on the neck and shoulders which protect against damage in fights (Packer 1983; Leader-Williams 1988). Similarly, among polygynous macropods, males commonly have thicker skin over the belly and neck to protect them against injury in fights (Jarman 1991, 2000) (see Chapter 13).

Across groups of related species, the development of male weaponry and the extent of sex differences are often associated with differences in the degree of polygyny and the intensity of reproductive competition. For example, in Old World primates, where a male's principal weapons are his canine teeth, the size of a male's canines relative to his body size is greater in species that live in harems or multi-male troops than in those that

Figure 18.3 Contrasting canines in male and female Chinese water deer. *Source*: © Tom Houslay.

live in monogamous pairs (Harvey *et al.* 1978a; Plavcan and van Schaik 1992; Plavcan 1998). Similar associations between polygyny and the relative development of sex differences in weaponry occur in other groups. In male deer, antler length relative to body size increases in polygynous species that form large breeding groups (Clutton-Brock *et al.* 1980) and an association between male horn size and group size is also present in bovids, whereas the size of female horns is not related to indices of sexual selection (Bro-Jørgensen 2007).

Sex differences in weaponry are also affected by the development of weapons in females. As in males, ecological factors appear to affect the development of weapons in females (Stockley and Bro-Jørgensen 2011; Uller 2012; Clutton-Brock and Huchard 2013a). For example, among bovids and cervids, females most frequently carry horns in bigger species that live in relatively large groups in open habitats (Roberts 1996; Stankowich and Caro 2009). The relative intensity of competition for resources may also be important: for example, in caribou, where females compete for access to feeding sites where they can obtain vegetation beneath the snow, the percentage of antlered females in different populations increases with mean snow depth in winter (Schaefer and Mahoney 2001).

Reproductive competition between females may also be important. For example, in lemurs, where direct competition between females is frequent, there are no consistent sex differences in relative canine size in most species (Kappeler 1996). Across primates, females of species that usually form monogamous pairs that defend breeding territories have relatively large canines: for example, in gibbons, where females play an active role in territory defence, they have large canines that resemble those of males. Females also have relatively large canines in species where they live in relatively large social groups, though (as in males) canine size is reduced where females form competitive coalitions (Harvey *et al.* 1978b; Plavcan 2004; Müller and Emery Thompson 2012).

An increase in female weaponry might also be expected in singular cooperative breeders where variance in female breeding success is large and competition between females is intense, but there is little evidence that this is the case. One possible explanation is that, in these species, females associate with close matrilineal relatives and that kin selection favours the resolution of conflict without resort to physical contests (Young and Bennett 2013). Alternatively, the absence of dimorphic weaponry may be associated with the tendency for breeding females to suppress or evict potential competitors before they reach a size and age where they become serious competitors or with their reliance on other group members to help them repel intruders (Clutton-Brock and Huchard 2013a).

18.4 Ornaments

Though sex differences in coloration are less common in mammals than in birds, some degree of sexual dichromatism is widespread in polygynous mammals. Sex differences in colouring are often associated with differences in coat colouring that accentuate contrasts in appearance between the sexes (Darwin 1859) and emphasise traits that are involved in fighting, such as the size of neck or temporal muscles (Geist 1971). As would be expected, the evolution of sex differences in colouring is less highly developed in nocturnal than diurnal species.

The distribution and extent of sex differences in coloration is often associated with the form of breeding systems and the intensity of competition between males. For example, in pinnipeds, sex differences in coloration occur in some land-breeding species where males defend large harems while they are reduced or absent in ice- or cave-breeding species where individual males seldom monopolise access to multiple females (Caro *et al.* 2012). Similarly, in some groups of antelope, sex differences in coat colour are more highly developed in species that breed on leks, where competition between males is intense and variance in breeding success is very high (see Chapter 10), than in species where males defend resource-based territories. For example, in the Reduncinae (African antelopes associated with flood plains), the extent of black markings in males (which are reduced or absent in females) is less pronounced in species where males defend resource-based mating territories, like reedbuck, puku and red lechwe, than in lek-breeding species like Uganda kob, white-eared kob, Kafue lechwe and Nile lechwe (Figure 18.4) and there appears to be a cline in male coloration from south to north.

Although many of the most pronounced sex differences in coat form and skin colour are found in polygynous species, sexual dichromatism also occurs in monogamous species, like some gibbons (Chivers 1977; Bartlett 2007). The selection pressures responsible for the evolution of dichromatism in these cases are unknown but, in many of them, pairs defend feeding territories against conspecifics and comparable studies of monogamous birds suggest that dichromatism may be associated with the contrasting roles played by the two sexes in territorial defence (Heinsohn *et al.* 2005). As in males, the displays of females often signal their age, size and condition (see Chapter 7) and selection operating on females has probably had an influence on the extent of sex differences.

Changes in skin or coat colour or in the development of manes or crests in males are often associated with variation in social or reproductive status. For example, in male lions, mane size and colour increase with age and play an important role in assessment by rivals and females (Box 18.1 and Figures 18.5 and 18.6). Similarly, the striking red coloration of male mandrills is associated with alpha status as well as with heightened testosterone levels and increased testicular size, developing when males attain high social rank and declining when they fall in rank (see Chapter 13). Males appear to use the relative brightness of rivals as an indicator of their likely competitive ability (Setchell and Wickings 2005) and females prefer bright males as social and mating partners (Setchell 2005). In some species, including mandrills and orangutans, the presence of dominant males is sufficient to suppress the development of sexually dimorphic ornaments and coloration in younger males (Setchell and Lee 2004).

In seasonal breeders, male ornaments and coloration often change throughout the year and are most striking in the mating season. For example, over the months preceding the autumn rut, male red deer develop a pronounced mane, their neck muscles enlarge, their coat darkens and changes in the structure of their vocal tract lower the pitch of their vocalisations (Lincoln 1971; Reby and McComb 2003).

18.5 Growth

Prenatal growth and development

In many sexually dimorphic animals, individual differences in adult size and reproductive success are associated with variation in early growth (Clutton-Brock 1991b). In sexually dimorphic species, early development often has a stronger effect on survival and breeding success in males than females. For example, the weight

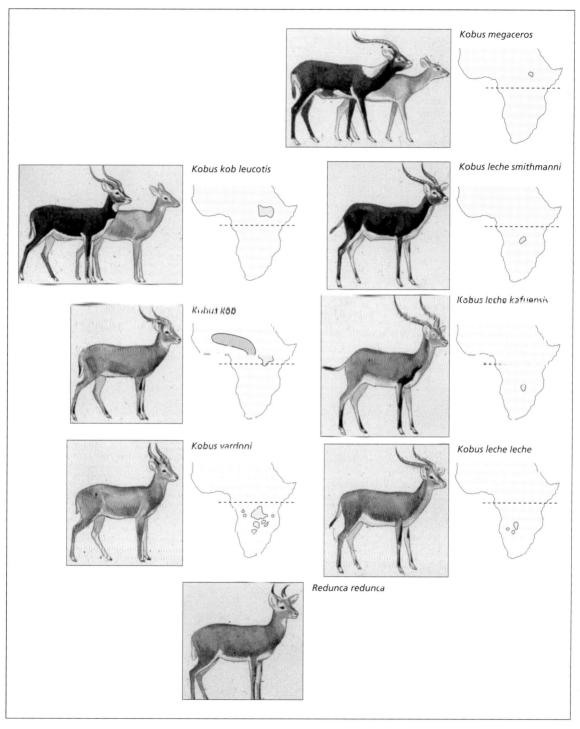

Figure 18.4 Sexually dimorphic male colouration in knob (left) and Lechwe (right) shows a south to north cline. Both groups of species may have evolved from an ancestor similar to the common reedbuck (*Redunca redunca*). Maps show the contrasting distributions of different species and subspecies. *Redunca redunca*, common reedbuck; *Kobus vardoni*, puku; *Kobus kob*, Uganda kob; *Kobus kob leucotis*, white-eared kob; *Kobus leche leche*, red lechwe; *Kobus leche kafuensis*, Kafue lechwe; *Kobus leche smithmanni*, black lechwe; *Kobus megaceros*, Nile lechwe. *Source*: © Dafila Scott.

Box 18.1 Sexual selection and the lion's mane

The manes of African lions develop in puberty and are variable in length and colour (Figure 18.5). Lions are the only existing felid where males possess sexually dimorphic manes and, in contrast to other felids, they live in multi-male, multi-female groups and have a polygynandrous mating system (see Chapter 11). Groups of males, who are often brothers, compete intensely to monopolise breeding access to groups of females. All resident males have potential mating access to females in their groups, there are usually large differences in mating success between individuals.

Figure 18.5 Short- and long-maned African lions. *Source*: © Dom Cram.

Manes may help to shield the neck and shoulders in fights or may have evolved in male lions either to signal their strength and condition to external competitors or to signal their phenotypic quality to females. The absence of manes in other felids suggests that protection cannot be their only function: using long-term records of individual life histories from wild lions in the Serengeti and neighbouring areas, West and Packer (2002) showed that individual differences in the length and darkness of manes were correlated with the male's age, testosterone levels and nutrition (Figure 18.6), and that mane length appears to reflect the individual's short-term health and fighting ability while mane darkness reflects their foraging success, nutrition and fighting ability as well as their likely breeding tenure in female groups. They predicted that both sexes should adjust their behaviour in social interactions on the basis of mane length and colour and showed that males attracted to life-size models (that differed in mane length and colour) by recordings of scavenging hyenas (see Chapter 15) were more likely to approach individuals with short or light manes while females were more likely to approach those with dark manes and did not differentiate between those with short or long manes. Dark manes are likely to increase heat load and so may have costs as well as benefits to males and, as expected, manes are darker in populations and individuals living in cooler areas and lighter in those living in hotter conditions (West and Packer 2002). Like many other male ornaments, the manes of lions appear to signal male quality both to rivals and to prospective mates.

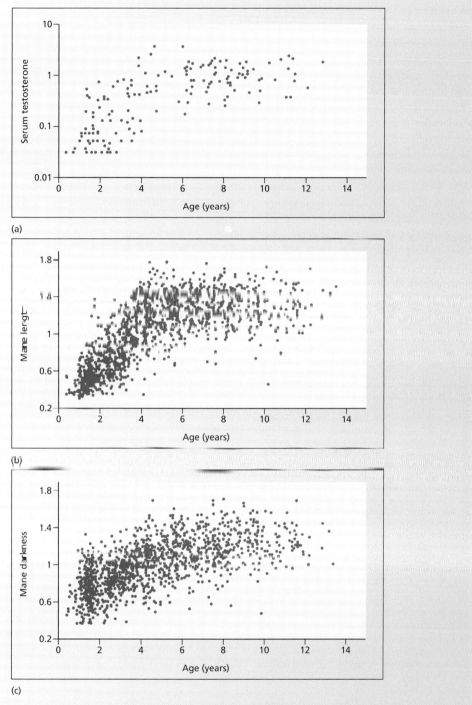

Figure 18.6 As lions increase in age, circulating levels of testosterone increase and their manes increase in length and darken in colour. The figures show age-related changes in (a) testosterone levels, (b) mane length and (c) mane darkness in wild lions. *Source*: From West and Packer (2002). Reproduced with permission from AAAS.

and breeding success of adult male rhesus macaques is correlated with their weight at 1 year while there is no evidence of similar relationships in females (Bercovitch *et al.* 2000). Similarly, in red deer, lifetime reproductive success is significantly correlated with birthweight in males but not in females (see Figure 14.8).

Where selection on birthweight is stronger in males than females, it can favour the evolution of sex differences in prenatal growth. For example, experiments with laboratory mice show that selecting on adult dimorphism increases sex differences in weight at birth (Krackow *et al.* 2003). Across species, sex differences in birthweight are common in sexually dimorphic mammals (Clutton-Brock *et al.* 1981; Bercovitch *et al.* 2000) and their magnitude is correlated with the extent of adult dimorphism in body mass (Figure 18.7). Sex differences in birthweight are usually small or absent in sexually monomorphic species as well as in those where adult females are larger than males (Jarvis *et al.* 1991; Smale *et al.* 1999).

In some species, there is also evidence that the costs of bearing males exceed those of bearing females. For example, in laboratory mice, genes on the Y chromosome can cause male embryos to accrue relatively more maternal resources at the expense of female embryos (Hurst 1994). Male embryos may also compete more intensely for resources than females: in several sexually dimorphic mammals, male fetuses grow more rapidly than females and male-biased litter sex ratios reduce prenatal growth and birthweight in female offspring, with downstream effects on their survival and breeding success (Clutton-Brock 1991b; Uller 2006). For example, in a naturally regulated population of Soay sheep, female lambs born co-twin with a male show reduced birthweight, lower first-year survival (relative to their birthweight) and lower fitness than female lambs born co-twin with a female, while the birthweight of males is unaffected by the sex of their twin (Korsten *et al.* 2009).

In litter-bearing mammals, the sex of neighbouring fetuses can also exert important effects on development. Studies of mice and rats show that females positioned between male fetuses in the uterus ('2M' females) have higher concentrations of testosterone in blood and amniotic fluid and show evidence of masculinisation, including relatively long anogenital distances compared with females positioned between two females ('0M' females) (vom Saal and Bronson 1980; Ryan and Vandenbergh 2002). These effects are associated with differences in subsequent reproductive performance: in laboratory populations of mice, 2M females show later ages at first breeding, produce fewer viable litters and relatively fewer sons, and show an earlier onset of infertility with increasing age than 0M females (vom Saal and Moyer 1985; Drickamer 1996). In some rodents (including gerbils), 2M males also show signs of increased masculinisation and greater mating success as adults than 0M males (Clark *et al.* 1992, 1998) though effects of intrauterine position on males appear to be more variable than those in females. Similar effects of intrauterine position on females have been demonstrated in free-ranging populations of rodents as well as in other polytocous mammals, including voles, ferrets and pigs (Drickamer *et al.* 1997; Ryan and Vandenbergh 2002). Several studies also suggest that intrauterine position may have important effects on ranging behaviour and dispersal. For example, in populations of wild mice introduced onto motorway islands, 2M females had larger home ranges than 0M females (Zielinski *et al.* 1992), while in grey-sided voles juveniles from

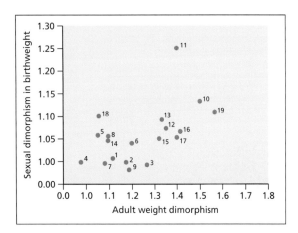

Figure 18.7 Sex differences in weight at birth are correlated with sex differences in adult weight. The figure shows sex differences in weight at or soon after birth plotted on adult weight dimorphism for a sample of 19 mammals: 1 and 2, *Onchomys*; 3, *Proechimys*; 4, coypu; 5, ringtail possum; 6, pig; 7, horse; 8, roe deer; 9, moose; 10, Chinese water deer; 11, mouflon; 12, Soay sheep; 13, pigtail macaque; 14, rhesus macaque; 15, fallow deer; 16, red deer; 17, wapiti; 18, reindeer/caribou; 19, white-tailed deer. *Source*: From Clutton-Brock (1991b). Reproduced with permission of Princeton University Press.

male-biased litters showed a greater tendency to disperse than those from female-biased litters (Ims 1989, 1990).

Postnatal growth and development

In many polygynous dimorphic mammals, there are consistent sex differences in the rate, timing and duration of postnatal growth (Clutton-Brock 1991a,b). Males commonly reach sexual maturity and adult size later than females and there are sex differences in average age at first breeding. Sexual bimaturism may often be a consequence of sex differences in reproductive competition. For example, among the great apes, sex differences in maturation timing are greater in gorillas and orang-utans, where one-on-one combat is the typical form of male mating competition, than in the two chimpanzee species, where multiple individuals are commonly involved and the size of individuals has less effect on fighting success (Watts and Pusey 1993).

Growth in dimorphic mammals often continues for longer in males than in females (Clutton-Brock 1991b) and in some species (including African elephants, sperm whales and some kangaroos), males continue to grow in size and weight throughout much of their lives (Jarman 2000; Whitehead and Weilgart 2000). In contrast, in some mammals where reproductive competition is more intense among females than males, growth continues for longer in females: for example, in meerkats, mass increases throughout the lifespan in females but not in males (see Chapter 17). Species also differ in the extent to which differences in growth rate (as against the duration of growth) contribute to sex differences in adult size. For example, among cercopithecine monkeys, sex differences in adult size are a consequence of differences in the duration of growth in talapoins but of differences in growth rate in blue monkeys, drills and some macaques (Leigh 1992; Martin et al. 1994).

Within species, growth is often more sensitive to environmental conditions in males than in females with the result that the extent of sex differences often varies with the quality and availability of resources (Clutton-Brock 1991b). For example, among captive reindeer maintained on a high level of nutrition, male calves gain weight faster than females, while in wild populations living on an inferior diet, there is little difference in growth between males and females during the first months of life (McEwan 1968). In some cases, variation in food availability can even reverse the direction of sex differences in growth. For example, though juvenile male wood rats grow faster than females when food is plentiful, in food-restricted litters female juveniles increase in weight more rapidly than males (McClure 1981). Similarly, while infant male pigtailed macaques reared in breeding colonies are (slightly) heavier than females of the same age, male infants rejected by their mothers and raised on an artificial diet grow more slowly than females (Sackett et al. 1975).

Variation in resource availability often has a particularly strong influence on the development of secondary sexual characters in males. For example, in deer, density-dependent changes in the antler length of juvenile males are correlated with variation in adult antler size and well-grown adults commonly show a relative increase in the size of secondary sexual characters (Huxley 1931). Individual differences in the relative size of secondary sexual characters are often correlated with other aspects of male performance and behaviour, including social dominance, disease resistance, body condition, sperm quality, signalling behaviour and dispersal timing (Geary 2015; Edelman 2011).

There may be several reasons for sex differences in the extent to which resource availability affects growth and the development of secondary sexual characters to different extents in males and females. The faster growth rates and greater energy demands of males in dimorphic species may cause growth in males to be more strongly affected by food shortage or adverse conditions than growth in females. In addition, males may be more likely to utilise muscle than females as a consequence of the effects of androgens during periods of starvation (Widdowson 1976). Finally, improved nutrition may reduce the adverse consequences of heightened testosterone levels, reduced immune response and increased parasite loads in males (see Section 18.9).

18.6 Nursing

Sex differences in nursing time and ingestion

Where juvenile males grow faster than females, they often have larger energetic requirements and males may show greater frequency or persistence in suckling attempts.

Juvenile males nurse more frequently or for longer periods in several sexually dimorphic mammals (Clutton-Brock 1991b; Hewison and Gaillard 1999), including Scottish red deer (Clutton-Brock and Albon 1982), goats (Pickering 1983) and Galapagos fur seals (Trillmich 1986) as well as in some relatively monomorphic species (White *et al.* 2007). Differences in the treatment of male and female pups have been documented in rodents (McGuire and Bemis 2007). For example, female gerbils raising all-male litters spend more time in the nest with young (Clark *et al.* 1990). Similarly, female Norway rats and house mice raising all-male litters invest more heavily in nursing and nest-building and lick the anogenital regions of male pups more than those of female pups (Moore and Morelli 1979; Alleva *et al.* 1989).

Since nursing time may not provide a reliable index of milk ingested (Mendl and Paul 1989; Cameron 1998), several recent studies have explored more direct measures of milk intake by male and female juveniles. In several cases, they confirm the results of observational studies. In red deer, where milk production by mothers of sons and daughters has been measured directly by milking captive females, comparisons confirm that mothers of sons produce more milk than mothers of daughters (Landete-Castillejos *et al.* 2005), while in fur seals indirect measures of milk transfer from mothers to pups have also shown that male pups ingest more milk than females (Costa and Gentry 1986; Trillmich 1986). In contrast, in Holstein cows, mothers of daughters produce significantly more milk across the lactation period than mothers of sons and the production and rearing of a daughter by primiparous females increases their milk production in the next breeding attempt (Hinde *et al.* 2014).

At the moment, it is still unclear whether sex differences in nursing behaviour are usually a consequence of differences in offspring demand or of contrasts in maternal responsiveness, or of a combination of the two. Several studies of species showing sex differences in nursing behaviour have failed to find differences in the proportion of suckling attempts by sons and daughters that are accepted by their mothers, suggesting that these differences are a consequence of contrasts in the behaviour of offspring rather than in that of their mothers (Clutton-Brock 1991a; Birgersson 1998). However, a recent study of white rhinos suggests that attempts to nurse by sons are more likely to be successful than those by daughters (White *et al.* 2007).

Rearing costs

Where males nurse more than females, rearing sons can have larger costs to the mother's subsequent survival or breeding success than rearing daughters. For example, in wild red deer, mothers that have reared a male calf are more likely to die the following winter (Figure 18.8a), though their relative chances of survival vary with their rank (Figure 18.8b). If male-rearing mothers survive, they are almost twice as likely to fail to produce a calf the following year compared with mothers that have reared female calves (Figure 18.8c) and, if they do breed, they conceive and give birth later in the season than female-rearing mothers (Figure 18.8d). Similar differences in subsequent reproductive performance or survival between females that have reared sons and those that have reared daughters have been found in a number of other sexually dimorphic mammals (Lee and Moss 1986; Hogg *et al.* 1992; Birgersson 1998).

Although sex differences in maternal investment during lactation are common in sexually dimorphic species, they have not been found in all species or populations. For example, there appear to be no sex differences in nursing behaviour or rearing costs in some sexually dimorphic ungulates (Byers and Moodie 1990; Weladji *et al.* 2003) and seals (McCann *et al.* 1989; Smiseth and Lorentsen 1995). One possible explanation is that sex differences in the costs of raising young are reduced where resource availability is high or mothers are in unusually good condition. For example, in red deer, no effects of offspring sex on the mother's subsequent breeding success or survival were found in a population living in a relatively productive environment (Bonenfant *et al.* 2003) and other studies have only detected significant sex differences in rearing costs when they controlled for the effects of maternal condition (Cameron and Linklater 2000).

While several studies of ungulates have shown that mothers invest more heavily in sons than daughters, studies of some primates suggest that raising daughters may depress the mother's subsequent reproductive performance more than raising sons (Simpson *et al.* 1981; Hiraiwa-Hasegawa 1993; Maestripieri 2001). A possible explanation is that in species where maternal rank is inherited by daughters, mothers maximise their fitness by investing more heavily in daughters than in sons (Gomendio *et al.* 1990; Brown and Silk 2002). Alternatively, female competition may affect the relative costs of

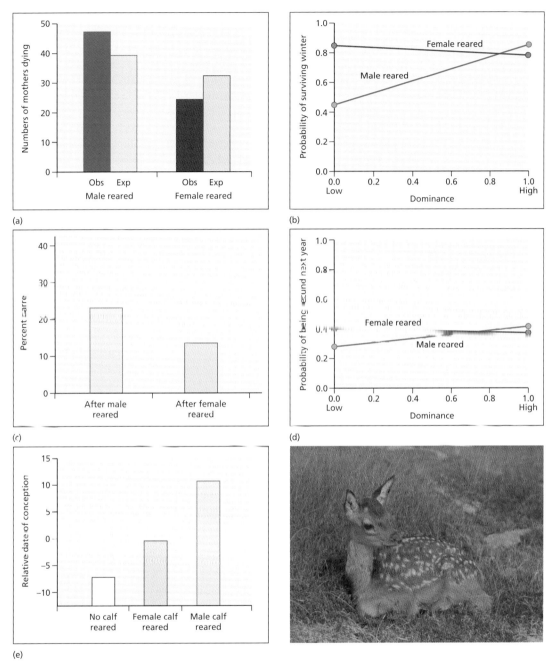

Figure 18.8 Relative cost of rearing sons and daughters to female red deer: (a) observed and expected numbers of mature females dying in winter after rearing males and females; (b) effects of maternal rank on the survival of females that reared sons and daughters; (c) probability of females failing to breed the next year (percent barren) after rearing a female calf and rearing a male calf; (d) effects of maternal rank on the probability that females would breed again the next year after rearing male and female calves; (e) relative conception dates (based on back-dating from known birth dates) of calves whose mothers had raised no calf, a female calf or a male calf the previous year. *Source*: Plots (a) and (c) from Clutton-Brock *et al.* (1981). Reproduced with permission of Nature Publishing Group. Plots (b) and (d) from Gomendio *et al.* (1990). Reproduced with permission of Nature Publishing Group. Plot (e) from Clutton-Brock (1991b). Reproduced with permission of Princeton University Press. *Photo source*: © Tim Clutton-Brock.

raising sons and daughters. For example, in some cerco-pithecine monkeys, daughters attract more aggression from females belonging to other matrilines than sons and subordinate mothers that have reared female infants do not conceive again as quickly as those that have reared males and are more likely to lose their next offspring (Simpson *et al.* 1981; Simpson and Simpson 1985; Maestripieri 2001). A possible cause of these effects is that heightened levels of aggression directed at female infants cause them to spend more time in contact with their mother and greater frequency of nipple stimulation raises levels of gonadotrophin and delays the next conception (McNeilly 1987).

Comparisons of nursing time and milk production may also be complicated by sex-specific differences in milk quality. In some species, milk quality differs between mothers of sons and mothers of daughters (Hinde 2007, 2009). For example, in red deer, a study of captive animals found that the percentage of protein was higher in milk produced by mothers of sons than in milk produced by mothers of daughters (Landete-Castillejos *et al.* 2005). However, these differences could be a consequence of contrasts in suckling frequency and milk yield. For example, in captive macaques (where daughters often spend more time in contact with their mothers than sons) subordinate mothers of daughters produce more milk than mothers of sons but the energy content of their milk is lower, so that the total amount of energy available to sons and daughters is similar (Hinde 2007, 2009). One difficulty in interpreting these results is that they involve captive animals maintained on generous nutritional regimens and both captivity and food intake may affect suckling frequency and milk quality. As yet, it is not clear whether sex differences in the quality of milk given to sons and daughters are common in natural populations or to what extent they affect estimates of relative investment in offspring of different sexes.

It is also important to appreciate that sex differences in the costs of raising sons and daughters are not necessarily limited to the period when offspring depend directly on their mothers for nutrition. Where daughters are philopatric and groups have reached the carrying capacity of their ranges, the continued presence of daughters can increase feeding competition and reduce the breeding success of their mothers (Clark 1978; Clutton-Brock 1991b). For example, in red deer, success in raising daughters may generate protracted costs to their mother's fitness which offset the additional costs of raising sons during the period of early development (Clutton-Brock and Albon 1982).

Sex differences in philopatry can also have important consequences for the duration of the effects of the mother's presence on the development of her offspring. Where mothers provide social support or protection for philopatric daughters, orphaning may have more prolonged effects on the subsequent fitness of daughters than sons. For example, in red deer, orphaning depresses the survival of sons and daughters during their first two years of life (the period for which males continue to associate with their mothers) but these effects only persist in daughters during the subsequent stages of their lives (Andres *et al.* 2013) (see Chapter 5). Similar differences in the effects of the mother's presence on the fitness of their sons and daughters also occur in some primates. For example, in Japanese macaques, where daughters usually remain in their natal group and associate closely with matrilineal kin, the presence of mothers has an important influence on the fitness of daughters (Pavelka *et al.* 2002). In contrast, in killer whales, where members of both sexes may remain in their natal group but mothers associate more consistently with sons (Baird and Whitehead 2000), the presence of mothers has a stronger influence on the fitness of sons (Foster *et al.* 2012). Similarly, in bonobos, where females commonly disperse to breed in other groups and mothers provide social support for sons, the mother's presence has an important influence on the rank and breeding success of her sons (Surbeck *et al.* 2011).

18.7 Social development

Maternal proximity

In some social mammals (including some ungulates and several primates), juvenile males spend more time further away from their mothers and show a more rapid development of independence than juvenile females and these differences can affect the number of other individuals they interact with (Clutton-Brock 1991b). For example, in wild chimpanzees, male infants interact with a larger number of other individuals (including adult males) than female infants (Lonsdorf *et al.* 2014).

Sex differences in infant behaviour may be associated with differences in the treatment of male and female infants by their mothers. For example, in rhesus macaques (where females are usually philopatric and males disperse to breed), mothers restrain male offspring less than females and are more protective of their daughters, rejecting them less frequently than their sons (Hinde and Spencer-Booth 1967; White and Hinde 1975; Meaney *et al.* 1985). Mothers also respond more strongly to separation–rejection vocalisations given by sons and this tendency is reduced if prenatal exposure of offspring to testosterone is blocked (Tomaszycki *et al.* 2001).

Relationships between mothers and sons may also involve more frequent exchanges of aggression: for example, in rhesus macaques, sons bite their mothers more frequently than do daughters (Mitchell 1968) while affiliative interactions can be more frequent between mothers and daughters. In vervet monkeys (where females are also philopatric), mothers allo-groom more frequently with their daughters than their sons (Fairbanks and McGuire 1985), while in baboons juvenile and sub-adult females reciprocate grooming received from their mothers to a greater extent than males of the same age (Pereira and Altmann 1985; Pereira 1988a). Whether these differences are associated with sex differences in philopatry and the likely duration of mother–offspring association is not yet clear but, in contrast to baboons, sex differences in the degree of reciprocity in grooming relationships between juveniles and their mothers do not occur in chimpanzees, where adolescent females usually disperse from their natal communities (Pusey 1983, 1990).

Sex differences are also common in social relationships between juveniles and adults other than their mother (Pereira 1988b; van Noordwijk *et al.* 1993). In several primates where females are philopatric, adolescent females show a tendency to form close spatial relationships with female kin: for example, in vervet monkeys, juvenile and adolescent females associate and interact more with adult females (especially high-ranking females and relatives) than with adult males, while adolescent males associate more frequently with adult males than adult females and show no tendency to approach high-ranking females more than low-ranking ones (Fairbanks 1993). In contrast, in chimpanzees (where males commonly remain in their natal group throughout their lives) immature males associate and interact increasingly often with adult

males as they grow older, gradually joining social networks in their communities (Pusey 1983; Watts and Pusey 1993) and a similar trend is found in muriquis, where many males also remain in their natal group throughout their lives (Strier 1993).

Aggression

Sex differences in intrasexual agonistic behaviour between adults are often associated with similar differences in aggressive behaviour among juveniles and sub-adults. In some polygynous mammals (including both monomorphic and sexually dimorphic species), juvenile and sub-adult males more frequently interact aggressively with each other than females of the same age (Watts and Pusey 1993; Fadem and Corbett 1997) and these differences can extend back to the first months of life. In contrast, in species where intrasexual competition is more intense between adult females than between males, intrasexual aggressive interactions can be more frequent between female juveniles and sub-adults. For example, in meerkats, aggression between female juveniles and sub-adults is more frequent than between males, female sub-adults show ritualised submission to each other more frequently than sub-adult males, and the frequency of aggression between females increases in relation to the number of sub-adult females in the group, while no similar relationship is present between male aggression and male numbers (Sharpe 2004; Clutton-Brock *et al.* 2010).

In some species, sex differences in aggression are evident immediately after birth. For example, in spotted hyenas (where female breeding success is closely related to rank and females are larger and more aggressive than males), females normally give birth to twins; although cubs initially remain below ground, they are precocial at birth, with open eyes and fully erupted canines and incisors (Frank *et al.* 1991; Frank 1996). Within minutes of the birth of a second infant, the first-born may attack it violently and, if maternal resources are insufficient to sustain two cubs, sibs fight repeatedly during the first hours or days of life until one establishes dominance over the other (Smale *et al.* 1999; Hofer and East 2008). In around half of all litters, aggression between littermates leads to the death of one twin before the end of the neonatal period. Conflict is particularly intense between sisters, and pairs of sisters seldom survive (Mills 1990; Frank *et al.* 1995; Hofer and East 1997).

Play

Sex differences in play are also common. In some species, immature males play more than females: for example, in wild Assamese macaques, immature males play more than immature females and show reduced rates of growth (Berghanel *et al.* 2015). In many social species, there are also marked sex differences in the form of play-fights which reflect sex differences in adult fighting techniques (Figure 18.9). For example, in horses, play-fights between females commonly involve kicking, while those between males involve biting, reflecting differences in fighting tactics that exist in adults (C. Feh, personal communication). Sex differences in aggression are also common during play between juveniles and adolescents (Pellis and Pellis 2007; Geary 2010): in polygynous species, juvenile males often engage in rough and tumble play and play-fighting more than females (Meaney *et al.* 1985) and also win intersexual play-fights more frequently (Jamieson and Armitage 1987; Biben 1998). These differences appear to be linked to the adult roles of the two sexes. For example, in spotted hyenas (where females compete intensely and are dominant to males), females engage in more play-fighting than males (Pedersen *et al.* 1990).

The development of sex differences in play may be affected by the social environment that individuals experience. For example, male rhesus monkeys reared in all-male groups show more rough and tumble play than those reared in normal mixed-sex groups or in isolation (Wallen 1996). These differences may be associated with prenatal exposure to gonadal hormones (Geary 2010): for example, in rats and rhesus monkeys

(where males typically play-fight more than females) females exposed to increased testosterone levels before birth engage in rough and tumble play as frequently as males (Goy and Phoenix 1971; Pellis 2002). High levels of androgens during the period of neural differentiation may promote the growth of sex-specific neural circuitry that mediates sex differences in play-fighting (Meaney *et al.* 1985).

In contrast to play-fighting, play-parenting is usually more common in female juveniles than in males (Geary 2010). In cercopithecine monkeys, juvenile females commonly handle, groom and play with sibling infants more than males (Cheney 1977, 1978; Walters and Seyfarth 1997). Similar sex differences occur in some New World monkeys: for example, in ursine howler monkeys, adolescent females help to carry younger siblings more often than males (Crockett and Pope 1993) and, in squirrel monkeys, female juveniles and sub-adults are more commonly involved in play directed at infants than males (Baldwin 1969). Male and female juveniles may also favour different objects to play with (Alexander and Hines 2002; Hassett *et al.* 2008).

The functional significance of sex differences in juvenile social behaviour probably varies. In some cases, sex differences in early social behaviour may help individuals to develop behavioural skills that are important in later life. For example, the greater involvement of female juveniles in allo-parental care may help them to develop similar behaviour and social skills in adulthood (Fairbanks 1990, 1993; Pryce 1993) while the greater involvement of males in play-fights may help to improve their competitive abilities or to establish dominance relationships. However, one of the few studies to investigate this possibility found no relationship between the frequency of play-fighting or the proportion of play-fights won and their eventual acquisition of dominant status (Sharpe 2005). Similarly, some studies have suggested that sex differences in spatial relationships between juveniles and their mothers in species where males disperse to breed are a consequence of maternal strategies adapted to increase the independence of sons (Meredith 2013). However, other explanations are possible in this case, too. For example, in some cercopithecine monkeys, females commonly direct aggression at juvenile females born into subordinate matrilines and, as a result, subordinate mothers with daughters may be more protective of their offspring than those with sons, with the result

Figure 18.9 Juvenile chacma baboons playing. *Source*: © Alecia Carter.

that female offspring develop closer spatial relationships with their mothers (see Chapter 8).

Sex differences in social behaviour among juveniles may also help individuals to establish social relationships that persist into later life (van Noordwijk *et al.* 1993; Watts and Pusey 1993; Walters and Seyfarth 1997). In both sexes, dominance relations established between juveniles may affect their relative rank and fitness as sub-adults and adults (Chapais and Schulman 1980; Chapais 1988; Fairbanks 1993; O'Brien and Robinson 1993). Affiliative relationships between juveniles and adults can also persist. For example, in vervet monkeys, relationships between female juveniles and particular adult females (whether they are kin or not) are often maintained into adulthood (Fairbanks 1993). Inter-specific differences in the pattern of sex differences in juvenile behaviour may reflect differences in the potential benefits that males and females derive from relationships with resident adults: for example, in several primates where female juveniles usually remain in their natal groups throughout their lives, they develop closer relationships with resident adult females than do males (Watts and Pusey 1993), whereas in species where males commonly remain in their natal groups throughout their lives, male juveniles and sub-adults develop closer relations with resident males than with females (Watts and Pusey 1993, see also Cürio 1993).

18.8 Feeding ecology

Sex differences in growth and body size are often associated with sex differences in feeding behaviour. In some species, the additional size or weight of males affects their access to particular resources. For example, male kudu and giraffe are able to feed at higher levels than females (Ginnett and Demment 1997; Mysterud 2000; du Toit 2005; Main and du Toit 2005), while in arboreal primates where males are substantially heavier than females, males may be unable to move safely in the terminal twigs of canopy trees and so may be restricted to the main branches or may be forced to adopt different foraging techniques (Clutton-Brock 1977; Grassi 2002). In some terrestrial carnivores, like fossas and African lions, the larger size of males and their greater strength allows males to hunt larger prey than females (Schaller 1972; Jones and Barmuta 1998, 2000; Lührs and Dammhahn 2010). For example, in African lions, males hunt large

prey, like buffalo or young elephants, more commonly than females, though both sexes usually prey on medium-sized ungulates like wildebeest and zebra (Funston *et al.* 1998, 2001; McComb *et al.* 2011). In some marine species, the larger size of males allows them to dive to greater depths and to use different foods. For example, in the non-breeding season, male elephant seals and fur seals travel to different feeding areas and feed at greater depths than females and on different prey (Michaud 2005; Staniland 2005).

In ungulates, the energetic consequences of increased body size in males can also lead to sex differences in habitat use (Clutton-Brock *et al.* 1982a; Ruckstuhl and Neuhaus 2000; Macfarlane and Coulson 2005; Main and du Toit 2005). In some species, males and females are partially or completely segregated for much of the year and males differ from females in their foraging behaviour and their habitat use. For example, in red deer, males spend much of the year segregated from females, use a wider range of habitats, feed less on heavily cropped swards and more in areas where vegetation is longer, and their diet has a higher fibre content than that of females (Staines and Crisp 1978; Clutton-Brock *et al.* 1982a). Habitat use in males is more strongly affected by adverse weather and males spend more time grazing in sheltered sites than females (Conradt *et al.* 2000). In addition, they graze more at night and, in winter, spend around 15% more time grazing over the 24 hours. Similarly, bull elephants spend much of the year separate from females, use a wider range of habitats, feed on a lower diversity of plants, are less selective of specific parts, and eat a higher proportion of woody vegetation (du Toit 2005; Lindsay 2011) (Figure 18.10). Sex differences in foraging behaviour and habitat use have also been found in other groups of herbivores, including African antelopes (Main and du Toit 2005) and macropods (Macfarlane and Coulson 2005), though their extent often varies between and within populations. In some cases, they are more pronounced at high population density than at low density, while in others they are affected by weather or by the risk of predation (Clutton-Brock *et al.* 1987; du Toit 2005). For example, male and female elk use different foods when wolves are absent but sex differences in diet are reduced or disappear when they are present (Christianson and Creel 2008).

The energetic reasons relationships between feeding behaviour and body size in herbivores were first identified by studies of African ungulates (Bell 1971; Geist 1974;

Figure 18.10 Male elephants feed more on woody matter than females and destroy more trees. *Source*: © Tim Clutton-Brock.

Jarman 1974) which argued that, as the total energetic requirements of individuals increase as approximately (body weight)$^{0.75}$, larger animals require smaller amounts of energy per unit body weight than smaller species and so can subsist on foods of lower energetic value. Jarman (1974) suggested that the occupation of open grassland habitats by ancestral ungulates was associated with increases in both the degree of polygyny and the extent of sexual dimorphism and phylogenetic analysis provides broad support for this suggestion (Pérez-Barbería *et al.* 2002) (Box 18.2).

While there is general agreement that differences in habitat use and feeding ecology among herbivores are associated with contrasts in body size and energetic requirements, the precise causes of these differences are still debated. Several extensions of Jarman's theory

have been suggested. Allometric relations between mouth size and bite size in herbivores may play some part: as bite size and rate of food intake on short swards is likely to be related to the breadth of the incisor arcade and to scale as approximately (body mass)$^{0.33}$ while energetic requirements scale as (body weight)$^{0.75}$, the (absolutely) greater energy requirements of larger animals may constrain them to feed on longer swards where bite size and intake rates are large enough to satisfy their energy requirements (Clutton-Brock *et al.* 1987). Another explanation is that, if gut volume increases isometrically with body mass, large animals have a high ratio of gut volume to metabolic requirements and so can afford to retain food in the rumen for longer, increasing their extraction efficiency and allowing them to use vegetation with a higher cellulose or lignin content (Demment and Van Soest 1985; Bowyer 1987, 2004; Illius and Gordon 1991).

Similar effects of body size on energy intake rates and energy requirements may contribute to sex differences in feeding behaviour in sexually dimorphic herbivores (Clutton-Brock *et al.* 1987; Pérez-Barbería and Gordon 1998). If so, this would suggest that, in strongly dimorphic species, competition with females may force males into secondary habitat and contribute to increased rates of male emigration and mortality. Though there is some evidence that this is the case, results are mixed and it is not yet clear whether sex differences in responses to scramble competition play an important role in the development of sex differences in habitat use and mortality in herbivores (Bowyer 2004; du Toit 2005).

Box 18.2 Body size, food selection and the Jarman–Bell hypothesis

Across species of herbivores, gut volume increases isometrically with body mass, while total energy requirements increase as approximately (body weight)$^{0.75}$ (Demment and Van Soest 1985) and larger herbivores consequently need to extract less energy from each unit of food than smaller ones in order to satisfy their energy requirements. The relevance of these relationships for species differences in food choice was independently recognised by Jarman (1968, 1974) and Bell (1969, 1971), who pointed out that they allow larger herbivores to select food with a lower energy content, thus avoiding competition with smaller species. An additional factor may be that larger herbivores can afford to process food more slowly, allowing them to extract relatively more energy from their diet or to select foods that require longer digestion times (Barboza and Bowyer 2000; Main and du Toit 2005).

Allometries in the rate of food intake may also contribute to relations between the body size of herbivores and the quality of their diets (Clutton-Brock and Harvey 1983; Pérez-Barbería and Gordon 1998). Where bite size increases with mouth breadth, which (like other linear measures) scales as approximately (body weight)$^{0.33}$ while metabolic requirements increase as approximately (body weight)$^{0.75}$, larger animals will ingest a smaller fraction of their metabolic requirements at each bite. As a result, they may be unable to feed economically in areas heavily used by smaller ones and so may tend to graze in less heavily used areas where standing crops are higher but food quality is lower. As expected, there is considerable evidence that smaller herbivores often reduce food availability to levels that cannot be tolerated by larger competitors, gradually excluding them from preferred habitats (Murray and Illius 2000; Main and du Toit 2005).

Sex differences in reproductive strategies may also contribute to sex differences in feeding behaviour and habitat use. The costs of gestation and lactation may cause female herbivores to feed on higher-quality swards than males, with the result that they use areas that are preferred by both sexes and spend more time in heavily grazed areas than males (Clutton-Brock *et al.* 1982a; du Toit 2005). For example, in red deer, comparisons of habitat use by mothers with and without dependent young show that females with calves use heavily grazed, herb-rich swards to a greater extent than those without (Clutton-Brock *et al.* 1982b). Similarly, in chimpanzees, pregnant and lactating females consume higher-quality foods than non-pregnant, non-lactating females and travel less far each day (Murray *et al.* 2009) The presence of dependent young may also limit female ranging behaviour (Main and Coblentz 1996, du Toit 2005): for example, in Grevy's zebra, lactating females spend the weeks following gestation close to water sources (Becker and Ginsberg 1990) In other cases, the risk of predation may cause females with dependent young to spend more time in thicker cover, even if resources are less abundant there (Main and du Toit 2005), or to move to areas where predators are scarce or where they can conceal their young in the weeks following their birth, and this can affect both the availability of food and the food species eaten (Bergerud *et al.* 1984; Main and du Toit 2005).

The reproductive strategies of males can also lead to sex differences in foraging behaviour and habitat use. Where males have larger ranges than females (see Chapter 10) this may cause them to use a wider range of habitats. Conversely, where mature males defend territories overlapping the ranges of several females, younger males may be forced to move into secondary habitat with consequences for their use of particular foods and habitats (Clutton-Brock *et al.* 1982a; Main and du Toit 2005). Mate guarding can also reduce foraging efficiency in males and may consequently affect food choice (see Chapter 13). Social relationships, too, may be important: for example in populations of chimpanzees where several individuals cooperate to hunt monkeys, hunts usually involve males (Stanford *et al.* 1994), though in Senegalese populations, where individuals use sharpened sticks to extract bushbabies from hollow trees, females are more commonly involved than males (Pruetz *et al.* 2015).

Where sex differences in feeding behaviour are pronounced in adults, similar differences are often found in sub-adults and juveniles and, sometimes, in infants too

(Terborgh 1983; Fragaszy 1986; Robinson 1986). For example, sex differences in the frequency of ant and termite eating among adult chimpanzees are associated with similar sex differences in frequency among juveniles (Hiraiwa-Hasegawa 1989) and females begin to 'fish' for termites at a younger age than males, and become more proficient at it at an earlier age. In addition, females adopt techniques similar to those of their mothers, while males do not (Lonsdorf *et al.* 2004). In contrast, sex differences in juvenile feeding behaviour are absent in several mammals where there are no obvious sex differences in the feeding behaviour of adults (van Noordwijk *et al.* 1993).

18.9 Mortality

Adults

Sex differences in growth, body size, feeding ecology and reproductive competition have important consequences for survival and longevity in the two sexes (Geary 2015). Higher annual rates of adult mortality, and reduced life expectancy in males compared with females are common in polygynous, sexually dimorphic species while they are often reduced or absent in relatively monomorphic species (Promislow 1992; Owens 2002; Clutton-Brock and Isvaran 2007), although there are some exceptions (Bocci *et al.* 2010). In many polygynous species, reduced longevity in males is associated with an earlier and more rapid decline in physical condition and survival with increasing age than in females (Clutton-Brock and Isvaran 2007; Beirne *et al.* 2015). Faster rates of ageing in males may in part be a direct consequence of the greater intensity of intrasexual competition in males (see Chapter 13) but the comparatively short effective breeding lifespans of males may also weaken selection for male traits that enhance longevity, including anatomical traits, like the size and structure of cheek teeth in ungulates (Carranza *et al.* 2004, 2008), physiological defences against parasites (Restif and Amos 2010) or oxidative damage (Selman *et al.* 2012) and genetic mechanisms that contribute to longevity (Barrett and Richardson 2011; Jemielity *et al.* 2007).

Several different processes probably contribute to sex differences in mortality. In many polygynous species, males are more likely to be injured or killed as a result of intraspecific fighting (see Chapter 13). In addition, where males defend individual territories, they may be more susceptible to predation because they spend much of the day on their own, and are more likely to be injured or in poor condition at the end of breeding seasons (see Chapter 10). The energetic costs of reproductive competition may also increase the risk of starvation in males and their greater energetic requirements (as a result of their larger body size) may affect their ability to compete with females for scarce resources. In many polygynous species, sex differences in mortality are most pronounced during periods of acute food shortage: for example, when a population of 6000 reindeer on St Matthew Island crashed to 42 in the course of a single winter, males of all ages were almost totally eliminated (Klein 1968) and similar sex differences have been recorded in die-offs of other herbivores, including white-tailed deer (Woolf and Harder 1979), wapiti (Flook 1970), mule deer (Taber and Dasmann 1954; Robinette *et al.* 1957), wildebeest (Child 1972) and Soay sheep (Grubb 1974) as well as in some human groups (see Chapter 19). Finally, more intense intrasexual competition between young males and associated increases in cortisol and testosterone levels may advance rates of ageing to a greater extent in males than in females, leading to earlier senescence and death in the second half of the lifespan: for example, in European badgers, increases in intrasexual competition among young adult males advance ageing rates in older adult males to a greater extent than in females (Beirne *et al.* 2015).

Sex differences in condition caused by contrasts in energetic expenditure during the breeding season may also cause males to be more susceptible to parasites (Promislow 1992; Owens 2002). In many polygynous mammals, males are more heavily parasitised than females and their increased parasite loads contribute to sex differences in mortality at times when resources are scarce (Schalk and Forbes 1997; Moore and Wilson 2002; Owens 2002; Wilson *et al.* 2003) (see Chapter 14): for example, experimental manipulation of parasite load in adolescent sheep shows that it has an important influence on overwinter mortality (Stevenson and Bancroft 1995). Across species, sex differences in parasite load increase with the degree of sexual dimorphism in adult size (Figure 18.11a) and are positively correlated with sex differences in mortality (Figure 18.11b), suggesting that sex differences in mortality may be a consequence of sex differences in immune responses associated with contrasts in the development and reproductive strategies of males and females (Moore and Wilson 2002; Wilson *et al.* 2004; French *et al.* 2007; Nunn *et al.* 2009). Whether

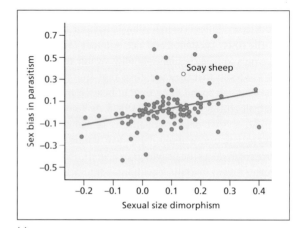

(a)

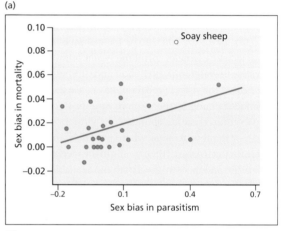

(b)

Figure 18.11 Sexual dimorphism and sex differences in parasite load in natural populations of mammals. The figure shows the sex bias in parasitism rate (male prevalence – female prevalence) in relation to (a) sexual size dimorphism in adults and (b) sex biases in mortality. The lines show the least-squares regression lines. *Source*: From Wilson *et al.* (2004). Reproduced with permission from Cambridge University Press.

these differences are reversed in species where competition is more intense among females is not yet known, but a recent study of meerkats (where dominant females are more frequently aggressive than other group members) has shown that levels of some parasites are highest and most strongly affected by group size in dominant females (Smyth and Drea 2015).

Juveniles and adolescents

Sex differences in growth and development can also have important consequences for survival in juveniles and

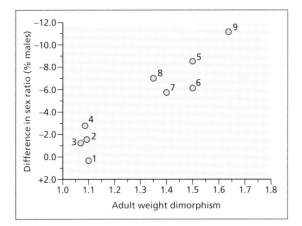

Figure 18.12 Sexual dimorphism and sex differences in juvenile mortality in ungulates. Differences in the sex ratio (percent males) between birth and the end of the first year of life in free-ranging populations of ungulates, showing different degrees of adult weight dimorphism (mean male weight/mean female weight). 1, zebra; 2, roe deer; 3, sable; 4, feral horse; 5, black tailed deer; 6, red deer; 7, wapiti; 8, Soay sheep; 9, reindeer. *Source*: From Clutton-Brock *et al.* (1985). Reproduced with permission of Nature Publishing Group.

adolescents. In many polygynous mammals, male juveniles are more likely to die before they disperse from their natal group than females (Clutton-Brock *et al.* 1985). While sex differences in neonatal mortality are often small or absent, mortality during the first and second winter of life is often biased towards males and sex differences increase with the degree of adult sexual dimorphism (Figure 18.12), suggesting that they may be a consequence of the energetic costs of faster growth rates and higher metabolic requirements of growing males. Like sex differences in adult mortality, they are most pronounced when resources are scarce or conditions are harsh and often rise as populations approach ecological carrying capacity (Figure 18.13). Although sex differences in mortality among juveniles and adolescents have been most extensively documented in ungulates, they also occur in other groups, including rodents and primates (Clutton-Brock 1991a; McGuire and Bemis 2007) as well as in humans (see Chapter 19) and birds (Clutton-Brock *et al.* 1985).

Although a tendency for juvenile and sub-adult males to show higher mortality than females is common, in some dimorphic species there are no obvious sex differences in mortality (Clutton-Brock 1991a). For example, some studies of white-tailed deer and mule deer have

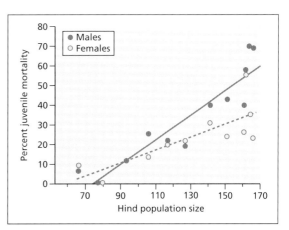

Figure 18.13 Density dependence in sex differences in juvenile mortality in red deer. The figure shows mortality during the first 2 years of life among males and females born between 1971 and 1982 in the red deer population of the North Block of Rum, plotted against a measure of population density (number of resident females). *Source*: From Clutton-Brock *et al.* (1985). Reproduced with permission of Nature Publishing Group. *Photo source*: © Martyn Baker.

found no evidence of sex differences in mortality, while others have found evidence either of male or of female biases (Mohler *et al.* 1951; Taber and Dasmann 1954; Robinette *et al.* 1957; Dapson *et al.* 1979; Woolf and Harder 1979). One possibility is that these differences are a consequence of variation in environmental factors, for both unusually favourable conditions and extremely harsh ones may reduce the extent of sex differences in survival: where food is abundant, the increased energetic needs of males may fail to generate sex differences in mortality; where food is very scarce, such a high proportion of all juveniles may die that sex differences in mortality are reduced. For example, in Soay sheep, where periods of acute starvation reduce the survival of lighter animals of both sexes disproportionately, sex differences in mortality are reduced among light juveniles compared to heavy ones because almost all light juveniles die in winter (Clutton-Brock *et al.* 1992).

In a number of mammals, male biases in juvenile mortality are larger among the offspring of subordinate mothers than among those of dominants (Clutton-Brock *et al.* 1985) and it has been suggested that some sex differences in mortality arise because mothers that cannot afford to invest adequately in sons terminate investment prematurely, generating increased mortality in males (Trivers and Willard 1973). However, there is little evidence that inferior mothers discriminate against sons and sex differences in mortality persist among juveniles reared away from their parents (McCance and Widdowson 1974; Widdowson 1976; Moses *et al.* 1998). For example, when lactating female wood rats were maintained on 70–90% of the maintenance food requirements of non-reproductive females of equivalent size, the sex ratio of their pups declined from around 50% males at birth to 29% by 20 days and in all but one of the litters, every male pup died before there were any deaths among females (McClure 1981). In addition, the most pronounced sex differences in mortality often occur after parental care has been terminated (Clutton-Brock 1991b).

In a few species, the usual pattern of differential mortality is reversed and females consistently show higher mortality than males. Several of the best-documented examples involve societies where competition between females is unusually intense. For example, in spotted hyenas, female infants are more aggressive than males and, when resources are in short supply, females are more likely than males to kill other infants of the same sex (Hofer and East 1997; James and Hofer 1999). In addition, in some primates where members of different matrilines compete intensely with each other, adult females are more commonly aggressive towards the daughters of subordinate females than towards their sons and females born to subordinate mothers are less likely to survive than males (Dittus 1979; Silk *et al.* 1981; van Schaik *et al.* 1983; Altmann *et al.* 1988) (see Chapter 8).

Like differences in adult mortality, sex differences in juvenile mortality probably have multiple causes. Sex differences in nutritional requirements associated with differences in growth, metabolism and fat deposition are often likely to be involved (Trivers 1972; Clutton-Brock 1991b). In addition, the immunosuppressive effects of testosterone and other androgens may lead to increased parasite loads and higher mortality among males: in several polygynous mammals, juvenile and sub-adult males have higher parasite loads than females of the same age (Festa-Bianchet 1991; Schalk and Forbes 1997). Experimental elimination of parasite load can remove sex differences in juvenile mortality: for example, in yearling Soay sheep, experimental treatment with anthelmintic drugs that kill intestinal worms removes sex differences in overwinter survival, which are otherwise pronounced (Stevenson and Bancroft 1995; Wilson *et al.* 2004).

Sex differences in mortality commonly extend into adolescence though it is often difficult to disentangle the consequences of sex differences in energy requirements from those of differential dispersal (see Chapter 12). One useful approach would be to compare sex differences in mortality among adolescents between species where males habitually disperse to breed and those where they remain in their natal groups, but no systematic comparison of this kind is yet available.

18.10 Sex ratios at birth

Population averages

While many mammals produce approximately equal numbers of male and female offspring, statistically significant biases in birth sex ratios are not uncommon and vary in both strength and direction (Clutton-Brock and Iason 1986; Sheldon and West 2004). In some cases, it is likely that these biases are caused by sex differences in mortality before birth, for harsh environmental conditions can be associated with reductions in the proportion of males born. For example, in red deer, the percentage of male calves born declines when population density is high or winter weather is unusually wet (Kruuk *et al.* 1999b) (Figure 18.14) and experimental studies of rodents show that sex ratios commonly decline when food availability is reduced or environmental conditions during gestation are harsh (Austad and Sunquist 1986; Clutton-Brock 1991a; Rosenfeld *et al.* 2003). Several different mechanisms may contribute to sex biases in fetal mortality, including sex differences in growth rates and energy needs, the greater susceptibility of males to infectious diseases and increased maternal immuno-intolerance of male fetuses as a result of the expression

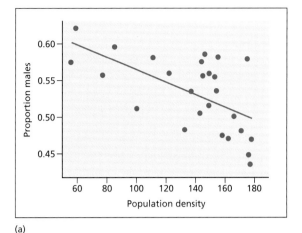

(a)

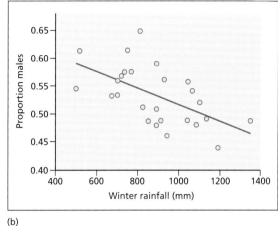

(b)

Figure 18.14 Effects of population density and winter rainfall on the sex ratio of red deer calves born to identified mothers on Rum. (a) Proportion of males born each year in relation to population density after correcting for effect of winter rainfall. Regression coefficient: −0.080 (±0.026 SE). (b) Proportion of males born each year in relation to winter rainfall (November to January, in mm) after correcting for the effect of density. Regression coefficient: −0.0128 (±0.046 SE). *Source*: From Kruuk *et al.* (1999b). Reproduced with permission from Macmillan Group Ltd. *Photo source*: © Martyn Baker.

of Y-linked genes generating antigens in females (Kirby *et al.* 1967).

Although sex differential survival during the course of gestation may often contribute to biases in birth sex ratios, evidence suggests that biases can also be present at conception or at least during the earliest stages of gestation. Although prenatal mortality usually appears to be higher in males than females, sex ratios at birth frequently show a small but significant male bias, suggesting that unless there is an initial loss of females fetuses, larger biases may be present at conception (Clutton-Brock and Iason 1986). In some species, aborted fetuses show a male bias (DiGiacomo and Shaughnessy 1979; McMillen 1979), while several studies of mammals killed at different stages of gestation have found that the sex ratio of fetuses (calculated as the percentage of males) shows a male bias that declines as gestation progresses (Clutton-Brock 1991b).

While average sex ratios at birth often show a small bias towards males, in some species these biases are more pronounced. For example, in some cooperative breeders where males commonly remain in their natal group and assist in their mother's subsequent breeding attempts, like African wild dogs, birth sex ratios are often relatively strongly biased towards males (Creel *et al.* 1998) (see Chapter 17). In other cases, male-biased sex ratios are associated with intense competition between females for resources: for example, in some populations of lemurs where local resource competition between females is unusually strong, mothers produce male-biased sex ratios (Richard *et al.* 2002). Other studies have reported birth sex ratios that show a consistent bias towards females: for example, in some marsupials, the average sex ratio of pouch young is biased towards females and a similar bias appears to be present at birth (Davison and Ward 1998).

While it is likely that some biases in the average sex ratio are associated with contrasts in the costs or benefits to parents of rearing sons or daughters, the reasons for the distribution of biases in the average sex ratio among vertebrates are not well understood (Clutton-Brock and Iason 1986; West 2009). Since selection is expected to favour equal total investment in male and female progeny (Fisher 1930), both sex differences in rearing costs and sex differences in mortality before the end of the period of parental investment would be expected to affect the evolution of birth sex ratios (see Chapter 1). However, in several sexually dimorphic species where males are more likely to die before or soon after birth, the sex ratio of juveniles is still biased towards males at the end of the period of parental care (Clutton-Brock 1991b). In addition, there is little indication that average birth sex ratios are biased towards females in species, like spotted hyenas, where female juveniles are more likely to die before the end of the period of parental care than males (Hofer and East 1997; James and Hofer 1999). One recent suggestion is that demographic processes that influence the reproductive value of males and females after the end of the period of parental investment may affect the evolution of sex ratio biases (Schindler *et al.* 2015).

Variation between females

Birth sex ratios can also differ between females and may vary with the mother's age, dominance rank and nutritional status (Clutton-Brock 1991b; Cameron 2004; Sheldon and West 2004; Robert and Schwanz 2011). For example, in red deer in an expanding population on the Scottish island of Rum, the proportion of males born to females during their lifespan varied from around 35% in subordinate mothers to around 70% among dominants (Clutton-Brock and Iason 1986). Similar correlations between maternal dominance or indices of maternal size or condition and birth sex ratios have been documented in a number of other dimorphic ungulates (Clutton-Brock 1991b; Sheldon and West 2004).

Several studies of rodents have also shown that the sex ratios of offspring produced by individual females varies with the mother's early development. For example, in golden hamsters, females whose access to food was restricted during their first 50 days of life and were subsequently replaced on ad lib diets produced smaller female-biased sex litters throughout their adult lives (Huck *et al.* 1986). The effects of food restriction during early development can span generations: in hamsters, the daughters of food-restricted females (themselves reared on ad lib diets) produced smaller litters and relatively fewer sons than the daughters of control females whose mothers were not food restricted (Huck *et al.* 1987). Exposure to varying levels of sex hormones during gestation may also be important. For example, in house mice, females that develop with male fetuses on either side of them produce more sons as adults than females that develop with female fetuses on either side, while those that were neighbours to one male and one female fetus produced intermediate sex ratios (Vandenbergh and Huggett 1994; Ryan and Vandenbergh 2002).

While there is substantial evidence of significant sex ratio biases, they have not been found in all species. For example, although meta-analysis of data for ungulates shows that, overall, there is a positive correlation between maternal rank or correlates of maternal condition and birth sex ratios (Sheldon and West 2004), there is no evidence of consistent trends in birth sex ratios in some species for which large sets of quality data are available (Hewison and Gaillard 1999; Lindström *et al.* 2002). In addition, the presence and strength of biases appears to vary between and within populations (Clutton-Brock 1991b; Cameron 2004; Komdeur 2012). For example, the positive correlation between maternal rank and the sex ratio in red deer on Rum disappeared when numbers reached carrying capacity and food supplies were reduced and erratic (Kruuk *et al.* 1999b).

One possible explanation of some of this variation is that multiple factors affect the sex ratio, complicating simple relationships between maternal rank or condition (Alonzo and Sheldon 2010). For example, in bighorn sheep, birth sex ratios vary with maternal age: while young females produce more sons than daughters, old ones do the reverse (Martin and Festa-Bianchet 2011). Similarly, in brush-tail possums, shortages of den sites correlated with high local densities of females are associated with male-biased sex ratios (Johnson *et al.* 2001). Another possibility is that the strength of direction of sex ratio trends varies with the point in time at which maternal characteristics are measured: trends are commonly stronger when maternal characteristics are measured at the time of conception than when they are measured at the time of birth (Cameron 2004; Sheldon and West 2004). In addition, relationships can become clearer if absolute estimates of maternal condition are replaced by measures of the extent to which maternal

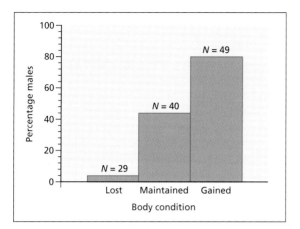

Figure 18.15 Extreme sex ratios in Kaimanawa horses. The figure shows the sex ratio (percentage males) produced by mothers that lost condition, maintained condition or gained condition between pre-conception (-20 ± 10 days) and post-conception ($+20 \pm 10$ days) for foals born in 1995, 1996 and 1997. Source: from Cameron and Linklater (2007). Reproduced with permission from the Royal Society.

condition is changing. For example, in one population of feral horses where mothers in superior condition tended to produce a small excess of sons, sex ratio biases increased dramatically when measures of absolute condition were replaced with measures of change in maternal condition: only 3% of mares that were losing condition at the time of conception subsequently gave birth to sons while 80% of females that were gaining condition at the time did so (Cameron and Linklater 2007) (Figure 18.15). A final possibility is that the relative benefits of producing sons and daughters are affected by variation in the relative reproductive value of males and females across the lifespan associated with variation in demographic conditions (Schindler *et al.* 2015).

Mechanisms

So what mechanisms might be responsible for variation in sex ratios at birth? Some biases are probably a consequence of the greater susceptibility of male fetuses to starvation as a result of their faster growth and larger metabolic requirements: associations between maternal size, dominance or condition that affect the mother's capacity to invest during gestation may affect the resources available to offspring and generate sex differences in prenatal survival (Clutton-Brock 1991b). However, effects of this kind do not provide a satisfactory

explanation of cases where biased birth sex ratios are not associated with any reduction in the number of offspring born and several lines of evidence suggest that biases can occur before male and female fetuses show measurable differences in growth (Clutton-Brock 1991b). For example, studies of natural abortions in women show that samples from the early months of gestation often show a strong bias towards males (McMillen 1979; Clutton-Brock 1991b).

Another possibility is that biases in the birth sex ratio are a consequence of sex differences in the susceptibility of male and female blastocysts to variation in glucose levels *in utero* (Cameron 2004). In mice, increases in glucose levels *in utero* favour development in male blastocysts during early cell division and low or declining levels may be responsible for a reduction in the number of males that develop successfully (Cameron 2004; Cameron *et al.* 2008) and this may help to explain a number of other sex ratio trends. For example, mice produce significantly more males when fed high-fat diets (which are likely to increase glucose levels) compared with low-fat diets with the same calorific value (Rosenfeld *et al.* 2003).

A third possibility is that sex ratio biases are a consequence of variation in the proportion of X- and Y-bearing sperm produced by males. The characteristics of sperm (including their velocity, concentration, morphology and viability) are known to vary between males and, in some cases, these differences are associated with the probability of sperm competition (Tourmente *et al.* 2011; Edwards and Cameron 2014). Recent advances that have facilitated the analysis of variation in the frequency of X- and Y-bearing sperm show that ratios can vary both between populations and between individuals and may be affected by a range of environmental contaminants as well as by heat stress, mating frequency and male age (Edwards and Cameron 2014). Many of the same variables are also associated with reductions in the proportion of males born, suggesting that variation in X/Y ratios or sperm characteristics may have an important influence on the birth sex ratio. For example, in captive red deer, more fertile males produce a higher proportion of male offspring than less fertile individuals (Gomendio *et al.* 2006).

Adaptive significance

In 1973, Trivers and Willard suggested that parents might be expected to adjust the sex ratio of their offspring in relation to their capacity for investment and its effects on the relative fitness of their sons and daughters. Their specific argument was that where offspring of one sex require higher levels of parental investment in order to survive and breed successfully, parents of high phenotypic quality (who can afford to invest heavily in their offspring) would be expected to produce a disproportionate number of the more expensive sex, while inferior parents might be expected to favour the less costly sex (Trivers and Willard 1973). This explanation is more likely to apply if sex ratio biases are a consequence of variation in conception ratios, for manipulation by differential mortality in the later stages of gestation is likely to have high fitness costs, especially in monotocous species.

Since 1973, a large number of variants or modifications of Trivers and Willard's hypothesis have been suggested (West 2009; Komdeur 2012). In polytocous species, there may be trade-offs between the sex and number of offspring (Williams 1979; Burley 1982; Gosling 1986): for example, mothers in very poor condition might produce a single offspring of the cheaper sex, those in slightly better condition a singleton of the more expensive sex, those in good condition twins of the cheaper sex, and those in very good condition twins of the more expensive sex (Figure 18.16). Alternatively, if the relative attractiveness (and hence the reproductive

value) of sons varies with the quality of their fathers, females mated to attractive males might bias the sex ratio of their offspring towards males, while females mated to unattractive partners might tend to produce daughters (Burley 1977, 1981, 1986; Ellegren *et al.* 1996; Sheldon 1998; Pen and Weissing 2000; West and Sheldon 2002; West 2009).

Contrasts in competition between parents and their daughters or sons might also be expected to affect birth sex ratios. For example, where daughters settle in their mother's range and compete with her for resources while sons disperse at adolescence, mothers might be expected to produce more sons than daughters, an idea that is usually referred to as *local resource competition* (Clark 1978). Or, if parents are able to pass on resources (such as territories or home ranges) to offspring of one sex and this affects their fitness, this, too, may affect the relative benefits of rearing sons and daughters and could generate biases in the average sex ratio as well as adaptive changes in the sex ratio of particular times: for example, ageing parents without resident 'heirs' might be expected to bias the sex ratio of their offspring towards the inheriting sex (Leimar 1996). Differences in the extent of cooperation between sibs could have similar effects: for example, where sons are more likely to cooperate with each other than daughters (for example, by dispersing together) parents might be expected to

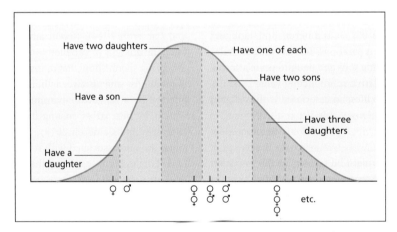

Figure 18.16 Litter sex ratio optimisation in a species that usually has two but sometimes one or three young per clutch. Where female quality is normally distributed and is related to the capacity of mothers to invest in their offspring and sons cost more to rear than daughters, low-quality females might be expected a produce a single daughter; higher-quality ones, a single son; even higher-quality females, two daughters; and females with the greatest ability to invest, two sons. *Source*: From Williams (1979). Reproduced with permission from the Royal Society.

produce more sons than daughters or to produce litters that are biased towards one sex or the other (Clutton-Brock 1991b; West 2009). In addition, in cooperative breeders where offspring of one sex are more likely to remain in their natal group and assist their parents, mothers might be expected to adjust the sex ratio of young to their need for helpers (see Chapter 17).

Males might also be expected to adjust the ratio of X- and Y-bearing sperm that they produce so as to maximise their fitness (Edwards and Cameron 2014). For example, attractive fathers whose sons are likely to inherit the same traits might be expected to produce an excess of Y-bearing sperm, which could explain why, in some species, more fertile males produce more sons (Gomendio et al. 2006). High mating rates could provide a rule of thumb for a male's attractiveness, leading to selection on successful males to produce an excess of Y-bearing sperm (Edwards and Cameron 2014).

So are most observed cases of sex ratio variation in mammals likely to be the product of differences in the susceptibility of the two sexes to starvation or are they the result of adaptive manipulation of the sex ratio by parents? One way of investigating whether particular sex ratio trends are likely to be adaptive is to investigate the effects of maternal characteristics associated with sex ratio variation on the relative fitness of sons and daughters. For example, in red deer, in an expanding population, the lifetime reproductive success of sons increased more rapidly with maternal rank than that of daughters (see Figure 14.9), indicating that a tendency for dominant females to produce male-biased sex ratios and for subordinate mothers to produce an excess of daughters probably increases the mother's fitness (Clutton-Brock et al. 1984). There is also some evidence that sex ratio trends may be reversed where maternal characteristics are likely to have a stronger influence on the fitness of daughters than sons. For example, in some cercopithecine primates (including some baboons and macaques) and some equids, maternal support for daughters may lead to situations where a mother's rank affects the fitness of her daughters more than that of her sons and negative correlations between birth sex ratios and maternal rank have been documented in some species (Simpson and Simpson 1985; Altmann et al. 1988; Lloyd and Rasa 1989). However, even data of this kind does not necessarily show that sex ratio trends are adaptive for the benefits of producing sons and daughters may be affected by variation in the reproductive value of the two sexes (Schindler et al. 2015).

Other studies have shown that birth sex ratios vary in ways that appear likely to be adaptive. For example, some studies suggest that stronger local resource competition in one sex can lead to changes in the birth sex ratio. For example, in some lemurs where females are philopatric and competition between them is intense, females kept in groups or experimentally exposed to urine from other females produce a higher proportion of sons than those caged individually (Perret 1996). Sex differences in cooperation may also affect the sex ratio: in wild dogs, where sons commonly assist their mother in rearing subsequent offspring, mothers may adjust the sex ratio of their offspring to their need for additional helpers (Creel et al. 1998; McNutt and Silk 2008); in Damaraland mole rats (where offspring of both sexes contribute equally to rearing young but males are more likely to disperse), the proportion of males born rises in relation to the number of female helpers already present in the group (Lutermann et al. 2014).

While these results suggest that females of some species may adjust the sex ratio of their offspring in an adaptive fashion, this does not indicate that all trends in birth sex ratios are adaptive. One problem in testing adaptive explanations of sex ratio variation is that their diversity means that few observed trends do not fit the predictions of at least one adaptive explanation. As a result, while it is possible to find an adaptive explanation for most trends, this does not necessarily indicate they are correct (Clutton-Brock 1991b). Determining which sex ratio trends are a consequence of adaptive parental manipulation may consequently remain problematic until we know substantially more about the processes responsible for sex ratio biases. Studies of species where it is possible to explore the effects of temporal changes in female condition, to measure variation in the relative frequency of X- and Y-bearing sperm or to manipulate sex ratios before birth are likely to have an important role to play (Robert and Schwanz 2011).

18.11 Adult sex ratios

Sex differences in development and survival have important consequences for adult sex ratios. In sexually dimorphic species, the effects of population density on the sex ratio at birth and on sex differences in mortality and emigration (see Chapter 12) can lead to adult sex ratios that are strongly biased towards females where populations are close to carrying capacity (Clutton-Brock *et al.* 1982a; Jorgenson *et al.* 1998). For example, in red deer, increases in the density of females are associated with larger increases in mortality and net emigration in males than in females and, as population density increases, adult sex ratios commonly show an increasing bias towards females (Clutton-Brock and Albon 1989). A demographic experiment in which female deer numbers were reduced by culling in one area and allowed to increase unchecked in another confirmed that reductions in female density led to increases in male numbers (Clutton-Brock *et al.* 2002).

Biases in the adult sex ratio have important consequences for both sexes. In females, they can lead to increased reproductive competition as well as to reductions in the capacity of males to protect them against predators or conspecifics. For example, in chacma baboons, adult sex ratios that are strongly skewed towards females are associated with increased intrasexual aggression among females, less stable social bonds and higher mortality rates in females (Cheney *et al.* 2012). Female-biased adult sex ratios may also delay conception and birth timing and reduce synchrony between females, with negative effects on juvenile survival, especially in males (Clutton-Brock and Parker 1992; Mysterud *et al.* 2002). In extreme cases, they may also reduce female fecundity, though males are generally capable of fertilising large numbers of females (Bergerud 1974; Haigh and Hudson 1993; Solberg *et al.* 2002) and the adult sex ratio is seldom so strongly biased that female fecundity is likely to be affected (Mysterud *et al.* 2002).

Female-biased adult sex ratios can also have important consequences for males by reducing the intensity of reproductive competition (Geist 1971). For example, in red deer populations that are strongly skewed towards females, dominant males do not monopolise as large a proportion of matings as in populations where the adult sex ratio is more equal, with the result that younger males are more likely to breed successfully (Clutton-Brock *et al.* 1997). Similarly, in Soay sheep, female-biased adult sex ratios are associated with both increased breeding by younger males and reduced time spent guarding by mature males (Stevenson 1994; Coltman *et al.* 1999).

Biased adult sex ratios can also have important implications for management, especially in populations where the main income is derived from the offtake of males (Clutton-Brock and Albon 1989). For example, in red deer populations that are managed for sport, female-biased adult sex ratios are likely to cause a reduction in the potential offtake of males as well as in income and models show that managers can maximise income by maintaining female numbers substantially below ecological carrying capacity (Clutton-Brock and Lonergan 1994; Buckland *et al.* 1996; Clutton-Brock *et al.* 2002) (Figure 18.17). Female densities that maximise the offtake of males can be relatively low: for example, calculations suggest that income from Scottish red deer populations may be maximised when female numbers are little more than 50% of the levels they would reach if they were not culled.

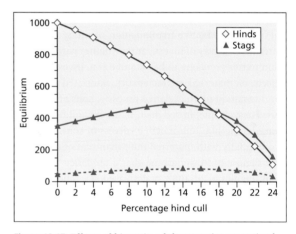

Figure 18.17 Effects of biases in adult sex ratios on optimal culling rates in red deer. The figure shows number of hinds (diamonds) and stags (triangles) of different age categories (hinds: solid line indicates all animals ≥1 year old; stags: solid line indicates ≥1 year old, dashed line 5 years old) in the population expected at equilibrium before the onset of the annual cull. The size of the unculled population is scaled to 1000 hinds. The model assumes that reproduction and survival change with the density of hinds alone and that hind culls vary (percentage hind cull). All stags are culled at age 5, giving the maximum sustainable yield for hind density. The highest potential offtake of males occurs when 12–16% of hinds are culled annually and the equilibrium number of hinds is around 58% of the number that would be present if they were not culled. *Source*: From Clutton-Brock and Lonergan (1994). Reproduced with permission of John Wiley & Sons.

SUMMARY

1. Sex differences in the strength of selection for traits that improve competitive ability or attractiveness to the opposite sex are common in polygynous and promiscuous mammals and are often associated with the greater development of body size and weaponry in one sex, usually males. However, competition between females can lead to the evolution of similar traits in females and reduce the magnitude of sex differences.

2. Sex differences in body size are often associated with sex differences in the rate and duration of growth, and in some polygynous species males continue to increase in size and weight throughout their lifespan. Faster growth rates in juvenile males are frequently associated with greater nutritional needs, with increases in the frequency and duration of nursing by males, and with increased energetic demands on their mothers. In some species, rearing sons has a stronger influence on the subsequent breeding success and survival of their mothers than rearing daughters.

3. Sex differences in adult size are also frequently associated with sex differences in habitat use and feeding behaviour: in many herbivores, males utilise habitats where food is more abundant but of lower nutritional value more than females.

4. Sex differences in growth and energetic requirements are also associated with sex differences in mortality, especially when conditions are harsh or food is scarce. In many polygynous species, average rates of mortality are higher in males than in females and males age faster and die earlier than females.

5. Sex differences in mortality can also occur before birth and are probably a common cause of variation in birth sex ratios, though there is also evidence of variation in the sex ratio at (or soon after) conception. While average birth sex ratios are close to parity in most mammals, some show consistent biases which vary in strength and direction and birth sex ratios also vary between mothers. Although some biases in birth sex ratios may be a consequence of adaptive adjustments in relation to variation in the costs or benefits of raising sons or daughters, others may be non-adaptive by-products of their sex differences.

References

Alexander, G.M. and Hines, M. (2002) Sex differences in response to children's toys in nonhuman primates (*Cercopithecus aethiops sabaeus*). *Evolution and Human Behavior* **23**:467–479.

Alexander, R.D., et al. (1979) Sexual dimorphism and breeding systems in pinnipeds, ungulates, primates and humans. In: *Evolutionary Biology and Human Social Behavior: An Anthropological Perspective* (eds N. A. Chapman and W. Irons). North Scituate, MA: Duxbury Press, 402–435.

Alleva, E., et al. (1989) Litter gender composition affects maternal behavior of the primiparous mouse dam (*Mus musculus*). *Journal of Comparative Psychology* **103**:83–87.

Alonzo, S.H. and Sheldon, B.C. (2010) Population density, social behaviour and sex allocation. In: *Social Behaviour: Genes, Ecology and Evolution* (eds T. Székely, A.J. Moore and J. Komdeur). Cambridge: Cambridge University Press, 474–488.

Altmann, J., et al. (1988) Determinants of reproductive success in savannah baboons, *Papio cynocephalus*. In: *Reproductive Success: Studies of Individual Variation in Contrasting Breeding Systems* (ed. T.H. Clutton-Brock). Chicago: University of Chicago Press, 403–418.

Andres, D., et al. (2013) Sex differences in the consequences of maternal loss in a long-lived mammal, the red deer (*Cervus elaphus*). *Behavioral Ecology and Sociobiology* **67**:1249–1258.

Austad, S.N. and Sunquist, M.E. (1986) Sex-ratio manipulation in the common opossum. *Nature* **324**:58–60.

Baird, R.W. and Whitehead, H. (2000) Social organisation of mammal-eating killer whales: group stability and dispersal patterns. *Canadian Journal of Zoology* **78**:2096–2015.

Baldwin, J.D. (1969) The ontogeny of social behavior of squirrel monkeys (*Saimiri sciureus*) in a seminatural environment. *Folia Primatologica* **11**:35–79.

Barboza, P.S. and Bowyer, R.T. (2000) Sexual segregation in dimorphic deer: a new gastrocentric hypothesis. *Journal of Mammalogy* **81**:473–489.

Barrett, E.L. and Richardson, D.S. (2011) Sex differences in telomeres and lifespan. *Aging Cell* **10**:913–921.

Bartlett, T.Q. (2007) The Hylobatidae: small apes of Asia. In: *Primates in Perspective* (eds C. Campbell, A. Fuentes and K. MacKinnon). Oxford: Oxford University Press, 274–289.

Becker, C.D. and Ginsberg, J.R. (1990) Mother-infant behaviour of wild Grevy's zebra: adaptations for survival in semidesert East Africa. *Animal Behaviour* **40**:1111–1118.

Beirne, C., et al. (2015) Sex differences in senescence: the role of intra-sexual competition in early adulthood. *Proceedings of the Royal Society of London. Series B: Biological Sciences* **282**:20151086.

Bell, R.H.V. (1969) The use of the herb layer by grazing ungulates in the Serengeti National Park, Tanzania. PhD thesis, University of Manchester.

Bell, R.H.V. (1971) A grazing ecosystem in the Serengeti. *Scientific American* **225**:86–93.

Bercovitch, F.B., et al. (2000) Maternal investment in rhesus macaques (*Macaca mulatta*): reproductive costs and consequences of raising sons. *Behavioral Ecology and Sociobiology* **48**:1–11.

Bergerud, A.T. (1974) Rutting behaviour of Newfoundland caribou. In: *The Behaviour of Ungulates and its Relation to Management* (eds V. Geist and F. Walther). Morges, Switzerland: IUCN 395–435.

Bergerud, A.T., *et al.* (1984) Antipredator tactics of calving caribou: dispersion in mountains. *Canadian Journal of Zoology* **62**:1566–1575.

Berghanel, A., *et al.* (2015) Locomotor play drives motor skill acquisition at the expense of growth: a life history trade-off. *Science Advances* **1** (7): e1500451.

Biben, M. (1998) Squirrel monkey playfighting: making the case for a cognitive training function for play. In: *Animal Play: Evolutionary, Comparative and Ecological Perspectives* (eds M. Bekoff and J.A. Byers). Cambridge: Cambridge University Press, 161–182.

Birgersson, B. (1998) Male-biased maternal expenditure and associated costs in fallow deer. *Behavioral Ecology and Sociobiology* **43**:87–93.

Bocci, A., *et al.* (2010) Even mortality patterns of the two sexes in a polygynous, near-monomorphic species: is there a flaw? *Journal of Zoology* **280**:379–386.

Bonenfant, C., *et al.* (2003) Sex-ratio variation and reproductive costs in relation to density in a forest-dwelling population of red deer (*Cervus elaphus*). *Behavioral Ecology* **14**:862–869.

Bouteiller-Reuter, C. and Perrin, N. (2005) Sex-specific selective pressures on body mass in the greater white-toothed shrew, *Crocidura russula*. *Journal of Evolutionary Biology* **18**:290–300.

Bowyer, R.T. (1987) Coyote group size relative to predation on mule deer. *Mammalia* **51**:515–526.

Bowyer, R.T. (2004) Sexual segregation in ruminants: definitions, hypotheses, and implications for conservation and management. *Journal of Mammalogy* **85**:1039–1052.

Bradley, B.J., *et al.* (in press) Non-human primates avoid the detrimental effects of prenatal androgen exposure in mixed-sex litters. *BMC Evolutionary Biology*.

Bro-Jørgensen, J. (2007) The intensity of sexual selection predicts weapon size in male bovids. *Evolution* **61**:1316–1326.

Brown, G.R. and Silk, J.B. (2002) Reconsidering the null hypothesis: is maternal rank associated with birth sex ratios in primate groups? *Proceedings of the National Academy of Sciences of the United States of America* **99**:11252–11255.

Buckland, S.T., *et al.* (1996) Estimating the minimum population size that allows a given annual number of mature red deer stags to be culled sustainably. *Journal of Applied Ecology* **33**:118–130.

Burley, N. (1977) Parental investment, mate choice and mate quality. *Proceedings of the National Academy of Sciences of the United States of America* **74**:3476–3479.

Burley, N. (1981) Sex ratio manipulation and selection for attractiveness. *Science* **211**:721–722.

Burley, N. (1982) Facultative sex-ratio manipulation. *American Naturalist* **120**:81–107.

Burley, N. (1986) Sex ratio manipulation in color-banded populations of zebra finches. *Evolution* **40**:1191–1206.

Byers, J.A. and Moodie, J.D. (1990) Sex-specific maternal investment in pronghorn, and the question of a limit on differential provisioning in ungulates. *Behavioral Ecology and Sociobiology* **26**:157–164.

Cameron, E.Z. (1998) Is suckling behaviour a useful predictor of milk intake? A review. *Animal Behaviour* **56**:521–532.

Cameron, E.Z. (2004) Facultative adjustment of mammalian sex ratios in support of the Trivers–Willard hypothesis: evidence for a mechanism. *Proceedings of the Royal Society of London. Series B: Biological Sciences* **271**:1723–1728.

Cameron, E.Z. and Linklater, W.L. (2000) Individual mares bias investment in sons and daughters in relation to their condition. *Animal Behaviour* **60**:359–367.

Cameron, E.Z. and Linklater, W.L. (2007) Extreme sex ratio variation in relation to change in condition around conception. *Biology Letters* **3**:395–397.

Cameron, E.Z., *et al.* (2008) Experimental alteration of litter sex ratios in a mammal. *Proceedings of the Royal Society of London. Series B: Biological Sciences* **275**:323–327.

Caro, T., *et al.* (2012) Pelage coloration in pinnipeds: functional consideratios. *Behavioral Ecology* **23**:765–774.

Carranza, J., *et al.* (2004) Disposable-soma senescence mediated by sexual selection in an ungulate. *Nature* **432**:215–218.

Carranza, J., *et al.* (2008) Sex-specific strategies of dentine depletion in red deer. *Biological Journal of the Linnean Society* **93**:487–497.

Chapais, B. (1988) Experimental matrilineal inheritance of rank in female Japanese monkeys. *Animal Behaviour* **36**:1025–1037.

Chapais, B. and Schulman, S.R. (1980) An evolutionary model of female dominance relations in primates. *Journal of Theoretical Biology* **82**:47–89.

Cheney, D.L. (1977) The acquisition of rank and the development of reciprocal alliances among free-ranging immature baboons. *Behavioral Ecology and Sociobiology* **2**:303–318.

Cheney, D.L. (1978) The play partners of immature baboons. *Animal Behaviour* **26**:1038–1050.

Cheney, D.L., *et al.* (2012) Evidence for intrasexual selection in wild female baboons. *Animal Behaviour* **84**:21–27.

Child, G. (1972) Observations on a wildebeest die-off in Botswana. *Arnoldia* **31**:1–13.

Chivers, D.J. (1977) The lesser apes. In: *Primate Conservation* (eds Prince Rainier of Monaco and G.H. Bourne). New York: Academic Press, 539–598.

Christianson, D. and Creel, S. (2008) Risk effects in elk: sex-specific responses in grazing and browsing due to predation risk from wolves. *Behavioral Ecology* **19**:1258–1266.

Clark, A.B. (1978) Sex ratio and local resource competition in a prosimian primate. *Science* **201**:163–165.

Clark, M.M., *et al.* (1990) Evidence of sex-biased postnatal maternal investment by Mongolian gerbils. *Animal Behaviour* **39**:735–744.

Clark, M.M., *et al.* (1992) Stud males and dud males: intra-uterine position effects on the reproductive success of male gerbils. *Animal Behaviour* **43**:215–221.

Clark, M.M., *et al.* (1998) Intrauterine position, parenting, and nest-site attachment in male Mongolian gerbils. *Developmental Psychobiology* **32**:177–181.

Clutton-Brock, T.H. (1977) Some aspects of intraspecific variation in feeding and ranging behaviour in primates. In: *Primate Ecology: Studies of Feeding and Ranging Behaviour in Lemurs, Monkeys and Apes* (ed. T.H. Clutton-Brock). London: Academic Press, 539–556.

Clutton-Brock, T.H. (1991a) The evolution of sex differences and the consequences of polygyny in mammals. In: *The Development and Integration of Behaviour: Essays in Honour of Robert Hinde* (ed. P. Bateson). Cambridge: Cambridge University Press, 229–253.

Clutton-Brock, T.H. (1991b) *The Evolution of Parental Care*. Princeton, NJ: Princeton University Press.

Clutton-Brock, T.H. (2007) Sexual selection in males and females. *Science* **318**:1882–1885.

Clutton-Brock, T.H. and Albon, S.D. (1982) Parental investment in male and female offspring in mammals. In: *Current Problems in Sociobiology* (eds Kings College Sociobiology Group). Cambridge: Cambridge University Press, 223–247.

Clutton-Brock, T.H. and Albon, S.D. (1989) *Red Deer in the Highlands*. Oxford: Blackwell Scientific Publications.

Clutton-Brock, T.H. and Harvey, P.H. (1983) The functional significance of variation in body size among mammals. In: *Advances in the Study of Mammalian Behavior* (eds J.F. Eisenberg and D.G. Kleiman). American Society of Mammalogists Special Publication No. 7. Stillwater, OK: American Society of Mammalogists, 632–663.

Clutton-Brock, T.H. and Huchard, E. (2013a) Social competition and its consequences in female mammals. *Journal of Zoology* **289**:151–171.

Clutton-Brock, T.H. and Huchard, E. (2013b) Social competition and selection in males and females. *Philosophical Transactions of the Royal Society B: Biological Sciences* **368**:20130074.

Clutton-Brock, T.H. and Iason, G.R. (1986) Sex ratio variation in mammals. *Quarterly Review of Biology* **61**:339–374.

Clutton-Brock, T.H. and Isvaran, K. (2007) Sex differences in ageing in natural populations of vertebrates. *Proceedings of the Royal Society of London. Series B: Biological Sciences* **274**:3097–3104.

Clutton-Brock, T.H. and Lonergan, M.E. (1994) Culling regimes and sex ratio biases in Highland red deer. *Journal of Applied Ecology* **31**:521–527.

Clutton-Brock, T.H. and Parker, G.A. (1992) Potential reproductive rates and the operation of sexual selection. *Quarterly Review of Biology* **67**:437–456.

Clutton-Brock, T.H., *et al.* (1977) Sexual dimorphism, socionomic sex ratio and body weight in primates. *Nature* **269**:797–800.

Clutton-Brock, T.H., *et al.* (1980) Antlers, body-size and breeding group size in the Cervidae. *Nature* **285**:565–567.

Clutton-Brock, T.H., *et al.* (1981) Parental investment in male and female offspring in polygynous mammals. *Nature* **289**:487–489.

Clutton-Brock, T.H., *et al.* (1982a) *Red Deer: The Behaviour and Ecology of Two Sexes*. Chicago: University of Chicago Press.

Clutton-Brock, T.H., *et al.* (1982b) Effects of lactation on feeding behaviour and habitat use in wild red deer hinds. *Journal of Zoology* **198**:227–236.

Clutton-Brock, T.H., *et al.* (1984) Maternal dominance, breeding success and birth sex ratios in red deer. *Nature* **308**:358–360.

Clutton-Brock, T.H., *et al.* (1985) Parental investment and sex differences in juvenile mortality in birds and mammals. *Nature* **313**:131–133.

Clutton-Brock, T.H., *et al.* (1987) Sexual segregation and density-related changes in habitat use in male and female red deer (*Cervus elaphus*). *Journal of Zoology* **211**:275–289.

Clutton-Brock, T.H., *et al.* (1992) Early development and population fluctuations in Soay sheep. *Journal of Animal Ecology* **61**:381–396.

Clutton-Brock, T.H., *et al.* (1997) Stability and instability in ungulate populations, an empirical analysis. *American Naturalist* **149**:195–219.

Clutton-Brock, T.H., *et al.* (2002) Sex differences in emigration and mortality affect optimal management of deer populations. *Nature* **415**:633–637.

Clutton-Brock, T.H., *et al.* (2006) Intrasexual competition and sexual selection in cooperative mammals. *Nature* **444**:1065–1068.

Clutton-Brock, T.H., *et al.* (2010) Adaptive suppression of subordinate reproduction in cooperative mammals. *American Naturalist* **176**:664–673.

Coltman, D.W., *et al.* (1999) Density-dependent variation in lifetime breeding success and in natural and sexual selection in Soay rams. *American Naturalist* **154**:730–746.

Conradt, L., *et al.* (2000) Sex differences in weather sensitivity can cause habitat segregation: red deer as an example. *Animal Behaviour* **59**:1049–1060.

Costa, D.P. and Gentry, R.L. (1986) Free-ranging energetics of northern fur seals. In: *Fur Seals: Maternal Strategies on Land and at Sea* (eds R.L. Gentry and G. Koogman). Princeton, NJ: Princeton University Press, 79–101.

Creel, S., *et al.* (1998) Birth order, estrogens and sex-ratio adaptation in African wild dogs (*Lycaon pictus*). *Animal Reproduction Science* **53**:315–320.

Crockett, C.M. and Pope, T.R. (1993) Consequences of sex differences in dispersal for juvenile red howler monkeys. In: *Juvenile Primates: Life History, Development, and Behavior* (eds M.E. Pereira and L.A. Fairbanks). New York: Oxford University Press, 104–118.

Dapson, R.W., *et al.* (1979) Demographic differences in contiguous populations of white-tailed deer. *Journal of Wildlife Management* **43**:889–898.

Darwin, C. (1859) *On The Origin of Species By Means of Natural Selection*. London: John Murray.

Davison, M.J. and Ward, S.J. (1998) Prenatal bias in sex ratios in a marsupial, *Antechinus agilis. Proceedings of the Royal Society of London. Series B: Biological Sciences* **265**:2095–2099.

Demment, M.W. and Van Soest, P.J. (1985) A nutritional explanation for body-size patterns of ruminant and non-ruminant herbivores. *American Naturalist* **125**:641–672.

DiGiacomo, R.F. and Shaughnessy, P.W. (1979) Fetal sex ratio in the rhesus (*Macaca mulatta*). *Folia Primatologica* **31**:246–250.

Dittus, W.P.J. (1979) The evolution of behavior regulating density and age-specific sex ratios in a primate population. *Behaviour* **69**:265–302.

Drickamer, L.C. (1996) Intrauterine position and anogenital distance in house mice: consequences under field conditions. *Animal Behaviour* **51**:925–934.

Drickamer, L.C., *et al.* (1997) Conception failure in swine: importance of the sex ratio of a female's birth litter and tests of other factors. *Journal of Animal Science* **75**:2192–2196.

du Toit, J.T. (2005) Sex differences in the foraging ecology of large mammalian herbivores. In: *Sexual Segregation in Vertebrates* (eds K.E. Ruckstuhl and P. Neuhaus). Cambridge: Cambridge University Press, 35–52.

Edelman, A.J. (2011) Sex-specific effects of size and condition on timing of natal dispersal in kangaroo rats. *Behavioral Ecology* **22**:776–783.

Edwards, A.M. and Cameron, E.Z. (2014) Forgotten fathers: paternal influences on mammalian sex allocation. *Trends in Ecology and Evolution* **29**:158–164.

Ellegren, H., *et al.* (1996) Sex ratio adjustment in relation to paternal attractiveness in a wild bird population. *Proceedings of the National Academy of Sciences of the United States of America* **93**:11723–11728.

Fadem, B.H. and Corbett, A. (1997) Sex differences and the development of social behavior in a marsupial, the gray short-tailed opossum (*Monodelphis domestica*). *Physiology and Behavior* **61**:857–861.

Fairbanks, L.A. (1990) Reciprocal benefits of allomothering for female vervet monkeys. *Animal Behaviour* **40**:553–562.

Fairbanks, L.A. (1993) Juvenile vervet monkeys: establishing relationships and practicing skills for the future. In: *Juvenile Primates: Life History, Development, and Behavior* (eds M.E. Pereira and L.A. Fairbanks). New York: Oxford University Press, 211–227.

Fairbanks, L.A. and McGuire, M.T. (1985) Relationships of vervet mothers with sons and daughters from one through three years of age. *Animal Behaviour* **33**:40–50.

Festa-Bianchet, M. (1991) Numbers of lungworm larvae in faeces of bighorn sheep: yearly changes, influence of host sex, and effects on host survival. *Canadian Journal of Zoology* **69**:547–554.

Fisher, R.A. (1930) *The Genetical Theory of Natural Selection*. Oxford: Clarendon Press.

Flook, D.R. (1970) A study of sex differential in the survival of wapiti. Canadian Wildlife Service Report Series No. 11. Ottawa: The Queen's Printer for Canada.

Foster, E.A., *et al.* (2012) Adaptive prolonged postreproductive life span in killer whales. *Science* **337**:1313.

Fragaszy, D.M. (1986) Time budgets and foraging behavior in wedge-capped capuchins (*Cebus olivaceus*): age and sex differences. In: *Current Perspectives in Primate Social Dynamics* (eds D. Taub and F. King). New York: Van Nostrand Reinhold, 159–174.

Frank, L.G. (1996) Female masculinization in the spotted hyena: endocrinology, behavioral ecology, and evolution. In: *Carnivore Behavior, Ecology and Evolution* (ed. J.L. Gittleman). Ithaca, NY: Cornell University Press, 78–131.

Frank, L.G., *et al.* (1991) Fatal sibling aggression, precocial development, and androgens in neonatal spotted hyenas. *Science* **252**:702–704.

Frank, L.G., *et al.* (1995) Dominance, demographics and reproductive success in female spotted hyenas: a long-term study. In: *Serengeti II: Dynamics, Management, and Conservation of an Ecosystem* (eds A.R.E. Sinclair and P. Arcese). Chicago: University of Chicago Press, 364–384.

French, S.S., *et al.* (2007) Trade-offs between the reproductive and immune systems: facultative responses to resources or obligate responses to reproduction? *American Naturalist* **170**:79–89.

Funston, P.J., *et al.* (1998) Hunting by male lions: ecological influences and socioecological implications. *Animal Behaviour* **56**:1333–1345.

Funston, P.J., *et al.* (2001) Factors affecting the hunting success of male and female lions in the Kruger National Park. *Journal of Zoology* **253**:419–431.

Geary, D.C. (2010) *Male, Female: The Evolution of Human Sex Differences*. Washington, DC: American Psychological Association.

Geary, D.C. (2015) *Evolution of vulnerability. Implications for Sex Differences in Health and Development*. Academic Press, Amsterdam.

Geist, V. (1971) *Mountain Sheep: A Study in Behavior and Evolution*. Chicago: University of Chicago Press.

Geist, V. (1974) On the relationship of social evolution and ecology in ungulates. *American Zoologist* **14**:205–220.

Ginnett, T.F. and Demment, M.W. (1997) Sex differences in giraffe foraging behavior at two spatial scales. *Oecologia* **110**:291–300.

Gomendio, M., *et al.* (1990) Mammalian sex ratios and variation in costs of rearing sons and daughters. *Nature* **343**:261–263.

Gomendio, M., *et al.* (2006) Male fertility and sex ratio at birth in red deer. *Science* **314**:1445–1447.

Gosling, L.M. (1986) Selective abortion of entire litters in the coypu: adaptive control of offspring production in relation to quality and sex. *American Naturalist* **127**:772–795.

Goy, R.W. and Phoenix, C.H. (1971) The effects of testosterone propionate administered before birth on the development of behavior in genetic female rhesus monkeys. In: *Steroid Hormones and Brain Function* (eds C.H. Sawyer and R.A. Gorski). Berkeley: University of California Press, 193–201.

Grassi, C. (2002) Sex differences in feeding, height, and space use in *Hapalemur griseus. International Journal of Primatology* **23**:677–693.

Grubb, P. (1974) Population dynamics of the Soay sheep. In: *Island Survivors: The Ecology of the Soay Sheep of St Kilda* (eds P.A. Jewell, C. Milner and J.M. Boyd). London: Athlone Press, 242–272.

Haigh, J.C. and Hudson, R.J. (1993) *Farming Wapiti and Red Deer*. St Louis, MO: Mosby.

Harvey, P.H., *et al.* (1978a) Sexual dimorphism in primate teeth. *Journal of Zoology* **186**:475–485.

Harvey, P.H., *et al.* (1978b) Canine tooth size in female primates. *Nature* **276**:817–818.

Hassett, J.M., *et al.* (2008) Sex differences in rhesus monkey toy preferences parallel those of children. *Hormones and Behavior* **54**:359–364.

Heinsohn, R., *et al.* (2005) Extreme reversed sexual dichromatism in a bird without sex role reversal. *Science* **309**:617–619.

Hewison, A.J.M. and Gaillard, J.M. (1999) Successful sons or advantaged daughters? The Trivers–Willard model and sex-biased maternal investment in ungulates. *Trends in Ecology and Evolution* **14**:229–234.

Hinde, K. (2007) First-time macaque mothers bias milk composition in favor of sons. *Current Biology* **17**:R958–R959.

Hinde, K. (2009) Richer milk for sons but more milk for daughters: sex-biased investment during lactation varies with maternal life history in rhesus macaques. *American Journal of Human Biology* **21**:512–519.

Hinde, K., *et al.* (2014) Holsteins favor heifers, not bulls: biased milk production programmed during pregnancy as a function of fetal sex. *PLOS ONE* **9** (2): e86169.

Hinde, R.A. and Spencer-Booth, Y. (1967) The behaviour of socially living rhesus monkeys in their first two and a half years. *Animal Behaviour* **15**:169–196.

Hiraiwa-Hasegawa, M. (1989) Sex differences in the behavioral development of chimpanzees at Mahale. In: *Understanding Chimpanzees* (eds P.G. Heltne and L.A. Marquardt). Cambridge, MA: Harvard University Press, 104–111.

Hiraiwa-Hasegawa, M. (1993) Skewed birth sex ratios in primates: should high-ranking mothers have sons or daughters? *Trends in Ecology and Evolution* **8**:395–400.

Hofer, H. and East, M.L. (1997) Skewed offspring sex ratios and sex composition of twin litters in Serengeti spotted hyaenas (*Crocuta crocuta*) are a consequence of siblicide. *Applied Animal Behaviour Science* **51**:307–316.

Hofer, H. and East, M.L. (2008) Siblicide in Serengeti spotted hyenas: a long-term study of maternal input and cub survival. *Behavioral Ecology and Sociobiology* **62**:341–351.

Hogg, J.T., *et al.* (1992) Sex-biased maternal expenditure in Rocky Mountain sheep. *Behavioral Ecology and Sociobiology* **31**:243–251.

Huck, U.W., *et al.* (1986) Food restricting young hamsters (*Mesocricetus auratus*) affects sex ratio and growth of subsequent offspring. *Biology of Reproduction* **35**:592–598.

Huck, U.W., *et al.* (1987) Food restricting first generation juvenile female hamsters (*Mesocricetus auratus*) affects sex ratio and growth of third generation offspring. *Biology of Reproduction* **37**:612–617.

Hurst, L.D. (1994) Embryonic growth and the evolution of the mammalian Y chromosome. I. The Y as an attractor for selfish growth factors. *Heredity* **73**:223–232.

Huxley, J.S. (1931) The relative size of antlers in deer. *Proceedings of the Zoological Society of London* **101**:819–864.

Illius, A.W. and Gordon, I.J. (1991) Prediction of intake and digestion in ruminants by a model of rumen kinetics integrating animal size and plant characteristics. *Journal of Agricultural Sciences* **116**:145–157.

Ims, R.A. (1989) Kinship and origin effects on dispersal and space sharing in *Clethrionomys rufocanus*. *Ecology* **70**:607–616.

Ims, R.A. (1990) Determinants of natal dispersal and space use in grey-sided voles, *Clethrionomys rufocanus*: a combined field and laboratory experiment. *Oikos* **57**:106–113.

James, W.H. and Hofer, H. (1999) A note on sex ratios and sex combinations in Serengeti spotted hyenas: siblicide and sub-binomial variance. *Applied Animal Behaviour Science* **65**:153–158.

Jamieson, S.H. and Armitage, K.B. (1987) Sex differences in the play behavior of yearling yellow-bellied marmots. *Ethology* **74**:237–253.

Jarman, P.J. (1968) The effects of the creation of Lake Kariba upon the terrestrial ecology of the Middle Zambezi Valley, with particular reference to the large mammals. PhD thesis, University of Manchester.

Jarman, P.J. (1974) The social organisation of antelope in relation to their ecology. *Behaviour* **48**:215–267.

Jarman, P.J. (1991) Social behaviour and organisation in the Macropodoidea. In: *Advances in the Study of Behavior*, Vol. **20** (eds P.J.B. Slater, J.S. Rosenblatt, C. Beer and M. Milinski). San Diego, CA: Academic Press, 1–50.

Jarman, P.J. (2000) Males in macropod society. In: *Primate Males: Causes and Consequences of Variation in Group Composition* (ed. P.M. Kappeler). Cambridge: Cambridge University Press, 21–33.

Jarvis, J.U.M., *et al.* (1991) Growth and factors affecting body size in naked mole-rats. In: *The Biology of the Naked Mole-rat* (eds P.W. Sherman, J.U.M. Jarvis and R.D. Alexander). Princeton, NJ: Princeton University Press, 358–383.

Jemielity, S. *et al.* (2007) Short telomeres in short-live males: what are the molecular and evolutionary causes? *Aging Cell* **6**:225–233.

Johnson, C.N., *et al.* (2001) Adjustment of offspring sex ratios in relation to the availability of resources for philopatric offspring in the common brushtail possum. *Proceedings of the Royal Society of London. Series B: Biological Sciences* **268**:2001–2005.

Jones, M.E. and Barmuta, L.A. (1998) Diet overlap and relative abundance of sympatric dasyurid carnivores: a hypothesis of competition. *Journal of Animal Ecology* **67**:410–421.

Jones, M.E. and Barmuta, L.A. (2000) Niche differentiation among sympatric Australian dasyurid carnivores. *Journal of Mammalogy* **81**:434–447.

Jorgenson, J.T., *et al.* (1998) Effects of population density on horn development in bighorn rams. *Journal of Wildlife Management* **62**:1011–1020.

Kappeler, P.M. (1990) The evolution of sexual size dimorphism in prosimian primates. *American Journal of Primatology* **21**:201–214.

Kappeler, P.M. (1996) Intrasexual selection and phylogenetic constraints in the evolution of sexual canine dimorphism in strepsirhine primates. *Journal of Evolutionary Biology* **9**:43–65.

Key, C. and Ross, C. (1999) Sex differences in energy expenditure in non-human primates. *Proceedings of the Royal Society of London. Series B: Biological Sciences* **266**:2479–2485.

Kirby, D.R., *et al.* (1967) A possible immunological influence on sex ratio. *Lancet* **ii**:139–140.

Klein, D.R. (1968) The introduction, increase, and crash of reindeer on St. Matthew Island. *Journal of Wildlife Management* **32**:350–367.

Komdeur, J. (2012) Sex allocation. In: *The Evolution of Parental Care* (eds N.J. Royle, P.T. Smiseth and M. Kölliker). Oxford: Oxford University Press, 171–188.

Korsten, P., *et al.* (2009) Sexual conflict in twins: male co-twins reduce fitness of female Soay sheep. *Biology Letters* **5**:663–666.

Krackow, S., *et al.* (2003) Sexual growth dimorphism affects birth sex ratio in house mice. *Proceedings of the Royal Society of London. Series B: Biological Sciences* **270**:943–947.

Kruuk, L.E.B., *et al.* (1999a) Early determinants of lifetime reproductive success differ between the sexes in red deer. *Proceedings of the Royal Society of London. Series B: Biological Sciences* **266**:1655–1661.

Kruuk, L.E.B., *et al.* (1999b) Population density affects sex ratio variation in red deer. *Nature* **399**:459–461.

Landete-Castillejos, T., *et al.* (2005) Maternal quality and differences in milk production and composition for male and female Iberian red deer calves (*Cervus elaphus hispanicus*). *Behavioral Ecology and Sociobiology* **57**:267–274.

Lawler, R.R., *et al.* (2005) Intrasexual selection in Verreaux's sifaka (*Propithecus verreauxi verreauxi*). *Journal of Human Evolution* **48**:259–277.

Lawson Handley, L.J. and Perrin, N. (2007) Advances in our understanding of mammalian sex-biased dispersal. *Molecular Ecology* **16**:1559–1578.

Leader-Williams, N. (1988) *Reindeer in South Georgia: The Ecology of an Introduced Population*. Cambridge: Cambridge University Press.

Lee, P.C. and Moss, C.J. (1986) Early maternal investment in male and female African elephant calves. *Behavioral Ecology and Sociobiology* **18**:353–361.

Leigh, S. (1992) Patterns of variation in the ontogeny of primate body size dimorphism. *Journal of Human Evolution* **23**:27–50.

Leimar, O. (1996) Life-history analysis of the Trivers and Willard sex-ratio problem. *Behavioral Ecology* **7**:316–325.

Lincoln, G.A. (1971) Puberty in a seasonally breeding male, the red deer stag (*Cervus elaphus*, L.). *Journal of Reproduction and Fertility* **25**:41–54.

Lindenfors, P., *et al.* (2007) Sexual size dimorphism in mammals. In: *Sex, Size and Gender Roles: Evolutionary Studies of Sexual Size Dimorphism* (eds D.J. Fairbairn, W.U. Blanckenhorn and T. Székely). Oxford: Oxford University Press, 16–26.

Lindsay, W.K. (2011) Habitat use, diet choice, and nutritional status in female and male Amboseli elephants. In: *The Amboseli Elephants: A Long-term Perspective on a Long-lived Mammal* (eds C.J. Moss, H. Croze and P.C. Lee). Chicago: University of Chicago Press, 51–73.

Lindström, J., *et al.* (2002) Sex-ratio variation in Soay sheep. *Behavioral Ecology and Sociobiology* **53**:25–30.

Lloyd, P.H. and Rasa, O.A.E. (1989) Status, reproductive success and fitness in Cape mountain zebra (*Equus zebra zebra*). *Behavioral Ecology and Sociobiology* **25**:411–420.

Lonsdorf, E.V., *et al.* (2004) Sex differences in learning in chimpanzees. *Nature* **428**:715–716.

Lonsdorf, E.V., *et al.* (2014) Boys will be boys: sex differences in wild infant chimpanzee social interactions. *Animal Behaviour* **88**:79–83.

Lührs, M.-L. and Dammhahn, M. (2010) An unusual case of cooperative hunting in a solitary carnivore. *Journal of Ethology* **28**:379–383.

Lutermann, H., *et al.* (2014) Sex ratio variation in a eusocial mammal, the Damaraland mole-rat, *Fukomys damarensis*. *Journal of Zoology* **294**:139–145.

McCance, R.A. and Widdowson, E.M. (1974) Review lecture: the determinants of growth and form. *Proceedings of the Royal Society of London. Series B: Biological Sciences* **185**:1–17.

McCann, T.S., *et al.* (1989) Parental investment in southern elephant seals, *Mirounga leonina*. *Behavioral Ecology and Sociobiology* **25**:81–87.

McClure, P.A. (1981) Sex-biased litter reduction in food-restricted wood rats (*Neotoma floridana*). *Science* **211**:1058–1060.

McComb, K., *et al.* (2011) Leadership in elephants: the adaptive value of age. *Proceedings of the Royal Society of London. Series B: Biological Sciences* **282**:3270–3276.

McEwan, E.H. (1968) Growth and development of the barren-ground caribou. II. Postnatal growth rates. *Canadian Journal of Zoology* **46**:1023–1029.

Macfarlane, A.M. and Coulson, G. (2005) Sexual segregation in Australian marsupials. In: *Sexual Segregation in Vertebrates* (eds K.E. Ruckstuhl and P. Neuhaus). Cambridge: Cambridge University Press, 254–279.

McGuire, B. and Bemis, W.E. (2007) Parental care. In: *Rodent Societies: An Ecological and Evolutionary Perspective* (eds J.O. Wolff and P.W. Sherman). Chicago: University of Chicago Press, 231–242.

McMillen, M.M. (1979) Differential mortality by sex in fetal and neonatal deaths. *Science* **204**:89–91.

McNeilly, A.S. (1987) Prolactin and the control of gonadotropin secretion. *Journal of Endocrinology* **115**:1–5.

McNutt, J.W. and Silk, J.B. (2008) Pup production, sex ratios, and survivorship in African wild dogs, *Lycaon pictus*. *Behavioral Ecology and Sociobiology* **62**:1061–1067.

Maestripieri, D. (2001) Female-biased maternal investment in rhesus macaques. *Folia Primatologica* **72**:44–47.

Main, M.B. and Coblentz, B.E. (1996) Sexual segregation in rocky mountain mule deer. *Journal of Wildlife Management* **60**:497–507.

Main, M.B. and du Toit, J.T. (2005) Sex differences in reproductive strategies affect habitat choice in ungulates. In: *Sexual Segregation in Vertebrates* (eds K.E. Ruckstuhl and P. Neuhaus). Cambridge: Cambridge University Press, 148–161.

Martin, J.G.A. and Festa-Bianchet, M. (2011) Age-independent and age-dependent decreases in reproduction of females. *Ecology Letters* **14**:576–581.

Martin, R.D., *et al.* (1994) The evolution of sexual dimorphism in primates. In: *The Differences Between the Sexes* (eds R.V. Short and E. Balaban). Cambridge: Cambridge University Press, 159–200.

Meaney, M.J., *et al.* (1985) Sex differences in social play: the socialization of sex roles. In: *Advances in the Study of Behavior*, Vol. **15** (ed. J.S. Rosenblatt). San Diego, CA: Academic Press, 1–58.

Mendl, M. and Paul, E.S. (1989) Observation of nursing and sucking behaviour as an indicator of milk transfer and parental investment. *Animal Behaviour* **37**:513–513.

Meredith, S.L. (2013) Identifying proximate and ultimate causation in the development of primate sex-typed social behavior. In: *Building Babies: Primate Development in Proximate and Ultimate Perspective* (eds K.B.H. Clancy, K. Hinde and J.N. Rutherford). New York: Springer, 411–434.

Michaud, R. (2005) Sociality and ecology of the odontocetes. In: *Sexual Segregation in Vertebrates* (eds K.E. Ruckstuhl and P. Neuhaus). Cambridge: Cambridge University Press, 303–326.

Mills, M.G.L. (1990) *Kalahari Hyaenas: Comparative Behavioural Ecology of Two Species*. London: Unwin Hyman.

Mitchell, G.D. (1968) Persistent behavior pathology in rhesus monkeys following early social isolation. *Folia Primatologica* **8**:132–147.

Mohler, L.L., *et al.* (1951) Mule deer in Nebraska national forest. *Journal of Wildlife Management* **15**:129–157.

Moore, C.L. and Morelli, G.A. (1979) Mother rats interact differently with male and female offspring. *Journal of Comparative and Physiological Psychology* **93**:677–684.

Moore, S.L. and Wilson, K. (2002) Parasites as a viability cost of sexual selection in natural populations of mammals. *Science* **297**:2015–2018.

Moses, R.A., *et al.* (1998) Sex-biased mortality in woodrats occurs in the absence of parental intervention. *Animal Behaviour* **55**:563–571.

Müller, M.N. and Emery Thompson, M. (2012) Mating, parenting and male reproductive strategies. In: *The Evolution of Primate Societies* (eds J.C. Mitani, J. Call, P.M. Kappeler, R.A. Palombit and J.B. Silk). Chicago: University of Chicago Press, 387–411.

Murray, C.M., *et al.* (2009) Reproductive energetics in free-living female chimpanzees (*Pan troglodytes schweinfurthii*). *Behavioral Ecology* **20**:1211–1216.

Murray, M.G. and Illius, A.W. (2000) Vegetation modification and resource competition in grazing ungulates. *Oikos* **89**:501–508.

Mysterud, A. (2000) The relationship between ecological segregation and sexual body size dimorphism in large herbivores. *Oecologia* **124**:40–54.

Mysterud, A., *et al.* (2002) The role of males in the dynamics of ungulate populations. *Journal of Animal Ecology* **71**:907–915.

Nunn, C.L., *et al.* (2009) On sexual dimorphism in immune function. *Philosophical Transactions of the Royal Society B: Biological Sciences* **364**:61–69.

O'Brien, T.G. and Robinson, J.G. (1993) Stability of social relationships in female wedge-capped capuchin monkeys. In: *Primate Societies* (eds M.E. Pereira and L.A. Fairbanks). Chicago: University of Chicago Press, 197–210.

Owens, I.P.F. (2002) Sex differences in mortality rate. *Science* **297**:2008–2009.

Packer, C. (1983) Sexual dimorphism: the horns of African antelopes. *Science* **221**:1191–1193.

Parker, G.A. (1979) Sexual selection and sexual conflict. In: *Sexual Selection and Reproductive Competition in Insects* (eds M.S. Blum and N.A. Blum). New York: Academic Press, 123–166.

Parkes, A.S. (1926) The mammalian sex-ratio. *Biological Reviews of the Cambridge Philosophical Society* **2**:1–51.

Pavelka, M.S.M., *et al.* (2002) Availability and adaptive value of reproductive and postreproductive Japanese macaque mothers and grandmothers. *Animal Behaviour* **64**:407–414.

Pedersen, J.M., *et al.* (1990) Sex differences in the play behavior of immature spotted hyenas (*Crocuta crocuta*). *Hormones and Behavior* **24**:403–420.

Pellis, S.M. (2002) Sex differences in play fighting revisited: traditional and nontraditional mechanisms of sexual differentiation in rats. *Archives of Sexual Behavior* **31**:17–26.

Pellis, S.M. and Pellis, V.C. (2007) Rough-and-tumble play and the development of the social brain. *Current Directions in Psychological Science* **16**:95–98.

Pen, I. and Weissing, F.J. (2000) Sex-ratio optimization with helpers at the nest. *Proceedings of the Royal Society of London. Series B: Biological Sciences* **267**:539–543.

Pereira, M.E. (1988a) Effects of age and sex on intra-group spacing behaviour in juvenile savannah baboons, *Papio cynocephalus cynocephalus*. *Animal Behaviour* **36**:184–204.

Pereira, M.E. (1988b) Agonistic interactions of juvenile savanna baboons. I. Fundamental features. *Ethology* **79**:195–217.

Pereira, M.E. and Altmann, J. (1985) Development of social behavior in free-living nonhuman primates. In: *Nonhuman Primate Models for Human Growth and Development* (ed. E.S. Watts). New York: John Wiley & Sons, 217–309.

Pérez-Barbería, F.J. and Gordon, I.J. (1998) The influence of molar occlusal surface area on the voluntary intake, digestion, chewing behaviour and diet selection of red deer (*Cervus elaphus*). *Journal of Zoology* **245**:307–316.

Pérez-Barbería, F.J., *et al.* (2002) The origins of sexual dimorphism in body size in ungulates. *Evolution* **56**:1276–1285.

Perret, M. (1996) Manipulation of sex ratio at birth by urinary cues in a prosimian primate. *Behavioral Ecology and Sociobiology* **38**:259–266.

Pickering, S.P.C. (1983) Aspects of the behavioural ecology of feral goats (Capra (domestic)). PhD thesis, University of Durham.

Plavcan, J.M. (1998) Correlated response, competition, and female canine size in primates. *American Journal of Physical Anthropology* **107**:410–416.

Plavcan, J.M. (2004) Sexual selection, measures of sexual selection and sexual dimorphism in primates. In: *Sexual Selection in Primates* (eds P.M. Kappeler and C.P. van Schaik). Cambridge: Cambridge University Press, 230–252.

Plavcan, J.M. and van Schaik, C.P. (1992) Intrasexual competition and canine dimorphism in anthropoid primates. *American Journal of Physical Anthropology* **87**:461–477.

Plavcan, J.M. and van Schaik, C.P. (1997) Intrasexual competition and body weight dimorphism in anthropoid primates. *American Journal of Physical Anthropology* **103**:37–67.

Poole, J.H., *et al.* (2011) Longevity, competition, and musth: a long-term perspective on male reproductive strategies. In: *The Amboseli Elephants: A Long-term Perspective on a Long-lived Mammal* (eds C.J. Moss, H. Croze and P.C. Lee). Chicago: University of Chicago Press, 272–290.

Promislow, D.E.L. (1992) Costs of sexual selection in natural populations of mammals. *Proceedings of the Royal Society of London. Series B: Biological Sciences* **247**:203–210.

Pruetz, J.D., *et al.* (2015) New evidence on the tool-assisted hunting exhibited by chimpanzees (*Pan troglodytes verus*) in a savannah habitat at Fongoli, Sénégal. *Royal Society Open Science* doi: 10.1098/rsos.140507.

Pryce, C.R. (1993) The regulation of maternal behaviour in marmosets and tamarins. *Behavioural Processes* **30**:201–224.

Pusey, A.E. (1983) Mother–offspring relationships in chimpanzees after weaning. *Animal Behaviour* **31**:363–377.

Pusey, A.E. (1990) Behavioural changes at adolescence in chimpanzees. *Behaviour* **115**:203–246.

Reby, D. and McComb, K. (2003) Vocal communication and reproduction in deer. In: *Advances in the Study of Behavior*, Vol. **33** (eds P.J.B. Slater, J.S. Rosenblatt, C.T. Snowdown, T.J. Roper and M. Naguib). San Diego, CA: Academic Press, 231–264.

Restif, O. and Amos, W. (2010) The evolution of sex-specific immune defences. *Proceedings of the Royal Society of London. Series B: Biological Sciences* **277**:2247–2255.

Richard, A.F., *et al.* (2002) Life in the slow lane? Demography and life histories of male and female sifaka (*Propithecus verreauxi verreauxi*). *Journal of Zoology* **256**:421–436.

Robert, K.A. and Schwanz, L.E. (2011) Emerging sex allocation research in mammals: marsupials and the pouch advantage. *Mammal Review* **41**:1–22.

Roberts, S.C. (1996) The evolution of hornedness in female ruminants. *Behaviour* **133**:399–442.

Robinette, W.L., *et al.* (1957) Differential mortality by sex and age among mule deer. *Journal of Wildlife Management* **21**:1–16.

Robinson, J.G. (1986) *Seasonal Variation in Use of Time and Space by the Wedge-capped Capuchin Monkey Cebus olivaceus: Implications for Foraging Theory.* Washington, DC: Smithsonian Institution Press.

Rosenfeld, C.S., *et al.* (2003) Striking variation in the sex ratio of pups born to mice according to whether maternal diet is high in fat or carbohydrate. *Proceedings of the National Academy of Sciences of the United States of America* **100**:4628–4632.

Rosvall, K.A. (2011) Intrasexual competition in females: evidence for sexual selection? *Behavioral Ecology* **22**:1131–1140.

Ruckstuhl, K.E. and Neuhaus, P. (2000) Sexual segregation in ungulates: a new approach. *Behaviour* **137**:361–377.

Ryan, B.C. and Vandenbergh, J.G. (2002) Intrauterine position effects. *Neuroscience and Biobehavioral Reviews* **26**:665–678.

Sackett, G.P., *et al.* (1975) Vulnerability for abnormal development: pregnancy outcomes and sex differences in macaque monkeys. In: *Aberrant Development in Infancy: Human and Animal Studies* (ed. N.R. Ellis). Hillsdale, NJ: Lawrence Erlbaum Associates, 59–76.

Schaefer, J.A. and Mahoney, S.P. (2001) Antlers on female caribou: biogeography of the bones of contention. *Ecology* **82**:3556–3560.

Schalk, G. and Forbes, M.R. (1997) Male biases in parasitism of mammals: effects of study type, host age, and parasite taxon. *Oikos* **78**:67–74.

Schaller, G.B. (1972) *The Serengeti Lion: A Study of Predator–Prey Relations.* Chicago: Chicago University Press.

Schmid, J. and Kappeler, P.M. (1998) Fluctuating sexual dimorphism and differential hibernation by sex in a primate, the gray mouse lemur (*Microcebus murinus*). *Behavioral Ecology and Sociobiology* **43**:125–132.

Schindler, S. *et al.* (2015) Sex-specific demography and generalization of the Trivers–Willard theory. *Nature* **526**:249–252.

Selman, C., *et al.* (2012) Oxidative damage, ageing and life-history evolution: where now. *Trends in Ecology and Evolution* **27**:570–577.

Setchell, J.M. (2005) Do female mandrills prefer brightly colored males? *International Journal of Primatology* **26**:715–735.

Setchell, J.M. and Dixson, A.F. (2001) Changes in the secondary adornments of male mandrills (*Mandrillus sphinx*) are associated with gain and loss of alpha status. *Hormones and Behavior* **39**:177–184.

Setchell, J.M. and Lee, P.C. (2004) Development and sexual selection in primates. In: *Sexual Selection in Primates: New and Comparative Perspectives* (eds P.M. Kappeler and C.P. van Schaik). Cambridge: Cambridge University Press, 175–195.

Setchell, J.M. and Wickings, E.J. (2005) Dominance, status signals and coloration in male mandrills (*Mandrillus sphinx*). *Ethology* **111**:25–50.

Setchell, J.M., *et al.* (2008) Social correlates of testosterone and ornamentation in male mandrills. *Hormones and Behavior* **54**:365–372.

Sharpe, L.L. (2004) Play and social relationships in the meerkat *Suricata suricatta.* PhD thesis, University of Stellenbosch.

Sharpe, L.L. (2005) Frequency of social play does not affect dispersal partnerships in wild meerkats. *Animal Behaviour* **70**:559–569.

Sheldon, B.C. (1998) Recent studies of avian sex ratios. *Heredity* **80**:397–402.

Sheldon, B.C. and West, S.A. (2004) Maternal dominance, maternal condition and offspring sex ratio in ungulate mammals. *American Naturalist* **163**:40–54.

Silk, J.B., *et al.* (1981) Differential reproductive success and facultative adjustment of sex ratios among captive female bonnet macaques (*Macaca radiata*). *Animal Behaviour* **29**:1106–1120.

Simpson, A.E. and Simpson, M.J.A. (1985) Short-term consequences of different breeding histories for captive rhesus macaque mothers and young. *Behavioral Ecology and Sociobiology* **18**:83–89.

Simpson, M.J.A., *et al.* (1981) Infant-related influences on birth intervals in rhesus monkeys. *Nature* **290**:49–51.

Smale, L., *et al.* (1999) Siblicide revisited in the spotted hyaena: does it conform to obligate or facultative models? *Animal Behaviour* **58**:545–551.

Smiseth, P.T. and Lorentsen, S.-H. (1995) Evidence of equal maternal investment in the sexes in the polygynous and sexually dimorphic grey seal (*Halichoerus grypus*). *Behavioral Ecology and Sociobiology* **36**:145–150.

Smyth, K.N. and Drea, C.M. (2015) Patterns of parasitism in the cooperatively breeding meerkat: a cost of dominance for females. *Behavioral Ecology* doi: 10.1093/beheco/arv132.

Solberg, E.J., *et al.* (2002) Biased adult sex ratio can affect fecundity in primiparous moose *Alces alces*. *Wildlife Biology* **8**:117–128.

Staines, B.W. and Crisp, J.M. (1978) Observations on food quality in Scottish red deer (*Cervus elaphus*) as determined by chemical analysis of rumen contents. *Journal of Zoology* **185**:253–259.

Stanford, C.B., *et al.* (1994) Hunting decisions in wild chimpanzees. *Behaviour* **131**:1–18

Stanland, I.J. (2005) Sexual segregation in seals. In: *Sexual Segregation in Vertebrates* (eds K.E. Ruckstuhl and P. Neuhaus). Cambridge: Cambridge University Press, 53–73.

Stankowich, T. and Caro, T. (2009) Evolution of weaponry in female bovids. *Proceedings of the Royal Society of London. Series B: Biological Sciences* **276**:4329–4334.

Stevenson, I.R. (1994) Male-biased mortality in Soay sheep. PhD thesis, University of Cambridge.

Stevenson, I.R. and Bancroft, D.R. (1995) Fluctuating trade-offs favour precocial maturity in male Soay sheep. *Proceedings of the Royal Society of London. Series B: Biological Sciences* **262**:267–275.

Stockley, P. and Bro-Jørgensen, J. (2011) Female competition and its evolutionary consequences in mammals. *Biological Reviews of the Cambridge Philosophical Society* **86**:341–366.

Strier, K.B. (1993) Development in a patrilocal monkey society: sex differences in the behavior of immature muriquis (*Brachyteles arachnoides*). In: *Juvenile Primates: Life History, Development, and Behavior* (eds M.E. Pereira and L.A. Fairbanks). New York: Oxford University Press.

Surbeck, M., *et al.* (2011) Mothers matter! Maternal support, dominance status and mating success in male bonobos (*Pan paniscus*). *Proceedings of the Royal Society of London. Series B: Biological Sciences* **278**:590–598.

Taber, R.D. and Dasmann, R.F. (1954) A sex difference in mortality in young Columbian black-tailed deer. *Journal of Wildlife Management* **18**:309–315.

Terborgh, J.W. (1983) *Five New World Primates: A Study in Comparative Ecology*. Princeton, NJ: Princeton University Press.

Tomaszycki, M.L., *et al.* (2001) Sex differences in infant rhesus macaque separation–rejection vocalizations and effects of prenatal androgens. *Hormones and Behavior* **39**:267–276.

Tourmente, M., *et al.* (2011) Sperm competition and the evolution of sperm design in mammals. *BMC Evolutionary Biology* **11**:12.

Trillmich, F. (1986) Maternal investment and sex-allocation in the Galapagos fur seal, *Arctocephalus galapagoensis*. *Behavioral Ecology and Sociobiology* **19**:157–164.

Trivers, R.L. (1972) Parental investment and sexual selection. In: *Sexual Selection and the Descent of Man, 1871–1971* (ed. B. Campbell). Chicago: Aldine, 136–179.

Trivers, R.L. and Willard, D.E. (1973) Natural selection of parental ability to vary the sex ratio of offspring. *Science* **179**:90–92.

Uller, T. (2006) Sex-specific sibling interactions and offspring fitness in vertebrates: patterns and implications for maternal sex ratios. *Biological Reviews of the Cambridge Philosophical Society* **81**:207–217.

Uller, T. (2012) Parental effects in development and evolution. In: *The Evolution of Parental Care* (eds N.J. Royle, P.T. Smiseth and M. Kölliker). Oxford: Oxford University Press, 247–266.

Vandenbergh, J.G. and Huggett, C.L. (1994) Mother's prior intrauterine position affects the sex ratio of her offspring in house mice. *Proceedings of the National Academy of Sciences of the United States of America* **91**:11055–11059.

van Noordwijk, M.A., *et al.* (1993) Spatial position and behavioral sex differences in juvenile long-tailed macaques. In: *Juvenile Primates: Life History, Development, and Behavior* (eds M.E. Pereira and L.A. Fairbanks). New York: Oxford University Press, 77–85.

van Schaik, C.P., *et al.* (1983) The effect of group size on time budgets and social behaviour in wild long tailed macaques (*Macaca fascicularis*). *Behavioral Ecology and Sociobiology* **13**:173–181.

vom Saal, F.S. and Bronson, F.H. (1980) Sexual characteristics of adult female mice are correlated with their blood testosterone levels during prenatal development. *Science* **208**:597–599.

vom Saal, F.S. and Moyer, C.L. (1985) Prenatal effects on reproductive capacity during aging in female mice. *Biology of Reproduction* **32**:1116–1126.

Wallen, K. (1996) Nature needs nurture: the interaction of hormonal and social influences on the development of behavioral sex differences in rhesus monkeys. *Hormones and Behavior* **30**:364–378.

Walters, J.R. and Seyfarth, R.M. (1997) Conflict and cooperation. In: *Primate Societies* (eds B.B. Smuts, D.L. Cheney, R.M. Seyfarth, R.W. Wrangham and T.T. Struhsaker). Chicago: University of Chicago Press, 306–317.

Watts, D.P. and Pusey, A.E. (1993) Behavior of juvenile and adolescent great apes. In: *Juvenile Primates: Life History, Development, and Behavior* (eds M.E. Pereira and L.A. Fairbanks). Chicago: University of Chicago Press, 148–167.

Weladji, R.B., *et al.* (2003) Sex-specific preweaning maternal care in reindeer (*Rangifer tarandus t.*). *Behavioral Ecology and Sociobiology* **53**:308–314.

West, P.M. and Packer, C. (2002) Sexual selection, temperature, and the lion's mane. *Science* **297**:1339–1343.

West, S.A. (2009) *Sex Allocation*. Princeton, NJ: Princeton University Press.

West, S.A. and Sheldon, B.C. (2002) Constraints in the evolution of sex ratio adjustment. *Science* **295**:1685–1688.

White, A.M., *et al.* (2007) Differential investment in sons and daughters: do white rhinoceros mothers favor sons? *Journal of Mammalogy* **88**:632–638.

White, L.E. and Hinde, R.A. (1975) Some factors affecting mother–infant relations in rhesus monkeys. *Animal Behaviour* **23**:527–542.

Whitehead, H. and Weilgart, L. (2000) The sperm whale: social females and roving males. In: *Cetacean Societies: Field Studies of Dolphins and Whales* (eds J. Mann, R. Connor, P. Tyack and H. Whitehead). Chicago: University of Chicago Press, 154–172.

Widdowson, E.M. (1976) The response of the sexes to nutritional stress. *Proceedings of the Nutrition Society* **35**: 175–180.

Williams, G.C. (1979) The question of adaptive sex ratio in outcrossed vertebrates. *Proceedings of the Royal Society of London. Series B: Biological Sciences* **205**:567–580.

Wilson, K., *et al.* (2003) Response to comment on Parasites as a viability cost of sexual selection in natural populations of mammals. *Science* **300**:55.

Wilson, K., *et al.* (2004) Parasites and their impact. In: *Soay Sheep: Dynamics and Selection in an Island Population* (eds T.H. Clutton-Brock and J.M. Pemberton). Cambridge: Cambridge University Press, 113–165.

Woolf, A. and Harder, J.D. (1979) Population dynamics of a captive white-tailed deer herd with emphasis on reproduction and mortality. *Wildlife Monographs* **67**:3–53.

Wright, P.C. (1999) Lemur traits and Madagascar ecology: coping with an island environment. *American Journal of Physical Anthropology* **110**:31–72.

Young, A.J. and Bennett, N.C. (2013) Intra-sexual selection in cooperative mammals and birds: why are females not bigger and better armed? *Philosophical Transactions of the Royal Society B: Biological Sciences* **368**:20130075.

Zielinski, W.J., *et al.* (1992) The effect of intrauterine position on the survival, reproduction and home range size of female house mice (*Mus musculus*). *Behavioral Ecology and Sociobiology* **30**:185–191.

CHAPTER 19
Hominins and humans

19.1 Introduction

Like many other social mammals, humans live in structured social groups, compete with other group members for resources and breeding opportunities, maintain close links with relatives and unrelated individuals and cooperate with each other. Dominance hierarchies and individual differences in power and influence are common in both sexes and individuals frequently attempt to manipulate each other to their own advantage. Relationships between groups in traditional societies can be tolerant or cooperative but, as in other social mammals, they can also be competitive, and hostilities between neighbouring groups are not uncommon in traditional societies and groups sometimes eradicate their neighbours or force them to leave the area.

While there are fundamental similarities in the structure and organisation of social groups between humans and other mammals, there are also both quantitative and qualitative differences. Improvements in cognitive ability associated with increases in brain size and structure and changes in function provide humans with an understanding of cause and effect that allows them to predict the consequences of their actions for themselves and others. Unlike other social mammals, humans can recognise the state of knowledge of others, assess the risks they face and interpret their intentions. Improvements in cognitive ability have allowed us to investigate the world we live in and have led to the development of modern science and to a systematic understanding of physical processes ranging from the structure of atoms to the structure of the cosmos and of biological processes ranging from the level of molecules to the level of ecosystems. In addition, the evolution of language allows humans to communicate in ways not available to other animals. As a result, humans can coordinate their activities precisely, form agreements and contracts, identify and enforce social norms and punish cheats and defectors. Exchange, interchange and barter are possible. Social endorsement can encourage and maintain costly forms of cooperation that are rare or absent in other animals and connections can be maintained with relatives after they have dispersed from their natal group, providing a basis for alliances between groups. Information, agreements and, in some cases, feuds can be transferred across generations.

The development of technology, in all its forms, has also had a profound impact on human evolution. Hominins have made stone tools for at least 3 million years and for at least the last 1.5 million years the range, sophistication and specificity of human tools has surpassed all forms of animal tool-making, enabling humans to access a wide range of resources unavailable to other animals. Following the development of agriculture, humans have also been able to achieve a degree of control over their food supplies and to increase their productivity, permitting rapid increases in population size that have led to the development of complex multi-layered societies. Agriculture has also allowed individuals to specialise in particular activities and to develop divergent skills to an extent that is not found in other mammals, improving technological expertise and increasing the degree of interdependence between group members.

The evolution of specific human characteristics and capabilities has been closely associated with the progressive development of cultural adaptations maintained through social learning and teaching. The increasing importance of culture has led to interactions between cultural traits and processes and natural selection (a process known as *gene–culture coevolution*) that has encouraged the emergence of cumulative adaptations and affected both the direction and intensity of selection and the rate of evolution.

Mammal Societies, First Edition. Tim Clutton-Brock.
© 2016 Tim Clutton-Brock. Published 2016 by John Wiley & Sons, Ltd.

My aim in this chapter and the next is to provide biologists interested in the evolution of societies in non-human animals with a brief introduction to related research on hominins and humans. Section 19.2 briefly describes human phylogeny and some of the changes in behaviour and technology that have occurred in the course of human evolution. Section 19.3 compares the life histories of humans with those of chimpanzees and speculates about the evolution of hominin breeding systems. Fossils rarely provide direct information about the nature of breeding systems, but as the extent of sex differences in growth and behaviour are associated with variation in mating systems (see Chapter 18), the magnitude of sex differences in hominin fossils and modern humans provides some indication of the past history of hominin breeding systems (section 19.4). Section 19.5 speculates about the social systems of hominins and compares traditional breeding systems and patterns of social organisation between hunter-gatherers, pastoralists and agricultural peoples. (Both these two sections focus on general patterns and comparisons and it is important to realise that there is also considerable variation between and within ethnic groups.) Finally, section 19.6 attempts to answer why it was an African ape that developed the cognitive abilities and technology that allowed it to dominate the planet.

19.2 Human evolution

Phylogeny

Both genetic and palaeontological evidence suggest that the hominin lineage diverged from a common ancestor in Africa shared with chimpanzees and bonobos between 6 and 8 million years ago (Mya). The earliest possible hominins are poorly known and diverse: *Sahelanthropus* (at 7 Mya) and *Ororrin* (at 6.0 Mya) both show potentially hominin features related to bipedalism, while *Ardipithecus* (at 4.4 Mya), which appears to have lived in wooded environments rather than in savannahs or grasslands, shows reduced canines and possible bipedalism. However, it is only after 4.2 Mya that a clear hominin line, represented by *Australopithecus*, appears, showing greater evidence for committed bipedalism (Stringer and Andrews 2005; Dixson 2009) (Figure 19.1). There appears to have been a separation of australopithecine lineages leading to larger heavier species and those

leading to more gracile species before 3 Mya. The latter may have been ancestral to early members of the genus *Homo*, characterised by larger brains and reduced prognathism, which appeared around 2.5 Mya.

The earliest australopithecines appear to have been associated with relatively mesic, wooded environments (Reed 1997). Their adaptations to walking upright, suggest that they were more terrestrial than chimpanzees and bonobos and were less reliant on trees to escape from predators. The relatively thick enamel of their teeth indicates a change in diet from the ancestors of the apes, possibly involving increased reliance on meat, either scavenged or hunted. In addition, the structure of human intestines suggests that meat may have been an important component.

Between 4 and 2.5 Mya, many African habitats became progressively arid and seasonal, and in both South and East Africa the first *Homo* species probably lived in open woodlands and grasslands (Stanley 1992; Reed 1997; Bobe and Behrensmeyer 2004). The appearance of *Homo erectus* after 1.8 Mya showed the evolution of a phenotype more closely resembling that of modern humans, with adaptations for full bipedalism, changes in body proportions, and larger brain size. Between 1.6 and 1.0 Mya, and probably again between 0.8 and 0.6 Mya, populations of *Homo erectus* or related species radiated from Africa into Europe and Asia, giving rise to *Homo heidelbergensis* and later to Neanderthals and possibly to other *Homo* species, too.

Modern humans (*Homo sapiens*) evolved in Africa before 195,000 years ago and dispersed out of Africa less than 100,000 years ago, reaching China by 80,000 years ago (Liu *et al.* 2015), Europe and Australia between 40,000 and 50,000 years ago and North America by around 13,000 years ago (Figure 19.2). Additional evidence of an African origin of modern human populations is provided by studies showing a progressive reduction in phenotypic and genetic diversity in human populations from west to east (Manica *et al.* 2007).

Brain size

Over the course of human evolution, there has been a progressive increase in body size and relative brain size and structure. Cranial capacity in the earliest hominids was between 400 and 500 cm^3, increasing to 600–700 cm^3 in the first hominins, to 1000–1200 cm^3 in later *Homo erectus* and to 1400 cm^3 in modern humans (Figure 19.3). The evolution of *Homo erectus* was also

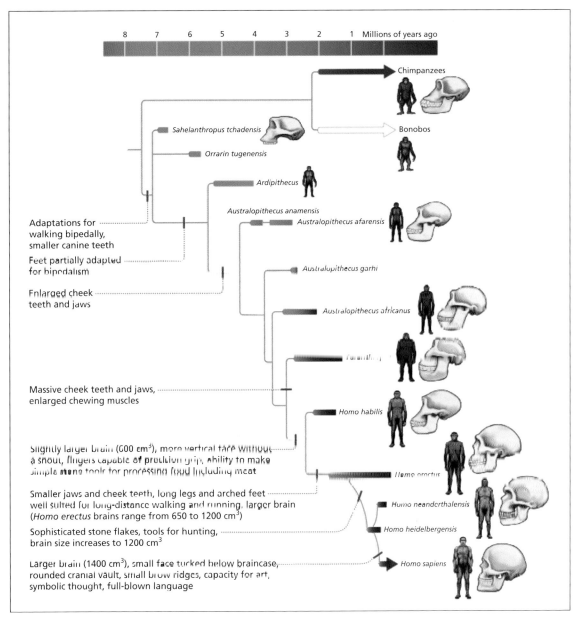

Figure 19.1 Chimpanzee, bonobo and hominin phylogeny, illustrating key evolutionary trends in hominin biology and behaviour. *Source*: Courtesy of Ben Roberts, Roberts and Company Publishing, reproduced from *The Tangled Bank*.

associated with increases in body size and changes in leg length and foot structure that indicate adaptation to long-distance walking or running, and some reductions in the size of molar teeth also occurred.

There has also been a decline in sex differences in body size (Figure 19.4). The reasons for these changes are not known: they could have been a consequence of a gradual transition from polygyny to social monogamy or they could reflect a gradual decline in the extent to which the competitive ability of males depended on their size and weight as male coalitions or artificial weaponry (and the skills associated with it) became increasingly important in

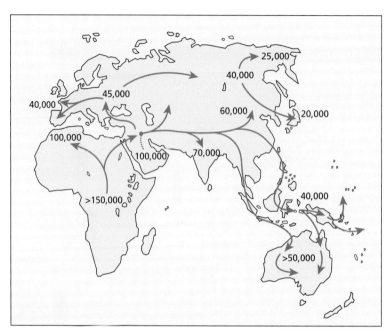

Figure 19.2 Modern humans originated in Africa more than 195,000 years ago, and by 100,000 years ago had begun to colonise the rest of the world. More recent discoveries suggest an earlier occupation of Asia. *Source*: From Dixson (2009), based on information in Stringer and Andrews (2005). Reproduced with permission from Oxford University Press.

deciding the outcome of competitive interactions and the strength of selection for male size declined. However, as increases in female size appear to have been responsible for a large part of the reduction in dimorphism, they could also have been a consequence of selection for increased strength or mobility in females (O'Connell *et al.* 1999).

Technology

Stone tools, made from sharpened flints or other rocks ('cores'), appear by 3.3 Mya, pre-dating the earliest known forms of *Homo* (de la Torre 2011; Harmand *et al.* 2015) (Figure 19.5a). Subsequently, core tools became progressively more standardised and sophisticated (Figure 19.5b) until, around 750,000 years ago, they were gradually replaced by tools made from sharpened flakes struck off cores and, eventually, by smaller and narrower flakes (blades) that were struck from pre-prepared cylindrical cores and which were probably used to make composite tools and projectile weapons (Stout 2011). Blade tools of this kind appear around 195,000 years ago in southern Africa in association with *Homo sapiens* and were made in western Europe by about 40,000 years ago (d'Errico and Stringer 2011): they show considerable geographical and temporal diversity and suggest that the evolution of modern cultures occurred through a complex process of asynchronous advances.

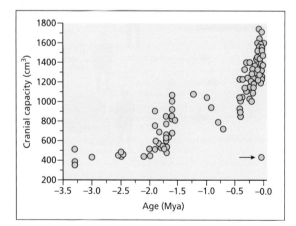

Figure 19.3 Progressive increases in brain size during hominin evolution. Cranial capacities of individual fossil hominin specimens have been plotted against time, beginning at 3.5 Mya. The arrow indicates the anomalous position of *Homo floresiensis*, given its recent date and small cranial capacity. *Source*: From Martin *et al.* (2006), based on data from Stanyon *et al.* (1993).

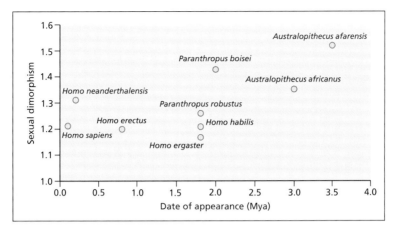

Figure 19.4 Estimates of sexual dimorphism in body weight plotted against date of appearance for different fossil hominins. *Source*: Data from Dixson (2009).

Neolithic agricultural cultures originated in western Asia before 12,000 years ago and spread throughout Asia, Africa and Europe, reaching north-western Europe around 6500 years ago (Johnson and Earle 2000). Similar processes of plant and animal domestication also took place at later dates in other parts of the world. Metalworking originated around 5000 years ago in western Asia and copper- or bronze-working then spread through Europe and Asia until they were replaced by the use of iron after 3200 years ago.

While it is usual to quote the dates of the first known development of different technological developments, it is important to appreciate that independent development in different parts of the world as well as variable rates of cultural diffusion means that, at any period of prehistory, different human populations relied on different technologies (Diamond 1997; Johnson and Earle 2000). For example, farming cultures did not reach western Europe until 6000–7000 years ago while the use of stone tools persisted in Highland New Guinea until recently.

The development of agriculture was associated with increases in population density and although total human numbers are thought to have remained low until around a thousand years ago (Figure 19.6) communities became larger and more settled and built permanent structures. Comparisons of population densities in modern hunter-gatherers and traditional agriculturalists show that average population densities are considerably lower in hunter-gatherers, though this difference may partly reflect the persistence of hunter-gatherers in relatively unfavourable environments.

Human traits

The timing of the evolution of other specifically human characteristics is also uncertain and is widely debated (Foley 1987; Kappeler 2011). Bipedal locomotion, hunting, tool-making, cooking, language, social learning, social organisation, investment by grandmothers, cooperative breeding and cooperative hunting have all been proposed as formative developments that stimulated the subsequent evolution of other human characteristics (Oakley 1957; O'Connell *et al.* 1999; Silk and Boyd 2006; Wrangham 2009; van Schaik and Burkart 2010). However, since the origins and sequence of their development is not known, arguments are necessarily speculative.

Both bipedal locomotion and tool use occur intermittently in chimpanzees and both evolved early in hominin evolution. As hominins colonised the open savannahs, they may also have relied increasingly on cooperative behaviour to provide protection from predators and rival groups and improved survival may have led to greater longevity (Hawkes 2003). The evolution of cooperative foraging or hunting and of extensive allo-parental care could also have originated with the occupation of grassland and savannah biomes to meet the threat of predation on juveniles and adolescents and may have facilitated the evolution of other forms of cooperation and prosocial behaviour (Burkart *et al.* 2014).

Reliance on hunting large game animals and cooking probably represent relatively late developments. The earliest evidence of the butchery of mammals by hominins involves *Australopithecus* species at sites in Ethiopia dated to around 2.5 Mya (Asfaw *et al.* 1999; de Heinzelin

(a)

(b)

Figure 19.5 (a) One of the earliest stone tools yet found, from Lomekwi 3 in northern Kenya, dated to 3.3 Mya. Scars on the stone indicate that it was a core from which flakes were struck. (b) Acheulean handaxe, also made by trimming a flint core. Similar tools first appear around 1.75 Mya but continued to be made as recently as 0.1 Mya in some areas. *Sources:* (a) Jason Lewis and Sonia Harmand. (b) © Zeresenay Alemseged/Science Photo Library.

et al. 1999). Specialised hunting of large herbivores was common by 50,000 years ago, though reliance on meat as an important component of the diet probably occurred much earlier (Stahl *et al.* 1984; Buss 2011).

There is some evidence of the use of fire by 1.0 Mya and it has been suggested that the early development of cooking may have facilitated increases in brain size in early *Homo* (Wrangham 2009). However, the early evidence of the controlled use of fire is limited and hard to interpret and the extensive use of fire may not have developed until after 500,000 years ago (Roebroeks and Villa 2011). By 400,000 years ago, it is clear that there was extensive use of fire which probably pre-dated the divergence of modern humans and Neanderthals and was presumably associated with cooking food (James *et al.* 1989; Pennisi 1999; Webb and Domanski 2009; Wrangham 2009). Cooking may in turn have had important energetic benefits and its development may have been associated with a wide range of changes in diet, habitat use, social organisation and niche breadth (Carmody and Wrangham 2009; Wrangham and Carmody 2010).

Language

Language is accepted as a human universal and it has been suggested that it is based on an innate propensity to structure words and symbols (Chomsky 1957, 1968; Pinker 1994). However, as more languages have been examined and the diversity of their structure and grammar has come to be better understood, the suggestion that language depends on a specific heritable capacity or *language acquisition device*, as Chomsky suggested, looks increasingly improbable and instead it seems more likely that its acquisition is gradual and piecemeal and depends on a protracted period of individual and social learning (Evans 2014).

Human language differs from vocal communication in other mammals in important ways (Cheney and Seyfarth 2010). In particular, human vocalisations are both more flexible and more variable than the vocalisations of non-human primates, and individuals continue to add to their vocabulary throughout life, generating far larger 'vocabularies' than those of other mammals. Humans also have a lexical syntax based on arbitrary sounds, and their vocalisations are arranged into complex sequences with recursive structures that affect their meaning as well as the responses of listeners. In contrast to other mammals (which show limited evidence of theory of mind; see Chapter 7), humans commonly take into account the knowledge and mental state of their audiences when they vocalise and the intentions of callers when they respond to them and language use is deeply sensitive to context.

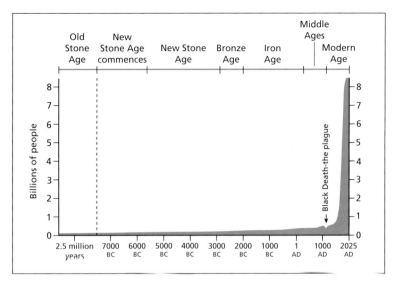

Figure 19.6 Changes in human population size from 2.5 Mya until the present. *Source*: From https://commons.wikimedia.org/wiki/File%3APopulation_curve.svg.

When language developed in the form that we know it today is uncertain. Evidence of decoration and the use of pigments by early African populations of *Homo sapiens* are commonly taken to suggest that it evolved before 100,000 years ago but some form of language probably had a much older origin and may have preceded the divergence of Neanderthals and modern humans around 300,000 years ago (Foley 2001; Foley and Gamble 2009).

The reasons for the evolution of language are a topic of continuing debate (Maynard Smith and Szathmáry 1995; Christiansen and Kirby 2003; Fitch 2010; Richerson and Boyd 2010). Explanations of the evolution of language range from suggestions that the development of bipedalism freed the hands for use in gesturing which in turn facilitated the evolution of complex vocal signals (Corballis 2009) to suggestions that the evolution of language was stimulated by the need for group members to exchange information about each other and to cooperate extensively (Dunbar 1998; Richerson and Boyd 2010). Overall, it seems likely that the occupation of grassland biomes by proto-hominins was associated with a need for greater coordination between group members, communication over longer distances and more effective transfer of information about resources out of sight or events that individuals had not witnessed. This could perhaps have led both to the gradual development of a *theory of mind* and to selection pressures favouring the development of flexible vocal production and a lexical syntax in order to provide greater specificity in communication while minimising the frequency of misunderstandings (Nowak and Krakauer 1999; Cheney and Seyfarth 2003). If so, long before hominins spoke in sentences, they probably had an understanding of each other's knowledge and mental states which tempered their signals and their responses. It would not be surprising if this was associated with selection for improved understanding of cause and effect, both in physical and social interactions and with increases in general cognitive abilities and brain development (see section 19.7) (Richerson and Boyd 2010).

19.3 Life histories

In the absence of direct records of the life histories of early hominins, it is useful to understand the similarities and differences between human life histories and those of the African apes. The principal differences in life-history traits between humans and chimpanzees lie in age at first breeding, breeding frequency, the duration of juvenile dependence on adults and the duration of post-reproductive female lifespans. In (wild) chimpanzees, females usually breed for the first time at 10–14 years, producing single infants after an 8-month gestation period (Emery Thompson *et al.* 2007; Stumpf 2007). Juveniles are nursed for their first 4 or 5 years of life and around 60% of infants

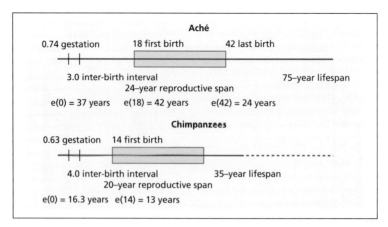

Figure 19.7 Contrasts in some important life-history variables between women (based on data from Aché hunter-gatherers) and chimpanzees. Points along the bar mark gestation period, inter-birth interval, age at first birth, reproductive span, age at last birth, and maximum lifespan (age at which 3% of all individuals born are still alive) in years. Below the bar are life expectancies (e) calculated at birth, at first reproduction and at last reproduction. *Source*: From Hill and Hurtado (1996) with additional data from Goodall (1986) and Nishida, *et al.* (1990).

survive to the age of 15 years (Emery Thompson *et al.* 2012) (Figure 19.7). By the time they are 4 or 5 years old, juvenile chimpanzees are no longer nutritionally dependent on their mothers. Inter-birth intervals after the birth of surviving offspring are at least 5 years and females continue to breed until shortly before their death. Few females survive for more than 40 years.

Among hunter-gatherers and subsistence farmers, women typically give birth for the first time when they are 18–20 years old: for example, average ages at first birth for Aché, !Kung and Yanomamö women are 19.5, 18.8 and 18.4 years respectively (Hill and Hurtado 1996). Children are born after a 9-month gestation period and are commonly nursed for 2–3 years in

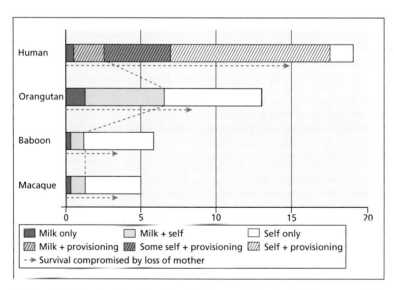

Figure 19.8 Duration of different periods of maternal investment and care and offspring development until female age at first birth in long-tailed macaques, savannah baboons, Bornean orangutans and human foragers. Arrows underneath the bars indicate until what age the presence of the mother has been shown to affect an offspring's fitness through a direct effect on survival or adult success, mediated by factors such as dominance rank. Extensive post-weaning maternal care affecting offspring reproductive success has been demonstrated in all four species. *Source*: From van Noordwijk *et al.* (2013). Reproduced with permission of Elsevier.

traditional societies. Inter-birth intervals are substantially shorter than in chimpanzees and are typically 3–4 years, though longer intervals occur in groups living in harsh environments, like the Kalahari (Blurton Jones 1986). Relative to the great apes, human infants are provisioned with food other than milk at a relatively early age. In addition, in hunter-gatherer societies, juveniles and adolescents rely partly on food provided by their parents or other group members for their first 20 years of life (Figure 19.8) and parents frequently support several offspring of different ages at the same time (Kaplan et al. 2000; Gurven and Kaplan 2008). However, comparisons suggest that the proportion of metabolised

energy diverted to reproduction over the lifespan in humans is similar to that for other mammals (Burger et al. 2010).

Women have longer breeding lifespans than female chimpanzees and substantially greater longevity. Though some women in hunter-gatherer societies continue to have children until they are more than 50 years old, mean age at last birth is usually lower, ranging from 34 years in the !Kung to 42 years in the Aché. Average life expectancy after their last child is substantially longer in women than in female chimpanzees, who typically die within a few years of rearing their last offspring (Figure 19.9), whereas women

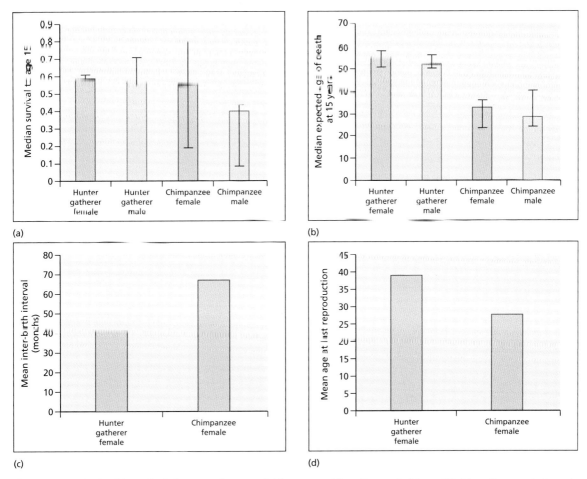

Figure 19.9 Survival and fecundity in hunter-gatherers and chimpanzees: (a) median survival to age 15; (b) median expected age at death at 15 years; (c) mean inter-birth interval (months); (d) mean age at last reproduction in females. Figures show medians or means and the range of values for each parameter from research on four hunter-gatherers (Aché, Hadza, Hiwi and !Kung) and five studies of chimpanzees (at Bossou, Gombe, Kibale, Mahale and Taï). Source: From Kaplan et al. (2000). Reproduced with permission of John Wiley & Sons.

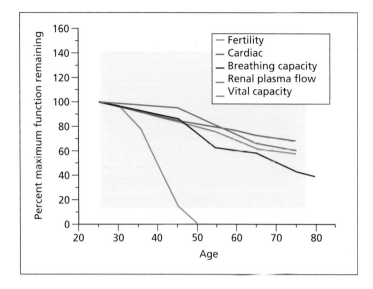

Figure 19.10 Age-related changes in physiological functions of human females. Most body functions senesce at about the same rate, but female reproductive function declines more rapidly and ends by age 50. *Source*: Adapted from Hill and Hurtado (1996). Reproduced with permission from Transaction Aldine with data from Mildvan and Strehler (1960) and from Wood (1990).

in many societies commonly live for 10–20 years after they cease to reproduce and play an important role in the care of their grandchildren (see Chapter 20). The longer post-breeding lifespan of women is not a consequence of a general decline in bodily functions, for fecundity declines earlier and more rapidly than other biological functions (Figure 19.10). Similar changes in female fecundity with increasing age do not occur in other primates for which data are available (Alberts *et al.* 2013).

Both in chimpanzees and in modern humans, males reach sexual maturity later than females but usually continue to be fertile throughout their lives. In chimpanzees, the reproductive success of males is relatively low until they are at least 15 years old, peaks between the ages of 16 and 25 years and then declines (Figure 19.11). In human societies (especially in polygynous populations) most men start to breed later than women but can continue to breed until their death (Borgerhoff Mulder 1988) and, in some societies, their reproductive success may increase with age as a result of the accumulation of wealth and wives. Generation times are shorter in chimpanzees than humans and longer in males than females in both species (Fenner 2005; Langergraber *et al.* 2007).

Contrasts in life-history patterns between humans and African apes are associated with a wide range of ecological differences, including a more diverse diet, greater reliance on extractive foods and hunting, and more extensive preparation and cooking of food (Silk and Boyd 2006; Wrangham 2009). Explanations of the initial evolution of contrasts in life history between modern humans and apes differ in the role they attribute to different changes in ecology. One suggestion is that changes in life histories were caused by a dietary shift towards high-quality but sparsely distributed foods (including an increased reliance on meat) associated with cooperative foraging and food sharing (Kaplan 1996) (see Chapter 20). Another is that the occupation of arid seasonal environments after 2 Mya led to selection on older females to invest in grand-offspring rather than offspring and to the evolution of an extended post-reproductive lifespan (Hawkes 2003). These two hypotheses are not exclusive and our knowledge of the timing of changes in life-history traits is still rudimentary. Reconstructions based on the developmental rates of teeth suggest that australopithecines had ape-like life histories and that the change towards modern life histories may have occurred with the emergence of *Homo* between 2.0 and 2.5 Mya (Foley 2001).

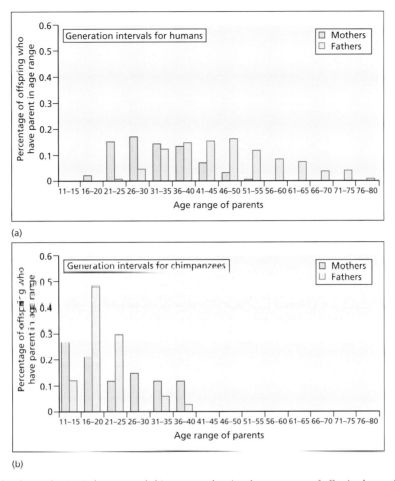

(a)

(b)

Figure 19.11 Age-related reproduction in humans and chimpanzees showing the percentage of offspring born with mothers and fathers of different ages. (a) Human agriculturalists in the Gambia. (b) Chimpanzees from Gombe. (*Source*: (a) From Ratcliffe *et al.* (2000). Reproduced with permission from WHO. (b) Data from Wroblewski *et al.* (2008). Reproduced with permission of Elsevier.

19.4 Sex differences

Size and growth

Since sex differences in size and growth among mammals are commonly associated with polygynous breeding systems (see Chapter 18), their presence and magnitude in hominins and modern humans provides some insight into the origins of human breeding systems (Geary 2003). It is frequently hard to sex early human fossils reliably but, with some possible exceptions (Reno *et al.* 2010), most palaeontologists are agreed that early hominins showed a substantial degree of sexual dimorphism in body size (McHenry 1994; Gordon *et al.* 2008; Foley and Gamble 2009). Sex differences in body size have

declined during hominin evolution and, in modern humans, the ratio of adult male-to-female body weight is around 1.1–1.2, though there is considerable variation in the extent of sex differences in height and weight between human populations (Figure 19.12). However, the extent of sex differences in body size is outside the range of existing monogamous primates (Dixson 2009).

Modern humans show a range of other sex differences typical of a mildly polygynous species. As in other sexually dimorphic mammals, sex differences in adult size are associated with small sex differences in birthweight and postnatal growth rates (Lummaa and Clutton-Brock 2002; Halileh *et al.* 2008) but whether these result in measurable differences in the costs of rearing sons and

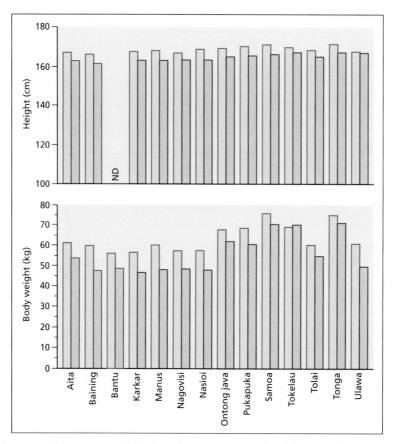

Figure 19.12 Sex differences in body weight and height in various human populations. Blue bars indicate men, red bars women. *Source*: From Dixson (2009). Data from Collard (2002) and adapted from Houghton (1996). Reproduced with permission of Oxford University Press.

daughters is uncertain. Some studies suggest that the cost of rearing sons can be higher than rearing daughters. For example, in pre-industrial Finnish populations, mothers that raised twin sons were subsequently more likely to cease breeding than mothers that raised twin daughters (Lummaa *et al.* 2001) and offspring whose immediately previous sibling was male showed reduced fecundity, survival and lifetime breeding success compared to those whose previous sibling was female (Rickard *et al.* 2007). However, the first effect could occur because sons are valued more highly and women that have reared twin males are more likely to delay breeding again than those that have raised twin daughters, while the second could occur if mothers continue to invest in sons for longer than daughters.

Men also reach sexual maturity later than women (Dixson 2009; Geary 2010). At adolescence, they become

leaner and more muscular, and develop larger hearts, larger lungs, higher systolic blood pressure and greater capacity for carrying oxygen in the blood (Tanner 1990; Bogin 1999). By adulthood, men have a higher percentage of muscle and a lower percentage of body fat than women (Clarys *et al.* 1984; Dixson 2009) (Figure 19.13). They typically have higher levels of circulating testosterone and higher resting metabolic rates and can run faster and throw objects further and with greater velocity (Geary 2010).

There are consistent differences between men and women in a range of other physical traits (Figure 19.14). Men have larger canine teeth than women, though sexual dimorphism in canine size appears to have declined over time and estimates for *Australopithecus afarensis* suggest that sex differences in canine size were substantially larger than those in

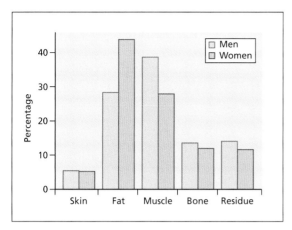

Figure 19.13 Sex differences in human body composition. In relation to their overall body composition, men show higher percentages of muscle, and women show higher percentages of fat. *Source*: From Dixson (2009). Originally from Clarys *et al.* (1984). Reproduced with permission of Oxford University Press.

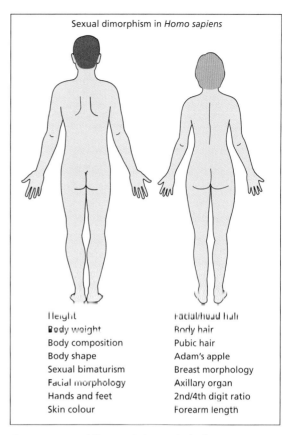

Figure 19.14 Sex differences in human body shape. Images depict a mesomorphic masculine somatotype and an hourglass feminine somatotype, with a weight-to-hip ratio of 0.7, together with a list of the morphological sex differences which occur in *Homo sapiens*. *Source*: From Dixson (2009). Reproduced with permission of Oxford University Press.

modern humans (Geary 2010). There are also consistent sex differences in face and body shape: men typically have larger jaws, chins and cheekbones, heavier brow ridges and narrower deeper-set eyes, while women have larger breasts, narrower waists, broader hips and, in some populations, much more protuberant buttocks (Dixson 2009). Men have more body hair than women as well as substantial facial hair, whose growth is stimulated by testosterone: in a well-known paper, a field worker on the island of Rum measured his daily beard growth and showed that this accelerated before visits to the mainland to see his girlfriend and declined after the resumption of sexual relationships (Anon 1970). Within populations, men tend to be darker skinned than women and these differences, too, are associated with contrasts in the effects of reproductive hormones: oestrogen promotes lightening of the skin in young women, while androgens increase deposition of melanin in the epidermis and increase the number of red blood cells in young men, leading to 'browner and ruddier' complexions (Frost 1994; Dixson 2009).

Like some other primates, humans show sex differences in vocal apparatus. During puberty, the male larynx enlarges and the vocal cords increase in length and the voice deepens (Titze 2000; Dixson 2009). Men have larger larynxes than women and have deeper voices and differences in the pitch of their voices is correlated with their height (Pisanski *et al.* 2014).

Individual differences in voice depth have also been shown to be associated both with variation in the ratio of testosterone to oestradiol (T/O) and with sexual activity (Dixson 2009): one early study compared T/O ratios and copulation frequency in tenor, baritone and bass singers and showed that bass singers had significantly higher T/O ratios as well as increased (reported) ejaculation frequencies (Nieschlag 1979).

Reproductive traits
Males
The structure of male genitalia also provides clues to the evolution of human breeding systems and suggests a history of polygyny. Though available data are limited, the evidence for modern human populations

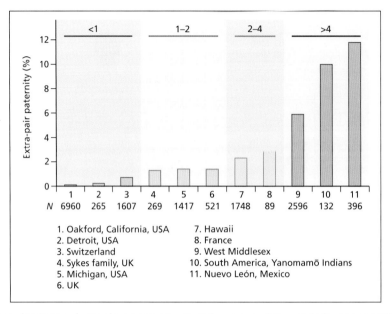

Figure 19.15 Frequency distribution of rates of extra-pair paternity in human populations. Data from Dixson (2009) and Simmons *et al.* (2004).

suggests that, as expected, the incidence of extra-pair paternity is relatively low in many human populations (Figure 19.15) (Anderson 2006; Dixson 2009). Across human populations, estimates of the combined weight of both testes range from 19.0 g (in a sample of men from Hong Kong) to 50.1 g (in a sample from Nigeria) (Dixson 2009): relative testes size is consistently smaller in samples from China, Japan and Korea compared with those from Europe, North America, Australia and Africa. Whether this is a consequence of the contrasts in the relative intensity of sexual selection in the past, of natural selection favouring reduced levels of testosterone, of countervailing selection on women or of genetic drift is uncertain (Dixson 2009). Compared to the other great apes, testis size relative to body size in men is substantially smaller than that of chimpanzees (Figure 19.16) and human testes are relatively larger than those of gorillas and approximately similar to those of orangutans (Harcourt *et al.* 1981) (see Chapter 13). Average values in humans lie outside the range of primate species with multi-male breeding systems, suggesting an ancestral breeding system that involved either unimale groups or monogamy (Dixson 2009).

The structure of human sperm is also consistent with low levels of sperm competition. Both among primates and across mammals, midpiece volume in sperm is correlated with relative testes size and is relatively low in humans (Dixson 2009). In addition, spermatogenesis is comparatively slow in humans and human males have relatively low numbers of sperm per unit

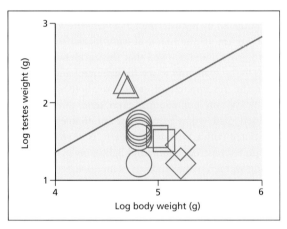

Figure 19.16 Relative testes sizes in human populations as compared to the great apes. Circles indicate *Homo*; diamonds indicate *Gorilla* (the western lowland gorilla is plotted separately and falls below the mountain gorilla on the graph); squares indicate orangutans (two species plotted); triangles indicate chimpanzees and bonobos (two species). *Source*: From Dixson (2009). Reproduced with permission of Oxford University Press.

volume of ejaculate, and relatively low storage of sperm in the tail of the epididymis, with the result that sperm counts decline relatively rapidly as a result of repeated ejaculations (Dixson 2009). Other characteristics of human reproductive anatomy associated with sperm transport or with production of seminal fluid (including the structure of the seminal vesicles and the prostate gland) also show similarities with other primates where the risk of sperm competition is low (Dixson 2009). For example, rates of seminal coagulation (which are relatively high in species, like chimpanzees, where sperm competition is frequent and ejaculates coagulate to form copulatory plugs) are relatively low in humans and are similar to those of gibbons and gorillas.

In contrast to the structure of testes, the size and shape of the human penis provides little information about the likely breeding system under which it evolved. Despite some claims, human penises are not substantially longer than those of chimpanzees or other primates when the effects of variation in body size are controlled, though they are relatively thick (Dixson 2009). Unlike the penises of many primates that live in multi-male groups, human penis structure is comparatively simple and is devoid of spines and bacula.

Females

Like men, women will engage in sexual activity throughout the reproductive cycle and neuroendocrine mechanisms do not control sexual receptivity as closely as they do in some other primates (Dixson 2009; Gray and Garcia 2013). Nevertheless, women show hormonally influenced peaks of sexual receptivity around mid-cycle when ovulation is most likely and are more likely to dress provocatively, wander further, initiate sexual interactions, show reduced selectivity of partners and have orgasms during sex over this period (Hrdy 1981) (Figure 19.17) Changes in the behaviour of women have generated disagreements over whether or not women should be regarded as having a period of oestrus: while some biologists and anthropologists regard the presence of continuous receptivity and the lack of close neuroendocrine control as indicating that women (and many other female primates) do not have a period of oestrus (Dixson 2009), others adopt a broader definition and regard the presence of consistent change in receptivity throughout the menstrual cycle as an indicator of oestrous behaviour (Hrdy 1981; Gangestad and Thornhill 2008).

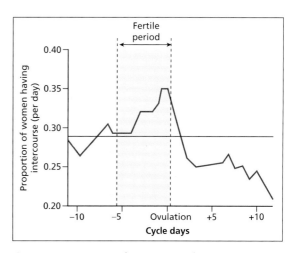

Figure 19.17 Variation in the proportion of women who report having intercourse at different phases of the menstrual cycle. Increases occur during the 'fertile phase' of the cycle (defined as the 6 days culminating in the probable day of ovulation). Source: From Dixson (2009), Adapted from Wilcox et al. (2004), data from Simmons et al. (2004). Reproduced with permission of Oxford University Press.

Several evolutionary explanations of the protracted receptivity of women have been suggested (Symons 1979; Strassmann 1981), including the possibility that it is associated with selection for pair-bonding (Morris 1967; Symons 1979) or control of fecundity (Burley 1979). By concealing ovulation, women may also reduce the risk of male infanticide and increase the chance that multiple males will contribute to caring for their offspring (Hrdy 2000).

Explanations of the absence of advertisements of temporal changes in receptivity and fertility may suggest that, at some stage of human evolution, some women mated with more than one partner in the course of the same reproductive cycle. Although extra-pair paternity is uncommon in many human populations (see Figure 19.15), it is relatively frequent in some, so it could have been more pervasive in the past. For example, in some societies, it is believed that paternity is shared ('partible paternity') and infidelity is accepted and in several of these societies, women that associate with multiple men benefit from increases in the number of individuals that supply them or their children with food or other gifts (Scelza 2013). As in some other mammals (see Chapter 16), this can lead to increases in the survival of offspring with multiple fathers (Hill and Hurtado 1996; Beckerman et al. 1998; Scelza 2013).

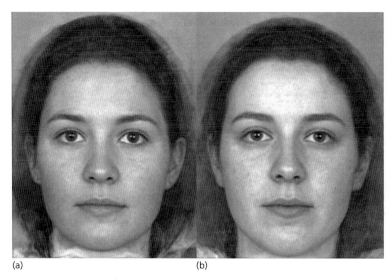

(a)   (b)

Figure 19.18 Composite photographs of (a) a woman with high follicular levels of oestradiol-17β and (b) a woman with lower oestradiol levels. *Source*: From Law Smith *et al.* (2006). Reproduced with permission from the Royal Society.

Many other primates that live in multi-male groups, women possess physical ornaments of various kinds that advertise their health and condition and play an important role in competition for breeding partner, including prominent breasts, constricted waists and, in some populations, prominent buttocks (Cant 1981; Low *et al.* 1987). Individuals with relatively large breasts and narrow waists show higher levels of salivary oestradiol at mid-cycle than those with relatively small breasts and relatively broad waists and have higher levels of fertility (Jasieńska *et al.* 2004). In addition, women with relatively high oestrogen levels tend to have rounder, more padded faces (Figure 19.18). Physical ornamentation in women is often supplemented by artificial ornaments that accentuate sexual characteristics (Morris 1967) (Figure 19.19). For example, in the latter part of

(a)   (b)

Figure 19.19 (a) A Paduang woman from Burma wearing rings that have elongated her neck. (b) Colourfully dressed and face-painted local tribes celebrating the traditional Sing Sing in the Highlands of Papua New Guinea. *Sourced*: (a) © Kevin Hellon/Alamy Stock Photo. (b) © Robert Harding World Imagery/Alamy Stock Photo.

the nineteenth century, European women commonly wore costumes that incorporated the elaborate plumage of male birds (Munro 2009) and artists of the period commonly associated female beauty with elaborate male ornaments borrowed from other species. However, artificial ornamentation is not confined to women and is also extensively used by men (see Figure 19.19b).

The presence of physical ornaments in women raises questions about their evolutionary origins. Darwin (1871) was puzzled by the reversal in humans of the usual pattern of sex differences in ornamentation and attributed it to the (supposedly) greater intellectual development of men, suggesting that it was a product of their greater selectivity of mates. A more likely explanation is that, as in other species where both sexes invest heavily in their offspring, preferences for characteristics that signal a healthy distribution of body fat and the capacity of individuals for parental investment developed in humans at an early stage in association with male mating preferences and led to a progressive elaboration of physical ornaments in females (Pond 1978; Cant 1981; Geary 2003) (see Chapter 20).

The role of female orgasm in cryptic female choice is also widely debated. Orgasm is not restricted to humans and similar responses to copulation occur in some other primates (Dixson 2009; Hrdy 1981). Although it has been suggested that orgasm might play a role in sperm transport and might help to draw sperm through the cervix and into the uterus (the inelegantly named 'upsuck' response) (Baker and Bellis 1993, 1995), there is little evidence that it affects the probability of conception (Lloyd 2005; Dixson 2009). An alternative view is that its function is to generate pleasurable sensations associated with mating and to enhance the receptivity of women or their bonds with regular partners (Hrdy 1981, 1999), while a third is that female orgasm is a non-functional homologue of orgasm in men (Symons 1979; Dixson 2009). One line of evidence that argues against the last explanation and suggests that female orgasm may have evolved to enhance female pleasure in sexual interactions is that female bonobos (who regularly engage in mutual genital stimulation with partners of both sexes) have an unusually large clitoris (Hrdy 1999).

Juvenile behaviour

Boys and girls show many of the same sex differences in social behaviour that are found in other sexually dimorphic mammals, including chimpanzees (Geary 2003; Gray and Garcia 2013; Lonsdorf *et al.* 2014). Both sexes tend to segregate during play and tend to form closer attachments to adults of the same sex as themselves (McIntyre and Edwards 2009). The sexes differ in the relative frequency with which they are involved in different types of play: boys are more frequently involved in rough and tumble play than girls; they are more frequently competitive and their interactions are more frequently directed at achieving hierarchical dominance (Geary 2003, 2010). Girls engage less frequently than boys in rough and tumble play and physical competition and girls exposed prenatally to high levels of androgens behave more like boys (Berenbaum and Snyder 1995; Geary 2003, 2010).

Boys also engage more frequently in object-oriented play than girls and learn to use tools more readily (Willingham and Cole 1997; Chen and Siegler 2000). Here too, girls exposed to high prenatal levels of androgens show patterns of play that resemble those of boys (Berenbaum and Hines 1992). Boys also have a wider range of play behaviour, more commonly build things and tend to be more exploratory than girls (Geary 2003). As they develop, boys in hunter-gatherer societies initially play at hunting animals and, later, start to do so in earnest, even though their success is usually low and they could obtain more calories by continuing foraging with the women (Blurton Jones *et al.* 1997). Like females in other primate species, girls engage more in play-parenting than boys and girls exposed to high levels of androgens before birth behave more like boys (Geary 2003).

Aggression

As in other social mammals, there are consistent sex differences in aggressive behaviour in adult humans. In both sexes, more frequent aggression is often associated with heightened levels of testosterone (Dabbs and Hargrove 1997; Mazur and Booth 1998) though, in most societies, men are more commonly involved in physical aggression with each other than women, often over access to women (Wrangham and Peterson 1996; Geary 2010; Hess *et al.* 2010). For example, in the Yanomamö, aggressive interactions between men belonging to the same community are common and involve intimidating displays, like chest pounding, wrestling or fights with clubs (Figure 19.20) and similar fights occur in the Aché (Hill and Hurtado 1996). The usual aim is not to kill opponents but to intimidate them or establish

Figure 19.20 Yanomamö men engage in wrestling and club fighting with other members of the same community. Though the aim is to establish status or respond to insults, injuries and occasional deaths are not uncommon. *Source*: © Napoleon Chagnon.

dominance, but deaths can occur and around 8% of Aché men die as a result of club fights (Hill and Hurtado 1996). In addition, men from the same community engage in synchronised dances that may help to consolidate social bonds and display the strength of the community (Figure 19.21). In many societies, rates of same-sex homicide are also substantially higher in men than women (Daly and Wilson 1997) and men are more likely to be involved in inter-group conflicts than women and rates of mortality in men are consequently substantially higher than those in women. For example, Figure 19.22 compares estimated rates of mortality resulting from raids or warfare in six traditional societies and, in all cases, estimated mortality rates for fighting are at least twice as high among men as among women.

Women may not engage in fights with other members of their group as frequently as men because the outcomes of fights are less likely to influence their reproductive success and potential costs may be higher (Campbell 1999). However, in traditional societies, they commonly compete for resources with other families belonging to the same social group and are more commonly involved in indirect aggression than men (Hess *et al.* 2010). In polygynous societies, women commonly compete with co-wives for household goods and their children's inheritance (Mace 2013) and, in some cases, these interactions can have lethal consequences (Geary *et al.* 2014). For example, among the Dogon (a polygynous West African society where sons share the land belonging to their father on his death) aggression directed by wives at the children of co-wives is common and children are sometimes poisoned by their father's co-wives (Geary 2010). The probability of premature death is substantially higher in children from polygynous marriages than in children from monogamous ones and sons are more than twice as likely to die as daughters. Murderous step-mothers also feature in the myths and fairy tales of many Western societies. However, lethal competition between women is relatively rare and women and girls more commonly engage in manipulative strategies and gossip, sometimes referred to as 'relational aggression' (Geary 2010). This is often associated with competition for men and physically attractive girls are often targeted (Campbell 1995; Leenaars *et al.* 2008).

While contrasts in aggression between the sexes are associated with hormonal differences which have a genetic basis, like other forms of social behaviour, they are strongly affected by cultural mechanisms. A good

(a)

(b)

Figure 19.21 Tribal dances, club fights and wrestling by Yanomamö men serve to consolidate the community and to emphasise its strength. *Source*: © Napoleon Chagnon.

example comes from studies of the reasons for the relatively high rates of violence in the southern USA versus the northern USA. Surveys show that southern white men are more willing to endorse violence when it is used for self-protection, to maintain 'honour' or to socialise children (Cohen and Nisbett 1994). Experiments in which males from the north or south of the USA were experimentally insulted showed that men from the south were more likely to think that their reputation had been damaged by insults and showed larger rises in circulating levels of cortisol and testosterone and more aggressive responses than those from the north (Cohen *et al.* 1996). Similar interactions

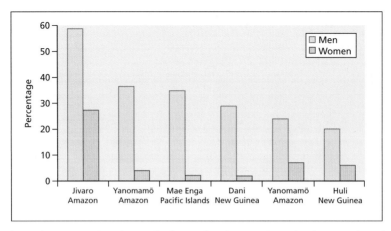

Figure 19.22 Estimated mortality rates resulting from ambushes, raids or larger-scale warfare for six traditional societies. The values for the Yanomamö are from two different groups, the Shamatari (left) and Namowei (right). *Source*: From Geary (2010). Reproduced with permission of the American Psychological Association, with data from Keeley (1996).

between biological and cultural factors are probably very common in many human societies.

Cooperation

In all human societies, men and women contribute extensively to cooperative activities, though in many societies there are qualitative differences in the cooperative activities that are involved, complicating attempts to compare the overall contributions of the two sexes to cooperative activities (Hrdy 1999, 2009). For example, women often cooperate in the collection of vegetable foods and in childcare while men cooperate in hunting and in aggressive interactions with neighbouring communities. The extent of cooperation and the roles of the two sexes are described in Chapter 20.

Cognition

Comparisons of the cognitive abilities of men and women have a long history (Gould 1981). Like many men of his time, Darwin believed that there was an inherent difference in the cognitive powers of the two sexes, which he attributed to male competition. Today, it is appreciated that sex differences in cognition are negligible and that there is no overall sex difference in IQ scores, though variance in scores may be slightly greater among men than among women (Colom *et al*. 2000; Geary *et al*. 2000; Seabright 2012). There are, however, small sex differences in cognitive performance in specific areas: for example, on average, females tend to perform slightly better than males on verbal memory tasks and to respond

more strongly to social stimuli, while males tend to perform better on tasks involving mental rotation in spatial abilities and respond more strongly to mechanical stimuli (Fairweather 1976; Connellan *et al*. 2000; Geary *et al*. 2000; Weiss 2003). As in some other mammals (Linn and Petersen 1985; Jacobs *et al*. 1990) these differences are present among young children, suggesting that they may have a heritable basis. There is also evidence that fetal testosterone levels may exert important effects on development, accentuating the expression of male traits and, in extreme cases, leading to autism (Knickmeyer *et al*. 2005).

Survival

Sex differences in survival are common among humans as they are in many other sexually dimorphic mammals. Rates of natural fetal mortality are often higher in males and stressful social or ecological conditions are commonly associated with reductions in the proportion of males born (Davis *et al*. 1998). Obstetric complications at birth and associated neonatal deaths are also more common in males (Grech *et al*. 2003) and childhood mortality rates are higher (Geary 2010).

Among adults, men are less likely than women to survive periods of acute starvation. For example, striking sex differences in survival occurred in several parties of pioneers in the USA that starved during their migration across America, like the Donner Party, which was stranded throughout the winter in the Sierra Nevada (Grayson 1990, 1993). Similarly, when the population of the Netherlands was subjected to acute famine during the last months of the Second World War, mortality among adult males was higher than among females (Lumey and van Poppel 1994). Extreme environmental hardship is particularly likely to affect older men: for example, in the Donner Party, the chances of a male of under 40 dying were rather less than twice as high as for a female, but among individuals over 40, the chances of dying were ten times higher in males than females (Grayson 1990). Among Donner men, survival was positively related to the number of kin present in the group. In most Western human populations, men age faster and die earlier than women with the result that there is a preponderance of women in older age categories (Figure 19.23).

Although there is a persistent tendency for mortality to be higher in males than females, this difference can be obscured or reversed by cultural practices that value sons more than daughters (Shenk *et al.* 2014). For example, in some societies, neonatal or infant mortality is higher in females as a result of sex-biased infanticide or selective investment by parents (Coales and Banister 1994). Similarly, despite faster rates of ageing in men, their reproductive success can extend for longer than that of females as a result of the accumulation of wealth and wives and can even increase in the later years of the lifespan (see section 19.3).

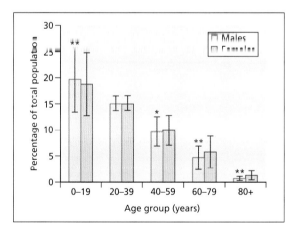

Figure 19.23 Sex differences in human longevity. Histogram shows age-related changes in percentage of males and females in the total population, in 69 countries worldwide. Data are mean ± SEM percentages of males and females of each age group in the total population of 69 countries, as listed in *Financial Time World Desk Reference 2004*. *, $P < 0.05$; **, $P < 0.01$. *Source*: Data from Dixson (2009) and the *Financial Time World Desk Reference 2004*.

19.5 Hominin and human societies

Early hominins

Both the prevalence of sexual size dimorphism in early hominins (see section 19.3) and the retention of sex differences in growth and physiology in modern humans suggest that hominins and humans have been polygynous throughout much of their evolutionary history (Dixson 2009), although it is also possible that some sexually dimorphic traits in humans may have evolved since their separation from the African apes (Plavcan 2012). Analyses of Y-chromosome diversity in humans suggest that the evolution of monogamy in humans may have been a relatively recent development (Dupanloup *et al.* 2003).

In contrast to most other mammals, females usually disperse to breed in neighbouring groups in chimpanzees, bonobos and gorillas while males may be philopatric in all three species, so that breeding groups commonly consist of unrelated females and males that are familiar with each other though they may not be closely related (see Chapter 12). As a result, it seems most likely that the earliest hominins also lived in similar groups (Ghiglieri 1987; Wrangham 1987; Chapais 2010), though not all anthropologists agree (Hawkes 2004). If so, most co-resident breeding females in groups of early hominins would probably have been unrelated and average coefficients of relatedness between group members would have been low.

The formation of multi-male groups in chimpanzees and humans could also be homologous and might have enhanced the ability of early hominins to defend themselves and their offspring against predators and pre-adapted them for the occupation of savannahs and, at a later stage, the evolution of cooperative hunting (Service 1962; Foley 2001). However the retention of fission–fusion groups in savannah-living hominins that were unable to escape predators by climbing trees seems unlikely and a more probable scenario is that most hominins living in savannahs and grasslands formed cohesive breeding groups.

Modern humans lack many of the traits characteristic of multi-male species where sperm competition is common (see section 19.4) suggesting that multi-male groups may have been replaced by single-male groups at a relatively early stage of hominin evolution. In single-male breeding groups where dominant males are continuously associated with small numbers of breeding females, they can usually control breeding access to them, so that the incidence of extra group paternity is low (see Chapter 13). As a result, the reproductive costs to males of associating with other breeding units are often likely to be low, too, and in many larger mammals that live in single-male breeding groups in open country, multiple groups aggregate in large, unstable herds or super-groups to reduce the risk of predation (see Chapter 10). If early hominins formed stable, single-male groups after they occupied open savannahs, they may well have aggregated in unstable super-groups when resource availability permitted and this could have facilitated both the recognition of kin (as a result of high levels of paternity certainty in one-male groups) and the maintenance of protracted bonds with kin after their dispersal to other breeding groups (as a consequence of the regular association of the same one-male groups in the same super-groups) (see Chapais 2008, 2013). A subsequent transition from one-male groups to monogamous pairs could have been a consequence either of increased costs of defending multiple mating partners (perhaps as a consequence of the use of weapons in disputes, frequent intervention by third parties or strong negative effects of disputes on cooperative relationships) or of increases in the needs of dependant offspring for paternal investment (perhaps because of increases in the period of nutritional dependence or of the occupation of habitats where food supplies were sparse and unpredictable). The stage at which this transition occurred is uncertain, but the pattern of sex differences in modern humans suggests a protracted period of polygynous breeding (see Section 19.4).

Cooperation involving non-kin as well as kin pervades all aspects of human societies (see Chapter 20) and the occupation of savannah and grassland ecosystems by Pleistocene hominins may have led to the progressive evolution of cooperative foraging and food sharing. Reproductive cooperation is also pervasive in humans and could also have developed at an early stage of human evolution (Hrdy 2009). One suggestion is that reproductive cooperation in hominins played a causal role in the evolution of cognitive development and increases in brain size by reducing or spreading the net energetic costs of breeding to mothers, either by allowing them to support the substantial energy costs of big brains without compromising expenditure on reproduction or by permitting the evolution of prolonged juvenile dependence and brain growth (Hrdy 2009; Burkart and van Schaik

2010; van Schaik and Burkart 2010; Burkart *et al.* 2014). While this is possible, the evolution of cooperative breeding involving non-breeding helpers in other mammals appears to be restricted to monogamous lineages where females produce multiple young so that it would be an unlikely development in African apes. In addition, cooperative breeders show little evidence of increased cognitive capacities or brain size in either mammals or birds so it is not clear that reproductive cooperation would necessarily have favoured the evolution of improved cognitive abilities or brain development. An alternative view of the evolution of reproductive cooperation in humans is consequently that it represents a relatively late development that either followed the evolution of cooperative foraging, language, improved cognitive abilities and protracted development in juveniles or occurred over the same period (see Silk and House 2016) (see Chapter 17).

Hunter-gatherers and foragers

Until around 10,000 years ago, all human societies depended on some combination of foraging for edible vegetable matter, fishing, scavenging and hunting. Studies of modern foragers or hunter-gatherers consequently provide insights into how early humans may have lived, though they have only persisted in a limited range of habitats and so are unlikely to reflect the original range of social structures and breeding systems. Unfortunately, detailed records of demography, reproduction and social relationships are available for relatively few hunter-gatherer societies: exceptions include the !Kung, Hadza and Aka in Africa and the Aché Indians of Paraguay (Lee 1972; Pennington and Harpending 1988; Hewlett 1991; Hill and Hurtado 1996; Marlowe 2010).

Mating systems and marriage practices are diverse and variable in hunter-gatherers and differences are often associated with local traditions and religious beliefs whose development may not have been closely constrained by ecological or genetic mechanisms. In contrast to the African apes (and probably to early hominins, too), the most common social unit in modern hunter-gatherers is a socially monogamous pair, with their dependent children and older relatives, though individuals of both sexes may change mates several times in the course of their reproductive careers (Marlowe 2003a, 2010). Simultaneous polygyny occurs in some settled communities but is seldom common in nomadic hunter-gatherers and does not reach the levels found in some agricultural and pastoral societies. It is not clear at what stage social monogamy developed. Some regard it as an early development in the hominin lineage that pre-dated (and may have permitted) the increase in brain capacity and prolonged infant dependence that occurred between 1.25 and 0.75 Mya, while others suggest that hominins have probably been polygynous for most of their history and emphasise the frequency of various forms of polygyny in modern populations and of sexually dimorphic traits (Dixson 2009; Kramer and Russell 2014). The prevalence of sex differences in size and growth among hominins and humans tends to favour the latter view (see section 19.3).

The reasons for the prevalence of social monogamy in hunter-gatherers are still debated. One possible explanation is that it is a consequence of reliance on hunting and the need for a sexual division of labour (Lee and DeVore 1968; Washburn and Lancaster 1968). Alternatively, the evolution of large brains and prolonged dependence of juveniles on their parents may increase the need for paternal investment and restricted most males to monogamy (Westermarck 1889; Kaplan *et al.* 2000). Energetic studies show that the acquisition of resources by men commonly requires advanced, slow-developing skills and that men make substantial energetic contributions to their offspring and their partners, who would be unable to rear children without their assistance (Box 19.1 and Figures 19.24 and 19.25; see section 20.5). A third possibility is that reduced asymmetries in the competitive abilities of males as a result of the development of weapons increased the costs to individuals of attempting to monopolise multiple

Box 19.1 Energetic constraints and male investment in hunter-gatherers

One interpretation of the evolution of human monogamy is that it is associated with the ecological niche occupied by humans and the slow acquisition of foraging skills (Kaplan and Gurven 2005). While many of the higher primates have relatively long periods of dependency on their mothers, by the time they are adolescents they are usually capable of obtaining the food they need. For example, chimpanzees are nutritionally dependent on their mothers until they are around 4 or 5 years old; from age 5 years to adulthood, net production (energy acquired independently less energy expended) is around zero but subsequently increases to a peak of around 250 calories/day, substantially lower than that of humans (see Figure 19.24). In contrast to humans, net production probably peaks between the ages of 20 and 25 years, and then declines after individuals are 35–40 years old.

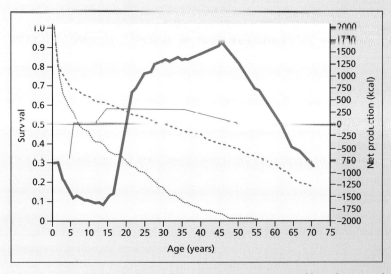

Figure 19.24 Net food production and survival in human foragers and chimpanzees. Dotted lines show age-related survival in chimpanzees, broken lines indicate human survival; thick solid line indicates net production in humans; thin solid lines indicate net production in chimpanzees. *Source*: Adapted from Kaplan *et al.* (2009). Reproduced with permission of John Wiley & Sons.

Human foragers rely on large, nutrient-dense, high-quality foods that are difficult to find and the skills and techniques used to acquire them take years to learn (Kaplan *et al.* 2000). Estimates of net production for hunter-gatherers and horticulturalist foragers show that individuals produce less energy than they consume until they are around 20 years old and so are reliant on others. Net production then increases gradually to a peak at around the age of 45, when they produce around 1750 calories/day; subsequently, men and women become net consumers, initially as a result of increases in the number of their dependants and later as a consequence of ageing (Kaplan and Gurven 2005).

Estimates of the relative contributions of different family members to provisioning young in some hunter-gatherers and horticulturalists emphasises the importance of the father's contributions. For example, in Tsimane forage-horticulturalists, fathers' contributions to feeding their offspring substantially exceed those of mothers or grandparents (see Figure 19.25). However, the relative importance of paternal contributions varies widely, both between communities and between tribes (Schacht and Borgerhoff Mulder 2015).

While net production in humans of both sexes is positive after the age of 20, there are substantial differences between men and women (Kaplan and Gurven 2005; Kaplan *et al.* 2009). Comparative data for ten foraging societies show that, on average, men acquired 68% of the calories and 88% of the protein and that, after their own consumption was subtracted, women supplied only 3% of the calories consumed by their offspring and men supplied almost all the protein eaten by offspring as well as the bulk of protein eaten by women. By comparison, sex differences in the acquisition of protein, starch and total calories by chimpanzees are small. In contrast to chimpanzees and other primates, the contributions of other group members also allow lactating human mothers to reduce time spent foraging and devote more time to the direct care of their dependent offspring (see Chapter 20).

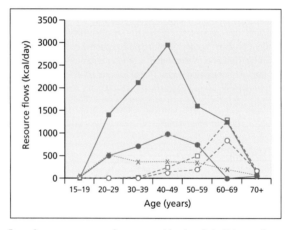

Figure 19.25 Net caloric resource flows from parents, grandparents and husbands in Tsimane forager-horticulturalists. Filled circles, mother to children; filled squares, father to children; open circles, grandmother to grandchildren; open squares, grandfather to grandchildren; crosses, husband to wife. *Source*: From Kaplan *et al.* (2009). Reproduced with permission from the Royal Society.

breeding partners (Mesnick 1997; Marlowe and Wetsman 2001; Chapais, 2010). These three processes are not exclusive and all of them could have contributed to the evolution of monogamy.

While the basic reproductive unit of many contemporary hunter-gatherers is the conjugal family, several families often associate with each other, forming bands with around thirty members (Service 1962; Johnson and Earle 2000; Hill *et al.* 2011). Band members typically forage or hunt in subgroups but are able to recognise each other and are aware of each other's social connections. They frequently share foraging or hunting rights to the same area, though in some societies particular families have priority of access to particular areas (see Chapter 20). Relationships between band members are relatively egalitarian, there is limited economic specialisation, resources (especially meat) are frequently shared, there is little or no formal authority and group

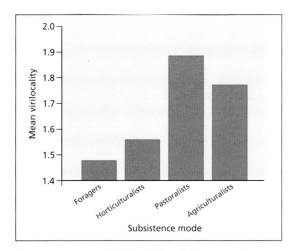

Figure 19.26 Degree of virilocality by subsistence mode
($N-186$) across different categories of societies based on the
Standard Cross-cultural Sample of Human societies. *Source*:
From Marlowe (2000). Reproduced with permission of Elsevier.

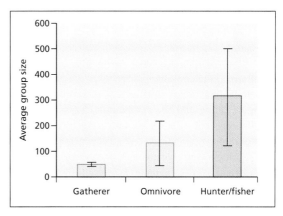

Figure 19.27 Average group size for hunter-gatherers
depending on contrasting resources. *Source*: Data from Kelly
(1983).

decisions are made democratically (Woodburn 1982;
Layton 1986).

In many hunter-gatherers, members of both sexes
frequently marry outside their natal band and may
transfer to their partner's band, so that average levels
of relatedness between band members are typically low
(Hill *et al.* 2011; Dyble *et al.* 2015). Although early
studies of hunter gatherers suggested that men were
more likely to remain in their natal community
throughout their lives than women, subsequent studies
suggest that bilocal systems (where couples may either
join the husband's kin or the wife's or may move
between the two) are usual (Figure 19.26) (Marlowe
2003b; Alvarez 2004; Hill *et al.* 2011). In contrast to
other mammals, social connections are commonly
maintained with offspring and other relatives that
have left their natal family, including matrilineal and
patrilineal kin, as well as with the kin of marriage
partners and most individuals (and their families) are
embedded in an extended network of kinship relations
(Lévi-Strauss 1969; Chapais 2010; Dyble *et al.* 2015).

The size of hunter-gatherer bands varies with population density, the abundance and distribution of the
resource base and the techniques used. Specialised hunters or fishermen, like the Indians of the north-west coast
of America that relied on annual salmon runs, commonly
live (or lived) in larger groups than communities that rely
on gathering (Kelly 1983) (Figure 19.27). The

movements of bands also depend on the distribution
and abundance of resources and are often strongly
affected by key sites, such as waterholes (Sealy *et al.*
2006) and the size of community ranges is typically larger
in communities living in savannah systems than in those
living in forest biomes or in coastal communities (Borrero
and Barberena 2006; Hamilton *et al.* 2007a). Larger
bands usually have larger ranges but the area used per
individual declines with increasing population size
(Hamilton *et al.* 2007b).

In many societies it is possible to identify nested aggregations of bands occupying progressively larger areas
(Hamilton *et al.* 2007a). Relationships between neighbouring bands vary and may be relaxed, peaceable and
cooperative (especially where there are well-developed
exchange networks) or hostile (especially where there is
competition over localised resources) (Cashdan *et al.*
1983; Layton 1986; Kelly 2005). In some hunter-gatherer groups, men and women commonly visit neighbouring groups and may acquire new technological skills
from their members that they can add to their existing
repertoire (Hill *et al.* 2011), though total strangers
(whether alone or in groups) are often regarded with
suspicion and may be attacked.

The origins of the formation of bands and higher-level
social connections can only be guessed at. After hominins
colonised savannah areas, the risk of predation could
either have favoured the retention of multi-male groups
or, after smaller breeding units had developed, the aggregation of breeding units in larger groups as in many non-
human mammals (see Chapter 2). As meat became a

more important component of their diet, the inclusion of women and children in hunting or scavenging forays would have presented problems and central-place foraging combined with a division of labour are likely to have developed (Washburn and Lancaster 1968). The risk of predation may have favoured the development of defended camps and the aggregation of breeding units into bands, while reliance on scarce unpredictably distributed foods may have favoured the development of food sharing between band members, as in vampire bats (see Chapter 9). For example, models of the benefits of food sharing in Aché groups suggest that its primary effect is to reduce the risk of days without food and that effects on mean harvest rates are smaller (Janssen and Hill 2014).

The need for defence against predators and neighbours may also have contributed to regular associations between bands (Durham 1976; Kelly 2005) and encouraged the development of exogamy for strategic reasons (Chapais 2008). As Edward Tylor suggested in 1889, the potential fate of human families may have been to 'marry out or die out'. Tylor's views formed the basis of subsequent treatments of kinship by anthropologists and similar arguments were developed by Lévi-Strauss (1969), who argued that reciprocal exchange of women represents the cornerstone of human societies.

Agriculturalists and pastoralists

Traditional agriculturalists and horticulturalists, like the Yanomamö, live in denser, more stratified societies than hunter-gatherers or 'foragers' though the size of their communities and the nature of the resources they rely on varies widely (Figure 19.28). Among horticulturalists who mix shifting agriculture with fishing or hunting, like the Yanomamö, monogamy or slight polygyny is common (Figure 19.29). In contrast, higher levels of polygyny are not uncommon in pastoralists and settled agriculturalists, where wealthy men often have multiple wives and variance in male reproductive success is relatively high (Flinn and Low 1986; Brown *et al.* 2009) (Figure 19.30).

Mating systems and marriage practices in traditional societies are diverse and variable, though relationships between ecological or economic factors and patterns of

(a)   (b)

Figure 19.28 (a) Yanomamö woman from Casiquiare River, Venezuela. (b) A monogamous Yanomamö man and wife and all their possessions, including their dog, preparing to go to a feast in a friendly village. She is painted. *Source*: (a) © Art Directors & TRIP / Alamy Stock Photo. (b) © Napoleon Chagnon.

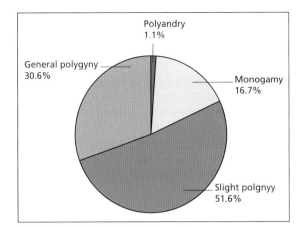

Figure 19.29 The relative frequency of different human mating systems based on the Standard Cross-cultural Sample of Human societies (*N* = 186). *Source*: From Marlowe (2000). Reproduced with permission of Elsevier.

social organisation are not obscured entirely. For example, in societies where individual men are able to monopolise a relatively large proportion of local resources, as in some pastoralists and agriculturalists, polygyny is common and wealthy or powerful men may have multiple wives, though polygynous families are usually in a minority (Borgerhoff Mulder and Caro 1983; Low 2007). For example, in the Kipsigi agropastoralists of Kenya, wealthy men are more likely to be married to multiple partners and

land ownership of their husbands is the primary determinant of the reproductive success of women (Borgerhoff Mulder 1990; Lawson *et al.* 2015). Similarly, among Gabra pastoralists, the reproductive success of men and women increases with the number of camels in their household herd (Mace 1996, 1998). Polygyny is widespread in many societies, but where it occurs polygynous families are usually in a minority (Figure 19.31).

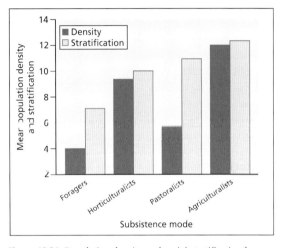

Figure 19.31 Population density and social stratification by subsistence mode across human societies, converted to same scale (*N* = 186). *Source*: From Marlowe (2000). Reproduced with permission of Elsevier.

(a)   (b)

Figure 19.30 Polygynous families: (a) Maasai; (b) Tuareg. *Sources*: (a) © Harry Hook/Getty Images; (b) © Anthony Ham/Lonely Planet Images/Getty Images.

At the opposite extreme are societies living in areas where land or other resources are very scarce and male investment is needed to support a nuclear family. Here, monogamy is normal and polyandrous marriages may also occur (Haddix 2001; Starkweather and Hames 2012). For example, in the Nyinba people of north-western Nepal, several men (who are usually brothers) may be married to the same wife (Levine 1988). Wives and their husbands both attempt to identify which off-spring are fathered by which husband and children inherit land holdings from their (presumed) father. As evolutionary theory would predict, Nyinba men married to women with a relatively large number of husbands and those that father relatively few children with their joint wife are more likely to abandon their marriages (Levine and Silk 1997). Contrary to expectation, mar-riage bonds are no more likely to persist if husbands are close relatives than if they are not (Levine and Silk 1997) but are often affected by changes in economic circum-stances (Haddix 2001).

Differences in the demographic structure of different communities within tribes or tribal groups may also affect mating practices and parental investment by males. For example, across eight communities of Macushi people in south Guyana, contrasts in the adult sex ratio are associ-ated with differences in male investment in parental care and in the duration of sexual relationships: in commu-nities where adult sex ratios are biased towards women, men more commonly engage in short-term, low-invest-ment relationships than in communities where the sex ratio is equal or male-biased (Schacht and Borgerhoff Mulder, 2015).

Like hunter-gatherers, families of agriculturalists and pastoralists maintain long-lasting connections with each other, forming nested multi-level groups within larger social units that may include several hundred individuals. Both exogamy and patrilocality are com-mon and genetic studies show that rates of migration in women are usually higher than in men (Wilkins and Marlowe 2006; Silk and Boyd 2006). While female dispersal in non-human mammals often appears to be a response to the risk of inbreeding with close relatives, unrelated males are usually available within local human communities and contrasts in matrilocality may be related to variation in the structure of social groups and the need to maintain or reinforce social connections rather than to the risk of inbreeding (Lévi-Strauss 1969; Chapais 2008).

As in hunter-gatherer bands, average levels of kinship are relatively low in most horticultural and agricultural communities. For example, Chagnon (2013) refers to Yanomamö villages as 'not so much a band of brothers as a band of brothers-in-law'. In many societies, kinship connections and ties are maintained and reinforced by customary marriage practices, including sororal marriage (the marriage of men to sisters of their wife), the levirate (marriages between men and the wives of their deceased brothers), cross-cousin marriage (marriages between cousins born to sibs of the opposite sex) and parallel-cousin marriage (marriages between cousins born to sibs of the same sex). Many of these practices may stem from natural patterns of association or proximity between individuals and similar patterns occur in societies where there are no formal prescriptions concerning marriages between cous-ins or in-laws (Kuper 2009).

Relations between groups among pastoralists and agri-culturalists vary widely, as they do among hunter-gath-erers (see Chapter 20). Groups that exchange members or have well-developed exchange networks or trading relations are often friendly and peaceable. However, competition between neighbouring communities for resources can lead to feuds, raiding for stock or women and retaliatory attacks that can persist across generations (Keeley 1996; Geary 2010; Chagnon 2013). In many traditional societies, raids and ambushes are commoner than pitched battles though the latter can also occur and associated mortality can be high (see Chapter 20).

Tribes, chiefdoms and states

Increases in population density after the development of agriculture and the domestication of animals were asso-ciated with increases in group size and social stratification as well as with increases in variance in breeding success among males (Johnson and Earle 2000; Marlowe 2000; Betzig 2012). Social institutions developed that encour-aged some forms of social behaviour and discouraged or punished others (Powers *et al.* 2016). Initially, local groups may have been organised into tribes consisting of up to a few hundred individuals whose structure resembled that of bands of hunter-gatherers and some groups of subsistence agriculturalists and horticultural-ists, like the Yanomamö of Venezuela or the Dani of New Guinea, live in groups of this kind today.

As numbers increased further, tribes became organised into chiefdoms by a centralisation of power (Johnson and Earle 2000). Economic specialisation was common and social hierarchies became institutionalised, sometimes leading to the formation of different castes (Bamshad *et al.* 2001). Chiefs collected tributes or taxes of food or labour and used them to maintain warriors, priests and craftsmen and attempted to monopolise the use of coercive tactics. They fostered belief in gods that enforced moral codes and social order (Norenzayan 2013), built religious and secular monuments and, in some cases, public goods, like irrigation systems (Peoples and Marlowe 2012). Since social units were too large and too widely dispersed for all individuals within them to recognise each other, hostile reactions to strangers were usually suppressed or controlled. However, competition for power or succession frequently led to violent interactions between potential heirs or political cliques.

States, consisting of multiple chiefdoms, often amalgamated by force and frequently including ethnically diverse populations, emerged around 5400 years ago (Johnson and Earle 2000; Diamond 2012). They were associated with increases in bureaucracy and centralisation, the formation of standing armies of specialised soldiers, increased economic specialisation both of regions and of individuals, the codification of laws and the enforcement of religious practices. Some developed as monarchies, others as theocracies or oligarchies and a few as democracies. Centralised power provides opportunities for rapid political change and revisions between the different types of society are still common. Like relationships between tribal societies, relationships between states range from cooperation to open hostilities. States have frequently formed defensive or offensive alliances, as in both the First and Second World Wars, and have intermittently attempted to agree and enforce ground rules governing their interactions with each other and the treatment of their citizens. However, these agreements are difficult to enforce and are frequently trumped by strategic or economic considerations (see Chapter 20).

19.6 Why us?

A question of central interest in human evolution is why it should have been primates, and African apes in particular, that produced a descendant whose development of cognition, culture and technology allowed them to dominate the planet (Richerson and Boyd 2005; Hrdy 2009; Whiten *et al.* 2011). In attempting to answer this question, it is important not to underestimate the role of serendipity or to overestimate human uniqueness: all animal species possess a unique array of adaptations whose assemblage is strongly affected by their evolutionary history (Foley 1987, 2001). Nevertheless, it is possible to identify some possible reasons why primates and, later, hominins produced descendants whose social, cognitive and technological developments allowed them to occupy such an unusual niche (Whiten *et al.* 2011).

In contrast to many groups of mammals, primates have relatively diverse diets that commonly involve the extraction of foods from protective shells and the application of considerable manual dexterity and their ecology, behaviour and anatomy and may have pre-adapted them for the subsequent development of extensive tool use (Reader *et al.* 2011). The life-history characteristics of primates may also have played an important role. Compared to other mammals, most higher primates have relatively long lifespans and produce single young at comparatively long intervals (Eisenberg 1981). The production of single young with long periods of dependence may have encouraged the evolution of reliance on social learning and prolonged parental investment. In addition, monotocy may have permitted the evolution of tolerance between breeding females and the formation of stable groups including several breeding females, associated with one or more males, for groups of this kind are relatively uncommon in polytocous species. A common feature of groups where multiple females breed and mothers produce single young at long intervals is that most group members are not close relatives, even if members of one sex are usually philopatric (see Chapter 2) and low relatedness between group members may intensify social competition, increase the frequency of aggressive interactions and encourage the development of differentiated supportive relationships both between kin and between non-kin, which are a distinctive feature of human societies (see Chapters 9 and 14).

The frugivorous African apes exemplify many of the unusual characteristics of primates. They live on foods whose acquisition often involves extractive skills and manual dexterity and possess relatively large brains and advanced cognitive skills (see Chapter 6). In addition, they form stable communities, where average levels of kinship between group members of both sexes are low and complex social relationships between individual group members, involving alliances and feuds, are common (see Chapters 8 and 14). Chimpanzees, in particular, show a number of characteristics that may have pre-adapted them for the occupation of seasonal woodland and savannah ecosystems. They occur in a wide range of habitats and are large enough for cooperative defence to deter most potential predators. They are adapted to terrestrial as well as arboreal locomotion, and have large ranges and communication systems adapted to information transfer over relatively long distances. They make extensive use of tools and often hunt smaller animals. And, while they normally forage in small unstable parties, their communities may have provided a basis for the formation of social groups large enough to provide effective defence against terrestrial carnivores.

The occupation of seasonal woodlands and savannah ecosystems by hominins in the early Pleistocene was probably associated with increased reliance on animal foods, increased range size, greater cohesion of groups to provide effective defence against predators, improved specificity of communication and increasing reliance on tools for extractive purposes. The regular construction of stone tools, which had developed by 3.3 Mya (see Figure 19.5) preceded by the regular use of tools of wood or bone. Between the origin of lithic technology in the early Pleistocene and the evolution of modern humans in Africa around 200,000 years ago, hominins showed a wide range of changes in diet, anatomy, behaviour, social organisation and physical culture (see section 19.2). These changes were probably asynchronous and may have occurred in a piecemeal fashion: some traits may have originated and spread rapidly throughout the whole human population but others may have appeared, died out and subsequently either reappeared or been permanently lost (d'Errico and Stringer 2011).

Although the relative timing of their evolution is still uncertain, three related developments are likely to have been of particular importance in the evolution of human societies. First, the gradual development of language improved and extended the specificity of communication and social learning and permitted the proliferation of reciprocal assistance and of exchanges and social contracts between group members (see section 19.2). These probably encouraged cooperation between group members of both sexes in protecting and provisioning dependent juveniles, allowing women to reduce the duration of inter-birth intervals. Second, the recognition of social norms and primitive social contracts may have helped to stabilise and extend cooperative relationships through third-party disapproval and punishment, increasing the effectiveness of cultural group selection and leading to increased complexity in cooperative relationships and, eventually, to functional specialisation within communities (Boyd and Richerson 1992; Gintis et al. 2003). And, third, increased reliance on tools combined with social learning is likely to have led to rapid improvements in technology as a result of cumulative experience and adaption as well as to interactions between cultural and genetic evolution (gene–culture coevolution) involving the modification of genetic evolution by cultural processes and modification of cultural evolution by genetic changes (Feldman and Cavalli-Sforza 1985; Boyd and Richerson 1989; Laland et al. 1995; Odling-Smee 1995; Gintis 2011). The clearest examples of these interactions are relatively specific traits, like the coevolution of agriculture and lactase persistence across Europe (Itan et al. 2009; Mace 2014) but similar processes may have played a crucial role in the evolution of many other human characteristics, including the evolution of language, cognitive skills neuronal mechanisms and brain development (Richerson and Boyd, 2005; Platt et al. 2016).

Following the evolution of modern humans with a sophisticated tool kit, advanced linguistic skills and sophisticated social organisation involving extensive cooperation between related and unrelated individuals, the basis for the rapid development of human culture and technology was established and the influence of both cultural group selection and gene–culture coevolution is likely to have increased (Richerson and Boyd 2005; Whiten et al. 2011). With the benefit of hindsight, many subsequent developments (including domestication and agriculture, the use of metal, the extraction and use of fossil fuels, the evolution of scientific enquiry and knowledge, the development of medicine and the construction of cities) can be seen as predictable consequences of this process. So, too, can the by-products of these developments, including the rise in human numbers, the evolution of industrialised warfare, improved living standards, increased per-capita use of resources and the progressive impact of human populations on other animals and on their environment (see section 20.6).

SUMMARY

1. While there are similarities between the social behaviour of humans and the other apes, the evolution of language and technology and the development of ethical systems, complex forms of social organisation and a sophisticated understanding of cause and effect create fundamental differences between humans and other animals that have a profound influence on evolutionary and ecological processes.

2. The first bipedal hominins appeared at least 4.2 Mya in Africa and inhabited relatively mesic wooded environments. The manufacture of primitive stone tools dates from at least 3.3 Mya. The first *Homo* species have been dated to around 2 Mya and were associated with drier, more open habitats, improved adaptations for walking and running, and primitive stone tools. Between 1.6 and 1.0 Mya, and probably again between 0.8 and 0.6 Mya, populations of *Homo* radiated from Africa into Europe and Asia. Modern humans (*Homo sapiens*) were present in southern Africa by 195,000 years ago and dispersed out of Africa around 70,000 years ago.

3. The life-history patterns of modern humans resemble those of chimpanzees, though women have longer lifespans, shorter inter-birth intervals and longer post-reproductive lifespans than female chimpanzees and both sexes are dependent on their parents and other group members for a much longer period.

4. Humans show many of the anatomical and physiological characteristics of a mildly polygynous primate, including sex differences in growth, body size, canine size and longevity while their reproductive anatomy and physiology suggest a relatively low incidence of sperm competition.

5. Both among hunter-gatherers and among pastoral and agricultural peoples, mating systems are diverse and variable, though the most common social unit is a monogamous pair with their dependent children and older relatives. Both sexes may change mates several times in the course of their reproductive careers. In most hunter-gatherers, several families form cohesive bands with traditional ranges and resources and a relatively egalitarian structure.

6. After the development of domestication and agriculture, progressive increases occurred in the size and structuring of human groups as well as in local population density and overall human numbers.

7. Many of the characteristics of higher primates may have facilitated the evolution of the unusual characteristics of humans, including their reliance on diverse diets requiring manual dexterity to extract resources, the production of single young with extended periods of development and dependence, and the formation of stable groups of breeding females, where average relatedness is low and competition between group members to raise young favours the evolution of extended maternal care and support and the evolution of complex social relationships and alliances.

References

Alberts, S.C., *et al.* (2013) Reproductive aging patterns in primates reveal that humans are distinct. *Proceedings of the National Academy of Sciences of the United States of America* **110**:13440–13445.

Alvarez, H.P. (2004) Residence groups among hunter-gatherers: a view of the claims and evidence for patrilocal bands. In: *Kinship and Behavior in Primates* (eds B. Chapais and C. Berman). Oxford: Oxford University Press, 420–441.

Anderson, K.G. (2006) How well does paternity confidence match actual paternity? *Current Anthropology* **47**:513–520.

Asfaw, B., *et al.* (1999) *Australopithecus garhi*: a new species of early hominid from Ethiopia. *Science* **284**:629–635.

Baker, R.R. and Bellis, M.A. (1993) Human sperm competition: ejaculate adjustment by females and a function for the female orgasm. *Animal Behaviour* **46**:887–909.

Baker, R.R. and Bellis, M.A. (1995) *Human Sperm Competition: Copulation, Competition and Infidelity*. London: Chapman & Hall.

Bamshad, M., *et al.* (2001) Genetic evidence on the origins of Indian caste populations. *Genome Research* **11**:994–1004.

Beckerman, S., *et al.* (1998) The Bari Partible Paternity Project: preliminary results. *Current Anthropology* **39**:164–167.

Berenbaum, S. and Hines, M. (1992) Early androgens are related to childhood sex-typed toy preferences. *Psychological Science* **3**:203–206.

Berenbaum, S. and Snyder, E. (1995) Early hormonal influences on childhood sex-typed toy preferences: implications for the development of sexual orientation. *Developmental Psychology* **31**:31–42.

Betzig, L. (2012) Means, variances, and ranges in reproductive success: comparative evidence. *Evolution and Human Behavior* **33**:309–317.

Blurton Jones, N.G. (1986) Bushman birth spacing: a test for optimal interbirth intervals. *Ethology and Sociobiology* **7**:91–105.

Blurton Jones, N.G., *et al.* (1997) Why do Hadza children forage? In: *Uniting Psychology and Biology: Integrative Perspectives on Human Development* (eds N. Segal, G. Weisfeld and C. Weisfeld). Washington, DC: American Psychological Association, chapter 11.

Bobe, R. and Behrensmeyer, A.K. (2004) The expansion of grassland ecosystems in Africa in relation to mammalian evolution and the origin of the genus *Homo. Palaeogeography, Palaeoclimatology, Palaeoecology* **207**:399–420.

Bogin, B. (1999) *Patterns of Human Growth.* Cambridge: Cambridge University Press.

Borgerhoff Mulder, M. (1988) Reproductive success in three Kipsigis cohorts. In: *Reproductive Success: Studies of Individual Variation in Contrasting Breeding Systems* (ed. T.H. Clutton-Brock). Chicago: University of Chicago Press, 419–435.

Borgerhoff Mulder, M. (1990) Kipsigis women's preferences for wealthy men: evidence for female choice in mammals. *Behavioral Ecology and Sociobiology* **27**:255–264.

Borgerhoff Mulder, M. and Caro, T. (1983) Polygyny: definition and application to human data. *Animal Behaviour* **31**:609–610.

Borrero, L.A. and Barberena, R. (2006) Hunter-gatherer home ranges and marine resources. *Current Anthropology* **47**:855–868.

Boyd, R. and Richerson, P.J. (1989) The evolution of indirect reciprocity. *Social Networks* **11**:213–236.

Boyd, R. and Richerson, P.J. (1992) Punishment allows the evolution of cooperation (or anything else) in sizable groups. *Ethology and Sociobiology* **13**:171–195.

Brown, G. R., *et al.* (2009) Bateman's principles and human sex roles. *Trends in Ecology and Evolution* **24**:297–304.

Burger, O., *et al.* (2010) Lifetime reproductive effort in humans. *Proceedings of the Royal Society of London. Series B: Biological Sciences* **277**:773–777.

Burkart, J.M. and van Schaik, C.P. (2010) Cognitive consequences of cooperative breeding in primates? *Animal Cognition* **13**:1–19.

Burkart, J.M., *et al.* (2014) The evolutionary origin of human hyper-cooperation. *Nature Communications* **5**:4747.

Burley, N. (1979) The evolution of concealed ovulation. *American Naturalist* **114**:835–858.

Buss, D.M. (2011) *Evolutionary Psychology: The New Science of the Mind.* Boston: Allyn and Bacon.

Campbell, A. (1995) A few good men: evolutionary psychology and female adolescent aggression. *Ethology and Sociobiology* **16**:99–123.

Campbell, A. (1999) Staying alive: evolution, culture, and women's intrasexual aggression. *Behavioral and Brain Sciences* **22**:203–214.

Cant, J.G.H. (1981) Hypothesis for the evolution of human breasts and buttocks. *American Naturalist* **117**:199–204.

Carmody, R.N. and Wrangham, R.W. (2009) The energetic significance of cooking. *Journal of Human Evolution* **57**:379–391.

Cashdan, E., *et al.* (1983) Territoriality among human foragers: ecological models and an application to four Bushman groups. *Current Anthropology* **24**:47–66.

Chagnon, N.A. (2013) *Noble Savages. My Life Among Two Dangerous Tribes: The Yanomano and the Anthropologists.* New York: Simon & Schuster.

Chapais, B. (2008) *Primeval Kinship. How Pair-Bonding Gave Birth to Human Society.* Cambridge, MA: Harvard University Press.

Chapais, B. (2010) The deep structure of human society: primate origins and evolution. In: *Mind the Gap: Tracing the Origin of Human Universals* (eds P.M. Kappeler and J.B. Silk). Berlin: Springer, 19–51.

Chapais, B. (2013) Monogamy, strongly bonded groups, and the evolution of human social structure. *Evolution Anthropology* **22**:52–65.

Chen, Z. and Siegler, R. (2000) Across the great divide: bridging the gap between understanding toddlers' and older childrens' thinking. *Monographs of the Society for Research in Child Development* **65**:1–96.

Cheney, D.L. and Seyfarth, R.M. (2005) Constraints and pre-adaptations in the earliest stages of language evolution. *The Linguistic Review* **22**:135–159.

Cheney, D.L. and Seyfarth, R.M. (2010) Primate communication and human language: continuities and discontinuities. In: *Mind the Gap: Tracing the Origin of Human Universals* (eds P.M. Kappeler and J.B. Silk). Berlin: Springer, 283–298.

Chomsky, N. (1957) *Syntactic Structures.* The Hague: Mouton.

Chomsky, N. (1968) *Language and Mind.* New York: Harcourt Brace and Jovanovich.

Christiansen, M.H. and Kirby, S. (2003) Language evolution: consensus and controversies. *Trends in Cognitive Sciences* **7**:300–307.

Clarys, J.P., *et al.* (1984) Gross tissue weights in the human body by cadavar dissection. *Human Biology* **56**:459–473.

Clutton-Brock, T.H. and Isvaran, K. (2007) Sex differences in ageing in natural populations of vertebrates. *Proceedings of the Royal Society of London. Series B: Biological Sciences* **274**:3097–3104.

Coales, A.J. and Banister, J. (1994) Five decades of missing females in China. *Demography* **31**:459–479.

Cohen, D. and Nisbett, R. (1994) Self-protection and the culture of honor: explaining southern violence. *Personality and Social Psychology Bulletin* **20**:551–567.

Cohen, D., *et al.* (1996) Insult, aggression, and the southern culture of honor: an 'experimental ethnography'. *International Relations and Group Processes* **70**:945–960.

Collard, M. (2002) Grades and transitions in human evolution. In: *The Speciation of Modern* Homo sapiens (ed. T.J. Crow). Oxford: Oxford University Press, 61–102.

Colom, R., *et al.* (2000) Negligible sex differences in general intelligence. *Intelligence* **28**:57–68.

Connellan, J., *et al.* (2000) Sex differences in human neonatal social perception. *Infant Behavior and Development* **23**:113–118.

Corballis, M.C. (2009) The evolution of language. *Annals of the New York Academy of Science* **1156**:19–43.

Dabbs, J.M. and Hargrove, M.F. (1997) Age, testosterone, and behavior among female prison inmates. *Psychosomatic Medicine* **59**:477–480.

Daly, M. and Wilson, M. (1997) Crime and conflict: homicide in evolutionary psychological perspective. *Crime and Justice* **22**:51–100.

Darwin, C. (1871) *The Descent of Man, and Selection in Relation to Sex.* London: John Murray.

Davis, D.L., *et al.* (1998) Reduced ratio of male to female births in several industrial countries: a sentinel health indicator? *JAMA* **279**:1018–1023.

de Heinzelin, J., *et al.* (1999) Environment and behavior of 2.5-million-year-old bouri hominids. *Science* **284**:625–629.

de la Torre, I. (2011) The origins of stone tool technology in Africa: a historical perspective. *Philosophical Transactions of the Royal Society B: Biological Sciences* **366**:1028–1037.

d'Errico, F. and Stringer, C.B. (2011) Evolution, revolution or saltation scenario for the emergence of modern cultures? *Philosophical Transactions of the Royal Society B: Biological Sciences* **366**:1060–1069.

Diamond, J. (1997) *Guns, Germs and Steel: The Fates of Human Societies.* New York: Norton & Co.

Diamond, J. (2012) *The World Until Yesterday: What Can We Learn from Traditional Societies?* New York: Viking Penguin.

Dixson, A.F. (2009) *Sexual Selection and the Origins of Human Mating Systems.* Oxford: Oxford University Press.

Dunbar, R.I.M. (1998) *Grooming, Gossip and the Evolution of Language.* Cambridge, MA: Harvard University Press.

Dupanloup, I., *et al.* (2003) A recent shift from polygyny to monogamy in humans is suggested by the analysis of worldwide Y-chromosome diversity. *Journal of Molecular Evolution* **57**:85–97.

Durham, W.H. (1976) Resource competition and human aggression, part I: a review of primitive war. *Quarterly Review of Biology* **51**:385–415.

Dyble, M *et al.* (2015) Sex equality can explain the unique social structure of hunter-gatherer bands. *Science* **348**:796–798.

Eisenberg, J.F. (1981) *The Mammalian Radiations: An Analysis of Trends in Evolution, Adaptation, and Behavior.* Chicago: University of Chicago Press.

Emery Thompson, M., *et al.* (2007) Aging and fertility patterns in wild chimpanzees provide insights into the evolution of menopause. *Current Biology* **17**:2150–2156.

Emery Thompson, M., *et al.* (2012) The energetics of lactation and the return to fecundity in wild chimpanzees. *Behavioral Ecology* **23**:1234–1241.

Evans, V. (2014) *The Language Myth: Why Language is not an Instinct.* Cambridge: Cambridge University Press.

Fairweather, H. (1976) Sex differences in cognition. *Cognition* **4**:231–280.

Fenner, J.N. (2005) Cross-cultural estimation of the human generation interval for use in genetics-based population divergence estimates. *American Journal of Physical Anthropology* **128**:415–423.

Feldman, M.W. *et al.* (1985) Gene-culture coevolution: Models for the evolution of altruism with cultural transmission. *Proceedings of the National Academy of Sciences of the United States of America* **82**:5814–5818.

Fitch, W.T. (2010) *The Evolution of Language.* Cambridge: Cambridge University Press.

Flinn, M.V. and Low, B.S. (1986) Resource distribution, social competition and mating patterns in human societies. In: *Ecological Aspects of Social Evolution: Birds and Mammals* (eds D.I. Rubenstein and R.W. Wrangham). Princeton, NJ: Princeton University Press, 217–243.

Foley, R.A. (1987) *Another Unique Species: Patterns in Human Evolutionary Ecology.* London: Longman.

Foley, R.A. (2001) Evolutionary perspectives on the origins of human social institutions. *Proceedings of the British Academy* **110**:171–195.

Foley, R.A. and Gamble, C. (2009) The ecology of social transitions in human evolution. *Philosophical Transactions of the Royal Society B: Biological Sciences* **364**:3267–3279.

Frost, P. (1994) Geographic distribution of human skin colour: a selective compromise between natural and sexual selection? *Human Evolution* **9**:141–153.

Gangestad, S.W. and Thornhill, R. (2008) Human oestrus. *Proceedings of the Royal Society of London. Series B: Biological Sciences* **275**:991–1000.

Geary, D.C. (2003) Sexual selection and human life history. In: *Advances in Child Development and Behavior*, Vol. **30** (ed. R. Kail). San Diego, CA: Academic Press, 41–101.

Geary, D.C. (2010) *Male, Female: The Evolution of Human Sex Differences.* Washington, DC: American Psychological Association.

Geary, D.C., *et al.* (2000) Sexual differences in spatial cognition, computational fluency, and arithmetical reasoning. *Journal of Experimental Child Psychology* **77**:337–353.

Geary, D.C., *et al.* (2014) Reflections on the evolution of human sex differences: social selection and the evolution of competition among women. In: *Evolutionary Perspectives on Human Sexual Psychology and Behavior* (eds V.A. Weekes-Shackelford and T.K. Shackelford). New York: Springer, 393–411.

Ghiglieri, M.P. (1987) Sociobiology of the great apes and the hominid ancestor. *Journal of Human Evolution* **16**:319–357.

Gintis, H. (2011) Gene–culture coevolution and the nature of human sociality. *Philosophical Transactions of the Royal Society B: Biological Sciences* **366**:878–888.

Gintis, H., *et al.* (2003) Explaining altruistic behavior in humans. *Evolution and Human Behavior* **24**:153–172.

Gordon, A.D., *et al.* (2008) Strong postcranial size dimorphism in *Australopithecus afarensis*: results from two new resampling methods for multivariate data sets with missing data. *American Journal of Physical Anthropology* **135**:311–328.

Gould, S.J. (1981) *The Mismeasure of Man.* New York: W.W. Norton and Company.

Gray, P.B. and Garcia, J.R. (2013) *Evolution and Human Sexual Behavior.* Cambridge, MA: Harvard University Press.

Grayson, D.K. (1990) Donner Party deaths: a demographic assessment. *Journal of Anthropological Research* **46**:223–242.

Grayson, D.K. (1993) Differential mortality and the Donner Party disaster. *Evolutionary Anthropology: Issues, News, and Reviews* **2**:151–159.

Grech, V., *et al.* (2003) Secular trends in sex ratios at birth in North America and Europe over the second half of the 20th century. *Journal of Epidemiology and Community Health* **57**:612–615.

Gurven, M. and Kaplan, H. (2008) Beyond the grandmother hypothesis: evolutionary models of human longevity. In: *Cultural Context of Aging: Worldwide Perspectives*, 3rd edn (ed. J. Sokolovosky). Westport, CT: Praeger, 53–65.

Haddix, K.A. (2001) Leaving your wife and your brothers: when polyandrous marriages fall apart. *Evolution and Human Behavior* **22**:47–60.

Halileh, S., *et al.* (2008) Determinants of birthweight: gender based analysis. *Maternal and Child Health Journal* **12**:606–612.

Hamilton, M.J., *et al.* (2007a) The complex structure of hunter-gatherer social networks. *Proceedings of the Royal Society of London. Series B: Biological Sciences* **274**:2195–2203.

Hamilton, M.J., *et al.* (2007b) Nonlinear scaling of space use in human hunter-gatherers. *Proceedings of the National Academy of Sciences of the United States of America* **104**:4765–4769.

Harcourt, A.H., *et al.* (1981) Testis weight, body weight and breeding system in primates. *Nature* **293**:55–57.

Harmand, S., *et al.* (2015) 3.3-million-year-old stone tools from Lomekwi 3, West Turkana, Kenya. *Nature* **521**:310–315.

Hawkes, K. (2003) Grandmothers and the evolution of human longevity. *American Journal of Human Biology* **15**:380–400.

Hawkes, K. (2004) Human longevity: the grandmother effect. *Nature* **428**:128–129.

Hess, N.H., *et al.* (2010) Interpersonal aggression among Aka hunter-gatherers of the Central African Republic: assessing the effects of sex, strength and anger. *Human Nature* **21**:330–354.

Hewlett, B.S. (1991) Demography and childcare in preindustrial societies. *Journal of Anthropological Research* **47**:1–37.

Hill, K.R. and Hurtado, A.M. (1996) *Aché Life History: The Ecology and Demography of a Foraging People*. New York: de Gruyter.

Hill, K.R., *et al.* (2011) Co-residence patterns in hunter-gatherer societies show unique human social structure. *Science* **331**:1286–1289.

Houghton, P. (1996) *People of the Great Ocean: Aspects of the Biology of the Early Pacific*. Cambridge: Cambridge University Press.

Hrdy, S.B. (1981) *The Woman That Never Evolved*. Cambridge, MA: Harvard University Press.

Hrdy, S.B. (1999) *Mother Nature: A History of Mothers, Infants, and Natural Selection*. New York: Random House.

Hrdy, S.B. (2000) The optimal number of fathers: evolution, demography, and history in the shaping of female mate preferences. *Annals of the New York Academy of Sciences* **907**:75–96.

Hrdy, S.B. (2009) *Mothers and Others: The Evolutionary Origins of Mutual Understanding*. Cambridge, MA: Harvard University Press.

Itan, Y., *et al.* (2009) The origins of lactase persistence in Europe. *PLOS Computational Biology* **5**(8): e1000491.

Jacobs, L.F., *et al.* (1990) Evolution of spatial cognition: sex-specific patterns of behavior predict hippocampal size. *Proceedings of the National Academy of Sciences of the United States of America* **87**:6349–6352.

James, S.R., *et al.* (1989) Hominid use of fire in the Lower and Middle Pleistocene: a review of the evidence. *Current Anthropology* **30**:1–26.

Janssen, M.A. and Hill, K. (2014) Benefits of grouping and cooperative hunting among Ache hunter-gatherers: insights from an agent-based foraging model. *Human Ecology* **42**:823–825.

Jasieńska, G., *et al.* (2004) Large breasts and narrow waists indicate high reproductive potential in women. *Proceedings of the Royal Society of London Series B: Biological Sciences* **271**:1213–1217.

Johnson, A.W. and Earle, T. (2000) *The Evolution of Human Societies: From Foraging Band to Agrarian State*. Stanford, CA: Standford University Press.

Kaplan, H. (1996) A theory of fertility and parental investment in traditional and modern societies. *Yearbook of Physical Anthropology* **39**:91–135.

Kaplan, H. and Gurven, M. (2005) The natural history of human food sharing and cooperation: a review and a new multi-individual approach to the negotiation of norms. In: *Moral Sentiments and Material Interests: The Foundations of Cooperation in Economic Life* (eds H. Gintis, S. Bowles, R. Boyd and E. Fehrs). Cambridge, MA: MIT Press, 75–113.

Kaplan, H., *et al.* (2000) A theory of human life history evolution: diet, intelligence, and longevity. *Evolutionary Anthropology* **9**:156–185.

Kaplan, H.S., *et al.* (2009) The evolutionary and ecological roots of human social organization. *Philosophical Transactions of the Royal Society B: Biological Sciences* **364**:3289–3300.

Kappeler, P. (2011) Our origins: how and why we do and do not differ from primates. In: *Essential Building Blocks of Human Nature* (eds U. Frey, C. Störmer and K. Wilführ). Berlin: Springer, 5–15.

Keeley, L.H. (1996) *War Before Civilization*. New York: Oxford University Press.

Kelly, R.C. (2005) The evolution of lethal intergroup violence. *Proceedings of the National Academy of Sciences of the United States of America* **102**:15294–15298.

Kelly, R.L. (1983) Hunter-gatherer mobility strategies. *Journal of Anthropological Research* **39**:277–306.

Knickmeyer, R., *et al.* (2005) Foetal testosterone, social relationships, and restricted interests in children. *Journal of Child Psychology and Psychiatry* **46**:198–210.

Kramer, K.L. and Russell, A.F. (2014) Kin-selected cooperation without lifetime monogamy: human insights and animal implications. *Trends in Ecology and Evolution* **29**:600–606.

Kuper, A. (2009) *Incest and Influence: The Private Life of Bourgeois England*. Cambridge, MA: Harvard University Press.

Laland, K.N., *et al.* (1995) Gene–culture coevolutionary theory: a test case. *Current Anthropology* **36**:131–156.

Langergraber, K.E., *et al.* (2007) The genetic signature of sex-biased migration in patrilocal chimpanzees and humans. *PLOS ONE* **2**(10): e973.

Law Smith, M.J.L., *et al.* (2006) Facial appearance is a cue to oestrogen levels in women. *Proceedings of the Royal Society of London. Series B: Biological Sciences* **273**:135–140.

Lawson, D.W., *et al.* (2015) No evidence that polygynous marriage is a harmful cultural practice in northern Tanzania. *Proceedings of the National Academy of Sciences of the United States of America* **112**:13827–13832.

Layton, R.H. (1986) Political and territorial structures among hunter-gatherers. *Man* **21**:18–33.

Lee, R.B. (1972) The !Kung bushmen of Botswanna. In: *Hunters and Gatherers Today* (ed. M.G. Bicchieri). New York: Holt, Rinehart and Winston.

Lee, R.B. and Devore, I. (1968) *Man the Hunter*. Chicago: Aldine.

Leenaars, L.S., *et al.* (2008) Evolutionary perspective on indirect victimization in adolescence: the role of attractiveness, dating and sexual behviaor. *Aggressive Behavior* **34**:404–415.

Levine, N.E. (1988) *The Dynamics of Polyandry: Kinship, Domesticity, and Population on the Tibetan Border*. Chicago: University of Chicago Press.

Levine, N.E. and Silk, J.B. (1997) Why polyandry fails: sources of instability in polyandrous marriages. *Current Anthropology* **38**:375–398.

Lévi-Strauss, C. (1969) *The Elementary Structures of Kinship*. Boston: Beacon Press. Translation of *Les Structures Elementaires de la Parente*, 1967.

Linn, M.C. and Petersen, A.C. (1985) Emergence and characterization of sex differences in spatial ability: a meta-analysis. *Child Development* **56**:1479–1498.

Liu, W., *et al.* (2015) The earliest unequivocally modern humans in southern China. *Nature* **526**:696–699.

Lloyd, E.A. (2005) *The Case of the Female Orgasm: Bias in the Science of Evolution*. Cambridge, MA: Harvard University Press.

Lonsdorf, E.V., *et al.* (2014) Boys will be boys: sex differences in wild infant chimpanzee social interactions. *Animal Behaviour* **88**:79–83.

Low, B.S. (2007) Ecological and socio-cultural impacts on mating and marriage systems. In: *The Oxford Handbook of Evolutionary Psychology* (eds R.I.M. Dunbar and L. Barrett). Oxford: Oxford University Press, 449–461.

Low, B.S., *et al.* (1987) Human hips, breasts and buttocks: is fat deceptive? *Ethology and Sociobiology* **8**:249–257.

Lumey, L.H. and van Poppel, F.W.A. (1994) The Dutch famine of 1944–45: mortality and morbidity in past and present generations. *Social History of Medicine* **7**:229–246.

Lummaa, V. and Clutton-Brock, T.H. (2002) Early development, survival and reproduction in humans. *Trends in Ecology and Evolution* **17**:141–147.

Lummaa, V., *et al.* (2001) Gender difference in benefits of twinning in pre-industrial humans: boys did not pay. *Journal of Animal Ecology* **70**:739–746.

Mace, R. (1996) Biased parental investment and reproductive success in Gabbra pastoralists. *Behavioral Ecology and Sociobiology* **38**:75–81.

Mace, R. (1998) The co-evolution of human fertility and wealth inheritance strategies. *Philosophical Transactions of the Royal Society B: Biological Sciences* **353**:389–397.

Mace, R. (2013) Cooperation and conflict between women in the family. *Evolutionary Anthropology* **22**:251–258.

Mace, R. (2014) Human behavioural ecology and its evil twin. *Behavioral Ecology* **25**:443–449.

McHenry, H.M. (1994) Behavioral ecological implications of early hominid body size. *Journal of Human Evolution* **27**:77–87.

McIntyre, M.H. and Edwards, C.P. (2009) The early development of gender differences. *Annual Review of Anthropology* **38**:83–97.

Manica, A., *et al.* (2007) The effect of ancient population bottlenecks on human phenotypic variation. *Nature* **448**:346–348.

Marlowe, F.W. (2000) Paternal investment and the human mating system. *Behavioural Processes* **51**:45–61.

Marlowe, F.W. (2003a) A critical period for provisioning by Hadza men: implications for pair bonding. *Evolution and Human Behavior* **24**:217–229.

Marlowe, F.W. (2003b) The mating system of foragers in the standard cross-cultural sample. *Cross-Cultural Research* **37**:282–306.

Marlowe, F.W. (2010) *The Hadza: Hunter-Gatherers of Tanzania*. Berkeley, CA: University of California Press.

Marlowe, F.W. and Wetsman, A. (2001) Preferred waist-to-hip ratio and ecology. *Personality and Individual Differences* **30**:481–489.

Martin, R.D. *et al.* (2006) Flores hominid: new species or microcephalic dwarf? *Anatomical Record Part A: Discoveries in Molecular, Cellular, and Evolutionary Biology* **288A**:1123–1145.

Maynard Smith, J. and Szathmáry, E. (1995) *The Major Transitions in Evolution*. Oxford: Oxford University Press.

Mazur, A. and Booth, A. (1998) Testosterone and dominance in men. *Behavioral and Brain Sciences* **21**:353–363.

Mesnick, S.L. (1997) Sexual alliances: evidence and evolutionary implications. In: *Feminism and Evolutionary Biology* (ed. P.A. Gowaty). New York: Chapman & Hall, 207–259.

Morris, D. (1967) *The Naked Ape. A Zoologist's Study of the Human Animal*. London: Jonathan Cape.

Munro, J. (2009) 'More like a work of art than of nature'. Darwin, beauty and sexual selection. In: *Endless Forms. Charles Darwin, Natural Sciences and the Visual Arts* (eds D. Donald and J. Munro). New Haven, CT: Yale University Press.

Nieschlag, E. (1979) The male climacteric. In: *Female and Male Climacteric* (eds P.A. Van Keep, D.M. Serr and R.B. Greenblatt). Lancaster: MTP Press.

Norenzayan, A. (2013) *Big Gods: How Religion Transformed Cooperation and Conflict*. Princeton, NJ: Princeton University Press.

Nowak, M.A. and Krakauer, D.C. (1999) The evolution of language. *Proceedings of the National Academy of Sciences of the United States of America* **96**:8028–8033.

Oakley, K.P. (1957) *Man the Toolmaker*. London: British Muesum.

O'Connell, J.F., *et al.* (1999) Grandmothering and the evolution of *Homo erectus*. *Journal of Human Evolution* **36**:461–485.

Odling-Smee, F.J. (1995) Niche construction, genetic evolution and cultural change. *Behavioural Processes* **35**:195–205.

Pennington, R. and Harpending, H. (1988) Fitness and fertility among Kalahari !Kung. *American Journal of Physical Anthropology* **77**:303–319.

Pennisi, E. (1999) Did cooked tubers spur the evolution of big brains. *Science* **283**:2004–2005.

Peoples, H. and Marlowe, F.W. (2012) Subsistence and the evolution of religion. *Human Nature* **23**:253–269.

Pinker, S. (1994) *The Language Instinct: The New Science of Language and Mind*. London: Penguin Books.

Pisanski, K., *et al.* (2014) Return to Oz: voice pitch facilitates assessments of men's body size. *Journal of Experimental Psychology: Human Perception and Performance* **40**:1316–1331.

Plavcan, J.M. (2012) Body size, size variation, and sexual dimorphism in early *Homo*. *Current Anthropology* **53**:S409–S423.

Platt, *et al.* (2016) Adaptations for social cognition in the primate brain. *Philosophical Transactions of the Royal Society B: Biological Sciences.* 371. DOI:10.1098/rstb.2015.0096 [Accessed 26 January 2016]

Powers, S.T. *et al.* (2016) How institutions shaped the last major evolutionary transition to large-scale human societies. *Philosophical Transactions of the Royal Society B: Biological Sciences.* 371. DOI:10.1098/rstb.2015.0098. [Accessed 26 January 2016]

Pond, C.M. (1978) Morphological aspects and the ecological and mechanical consequences of fat deposition in wild vertebrates. *Annual Review of Ecology and Systematics* **9**:519–570.

Ratcliffe, A.A., *et al.* (2000) Separate lives, different interests: male and female reproduction in the Gambia. *Bulletin of the World Health Organization* **78**:570–579.

Reader, S.M., *et al.* (2011) The evolution of primate general and cultural intelligence. *Philosophical Transactions of the Royal Society B: Biological Sciences* **366**:1017–1027.

Reed, K.E. (1997) Early hominid evolution and ecological change through the African Plio-Pleistocene. *Journal of Human Evolution* **32**:289–322.

Reno, P.L., *et al.* (2010) An enlarged postcranial sample confirms *Australopithecus afarensis* dimorphism was similar to modern humans. *Philosophical Transactions of the Royal Society B: Biological Sciences* **365**:3355–3363.

Richerson, P.J. and Boyd, R. (2005) *Not by Genes Alone: How Culture Transformed Human Evolution*. Chicago: University of Chicago Press.

Richerson, P.J. and Boyd, R. (2010) Why possibly language evolved. *Biolinguistics* **4**:289–306.

Rickard, I.J., *et al.* (2007) Producing sons reduces lifetime reproductive success of subsequent offspring in pre-industrial Finns. *Proceedings of the Royal Society of London. Series B: Biological Sciences* **274**:2981–2988.

Roebroeks, W. and Villa, P. (2011) On the earliest evidence for habitual use of fire in Europe. *Proceedings of the National Academy of Sciences of the United States of America* **108**:5209–5214.

Scelza, B.A. (2013) Choosy but not chaste: multiple mating in human females. *Evolutionary Anthropology* **22**:259–269.

Schacht, R. and Borgerhoff Mulder, M. (2015) Sex ratio effects on reproductive strategies in humans. *Royal Society Open Science* **2**:140402.

Seabright, P. (2012) *The War of the Sexes: How Conflict and Cooperation Have Shaped Men and Women from Prehistory to the Present*. Princeton, NJ: Princeton University Press.

Sealy, J., *et al.* (2006) Diet, mobility, and settlement pattern among Holocene hunter-gatherers in southernmost Africa. *Current Anthropology* **47**:569–595.

Service, E.R. (1962) *Primitive Social Organisation: An Evolutionary Perspective*. New York: Random House.

Shenk, M.K., *et al.* (2014) The evolutionary demography of sex ratios in rural Bangladesh. In: *Applied Evolutionary Anthropology: Darwinian Approaches to Contemporary World Issues* (eds M.A. Gibson and D.W. Lawson). New York: Springer, 141–173.

Silk, J. and Boyd, R. (2006) *How Humans Evolved*. New York: W.W. Norton & Company.

Silk, J.B. and House, B.R. (2016) The evolution of altruistic social preferences in human groups. *Philosophical Transactions of the Royal Society B: Biological Sciences.* 371 DOI:10.1098/rstb.2015.0097. [Accessed 26 January 2016]

Simmons, L.W., *et al.* (2004) Human sperm competition: testis size, sperm production and rates of extrapair copulations. *Animal Behaviour* **68**:297–302.

Stahl, A.B., *et al.* (1984) Hominid dietary selection before fire. *Current Anthropology* **25**:151–168.

Stanley, S.M. (1992) An ecological theory for the origin of *Homo*. *Paleobiology* **18**:237–257.

Starkweather, K.E. and Hames, R. (2012) A survey of non-classical polyandry. *Human Nature* **23**:149–172.

Stout, D. (2011) Stone toolmaking and the evolution of human culture and cognition. *Philosophical Transactions of the Royal Society B: Biological Sciences* **366**:1050–1059.

Strassmann, B. (1981) Sexual selection, paternal care and concealed ovulation in humans. *Ethology and Sociobiology* **2**:31–40.

Stringer, C. and Andrews, P. (2005) *The Complete World of Human Evolution*. London: Thames and Hudson.

Stumpf, R. (2007) Chimpanzees and bonobos: diversity within and between species. In: *Primates in Perspective* (eds C.J. Campbell, A. Fuentes, K.C. MacKinnon, N. Panger and S.K. Bearder). Oxford: Oxford University Press, 321–344.

Symons, D. (1979) *The Evolution of Human Sexuality*. Oxford: Oxford University Press.

Tanner, J.M. (1990) *Fetus Into Man: Physical Growth from Conception to Maturity*. Cambridge, MA: Harvard University Press.

Titze, I.R. (2000) *Principles of Voice Production*. Iowa City, IA: National Center for Voice and Speech.

van Noordwijk, M.A., *et al.* (2013) The evolution of the patterning of human lactation: a comparative persective. *Evolutionary Anthropology* **22**:202–212.

van Schaik, C. and Burkart, J.M. (2010) Mind the gap: cooperative breeding and the evolution of our unique features. In: *Mind The Gap: Tracing the Origins of Human Universals* (eds P.M. Kappeler and J.B. Silk). Berlin: Springer, 477–495.

Washburn, S.L. and Lancaster, C. (1968) Hunting and human evolution. In: *Man the Hunter* (eds R. Lee and I. Devore). Chicago: Aldine.

Webb, J. and Domanski, M. (2009) Fire and stone. *Science* **325**:820–821.

Weiss, E.M. (2003) Sex differences in cognitive functions. *Personality and Individual Differences* **35**:863–875.

Westermarck, E.A. (1889) *The Origin of Human Marriage.* Frenckell.

Whiten, A., *et al.* (2011) Culture evolves. *Philosophical Transactions of the Royal Society B: Biological Sciences* **366**:938–948.

Wilkins, J.F. and Marlowe, F.W. (2006) Sex-biased migration in humans: what should we expect from genetic data? *Bioessays* **28**:290–300.

Willingham, W. and Cole, N. (1997) *Gender and Fair Assessment.* Mahwah, NJ: Erlbaum.

Woodburn, J. (1982) Egalitarian societies. *Man* **17**:431–451.

Wrangham, R.W. (1987) Evolution of social structure. In: *Primate Societies* (eds B.B. Smuts, D.L. Cheney, R.M. Seyfarth, R.W. Wrangham and T.T. Struhsaker). Chicago: University of Chicago Press, 282–296.

Wrangham, R.W. (2009) *Catching Fire: How Cooking Made Us Human.* New York: Basic Books.

Wrangham, R.W. and Carmody, R.N. (2010) Human adaptation to the control of fire. *Evolutionary Anthropology* **19**:187–199.

Wrangham, R.W. and Peterson, D. (1996) *Demonic Males: Apes and the Origins of Human Violence.* New York: Houghton Mifflin Harcourt.

Wroblewski, E.E. (2010) *Paternity and father–offspring relationships in wild chimpanzees,* Pan troglodytes schweinfurthii. PhD thesis, University of Minnesota.

CHAPTER 20

Human behaviour

20.1 Introduction

In the course of their lives, humans have to make many decisions similar to those made by animals, including choosing mates, determining what proportion of their resources to allocate to their offspring and deciding whether and to what extent they should assist other members of their family or their community. However, the processes involved differ in important ways from those operating in other mammals, for the evolution of language and the development of culture, technology and agriculture have had far-reaching consequences for human evolution (see Chapter 19). In particular, the evolution of language makes it possible for individuals to specify their intentions in a way that is impossible in other mammals and, as a result, individuals can make agreements about exchanges of goods and services and establish informal contracts concerning the maintenance of social norms that have no close parallels in other animals, permitting the evolution of extensive cooperation with unrelated unfamiliar individuals. In addition, the development and diversity of human culture has a profound influence on the costs and benefits of human decisions, with the result that variation in human behaviour often has no simple relationship to the environmental factors that predict variation in behaviour among non-human animals. Nevertheless, there are few human decisions that are not affected by ecological (or economic) factors, as well as by biological mechanisms, and there is consequently a need for interpretations of human behaviour that combine cultural, psychological, ecological and evolutionary approaches.

Over the last 40 years, research in human ecology, economics, sociobiology, behavioural ecology, cultural and biological anthropology, and evolutionary and social psychology has done much to explore the common ground between these disciplines and to build an integrated approach to understanding human reproduction and behaviour based on a framework of ecological and evolutionary theory (Hrdy, 1981, 1999; Barrett *et al.* 2002; Mesoudi *et al.* 2006; Borgerhoff Mulder and Schracht 2012; Nettle *et al.* 2013; Mace 2014). However the process is far from complete and there are still voices on both sides that argue against the feasibility (and, in some cases, the desirability) of developing these connections, so that producing a balanced assessment of the evolution of human social behaviour is still fraught with difficulty.

This chapter examines a selection of the reproductive decisions and social behaviour of humans. Section 20.2 describes mating preferences in women and men while section 20.3 reviews studies of maternal and paternal care and their consequences. Subsequently, section 20.4 explores allo-parental care, the role of grandmothers and the evolution of the menopause. Section 20.5 examines cooperation and prosocial behaviour in human societies, focusing mainly on cooperation between individuals in three contexts (defence, food sharing and reproduction). Finally, section 20.6 examines ideas about human prosociality while section 20.7 briefly reviews our understanding of the human condition.

Three important caveats regarding evolutionary studies of human behaviour deserve to be emphasised. First, many studies of human behaviour have involved 'WEIRD' subjects (Western, educated, industrialised, rich and democratic), whereas WEIRD humans represent a small proportion of total human numbers (Henrich *et al.* 2010; Mace 2014). The extent to which the results of studies of WEIRD subjects reflect the situation in human populations where numbers are regulated by natural processes similar to those operating in other mammals is open to question. Second, despite extensive interest in comparing human societies categorised by continent, latitude, mode of subsistence or mating system, detailed data on human life histories and behaviour of the kind that is available for a growing number of other mammals (see Chapter 1) is available for very few human societies. As a result, comparisons are often based either on individual societies or on data derived from databases that are often variable in quality. And, third, as it is frequently difficult to interpret the direction of causation underlying correlations between social or environmental factors and human behaviour since ethical considerations limit experiments involving humans.

20.2 Mate choice

Incest and inbreeding avoidance

One of the most extensively debated aspects of human mate choice concerns the influence of kinship. Although both psychologists and anthropologists initially suggested that humans commonly have a deep-rooted desire to mate with close family members (Freud 1950; Lévi-Strauss 1969), a large body of evidence suggests otherwise and that, like other mammals, most humans are seldom sexually attracted to familiar first-order relatives and avoid breeding with them.

With a very small number of exceptions (Scheidel 1996), mating between first-order relatives (parent and offspring or full siblings) is very rare in human populations and there is not a single documented case of a society where marriages regularly occur between brothers and sisters or in which parents regularly mate with their own children (Sillé and Boyd 2006). In the late nineteenth century the Finnish anthropologist Edward Westermarck (1889) suggested that children of the opposite sex that are reared together seldom find each other attractive as sexual partners and several studies of marital patterns support this suggestion. For example, studies of children reared in communal nurseries in Israeli kibbutz show that marriages between individuals reared in the same communal nursery are uncommon (Shepher 1971; Bevc and Silverman 1993; Shor and Simchai 2009). In China during the last century, some families adopted prospective brides as children and raised them with their future husband, treating them much like sibs. Sexual attraction between couples reared together as children was weak, they produced few children and divorce was common (Wolf 1966, 1970; Lieberman 2009). Similarly, American college students invited to consider consensual mating between adult siblings showed stronger evidence of aversion if they had been co-socialised with one or

more individuals of the opposite sex as children and this effect was stronger in women than men (Lieberman *et al.* 2003; Fessler and Navarrete 2004).

One further line of evidence may suggest that there could be biological adaptations that reduce the risk of mating between first-order relatives. As Chapter 4 describes, in a number of mammals the presence of fathers in the same social group inhibits the sexual development of their daughters while immigration by unrelated males has an opposite effect. In several human societies, the presence of their fathers is associated with delays in the age of menarche in girls, while the absence of fathers and the presence of stepfathers can advance it (Belsky *et al.* 1991; Hulanicka 1999; Quinlan 2003; Bogaert 2008), though this is not the case in all societies, possibly as a result of variation in nutrition (Sheppard *et al.* 2014). Some studies have also shown that the presence of stepfathers is associated with earlier reproductive development after the effects of the father's absence have been allowed for (Quinlan 2003). In addition to earlier ages at menarche, girls whose fathers are absent also show greater attraction to visual stimuli from infants and are characterised by an earlier readiness for reproduction (Maestripieri *et al.* 2004). However, avoidance of the risk of inbreeding is not the only possible explanation of these trends and several others have been suggested, including the possibility that advances in age at menarche are a direct consequence of stressful social circumstances (Hulanicka 1999; Tither and Ellis 2008) or of associations between genetic factors affecting the duration of relationships and the age at first breeding (Comings *et al.* 2002; Mendle *et al.* 2006) or that uncertain environments favour reductions in age at first breeding (Ellis and Garber 2000).

While there is suggestive evidence that there may be biological mechanisms affecting sexual attraction that reduce the risk of close inbreeding, beliefs acquired through early social learning also play a role and it is difficult to separate their effects. As a result, there is still disagreement about the relative role of adaptive strategies and cultural beliefs in discouraging breeding between close kin in humans (Leavitt 1990; Moore 1992; Ingham and Spain 2005). It is also important to recognise that evidence suggesting that the avoidance of inbreeding with close relatives has a biological basis does not suggest that all incest taboos do so, and taboos relating to sexual relations between more distant kin vary widely and are commonly affected by economic or political considerations or by symbolical associations. For example, prescriptions concerning marriage between cousins differ between societies and cousin marriage (especially marriages between a parent's offspring and the offspring of its siblings of the opposite sex) is a common practice in some cultures while intermarriage between cousins is forbidden in others (Lévi-Strauss 1969; Chapais 2008). Practices have varied over time and a range of studies have shown that they are influenced by social, economic and political rather than biological considerations: for example, marriages between relatives (including first cousins) are frequently used to maintain wealth in the family or to avoid sharing liability with unrelated individuals (Kuper 2009). One case where genetic considerations can be discounted involves taboos on marriages between widowers or widows and the relatives of their dead partners, which vary widely and are closely linked to religious beliefs and practices (Box 20.1).

Female preferences

Evolutionary theory suggests that women should often be expected to be more choosey of their mating partners than men, though the extent (and in some cases the direction) of sex differences in choosiness varies (see Chapter 14). There is evidence from Western societies that the physical characteristics of men affect their desirability as partners, though preferences are often context-specific and vary between societies. Preferred traits are often ones likely to be involved in intersexual signalling (Miller 2000) and include relative masculinity, facial attractiveness and facial symmetry (Thornhill and Gangestad 1999; Johnston et al. 2001). Women generally prefer men with relatively masculine faces, characterised by wider jaws, longer chins and narrower (less open) eyes (Figure 20.1). Their preferences for male masculinity vary throughout the menstrual cycle and are greatest at times when conception is most likely (Barrett et al. 2002; Geary 2010).

Facial characteristics are often perceived as indicators of personality and also reflect the priorities of women for partners that are kind, understanding, intelligent, dependable and healthy (Buss 2006; Geary 2010). Vocal characteristics may be important, too: studies of Western men have shown that deeper voices are associated with physical and social dominance in men and are preferred by women (Feinberg et al. 2005; Puts et al. 2006). In addition, vocal attractiveness has been shown to be correlated with visual cues and to be associated with heightened levels of sexual activity (Hughes et al. 2004). The importance of vocal characteristics is not confined to

Box 20.1 Marriage to in-laws

Many societies have prohibitions or prescriptions concerning the marriage of widowers or widows to relatives of their dead partners which have no connection to genetic benefits. For example, in some societies, widowers are expected to marry any available sisters of their dead wife while, in others, such marriages are prescribed as incestuous (Kuper 2009; Chapais 2008). Similarly, in some cultures, men whose brothers die are expected to marry their wives (a practice known as the levirate), while in others this, too, is forbidden. In many Western societies, official attitudes to marriage with in-laws have varied over time, often provoking heated debates.

In England, debates about sororal marriage persisted until the early twentieth century. Leviticus (18, 6–18) prohibits sexual intercourse between a man and a woman previously married to his father, son or brother and the Catholic church adopted (and extended) the prohibition, arguing that, as a consequence of sexual intercourse, a wife became a part of her husband's body, so that subsequent marriages between widowers and sisters of their deceased wives represented unions between close kin and were consequently forbidden (Kuper 2009). After the Reformation, the Anglican church inherited the same doctrine and reform was resisted until the early twentieth century: for example, as late as 1903, Winston Churchill opposed the legalisation of sororal marriage, arguing to the House of Commons that it should maintain the prohibition on the grounds that 'when a man and a woman were married they became as one' so that 'any person the man could not marry by reason of consanguinity to himself, he could also not marry if similarly related to his wife' (Kuper 2009). Sororal marriage was finally legalised in England by the Deceased Wife's Sister's Marriage Act 1907, though it was not until 1921 that a similar law was passed legalising marriages between widows and brothers of their deceased husbands.

While marriages to in-laws are unlikely to generate genetic benefits to offspring, they could have other advantages. For example, they may help to ensure that related dependants will be provided for. In addition, depending on patterns of inheritance, they may help to retain wealth within the family. A likely origin of many similar practices is that social connections between individuals and their in-laws are close, increasing the probability of remarriage, irrespective of any benefits incurred (Kuper 2009).

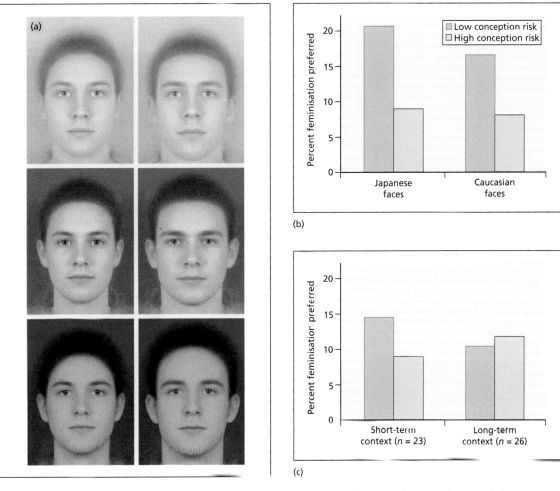

Figure 20.1 Female preferences for more masculine facial characteristics, (a, right) change as a function of menstrual phase at the time of testing (b) and differ between short and long term relationships (c). Composite faces have been 50% feminised (left) and 50% masculinised (right). *Source*: From Penton-Voak *et al*. (1999). Reproduced with permission of Macmillan Publishers Ltd.

Western men: for example, among the Hadza, men with deeper voices show higher reproductive success (Apicella *et al*. 2007).

Other studies have demonstrated preferences for facial and physical symmetry (Figure 20.2) in prospective partners and contrasts in symmetry have been shown to be correlated with testosterone levels during pubertal development as well as in adulthood (Gangestad and Thornill 2003). However, facial symmetry is not consistently correlated with masculinity (Penton-Voak and Perrett 2001) and may reflect differences in developmental stability and immunocompetence which, in turn, may

permit the development of high testosterone levels despite their immunosuppressive effects. Physical symmetry among men may also be associated with olfactory signals used by women in assessing potential partners. For example, Western women asked to rate the attractiveness of the smell of T-shirts worn by male subjects for three consecutive nights under controlled conditions preferred those worn by individuals with high levels of facial symmetry during the most fertile phase of their menstrual cycle and, in a sample of American women, individuals with relatively symmetrical partners reported significantly more copulatory orgasms (Thornhill *et al*.

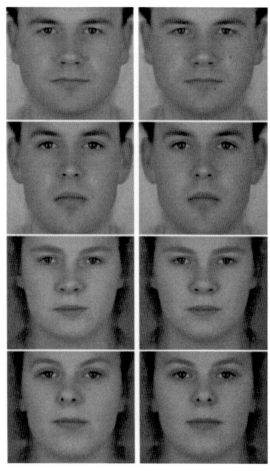

Figure 20.2 Facial symmetry correlates with ratings of attractiveness in prospective partners. Participants were asked to rate the attractiveness of normal images of human faces (left) and more symmetric shaped versions (right). Images were produced by warping and blending 15 faces and their mirror images in the original asymmetric face shapes (left) and the symmetric shapes (right). While facial symmetry was manipulated, average skin texture was kept constant. *Source*: From Perrett *et al.* (1999). Reproduced with permission of Elsevier.

1995). However, it is important to remember that none of these correlations necessarily have a causal basis.

Some evidence also suggests that women may mate disassortatively for MHC genotype, perhaps because MHC heterozygosity is positively correlated with immunocompetence (Thornhill *et al.* 2003). MHC genotype affects body odours and several studies show that women prefer the odours of men with MHC genotypes different

from their own and that, like preferences for facial characteristics, these vary in strength throughout the menstrual cycle and are reduced by taking oral contraceptives (Thornhill *et al.* 1995; Milinski and Wedekind 2001). Similarly, comparisons of human leukocyte antigen haplotypes (which are paternally inherited and affect the ability of women to discriminate and choose odours) in Hutterite communities show that marriage between individuals that share the same HLA haplotype are significantly under-represented (Ober *et al.* 1997; Jacob *et al.* 2002).

An additional factor affecting mate choice in Western women is their perception of their own relative attractiveness and potential as partners (Little *et al.* 2001). A two-part questionnaire completed by a large sample of heterosexual couples invited participants to rate the extent to which they would select partners on four groups of attributes reflecting their wealth and status, family commitment, physical appearance and sexual fidelity and then to rate themselves on each attribute (Buston and Emlen, 2003). The largest amount of variation in the selectivity of mate preferences in each category was explained by the ratings which individuals gave themselves, indicating that individuals of both sexes are attracted to partners whose relative attractiveness resembles their own.

Economic factors also have an important influence on the preferences of women. In many traditional societies, the reproductive success of women increases with the wealth of their partners and wealthy or powerful men are usually preferred as marriage partners. For example, in Kipsigis, the reproductive success of women increases with the wealth of their husbands and polygynous marriages to landowners are frequently preferred to monogamous marriages with poorer men since polygyny does not necessarily compromise the well-being of women or their children (Borgerhoff Mulder 1990; Lawson *et al.* 2015). The partner preferences of Western women are also related both to the earning potential of men and to factors likely to affect the duration of bonds (Pawłowski and Dunbar 1999) and Western women commonly show a preference for partners that have good financial prospects (Figure 20.3). As would be expected, the relevance of economic factors varies between societies and is related to environmental parameters, including population density and resource availability (McGraw 2002; Little *et al.* 2007). Peer opinion and mate choice copying may also be important: for example, Western women

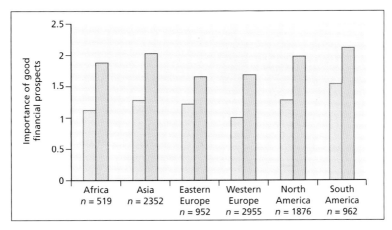

Figure 20.3 The importance of 'good financial prospects' of a prospective marriage partner on a 0 (unimportant) to 3 (indispensible) scale for men (blue bars) and women (pink bars). *Source*: From Geary (2010); Reproduced with permission from the American Psychological Association with data from Buss (1989).

given the choice between a male face paired with a smiling female face and one paired with a neutral female face consistently preferred the one paired with a smiling female face (Jones *et al.* 2007) and witnessing a rival showing interest in a member of the opposite sex causes both men and women to rate them as more desirable partners (Bowers *et al.* 2011).

Male preferences

Like males of other mammals, men commonly favour females that are likely to be fecund (Pawłowski and Dunbar 1999; Geary *et al.* 2004; Mandal 2012). Compared to women, men usually favour younger sexual partners and their preferences are often associated with physical characteristics correlated with youth, health, hormonal status and fecundity, including symmetry, facial neoteny, large breast size, protruding buttocks, low waist/hip ratios and low volume/height ratios (Figure 20.4) (Jasieńska *et al.* 2004, 2006; Furnham and Reeves 2006; Mandal 2012). Although there are arguments about the reliability of different measures as indicators of fecundity as well as about their independence and their relative contributions to attractiveness (Tovée *et al.* 1999; Marlowe *et al.* 2005; Furnham and Reeves 2006; Swami *et al.* 2007), cross-cultural studies have often found similar preferences in different populations. For example, comparisons of Kenyan and British men showed that both groups favoured lighter women and individuals with relatively low waist/hip ratios

(Furnham *et al.* 2003). Similarly, comparisons of preferences in men from Cameroon, Indonesia, Samoa and New Zealand found a consistent tendency for men to rate women with low waist/hip ratios as attractive after effects of variation in body mass index (BMI) had been controlled (Singh *et al.* 2010). Vocal and olfactory cues can also be important (Hughes *et al.* 2004): for example, a study of British men found that higher-frequency female voices were judged to be associated with youth and were rated as more attractive than lower-frequency ones and blind tests showed that ratings of individual differences in the facial attractiveness of women were correlated with their rating of the relative attractiveness of their voices (Collins and Missing 2003).

The preferences of men, like those of women, are also strongly influenced by cultural factors. For example, Hadza men place greater emphasis on women being hard-working and favour women with higher waist-to-hip ratios than men in Western populations, reflecting a preference for heavier women (Wetsman and Marlowe 1999; Marlowe 2004). Similarly, variation in waist-to-hip ratio has a stronger influence on attractiveness ratings given by Spanish and Portuguese men than on ratings given by British men (Swami *et al.* 2007).

Homosexuality

Homosexual preferences are widespread in humans and are not confined to Western societies (Roughgarden 2013). Most surveys suggest that between 10 and 25%

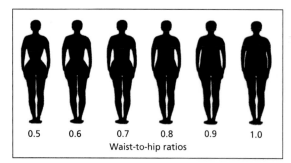

(a)

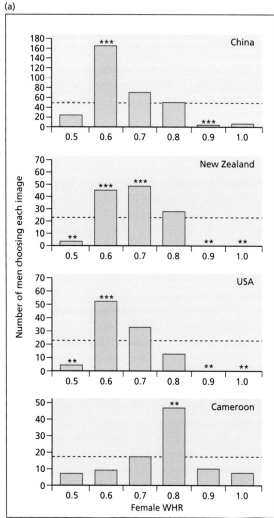

(b)

Figure 20.4 (a) Backposed images of women used in cross-cultural preferences for waist-to-hip ratios. (b) Numbers of men from different cultural groups preferring different levels of waist-to-hip ratio. *Source*: From Dixson (2009); data from Dixson (2007a,b). Reproduced with permission from Oxford University Press.

of adults have had at least one homosexual experience in their lives while 2–6% of adults identify themselves as gay, lesbian or bisexual. There are few reliable figures for other mammals, but these suggest that the incidence of individuals with consistent preferences for sexual interactions with members of the same sex is low in most other mammals for which data are available (see Chapter 15).

Both the mechanisms affecting the development of homosexuality and the evolutionary processes involved are widely debated and contentious (Muscarella *et al.* 2001; Poiani 2010). Homosexual preferences in men are associated with later fraternal birth order, possibly suggesting that a maternal immunisation response affects the sexual development of offspring (Rahman 2005). Genetic factors are also implicated and surveys have identified two regions of linkage with homosexual tendencies (Sanders *et al.* 2014) while other studies suggest that homosexual tendencies may be maternally related (Risch *et al.* 1993; Camperio-Ciani *et al.* 2004).

Several different evolutionary mechanisms that might help to maintain homosexual preferences in humans have been suggested but their relative importance is unknown. For example, homosexual preferences might be maintained through indirect fitness benefits if homosexual individuals assist close kin though there is little empirical evidence that assistance from homosexual individuals has an important influence on the fitness of their relatives (Kirkpatrick 2000). Alternatively, homosexual preferences may be a by-product of selection for tolerance, cooperation or conflict avoidance between members of the same sex (Kirkpatrick 2000). Yet again, they may be a consequence of selection operating on traits in the opposite sex: for example, there is some evidence that the female relatives of homosexual men have higher fecundity than those of heterosexual men (Camperio-Ciani *et al.* 2004, 2009, 2012).

Coercion

Although many of the forms of male coercive behaviour that are common in non-human animals (see Chapter 15) are discouraged or prohibited in many human societies, they continue to occur (Müller and Wrangham 2009), and male mating tactics ranging from harassment and punishment to the direct use of force are not uncommon in humans (Wilson and Daly 2009; Emery Thompson *et al.* 2012). In some cases, these can be associated with

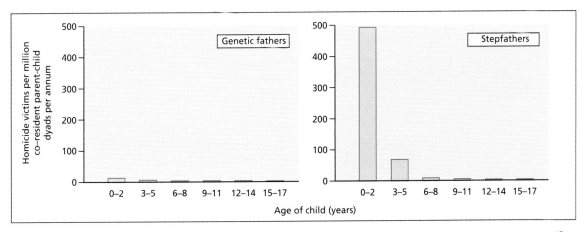

Figure 20.5 Estimated rates of homicide by genetic fathers versus stepfathers in Canada, 1974–1990. Denominators are age-specific estimates of the numbers of Canadian children residing with each type of father, based on census information on the numbers of Canadian children in each age class in each year and age-specific proportions living with genetic fathers versus stepfathers averaged across two national surveys, conducted in 1984. Homicide rates for stepfathers are probably underestimates. *Source*: From Daly and Wilson (1996). Reproduced with permission of Sage Publications.

homicidal attacks by men on rivals or on women that do not comply with their wishes (Chagnon 1988; Daly and Wilson 1997; Macfarlan *et al.* 2014).

Male infanticide also occurs and, with some exceptions (Temrin *et al.* 2004), stepchildren are particularly vulnerable (Daly and Wilson 1988; Harris *et al.* 2007). For example, comparison of age-specific rates of homicide of Canadian children by genetic fathers versus stepfathers shows a substantial increase in the risks that children under the age of five face from stepfathers (Figure 20.5). In addition, there is evidence that mothers with resident minor children fathered by other men are substantially more likely to be killed by their current partners than women without dependent children fathered by other men (Daly *et al.* 1997). Like females in non-human mammals, women avoid violent partners and defend themselves and their children against attacks and a substantial proportion of homicides perpetrated by women involve attacks on current or previous sexual partners.

While parallels have been drawn between male infanticide in humans and male infanticide in non-human mammals (see Chapter 15), there are important differences. Male infanticide in humans is far less common than in many of the other mammals where it occurs. And, while immigrant males in non-human animals systematically attempt to eliminate dependent offspring, homicidal attacks by stepfathers often appear to be a consequence of intolerance of the demands of young children and may be a product of a temporary loss of control by stepfathers who lack the hormonal changes associated with paternal care.

20.3 Parental care

Maternal care

Compared with other mammals, parental care in humans has several unusual characteristics. Both parents commonly invest in their offspring throughout an unusually long period of dependence and parents also receive substantial assistance from other members of their community, including close kin and unrelated individuals (see section 19.5). However, the neuro-endocrine basis of maternal care is broadly similar to that in other primates and differences in maternal responsiveness to infants are correlated with variation in oestradiol, cortisol and oxytocin levels (Maestripieri 1999). As in other mammals, individual differences in parenting styles and in the treatment of older infants are often related to variation in the experience of females as infants and juveniles, the temperaments of their mothers and the socio-demographic circumstances they are reared in (Fernandez-Duque *et al.* 2009).

The energetic consequences of reproduction in humans also resemble those found in other mammals.

Figure 20.6 A !Kung San mother nurses her infant.
Source: © Robert Estall Photo Agency/Alamy Stock Photo.

The energy costs of gestation rise sharply during the second half of pregnancy, leading to reductions in maternal condition (Ellison 2001) and the high metabolic costs of lactation increase the energetic needs of human mothers and limit their capacity to collect resources (Figure 20.6). For example, in the Hadza, the presence of infants constrains the foraging opportunities of their mothers (Hawkes 2003). Although it was previously thought that increases in prolactin triggered directly by nursing controlled the duration of post-lactational amenorrhoea, subsequent studies suggest that increases in relative metabolic load and associated reductions in condition determine the duration of both amenorrhoea and inter-birth intervals and that insulin production may play an important role (Willis *et al.* 1996; Ellison and Valeggia 2003).

The duration of inter-birth intervals in humans varies widely and is often related to ecological variables that affect food availability or energetic demands on mothers (Ellison 1990; Rosetta 1993; Ulijaszek 1993). Both in traditional and in industrial societies, adverse

environments that constrain the resources available to mothers with dependent young commonly affect the growth and development of surviving offspring and sometimes have trans-generational effects (Barker 1998; Lummaa 2003; Singhal and Lucas 2004). For example, as in non-human mammals, the birthweights of mothers are often correlated with their weight as adults, as well as with the birthweight of their offspring (Lummaa and Clutton-Brock 2002). Restrictions on early growth can also have protracted effects on the health of offspring and have been shown to be correlated with the risk of cardiovascular disease (Barker *et al.* 1993), diabetes (Hales and Barker 1992) and cancer (Michels *et al.* 1996). As in other mammals, one interpretation of reductions in the birthweight and subsequent development of neonates is that they represent adaptations that adjust the development of individuals to the environmental conditions they are likely to face and increase the fitness of individuals exposed to similar challenges in later life (the *thrifty phenotype* hypothesis), while another is that they are by-products of constraints on growth imposed by resource shortages and that individuals that grow fast during the early stages of their lives are subsequently more likely to survive challenging circumstances than those that grow more slowly (the *silver spoon* effect). In one of the few studies to test whether human responses to adversity were consistent with 'thrifty phenotype' or 'silver spoon' effects, Hayward and Lummaa showed that pre-industrial Finnish agriculturalists that experienced poor nutrition in early life showed lower survival and fertility under famine conditions in later life than individuals that had experienced more plentiful conditions as children (Hayward and Lummaa 2013; Hayward *et al.* 2013) and studies of other primates have produced similar results (Lea *et al.* 2015).

The mother's perception of the likely costs of raising offspring is also important. In traditional and pre-industrial societies, parents commonly appear to balance family size against offspring survival and development (Lawson and Mace 2011). The available data suggest that parental responses may often reduce their lifetime reproductive output, though it is difficult to assess their effects on fitness in subsequent generations. In contrast to most other higher primates (with the exception of callitrichids), human mothers that are unlikely or unable to support their infants may abandon or kill them, especially if they were conceived outside marriage, usually within 72 hours of birth (Friedman *et al.* 2005; Bourget *et al.* 2007; Hrdy 1999; Aengst 2014). In traditional societies, female infanticide is commonly associated with harsh environmental conditions or food shortage and is particularly likely to occur where mothers are still nursing an older sib whose survival may be jeopardised by the demands of a younger sibling. The death of the father can also increase the chance that subsequent children will be killed: for example, Aché mothers sometimes kill their infants if their father has died or they have divorced (Hill and Hurtado 1996).

In some South American hunter-gatherer societies where band members rely on shared resources, dependent children whose parents die may also be killed by other group members. In the Aché, children of less than a year old may be killed by other group members if their mother dies and older children of up to 13 years can also be at risk. Though their relatives sometimes intervene to prevent them being killed (Hill and Hurtado 1996). In all, Hill estimates that 14% of male children and 23% of all female children among the Aché were victims of infanticide before the age of 10 years. Females were killed at 1.67 times the rate of males and children without mothers were 4.5 times as likely to be killed in each year of their childhood as those with mothers. As one Aché told an interviewer, 'we really hate orphans' (Hill and Hurtado 1996). In contrast, infanticide appears to be relatively uncommon in African hunter gatherers.

Paternal care

In contrast to the African apes, men commonly contribute to the maintenance of women and their children and, in many human societies, they are also directly involved in the care of their own children (Anderson *et al.* 1999; Geary 2000; Anderson 2006), though the involvement of fathers in direct forms of care is generally lower than that of mothers (Gray and Anderson 2010; Bridescac *et al.* 2012). In some societies, men adjust their investment in relation to their impression of paternity, sometimes using cues based on physical similarity (Apicella and Marlowe 2004). For example, in a polygynous society in Senegal, paternal investment increased with the degree of similarity in facial characteristics and odours between fathers and their children and was positively correlated with the growth and nutritional status of children (Alvergne *et al.* 2009).

Variation in paternal care between societies may be related both to mating systems and to the mode of subsistence. For example, comparisons based on the Standard Cross-cultural Sample of societies suggest that proximity between fathers and infants tends to be highest in polyandrous societies, intermediate in monogamous ones and those showing occasional polygyny, and lowest in societies where polygyny is highly developed (Figure 20.7) and that fathers spend more time close to their offspring in hunter-gatherers and foragers and less in agriculturalists, though this varies with the age of

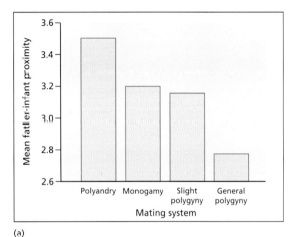

(a)

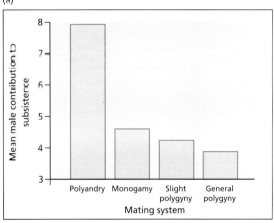

(b)

Figure 20.7 Differences with mating system of (a) mean time spent by fathers close to their infant offspring and (b) male contributions to subsistence. Comparisons are based on large samples of comparative data extracted from the Standard Cross-Cultural Sample of Human Societies. *Source*: From Marlowe (2000). Reproduced with permission of Elsevier.

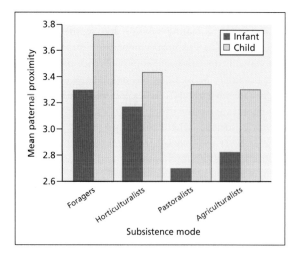

Figure 20.8 Effects of subsistence mode on mean time spent by fathers close to their infant and older offspring, based on the Standard Cross-cultural Sample of Human societies. *Source*: From Marlowe (2000). Reproduced with permission of Elsevier.

offspring (Figure 20.8). However, there are also large differences between individual societies. For example, among the Aka, who are net hunters, families frequently join in the hunt and men spend a relatively large

amount of time holding infants (Hewlett *et al.* 1998) (Figure 20.9).

In all societies, there are also substantial differences between individuals and families in direct paternal contributions to care. In traditional societies, these are often correlated with variation in the same social and ecological factors that are associated with contrasts between societies: for example, among the Yanamamö, monogamous men provide more care than polygynous ones (Hames 1992). In Western societies, differences in paternal care are often associated with variation in wealth or socioeconomic status or with contrasts in relations between spouses.

Differences in the personality of parents also affect the extent and quality of parental care and are frequently correlated with variation in hormone levels, including differences in levels of prolactin, oxytocin, vasopressin and testosterone (Fernandez-Duque *et al.* 2009; Müller *et al.* 2009). Oxytocin is implicated in the expression of maternal behaviour, while its effects in males are variable and vasopressin may play a more important role (Carter *et al.* 2008; Fernandez-Duque *et al.* 2009). Low testosterone levels may also facilitate the development of paternal care and brain activity associated with nurturing has been shown to be negatively related to testes volume

Figure 20.9 The families of Aka men, who are net hunters, often accompany the men on hunts. The men spend over 20% of daytime holding their children. *Source*: © Martin Harvey/CORBIS.

among fathers (Mascaro *et al.* 2013): experiments in which the cries of infants were played to samples of fathers with and without children of similar age showed that fathers responded more strongly to the cries of children. Both among fathers and among non-fathers, individuals with relatively low testosterone responded more strongly than individuals with high testosterone and experienced fathers responded more strongly than inexperienced ones (Fleming *et al.* 2002). Testosterone levels may in turn be affected by the stability of sexual relationships: several studies of Western men have shown that individual males in long-established monogamous relationships have lower levels of circulating testosterone than unpaired men or those in recently established relationships and there is a tendency for testosterone levels to be relatively low in men with young children (Gray *et al.* 2002, 2006; Burnham *et al.* 2003) (Figure 20.10). Although more than one causal relationship could underlie these associations, a likely one is that established relationships reduce conflict and so are associated with relatively low testosterone levels, facilitating male care (Gray *et al.* 2002).

Less is known about the consequences of paternal than maternal care. Estimates suggest that the energetic costs of raising a human infant from birth to nutritional independence are around 13 million kilocalories and

commonly exceed the level that mothers can produce on their own, suggesting that male contributions to supporting their wives and dependent offspring probably represent a significant proportion of the costs of raising young (Kaplan *et al.* 2000; Wood and Marlowe 2013). Several studies of traditional societies provide evidence of the importance of male contributions. The amount of food provided by men to their families varies widely: for example, among the Hadza, the best hunters bring three to four times as much food to their families as poor hunters (Wood and Marlowe 2013). Across societies, the nutritional contributions of men are correlated with the fertility of their partners and their total reproductive success (Marlowe 2001) (Figure 20.11).

The amount of food provided by men may also help to increase their attractiveness as partners (Hawkes 1991) and it is sometimes suggested that this represents the principal benefit of care to men. However, in several societies there is strong evidence that the food provided by men in groups of hunter-gatherers represents an important component of the diet of their families (Smith 2004). In addition, in some Western societies, there are correlations between paternal investment and the development of children, including measures of their cognitive competence, psychological adjustment and career development (Sarkadi *et al.* 2008; Kentner *et al.* 2010; Scelza 2010), and where offspring commonly inherit resources from their fathers, this can increase their reproductive success and the longevity of their children (Gray and Anderson 2010). Conversely, both the absence of fathers and harsh or disruptive parenting can lead to increased levels of aggression and violence in pre-adolescent males, though it does not always do so (Ember and Ember 1994; Cassar *et al.* 2003).

Although paternal behaviour is likely to have important costs to fathers, these have seldom been quantified. Among hunter-gatherers, they may include costs to the father's condition and survival as well as opportunity costs to his mating success (Gray and Crittenden 2014). In Western societies, the presence of children can also lead to reductions in circulating levels of testosterone in men (see Figure 21.10) and to reductions in the sexual function of fathers (von Sydow 1999; Gray and Garcia 2013) as well as in the quality of marital relationships (Twenge *et al.* 2003).

As in other biparental mammals (see Chapter 16), there are likely to be conflicts of interest between parents over the frequency of breeding attempts as well as over

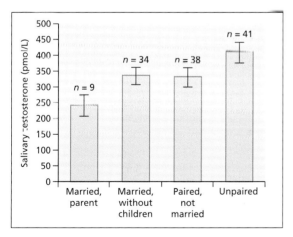

Figure 20.10 American men in relationships, especially fathers, have lower testosterone levels than unpaired men. The figure shows salivary testosterone levels for samples of male Harvard Business School students in contrasting relationships. *Source*: From Burnham *et al.* (2003). Reproduced with permission of Elsevier.

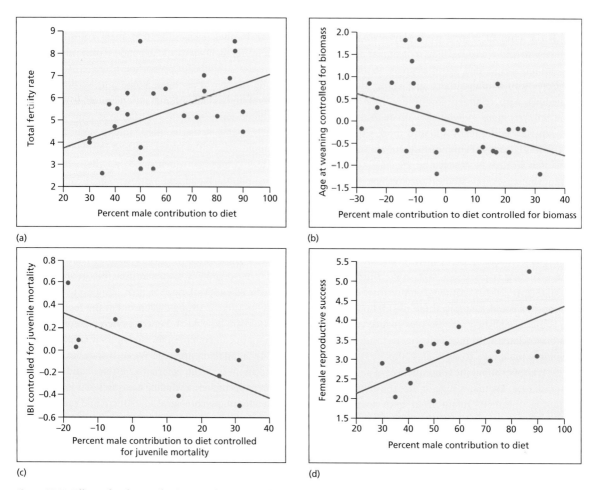

Figure 20.11 Effects of male contributions to subsistence in foraging societies on (a) total fertility rate in women; (b) age of children at weaning, controlling for variation in the biomass of the habitat occupied; (c) inter-birth interval (IBI), allowing for juvenile mortality; and (d) total reproductive success of women, based on the Standard Cross-cultural Sample of Human Societies. *Source*: From Marlowe (2001). Reproduced with permission of the University of Chicago Press.

their relative contributions to the care of offspring. For example, wives and husbands may differ in their view of the optimal number of children: in many traditional societies, men may frequently favour larger families than their wives, since their energetic contributions to rearing children are generally smaller and they do not experience the costs of short inter-birth intervals (Mason and Taj 1987). Frequent reproduction is also less likely to affect the longevity of men, and in the event that it reduces the lifespan of their wife, they may be in a position to remarry with a younger partner, so that their fertility may either be unaffected or may increase (Borgerhoff Mulder and Rauch 2009).

Conflicts of interest between partners may involve other family members, too. For example, if there is a trade-off between fertility and maternal survival, a woman's maternal kin may favour her survival over that of her children while her husband's kin may favour the production of additional children at the expense of the survival of the mother, to whom they are unrelated (Mace and Sear 2005). Other conflicts between partners involve fidelity and remarriage as well as relative contributions to childcare (Borgerhoff Mulder and Rauch 2009). In some societies there are also intense conflicts of interest over the treatment of particular children. For example, in the Dogon, where sons inherit land from their fathers, there are historical reports of frequent attempts by co-wives to poison each other's sons (Geary 2010) (see section 19.4).

20.4 Allo-parental care

Grandmothers and others

In many human societies (including many hunter-gatherers), mothers receive assistance in caring for and provisioning their offspring from male and female group members, including their parents and siblings, cousins, aunts, uncles and grandmothers (Hrdy 2009). Grandmothers, in particular, often play an important role in caring for their grandchildren and several studies have shown that their presence is associated with improved growth and survival among their grand-offspring. For example, in pre modern populations of Finnish women (Figure 20.12), children whose grandmothers were still alive were more likely to survive than those whose grandmothers had died (Lahdenperä *et al.* 2004) (Figure 20.13). Positive relationships between the presence of grandmothers and the reproductive performance of their daughters and the survival of their grand-offspring have now been shown in several other populations (Leonetti *et al.* 2005; Mace and Sear, 2005).

While correlations between the presence of grandmothers and the growth or survival of their grandchildren are frequently interpreted as the result of causal relationships, other interpretations are possible. In particular, there is a risk that relationships between the presence of grandmothers and the survival of their grand-offspring are generated by independent effects of variation in wealth, health or genetic quality on the survival of infants and older group members, for both young and old are often more strongly impacted by hardship than adults in their prime. However, analysis of temporal changes in the fitness of grand-offspring in relation to the presence and absence of their grandmothers suggests that these associations have a causal basis. For example, analysis of data for tribal agriculturalists in Gambia show that the loss of maternal grandmothers and other maternal relatives is associated with reductions in the survival of children below the age of 5 years (Sear *et al.* 2000, 2002) and other studies have shown that variation in the grandmother's contributions to family labour is correlated with the growth of her grandchildren, especially during times of food shortage (Hawkes *et al.* 1989; Leonetti *et al.* 2005). In addition, there is little indication that the presence of paternal grandmothers has similar effects and some that it can have negative ones, possibly because they encourage

Figure 20.12 A multi-generational Finnish family. *Source:* © Dr V. Lummaa.

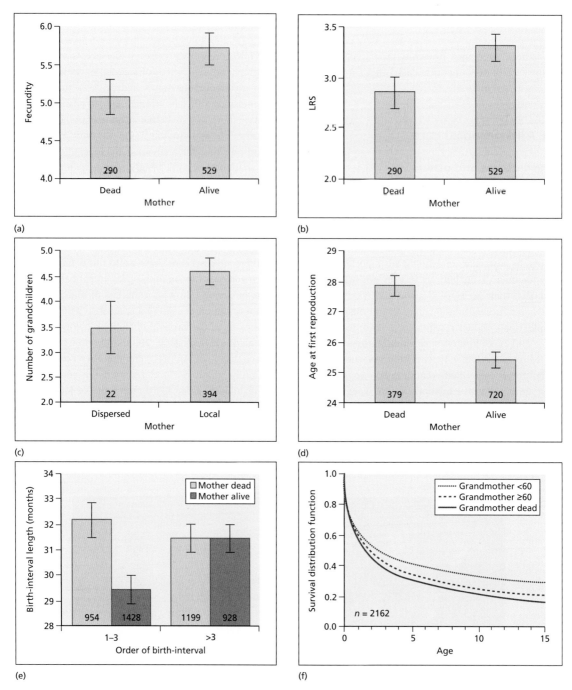

Figure 20.13 Effects of the presence of a post-reproductive mother at the time that each offspring of both sexes began to reproduce and correlates of each offspring's fitness and key life-history traits in pre-industrial Finland. (a–c) Offspring fitness; (d–f) key life-history traits. Data show that having a living post-reproductive mother is associated with (a) increased lifetime fecundity of offspring; (b) increased lifetime reproductive success (LRS) of offspring; (c) increased numbers of grandchildren produced when the post-reproductive mother was living in the same village (local) versus elsewhere (dispersed); (d) reduced age at first reproduction of offspring; (e) reduced first three inter-birth intervals, but not those thereafter; and (f) increased survival of grandchildren until age 15. Panels (a–e) show predicted means from general linear mixed models (GLMMs) after controlling for the effects of offspring socioeconomic status, geographic and temporal differences in living conditions, birth order, sex, number of siblings, mother's age and repeated measures of mother, where appropriate. Panel (f) shows Kaplan–Meier survival curves for grandchildren depending on the age and presence of a post-reproductive grandmother. *Source*: From Lahdenperä *et al.* (2004). Reprinted by permission of Macmillan Publishers Ltd.

their sons (and daughters-in-law) to increase the number of breeding attempts or because they favour increased investment in male grandchildren at the expense of their sisters (Hrdy 2009). It can also be difficult to distinguish between the contribution of grandmothers to the fitness of their grandchildren and the contributions of other relatives: in some societies, the presence of older sisters also enhances the survival of younger siblings (Crognier *et al.* 2002; Sear *et al.* 2002) while the presence of other non-reproductive relatives has little or no effect (Nitsch *et al.* 2014).

While maternal grandmothers can play an important role in the care of their grand-offspring, many women start to breed after their mothers have died. In addition, most would have one or more older siblings or other maternal or paternal relatives. This could suggest that selection favouring the formation of networks of supportive relationships with younger individuals than grandmothers has been strong and that the average influence of these relationships on female fitness may be as large as (or larger than) that of grandmothers (Hrdy 2009). Moreover, the effects of grandmothers and other relatives on female fitness probably vary widely and may not always be positive. For example, in Pimbwe horticulturalists, the presence of kin in the same village dilutes the ability of parents to invest in their children and (after effects of other social and environmental factors have been allowed for) is associated with increased mortality among their children (Borgerhoff Mulder and Beheim, 2011).

Contributions to the care of offspring by female relatives could have had an important influence on the evolution of human life-histories. The establishment of supportive relationships between women may have raised rates of survival in dependent young and their mothers, permitting the evolution of prolonged development of juveniles, and the overlapping dependence of successive offspring and extended lifespans (Hawkes 2003; Hrdy 2009). These changes, combined with selection on juveniles to extract resources from multiple adults, may have permitted and encouraged improvements in cognitive abilities and led to a more prolonged period of brain development and to increases in eventual brain size (Hrdy 2009; Isler and van Schaik 2009). An alternative interpretation is that social and technological changes associated with the occupation of savannah ecosystems and dependence on scavenging and hunting led to the initial development of brain size and cognitive

abilities in hominins (see Chapter 19) and reproductive cooperation arose at a comparatively late stage in human evolution.

While grandmothers may work hard for the benefit of their grandchildren because investment in their grand-offspring represents the most effective use of their remaining resources once they have reached the menopause, there may also be other reasons, too why they benefit by helping other members of their group for, in some societies, orphans are not the only individuals that are killed by other group members if they are likely to become a burden (Hill and Hurtado 1996). An old Aché man described how, when he was younger, he would stalk old women who were a drain on the group when they came down to the river to get water and attack them with an axe: 'I would step on them they all died there by the big river . . . I didn't used to wait until they were completely dead to bury them. When they were still moving I would [break their backs and necks]'. As Hrdy (2009) points out, it may benefit grandmothers to keep active.

The evolution of post-reproductive lifespans in women

The needs of mothers for assistance from maternal relatives may have been responsible for the evolution of the unusual post-reproductive lifespans in women. As section 19.3 describes, women typically cease breeding by the age of 50 when their egg supplies have become depleted. A similar decline in fertility occurs at a slightly earlier age in female chimpanzees (Hrdy 1999) but in contrast to female chimpanzees, many women live for at least a further decade after ceasing to be fertile. Although some females survive for several years after ceasing to breed in many other mammals (Promislow 1991; Packer *et al.* 1998; Cohen 2004), the average post-reproductive lifespans of most women are substantially longer than those of species, with the possible exception of some cetaceans.

The evolution of extended post-reproductive lifespans in women consequently requires an explanation (Sherman 1998). One of the first suggestions (sometimes called the *mother hypothesis*) was that, by ceasing to breed at a relatively early stage of their lifespan, mothers reduce the risk that they will produce offspring they are unlikely to be able to rear before their own death and that an early cessation of fertility allows them to channel their remaining resources to their last-born offspring and so maximises their fitness (Williams 1957; Nesse and Williams 1995). Effects of this kind might, it was suggested, be particularly powerful if the chances that mothers would die in childbirth increased towards the end of the lifespan. Subsequently, evidence that the presence of grandmothers was associated with increased reproductive success in their offspring and increased numbers of grandchildren (Lahdenperä *et al.* 2004) led to suggestions that an extension of post-reproductive lifespans allowed older females to invest in their grand-offspring directly, an idea that came to be known as the *grandmother hypothesis* (Hawkes *et al.* 1998; Peccei 2001; Hawkes 2003; Hrdy 2009). One difference between these two explanations is that the grandmother hypothesis assumes that selection has favoured an extension of female longevity while the mother hypothesis suggests that the age at which females cease to breed has either advanced or has remained constant as female longevity has increased.

When the grandmother hypothesis was first suggested, it faced the objection that both female chimpanzees and female hunter-gatherers usually leave their natal communities to breed, suggesting that female hominins would have been unlikely to have had an opportunity to invest in their grand-offspring. However, both these generalisations have now weakened: some chimpanzees remain and breed in their natal community (Pusey *et al.* 1997; Hrdy 2009), while in many hunter-gatherer societies women often breed for the first time in their natal community and their societies are more accurately categorised as bilocal rather than patrilocal (Alvarez 2004; Hrdy 2009). In addition, there is little increase in mortality during childbirth with increasing maternal age (Lahdenperä *et al.* 2011) and recent genetic research has confirmed the existence of genes that can play an important role in extending lifespan (Perls and Fretts, 2001).

Despite these results, the grandmother hypothesis still faces a number of difficulties. Grandmothers do not make a major contribution to provisioning their grandchildren in all societies and their presence is not always associated with improvements in the growth or survival of grandchildren in all hunter-gatherers (Hill and Hurtado 1996; Borgerhoff Mulder and Beheim 2011). In addition, it is reasonable to suppose that increases in longevity are likely to delay age at first breeding and reduce fecundity, even if negative phenotypic correlations between longevity and fecundity are not evident in all cases (Kaplan 1997; Helle *et al.* 2005; Kachel *et al.* 2011), and as only a proportion of females reach old age, the benefits of an early cessation of reproduction need to be large to compensate for reductions in early fecundity as well as for fecundity in later years (Hill and Hurtado 1991; Rogers 1993). A combination of benefits to the fitness of children and grandchildren could be sufficient to favour an early termination of breeding (Shanley and Kirkwood 2001; Kirkwood and Shanley 2010) or the extension of female lifespans but whether this is likely is debatable (and is still debated).

As a result of these problems, other explanations of the evolution of post-reproductive lifespans in women have been explored. For example, Cant and Johnstone point out that humans differ from other primates in showing little generational overlap in reproduction (Figure 20.14) and argue that this is because there is likely to be competition between co-resident women for resources and breeding opportunities (Cant and Johnstone 2008; Johnstone and Cant 2010). As their argument suggests, in pre-modern Finnish populations, reproductive overlap between women and their daughters-in-law is associated with reduced survival of offspring born either to older women or to their daughters-in-law (Lahdenperä *et al.* 2012) (Figure 20.15) a result that has close parallels in other social mammals (see Chapter 17). In theory, these conflicts could lead either to an early cessation of breeding in mothers-in-law or to deferred breeding by their daughters-in-law. However, where females disperse and males remain and breed in their natal group, mothers of mature sons will be more closely related to offspring produced by their daughters-in-law than their daughters-in-law will be to any offspring that their mothers-in-law produce, with the result that daughters-in-law might be expected to win competitive encounters over breeding with their mothers-in-law (Johnstone and Cant 2010). However, attempts to test these 'conflict' theories with empirical data have so far produced little supporting evidence: for example, comparisons of the timing of the menopause between patrilocal and matrilocal societies have found no evidence that it occurs earlier in patrilocal ones (Snopkowski *et al.* 2014).

Other conflicts may also be involved and the interests of males may need to be considered. Where males provide resources that are shared with close relatives, the sons of ageing mothers (who will, on average, usually be more closely related to their own offspring than to subsequent siblings) may ensure that their own offspring have priority to resources over subsequent siblings, making it unprofitable for their mothers to continue to breed after resident sons have started to do so. Resident sons

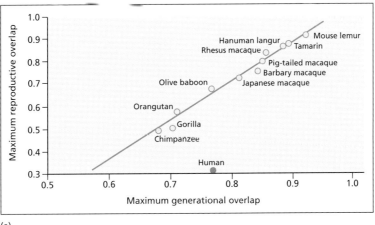

(a)

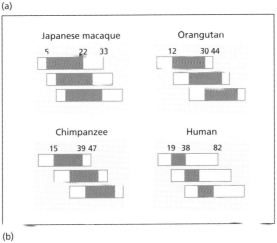

(b)

Figure 20.14 Generational overlap in primates. (a) Maximum reproductive versus maximum generational overlap in 12 primate species. Maximum generational overlap is defined as (MLS − AFB)/MLS, where AFB is average age at first birth and MLS is maximum recorded lifespan. Maximum reproductive overlap is defined as (MRS − AFB)/MRS, where MRS is the maximum reproductive span, calculated as maximum age at last birth minus AFB. (b) Pattern of overlap for three non-human primates and for humans. For each species, horizontal bars represent the maximum lifespans of three successive generations, scaled to a standard length and offset in accordance with the value of AFB relative to MLS, with mean reproductive spans shaded. *Source*: From Cant and Johnstone (2008). © 2008 National Academy of Sciences, USA.

might be expected to become progressively more reluctant to make reproductive concessions to their mothers as they both age if the probability that offspring produced by their mother will be full sibs declines with time. This illustrates the point that a range of conflict models can be constructed and that models based on divergent assumptions can generate similar predictions. As a result, a match between the predictions of particular models and empirical data does not necessarily provide strong evidence that the evolutionary mechanisms assumed by the model are involved.

Finally, there are other possible processes that could contribute to the extension of female lifespans after the age at which breeding ceases (Promislow 1991; Cohen 2004). Perhaps the maintenance of fertility has higher costs than the maintenance of soma, possibly because large brains act as a homeostatic mechanism and delay senescence (Sacher 1978). Or perhaps phenotypic selection operating over the course of female lifespans progressively removes females that invest heavily in breeding at a high cost to their own soma, so that mothers that show low fecundity are disproportionately

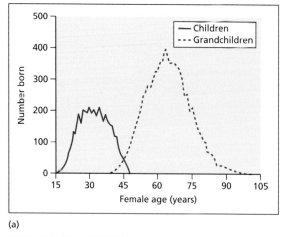

(a)

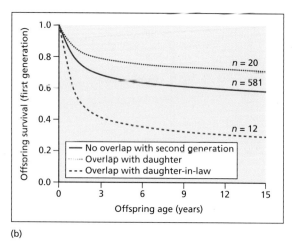

(b)

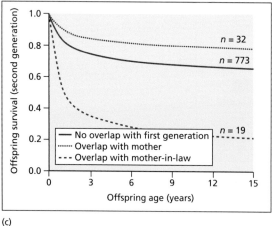

(c)

Figure 20.15 Effects of reproductive overlap on child survival in pre-industrial Finnish families. (a) Age-specific fertility distribution illustrating the ages over which mothers and offspring overlapped in their reproductive periods. Reproductive overlap significantly reduced the survival of offspring from (b) first-generation mothers and (c) second-generation mothers when the reproductive overlap occurred between unrelated women (mothers-in-law and daughters-in-law), but had less effect when mothers and daughters reproduced simultaneously. Figures show predicted relationships from Cox regressions. Gen refers to generation. *Source*: From Lahdenpera *et al.* (2012). Reproduced with permission of John Wiley & Sons.

represented in the oldest age classes (Cohen 2004). Debates over the origin and function of the human menopause and the extension of female lifespans look set to continue.

20.5 Cooperation

The evolution of human cooperation

Cooperation pervades all aspects of modern human society, affecting the distribution of labour, the division of food, the organisation and defence of land tenure, the distribution of breeding partners and the care of dependent young. It is also involved in all trading relationships and all local alliances between groups or states (Bowles and Gintis 2011; Nowak 2012). There are many similarities between human cooperation and cooperation in other social mammals: individuals often cooperate preferentially with relatives and with familiar individuals; cooperators often gain mutualistic benefits from assisting each other and competition within groups may reduce the extent of cooperation (Barker *et al.* 2012). As in other social mammals the readiness of individuals to assist particular partners is often related to the assistance

In contrast to other animals, humans occasionally provide unrelated and unfamiliar individuals with assistance at substantial cost to their own fitness (Oliner 2005). Extreme examples of self-sacrifice of this kind occur regularly in wartime but are not confined to military action. In Britain, the operations of lifeboat crews provide a good example. Lifeboats owned by a national charity (the Royal National Lifeboat Institution or RNLI) are stationed around the coast (Figure 20.16). They are manned by (mostly) unpaid local volunteers who routinely attempt to rescue the crews of ships in danger under the worst weather conditions imaginable. When the RNLI was founded in 1825, its boats were mostly open and were powered by oar or sail, so that rescues in extreme weather were very hazardous and entire crews were frequently lost. Even in the twentieth century, when motor power and advances in lifeboat design reduced risks, they were still high: for example, between 1900 and 1981, twenty-eight lifeboats capsized or were wrecked, frequently with the loss of the entire crew and over 150 lifeboat-men drowned. In virtually all cases, crews were attempting to rescue sailors with whom they had no connection, many of whom were not British nationals, and few crew-members received any form of public recognition. While individuals may sacrifice themselves for the benefit of the colony in some social animals where colony members are close relatives, extreme cooperation involving such high risks or costs to fitness does not occur between unrelated individuals on a regular basis in non-human animals and, in humans, is presumably maintained by ethical and religious beliefs or commitments rather than by any form of selection. A common theme in studies of extreme cooperation and altruism is that, like acts of extreme cruelty, they commonly involve ordinary people doing extraordinary things (Oliner 2002).

Figure 20.16 The Royal National Lifeboat Institution's lifeboat, the *Solomon Browne* of Penlee in Cornwall. In December 1981, the *Solomon Browne*, with a crew of eight, launched in hurricane-force winds in an attempt to rescue the crew of a Danish trawler, in trouble off the Wolf Rock. After successfully rescuing four of the trawler's crew, the lifeboat tried to take off the remaining crew members and capsized or was crushed with the loss of the entire crew and those that they had rescued. Similar accidents involving the crews of RNLI lifeboats occurred throughout the twentieth century but had little effect on the numbers of volunteers
Source: © Lalouette Photographers.

that they have received from them (Wilson 1978; Taborsky *et al.* 2016). However, cooperative interactions between humans differ from those in other mammals in fundamental ways. They commonly involve individuals that are both unrelated and unfamiliar to each other and the provision of assistance can be associated with substantial risks or costs (Box 20.2 and Figure 20.16).

Cooperation is frequently based on implicit or explicit agreements or contracts between dyads or group members that link the actions of different parties to each other, creating specific expectations about the responses of partners (Bowles and Gintis 2011) that generate disapproval, anger or punishing reactions when they are not met (see Box 20.3). Their reactions thus differ from those of non-human cooperators who may cease to assist or associate with individuals that fail to reciprocate but seldom attempt to coerce cooperation or to punish 'defectors' (Riehl and Fredrickson 2016; Taborsky *et al.* 2016). Moreover, in all human societies, agreements and contracts (and the norms of behaviour associated with them) are often supported or enforced by the actions of third parties, including various forms of disapproval, intervention and punishment by designated individuals ('policemen'), by groups of elders (Boyd and Richerson 2009) or individual leaders (von Rueden *et al.* 2015; Glowacki and von Rueden 2015) or their peers. The risk of social disapproval from peers can also be sufficient to enforce normative behaviour. For example, a recent experiment in which some honesty boxes provided for payments for coffee and other small purchases were paired with photographs of eyes showed that this increased takings substantially (Figure 20.17). In

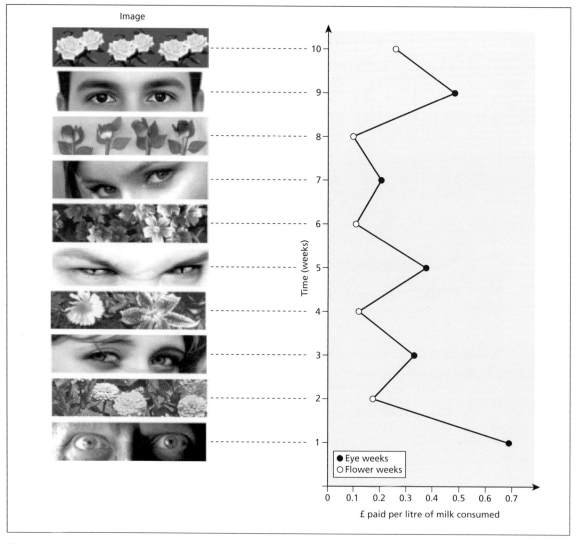

Figure 20.17 Pictures of human eyes increase the tendency for individuals to cooperate. In this experiment, individuals contributed nearly three times as much to an honesty box in weeks when photographs of eyes were displayed nearby than in weeks when they were replaced by pictures of flowers. *Source*: From Bateson *et al.* (2006). Reproduced with permission from the Royal Society.

addition, in some societies, the role of public watchdog is attributed to morally concerned, all-seeing deities who are often depicted as keeping a close eye on the actions of individuals and are expected to punish deviations from accepted social norms (Hinde 2009; Norenzayan 2013) (Figure 20.18). Belief in morally concerned deities is commonly associated with relatively large human groups and experiments show that religious thoughts can

(a)

(b)

(c)

(d)

Figure 20.18 Watching eyes in images associated with personal religions. (a) Eye of Horus from ancient Egypt, late sixth to fourth centuries BCE. (b) Buddha eyes on a stupa in Swayambhunath temple, Kathmandu, Nepal. (c) Stone carving of Viracocha, the chief God of the Inca Empire, from Tiahuanaco, Bolivia. (d) Eye in the Sky, an alchemical woodcut attributed to a European text from the sixteenth century. It is often depicted with the motto *Quo Modo Deum*, Latin for 'this is the way of God'. *Source*: (a) © Marie-Lan Nguyen. (b) © Ara Norenzayan. (c) © Martin Gray/Getty images.

encourage cooperation between strangers and reduce cheating (Norenzayan and Shariff 2008).

Disapproval or punishment of deviance from social norms has an important influence on evolutionary processes maintaining cooperation. As a result of third-party intervention in support of social norms and reputational benefits, the potential benefits of cheating strategies can be minimised, increasing the potential for group selection to maintain behaviour that maximises the average fitness of group members even if it reduces the fitness of some (Boyd and Richerson 2009; Nowak 2012; Macfarlan and Lyle 2015). Where cultural mechanisms that have developed through social learning impose substantial costs on cheats, mutualistic cooperation (or public goods games) may be preserved from erosion by cheating strategies and cooperation may be more likely to be maintained (Sigmund *et al.* 2010; Norenzayan 2013; Schonmann and Boyd 2016). In democratic societies, this could extend to the conservation of resources for future generation (Hauser *et al.* 2014). Where cultural and religious factors play an important role in controlling cheating strategies, they may interact with selection operating on genes, generating coevolutionary processes that affect both cultural practices, gene frequencies and neuronal development (Boyd and Richerson 2009) (see Chapter 19).

The evolution of human cooperation has far-reaching consequences and some biologists have drawn parallels between the success of human societies and that of social insects (Crespi 2014). In conjunction with improved technology, cooperation has allowed humans to occupy areas and habitats inaccessible to other similar mammals. While many forms of cooperation may have had an important influence on human evolution, cooperative defence and aggression, cooperative foraging and reproductive cooperation have probably played particularly important roles in human evolution. The following sections describe each of them in turn.

Cooperative defence and warfare

Early hominins probably relied on cooperative defence to protect themselves against predators (see Chapter 19) and the need for individuals to aggregate and cooperate to provide effective defence probably persisted throughout much of human evolution. In many populations, the construction of defensive structures around camps and settlements may have originated as an adaptation to this need (Figure 20.19a). Cooperative defence or aggression against neighbouring groups is also likely to have been important in hominin societies. In many social animals, from ants (Hölldobler and Wilson 1990) to chimpanzees (Wilson 2013), relationships between neighbouring groups are usually hostile, neighbours or intruders are likely to be attacked and may be killed, and bigger groups sometimes evict or extinguish smaller ones (see Chapter 2). However, the intensity of conflict between social groups varies widely between and within species: for example, while relationships between males from neighbouring chimpanzee communities are often intensely hostile, bonobo groups appear to avoid each other and there is little evidence of regular aggression (White *et al.* 2013; Wilson 2013). Within species, too, the frequency and intensity of hostile interactions between groups vary widely and are often affected by population density, range size and the rate of group encounters, as well as by contrasts in traditions resulting from social learning (Sapolsky and Share 2004; Wilson 2013).

The intensity of inter-group aggression in hominins and humans is a subject of extensive debate (Fry 2006, 2013). There is no direct way to estimate the frequency of inter-group aggression in early hominins and arguments concerning the importance of aggression between competing groups are often based on the questionable assumption that inter-group relations resembled those in modern chimpanzees. Although there are claims that archaeological evidence from some Upper Palaeolithic sites indicates the presence of interpersonal or inter-group violence as early as 35,000 years ago (Keeley 1996; Otterbein 2004) or even earlier, the available evidence consists largely of bone fractures that cannot reliably be attributed to interpersonal violence or warfare (Kelly 2013). One of the first convincing examples of inter-group violence comes from a 12,000–14,000 year old cemetery in the northern Sudan, where around two dozen people (including men, women and children) apparently met violent deaths and some had points still embedded in their bones (Wendorf 1968; Haas and Pisiticelli 2013).

So how common is violence among modern and historical communities of hunter-gatherers and traditional agriculturalists? Interpersonal violence is not uncommon within hunter-gatherer communities (Durham 1976; Kelly 2005) but inter-group hostility appears to be unusual, especially in communities living in harsh and unpredictable conditions (Butovskaya 2013; Endicott 2013), though it may be more frequent in more settled communities that exploit more abundant resources (Dye 2013). Hospitality between neighbouring groups appears to be more frequent in some horticultural

(a)

(b)

Figure 20.19 (a) Tuareg encampment. (b) The Norman keep of Cardiff castle. *Source:* (a) © De Agostini/N. Cirani/De Agostini Picture Library/Getty Images. (b) © VisitBritain/Britain on View/Getty Images.

and agricultural communities where population density is relatively high as well as in pastoralists, where boundaries are frequently disputed and stock theft is common (Chagnon 1988; Leff 2009; Hundie 2010) and can be an important cause of mortality in men (see Figure 19.22). For example, in the Yanomamö, men of similar age, who are not necessarily close relatives, form coalitions that often involve co-residence and marriage ties which they use in defence and in attacks on rival communities (Macfarlan *et al.* 2014). In exceptional cases, feuding can lead to pitched battles and violence can be a common cause of death in men. As in non-human animals, the relative size of competing groups can have an important influence on the outcome of inter-group conflicts (see Chapter 2): for example, among the Yanomamö, large villages commonly raid smaller ones and may eventually displace or annihilate them (Chagnon 1997). While hostile relations between social groups are not uncommon in many human societies, popular accounts have often exaggerated their frequency (Ferguson 2013). Many human societies have devised ways of resolving conflicts (Boehm 2013), members of neighbouring groups (especially older individuals) often seek to avoid hostilities where possible, and peaceful relations between neighbouring groups are the norm in many populations (Fry 2013).

After improvements in technology and agriculture led to the storage of food surpluses and to increases in population density and the size of social units, progressive specialisation by group members was associated with increasing sophistication of military expertise and organisation, pitched battles became more frequent and defensive structures became more elaborate and more permanent (Figure 20.19b). Though the outcomes of military encounters can be influenced by multiple factors, including experience of combat, morale and military technology, the size of opposing forces and their capacity to sustain losses can still be crucial (Weiss 1966).

For example, during the First World War, military theoreticians developed models predicting the outcome of battles either between foot soldiers that cannot fight more than one individual at a time (Lanchester's linear law) or between soldiers with firearms that can engage multiple targets (Lanchester's square law) (Lanchester 1956). Both models have been applied to a wide range of human conflicts as well as to conflicts between animals and their results confirm the importance of numbers when other factors are allowed for (Weiss 1966; Lepingwell 1987; Wilson *et al.* 2002; Plowes and Adams 2005).

Underlying many discussions of violence and warfare in hominins and humans is the wish to ask whether humans have an innate tendency to direct aggression at neighbouring groups. While it is clear that competition for resources and associated hostilities between social groups can occur in traditional human societies, as they do in many other social animals, human aggression is typically opportunistic, its frequency varies in relation to the costs and benefits of winning or losing competitive encounters and it is strongly influenced by social learning and social norms (Kokko 2013; Sussman 2013). Offered a free choice, older members of local groups in traditional societies commonly discourage attacks on neighbours and democracies relatively seldom make war on each other (Hughbank and Grossman 2013). Although theoretical treatments suggest that cultural processes have the capacity to interact with selection operating through inter-group competition to maintain genes favouring cooperative aggression towards rival groups (Bowles 2008; Zefferman and Mathew 2015), there is plentiful evidence that social contracts and the enforcement of group norms through shaming and punishment can, on their own, encourage or enforce the involvement of individuals in aggression and warfare. As a result, there is little need to suppose that humans have evolved a heritable tendency to attack their neighbours.

Food sharing

Cooperative foraging and hunting probably evolved at an early stage of hominin evolution (see Chapter 19) and are likely to have had a profound influence on the course of social evolution, permitting the extended development and dependence and favouring improvements in communication that led ultimately to the evolution of language. Studies of modern hunter-gatherers have documented the extent to which meat is shared between group members, while other foraged resources (including berries, nuts and roots) are usually either retained by their collectors or are shared less widely (Kaplan and Hill 1985; Hill and Hurtado 1996). In many hunter-gatherer communities, hunting parties consist exclusively of men (Figure 20.20). If hunters are successful, their prey is usually dismembered and brought back to camp (Figure 20.21) to be shared with other members of their band, who then pass parts of their share to children (Kaplan and Hill 1985). Cooperative fishing is also common in hunter-gatherers and horticulturalists and often involves both men and women (Figure 20.22).

Figure 20.20 (a) A group of Gwi San men hunting, and (b) their eventual kill. *Sources*: (a) © John Warburton-Lee Photography/ Alamy Stock Photo; (b) © Greatstock Photographic Library/Alamy Stock Photo.

Figure 20.21 A Hadza man returning to camp with meat to be shared with the community. *Source*: © Travel Africa/Alamy Stock Photo.

Women forage for a range of other foods, including fruit, roots and insects and their larvae (Figure 20.23).

The relative contributions of men to subsistence resources used by women and children are highest in pastoralists (where males usually own and control

Figure 20.22 Yanomamö women collecting fish from a stream that has been poisoned by men. *Source*: © Napoleon Chagnon.

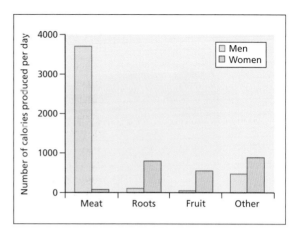

Figure 20.23 Comparisons of the number of calories produced per day from different sources by men and women in three foraging societies (Ache, Hadza and Hiwi). Men and women specialise in different foraging tasks: men hunt, while women specialise in extractive foraging. *Source*: From Boyd and Silk (2000). Reproduced with permission from W. W. Norton.

livestock), intermediate in hunter-gatherers and agricultural societies and lowest in horticulturalists, where women commonly play an important role in producing food. Among hunter-gatherers, male contributions tend to increase at higher latitudes and in colder climates, where the availability of edible plant food is reduced so that women and their offspring are almost totally dependent on resources contributed by men for a substantial proportion of the year (Hiatt 1974; Kelly 1995). The proportion of total calories used by women and children that are provided by men varies widely and is influenced both by ecological factors and by social expectations, customs and laws (Geary 2000). For example, among Hadza hunter-gatherers, total calories collected by foraging women equal or exceed those provided by men, though food provided by men may be of greater nutritional value and can represent an important component of the diet at particular times of year or when women are unable to forage effectively (Lee 1979; Hawkes and O'Connell 1997; Marlowe 2010). In contrast, in some Amerindian societies, including the Aché, total calories generated by men (mostly from meat and honey) exceed those generated by women (Hill and Hawkes 1983; Hill and Hurtado 1996; Kaplan and Gurven 2005). Whether these differences represent a contrast between African and Amerindian hunter-gatherers is still uncertain and broader comparative studies are needed.

In many hunter-gatherer societies, resources provided by other group members are of crucial importance for categories of individuals whose capacity to collect food themselves does not match their needs, including lactating mothers, children and adolescents and older group members. For example, as Aché parents get older and the number of their dependent offspring rises, the energetic burden on them increases. Empirical studies of energy allocation show that energy demands peak in parents of between 40 and 50 years old and often persist at a high level until parents are in their sixties and their own capacity to acquire resources has begun to decline (Figure 20.24). As a result, more successful parents with larger numbers of dependants are reliant on nutritional contributions from other group members (Kaplan and Gurven 2005).

Meat sharing may have direct energetic advantages to hunters and their families because it reduces temporal variance in resources. In most hunter-gatherers, hunting success varies widely, often providing nothing and occasionally generating more meat than one family can utilise (see Figure 20.20). For example, Aché hunters fail to make kills on 40% of the days that they hunt but sometimes make kills that greatly exceed their own

needs or those of their immediate families (Hill and Hawkes 1983); by sharing meat, they are likely to reduce the fitness costs of failure without generating similar reductions in the benefits of success (Janssen and Hill 2014). Cost-counting reciprocity or generalised reciprocity may also be involved, especially in the sharing of foraged food: in some societies, individuals or families share what they find with other families with similar rates of success and individuals that avoid sharing may be excluded (Kaplan and Gurven 2005).

Meat sharing may also generate indirect fitness benefits to the kin of successful hunters. In many societies, there is a component of nepotism in the distribution of meat, families of successful hunters benefit more than others, and networks of sharing meat and other foods link related families and their associates (Kaplan and Gurven 2005). For example, in the Lamalera whale-hunters of Indonesia, hunters own specific parts of the animals they kill and their families benefit disproportionately (Alvard 2002) (Figure 20.25). In some cases, food sharing may also be a consequence of harassment or scrounging, as is sometimes the case in chimpanzees (Gilby 2006). For example, in Mikea agriculturalists of Madagascar, voluntary sharing of

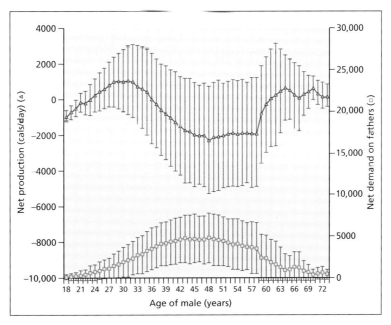

Figure 20.24 Age-related changes in the capacity of parents to feed their families in Aché foragers. The upper plot shows age-related changes in net production, the lower one the net demand for resources produced by fathers. *Source*: From Gintia *et al.* (2005). Reproduced with permission from Massachusetts Institute of Technology.

Figure 20.25 Lamalera whale hunters on Lesser Sunda Island, with a whale boat and dead pilot whale. *Source*: © Fred Bruemmer/ Photolibrary/Getty Images.

food is uncommon but scrounging is common (Tucker 2004).

Cooperative hunting and food sharing may also have reputational benefits (Kaplan and Gurven 2005). In some cases, the pursuit of foods that are difficult to acquire appears to generate suboptimal energetic returns (Bird *et al.* 2001; Kaplan and Gurven 2005) and men may increase their own breeding success by hunting and sharing food because it increases their reputation and enhances their desirability as mates, in-laws or allies (Hawkes *et al.* 2001) and hunting (and subsequent meat sharing) may represent an example of costly signalling, reflecting the quality of signallers (Hawkes *et al.* 2001; Patton 2005).

Reproductive cooperation

Reproductive cooperation involving shared care and provisioning of dependent children between women is also widespread in human societies (Hrdy 2009) (Figure 20.26) as section 19.3 describes. Like cooperative foraging this, too, could have an ancient origin and could have played an important role in the subsequent evolution of human cooperative behaviour and life-history patterns (Burkart and van Schaik 2010; Kramer and Russell 2014). Assistance from other group members in feeding and caring for dependent infants and juveniles might have allowed women to reduce their inter-birth intervals and to support several dependants of different ages at the same time (Hawkes 2003). However, we have no firm evidence of the origin of alloparental care or current human life-history patterns and an alternative interpretation is that extensive reproductive cooperation between women is a relatively recent phenomenon associated with the evolution of language and the development of extensive reciprocity.

Evidence of the extent to which human mothers rely on other caregivers (including husbands, grandmothers and older sisters) and of the effects of their presence on the survival of infants has led to the suggestion that humans should be classified as cooperative breeders (Hrdy 2009; Kramer 2010). Most proponents of this view usually do not distinguish between cooperative and communal breeders: for example, Hrdy defines cooperative breeders as 'any species with alloparental assistance in both care and provisioning of young', a definition that includes many communal breeders, like house mice, as well as cooperative breeders, where young are reared by non-breeding helpers (see Chapter 17). Definitions of cooperative breeding have

Figure 20.26 A !Kung grandmother and child, Namibia.
Source: © Martin Harvey/Photolibrary/Getty Images.

varied as our understanding of them has developed but, as Chapter 17 argues, it is useful to distinguish between communal and cooperative breeding since communal breeding can be maintained by direct mutualistic benefits and it can evolve where relatedness between group members is relatively low, while the evolution of cooperative breeding systems involving non-breeding helpers appears to involve kin selection and is associated with relatively high levels of relatedness between non-breeding helpers and the individuals that they assist has been confined to polytocous species.

So should humans be regarded as cooperative or communal breeders? Contemporary human breeding systems show some of the characteristics of both systems, though in most respects they resemble communal breeders more closely than cooperative breeders. Adult females are usually tolerant of each other's breeding attempts and breeding females often assist each other

(Hrdy 2009); groups rarely include adult females whose sexual development has been suppressed, as they do in non-human cooperative breeders; and, like some communal breeders (and unlike most cooperative breeders), the ancestral breeding system of hominins and humans probably involved some form of polygyny (see section 19.3). Finally in many human societies, parents derive assistance in raising their offspring from unrelated or distantly related group members as well as from close relatives, providing additional evidence that mutualistic benefits played an important role in the evolution of human cooperation (Tomasello *et al.* 2012). In other ways, humans differ from cooperative and mammals. In particular, they are monotocous – though the evolution of overlapping periods of dependence of successive offspring could have generated selection pressures similar to those in polytocous species – and the relative importance of males as providers of shared food differs from their role in most communal and cooperative mammals.

Prosociality

Arguments that the evolution of human life histories has required extensive cooperation combined with the prevalence of cooperation in hunter-gatherer societies and its benefits to breeding success have also led to suggestions that humans may have an innate tendency to cooperate with each other and show prosocial behaviour. Three lines of evidence are commonly cited in support of this idea. First, humans commonly show a predisposition to behave cooperatively in economic games that explore the probability that individuals will behave cooperatively or selfishly in 'one-shot' encounters where they cannot benefit from cooperating since they change partners after each interaction and identities are not revealed (Fehr and Gächter 2000) (Box 20.3). Unconditional cooperation of this kind is known as 'strong reciprocity' and a substantial number of similar experiments performed with different groups show that it is widespread (Gintis *et al.* 2003; Henrich *et al.* 2006). Humans also show a strong tendency to punish individuals that do not behave cooperatively and individuals that are exposed to uncooperative behaviour in one-shot encounters tend to punish their partner, even if this involves costs and generates no benefits since their next interaction will be with a novel partner (Fehr and Gächter 2002; Henrich *et al.* 2006).

Second, psychological research on humans has demonstrated correlations between individual differences in

Box 20.3 Cooperation and punishment in economic games

Experimental economists have explored the tendency of humans to cooperate with each other in one-shot games where partners change after each encounter and the identity of individuals is suppressed, so that players cannot gain from behaving cooperatively (Fehr and Gächter 2000). Despite this, they commonly cooperate. For example, in one experiment designed to imitate labour markets, participants were randomly allocated to roles as employers or workers. 'Employers' could either pay a minimum wage or could pay a higher wage and 'workers' could vary the amount of work they did. After each interaction, both individuals maintained their roles but played with other partners who knew nothing of their previous actions. 'Workers' who were treated generously in one round were more likely to work harder in the next, though this gave them no benefit since they were playing with a new partner who was unaware of their previous behaviour (Fehr and Gächter 2000).

Humans also show a strong tendency to punish individuals who do not behave cooperatively, even when there are costs to punishing others and no direct benefits. For example, in an experiment involving two subjects, one was asked to decide on how a sum of money should be divided between them and their partner and inform their partner of their decisions. The partner could then either accept the division (in which case the sum was divided in the agreed fashion) or could reject the proposal, in which case neither partner received anything. Once again, partners were subsequently re-paired with anonymous individuals after each interaction. In these games (known as 'ultimatum' games), partners offered a relatively small share commonly reject the proposal, even though this means that they gain nothing themselves and punishing their opponents by rejecting their offer has no effect on how they will be treated in subsequent interactions (since these are with new partners). The tendency for individuals to punish non-cooperators is not confined to Western populations and occurs widely in other populations, though the readiness of individuals to punish varies between societies (Henrich *et al.* 2006). Additional studies have also shown that individuals who punish partners that they perceive as having offered unfair (i.e. unequal) offers show heightened levels of activity in the caudate nucleus, an area of the brain that responds to pleasurable sensations (Fehr and Rockenbach 2004).

The evolutionary mechanisms that maintain strong reciprocity and 'altruistic' punishment are a topic of debate. One suggestion is that they are maintained by some form of group or multi-level selection favouring cooperation and the punishment of non-cooperators (Fehr and Fischbacher 2003). However, another interpretation is that although individuals have been told that they are involved in anonymous one-shot games, they are used to responding repeatedly with the same group of familiar individuals and continue to act as if they were doing so.

behaviour and variation in brain structure (Bickert 2012; Kanai 2012). Humans show an unusual degree of empathy between individuals and unusual understanding of the mental states and motives of others that allows individuals to anticipate each other's actions. These processes are associated with specific neural substrates and neuroimaging studies have documented circuits that respond to the perception of distress in other individuals (Lamm *et al.* 2007; Decety 2011). There also appear to be contrasts in the neuroendocrine systems underlying social behaviour between humans and other social mammals, as well as in cognition and in patterns of development (Bjorklund *et al.* 2010; Gallagher and Skuse 2010). For example, detailed comparisons of the development of cognitive abilities in human infants and infant great apes show that skills used in dealing with social interactions (including imitative learning, gestural communication and understanding intentions) are more highly developed in humans than in other animals (Herrmann *et al.* 2007, 2009). In conjunction, these comparisons suggest that the contrast between humans and the other great apes lies not so much in the greater computational power of their brains as in their capacity to apply their cognitive abilities to social interactions and issues (Tomasello and Moll 2010). These differences are amplified among older juveniles: human children are both more likely to play cooperative games and are more adept at solving problems that require cooperation than apes of the same age (Warneken and Tomasello 2006) and they show evidence of greater awareness of social signals as well as of more sophisticated social communication than the other apes (Tomasello and Moll 2010).

Third, several studies show that, even under urban conditions, measures of social connectedness and involvement of individuals in positive interactions with other humans are correlated both with variation in brain structure (Bickart *et al.* 2012; Kanai *et al.* 2012) and with physical and psychological health and longevity (House *et al.* 1988; Holt-Lunstad *et al.* 2010). Similar correlations between social behaviour and health also occur across communities. For example, across the USA, estimates of happiness increase with social connectedness and measures of 'social capital' (an index that incorporates measures of connectedness and contributions to public goods) are positively related to variation in health and negatively to age-specific mortality (Figure 20.27) (Putnam 2000). Measures of social connectedness and social capital in the USA have declined over the last 40 years and so, too, have estimates of happiness. Further evidence of the positive effects of social interactions comes from studies of relationships between humans and their pets, which show that interactions with pets help to reduce levels of perceived stress, blood pressure and antisocial behaviour (Furst 2006).

(a)

(b)

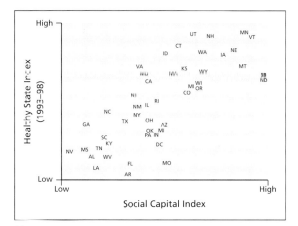

(a)

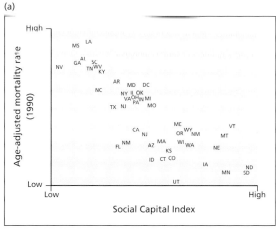

(b)

Figure 20.27 Variation in measures of (a) public health and (b) age-adjusted mortality rates across states in the USA between 1990 and 1998 are correlated with contrasts in Social Capital Index (a measure which estimates individual connectedness and contributions to public goods).
Source: Adapted from Putnam (2000). Reproduced with permission of Simon & Schuster.

Figure 20.28 (a) Holocaust victims at Lager Nordhausen (12 April 1945), where around 20,000 inmates were killed by the Nazis. (b) Skulls on display at the Tuol Sleng genocide museum, Cambodia. *Sources*: (a) © James E Myers. https://commons.wikimedia.org/wiki/File%3ARows_of_bodies_of_dead_inmates_fill_the_yard_of_Lager_Nordhausen%2C_a_Gestapo_concentration_camp.jpg. (b) © Pacifica/The Image Bank/Getty Images.

While it is possible that humans have a heritable tendency towards tolerance, sociality and social awareness, the available evidence falls some way short of providing a strong indication that this is the case. An alternative interpretation is that humans possess adaptations to social life that facilitate recognition of social signals and opportunities to modify their actions so as to derive benefits from social encounters and avoid costs. Their social responses are often strongly influenced by individual experience, social learning and cultural mechanisms as is the development of the structures and processes that control them (Maguire *et al.* 2000) and, it maybe possible that they have no innate tendency either to be helpful or to be cruel. As this second interpretation would suggest, prosocial tendencies differ widely between individuals and are strongly affected by early experience as well as by cultural differences (Penner *et al.* 2005; Dovidio *et al.* 2006; Norenzayan 2013). In most societies, cooperative or affiliative behaviour is most commonly directed at familiar individuals whereas strangers or outsiders may be treated in a fashion that is difficult to reconcile with any innate commitment to empathy or universal cooperation (Figure 20.28). Like many examples of extreme altruism, torture and genocidal killing are often perpetrated not by psychopaths but by ordinary people, whose social or political context has encouraged them to distance themselves from their victims (Fiske *et al.* 2004; Waller 2007; Travis 2010).

Nor is it the case that distancing of this kind is uncommon:

> Father and Mother, and Me,
> Sister and Auntie say
> All the people like us are We,
> And every one else is They.
> And They live over the sea,
> While We live over the way,
> But – would you believe it? – They look upon We
> As only a sort of They!

We and They by Rudyard Kipling

It is also important to appreciate that cooperation between groups of humans (including tribes, chiefdoms and states) is typically more limited, less reliable and less generous than cooperation between individuals. Although political units frequently make agreements, treaties and alliances and some deliver important benefits (Falger 1992), states commonly pursue their own interests and the terms of agreements and treaties and the obligations involved are frequently modified or ignored (Singer and Small 1966; Leeds 2003). Historical examples are numerous, but relationships between the Anglo Saxon populations of eastern England and the Norse invaders provide a good example that illustrates both the fickleness of inter-state agreements and the ineffectiveness of attempts to control competitors through rewards (Box 20.4). There is little indication that treaties between democracies are more effective or more stable than those between other kinds of

Box 20.4 Danegeld

One of the best-known examples of the limitations of attempts to control competitors or exploiters through reward systems is provided by payments made by the Anglo-Saxon kings of England to Norse invaders between 991 and 1016 to leave them alone. After losing the battle of Maldon in 991, Æthelred, the Anglo-Saxon king of England, paid Viking invaders 10,000 Roman pounds (3300 kg of silver) to go back to Scandinavia. Three years later they returned, demanding a larger payment, and Viking invasions, followed by payments of 'Danegeld' of rapidly escalating size, were made every few years until 1016, when a Viking leader, Knut, eventually became King of England.

Kipling neatly encapsulates the fallacy of attempts to control exploiters by rewards or bribes:

> It is always a temptation to an armed and agile nation
> To call upon a neighbour and to say:
> 'We invaded you last night – we are quite prepared to fight,
> Unless you pay us cash to go away.'

> And that is called paying the Dane-geld;
> But we've proved it again and again,
> That if once you have paid him the Dane-geld
> You never get rid of the Dane.

Dane-geld by Rudyard Kipling

governments and the need for democratic politicians to court public opinion and the relatively high frequency with which governments change may reduce their reliability as allies (Gaubatz 1996; Gartzke and Gleditsch 2003).

The limited effectiveness and duration of agreements between independent states reflects the problems of policing and enforcing agreements at an international level. Unlike individuals within close social groups, states are frequently able to disguise their actions and pursue their own interests when these differ from agreements made with competitors, neighbours or partners and their representatives are carefully trained in the control and manipulation of information (Dearth 2002; Reid 2002). Unilateral or multilateral attempts to enforce international agreements are often likely to trigger defensive responses that can have high costs to enforcers. As a result, agreements between states are particularly susceptible to cheating at all levels and overt or covert defection is common.

20.6 The human condition

What conclusions can be drawn from this brief review of human behaviour? First, that humans evolved from a polygynous ape whose occupation of a partially terrestrial niche encouraged the development of advanced cooperative behaviour, which led in turn to the evolution of energetic interdependence between group members, increasingly complex social relationships involving cooperation and manipulation, advances in social awareness and communication, and associated increases in brain structure and size. Improvements in communication and the eventual development of language may have allowed early humans to make (and enforce) agreements and informal contracts of a kind not found among non-human animals, permitting the evolution of new forms of cooperation between unfamiliar as well as familiar individuals. The sharing of resources in many modern hunter-gatherer communities suggests that energetic interdependence of group members probably persisted in most societies until the development of agriculture.

So how similar are humans, and human societies, to those of other mammals? Similarities between human behaviour and the behaviour of non-human mammals are widespread and many are a consequence of shared physiological and developmental processes. There are also important similarities in the benefits of social living

and the structure of relationships. As in many other mammals, groups compete with each other, sociality provides protection against predators and neighbours and larger groups provide more effective protection than small ones. Within groups, social competition, alliances and coalitions and dominance hierarchies are all common and a combination of punishments and rewards are used to manipulate others. Family ties are pervasive and often form a basis for cooperation and support.

There are also similarities in reproductive behaviour. Like other female mammals, women are seldom strongly attracted to men with whom they were reared as children and show preferences for partners with relatively masculine features, while men tend to prefer partners whose shape and appearance are indicators of heightened oestrogen levels and fecundity. Birth and the presence of young infants trigger similar hormonal changes in humans and other mammals, while conflicts generate rises in cortisol and testosterone that are often associated with increased levels of aggression. As in many sexually dimorphic species, heightened levels of testosterone and increased rates of growth in males raise their metabolic needs, and they are commonly more susceptible to food shortage or other forms of environmental hardship than females. As in other mammals, the effects of variation in hormone levels differ between individuals and are likely to have some genetic basis.

But there are important differences, too. Kinship ties among humans are maintained with dispersed kin as well as with the kin of marriage partners, so that kinship networks are more extensive and provide a broader basis for alliances between social groups. Language allows individuals and groups to make agreements and to exchange resources or services. It also permits group members to establish and enforce social contracts and to develop ethical systems and religious codes that structure social processes. These can, in turn, reinforce cooperation within groups and can lead to extreme forms of cooperation that have no parallel in other animals (see Box 20.2). The evolution of language is also likely to have played an important role in the development and effectiveness of social learning and to have facilitated the transfer and sharing of increasingly specific and complex technological information between individuals. This in turn may have resulted in the cumulative development and improvement of technology over time and stimulated successive improvements in established techniques and novel solutions to pre-existing problems (see Chapter 19).

Changes in social and technological culture must have modified the selection pressures operating on individuals. In some cases, they may have allowed populations to remain in changing habitats or to occupy novel ones until selection pressure led to the evolution of heritable adaptations (Crispo 2007). In others, cultural changes may have generated novel selection pressures and stimulated coevolutionary processes involving cultural and genetic adaptations (Richerson and Boyd 2008). Similar processes occur in non-human animals, too, but the greater technological development of humans has increased their extent and intensity.

The spread of agriculture and the development of human technology have had profound demographic and ecological consequences. They have led to increases in group size and population density as well as in social stratification and variance in breeding success (Johnson and Earle 2000; Marlowe 2000; Betzig 2012). Human populations have continued to increase. A world population of an estimated 35,000 hunter-gatherers is thought to have increased to around 15 million by the time agriculture was invented around 12,000 years ago. The world population is thought to have risen to around 1 billion by 1800, to 6 billion by 2000, and now (2015) exceeds 7 billion (Figure 19.6). United Nations' projections suggest that human numbers are likely to rise to between 8 and 11 billion by 2050 and may then stabilise. This spectacular increase has been associated with a progressive increase in the proportion of the population living in cities, where heightened levels of social stress are associated with increased incidence of mood and anxiety disorders as well as schizophrenia (Lederbogen *et al.* 2011).

Increases in human numbers have also had profound consequences for other terrestrial life forms (Barnosky *et al.* 2004) and marine ecosystems (Jackson *et al.* 2001) as well as for the natural resources of the planet, including air and water (Blois *et al.* 2013). The consequences of human impact are now affecting global climates and are likely to have increasingly disadvantageous effects on many human populations, as well as for other animals (Blois *et al.* 2013; Diffenbaugh and Field 2013; Kubiszewski *et al.* 2013; Wheeler and von Braun 2013). While a combination of improvements in agricultural and medical technology and an extension of international aid has limited the consequences of increasing numbers and improved living standards on human welfare so far,

this cannot be sustained indefinitely and effective global agreements will be necessary to reduce the rate of increase and modify the impact of human populations on the natural resources of the planet.

A final contrast between humans and other animals is that the evolution of human cognitive abilities and an understanding of cause and effect allows us to identify and anticipate the social and environmental consequences of further increases in population size and living standards while our technological development provides the tools necessary to control population growth and to reduce the effects of increasing living standards on natural ecosystems, and to arrest environmental deterioration. However, their effective deployment will require individuals and states to accept substantial short-term costs in order to access larger long-term benefits shared with unknown individuals in other populations. Mutualistic cooperation on this scale would be a novel development in mammal societies and is likely to be particularly susceptible to cheating. It remains to be seen whether humans will be able to meet this challenge and to mobilise the ethical, political and economic motivation to counter the threats generated by their own success.

References

Aengst, J. (2014) Silences and moral narratives: infanticide as reproductive disruption. *Medical Anthropology* **33**:411–427.

Alvard, M.S. (2002) Carcass ownership and meat distribution by big-game cooperative hunters. *Research in Economic Anthropology* **21**:99–131.

Alvarez, H. (2004) Residence groups among hunter-gatherers: a view of the claims and evidence for patrilocal bands. In: *Kinship and Behavior in Primates* (eds B. Chapais and C. Berman). Oxford: Oxford University Press, 420–441.

Alvergne, A., *et al.* (2009) Father–offspring resemblance predicts paternal investment in humans. *Animal Behaviour* **78**:61–69.

Anderson, K.G. (2006) How well does paternity confidence match actual paternity? *Current Anthropology* **47**:513–520.

Anderson, K.G., *et al.* (1999) Paternal care by genetic fathers and stepfathers. I: reports from Albuquerque men. *Evolution and Human Behavior* **20**:405–431.

Apicella, C.L. and Marlowe, F.W. (2004) Perceived male fidelity and paternal resemblance predict men's investment in children. *Evolution and Human Behavior* **25**:371–378.

Apicella, C.L., *et al.* (2007) Voice pitch predicts reproductive success in male hunter-gatherers. *Biology Letters* **3**:682–684.

Barker, D.J. (1998) *Mothers, Babies and Health in Later Life.* Edinburgh: Churchill Livingstone.

Barker, D.J., *et al.* (1993) Fetal nutrition and cardiovascular disease in adult life. *Lancet* **341**:938–941.

Barker, J.L., *et al.* (2012) Within-group competition reduces cooperation and payoffs in human groups. *Behavioral Ecology* **23**:735–741.

Barnosky, A.D., *et al.* (2004) Assessing the causes of late Pleistocene extinctions on the continents. *Science* **306**:70–75.

Barrett, L., *et al.* (2002) *Human Evolutionary Psychology*. Princeton, NJ: Princeton University Press.

Bateson, M., *et al.* (2006) Cues of being watched enhance cooperation in a real-world setting. *Biology Letters* **2**: 412–414.

Belsky, J., *et al.* (1991) Childhood experience, interpersonal development, and reproductive strategy: and evolutionary theory of socialization. *Child Development* **62**:647–670.

Betzig, L. (2012) Means, variances, and ranges in reproductive success: comparative evidence. *Evolution and Human Behavior* **33**:309–317.

Bevc, I. and Silverman, I. (1993) Early proximity and intimacy between siblings and incestuous behavior: a test of the Westermarck theory. *Ethology and Sociobiology* **14**:171–181.

Bickart, K.C., *et al.* (2012) Intrinsic Amygdala–Cortical Functional Connectivity Predicts Social Network Size in Humans. *Journal of Neuroscience* **32**:14729–14741.

Bird, R.B., *et al.* (2001) The hunting handicap: costly signaling in human foraging strategies. *Behavioral Ecology and Sociobiology* **50**:9–19.

Bjorklund, D.F., *et al.* (2010) The evolution and development of human social cognition. In: *Mind the Gap: Tracing the Origin of Human Universals* (eds P. M. Kappeler and J.B. Silk). Berlin: Springer, 351–371.

Blois, J., *et al.* (2013) Climate change and the past, present, and future of biotic interactions. *Science* **341**:499–503.

Boehm, C. (2013) The biocultural evolution of conflict resolution between groups. In: *War, Peace, and Human Nature: The Convergence of Evolutionary and Cultural Views* (ed. D.P. Fry). Oxford: Oxford University Press, 315–340.

Bogaert, A.F. (2008) Menarche and father absence in a national probability sample. *Journal of Biosocial Science* **40**:623–636.

Borgerhoff Mulder, M. (1990) Kipsigis women's preferences for wealthy men: evidence for female choice in mammals. *Behavioral Ecology and Sociobiology* **27**:255–264.

Borgerhoff Mulder, M. and Beheim, B.A. (2011) Understanding the nature of wealth and its effects on human fitness. *Philosophical Transactions of the Royal Society B: Biological Sciences* **366**:344–356.

Borgerhoff Mulder, M. and Rauch, K.L. (2009) Sexual conflict in humans: variation and solutions. *Evolutionary Anthropology* **18**:201–214.

Borgerhoff Mulder, M. and Schracht, R. (2012) Human behavioural ecology. In: *eLS*. Chichester: John Wiley & Sons Ltd. http://www.els.net [doi: 10.1002/9780470015902.a0003671. pub2] .

Bourget, D., *et al.* (2007) A review of maternal and paternal filicide. *Journal of the American Academy of Psychiatry and the Law* **35**:74–82.

Bowers, R.I., *et al.* (2011) Generalization in mate-choice copying in humans. *Behavioral Ecology* **23**:112–124.

Bowles, S. (2008) Being human: conflict: altruism's midwife. *Nature* **456**:326–327.

Bowles, S. and Gintis, H. (2011) *A Cooperative Species: Human Reciprocity and Its Evolution*. Princeton, NJ: Princeton University Press.

Boyd, R. and Richerson, P.J. (2009) Culture and the evolution of human cooperation. *Philosophical Transactions of the Royal Society B: Biological Sciences* **364**:3281–3288.

Boyd, R. and Silk, J.B. (2000) *How Humans Evolved*. New York: W.W. Norton and Company.

Bridiescac, R.G., *et al.* (2012) Male life history, reproductive effort, and the evolution of the genus *Homo*: new directions and perspectives. *Current Anthropology* **53**:S424–S435.

Burkart, J.M. and van Schaik, C.P. (2010) Cognitive consequences of cooperative breeding in primates? *Animal Cognition* **13**:1–19.

Burnham, T.C., *et al.* (2003) Men in committed romantic relationships have lower testosterone. *Hormones and Behavior* **44**:119–122.

Buss, D.M. (2006) Strategies of human mating. *Psychological Topics* **15**:239–260.

Buston, P.M. and Emlen, S.T. (2003) Cognitive processes underlying human mate choice: the relationship between self-perception and mate preference in Western society. *Proceedings of the National Academy of Sciences of the United States of America* **100**:8805–8810.

Butovskaya, M.L. (2013) Aggression and conflict resolution among the nomadic Hadza of Tanzania compared with their pastoralist neighbors. In: *War, Peace, and Human Nature: The Convergence of Evolutionary and Cultural Views* (ed. D.P. Fry). Oxford: Oxford University Press, 278–296.

Camperio-Ciani, A., *et al.* (2004) Evidence for maternally inherited factors favouring male homosexuality and promoting female fecundity. *Proceedings of the Royal Society of London. Series B: Biological Sciences* **271**:2217–2221.

Camperio-Ciani, A., *et al.* (2009) Genetic factors increase fecundity in female maternal relatives of bisexual men as in homosexuals. *Journal of Sexual Medicine* **6**:449–455.

Camperio-Ciani, A., *et al.* (2012) Factors associated with higher fecundity in female maternal relatives of homosexual men. *Journal of Sexual Medicine* **9**:2878–2887.

Cant, M.A. and Johnstone, R.A. (2008) Reproductive conflict and the separation of reproductive generations in humans. *Proceedings of the National Academy of Sciences of the United States of America* **105**:5332–5335.

Carter, C.S., *et al.* (2008) Oxytocin, vasopressin and sociality. *Progress in Brain Research* **170**:331–336.

Cassar, E., *et al.* (2003) A descriptive model of the homicide process. *Behaviour Change* **20**:76–93.

Chagnon, N. A. (1988) Life histories, blood revenge and warfare in a tribal population. *Science* **239**:985–992.

Chagnon, N.A. (1997) *Yanomamö*. Fort Worth, TX: Harcourt Brace.

Chapais, B. (2008) *Primeval Kinship. How Pair-Bonding Gave Birth to Human Society*. Cambridge, MA: Harvard University Press.

Cohen, A.A. (2004) Female post-reproductive lifespan: a general mammalian trait. *Biological Reviews of the Cambridge Philosophical Society* **79**:733–750.

Collins, S.A. and Missing, C. (2003) Vocal and visual attractiveness are related in women. *Animal Behaviour* **65**: 997–1004.

Comings, D.E., *et al.* (2002) Parent–daughter transmission of the androgen receptor gene as an explanation of the effect of father absence on age of menarche. *Child Development* **73**:1046–1051.

Crespi, B.J. (2014) The insectan apes. *Human Nature* **25**: 6–27.

Crispo, E. (2007) The Baldwin effect and genetic assimilation: revisiting two mechanisms of evolutionary change mediated by phenotypic plasticity. *Evolution* **61**:2469–2479.

Crognier, E., *et al.* (2002) Helping patterns and reproductive success in Aymara communities. *American Journal of Human Biology* **14**:372–379.

Daly, M. and Wilson, M. (1988) Evolutionary social psychology and family homicide. *Science* **242**:519–524.

Daly, M. and Wilson, M.I. (1996) Violence against stepchildren. *Current Directions in Psychological Science* **5**:77–81.

Daly, M. and Wilson, M. (1997) Crime and conflict: homicide in evolutionary psychological perspective. *Crime and Justice* **22**:51–100.

Daly, M., *et al.* (1997) Women with children sired by previous partners incur excess risk of uxoricide. *Homicide Studies* **1**:61–71.

Dearth, D.H. (2002) Shaping the information space. *Journal of Information Warfare* **1**:1–15.

Decety, J. (2011) The neuroevolution of empathy. *Annals of the New York Academy of Sciences* **1231**:35–45.

de Waal, F.B.M. (1996) *Good Natured: The Origins of Rich and Wrong in Humans and Other Animals*. Cambridge, MA: Harvard University Press.

Diffenbaugh, N. and Field, C. (2013) Changes in ecologically critical terrestrial climate conditions. *Science* **341**:486–492.

Dovidio, J.F., *et al.* (2006) *The Social Psychology of Prosocial Behavior*. Hillsdale, NJ: Lawrence Erlbaum Associates.

Durham, W.H. (1976) Resource competition and human aggression, part I: a review of primitive war. *Quarterly Review of Biology* **51**:385–415.

Dye, D.H. (2013) Trends in cooperation and conflict in native eastern North America. In: *War, Peace, and Human Nature: The Convergence of Evolutionary and Cultural Views* (ed. D.P. Fry). Oxford: Oxford University Press, 132–150.

Ellis, B.J. and Garber, J. (2000) Psychosocial antecedents of variation in girls' pubertal timing: maternal depression, stepfather presence, and marital and family stress. *Child Development* **71**:485–501.

Ellison, P.T. (1990) Human ovarian function and reproductive ecology: new hypotheses. *American Anthropologist* **92**: 191–195.

Ellison, P.T. (2001) *On Fertile Ground: Natural History of Human Reproduction*. Cambridge, MA: Harvard University Press.

Ellison, P.T. and Valeggia, C.R. (2003) C-peptide levels and lactational amenorrhea among Toba women of northern Argentina. *Fertility and Sterility* **80**:1279–1280.

Ember, C.R. and Ember, M. (1994) War, socialization, and interpersonal violence: a cross-cultural study. *Journal of Conflict Resolution* **38**:620–646.

Emery Thompson, M., *et al.* (2012) The energetics of lactation and the return to fecundity in wild chimpanzees. *Behavioral Ecology* **23**:1234–1241.

Endicott, K. (2013) Peaceful foragers: the significance of the Batek and Moriori for the questions of innate human violence. In: *War, Peace, and Human Nature: The Convergence of Evolutionary and Cultural Views* (ed. D.P. Fry). Oxford: Oxford University Press, 243–261.

Falger, V.S.E. (1992) Cooperation in conflict: alliances in international politics. In: *Coalitions and Alliances in Humans and Other Animals* (eds A.H. Harcourt and F.B.M. de Waal). Oxford: Oxford University Press.

Fehr, E. and Fischbacher, U. (2003) The nature of human altruism. *Nature* **425**:785–791.

Fehr, E. and Gächter, S. (2000) Fairness and retaliation: the economics of reciprocity. *Journal of Economic Perspectives* **14**:159–181.

Fehr, E. and Gächter, S. (2002) Altruistic punishment in humans. *Nature* **415**:137–140.

Fehr, E. and Rockenbach, B. (2004) Human altruism: economic, neural, and evolutionary perspectives. *Current Opinion in Neurobiology* **14**:784–790.

Feinberg, D.R., *et al.* (2005) Manipulations of fundamental and formant frequencies influence the attractiveness of human male voices. *Animal Behaviour* **69**:561–568.

Ferguson, R.B. (2013) Pinker's list: exaggerating prehistoric war mortality. In: *War, Peace, and Human Nature: The Convergence of Evolutionary and Cultural Views* (ed. D.P. Fry). Oxford: Oxford University Press, 112–131.

Fernandez-Duque, E., *et al.* (2009) The biology of paternal care in human and nonhuman primates. *Annual Review of Anthropology* **38**:115–130.

Fessler, D.M.T. and Navarrete, C.D. (2004) Third-party attitudes toward sibling incest: evidence for Westermarck's hypotheses. *Evolution and Human Behavior* **25**:277–294.

Fiske, S.T., *et al.* (2004) Policy forum: Why ordinary people torture enemy prisoners. *Science* **306**:1482–1483.

Fleming, A.S., *et al.* (2002) Testosterone and prolactin are associated with emotional responses to infant cries in new fathers. *Hormones and Behavior* **42**:399–413.

Freud, S. (1950) *Totem and Taboo: Some Points of Agreement Between the Mental Lives of Savages and Neurotics*. London: Routledge.

Friedman, S H., et al. (2005) Child murder by mothers: a critical analysis of the current state of knowledge and a research agenda. *American Journal of Psychiatry* **162**:1578–1587.

Fry, D.P. (2006) *The Human Potential For Peace: An Anthropological Challenge to Assumptions About War and Violence*. New York: Oxford University Press.

Fry, D.P. (2013) War, peace, and human nature: the challenge of achieving scientific objectivity. In: *War, Peace, and Human Nature: The Convergence of Evolutionary and Cultural Views* (ed. D.P. Fry). Oxford: Oxford University Press, 1–22.

Furnham, A. and Reeves, E. (2006) The relative influence of facial neoteny and waist-to-hip ratio on judgements of female attractiveness and fecundity. *Psychology, Health and Medicine* **11**:129–141.

Furnham, A., et al. (2003) A cross-cultural comparison of ratings of perceived fecundity and sexual attractiveness as a function of body weight and waist-to-hip ratio. *Psychology, Health and Medicine* **8**:219–230.

Furst, G. (2006) Prison-based animal programs: a national survey. *The Prison Journal* **86**:407–430.

Gallagher, I. and Skuse, D. (2010) Molecular and genetic influences on the neural substrate of social cognition in humans. In: *Social Behaviour: Genes, Ecology and Evolution* (eds T. Székely, A.J. Moore and J. Komdeur). Cambridge: Cambridge University Press, 446–469.

Gangestad, S.W. and Thornill, R. (2003) Facial masculinity and fluctuating asymmetry. *Evolution and Human Behavior* **24**:231–241.

Gartzke, E. and Gleditsch, K. (2003) Why democracies may actually be less reliable allies. *American Journal of Political Science* **48**:775–795.

Gaubatz, K. (1996) Democratic states and commitment in international relations. *International Organization* **50**: 109–139.

Geary, D.C. (2000) Evolution and proximate expression of human paternal investment. *Psychological Bulletin* **126**:55–77.

Geary, D.C. (2010) *Male, Female: The Evolution of Human Sex Differences*. Washington, DC: American Psychological Association.

Geary, D.C., et al. (2004) Evolution of human mate choice. *Journal of Sex Research* **41**:27–42.

Gettler, L.T., et al. (2011) Longitudinal evidence that fatherhood decreases testosterone in human males. *Proceedings of the National Academy of Sciences of the United States of America* **208**:16194–16199.

Gilby, I.C. (2006) Meat sharing among the Gombe chimpanzees: harassment and reciprocal exchange. *Animal Behaviour* **71**:953–963.

Gintis, H., et al. (2003) Explaining altruistic behavior in humans. *Evolution and Human Behavior* **24**:153–172.

Glowacki, L. and von Rueden, C. (2015) Leadership solves collective action problems in small-scale societies. *Philosophical Transactions of the Royal Society B: Biological Sciences*. DOI: 10.1098/rstb.2015.0010. [Accessed 27 January 2016]

Gray, P.B. and Anderson, K.G. (2010) *Fatherhood: Evolution and Human Paternal Behavior*. Cambridge, MA: Harvard University Press.

Gray, P.B. and Crittenden, A.N. (2014) Father Darwin: effects of children on men, viewed from an evolutionary perspective. *Fathering* **12**:121–142.

Gray, P.B. and Garcia, J.R. (2013) *Evolution and Human Sexual Behavior*. Cambridge, MA: Harvard University Press.

Gray, P.B., et al. (2002) Marriage and fatherhood are associated with lower testosterone in males. *Evolution and Human Behavior* **23**:193–201.

Gray, P.B., et al. (2006) Fathers have lower salivary testosterone levels than unmarried men and married non-fathers in Beijing, China. *Proceedings of the Royal Society of London. Series B: Biological Sciences* **273**:333–339.

Haas, J. and Piscitelli, M. (2013) The prehistory of warfare: misled by ethnography. In: *War, Peace, and Human Nature: The Convergence of Evolutionary and Cultural Views* (ed. D.P. Fry). Oxford: Oxford University Press, 168–190.

Hales, C.N. and Barker, D.J. (1992) Type 2 (non-insulin-dependent) diabetes mellitus: the thrifty phenotype hypothesis. *Diabetologia* **35**:595–601.

Hames, R. (1992) Variation in paternal care among the Yanomamö. In: *Father–Child Relations: Cultural and Biosocial Contexts* (ed. B.S. Hewlett). New York: Walter de Gruyter.

Harris, G.T., et al. (2007) Children killed by genetic parents versus stepparents. *Evolution and Human Behavior* **28**:85–95.

Hauser, O.P., et al. (2014) Cooperating with the future. *Nature* **511**:220–223.

Hawkes, K. (1991) Showing off: tests of an hypothesis about men's foraging goals. *Ethology and Sociobiology* **12**:29–54.

Hawkes, K. (2003) Grandmothers and the evolution of human longevity. *American Journal of Human Biology* **15**:380–400.

Hawkes, K. and O'Connell, J.F. (1997) Hadza women's time allocation, offspring provisioning, and the evolution of long postmenopausal life spans. *Current Anthropology* **38**:551–577.

Hawkes, K., et al. (1989) Hardworking Hadza grandmothers. In: *Comparative Socioecology: The Behavioural Ecology of Humans and Other Mammals* (eds V. Standen and R. A. Foley). Oxford: Blackwell Scientific Publications.

Hawkes, K., et al. (1998) Grandmothering, menopause, and the evolution of human life histories. *Proceedings of the National Academy of Sciences of the United States of America* **95**:1336–1339.

Hawkes, K., et al. (2001) Hadza meat sharing. *Evolution and Human Behavior* **22**:113–142.

Hayward, A.D. and Lummaa, V. (2013) Testing the evolutionary basis of the predictive adaptive response hypothesis in a preindustrial human population. *Evolution, Medicine and Public Health* **2013**:106–117.

Hayward, A.D., et al. (2013) Influence of early-life nutrition on mortality and reproductive success during a subsequent famine in a preindustrial population. *Proceedings of the National Academy of Sciences of the United States of America* **110**:13886–13891.

Helle, S., *et al.* (2005) Are reproductive and somatic senescence coupled in humans? Late, but not early, reproduction correlated with longevity in historical Sami women. *Proceedings of the Royal Society of London. Series B. Biological Sciences* **272**:29–37.

Henrich, J., *et al.* (2006) Costly punishment across human societies. *Science* **312**:1767–1770.

Henrich, J., *et al.* (2010) Most people are not WEIRD. *Nature* **466**:29

Herrmann, E., *et al.* (2007) Humans have evolved specialized skills of social cognition: the culture intelligence hypothesis. *Science* **317**:1360–1366.

Herrmann, E., *et al.* (2009) The structure of individual differences in the cognitive abilities of children and chimpanzees. *Psychological Science* **21**:102–110.

Hewlett, B.S., *et al.* (1998) Culture and early infancy among Central African foragers and farmers. *Developmental Psychobiology* **34**:653–661.

Hiatt, B. (1974) Woman the gatherer. In: *Woman's Role in Aboriginal Society* (ed. F. Gale). Canberra, ACT: Australian Institute of Aboriginal Studies.

Hill, K. and Hawkes, K. (1983) Neotropical hunting among the Ache of Eastern Paraguay. In: *Adaptive Responses of Native Amazonians* (eds R. Hames and W. Vickers). New York: Academic Press, 139–187.

Hill, K. and Hurtado, A.M. (1991) The evolution of premature reproductive senescence and menopause in human females. *Human Nature* **2**:313–350.

Hill, K.R. and Hurtado, A.M. (1996) *Aché Life History: The Ecology and Demography of a Foraging People*. New York: de Gruyter.

Hinde, R.A. (2009) *Why Gods Persist: A Scientific Approach to Religion*. London and New York: Routledge.

Hölldobler, B. and Wilson, E.O. (1990) *The Ants*. Cambridge, MA: Harvard University Press.

Holt-Lunstad, J., *et al.* (2010) Social relationships and mortality risk: a meta-analytic review. *PLOS Medicine* **7** (7): e1000316.

House, J.S., *et al.* (1988) Social relationships and health. *Science* **241**:540–545.

Hrdy, S.B. (1981) *The Woman That Never Evolved*. Cambridge, MA: Harvard University Press.

Hrdy, S.B. (1999) *Mother Nature: A History of Mothers, Infants and Natural Selection*. New York: Random House.

Hrdy, S.B. (2005) Evolutionary context of human development: the cooperative breeding model. In: *Attachment and Bonding: A New Synthesis* (eds C.S. Carter *et al.*). Cambridge, MA: MIT Press, 9–31.

Hrdy, S.B. (2009) *Mothers and Others: The Evolutionary Origins of Mutual Understanding*. Cambridge, MA: Harvard University Press.

Hughbank, R.J. and Grossman, D. (2013) The challenge of getting men to kill: a view from military science. In: *War,*

Peace, and Human Nature: The Convergence of Evolutionary and Cultural Views (ed. D.P. Fry). Oxford: Oxford University Press, 495–513.

Hughes, S.M., *et al.* (2004) Ratings of voice attractiveness predict sexual behavior and body configuration. *Evolution and Human Behavior* **25**:295–304.

Hulanicka, B. (1999) Acceleration of menarcheal age of girls from dysfunctional families. *Journal of Reproductive and Infant Psychology* **17**:119–132.

Hundie, B. (2010) Conflicts between Afar pastoralists and their neighbors: triggers and motivations. *International Journal of Conflict and Violence* **4**:134–148.

Ingham, J.M. and Spain, D.H. (2005) Sensual attachment and incest avoidance in human evolution and child development. *Journal of the Royal Anthropological Institute* **11**:677–701.

Isler, K. and van Schaik, C.P. (2009) The expensive brain: a framework for explaining evolutionary changes in brain size. *Journal of Human Evolution* **57**:392–400.

Jackson, J.B., *et al.* (2001) Historical overfishing and the recent collapse of coastal ecosystems. *Science* **293**:629–637.

Jacob, S., *et al.* (2002) Paternally inherited HLA alleles are associated with women's choice of male odor. *Nature Genetics* **30**:175–179.

Janssen, M.A. and Hill, K. (2014) Benefits of grouping and cooperative hunting among Ache hunter-gatherers: insights from an agent-based foraging model. *Human Ecology* **42**:823–825.

Jasieńska, G., *et al.* (2004) Large breasts and narrow waists indicate high reproductive potential in women. *Proceedings of the Royal Society of London. Series B: Biological Sciences* **271**:1213–1217.

Jasieńska, G., *et al.* (2006) Symmetrical women have higher potential fertility. *Evolution and Human Behavior* **27**:390–400.

Johnson, A.W. and Earle, T. (2000) *The Evolution of Human Societies: From Foraging Band to Agrarian State*. Stanford, CA: Stanford University Press.

Johnston, V.S., *et al.* (2001) Male facial attractiveness: evidence for hormone-mediated adaptive design. *Evolution and Human Behavior* **22**:251–267.

Johnstone, R.A. and Cant, M.A. (2010) The evolution of menopause in cetaceans and humans: the role of demography. *Proceedings of the Royal Society of London. Series B: Biological Sciences* **277**:3765–3771.

Jones, B.C., *et al.* (2007) Social transmission of face preferences among humans. *Proceedings of the Royal Society of London. Series B: Biological Sciences* **274**:899–903.

Kachel, A.F., *et al.* (2011) Grandmothering and natural selection revisited. *Proceedings of the Royal Society of London. Series B: Biological Sciences* **278**:1939–1941.

Kanai, R., *et al.* (2012) Online social network size is reflected in human brain structure. *Philosophical Transactions of the Royal Society B: Biological Sciences*. DOI: 10.1098/rspb.2011.1959. [Accessed 27 January 2016]

Kaplan, H. (1997) The evolution of the human life course. In: *Between Zeus and the Salmon: The Biodemography of Longevity* (eds

K. Wachter and C.E. Finch). Washington, DC: National Academy of Sciences, 175–211.

Kaplan, H. and Gurven, M. (2005) The natural history of human food sharing and cooperation: a review and a new multi-individual approach to the negotiation of norms. In: *Moral Sentiments and Material Interests: The Foundations of Cooperation in Economic Life* (eds H. Gintis, S. Bowles, R. Boyd and E. Fehr). Cambridge, MA: MIT Press, 75–113.

Kaplan, H. and Hill, K. (1985) Food sharing among Ache foragers: tests of explanatory hypotheses. *Current Anthropology* **26**:223–245.

Kaplan, H., *et al.* (2000) A theory of human life history evolution: diet, intelligence, and longevity. *Evolutionary Anthropology* **9**:156–185.

Keeley, L. H. (1996) *War Before Civilization*. New York: Oxford University Press.

Kelly, R.C. (2005) The evolution of lethal intergroup violence. *Proceedings of the National Academy of Sciences of the United States of America* **102**:15294–15298.

Kelly, R.L. (1995) *The Foraging Spectrum. Diversity in Hunter-gatherer Lifeways*. Washington, DC: Smithsonian Institution Press

Kelly, R.L. (2013) From the peaceful to the warlike: ethnographic and archaeological insights into hunter-gatherer warfare and homicide. In: *War, Peace, and Human Nature: The Convergence of Evolutionary and Cultural Views* (ed. D.P. Fry). Oxford: Oxford University Press, 151–167.

Kentner, A.C., *et al.* (2010) Modeling dad: animal models of paternal behavior. *Neuroscience and Biobehavioral Reviews* **34**:438–451.

Kirkpatrick, R.C. (2000) The evolution of human homosexual behavior. *Current Anthropology* **41**:385–413.

Kirkwood, T.B.L. and Shanley, D.P. (2010) The connections between general and reproductive senescence and the evolutionary basis of menopause. *Annals of the New York Academy of Science* **1204**:21–29.

Kokko, H. (2013) Conflict and restraint in animal species: implications for war and peace. In: *War, Peace, and Human Nature: The Convergence of Evolutionary and Cultural Views* (ed. D.P. Fry). Oxford: Oxford University Press, 38–53.

Kramer, K.L. (2010) Cooperative breeding and its significance to the demographic success of humans. *Annual Review of Anthropology* **39**:417–436.

Kramer, K.L. and Russell, A.F. (2014) Kin-selected cooperation without lifetime monogamy: human insights and animal implications. *Trends in Ecology and Evolution* **29**:600–606.

Kubiszewski, I., *et al.* (2013) Beyond GDP: measuring and achieving global genuine progress. *Ecological Economics* **93**:57–68.

Kuper, A. (2009) *Incest and Influence: The Private Life of Bourgeois England*. Cambridge, MA: Harvard University Press.

Lahdenperä, M., *et al.* (2004) Fitness benefits of prolonged postreproductive lifespan in women. *Nature* **428**:178–181.

Lahdenperä, M., *et al.* (2011) Selection on menopause in two premodern human populations: no evidence for the Mother hypothesis. *Evolution* **65**:476–489.

Lahdenperä, M., *et al.* (2012) Severe intergenerational conflict and the evolution of menopause. *Ecology Letters* **15**:1283–1290.

Lamm, C., *et al.* (2007) The neural substrate of human empathy: effects of perspective-taking and cognitive appraisal. *Journal of Cognitive Neuroscience* **19**:42–58.

Lanchester, F. (1956) Mathematics in warfare. In: *The World of Mathematics* (ed. J.R. Newman). New York: Simon & Schuster.

Lawson, D.W. and Mace, R. (2011) Parental investment and the optimization of human family size. *Philosophical Transactions of the Royal Society B: Biological Sciences* **366**:333–343.

Lawson, D.W., *et al.* (2015) No evidence that polygynous marriage is a harmful cultural practice in northern Tanzania. *Proceedings of the National Academy of Sciences of the United States of America* **112**:13827–13832.

Lea, A.J., *et al.* (2015) Developmental constraints in a wild primate. *American Naturalist* **185**:809–821.

Leavitt, G.C. (1990) Sociobiological explanations of incest avoidance: a critical review of evidential claims. *American Anthropologist* **92**:971–993.

Lederbogen, F., *et al.* (2011) City living and urban upbringing affect neural social stress processing in humans. *Nature* **474**:498–501.

Lee, R. (1979) *The !Kung San: Men Women and Work in a Foraging Society*. Cambridge: Cambridge University Press.

Leeds B. (2003) Alliance reliability in times of war: explaining state decisions to violate treaties. *International Organization* **57**:801–827.

Leff, J. (2009) Pastoralists at war: violence and security in the Kenya–Sudan–Uganda border region. *International Journal of Conflict and Violence* **3**:188–203.

Leonetti, D.L., *et al.* (2005) Kinship organization and the impact of grandmothers on reproductive success among the matrilineal Khasi and patrilineal Bengali of Northeast India. In: *Grandmotherhood: The Evolutionary Significance of the Second Half of Female Life* (eds E. Voland, A. Chasiotis and W. Schiefenhöevel). Piscataway, NJ: Rutgers University Press, 194–214.

Lepingwell, J.W.R. (1987) The law of combat? Lanchester reexamined. *International Security* **12**.89–134.

Lévi-Strauss, C. (1969) *Elementary Forms of Kinship*. Boston: Beacon Press.

Lieberman, D. (2009) Rethinking the Taiwanese minor marriage data: evidence the mind uses multiple kinship cues to regulate inbreeding avoidance. *Evolution and Human Behavior* **30**:153–160.

Lieberman, D., *et al.* (2003) Does morality have a biological basis? An empirical test of the factors governing moral sentiments relating to incest. *Proceedings of the Royal Society of London. Series B: Biological Sciences* **270**:819–826.

Little, A.C., *et al.* (2001) Self-perceived attractiveness influences human female preferences for sexual dimorphism and symmetry in male faces. *Proceedings of the Royal Society of London. Series B: Biological Sciences* **268**:39–44.

Little, A.C., *et al.* (2007) Human preferences for facial masculinity change with relationship type and environmental harshness. *Behavioral Ecology and Sociobiology* **61**:967–973.

Lummaa, V. (2003) Early developmental conditions and reproductive success in humans: downstream effects of prenatal famine, birthweight, and timing of birth. *American Journal of Human Biology* **15**:370–379.

Lummaa, V. and Clutton-Brock, T.H. (2002) Early development, survival and reproduction in humans. *Trends in Ecology and Evolution* **17**:141–147.

Mace, R. (2014) Human behavioural ecology and its evil twin. *Behavioral Ecology* **25**:443–449.

Mace, R. and Sear, R. (2005) Are humans cooperative breeders? In: *Grandmotherhood: The Evolutionary Significance of the Second Half of Female Life* (eds E. Voland, A. Chasiotis and W. Schiefenhöevel). Piscatawy, NJ: Rutgers University Press, 143–159.

Macfarlan, S.J., *et al.* (2014) Lethal coalitionary aggression and long-term alliance formation among Yanomamö men. *Proceedings of the National Academy of Sciences of the United States of America* **111**:16662–16669.

Macfarlan, S.J. and Lyle, H.F. (2015) Multiple reputation domains and cooperative behaviour in two Latin American communities. *Philosophical Transactions of the Royal Society B: Biological Sciences*. DOI: 10.1098/rstb.2015.0009. [Accessed 27 January 2016]

McGraw, K.J. (2002) Environmental predictors of geographic variation in human mating preferences. *Ethnology* **108**:303–317.

Maestripieri, D. (1999) The biology of human parenting: insights from nonhuman primates. *Neuroscience and Biobehavioral Reviews* **23**:411–422.

Maestripieri, D., *et al.* (2004) Father absence, menarche and interest in infants among adolescent girls. *Developmental Science* **7**:560–566.

Mandal, F.B. (2012) Mate choice in humans. *International Journal of Psychology and Behavioral Sciences* **2**:51–56.

Marlowe, F.W. (2000) Paternal investment and the human mating system. *Behavioural Processes* **51**:45–61.

Marlowe, F.W. (2001) Male contribution to diet and female reproductive success among foragers. *Current Anthropology* **42**:755–759.

Marlowe, F.W. (2004) Mate preferences among Hadza hunter-gatherers. *Human Nature* **15**:365–376.

Marlowe, F.W. (2010) *The Hadza: Hunter-Gatherers of Tanzania*. Berkeley: University of California Press.

Marlowe, F.W., *et al.* (2005) Men's preferences for women's profile waist-to-hip ratio in two societies. *Evolution and Human Behavior* **26**:458–468.

Mascaro, J.S., *et al.* (2013) Testicular volume is inversely correlated with nurturing-related brain activity in human fathers.

Proceedings of the National Academy of Sciences of the United States of America **110**:15746–15751.

Mason, K.O. and Taj, A.M. (1987) Differences between women's and men's reproductive goals in developing countries. *Population and Development Review* **13**:611–638.

Mendle, J., *et al.* (2006) Family structure and age at menarche: a children-of-twins approach. *Developmental Psychology* **42**:533–542.

Mesoudi, A., *et al.* (2006) Towards a unified science of cultural evolution. *Behavioral and Brain Sciences* **29**:329–347.

Michels, K.B., *et al.* (1996) Birthweight as a risk factor for breast cancer. *Lancet* **348**:1542–1546.

Milinski, M. and Wedekind, C. (2001). Evidence for MHC-correlated perfume preferences in humans. *Behavioral Ecology* **12**:140–149.

Miller, G. (2000) Sexual selection for indicators of intelligence. In: *The Nature of Intelligence: Novartis Foundation Symposium* (eds G.R. Bock, J.A. Goode and K. Webb). Chichester: John Wiley & Sons Ltd, 260–269.

Moore, J. (1992) Sociobiology and incest avoidance: a critical look at a critical review. *American Anthropologist* **94**:929–932.

Müller, M.N. and Wrangham, R. (eds), (2009) *Sexual Coercion in Primates and Humans: An Evolutionary Perspective on Male Aggression Against Females*. Cambridge, MA: Harvard University Press.

Müller, M.N., *et al.* (2009) Testosterone and paternal care in East African foragers and pastoralists. *Proceedings of the Royal Society of London. Series B: Biological Sciences* **276**:347–354.

Muscarella, F., *et al.* (2001) Homosexual orientation in males: evolutionary and ethological aspects. *Neuroendocrinology Letters* **22**:393–400.

Nesse, R. and Williams, G.C. (1995) *Why We Get Sick*. New York: Random House.

Nettle, D., *et al.* (2013) Human behavioural ecology: current research and future prospects. *Behavioral Ecology* **24**:1031–1040.

Nitsch, A., *et al.* (2014) Alloparenting in humans: fitness consequences of aunts and uncles on survival in historical Finland. *Behavioral Ecology* **25**:424–433.

Norenzayan, A. (2013) *Big Gods: How Religion Transformed Cooperation and Conflict*. Princeton, NJ: Princeton University Press.

Norenzayan, A. and Shariff, A.F. (2008) The origin and evolution of religious prosociality. *Science* **322**:58–62.

Nowak, M.A. (2012) Evolving cooperation. *Journal of Theoretical Biology* **299**:1–8.

Ober, C., *et al.* (1997) HLA and mate choice in humans. *American Journal of Human Genetics* **61**:497–504.

Oliner, S.P. (2002) Extraordinary acts of ordinary people: faces of heroism and altruism. In: *Altruism and Altruistic Love: Science, Philosophy and Religion in Dialogue* (eds S.G. Post, L.G. Underwood, J.P. Schloss and W.B. Hurlbut). New York: Oxford University Press, 123–139.

Oliner, S.P. (2005) Altruism, forgiveness, empathy and intergroup apology. *Humboldt Journal of Social Relations* **29**:8–39.

Otterbein, K.F. (2004) *How War Began*. College Station, Texas: A&M University Press.

Packer, C., *et al.* (1998) Reproductive cessation in female mammals. *Nature* **392**:807–811.

Patton, J.Q. (2005) Meat sharing for coalitional support. *Evolution and Human Behavior* **26**:137–157.

Pawłowski, B. and Dunbar, R.I.M. (1999) Impact of market value on human mate choice decisions. *Proceedings of the Royal Society of London Series B: Biological Sciences* **266**: 281–285.

Peccei, J.S. (2001) Menopause: adaptation or epiphenomenon? *Evolutionary Anthropology* **10**:43–57.

Penner, L., *et al.* (2005) Prosocial behavior: multilevel perspectives. *Annual Review of Psychology* **56**:365–392.

Penton-Voak, I.S. and Perrett, D.I. (2001) Male facial attractiveness: perceived personality and shifting female preferences for male traits across the menstrual cycle. In: *Advances in the Study of Behavior*, Vol. **30** (eds P.J.B. Slater, J.S. Rosenblatt, C.T. Snowdon and T.J. Roper). San Diego, CA: Academic Press, 219–259.

Penton-Voak, I.S. *et al.* (1999) Menstrual cycle alters face preference. *Nature* **399**:741–742.

Perls, T.T. and Fretts, R.C. (2001) The evolution of menopause and human life span. *Annals of Human Biology* **28**: 237–245.

Perrett, D., *et al.* (1999) Symmetry and human facial attractiveness. *Evolution and Human Behavior* **20**:295–307.

Plowes, N.J.R. and Adams, E.S. (2005) An empirical test of Lanchester's square law: mortality during battles of the fire ant *Solenopsis invicta*. *Proceedings of the Royal Society of London Series B: Biological Sciences* **272**:1809–1814.

Poiani, A. (2010) *Animal Homosexuality: A Biosocial Perspective*. Cambridge: Cambridge University Press.

Promislow, D.E.L. (1991) Senescence in natural populations of mammals: a comparative study. *Evolution* **45**: 1869–1887.

Pusey, A.L., *et al.* (1997) The influence of dominance rank on the reproductive success of female chimpanzees. *Science* **277**:828–831.

Putnam, R.D. (2000) *Bowling Alone*. New York: Simon & Schuster.

Puts, D.A., *et al.* (2006) Dominance and the evolution of sexual dimorphism in human voice pitch. *Evolution and Human Behavior* **27**:283–296.

Quinlan, R.J. (2003) Father absence, parental care, and female reproductive development. *Evolution and Human Behavior* **24**:376–390.

Rahman, Q. (2005) The association between the fraternal birth order effect in male homosexuality and other markers of human sexual orientation. *Biology Letters* **1**:393–395.

Reid, R.P. (2002) Waging public relations: a cornerstone of fourth-generation warfare. *Journal of Information Warfare* **1**:51–64.

Richerson, P.J. and Boyd, R. (2008) *Not By Genes Alone: How Culture Transformed Human Evolution*. Chicago: University of Chicago Press.

Riehl, C. and Frederickson, M.E. (2016) Cheating and punishment in cooperation animal societies. *Philosophical Transactions of the Royal Society B: Biological Sciences*. DOI: 10.1098/rstb.2015.0090. [Accessed 27 January 2016]

Risch, N.J., *et al.* (1993) Male sexual orientation and genetic evidence. *Science* **262**:2063–2065.

Rogers, A.R. (1993) Why menopause? *Evolutionary Ecology* **7**:406–420.

Rosetta, L. (1993) Seasonality and fertility. In: *Seasonality and Human Ecology* (eds S.J. Ulijaszek and S.S. Strickland). Cambridge: Cambridge University Press, 65–75.

Roughgarden, J. (2013) *Evolution's Rainbow: Diversity, Gender and Sexuality in Nature and People*. Berkeley: University of California Press.

Sacher, G.A. (1978) Longevity and aging in vertebrate evolution. *BioScience* **28**:497–501.

Sanders, A.R., *et al.* (2014) Genome-wide scan demonstrates significant linkage for male sexual orientation. *Psychological Medicine* **45**:1379–1388.

Sapolsky, R.M. and Share, L.J. (2004) A pacific culture among wild baboons: its emergence and transmission. *PLOS Biology* **2**: e106.

Sarkadi, A., *et al.* (2008) Fathers' involvment and children's developmental outcomes: a systematic review of longitudinal studies. *Acta Paediatrica* **97**:153–158.

Scelza, B.A. (2010) Fathers' presence speeds the social and reproductive careers of sons. *Current Anthropology* **51**:295–303.

Scheidel, W. (1996) Brother–sister and parent–child marriage outside royal families in ancient Egypt and Iran: a challenge to the sociobiological view of incest avoidance? *Ethology and Sociobiology* **17**:319–340.

Schonmann, R.H. and Boyd, R. (2016) A simple rule for the evolution of contingent cooperation in large groups. *Philosophical Transactions of the Royal Society B: Biological Sciences*. DOI: 10.1098/rstb.2015.0099. [Accessed 27 January 2016]

Sear, R., *et al.* (2000) Maternal grandmothers improve the nutritional status and survival of children in rural Gambia. *Proceedings of the Royal Society of London Series B: Biological Sciences* **276**:461–467.

Sear, R., *et al.* (2002) The effects of kin on child mortality in rural Gambia. *Demography* **39**:43–63.

Shanley, D.P. and Kirkwood, T.B.L. (2001) Evolution of the human menopause. *BioEssays* **23**:282–287.

Shepher, J. (1971) Mate selection among second generation kibbutz adolescents and adults: incest avoidance and negative imprinting. *Archives of Sexual Behavior* **1**:293–307.

Sheppard, P., *et al.* (2014) Father absence and reproduction-related outcomes in Malaysia, a transitional fertility population. *Human Nature* **25**:213–234.

Sherman, P.W. (1998) Animal behaviour: the evolution of menopause. *Nature* **392**:759–761.

Shor, E. and Simchai, D. (2009) Incest avoidance, the incest taboo, and social cohesion: revisiting Westermarck and the case of the Israeli kibbutzim. *American Journal of Sociology* **114**:1803–1842.

Sigmund, K., *et al.* (2010) Social learning promotes institutions for governing the commons. *Nature* **466**:861–863.

Silk, J. and Boyd, R. (2006) *How Humans Evolved* New York: W. W. Norton and Company.

Singer, D. and Small, M. (1966) Formal alliances, 1815–1939. *Journal of Peace Research* **3**:1–31.

Singh, D., *et al.* (2010) Cross-cultural consensus for waist–hip ratio and women's attractiveness. *Evolution and Human Behavior* **31**:176–181.

Singhal, A. and Lucas, A. (2004) Early origins of cardiovascular disease: is there a unifying hypothesis? *Lancet* **363**:1642–1645.

Smith, E.A. (2004) Why do good hunters have higher reproductive success? *Human Nature* **15**:343–364.

Snopkowski, K., *et al.* (2014) A test of the intergenerational conflict in Indonesia shows no evidence of earlier menopause in female-dispersing groups. *Proceedings of the Royal Society of London Series B: Biological Sciences* **281**:20140580.

Sussman, R.W. (2013) Why the legend of the killer ape never dies: the enduring power of cultural beliefs to distort our view of human nature. In: *War, Peace, and Human Nature: The Convergence of Evolutionary and Cultural Views* (ed. D.P. Fry). Oxford: Oxford University Press, 97–111.

Swami, V., *et al.* (2007) Preferences for female body weight and shape in three European countries. *European Psychologist* **12**:220–228.

Taborsky, M., *et al.* (2016) The evolution of cooperation based on direct fitness benefits. Special issue of *Philosophical Transactions of the Royal Society B: Biological Sciences*. **371**:1–173.

Temrin, H., *et al.* (2004) Are stepchildren over-represented as victims of lethal parental violence in Sweden? *Proceedings of the Royal Society of London Series B: Biological Sciences* **271**: S124–S126.

Thornhill, R. and Gangestad, S.W. (1999) Facial attractiveness. *Trends in Ecology and Evolution* **3**:452–460.

Thornhill, R., *et al.* (1995) Human female orgasm and mate fluctuating asymmetry. *Animal Behaviour* **50**:1601–1615.

Thornhill, R., *et al.* (2003) Major histocompatibility complex genes, symmetry, and body scent attractiveness in men and women. *Behavioral Ecology* **14**:668–678.

Tither, J.M. and Ellis, B.J. (2008) Impact of fathers on daughters age at menarche: a genetically and environmentally controlled sibling study. *Developmental Psychobiology* **44**:1409–1420.

Tomasello, M. and Moll, H. (2010) The gap is social: human shared intentionality and culture. In: *Mind the Gap: Tracing the Origin of Human Universals* (eds P.M. Kappeler and J.B. Silk). Berlin: Springer, 331–349.

Tomasello, M., *et al.* (2012) Two key steps in the evolution of human cooperation. *Current Anthropology* **53**:673–692.

Tovée, M.J., *et al.* (1999) Visual cues to female physical attractiveness. *Proceedings of the Royal Society of London Series B: Biological Sciences* **266**:211–218.

Travis, H. (2010) *Genocide in the Middle East: The Ottoman Empire, Iraq and Sudan*. Durham, NC: Carolina Academic Press.

Tucker, B. (2004) Giving, scrounging, hiding, and selling: minimal food sharing among Mikea of Madagascar. *Research in Economic Anthropology* **23**:45–68.

Twenge, J.M., *et al.* (2003) Parenthood and marital satisfaction: a meta-analytic review. *Journal of Marriage and Family* **65**:574–583.

Ulijaszek, S.J. (1993) Influence of birth interval and child labour on family energy requirements and dependency ratios in two traditional subsistence economies in Africa. *Journal of Biosocial Science* **25**:79–86.

von Rueder, C. *et al.* (2015) Solving the puzzle of collective action through inter-individual differences. *Philosophical Transactions of the Royal Society B: Biological Sciences*. DOI: 10.1098/rstb.2015.0002. [Accessed 27 January 2016]

von Sydow, K. (1999) Sexuality during pregnancy and after childbirth: a metacontent analysis of 59 studies. *Journal of Psychosomatic Research* **47**:27–49.

Waller, J. (2007) *Becoming Evil: How Ordinary People Commit Genocide and Mass Killing*. New York: Oxford University Press.

Warneken, F. and Tomasello, M. (2006) Altruistic helping in human infants and young chimpanzees. *Science* **311**: 1301–1303.

Weiss, H.K. (1966) Combat models and historical data: the U.S. Civil War. *Operations Research* **14**:759–790.

Wendorf, F. (1968) *The Prehistory of Nubia*. Dallas, TX: Southern Methodist University Press.

Westermarck, E.A. (1889) *The Origin of Human Marriage*. Frenckell.

Wetsman, A. and Marlowe, F.W. (1999) How universal are preferences for female waist-to-hip ratios? Evidence from the Hadza of Tanzania. *Evolution and Human Behavior* **20**:219–228.

Wheeler, T. and von Braun, J. (2013) Climate change impacts on global food security. *Science* **341**:508–513.

White, F.J., *et al.* (2013) Evolution of primate peace. In: *War, Peace, and Human Nature: The Convergence of Evolutionary and Cultural Views* (ed. D.P. Fry). Oxford: Oxford University Press, 389–405.

Williams, G.C. (1957) Pleiotropy, natural selection, and the evolution of senescence. *Evolution* **11**:398–411.

Willis, D., *et al.* (1996) Modulation by insulin of follicle-stimulating hormone and luteinizing hormone actions in human granulosa cells of normal and polycystic ovaries. *Journal of Clinical Endocrinology and Metabolism* **81**:302–309.

Wilson, E.O. (1978) *On Human Nature*. Cambridge, MA: Harvard University Press.

Wilson, M.L. (2013) Chimpanzees, warfare, and the invention of peace. In: *War, Peace, and Human Nature: The Convergence of Evolutionary and Cultural Views* (ed. D.P. Fry). Oxford: Oxford University Press, 361–388.

Wilson, M. and Daly, M. (2009) Coercive violence by human males against their female partners. In: *Sexual Coercion in Primates and Humans: An Evolutionary Perspective on Male Aggression* (eds M.N. Müller and R. Wrangham). Cambridge, MA: Harvard University Press, 271–291.

Wilson, M.L., *et al.* (2002) Chimpanzees and the mathematics of battle. *Proceedings of the Royal Society of London. Series B: Biological Sciences* **269**:1107–1112.

Wolf, A.P. (1966) Childhood association, sexual attraction, and the incest taboo. a Chinese case. *American Anthropologist* **68**:883–898.

Wolf, A.P. (1970) Childhood association and sexual attraction: a further test of the Westermarck hypothesis. *American Anthropologist* **72**:503–515.

Wood, B.M. and Marlowe, F.W. (2013) Household and kin provisioning by Hadza men. *Human Nature* **24**:280–317.

Zefferman, M.R. and Mathew, S. (2015) An evolutionary theory of large-scale human warfare: group-structured cultural selection. *Evolutionary Anthropology* **24**:50–61.

Index